CAMBRIDGE LIBRARY COLLECTION

Books of enduring scholarly value

Botany and Horticulture

Until the nineteenth century, the investigation of natural phenomena, plants and animals
was considered either the preserve of elite scholars or a pastime for the leisured upper
classes. As increasing academic rigour and systematisation was brought to the study of
'natural history', its subdisciplines were adopted into university curricula, and learned
societies (such as the Royal Horticultural Society, founded in 1804) were established to
support research in these areas. A related development was strong enthusiasm for exotic
garden plants, which resulted in plant collecting expeditions to every corner of the globe,
sometimes with tragic consequences. This series includes accounts of some of those
expeditions, detailed reference works on the flora of different regions, and practical advice
for amateur and professional gardeners.

A Dictionary of the Economic Products of India

A Scottish doctor and botanist, George Watt (1851–1930) had studied the flora of India
for more than a decade before he took on the task of compiling this monumental work.
Assisted by numerous contributors, he set about organising vast amounts of information
on India's commercial plants and produce, including scientific and vernacular names,
properties, domestic and medical uses, trade statistics, and published sources. Watt hoped
that the dictionary, 'though not a strictly scientific publication', would be found 'sufficiently
accurate in its scientific details for all practical and commercial purposes'. First published
in six volumes between 1889 and 1893, with an index volume completed in 1896, the whole
work is now reissued in nine separate parts. Volume 6, Part 1 (1892) contains entries from
Pachyrhizus angulatus (a large climbing herb) to *rye* (not indigenous to India).

A Dictionary of the Economic Products of India

VOLUME 6 – PART 1: PACHYRHIZUS TO RYE

GEORGE WATT

CAMBRIDGE
UNIVERSITY PRESS

CAMBRIDGE
UNIVERSITY PRESS

University Printing House, Cambridge, CB2 8BS, United Kingdom

Published in the United States of America by Cambridge University Press, New York

Cambridge University Press is part of the University of Cambridge.

It furthers the University's mission by disseminating knowledge in the pursuit of
education, learning and research at the highest international levels of excellence.

www.cambridge.org
Information on this title: www.cambridge.org/9781108068789

© in this compilation Cambridge University Press 2014

This edition first published 1892
This digitally printed version 2014

ISBN 978-1-108-06878-9 Paperback

A

DICTIONARY

OF

THE ECONOMIC PRODUCTS OF INDIA.

BY

GEORGE WATT, M.B., C.M., C.I.E.,

REPORTER ON ECONOMIC PRODUCTS WITH THE GOVERNMENT OF INDIA.
OFFICIER D'ACADEMIE; FELLOW OF THE LINNEAN SOCIETY; CORRESPONDING MEMBER OF THE
ROYAL HORTICULTURAL SOCIETY, &C., &C.

(ASSISTED BY NUMEROUS CONTRIBUTORS.)
IN SIX VOLUMES.

———

VOLUME VI, PART I.

[Pachyrhizus to Rye.]

———

Published under the Authority of the Government of India,
Department of Revenue and Agriculture.

LONDON:
W. H. ALLEN & Co., 13, WATERLOO PLACE, S.W., PUBLISHERS TO THE
INDIA OFFICE.

CALCUTTA:
OFFICE OF THE SUPERINTENDENT OF GOVERNMENT PRINTING, INDIA,
8, HASTINGS STREET.

———

1892.

CALCUTTA
GOVERNMENT OF INDIA CENTRAL PRINTIN OFFICE,
8, HASTINGS STREET.

PREFACE to Vol. VI.

OWING to the important character of many of the Economic Products dealt with in the sixth volume (*e.g.*, Sugar, Silk, Tea, Wool, and Wheat), it has been found necessary to expand the volume into four parts which are designated as VI, Parts i, ii, iii, and iv, respectively. The adoption of this numbering, in preference to the designation of the last three volumes as VII, VIII, IX, has been made necessary by the circumstance that, in the first five volumes, Economic Products falling under the concluding letters of the alphabet have been referred to as described in Volume VI.

The price of the Dictionary will not be affected by the expansion of the sixth volume into four parts.

The opportunity is taken to state that an index is under preparation, and will be published soon after the completion of the sixth volume, in which vernacular names of products used or quoted in the Dictionary are brought together in geographical groups.

With the aid of the index, the article describing any product, of which the vernacular name only is known, can be readily traced, provided that the vernacular term was on record at the time when the Dictionary was compiled.

The occasion of issuing the final volume is taken to acknowledge the services rendered to the Editor of the Dictionary by Mr. F. M. W. Schofield, Superintendent of the Statistical Branch of the Revenue and Agricultural Department of the Government of India, and by Mr. A. R. Tucker, Registrar of the same Department, who superintended the administrative arrangements under which the records in the Office of the Department were provided for the use of the Editor.

Finally the Governor General in Council desires to place on record his high appreciation of the industry, ability and research of the Editor, Dr. George Watt, C.I.E. ; of whose merits the Dictionary itself affords the best testimony, and to thank his official collaborateurs Mr. J. F. Duthie and Surgeons-Captain J. Murray and W. R. Clark for their valuable assistance.

E. C. BUCK,

Secy. to the Govt. of India.

REVENUE AND AGRICULTURAL DEPARTMENT :

Simla, the 6th August 1892.

DICTIONARY

OF

THE ECONOMIC PRODUCTS OF INDIA.

PACHYRHIZUS, *Rich. ; Gen. Pl., I., 540.*

Pachyrhizus angulatus, *Rich. ; Fl. Br. Ind., II., 207 ;* LEGUMINOSÆ.   I
    YAKA or WAYAKA.

    Syn.—DOLICHOS BULBOSUS, *Linn. ;* PACHYRHIZUS TRILOBUS, *DC.*

    Vern.—*Sánkálu,* BENG.; *Sankhálu,* SANS.

    Habitat.—A large climbing herb, cultivated throughout India, but not known in a wild state.

    Food.—The tuberous ROOT, which is from 6 to 8 feet in length and as thick as a man's thigh, resembles a turnip in taste and consistence. It is eaten both raw and boiled. When cooked it has a dirty white colour, and an insipid flavour, but is palatable enough and is considerably used in times of scarcity.

FOOD.
Root,
2

PACKING CASES.

3

TEA, COFFEE, INDIGO, TOBACCO, OPIUM, AND CINCHONA CHESTS, CASKS, AND PACKING CASES, Timbers used for—

Abies Webbiana ; CONIFERÆ ; packing cases.
Acer cæsium ;
A. Cambellii ; } SAPINDACEÆ ; tea chests.
A. lævigatum ;
Acrocarpus fraxinifolius ; LEGUMINOSÆ ; tea-chests.
Adina cordifolia ; RUBIACEÆ ; opium boxes.
Æsculus indica ; SAPINDACEÆ ; tea chests, packing cases.
Albizzia procera ; } LEGUMINOSÆ ; tea chests.
A. stipulata ;
Alnus nepalensis ; BETULACEÆ ; tea chests.
Alstonia scholaris ; APOCYNACEÆ ; tea chests, coffins.
Anacardium occidentale ; ANACARDIACEÆ ; packing cases.
Anthocephalus Cadamba ; RUBIACEÆ ; tea chests.
Artocarpus Chaplasha ; URTICACEÆ ; tea chests.
Bassia latifolia ; SAPOTACEÆ ; tea chests.
Beilschmiedia Roxburghiana ; LAURINEÆ ; tea chests.
Bœhmeria rugulosa ; URTICACEÆ ; tobacco boxes.
Bombax malabaricum ; MALVACEÆ; tea chests, packing cases, coffins.
Canarium bengalense ; BURSERACEÆ ; tea chests.

PÆDERIA foetida.	Timbers for Tea; Coffee, Indigo, and other Chests.

Casearia glomerata; CAMYDACEÆ; tea chests.
Cedrela Toona; MELIACEÆ; tea & indigo chests, also cigar boxes.
Cordia Myxa; BORAGINEÆ; tea chests.
Daphnidium pulcherrimum; LAURINEÆ; tea chests.
Derris robusta; LEGUMINOSÆ; tea chests.
Duabanga sonneratoides; LYTHRACEÆ; tea chests.
Echinocarpus dasycarpus; TILIACEÆ; tea chests.
Ehretia Wallichiana; BORAGINEÆ; tea chests.
Elæocarpus lanceæfolius; TILIACEÆ; tea chests.
Engelhardtia spicata; JUGLANDEÆ; tea chests.
Erythrina indica; LEGUMINOSÆ; tea chests.
Ficus religiosa; URTICACEÆ; packing cases.
Gmelina arborea; VERBENACEÆ; tea chests, packing cases.
Guazuma tomentosa; STERCULIACEÆ; packing cases.
Hymenodictyon thyrsiflorum; RUBIACEÆ; packing cases.
Lagerstrœmia Flos-Reginæ; LYTHRACEÆ; casks.
L. microcarpa; coffee cases.
Machilus odoratissima; LAURINEÆ; tea chests.
Mangifera indica; ANACARDIACEÆ; indigo & tea chests, packing cases.
M. sylvatica; tea chests.
Melia Azadirachta; MELIACEÆ; trunks, packing cases.
M. dubia; tea chests.
Michelia Cathcartii; MAGNOLIACEÆ; tea chests.
Morus indica; URTICACEÆ; tea chests.
Pentace burmanica; TILIACEÆ; tea chests.
Phœbe attenuata; LAURINEÆ; tea chests.
Picea morinda (Abies Smithiana); CONIFERÆ; packing cases.
Pinus excelsa; CONIFERÆ; tea chests.
P. longifolia; tea chests.
Polyalthia Jenkinsii; ANONACEÆ; tea chests.
P. longifolia; tea chests.
P. simiarum; tea chests.
Prunus communis; ROSACEÆ; papier-maché boxes.
Pyrus lanata; ROSACEÆ; tea chests.
Sonneratia apetala; LYTHRACEÆ; packing cases.
Sterculia villosa; STERCULIACEÆ; tea chests.
Stereospermum chelonoides; BIGNONIACEÆ; tea chests.
Terminalia belerica; COMBRETACEÆ; coffee cases, packing cases.
T. myriocarpa; tea chests.
Tetrameles nudiflora; DATISCEÆ; tea chests.
Vateria indica; DIPTEROCARPEÆ; coffins.

For further information regarding these timbers, the reader is referred to the description of each in its respective alphabetical position in this work.

Paddy, see Oryza sativa, *Linn.;* GRAMINEÆ; Vol. V., 502.

PÆDERIA, *Linn.; Gen. Pl., II., 133.*

Pædería fœtida, *Linn.; Fl. Br. Ind., III., 195;* RUBIACEÆ.

    Syn.—P. OVATA, *Miq.;* P. SESSILIFLORA, *DC.*

    Var. microcarpa, *Kurz.*

    Vern.—*Khip, gandháli, so marájí, sómráj, bacuchi,* HIND.; *Gandhabhádulia,* BENG.; *Gandali,* URIYA; *Bedolí sutta,* ASSAM; *Padebiri,* NEPAL; *Takpœdrik,* LEPCHA; *Paedebiri,* PAHARIYA; *Prasáram,* BOMB.; *Hiranvel,* MAR.; *Gandhana,* GUZ.; *Savirela,* TEL.; *Prasárani,* SANS.

4

The Pæony Rose. (*J. Murray.*)

References.—*Roxb., Fl. Ind., Ed. C.B.C., 229 ; Voigt, Hort. Sub. Cal., 388 ; Kurz, For. Fl. Burm., II., 26 ; Gamble, Man. Timb., 219 ; Elliot, Fl. Andhr., 167 ; Sir W. Jones, Treat. Pl. Ind., V., 94 ; U. C. Dutt, Mat. Med. Hind., 178, 314 ; Dymock, Mat. Med. W. Ind., 2nd Ed., 401 ; Dymock, Warden & Hooper, Pharmacog. Ind., II., 229 ; Drury, U. Pl., 325.*

Habitat.—An extensive climber, met with in the Central and Eastern Himálaya, up to 5,000 feet, and extending southward to Malacca and westward to Bengal and Assam.

Fibre.—The PLANT yields a fibre which has recently attracted considerable attention. It is strong and flexible, and has a silk-like appearance. The best quality is obtained from plants growing on alluvial deposits, such as those on the banks of the Bráhmapútra.

FIBRE.
Plant.
5

Medicine.—The LEAVES and ROOT were considered of medicinal value by the Sanskrit writers. The former, boiled and made into a soup, are considered wholesome and a suitable diet for the sick and convalescent. The entire plant, including the STEM, leaves, and root, is used both internally and externally in rheumatic affections, for which it is regarded as a specific (*U. C. Dutt*). Roxburgh remarks that the Hindus employ the root as an emetic. Dymock states that the properties of this plant do not appear to be known to Muhammadan writers.

MEDICINE.
Leaves.
6
Root.
7
Stem.
8

Chemical Composition.—The plant has recently been analysed by the authors of the *Pharmacographia Indica*, who state that a volatile oil is obtained by distillation with water, which has the highly offensive bisulphide of carbon odour of the fresh drug. Evidence was also obtained of the presence of at least two alkaloids, one soluble in ether, from which it could be deposited as minute needles, the other apparently non-crystalline, and only slightly soluble in amylic alcohol, chloroform, and benzene. For these principles the names of α. and β. *pæderine* have been provisionally proposed.

Chemistry.
9

Special Opinions.—§ "The leaves are much used in Bengal as a constituent of stews and curries in convalescence from all kinds of disease. They are certainly wholesome and tonic" (*Surgeon R. L. Dutt, M.D., Pubna*). "Soup prepared from the leaves is wholesome and is frequently given in convalescence from fevers" (*Assistant Surgeon S. C. Bhattacharji, Chanda*). "The juice of the leaves is considered astringent, and is given to children when suffering from diarrhœa; dose 1 drachm" (*Surgeon A. C. Mukerji, Noakhally*).

Food.—The LEAVES are used as an article of invalid diet.

FOOD.
Leaves.
10

PÆONIA, *Linn.; Gen. Pl., I., 10.*

Pæonia emodi, *Wall.; Fl. Br. Ind., I., 30 ;* RANUNCULACEÆ.

II

THE PÆONY ROSE.

Syn.—P. OFFICINALIS, *H. f. & T.*

Var., 1 emodi, *proper ;* follicles strigose.

Var., 2 **glabrata ;** follicles glabrous.

Vern.—*Ud-sálap,* HIND. ; *Bhuma-madiya, yet ghás,* BHUT.; *Mamekh,* PB.; *Chandra* (the plant), *sújuniya* (the young edible shoots), N.-W. P.

References.—*Gamble, Man. Timb., 1 ; Stewart, Pb. Pl., 4; Pharmacog. Ind., I., 29, 37; Dymock, Mat. Med. W. Ind., 2nd Ed., 17 ; Murray, Pl. & Drugs, Sind, 75 ; Year-Book of Pharmacy, 1875, 622 ; Atkinson, Him. Dist., 304, 745 ; Smith, Dic., 317 ; Gazetteer, Peshawar, 28 ; Ind. Forester, IX., 197.*

Habitat.—An erect, stout, leafy herb, found in the west Temperate Himálaya, from Kumáon and Hazára, at altitudes of 5,000 to 10,000 feet.

| PANDANUS furcatus. | The Pandanus Fibre. |

MEDICINE.
Tubers.
12

Medicine.—The TUBERS of this plant, like those of **P. officinalis**, *L.*, are highly esteemed as a medicine for uterine diseases, colic, bilious obstructions, dropsy, epilepsy, convulsions, and hysteria. *Ud-sálap* is generally given to children as a blood-purifier. It was a common belief in ancient times, and is so even now among the peasantry of Europe, that pæony root, if worn by children round their necks, has the power of preventing epileptic attacks. If taken in full doses, the drug produces head-ache, noise in the ears, confused vision and vomiting (*Dymock*).

Flowers.
13
Seeds.
14
Root.
15
FOOD.
Shoots.
16

The infusion of the dried FLOWERS is highly valued as a remedy for diarrhœa. The SEEDS are emetic and cathartic. Madden mentions that the long ROOT may furnish one of the kinds of *bikh* used in Kumáon (generally Aconite, *q. v.*, Vol. I., 87).

Food.—The young SHOOTS are eaten in Kumáon as a vegetable.

PAJANELIA, *DC.; Gen. Pl., II., 1047.*

[*1343-44;* BIGNONIACEÆ.

17

Pajanelia Rheedii, *DC.; Fl. Br. Ind., IV., 384; Wight, Ic., tt.*

Syn.—BIGNONIA INDICA, VAR. β, *Linn.*; B. PAJANELIA, *Ham.*; B. LONGIFOLIA, *Willd.*; B. MULTIJUGA, *Wall.*; PAYANELIA MULTIJUGA, *Kurz.*
Vern.—*Kyoungdouk, kingalun,* BURM.; *Kaukonda,* AND.
References.—*Gamble, Man. Timb., 279; Kurz, For. Fl., II., 237; Gazetteer, N.-W. P., I., 82.*

TIMBER.
18

Habitat.—A large evergreen tree of Burma and the Andaman Islands.
Structure of the Wood.—Orange-brown, very hard, close-grained; weight 52℔ per cubic foot. A fine timber which is used for canoes by the Andamanese.

Pala indigo, see Wrightia tinctoria, *Br.*; APOCYNACEÆ, Vol. VI., Pt. II.

Pale catechu, see Uncaria Gambier, *Roxb.*; RUBIACEÆ, Vol. VI., Pt. II.

Palma christi, see Ricinus communis, *Linn.*; EUPHORBIACEÆ.

Palm Wine or **Toddy**, see Borassus flabelliformis, *Linn.*; PALMÆ, Vol. I., 495; see Caryota urens, *Linn.*, Vol. II., 208; Cocos nucifera, *Linn.*, Vol. II., 451; Phœnix sylvestris, *Roxb.*; also Spirits under Narcotics, Vol. V., 332.

Palmyra Palm, see Borassus flabelliformis, *Linn.*; PALMÆ, Vol. I., 495.

PANDANUS, *Linn.; Gen. Pl., III., 949.*

[PANDANEÆ.

19

Pandanus andamanensium, *Kurz; For. Fl. Burm., II., 508;*

Habitat.—An evergreen tree, frequent in the tropical forests of the Andamans, especially in the neighbourhood of the sea.

FIBRE.
20

Fibre.—It yields a fibre which is employed in the Andamans for making various articles of apparel, such as the " tails " worn by the women.

21

P, furcatus, *Roxb.; Kurz, For. Fl. Burm., II., 507.*

Vern.—*Jarika,* NEPAL; *Bor,* LEPCHA; *Thabaw, tagyet,* BURM.; *Okaiyega,* SING.
References.—*Roxb., Fl. Ind., Ed. C.B.C., 708; Gamble, Man. Timb., 425; List of Trees, Shrubs, &c., of Darjeeling, 86; Beddome, For. Man., ccxxviii.; Mason, Burma & Its People, 521, 815; Gazetteer, Bomb., XV., Pt. I., 445; Ind. Forester:—IV., 241; VI., 237; XIV., 341.*

Habitat.—A palm-like tree of Northern and Eastern Bengal, Western India, and Burma.

| The Keora. | (*J. Murray.*) | PANDANUS odoratissimus. |

Fibre.—The LEAVES are used in Burma for making mats. Mason states that the leaves of a species of **Pandanus** (probably this plant) are by the Burmans sewn together to make sails. | FIBRE. Leaves. 22

Structure of the Wood.—Outer wood moderately hard, containing satiny vascular bundles, inner wood soft and spongy; weight 20℔ per cubic foot; employed in Burma to make floats for fishing nets. | TIMBER. 23

Pandanus Mellori.
24

A species of PANDANUS (described under this name in "*A vocabulary of the Dialects spoken in the Nicobar and Andaman Islands*") is said to yield a FRUIT from which bread is made. "It grows to perfection in this warm, moist climate and in half swampy soil. It requires, however, great labour to extract and cook the edible farinaceous part." | Fruit. 25

P. odoratissimus, *Willd.; Kurz, For. Fl. Burm., II., 508.*
26

Vern.—*Keorá, ketgi, gagandhul,* HIND.; *Keyá, ketuki, ketki-keya, keori,* BENG.; *Kedgi,* DEC.; *Kenda, keur,* BOMB.; *Keora,* MAR.; *Kewoda,* GUZ.; *Talum, tazhai, thalay,* TAM.; *Mugali, gájangi, gédanvi, kétaki,* TEL.; *Mogalisandlu, mogili nara,* NELLORE; *Tale mara, kyad-age-gida, ketaki,* KAN.; *Pandang, kaida, thala,* MALAY.; *Tsat-tha-pu,* BURM.; *Wœta keyiya, mudu keyiya,* SING.; *Leram,* NICOBAR; *Ketaki, ketaka,* SANS.; *Cadhi, kasi, kadar,* ARAB.; *Kadi,* PERS.

References.—*Roxb., Fl. Ind., Ed. C.B.C., 707; Beddome, For. Man., ccxxviii.; Kurz, For. Fl. Burm., II., 508; Gamble, Man. Timb., 425; Thwaites, En. Ceylon Pl., 327; Cleghorn, Plants of Southern Ind., 145; Elliot, Fl. Andhr., 56, 59, 61, 90, 116; U. C. Dutt, Mat. Med. Hind., 304; S. Arjun, Bomb. Drugs, 149; Lisboa, U. Pl. Bomb., 207, 225, 237; Bird-wood, Bomb. Pr., 185; Royle, Ill. Him. Bot., 408; Royle, Fib. Pl., 36; Liotard, Paper-making Mat., 7, 11, 15, 20; Butler, Top. & Statis., Oudh & Sultanpur, 33; Smith, Dic., 372; Kew Off. Guide to Bot. Gardens & Arboretum, 39, 40; Gazetteers:—Mysore & Coorg, I., 66; II., 7; III., 24; Bomb., V., 28; XV., 445; N.-W. P., I., 84; IV., lxxviii.; Ind. Forester:—III., 46, 237; V., 329; X., 175; Man. Mad. Adm., I., 359; II., 85; W. W. Hunter, Orissa, II., 133, 179; District Manuals: Trichinopoly, 72; Nellore, 117, 1128; Agri.-Horti., Soc. Jour. (Old Series), II., 93; IV., 232; IX., 176, 177; X., 353, 354.*

Habitat.—A common, much-branched shrub, frequently planted on account of the powerful fragrance of its flowers, but wild on the coasts of South India, Burma, and the Andamans. It is found abundantly in Bengal, Madras, the Straits Settlements, and the South Sea Islands.

Fibre.—The LEAVES are composed of tough, longitudinal, white, glossy fibres, which are employed for covering huts, making matting, for cordage, and in South India for the larger kinds of hunting nets, and for the drag-ropes of large fishing nets. In Mauritius the fibre is employed for making sacks for coffee, sugar, and grain. The reader is referred for a full description of the method of preparation of the fibre in Mauritius to a paper by **Mr. Henley** in the *Agri.-Horti. Soc. Ind. Jour., II., 93*. It is said to be a good paper-making material. The leaves are cut every second year, and each plant yields sufficient fibre to make two large bags. The ROOTS also are fibrous and are used by basket-makers for binding. When cut into lengths and beaten out, they are very commonly used as brushes for painting and white-washing. It is possible that this root-fibre might be found suitable for brush-making as a substitute for bristle, a form of fibre which is now in great demand. | FIBRE. Leaves. 27 · Roots. 28 · OIL. Bracts. 29

Oil.—An oil called KEORA OIL is obtained by distillation from the fragrant BRACTS. An aromatic water or otto, *keora-árak,* is also distilled from the bracts. Both are employed medicinally. | MEDICINE. Oil. 30

Medicine.—The OIL and OTTO, obtained from the bracts, are considered stimulant and antispasmodic, and are administered for headache and | Otto. 31

P. 31

PANICUM	Fodder Grasses.

MEDICINE.
Roots.
32
Aerial Root.
33
FOOD.
Leaves.
34
Drupe.
35
Flowers.
36
DOMESTIC.
Blossoms.
37
Fruit.
38
Roots.
39
Leaves.
40

41

rheumatism. A medicinal oil is prepared from the ROOTS. The AERIAL ROOT is used medicinally by the Sinhalese.

Food.—The tender FLORAL LEAVES are eaten raw, or cooked with various condiments; the pulp contained in the lower part of the DRUPE is also edible, and is eaten by the Natives of Southern India in times of scarcity. The FLOWERS, together with catechu and certain spices, form a substance known as *keá khoir*, which is used in *pán.*

Domestic & Sacred.—Strettell states that the women of Upper Burma adorn their hair with the fragrant BLOSSOMS, and use the FRUIT for hackling thread. The thick aerial rope-like ROOTS, cut into lengths, are beaten out at one end to form brushes, and are also employed as a substitute for corks. In Southern India the LEAVES are used to make umbrellas. The plant is recommended by Cleghorn as " a very strong sand-binder." Lisboa writes, " *Chapter III. of Kartik Máhatma* relates that this plant is cursed by *Shiv* for telling a lie and giving false evidence on the occasion of a dispute between *Shiv* and *Vishnu.* But he again took pity on it, and ordered that he (*Shiv*) should be worshipped with this plant on the day of *Shivratri.* The plant is also employed in the worship of many other gods."

(*J. F. Duthie.*)

PANICUM, *Linn.; Gen. Pl., III., 1100.*

This, the largest genus of the grass family, contains nearly 300 species. Of the Indian species several are much valued as fodder-grasses, owing to their abundant yield of grain ; and on this account many kinds, such as *chena, sanwák, kutki,* &c., known as minor millets, are cultivated in various parts of India, and afford a large amount of food for the lower classes. In addition to the more important kinds about to be described, the following deserve to be briefly mentioned :—
P. cimicinum, *Retz.* , SYN.—CORIDOCHLOA FIMBRIATA, *Nees.,* a native of the hilly parts of Northern and Central India, abundant along the base of the Himálayan range ; this grass, when drying, emits a strong perfume resembling that of the sweet-scented Anthoxanthum of Europe, and might be used with advantage in hay mixtures. P. distachyum, *Linn.,* VERN.—*Motia* (Doáb); as a fodder grass this is probably as nutritious as any of the other indigenous species of Panicum, but it is less plentiful ; in Australia it is largely used as hay. P. erucæforme, *Sibth. & Sm.,* SYN.—P. caucasicum, *Trin.,* VERN.— *Tiliya, chinwári* (Bundelkhand); *Guhria, loidan-siput, sarpot, sarput* (C. Prov.); abundant on cultivated ground, especially on the black soil of Central India, it yields a very large amount of grain. P. fluitans, *Retz.,* SYN.—P. brizoides, *Retz.* (not of *Linn.*), VERN.—*Petinar* (Beng.); *Dusa* (Tel.); an aquatic species allied to P. flavidum, produces an abundant quantity of grain. P. humile, *Nees.,* VERN.—*Katki, urdiya* (C. Prov.); common in Northern and Central India, generally on cultivated ground, and reckoned a good fodder grass. P. pabulare, *Aitch. & Hensl. in Journ. Linn. Soc., XVIII.* (*1880*), *p. 190,* was discovered by Dr. Aitchison in the Kuram Valley, and is considered to be the best fodder grass for cattle in that district. P. paludosum, *Roxb..* SYN.—P. decompositum, *R. Br.,* VERN.—*Boruti, kulus-nar* (Beng.); *Soda* (Tel.); an aquatic species, but not common; it is valued in Australia for fodder, and the grain is made into cakes by the aborigines. P. Petiverii, *Trin.,* VERN.—*Chápar, chaprur, chaprura* (N.-W. India); abundant in Northern India, and valuable as green fodder. P. psilopodium, *Trin.,* allied to P. miliare, of which it may be the wild form; a good fodder grass, abundant in Northern India and up to moderate elevations on the Himálaya. P. tenuiflorum, *R. Br.,* SYN.—Paspalum brevifolium, *Flugge;* not uncommon in Northern India. It is used for fodder in Australia, and yields plenty of grain. P. turgidum, *Forsk.,* a coarse perennial species found in Sind and in Central India, extending to Arabia and Egypt, where the grain is used for making a kind of bread.

[GRAMINEÆ.

Panicum antidotale, *Retz.; Duthie, Fodder Grasses, 4, Pl. iii.;*   **42**
  Syn.—P. SUBALBIDUM, *Kunth.*
  Vern.—*Gunara,* HIND.; *Layo-gundli,* SANTAL; *Gamur, ghamor,* N.-W.
    P.; *Gharam, ghamur, ghamrur, ghirri, garm, girúi, baru, mangrúr,*
    PB.; *Male, shamukha,* PUSHTU; *Barwári, bari gagli,* RAJ.; *Krimisastru,*
    SINH.
  References.—*Stewart, Pb. Pl., 258; Aitchison, Cat. Pb. & Sind Pl., 158;*
    Coldstream, Grasses of S. Panjáb, Nos. 1 & 1a.; Boiss., Fl. Or., V., 440;
    U. C. Dutt, Mat. Med. Hind., 299; Kew Report, 1880, 16; Gazetteers,
    N.-W. P., I., 85; IV., lxxix.
  Habitat.—A rather tall coarse grass, usually growing in clumps, or in
the shelter of bushes and hedges. Abundant in the plains of Northern
India and extends to Baluchistán and Afghánistán. It occurs also in
Africa and in Australia.
  Medicine.—Dr. Stewart says that the smoke of the burning PLANT is used   **MEDICINE.**
for fumigating wounds, also as a disinfectant in small-pox. In Madras it   **Plant.**
is said to be employed medicinally in throat affections; it is stated by   **43**
Thwaites to be used medicinally by the Sinhalese.
  Fodder.—Opinions differ as to the utility of this grass for fodder pur-   **FODDER.**
poses. **Mr. Coldstream,** writing from Hissar, says that it is grazed only   **44**
when quite young, as it afterwards acquires a bitter or saltish taste. In the
Sirsa Settlement Report it is mentioned that cattle feed only on the dry
grass, for if they eat it when it is green and young, poisonous, and sometimes
fatal, results follow. **Captain Wingate** informs us that it is very abundant
at Changa Manga near Lahore, where more than three-fourths of the grass
herbage, in the actual forest plantation, is composed of this species, and
that Natives feed their cattle on the green fodder. Like *dáb* (**Eragrostis
cynosuroides**) and other coarse deep-rooting grasses, its real value probably
commences at those periods when the better class of fodder grasses fail.

P. colonum, *Linn.; Duthie, Fodder Grasses, 4, Pl. iv.*   **45**
  Syn.—P. BRIZOIDES, *Linn.;* OPLISMENUS COLONUS, *Kunth.;* ECHINOCH-
    LOA COLONA, *Kunth.*
  Vern.—*Sawánk,* HIND.; *Shama,* BENG.; *Sáma,* SANTAL.; *Sawan, jangli
    sáwank, sarwak, shamak,* N.-W. P.; *Sivaen,* OUDH; *Sáwank saun,
    jangli mandira, ganára,* KUMAON; *Sánwak, jangli sámak, sirmakar,
    chatta,* PB.; *Sawuk, jangli sawuk,* SIND; *Shámakh,* C. P.; *Saweli,*
    BERAR; *Borrur,* BOMB.; *Wundu,* TEL.
  References.—*Roxb., Fl. Ind., Ed.C.B.C., 99; Dals. & Gibs., Bomb. Fl., 291;
    Stewart, Pb. Pl., 258; Aitchison, Cat. Pb. & Sind Pl., 161; Boiss.,
    Fl. Or., V., 435; Coldstream, Fodder Grasses of S. Panjáb, No. 2;
    Baden Powell, Pb. Pr., 245; Lisboa, U. Pl. Bomb., 276; Royle, Ill. Him.
    Bot., 422; Church, Food Grains of India, 50; Econ. Prod. N.-W.
    Prov., Pt. V., 99; Gazetteers:—N.-W. P., I., 85; Gurgaon, 17; Karnal,
    19; Ind. Forester, XII., App., 22; Settle. Rept., Montgomery, 22; Land
    Revenue Settle. Rept., Nursingpore, 52.*
  Habitat.—An annual grass, abundant throughout the plains, ascending
to moderate elevations on the Himálaya. It prefers a rich soil, and is
consequently often met with as a weed of cultivation. In parts of the
Panjáb it is cultivated for its grain. Out of India it occurs in South
Europe, North Africa, and North America.
  Food & Fodder.—The GRAIN is collected for food by the poorer people,   **FOOD &**
in many parts of Northern India. In the Montgomery district it is made   **FODDER.**
into a paste, called *bát* or *phat,* and eaten with milk. **Stewart** mentions   **Grain.**
that this preparation constituted the chief food of the Akhund of Swát,   **46**
and that the plant was cultivated for that purpose. Another preparation
called *khir,* made by boiling the grain in milk, is eaten by Hindús on
their fast days.

PANICUM flavidum.	Fodder Grasses.

The following analysis of the husked grain is given by Professor Church, in his *Food Grains of India* :—

	In 100 parts.
Water	12˙0
Albumenoids	9˙6
Starch	74˙3
Oil	˙6
Fibre	1˙5
Ash	2˙0

The nutrient ratio being 1 : 8, and the nutrient value 85.

Grass.
47

This GRASS is generally considered one of the best fodder grasses in India. It is greedily eaten by all kinds of cattle, both before and after it has flowered, the abundant crop of grain which it yields adding materially to its nutritive value.

48

Panicum Crus-Galli, *Linn. ; Duthie, Fodder Grasses, 5, Pl. v.*

Syn.—P. CRUS-CORVI, *Linn.*; OPLISMENUS CRUS-GALLI, *Kunth.*; ECHI-NOCHLOA CRUS-GALLI, *Beauv.*; ORTHOPOGON CRUS-GALLI, *Spreng.*; P. STAGNINUM, *Linn.*; P. HISPIDULUM, *Roxb.*; P. TOMENTOSUM, *Roxb.*

Vern.—*Sánwak,* HIND. ; *Bura shama, dul,* BENG. ; *Dhand, jal-sawank,* N.-W. P. ; *Bharti, bara sánwak, jarotha, dhand,* PB. ; *Sama, horma,* RAJPUTANA ; *Bari bhodore, bharta, datia, kunda buttam godi,* C. P. ; *Pedda wúndú,* TEL.

References.—*Roxb., Fl. Ind., Ed. C.B.C., 99; Dals. & Gibs., Bomb. Fl., 292; Royle, Ill. Him. Bot., 420; Coldstream, Fodder Grasses of S. Pánj., No. 3; F. von Mueller, Select Extra-Trop. Pl., 272.*

Habitat.—Some of the forms of this variable species approach **P. colonum,** but it is usually a larger and coarser plant, and prefers wet ground, such as the borders of *jhils* and edges of streams. It is abundant all over India, and is found on the Himálaya up to 6,000 feet. It is distributed throughout the world except in the Arctic regions.

Food & Fodder.—The GRAIN is collected and used as food by the lower classes, and is also made into *khir.* It is said to be cultivated in parts of the Lahore district for its grain.

When young, this grass is much relished by cattle, especially by buffaloes. In America, where it is known under the name of "Barn-yard grass," it is highly valued as hay.

50

P. flavidum, *Retz. ; Duthie, Fodder Grasses, 7.*

Syn.—P. BRIZOIDES, *Jacq.*

Vern.—*Sánka* (Dehra-Dun), *dhanera,baunri,* N.-W. P. ; *Sathiya, sitiya,* OUDH ; *Kangna, pálon, bharti,* PB. ; *Chida sáwán, homa* (Abu), RAJ. ; *Paddatunga, kura-tuka, chichwi, sama,* C. P. ; *Barti, bauti,* BOMB. ; *Barti,* MAR. ; *Banti,* GUZ. ; *Uda-gaddí, vúda gaddí,* TEL.

References.—*Roxb., Fl. Ind., Ed. C.B.C., 98 ; Dals. & Gibs., Bomb. Fl., 290 ; Elliot, Fl. Andhr., 185, 193 ; Baden Powell, Pb. Pr., 237 ; Birdwood, Bomb. Pr., 110 ; Royle, Ill. Him. Bot., 420 ; Gazetteers:—Bombay, VIII., 182; N.-W. P., I., 85; Ind. Forester, XII., App., 22; Bombay Man., Rev. Accts., 101 ; Bulletin, R. Gardens, Kew (1887), No. 12 ; F. von Mueller, Select Extra-Trop. Pl., 273.*

Habitat.—Common throughout the plains and up to moderate elevations on the Himálaya. It is cultivated as a rainy season crop in some parts of India for its grain. It is found also in Australia.

Cultivation.
51

CULTIVATION.—In certain districts of the Bombay Presidency this grass is regularly grown to a limited extent. The returns for 1888-89 showed an area of 45,613 acres. Comparing this with the acreage given for the three preceding years, the cultivation of this crop appears to be rapidly extending. The Agra Canal returns for 1878-79 gave 80 acres under *bharti* in the Gurgaon district of the Panjáb. It is said to be the cheapest grain grown, and will keep many years without being eaten by insects; for this reason it is stored up as a provision against years of scarcity and famine.

Food & Fodder.—An analysis, lately made by Professor Church, shows that the GRAIN of this grass contains much more indigestible fibre than that of any species yet examined, but that it is exceptionally rich in oil or fat since it possesses nearly twice as much of that constituent as any other kind.

Affords excellent fodder both for bullocks and horses.

FOOD &
FODDER.
Grain.
52

Panicum frumentaceum, *Roxb.; Duthie, Fodder Grasses, 8.*

53

Syn.—OPLISMENUS FRUMENTACEUS, *Kunth;* ECHINOCHLOA FRUMENTACEA, *Link.*

Vern.—*Shamula, sáwa, sánwa, shama, samak,* HIND.; *Shamula, sanwa, saon, samra-shama, syámá dhán,* BENG.; *Samu,* URIYA; *Sánwán, sáwán, sáma, sama,* BIHAR; *Sáwan, sánwan, sáwan-bhadeha, sama, samei,* OUDH; *Jhúngara, sáman, sáwan,* N.-W. P.; *Mandira, jhangora, koni, kungni,* KUMAON; *Jhungara,* GARHWAL; *Soák, karin,* KASHMIR; *Samuka, samá, sánwak, sawank, soák, chandra,* PB.; *Eaon saron,* SIND; *Sawa, sema,* C. P.; *Bávto,* BOMB.; *Shamula, kathi, kangra, sawa, sanwa, saon, shama, kathli,* DEC.; *Bonta-shama, bonth-shama, sawa, bonta chámalu, chámalu, chama,* TEL.; *Sáme, sáve,* KAN.; *Welmarukku,* SING.; *Syámák, shyámaka,* SANS.; *Bajri,* PERS.

References—*Roxb., Fl. Ind., Ed. C.B.C., 102; Dals. & Gibs., Bomb. Fl., Suppl., 98; Stewart, Pb. Pl., 257; Aitchison, Cat. Pb. & Sind Pl., 161; Elliot, Fl Andhr., 30, 33; U. C. Dutt, Mat. Med. Hind., 320; Baden Powell, Pb. Pr., 237; Atkinson, Him. Dist., 689; Duthie & Fuller, Field & Garden Crops, pt. II., 3; Lisboa, U. Pl. Bomb., 184, 276; Birdwood, Bomb. Pr., 111; Royle, Ill. Him. Bot., 421; Church, Food Grains of India, 49; Balfour, Cyclop., III., 111; Rep. Agri. Station, Cawnpore, 1884, 7; Gazetteers:—Mysore & Coorg, II., 11; C. P., 50; Gujrát, 79; Musaffargarh, 93; Hoshiarpur, 94; Meerut, 225; Settle. Repts:—Upper Godávery, 35; Bareilly, 82; Kángra, 25; Azamgarh, 114; Chánda, 81; Betul, 63; Lahore, 10; Grierson, Bihar Peasant Life, 226 (under P. miliaceum); F. von Mueller, Select Extra-Trop. Pl., 274.*

Habitat.—Extensively cultivated as a rainy season crop over the greater part of India, and on the Himálayan range up to 6,500 feet.

CULTIVATION.—This is the quickest growing of all the millets, and in some localities can be harvested within six weeks after being sown. It thrives best in light sandy soils, and is often cultivated, when the rains are over, on the banks of rich silt deposited by rivers. According to Roxburgh it yields fifty-fold in a good soil. In the North-Western Provinces it is generally sown at the commencement of the rainy season, and a spring crop usually follows it. It is grown most extensively in the Rohilkhand and Benares Divisions in the tracts bordering the hills. The seed is sown at the rate of 10℔ to the acre, and the young plants require at least two weedings. In the drier districts of the Doáb it is oftener grown as a subordinate crop along with *juár*. The crop is liable to damage from excessive rains, and suffers from blight, but is, as a rule, a generous one, producing from 8 to 10 maunds of grain to the acre (*Field & Garden Crops*). Mr. Atkinson (*Him. Districts*) says that on the Himálaya it is sown in July and gathered in September. It is not much grown in the Panjáb, except near the hills, in the eastern districts. In Gujrát it is cultivated as a subordinate crop along with maize, on well irrigated land. In the Muzaffargarh district it is sown in August and September on the large flats of quicksand which border the rivers; and as the mud dries, the plants spring up and produce grain in October. In the Upper Godáveri district of the Central Provinces it is sown in forest

Cultivation.
54

FOOD &
FODDER.
Grain.
55

clearings ; also on river banks, the seed being dropped into the water by men in canoes.

Food & Fodder.—The GRAIN is consumed chiefly by the poorer classes of people, for whom it has a special utility in its ripening early and affording a cheap article of food before *bájra* and other millets. It is eaten either boiled in milk, or parched. It is said to be heating, but becomes more wholesome by keeping. Hindus consider it to be a very pure grain, and it is preferred by them for religious offerings. An analysis of the grain (with husk) made by Professor Church gave the following result :—

			In 100 parts.
Water	.	.	12·0
Albumenoids	.	.	8·4
Starch	.	.	72·5
Oil	.	.	3·0
Fibre	.	.	2·2
Ash	.	.	1·9

Straw.
56

The nutrient ratio is here 1 : 9·5, and the nutrient value 88.

The STRAW is much used in the Madras Presidency and in Mysore as cattle fodder, though considered to be inferior to that of *rági* or rice. In the Meerut district it is sometimes grown as a fodder crop. On the Himálaya the ears are first cut, and the stalks are afterwards given to cattle.

57

Panicum Helopus, *Trin. ; Duthie, Fodder Grasses, 8.*

Syn.—P. SETIGERUM, *Retz.* ; P. HIRSUTUM, *Kœn.* ; P. KŒNIGII, *Spreng.* ; UROCHLOA PUBESCENS, *Beauv.*

Vern.—*Kuri, kuriya,* HIND. ; *Bara jal-ganti, jal-ganti,* BENG. ; *Basaunta,* N.-W. P. ; *Chatia, kowin, thun,* PB. ; *Kuri,* BOMB. ; *Salla-wúdú,* TEL.

References.—*Roxb., Fl. Ind., Ed. C.B.C., 100, 101 ; Voigt, Hort. Sub. Cal., 701 ; Royle, Ill. Him. Bot., 420.*

Habitat.—Common in the plains on rich or cultivated land, and on the Himálaya up to about 5,000 feet. The short awn-like point terminating the fruiting glume is said to be the best character by which it may be distinguished from other species.

FOOD &
FODDER.
Grain.
58

Food & Fodder.—Royle mentions this species as yielding a GRAIN eaten by the poorer classes. In the Baroda district of Bombay *Kuri* is said to be grown as a rainy season crop. It is reckoned as one of the best of Indian fodder-grasses for both horses and cattle.

59

P. jumentorum, *Pers. ; Duthie, Fodder Grasses, 9.*

GUINEA GRASS.

Syn.—P. MAXIMUM, *Jacq.*

Vern.—*Gini gawat,* MAR. ; *Gini ghás,* GUZ. ; *Geneo-pullu,* TAM. ; *Gini hullu,* KAN. ; *Nau-ka-thau-hau,* BURM. ; *Rata-tana,* SING.

References.—*Voigt, Hort. Sub. Cal., 700 ; Thwaites, En. Ceylon Pl., 361 ; Dals. & Gibs., Bomb. Fl., Suppl., 98 ; DC., Origin Cult. Pl., 115 ; Mason, Burma & its People, 477, 816 ; Atkinson, Him. Dist., 320 ; F. von-Müeller, Select Extra Trop. Pl., 275 ; Balfour, Cyclop., III., 112 ; Smith, Dic., 202 ; Rep. Agri. Dept. Madras, 1878, p. 163 ; 1879, p. 105 ; Assam, 1885-86, p. 19 ; Saidapet Farm Man. and Guide (1879), 47 ; Rep. Madras Exper. Farm, 1871, p. 12 ; 1872, p. 21 ; 1875, pp. 29, 30 ; 1877, p. 95 ; 1878, p. 23 ; 1879, pp. 41, 105 ; Rep. Nagpur Exper. Farm, 1885-86, p. 4 ; Rep. Exper. Farm, Sindh, 1887-88, p. 11 ; Rep. R. Gardens, Kew (1879), p. 16 ; Ind. Forester, X., 111, 292 ; Journ. Agri-Horti. Soc., VI. (Old Series), Pt. II., 173 ; VII. (New Series), 111.*

Habitat.—A tall perennial grass, native of Tropical Africa, and now cultivated in most tropical countries.

HISTORY.
60

History.—Guinea grass appears to have been first cultivated in the West Indies, having been grown in Jamaica since the middle of the last

Guinea Grass. (*J. F. Duthie.*) **PANICUM jumentorum.**

century. It was recommended for cultivation in India as early as 1793 by Mr. W. Fitzmaurice in a *"Treatise on the Cultivation of Sugarcane and the Manufacture of Sugar,"* a work published in Calcutta and dedicated to Sir W. Jones. The extract relating to guinea grass will be found in the *Agri.-Horticultural Society's Journal, VI., loc. cit.* As this grass is not mentioned in Roxburgh's works, it could not have been cultivated to any extent before his time. Mr. J. Bell, who was Secretary to the Agri.-Horticultural Society, is reported to have grown it successfully in 1831; and prizes were offered by the Society in 1837 for the best results. The numerous experiments undertaken at Government gardens and model farms, in different parts of India, as well as those carried out by private individuals, sufficiently confirm the general opinion as to the excellent feeding qualities of this useful fodder grass.

Cultivation. 61

CULTIVATION.—Although in India this grass seeds freely in the plains, it is best propagated by root cuttings. A sandy soil is found to be most suitable. After the ground has been prepared in the ordinary way, the roots should be planted out about 2 feet apart on parallel ridges, at the commencement of the rains, care being taken to arrange the plants so that they will form lines at right angles to the ridges. This arrangement will facilitate weeding until the plants have established themselves. As soon as the roots have fairly taken hold of the ground very little watering will be necessary; indeed, the capabilities of this grass for resisting the effects of the severest droughts have been fully tested. For instance, in the *Report of the Madras Agricultural Department for 1878-79*, it is stated:—"A field of guinea grass about 2 acres in extent, planted in September 1875, was, towards the beginning of May, as brown and as dry as if it were totally devoid of life; the heavy rains in that month, however, led to its immediate revival, and before the third day of rain was over, green shoots had appeared throughout the field, which grew on and produced a good crop of fodder in two months." "Not one per cent of the tufts failed to throw out shoots after the rain referred to, thus showing that the plant is capable of withstanding our longest droughts."

The writer of the *Saidapet Farm Manual* makes the following remarks:—"It is perfectly true that if guinea grass is planted in the hot season or during dry weather, the plants will fail completely if not watered a few times; but there seldom can be any necessity for planting under such circumstances. In the driest parts of the country there is always a time when the sky is cloudy and the weather showery, and if such opportunities are properly utilised for planting, the roots can be established sufficiently well without the aid of irrigation. Irrigation, of course, enables more cuttings to be obtained in a year, but is never necessary for the maintenance of the crop. It is, however, most necessary to manure land for guinea grass well." "It is advisable to pass a plough or cultivator occasionally through and across the crop, as the absorptive powers of the soil can in this way be kept up. This ploughing or cultivating should be repeated at any rate once after the removal of each cutting. Before the end of the second year the plants from frequent cutting will have formed large tussucks; these should be reduced by simple chopping with a spade or hoe. It has been found best to make two cuts across the tussucks at right angles to each other, thus dividing it into four parts. Of these, three may be removed and they will form excellent bedding for cattle stalls, the fourth remaining to perpetuate the crop. In this way, there is no necessity to remove the plants to other ground, but care must be exercised to see that the soil is properly manured, as a crop which yields such large returns necessarily makes large demands

PANICM miliaceum.	The Common Millet.

CULTIVATION

on the soil." During the cold weather of Northern India the plants dry up, and remain dormant until the approach of spring. In places where frost is prevalent manure should be applied to the roots at the commencement of the cold season.

FODDER.
62

Fodder.—Guinea grass, when carefully cultivated, yields during the year several cuttings of good fodder suitable for all kinds of stock. "At first," says the writer of the *Saidapet Experimental Farm Manual,* "it seems to disturb the digestive organs of some animals, but this is only temporary; cattle and sheep have been fed on it exclusively for months with the most satisfactory results. A guinea grass field is a capital place in which to graze working cattle during the hot season, while for ewes with young lambs better pasture could scarcely be discovered. It produces an abundant flow of milk in the ewes, without, what is common in such cases, disturbing the health of either mother or lamb. Care must, however, be always observed never to graze guinea grass too closely." In favourable seasons, and when the climate is suitable, it has been found possible to obtain five or six cuttings during the year, and excellent hay may be prepared from the crop which matures at the close of the rainy season. In Assam, according to the report of the Agricultural Department for 1885-86, the weight of six cuttings was equivalent to 500 maunds per acre, and the cost of cultivating one acre was estimated at R54. In the Coimbatore Municipal Garden the cuttings, obtained during the year 1876, were reported to have averaged 960 maunds per acre. In 1881, in the Government Garden at Saharunpur, the yield was at the rate of 600 maunds per acre. Mr. Fuller in his report on the *kharif* season of 1880, at the Cawnpore Model Farm, says that a single cutting will yield as much as 180 maunds of green fodder to the *pakha bíghá.*

63

Panicum miliaceum, *Linn. ; Duthie, Fodder Grasses, 9, Pl. A., Fig. 2.*
COMMON MILLET.

Syn.—P. MILIUM, *Pers.* ; P. ASPERIMUM, *Lag.*

Vern.—*Chena, chin,* HIND.; *Chiná,* BENG.; *Chinna, china, chinh,* BIHAR; *Phikai, rali, bansi,* BUNDEL.; *Chehna, chinwa, chirwa, sáwan-chaitwa, sáwan-jethwa, kuri,* N.-W.P.; *Chinwa,* KASHMIR; *China, chini, chena, anne, sálan, sálar,* PB.; *Tsedse,* LADAK; *Chinu,* SIND.; *Wadi, sáwa, varísáva, chino, chenah, sáma, varika-anu, vari,* BOMB.; *Vari, sáva, barag,* MAR.; *Vari, chino, sámli,* GUZ.; *Sáva, sawi, wári, shamakh,* DEC.; *Katakanai, varagu,* TAM.; *Worga, varagalu,* TEL.; *Bili baragu, karibaragu, sáve, baragu,* KAN.; *Mainairi,* SING.; *China, chini, rad, vrihib-heda, anu,* SANS.; *Worga, worglo, dokhu,* ARAB.; *Arsan,* PERS.

References.—*Roxb., Fl. Ind., Ed. C.B.C., 104 ; Stewart, Pb. Pl., 258 ; Aitchison, Cat. Pb. & Sind Pl., 159 ; Bot. Afghan Del. Com., 122 ; DC., Origin Cult. Pl., 376 ; U. C. Dutt, Mat. Med. Hind., 295 ; Bidie, Cat. Raw Pr., Paris Exh., 71 ; Baden Powell, Pb. Pr., 237 ; Atkinson, Him. Dist., 688 ; Duthie & Fuller, Field & Garden Crops, Pt. II., 1 ; Lisboa, U. Pl. Bomb., 184, 276 ; Birdwood, Bomb. Pr., 111 ; Royle, Ill-Him. Bot., 420 ; Church, Food Grains of India, 42 ; F. von-Mueller, Select. Extra-Trop. Pl., 275 ; Balfour, Cyclop., III., 112 ; Gazetteers :—Mysore & Coorg, I, 68 ; II., 11 ; Bombay, IV., 53 ; VII., 30 ; XIII., 289 ; XVI., 91 ; N.-W. P., I., 85 ; III., 225 ; IV., lxxix ; Gurgáon, 86 ; Hoshiárpur, 94 ; Simla, 57 ; Gujerát, 79 ; Kángra, I., 158 ; Montgomery, 115 ; Jhang, 115 ; Settle. Repts. :—Simla, App. xxxix. ; Jhang, 85, 94 ; Lahore, 10 ; Bareilly, 82 ; Montgomery, 107 ; Chánda, 81 ; Moore, Man., Trichinopoly, 71 ; Nicholson, Man., Coimbatore, 221 ; Buchanan, Journey through Mysore and Canara, I., 290, 371, 381, 411 ; Grierson, Bihár Peasant Life, 227 (under P. frumentosum).*

Habitat.—This small millet is easily recognised by its bright green foliage, and spikelets of flowers arranged in lax drooping panicles. It is supposed to have been introduced originally from Egypt or Arabia, but

| The Common Millet. | (*J. F. Duthie*.) | **PANICUM miliare.** |

its cultivation in India probably dates from a very early period. It is grown in various parts of the country, even up to 10,000 feet on the Himálaya, but nowhere to any great extent.

CULTIVATION.—In the plains of the North-Western Provinces *Chehna* is almost invariably grown as a hot weather crop and is irrigated from wells. It is sown in March at the rate of about 10℔ to the acre, and the crop ripens towards the end of May, during which period it may have required as many as fourteen waterings. It is liable to be damaged by the hot winds which are prevalent at this season, the grain being quickly scattered when fully ripe. Six to eight maunds of grain per acre is considered a fair average yield. In Bundelkhand there are two varieties, called *phikai* and *rali*, both of which are grown as rainy season crops. The former is sown a little earlier than the other and yields a heavier outturn (*Field & Garden Crops, II., 1*). On the Himálaya it is cultivated near villages as a summer crop, the grain being used for home consumption. In the Panjáb it is chiefly grown in the northern districts, both as a spring and as an autumn crop. The latter season is usually preferred, there being less wind at that time. In many districts the crop is cut as green fodder before it ripens. In the Baroda district it is sown on irrigated land in January. It takes 45 days to ripen, and requires as many as fifteen waterings. In the Panjáb the crop is sometimes attacked by *tela* and *kummi*, the latter being caused by a small red, tortoise-shaped insect, which feeds on the pollen.

Food & Fodder.—The GRAIN is considered to be digestible and nutritious, and in many places is eaten unground and cooked like rice. In Bihár, it is called, when boiled and parched, *marha*, *mánhra* or *már*. Prepared with milk and sugar it constitutes a favourite food at marriage ceremonies; it is also eaten by Hindus during their religious fasts. In the neighbourhood of Simla it is sometimes used as bread in the form of chapatties, called *chinatti*. The following is given by **Professor Church** as the chemical composition of the grain :—

	In 100 parts.
Water	12·0
Albumenoids	12·6
Starch	69·4
Oil	3·6
Fibre	1·0
Ash	1·4

The nutrient ratio is here 1 : 6, and the nutrient value 89.

This millet in the green state affords excellent fodder for cattle and horses, and in parts of the Panjáb is not unfrequently grown for this purpose only; or it has occasionally to be used as such should the rains fail, and grass fodder become scarce.

The dry STRAW, called *prál* or *práli* in the Montgomery district, is sometimes given to cattle, but is considered heating. In Northern India, as a rule, it is either used as bedding, or as a contribution to the manure heap. In Mysore and in the Madras Presidency, however, the straw is regarded as being nearly equal in value to that of rice.

Panicum miliare, *Lamk.; Duthie, Fodder Grasses, 10, Pl. xlvi.*

LITTLE MILLET.

Vern.—*Kungu, kutki,* HIND.; *Gondula,* BENG.; *Gundli,* SANTAL ; *Kutki, mighri,* N.-W. P.; *Kútki,* PB.; *Chika,* C.P.; *Warai,* BOMB.; *Shamai, samai, chámai,* TAM.; *Nella-shama, nella-shamalu, nalla-chámalu, saumai,* TEL.; *Menéri,* SING.

References.—*Roxb. Fl. Ind., Ed. C.B.C., 104, Stewart, Pb. Pl., 259; Aitchison, Cat. Pb. & Sind Pl., 159; Elliot, Fl. Andhr., 124; Bidie, Cat. Raw Pr., Paris Exh., 71; Atkinson, Him. Dist., 320; Duthie &*

PANICUM miliare.	**The Little Millet.**

Fuller, Field & Garden Crops, 7 (under P. psilopodium) ; *Birdwood, Bomb. Pr., 111* ; *Royle, Ill. Him. Bot., 421* ; *Church, Food-Grains of India, 44* ; *Balfour, Cyclop., III., 112* ; *Gazetteers :—Bombay, VII., 30; X., 147 ; XI., 96 ; XVI.. 91 ; N.-W.P., I., 85 ; IV., lxxix. ; Settlement Rep., Betul, 63 ; Madras Man., Administration, I, 288 ; Nicholson, Man., Coimbatore, 221 ; Buchanan, Journey through Mysore and Canara, I., 106, 287, 376 408 ; II., 105, 223, 452 ; III., 440.*

Habitat.—This is another of the minor millets, smaller in all its parts than *chena* (**P. miliaceum**). It is cultivated to some extent by the poorer people and aboriginal tribes of Central India and the Central Provinces. A nearly allied species, **P. psilopodium,** *Trin.,* grows wild in Northern India, and up to moderate elevations on the Himálaya. It is cultivated in the same way as **P. miliare,** under the name of *mijhri,* and may possibly constitute one of its so-called varieties.

Cultivation.
68

CULTIVATION.—This crop is usually sown at the commencement of the rainy season in the poorer soils of the elevated tracts of Central and Southern India. In Mysore it is sown in the worst descriptions of soils, and manure is given only when it can be spared. It ripens in three months, is then cut close by the ground, and stacked, and after five or six days is spread out and the grain trodden out by oxen. There are said to be three distinct varieties in Mysore, which are never inter-mixed.

In Coimbatore, according to **Mr. Nicholson,** *samei* is grown mainly as a dry crop, and is sown usually in August-September, but in some places May-June is the sowing season. It is either grown by itself or mixed with various pulses, and is reaped before the latter ripen. There are several varieties, one of which, known as *pillu-samei,* has a superior grain, equal, it is said, to the best rice in delicacy. (This is probably **P. miliaceum** or some other distinct species.) *Samei* is weeded, but not interploughed, about four weeks after sowing. The maximum yield is said to be 400 measures for a garden crop, and somewhat less for a dry crop. Only two or three waterings are required, and when grown in gardens manure is freely applied.

In the *Report of the Madras Agricultural Department, 1871,* it is described as a useful dry crop for sandy soils. It is sown, 10℔ to the acre, in October and harvested in February. The average yield per acre is given as 270℔, and its value about 20℔ per rupee. In the North-West Provinces its cultivation is chiefly confined to the southern hilly districts. It is sown in June and reaped in October, forming, together with *kodon,* the crop which is generally taken from the poorest land in the village. Indeed it is often grown on soils, which could hardly produce a crop of *kodon* (*Field & Garden Crops*).

FOOD & FODDER. Grain.
69

Food & Fodder.—The GRAIN is of an inferior kind, and is mostly consumed by the poorer classes of people. It is sometimes boiled whole like rice or ground into flour for cakes (*Buchanan, Journey, I., 106*). An analysis by **Professor Church** gave the following results :—

Water	10·2
Albumenoids	9·1
Starch	69·0
Oil	3·6
Fibre	4·6
Ash	3·5
	100·0

Straw.
70

The nutrient ratio is 1 : 8·4, and the nutrient value 85.

Cattle are very fond of the STRAW, which is largely used in Southern India as fodder. It is considered, however, inferior to that of rice and *rági.*

P. 70

Domestic.—Buchanan remarks that in Mysore the STRAW of this grass is considered to be the best for stuffing pack-saddles.

Panicum prostratum, *Lamk. ; Duthie, Fodder Grasses, 11, Pl. xlv.*
> **Syn.**—P. PROCUMBENS, *Nees.*
> **Vern.**—*Chaurila,* BUNDEL. ; *Choti semai. sarpur,* C. P.
> **References.**—*Dals. & Gibs., Bomb. Fl., 290 ; Aitchison, Cat. Pb & Sind Pl., 160 ; F. von Mueller, Select Extra-Trop. Pl., 216.*
> **Habitat.**—A perennial with creeping stems which root at the nodes; common all over the plains. It assumes a more erect habit of growth in rich cultivated ground.
> **Food & Fodder.**—The GRAIN is collected as food in times of scarcity, while the PLANT is one of the most nutritious of Indian fodder grasses.

P. repens, *Linn. ; Duthie, Fodder Grasses, 11.*
> **Vern.**—*Etora,* SING.
> **References.**—*Roxb., Fl. Ind., Ed. C.B.C., 110 ; Thwaites, En. Ceyl. Pl., 360 ; Trimen, Syst. Cat. Ceyl. Pl., 107 ; Royle, Ill, Him. Bot., 422 ; F. von Mueller, Select Extra-Trop. Pl., 276.*
> **Habitat.**—A perennial species with glaucous-tinted foliage, not uncommon in wet places in the plains. It is found also in South Europe, North Africa, and Australia, as well as on the coast of Brazil.
> **Fodder.**—Both Roxburgh and Royle have stated that cattle are fond of this grass. In Ceylon it is said to be much valued as a good fodder for cattle.

P. sanguinale, *Linn. ; Duthie, Fodder Grasses, 12.*
> **Syn.**—P. ÆGYPTIACUM, *Retz. ;* DIGITARIA SANGUINALIS, *Scop. ;* DACTYLON SANGUINALE, *Vill. ;* PASPALUM SANGUINALE, *DC.*
> **Vern.**—*Takri takria,* HIND. ; *Kewai, charmara,* N.-W. P. ; *Khurásh* (Trans-Indus), *moti khabbal, takkri, farw, dubra, bara takria,* PB. ; *Hen,* RAJ. ; *Korkol-jodi,* C.P. ; *Chikhari,* BERAR.
> **References.**—*Roxb., Fl. Ind., Ed. C.B.C., 97 ; Dals. & Gibs., Bomb. Fl., 291 ; Stewart, Pb. Pl., 254 ; Aitchison, Cat. Pb. & Sind Pl., 161 ; F. von Mueller, Select Extra-Trop., Pl., 277 ; Ind. Forester, XII., App., 23.*
> **Habitat.**—A very common grass in India, and in many other countries. Easily recognised by its finger-like spikes of flowers terminating the long peduncle. Its general resemblance to a magnified form of *dúb* is indicated in some of the vernacular names. Its typical form is confined chiefly to the Himálaya.
> **Fodder.**—Universally regarded as an excellent fodder GRASS, whether green or dry, both for horses and cattle. In the United States of America it is also highly valued under the name of "brab grass."

Var. ciliare.
> **Syn.**—P CILIARE, *Retz.*
> **Vern.**—*Makur-jali,* BENG. ; *Sinri, kabdai, kewai,* N.-W. P. ; *Dobra,* PB. ; *Chopar,* BALUCH. ; *Chhinke,* RAJ. ; *Mandiya, ráha,* C. P. ; *Shangali-gaddi,* TEL.

In the plains of India this variety is more abundant than the typical form. It differs by having the lateral nerves and margins of the inner glumes more or less densely clothed with white silky hairs.

Pan-leaf, see **Piper betel** in this volume.
(G. Watt.)

PAPAVER, Linn. ; Gen. Pl., I., 51.

Papaver is the Latin classical name for the Poppy, the equivalent of the Greek μήχων.

DOMESTIC. Straw.
71
72
FOOD & FODDER. Grain.
73
Plant.
74
75
FODDER.
76
77
FODDER. Grass.
78
79
80

P. 80

PAPAVER Rhœas.	The Red Poppy.

81

Papaver dubium, *Linn.; Fl. Br. Ind., I., 117;* PAPAVERACEÆ.

References.—*Aitchison, Afghan Delim. Com. Report; Stewart, Journal of a Tour in Hazára; Atkinson, Him. Dist., 304; Royle, Ill. Him. Bot., 67.*

Habitat.—A species more frequently met with in India than the allied form P. Rhœas, though occasional in the Western Himálaya from Garhwál and Kumáon to Hazára, Afghánistán, Baluchistán, Persia, and Europe. It is, in fact, separable from P. Rhœas by minor characters, such as its more glabrous nature, narrower leaf segments, the hairs on the scape being appressed, and the capsule sessile.

It is fairly abundant in the wheat fields of the temperate tracts of India, and is even found as a cold season weed in the plains. It nowhere, however, attains such a degree of prevalence as in Europe, and the flowers are often not more than half an inch in diameter, the plant being correspondingly diminutive.

82

P. Rhœas, *Linn.; Fl. Br. Ind., I., 117.*

RED POPPY OR CORN ROSE.

The classical name of this species (ῥοίας) gave origin to its botanical and to most of the vernaculars by which it is distinguished by popular writers. It is probably also the μήκων ῥοίας of Dioscorides. The red poppy is recognised as distinct from the opium-yielding species by even the most primitive of savage races and bears its own distinctive names. In the plains of India, these literally mean " red poppy." It is, however, doubtful how far P. Rhœas is separately recognised from P. dubium, and of the two forms the latter is undoubtedly the more prevalent in India; indeed, the former can only be said to exist in a wild state in Kashmír, while the latter occurs throughout the temperate tracts in wheat and barley fields, but nowhere to such an extent, nor so prolific, as the red-poppy of Europe.

Vern.—*Lálá, lál-póst,* HIND.; *Lál-póshta, lál-poshtér-gáchh,* BENG.; *Jang-li-mudrika,* BOMB.; *Támbada-khasa-khasá-cha-jháda,* MAR.; *Lálá, lál-khas-khas-nu-jháda,* GUZ.; *Lál-khash-khash-ká-jhár,* DEC.; *Shivappu-gasha-gashá-chedi, shigappu-póstaká-chedi,* TAM.; *Erra-gasa-gasala-chethe, erra-posta-káya-chethe,* TEL.; *Kempu-khasa-khasi-gída,* KAN.; *Chovanna-kasha-kashach-cheti,* MALAY.; *Bhin-bin-ami,* or *bh-ain-bin-ami,* BURM.; *Rakta-pósta-vrikshaha,* SANS.; *Nabatúl-khash-khashul-ahmar,* ARAB.; *Kóknáre-surkh,* PERS.

References.— *O'Shaughnessy, Beng. Dispens., 171; Moodeen Sheriff, Supp. Pharm. Ind., 192; Pharmacog. Indica, I., 108; Dymock, Mat. Med. W. Ind., 2nd Ed., 48-49; Flück. & Hanb., Pharmacog., 39; U. S. Dispens., 15th Ed., 1243; Bent. & Trim., Med. Pl., 19; Year-Book Pharm., 1873, 493, 495; 1874, 622; 1878, 288; Kew Off. Guide to the Mus. of Ec. Bot., 12; Agri.-Hort. Soc. Ind., Transactions, VII., 45; Gazetteer, Mysore & Coorg, I., 57.*

Habitat.—Met with in Kashmír and Pangi, but frequently cultivated in gardens throughout the plains of India, and often, as at Patna in Behar, seen as an escape from gardens in the neighbouring wheat fields.

MEDICINE.

Medicine.—The red poppy seems to be the *Khashkhásh-i-Mansúr* of the Arabs and Persians, which gets the name *mansúr* from the ease with which it sheds its petals. It is by the Indian Muhammadans viewed as the *Lálá* of their Persian poets, and by the Hindus the capsules are often designated *Mudrika,* from their resemblance to the *Mudra,* a seal used to give impressions on the forehead. All these characters might, however, apply with equal force to P. dubium, a plant which is plentiful in Afghánistán and Persia, where P. Rhœas only doubtfully occurs. The red poppy is, however, frequently cultivated in gardens, and in India the PETALS are sometimes collected and used in colouring drugs, much after the fashion of their former employment in Europe. The MILK from the CAPSULES is narcotic and has slightly sedative properties.

Petals.
83
Milk.
84
Capsules.
85

P. 85

The Opium or White Poppy.	(G. Watt.)	PAPAVER somniferum.

CHEMISTRY.—The milky sap from the capsules has, in a weak form, the same properties as opium, and M. Filhol even held that it contained morphine in exceedingly minute proportions, and the same alkaloid, Chevallier thought he had detected in an extract prepared from the petals. This statement has, however, been denied by Attfield, who could not detect the alkaloid in an extract prepared from a pound of the petals. The colouring principle is that for which the petals are chiefly valued, and, according to Meier, this is due to two acids, which he denominates *rhœadic* and *papaveric*.

CHEMISTRY.
86

O. Hesse has, however, devoted the greatest attention to the chemistry of this poppy, and has discovered an alkaloid in it for which he assigns the name *rhœadine*. It was found to pervade all the parts of the plant. It may be isolated in small white prismatic crystals, tasteless, fusible at 232·2°C. It is almost insoluble in water, alcohol, ether, chloroform, benzol, ammonia, solution of carbonate of sodium, and lime water, but is dissolved by dilute acids which produce from it colourless solutions. It may be represented by the formula of $C_{21}H_{21}NO_6$. It does not appear to be poisonous. It may be detected by its reduction with hydrochloric and sulphuric acids moderately concentrated, giving a deep purple colour to water even when diluted to 800,000 parts. By this means traces of it may be detected in all parts of P. Rhœas, in the ripe capsules of the opium poppy, and in opium itself. According to Hesse, the milky sap of P. Rhœas also contains *Meconic Acid*.

SPECIAL OPINION.—§ " Profuse in Kashmír, where it is collected. When cows are fed on it, their milk becomes flavoured with opium " (*Surgeon-Major J. E. T. Aitchison, Simla*).

Papaver somniferum, *Linn.; Fl. Br. Ind., I.,* 117.

THE OPIUM, OR WHITE POPPY.

87

Vern.—*Afiyun, afim, kashkásh, post,* HIND.; *Post,* BENG.; *Aphim,* NEPAL; *Posta,* OUDH; *Posht,* KUMÁON; *Khash-khash, post, doda, afim, khish-khash,* PB.; *Aphim, appo, khaskhas, post,* BOMB.; *Aphu, posta, khuskhus (afú-ke-thar),* MAR.; *Aphina, posta, khuskhus,* GUZ.; *Afim, khashkhash-ké-bóndé, khashkhash,* DEC.; *Abini, gashagasha, postakatol, gashagasha-tól, casa-casa,* TAM.; *Abhini, gasagasála-tólu, gasagasálú, casa-casa,* TEL.; *Khasakhasi, gasagase, afim,* KAN.; *Kashakasha-karuppa, kashkashat-tól, kasha-kashak kuru, afiun,* MALAY.; *Bhain, bhain-si,* BURM.; *Abin,* SING.; *Ahiphena (c'hosa,* according to Fleming), *(khaskhasa, póstubéjam?),* SANS.; *Afiun, qishrul-khashkhásh, bisrul-khashkhash, abunom,* ARAB.; *Khashkhásh, afiún, póstékóknár, tukhme-koknar,* PERS.

NOTE.—The distinction into Opium, Poppy seeds, Poppy capsules, Poppy plant, and beverage from Poppy, has not been attempted in the above enumeration of vernacular names.

References.—*Boiss., Fl. Orient., I.,* 116 *; Roxb., Fl. Ind., Ed. C.B.C.,* 426 *; Voigt, Hort. Sub. Cal.,* 5 *; Dalz. & Gibs., Bomb. Fl., Supp.,* 3 *; Stewart, Pb. Pl.,* 10 *; Aitchison, Cat. Pb. & Sind Pl.,* 4 *; Kuram Valley Rept., Pt. I.,* 9 *; Rept. Pl. Coll. Afgh. Del. Com.,* 32 *; DC., Orig. Cult. Pl.,* 397 *; Graham, Cat. Bomb. Pl.,* 6 *; Mason, Burma & Its People,* 487 *; Enumeration Plants of China Proper by Forbes & Hemsley (Linn. Soc. Jour., Vol. XXIII.,* 34) *; Hooker, Himálayan Journals, I.,* 86 *; Pharm. Ind.,* 13, 15-23 *; British Pharm.* (Ed. 1885), 295-297, 299 *; Flück. & Hanb., Pharmacog.,* 40-64 *; U. S. Dispens., 15th Ed.,* 977, 1054-1078, 1082 *; Martindale, Extra Pharm.,* 20, 151, 236-241, 244, 259 *; Fleming, Med. Pl. & Drugs (Asiatic Res., XI.),* 174 *; Ainslie, Mat. Ind., I.,* 271, 326 *; II.,* 339 *; O'Shaughnessy, Beng. Dispens.,* 170-183 *; Irvine, Mat. Med. Patna,* 6, 77 *; Moodeen Sheriff, Supp. Pharm. Ind.,* 87, 190, 194, 358 *; Mat. Med. S. Ind. (in MSS.),* 18 *; U. C. Dutt, Mat. Med. Hindus,* 112, 289 *; Sakharam Arjun, Cat. Bomb. Drugs,* 9 *; K. L. De, Indig. Drugs, Ind.,* 82-83 *; Murray, Pl. & Drugs,*

Sind, 75 ; *Bidie, Cat. Raw Pr., Paris Exh., 2, 84, 103 ; Waring, Bazar Med., 104 ; Bent. & Trim., Med. Pl., t. 18 ; Dymock, Mat. Med. W. Ind., 2nd Ed., 39-48 ; Dymock, Warden & Hooper, Pharmacog. Ind., Vol. I., 73-108 ; Official Corresp. on Proposed New Pharm. Ind., numerous references ; Watts, Dic. Chemistry, IV., 207-210, 338 ; VI., 882, 896 ; VII., 875-877 ; VIII., 1437 ; Johnston (Church Ed.), Chemistry of Common Life, 309-331 ; Report of Opium Commission, Part IV., Rules, &c., for Guidance of Officers of Patna & Ghasipur Opium Factories, 1883 ; J. Scott, Man. Opium Husbandry ; Selections, Records, Bengal Govt., No. 1 (1851) ; The Opium Poppy by Dr. Eatwell ; also No. XXV. of 1857 ; Notes by Dr. Lyall ; Gennoe, Notes on Poppy Cultivation ; Selections, Records, Financial Department, 1871 ; Blight in Poppy Crop in the Behar & Benares Agencies ; Selections, Records, N.-W. P. Govt., Vol. VI. (1873), 429 ; Birdwood, Bomb. Prod., 6, 18, 125, 198, 275 ; Baden Powell, Pb. Pr., 293 ; Drury, U. Pl. Ind., 327 ; Atkinson, Him. Dist. (Vol. X., N.-W. P. Gas.), 304, 704, 757 ; Duthie & Fuller, Field & Garden Crops, 64-68 ; Useful Pl. Bomb. (Vol. XXV., Bomb. Gas.), 144 ; Econ. Prod. N.-W. Prov., Pt. V. (Vegetables, Spices, & Fruits), 27 ; Royle, Prod. Res., 102-107 ; Cooke, Oils & Oilseeds, 63-64 ; Hawkes, Report on the Oils of South India, 19, 20, 31 ; Tropical Agriculture, 422, 430 ; Ayeen Akbary, Gladwin's Trans., Vol. I., 82 ; II., 44 ; Ain-i-Akbari, Blochmann's Trans., Vol. I., 378, 384 ; Linschoten, Voyage to East Indies (Ed. Burnell, Tiele & Yule), Vol. II., 112 ; Milburn, Oriental Commerce (1825), 293-4 ; Hove's Journal of a Tour in Bombay in 1787, 59-60, 111 ; Man. Madras Adm., Vol. I., 456-59 ; II., 291 ; Nicholson, Man. Coimbatore, 8 ; Boswell, Man. Nellore, 305 ; Madden, Official Rept., Kumáon, 279 ; Bomb. Man., Rev. Accts., 102 ; Quarterly Jour., Agri., X., 37 ; XI., 120 ; Adm. Rept., C. P., 1866-67, 69 ; Settlement Reports :—Panjáb, Dera Ismail Khan, 75 ; Jhang, 85 ; Kangra, 25, 28 ; Montgomery, 107 ; Dera Ghasi Khan, 10 ; Lahore, 8 ; Hasára, 93 ; N.-W. P., Hardoi, 15 ; Kumáon, App., 33 ; Asamgarh, 135 ; Shahjehanpur, 15 ; Allahabad, 35 ; Central Provinces, Nimar, 199 ; Baitool, 68 ; Chhindwárá, 28 ; Upper Godavery, 36 ; Nagpur, 274 ; Wardha, 60 ; Administration Report, Bombay, 1889-90, 143-144 (instructive statement) ; Administration Report, Bengal, 1889-90, 34-36 with three maps showing produce per bighá of land irrigated, and also percentage of cultivation, 331 ; Gazetteers :—Bengal, VII., 307 ; XI., 145, 287-292 ; XIII., 92-98, 269-271 ; XV., 93-101 ; XVI., 350-352 ; Bombay, VII., 97, 152, 430 ; XII., 167 ; Panjáb, Montgomery, 19 ; Shahpur, 65 ; Dera Ghasi Khan, 82 ; Simla, 56 ; N.-W. P., I., 79 ; II., 505 ; IV., 525, lxvii. ; Oudh, I., 163, 235, 415 ; III., 81, 419 ; Central Provinces, 50, 290, 328, 385 ; Mysore & Coorg, I., 57 ; II., 11 ; Rajputana, 98, 109, 151, 227, 269 ; Agri.-Horti. Soc. Ind. :—Trans., I., 50, 51, 54-58 ; II., 226, App. 314 ; V., 1119 ; VI., 1-3 ; VII., 45, Pro., 95, VIII., 9 ; Journals (Old Series), II., 319-322, Sel., 465 ; IV., 199, Sel., 151, 211 ; VI., 34 ; VIII., 143 ; XII., 354, 355, 413-417, Pro., 46-48 ; XIII., 390, Pro. (1862), 2, 17 (1863), 9, 10 ; XIV., 113-119 ; (New Series), II., Pro. (1870), 10 ; V., Pro. (1875), 28 ; Ind. Forester, II., 292 ; Spons' Encycl., II., 1308-1324 ; Encyclop. Brit., XVII., 787-794 ; Balfour, Cyclop. Ind., III., 28-39 ; Ure, Dict. Ind., Arts. & Man., III., 312 ; Smith, Ec. Dict., 335, &c., &c.*

Habitat.—DeCandolle says that botanists are agreed in regarding the opium-yielding poppy to be a cultivated state of **Papaver setigerum**,—a species which is wild on the shores of the Mediterranean, notably in Spain, Algeria, Corsica, Sicily, Greece, and the Island of Cyprus. He adds, "Its cultivation must have begun in Europe or in the north of Africa. In support of this theory we find that the Swiss lake-dwellers of the stone age cultivated a poppy which is nearer to **P. setigerum** than to **P. somniferum**. Heer has not been able to find any of the leaves, but the capsule is surmounted by eight stigmas, as in **P. setigerum**, and not by 10 or 12, as in the cultivated poppy. This latter form, unknown in nature, seems therefore to have been developed within historic times. **P. setigerum** is still cultivated in the north of France, together with **P. somniferum**, for the sake of its oil-seed."

| The Opium or White Poppy. | (*G. Watt.*) | **PAPAVER somniferum.** |

In India the opium-poppy is grown here and there throughout the country, especially so in Native States, but its cultivation is mainly confined to three centres, which afford the following opiums :—"Patna Opium" in Behar, "Benares Opium" in the North-West Provinces, and "Malwa Opium" in Central India and certain parts of Rájputana. It nowhere manifests in India a tendency to escape from cultivation, still less to become naturalised; it thus confirms so far the historic facts which point to its introduction within comparatively recent times, though long anterior to the British conquests and influence.

Oil.—In France a form of poppy is specially grown on account of its oil-yielding SEED. In India this is very rarely the case, permission to grow the plant being, in British India, governed by the regulations regarding opium. The seed obtained from the ripe capsules (after the removal of the drug) constitutes, however, a regular article of trade. It is also to some extent eaten or employed in the preparation of certain curry-powders. At the same time poppy-seed oil is edible or is employed for illumination, and the cake left after the expression of the oil is an article of diet amongst the poor or is used for feeding cattle. The yield of oil is said to be proportioned to the freshness of the seed, and as much as 14 ozs. are sometimes obtained from 4℔. The oil readily bleaches by exposure to the sun, and becomes transparent and almost tasteless. **Mr. Bingham** wrote an account of the oil and seed in connection with Sasseram (*Jour. Agri.-Hort. Soc. Ind., XII., 345*); he there states that the yield of oil is about 30 per cent. It has no intoxicating properties; the seed has a sweet taste and is eaten parched, and largely employed by the sweetmeat-makers as an addition in their wares. In 1862 oil sold at about 5 seers to the rupee. "It appears to me that this oil, if properly prepared, would, from its thin and limpid character, be admirably adapted to supersede many of the purposes, if not all, where the more expensive olive oils of Southern France and Italy are now used, and would be an admirable watch-maker's oil." Hawkes furnished for 1855 more detailed information regarding the Malwa source of this oil :—" There are three lakhs of *bíghás* under poppy cultivation in Malwa, the average produce of seed per *bíghá* being 2 maunds, each of 40 seers, or 82℔, which gives a total of 6,00,000 maunds of seed, worth from R1 to 1¼ per maund. From these six lakhs, deduct 1¼ seers per *bíghá* required for seed, and there remains 5,90,623 maunds of seed, for oil. The oil extracted from one maund of seed being about 13 seers, the above quantity of seed would yield a total of 1,91,952 maunds of oil, which sells at from R4-8-8 to R5 per maund, or £40-6 per ton. The whole of the oil at present made here appears to be consumed in Malwa. **Mr. Anson,** the First Assistant to the Governor General and Superintendent of the Opium Department at Indore, concludes that 5,000 maunds might be available for export at Indore alone" Hawkes then discussed the advantages of the two possible routes by which the seed or oil could pass from the "Malwa" poppy fields, namely, to Bombay, or to Calcutta; and he concluded his account of this oil in the following words :—" By simple exposure to the rays of the sun, in shallow vessels, this oil is rendered perfectly colourless. It is supposed by the Natives to produce sleep and to strengthen the brain. Poppy oil is peculiarly suitable for mixing with paints; 'with white lead it leaves a beautiful surface which does not afterwards change by the action of light into a dirty yellow'" (*Hawkes,* 31-32). "In Oudh, each *rayat* sows from two to four *bíghás* in the month of October. The oil is extracted by the common native press. The cost of the seed is 10 seers for the rupee, and the oil sells for 3 seers for the rupee; two-fifths of the weight of the seed employed is about the proportion of oil yielded by the native process. The poppy seed is eaten by the

OIL.
Seed.
88

2 A

PAPAVER
somniferum.   Trade in Poppy Seed.

OIL.

Natives made into sweetmeats, provided the opium has been extracted from the seed vessels, otherwise, it is bitter and narcotic, and under these circumstances the oil extracted is also bitter" (*Simmonds, Trop. Agri., 423*).

One or two similar brief passages occur in Indian works on this subject, but for the past forty years no writer of any note appears to have dealt with poppy seed or oil. All that has appeared on the subject might, in fact, be characterised as a republication of the statements made by **Hawkes** and **Bingham**, in which the oil is commended as one likely to find a place in European commerce. Silently it has, however, advanced, and, at the present moment, it holds a distinct place among the oil-seeds of India.

TRADE
in
Seed.
89

TRADE IN POPPY SEED.

The following table exhibits the FOREIGN IMPORTS and EXPORTS to and from India in poppy seed for the past ten years :—

YEARS.					Imports.		Exports.	
					Cwt.	R	Cwt.	R
1880-81	.	.	.	.	892	6,526	579,544	39,76,254
1881-82	.	.	.	.	430	3,096	603,289	39,04,065
1882-83	.	.	.	.	1,201	9,768	571,542	30,26,401
1883-84	.	.	.	.	109	723	514,228	30,30,184
1884-85	.	.	.	.	709	4,348	659,900	40,91,595
1885-86	.	.	.	.	320	1,857	695,097	40,28,685
1886-87	.	.	.	.	583	3,609	612,654	34,10,452
1887-88	.	.	.	.	246	1,769	456,308	27,76,571
1888-89	.	.	.	.	1,198	8,464	730,455	47,61,731
1889-90	.	.	.	.	2,191	16,059	482,893	36,79,626

The fact that India imports from foreign countries any poppy seed at all seems surprising. Taking the last of the above series of years for the subject of special enquiry, it is found that the major portion of the imports came from Persia (*viz.*, 2,161 cwt.) and was received by Bombay (*viz.*, 2,073 cwt.), with the balance to Sind and Madras. It is not known whether any special merit is attributed to the Persian seed, but, **Mr. J. E. O'Conor,** in a letter to the *Editor* on this subject, says that it is believed the Persian seed is imported, as the supply from which much of the Malwa Opium is raised. If this be so the fact is of some interest in the light it throws on the difference between Bengal and Malwa Opiums, and on the discussion, which will be found below, on the botanical nature of the forms of poppy grown in these centres. It is probable that these imports are to some extent, however, the expression only of one of those abnormal but conservative features of trade which adjust the balance of exchange in other commodities. Of the exports for 1889-90 by far the larger quantity went to Belgium (*viz.*, 220,849 cwt.); next in importance came France (192,929 cwt.), followed by the United Kingdom (64,839 cwt.). The largest exporting province appears to be Bombay, which issued 279,981 cwt., and was followed by Bengal as exporting 202,116 cwt. This analysis of the returns of any one of the above years would express very nearly the features of the present trade.

Malwa Seed.

Coastwise.
90

OF THE COASTWISE TRANSACTION—No separate record appears in the trade returns. The interchanges between the provinces along the coast must, therefore, be unimportant and included in the collective list of " Other Seeds " which is supplementary to those separately dealt with.

Internal.
91

OF THE INTERNAL TRADE BY LAND ROUTES—The North-West Provinces are shown to export the largest quantity. In 1888-89, these Provinces consigned to Bombay port 4,98,223 maunds (say 355,873 cwt.), and to

Calcutta 2,10,328 maunds (150,234 cwt.). The internal trade returns for 1889-90 have not as yet been published, so that the imports into Bombay port and Calcutta, given above, were not the sources from which the foreign exports were drawn in 1889-90. An inspection of the returns of foreign trade for 1888-89 reveals the fact, however, that Bombay exported 428,794 cwt., so that by far the larger portion of the Bombay foreign exports in poppy seed were in that year drawn from the North-West Provinces; only a small amount, if any, of the Malwa poppy seed was exported. Rájputana and Central India are, however, shown to have exported a considerable amount of poppy seed to Bombay port. In 1888-89 these exports amounted to 83,083 maunds. Here, again, then the nature of the poppy seed exported from Bombay is of some interest, in connection with the discussion which will be found below, as to the Malwa poppy being possibly the dark-seeded plant, **Papaver somniferum,** *var.* glabrum. But the writer can find no direct evidence on this point, and only offers the suggestion as worthy of consideration. That the poppy seed exported from India is mainly, if not exclusively, drawn from the Patna and Benares Agencies there would seem to be no doubt. Bengal, however, figures second on the list of internal trade in poppy seed. The Province consigned its supplies (available for export) almost exclusively to Calcutta.

POPPY OIL.—There would appear to be little or no export in this substance from India. The oil obtained from the Bengal seed seems to be of superior quality to that from the Malwa plant, since the former is more generally eaten than the latter, Malwa poppy oil is, on the other hand, apparently more largely used for illumination purposes than the Bengal. In the article on **Oils** (in this work) it will be found, however, that poppy oil is recorded under the sections of oils suitable for candles, soap, illumination, food, and medicine. In Europe it appears to be largely employed in adulteration of olive oil and as a substitute for linseed and other oils in the preparation of paint, especially artists colours.

Medicine.—Opium is so well known to the medical profession that it would be quite beside the character of this work to deal with its therapeutic properties. As stated in the special chapter under OPIUM below, the medicinal drug of Europe is mainly derived from Asia Minor and is the produce of P. **somniferum** *var.* glabrum. Egyptian opium, though now-a-days little used, is still met with in Europe. In India a specially prepared article is issued from the Government Agencies to the medical department, and this takes the place, in this country, of the European form of the drug. The alkaloids are also specially prepared and issued from the Government factories, so that, as far as the supply of this drug or its alkaloids are concerned, India is quite independent of Asia Minor.

In the words of the *Pharmacopœia of India* it may be said that opium, in its primary effects medicinally, is stimulant, and its secondary action narcotic, anodyne, and antispasmodic. "It operates chiefly on the cerebro-spinal system, and, through the nerves arising therefrom, it affects more or less every organ of the body. It tends to diminish every secretion excepting that of the skin, which increases under its use. In overdoses it is a powerful poison." "In inflammation, especially of serous membranes, it has been employed extensively, either alone, or in combination with calomel, antimony, and other remedies. As a general rule it is less applicable to inflammatory and other diseases, in which the tendency to death is by coma or by apnœa, than to those which produce death by asthenia. In the latter, as in peritonitis, it holds the first place as a remedial agent. In fevers, especially in the advanced stages, it is of the highest value, either alone, or in combination with antimony, camphor, &c., allaying vascular and nervous excitement, and

**PAPAVER
  somniferum.**   Medicinal Properties of Opium.

MEDICINE.

procuring sleep; but, as a general rule, it is inadmissible when contraction of the pupil is strongly marked. In painful spasmodic affections, opium in large and repeated doses often affords immediate relief. In various morbid states of the abdominal viscera, *e.g.*, simple or cancerous ulceration of the stomach, chronic gastritis, gastrodynia, nervous and sympathetic vomiting, diarrhœa, dysentery, colica pictonum, strangulated hernia, visceral obstructions, &c., it is often given with the best results. In diseases of the genito-urinary system, *e g.*, cystitis, cystirrhæa, spasmodic stricture of the urethra, also in menorrhagia, dysmenorrhæa, irritable states of the uterus, metritis, &c., it is a remedy of the highest value. Tetanus, acute rheumatism, and delirium tremens are amongst the other diseases in which opium has been employed as a sheet anchor. In cholera it has been used but with doubtful results. In cancerous and gangrenous ulceration, opium, by allaying constitutional disturbances, often exercises a favourable influence on the local symptoms. As an external application, opium proves valuable in various rheumatic, neuralgic, and other painful affections; also in ophthalmia and other diseases of the eye."

"DOSE.—From a quarter of a grain to two grains or more according to circumstances. In many painful affections the sole criterion for the regulation of the dose is the amount of relief afforded."

**Habitual
  Use.**

Opinions differ very materially as to the effect of habitual use of small doses of opium as a luxury or indulgence. The bulk of the medical evidence goes to support the verdict that it is not more injurious than the moderate use of alcohol, and that even its abusive use is less destructive to the victim and his friends than intemperance. In tropical countries the protection it affords against fever and other depressing influences is contended to justify its use as almost a necessity of life. The danger lies in its greater power to retain its victims than prevails with alcohol. But to the people of India this feature of opium seems far less serious than in Europe. A very large community are known to take a small amount daily and to continue to do so throughout life, without ever showing any tendency or desire to increase the daily allowance. It is even contended that in China, where opium indulgence attains its greatest proportions, the people are strong, healthy, active, and lose none of their national characteristics through the daily use of opium. That, on the contrary, the majority of the working classes manifest far less evidence of demoralisation and physical degeneracy than occurs with the labouring classes of Europe, who participate daily in an allowance of spirits. The moral and physiological aspects of the opium question are, however, entirely beyond the scope of this article, which has to deal with its production, uses, and traffic, apart altogether from party opinions as to its value or otherwise to the human race.

**Chemistry.
  94**

CHEMISTRY.—The chemistry of opium will be found discussed in a further paragraph in connection with the process of purification of the crude drug and manufacture of the alkaloids.

FOOD.

Food.—Throughout the pages devoted below to the subject of OPIUM, frequent mention will be found to the manner in which the poppy plant, its seeds and narcotic, are eaten. Briefly it may be here said that sometimes

**Plant.
  95
Seeds.
  96
Oil.
  97
Opium.
  98**

the PLANT is grown as a salad and eaten in a young state like lettuce. In other cases the young plants weeded from the opium fields are so eaten. The SEEDS form a valuable article of food in the districts where poppy cultivation occurs. They are also exported from these districts all over India, and are extensively employed by the sweetmeat-makers or in the manufacture of certain curries. The OIL obtained from the seeds is also largely eaten or used for culinary purposes. The narcotic, OPIUM, is eaten in many parts of India, or a liquid is prepared by maceration in water and

FOOD.
Capsules.
99

then drunk. A beverage called *post* or *koknar* is also prepared from the CAPSULES and consumed in many parts of India. But opium-smoking is by no means uncommon, two special preparations being used for this purpose, *vis.*, *Maduk* and *Chandu*. Though the expression opium-smoking-shop raises at once the picture of Chinese demoralization, which for years past has been flaunted before the eyes of a sympathetic public, yet were the victims and materials of the indulgence changed for those of a drinking den in London or any other European town, the picture of the one scene might be substituted for the other. So general is the moderate use of opium, in many parts of India, that the comparison is practically legitimate between the people of this country and the inhabitants of Europe in their use of beers, wines, and spirits. It is, as already remarked, almost a necessity of life, and so much faith do the people place in the drug that they by no means infrequently give it to their horses when an exceptionally heavy task has to be rendered. Barnes alludes to this practice as specially prevalent with the Cutchees. Indeed, some difficulty is experienced in checking a certain amount of illicit importation of the drug through the habit of Afghan and Shan horse-dealers bringing opium with them for their horses, the unused surplus of which they sell in India, or cover absolute contraband traffic under pretence of the surplus provision for their horses. The Rájputs and other Hindus also frequently present opium at ceremonial visits and religious entertainments with the same familiarity, Forbes remarks, as the snuff-box was formerly handed round the friendly circle in Europe.

OPIUM.

100

An inspissated juice obtained by scratching the unripe capsules of **Papaver somniferum**, *Linn.*, and allowing the milky sap, which exudes thereby, to dry spontaneously. The bulk of the medicinal article is obtained from Asia Minor, at the present day, as it has been for nearly the past 2,000 years. Indian Opium is, however, the most important form commercially, being that which is smoked, eaten, or drunk, in various preparations, by the inhabitants of Asiatic countries, chiefly the Chinese.

HISTORY OF OPIUM, OF OPIUM-EATING, AND OF SMOKING.

HISTORY.
101

Historically there would appear to have been several phases in the growth of our knowledge of this substance. Various species of poppy are mentioned by early writers as ornamental garden plants: the merits of the seed, as affording oil and as being medicinal and edible, were early recognised, perhaps before the discovery of the somniferous property of the capsules, and certainly anterior to the recognition of the value of their milky sap: the plant was grown, as it is at the present day, in most parts of India and Central Asia, as a vegetable, similar in its action to lettuce: the capsules were employed in the preparation of soporific drugs or in the manufacture of a stimulating and soothing beverage: and lastly came the discovery of the more potent nature of the inspissated sap. It seems probable these phases were evolved in the order named, and that **Papaver somniferum** was grown in Asia Minor for its capsules which the Arabs carried all over the East even to China, some time before the Greek discovery of the value of the juice. Opium was probably originally detected in **Papaver setigerum**, though that species must have been rapidly displaced by **P. somniferum**, a form which took its origin apparently in Asia Minor and was itself very likely but a sport from **P. setigerum**.

In Classic
periods.
102

We learn, at all events, that the poppy was grown as a garden plant, and its medicinal properties understood during the early classic periods of Greece and Rome. Homer gives us the simile of the exhausted warrior hanging his heavily helmetted head, like the drooping poppy flower buds.

P. 102

PAPAVER
somniferum.

HISTORY :
In Classic
periods.

Livy tells the story of one of the last of the Roman Kings sending the message to his son, **Sextus Tarquinius,** to remove the headmen of one of the Etrurian cities. That message was symbolically conveyed by the king striking off the capsules from the tallest poppy plants in his garden. **Theophrastus,** who lived in the beginning of the third century, was acquainted with a preparation from the poppy which he called μήκώνιον or the juice (ὀπός) of the μήκωι plant. From the fact that all nations of the world have adopted some form of the word ὀπός or ὄπιον for the inspissated juice, while they often possess indigenous names for the plant, its seed or medicinal capsules, it may safely be inferred that the discovery of the isolation of opium proceeded from the Greeks. Subsequent writers certainly dwell with greater significance on the soporific properties of the poppy (after the date of **Theophrastus**), than the earlier authors thought it necessary to do, so that the discovery whether made by Greeks or Arabs took place somewhere about that date. By the beginning of the Christian Era the knowledge in opium may be said to have become universal. Virgil speaks of the lethean virtue of the plant ("*Lethæo perfusa papavera somno," Georg. I., 78*). **Unger** failed to find any record of the poppy or of opium in connection with ancient Egypt, but the Egyptians of the first century A.D., used the capsules, if not the opium, as a medicament. **Pliny** (*Lib. XX., c. 76*) pays special attention to the medicinal value of *opion*. Amongst his contemporaries **Dioscorides** narrates with the minutest detail the process of extracting the ὀπός or ὄπιον, and distinguishes it from μηκώνειον. From his remarks it may be inferred that in his time the art of extracting the drug constituted an important industry in Asia Minor. Indeed during the classic period of the Roman Empire, as also the early Middle Ages, the only sort of opium known was that of Asia Minor. In the thirteenth century **Simon Januensis,** physician to Pope Nicholas IV., refers to *Opium thebaicum,* although *Meconium* was even then still in use.

India made
acquainted
with
Opium.
103

But if to the Greeks be due the discovery of opium, the Arabs most undoubtedly carried, to the utmost corners of the Eastern countries, the knowledge of that drug. In the first instance, they made it known to Persia and subsequently to India and China. That the followers of the faith of Islám proclaimed the properties of opium to the people of India there can be no doubt, since the Sanskrit and all the vernacular names in this country are traceable to the Semitic corruption of ὀπος or

Names
derived from
the Semitic.
104

ὄπιον into *Afyún.* Thus its best known Sanskrit name is *Ahiphena* and its Hindí *Afím.* But direct historic records exist which also leave no room for doubt that the poppy was not known to the people of India prior to the Arab influence. The early Chinese works tell of the Arab traders exchanging poppy capsules for the merchandise of China, and its Arabic name became in the Chinese tongue *Ya-pien.* In the time of the Caliphs, the Arabs certainly visited India and China, especially after the founding of Baghdad (A. D. 763) ; previous to the Táng dynasty the poppy was apparently not known to the Chinese. When first shown to them its urn-shaped capsules, full of millet-like seeds, suggested

Soporific
beverage of
capsules.
105

the names *Mi-nang* (=millet vessels) and *Ying-su* (=jar-millet). The Arab doctors directed the Chinese to prepare of these capsules a soporific beverage and medicine. Hence it may be stated that the introduction of the *ya-pien* followed the *mi-nang.* Interest in the oil-yielding seeds and medicinal capsules had thus been fully aroused in Asiatic countries, long anterior to the introduction of opium. It is, therefore, not difficult to understand the existence of names for these parts of the plant that are undoubtedly more ancient than the word opium ; some of which, indeed, even seem, if they are not actually, of an indigenous character in the countries where used. Thus, for example, as has been stated, the Greeks were acquainted with the value of the seeds and capsules prior to their knowledge of the inspissated juice. Virgil commemorates not only the poppy as possessing, like the river Lethe, the power of producing forgetfulness, but he connects the opium plant with the goddess Ceres. The contention here advanced (of the discovery of opium being intimately associated with the Greeks) by no means, however, precludes the inhabitants of Asiatic countries, where species of **Papaver** were found wild, from having contemporaneously, or even at an earlier date than the classic records of Greece and Rome, recognised the value of these plants as vegetables, as sources of oil, or as medicines. Some of the Asiatic names for these parts of the plant may, therefore, be spontaneous; that is to say, they were not necessarily imported with a communication of the value of opium. Several species of **Papaver** are met with in India— some very plentifully. In Kashmir **P. Rhœas** is perhaps as abundant as in Europe. **P. dubium, P. cornigerum,** and **P. pavoninum,** are described as wild species found in Afghánistán and Balúchistán and in considerable abundance. The first mentioned species is in fact plentiful in fields throughout the northern plains of India, during the winter months, and in summer it is common in similar situations on the temperate tracts. **P. hybridum** also is fairly common on the plains of the Panjáb, and is distributed to Afghánistán. **B**ois-sier describes, in his *Flora Orientalis,* some 38 wild species, including **P. setigerum,** besides the cultivated conditions of **P. somniferum.** Aitchison (*Notes on the Products of Western Afghánistán and of North-Eastern Persia*) gives an instructive account of opium cultivation accompanied with the vernacular names for the various parts of the plant and the apparatus used in the preparation of opium. "The plant," he says, "is known as *kóknár;* the capsules, *góza, khól-a-kóknár, póst-a-kóknár;* the seed, *tukhm-i-khash-khash;* the milky juice *shíra;* the needles set in wax which are used in scratching the capsules, *sanesh;* the inspissated juice opium, *afíun, tariák;* and the oil of the seeds, *roghan-i-khash-khash.*" But it is perhaps questionable as to the value of even so complete a vocabulary of vernacular names as that given by **Aitchison,** in proving the antiquity of opium cultivation. It is a highly remunerative crop, and hence specially adapted or newly coined names (the original meanings of which may be now lost to philology) would be necessitated for every state and process of cultivation and manufacture. Thus, for example, **Mr. Grierson,** in his *Behar Peasant Life,* gives an infinitely more complete enumeration, than **Aitchison's,** for one of the chief opium-producing provinces of India (Behar), but of which there can be no possible doubt the cultivation does not date further back than a couple of centuries.

Moorcroft states that in Garhwál the young plant is used as a vegetable, raw or cooked with buttermilk. This same fact has been observed by the writer in connection with many parts of the Panjáb Himálaya, where white, red, or white and red variegated, forms are grown. Throughout the opium-producing provinces of the plains the young plants obtained from the early weedings are universally eaten as a salad. The plant is, however, never exclusively cultivated as a vegetable, but the superabundant seedlings are weeded out from time to time and eaten, thus leaving the healthiest to mature their capsules. In many parts of the Himálaya these are not scratched for their opium though that property is fully known. Thus **Mr. Baden Powell** informs us that this is the case in Jálandhar where the poppy is grown for its capsules which are used in making *post,* an infusion consumed as a beverage. The *post* of the Panjáb appears to closely resemble the *kuknár* which was a luxury among the Muhammadans in the time of Akbar. **Bontius,** writing of Batavia in 1658, tells us that the

HISTORY :

In Classic periods.

Indians were there divided popularly into two sections—those called *pusti·* who smoke an inferior opium prepared from the leaves, stems, and the capsules of the poppy, and those called *afyuni*, who could afford to purchase ordinary opium. Throughout India the seeds of the poppy are cooked in various sweetmeats or are employed in the extraction of an oil which is used for culinary and other purposes. It need not, therefore, be a subject of much surprise that in the Persian, Arabic, and Indian languages there should exist names (some of which became like *post* or *pust*) widely diffused, and which appear quite independent of Greek literature. Indeed, had the cultivation of the poppy proceeded exclusively from the Greeks and diffused eastward, prior to the Aryan invasion of India, the people of Egypt and Palestine might have been expected to have possessed an earlier knowledge of it than the Roman epoch. Far from this being so, it can hardly be said to be mentioned in Hebrew literature, since the Jerusalem Talmud (seventh century) is the earliest work that alludes to *ophyon*—a poisonous drug—and Pliny in the first century is the earliest author who refers to Egyptian opium. It seems, therefore, highly probable that the merits of the poppy seed and poppy capsules were known to the Arabs and to the people of large tracts of Central Asia and even of Upper India anterior to the supposed Greek discovery of opium. But the silence of the early Sanskrit authors leaves no room for doubt that right down to the commencement of Muhammadan influence in India the various wild poppies of the country attracted very little attention. We have no evidence that **Papaver somniferum** ever existed in any part of India, Balúchistán, Afghánistán, or even Persia, in a wild state, and even now, after at least a thousand years of cultivation, it has nowhere manifested any tendency to become naturalized. It thus seems likely that for many years the early Arab traders brought by land and sea the capsules of the poppy just as they carried them to China, and it need only be here added that long after. poppy cultivation had been started in India, the Arabs continued to bring to this country Arabian opium in competition with that of Cambay.

Capsules Imported into India.
109

CHINESE INDULGENCE.
110

TANG DYNASTY.
8th century A.D.
111

CHINESE TRADE.—A Chinese writer (**Ch'en Ts'ang-ch'i**) of the eighth century quotes **Sung Yang-tzu** as stating that the poppy has four petals. It is white and red. Above is a pale red rim. This last expression probably denotes the poppy often seen on the Himálaya with petals white below but with a purple band on their extremities. **Dr. Edkins** in his *Historical Note on the Poppy in China* comments on the above statement that "at this time, the Arabs had been trading with China for at least a century, for Mahomet's death occurred A.D. 632 and that of his uncle" (an immigrant to China), "not long afterwards. It was easy for the poppy to be cultivated with the jasmine and the rose everywhere throughout the country." We know, indeed, from the *Nan-fang·ts'ao-mu-chuang*, a work which dates from the beginning of the fourth century, that the jasmine and the *henna* plants, which must have come with the Arabian commerce, were already in China when that book was written. Though there are thus early indications of Arab influence in China, the first distinct mention of the poppy is in the work of **Ch'ên Ts'ang-ch'i** quoted above. A poem written at the close of the T'ang dynasty shows that at that time the poppy was cultivated at Ch'êng-tu-fu in Szechwan.

632 A.D.

Early Cultivation in Szechwan.
112
973 A.D.

The Emperor **Sung T'ai-tsu** in 973 A.D. gave orders that the medical work *K'ai-pao-pên-ts'ao* should be prepared. In that work the poppy is called *ying-tsŭ-su*, and the seeds have medicinal virtues given them. The name *ying-su* means "jar-millet." About this same period a poet speaks of the poppy as being eaten like the vegetables of spring. Its seeds are like autumn millet. When ground they yield a sap like cow's milk; when boiled they become a drink fit for Buddha.

Ying-su or Jar-Millet.
113

The Emperor **Jên Tsung** ordered the compilation of a second great medical work, and for this purpose the magistrates were directed to furnish the compiler **Su Sung** with all information regarding drugs. In that work (which appeared in 1057) it is stated "the poppy is found everywhere. Many persons cultivate it as an ornamental flower. There are two kinds, one with red flowers and another with white." As pointed out by **Dr. Edkins** it is safe to infer from that statement that, though the name by which opium was subsequently known to the Chinese, had not then found its way to that country, the two chief forms of the plant (red and white), which afford the drug, were widely cultivated in the eleventh century. **Lin Hung** in the twelfth century is the first Chinese author who definitely alludes to the preparations of the milk from poppy-heads. A poem of **Hsieh K'o** written in the Sung dynasty speaks of the white flowers as resembling snow, but that author says the seeds are black. It is customary to read, at the present day, that the white-flowered form has white, and the red-flowered black, seeds. Most of the Sung dynasty medical writers and from them downwards extol the merits of poppy capsules in the treatment of dysentery, especially when combined with astringent drugs. Thus **Wang Shih** in his work *I-chien-fang* says their action in dysentery is simply magical. In the Yuan dynasty a medical author alludes to the preparations of the poppy capsule as "killing like a knife," hence the true opium-yielding plant must have been employed.

During the Ming dynasty, which lasted during the fifteenth, sixteenth, and part of the seventeenth centuries, the Chinese trade with India, Arabia, and the islands of the Eastern Archipelago greatly increased, and the extension of the Mongol Empire to Persia also helped to facilitate intercourse with China. Commenting on this subject **Dr. Edkins** says, "**Cheng Ho**, who was sent on a diplomatic mission to all important seaports from Canton to Aden, succeeded so well on his first voyage that he was repeatedly despatched afterwards, and brought back a fairly minute account of the places he visited. He was in diplomatic communication with the chief persons in authority in Aden and some other Arabian ports, in Hormu on the Persian Gulf, in several cities of India, such as Goa, Cochin, Quilon, and Calicut, as well as other centres of trade nearer home. Can we wonder that all the principal exports in those countries became known to the merchants of Canton and Amoy? They were then probably, next to the Arabs, the chief traders in the Indian seas.

"When the Portuguese appeared unexpectedly at Cochin in 1498, they commenced at once a career of conquest, and quickly made themselves masters of Aden, Hormuz, Goa, Cochin, Calicut, Malacca, and many other cities. With military prestige they joined great activity in commerce, and became the chief merchants in the East. At this time, as we learn from **Barbosa**, opium was among the articles brought to Malacca by Arabs and Gentile merchants, to exchange for the cargoes of Chinese junks. He also states that opium was taken from Arabia to Calicut, and from Cambay to the same place, the Arabian being one-third higher in price than the Cambay. The opium exported from this seaport may be assumed to have been manufactured in Malwa, which lies quite near to it.

"The Arabs, then, had begun to grow opium in India in the sixteenth century. In addition to this we are told that from places on the Coromandel Coast opium was exported to Siam and Pegu. Here we also find clear indications of the activity of Arab traders in extending the cultivation of the poppy in India. The Chinese also at this time imported opium themselves, to be used medicinally. It is important to note this for the proper understanding of the history of Opium in China."

HISTORY :
Chinese.

SUNG. DYNASTY 1057 A.D.

Red and White Flowered.
114

12th century. Preparation from Poppies.
115

YUAN DYNASTY (13th century).
116

MING DYNASTY (15th, 16th, & part of 17th centuries).
117

Chinese Traders.
118

Advent of the Portuguese, 1498 A.D.
119

The Chinese import Opium.
120

Poppy grown in India in the 16th century.
121

P. 121

PAPAVER somniferum.	History of Opium-eating

HISTORY :

Chinese.

First knowledge of, in China, 15th century.
122

CULTIVATION IN CHINA of Poppy and preparation of Opium ; 15th century.
123

Description, Method, Extraction.
124

Imposition of Import Duty.
125

Native & Foreign Opiums in China (15th century).
126

Introduction of Tobacco into China : 1620 A.D.
127

From about the fifteenth century, therefore, interest in poppy capsules is in Chinese works diverted to opium. **Wang Hsi**, who died A.D. 1488, says it is obtained in Arabia from a poppy with red flowers. He gives a minute description of its use in dysentery, and as he was governor of the province of Kansuh for 20 years and had thus to look after a large Muhammadan community, his knowledge of opium must have proceeded from more than casual report. He died ten years before the arrival of **Vasco de Gama** in India. But not only does **Wang Hsi** describe the uses of opium, but also the cultivation of the plant and the extraction of the narcotic. Thus, there can be no doubt that the Chinese had commenced the cultivation and the preparation of opium as early as the fifteenth century. Several of the Ming dynasty writers describe the cultivation and extraction of opium, but only for medicinal purposes. The prohibition against foreigners trading with China (1523), consequent on the raids made by the Japanese on the coast, caused foreign drugs to become scarce, hence directions were actually issued for the more careful cultivation of opium in China itself, and the value of the poppy, as an agricultural crop, was thereby enhanced and greatly extended.

In the year 1589 an import duty was imposed on opium in common with Myrrh, Olibanum, Asafœtida, and other goods. This was modified in 1615, opium being rated at $1\frac{78}{100}$ *mace* for each 10 *catties*.

" 'In the work called *Wu-li-hsiao-shih* written at the end of the Ming dynasty and the beginning of the present, it is stated of the poppy that it is sown in the middle month of autumn, at noon. After flowering, the seed vessel grows into the shape of a vase. The tiny seeds can be eaten as porridge. Oil is also obtained from them, and the capsules are useful in medicine ; they are powerfully astringent. When the capsules are still green, if a needle be used to puncture them in ten or fifteen places, the sap will come out. This should be received into an earthenware cup, which may be covered carefully with paper pasted round the edge. Let the cup be exposed to the sun for fourteen days ; it is then opium, ready for use as an astringent, and restrains reproduction most powerfully.'

" Carefully weighing what is said in the passages preceding, it appears plain that from the latter part of the fifteenth century the manufacture of Native opium has existed in China ; it is not only in recent years that there has been both Native and Foreign opium in this country. Let the reader examine the various accounts of the manipulation by four different authors. **Wang Hsi's** book cannot be now procured, but judging by what is quoted from him in **Li Shih-chen's** work, he meant to describe the method of Poppy culture in Arabia, and spoke particularly of a kind which yielded, the opium sap in the seventh and eighth months or later. When, however, he speaks, as in the passage translated from the *Tung-ı-pao-chien*, of obstinate diarrhœa needing opium to cure it, and advises the physician to make opium direct from the Poppy, in a way which he describes, he must be speaking of a Chinese article. **Li T'ing's** account differs in too many points from that of **Wang Hsi** to be regarded as a second-hand statement based exclusively upon it. If so, then **Li T'ing** is a third and independent witness on this subject, the fourth being the author of the work *Wu-li-hsiao-shih*" (*Dr. Edkins, Note on Opium*).

The brief extracts given above, of the chief historic facts which bear on the introduction of opium into China, will be found to correspond very closely with the sketch written by **Dr. F. Porter Smith** (*Mat. Med. & Natural History of China, pp. 162-164*) and with the learned discoveries made known from time to time by **Dr. Bretschneider**—the indefatigable student of Chinese classic literature. **Dr. Edkins**, in the further pages of his admirable paper, deals with the growth of the indulgence of opium-smoking. To-

wards the end of the Ming dynasty, he says, the legitimate practice of taking opium medicinally was destined to soon change for the habit of smoking
it. This new phase of the opium question was intimately associated with
the introduction of tobacco-smoking. Tobacco was introduced into China
from the Philippine Islands. On these Islands the Spaniards had settled,
and being in intimate relations with America, they conveyed the tobacco
plant to the Philippines : from these Islands it soon after crossed to Amoy,
being conveyed there by Fuhkien sailors trading with Manilla.

It thus reached China about 1620 A.D., just at the time King
James I. published his *Counterblast to Tobacco*. Dr. Edkins remarks that
though tobacco-smoking was prohibited by the last Ming Emperor, the
habit rapidly spread, and in the provinces of China various substances
were mixed with it in the preparation of smoking-mixtures. Amongst
these were opium and arsenic. The first of the Manchu Emperors in 1641
accused his nobility of neglecting to set the soldiery and populace a good
example, instead of leading them into the vicious practice of tobacco-smoking. But the edicts against tobacco were just as ineffectual as were those
framed in far more stringent terms against opium-smoking. " The immense
popularity of tobacco-smoking was an indication of the readiness of the
Chinese nation to adopt the use of narcotics. The same thing which took
place in the nineteenth century with opium-smoking occurred in the seventeenth century with tobacco-smoking. The Confucian mind was shocked,
the sense of propriety was wounded ; but this did not prevent the rapid
spread of both of these modes of indulgence in all circles. Prohibitory edicts
were issued in vain by Emperors animated by paternal affection for their
people. Tobacco was a less evil than they supposed ; opium-smoking
was a far greater evil than they feared. In both cases the Emperor was
powerless. The Emperor **Ch'éng Tsung**, as we should call him, but who
is better known as **Tao Kuang**, is much to be respected for his strong
moral convictions on the subject of opium. He made really great efforts
to cope with this evil, but it was in vain. The fondness of the people for
inhaling a narcotic was too strong for him to overcome. He failed utterly
in the attempt to put down opium-smoking even in the city of Peking."
" The habit of tobacco-smoking became national, and went on extending
itself for a century, till soon after the close of the long reign of **Kang Hsi**
the attention of the Government was drawn to opium-smoking as a new vice
in Formosa and at Amoy." The first edict against opium-smoking was
issued about 1729, and by 1746 the practice had become firmly established
in Formosa. Several writers about that period mention that this form of
indulgence came from Java. It thus took its birth in the country where
tobacco-smoking had been indulged in for the longest time. Dr. Bretschneider (in a letter to M. deCandolle) says that according to Don
Sinibaldo di Mas, the Chinese took the practice of opium-smoking from
the people of Assam " where the custom had long existed." This is very
probably a mistake ; at all events, there is no sufficient grounds for supposing that the people of India anywhere smoked opium prior to the Chinese,
and it seems much more natural to suppose that it came to China from the
same source as the earlier indulgence—tobacco-smoking.

Dr. Edkins, who has been placed under contribution for the chief
facts given above, regarding China, turns from the ancient writings of
that country to the journal of the great Westphalian traveller **Kampfer**.
That indefatigable explorer was born in 1651 ; visited Persia in 1683—
85 ; India, Java, and Siam in 1690 ; Japan in 1690—92 ; returned to
Leiden in 1694, and died at Lemgo in 1716. His great work *Amœnitates
Exoticœ* was published in 1712. In that publication he deals with tobacco,
hemp, and opium, and gives the first authentic description of an opium-

HISTORY :

Chinese.

1620 A.D.

MANCHU
DYNASTY.
(17th cen-
tury).
128

Predisposi-
tion of
Chinese to
Narcotism.
129

Introduction
of Opium-
smoking
(18th cen-
tury).
130
Appeared in
Formosa and
Amoy.
131

Assam.
132

PAPAVER somniferum.

HISTORY :

Chinese.

First mention of an opium-smoking shop, *1712 A.D.*

133

Chinese opium-smoking preparation.

134

Punishments against Traffic in Opium.

135

Local Production greater than Imports.

136

Sale prohibited but drug admitted.

137

Early British Transactions in Chinese Opium, *1781 A. D.*

138

smoking shop. In his account of the method of smoking opium, in which tobacco was saturated with a dilute infusion of the narcotic, he uses the term *haurio.* That the tobacco impregnated with opium was actually smoked there can be no doubt, and indeed the expression of "drinking, tobacco" was at one time in use in England just as it is at the present day with the people of India. Kampfer, in speaking of Persia, says the best sorts of opium were flavoured with spices and were called *theriaka* preparations; these were held in great esteem during the Middle Ages. Dr. Edkins in his concluding pages deals with the various Chinese accounts of the special preparations used in opium-smoking from the year 1723 down to the present date. He is strongly of opinion that the practice came from Java to Amoy, and Formosa, because of the fact that the earliest Chinese works describe the system of smoking in almost the same words as those employed by Kampfer when speaking of the method of indulgence which he saw in Java. Dr. Bretschneider is of opinion that opium-smoking is a Chinese invention and modern. "Nothing," he says, "proves that the Chinese smoked opium before the middle of the last century."

Dr. Edkins, however, reviews the various edicts that were issued (from 1729 to 1796) against opium-smoking, and shows that the connivance of the Chinese authorities, from the highest to the lowest, served to render repressive measures futile. Every person, except the opium-smoker, was, by the laws of China, to be meted out the severest punishment, even death, for participation in the opium traffic. The people were stronger, however, than their lawgivers and found the means of establishing with impunity, under the very eyes of the authorities, the agencies and shops necessary for the import traffic, and we know that they also rapidly began to partake in the profits of production, for the cultivation of the poppy was soon extended throughout the empire until the home manufacture of opium far exceeded the quantity imported. While discussing the spread of opium-smoking in the eighteenth century, Dr. Edkins says : " It sprang up in a lawless locality at a great distance from Peking ; there was, therefore, no inclination to leniency from the fear of offending persons or classes whom the Government would not like to offend. The law was in consequence promptly made decided in tone, and severe in detail. Was this law acted upon ? No allusion was made to it by the Jesuit Missionaries in the *Lettres édifiantes* or in the *Mémoires concernant les Chinois.* The habit of opium-smoking is not mentioned in these works. The trade in opium certainly remained as before. Two hundred chests a year continued to be imported, and in 1767 that quantity had gradually increased to 1,000 chests. The duty was Taels 3 a chest. It would appear, then, that the old tariff of the Ming dynasty was still followed in the main. The sale of opium was prohibited by Statute, but we do not find proof that it was refused as a drug at the Custom Houses of Amoy and Canton. The import steadily increased during the time it was in the hands of the Portuguese, till English merchants took it up in 1773, after the conquest of Bengal by Clive. The East India Company took the opium trade into its hands in 1781. At that time the minor portion only of the imported opium was devoted to opium-smoking,—at least we may assume this. The Superintendent of Customs in those days would continue to take the duty on opium as a drug. What was contraband they would say was *ya-pien-yen,* which means opium for smoking : the drug *ya-pien* would still pass the Customs as a medicine." Thus the opium monopoly was a legacy from the Muhammadan rulers of India to the British conquerors at Plassey in 1757, but the external traffic in it was not taken up until some time after when its direct control became a necessity in the administration of India.

The habit of opium-smoking had reached Peking in 1790 and assumed there so alarming proportions as to call forth fresh edicts against it. At about the same time the local authorities at Canton began to complain of rapid increase in the trade in opium. In 1800 there was an edict issued prohibiting opium from being brought to China in any ship. It was from this time that the more distinctly smuggling period commenced. It was a contraband trade, but connived at by Viceroys and Governors; they felt a difficulty, and concluded not to touch the evil with any firm intention to heal. How to treat it they knew not. The evil grew beyond their power of control. They regarded it as the " vile dirt of foreign countries;" they feared it would spread among all the people of the inner land, wasting their time and destroying their property; they advocated the prohibition of the trade, and the Government consented to their advice, and frequently issued prohibitory edicts, but too often some of the officials themselves smoked, or their nearest friends smoked, and so the hand of interference was paralysed, and the demand for opium continuing, the import was never seriously checked till the time of **Lin Tsê-hsü**, and the war of 1841."

The unconstitutional and unauthorised action of the provincial Commissioner Lin in destroying £2,000,000 sterling worth of opium, the property of British traders, while the Imperial negotiations were proceeding which were intended to meet the wishes of the Chinese Government, could have resulted in nothing but war. It was clearly not the duty of external Governments to superintend the commerce of their subjects, on the standpoint of what *they* conceived should be the moral character of the Chinese people, even if the perniciousness of opium-smoking be admitted. It would have been the duty of these Governments, however, to see that their traders did not violate the treaties made with China. Had the Chinese Government taken the course open to it, and that too without arbitrary injury to a trade of large proportions which had been fostered by the governing commercial principle of demand, to impose a gradually increasing taxation on imported opium; had it exercised also the power, which it should have possessed if it did not do so, of restricting or prohibiting the cultivation of the poppy within its own territory, little would have been heard of the perplexing opium question of the present day. But that the Chinese Government was either helpless against the popular desires of its people or never earnestly intended to combat the evil there seems no doubt. Matters were therefore left to drift along in a state of lawlessness, and the conflicting interests of the Chinese Imperial and Provincial Governments encouraged rather than prevented the spirit of high-handedness and robbery which culminated in a violation of international law. The Chinese Government had thus to pay 21,000,000 dollars towards the war which its vacillation and weakness entailed, as well as the further sum of £1,250,000 in compensation to the British merchants whose property had been confiscated or destroyed. But even losses of such severity did not teach the Chinese Government that the suppression of the opium traffic (if that was actually its object) depended on a reform in its own fiscal administration. In spite of all the prohibitory measures which had been taken, the habit of opium-smoking became firmly established, and that, too, long before British commerce had anything to say to it. But the foreign supply, at no period since the introduction of opium-smoking, has been able to keep pace with the demand. Local production has accordingly grown into infinitely greater importance. Had the Indian Government desired, did it desire now, to repress Chinese production, and competition with Indian opium, a reduction of the export duty would very probably have the immediate effect, provided (and this is the keynote to the whole question) the Chinese Government refrained from imposing a correspondingly increased

HISTORY:
Chinese.

Opium-smoking in Peking, *1790 A.D.*
139

Prohibition against import of Opium, 1800.
140

Chinese war, *1841 A.D.*
141

Power to control Opium Traffic.
142

Cultivation in China.
143

PAPAVER somniferum.	Controlling factor in Indian Exports.

HISTORY :
Chinese.

Indian
Opium
a
Luxury.
144

import duty. The limits of possibility in Indian production are not the controlling factors in the exports to China. The Indian duty is such as to make our opium the luxury of the rich, and a reduction in the export duties from India would but transfer a larger share in the profits of the traffic from India to China without either lowering the price to the Chinese-smoker or restricting the amount consumed. That the Chinese Government are fully alive to their own interest and do not now and never have desired to lose their opium revenue is a fact that no student of the question can be ignorant of. " It appears thus to be certain that, while the Court at Peking was endeavouring to suppress the foreign trade in opium from 1796 to 1840, it did not or could not put a stop to home cultivation of the drug, since a Chinese Censor in 1830 represented to the throne that the poppy was grown over one half of the province of Chekeang, and in 1836 another (Cho Tsun) stated that the annual production of opium in Yunnan could not be less than several thousand piculs " (*Encycl. Brit.*).

Magnitude
of
Native Opium
Production.
145

But on tracing out the expansion of Chinese cultivation to more recent times it is seen to have attained, as might have been expected, alarming proportions. The following brief passages exhibit this fact : —" Mr. Donald Spence, British Consul at Ch'ung-ch'ing-fu, in Szechwan, in the year 1881, made enquiries into the amount of opium produced at that time in the four south-western provinces. He states that in Szechwan the consumption of Native opium within the province amounts to 54,000 piculs, while 123,000 piculs are sent to other provinces ; of these 70,000 piculs are exported in an easterly direction, 40,000 piculs paying duty, and 30,000 piculs being smuggled. Yunnan produces annually 35,000 piculs, and Kweichow 10,000 piculs, while Hupeh supplies to the market not more than 2,000 piculs. In all, the production of Native opium amounts to 224,000. Mr. Spence's Report on the Native production of Opium was forwarded to the Foreign Office of the British Government, and subsequently presented to Parliament and printed. If a comparison be made of the amount of opium produced in the four above-mentioned provinces, *viz.*, 224,000 piculs, with the quantity of Foreign opium imported in 1882, *viz.*, 66,000 piculs, it will be seen that the opium of Native production is more than three times as much in quantity as that introduced from India and elsewhere."

Growth
of
Chinese
Production.
146

On the subject of the growth of the Chinese production and importation of Indian opium, there appeared in the *Gazette of the Government of India for February 23rd, 1871,* a long and instructive historic sketch of the trade. In that very able and detailed statement the progress made in the cultivation of the drug in China and its effect on the Indian market was fully dealt with, from 1848 to the year of publication. For early information on this subject, therefore, the reader should consult the *Gazette* in question, as also the report issued in 1869 on the Trade of the Treaty Ports of China, to which is appended a special report drawn up by a native of Hupeh regarding the cultivation in Szechwan. In the Consular Reports of China will also be found the figures, year by year, of the continued expansion of poppy cultivation.

Cultivation
prohibited
in
China.
147

The cultivation and production of opium was at first prohibited in China by equally severe edicts to those published regarding the importation of the drug. But the one series of laws were as futile as the other. The demand for opium which from the fifteenth to the nineteenth century had steadily increased was not to be repressed by law. A large community could not do, or did not care to do, without the drug. The foreign supply obtained from India, Arabia, Egypt, and Persia was not sufficient. It naturally followed that just as the difficulties of importation were overcome by the corruption of the officials, the same loophole allowed the cultivators in the most rural districts to not only produce opium but to have it conveyed throughout the length and breadth of the land. Awakened at last to the reality of the case, the Chinese Government saw that the supply of opium was a national demand, similar to that for tobacco and spirits in other countries. That the attempts to stifle it by the high hand of law had but led to a deeper moral corruption which stultified the humane desires of the refined and cultured reformers. All restriction was therefore removed. Importation and cultivation were legalised, and a fiscal revenue, which could be made to control the traffic far more effectually, was thus drawn into the Imperial treasury, instead of being swallowed up by unscrupulous officials in the form of blackmail. For some years past the influences of this wise decision have been felt powerfully by all opium-producing countries. The Malwa opium of India, which resembles most that produced in China, has proved a less and less remunerative enterprise. The competition which has, for some years, been carried on between the Chinese and Indian drug, appears at length to have appreciably affected the latter, especially the Malwa description. Prices have fallen heavily, and the Government of India has found it expedient, in view of representations from the traders of increasing losses, to reduce the transit duty to the level at which it stood before 1877. The duty has been levied at the following rates since 1845 :—

	R			R
1845	 300	1862	. . . .	600
1847	 400	1877	. . . .	650
1859	 500	1879	. . . .	700
1860	 600	1882	. . . .	650
1861	 700	1890	. . . .	600

Although the price of Bengal opium has also fallen seriously, its superior quality has largely saved it from feeling the full effects of the Chinese production, and it seems probable that, as an expensive luxury and a medicine, it will for some years to come be able to hold its own against China.

INDIAN TRADE.—Turning now to the Indian side of the HISTORY OF OPIUM it has to be repeated that none of the Sanskrit authors deal with the subject. Dr. U. C. Dutt, in his Materia Medica of the Sanskrit medical writers, says :—"Opium appears to have been brought to India by the Mussulmans, as its Sanskrit name is evidently derived from the Arabic *Afyún*, and as it is not mentioned by the older Sanskrit writers. The capsules of the poppy are called *khákhas*, and the seeds *khasatila*, in Sanskrit." DeCandolle, on the other hand (on what is probably very doubtful authority), seems to regard the Sanskrit and Persian names as denoting an ancient cultivation. After pointing to the defect of knowledge indicated in the Hebrew literature he says : "On the other hand, there are one or two Sanskrit names. Piddington gives *chosa*, and Adolphe Picket, *khaskhasa*, which recurs in the Persian *chash-chásh*, the Armenian *chash-chash*, and in the Arabic. Another Persian name (given by Ainslie) is *kouknar*. These names, and others I could quote, very different from the *maikon* (μήχων) of the Greeks, are an indication of an ancient cultivation in Europe and Western Asia." It need scarcely be repeated that these are names that point to a knowledge in the capsules and seeds, not to opium, and were very probably originally derived from the Arabic, certainly not from the Sanskrit. Indeed, the two words given by DeCandolle as Sanskrit do not occur in standard works on the subject and should therefore be set aside. Thus, for example, Ainslie remarks on *chasa*: "This is more properly the Sanskrit name of the poppy plant; but, for either it is not given with much confidence." The only word in Sanskrit, which distinctly refers to Opium, is *Ahiphena* from ὀπός, through the Arabic *Afyún*. Thus the literature of the Hindu physicians throws little or no light on the

HISTORY:
Chinese.

China recognises true character Opium Traffic.
148

INDIAN TRADE.
149

Absence of Sanskrit Records.
150

PAPAVER
somniferum.   History of Opium in India.

HISTORY :
Indian.

Cultivation
in
India during
15th
century.
151

Opium—
a State
Monopoly
during
Mogul
Dynasty.
152

subject of the Indian acquaintance with opium, while that of the Muhammadans is unfortunately of a comparatively speaking modern character. Prior to the Emperor Akbar, the Arabian, Persian, and Afghán invaders and conquerors of India had enough to do in holding the country and enforcing tribute, as a military right, without thinking of organised systems of Government and Revenue Administration. Hence it followed that the Arab trading influence did not extend far beyond the coast, more than did that of the early European settlers. We know, however, that they had succeeded during the fifteenth century in introducing poppy cultivation into Cambay and Malwa, and that when Akbar reduced the lawless Muhammadan oppressors and fugitive Hindu chieftains to an Empire, he found Malwa opium a characteristic feature of that country. Abul Fazl specially states, however, that the poppy culture was chiefly practised in Fatehpur, Allahabad, and Ghazipur. We learn that the founder of the Mogul civil dynasty and his successors regarded opium as of necessity a State monopoly. They found it, however, at once the most convenient and successful course to farm out the right to manufacture and to sell the drug, and they imposed a very considerably heavier land tax on poppy cultivation than on any other crop.

In the Records of the Government of India much interesting information exists on the early history of opium culture. From these sources of information many of the facts here given have been culled. In Abul Fazl's time, all the districts from which the Benares Agency Opium is now provided were included in the *Subahs* of Agra, Oudh, and Allahabad. The *Subah* of Allahabad comprised nearly the whole of the present districts of Fatehpur, Allahabad, Benares, Mirzapur, Jaunpur, Ghazipur, and Ballia; also parts of the districts of Jalaun, Hamirpur, Banda, Cawnpore, Partabgarh, and Azamgarh. The seasons were supposed to undergo a complete revolution in the course of nineteen years—a cycle of the moon. In the tables of nineteen years' rates for the *Subahs* noted in the text, the highest and lowest rates for land under wheat and under poppy are thus stated :—

Taxation
of
Opium &
Wheat
Cultivation
in
16th century.
153

| SUBAH. | WHEAT. | | | | POPPY. | | | |
| | HIGHEST. | | LOWEST. | | HIGHEST. | | LOWEST. | |
	Period.	Rent-rate per bigha.	Period.	Rent-rate per bigha.	Period.	Rent-rate per bigha.	Period.	Rent-rate per bigha.
		R a.		*R a.*		*R a.*		*R a.*
Agra	1st to 4th year.	2 4	10th year .	1 3	1st to 4th year.	4 0	10th to 19th year.	3 4
Allahabad	Ditto	2 4	9th ,,	1 9	Ditto	4 0	9th year .	3 4
Oudh	Ditto	2 4	14th ,,	1 1	Ditto	4 0	9th ,,	3 4
Delhi	Ditto	2 4	14th ,,	1 0	6th to 9th year.	3 8	1st to 5th year.	2 11
Lahore	1st and 2nd year.	2 4	13th ,,	0 10	1st to 4th year.	4 0	11th year .	2 8
Multan	10th and 11th year.	1 5	13th ,,	0 10	10th year .	3 4	12th to 14th year.	2 8
Malwa	1st to 14th year.	1 4	15th to 19th year.	1 2	15th to 19th year.	2 8	1st to 9th year.	1 4

NOTE.—The *bighá* in Akbar's time was 3,025 square yards, and thus equivalent to the present opium *bighá* or ⅝th of an acre.

Opium-smoking & Opium-eating.   (*G. Watt.*)   PAPAVER somniferum.

There is, however, no mention of the poppy being grown in Behar, but, as an instance of how agricultural systems have changed in India, it may be added that Abul Fazl (the historian of Akbar's reign) specially points out that *juár* and *bájrá* were not grown in Allahabad. At the present day these might be described as the chief autumn crops of that district.

Thus the State Monopoly in Opium and the policy which is practically that pursued at the present day, was a hereditary gift to the British successors of the great Mogul Emperors. The extent to which opium was in use by the people of India (from the peasant to the noble) may be judged of from one or two passages from the *Ain*. In Gladwin's Translation we learn that the people of Mahoa "gave their children opium to eat till they were three years old." Abul Fazl, the intimate adviser and historian of Akbar, feels no sense of odium attached to the confession that "whenever His Majesty is inclined to drink wine, or take opium, or *kuknár*, trays of fruit are set before him. He eats a few, but the greatest part is distributed amongst the attendants and nobility." The *kuknár* alluded to was, no doubt, a beverage prepared from poppy capsules, and the word rendered "opium" in the translation shows that both the inspissated juice and the poppy capsules were in use in Akbar's time. The dissipation of the grandees at the courts of Akbar and his son Jahangir is, perhaps, one of the most striking features of the annals of that period, when it is remembered that the tenets of Islám forbid the use of wine. In Blochmann's biographical notes of these noblemen, the remark, in varying form, "that he fell a victim to the vice of the age and died from excessive wine-drinking," will be found some twenty to thirty times, while, not infrequently, the addition of some reference to opium occurs. Thus of Shabeg Khan it is stated, " he was much given to wine-drinking. He drank, in fact, wine, hemp, opium, and *kuknár*, mixed together, and called his beverage of four ingredients *Chár Bughra*." Jalal-uddin Mas'ud, referred to in the *Tuzuk*, is said to have "ate opium like cheese out of the hands of his mother." Barkhurdar (Mirza Khan Alam), Governor of Bihar, was much "given to *kuknár* (opium and hemp) and neglected his duties." But it is, perhaps, unnecessary to quote other indications of the extent to which the habit of eating opium or drinking a beverage made from poppy capsules prevailed in India during the sixteenth century; suffice it to add that, in none of these passages does any mention occur of smoking opium in the form of *madak* or *chandu*, the special preparations which are in use at the present day.

Reference has already been made to the Portuguese traveller Barbosa as the earliest European writer who deals with the subject of Indian opium. He wrote of the years 1511 to 1516. The reference to Barbosa, given in Crawfurd's *Descriptive Dictionary of the Indian Islands, p. 312*, may be here reproduced :—" The earliest account we have of the use of Opium, not only from the Archipelago, but also from India and China, is by the faithful, intelligent Barbosa. He rates it among the articles brought by the Moorish and Gentile merchants of Western India, to exchange for the cargoes of Chinese junks." Garcia D'Orta, who wrote half a century later (1574), tells us that the opium exported from Cambay was chiefly obtained from Malwa ; it was soft and yellowish. John Huyghen van Linschoten, the account of whose voyage to the East Indies was published in 1596, says, " *Amfion*, so called by the Portingales, is by the Arabians, Mores and Indians called *Affion*, in Latin *opio* or *opium*. It commeth out of Cairo in Egypt and out of Aden, upon the coast of Arabia, which is the point of the land entering into the Red Sea, sometimes belonging to the Portingales ; but most part out of Cambaia and from Deccan. That of Cairo is whitish and is called *mecerii* " (*i.e.*, the Arabic *misri*=Egypt) ; " that of Aden and

Marginal notes

HISTORY:
Indian.

State Monopoly a gift to the British.
154

A Beverage prepared from Poppy Capsules.
155

Mentioned by Early European Travellers.
156

PAPAVER
somniferum.   Early Traffic from India.

the places bordering upon the mouth of the Red Sea is blackish and hard. That which commeth from Cambaia and Deccan is softer and reddish. *Amfion* is made of sleepe balls" (*i.e.*, poppy-heads) "and is the gumme which commeth forth of the same to ye which end it is cut up and opened. The Indians use much to eat *Amfion*, specially the Malabares, and thether it is brought by those of Cambaia and other places in great abundance. Hee that useth to eate it, must eate it daylie, otherwise he dieth and consumeth himselfe. When they begin to eate it, and are used unto it they eate at the least twenty or thirty grains in weight everie day, sometimes more : but if for foure or five dayes hee chanceth to leave it, he dieth without faile : likewise hee that hath never eaten it, and will venture at the first to eate as much as those that dayley use it, it will surely kill him : for I certainly believe it is a kind of poyson."

Though opium had been made known to Europe long anterior to Linschoten's time, still his account was far more complete than that of his predecessors in Indian exploration. His remarks leave no room for doubt but that in the sixteenth century poppy cultivation and opium indulgence had attained considerable proportions in India. He makes no mention of opium-smoking, but he confirms the statement, already recorded, that eating opium was by no means confined to the Muhammadan nobility but had taken a powerful hold on the people of India. Indeed, the practice of tobacco-smoking was not introduced into India until the sixteenth century, and it seems highly probable that as in China so in India, opium-smoking was an outcome of the inhalation of tobacco fumes. Tavernier, the great French traveller who sojourned in India about the middle of the seventeenth century, was apparently not brought prominently into contact with poppy cultivation, since he specializes all the crops he had studied on his journey through India, but says little or nothing of the poppy.

Dutch &
Portuguese
Traffic in
Indian
Opium
during
16th & 17th
centuries.
158

He, however, saw tobacco, and remarks of Burhánpur : "In certain years I have known the people to neglect saving it because they had too much." Of opium he simply remarks (and his words are significant, as they imply a well-established trade which scarce deserved, in his eyes, more than a passing notice) : "Opium comes from Burhánpur, a good mercantile town between Surat and Agra. The Dutch go there for it, and exchange their pepper for it." Numerous other records show that the traffic in opium from India had assumed by the sixteenth and seventeenth centuries an important position in the Portuguese and Dutch trade prior to the advent of British merchants. Thus, an Italian writer in 1511 says, that General Albuquerque had seized "8 Guzzarate ships" "because they were enemies of the king of Portugal ; and that these had many rich stuffs and much merchandize, and *arfiun* (for so they call *opio-tebaico*) which they eat to cool themselves ; all which he would sell to the king for 300,000 ducats worth of goods, cheaper than they could buy it from the Moors." Ramusio in 1668 alludes to a purchase of 60 parcels of opium from Cambaya. Valentijn deals also with the export traffic of opium from India. "It will hardly be believed, he says, that Java alone consumes monthly 350 packs of opium, each being of 136 *cattis*, though the East India Company make 145 *cattis* out of it" (Vol. IV., 61, published 1726).

Incidentally allusion has been made to opium being conveyed to Batavia from the Coromandel Coast of India. Roxburgh, Elliot, and Ainslie, who wrote of that region and of Madras generally, make no mention of opium cultivation in South India. That it was once on a time cultivated there, as well as in Orissa and Bengal generally, there can be no doubt, for we have the East India Company orders that its cultivation should be restricted to Patna and Benares. Doubtless the danger to the community and the difficulty in preventing illicit transactions, with a widespread, almost promiscuous, cultivation, must have forced on the Directors of the Honourable Company the necessity for confining the traffic within narrow limits, where fiscal restrictions could be brought directly to bear on it. Prior to that arrangement there were no doubt frequent cases of contraband shipments leaving the smaller port towns of South India. At all events we read of Siam and Pegu obtaining their opium in the sixteenth century from Negapatam and Meliapur. The statement that opium was unknown in Burma prior to British annexation is thus quite incorrect. The Arabs introduced the drug into Burma and succeeded in making its use very general.

Kampfer in 1688 saw the people of Java smoking tobacco impregnated with opium which, he says, the Batavians brought from Bengal and Coromandel. In 1516 Pyres referred to opium as a production of the kingdom of Cous (Kuch) Behar and also of Malwa. Perhaps the earliest direct reference to the opium of Bengal, however, occurs in **Captain A. Hamilton's** *New Account of the East Indies* (*I.*, *315*) published 1727, where he says : "The chiefs of Calicut for many years had vended between 500 and 1,000 chests of *Bengal ophium* yearly up in the inland countries, where it is very much used." A little later (1770) the **Abbe Raynal** (*Histoire Philosophique, &c.*, *I.*, *424*) wrote, " Patna . . . is the most celebrated place in the world for the cultivation of *opium*. Besides what is carried into the inland parts, there are annually 3,000 or 4,000 chests exported, each weighing 300lb. . . . . An excessive fondness for opium prevails in all the countries to the east of India." About the same period (1787) **Dr. Hove**— a Polish savant—was devoting himself to the study of the natural productions of Western India. He noted the fact that the Moguls "chew opium constantly" and extract the juice of **Papaver somniferum** from the stalks "and make it their common beverage." Then, again, he says, "After supper the Rajah of Ahmood introduced a beverage of opium diluted in water which was the horridest treat a man could invent. They, however, drank it as we do wine (in much smaller quantities) and, to my great surprise, it seemed that it had but little effect on them. All these opium drinkers are, at 40, the most miserable creatures the world supports. They are either hectic, or so nervous and spasmodic that they are hardly able to out-crawl a snail. They have, however, a convinced idea of its bad qualities, and in consequence said ' we would leave it off if we could, but on cessation for a little while we found ourselves so much reduced, and should perish if we had not recourse to it again'."

By the middle of the eighteenth century Behar had become the province in which opium of the best quality, and in the greatest quantity, was procured. In the anarchy of that period, the Government monopoly had fallen into abeyance. The system under which, in the early part of the century, business in opium was conducted in that part of India is described in a memoir by **Ram Chand Pandit**. The authority of that writer is unexceptionable. In the first year of British monopoly (1773) he was one of the joint-contractors of the opium provision (see the Selections from the Duncan Records, Vol. II., pp. 159—166).

Ram Chand thus describes the system of independent right to cultivate and manufacture opium that prevailed in Patna prior to the British monopoly :—" There was a body of native merchants, then resident at Patna, known under the appellation of the opium-dealers, who, from the time of sowing it in the month of *Assin*, made advances to the cultivators of the poppy, under a stipulation to get interest for the amount thereof, and to receive their opium, in consideration of their thus assisting the *rayats* with advances, with a small rate in their (the merchant's) favour, over and

PAPAVER
somniferum.   History of British Indian Monopoly.

above what would be the common selling rate of the subsequent month of *Bysakh*, as it was in that month that the rate or price of the opium was fixed, according to the smaller or greater quantity of the produce, being first settled at about 2 rupees per seer of 80 sicca weight : and the Patna merchants at this rate, receiving according to their stipulations, with interest, &c., the commodity for the advances they had made, they brought it into town and prepared the same carefully in their houses, so that it might suit the European purchasers. There were also smaller dealers, known by the names of *Pykars*, resident throughout the country at large who made purchases and prepared the opium in their houses from 10 to 50 maunds, but the commodity thus received from the *Pykars* was never genuine. After the opium-growers had delivered to the Patna opium merchants as much of the commodity as liquidated their advances, they then heightened their prices. In the month of October, the opium being prepared, the merchant used first to offer it for sale to the Dutch, and all the merchants joining together fixed a price and agreement with the Dutch Chief, a merchant having 500 maunds in his house, contracting to deliver 200 maunds, and receiving payment thereof to that amount. After thus settling with the Dutch, the merchants used to make further contracts with the English, but with an enhanced price above what the Dutch had agreed for ; and thereafter they went and contracted with the French at a still higher price than they had stipulated for with the English. Thereafter, in the month of November, the Dutch made a second contract with the merchants at a higher price than they had at first agreed for, and the commodity began to be delivered ; and the mofussil *Pykars* taking something by way of *Arat* or brokerage, disposed of their goods to the Patna merchants ; but there was very little thereof found of such a quality as to be accepted, it being, on the contrary, for the most part rejected. If in any year the demand for opium was more urgent than usual, the *Pykars*, by collecting the article from Shahabad, &c., made it sink very much in quality by which means the commodity became in general very much adulterated ; the purchasing price of the Europeans ran then from 100 to 150 rupees per maund, though, in some years, by reason of the urgency of the demand and the competition among the gentlemen, the price has now and then risen to 200 rupees, whereby many native merchants made fortunes, whilst, on the other hand, by reason of European wars and other causes, the price has fallen as low as 70 or 75 rupees, in which time both the merchants and *Pykars* were great sufferers. Thus the purchasing price of opium was various, but that variety never affected the revenue rate of the poppy lands."

In the official papers which the writer has had the privilege of consulting, the historic facts of the period of which Ram Chand wrote down to the establishment of the Honourable the East India Company's monopoly has been ably reviewed. A few passages from these records may be here reproduced :—

" The triumphs of Suraj-ud-Daula over the European Companies in 1756 brought ruin to Patna opium-dealers, who, in default of competition from the English Company, were compelled to dispose of their opium to the Dutch at R70 the maund. The dealers being impoverished were unable to make advances to the cultivators ; and on the restoration of peace in 1765, so little opium was to be had, that even the produce of former seasons, which had lain by in the dealer's houses, fetched as much as R200 the maund. The rise in price stimulated the cultivation, and the servants of the Companies began to send their own agents into the districts to purchase direct from the cultivators. Quarrels ensued between these agents, and in 1767 the Companies made a joint concern of the trade, with one general agent, for all the opium produced. The servants, however,

History of British Indian Monopoly.   (*G. Watt.*)   **PAPAVER somniferum.**

of the Companies continued to trade clandestinely on their own account in the drug, with the result that a large quantity of adulterated opium was brought into market. At length, in 1773, an end was put to all disputes by the Governor of Bengal (**Warren Hastings**), who assumed, on behalf of the English East India Company, a monopoly of all the opium produced in Bengal, Behar, and Orissa, promising to the Danes, the Dutch, and the French, a certain quantity of opium annually, to be received by them from the English Company's agent." " The arrangement with the French and the Dutch was terminated by the taking of all the possessions in India and Ceylon, of those nations, in 1793 and 1795. That with the Danes continued in force for some years later. But the arrangement with the French was renewed in 1815 by Article 6 of a Convention executed in London on the 7th of March of that year." By that article the French were to receive upon requisition duly made opium to the maximum extent of 300 chests. The right thus conveyed is still exercised by the French Administration in India. Delivery of the opium is never taken for sale or consumption in the French settlements ; but on the issue of the notification of opium sales for the ensuing year, the privilege to receive the 300 chests is put up for sale to the highest bidder. The purchaser is by means of it able to keep back from sale part of the whole of the reserved chests, and thus raises the price (a form of opium gambling) at a given sale. Delivery is rarely taken, and the loss on re-sale (often considerable) falls upon the British Government. From the remarks below on the opium Fiscal Administration it will be found that for some years past the right to this claim, 300 chests, has been sold to the Indian Government for an annual sum, so the difficulty indicated above has for the present been overcome.

But to return to the historic facts which bear on the Indian Opium monopoly, it may be said that from 1773 to 1797 the right to the exclusive manufacture of opium on account of the Company was sold annually in the first instance, but from the year 1781 by four-year contracts (*see 9th Report of a Select Committee of the House of Commons, 1783: also Colebrooke's Digest of the Regulations, &c. (Calcutta, 1807), pp. 396 to 423).* " The contracts were disposed of by private bargain and favour till the year 1785, when they were opened to public competition, and were assigned to the highest bidder. Definite stipulations were thenceforward exacted for protecting the cultivators from being compelled to grow the poppy, for securing to them freedom from vexatious imposts, and a fair price for the raw-juice delivered by them to the contractors, and for the payment by the contractors of certain charitable allowances. The cultivators, on the other hand, were made liable to penalties for keeping back the opium produced by them, and to a deduction of *batta* (to be assessed by arbitrators) upon opium with which water had been mixed. Opium delivered by the growers adulterated otherwise than with water was made liable to confiscation ; and, under certain conditions, became the property of the contractors, who were allowed to dispose of it as they pleased, except as part of the Company's provision. Dealing in opium, contrary to the monopoly, was made punishable, in the case of British subjects, by forfeiture of the Company's protection and deportation to Europe ; and in the case of other persons, by fine recoverable ' by the mode of process laid down in the judicial regulations for enforcement of decrees.' " (*Official papers in the Department of Finance & Commerce.*) Colebrooke (*Remarks on the Husbandry of Bengal, Calcutta, 1804, and London, 1806*) alludes to the system of enforced cultivation of poppy that seems to have been engendered by the contract system. The Provincial Council of Revenue also, in a notification on this subject, expressed their strongest displeasure (1st August 1777) at a high-

HISTORY :
Indian.

Birth of the British Monopoly in 1773.
170

Right to manufacture farmed.
171

Evils of the Contract system.
172

**HISTORY :
Indian.**

handed proceeding which had come to their notice in which " a considerable tract of land in the neighbourhood of Ghya was covered with green corn, and which would have been fit to cut in about a month or six weeks: this corn was suddenly cut down, in order that the land might be prepared for the immediate cultivation of poppy." Frequent mention occurs in the records of the period that the farmers of the opium monopoly insisted that when once a cultivator had produced opium on a field he must, year after year, continue to do so. That their vested rights were infringed by a spasmodic cultivation. That if a *rayat* did not wish to continue the opium cultivation, he should forfeit his right to the land and allow the contractors to find a *rayat* who would be willing to continue the cultivation. The instances of abuse in spite of the most stringent injunctions made by the Company with the contractors became so rife and flagrant, that the Board of Directors were forced to seek some other plan of operations than that of

**Necessity for
Government
Agencies
(1797).
173**

farming out the right to produce a drug of such potency. Moreover, the opening of the contracts to competition compelled the contractors to lower their prices and to adopt false contrivances to secure their profit. Amongst others they incited adulteration or themselves directly adulterated the opium, and thus coming into possession of large quantities that the Company would not take from them they became retail vendors to an alarming extent. Year by year the provision opium became worse and worse, the revenue fell in consequence until, in the early part of 1797, it was decided to abandon the contract system and to open direct Government agencies.

**Profits of
Opium
credited to
Revenue.
174**

In 1775, the issue was raised as to whether the profits of the opium traffic shall be credited to Commerce or Revenue. Ultimately, much to the honour of the great trading Company, the question was decided in favour of the latter. It thus became all the more necessary that no intermediary profits should be tolerated, but that the country at large should get the full benefit of the savings which could be effected by a proper administration of the Opium Department. The vested interests of the community at large were thus early recognised as having a direct claim on the proper control and profits derived from opium. This change was highly commended by Colebrooke (*Remarks on Husbandry, &c.*), and even Ram Chand Pundit, who could contrast the old free-trade system, in which he formerly participated, with the subsequent contract system, after discussing the grievances of the *rayats*, amid bribery and corruption, the enforced cultivation and dishonest weighments of their produce, admitted ' If the grievances above stated, supposing them still to exist, were fairly obviated, it appears to me that the carrying on the opium business as a monopoly is the best mode, as well for the *rayat* as for the preservation of the quality and the good of the country at large, including the advantage of the Company." After discussing the many recommendations of the new system, *viz.*, a constant market, liberal advances, freedom from all compulsion to cultivate, he adds : " The *rayats* should never be let off without punishment whenever detected either in delivering the juice of the poppy in an adulterated state or in disposing of it elsewhere ; and finally, after its receipt into the warehouse, it should be made up with the greatest care, that to whatever region it may be exported its good quality may redound to the honour of the agent." It will thus be seen that the free trade, as also the contract systems, had been found by the East India Company, as well as by the community at large, defective and pernicious, and that direct control was hailed by the

**Consumption
of Opium in
India
restricted.
175**

enlightened people of India as a greatly needed and valuable reform.

It is to be regretted that definite statistics do not exist of the consumption of opium in India for more than the past twenty or thirty years. From the general terms in which it is spoken of, however, by the seventeenth and

Introduction of Opium-smoking in India. (*G. Watt.*) **PAPAVER somniferum.**

eighteenth century writers, there would appear no doubt that it was more prevalent formerly than at the present day. One of the direct results of the control of Government has been to restrict cultivation to narrow limits and thus to remove temptation from the door of the peasant. This would appear to have altered the form of indulgence, for the capsules and plant could not so readily be obtained as formerly, and, in many parts of the country, the beverage *post* or *kuknár* rapidly became unknown. The consumption of opium itself by eating it or making a water decoction from it, to be drunk in place of *kuknár*, seems to have been unaffected by the change. Down to the beginning of the present century no writer appears to have recorded the existence in India of the Chinese special form of indulgence, namely, the smoking of the drug. To the ordinary student of Indian questions of political economy, the fact of opium being actually smoked in this country appears to have escaped observation—probably owing to the returns of the traffic appearing under the names of the two chief smoking preparations, *madak* and *chandu.* How or when opium-smoking was commenced in India it is difficult to say. The practice is most undoubtedly of a modern nature—very considerably after the date of its appearance in China. It has fortunately, however, by no means assumed so formidable a character, nor indeed does it show such a tendency, as in China, to become widely popular with the people of India. But it would be quite incorrect to continue the statement, current in works on this subject, that the people of India *eat and drink opium but do not smoke it.* They most undoubtedly do smoke it, and in some parts of the country to an almost equal extent with the practice of eating and drinking it. The majority of opium-smokers, however, take the drug only to a very slight extent and continue to use that limited amount for years without ever showing any desire to increase their daily consumption. It is chiefly used as a flavouring ingredient along with other substances in the cakes of prepared tobacco smoked in the *hukah.* The habituee resorts to *madak* or *chandu,* and he belongs to the lowest and most despised of the Indian community. The picture of an Indian smoking den, however, might be transcribed from the literature of the demoralization of China which is kept ever before the eyes of a sympathetic community, as if Europe did not, under other names and in other disguises, possess moral corruptions and depravities that call in even louder terms for the saving hand of the reformer than do the opium-smoking dens of China and India.

The subject of the introduction of opium-smoking into India may be said to be involved in an obscurity far more unaccountable than that of *ganja*-smoking—a vice apparently more injurious than that of opium; for the *ganja* inebriate cannot always be said to be in a state of "harmless irrationality and helpless intoxication." Opium-smoking has insidiously appeared and become diffused throughout the country, here and there, but no one seems to have considered from where or how. The whole volume of evidence consulted by the writer proves conclusively, however, that far from having expanded, since the advent of British rule, the total consumption of opium by all methods collectively has vastly declined. It is still eaten and, in many parts of the country, certain writers hold that to the degree to which indulged in, it is, far from being pernicious, a much less injurious luxury than the beer and spirits of England. Be that as it may, the total consumption in India bears little or no comparison to that in China and other eastern countries. This may be shown by one fact (brought out in a Despatch of the Government of India in 1882), namely, that the opium excise revenue of this country amounted, for the average of the three years prior to 1882, to only £381,000, while the corresponding average profit on the foreign exports (chiefly to China) was £5,000,000.

HISTORY :

Indian.

Form of Consumption changed.
176

Opium-smoking in India.
177

P. 177

PAPAVER somniferum.	Attitude of Indian Government

THE ATTITUDE OF THE INDIAN GOVERNMENT ON THE CHINESE OPIUM QUESTION.

It would be foreign to the purpose of the present article to enter fully on the arguments that have been advanced in support of, or against, the attitude taken up by the Indian Government in the opium question. The controversy may be said to have assumed a voluminous form with the details of which the public are already conversant. The despatch (alluded to above) which issued from the Marquis of Ripon's Government and was presented to both Houses of Parliament by Command of Her Majesty, sets forth the chief difficulties that exist. In that despatch the following succinct passages occur : " The economic objections to the manner in which the opium revenue is raised, whether in Bengal or Bombay, may be admitted to be considerable. In the former case, the Government itself engages in private trade, a course which is open to obvious objections. In the second case, a very heavy export duty is imposed. In both cases the course adopted interferes with and restricts the free production of and trade in opium. It cannot be doubted that it would be highly profitable to any private trader to pay for crude opium a very much higher sum than is now paid by the Government to the cultivators of Bengal. If, therefore, supposing such a thing to be possible, no restriction were placed upon the cultivation of the poppy, and if, at the same time, the export duty were taken off, it is certain that an immense stimulus would be given to the production of opium, and that China would be flooded with the Indian drug. Thus, in direct proportion to the removal of the economic objections, the moral objections would be intensified in degree. So long, therefore, as the plea of the Anti-Opium Society is confined to the contention that the Indian Government should cease its direct connection with the opium trade, it may be said, with perfect truth, that their policy is based purely on theory. Not only can it effect no practical good, but it almost certainly would do a great deal of harm. It would increase the consumption of opium in China. It would, by cheapening the price of the Indian drug, cause the poorer classes of Chinamen, who now smoke native opium, to substitute Indian opium in its place. It would, moreover, encourage the use of opium amongst the native population of India, some of whom, notably the Sikhs, are already addicted to the practice; and it would result in a diminution of the food-supply of India by reason of the cultivation of the poppy over tracts where cereals are now grown. If, therefore, the policy is to be not merely theoretical, but is to be productive of some practical good, it must aim, not only at the disconnection of the Indian Government with the opium trade, but at the total suppression of the trade itself.

" As to whether it be more immoral for the Government to be directly connected with the manufacture and sale of opium than merely to derive a revenue from the manufacture and sale of the drug by others, that is a point on which, without doubt, much difference of opinion may exist. We do not think that any useful object would be gained by a discussion of this point, or of the cognate question of whether, in Mr. Fawcett's words (*Times*, 5th June 1880), there is ' much difference between raising revenue from opium and raising 26 millions as we do in this country (England) to a great extent out of the intemperance, improvidence, and vice of the people."

" We turn to the second point, in respect to which the position of the Government of India is especially open to attack, namely, the policy pursued towards the Chinese Government in relation to the opium trade. This branch of the question requires ample treatment. It is stated that the treaty under which opium is admitted into China was extorted from the Chinese Government ; that the Chinese Government are now forced to admit opium, and that they are both able and willing to put a stop to the consumption of opium in China, if the foreign import trade were stopped.

"These are grave accusations. We think, however, that we can show that, whatever opinions may be entertained as regards the policy of the various wars waged with China, it is wholly incorrect to say that the article in the Treaty of Tientsin, which provided for the admission of opium, was extorted from the Chinese Government; that Indian opium is in no way forced upon China, but that, on the contrary, the Chinese Government derives a large revenue from the import trade, which it is very unwilling to sacrifice ; that the statement that the Chinese Government is willing to put down the use of opium must be received with great qualifications, and that irresistible evidence is forthcoming to show that they are not able to stop it, even if their willingness to do so be freely admitted.

"We need not go into the history of the various wars which have from time to time been waged with China, nor examine whether the policy of those wars was justifiable or the reverse. It may be very true that the Chinese were with difficulty got to assent to the terms of the Treaty of Tientsin. There is nothing very remarkable in the fact that, at the close of a war, the vanquished party should accept, with great reluctance, the terms on which the victors insist. But as regards the particular point which we are now discussing, namely, the admission of opium into China, we have excellent evidence to show that it is wholly incorrect to say that this provision of the treaty was extorted from the Chinese. **Mr. Lay,** who was Chinese Secretary to **Lord Elgin's** Mission, and who personally conducted the tariff negotiations, has specially stated that 'the Chinese Government admitted opium as a legal article of import, not under constraint, but of their own free will deliberately.' **Mr. Laurence Oliphant,** who was Secretary to the Mission, has confirmed **Mr. Lay's** statement, and has added that he informed the Chinese Commissioner that he had 'received instructions from **Lord Elgin** not to insist on the insertion of the drug in the tariff, should the Chinese Government wish to omit it. This he declined to do. I then proposed that the duty should be increased beyond the figure suggested in the tariff, but to this he objected, on the ground that it would increase the inducements to smuggling'. **Mr. Lay's** and **Mr. Oliphant's** statements appeared in the *Times* of the 22nd October and 25th October 1880. Their evidence, we venture to think, is conclusive answer to the charge that the clause of the Tientsin Treaty, under which opium was admitted into China, was extorted from the Chinese.

"In point of fact, it is quite clear that a constant struggle is going on in China between the Imperial and the local authorities. The former endeavour to check the consumption of native opium, and are not so much averse to the use of the foreign drug, inasmuch as they derive more revenue from the latter than from the former. The interests of the local authorities are in a totally different direction. To them the use of the native is far more lucrative than that of the foreign drug."

The Government of India in dealing with the question of the power of the Chinese Government—its ability and willingness—to check the use of opium wrote :—"Nominally (the despatch continues) the laws against the use of opium in China are exceedingly stringent." "In issuing these decrees, the Chinese Government may be credited with a certain amount of sincerity. Without doubt, the Emperor of China, and his Ministers, and the most enlightened portion of the population of China, deplore the extensive use of opium. If they could afford the loss of imperial revenue consequent on the importation of foreign opium,—if they could exercise any real control over the numberless corrupt officials who earn a livelihood from the use of

HISTORY:

Attitude of Indian Government.

the native drug,—if they thought it were possible to deal with a great social evil of this sort by legislation, and to coerce a large part of the population of a vast empire into a groove contrary to their inveterate habits and the current of their every-day life,—it is not improbable that they would gladly see the use of opium abandoned. **Mr. Fraser,** the Chargé d'Affaires at Pekin, writing on the 19th October 1878, says:—'Although the Imperial Government must undoubtedly have fixed principles, it is apt to be wanting in definite policy; and there is no question upon which its attitude is so uncertain as the opium question. It is impossible not to believe that the Emperor's Ministers are convinced of the expediency of checking the consumption of the drug by every available means; but it is certain that they have not been able to satisfy the interests involved, and enforce their conviction in the provinces, for more and more opium appears to be produced and used each year in China.' But, whatever be the views which the Chinese Ministers entertain on this question in the abstract, nothing is more certain than that they, equally with the Indian Government, are embarrassed by the loss of revenue which would be caused by a cessation of the opium trade; that, save on rare occasions, when some specially energetic official may have produced a temporary effect, they have up to the present time never earnestly endeavoured to check the use of the drug, and that they recognise their complete inability to do so."

The despatch from which the above passages have been taken deals so fully with every feature of the opium controversy, that it is difficult to extract a limited number of sentences with any hope of conveying even the leading features. Adverting to the argument advanced by some writers that India should cease to cultivate the poppy entirely, the Government of India state in reply to **Mr. Pease** (who had instituted a comparison between the suppression of the opium trade with the abolition of slavery) that he "appears to have thought this was merely a question of money, and that if England were again prepared to pay, the use of opium in China might be suppressed. We submit that there is but a very slight analogy between the suppression of the opium traffic and the abolition of the slave trade."

Compensation to India for loss of Opium Revenue.
180
(*Conf. with pp. 97-98.*)

"Even supposing England were prepared to compensate India for the loss of her opium revenue,—and to do so in any adequate measure, would, we may observe, cost a great deal more than £20,000,000,—and that, in consequence, the cultivation of the poppy in India were to be altogether suppressed, the result would, indeed, be that a connexion, which is by some regarded as involving a moral stigma, would be terminated, but the cessation of opium-smoking in China would be as far off as ever. India would suffer, and China would not gain. The use of opium has taken deep root amongst the Chinese; and we venture to think that the idea that the habits of the Chinese people can be changed by any other means than the slow process of education and moral training, which shall impart to those who are vicious a power of self-control to resist their favourite vice, is illusory and chimerical." (For a further elaboration of the policy of the Government of India, the reader should consult **Sir E. Baring's** Financial Statement for 1882-83.)

"*India*" by **Sir John Strachey,** though in no way an official publication, may be here quoted, since the utterances of that distinguished Statesman seem to be cast very much on the same lines as the declared policy of the Indian Government. In the work mentioned **Sir John** gives a brief but very pointed statement of the opium question. "The first thing," he writes, "to be learned is this, that, although the finest opium consumed in China is Indian, China does not depend on India for her supply. It is a common but complete mistake to suppose that the prohibition of the export of opium from India would have the result of putting a stop to

P. 180

opium-smoking in China. If the supply of opium from India were to cease, the richer classes in China would be deprived of a luxury which they prize, but, so far as the general population was concerned, the consumption of opium would remain much as it was before. Long before Indian opium went to China, opium was consumed there; no one can say how long the custom of opium-smoking has prevailed. A single province of Western China produces more opium than the whole of India; the cultivation is carried on, so far as the Chinese Government is concerned, with perfect freedom, and it is constantly and rapidly increasing. The population of China practically depends for its supply on the opium produced at home.

"If, therefore, all that is said about the ruin of the Chinese by opium were true, the prohibition of imports from India would afford no remedy. But it is certainly not true. Excess in opium, so far as the individual consumer is concerned, may probably be as bad as excess in alcohol; it cannot be worse, and its effects upon his neighbours are comparatively harmless. Used in moderation, as the vast majority of Chinese smokers use it, there is no reason to believe that opium is injurious. I do not doubt that the people of France, and Italy, and Spain are, on the whole, better for their wine, and that the people of England and Germany are better for their beer. Neither do I doubt that, on the whole, the Chinese are better for their opium.

"It is often said that the Chinese Government views the opium trade with dislike and desires its abolition. Whatever may once have been the case, it undoubtedly now desires that the trade should flourish, because it derives from duties on Indian opium a large and highly prized revenue. It has officially disclaimed any wish to see the imports from India diminished. Its real and reasonable object, for which it has long been striving, has been to obtain for itself a larger share of the profit derived by a foreign State from the consumption of opium by Chinese subjects."

One of the chief difficulties in the Chinese opium question is that which has been briefly touched on above, in the remarks regarding the complication of the Chinese Imperial and Provincial interests. By the Treaty of Tientsin (1860) the Imperial Government of China received 30 taels (R95) on every chest imported. But besides that duty opium has always been made to pay a far heavier Provincial charge known as the *likin* dues. These dues were, till within the past few months, imposed by the Provincial Governments at rates fixed by them not only on imported opium but also on what was produced in the country. The latter being greater in volume, and that volume being increased by the decrease of the foreign imports, the Provincial authorities naturally put greater value on local production. Their interests were thus diametrically opposed to the Imperial Government, since the latter obtained no revenue from local production. But these *likin* dues were a never-failing source of annoyance both to the people and to the administration, from the difficulty of collection and the countless facilities for blackmail, over and above the authorised charges, rendered possible in a trade which by Imperial legislation fell under the severest bane of the law. Formerly the *likin* charge on imported opium was not a constant one but often indeed assumed the form of an accumulative transit charge. By the Convention of Chefoo, finally ratified by the British Government February 1887, the *likin* taxes on imported opium were consolidated with the Imperial import duty. All imported opium by present arrangement is deposited in bond, from which it is removed only on payment of the entire duty due to the Government. Finding that the reform thus set on foot was by no means sufficient, the Imperial Government of China has recently advanced further and placed the entire traffic in opium and poppy cultivation under Imperial supervision.

HISTORY :

Attitude of Indian Government.

Effects of Opium Indulgence.
181

Imperial and Provincial Dues.
182

Convention of Chefoo in 1887.
183

P. 183

PAPAVER
　somniferum.　　　　　　　　Poppy Cultivation；

HISTORY :

Attitude of
Indian
Government.

Opium
a
State
Monopoly
in
China.
184

The *likin* dues are now collected by the Provincial Governments and credited to the Imperial revenue. Opium is thus in China as in India a *State Monopoly*. Of imported opium, however, it is currently reported that the working of this new arrangement is not giving satisfaction to the traders. The chances of contraband traffic and of escaping the *likin* charges through the easy bribery of the humbler officials has been greatly minimised, and that being so the profits are much less to the smaller native dealers, hence their purchasing limit has been considerably lowered. There seems little doubt, however, that as far as the Chinese Government is concerned, the traffic, by being thus legalised, is at last placed on a more certain footing, and one from which such restrictions, as may be found necessary to the expansion of consumption, will be within the power of the Government to effect on perfectly constitutional grounds. At present the revenue realised by China on its imported opium alone cannot be far short of £1,500,000; the Indian Government realises on its exports (mainly to China) a net revenue of, say, £5,500,000.* These two sums are, therefore, additional charges to the cost of production of the article, and though they do not very probably express much more than half the total profits made from the consumer of Indian opium, they show conclusively that if taxation can repress luxurious indulgences, opium is being fully brought under that influence. Opium, in fact, pays quite as heavy a tax as the corresponding indulgence of the working classes of Great Britain—spirits—which gives to the Government a little over £14,000,000 sterling a year. The actual revenue from opium in China is of course very considerably greater than here shown, since to the figures given have to be added the excise dues and license fees imposed both on Chinese-grown opium as well as on the imported drug.

CULTIVATION
185

CULTIVATION OF THE POPPY.

In the historic chapter above, reference will be found to the earliest mention of poppy culture in India. From the facts there brought out it will have been seen that the plant was apparently first introduced into India by the Arab traders at Cambay and Malwa, and that its cultivation spread all over the peninsula, chiefly on the higher tracts of the tableland, but ultimately, as a cold season annual, to the plains. On the subject of the cultivation of the poppy in India numerous valuable papers have appeared. The literature of the subject is, in fact, so accessible to the members of the Opium Department and others interested in the subject that it is unnecessary to give in this work more than the merest outline, sufficient to convey a general impression, to those outside the limits of the country. For full details the reader should consult the *Transactions and Journals of the Agri.-Horticultural Society of India;* Dr. W. C. B. Eatwell's *Poppy Cultivation and the Benares Opium Agency, 1851;* Dr. R. Lyell's *Notes on the Patna Opium Agency, 1857;* Mr. T. A. M. Gennoe's *Notes on the Cultivation of the Poppy, 1861;* and Mr. J. Scott's *Manual of Opium Husbandry, 1877.* The late Mr. Scott's work deals with the subject in an able manner, and though it occasionally enlarges on side issues, general agricultural principles, and even speculative problems of plant life, it embodies the results of patient and careful study by one of the foremost horticulturists who has ever lived in this country. It may in passing be said that the chief lessons taught by Mr. Scott's deputation to study the opium plant are that, with judicious cultivation, more manure and deeper ploughing, combined with careful selection of seed, the yield and the quality of the drug can be considerably improved. On the importance of a careful selection of the forms of the plant, and the discovery of the special suitability of these to certain soils and climates, Mr. Scott has much to say of great

---

* For actual figures of Indian Revenue see table at page 97.

Products of India. 47

Varieties of the Plant. (*G. Watt.*) **PAPAVER somniferum.**

CULTIVATION

value. Thus in one part of his work while commenting on the analysis of opium produced on certain soils, he remarks :—" The opium was the produce of one variety ; different varieties vary considerably in their relative richness in alkaloids. I should state," he continues, " that when these experimental analyses were made, I had not then observed the remarkable differences in the drug-producing functions of different individuals,—capsules of certain plants producing scarcely *one grain* of drug, while others yielded from *18 to 30 grains,*—and I had not thus adopted any selective system with a view to the elimination of the scanty drug-yielding plants and the multiplication of the most copious drug-producers only. With such individual differences then, it is easy to see that the relative drug-produce of the respective plots can afford us no criterion of the comparative fertilities of the soil, as being largely, indeed wholly, dependent on the proportion of scanty and copious drug-producing plants on each plot." Referring to this same subject **Dr. Lyell** (many years before the date of **Scott's** discoveries) wrote :—" Only the white variety of poppy is grown in the Patna and Benares Agencies,— a circumstance which probably affects in no slight degree the quality of the drug, for trials made several years ago in Germany tend to show that of the three varieties,—the white, red, and purple,—the white poppy yields the least opium and the purple the most of all, and the opium from the last kind contains nearly three times as much morphia as the white, but only an eighth part of its narcotine, while the opium from the red poppy is intermediate between the two. The reason, according to **Mr. Fleming**, who was for a long period Opium Agent at Patna, why the white variety is preferred is because it is known to suit the climate best. The purple flourishes, however, luxuriantly in Malwa." But in this connection it may be added that **Mr. Scott** found that there were vast differences in the forms of even the white poppy.

Difference in yield of Alkaloids of Different Forms. 186

There would seem to underlie these opinions a far greater importance than the authorities of the Opium Department appear to have given **Mr. Scott** credit for. If his views have been, or can be, confirmed by other investigators, the first step towards improvement would seem naturally to be the distribution of seed of the peculiar form or race that yields opium of the desired quality when considered in relation to its quantitative production. Taking this view of the case the writer feels it incumbent on him to republish the more important facts brought out by **Mr. Scott**, regarding the Indian forms of the opium poppy. (*Conf.* with the concluding remarks on the Varieties and Races of the Poppy, p. 55.)

Value of Mr. Scott's Investigations. 187

CULTIVATED VARIETIES & RACES OF THE POPPY.

VARIETIES. 188

Varieties.—In support of the theory that **Papaver somniferum** has been derived from **P. setigerum** by cultivation, it may be stated that **Mr. John Scott**, who was specially deputed to study the various forms of the poppy grown at the Government Agencies, discovered in a field a solitary plant with bristly or setigerous flower-stalks. He does not expressly state that he regarded the sport alluded to as **P. setigerum** of botanists, but shortly afterwards he obtained from Spain, seed of a poppy which, when reared, he found to closely resemble the sport above mentioned and which he had ever since continued to cultivate under its vernacular name of *darhi.* According to **Boissier** there are three varieties of **P. somniferum.** These may be briefly defined :

Var. 1st, setigerum—Leaves rather acutely denticulate and setose on the terminal lobe ; peduncles and calyx sparsely setigerous; stigma 7 to 8, with openings below.

189

A spontaneous form met with both wild and cultivated in the Mediterranean region.

P. 189

PAPAVER somniferum.	Poppy Cultivation ;

<table>
<tr><td>CULTIVATION
190</td><td>Var. 2nd, glabrum—A rather more robust, often profusely branched glabrous plant, having the leaves less densely lobed than in the preceding; flowers red or purple or variegated and the petals often deeply laciniated ; capsules almost spherical, stigmas 10 to 12, dehiscing with openings below the stigmatic lobes.
Cultivated for its opium in Asia Minor, Egypt, and Persia.</td></tr>
<tr><td>191</td><td>Var. 3rd, album—A glabrous plant with ovate globose capsules, devoid of the openings below the stigmatic surface which are met with in the other forms. It seems probable that red and even variegated flowers occur in this variety, though of course less frequently than in the second form mentioned above.
Cultivated for its opium in India and parts of Persia.</td></tr>
<tr><td>RACES.
192</td><td>Races.—In the Bengal plantations (an expression which will be understood to cover the area from which Patna and Benares Agencies derive their supplies) the plant generally grown is a white-flowered form of Papaver somniferum with pale-coloured capsules, safaid dherri (i e., white capsuled). As already remarked, Mr. Scott devoted much careful study to the subject of the races of Indian cultivated poppies. He describes three sets of forms : first, three depauperated states (escapes from careful cultivation) which he speaks of as " degenerated or wild forms which are so often seen under neglected husbandry ; " second, three cross-bred forms, which he produced from one of his so-called wild species with certain well-known Indian races ; and, third, some 20 cultivated states, known to the opium-growers, of which four had been obtained from Malwa. These 20 forms might be viewed as the Indian produced races of the plant. Before proceeding to deal with these it may perhaps be as well to state here that although Mr. Scott devoted his well-known energies to the subject, all that has as yet been accomplished has been the establishment of the relative value of the various races grown. Much still remains to be done both of a practical and scientific character. It may, in fact, be said that absolutely no progress has been made towards a scientific classification of the cultivated poppies of India. And it need scarcely be remarked that such a classification should naturally precede practical generalisations. Far from this being so, the greatest ambiguity and confusion prevails in works on the subject. Thus the authors of the Pharmacographia Indica say, " The poppy generally cultivated in India is the P. somniferum var. album, with white flowers and white seeds ; but a red-flowered and black seeded variety is met with in the Himálayas."</td></tr>
<tr><td>Bengal.
193</td><td>Speaking of the Malwa poppy these authors add, " It is stated that the variety grown there is the P. glabrum." Messrs. Duthie & Fuller (Field & Garden Crops), on the other hand, inform us that the varieties grown in the North-West Provinces " are all of the white flowered kind, which is found better suited to the climate than the red or purple flowered kind,</td></tr>
<tr><td>Malwa.
194</td><td>extensively grown in Malwa." In the passage quoted above, Dr. Lyell makes mention of the Malwa poppy being the purple plant, a form probably quite distinct from what is generally spoken of as the red. In the Himálaya a red flowered plant is certainly by no means so prevalent as the white, and, indeed, a parti-coloured stock is sometimes seen as more common than the pure coloured forms. If, therefore, any part of India could be specialised as having a red poppy, it would be Malwa, but whether that is actually P. somniferum var. glabrum does not appear to have been determined by any competent botanist. Scott, who cultivated at Patna four forms of the Malwa poppy, viz., lukria, lila, gungajuli, and uggarya, while he says their flowers were coloured or partly coloured, does not say that they were forms of the variety glabrum. He was perhaps not botanist enough to have been able to recognise the varieties of the species, but as he furnishes careful descriptions of the numerous plants grown by him, it would scarcely be justifiable to assume that he failed to recognise that the capsules of the Malwa poppy opened by small pores below the stigmatic folds, while the Bengal poppies were indehiscent, if that was actually characteristic of them. The fact, therefore, of his not observing</td></tr>
</table>

a peculiar dehiscence in his Malwa poppies argues against the possibility of their having belonged to *var.* glabrum. The definition of **P. somniferum,** however, as given by most botanists, provides for a considerable variability in the shape of the leaves and the form and colour of the petals. It is, therefore, by no means a matter of necessity that the red coloured poppies of Malwa belong to the variety glabrum. This discussion, as to the possible botanical nature of the opium poppies of India, founded on the peculiar expressions used by Indian authors, has been forced on the writer, owing to his not having had the opportunity of investigating the subject personally. It will, therefore, perhaps, be readily admitted as desirable that we should not much longer remain ignorant on so all-important a feature of the poppy industry. The subject is by no means one of purely botanical interest. The recognised merits of the opium of this country and of that depend upon a number of circumstances, such as climate, soil, systems of agriculture, methods of preparation and purity; but it cannot be denied that into that category must also be placed (and perhaps foremost of all other considerations) the peculiar merits of the variety or race of poppy grown. Even when cultivated side by side in the garden, Mr. Scott noted that the various races which he found in the hands of the cultivators, or which he was able to pick up as mere sports, or to produce by cross-breeding, yielded materially different results in quantity and quality of the drug. It is freely admitted that the highly valued medicinal form of opium obtained from Asia Minor is the produce of the variety glabrum. At the present moment we possess no proof as to whether or not that plant is grown in India. But to return to the forms of the Indian poppy, described by Mr. Scott, it would seem that many were, as he states, mere degenerations through neglect or from poverty of soil. Others were hybrids, if they may be so called (between recognised races), which he had himself produced, while a third class were sports, found in the fields and which he had cultivated in the hope of discovering more profitable strains than already existed. The writer has very carefully considered all the statements made by Mr. Scott, and has come to the conclusion that the following forms described by him are probably recognisable races, most of which seem to be known to the opium growers of India. A selection of Mr. Scott's forms must of necessity, however, be made with considerable diffidence, since that painstaking observer is the only writer who has dealt with the subject, and any such selection may tend to detract from the value of his work. At the same time it seems desirable to separate Mr. Scott's experiments in the production of what he calls "Mongrels" from old established cultivated races.

1st, Safaid-dherri.—"The white capsuled poppy is the most common of the varieties, and is chiefly distinguished by its roundish, oblong, and glaucous, or powdery—*i.e*, covered with a fine bloom or white powdery substance—capsules. Mixed with it, less or more, however, we find varieties with rather small sized, round or oblate, and powdery capsules, which as a rule are very scanty drug-producers. So also another sort with roundish oblong capsules of a dull olive green colour, and but slightly, if at all, powdery. The rind of this variety gets dry, hard, and almost woody before the capsule is full grown, and rarely affords more than one drug incision In selecting seed, therefore, it is well to avoid either of those, and especially the latter. The true white-capsuled variety, as indicated above, is, as a drug-producer, with the exception of the *teyleah,* one of our best." It gives a discharge of drug with from four to nine or more incisions with ordinary 4-bladed *nashtar* or lancet. "The drug is also of excellent quality, and, in the sample analysed for me by **Dr. Durant,** contained in nearly equal proportions, a total of 7·04 per cent. in *morphine* and *narcotine.*"

CULTIVATION Races.

Races. 195

Hybrids. 196 Sports. 197

Bengal Forms. *Safaid-dherri.* **198**

PAPAVER **somniferum.**	Poppy Cultivation;

CULTIVATION
Races.
Kalodanthi.
199

Monaria.
200

Teyleah.
201

Kutila.
202

Sabza kutila.
203

2nd, "Kalodanthi," or Black-stalked poppy.—"This is a compact branchy plant of a somewhat dwarfer habit than any of our other local varieties. It is readily distinguished from these by the stem and flower stalks acquiring a purply-black colour shortly after the fall of the flower, and before the maturation of the capsule. The latter are rather small, globose or roundish-oblong, glaucous, and from 1¾ to 2 inches long by 1½ to 2 in diameter. It is an excellent variety, though not so copious a drug-producer—on the lands here at least—as the *teyleah* variety. It would also appear to produce a drug less rich in alkaloids than some of the others, as in a sample analysed for me by Dr. Durant the total alkaloids at 70° consistence, was only 6·38, of which 2·98, however, was *morphine*."

3rd, "Monaria."—"A local variety so-called, and sent to me in the season of 1873-74 by the Sub-Deputy Opium Agent of Patna. It is distinguished by its large, round or spherical, *i.e.*, vertically depressed, capsules (from 2 to 2¼ inches by 2¼ to 2½ in diameter), covered with a fine white powdery matter. The better capsules afford a fairly copious discharge of drug to from five to nine incisions; but crops from the ordinary unselected seeds of the district cover a very large percentage of such as are wholly exhausted of their drug by the third or fourth incision. It would appear, according to the analysis of the sample sent by me to Dr. Durant, to be less than any of our other varieties suitable for officinal use, as deficient in alkaloid matters, the total *narcotine* and *morphine* being only 6·19 per cent. Of this, however, 2·88 consist of the latter."

4th, Teyleah, or Sabza-dherri.—"This is perhaps the most copious drug-producing variety of the opium poppy. Unfortunately it is far less generally cultivated than some of the other varieties, especially the *sufaid-dherri*: No. 6 of this list. In habit it is very similar to the latter variety, but the capsules are ovate-oblong (from 2 to 2½ inches long by 1½ to 2 inches broad) and of a dull-green colour, destitute of the fine white powdery matter of the above. It appears, however, to produce a drug less rich in alkaloids than that of some of the other varieties, as the sample analysed for me by Dr. Durant contained only 6·57 per cent., of which 3·16 was morphia. It has all the physical characters, however, of a first-class commercial article, and its comparative paucity in alkaloids does not at all lessen its value in the China market."

5th, "Kutila," or "kat-patta."—"This is a very distinct variety, and readily distinguished by its foliage being deeply cut into less or more narrow segments. The plant is of robust habit, the main stem giving off upwards a few simple branches; the leaves thick-textured, of a glaucous or pale sea-green colour, oblong-ovate, from 5 to 9 inches long by 4 to 5 inches broad, and bipinnately cut into narrow and bluntish segments; the capsules 2 to 2½ inches long by 1½ to 2 inches broad, oblong or ovate-oblong, and covered with a fine white powdery matter. This variety is decidedly worthy of a more extended cultivation. As I will subsequently show, it is less liable to be seriously infected with the poppy-mould than any of the other varieties commonly cultivated. Again, the thick and comparatively firm texture of the leaves, divided as they are (nearly to the mid-rib and primary veins) into narrow segments, enables it to withstand hailstorms of force sufficient to wholly mince the broad and but slightly divided foliage of the other varieties. It is a fairly copious drug-producer, and in the sample analysed for me by Dr. Durant the total alkaloids was 7·21 per cent., of which 3·56, or nearly half, was morphia."

6th, "Sabza kutila."—"This is a very rare variety as yet, but, like the other green-capsuled sorts, promises to be a copious drug-producer. It is only distinguished from the preceding variety by its capsules, which are of a dull-green colour, entirely void of the fine powdery matter which covers the capsules of that variety. I first observed a single slender specimen of this variety in one of my plots of *kutila* in the season 1873-74. I sowed the seed, but unfortunately none germinated. Again, in the following season. I found another plant which has this season afforded me a few more. I have also unexpectedly got an addition to these from a mongrel progeny raised first in the season 1874-75. The cross was with *kutila* as the female and the *safaid-dherri*, or common glaucous capsuled poppy, as male. The effects of the cross were clear on the first season's progeny ; as with many of the *kutila*-foliaged variety, there was also a considerable percentage with the leaves of the male parent— the remainder intermediate. This season's progeny presents a very small percentage of the male variety, nearly all having the *kutila* foliage, and considerably the major part of the remainder differing chiefly from it in having more entire leaves. Curiously enough, this season I observed that not a few of the *kutila*-foliaged plants had capsules of a dull-green colour, quite destitute of powdery matter. This is also the case, and more generally, with the more entire-foliaged forms. I was surprised to observe this, as in the previous season, without an exception, every plant bore glaucous capsules."

7th, "*Dusra kutila.*"—"This is another very distinct variety of more robust habit than the *kutila* : plants rather branchy ; leaves oblong-ovate, from 6 to 12 inches long by 5 to 7 broad, rather deeply cut from about the middle upwards into from 4 to 6 broad, coarsely toothed, blunt, and wavy-margined lobes ; texture thick and firm ; the colour a pale sea-green, the capsules roundish, about two inches in both diameters, covered with a fine white powdery matter. It affords, as a rule, a fairly copious discharge of drug to from five to eight incisions. I have, however, as yet only a very few plants of this variety."

8th, *Chaura,* or "*Tisra kutila.*"—"A robust but sparingly branched variety, with oblong, irregularly lobed leaves of a dull sea-green colour, from 12 to 20 inches long by 4 to 7 broad ; the lobes acute or tapering ; the margins coarsely toothed and wavy. The capsules are of an oblong shape, from 2 to 2$\frac{3}{8}$ inches long by 1$\frac{5}{8}$ to 2$\frac{1}{4}$ in diameter, glaucous, and fairly copious drug-producers. The opium, as judged by the sample analysed for me by Dr. Durant, is somewhat richer in alkaloids than the true *kutila*, yielding in *equal* proportions 7·42 per cent. of morphia and narcotine."

Of the other forms described by Mr. Scott two are striking sports, the others are admittedly hybrids. The special sports may be here dealt with :—

9th, "*Darhi, or Muedardanthi.*"—"This is a well-marked new variety, with bristly flower-stalks which I first observed in one of the poppy plots here in the season 1873-74. (A Spanish variety of the opium poppy which I had this season from the Board of Revenue, and directly imported from that country, presents the same peculiarities in the flower-stalks, but is very distinct in habit and foliage.) There was but a single plant of it. I collected the seed, and in the following season raised from it a similarly characterised progeny. The plant is of a robust, somewhat branchy habit, and readily distinguished by its strongly *setigerous flower-stalk*. The capsules are large, glaucous, and very similar in size and shape to the preceding variety. It has hitherto proved a less copious drug-producer than that variety, but this will no doubt be remedied by the selection of seed from the best producing capsules. The drug, however, is of excellent quality, being richer in total alkaloids by 1·03 per cent. (thus in morphine 3 48, narcotine 5·03) than any of our other cultivated sorts : another important quality is its comparative immunity from the mould infection. During the season of 1874-75, when the mould was so mischievously prevalent on all the other common varieties in the gardens here, this plant was but very slightly affected. Indeed, for weeks after the sorts on the

CULTIVATION
Races.

Dusra kutila.
204

Chaura.
205

Darhi.
206

PAPAVER
  somniferum.   Poppy Cultivation ;

CULTIVATION   adjoining plots had been overrun and destroyed by the mould, this variety
  Races.   resisted it very effectively, retaining to the last its normal functions for the
              secretion of drug.   This season (1875-76), which proved so very unfavour-
              able to the germination and health of the early-sown plant, thinned con-
              siderably my plot of this variety, but fortunately a sufficiency of plants
              escaped to afford me a few seers of seed.   The plant retains its robust
              habit, *setigerous* flower-stalks, with the large spherical capsules."
                 This is the form alluded to above as approximating to the type of
              Papaver setigerum.   Mr. Scott does not say that its capsules were
              dehiscent, nor did he apparently record the number of its stigmatic folds.
              (*Conf.* with the definition of *var.*setigerum, p. 47.)
  Gunagun-   *10th, Gunagun-posta, or Sufaid-patta.*—" This variety is distinguished
  posta.   by the variegation of its leaves, stalks, and capsules, all of which are more
  207   or less streaked and blotched with white.   I first observed a few specimens
              of it in the poppy plots here in 1873 and 1874.   Like the preceding, it is
              also but little liable to the poppy murrain.   In my annnal reports for 1873
              and 1874 to the Board of Revenue, I suggested what appeared to me to be
              the cause for this comparative immunity of the variegated   plant.   Subse-
              quent observations confirm that view, which will be explained in treating
              of the diseases and injuries of the poppy.   In the meantime I may state
              that last season (1874-75), when all the generally cultivated varieties, with
              the exception of the *kutila*, were seriously affected with the poppy mould,
              this variety was scarcely at all injured by it, the extension of the mould-
              spawn in the leaf being wholly confined to the green streaks and blotches;
              in no instance did I observe it cause any injury to the white parts.   Un-
              fortunately there is at present one. objection to the multiplication and
              extension of this variety, and that is, its scanty drug-producing quality.
              Thus the capsules, though of large size, about $2\frac{1}{4}$ inches in both diameters,
              as being round and vertically compressed, rarely afford more than three
              or four drug-yielding incisions.   This will no doubt be obviated, however,
              by continued selection of seed from the most copious drug-producing cap-
              sules.   The opium has also all the physical qualities of a good commercial
              article and yields 6·57 per cent. of alkaloids."
                 The above ten special forms are those of the Bengal Agencies which
              seem deserving of special notice.   It may be as well, however, to give here
  Malwa   Mr. Scott's four forms of Malwa Poppies. He writes of these :—
  Forms.   " I will briefly notice here four of the Malwa varieties which were intro-
  208   duced to this district in the season 1871-72, but all were clearly unsuited
              to the climate and soil.   They grew up spare and sickly, and Mr. Aber-
              crombie remarks in his report to the Board that they 'did not in any in-
              stance produce more than a *seer* of opium per *bíghá*, and generally only
              a few *chittacks*, while local seed on adjacent lands produced in some cases
              from eight to ten *seers*.'   The cultivation of the Malwa varieties by the
              Assamese was at once discontinued, and the seed they had collected pur-
              chased from them, to prevent them making use of it for the next season's
              sowings.   A small portion has been tried by some of the sub-divisional
              officers in their gardens to ascertain whether the acclimatised seed will be
              increasingly productive.   In the following season (1872-73) there was but
              a very slight increase in the opium produce, and I rather think the sub-
              divisional officers very generally discontinued their cultivation.   I, that
              season having been specially deputed by Government to the opium dis-
              tricts (to ascertain the nature of the poppy blights, and suggest, if possible
              mitigative or remedial measures for them, as also to have regard to the
              modes of improving the general husbandry of the plant), had likewise
              seeds of the different Malwa sorts for trial in the gardens which I had
              opened at Bankipore.   The plants, however, grew up again poor and sick-
              ly, and gave a very small quantity of drug.   In 1873-74 I repeated the

CULTIVATION
Races.

experiment on the Deegah land. The plants generally acquired a some-what more robust habit, but they again proved miserably poor drug-pro-ducers, the maximum produce being about 4½ seers per *bighá*. It is to be observed, however, that even this is a considerable increase on the first season's results. Again in 1874-75 I gave them a third season's trial. They all suffered, however, so seriously with the common poppy mould, that they gave considerably less opium than they had done the previous season. This blight injury, however, they suffered from in common with nearly all the local varieties. I will now describe the Malwa varieties from the specimens grown in the gardens last season : having now been grown in this district for *five* successive seasons, a period sufficient, as one might naturally anticipate, for the perfect acclimatisation of an annual herb in a district, where others of its own kin are so extensively and successfully cultivated."

Lukria. **209**

11th, Lukria.—"This variety has increased much in vigor of habit, and is, so to speak, more hardy, *i.e.*, better suited to the climate, &c., than it has hitherto proved. It has an erect, slightly branched stem from 4 to 5 feet high ; the leaves oblong-ovate, from 12 to 16 inches long by 5 to 8 inches broad, the *margins* sinuous, irregularly lobed, coarsely and bluntly toothed ; texture thin, papery ; the colour a pale sea-green ; the flowers, large, white, usually with rose, lilac, or pink margins, and deeply fringed ; the capsules oblong, from 2 to 2½ inches long by 1¾ to 2 broad, and of a dull glasseous green. It has now considerably improved in its drug-yielding qualities, having this season yielded 6s. 9c. per *bighá* of excellent drug."

"The rains being unusually light last season, not a few of all the sorts sprung up in the plots from seed which had been left on the soil in the harvesting of that season's crop. This was the case with the *lukria*. The plot was sown up this season with a variety of black-seeded poppy from Turkey. Not a single seed, however, germinated of it on this plot, nor on two other plots on which it was sown. The few young plants that did spring up on the first plot ultimately proved to be of the *lukria* variety, cultivated thereon last season. The plants are remarkably vigor-ous in habit, more resembling the more robust of the local sorts than the hitherto spare forms of the Malwa kinds. They rise up with stout, branchy stems to a height of from 4 to 5 feet ; the leaves large, oblong, with slightly lobed, irregularly toothed, and wavy margins, the lower leaves being from 16 to 21 inches long by 8 to 9 inches broad, of a thick, but soft texture, and of a dull sea-green colour. A very few have coloured flowers ; the others large, white, and deeply fringed. The capsules large, from 2½ to 3¼ inches long by 2 to 2½ in diameter. They have also proved a much more copious drug-producing race than any of the other Malwa kinds sown in the usual way. Thus from the 125 plants on the plot I collected 1¼c. of opium, so that allowing 14,520-12 inches apart, equal to the *bighá* (a low rather than a high estimate), we have a return of about 9s. 1c. of opium per *bighá*. We thus see that the process of acclimatisation has been greatly accelerated by a purely natural selective agency, the more vigorous germs, or those best adapted to the conditions of climate, soil, &c, surviving ; the weaker we may assume to have been destroyed. This, as I think, well illustrates the much disputed phenomena of accli-matisation. In the Malwa and local poppies we have clearly varieties endowed with different constitutions best adapting them to their respective localities, while by a few years' cultivation we find this variety acquiring increased vigor of habit, yielding more drug, and ultimately equalling the local sort."

Leela. **210**

12th, Leela.—"The seed of this sort was much mixed with that of the preceding in the sample originally received by me from the Agent. It presented, however, a considerable percentage of two well-marked

CULTIVATION
Races.

sorts which I have this season separated from the others. They are of vigorous but somewhat dwar habit, 3½ to 4 feet high, rather branchy, the leaves ovately oblong, 9 to 12 inches long by 5 to 7 broad, the margins but slightly lobed and coarsely toothed; the texture rather thick and firm; the colour a dull sea-green; the flowers large, white, with entire margins; the capsules roundish-oblong (2½ to 3 nches long by 2¼ to 2⅜ broad) and glaucous in the one variety—*leela*; in the other—*sabsa leela*—rather globose, about 2½ inches in both diameters, of a dull-green colour, and void of the white powdery matter coating the rind of the other. This season (1875-76) they have yielded an average of 5*s.* ½*c.* of opium per *bíghá*."

Gungajulee.
211

13th, Gungajulee.—"This sort is very like *lukria* in habit and growth. The leaves oblong to ovate-oblong, from 10 to 16 inches long by 5 to 8 broad, of a pale-green colour, the margins less or more lobed and coarsely toothed; flowers large, white, with or without rosy or pink margins, and deeply fringed; the capsules large, oblong, 3 to 3½ inches long by 2⅓ to 3 broad, less or¹ more furrowed longitudinally, and covered with a fine white powdery matter. The drug-produce this season averages 8*s.* 12¾*c.* per *bíghá*."

Uggarya.
212

14th, Uggarya.—"This is the least robust of all the Malwa kinds above described. The stems slender, and nearly simple, from 3 to 4 feet high; the leaves oblong, from 9 to 12 inches long by 5 to 7 broad, of a pale-green colour; the margins sinuous and toothed; the flowers smaller than those of the other varieties; the base of the cup stained with dull purple, the margins carmine-red or lilac, and generally deeply fringed; the capsules oblong or ovate-oblong, from 2 to 2½ inches long by 1¼ to 2½ in diameter, glaucous, and containing seeds of a pale purply-grey colour, very similar to the maw-seed. The drug-produce of this variety averaged only 3*s.* 14½*c.* per *bíghá*."

"It has been a matter of surprise to many that the Malwa varieties should have proved so ill-suited to the climate or soil of the opium tracts in Behar and Benares. I have already remarked that four of the most productive of the Malwa poppies have now been cultivated for *five* successive seasons here, and it must be confessed they have not yet attained their normal standard of productiveness in Malwa, though now exceeding the average produce of the local varieties in the Behar Agency for the past ten seasons, the highest average being that of 1863-64, which was 6*s.* ⅜*c.* per *bíghá*. The local poppy of Behar succeeded much better in Malwa, the minimum produce of the directly imported seed, according to the Superintendent of Rutlam, being 3*s* 5*c.* per *bíghá*, the maximum 5⅓*s.*, thus affording an average of 3*s.* 7*c.* per *bíghá*. This, I may remark, exceeds the average produce of the local poppy in *four* of the *twelve* divisions of the Behar Agency in 1870-71, by 6*c.* per *bíghá*, and indeed is only 5*c.* under the general average. A consideration of the climatic conditions of the two provinces affords, as I think, an explanation of the indifferent success of the Malwa plant in Behar. Malwa has a very uniformly mild and humid temperature, the cold season being extremely agreeable, and many of the products of the garden in temperate latitudes, which are destroyed by the heat of the Lower Provinces, succeed well on the table-land. The soil is extremely fertile, very generally consisting of a black vegetable loam, producing fine crops of cotton, indigo, opium, tobacco, wheat, &c. The cold season in Malwa, it is to be observed, differs much from that of Behar. It is never subject to the dry westerly winds so prevalent in the latter, nor is there those extremes between the day and night temperature. These with the atmospheric aridity are, as I think, the main cause of the indifferent success of the Malwa poppy in Behar. The foliage of the plant is of a much thinner and delicate texture than that of the local poppy, and it

CULTIVATION Races.

is thus exceedingly susceptible, in the early stages of its growth, especially, to the cooling of the soil by the rapid and incessant evaporation of moisture. As illustrating this, I may observe that some of the most healthy of my Malwa plants have always been those sheltered less or more from the westerly winds. I have observed also that under a soft and moist easterly wind they quickly acquire a more healthy appearance. Again, if slightly overshadowed by a tree, they are also greatly more vigorous than when fully exposed : this can be only due to decreased radiation and its consequent chilling effects, and some little perhaps to the plants being slightly screened from the direct rays of the sun. The chilling effects, however, of radiation is, as I believe, the main cause of the hitherto indifferent success of the Malwa plant here, and only when it has acquired a thicker and firmer textured foliage (as it now evidently has) will it afford a remunerative crop " (*Man. Opium Husbandry, pp. 44-52*).

Possible Improvement. (*Conf. with p. 73.*)

In a further page it will be found that Mr. Scott, while dealing with the subject of the fungoid diseases to which the poppy plant is liable, pointed out that an even greater interest centres around his discovery of the existence of distinct varieties and cultivated races, than the fact of these affording varying proportions of the drug or of its separate alkaloids. As with the potato and many other cultivated plants so with the poppy, Mr. Scott noted, that while one race was invariably severely injured by disease others enjoyed a comparative or even complete immunity from it. It may thus be inferred that one of the most hopeful courses of improvement would be to foster the cultivation of the races of poppy which yield the maximum of the drug commensurate with the minimum tendency to disease.

AREA OF POPPY CULTIVATION & YIELD.

Area & Production. 213

In modern language "Bengal opium" means opium manufactured at Patna and Ghazipur as distinct from Turkey, Malwa, and other opiums. The area and yield of Bengal opium are annually known with the utmost accuracy, but that which falls under the definition of " Malwa " and " Other Indian Opiums " can only be learned from the reports furnished by the chief producing Native States. In the Panjáb, opium is grown in almost every district, more or less, but the area is regularly declared. The permission to grow the plant and to manufacture the drug in the Panjáb is looked upon as too unimportant to call for special interference more than is done in the case of opium cultivation in certain mountainous tracts and petty Native States within British districts. The Panjáb opium is consumed locally, and the fact that it does not prove sufficient to meet the demand seems satisfactory evidence that the supervision and restriction imposed are sufficient to keep the traffic within legitimate bounds. Throughout the length of the Himálaya the poppy is grown, but little or no control is exercised, since nothing has as yet transpired to show that there is any export from these tracts. It may thus be said that wherever semi-independent Native States and wild mountainous country occur, a limited poppy cultivation is permitted, through the habit of non-interference, but that, with the exception of the Panjáb, no other British province is allowed to grow the plant save in the tracts that have been brought under the direct supervision of the Agencies of Patna and Benares. The accuracy of the returns of area furnished by the chief Native States that produce the Malwa opium, is doubtless open to suspicion. Further, the amounts registered at the weighment stations cannot be used as a factor to check these returns, since it is well known that, in spite of all precautions, considerable quantities do actually percolate illicitly into British districts. At the same time the local consumption within Native States is not known, so that the returns, such as they are, have to be accepted. Speaking generally,

PAPAVER somniferum. — **Area & Production.**

CULTIVATION Area & Production.

therefore, it may be said that the opium-producing area of British India lies in the Gangetic basin and follows the natural configuration of the country from Monghyr (87° E. long.) to Agra, Muttra, and almost Delhi (78° E. long.), and between a line on the south from Gya, Allahabad, Banda, Jalaon, and Agra (24° to 26° N. lat.) to a similar line on the north from Champarun (Moteeharee), and Keri to Moradabad (28° N. lat.). Beyond these limits a casual or insignificant cultivation of course exists from Kuch Behar in the east to Peshawar and Dera Ismail Khan in the west, but the British districts of Bengal, east of Monghyr, which formerly were permitted to grow the drug, have been prohibited from doing so since 1840. The area defined approximately embraces the region that produces the Bengal opium.

Bengal opium. **214**

The following three tables exhibit certain important features of the Bengal opium traffic:—

YEARS.	Poppy area in acres.	Opium produced in maunds.	Number of Abkari chests made.	Number of Provision chests made.	PRODUCE PER BIGHA. Behar.	Benar s.
					Sr. Ch.	Sr. Ch.
1860-61 . .	272,086	58,168	3,107⅓	29,398	4 12¼	6 6¼
1861-62 . .	388,228	75,044	3,019⅔	39,656	4 9	5 4½
1862-63 . .	445,221	93,583	3,190	49,727	6 1⅓	5 6¾
1863-64 . .	505,408	1,19,517	2,622	64,269	6 0	5 11¾
1864-65 . .	478,241	86,276	2,384	47,785	4 8	4 8¼
1865-66 . .	398,644	81,327	4,157	40,901	4 14¼	5 12¾
1866-67 . .	438,798	93,136	4,596	48,895	5 3¾	5 6¾
1867-68 . .	453,159	83,750	5,277	43,610	4 6¾	4 15
1868-69 . .	430,863	86,019	4,458	46,894⅖	4 11¾	5 5¾
1869-70 . .	485,016	99,124	2,579	54,072 7/40	5 5¼	4 12½
1870-71 . .	515,851	76,739	3,114	40,981⅕	3 12	3 10¾
1871-72 . .	535,452	81,431	3,431	42,975	4 2¾	3 4½
1872-73 . .	509,053	88,104	4,016½	45,770	4 10½	3 14
1873-74 . .	456,069	1,03,862	3,637½	54,716	5 15	5 6¼
1874-75 . .	543,541	98,178	3,892¼	51,754	4 10	4 5¼
1875-76 . .	530,730	1,28,817	3,972½	68,051	5 13¼	6 5¾
1876-77 . .	556,013	1,25,256	4,326	67,167	4 14¾	6 9¾
1877-78 . .	467,665	79,383	4,261	43,140	3 5½	5 4¾
1878-79 . .	506,943	97,905	5,605½	49,961	3 15½	5 11½
1879-80 . .	562,261	99,070	4,221	52,969	4 10¼	4 2¼
1880-81 . .	536,017	94,798	4,378	49,733	4 8½	4 4¼
1881-82 . .	531,275	1,00,889	3,959½	54,039	4 4¾	5 4½
1882 83 . .	495,740	69,287	591⅙	38,2'4	2 12	4 3¾
1883-84 . .	505,843	1,26,597	2,146⅓	65,993	5 15¼	6 8½
1884-85 . .	565,246	1,33,803	4,32'5/12	64,930	5 2¼	6 10
1885-86 . .	594,921	1,21,500	2,291	64,500	5 4¼	4 15¼
1886-87 . .	562,052	1,07,577	2,831½	57,500	4 5¼	5 4¼
1887-88 . .	536,607	1,23,043	3,367	69,500	5 6	6 2¾
1888-89 . .	459,864	67,732	3,485	38,305	3 2¾	4 3¾
1889-90 . .	482,557	95,868	6,320	44,760	4 3½	5 9½

It will thus be seen that during the past 30 years the average annual area under the crop in the Government Agencies, has been 490,000 acres (approximately), and that it has never attained 600,000 acres. From that acreage the annual production, during the period shown, has been about 96,000 maunds, of which less than 2,500 maunds have been consumed in India. The balance on total production would be approximately the amount which has been consigned to China.

The totals shown in the above table, for the past seven years, may now be analysed under the headings of the two Government Agencies:—

Area & Production. (G. Watt.) **PAPAVER somniferum.**

CULTIVATION Area & Production.

Extent of Cultivation and Total Produce in the Patna Agency.

OPIUM SUB-AGENCIES.	Bighás CULTIVATED WITH POPPY (OPIUM bighá = ⅛th OF AN ACRE).							PRODUCE IN maunds.							Average yield per bighá for the years 1883 to 1890.	
	1883-84.	1884-85.	1885-86.	1886-87.	1887-88.	1888-89.	1889-90.	1883-84.	1884-85.	1885-86.	1886-87.	1887-88.	1888-89.	1889-90.	Srs.	Ch.
Tirhoot	17,255	19,048	22,199	26,765	25,638	21,142	24,367	1,686	2,033	2,318	2,443	2,410	1,285	1,550	3	15
Hajeepore	13,138	11,256	15,338	17,444	13,227	14,119	11,507	1,538	1,321	2,009	1,871	1,721	1,160	925	4	9
Chupra	28,094	29,923	26,425	31,103	29,883	27,969	27,701	4,112	3,484	4,495	3,518	4,470	2,637	2,782	5	9
Alligunge	41,155	42,640	42,612	44,031	43,286	40,774	38,728	6,153	6,409	7,325	6,465	7,455	4,336	4,516	6	5
Moteeharee	55,208	71,232	74,343	74,142	74,300	57,787	63,041	4,842	6,683	6,970	6,968	6,534	3,354	5,021	3	10
Betteah	33,983	39,512	41,430	40,564	37,488	35,456	33,311	3,160	4,515	4,141	3,799	3,905	2,474	2,553	4	0
Shahabad	35,198	35,807	36,623	35,711	34,556	32,239	29,384	6,992	5,751	5,621	4,376	6,050	3,001	4,361	6	7
Gya	63,420	65,906	69,117	65,596	66,044	62,832	66,238	10,992	8,725	9,814	7,066	9,969	5,005	7,802	5	10
Tehtah	42,986	45,036	46,972	45,741	46,178	42,455	43,152	8,231	6,555	6,466	5,43	6,843	3,809	5,812	5	15
Patna	39,147	39,469	43,038	42,396	41,846	40,754	38,445	7,079	6,151	6,430	4,472	6,444	2,964	4,738	5	15
Monghyr	29,934	33,332	35,417	34,673	35,313	36,339	28,356	4,775	4,175	4,277	3,16	4,391	2,158	3,057	4	15
TOTAL { Bighás	399,518	433,161	453,514	458,266	447,759	405,866	398,230	59,560	55,802	59,866	49,582	50,192	32,233	43,147	5	2
TOTAL { Acres	249,699	270,726	283,446	286,416	279,849	253,666	248,893									

PAPAVER somniferum.   Area & Production.

CULTIVATION
Area & production.

Extent of Cultivation and Total Produce in the Benares Agency.

OPIUM SUB-AGENCIES	Bighás cultivated with Poppy (bighá = ⅚th acre)							Produce in maunds							Average yield per bighá for the years 1883 to 1890		Remarks
	1883-84	1884-85	1885-86	1886-87	1887-88	1888-89	1889-90	1883-84	1884-85	1885-86	1886-87	1887-88	1888-89	1889-90	Srs.	Ch.	
Ghazipur	36,133	39,214	39,836	39,856	38,471	21,139	20,995	6,092	5,820	6,150	4,549	6,248	2,316	2,864	5	15	
Mirzapur	...	...	...	...	...	16,690	18,421	...	...	...	...	...	1,728	3,021	...	...	
Azimgurh	15,377	17,677	20,574	20,263	19,028	15,531	17,231	2,759	3,186	3,252	2,674	3,059	1,691	2,218	6	7	
Gorakhpur	29,118	30,781	30,720	31,477	30,550	27,925	29,478	4,361	5,008	4,707	4,501	5,134	3,113	3,598	6	3	
Bustee	36,054	38,434	40,235	34,983	31,388	29,425	25,215	4,810	6,181	5,406	4,625	4,765	2,926	3,515	5	11	
Allahabad	22,505	23,983	24,418	23,349	23,074	13,050	14,421	4,937	4,594	3,312	3,907	4,059	1,186	2,085	7	1	
Etawah	...	...	...	...	...	11,364	13,103	...	...	...	...	...	832	1,218	...	...	
Cawnpore	30,450	11,237	12,349	10,928	10,817	...	...	3,818	1,057	862	764	814	...	...	3	7	
Futtehgarh	45,065	34,192	36,433	29,965	26,076	19,056	20,735	7,042	4,386	3,106	2,804	2,935	1,695	2,555	4	9	
Mainpuri	...	33,449	31,453	33,606	28,842	14,174	15,959	...	4,436	3,397	3,002	3,157	1,343	2,276	4	3	Average for four years only.
Rohilkhand	29,696	40,668	47,741	32,689	29,969	23,171	27,635	5,117	7,382	4,184	3,822	4,312	2,643	4,271	5	9	
Aligarh	6,540	9,098	11,310	8,976	8,093	...	...	937	1,156	743	897	895	...	...	3	9	
Lucknow	59,056	56,118	57,969	50,743	48,033	24,889	31,241	9,949	10,283	7,925	7,759	8,052	3,331	5,133	6	7	For four years only.
Sitapur	18,878	18,878	19,300	16,105	15,347	18,718	21,353	9,949	10,283	2,135	2,109	2,188	2,442	3,533	5	9	
Faizabad	38,030	43,809	46,552	40,142	36,812	18,159	21,639	5,360	7,365	6,178	5,902	5,874	2,285	3,502	5	15	
Gonda	...	...	...	...	...	18,708	27,023	...	...	...	...	...	1,916	3,186	...	...	
Pertabgarh	...	...	...	...	28,663	25,916	31,752	...	...	...	...	5,116	2,611	4,338	7	2	For 1887-88 only.
Rai Bareilly	...	39,910	41,053	38,263	35,650	32,043	36,661	...	7,517	5,810	5,917	6,233	3,441	5,407	6	9	For four years only.
Sultanpur	61,747	33,784	32,427	29,673	...	...	...	11,875	5,839	4,467	4,763	...	...	...	6	10	
TOTAL { Bighás	409,831	471,232	498,350	441,018	410,813	339,917	373,862	67,037	78,001	61,634	57,995	62,851	35,499	52,721	5	11	
{ Acres	256,144	294,520	311,475	275,636	356,758	206,198	233,664										

P. 214

From these tables it will thus be seen that the area under the poppy, the production of opium, and the consumption of the drug in India have been practically stationary, during the past thirty years, as far as the Patna and Benares Agencies are concerned.

CULTIVATION

Area & Production.

Turning now to the other Indian opiums it may be said that the Panjáb has on an average a little over 13,000 acres under the poppy (and a production of about 1,400 maunds); the Rajputana States, 178,757 acres (and a produce of 40,000 maunds) ; Ajmír-Merwara, 2,854 acres ; Central Indian States, 243,494 acres (and a produce of 65,500 maunds); and the Native States of Bombay, the Central Provinces, and Madras a small annual acreage of which at present no definite information exists. In the *Bombay Gazetteer*, for Baroda, for example, it is stated that prior to the establishment of a State monopoly by the Gaikwár's Administration the area under opium was 8,301 acres in 1878, but that it fell off to 1,376 acres and has since recovered (1880) to 5,936 acres. Cultivation on a small scale exists in Mysore, in the Bangalore and Kolar Districts, and formerly there was a little also by the hill tribes of Madras (Nilghiri hills chiefly). This has been since prohibited, and the opium required by Madras is imported from Bombay. Reference has been made to the fact that throughout the Himálaya a limited cultivation exists, but the same may be said of Assam and Burma in tracts outside the limits of the British possessions. In Assam proper, cultivation is not only prohibited but severely punished, so that it is met with only in the mountainous country inhabited by the aboriginal tribes, such as the Dufflas, Miris, Abors, Mishmis, Singphos, Manipuris, &c. In this connection it may be said that, with the exception of the Miris, the hill tribes of Assam, like the people of the valley, eat or drink the drug, the consumers of it being known as *kanias*. The Miris appear to smoke it, but the habit of opium-smoking is so little indulged in by the people of the eastern frontier of India at the present day as to throw grave doubt on the accuracy of the statement, quoted in the historic chapter, that the inhabitants of China learned the practice from the people of Assam. With the Nepalese it may be said that opium cultivation (though very extensive at the present day, and smuggling into British territory often carried on to a serious extent) is comparatively a modern branch of agricultural enterprise. It does not appear to have been seen by Dr. Buchanan-Hamilton during his residence in that country which led to the publication (in 1819) of his "Account of the Kingdom of Nepal."

Malwa & other Indian Opiums. 215

MODIFICATIONS EFFECTED IN THE AREA OF "BENGAL OPIUM."—By Regulation XIII. (section 6) of 1816 the production of opium in the districts of Bengal proper was legalised and a Commercial Resident was appointed (in special charge of opium) at Rungpore. But so inferior was the quality of the drug that it was determined to prohibit its exportation to China, the opium being reserved for Indian consumption. After a trial of four years the Agency was abolished and opium cultivation in Bengal prohibited. The keen competition of Malwa and Turkey opiums, however, induced the authorities to endeavour to lower the price, by extending cultivation, and in 1830-34 the present opium districts of Bhagulpore and Monghyr were accordingly added to the Behar Agency. The demand for opium continuing to increase, poppy culture was once again extended over the Bengal Districts (but this time under the Behar Agency), *viz.*, to Purneah, Dinajpur, South Rungpore, Hazaribagh, and Lohardugga. Similarly the Benares Agency was expanded to Jalaun, Banda, Hamírpur, Cawnpore, Fatehgarh, Etawah, Mainpuri, and Meerut. At the same time it must have also reached Nepal, since we read of permission being granted to the Agents to purchase the drug obtained from that kingdom. In 1839-40 came the collapse due to over-production, for instead of stamping out the Malwa

Modifications in area. 216

PAPAVER somniferum.	Area & Production.

CULTIVATION

Area & Production.

opium, the latter had similarly increased and the doom of the opium trade seemed then not far off. The less productive districts, and those which yielded inferior opium, were once again barred from cultivation. Affairs were wound up in Hazaribagh, Lohardugga, Rungpore, Dinajpur, Meerut, Allahabad, Jalaun, Etawah, Mainpuri, and Banda. But at the close of the Chinese war in 1842 prospects once more revived, but it was then resolved to extend cultivation within existing districts and not again to open up those lately abandoned. The Behar Agency, with the exception of the discontinuance in 1848 of Purneah and the re-introduction of the poppy into Chutia Nagpur in 1869-70 and its re-abandonment in 1877-78, has remained stationary since 1840. It is thus quite incorrect to say (as in the *Pharmacographia Indica*) that the eastern limits of opium production of the present day is Dinajpur and its southern Hazaribagh.

In the Benares Agency changes have been more frequent and the expansion larger. From 1840 to 1861 the only addition was the re-cultivation of the poppy in Mainpuri and in Banda. But after the disturbances of 1857 and the annexation of Oudh, opium-producing districts had either to be permitted to continue as they had been doing during Native rule or be prohibited. The conditions and necessities of the country were vastly changed and opium cultivation was accordingly permitted (in 1860 to 1864) in many new districts or restored in others, such as Saharanpur, Dehra Dun, Meerut, Muzaffarnagar, Aligarh, Bulandshahr, Muttra, Agra, Budaun, Moradabad, Bareilly, Fyzabad, Bara Banki, Unao, Lucknow, Gonda, Bahraich, Hardoi, Sitapur, Sultanpur, Partabgarh, and Rae Bareli. But as it became impossible to exercise proper supervision and control over the opium production and consumption, Meerut, Agra, Muttra, Muzaffarnagar, Bulandshahr, and Aligarh were given up and Etah added in 1864-65 to the North-West opium-growing tract. Moradabad, Dehra Dun, and Saharanpur were abandoned, while Kheri and Shahjahanpur were added in 1871-73. In 1881-82 a fresh re-arrangement and slight expansion of the area took place, but not such as to materially increase the total of all India, and, indeed, some contraction has since taken place in the North-West area. The changes that have occurred were governed by many considerations, chief among which appear to have been the administrative necessities of the country, and the discovery that one locality produced a better quality of the drug than another.

Total area.

217

In concluding this brief review, of the opium-producing area of India, it is perhaps safe to say that it does not now and has never under British domination in India materially exceeded 1,000,000 acres. The total annual production of opium might, therefore, be put at about 1,50,000 maunds. This statement assumes an average yield of 10 seers the acre. Duthie & Fuller in their *Field & Garden Crops* think, that, while the average yield recorded at Ghazipur, for the North-West Provinces, during 16 years prior to their estimate was 9 seers, the well-known fact that the consumption of Government opium in poppy-growing, is much lower than in non-poppy growing districts, proves, these authors contend, that the cultivators always retain a portion of the produce. On that account they regard a yield of 10 seers as more nearly correct. In most parts of the Behar Agency the yield is considerably less than in the North-West Provinces, but in Malwa it is greater, so that a yield of 10 seers is very probably not far from the average production for all India. The figure arrived at by such an estimate corresponds to the actual returns *plus* an allowance for illicit consumption. In conclusion, it may be said that the number of licensed cultivators in the Government Agencies is, as a rule, a little over 1,000,000.

P. 217

Preparation of the Land & Methods of Cultivation.

1.—PATNA AND BENARES AGENCIES.

In these agencies the opium year is considered to commence in September. It is customary to follow poppy cultivation after Indian-corn as a *kharif* crop, the soil being at once prepared after the collection of the corn. This is done by passing the plough and *hingah* (clod-crusher) over it, at intervals of a week or ten days, till the middle of October, when the earliest sowings commence. The sowing may, however, be extended to the middle of November. Land in the immediate vicinity of the villages is generally selected for the poppy on account of its being more highly manured. A soil which might be described as a sandy-loam is considered the best. The seed is generally that saved from last year's crop, but care is taken to select for this purpose the healthiest and most productive capsules. In some cases the *rayats*, however, recognise the advantage of getting seed from districts remote from their own, and indeed the seed from certain parts of the country they recognise as of special merit.

Manure.—In the words used by **Dr. Eatwell** it may be said that the poppy lands are "manured to the extent which the means of the cultivator will admit." **Mr. Gennoe** suggests as one of the best of manures that goats and sheep should be penned in the fields, "as the manure thus obtained operates favourably on and is peculiarly invigorating for the soil." He also speaks of the poppy lands being highly manured. **Mr. Scott** devotes Chapters V., VI , and VII. (*pp. 14 to 41*) of his *Opium Husbandry* to the subject of manures. He there furnishes a useful essay on the principles of scientific manuring and the different kinds of manure, but has little to say either as to the native methods of manuring the poppy fields or the system which he regards as the best. He occasionally speaks of certain manures as useful. For example, of green manures he urges the superior merits of *sani* (**Crotolaria juncea**), and he commends the value of the water in which both *san* hemp and jute have been retted. The poppy stalks, he says, would naturally be expected to prove valuable, but unfortunately they are far too often used as fuel and only return to the ground in the form of ashes. Of oil-cakes, he says, castor oil has perhaps a more immediate action than any of the others. Ashes (*rakh*), he remarks, are peculiarly valuable as a poppy land manure, provided the potash has not been allowed to be washed out of them through carelessness. Nitrate of potash (*shorah*) in **Mr. Scott's** opinion is one of the best mineral manures for poppy and may either be applied as a top-dressing at different stages of the crop or scattered over the field after the sowing and first watering of the seed. It may be applied at from *one* to *four* maunds per *bíghá*. Commenting on top-dressing he says :—"It is to be observed that the main object in view in the manurial top-dressing of the opium plant is to supply direct to the leaves certain compounds which, under atmospheric action, may be induced to afford, in a readily available form, materials likely to promote the secretion and enrich the quality of the drug in the plant." For this purpose he suggests the following as the best top dressings :—

(*1st*) *Oil-cakes* —These should be applied as a first top-dressing when the plants have been finally thinned. Cowdung, he adds, in a fresh state dried and broken up with ashes, makes an excellent top-dressing.

(*2nd*) *Shorah* 40 seers, lime 160 seers, with 20 maunds of *nonimattee*, to be applied after the flower-buds make their appearance.

(*3rd*) *Khareenoon* 40 seers, lime 160 seers, mixed with 20 maunds of *nonimattee* and applied as in No. 2.

CULTIVATION
in
Bengal.
218

Manure.
219

P. 219

**CULTIVATION
in
Bengal.**

Manure.

(*4th*) *Lime* 6 maunds, mixed with 3 maunds charcoal dust, to be applied at the early stages of growth in place of No. 1.

(*5th*) *Shorah* 4 maunds, with 4 maunds charcoal dust, all well pulverized and mixed, to be applied in a more advanced stage, say with the first appearance of the flower-buds.

(*6th*) *Common salt* 20 seers, *shorah* 40 seers, mixed with 160 seers of lime, to be applied as in No. 5.

Sowing.—Three seers of seed are sufficient for a *bighá* of land (3,025 square yards) and the seed is generally, though not always, steeped in water the evening before being sown. Scott discusses the merits of the various manurial substances recommended as advantageous if added to the water in which the seeds are steeped (*see Chapter VII., pp. 31 to 37 of Opium Husbandry*). " It will be observed," he says, " that the ammoniacal liquor and iron sulphate steeps proved the most active stimulants. The camphorated water is singularly active, the germination commencing on the fourth day, and copious on the sixth. The young plants were also extremely healthy and of a darker colour than the generality of those otherwise treated. Seeds steeped in the saturated solution of alcohol and camphor germinated tardily and sparely, the plants being all of a pale, sickly green colour, and in very few cases survived the cotylidonary stage." . . . . " Besides accelerating germination and invigorating the young plant, these camphor steeps appear to be a very effective antidote to the mould-blight, which periodically causes so much damage to the poppy." In this connection the writer would offer the suggestion that as a steeping material more convenient to the opium cultivator (since the plant grows on every road-side adjoining his fields), a decoction of **Adhatoda Vasica** should be tried. The reader will find full particulars of that substance under the articles*"Manure" and " **Oryza sativa**—Rice." In connection with the latter subject it will be found that the decoction here recommended is actually used to hasten germination in Afghánistán. If the fungoid pest that attacks the poppy exists, in its early stage, in the seed (as maintained by Scott), it doubtless would be more or less destroyed by some of the steeping agents (notably camphor) employed by him, but by none very probably of those so effectually as by the decoction here recommended. On this subject, however, the writer can offer no personal opinions, since he has not had the opportunity of studying the poppy pests (see the paragraph below on Pests).

**Irrigation.
220**

Irrigation.—As soon as the seed begins to germinate, which is about a week after it is sown, the field is divided by a cross series of ridges into rectangular compartments, about eight feet in length by four feet in breadth, the alternate ridges being made somewhat broader than the others, to form the water channels for the irrigation of the plant. The field should be watered as soon as the plants appear above ground, and seed again sown in those places in which germination has failed. Irrigation must be continued at regular intervals till the crop is matured, but care should be taken not to allow the entire seedlings to be under water. Upon unirrigated lands it is of course unnecessary to divide the field into compartments. Gennoe recommends that the beds should be parallel, so that they may be easily irrigated. A drain or outlet should intervene between every two beds for the passage of water. Lands bordering on rivers and *jhíls*, he adds, as they retain their moisture till December, do not require to be laid out in beds. Wells are essentially necessary for poppy fields, and hence every assistance and encouragement for their construction should be afforded.

**Weeding.
221**

Thinning and Weeding.—When the young plants are about two to three inches high they are thinned out. This is repeated perhaps two or

* Vol. V., 174 & 615-619.

three times, the sickly and superfluous plants being got rid of and the
vigorous plants left at a distance of seven to eight inches from each other.
Dr. Lyell says of Behar that the young plants which are removed at the
first weedings "are sold and eaten as a salad, those of second and third
thinnings possess narcotic properties and are seldom used."

Flowering and Period of Collection of the Opium.—The plants take
from 75 to 80 days after germination to reach the flowering condition.
The petals, which are four in number, are gently removed when fully
matured, which is on the third day after expansion. These petals constitute the "Flower Leaves" of the manufacturers and are employed in the
formation of the outer casing of the opium-cakes. In the course of another
eight or ten days the capsules are sufficiently advanced for the extraction of
the drug. The earliest collections in Behar may thus take place at the
end of January, but operations are continued throughout February and
even up to the middle of March. In the Benares Agency the season of
collection is a little later. The plants do not flower as a rule in the North-
West Provinces till March, and the collection of the drug is accordingly
sometimes extended to the second week in April.

Subsequent Treatment of the Land.—After the poppy is off the soil, the
land is usually allowed to lie fallow till the *kharif* sowings begin. In
some cases *chéna* is sown at once, then followed by a *kharif* crop, or sometimes the poppy is succeeded by indigo. In rare instances the fields are
specially reserved for poppy culture and left fallow between each crop.

II.—North-West Provinces & Oudh.

These remarks apply to the average opium cultivation under the Government Agencies. As exhibiting, however, special features and amplifying
the facts given, the following passages may be taken from the *Field &*
Garden Crops of the North-West Provinces :—

Permission to cultivate.—"The system on which opium is grown for
Government is not unlike that on which *badni* indigo is grown for an
indigo factory. Every cultivator wishing to grow that plant must obtain
a written license to do so, and receives at the same time an advance in
cash of from R12 to R13 an acre, paid in two instalments, one, two
months before the poppy is sown, and the second, one month after sowing. The whole of the produce is purchased by Government, at a rate
varying between R4-8 and R6 a seer. This is the only form of cultivation in which the peasant can get his seed without paying interest to the
money-lender. The price paid by Government for the drug alone is, in
districts where the soil is good and the cultivation moderately careful, not
much less than R50 per acre. Something more is realised from the
seed and dried capsules. While the profits per acre are at least as
large as they are from wheat or any other ordinary crop not requiring
unusual care and expense for its production, the market is absolutely
certain, and the cultivator has to fear none of the sudden changes in price
which a diminished demand makes not infrequent in the case of other
products. In times of scarcity and famine the advantages of this form of
cultivation are very great. The crop is almost invariably grown close to a
well, and is therefore less dependent than others on the vicissitudes of the
season. This certainty enables the Opium Department to make further
advances on the crop besides those that had originally been given for seed,
and supplies the cultivators with money, which could not otherwise have
been obtained at all, or except at an exorbitant rate of interest. Under
these circumstances it is not surprising that this crop is extremely popular
with all classes of cultivators, though in a few districts the landlords have
shown at its first introduction some signs of the hostility which is aroused

CULTIVATION
in
Bengal.

Collection
of
Opium.
222

N.-W. P. &
OUDH.
223

Permission.
224

**CULTIVATION
in
N.-W. P.
& Oudh.**

in every case by the direct dealings of a Government department with the people on their lands, a feeling which has at other times, with even less reason, led them to oppose the making of new canals. The ultimate gain to the landlord from opium cultivation differs only in degree from what he owes to canals. He is able to charge the highest rate of rent on the land devoted to it, and has little to fear from irrecoverable balances."

Opium Cultivators.—"The Kachi was formerly the opium cultivator *par excellence*, and owes his very name to the process of scraping the juice off the capsules (*kachna*), which is one of the most distinctive features in opium-growing. The cultivation has now, however, extended to Kurmis and Lodhas in equal proportion with the Kachis, and is gradually spreading to the higher castes, even Thakurs and Brahmins occasionally taking to it."

**Season.
225**

Seasons.—"In the hills of Jaunsar the opium season is from February to June, but in the plains it is from October to March and the poppy may therefore be classed as a *rabi* crop. If grown on very highly manured land, it often follows a crop of maize or millet in the preceding *kharif*, which, by exhausting some of the richness of the soil, prevents all risk of the poppy running unduly to stalk and leaf. It is most commonly grown alone, but occasionally the cultivator leaves the maize to be of service in attracting the attacks of insects which might otherwise injure the poppy plants."

**Manure.
226**

Soils and Manuring.—"A strong loam is preferred, and the field invariably lies in the highly manured circle round the village, known as *goind* or *gauhani*, receiving from 150 to 200 maunds of cattle dung to the acre each year. In Fatehpur the land is often manured by herding sheep or goats on it, the dung of which is supposed to be of peculiar value. The plants benefit greatly if they are irrigated from a well the water of which is impregnated with nitrates (*khari*), and, as in the case of tobacco, the selection of a field for opium cultivation is greatly dependent on the accessibility of any well of this description. Earth impregnated with saltpetre (*nonamitti*) is also extensively used, chiefly as a top-dressing after the plants have come up (Mainpuri), in which manner, too, well-rotted cowdung and ashes are often applied."

Tillage.—"A finely powdered tilth is absolutely essential, and opium land is ploughed as many times as the cultivator has leisure for."

Sowing.—"The seed is sown in October, broadcast, at the rate of about 3℔ to the acre, having been mixed with earth to assist in its even distribution and the log clod-crusher is then run over the ground."

**Irrigation.
227**

Irrigation.—"The ground is almost always prepared for sowing by a watering, and in the drier portions of the Provinces the plants are irrigated once in every fortnight or three weeks between germination and harvest time. In the Benares Division four or five waterings are generally sufficient."

"If the crop can be irrigated with water containing nitrates so much the better, but it is above all things important that the watering should be timed exactly to the requirements of the plants, and opium cultivators in consequence are shy of the canal. The field is kept scrupulously free from weeds, at least three weedings being as a rule given."

**Cost.
228**

Cost of Cultivation.—"The cost of cultivating an acre of poppy is given below :—

	R	a.	p.
Ploughing (eight times)	6	0	0
Clod-crushing	0	4	0
Seed	0	2	0
Sowing	0	3	0

Products of India. 65

N.-W. Provinces & Oudh. (*G. Watt.*) PAPAVER somniferum.

CULTIVATION in N.-W. P. & Oudh.

	R	*a.*	*p.*
Making water beds	0	3	0
Watering (six times)	9	8	0
Weeding (four times)	8	0	0
Harvesting (8 coolies at 2 annas a day for 15 days) . .	15	0	0
* Manure (200 maunds ; ⅔rds of cost) , . . .	4	0	0
	38	4	0
* Rent (⅔rds of annual)	10	0	0
TOTAL .	48	4	0 ”

* Only ⅔rds of the cost of manure and of the annual rent are charged, since at least ⅓rd of each must be debited to the crop of maize which nearly always precedes opium in the *kharif*.

III.—MALWA & OTHER NATIVE STATES.

MALWA. 229

Dr. Impey who resided for some years in Malwa furnishes useful information on the system of poppy cultivation which he witnessed. He writes :—" For the successful cultivation of opium, a mild climate, plentiful irrigation, a rich soil, and diligent husbandry, are indispensable. In reference to the first of these, Malwa is placed most favourably. The country is, in general, from 1,300 to 2,000 feet above the level of the sea ; the mean temperature is moderate, and range of the thermometer small. Opium is always cultivated in ground near a tank or running stream, so as to be insured at all times of an abundant supply of water. The rich black loam supposed to be produced by the decomposition of trap, and known by the name of ' Cotton soil,' is preferred for opium ; though fertile and rich enough to produce thirty successive crops of wheat without fallowing, it is not sufficiently rich for the growth of the poppy *until* well manured ; there is, in fact, no crop known to the agriculturist, unless sugar-cane, that requires so much care and labour as the poppy. The ground is first four times ploughed on four successive days, then carefully harrowed, when manure, at the rate of from eight to ten cart-loads an acre, is applied to it ; this is scarcely half what is allowed to a turnip crop in Britain. The crop is after this watered once every eight or ten days, the total number of waterings never exceeding nine in all. One *bíghá* takes two days to soak thoroughly in the cold weather, and four as the hot season approaches. Water applied after the petals drop from the flowers causes the whole to wither and decay. When the plants are six inches high, they are weeded and thinned, leaving about a foot and a half betwixt each plant ; in three months they reach maturity, and are then about four feet in height if well cultivated. The full-grown seed-pod measures three and a half inches vertically, and two and a half in horizontal diameter."

IV.—BARODA.

BARODA. 230

" The poppy is grown in all the Sub-Divisions except Dehgam. Land intended for it is, as a rule, left fallow for about four months and ploughed ten times before the seed is sown. But in some Sub-Divisions, it is usual to take a crop of *bájrí* before the land is utilised for poppy. In such cases immediately after the *bájrí* is removed, the ground is ploughed three times and saturated with water, and when completely dry, is again ploughed three times. The yield from fallow land is the greater, and the opium of a lighter colour. Manure is carried to the field in the month of June, and applied after the first rain at the rate of 1,400℔ or more per *bíghá* every third year. Cattle manure is most used, mixed with alluvial deposits when available. Poppy is sown in small rectangular beds of from five to eight feet each, nearly square, and so made as to allow water to spread evenly. The seed is thrown broadcast over

CULTIVATION
in
Baroda.

these beds, the earth is then turned over with an implement called the *khandi*, and is watered immediately after. Care is required in selecting the seed : it should be a year old and free from damp, and the cultivators, as a rule, preserve a stock from the crop of the previous year. The proportion sown is about 2℔ to the *bighá*. In poppy cultivation irrigation requires attention, channels having to be made for the even watering of the beds. Poppy fields are watered seven times : the first watering commences with the sowing, the second four days after, and others follow at intervals of about a fortnight, and after the flowers appear there is one watering. Weeding is a laborious process. It has to be gone through three times, and on each occasion, besides removing strange growth, the poppy plants are thinned until they are left at a distance of eight inches from one another. The first weeding is done twenty days after the plants appear, the other two at intervals of a fortnight. Withered or *jogida* plants are removed. But barren or *vánjia* plants are kept for the seed, although they produce no juice. The plants, when tender, are used as a vegetable by many classes. The earlier sowing is in flower in January, and the later in February. The poppy-heads are considered ready for scarification when they present a coating of a light brown colour and do not yield easily to the touch. The process of scarification commences in February or March according to the date of the sowing " (*Gazetteer, p. 98*).

PANJAB.
231

V.—PANJAB.

"There is no district in the Panjáb that produces more of this drug than Sháhpur. The poppy plant requires a rich soil and abundance of moisture. The mode of culture is this ; the land which it is proposed to sow with this crop is allowed to lie fallow for one season at least. During the rains it is repeatedly ploughed and well manured. It then remains untouched till the beginning of November, when it is prepared to receive the seed, which, at the rate of half a *seer* to the acre, is sown broadcast, mixed with equal parts of sand to ensure equal distribution. Water is supplied as often as the surface shows signs of dryness. The young plants begin to show themselves about the twelfth day, and from this time, till the pods begin to ripen, the successful cultivation of the crop depends on the attention paid to watering, weeding, and manuring. The pods begin to swell in March, and towards the end of this month, an estimate can be framed of the probable yield of opium. Traders then come forward and buy the standing crop, after which the cultivator has nothing to do but supply water as required. The drug is obtained by making incisions in the pod with a three-bladed lancet. The incisions are made vertically, about half an inch in length, in the centre of the pod. Three strokes are made with the instrument each time, making nine cuts, and this is repeated four times at intervals of as many days, making 36 incisions in all, the whole operation extending over about a fortnight. The work is carried on during the middle of the day, as it is found that the heat assists the exudation of the juice. The morning following the making of each set of incisions, the juice which has exuded from the cuts is scooped off with shells, and collected in cups made of the leaves of the plant itself. It is estimated that one man (women and children are not much employed in this work) can, on an average, incise the pods, and collect the juice of about 10 *marlas* ($\frac{1}{16}$ acre) of the crop in a day; and as this is repeated four times, and the labourers are paid from two to four annas a day, the cost of extraction varies from eight to sixteen rupees an acre. The produce of an acre is from four to eight seers, the selling price from eight to twelve rupees. In the process of drying, the extract loses about a fourth of its weight. In 1881-82 the area under poppy cultivation was little below

3,500 acres, the produce of which, at an average of six seers per acre, amounts to 525 maunds. Even reducing this by a fourth to allow for loss in drying, we have still the large quantity of 394 maunds, which, at ten rupees a seer represent no less a sum than R1,57,500. Careful enquiry has shown that, of the produce of the district, all but a few maunds leave it, the destination of by far the greater part being the great Sikh centres of Lahore and Amritsar" (*Panjab Gazetteer, pp. 65 & 66*).

CULTIVATION in Punjab.

VI.—CENTRAL PROVINCES.

Formerly poppy cultivation appears to have been fairly abundant in these provinces. Some few years ago it was prohibited, but the following passage from one of the Settlement Reports conveys an idea of the system of culture which was pursued and the methods of extracting the crude drug and of disposing of the same to the dealers :—

CENTRAL PROVINCES.

232

"Opium cultivation is chiefly pursued in the Mooltye pergunnah. It can be grown upon almost any soil by means of manure, but generally the *kharee* land near the villages is preferred. The land is prepared in much the same way as sugarcane, only that less ploughing is required. Similarly, the ground is divided into little beds in which the seed is sown broadcast. The sowing generally begins in November. In about ten days it springs up and soon afterwards is weeded and thinned out. Each part of the field is watered once in about four or five days and weeding is again required in the intervals ; in February the plant flowers, and when the pods are fully formed, irrigation is stopped for about a fortnight when the plant is supposed to be ripe. Small incisions are then made in the pods, and the juice that exudes is collected the day following. This operation is repeated three times, when the whole of the juice is supposed to be extracted. It is deposited in earthen vessels and is called *chik* by the people; it is exported in this state to Indore, where it is manufactured into opium and mixed with that of Malwa which is a superior article.

"The whole almost of the 'chik' produced in the district is bought up by Mahajuns in Mooltye who export it on bullocks.

"Much opium is grown under the system of *lawanee*, as it is termed, by which the cultivator receives an advance of money from the Mahajun and binds himself to deliver his produce to him in payment at a certain rate. The Mahajun of course derives a large profit on this transaction, and I am happy to say that the system is going out owing to the growth of intelligence and capital among the cultivating classes.

"I append an index of the cost of production and profits of an opium field of from two to three acres :—

Expenses.   R.

	R.
Seed	1
Wages of 2 men for 4 months	24
Leathern bag for well	5
Repairs, &c.	1
Weeding four times	8
Hire of bullocks for 4 months	25
Rent of land	4
TOTAL	68

Produce.   R

	R
20 seers *chik* at R7½*	82

"Over and above, however, the raw juice there is produce—*khushkhus* oil, *manooa*, and poppy seed, worth not less than R10, and in the same

* This appears to be a misprint.—*Ed.*

PAPAVER somniferum.	Diseases of the Poppy Plant.

field is also sown a little garlic or patches of barley. There are about 2,500 acres under opium cultivation, yielding an average of 180 maunds of *chik* annually.

" The chief inducement to grow opium is that it requires so much less labour than sugarcane, and being only four months in the ground requires but comparatively little water, and many streams which run quite dry during the hot weather are available for the irrigation of it, ripening as it does by the end of February" (*Report on the Land Revenue Settlement of the Baitool District, Nerbudda Division, Central Provinces, by W. Ramsay, pp. 68-70*).

DISEASES.
233

DISEASES OF THE POPPY PLANT.

The late **Mr. Scott** gave this subject much patient study, and the able chapter which he devoted to it in his *Opium Husbandry* should be consulted by all interested in the matter. It is impossible to devote to the diseases of the poppy the space in this work which its importance demands. It may be said at the outset that the life-history of the most injurious parasitic fungus has not apparently been as yet worked out to its final issue. **Mr. Scott** presumed that the recuperative germs that carry on the growth exist in the seed and is thus sown with the crop instead of coming to it from some other plant. On this supposition he explained the observation that seeds saturated in camphorated water were less attacked owing, as he thought, to the germs of the fungus having been killed. The writer has alluded to this subject above in the paragraph devoted to the manurial materials employed in steeping the seeds.

Scott discusses the diseases of the poppy under the following headings : (1) Specific diseases of a functional or organic nature, caused by an excess or deficiency of those agents which are necessary for the vigorous growth of plant, such as *soil, light, heat,* and *moisture* ; (2) Injuries caused by parasitic plants (*a*) of the higher or flowering plants, *e.g.,* the broom-rapes and (*b*) of the lower or so-called flowerless class, limited to the natural order of Fungi ; (3) Mechanical injuries of various kinds, chiefly the attacks of insects.

1st, Specific Diseases.—Mr. **Scott** describes under this heading the following :—

Root-canker.
234

(*a*) *Root-canker.*—In stiff clayey soils, in which there is an excess of the oxides of iron or manganese, the poppy is frequently seriously affected with the above disease. It also prevails even on the lighter class of soils when there is an excess of alkaline salts. Plants subject to this disease have the root variously affected ; very frequently it breaks out in isolated patches, or again it extends less or more over the whole tap-root from the point upwards. The following, **Mr. Scott** says, are the more characteristic symptoms :—The leaves gradually wither from below upwards, and individually from the apex and circumference to the base, the central parts thus being fresh and green, while the margins are quite withered. All ultimately assume a pale-brownish yellow with veins of a deeper colour, while the lower surface is generally less or more freely speckled with black resiny exudations.

A second form of root disease is also mentioned by **Mr. Scott,** in which the root is *ringed* immediately below the soil. The softer outer tissues are abruptly corroded and this not unfrequently to the central woody cord. The cause of this ringing, he adds, appears due in many cases to abrasion of the softer outer tissues in a dry and baked soil. This may be a result either of an unnatural shrinking of the soil or to the irrigation of the crop in windy weather.

Sun-burning.
235

(*b*) *Sun-burning.*—This is the *moorka* or *jookka* of the cultivators (the *assamees* as they are called), which occurs on light sandy soils in hot sea-

CULTIVATION

Diseases.

sons. The symptoms are a general drying up of the upper or younger parts of the plant, the leaves becoming shrivelled and variously contorted, with the veins of a brownish or purple-black colour, while a longitudinal section of the plant shows a dry and shrivelled apex, and the pith less or more decayed from above downwards. As showing that this injury is caused purely by heat, it is frequently observed that the stem throws off lateral shoots immediately after irrigation.

Sclerosis.
236

(*c*) *Sclerosis.*—The symptoms of this affection, says **Mr. Scott**, are a gradual drying and hardening of the whole tissues of the plant; it is evidently a constitutional disease. It appears to be rather a prevalent form and occurs indifferently in crops on irrigated and unirrigated lands, on dry and poor soils, as well as on those which lack neither manurial conditions nor moisture. It would appear to be due to a partial arrestment of the vegetative functions as the plant attains the flowering stage. The first symptom of it is ordinarily the slight persistence of the sepals, imperfect expansion of the petals, and less or more complete persistence of the stamens which are often found even in the ripe capsule as a fringe around its base. The capsules are generally of a small size, and very numerous (from 12 to 20, or 30 on each plant) and rarely yield the drug to more than two incisions and even that scantily. Where prevalent it thus seriously damages the crop. **Mr. Scott** found there was a strong tendency for this disease to become hereditary. It thus becomes necessary that the cultivators should be taught to avoid collecting seed from such diseased stock.

Petechia.
237

(*d*) *Petechia.*—In this disease the cuticle of the leaves becomes more or less densely covered with specks or spots, somewhat resembling those caused by the bite of insects. In moist weather plants thus affected are far from uncommon in our poppy-fields. The first sign of the disease is shown by the leaves acquiring a pale, dull, yellowish-green tinge, and soon after the whole surface becomes studded with specks and patches of purply-black. The affection is much less developed on the lower than on the upper surface of the leaf; the latter ultimately acquires a nearly uniform purply-black colour. The discoloration is mainly confined to the leaves, and very rarely accompanied by any external exudations from the tissues, though all parts of the plants are soft and juicy; indeed, in a *quasi*-dropsical condition. In many cases this disease seems largely attributable to over-crowding of the plants.

Gangrene.
238

(*e*) *Gangrene.*—The soft herbaceous parts of the plant are most affected by this disease. It frequently breaks out on injured parts of the plant but is more or less local. But when due to an excess of moisture and low temperature, the whole plant may be affected. Indeed, cold and damp seem frequent causes of this disease. The affected parts are, **Mr. Scott** says, easily recognised by their curved or depressed shoots and drooping and even revolute leaves. The flowers expand imperfectly and the stamens are always more or less persistent, forming a shrivelled fringe around the base of the capsule. The latter is also very generally arrested in an early stage of its growth, and very rarely are the seeds perfected. As the disease progresses, the stems, leaves, and capsules are spotted with black resiny exudations. If healthy plants are inoculated with this disease, it spreads upwards from the point of inoculation.

(*f*) *Parasitic flowering Plants.*
239

2nd, **Injuries Caused by Parasites.**—These have been already explained. **Mr. Scott** refers to two sections, the higher flowering parasitic plant and the lower or flowerless plants—fungi. Of the former he noted two members of the same genus—**Orobanche indica** and **O. cernua**—the broom-rapes. These occur parasitically on a great many other plants, such as tobacco, brinjal, rape, and mustard; indeed, the last-mentioned species affects the tobacco

CULTIVATION

Diseases.

Fungoid:
Peronospora.
240

more than the poppy. They grow up from the roots of the crop sending above ground a tuft of beautiful spikes of bluish flowers. When very prevalent they do considerable injury to the crop, but as they are large and easily seen can be exterminated. Indeed, unless when very prevalent, they rarely do any appreciable injury.

(g) *Peronospora.*—The fungoid pests are far more destructive. **Mr. Scott** devotes, however, so many pages to this subject that it is impossible to produce a brief abstract of his views. The chief disease is **Peronospora arborescens**—the poppy-mould. This appeared the first year (1872-73), as seen by **Mr. Scott**, to so slight an extent that he was disposed to view it as of no great moment. In the following season it was manifested to an almost alarming extent. But **Mr. Scott** may be allowed to describe this disease in his own words. Speaking of the disease as seen in 1873-74 he says :—" This was quite a periodic phenomenon, and extended from about the middle of December to the middle of February. At the latter period fully more than half the leaf surface of every plant was invaded and discoloured by the mould. The invaded portions of the leaf are readily distinguished by their dry, escharotic appearance, and dull-brownish colour. On examining the lower surface of these in February 1874, I found every patch crowded with spore-bearing filaments. The mould had apparently exhausted itself in this high reproductive fertility : it made no further extension, and the invaded parts dried up and withered, without causing rot or further injury to the uninvaded parts of the leaf. From the result of that season's observations of the mould, I had concluded that it was not at all likely to prove a serious enemy to the poppy. The less succulent structure, as compared with that of the potato, as I then wrote, ' is evidently unfavourable to any rapid or general extension of the *mycelial processes,* and out of the many thousands of poppy plants which I have now seen, with their whole leaf surface teeming more or less with the *fertile* shoots of the mould, I have not seen a single plant succumb to it : all have flowered, yielded drug,—*I am indeed not prepared to say in undiminished quantity,* and borne their normal quantity of good seeds.' Well-founded, apparently, though the above view was, the following season's observations proved it to be anything but true ; and the escape of the crops from a virulent blight, proved simply to be due to the mode of development, as I will now attempt to show in the following *resumé,* of the mould's extension and morbific action on the poppy crops in the season 1874-75.

" The first appearance of the mould on the poppy in the experimental gardens for the above season was towards the close of November 1874, and, as in the previous season, I found it entirely confined to the lower and mature leaves. It had again a periodic extension from the lower to the upper cycle of leaves, and by the middle of February the whole leaf series presented patches and streaks of a dark-brown colour, indicative of mould invasion. Hitherto it had given me no anxiety, my observations on extensively-affected specimens in the preceding season having afforded me grounds for the belief that the drug-yielding qualities of the poppy would not be sensibly impaired by the infection, nor did it appear in any marked way to have weakened the plants, as even those with the older leaves very extensively affected (the *mycelium,* as I should here observe, being *fertile* or *spore-bearing*) continued in good health. . . . Each mould-patch *fructified copiously,* and thus, as it would appear, utterly exhausted the *mycelium* or *spawn* in the inner tissues ; so that the latter and its fostering matrix simply dried up, without communicating disease or rot to any uninvaded portion of the leaf-tissues. On the other hand, in the season 1874-75, the inner tissues of the leaves appear to have been wholly inter-

CULTIVATION

Diseases.

Peronospora.

woven by a *sterile mycelium*, the decay of which caused their discoloration and death. Again, the fungal discolorations and disease of the leaves were communicated to the *nodes*,—*i.e.*, the points of union of the leaf and the stem,—thence extending upwards and downwards from each as a centre, until the whole stem not unfrequently assumed a nearly uniform glaucous bluish-black. At this stage all subsequent growth was arrested, and the plant sooner or later died. The cuticle, or skin of the stem, was seen to shrink and dry, but, unlike that of mould-affected potato-stalks, it continued firmly adherent to the dry woody layers beneath. This close adhesion of the epidermal layers to the woody zone tends, as I think, to show that neither the cambium layer nor liber, or inner bark, has been actually invaded by the mould; otherwise one might have expected a separation of the epidermal or upper layers on the death of the *mycelium* and softer tissues investing the wood. . . . The pith of plants affected with mould acquires a brownish-green colour, and in the earlier stages of the disease becomes gorged with sap. As this is exhausted—apparently in the supply of the capsules, which frequently continue green and soft, and even mature their seeds after the leaves of the plants are dry and withered—the pith also dries up, and the whole plant dies. . . . The weather during the period of this mould-invasion was upon the whole dry, with only light, occasional showers of rain. I do not doubt that with cold, foggy, or rainy weather those mould-blighted plants would have become so surcharged with sap, that all their parts would have become soft, and instead of withering and drying as they did, would have become a pulpy, rotten mass. This I now strongly suspect has been the case in the poppy blight of 1871-72, as it appears to have been everywhere preceded by cold, foggy, and rainy weather. It thus appears that, under certain unknown conditions, there may occur either a merely vegetative extension of the *spawn* or *mycelium* in the tissues of the poppy leaves, or, on the other hand, vegetative development may be in abeyance, and the reproductive correspondingly exalted, to the complete exhaustion of the mycelial processes. Now, in the latter case, we have a comparatively *harmless parasitism*, the injury to the tissues of the foster parent being local, and limited to those in immediate contact with the *mycelium*; whereas in the other case, with the reproductive function in abeyance, and a purely vegetative development in force, the results are most mischievous: this, as it appears to me, is due to the decay of the *unexhausted mycelium* in the tissues of its nurse. In all circumstances the spawn is but short-lived, decay fast following development; and it is thus that the soft, almost sebaceous tissues of the *mycelium*, full as they are of protoplasmic matter and juices, cause alterations in the cellulose and cell contents of the surrounding tissues, and generate a *quasi*-fermentive process, which being communicated to the surrounding tissues, the whole organ, or indeed the entire plant, undergoes decay. Now, it is easy to see that if this should occur in cold, wet, and cloudy weather, when transpiration is checked, and the tissues consequently less or more gorged with juices, the plant will present all the symptomatic appearances of *moist gangrene*: the leaves becoming soft, flaccid, and drooping with the first break of the sun, while the softer tissues of the stem will speedily rot. In a dry and hot atmosphere, on the other hand, while transpiration is increased and the absorption of moisture from the soil impeded, by the decay of the mould in the tissues, and the functional derangements thereby occasioned, the fluid contents of the cells being nearly exhausted, the several parts of the plant, as might be anticipated, do simply wither and die. Thus in the first instance, we have one of those *virulent rot-blights* which ruined the poppy-crops in 1871-72; and in the second it assumes the dry and less

CULTIVATION

Diseases.

Peronospora.

striking form as exemplified in those of 1874-75. The latter, its development and effects, we will now briefly notice."

Mr. Scott next deals with the manner in which the mycelial disease progresses. Starting from the first leaf on the seedling, he says, it ascends (never descends) attacking in succession each cycle of leaves until the whole plant is involved, the onward progress being temporarily arrested between each new cycle.

"From the above summary of a series of careful observations, there can, I think, be no question as to the correlation of periodicity in the development of the mould and its foster-parent, each periodic invasion of the former being strictly limited to a single leaf cycle of the latter. In its limitation to the fully mature leaves, and in presenting well-marked periods of rest and activity, the poppy-mould appears to differ altogether from its ally, the potato-mould, in which, so far as I am aware, no such phenomena have been observed. It is also important to note that the poppy-mould does not necessarily extend wherever a green leaf matrix presents itself. This, as it appears, is a clear proof that atmospheric influences have really little to do with the extension of the mould, but that, on the other hand, it is largely, if not wholly, dependent on the degree of maturation of the matrix; that is to say, on certain peculiar chemical changes which occur in the juices of the nurse plant at that period, and that these are alone favourable to any high degree of vital activity in the mould. This is a matter of high importance. Hitherto, it has been customary to correlate the sudden appearance and rapid extension of various fungoid-blights to certain peculiar atmospheric conditions. Now, there may be some such relation in the spread and extension of the truly terrestrial or even epiphytal species; but I do think that careful observations will show that this scarcely, if at all, holds in those of a truly parasitic character, and especially such as the poppy and potato-moulds, which are wholly dependent for their development on a supply of the juices afforded by these plants in the period of their vegetation. As regards the poppy-mould especially, I am convinced that its development and extension are largely dependent on certain changes in the juices of the maturing leaves, though I admit that there are grounds for the belief that the particular mode of development—that is, the dominance of the reproductive or vegetative function—may be partly due to certain obscure atmospheric conditions. This, of course, is an important influence, inducing, as I have shown, in the one case a comparatively harmless, in the other a most mischievous, parasitism."

The above passages from Mr. Scott's account have been reproduced since, though his conclusions may be subject to modification, his practical observations of the progression of the disease are doubtless of much practical value. Subsequent to the publication of Mr. Scott's *Opium Manual*, Surgeon-Major D. D. Cunningham took up the subject of the *Opium Blight* (**Peronospora arborescens**), and through his careful microscopic studies greatly advanced our knowledge of the structural peculiarities of the fungus and its methods of penetration and destruction of the poppy tissues. Dr. Cunningham, however, admits of his studies that "They refer solely to the appearances presented by the parasite and the lesions resulting from its presence as they occur in the leaves, as, partly owing to the difficulty of procuring perfectly fresh plants for study and partly to want of time, the examinations were almost entirely devoted to those points; and, even as regards them, are confessedly imperfect, experiments on the infection of previously healthy tissues not having been carried out for want of growing plants."

Dr. Cunningham found, however, that all the specimens examined by

CULTIVATION
Diseases.
Peronospora.

him, from whatever locality obtained, agreed in showing the presence of **Peronospora arborescens**, a fungus closely allied to that which occurs in potato disease. "The fungi," he says, "which belong to this genus, besides the two species mentioned above, include various others affecting other host-plants, are true entophytes, their mycelium or vegetative apparatus being developed within the tissues of the plants which play the part of hosts, whilst their reproductive apparatus consists partly of filaments bearing the so-called spores or *conidia* which emerge from the substance of the leaves and project on the surface as a whitish mouldy layer, and partly of larger reproductive bodies termed *Oogonia* which are developed within the substance of the affected tissues. These consist of a sac containing a mass of living protoplasm which subsequently forms the *Oospore*, and is ultimately invested with a thickened membrane, frequently marked with projecting tubercles, ridges or reticulations. The former class of reproductive bodies, the conidia, are produced in extreme abundance and serve for the rapid diffusion of the parasite from plant to plant, whilst the latter, wrapped up in the tissues of the host and additionally protected by the thickened character of their envelopes, are supposed to be capable of retaining their vitality under conditions destructive to that of the mycelium and conidia, and to be thus endowed with a capacity for germination after prolonged periods of rest. This process has, however, never been actually observed to occur in them, but is only inferred to exist from what has been observed in regard to the corresponding bodies in the allied genus **Cystopus**. It is by means of such *Oospores* that the species of **Peronospora** affecting various plants at home are supposed to survive the winter and to be enabled to appear year after year." Dr. Cunningham next alludes to a method of survival which pertains in perennial host-plants. "For example, *Oogonia* and their contained *Oospores* have never been detected on the **Peronospora infestans**, as that parasite exists in Europe on the potato plant, and yet it survives the rigours of the winter and appears year after year because portions of its mycelium affecting the tissues of the tubers remain protected and dormant, wrapped up in the substance of the latter and ready to grow and spread outwards with the young shoots arising from the tubers when they are planted." In the case of the potato disease Dr. Cunningham says that so far as the evidence goes it would seem as if certain breeds of the potato had a hereditary predisposition to attack or exemption, co-existant with perfect health in the host. This well-accepted view, regarding the potato disease and some other parasitic diseases, makes it very probable that **Mr. Scott's** observation is correct, namely, that certain forms of the poppy plant are less liable to the attacks of the disease than others. In support of the idea it may be added that Dr. Cunningham noted that the disease presented different appearances according to the nature of the leaves in which it was found. In his microscopical investigations also he failed to detect any *haustoria* or penetrating portions of the mycelium. The filaments were seen to be closely adapted to their containing intercellular spaces and to exhibit projections and dilatations corresponding with the irregularities of the intercellular spaces of the leaf-tissue within which they freely anastomosed, but were never seen to penetrate the cells of the host. The filaments were also observed to be devoid of any septa. "The fertile filaments are erect, strong, and persistent, many remaining recognisable adherent to dried fragments of leaves attached to dead plants, or obtained from the neighbouring soil. When quite fresh, they are perfectly colourless and full of finely molecular protoplasm; but on keeping, the latter shrinks into granular masses of bright yellow colour and remains without further change for many months in dried specimens." "The conidia appear very rapidly to lose the power of germinating. In

Conf. with p. 55.

PAPAVER somniferum.	Diseases of the Poppy Plant.

CULTIVATION.

Diseases.

Peronospora.

several instances in which fresh specimens were sown in water, germination was observed to take place, the method, curiously enough, being in general somewhat different from any of those described by **De Bary** as characteristic of the genus. The protoplasm at first became divided into several, generally four, distinct masses, as though a formation of zoospores were about to occur, but in place of this each mass gave origin to a simple germinal tube. The latter elongated considerably, but remained undivided, and showed no further development in water, and, owing to the want of fresh-growing plants, experiments on their behaviour towards the living tissues of the host could not be carried out."

" Prolonged and careful search was made for *Oogonia* and *Oospores,* but entirely without avail, for, although fresh leaves in every stage of the disease and withered ones which adhered to blighted plants or had fallen on the surrounding soil, as well as specimens of the soil itself, were carefully examined, in no instance could any such bodies be detected. This absence of *Oospores,* however, may have been a peculiarity in the leaves examined, or may be dependent on their development being confined to some part or parts of the host not submitted to examination, such as the capsules, cotyledons, first developed true leaves, &c." Dr. **Cunningham** also examined specimens of the common weed (allied to the poppy and met with throughout the plains of India) **Argemone mexicana,** in the hope of detecting the Oospores of **Peronospora** on that plant, but in this also he failed. " The absence of *Oospores,*" he adds, " in all the specimens submitted to examination is remarkable, as there can be little doubt that they must be produced, the recurrence of the disease from year to year being otherwise almost inexplicable, as the conidia are, in so far as observation goes, incapable of retaining their vitality when dried, or of germination after any length of time. The possibility of their development being confined to special parts of the host has been already alluded to, and it is also possible that the germs of the mycelium may be contained in the seed." This opinion was held by the late **Mr. Scott,** *viz.,* that reproduction must pass through the seed. The only other likely method would be by the production of the *Oospores* on some altogether different plant in the vicinity of the poppy fields. This suggestion is of course at variance with the accepted life-history of the better known PERONOSPORÆ, but it is within the bounds of possibility that some such migration from one host to another is specially adapted to bridge over the period during which the poppy is off the fields. Dr. **Cunningham** suggests at least a safe precaution that no diseased parts of the plant should be left on the field from one season to another. It might further be recommended that the Bombay system of burning the surface soil (*rábbing*) might be tried as a prevention against the possible survival of Oospores in the soil or attached to the roots of the plant.

The above detailed extract from **Dr. Cunningham's** valuable paper has been given in the hope of its proving useful to persons interested in the subject who may not have access to the Annual Report of the Sanitary Commissioner with the Government of India (1873) in which the original paper occurs. The life-history, so far elaborated by **Dr. Cunningham,** gives the key to the natural course of future investigation by which the concluding admission of failure made by **Dr. Cunningham** may be removed from a subject of vast importance to Indian agriculture and an invaluable source of revenue to this country, namely, the failure to detect the location of the *Oospores* or to discover the existence of any other arrangement which could provide for the recurrence of the *Opium Blight.*

In summing up his views, **Mr. Scott** (a practical horticulturist) offers

CULTIVATION

Diseases.

Peronospora.

certain further remarks regarding poppy mould and other fungi seen on the opium plant.

"From these personal observations I was thus rather prepared for the periodicity in the mould attack, of which we are afforded unmistakable evidence in the Gya and Tehta returns, for the last quarter of a century or so. In conjunction with my own observations, these crop statistics indicate intervals of from three to four seasons in the appearance of the mould in a more or less severe blight form. Admitting, then, the operations of a law of periodicity, I would explain the apparent *triennial or quadrennial recurrence of mould in a virulent form*, thus :— *First*, I am led to assume, on a careful reconsideration of the results of all my observations and experiments, that the degree of virulence, or at least the particular form—*i.e.*, the *dry*, or *humid*, and *rotting*—is to a certain extent modified by prevailing atmospherical conditions during the latter stages of the mould's maturity, which, as a rule, is from about the close of February until the middle of March. Again, however, in seasons when the mould has attained a very extensive development on the crops, it, irrespective of the weather, tends to induce certain obscure chemical changes in the cell-contents and juices of its foster-parent, which would appear to entail an active resumption of the *vegetative function* and the *complete abeyance of the reproductive.*"

Some of the moulds observed by Mr. Scott are common **Saprophytes**, by no means confined to the Opium plant. It may, however, be found of some service to complete very briefly his enumeration.

"(*h*) *Sporocybe.*—A genus affecting the decaying parts of plants, and usually forming a blackish stratum thereon. The species have erect, simple, or branched peduncles, terminating in a capitate head studded with spores. The species found on the poppy much resembles the S. **bulbosa**, chiefly distinguished by its rigid, erect peduncles, slightly inflated at the base, and ending in a globose head of brownish-coloured spores. As affecting only the dead organs of the poppy, it is, of course, quite innocuous, and only alluded to here as being of common occurrence on the decaying parts.

Sporocybe.
241

"(*i*) *Helminthosporium.*—The species of this rather extensive genus of dark threaded moulds affect rotten wood and the decaying parts of plants generally. They have a gelatinous or indistinct mycelium, from which spring up erect, rigid, fertile filaments, terminating in large less or more elongated, often club-shaped, many-septate spores of a brownish-black colour. The species occurring on the poppy has erect septate filaments, with oblong and club shaped, 6 to 10 septate spores.

Helminthosporium.
242

"(*j*) *Macrosporium.*—A genus affecting similar habitats to the preceding, and distinguished from it by its large, transversely and vertically septate spores. Species of this genus are very abundant in India, but are perfectly innocuous to vegetable life, springing up only on the rotten or decaying parts of plants. The spores of this and the preceding genus are among the most abundant of the organic products found in the examination of the air at Calcutta.

Macrosporium.
243

"(*k*) *Cladosporium.*—A genus affecting both decaying animal and vegetable substances, and found in nearly every habitable part of the globe. **C. herbarum** is the species found on the poppy, forming thereon olive-green patches, which consist of jointed microscopic threads more or less varicose, simple, or branched, and bearing on their sides oblong or elliptic spores of an olive-colour, with one or two transverse divisions. It is very common on the older gummy exudations of the poppy, and also frequent on the injured capsules ; indeed, on all injured or decaying parts.

Cladosporium.
244

"(*l*) *Aspergillus.*—A genus found everywhere on decaying substances and accelerating decomposition. A well-known species is the **A. glaucus**,

Aspergillus.
245

**PAPAVER
  somniferum.**   Diseases of the Poppy Plant.

CULTIVATION

Diseases.

Dactylium.

246

which forms the *blue mouldiness* on cheese. It is distinguished by its necklaces of spores, rising from a less or more globular head, and supported by erect, septate, microscopic threads.

"(*m*) *Dactylium.*—A genus of thread-moulds found on the decaying or living tissues of plants and animals. The species are distinguished by their erect, jointed, and branched hyaline threads, bearing at their tips septate spores. D. roseum, the species found on the poppy, is thus described by the Rev. Mr. Berkeley :—" It consists of a creeping mycelium, from which arise short erect threads, crowned above with a few obovate uniseptate spores. The mass is first white, but at length acquires a pale rose colour, by which it is readily distinguished. The plant grows very abundantly on various objects, whether dead or living, and is sometimes highly destructive to cucumber plants, forming broad patches on the leaves and stems. It occurs also not unfrequently in closed cavities, as in nuts, to which it must have made its way from without through the tissues. A solution of *bisulphide of soda,* or indeed anything which contains sulphuric acid, will facilitate the destruction of this mould when requisite." The pale rose-coloured patches of this mould are very common on the poppy in moist, warm weather, occurring alike on the leaves and stems, and in the interior of the capsules, then densely webbing the seeds. It is very common in capsules which have been injured by grubs. In the opium go-downs, indeed, as on opium everywhere, when exposed to a moist and high temperature, it evidently finds a favourite *nidus,* covering the surface with its rosy web, when left for any time undisturbed. Dr. Durant, the present Principal Assistant in the Behar Agency however, informs me that it does really no injury to the opium. His only fungoid dread is the black-mould – *Mucor mucedo* —which will be subsequently described.

Sporotrichum.

247

"(*n*) *Sporotrichum.* — A genus of obscure thread-moulds, affecting decaying vegetable substances. They have tufted, jointed, ascending threads, with spores at first concealed, simple, and scattered. A species apparently belonging to this genus is found on the roots of the poppy. It occurs in oblong or roundish patches of a strawy colour, the threads forming a closely woven web, and bearing numerous minute subglobose spores. Plants with their roots affected by this mould have been sent me by the Sub-Deputy Opium Agent of Chuprah. The plants had withered in the ground ; the roots were covered with the above mould, and had assumed quite a dry and woody texture. I suspect this mould had been the main cause of the destruction of the plants.

Trichoderma.

248

"(*o*) *Trichoderma.*—A genus found on decaying wood and other parts of plants. It is characterised by a roundish peridium, composed of interwoven, branched, and jointed filaments, at length disappearing in the centre ; spores minute and spread over the disc. T. viride is not at all unfrequent on the base of the stem and roots of the poppy, but generally confined to diseased or injured parts. It forms roundish, intertangled tufts of snowy-white threads, on which rest numerous globose spores of a dusky-green colour.

Mucor.

249

"(*p*) *Mucor.*—The typical genus of the mucorinous moulds, characterised by a globose sporangium, into which the tip of the stem often enters in the guise of a clavate columella, and indefinite sporidia produced irregularly in the cavity : the spores are mostly elliptic—(*Berkeley*) . . . .
M. mucedo is the blackish-coloured mould so frequent on the opium shells during moist weather. It has a dense cobweb-like mycelium, consisting of tubular, jointed, simple threads, which bear at their apex a roundish membranous spore case, at first white, but which ultimately with the spores assumes a brownish black colour. Dr. Durant, the Principal Assistant in the Behar Agency, refers to it as the only serious cause of

| Diseases of the Poppy Plant. | (*G. Watt.*) | PAPAVER somniferum. |

injury to which the China opium cakes are liable. It appears to destroy the aroma and other physical qualities of the opium if allowed to affect it seriously. With regard to the conditions favourable or otherwise to this mould on the cakes, **Dr. Durant** remarks :—'I am much inclined to attribute it to the use of an undue proportion of *lewa*, as it is a well known fact that all adulterated opium of the kind and quality generally confiscated as unfit for use, if kept for a short time, will generate this fungus to any extent. This fungus also may be seen to perfection during the rainy season on the surface of all contraband opium classed as fourth quality, which, as a rule, is chiefly composed of foreign extracts, and is kept in large wooden boxes until destroyed. The only remedy for infected cakes is to strip off at once all affected parts of the shell, and to replace it by a new one well smeared with fresh *lewa*.' A moist and hot atmosphere is peculiarly favourable to its development and rapid extension, and really the only practical mode of suppressing it, for example, in our opium godowns, as in other closed structures, is to supplement free ventilation with a perfectly dry atmosphere—conditions which, throughout the rainy season at least, could only be secured by introducing hot air pipes or other heating apparatus into the godowns. The black mould is of that class technically called *hysterophytes*, which chiefly affects dead or decaying matters, and like many others characterised by a high selective power, absorbing and assimilating nutritious and rejecting noxious matters, so as to flourish amongst the most poisonous of chemical substances, if not of a corrosive nature It is thus useless to attempt its suppression by the artificial application of mineral salts, &c., to its *nidus*."

3rd, Mechanical Injuries.—Mr. **Scott** devotes some six or seven pages of his *Opium Husbandry* to this subject. Space cannot be afforded to deal with all the insects, birds, &c., which he alludes to. Suffice it to say that several caterpillars are very destructive, but that these are greedily devoured by the crow and the *mina*. Crickets often do considerable damage to the young poppy plants. A hemipterous insect known to the cultivators as *Lhi* is often destructive during dry weather. Similarly, during dry weather a mite does harm by spinning its web over the surface of the poppy leaves. The seed reserved for next year's crop is also attacked by various species of mites, beetles, and grubs; for these pests **Mr. Scott** recommends the use of camphor in the vessels used for storing the seed. (For further information on Insects destructive to the Poppy crop and to Opium, see the article **Pests** below.)

Of birds **Mr. Scott** specially mentions the parrakeet (*totha* or *soogha*) which feeds on the seeds in the capsules. To get at these it slits open the capsules and does much harm to the opium crop. If disturbed, it dexterously darts down from a height and nips off a capsule, the seeds of which are devoured leisurely on a neighbouring tree. **Mr. Scott** never observed the white-ants to eat opium, and thus disbelieves in the common report that they do so.

In many of the reports which the writer has consulted a further enemy, to the poppy crop, is said to exist in monkeys, who eat the leaves but do more damage by breaking down the plants than by their removal of the leaves.

EXTRACTION OF CRUDE DRUG FROM THE POPPY, AND MANUFACTURE OF OPIUM, MORPHIA, &c.

The cultivation of the poppy yields to the *rayat* several substances, from each of which he obtains a pecuniary return for his labours. These may briefly be said to be the (*a*) inspissated sap of the unripe capsules—opium—in its crude form ; (*b*) the moisture and soluble ingredients known as *pasewa* which drain from the opium ; (*c*) poppy petals (technically known as

MANUFAC-
TURE:

Extraction.
252

"leaves"); (*d*) "trash" or a powder prepared from the dried stems and leaves; (*e*) capsules; and (*f*) seed. The first mentioned three substances are those alone with which the Opium Department is concerned. They will, therefore, be dealt with in some detail in this place.

Extraction of the Drug.—The method of extracting the drug is described at page 74 of **Mr. Scott's Manual.** The capsules are lanced in the afternoon with an instrument known as the *nashtar*; this consists of four sharp blades tied together with cotton, the cotton being passed between the blades so as to keep them about $\frac{1}{10}$th of an inch apart. The incisions are drawn from below upwards, in perpendicular lines, but much care and great practice are necessary to ensure their being made of exactly the right depth. A merely superficial scratch is of no use, but, on the other hand, the walls of the capsule must not be cut through. Each capsule is usually lanced in this manner three or four times at intervals of two or three days, but sometimes a single incision exhausts the drug, while occasionally a productive capsule will give five, eight, or even ten discharges. The opium from the early sowings is thinner and more plentiful; that from the later sowings is more scanty, but of higher consistence. In the operation of lancing, a whole field is not taken in hand at once; it is divided into two or three portions, which are taken up on successive days. The work is generally performed, not by hired labour but by the *rayat* himself and his family. The object of the division of the fields is to make the labour continuous, by providing employment for each day.

The drug is collected in the early morning of the day following that on which the capsule has been lanced. The juice which has exuded from the incisions is scraped off with a small trowel-shaped scoop of thin iron, called a *setwah* (or *sutwa*). As the scoop is filled, the opium is transferred to a metal or earthen vessel, and is taken to the *rayat's* house for further manipulation. The subsequent treatment of the drug differs somewhat in the two Agencies. The Benares standard of consistence being 70 per cent. opium to 30 moisture, the Benares *rayat* keeps his opium in an earthen pan, which is slightly tilted to allow the *pasewa* to drain off. The *pasewa* is poured away into a separate vessel, and the opium is turned over by hand from time to time, at intervals of not more than a week. But as the standard in Behar is higher than that of Benares, *viz.*, 75, it is the object of the Behar *rayat* to raise the consistence of his opium more than this process would enable him to do. Accordingly after the *pasewa* has drained off, he puts the opium on a cloth tied over an earthen dish. This absorbs a good deal of the moisture, but a portion of the opium necessarily adheres to the cloth. This cloth is technically known as *kuffa* or *kaffá*, and is purchased by Government at a price proportioned to the quantity of opium it contains. The term *kaffá* or *kaphá* is also given to the special preparation of opium used in the manufacture of *madak* (*Conf.* with **Narcotics, V., 328**). *Pasewa* is purchased by Government at a fixed rate of R3-8 per seer. Fresh opium as collected contains about 50 per cent. of moisture. The average yield for each scarification is about 10 grains, but a healthy plant may yield 75 grains in all, *viz.*, with five to eight scarifications.

The above notes (compiled from official papers and from **Mr. Scott's** Manual) may be said to represent the chief facts of the extraction of crude opium, as practised in Behar and Benares. The following passages from **Dr. Impey's** account of Malwa convey the leading peculiarities of the methods pursued in Native States :—" Early in February and March the bleeding process commences. Three small lancet-shaped pieces of iron are bound together with cotton, about one-twelfth of an inch alone protruding, so that no discretion as to the depth of the wound to be inflicted shall be left to the operator; and this is drawn sharply up from the top of the stalk

MANUFAC-
TURE:

at the base to the summit of the pod. Three sets of people are so arranged that each plant is bled all over once every three or four days, the bleed-
ings being three or four times repeated on each plant. This operation
always begins to be performed about three or four o'clock in the afternoon,
the hottest part of the day. The juice appears almost immediately on the
wound being inflicted, in the shape of a thick, gummy milk, which is
soon thickly covered with a brown pellicle. The exudation is greatest
over-night, when the incisions are washed and kept open by the dew. The
opium thus derived is scraped off next morning with a blunt iron tool resembling a cleaver in miniature. Here the work of adulteration begins —
the scraper being passed heavily over the seed pod, so as to carry with it
a considerable portion of the beard, or pubescence, which contaminates
the drug and increases its apparent quantity The work of scraping
begins at dawn, and must be continued till 10 o'clock; during this time a
workman will collect 7 or 8 ounces of what is called *chick*. The drug is
next thrown into an earthen vessel, and covered over or drowned in linseed
oil at the rate of two parts of oil to one of *chick*, so as to prevent evaporation. This is the second process of adulteration, the *rayat* desiring to sell
the drug as much drenched with oil as possible, the retailers at the same
time refusing to purchase that which is thinner than half-dried glue. One
acre of well cultivated ground will yield from 70 to 100 pounds of *chick*.
The price of *chick* varies from 3 to 6 rupees a pound, so that an acre will
yield from 200 to 600 rupees' worth of opium at one crop. Three pounds
of *chick* will produce about two pounds of opium, from the third to the
fifth of the weight being lost in evaporation. It now passes into the hands
of the *bunneah*, who prepares it and brings it to market. From 25 to 50
pounds having been collected is tied up in parcels in double bags of sheet-
ing cloth, which are suspended from the ceiling so as to avoid air and
light, while the spare linseed oil is allowed to drop through. This operation is completed in a week or ten days, but the bags are allowed to
remain for a month or six weeks, during which period the last of the oil
which can be separated comes away, the rest probably absorbs oxygen
and becomes thicker, as in paint. This process occupies from April to June
or July, when the rains begin. The bags are next taken down, and their
contents carefully emptied into large vats from 10 to 15 feet in diameter
and six or eight inches deep. Here it is mixed together and worked up
with the hands five to six hours, until it has acquired an uniform colour and
consistence throughout, and become tough and capable of being formed
into masses. This process is peculiar to Malwa. It is now made up into
balls of from 8 to 10 ounces each, these being thrown, as formed, into a
basket full of the chaff of the seed pods. It is next spread on ground
previously covered with leaves and stalks of the poppy; here it remains
for a week or so, when it is turned over and left to consolidate, until hard
enough to bear packing; it is ready for weighing in October or November, and is then sent to market. It is next packed in chests of 150 cakes,
the total cost of the manufacture at the place of production being about
R 14 per chest."

Pasewa or Pasewha.—This is the dark coffee-coloured fluid which col-
lects in the bottom of the vessels in which the freshly gathered juice of the
capsules is placed by the cultivators when brought home. The shallow vessels are tilted to such a degree that the *pasewa* can drain off and be collected
and sent in separately for weighment. The reason for its separation is that
pasewa, if allowed to remain in opium, injures the physical characters of
the drug, causing it to look black and liquid, whilst at the same time it
gives the drug an unduly high assay when tested by evaporation on the
steam table. *Pasewa* consists of the most soluble of the principles of opium

Pasewa.
253

**PAPAVER
somniferum.**   Extraction of the Crude Drug from Poppy,

MANUFAC-
TURE :

dissolved in dew or in moisture absorbed from the atmosphere; it has peculiar smell, is strongly acid in reaction, and contains meconic acid, resin, morphia, and narcotine. *Pasewa* does not appear to be always present in opium, but is produced under certain atmospheric conditions. It is said to be never present, for example, in opium collected during strong westerly winds or in the absence of dew.

Leaves.
254

Leaves.—The flowering period of the poppy crop is from about the middle or end of January until the beginning of March. The formation of the petals is carefully watched by the cultivators. When ready to fall they are collected in the following manner. The base of the flower is encircled by the operator's forefinger and thumb ; the hand is drawn gently upwards and the matured petals at once detached without any injury to the capsule. Care is taken not to *pluck* the petals before they are mature, as that would cause a loss of juice from the wounded surface; nor are the petals gathered when moist, as they are then apt to become discoloured and to partially ferment. The petals are then made into what is departmentally known as "leaves." For this purpose an earthen plate is placed over a slow fire and on it is spread a handful of petals. These are covered with a moist cloth, above which is pressed a damp cloth pad until the steam from the cloth, acting upon the resinous matter contained in the petals, cause them to adhere together. The thin cake of petals thus formed is turned over in the earthen plate, and the process of pressing and consolidation repeated on the reverse side. These thin sheets pasted together with *lewa* form the shell or outer casing of the "provision" or Chinese opium. When fresh they have a pleasant fragrant smell which is supposed to impart to the Bengal drug its superior aroma. Cloth, silk, wax, and many other substances have been recommended and even tried, though unsuccessfully, as substitutes for the tissue of cemented petals. For cake-making three qualities of petal tissue are recognised. The first quality is known as *chandi* : it includes the silver-like petals of fine texture and colour : the second quality is known as *dawan* : the third quality as *gatha* or the thicker and coarser kinds. The terms *dawan* and *gatha* are chiefly in use in the Benares Agency : in Behar the terms are first, second, and third qualities of *chandi*.

The collection of the petals and the preparation of the "leaves" is a favourite task of the females of the cultivator's family, who generally appropriate the proceeds. The females of high caste cultivators, however, who are not accustomed to work in the fields (such as Brahmins, Rajputs, &c.) do not make "leaves." Hence these cultivators having to employ hired labour regard the preparation of "leaves" as a hardship. Where fuel is also scarce the preparation of leaves is looked upon with considerable disfavour, and for these two reasons it would appear to be more difficult to procure "leaves" in Behar than in Benares. The "leaves" are packed in baskets and brought into the Agencies by the cultivators on dates fixed by the departmental officers usually commencing in the Behar Agency about the 10th or 15th March, and in the Benares Agency a little later. Each cultivator's leaves are separately sorted and weighed, after which their value is placed to his credit in the accounts at the prescribed rates of R10, R7, and R5 for leaves of the first, second, and third quality respectively. The weighments of "leaves" begin earlier than that for opium, and are usually over in 15 days. The sum paid for leaves is generally about from R70,000 to R1,40,000 in each of the Agencies.

Trash.
255

Trash.—The pounded stalks and leaves of the poppy plant are known technically by this name. The substance is employed in packing the cakes. To prepare "trash" the plants are left standing in the fields till dry. The leaves and thinner parts of the stem are then collected and

P. 255

	PAPAVER
and the Manufacture of Opium. (*G. Watt.*)	somniferum.

broken up for " trash, " and the thicker parts of the stem are generally used for fuel or thatching. The quantity of "trash" required annually is regulated by the amount of opium manufactured.

The "trash" is usually delivered at the Sub-Agencies in bags of about 1 maund each. On arrival at the Agencies these bags are again weighed. One or more bags in each consignment are opened, and all foreign matter, such as earth and sticks, removed and separately weighed: 10 per cent. of foreign matter is allowed to pass, but, for any quantity in excess of that, a proportionate reduction is made from the weighment of the whole despatch. At each of the Agencies from R10,000 to R15,000 are paid annually for "trash."

Lewa—Is the paste by which the layers of the shell of the 'provision opium' cake are held together. In the preparation of *lewa* all the opium of inferior quality, unfit for the central mass of a cake by reason of its consistence or the presence in undue quantity of *pasewa*, is used, supplemented if necessary by good opium. The *pasewa* delivered is also employed to the extent of 5 per cent. in making *lewa*, to give greater adhesiveness to the paste.

Seed & Oil.—In another part of this article full particulars will be found of poppy seed and oil. The best seed is yielded by capsules which have not been lanced. The bland oil obtained from them is used at the opium factories in the manufacture of excise opium, a few drops being smeared on the inner paper wrapper to prevent this adhering to the Abkari cake.

ESTIMATION OF THE VALUE OF OPIUM.

Mr. Gregory gives the following practical notes on the subject of estimating the value of the crude opium presented by the *rayats* at the Government Agencies :—

"The points that an Opium Examiner keeps before him, and that intuitively pass through his mind, in the physical examination of the drug, are—

(*a*) consistence,	(*c*) texture,
(*b*) colour,	(*d*) aroma.

" Each one of the above points gives him some indication as to the quality of the drug and its ultimate appraisement, and also as to its disposal for factory uses.

" CONSISTENCE.—By this term we mean the actual percentage of solid and non-volatile matter in any given sample of the drug, if it were subjected to evaporation and reduced to dryness at a temperature of 200° Fahr.

" Pure opium being paid for by Government at a fixed rate for a certain standard of consistence, and being subject to a *pro ratâ* increase or decrease in price according as it is above or below that standard, it will be readily seen that the importance of arriving at the true consistence of any given parcel of the drug stands second to none of the many duties devolving on the Opium Examiner.

" By the help of sensitive balances and metallic tables heated by steam accurate results in the estimation of consistence can be relied on, and the mechanical method pursued at the present day has already been noticed. Such a delicate operation, however, as the 'assaying' of opium (as the estimation of the true consistence by steam tables is termed) can be applied to a very limited portion of the many thousand tons of the drug that pass through the factories. Every 100 grains of the drug, therefore, that is placed on the steam table is a representative sample of a large bulk that

Margin notes:
MANUFAC-
TURE :

Lewa.
256

Seed.
257
Oil.
258

VALUATION.
259

Consistence.

**PAPAVER
somniferum.**   Estimation of value of Opium.

MANUFAC-
TURE :

Valuation.

has been adjudged of nearly equal consistence by the remarkable power
of hand estimation practised at the factories, a power that is gained only
by years of experience in the examination of the drug.

"It would be difficult therefore—nay impossible—to lay down rules for
arriving at results that can be satisfactorily obtained only by practice. A
few guiding principles will, however, be touched on here.

"As a rule the consistence of opium freshly collected from the capsule
varies considerably, according to peculiarities of soil and weather, ranging
from 30° to 50°, that is, it contains from 30 to 50 per cent. of solid
matter.

"Between the time of collection, and of weighment and examination of
the drug at the Government scales, there is generally an interval of from
one to even three months, and during this period it is within the power of
the cultivator so to manipulate his drug as to raise it to any standard of
spissitude. Experience, however, shows that the cultivator is not so easily
schooled into turning out an article exactly suitable to the requirements of
our factories, and it is no uncommon thing to find in one season two jars
lying side by side, one of which contains opium yielding a clean section if
cut with a spatula, the other containing a drug so fluid as to be poured
out of the jar by tilting it over.

"The practical impossibility of guessing with certainty to a degree the
consistence of any given sample of opium has given rise to the ' classes '
of opium now obtaining at the two Agencies. Each class includes in it
a range of three degrees of consistence, and between the first and the last
class is included all the opium that is ordinarily brought to the Government scales.

"The following is the classification table adopted for good opium at the
two factories at Patna and Ghazipur, together with the distinctive mark of
each class :—

CLASS.							DISTINCTIVE MARK.		Degrees included in each Class.
							At Patna.	At Ghazipur.	
Báshi bála darawal	.	.	.	.	.		X̿	XXX̊	79, 80, 81
Bála darawal	.	.	.	.	.		X̄	XX	76, 77, 78
Darawal	.	.	.	.	.	.	X	X	73, 74, 75
Awal	.	.	.	.	.	.	I	I	70, 71, 72
Duyum	.	.	.	.	.	.	II	II	67, 68, 69
Siyum	.	.	.	.	.	.	III	III	64, 65, 66
Chaharum	.	.	.	.	.	.	...	IV	61, 62, 63
Panjum	.	.	.	.	.	.	...	V	58, 59, 60
Shishum	.	.	.	.	.	.	...	VI	55, 56, 57
Haftum	.	.	.	.	.	.	...	VII	52, 53, 54

"For purposes of district classification the above table answers admirably, and it is also adhered to at the factories when re-classifying by touch
the classification of district officers, prior to the ultimate appraisement of
the opium by the help of steam tables. During this final classification,
however, when the object at the factories is to arrive at the true consistence
of every parcel of opium, drug of a spissitude estimated by touch to be
above the highest or below the lowest class is assayed separately on the
steam table, and its true consistence adjudged.

"We have thus seen that there are two methods practised at the Agencies
for estimating consistence, (a) by steam tables, (b) by touch. The second is a

P. 259

MANUFACTURE :

Valuation.

rough-and-ready method of assigning into one class masses of opium the true average consistence of which is finally settled by the first method.

"For the determination of consistence which is dependent only on the quantity of moisture contained in the drug the mode of procedure is a simple one and the results satisfactory. In practice, however, disturbing elements are very often introduced, and one of these is *pasewha*. Opium with an admixture of *pasewha* is deceptive to the touch.

"In drug free from *pasewha* the granular texture appears to maintain cohesion between the particles which, as it were, support each other and offer a certain amount of resistance to pressure. In drug with a copious admixture of *pasewha* the granular texture is destroyed by the gradual merging of the tears into each other through the medium of the tenaceous and shiny *pasewha*, the cohesion existing is thus lessened, but the tenacity of the drug is increased.

"Where the bulk of the produce at the factories lies somewhere intermediate, with regard to the admixture of *pasewha* between the two descriptions of drug given above, the sense of touch is regulated by what comes most in its way. When dealing, therefore, with varieties bordering on the two extremes of the drug we are apt to go astray, and we are thus able to account in a large number of cases for what is known as being 'out in parakh' (judgment). We have thus prepared for ourselves an arbitrary and indefinable standard of 'touch;' it is, nevertheless, a standard so generally accepted by all examiners of opium in the Agencies that it is practically a fixed one, and it is a recognised maxim that opium entirely free from *pasewha* will assay lower than this our accepted standard of touch, and that opium with a copious admixture of that substance will assay correspondingly higher. A good 'parkhia' (examiner) will always, therefore, make due allowance for the absence or presence of *pasewha* in any sample of the drug that is being subjected to examination for consistence. The remarks made here refer entirely to good opium.

"Another disturbing element in estimating consistence is heat, particularly on drug charged with *pasewha*. Drug of this character, under the influence of heat, undergoes liquefaction to a moderate extent in the process of drying. Opium to be examined for consistence by touch should invariably be placed, therefore, in shaded and cooled verandahs, and the examination should be concluded by 9 or 10 o'clock in the morning, and before the sun gets hot. When, for want of accommodation, jars have to be placed in open yards, their examination should invariably be undertaken first, and in the early morning. The examination by touch, for consistence, of opium that is lying exposed to the sun's rays in the months of April, May, June, and July, when all the examination at the factories is conducted, must always be faulty and conjectural, and should never be attempted.

Colour.

"COLOUR.—The natural colour of the drug runs through infinite shades of brown, from a dull or even bright chestnut to a reddish-brown, and from a dark mahogany to a blackish-brown. It even appears black at times when viewed in bulk.

"These variations are due to causes with which we have no concern here, suffice it to say that they are natural, and to a practised eye easily discernible as the true colours of opium. Age and exposure may darken the colour of the drug but cannot alter its characteristics; and where an alteration appears it may be accepted as a sure indication of adulteration or sophistication of some sort, although, again, sophistication of the drug is possible without any perceptible alteration of colour.

"The true colour of opium is clearly seen when the drug is viewed in a very thin film; this is best accomplished by pressing a small portion between two glass slips against the light, or by rubbing it down with the

**MANUFAC-
TURE:
Valuation.**

finger on a white earthenware plate. Here it is that we see clearly the various shades of chestnut, reddish-brown, dark brown or mahogany, but never black. When rubbed between the fingers opium displays a shining surface and a waxy lustre.

"The colour of opium is a valuable indication as to its purity.

Texture.

"TEXTURE.—Like consistence and colour the drug delivered at the Government Factories may be said to differ, one sample from another, in texture. At the two extreme poles of variation there are the distinctly granular, and the perfectly homogeneous, and the bulk of the produce lies, as to texture, somewhere intermediate between those extremes.

"The primary causes of variation, into which our enquiry does not extend, are undoubtedly due to differences in soil, and to conditions of weather obtaining at the time of collecting the drug; they are also due, to some extent, to manipulation of the drug after collection. A light-coloured, chestnut or reddish-brown variety of the drug, which is free from *pasewha*, will, as a rule, be found to be distinctly granular, while the dark, or blackish-brown variety, which has more or less of *pasewha* in its composition, or an excess of moisture, will, on the other hand, tend to the homogeneous type.

"Ordinary manipulation, without the aid of sophistication, has little effect on texture, but long-continued manipulation will affect it materially. The presence of *pasewha*, again, affects it in a very marked degree, and so does an excess of moisture.

"As already explained under the head " consistence " to the presence of *pasewha* in varying quantities is due the merging, more or less, of the tears into each other whereby the granular nature of the drug passes by imperceptible gradations to the homogeneous. The presence of *pasewha* also alters the dull waxy appearance of the drug to one that is more or less smooth and shiny, adding to it tenacity, and making it more glutinous. Ordinarily, opium, free from *pasewha*, is moderately ductile, but the presence of *pasewha*, by adding tenacity, increases also the ductility of the drug. This is seen by drawing out with both hands opium of high consistence. If free from *pasewha* it will be found to be ductile to an extent varying according to consistence, with a uniform and minutely granular texture. When there is *pasewha* present this ductility is increased, while the granular texture is less marked, according to the proportion of *pasewha* present. The drug when thus drawn out breaks with an irregular fracture; it adheres to the fingers, is viscid, and of a plastic nature. The texture of the drug is also well seen in high consistence opium when a section is exposed with a spatula.

"Opium of the lower consistences—below about 66°—being in a somewhat fluid state, will not draw out at all but breaks off with ragged edges. Its texture is subject to change, under the same conditions, as in opium of higher consistences.

"The texture of any given sample of pure drug is always uniform. A practised eye can at once detect any irregularity, and where such exists it betrays the presence of a foreign substance in the composition of the drug.

Aroma.

"AROMA.—Chemistry has not yet isolated the volatile odorous principles of opium. Its aroma, however, is peculiar and characteristic. Some consider it not unpleasant, while others relegate it to the class of disagreeable odours. In well-prepared, fresh drug the aroma is decidedly fruity, but it varies with age, and is even said to vary somewhat with the description of soil on which the plant is grown, and with the manure used.

"Careless preparation of the drug, such as its collection or manipulation in plates not scrupulously clean, or allowing it to come in contact with animal substances, such as bladders for storing it away in, or keeping it in

P. 259

	MANUFAC-TURE:
	Valuation.

ill-ventilated and smoky closets, or shutting it up for security in small, close receptacles, will dissipate and destroy the aroma in drug that is otherwise intrinsically good, and will even give it an offensive odour.

"The aroma of the drug is one of its chief commercial criterions, and as such should be carefully guarded by the cultivator. To the Opium Examiner it gives a very important indication as to the suitability of the drug for the various factory purposes. It is only by chemical tests that the Examiner can be certain that opium that is devoid of aroma or offensive to the smell, although apparently good as to texture and colour, has not also a foreign substance in its composition, assuming that the foreign substance, if present, has not given the clue by its own specific odour. Under any circumstances, opium deteriorated in its aroma, although it may be otherwise pure, should be set aside and utilised for other than the main factory purpose, that is, amalgamation with drug intended for the central mass of cakes, otherwise there will be risk of the deteriorated drug tainting a much larger mass of good opium" (*Pharmacog. Indica, I., 101 to 107*).

Adulteration.
260

Adulteration of Opium.—The late Surgeon-Major Sheppard, Principal Assistant Opium Agent, Benares, drew up the following enumeration of the adulterants generally used by the cultivators:—

"1. Adulteration with fresh green parts of the poppy plant, including watery extracts.

"2. Adulteration with foreign extractives, and vegetable matter, such as the inspissated juice of the **Opuntia Dillenii** and **Calotropis gigantea**, extracts of the tobacco plant, datura, and hemp.

"3. Gums and resinous matters. A gum resin derived from different species of **Ficus**, and called *lassa*. The resin of **Shorea robusta** (*sál*), pulp of Bael fruit, gum from seeds of *Tal-i-makhana* (**Hygrophylla spinosa**), tamarind pulp, gum from **Acacia arabica**.

"4. Farinaceous admixtures, including linseed, poppy seed, seeds of leguminous plants, and esculent tubers and roots. The starchy matter is often heated for a long time before being used ; hence iodine reaction may fail.

"5. Vegetable substances containing tannin and colouring matters. Catechu, *gáb* (**Diospyros Embryopteris**, juice of fruit), turmeric, flowers of **Bassia latifolia**, betel-nut, extract of pomegranate bark.

"6. Saccharine matter; vegetable oils and *ghí;* soot, charcoal, and semi-burnt opium; cotton and paper.; cowdung; earthy and siliceous matter; pounded burnt bricks; impure carbonate of soda, &c. "

Scott in his *Opium Husbandry* gives an instructive review of the adulterants employed in opium, see pp. 121 to 127. He regards those alluded to in paragraph 4 above as the most important since the materials are at hand to every cultivator.

PROCESS OF MANUFACTURE AT THE GOVERNMENT AGENCIES.

OPIUM FACTORY.
261

The earliest and to the present day the best detailed account of the various stages and manipulations in the manufacture of opium, is given by Dr. Eatwell (in 1851). Most subsequent writers have satisfied themselves by reproducing Dr. Eatwell's account and have altered his expressions and sentences only so far as seemed necessary to lay claim to originality. Perhaps the most concise version of Dr. Eatwell's description is that given in *Spons' Encyclopædia*, which, from its being rendered in a less technical form than the original, may preferentially be reproduced here:— "When weighed into store, the opium is kept in large wooden boxes, holding about 10 cwt. ; if below the standard, it is occasionally stirred up, to favour its thickening; and if very low is placed in shallow wooden

MANUFAC-
  TURE:
  Opium
  Factory.

drawers, and constantly turned over. Whilst keeping, it becomes coated with a thin blackish crust, and deepens in colour according to the degree of exposure to the air and light. From this general store (*malkhana*), it is taken daily, in quantities of about 250 maunds (of 82⅔℔), for manufacture into 'cakes.' Various portions are selected (by test assay) so as to ensure the mass being of the standard consistence, and these, weighing exactly 10 seers (21℔) each, are thrown promiscuously into shallow drawers, and rapidly and thoroughly kneaded up together. The mass is filled into boxes, all of one size, from each of which a specimen is drawn and assayed. The mean is taken as the average. Before evening, these boxes are emptied into wooden vats, 20 feet long, 3½ feet wide, and 1¼ feet deep, and the opium is further kneaded and mixed by men wading knee-deep through it from end to end, till the consistence appears uniform.

"Next morning the manufacture of the cakes commences. Each cake-maker sits on a wooden stand, and is provided with a brass cup and a graduated tin vessel. The 'leaves' for forming the shells of the cakes are weighed out over-night, tied in bundles, and damped to make them supple; and boxes are provided containing *lewa*, for agglutinating the 'leaves' to form the shells of the cakes. This *lewa* consists of an admixture of inferior opium, *pasewa*, and the washings of vessels that have contained good opium, forming a semi-fluid paste of such a consistence that 100 gr. evaporated to dryness at 93½° (200° F.) leave 53 gr. residue. The *lewa*, 'leaves,' and opium are accurately weighed out for each cake. In his brass cup, the operator rapidly forms the lower segment of the shell of the cake, pasting 'leaf' over 'leaf,' till a thickness of ⅛ inch is reached, and allowing a certain free portion of the most external 'leaves' to hang down around and over the sides of the cup. The cake of opium brought from the scales is now inserted, and held away from the sides with the left hand, while one 'leaf' after another is tucked in, well smeared with *lewa*, and imbricated one over the other, till the circle is complete. The free portions of the 'leaves' left hanging over are drawn up tightly, and the opium cake is well compressed within the casing. A small aperture remains at the top; this is closed by adding more 'leaves,' and finally the cake is completed by applying a single large 'leaf' to the entire exposed half. The finished ball or 'cake' resembles a Dutch cheese in size and shape. It is rolled in a little finely powdered poppy trash, which adheres to its surface, is at once placed in a small earthen cup of the same dimensions as the brass cup used in shaping it, and is carried out and exposed in small dishes to the sun. It is so exposed for three days, and is meantime constantly turned and examined; should it become distended, it is opened to liberate the gas, and again tightly closed. On the third evening, still in their earthen cups, the cakes are placed on the 'frames,'—open battens allowing free circulation of air. The operation thus far is terminated by the end of July. The constituents of the average perfect cake are :—

Standard opium	.	.	.	.	.	1 seer 7·50 chittacks	
Lewa	.	.	.	.	.	3·75	,,
'Leaves' (poppy petals)		.	.	.	.	5·43	,,
Poppy trash	.	.	.	.	.	0·50	,,

                            2 seers 1.18 chks.=4℔ 3½oz.

"The number of cakes made by one man in a single day is about 70, though some can turn out 90—100. After manufacture, the cakes require much attention, and constant turning, on account of the mildew which attacks them. This is removed by rolling and rubbing in dry poppy-trash. Weak places are ? so looked for, and strengthened with **extra**

Manufacture at Government Agencies. (*G. Watt.*) **PAPAVER somniferum.**

'leaves' By October, the cakes are perfectly dry to the touch, and fairly solid; they are then packed in chests, furnished with a double tier of wooden partitions, each tier with twenty square compartments for the reception of so many cakes, which latter are steadied and packed around with loose poppy-trash. The chests must be most carefully kept from damp for a length of time; but ultimately the opium in the cake ceases to yield any more moisture to the shell, and the latter acquires extreme solidity. Each case contains 120 *catties* (about 160℔). The foregoing remarks refer exclusively to the Government prepared opium for the Chinese market, which includes the great bulk of the entire product.

" Bengal opium intended for internal consumption, and known as *abkari* opium, is prepared in the following manner:—It is inspissated by exposure to direct sun-heat till its standard is 90 per cent., and its consistence resembles wax. It is then moulded into square bricks weighing one *seer* each, which are wrapped in oiled Nepal paper, and packed in boxes furnished with compartments for their reception. Put up in this way, it has not the powerful aroma of the 'cake' drug, but it is more concentrated, and more easily packed. It is sometimes also made into flat square tablets."

CHEMISTRY OF OPIUM.

The complex nature chemically of the inspissated sap of poppy capsules has been dealt with so fully by writers on the subject that it is perhaps scarcely necessary to review here the history of the development of our knowledge of the subject. The Journals and Transactions of the Royal Pharmaceutical Society of Great Britain teem with papers on the chemistry of opium, and a large series of special works exist which are accessible to the student. The recent publication by Dymock, Warden, & Hooper (*Pharmacographia Indica*), however, contains much of a practical and specially Indian nature on the chemistry of opium It will perhaps, therefore, serve the purposes of the present publication better to reprint the special chemical chapter of that work rather than to attempt a separate compilation from the voluminous literature that exists:—

" The alkaloids which have been separated from opium are Hydrocotarnine, $C_{12} H_{15} NO_3$; Morphine, $C_{17} H_{19} NO_5$; Pseudomorphine, $C_{17} H_{19} NO_4$; Codeine, $C_{18} H_{21} NO_3$; Thebaine, $C_{19} H_{21} NO_3$; Protopine, $C_{20} H_{19} NO_5$; Laudanine, $C_{20} H_{25} NO_4$; Codamine, $C_{20} H_{25} NO_4$; Papaverine, $C_{20} H_{21} NO_4$; Rhœadine, $C_{21} H_{21} NO_6$; Opianine, $C_{21} H_{21} NO_7$; Meconidine, $C_{21} H_{23} NO_4$; Cryptopine, $C_{21} H_{23} NO_5$; Laudanosine, $C_{21} H_{27} NO_4$; Narcotine, $C_{22} H_{23} NO_7$; Lanthopine, $C_{23} H_{25} NO_4$; Narceine, $C_{23} H_{29} NO_9$; Gnoscopine, $C_{34} H_{36} N_2 O_{11}$. A bitter principle, Meconin, $C_{10} H_{10} O_4$, is also present in opium, accompanied by Meconic acid, $C_7 H_4 O_7$.

Porphyroxin, first described by Merck, occurs in East Indian, in Smyrna, and probably other opiums. The principle is of interest, because it has the property of being reddened by hydrochloric acid, a reaction which has been utilised for many years in testing for opium in medico-legal analysis in the Bengal Chemical Examiner's Department. In testing viscera for opium the ethereal extract obtained by Stas's process is evaporated in a po celain capsule, and the dry residue moistened with dilute hydrochloric acid; on the application of a gentle heat a red coloration is developed should opium be present. A good plan of applying the test is to place on the bottom of the capsule containing the dry ether extract, a very small watch glass moistened with a few drops of concentrated hydrochloric acid; the capsule is then covered with a glass plate; after standing some time a red or violet reddish coloration appears on the sides of the capsule should porphyroxin be present. The application of heat is unnecessary when the test is

PAPAVER somniferum.   **Chemistry of Opium.**

MANUFAC-
TURE :
Chemistry.

applied in this manner.* The chemical composition of porphyroxin appears to be a matter of some uncertainty; according to O. Hesse it is a mixture of several distinct principles. Pedler & Warden isolated in 1886 from Bengal opium a neutral principle insoluble in water, but dissolving in ether, chloroform, benzol, &c., and yielding solutions which exhibited a magnificent blue fluorescence. The morphine in opium is combined with meconic acid. The nature of these two substances was made known by Sertürner in 1816, who at the same time pointed out the difference between morphia and narcotine, a substance which had been discovered in opium by Derosne in 1803 and also by Seguin. There can be no doubt that these two chemists also obtained morphine, but failed to distinguish it from narcotine. Warden (*Chem. News, 38, 146*) has examined the ash of Behar opium. It was of a light grey colour and contained ·857 per cent. of charcoal, which was deducted before calculating the percentage composition, which is as follows :—$Fe_2 O_3$, 1·983; $Ca O$, 7·134; $Mg O$, 2·310; $K_2 O$, 37·240; $Na_2 O$, 1·700; SO_3, 23·141; $P_2 O_5$, 10·902; $Si O_2$, 15·274. There were also traces of alumina, manganese, carbon dioxide and chlorine present.

The examinations of various kinds of Indian opium conducted by Dr. Buri in Professor Flückiger's laboratory (*Pharm. Journ., April 24, 1875*) gave the following results :—

	Patna garden opium, 1838.	Indian medical opium, 1852.	Abkari provision opium.	Garden Behar opium.	Malwa opium, flat cake.	Sind opium.	Hyderabad, Sind.	Khandesh.	Persian, 1872.
a—Ethereal extract, *i.e.*, residue dried after the evaporation of the ether.	24·2	21·7	22·0	20·6	14·1	17·4	20·4	...	25·0
b—Crude narcotine	10·0	9·0	8·5	7·6	7·6	8·0	9·7	...	10·2
c—Wax: difference between *a* & *b*	14·2	12·7	13·5	13·0	6·5	9·4	10·7	...	14·8
d—Purified narcotine	4·0	6·1	5·5	4·5	4·7	3·1	5·4	7·7	6·4
e—Crude morphine	11·2	11·2	14·1	10·6	14·4	...	...	...	...
f—Purified morphine	8·6	4·3	3·5	4·6	6·1	3·8	3·2	6·07	7·1

Professor Flückiger remarks :—

" The process for the estimation of narcotine and morphine was that described in the *Pharmacographia, p. 59.* The extract *a* of the above table is that afforded by means of boiling ether, with which the powdered opium had almost absolutely been exhausted by repeating the treatment with ether from about twenty to thirty times. The extract remaining after the evaporation of the ether was boiled with acetic acid, 1·04 sp. gr. This liquid, after the acid had been driven off, yielded *b*, crude narcotine, as a crystalline brownish mass. It was washed with ether, and then afforded *d*, purified narcotine. Under *c* the difference between *a* and *b*, representing the amount of waxy matter, is calculated. It includes also the oily matter with which the Persian opium is impregnated, as well as a little wax in the case of sample I.

" In exhausting the opium with ether, a slightly yellowish fluid is obtained, which displays a bluish fluorescence, due to an unknown constituent of the drug.

" Before precipitating the morphine, the aqueous solution was concentrated in order to get a smaller volume.

* It is necessary to note that this test is only employed as a corroborative one for the presence of opium.

MANUFAC-
TURE:
Chemistry.

" It afforded *e*, the crude, dried morphine, which, after twice or three times repeated recrystallisation, finally furnished *f*, purified morphine. This purification of morphine cannot be performed without a loss of morphine; the real practical percentage of that alkaloid may, therefore, more correctly be regarded as somewhat superior to the figure *f*. It would be desirable to apply a process furnishing the exact percentage; yet there is, as far as I know, no such method thoroughly satisfactory. I have been struck with the very large discrepancy, in the Indian opium, of the figures under *e* and *f*, which, I think, is larger than in opium from Asia Minor. Another fact well worth considering is the usually low percentage of morphine of Indian opium, narcotine being frequently present to a larger amount. This has already been pointed out in the *Pharmacographia, p. 57.* It would appear, however, that this is of no consequence for the Chinese consumption, yet, possibly, it will be so some day if the home production of the Chinese further increases. Perhaps a more careful preparation of the Indian opium would at least prove of importance, not so much with regard to the smokers of the drug as to the possibility of extracting morphine from Indian opium profitably. It is not needful to point out that this would be highly desirable."

In the following table is shown the analyses of samples of Patna and Behar provision opium, Malwa opium and *pasewa*. These analyses are interesting, as they indicate the amount of extractive obtained by the action of cold and hot water on the drug. The amount of extractive as well as the alkaloidal content varies within narrow limits from year to year. Analyses of Behar and Patna provision opium, arranged as shown in this table, are yearly placed before the merchants at the annual inspection of opium, which takes place before the first sale of the season :—

VARIETY OF OPIUM.	Moisture at 100° C.	Cold-water extract on anhydrous opium at 100° C.	Hot-water extract on anhydrous opium at 100° C.	Narco-tine on anhy-drous opium.	Morphia on anhy-drous opium.	Total alka-loids.
Behar cake No. 1, manufac-tured 26th May 1883 .	26·43	64·25	65·54	5·91	3·86	...
Benares cake No. 4, manu-factured 1st January 1883	29·97	63·8	64·57	5·91	4·58	...
Malwa opium, 31st March 1883	8·56	65·90	68·58	6·81	4·92	11·73
Pasewa from Benares District, 1886 . .	19·75	66·31	70·46	5·04	3·19	8·23
Pasewa from Benares District, 1888 . .	22·00	72·05	76·16	4·10	·85	4·95

Regarding the amount of morphia in Malwa opium, according to Dr. Smyttan, formerly Opium Inspector, Bombay, the best Malwa opium yields 8 per cent., Flückiger's analysis gives 14·4 per cent. of crude and 6·1 per cent. of purified morphia, a larger yield than that obtained by Mr. Gregory, who only found 4·92 per cent. On the other hand, while Flückiger found only 4·7 per cent of narcotine, the Opium Factory analysis affords 6·81 per cent. Flückiger's analysis of Patna garden opium, in which 8·6 per cent. is given on the content of purified morphia, is an exceptional yield of the alkaloid for Bengal opium. The analyses of *pasewa* are of special interest as indicating the very wide differences which may occur in its composition.

PAPAVER
somniferum.

Manufacture of Alkaloids.

MANUFAC-
TURE :
Alkaloids.
263

MANUFACTURE OF ALKALOIDS.

In the *Pharmacographia Indica* the following account of the manufacture of the alkaloids as practised at Benares is given, from the pen of Mr. Gregory, Officiating Principal Assistant Opium Agent :—

"The opium used at Ghazipore for manufacture of alkaloids, consists of confiscated opium so adulterated as to be unfit for provision or *ábkári* purposes; adulterated contraband opium and *dhoi.* The average amount of opium used, taking the figures for three years from 1886 to 1888, amounts to about 16,626℔ annually.

The yield of alkaloids during 1887-88 was as follows :—

Hydrochlorate of Morphia		.	.	.	.	.	242℔	14½ oz.
Acetate	,,	,,	.	.	.	.	34 ,,	11 ,,
Sulphate	,,	,,	.	.	.	.	19 ,,	10½ ,,
Codeine	.	.	.	.	.	.	.	30 ,,	10¼ ,,

No narcotine has been manufactured since 1881-82, there being no demand for it. In 1879-80, the yield was 188℔.

Morphia is manufactured by the **Gregory-Robertson** system modified in a few minor details. The opium is steeped in small vats with water, and the liquor passed through blanket filters : the maceration of the residue is repeated until the filtrate is colourless. The mixed filtrates are evaporated by steam to a thin syrupy consistence. Chloride of calcium is then added in the proportion of about 5 per cent. of the weight of the opium used, and the mixture evaporated until it solidifies on cooling. The crystalline magma is then powerfully pressed. The dry cake is dissolved in boiling distilled water, filtered, and the filtrate evaporated until it solidifies on cooling. Pressure is again applied to the magma ; and the resulting cake again dissolved in water, and this process is repeated perhaps a dozen times, until the cake is almost white. The expressed mother liquors are again worked up for morphia. The nearly white cake is finally dissolved in boiling distilled water, and ammonia in slight excess added. The precipitate is collected and worked with cold distilled water, until it ceases to give the reaction for chlorides. The precipitated morphia is then neutralised with hydrochloric acid, and the solution crystallised. The crystals are pressed, and mixed with twice their weight of water, and wood charcoal added in the proportion of 2 oz. to each ℔ of the mass. This mixture is heated to 200° F. for about twenty minutes, and then filtered. On cooling, the hydrochlorate of morphia separates in crystals. Codeia is obtained from the mother liquor, left after the precipitation of the morphia by ammonia. The liquor is concentrated to a moist mass and strongly pressed ; the cake is moistened with water and again pressed, and this is repeated until the alkaloid is nearly white. The cake is broken up in water, and caustic potash added in considerable excess. The codeia separates in crystals slightly coloured. It is finally purified by crystallisation from alcohol. Narcotine is obtained by digesting with hydrochloric acid the insoluble residue left by the action of water on opium, and precipitating with ammonia. The impure narcotine is purified by repeated solution and crystallisation from acohol, and decolorised by charcoal."

The authors of the *Pharmacographia Indica* give the following remarks regarding the Indian prepared alkaloids of opium :—

Indian
Alkaloids.
264

"INDIAN MANUFACTURED ALKALOIDS.—We have examined morphia, codeine and narcotine manufactured at Ghazipore.

The morphia hydrochlorate was in white acicular prisms of silky lustre and free from odour. Dried at 100°C., the crystals lost 12·74 per cent. The hydrochlorate is usually stated to contain three molecules of water which would be equal to 14·38 per cent. The chlorine calculated as H. C. amounted to 9·42 per cent., the sample was consequently deficient in combined acid to the extent of ·3 per cent. The ash amounted to ·063 per cent. By the action of chloroform ·812 per cent. of extractive was obtained. The precise nature of this extractive was not determined; it probably contained a trace of morphia; it was tested specially for narcotine with negative results. Uncombined morphia to the extent of ·828 per cent. was detected in the sample.

The codeine was a perfectly white powder, and free from odour. Dried at 100° C., it lost 5·16 per cent. The ash amounted to ·056 per cent. The saturating power of the alkaloid for standard acid corresponded closely with that acquired by theory.

The narcotine was in faintly yellowish crystals. It contained only a minute trace of ash, and was free from morphia."

MANUFAC-TURE: Alkaloids.

FISCAL ADMINISTRATION OF OPIUM IN INDIA AND THE REVENUE THEREFROM.

Having now worked out the leading historic facts that have a bearing on the Indian and Chinese Opium question, it may not be out of place to record here the chief facts regarding the Indian localities where opium is produced; to exhibit the enactments by which the cultivation and manufacture are controlled; to explain the degree to which the Indian Government is directly in the position of a private individual who holds the sole right to cultivate and sell an article of commerce; and to show the revenue derived by India from the drug. These objects cannot be better met than by republishing a memorandum recently issued by the Department of Finance and Commerce :—

ADMINISTRA-TION. 265

BENGAL OPIUM.

"In British India the manufacture of opium for export to the Straits Settlements and China is restricted to two agencies in Behar and Benares in the Bengal Presidency. These are under the administration of the Government of Bengal, though the so-called Benares Agency includes sub-divisions which are partly situated in Oudh and are established in territory under the Government of the North-Western Provinces and Oudh.

"The cultivation of poppy and the manufacture of opium in these two Agencies are regulated by Act XIII. of 1857, under the general control of the Government of Bengal and the Board of Revenue, and the immediate supervision of the Opium Agents in charge of the Agencies. The possession, transport, import and export of opium are regulated by the provisions of the Rules passed under section 5 of the Opium Act of 1878. The Collectors of Land Revenue in the districts within the range of the Agencies are Deputy Agents for the enforcement of some of the provisions of Act XIII. of 1857, but they do not otherwise take part in the details of supervision which are conducted by Sub-Deputy Opium Agents acting under the orders of Opium Agents.

"The *extent* of cultivation and the quantity of opium to be annually brought forward for sale are regulated by the Government of India in communication with the Government of Bengal, while the limits of cultivation within each sub-division, and the persons by whom the poppy may be cultivated, are arranged annually by the Opium Department. The cultivators, though free to decline the cultivation, are restrained from selling their produce to any but the Opium Department.

Bengal Agencies. 266

Cultivation and Manufacture regulated by Act XIII. of 1857. 267

Possession of opium, right to sale : & transport, import, & export regulated by Opium Act of 1878. 268

Restriction to Cultivation. 269

P. 269

ADMINISTRA-
TION:
Purchase of
Produce.
270

"The *produce* is purchased, in the form of crude opium, at a price, per seer of a standard consistence of 70 degrees, which is fixed from time to time by the Government of India, any change in the price being notified to the cultivators at the time of their entering into the annual engagements with the Department.

"*Settlements* are made with the cultivators or their accredited agents at the end of July or early in August, when the license to cultivate is granted by the Sub-Deputy Agent or his Assistant, and after the details are settled an advance is paid to the *lumberdar* by the *gomashta* in their presence.

Method
of
Payment.
271

"The number of advances made during the year in the Behar Agency are not to exceed five, as follow :—

"*1st*—In September, at a rate not exceeding R5 per *bíghá* excepting when the Agent permits a higher rate.

" *2nd*—At the end of December or beginning of January, after the crops are above ground and the prospects are favourable, at a rate not exceeding R4 per *bíghá*, including any advance for poppy leaves.

" *3rd*—At the end of March, after the chief part of the crop has been gathered and the *taidad* or estimate of the probable outturn of the season has been completed, at a rate not exceeding R3 per *bíghá*.

" *4th*—At the time of weighment, the rate of the advance being regulated by the quality of the drug delivered at the rates here specified :—

								Behar.		Benares.	
								R	a.	R	a.
For class	I and above up to	.	.	.	.	.	5	0	5	0
Ditto	II	ditto	.	.	.	.	.	4	8	4	12
Ditto	III	ditto	.	.	.	.	.	4	0	4	8

And other sorts proportionately.

" *5th*—On obtaining from the Agent the godown receipts.

"The number of advances during the year in the Benares Agency are not to exceed four, as follow :—

" *1st*—In September, at the rate of R4 per *bíghá*, a higher rate being granted to parties living at a long distance to obviate the necessity of a second journey to receive the second advance.

" *2nd*—In January or February, after the crop is above ground and the prospects are favourable, at the same rate as the first.

" *3rd*—At the time of weighment, according to the quality of the drug and the rates specified above.

" *4th*—On obtaining from the Agent the godown receipts.

"The different rates of advances are not to be exceeded except with the permission of the Agent.

Manufacture.
272

"By the end of July the manufacture is finished, but the airing and drying are continued until October, by which time the balls are ready for packing.

I.
PROVISION
OPIUM.
273

"Forty balls are allotted to each chest of provision opium. The *manufactured opium* is classed as Provision Opium for export to China, and ABKARI or EXCISE OPIUM, for consumption in the country. The provision opium is brought down to Calcutta, where it is *sold by auction to the highest bidder*."

Direct con-
nection of
Government
with the
Opium ter-
minates with
the auction
sales.
274

"By a convention with the French Government, dated 7th March 1815, the authorities at Chandernagore are entitled to demand a quantity not exceeding in the aggregate 300 chests in each year, the price of the quantity claimed at any one of the periodical sales during the year being determined by the average price at which the rest of the opium is sold at such sale, and the requisition for such opium being " addressed to the Governor General at Calcutta within 30 days after the notice of the intended sales shall have been published in the Government Gazette." The French

P. 274

| Bengal Opium. | (*G. Watt.*) | PAPAVER somniferum. |

Government having proposed to commute for a period of five years their opium rights for a money payment, it was agreed by a convention executed between the Government of India and the French authorities at Chandernagore on the 16th July 1884 to fix the annual commutation payment at R3,000. The convention terminated at the end of 1888 and was renewed for a further period of five years. By the new convention the annual commutation payment has been raised to R5,000, conditionally on the French administration undertaking that no opium, except opium provided from the Hooghly Treasury, shall be admitted for consumption or for any other purpose into Chandernagore, and that the French officials endeavour to prevent contraband dealing in opium."

"The year's provision is not sold at once. Originally there were five sales in a year, then nine, but since 1848 they have been monthly. Of late years to check speculation, the quantity to be sold in any calendar year was, if the outturn was sufficient, notified in the previous year, and, as a further means of steadying prices, it was determined to accumulate a sufficient reserve for supplying a deficiency of out-turn in bad seasons.

"The **Abkari** or **Excise Opium** manufactured in the two Agencies is supplied at R7-4 a seer to the Governments of Bengal and the North-Western Provinces, and the Administrations of Burma, Assam, and Central Provinces, who permit its sale within their respective jurisdictions in accordance with rules framed under the Opium Act, I of 1878. Excise opium is issued to licensed dealers from the different District Treasuries in quantities of not less than one seer at a time, and at prices which are fixed by the Local Governments at their discretion, *but never so low as to encourage the exportation of Abkari Opium to China* in preference to provision opium.

"The *consistence* of the manufactured opium differs in the two Agencies. The moisture in 100 grains of crude opium being thoroughly evaporated, the residue is weighed, each grain after dryage being taken at 1°. If the cultivator delivers his drug of standard consistence, he receives for it the regulated price, otherwise a *pro ratâ* increase or deduction is made according as the drug is above or below standard.

"The *gross and net weight of a chest of opium*, and the consistence of provision and abkari opium, are as follows:—

	Consistence.	Gross weight of chest.		Net weight or quantity of opium in each chest.			Weight or quantity of fine opium in each chest.	One chest of excise opium equal to	
		Mds.	Srs.	Mds.	Srs.	Chs.	Srs.		
Behar Provision Opium.	75°	3	20	1	28	2	51·09375	1·05688	⎰ 1 md. 28 srs. 2 chks. of Benares opium at 70°=1 md. 23 srs. 9½ chks. when raised to the consistence of Behar.
Benares	70°	3	20	1	28	2	47·6875	1·132372	⎱
Malwa Opium*	90°	abt. 2	36½	1	28	2	61·3125	...	
Opium for consumption in India.	90°	2	10	1	20	0	...	...	
Persian Opium.	80°	Fair quality about 10½ to 11 shamans, 13½℔ avoir.=1 shaman.							

* It is assumed that the consistence of Malwa opium exported from Bombay is about 90 to 95 per cent. At the lower of these two consistencies a chest of Malwa contains 126 to 128℔ pure opium. The duty on Malwa opium is levied without reference to quality or consistence.

Marginal notes (right column):

ADMINISTRATION:

Regulations to prevent Contraband Traffic with French Possessions in India.
275

Reserve Stock of Provision Opium.
276

II.
EXCISE OPIUM.
Sale in India.
277

Consistence of Indian Opium.
278

Weight of Chest of Opium.
279

MALWA OPIUM.

**ADMINISTRA-
TION:**

**Malwa
Opium.
Revenue.
280**

"The *revenue* derived from the trade in opium known as Malwa Opium, which is produced in Central India and parts of Rájputana, consists chiefly of the duty charged for passes granted at the scales to cover the transit of the opium through the British territories to Bombay for exportation thence to China, or for consumption in India. The export is now regu-

**Regulated by
Act I. of 1878.
281**

lated by rules framed under the Opium Act, I. of 1878. The Act and the rules framed under sections 5 and 13 thereof were introduced into the Bombay Presidency on the 1st April 1878. Previous to the year 1831 the British Government reserved to itself a monopoly of Malwa opium which was purchased by the British Resident at Indore and sold by auction either at Bombay or at Calcutta. But in that year it was deemed advisable, chiefly on account of the large quantity of opium smuggled to the Portuguese Settlements of Damaun, &c., on the Coast, to relinquish the monopoly, to open the trade to the operations of private enterprise, and to substitute as a source of revenue, in place of the abandoned system, the grant, at a specified rate, of passes to cover the transit of opium through the Company's territories to Bombay. In determining the amount of transit duty it was proposed to be guided by a comparison of the cost of transit direct to Bombay with that of the transmission of the drug to the Coast by the cheapest of the more circuitous routes through the territories of Native States; and on the basis of such a comparison it was fixed at R175 per chest of 140℔ each. In 1835 the results of the preceding official year

**Difficulties
arising from
Contraband
Trade
with
Portuguese
Possessions.
282**

having been unfavourable and the shipments of opium from Bombay having largely declined, while those from Damaun had greatly increased, the rate was reduced to R125 per chest.

"The annexation of Sind facilitated the levy of a higher rate. Until then a large portion of the opium of Malwa had been conveyed through Sind to Karáchi, and thence onward to the Portuguese ports of Diu and Damaun. That route being now closed, it was reasonably expected that an advance might be made in the charge for passes, without risk of loss

**Increased
Duty.
283**

to the revenue from a diminished demand. The rate was accordingly increased in October 1843 from R125 to R200 per chest. Upon the principle that the duty should be fixed at the highest amount which could be levied without forcing the trade into other channels, a further increase was made in 1845, when it was determined that the charge should be R300 per chest. For the same reason it was, on 1st June 1847, raised to R400 per chest and subsequently as follows :—

**Decline.
284**

								R
1st July 1859	.	.	.	.	.	.	.	500
1st September 1860	.	.	.	.	.	.	.	600
1st October 1861	.	.	.	.	.	.	.	700
1st „ 1862	.	.	.	.	.	.	.	600
16th July 1877	.	.	.	.	.	.	.	650
16th September 1879	.	.	.	.	.	.	.	700
28th June 1882	.	.	.	.	.	.	.	675, if weighed at Ajmere for export by sea from Bombay; 650, if weighed elsewhere for export by sea from Bombay.
5th July 1890	.	.	.	.	.	.	.	R625 and R600.

**Sale of Malwa
Opium
in
Bombay.
285**

"Opium imported into Bombay for the declared purpose of export into China, may, on arrival, *be sold for home consumption.* The pass duty on opium consumed in India is at present R725 per chest if weighed at Ajmere, and R700 if weighed elsewhere. The pass duty on all opium consumed in India is credited to Provincial Revenue, and that on opium exported to China to Imperial Revenue. Under orders issued in August 1889 the duty levied on Malwa opium imported to the Native States of the Panjáb is now assigned to those States: the concession was granted with the object of interesting the States in an excise system and the suppression of opium smuggled from Rájputana.

"The poppy is sown in Malwa in November; the plants are in flower in the beginning of February, and by the end of March the whole of the opium is collected by the cultivators and ready for sale. The village bankers, who get possession of the raw opium, retain it till the end of April, and during May and June it is bought by the large dealers who make it up into cakes of 12 ounces each, and expose it in store-houses to dry for the next two months, after which it is ready for the scales. Generally the opium is ready for export in September, but as considerable dryage takes place in its transport to Bombay while new, it is usually kept until October, unless an expected increase of duty or pecuniary difficulties of the dealer compel the owners to bring the drug earlier to the scales. Before September 1,866 chests were sent to Bombay from Indore *via* Manpur, Scindwa, and Manmad, the transit occupying twenty days on an average. In that month a new route was adopted *via* Simrol and Barwari to the Great Indian Peninsula Railway at Khandwa which reduced the time of transit to eight days. The only authorised routes by which opium can now be imported by land into Bombay (after leaving the scales) are—(*a*) direct into the town of Ahmedabad from any place on the frontier of the Kadi division in the territory of His Highness the Gaekwar, which the local Government may from time to time appoint; (*b*) by rail from Khandwa or *via* Palanpur and Ahmedabad; (*c*) and in the case of opium belonging to His Highness the Maharaval of Dungarpur, from Kherwada, by Vichwada, Samera, Sambagi, Tintoi, Bakrol, Lembhoi, Dhakrol, Madhuka, Harsol, Ujdia, Dehgam, Naroda, to Ahmedahad under an escort furnished by His Highness and subject to examination as to number, weight, and tampering in transit at any of the above places which the Local Government may appoint in this behalf.

"Passports have been granted since January 1858 at Ahmedabad where the scales are superintended by an officer of the Opium Department, Bombay, and, since February 1877, at Ajmere. Passports are now granted by the Government Opium Agent at Indore and Ajmere (or by an officer authorised by one of these Agents) or Ahmedabad, or by some other officer duly authorised in that behalf by the Governor General in Council or the Local Government. The Ajmere scales were made use of for the first time in November 1877. A chest of Malwa opium contains 140℔ net weight, to which an allowance of 4 oz. is added for leaf and dust, making a total of 140¼℔.

"The rules in force at the Malwa Agency (and the same are applicable at Ahmedabad and Ajmere) for *the weighment of opium and the grant of passports* are as follow:—The opium ready packed in half chests (for the convenience of carriage), is brought to the Government godown by the merchants and brokers, who tender for the duty *hundis* bearing interest at five per cent. per annum, drawn on some trustworthy firm in Bombay, and payable at six days' sight to the General Treasury at the Bank of Bombay. These, on being approved, registered, and numbered, are forwarded with the register to the Accountant General, Bombay. At the Ahmedabad Agency the duty (R650) may be paid cash or by *hundi*, payable at sight at the Bank of Bombay, and at the Ajmere Agency also by cash or by *kundi*, drawn on some trustworthy firm at Bombay and payable at sight.

ADMINISTRATION:

Encouragement given to Native States to a more careful excise system. **286**

Cultivation of Malwa Poppy. **287**

Period of Export. **288**

Routes of Transit. **289**

Issue of Passports. **290**

Weight of Chest. **291**

Weighment & Duty. **292**

PAPAVER somniferum	Fiscal Administration of

ADMINISTRA-TION:

Weighment & Duty.

The *Government duty* on opium weighed at Ajmere for export by sea from Bombay is at present R675 per chest containing 140¼℔ net weight including dust and leaf. The duty on opium weighed at Ajmere was raised from R625 per chest to R675 on the 16th July 1877, and to R725 on the 15th September 1879, and again reduced to R675 per chest from 28th June 1882. The *hundis* tendered having been received and approved, the half chests are arranged in the godown and numbered consecutively in English. The merchants provide and pay for porters, carpenters, and men for screwing up the chests opened for weighment. On the arrival of the officer who makes the weighment, he selects at hazard two half-chests out of every ten of each consignment brought to the scales. From these an average is struck; when the average falls short of the allowance made by Government, *vis.*, 70℔ 20z. per half-chest, the deficiency is allowed to be made good; when it is in excess, the number of pounds in excess is withdrawn. This is called *net weighment*.

Guards for Escort of Consignments. 293

" When the *net weighment* is completed, the boxes are nailed up, and repacked and the gross weighments are proceeded with—that is, the chests as ready for conveyance, are weighed and marked in English. Full particulars are then entered in the passport granted for each weighment. The gross weight exceeds the net by about 50℔ in each half-chest. To prevent molestation on the road, a vernacular pass is granted to the person appointed by the merchant to accompany the despatch, in which the particulars of the consignment are inserted, and the weekly register of the passports is sent to the Commissioner of Customs, Opium and Abkari, Bombay. The passport is valid for two years from the date of its issue. *The guards necessary for the escort of the sonsignments* are supplied or paid for by the merchants or insurers. Advice of each despatch is forwarded to the Commissioner of Customs, Salt and Opium, Bombay, the day after it occurs. There are seven subordinate offices attached to the Malwa Agency (Indore),—Ujjain, Rutlam, Dhar, Udaipur, Joara, Mandesar, and Bhopal, the officers in charge of which forward their *hundis* to the head office, and in return receive passports bearing the head office registered number.

Receipt in Bombay. 294

" At Bombay the process observed with the pass fees levied on opium is as follows :—The Malwa-Ahmedabad or Ajmere Agent, as the case may be, grants the pass, receives a *hundi* in payment of the pass fee, and hands it on to the Accountant-General. The Accountant-General sends the *hundi* to the General Treasury to be realised, and informs the Commissioner of Customs that he has done so. The Commissioner of Customs at once records it as a debt due by him and takes care that the amount is recovered. The Treasury receives the amount of the *hundi*, and informs the Commissioner of Customs to that effect. The Commissioner of Customs, on the arrival of the opium at Bombay, receives it, and, *as a check against smuggling on the way,* sees that the chest corresponds with the pass, that the weighment also corresponds, less a certain amount of dryage, and that the pass fees have been duly recovered. If the pass fees have not been paid, he seizes the opium and recovers the amount of the *hundi* and the interest thereon. Before the export of the opium to China from Bombay, the Commissioner of Customs weighs it, examines the pass under which it was imported, and checks off therefrom the quantity to be exported."

It will thus be seen from the above brief sketch of the Opium Policy, that at every stage in the cultivation, manufacture, sale and export, the Government of India exercises the most direct supervision that is possible. It thus holds the traffic in its power, and, by the imposition of heavy taxes, restricts to the utmost the consumption in India by retaining the price at

P. 294

Malwa Opium. (*G. Watt.*) PAPAVER somniferum.

which opium is procurable in this country, at a figure considerably above that at which it can be obtained in China and other parts of the world. This not only saves contraband exportation in excise opium, but makes the governing price in the country with greatest demand and highest ruling price determine the Indian consumption. It need only be necessary, therefore, in concluding this chapter to furnish a table of the revenue derived by India from the drug.

REVENUE DERIVED FROM OPIUM.

The following table analyses the official returns of this subject for a period of 23 years, the figures being given in the conventional Rx. as equivalent to one pound sterling. The figures for Net Revenue are approximately as shown below, a few minor details of revenue and expenditure having been purposely left out of account :—

Revenue
from export-
ed Opium.
295

OFFICIAL YEARS.	Cost of Opium, being amount chiefly paid to the Indian *Rayat.*	Total Charges on account of Bengal sales and Bombay pass duty.	Net Revenue from provision opium.	Net Revenue approximately from Excise Opium (*i.e.*, consumption in India).	Total Indian Net Opium Revenue approximately.
	Rx.	Rx.	Rx.	Rx.	Rx.
1867-68	1,745,193	1,874,121	7,048,065	345,918	7,493,983
1868-69	1,550,207	1,720,111	6,731,330	352,930	7,084,260
1869-70	1,678,242	1,820,683	6,130,872	332,406	6,463,278
1870-71	1,861,331	2,014,425	6,031,034	356,175	6,387,209
1871-72	1,452,125	1,596,646	7,657,213	256,910	7,914,123
1872-73	1,664,824	1,814,268	6,870,423	413,178	7,283,601
1873-74	1,851,268	2,001,280	6,323,599	409,269	6,732,868
1874-75	2,178,443	2,341,282	6,215,046	468,780	6,683,826
1875-76	2,053,921	2,217,851	6,252,740	503,507	6,756,247
1876-77	2,664,750	2,841,644	6,280,784	556,174	6,836,958
1877-78	2,467,7451	2,659,504	6,523,099	585,896	7,108,995
1878-79	1,548,890	1,697,792	7,699,970	660,043	8,360,013
1879-80	1,912,635	2,067,336	8,249,964	702,677	8,952,641
1880-81	1,869,903	2,028,669	8,451,273	810,963	9,262,236
1881-82	1,895,489	2,056,273	7,804,282	800,490	8,604,772
1882-83	2,110,313	2,281,231	7,218,033	819,517	8,037,550
1883-84	1,681,964	1,853,410	7,702,380	845,634	8,548,014
1884-85	2,766,468	2,962,532	5,853,548	844,654	6,698,202
1885-86	2,850,971	3,051,481	5,890,818	844,177	6,734,995
1886-87	2,537,775	2,726,512	6,216,396	863,073	7,079,469
1887-88	2,237,923	2,423,238	6,092,095	877,377	6,969,472
1888-89	2,413,603	2,596,772	5,965,498	919,980	6,885,478
1889-90	1,447,849	1,603,596	6,979,394	928,928	7,908,322

When it is recollected that the amounts shown in the first column of the above table are practically the sums paid in each year to the cultivators in the Behar and Benares Agencies, for the produce of a little under 526,000 acres (the average of the past five years' returns), some idea is obtained of the importance of the crop and of the vested interests concerned in the production of opium. But the cultivators are not the only persons in India who have to be considered as having the right of centuries in justification of their claims on the opium trade. From the time of Akbar to the present day that traffic has been authorised and protected by Government. The vested mercantile interests concentrated on it are very

7

PAPAVER somniferum.	Excise Regulations, and

extensive, far more so than with any other single article of Indian Commerce. Should occasion ever arise, therefore, to prohibit that traffic, the loss to the cultivators would be the least important item that would have to be considered ; for no administration would be justified in bringing ruin to a large mercantile community by any sudden and arbitrary interference with a trade which it has hitherto protected and legalised. The amounts of Malwa and other opiums purchased by Government are very small, so that it is fairly safe to say that Government paid, for the past five years, an average of, say, 2½ million pounds sterling for the opium grown in British districts. The value of the opium crop to the cultivators within our territory is thus by no means inconsiderable, but the saving in taxation effected by the profits of the trade is by far its most important feature. But there are other considerations. By taking the traffic under direct supervision the Government is enabled most effectually to check the abuses that would otherwise prevail and have prevailed. A pure article is furnished in place of an adulterated one. To the cultivators the crop is by no means unimportant. Its value is not represented by dividing the amounts shown as annually paid by Government by the returned acreage. The crop occupies the field for but a short period. The system of poppy cultivation enriches the soil so that another crop may be taken off it at a nominal cost. The cultivator has the poppy seed and other bye-products of the opium crop as an additional source of income. Opium production is therefore highly remunerative, and the advances made on the crops come opportunely.

EXCISE REGULATIONS AND THE CONSUMPTION OF OPIUM IN INDIA.

Repeated reference has been made in the foregoing chapters to the consumption of opium in India. In connection with the HISTORY of the drug, it has been shown that, long before the British conquests, the practice of using opium was widely diffused throughout the country. It would be beside the scope of this work to deal with the moral aspects of the traffic or the perniciousness or otherwise of the habit of opium indulgence. The fact that opium-eating, drinking, or smoking existed in the country, necessitated supervision and control. For many years now, it has been the habit to vest local Governments with the management of all forms of Excise Administration, and opium is, therefore, a Provincial revenue. A large proportion of the opium used in India is specially prepared at Patna and Benares under the name of "Excise Opium" ; it is issued at cost price to the local Governments and Administrations. From the district authorities it is issued to the licensed dealers on payment of the excise dues. But in certain parts of India locally-grown opium is used, or specially selected Malwa opium takes the place of the "Excise Opium" of the Bengal Agencies. The table given at page 97 shows the total excise net revenue from opium, but, as exhibiting in some degree the relative consumption in each province, the analysis of the returns for ten years, given on the next page, may be cited. That table will be observed to be an elaboration of the information furnished at page 56. The three most important consuming provinces of Excise Opium may be said to be Burma, Assam, and Bengal. Perhaps the most noticeable feature of the Indian provincial consumption is the increasing traffic with Burma, a fact which is doubtless largely due to Chinese influence and Chinese immigration. It would, however, appear from other than purely statistical evidence that the Sikhs are the community of India most addicted to the use of opium, and they are certainly not an effeminate and emaciate race. The moderate indulgence authorised by religion and habit does not appear to be more injurious to them than the moderate use of wine or tobacco is in Europe.

Consumption of Opium in India.   (*G. Watt.*)   **PAPAVER somniferum.**

ADMINISTRATION: Revenue.

Revenue on Consumption of Opium in India.

PROVINCES.	1878-79.	1879-80.	1880-81.	1881-82.	1882-83.	1883-84.	1884-85.	1885-86.	1886-87.	1887-88.	1888-89.	1889-90.
	Rx.	Rx.	Rx.	Rx.	Rx.	Rx.	Rx.	Rx.	Rx.	Rx.	Rx.	Rx.
India, General	918	'522	319	381	650	1,610	2,902	2,434	2,799	2,720	3,174	5,012
Oudh	8,515	12,090	13,512	13,486	14,448	14,618	14,203	15,221	15,808	15,189	18,596	19,899
Central Provinces	64,239	75,946	83,447	89,511	92,938	94,922	94,624	95,441	93,709	88,129	90,331	93,824
Burma	151,296	162,644	180,593	177,035	179,982	181,554	176,769	164,383	197,340	209,106	226,009	229,463
Assam	154,813	156,760	163,740	155,351	164,703	165,059	169,232	166,700	166,594	162,746	163,550	173,681
Bengal	148,294	155,358	164,398	174,760	179,592	190,548	188,393	185,798	190,575	198,491	200,813	195,948
North-West Provinces	67,244	72,155	84,809	80,370	83,384	81,137	89,131	80,617	80,838	79,749	87,340	89,142
Panjab	32,863	32,511	133,171	47,818	43,344	42,589	41,668	36,333	36,722	37,119	38,183	38,521
Madras	Not recorded.	63,137	63,137	47,018	51,406	63,973	66,524	76,014	67,632	65,281	68,387	64,022
Bombay	31,861	34,691	23,837	14,760	9,070	9,624	10,208	21,236	11,056	18,847	23,587	19,416
TOTAL Rx.	660,043	702,677	810,963	800,490	819,517	845,634	844,654	844,177	863,073	877,377	919,980	928,928
Number of Abkari Chests manufactured for consumption in India	5,605	4,221	4,378	3,959	591	2,146	4,321	2,291	2,831	3,367	3,485	6,320

A

P. 297

| PAPAVER somniferum. | Fiscal Administration of Opium in India, |

ADMINISTRA-TION:

Revenue.

The above provincial shares of the Excise Net Revenue from Opium cannot, however, be accepted as exhibiting the amounts actually consumed, since the degree of taxation is not uniform in the provinces. Nor can the number of chests of Government Abkari Opium be taken as an absolute statement of the amount of opium used up in India. The figures of net revenue include the proceeds derived from all opium (*i.e.*, Government Abkari, Malwa, locally-grown, &c.), but it can safely be said that the Abkari opium is by far the most important item in Indian consumption. Additional information on this subject will be found in the chapter below on the TRADE IN OPIUM, more especially the sections COASTWISE and INTERNAL.

MALWA.

298

The following instructive account of the Bombay opium traffic (taken from the Administration Report for 1889-90) deals with the subject of Malwa opium whether consumed in India or exported to China. It may, therefore, fitly conclude these remarks, since it links together what has been said, under FISCAL ADMINISTRATION, with the sketch given of the EXCISE OPIUM of India:—

Scales and rates of duty.

299

"No change was made during the year in the number or location of the scales for the weighment of opium. The rate of duty on opium intended for export to China and other foreign countries remained at R650 per chest up to 4th July 1890 inclusive, after which date it was reduced to R600. The rate on opium imported for local consumption in the Bombay Presidency remained at R7co per chest. The rates levied at the scales at Ajmere were as usual R25 per chest higher in each case.

Imports and Exports.

300

"The following table shows the number of opium chests imported from Málwa, Meywár, Márwar, and the territories of His Highness the Gáekwár into Bombay, the amount of pass-duty realised, and the number of chests exported to China and otherwise disposed of in the year 1889-90, as compared with corresponding particulars for the two previous years:—

YEARS.	Number of chests imported into Bombay.	Amount of pass fee actually realised and brought to account.	Number of chests exported to China.	Number of chests bought for home consumption.	Number of chests otherwise disposed of.	TOTAL.
		R				
1887-88 .	38,716½	2,51,31,300	36,353	1,575	119½	38,047½
1888-89 .	31,617	2,05,67,800	29,777½	1,571½	91½	31,440½
1889-90 .	28,835½	1,86,56,125	28,372	1,741½	78	30,191½

"The quantity of opium exported during the year to China was less by 1,405½ chests than the quantity exported in 1888-89. Fifteen-and a-half chests were exported to African ports, 60½ chests to Cochin and Allepy, and 2 to London.

Prices.

301

"The average prices of opium in Bombay during the year under review were lower than those prevailing in the previous year, being R1,165 per chest of one year old opium, and R1,226 per chest of two and three years old opium, against R1,323 per chest of the former, and R1,380 per chest of the latter in 1888-89.

Consumption in the Presidency.

302

"Opium was issued from the Government depôts at the rate of R10 per ℔ throughout the year under review. From the Government depôts 61,702℔ of opium were issued to British retail vendors or to Native Chiefs. Besides this quantity, 955½ chests, equal to 130,054 ℔, were bought in the Bombay market, and 440½ chests, equal to 59,957℔, were imported

from Málwa by Chiefs and vendors permitted on payment of duty to supply themselves directly with opium for retail distribution.

"The licit sales of the year under review exceeded those of the preceding year by 9,664℔. The increase was noticeable in the districts of Kaira,
Broach, and Khandesh, while there was a considerable decrease in the
district of Ahmedabad. In the Native States of the Presidency the licit
sales amounted to 154,555℔, of which 62,844℔ were sold in the dominions
of the Gáekwár of Baroda, 10,880℔ in Cutch, 46,863℔ in Káthiáwar, and
31,375℔ in other States in Gujarat. The opium required by Native Chiefs
in the Konkan, Deccan, and the Southern Maratha Country amounted to
2,593℔ and was obtained by the Chiefs from the nearest British depôt or
from the Collector of Customs, Bombay.

"The total pass-fee realisations from opium amounted to R1,87,35,185,
of which R12,61,260, representing duty on excised opium bought for home
consumption on account of Government by British licensed vendors and
by Native Chiefs in the Bombay and Madras Presidencies, were transferred
to the account head of "Excise," leaving R1,74,73,925 to the credit of
Imperial revenues. Adding to the latter sum the fees levied on account of
the transhipment of opium chests in the Bombay Harbour and miscellaneous receipts, the total receipts for the year credited to imperial revenues
amounted to R1,75,01,507 or R20,09,559 less than in the previous year.
The decrease was wholly due to smaller imports into Bombay of opium
from Málwa for export to China. The total excise realisations from opium,
allowing for refunds of duty to Native States, were R11,68,269 against
R12,13,858 in the previous year. They included the pass-fee on opium
consumed in British territory, two-thirds of the pass-duty on opium
consumed in Kathiáwar and Cutch, nine-tenths of the pass-duty on opium
consumed in the Sátára Jághirs, and four-fifths of the pass-duty supplied
to other Native Chiefs except those in Rewa Kántha, Mahi Kántha,
Pálanpur, and Cambay. The last-mentioned States are entitled to receive
opium free of duty. In Baroda consumption was confined to the opium
grown in the State and no duty was paid to Government.

"The expenditure on account of a portion of the opium establishment
employed in Bombay and that employed at the Ahmedabad opium
agency, which is debited to Imperial services under the head 'Opium,'
amounted to R25,233 against R25,021 in the previous year. The provincial expenditure of the Opium Department is included in that of the Abkári
Department and cannot be accurately distinguished, but may be estimated
approximately at R30,000."

TRADE IN OPIUM.

In the chapter on HISTORY above, some of the leading ideas regarding
the early opium trade have already been dealt with. In the succeeding
chapters on CULTIVATION and FISCAL ADMINISTRATION, additional facts
have been brought out, as to area and yield, manufacture, purchase of the
drug, and the system under which it is sold to the highest bidder. Very
little remains, therefore, to be discussed in this place. The table at page
56 shows the number of chests of Provision Opium manufactured at the
Government Agencies since 1860, and the table at page 57 shows the number of maunds of opium produced at the Patna (Behar) Agency during the
past five years, while the table at page 58 gives the same information for
the Benares Agency. As exhibiting certain facts of historic interest, the
following brief account of opium may be taken from **Milburn's** *Oriental
Commerce* (published in 1825, pp. 294-295). It shows the constancy of
the policy pursued in the traffic, and also gives certain figures of early
transactions :—"The monopoly in the trade of opium, or the cultivation

Margin notes (right column):

ADMINISTRA-
TION :

Revenue.

Malwa.

Pass Fees.
303

*Conf. with
remarks on
Revenue,
p. 97.*

TRADE.
304

TRADE.

of the poppy, may be traced at least as far back as the commencement of the British influence in Bengal. The advantages resulting from it were for several years merely considered as a part of the emoluments of certain officers under the Government. In the year 1773, it was taken out of their hands, and the profit of the trade assumed for the benefit of the Company. The provision of the article was for many years let out upon contract. The opium concern continued under the direction of the Board of Revenue till 1793, when it was transferred to the Board of Trade. On the expiration of the contracts in 1797, the cultivation of opium was restricted to Behar and Benares, and discontinued in Bengal; the mode of provision by agency was resorted to, and still continues in practice. In July 1799, some regulations were published 'for the guidance of all persons concerned in the provision of opium on the part of Government, and for preventing the illicit cultivation of the poppy, and the illicit importation or traffic in the article of opium.' Under these regulations, which were further modified in 1807, the cultivation of the poppy, except on account of Government, is expressly prohibited, but it is left entirely at the option of the cultivator to enter into engagements on account of Government at a settled price, or to decline it altogether. The quantity grown, which is limited, is sold by public auction at two annual sales at Calcutta, in December and February. It is usually about 4,000 chests. The trade in opium is liable to be affected by many contingencies, not only from adverse seasons, but by the state of the markets to the eastward, which fluctuate considerably. The superior advantages of the agency system, and the measures resorted to for securing the provision of the drug pure and unadulterated, have proved of essential service."

"The Bengal opium is distinguished in commerce into two kinds, Patna and Benares; the former is most esteemed. There is another kind of India opium, less esteemed, produced in considerable quantity in the province of Malwa; it usually fetches little more than half the price of Patna opium.

"The quantity of opium exported by sea from Calcutta, in 1821, was 4,337 chests, of which 3,137 were to China and Macao, 632 to Penang and eastward, 420 to Java, the rest in small quantities to various parts."

It is scarcely necessary to trace out the growth of the trade from 1821 up to the present time. Suffice it to say that it manifests no period of spasmodic expansion. Year by year it gradually increased, being little affected even by the troubles of 1840 or of 1857. In 1831, the total exports were 11,726 chests, in 1841, 29,432 chests, and in 1851, 52,040 chests. The following table furnishes a complete statement of the traffic from 1860 to 1890; it distinguishes the Bengal from the Malwa and shows the proportion of each kind sent to the chief consuming countries. One of the most striking features of the trade will be seen to be the immense importance of the Indore exports. Should occasion arise for the Government of India to abandon its opium supervision and to discontinue the Bengal traffic, the difficulty in prohibiting the cultivation of opium in the Native States of Central and Western India would be almost unsurmountable. If cultivated it would find its way all over India and even to China, in spite of every regulation that might be passed against the traffic. Such illicit traffic is at present checked powerfully by the natural effects of competition in trade, but were Malwa Opium to enjoy a monopoly no such restrictions would then exist.

P. 304

TRADE.

Tabular Statement of the number of Chests of Bengal and other Indian Opiums Exported to Foreign Countries.

Official Years.	Bengal Chests sold.	Malwa Opium passed on paying duty.			Total of Bengal and Malwa.	Exported from India.							Grand Total of Chests Exported.
						Bengal Opium.				Malwa Opium.			
		Indore.	Ahmedabad.	Ajmír.		To China.	To Straits.	To other countries.	Total.	To China.	To other countries.	Total.	
	1	2	3	4	5	6	7	8	9	10	11	12	13
1859-60	25,253	30,488	2,416	The Ajmír Agency was established in February 1877, but the scales were made use of for the first time in November 1877.	58,157	22,329	3,483	148	25,950	32,534	197½	32,731½	58,681½
1860-61	21,363	43,057	3,074		67,494	15,688	3,810	10	19,508	43,691	265	43,956	63,464
1861-62	24,063	33,719	3,328		61,110	21,332	5,192	48	26,572	35,680	110½	35,790½	62,362½
1862-63	32,833	46,875	4,353		84,061	25,846	6,784	31	32,661	49,485½	70½	49,556	82,217
1863-64	42,620	23,351	2,324		68,295	33,815	8,766	40	42,621	28,210½	2½	28,213	70,834
1864-65	54,486	31,932	690		87,108	40,915	9,288	74	50,277	34,213½	½	34,214	84,491
1865-66	56,011	34,193	1,927		92,131	42,697	11,549	27	54,273	34,166½	...	34,166½	88,439½
1866-67	38,680	29,260	1,332		69,272	37,279	4,444	34	41,757	33,081	17	33,098	74,855
1867-68	47,999	36,101	2,999		87,099	41,672	6,352	22	48,046	38,883	1½	38,884½	86,930½
1868-69	47,235	29,787	1,185		78,207	37,985	6,167	103	44,255	30,690	10½	30,700½	74,955½
1869-70	45,680	35,828	3,543		85,051	43,054	6,680	228	49,962	38,716	7	38,723	88,685
1870-71	49,030	37,608	1,898		88,536	39,669	8,054	292	48,015	40,992½	67	41,059½	89,074½
1871-72	49,695	37,591	1,163		88,449	41,569	7,843	43	49,455	39,334½	7½	39,342	88,797
1872-73	42,675	42,688	1,315		86,678	34,009	6,438	38	40,485	42,369	80	42,449	82,934
1873-74	42,750	42,112	2,845		87,707	34,820	8,464	53	43,337	45,301	86½	45,387½	88,724½
1874-75	45,000	47,982	1,579		94,561	36,678	8,939	91	45,704	48,845½	188	49,033½	94,737½
1875-76	45,510	38,753	3,051		87,314	35,673	11,005	90	46,768	41,517	9	41,526	88,354
1876-77	47,240	48,018	1,768		97,026	37,426	9,701	194	47,321	49,430½	121	49,551½	96,872½
1877-78	49,500	41,617	1,493	171	92,781	40,234	9,357	115	49,706	43,083½	30½	43,114	92,820
1878-79	55,500	32,897½	3,771½	211	92,380	45,891	8,881	159	54,931	36,262	10	36,272	91,203
1879-80	59,100	46,204½	68	459½	105,832	48,722	10,586	79	59,387	46,114½	6½	46,121	105,508
1880-81	56,400	36,649	1	555	93,605	46,767	9,622	141	56,530	35,626	36½	35,662½	92,192½
1881-82	56,400	32,009½	...	380	88,789½	45,731	10,005	759	56,495	32,825½	20	32,845½	89,340½
1882-83	56,400	36,103½	1,185½	315	94,004	44,961	10,316	768	56,045	35,736½	19	35,755½	91,800½
1883-84	54,400	38,717½	882½	177½	94,177½	45,498	7,320	901	53,719	38,231	14½	38,245½	91,964½
1884-85	46,698	39,013½	350½	136	86,198	36,714	10,134	1,046	47,894	38,679½	6½	38,686	86,580
1885-86	50,994	38,967	...	...	89,961	39,223	10,728	1,103	51,054	36,892½	9	36,901½	87,955½
1886-87	54,750	40,839	205½	...	95,794½	41,920	11,396	1,300	54,616	41,204½	18	41,222½	95,838½
1887-88	57,000	36,799	415	...	94,214	43,128	12,368	889	56,385	33,687	24	33,711	90,096
1888-89	57,000	30,315	29	14	87,358	42,493	13,995	870	57,358	30,398	33	30,431	87,789
1889-90	57,000	30,564	...	...	87,564	40,942	13,749	1,294	55,985	29,160	21	29,181	85,166

| PAPAVER
somniferum. | Trade in Opium. |

TRADE.

IMPORTS.—India imports but a very small amount of opium from foreign countries; during the past five years the imports have amounted to about 250℔, valued at between R3,500 and R4,100. Of these foreign imports none appears to be re-exported.

Foreign.
305

EXPORTS.—Of the exports to foreign countries of Indian opium the following table may be given for the traffic during the past five years:—

	1885-86.	1886-87.	1887-88.	1888-89.	1889-90.
Chests	87,956	95,839	90,096	87,789	85,166
Weighing Cwt.	120,995	131,630	125,871	122,160	118,598
Value in Rupees	10,73,55,180	11,07,76,689	10,06,77,636	10,50,80,808	10,11,59,362

The table in the previous page exhibits the countries to which the above opium was consigned, and the present table has been given to show the weight of the chests and their declared values.

Coastwise.
306

COASTWISE TRAFFIC.—The chief items of this trade in the year 1889-90 were the exports from Bombay to Madras and from Bombay to the Native States of Cutch, Kathywar, and Travancore. Last year, for example, Bombay sent to Madras opium to the value of R56,591, to Cutch R1,26,875, to Kathywar R655, and to Travancore R30,500. These are apparently commercial transactions. Under the heading 'Government Stores' entries also occur in the *Annual Statements of the Trade and Navigation of British India Coastwise.* The following were the chief transactions:—Bengal exported opium to the value of R15,57,668, of which Burma took R11,20,560, and ports within Bengal itself the remainder. Bombay exported opium to the value of R1,96,330, of which R1,90,638 went to Sind, and the balance to ports within the Presidency. The supplies for Madras do not appear to be drawn from Government Stores; they have already been recorded under general mercantile coastwise transactions.

Inland.
307

INLAND TRADE.—During the year 1888-89 (the most recent year for which returns are available) Bengal imported from the North-West Provinces and Oudh opium to the value of R3,16,100, the North-West Provinces imported from Bengal R9,240, the Central Provinces imported R2,24,750 from the North-West Provinces, and Assam imported R1,08,900 worth from Calcutta. These were the chief transactions by land routes between British Provinces, but from Rájputana and Central India the following amounts of opium were consigned to British Provinces:—Madras R6,112 worth, Bombay R4,24,532 worth, the North-West Provinces R6,960 worth, the Central Provinces R1,920 worth, and Berar R2,21,318 worth. The imports into the chief seaports furnish the supply from which the foreign exports are mainly drawn. Calcutta received from Bengal opium to the value of R1,89,57,620, and from the North-West Provinces R2,07,32,970 worth. Bombay drew its supplies mainly from Rájputana and Central India, namely, opium to the value of R4,65,81,524.

The movements of opium by land routes and coastwise convey a fairly accurate conception of the Indian consumption. The amounts recorded above in money value, less the imports into the chief port towns (which met foreign marts chiefly), were the interprovincial adjustments to meet the Indian consumption.

Trans-frontier.
308

TRANS-FRONTIER LAND TRAFFIC.—This is comparatively unimportant, though it gives considerable trouble, since large quantities are constantly

P. 308

imported illicitly, especially from Nepal. The following table exhibits
the traffic for the past three years :—

	IMPORTS INTO BRITISH INDIA.			EXPORTS FROM BRITISH INDIA.		
	COUNTRIES FROM WHENCE—KASHMIR AND NEPAL.			COUNTRIES TO WHICH— NAGA AND MISHMI HILLS.		
	1887-88.	1888-89.	1889-90.	1887-88.	1888-89.	1889-90.
Weight in Cwt. . . .	167	306	200	4	1	4
Value in Rupees . . .	2,68,650	4,20,088	2,38,490	8,309	2,102	8,695

The exports from Nepal (which represent the major portion of the imports into British India) are received by the province of Bengal. The exports from British India go from the province of Assam to the frontier
Hill States.

(*J. Murray.*)

PAPER & PAPER FIBRES.

309

Paper & Paper Fibres.

The first authentic account of paper-making in India dates from the
time of the Emperor Akbar, when the art is said to have been introduced
into Kashmír. It spread rapidly all over India, displacing the birch-
bark used by the hill tribes and the palm-leaves in the case of the people
of the plains. It is probable, however, that the art of Nepál paper-making
came from China to the inhabitants of the Eastern Himálaya long anterior to the introduction of the paper industry into Kashmír by Akbar.
The art is one that has advanced little, save under immediate European
supervision, and even at the present day the paper-making of the Natives
of India is very much like their glass-blowing, since it largely consists in
remaking from waste manufactured material.

In a country teeming with fibres, it is a surprising fact that the question of a paper-fibre should still be under consideration. European writers
seem always to forget the immense size of the continent of India, and to
assume that a few hundred miles are of little consideration in most Indian questions, but, when this distance entails a heavy railway freight
upon a bulky article, it becomes prohibitive, even to industries that can
afford to pay more than paper-making. To cultivate fibre specially for
paper has many drawbacks, of which but one need be mentioned—the cost
of land near commercial centres is so high as to preclude the idea as a
practicable consideration. Moreover, the paper fibre must, like esparto,
be fit almost for immediate immersion in the vats, for paper-making
can never pay the cost of separation or preparation however simple and
inexpensive. No fibre known to commerce can compete with jute in point
of cheapness, yet the paper-maker can afford to purchase jute waste and
jute cuttings only, so that but for the demand for an altogether different
purpose, the paper-maker could never procure jute.

The two most important indigenous paper grass-fibres in India are
munj and *bhabar.* These are now being largely used by our Indian
paper mills, the supply at a remunerative price being the chief obstacle
to their further utilisation. The roots and lower stems of rice have been
suggested as a paper material, and if it be proved that these are worth the
trouble of collecting, the supply might be practically unlimited ; but it is
doubtful whether the paper-maker could pay sufficient to cover freight and

PAPER
MILLS.
310

TRADE.
311

PAPER-MAK-
ING FIBRES.
312

expense of collection. It has been demonstrated that bamboo affords excellent paper, but, in this case also, practical difficulties exist, which have, for the present at least, dispelled the hopes once entertained of the immense tracts of bamboo forest becoming of commercial value. The most valuable paper materials in India are old rags, waste gunny bags and jute in other forms, *sunn* ropes, &c., &c.

Paper Mills.—Mr. O'Conor, in the Fourteenth Issue of his *Statistical Tables,* enumerates nine paper mills as existing in India in 1888, of which five are in Bombay, two in Bengal, one in the North-West Provinces, and one in Central India. The total nominal capital of six of these, which belong to Joint-Stock Companies, amounts to nearly 39 lakhs of rupees, the average daily number of hands employed aggregate 1,687, and the yearly outturn is 18,363,665℔, valued at R26,84,431. The descriptions of paper, said to be manufactured, include;—writing, printing, blotting, cartridge, *badami,* coloured, and brown papers. The quality of the finer papers produced is improving year by year, as will be seen by the returns of Government imports ; the yearly decrease in these manifests the fact that stationery made by Indian mills is ousting European-made paper from office use.

Trade.—Returns of trade connected with paper show that India exports no manufactured goods, but supplies a considerable and increasing amount of paper-making materials. The imports, on the other hand, consist entirely of paper and paste-board, and, notwithstanding the increased number and outturn of Indian paper mills, these imports have largely increased during the past few years. A considerable decrease is, however, apparent (as already remarked) in the imports on account of Government, a decrease due to the large extent to which office stationery is now supplied by Indian mills. The exports of rags and other paper materials for the past five years were—

						Cwt.	R
1884-85	.	.	.	.	.	96,408	3,70,533
1885-86	.	.	.	.	.	107,531	3,98,269
1886-87	.	.	.	.	.	89,522	3,45,697·
1887-88	.	.	.	.	.	125,988	4,72,916
1888-89	.	.	.	.	.	135,969	5,64,871

The following analysis may be given of the exports for the year 1888 89 :—

Presidency from which exported.	Value in Rupees.	Country to which exported.	Value in Rupees.
Bengal	4,75,816	United Kingdom . .	93,614
Bombay	81,521	United States . . .	4,70,687
Sind	7,524	Other Countries . .	570
Madras	10		

The imports of paper and pasteboard for the past five years were—

								General.	Government.	Total.
								R	R	R
1884-85	.	.	.	.	.	.	.	35,21,026	13,71,095	48,92,121
1885-86	.	.	.	.	.	.	.	33,36,753	10,01,247	43,38,000
1886-87	.	.	.	.	.	.	.	31,66,767	7.59,448	39,26,215
1887-88	.	.	.	.	.	.	.	41,83,070	10,79,993	52,63,063
1888-89	.	.	.	.	.	.	.	41,05,349	7,84,574	48,89,923

It is perhaps needless to attempt to specialise all the fibres of India which might be used in paper-making. It has already been indicated that the only obstacle is one of price. Any fibre may be used for paper, but the following are a few of those most deserving of mention :—

Adansonia digitata, *Linn.;* MALVACEÆ.

THE BAOBAB TREE or the MONKEY BREAD TREE OF AFRICA.

Cultivated in some parts of India to a small extent. The bark yields a fibre, which seems likely to come into use in paper manufacture.

Agave americana, *Linn.;* AMARYLLIDEÆ. THE AMERICAN ALOE.

A native of America, now naturalised in many parts of India.

The leaves and the root yield an excellent fibre, which might easily become an important paper material but for its high price. During experiments performed at the Bally Paper Mills, it was discovered that one of the greatest difficulties in the way of Agave fibre for paper manufacture was the fact that the young leaves yielded too fine a pulp; the best leaves were those three years old. A mixture was found injurious, and therefore a difficulty exists in getting uniformity.

Antiaris toxicaria, *Lesch.;* URTICACEÆ. THE TRAVANCORE SACKING TREE.

A gigantic tree of the evergreen forests of Burma, Western Gháts, and Ceylon. It seems probable that the bark of this tree may come into use as a paper fibre.

Bambuseæ; GRAMINEÆ. THE VARIOUS SPECIES OF BAMBOO.

Mr. Routledge, in his pamphlet on *"Bamboo as a paper-making material,"* makes the following remarks :—"Of all the fibre-yielding plants known to Botanical science, there is not one so well calculated to meet the pressing requirements of the paper-trade as bamboo, both as regards facility and economy of production, as well as the quality of the 'paper-stock' which can be manufactured therefrom." While this is doubtless correct, there are practical difficulties which render it extremely improbable that bamboo will ever come into use as a paper material. The difficulty of getting a continuous supply of young shoots without injuring, if not killing, the clumps, after two or three years, the heavy charges likely to be incurred in conveying the material from malarious jungles to the mill, and the fact of the scales and young stems being covered with hairs which cannot be removed, render the bamboo practically unsuitable for paper-making. Experiments recently conducted have proved that a clump of bamboo is killed after a very few years if a large percentage of the shoots be systematically removed. The natives of India, and of China, however, do make paper from bamboo, but they do not appear to regard it as absolutely necessary to use the young shoots only.

Broussonetia papyrifera, *Vent.;* URTICACEÆ.

A small tree, said to be wild in the Martaban hills. The Japanese make paper from the bark. There is, perhaps, no plant that offers a better future than this as a source of paper, and if extensively cultivated by the villagers as a hedge, it would undoubtedly become a distinct source of revenue. Samples of native-made **Broussonetia** paper were exhibited at the Colonial & Indian Exhibition.

Corchorus olitorius, *Linn.;* and **C. capsularis,** *Linn.;* TILIACEÆ. JUTE.

These plants are extensively cultivated for their fibre. The rejections and cuttings are largely employed in paper-making both in India and Europe.

Crotolaria juncea, *Linn.;* LEGUMINOSÆ. SUNN.

The plant is extensively cultivated all over India for its fibre. The waste is utilised in paper manufacture, but it would never pay to cultivate the plant solely for paper, and the uses of *sunn* are limited, so that the amount of waste available is comparatively small.

**PAPER-
MAKING
FIBRES.**

[Thymelæaceæ.

Daphne longifolia, *Meissn.*; **D. papyracea,** *Wall.*; and **D Wallichii,** *Meissn.*;

The barks of the above plants are used in the manufacture of paper. The so-called Nepál paper is prepared from the bark of the first two. As these plants, however, require a warm temperate climate, the cultivation would have to be conducted on the hills and therefore at a distance from commercial centres and paper mills, so that, although the fibre affords a remarkably strong paper, it does not seem likely to ever become of much importance.

Edgeworthia Gardeneri, *Meissn.*; Thymelæaceæ.

A large, elegant bush, found along the Himálaya from Nepál to Sikkim and Bhután, and recently found plentifully on the mountains of Manipur, extending to the northern frontier of Burma.

The finest Nepál paper is said to be made from the fibre of this plant, and to be purer and cleaner than the paper from **Daphne papyracea.** The same remark as already made, with regard to a possible development of the trade in **Daphne** fibre, is also applicable to this plant.

Helicteres Isora, *Linn.*; Sterculiaceæ.

A large shrub in the tropical and sub-tropical regions of India. The fibre from the bark might become serviceable as a paper material were it not for the fact that it will scarcely pay to cultivate any fibre that requires to be extracted for paper alone. The plant could be cultivated to any extent throughout India as a hedge.

Hibiscus cannabinus, *Linn.*; Malvaceæ. The Hemp-leaved Hibiscus.

A small, herbaceous shrub, apparently wild on the Northern Gháts, largely cultivated for its fibre, especially in the North-West Provinces and the Panjáb. It yields the chief fibre used in the manufacture of paper in the Dacca district, Bengal, and is also largely employed in the Madras Presidency. The fibre is too valuable, however, to be cultivated on a commercial scale for the purpose of paper alone.   [Grass.

Ischœmum angustifolium, *Hack.*; Gramineæ. The Bhabar or Bhaboi

A grass, met with throughout the central table-land of India from Bengal to Madras, the North-West Provinces and Central India; in some parts of the country extremely abundant. As a paper material this grass has been reported as little inferior to esparto, and the paper made from it as of good quality. The great drawback to the fibre is that it would not pay to cultivate the grass purely for the purpose of a paper supply, and the wild plant has to be collected over wide and distant areas. As already stated, the question of freight is in India one of most serious consequence, for if a bulky substance has to be carried for great distances by railway the price of the article is enhanced to a prohibitive extent. The Indian paper mills, however, use *bhabar* grass largely, and samples of paper and paper-stuff made of it were exhibited by the Bally Paper Mills at the Colonial & Indian Exhibition.

Musa sapientum, *Linn.*; Scitamineæ. The Banana or Plantain.

Considerable attention has of late years been attracted to the subject of plantain fibre as affording a paper material. It has been stated that a great future is before the paper industry of India, and that the enormous quantities of plantain stems which are annually thrown away as useless may yet come to be greedily purchased for paper manufacture. The principal objection to this idea is that unless cultivated within a definite area, it would never pay to collect the stems all over the country, unless some simple process could be suggested by which the people could reduce the stems to fibre, and thus to lessen the charges of freight.

P. 312

Opuntia Dillenii, *Haw.;* CACTEÆ. THE PRICKLY PEAR.

An erect, fleshy, thorny shrub common all over the arid and dry zones of India; originally brought from America but now quite naturalised. **Dr. Bidie,** speaking of this plant in the Madras Presidency, says: "This abounds in every part of the country, and has become such a nuisance that large sums are expended annually in cutting it down and burying it, on sanitary grounds." If found suitable as a paper material, this nuisance might be converted into a source of wealth.

Saccharum Sara, *Roxb.;* GRAMINEÆ. THE MUNJ GRASS.

The grass is common from Bengal to North India. *Munj* and several other species of **Saccharum** are largely used in the Upper India Paper Mill near Lucknow. Indeed, this is one of the most valuable of Indian paper materials, in some respects being even superior to *bhabar* (Ischœmum) grass. The great obstacle to it is the difficulty of procuring a large and constant supply, and the fact that from greed the collectors reap the grass too near the ground, injuring materially the quality of paper through the amount of hard stems mixed with the leaves.

Sansevieria zeylanica, *Willd.;* HÆMODORACEÆ. THE BOWSTRING HEMP.

A stemless bush found on the coast of Bengal extending to the Madras Presidency, and common on the Coromandel Coast. The fibre from the leaves makes a very superior paper, but the price precludes an extended use.

Yucca gloriosa, *Linn.;* LILIACEÆ. ADAM'S NEEDLE.

A native of America introduced into India. It yields, like **Agave,** an excellent fibre suitable for paper manufacture, but too valuable to be put to paper manufacture alone. The fibre is not employed in the textile industries to any appreciable extent, hence it is impossible to obtain a large supply of waste.

For further information regarding these fibre-yielding plants, the reader is referred to the article on each in its respective alphabetical position.

PARABAIK.

313

Parabaik.—This is the Burmese name for a description of paper tablet or slate, made from the fibrous materials of the bamboo or from the bark of **Broussonetia papyrifera** (*q.v., Vol. I., 538*). These tablets are either black or white, the former being coloured with a preparation of starch and teak-wood charcoal, made into a paint with the juice of Cyamopsis psoralioides, *DC.* To prepare the tablets the bamboo is cut into short lengths, the knots discarded, and the outer bark stripped off. The pieces are then soaked in lime water until the fibre separates. This is reduced to a pulp by pounding and again soaked in lime water for a week. It is then transferred to clean vessels and is ready for moulding. The mould consists of a frame with a bottom formed of cloth, upon which the pulp is floated layer upon layer until the desired thickness is obtained. The preparation of tablets from **Broussonetia** differs in no essential point from that of bamboo, and paper prepared from these fibres in Burma differs, from the tablets described, in thickness only.

PARAMERIA, *Benth.; Gen. Pl., II., 715.*

314

[*Ic., t. 1307;* APOCYNACEÆ.

Parameria glandulifera, *Benth.; Fl. Br. Ind., III., 660; Wight,*
     Syn.—ECDYSANTHERA GLANDULIFERA, *A. DC.;* E. BARBATA, *Miq.;*
          ECHITES GLANDULIFERA & MONILIFERA, *Wall.;* PARSONSIA BARBATA,
          Bl.
     Vern.—*Taline-no-thee,* BURM.

PARKINSONIA
aculeata.   The Jerusalem Thorn.

GUM.
315

Habitat.—An extensive climber, in the tidal forests of Burma, extending to Malacca, Singapore, and the Andaman Islands; distributed to Java and Borneo.

Gum.—This plant has recently attracted considerable attention as a source of India-rubber. See **India-rubber**, Vol. IV., 361, 363.

PARAMIGNYA, *Wight; Gen. Pl., I., 305.*

316

[*I., 510;* RUTACEÆ.

Paramignya monophylla, *Wight, Ill., I., 109, t. 42; Fl. Br. Ind.,*
   Vern.—*Natkanta*, NEP.; *Jhunok*, LEPCHA; *Kurwi wágeti, kari-wágeti,*
      BOMB. & GOA.
   References.—*Gamble, Man. Timb., 59; Kurz, For. Fl. Burm., I., 193; Dymock, Mat. Med. W. Ind., 133; Dymock, Warden, & Hooper, Pharmacographia Ind., I., 268.*
   Habitat.—A stout, climbing, evergreen shrub, native of the Sikkim Himálaya and the Khásia mountains, from 2,000 to 5,000 feet; also met with in Bhutan, the Western Peninsula, and Ceylon.

MEDICINE.
Root.
317

   Medicine.—Dymock writes, "In the Konkan, the ROOT is given to cattle suffering from bloody urine, or bloody fluxes, from the abdomen. When on a visit to Goa I observed that the country people made use of the root as an alterative tonic." The part used has a scabrous brown bark; it possesses a bitter saline taste, and abounds in large crystals of oxalate of lime.

TIMBER.
318

   Structure of the Wood.—White, hard, close-grained.

Pareira or **Pari**, see **Cissampelos Pareira**, *Linn.; Vol. II., 327.*

Paritium tiliaceum, see **Hibiscus tiliaceus**, *Linn.; Vol. IV., 247.*

PARKIA, *Br.; Gen. Pl., I., 588.*

319

Parkia insignis, *Kurz; Fl. Br. Ind., II., 290;* LEGUMINOSÆ.
   Vern.—*Myouk-ta-gnyek*, BURM.
   References.—*Kurz, For. Fl. Burm., I., 134; Gazetteer, Burm., I., 134.*
   Habitat.—An erect tree from 80 to 100 feet high, not infrequent in the tropical forests of Martaban, east of Tounghoo (*Kurz*).

RESIN.
320

   Resin.—It exudes a red resin, about which no information is obtainable.

TIMBER.
321

   Structure of the Wood.—Yellowish, turning pale brown, rather heavy, of a somewhat unequal coarse fibre, soon attacked by xylophages (*Kurz*).

PARKINSONIA, *Linn.; Gen. Pl., I., 570.*

22

Parkinsonia aculeata, *Linn.; Fl. Br. Ind., II., 260;*
   THE JERUSALEM THORN.   [LEGUMINOSÆ.
   Vern.— *Wilayáti kikar*, PB.; *Vilayti babúl* or *kikár*, SIND; *Devi bábul, vilayti kikar*, BOMB.; *Sima jíluga*, TEL.
   References.—*Kurz, For. Fl. Burm., I., 403; Beddome, Fl. Sylv., 91; Gamble, Man. Timb., 124; Stewart, Pb. Pl., 73; Cleghorn, Plants of Southern India, 145; Elliot, Flora Andhr., 168; Murray, Pl. & Drugs, Sind, 129; Baden Powell, Pb. Pr., 587; Lisboa, U. Pl. Bomb., 278, 392; Royle, Fib. Pl., 298, 391; Cross, Bevan, & King, Rep. on Indian Fibres, 10, 55; Gazetteers:—Mysore & Coorg, I., 59; Bombay, XV., 74; XVII., 23; N.-W. P., I., 80; IV., lxxi.; Shahpur, 70; Raj., 27; Ind. Forester, VI., 240; XII., App., 27.*
   Habitat.—An introduced shrub, or small tree, native of Troipcal America, now almost naturalised in India, especially in the hotter regions where it is grown as a hedge plant.

Rope Bridges.	(*J. Murray.*)	PASPALUM scrobiculatum.

Fibre.—It yields a fibre of a beautiful white colour, but short, and somewhat brittle. It is considered to be wanting in strength, though it might be made useful for paper-making by mixing with other fibrous substances and beating into a half stuff (*Royle*). Examined by **Messrs. Cross, Bevan, & King** it was found to yield 65·2 per cent. of cellulose and 12·5 per cent. of moisture.

Fodder.—In the Panjáb it is lopped of its smaller BRANCHES, which are given to goats as fodder.

Structure of the Wood. Whitish, close-grained, used for fuel; it also makes good charcoal.

Domestic, &c.—The plant is well adapted for fences, and is largely planted for that purpose.

FIBRE.
323

FODDER.
Branches.
324
TIMBER.
325
DOMESTIC.
326

Parmelia, see Lichens, Vol. IV., 635-639.

PARROTTIA, *C. A. Mey.; Gen. Pl., I., 666.*

Parrottia Jacquemontiana, *Dcne.; Fl. Br. Ind., II.. 426;*

327

[HAMAMELIDEÆ.

Syn.—FOTHERGILLA INVOLUCRATA, *Falc.*
Vern.—*Psher, pishor, páhú, po, killar, kírrú, páre, shá, spilecha,* PB.; *Spilecha,* PUSHTU.
References.—*Brandis, For. Fl., 216, t. xxviii.; Gamble, Man. Timb., 174, Stewart, Pb. Pl., 110; Aitchison, Flora of the Kuram Valley, 59; Baden Powell, Pb. Pr., 587; Gazetteers:—Bannu, 23; Gurdáspur, 55; Dera Ismail Khán, 19; Rawalpindi, 15; Ind. Forester, IV., 345; V., 184; VIII., 37, 38; XIV., 99.*

Habitat.—A large, deciduous shrub or small tree of the North-West Himálaya from the Indus to the Ravi, between 2,800 and 8,500 feet.

Fibre.—The strong fibrous TWIGS are used in the Panjáb for binding loads, making baskets, and very largely for constructing the rope or twig-bridges of the Himálayan rivers. These bridges or *jhulás* are generally constructed entirely of the twigs of this shrub, but occasionally the branches of **Cotoneaster, Olea,** and **Indigofera heteranthera** (*q. v.*) are mixed with them. **Stewart** writes, "For the bridges, &c., **Parrottia** is cut at all seasons, and is not very lasting, requiring frequent piece-meal renewal. Willow is said to be employed in Spiti, Zanskar, Ladak, &c. Near Mozaffarabad, there are several bridges of the same construction, (*viz.,* one longitudinal rope to walk on, and two lateral ones to hold by, connected with the former by thinner ropes), but made of twisted hide, and one is mentioned by Hutton in Kunawar made of Yak's hair."

Fodder.—The LEAVES are said to be browsed by cattle (*Stewart*).

Structure of the Wood.—Light pinkish red, hard, heavy, very close-grained, growth slow, weight 56℔ per cubic foot. Highly esteemed for walking-sticks, tent pegs, *charpoys,* and rice pestles, also for native bows used in throwing pellets.

FIBRE.
Twigs.
328

FODDER
Leaves.
329
TIMBER.
330

Pashm, pashmina, see Wool; Vol. VI.

(*J. F. Duthie.*)

PASPALUM, *Linn.; Gen. Pl., III., 1097.*

331

This genus is largely represented in America, where several of the species are highly valued both for grazing and stacking.

[GRAMINEÆ.

Paspalum scrobiculatum, *Linn.; Duthie, Fodder Grasses, 1;*

332

Vern.—*Kodo, kodaka,* HIND.; *Kodoá dhán,* BENG.; *Koda, kodai,* BIHAR; *Janhe,* SANTAL; *Kodon, koda, kodrám,* N.-W.P.; *Kodo, kodra, kodrám,* KUMAON; *Kodra, kodon,* PB.; *Kodo, kodie,* C. P.; *Kodra, kodri, harik, pakodi, pakod, kodroa-kora, kodro,* BOMB.; *Kodra, kodru, harik,* MAR.;

P. 332

The Kodo Millet.

Menya, kodra, Guz.; *Kiraruga, aruga, alu, állu, arikalu, árike, páta-
arige,* Tel.; *Harik,* Kan.; *Wal-amu,* Sing.; *Kodrava, korádusha,*
Sans.

References.—*Roxb., Fl. Ind., Ed. C.B.C., 93; Thwaites, En. Ceylon Pl.,
358; Dals. & Gibs., Bomb. Fl. Suppl., 97; Stewart, Pb. Pl., 259;
Aitchison, Cat. Pb. & Sind Pl., 158; Sir W. Elliot, Fl. Andhr., 13, 16,
146, 189; U. C. Dutt, Mat. Med. Hindus, 305; Birdwood, Bomb. Prod.,
112: Baden Powell, Pb. Pr., 238; Atkinson, Him. Dist. (X., N.-W.
P. Gas.), 688; Duthie & Fuller, Field & Garden Crops, II., 8;
Useful Pl. Bomb. (XXV., Bomb. Gas.), 141, 184, 270, 276; Royle, Ill.
Him. Bot., 420; Church, Food-Grains Ind., 39; Tropical Agriculture,
342; Mueller, Select Extra Trop. Pl., 282; Buchanan, Journey through
Mysore & Canara, &c., I., 106, 287, 379; II., 223; III., 352; Grier-
son, Bihar Peasant Life, 227; Settlement Reports:—Panjab, Kángra,
25; N.-W. P., Banda, 49; Azamgarh, 115; Central Provinces, Upper
Godávari, 35; Betul, 63, 64; Chanda, 81; Land Rev. Settle. Rept., Hos-
hungabad, 281; Rep. Agri. Station Cownpur, 1884, 7; Gazetteers:—
Bombay, II., 66; VI., 39; X., 146; XI., 97; XIII., Pt. I., 289; N.-W. P.,
I.. 85; IV., lxxix.; Ind. Forester, XII., App. 22; Balfour, Cyclop. Ind.,
III., 155.*

Habitat.—A native of India, and extensively cultivated in many parts
of the country during the rainy season. It is an erect-growing annual
grass, the stems averaging two feet in height.

Cultivation.—In the North-West Provinces it is grown far more extensively than any of the other minor millets, and over a large portion of the
Province it is the favourite crop for inferior outlying land. This is, however, on account of the readiness with which it grows on the poorest soil,
and not by reason of the quality of its grain, which is by no means a popular
article of food. It is cultivated chiefly in the districts south of the Jumna,
the Allahabad and Lalitpur districts comprising the largest area. It is
sown, as a rule, at the commencement of the rains, at the rate of from 12
to 20℔, to the acre, and is cut in October. It is either grown alone or (in
the Doáb) is often mixed with cotton, and in the Benares Division with
arhar. It is never succeeded by a spring crop, partly because it ripens
too late to be off the ground in time, and also because the soil on which
it is grown is too poor to bear two crops within the year. Careful weeding
is needed to secure a good outturn, which is estimated at from 10 to 12
maunds of grain to the acre. The plants have to lie for a week or so
after being reaped, in order to loosen the grain, which is even then not
thrashed out without a good deal of trouble (*Field & Garden Crops, Part
II., 8*).

In the Bombay Presidency several varieties are recognised by the
natives, the differences probably resulting from the nature of the soil, the
method of cultivation, &c. Two sorts are, however, well known, *viz.*, the
wholesome and the unwholesome. The grain of the former is smaller and
paler, and is known by the name of *pechadi* or *harkin* in the Konkan. In
Goa it is called *pakod*. The unwholesome variety is called *dhone* or
mojari-harik in the Konkan, and *mana kodra* in Gujarát. It is the
Sanskrit *kodrava* (injurious). The grain is the only poisonous part of the
plant. All the varieties are, as a matter of fact, more or less poisonous
(*Useful Pl. Bomb., 141*). *Harik* is one of the principal crops of the
Ratnagiri District. It is grown chiefly on the more elevated tracts, or on
steep hill slopes. The seed is sown broadcast soon after the rains have
commenced. The crop is once weeded, and ripens about the end of October or beginning of November. In the Kaira District this millet forms the
staple food of the lower classes along with **Pennisetum typhoideum**, to
which they consider it as next in importance.

In the Central Provinces the area occupied by this crop is greater than
in any other part of India, constituting in many districts the staple

CULTIVATION

food of a large portion of the poorer classes. It is much appreciated for its prolific yield and satisfying properties; also because it does not require so much water as rice. In the Biláspur District it is rarely grown for more than two years on the same land (*Gaz., Cent. Prov.*

This crop is said to suffer considerably from the attacks of insects in the North-West Provinces; in the Bombay Presidency, however, it is stated to be proof against all insect attacks. **Mr. Benson** mentions the parasitic growth of a species of **Striga** as being a cause of injury in the Kurnús District of Madras.

FOOD & FODDER. Grain. 334

Food & Fodder.—Though used as food by a very large number of people in India, the GRAIN can by no means be considered a wholesome article of diet; and unless special precautions are taken, it is liable to act as a narcotic poison. Chemical analysis has not as yet revealed the poisonous principle; but whatever it may prove to be, the deleterious effects are well known in the various parts of India where this crop is grown. **Dr. Stewart,** in his account of the Bijnor District, remarks, "That *kodon* at times produces delirium and vomiting is universally believed in the Bijnor District, and I find that various authors mention a similar phenomenon as occasionally occurring in many parts of all the three Presidencies. The Natives generally hold that with and among the ordinary *kodon*, and undistinguishable in appearance from it, grows a kind which they call *majna* or *majui*, and which produces the above effects; but it has been suggested, with greater probability, that these depend on the use of the new grain under certain conditions." Regarding the variety known in the Bombay Presidency as "the black *kodra*," the following account is taken from the *Bombay Gazetteer, Vol. XXV.* :—" The outer coat or husk has, according to **Surgeon-Major Pyrie**, a dark outline of a fungus-like character, and on the internal surface it appears to consist of minute roundish cells containing dark sporules. Several authorities have failed to recognise this fungus-like character, in which is supposed to reside the poisonous principle. The fact, however, that *kodra* grain, freshly reaped, if left unstacked in the fields for some days when it was rainy and wet, had become possessed of decidedly more poisonous properties than grain from the same field harvested and stacked when the weather was dry, together with the generally acknowledged truth that a very poisonous seed has, under peculiarities of soil and cultivation, yielded a comparatively harmless grain, seems to bear out the fungus theory." " Though every part of the grain is poisonous, the husk and testa are more so; hence the natives take good care to separate the light grain, by means of water in which it floats, from the heavy and less injurious one. *Kodra* grain is a common article of food with all the poor people in India. They prepare it by macerating it for three or four hours or more in a watery solution of cowdung, when the scum and hollow grain which rise to the surface are separated, and the good grain removed and spread out in the sun to dry. This process is repeated so long as any poison is suspected to remain in the grain. Boiling does not entirely destroy the poison, but if the grain is kept for a number of years, its poisonous properties are found to diminish. When required for use it is ground in earthen mills, which remove the pericarp; it is then pounded and winnowed, and the grain is fit for use." *Kodra* is much used in the Konkan, eaten with whey, which is supposed to have the property of neutralising the poison.

The symptoms of *kodra* poisoning, as described by **Dr. Pyrie**, are:—" Unconsciousness, delirium with violent tremors of the voluntary muscles, pupils dilated, pulse weak, skin cold and covered with profuse perspiration, and difficulty in swallowing." Animals appear to suffer much more than men from the evil effects of unwholesome *kodra*, especially in the case

8

| PAVETTA indica. | The Kodo Millet. |

FOOD & FOODER.

of those which are unable to vomit; and it should also be borne in mind that animals usually eat husk with the grain. Some native antidotes for *kodra* poisoning are :—Gruel made of the flour of *urd* (**Phaseolus radiatus**); the juice of the plantain stem; the astringent juice of the guava, or the leaves of **Nyctanthes Arbor-tristis.**

The following analysis by **Professor Church** shows the chemical composition of the husked grain in 100 parts : —

Water	11'7
Albumenoids	7'0
Starch	77'2
Oil	2'1
Fibre	0'7
Ash	1'3

Grass.
335
Straw.
336

The nutrient ratio being 1 : 11'7, and the nutrient value 89.

Cattle, and especially buffaloes, eat the GRASS readily when it is young. The STRAW is occasionally used as fodder, but on no account should cattle be allowed to stray into fields of this crop when it is ripening.

(*J. Murray.*)

337

PASSIFLORA, *Linn.; Gen. Pl., I., 810.*

A genus of twining shrubs which belongs to the Natural Order PASSIFLOREÆ, and comprises a large number of species, mostly natives of Tropical and Subtropical America. Two, **M. Leschenaultii,** *DC.,* and **M. nepalensis,** *Wall.,* are natives of India, but neither appears to be of economic value. **M. quadrangularis,** the Square-stalked Passion Flower, or Common Granadilla, is cultivated in the hotter parts of India and Burma on account of its flower and its sweet acid fruit, which is much esteemed, especially in the hot weather. **Mason** states that it has been introduced among the Karens in Burma, who highly appreciate the fruit.

Pastures & Pasture Lands, see Fodder, Vol. III., 407.

Patchouli oil, see **Plectranthus Patchouli,** *Clarke,* p. 291; see **Pogoste-**
[mon **Patchouli,** *Pelle.,* p. 307.

PAVETTA, *Linn.; Gen. Pl., II, 114.*

[*148, 186, 1065;* RUBIACEÆ.

338   **Pavetta indica,** *Linn.; Fl. Br. Ind., III., 150; Wight, Ic., tt.*

Var. indica proper,—P. INDICA, *Linn.;* P. FINLAYSONIANA, *Wall.;* P. GRACILIFLORA, *Wall.;* P. ALBA, *Vahl.;* P. PETIOLARIS, *Wall.;* IXORA PANICULATA, *Lamk.;* I. PAVETTA, *Roxb.*

Var. polyantha, *Wall.* (sp.),—P. INDICA, *Wall.;* P. ROTHIANA, *DC.;* P. VILLOSA, *Heyne, not of Vahl.;* IXORA TOMENTOSA, *var.* GLABRESCENS, *Kurz.*

Var. tomentosa, *Roxb.* (sp.),—IXORA TOMENTOSA, *DC.;* PAVETTA BRUNONIS, *Wight, Ic., t. 1065;* P. VELUTINA, MOLLIS, and CANESCENS, *Wall.*

Var. montana, *Thw. Mss.* (sp.).

Var. minor.

Vern.—*Kankra, carnicára, canchra, cat'hachampa, pápari,* HIND.; *Kúkúra-chúra, var.* tomentosa=*júi,* BENG.; *Sikriba sikérúp,* KÓL.; *Buri,* KHARWAR; *Budhi tilai, budhi ghasit,* SANTAL; *var.* tomentosa=*Sun dók,* LEPCHA; *var.* tomentosa=*Padera,* N.-W. P.; *Papát,* BOMB.; *Pápadi,* MAR; *Pavut-tay-vayr,* TAM.; *Pápata, núne-pápata, páputtavayrú,* TEL.; *Pavati, pappadi,* KAN.; *Pawetta,* SING.; *Páppána, carnicára,* SANS.

References.—*Roxb., Fl. Ind., Ed. C.B.C., 129; Brandis, For. Fl., 275; Kurz, For. Fl. Burm., II., 18, 19; Thwaites, En. Ceylon Pl., 155; Dals.*

P. 338

| The Pavonia Fibre. | (*J. Murray.*) | PAVONIA odorata. |

& Gibs., *Bomb. Fl., 112*; *Elliot, Flora Andhr., 138, 144, 178*; *Sir W. Jones, Treat. Pl. Ind., V., 81*; *Ainslie, Mat. Ind., II., 289*; *Dymock, Mat. Med. W. Ind , 2nd Ed., 400*; *Dymock, Warden & Hooper, Pharmacog. Ind., II.; 211*; *Bidie, Cat. Raw Pr., Paris Exh., 54*; *Drury, U. Pl., 333*; *Lisboa, U. Pl. Bomb., 162*; *Gazetteers :—Mysore & Coorg, I., 70*; *Bomb., XV., 436*; *Dist. Manual, Nellore, 133*; *Ind. Forester, III., 203.*

Habitat.—A common and very variable bush or small tree, found throughout India from the Western Himálaya in Gárhwal (**var.** tomentosa) ascending to 4,000 feet, to Bhután and Burma, also southwards to Ceylon, Malacca, and Penang.

MEDICINE.

Medicine.—Ainslie, who appears to have been the first English writer to notice the medicinal properties of this plant, writes :—" This is a bitter but not unpleasant tasted ROOT, possessing at the same time aperient qualities, and is one of those medicines commonly prescribed by the Native doctors in visceral obstructions; given in powder to children, the dose about a drachm or more." Bidie states that " in Madras the root is considered diuretic and given in dropsy. It is also said to be purgative. An infusion of the LEAVES is used externally as a local application in certain complaints."

Root.
339

Leaves.
340

CHEMICAL COMPOSITION.—The authors of the *Pharmacographia Indica* state that the root contains a bitter glucoside, closely related to salicin, but differing from that substance in its optical inactivity and its greater solubility in water.

Chemistry.
341

Food.—According to Lisboa the FRUIT is pickled and eaten in some parts of Madras; and the FLOWERS are used as food by the hill people of Mátherán.

**FOOD.
Fruit.**
342
Flowers.
343

Pavia indica, see *Æsculus indica, Colebr.*; Vol. I., 126.

PAVONIA, *Cav.; Gen. Pl., I., 205.*

Pavonia odorata, *Willd.; Fl. Br. Ind., I., 331*; MALVACEÆ.

344

Syn.—HIBISCUS ODORATUS, *Roxb.*; PAVONIA SIDOIDES, *Horn.*; P. ROSEA, *Wall.*; P. ROMBORUA, *Wall.*

Vern.—*Sugandha-bálá, bálá,* HIND.; *Bálá, súgandh-bala,* BENG.; *Bálú,* BOMB.; *Kálá-válá,* MAR.; *Perámútiver, paramutty, mudda pulagam,* TAM.; *Erra kúti,* TEL.; *Bálarakkasi-gida,* KAN.; *Bálá, hrivera,* SANS.

References.—*Roxb., Fl. Ind., Ed. C.B.C., 530*; *Voigt, Hort. Sub. Cal., 115*; *Thwaites, En. Ceylon Pl., 26*; *Elliot, Fl. Andhr., 52*; *Ainslie, Mat. Ind., II., 297*; *U. C. Dutt, Mat. Med. Hind., 123, 293, 300*; *Dymock, Mat. Med. W. Ind., 2nd Ed., 109*; *Dymock, Warden, & Hooper, Pharmacog. Indica, I., 224*; *Murray, Pl. & Drugs, Sind, 61*; *Irvine, Mat. Med., Patna, 103*; *Taylor, Topography of Dacca, 55*; *Drury, U. Pl., 334*; *Watt, Selections from Records of Govt. of India, R. & A. Dept., I., 179*; *Cross, Bevan, & King, Rep. on Indian Fibres, 9, 41*; *Gazetteers :—Mysore & Coorg, I., 57*; *N.-W. P.. IV., lxviii.*; *Nellore Manual, 128*; *Ind. Forester, XII., App., 7*; *XIV., 269.*

Habitat.—An erect herb, covered with sticky hairs; found in the North-West Provinces, Sind, and Banda, the Western Peninsula, Burma, and Ceylon; also considerably cultivated in gardens for its fragrant flowers.

Fibre.—This species, and P. zeylanica, *Cav.*, both yield a FIBRE of excellent quality, which somewhat resembles that obtained from the fibre-yielding species of Hibiscus. It is, if anything, of a finer texture, softer and whiter, and stands a good chance of coming into commercial use as a substitute for Hibiscus, and even for jute (*Watt*). Both species deserve to be more carefully cultivated and their fibres thoroughly tested. Specimens examined by Messrs. Cross, Bevan, & King were found to contain—Moisture, 8·8 per cent.; ash, 3·4; and cellulose, 74·7 per cent. The loss by hydrolysis on boiling with caustic soda for five minutes was 11 per cent., when boiled for

FIBRE.
345

8 A

PEACOCK;
Pavo cristatus.   The Peacock.

MEDICINE.
Root.
346

one hour, 13·5 per cent. Though the chemical composition indicates high excellence, the short length of the ultimate fibre, 1—1·5 mm., places it low in the scale of commercial value.

Medicine.—The fragrant ROOT is esteemed in Hindu medicine, being considered aromatic, cooling, and stomachic. It is much used, in combination with other drugs, as a remedy for fever, inflammation, and hæmorrhage from internal organs, and enters into the composition of a well-known fever drink called *shadanga pániya* (*U. C. Dutt*). Ainslie notices the utilisation of the plant by the Hindus, but expresses no opinion as to its medicinal virtues. Taylor describes a preparation of *bálá* with other carminatives and *bél* fruit. This, he says, is frequently prescribed as an astringent and tonic in cases of dysentery. Dymock states that in Bombay, serpentary root, imported from Europe, is universally substituted for *bálá*. The therapeutic properties of the root are probably due to the carminative quality of the odorous matter it contains, together with the mucilaginous character commonly met with in members of the Natural Order MALVACEÆ.

347

Pavonia zeylanica, *Cav. ; Fl. Br Ind., I., 331.*
   Syn.—HIBISCUS ZEYLANICUS, *Linn.*
   Vern.—*Pera mutti*, TEL. ; *Chittamutti, chinna muddapulagam*, TAM.
   References.—*Roxb., Fl. Ind., Ed. C.B.C.. 530; Thwaites, En. Ceylon Pl., 401; Dals. & Gibs., Bomb. Fl., 21; Burm. Fl. Ind., t. 48, f. 3; Elliot, Fl. Andhr., 151; Murray, Pl. & Drugs of Sind, 61; Cross, Bevan, & King, Rep. on Ind. Fibres, 9, 41; Watt, Sel. from Rec. of Govt. of India, R. & A. Dept., I., 179; Gazetteer, N.-W. P., IV., lxviii.; Nellore Manual, 121.*
   Habitat.—A much branched perennial, fairly abundant in the North-West Provinces, Sind, Bombay, Madras, and Ceylon.

FIBRE.
348

   Fibre.—See the paragraph "Fibre" under the preceding species.

349

PAYENA, *A. DC.; Gen. Pl., II., 659.*

A genus of trees which belongs to the Natural Order SAPOTACEÆ, and comprises from eight to ten species, natives of the Malay Peninsula and islands   Of these at least two, *viz.* P. **Maingayi,** *Clarke,* and P. **lucida,** *A. DC.* (*Fl. Br. Ind., III., 547*), are stated to yield Guttapercha (see Vol. III., 108)   The latter is also said to have a hard red wood, weighing 45℔ per cubic foot, which is used for planking (*Gamble, Man. Timb., 245*).

Pea, see Pisum sativum, *Linn.*, p. 277.

Peach, see Fruits, Vol. III., 449; see **Prunus persica,** *Benth. & Hook.*

PEACOCK.

350

Peacock ; Pavo cristatus, *Linn.*
   PFAU, *Germ.* ; PAON, *Fr.* ; PAUÚW, *Dutch.* ; PAVONE, *Ital.*
   Vern.—*Múr, mohr*, HIND. ; *Mab-ja*, BHOT. ; *Mong-yung*, LEPCHA ; *Mór*, MAR. ; *Mail, maylú*, TAM. ; *Némilie*, TEL. ; *Navelu*, KAN. ; *Mirrik, marak, to-gei*, MALAY. ; *Tokei*, SING. ; *Nilkantha, mayúra, varhi*, SANS. ; *Táous, taon*, ARAB. : *Maragh*, PERS.
   References.—*Forbes Watson, Ind. Survey, II., 392; Ainslie, Mat. Ind., I., 290; Balfour, Cyclop. of India, III., 165, 167; Tennant, Natural History of Ceylon, 2447.*
   Habitat.—The common pea-fowl is widely spread over India and Ceylon, of both of which countries it is a native. It is readily domesticated, and is bred in large numbers in the neighbourhood of many Hindu temples. The domestic pea-fowl of Europe is identical with that of India, and must

| The Peacock. | (*J. Murray.*) | **PEARLS.** |

have been introduced from the latter country in very remote times. Setting aside its importation to Palestine by Solomon (*I. Kings, x., 22; II. Chronicles, ix., 21*) its assignment in classical mythology as the favourite bird of Hera or Juno, testifies to the early acquaintance the Greeks must have had of it. Notwithstanding this, however, and the fact that it is mentioned by **Aristophanes** and other older writers, it is probable that the bird was little known in Europe till after the conquest of India by Alexander. Besides the common **P. cristatus**, *Linn.*, another species, **P. muticus**, *Linn.*, occurs in Assam, Burma, and the Malay Peninsula, but is of no interest from an economic point of view.

Medicine.—The FLESH of the pea-fowl is amongst the medicines of the Hindus, and, according to **Ainslie**, is prescribed by Tamil medical writers as a valuable remedy in cases of contracted limbs. Peacock-grease has also long been esteemed as an article of therapeutic value.

MEDICINE.
Flesh.
351

Food.—The peacock has been an esteemed dish from classical times. The first to present it at table is said to have been **Hortensius**, the orator. It subsequently became a famous dish towards the decline of the Roman republic, and was highly recommended as a wholesome flesh by **Celsus** and **Pliny**. In mediæval bills of fare on State occasions the peacock—which seems to have been served up garnished with its gaudy plumage—was a constant feature, and one of the most solemn oaths of the days of chivalry was taken "on the Peacock." It is not an uncommon article of food amongst Europeans in India to the present day, though the religious objections of the Hindus render it less commonly eaten than it otherwise might be. The young bird, at certain seasons, very much resembles turkey in flavour, and is in no way inferior; but if not in proper season it is tasteless and uninviting.

FOOD.
352

Domestic & Sacred.—The FEATHERS are much used in India for making fans and hand *pankáhs*. They are collected into the desired form, and sewn together with pieces of prepared quills and *shola*, the colour generally being lac dye. A wooden handle is then attached, by means of lac, at one of the corners. Large hand *pankáhs* thus made are used for fanning idols and people of importance on ceremonial occasions. The brilliantly coloured tail-feather of the peacock is also used to decorate the turban of the Rájput. A bunch of feathers is still the implement of conjuring, and is carried by mendicants who pretend to skill in magic, especially by Jaina vagrants.

DOMESTIC.
Feathers.
353

The peacock enters largely into Hindu mythology, and is considered sacred or semi-sacred.

Pear, see Fruits, Vol. III., 449; see **Pyrus communis,** *Linn* 374.

Pearl Barley, see **Hordeum vulgare,** *Linn.;* Vol. IV., p. 283.

PEARLS.

Pearls are found in several molluscs which inhabit shallow seas and sandbanks in the Old and New World; the most productive is the so-called "Pearl Oyster." This name is, strictly speaking a misnomer, since the so-called "Oyster" is in reality a mussel—**Meleagrina margaritifera,** *Lam.* It is a matter of contention whether it should remain **Meleagrina** as **Lamarck** separated it, or whether it should be relegated to the genus **Avicula.** Dr. **Kelaart,** who made valuable reports on the fisheries of Southern India in 1857, 1858, and 1859, preferred **Lamarck's** name, and as a consequence much of Indian literature of the subject retains his nomenclature. But modern classification favours its reduction to **Avicula.** Thus in **Tryon's** Conchology, 1884, the pearl-bearing mussel is named **Avicula (Meleagrina) margaritifera;** older synonyms are **Mytilus marga-**

354

ritiferus, *Linn.*, **Margarita sinensis**, *Leach*, **Avicula meleagrina**, *Blain*, and **Avicula radiata**, *Leach*.

Mr. E. Smith has recently separated the Ceylon pearl oyster from **Avicula (Meleagrina) margaritifera**, on the ground that it is more convex than that species; also that it does not attain the same size, and does not yield "mother of pearl." He names it **Avicula (Meleagrina) fucata**, *Gould*.

[Fisheries, 1884.

355   **Pearls & Pearl Fisheries**, *Thomas, Rep. on Pearl & Chank*

Vern.—*Moti*, HIND.; *Mutti*, GUZ.; *Muthu, muthu-chippi*, TAM.; *Muti-amu*, TEL.; *Muti, mutya, mutiyara lulu*, MALAY.; *Muthu-chippi*, SING.; *Muktá, marakata*, SANS.; *Lúlú, júhar,* gamán, shazzir*, ARAB.; *Marwaríd*, PERS.

* A common name for all precious stones, but more particularly applied to the pearl.

References.—*Forbes Watson, Ind. Survev, I., 368; Dr. Kelaart, Reports on Pearl Oyster & Pearl Fisheries, 1857, 1858, 1859; J. H. Van Linschoten, Voyage to the East Indies, 1596, Ed. Burnell, Tiele, & Yule, I., 45, 125, 128; II., 133; Tavernier, Travels in India (1640—1667), Ed. by Ball (1889), I., 137, 259, 287, 305, 397; II., 107, 122, 130-131; Ainslie, Mat. Ind., I., 292; Irvine, Mat. Med., Patna, 68; Med. Top. Ajmir, 146; Dutt, Mat. Med., Hindus, 93; Balfour, Cyclop., III., 168; Trop. Agriculturist, April, 1888, 667; February, 1889, 553; Gazetteers:— Bombay, VIII., 93; XIII., 55; District Manuals:—Tinnevelly, 73, 80, 154; Sel. Records, Bomb. Govt., XVII. (1847).*

Formation.

356

Formation.— It is believed that most pearls are formed by the presence of some foreign substance, which becomes imbedded in the interstices of the mantle, and forms a source of irritation. This irritation causes the deposition of nacreous matter (similar to that lining the interior of the shell), in concentric layers, until the substance becomes completely encysted. The popular notion that the foreign matter is generally a grain of sand is untenable. According to several eminent conchologists it is in most cases a minute parasite, but **Dr. Kelaart** believes the nucleus to be, in most cases at least, an ovum, or ova escaped through the distended coats of an overgrown ovary, and become imbedded in the interstices of the mantle. "I have repeatedly examined seed or young pearl," he writes, "in process of formation, and with a magnifying power of $\frac{1}{8}$th inch lens I was able to see distinctly the outlines of two or three ova through the first or superficial layer of nacre, surrounded by groups of ova." His theory is further supported by the fact that pearls are most frequently found imbedded in the mantle near the hinge (the place where the ovarium is most likely to be liable to rupture), and by the fact that, with careful examination he was generally able to find, when the pearls were not actually found in the interstices of the mantle near that locality, cicatrices on the structure where they once existed.

Though it is generally admitted that pearls originate in the mantle, they nevertheless, when large, frequently work their way out and lie loose between it and the shell, or become attached by subsequent nacreous deposit to the "mother of pearl" surface of the latter. In this position a pearl may become in time so covered up as to form a mass of hemispherical shape, which, when cut from the shell, forms what is known by jewellers as a *"perle bouton."* Occasionally, by a continuation of the process of deposit, a pearl may become completely hidden. It has thus happened that fine pearls have been occasionally brought to light in cutting up mother of pearl.

The hollow wartty pearl, known as "blister pearl," is supposed to be produced by a deposit of nacreous matter at the point of invasion of a

**PEARLS &
Pearl Fisheries.**

FORMATION.

boring parasite. Solid pearls of irregular form frequently occur, which are termed "pearles baroques," or " barrok pearls " in the trade. These are supposed to be formed by deposition on rough and irregular substances, but it seems to be possible that they may originate similarly to round pearls, *viz.*, by deposit on an irregular group of ova.

Structure.
357

Structure of the Pearl.—The substance of a pearl is, as will be gathered from the above description of its formation, essentially similar to that of the nacre, or "mother of pearl," lining the shell. Both owe their iridescence to the interference in the rays of light caused by the microscopic corrugations on their surfaces.

A pearl of the first water should be of delicate texture, free from speck or flaw, and of clear, almost translucent white colour, with a subdued iridescent sheen. It should also be perfectly spherical, or, if not, of a symmetrical pear shape. On removing the outer layer the underlying surface is generally dull, but it sometimes happens that a poor pearl encloses a "lively kernel," and may therefore be improved by careful peeling. Pearls of a bluish tinge are less valued, and when yellow they are still less esteemed. Tavernier wrote, however, that the pearls of Behren and El Katif, though of a yellowish tinge, were quite as highly esteemed in the East as the gems of purer water obtained from Manar, because the former were supposed to be mature or ripe and never to change colour.

Indian Pearl Fisheries.—The pearl fisheries of India have been famous from the remotest times; indeed, the ancients obtained their pearls almost entirely from India and the Persian Gulf. In the latter locality the industry has been carried on since the days of the Macedonians. Oyster beds are said to extend along the entire Arabian coast of the Gulf, but the most important are on sandbanks off the island of Bahren. These were visited (*1596*) and described by Linschoten, who writes:—" The principall and the best that are found in all the Orientall Countries, and the right orientall pearles, are [some] between Ormus and Bassora in the straights or Sinus Persicus, in the places called Bareyn, Catiffa, Julfar, Camaron, and other places in the said Sinus Persicus, from whence they are brought into Ormus. The King of Portingale hath also his factor in Bareyn, that stayeth there onlie for the fishing of pearles. There is great trafficke used with them, as well in Ormus as in Goa " Tavernier visited the Persian pearl fisheries nearly a century later, when they had been regained by the Persians. He writes:—"In the first place there is a pearl fishery round the island of Behren in the Persian Gulf. It belongs to the King of Persia, and there is a good fortress there with a garrison of 300 men." " While the Portuguese held Hermuz and Muscat each *terrate* or boat which used to fish, was obliged to take out a license from them, which cost 15 abassis (£1-2-6) and many brigantines were maintained there, to sink those who were unwilling to take out licenses. But since the Arabs have retaken Muscat, and the Portuguese are no longer supreme in the Gulf, every man who fishes pays to the King of Persia only 5 abassis, whether his fishing is successful or not. The merchant also pays the King something small for every 100 oysters.

Pearl
Fisheries.
358

" The second pearl fishery is opposite Bahren, on the coast of Arabia-Felix, close to the town of El Katif, which, with all the neighbouring country, belongs to an Arab prince."

The pearls fished in these places, Tavernier states, were nearly all sold in India, but some were also taken to Bassora, and some to Russia. In connection with these fisheries he mentions a very fine pearl which was then in the possession of the Emir of Vodana. Though small, weighing only $12\frac{1}{16}$ carats, it was perfect in shape and "so clear and so transparent

P. 358

PEARL
FISHERIES.

that you can almost see the light through it." The value of this gem is stated to have been £32,000.

The best beds in the Gulf are said to consist of coral with beds of white sand lying in clear water. According to Colonel Pelly's report there were in 1863, 1,500 boats in Bahren alone, and the annual profit from the whole fishery amounted to about £400,000. In 1879 the value of the pearls taken in the Persian Gulf was estimated at about £300,000. In the same year about 7,000,000 were obtained, and it was believed that but for frequent interruption by weather, 2,000,000 more might have been lifted. From these figures it would appear that Colonel Pelly's statement of an annual value of £400,000 must be very nearly correct.

The chief centre of the pearl trade in the Persian Gulf is Lingah. Most of the products of this fishery are known as "Bombay pearls," from the fact that many of the best are sold there. As already stated, the pearls have frequently a distinctly yellow water. These are chiefly sent to Bombay, since they have now, as in Tavernier's time, a ready sale in India. The whitest and purest gems go to Bagdad.

The pearl fisheries on the Gulf of Manar are, however, the most important in India. On the Ceylon side the oyster banks lie from 6 to 8 miles off the western shore a little to the south of the island of Manar, while those on the Madras side are situated off Tinnevelli and Madura. Fishing has been conducted for pearls in the Gulf from time immemorial. Thus in the earliest writings of Europeans on the East we find frequent reference made to the fisheries. The head-quarters of the fishery appears to have been the Koru of Ptolemy, the Kolkhi of Aryan, the Coli or Chayl of the travellers of the Middle Ages, the Ramana-Koil (temple of Rama) of the natives. This place is the sandy promontory of Ramnad, which sends off the reef of rocks known as Adam's Bridge towards Ceylon.

According to Friar Jordanus, 8,000 boats were engaged in the Manar Gulf in 1330. Linschoten writes in the sixteenth century : "There are also other fishings for pearle, as betweene the Island of Seylon, and the Cape de Comriin, where great numbers are yearlie found, for that the Kinge of Portingale hath a captaine there with [certaine] soldiers that looketh unto it ; they have yearlie at the lest above three or four thousand duckers, yt live onlie by fishing pearles, and so maintaine themselves." He then fully describes the method of fishing which appears to have been very similar to that of the present time, and adds some interesting remarks concerning the Portuguese method of working the fisheries : —"When they have made an end of the day's fishing," he writes, " all the fishers with the captaine soldiers, labourers and watchmen for the King, goe together, and taking all the pearles that are caught that day, they divide them into certaine heapes, that is, one part for the King, another part for the captaine and soldiers, the third part for the Jesuits, because they have their Cloyster in that place, and brought the countrie first into the Christian Faith, and the last part for the Fishers, which is done with great Justice and Equalitie. This fishing is done in the summer tyme, and there passeth not any yeare but that divers Fishers are drowned by the Cape de Comriin (which is called the King's Fishing) and manie devoured by fishes, so that when the fishing is done there is great and pitifull noyse and cry of women and children heard [upon the land, for the losse of their husbands and friends]. Yet the next yeare they must do the same work againe, for that they have no other meanes to live, as also for that they are partlie compelled thereunto by the Portingales. but most part (are content to doe it,) because of the gaine [they get thereby after all the danger is past]." The observant traveller then goes on to state that the pearls were sold according to their size, being passed through sieves, each sifting having its constant value. He also remarks that the pearls were rubbed with rice beaten with a little salt to give them "a faire colour," and that they then became "as faire and cleane as christall, and so continue."

Tavernier also notices these fisheries a hundred years later, remarking, "The pearls found there are the most beautiful, both as regards water and roundness of all the fisheries; but one is rarely found which exceeds 3 or 4 carats in weight."

In the seventeenth century, with the growth of Dutch commercial power in the south of India, the fisheries fell into the hands of Holland, and were conducted for many years by that power. In 1794, the Madras Government undertook the management of the pearl fisheries on the southeast coast of the Peninsula, and in 1796 the Ceylon pearl fisheries also came under the British Government. During the first 83 years the former realised about 12 lakhs of rupees, their annual expenditure being about R6,000. The Ceylon fisheries have proved much more valuable. In the first 40 years, namely, from 1796 to 1837, the total receipts amounted, according to one computation, to £964,803, and the expenditure to £51,752; according to another the total net revenue for that period was only £524,521. In the past fifty years (from 1838 to 1888), the total revenue has amounted to £437,110, and the expenditure to £105,656, or a net average annual profit, inclusive of many years when there was no fishing, of upwards of £6,600. The average number of oysters annually obtained during the same period amounted to about 3,575,600. In the year 1880, according to the Colonial Exhibition Handbook of Ceylon, 25,000,000 oysters sold for only £20,000, whereas in 1881, 18,000,000 sold for £59,000.

The revenue derived from pearl-fisheries is necessarily of a very precarious nature, since the molluscs possess locomotive powers and frequently disappear from certain banks and migrate to more favourable situations. New beds are thus formed from time to time, which can only be found at the expense of considerable time and money. The Dutch, probably owing to this reason, had no fishery from 1732 to 1746, and again from 1768 to 1796, while from 1820 to 1828, and 1838 to 1854 there was similarly no fishery on the Madras coast. Mr. Thomas states that again, between 1860-61 and 1884, fishing operations were suspended in Madras. The Ceylon and Madras fisheries are each under the charge of a Government officer, who spends a certain part of each year on inspection duty, and keeps an accurate record of the condition, at different times, of the various banks. These officers regulate the fisheries, and permit fishery only when they consider the banks to be in a satisfactory condition.

METHOD.—Off the coast of Ceylon the fishery season is inaugurated by numerous ceremonies, and the fleet of boats then puts to sea. Fishing, when allowed, generally commences in the second week of March, and lasts from four to six weeks, according to the season. The boats are grouped in fleets of from sixty to seventy, and usually start at midnight so as to reach the oyster-banks at sun-rise. Each boat generally carries the divers, and is manned by ten rowers, a steersman, and a shark-charmer (*Pillal karras.*). On reaching the bank a signal-gun is fired and diving commences. To facilitate the descent of the diver, a stone of granite, weighing about 40℔, is attached to the cord by which he is let down. The men work in pairs, one going down, the other staying on board and watching the signal-cord. When this is pulled the latter draws up the sink-stone first, then the basket of "oysters," and last of all the diver himself. Divers remain below on an average for from fifty to eighty seconds, though some can endure a much longer submergence. Exceptional cases, indeed, are recorded of men who could remain as long as six minutes under water.

Method.
359

PEAT.	Pearls used in Medicine.

Each man makes forty to fifty descents daily, and when exhausted, is re-lieved by the watcher. Notwithstanding the fact that the water is infested with sharks, accidents almost never occur. The divers, however, always arm themselves with a short pointed piece of iron wood for defence, and place implicit faith in the shark-charmers. The constant life of exposure, excite-ment, and exhaustion in a tropical climate is necessarily a very trying one, consequently the diver is generally a short-lived man. Work is continued from about sun-rise to noon, at which time a gun is fired and operations stop. On the arrival of the fleet at the shore, the oysters are thrown on the beach to putrify, or are carried to the Government *kottus* at once, where they are divided into four heaps. The divers then receive their share (one heap), and the remaining three are sold for Government to the highest bidder. The oysters generally fetch an average of from R30 to R40 per thousand. After removal from the dead oysters, the pearls are classed by being passed through a number of small brass cullenders, known as "baskets," the holes in the successive vessels being smaller and smaller. Having been sized in this way they are sorted as to colour, weighed, and valued.

MEDICINE.
360

Medicine.—Pearls have been used in medicine from a very ancient period, and are mentioned in Sanskrit literature by **Susruta**. They are puri-fied for use by being boiled with the juice of the leaves of **Sesbania aculeata** (*jayanti*), or of the flowers of **Sesbania grandiflora** (*vaka*). They are then calcined in covered crucibles and reduced to powder. The powder thus formed is believed to be similar in properties to coral, and is generally used in combination with that substance. It is esteemed in urinary diseases, consumption, &c., and is said to increase the strength, nutrition, and energy of weak patients. U. C. **Dutt** describes in full, two preparations of pearl, one containing nutmegs, mace, *jatámánsi* root, *kushta* root, leaves of **Pinus Webbiana**, aconite, iron pyrites, iron, talc, and realgar; the other gold, camphor, tin, lead, iron, and talc in addition to the pearl. The latter compound is supposed to be useful in diseases caused by "deranged bile," such as dyspepsia, jaundice, &c., the former is considered a valuable alter-ative tonic in chronic gonorrhœa and spermatorrhœa (*Mat. Med. Hind.*). **Ainslie** writes, "Arabian physicians suppose the powder of the pearl to have virtues in weak eyes; they also consider it as having efficacy when administered in palpitations, nervous tremors, atrabilarious affections, and hæmorrhage. They have, besides, this strange notion, that when applied externally, while in its embryo state in the shell, it cures leprosy." "Indian practitioners, especially the *Vytians*, recommend it as an affluent, calcined in cases of lientery, heart-burn and 'typhus fever.' The Arabians place it amongst their 'Cardiaca.'"

The only virtue possessed by the gem is doubtless that of an antacid, a property for which it was used at one time in European medicine, and even held a position in the British Pharmacopœia. Thus **Linschoten** writes in the sixteenth century of "seed-pearls":—"They are sold by the ounce, and used by Potticaries and by Physitions, and to that end many of them are carried into Portingall and Venice, and are very good and cheap."

PEAT.

361

Peat, *Ball, in Man. Geol. Ind., III., 120.*

True peat is to be found in several localities in India, but in most places it appears to be too impure to be of much economic value. According to **Ball** it is to be found on the Nilghiris in Madras; in several parts of Lower Bengal; in Pertabgurh in Oudh; in Kashmír; at Bhím Tal, near Naini Tal, North-West Provinces; in Nepál, in Assam, and in Burma. Of the localities mentioned the peat likely, however, to prove of value is that found on the Nilghiris. Propositions have frequently been made to employ the deposit of that locality as fuel for Indian railways. It is largely used locally for burning, but no data are available as to its working power in locomotives. According to **Ball**, the cost of the peat at the Nilghiri depôt might be about 10s. a ton, but its price would be increased by about 30s. by its conveyance from Beypur to Bombay ; the total cost when landed being £2 per ton. As the cost of coal at Bombay is now much less than this, there would appear to be very little chance of the Nilghiri peat ever competing with it as a fuel.

PEA STAKES; PAN HOUSES; WATTLE IN HUT CONSTRUCTION, Timbers used for—

362

Acacia arabica, Catechu, &c.	Dodonœa viscosa.
Ægiceras majus (native huts).	Evodia fraxinifolia (posts of huts).
Albizzia procera (used by tea-planters for stakes and by natives for huts).	Helicteres Isora.
	Leea robusta (stakes, huts), &c.
	Macaranga denticulata (huts).
Areca gracilis (native huts).	M. gummiflua.
Bamboo, various species.	Salix daphnoides.
Calligonum polygonoides (walls and roofs of huts).	Sesbania ægpytiaca.
	S. grandiflora (posts for native houses).
Corchorus capsularis (pea stakes).	Vitex Negundo (wattle work).
Cynometra ramiflora (posts).	&c., &c.

Pebbles, see Carnelian, &c., Vol. II., 167 ; see Flint, Vol. III., 404.

PEDALIUM, *Linn.* ; *Gen. Pl., II., 1056.*

[*1615 ;* PEDALINEÆ.

Pedalium Murex, *Linn.* ; *Fl. Br. Ind., IV., 386 ; Wight, Ic., t.*

363

Vern.—*Faríd-búti,** barà-gókhrú,* HIND. ; *Bara-ghókru,* BENG. ; *Gokshurá,* URIYA ; *Gokrú kalán,* PB. ; *Barà-ghókrú, haíti-ghókrú,* DEC. ; *Karonta,* GO ; *Hattí-charátté, mothe gokharu,* MAR. ; *Motto-ghókru, mothan gokharu,* GUZ. ; *Peru-nerunji, ánai-nerunti, ana-neringie,* TAM. ; *Enu-ga-palléru-mullu, yea-nugapulléru, pedda-palléru, yenuga palleru,* TEL. ; *Anne-galu-gidá,* KAN. ; *Káthe-nerinnil, ána-nerinnil, kákka-múllu,* MALAY. ; *Sule-gi,* BURM. ; *Ati-neranchi,* SING. ; *Khasake-kabir,* ARAB. ; *Khasake-kalán,* PERS.

* This name, according to certain authorities, is applied to **Cocculus villosus,** *DC.* ; see Vol. II., 397.

References.—*Roxb., Fl. Ind , Ed. C.B.C., 487, 495, 529 ; Rheede, Hort. Malab., X., 143, t. 72 ; Dals. & Gibs., Bomb. Fl., 162 ; Elliot, Flora Andhr.,51, 149 ; Pharm. Ind., 151 ; Ainslie, Mat. Ind., II., 15 ; O'Shaugh-nessy, Beng. Dispens., 479, 480 ; Dymock, Mat. Med. W. Ind., 2nd Ed., 553 ; S. Arjun, Bomb. Drugs, 92 ; Murray, Pl. & Drugs, Sind, 176 ; Baden Powell, Pb. Pr., 364 ; Drury, U. Pl ,334 ; Birdwood, Bomb. Pr., 57 ; Gazetteers :—Mysore & Coorg, I., 63 ; Bomb., V., 27 ; N.-W. P., I., 82 ; IV., lxxiv. ; Agri.-Horti. Soc., Ind., Journals (Old Series), IX., 177.*

Habitat.—A common plant in many parts of the Deccan Peninsula, especially near the sea ; abundant in Kathiawár and Guzerat ; also met with in Ceylon.

Medicine.—The LEAVES, or whole HERB, and FRUIT have been valued from remote times by the Natives of Southern India. When agitated with water they render it very mucilaginous without materially altering the taste, colour, or smell of the liquid. Though Muhammadan writers do not appear to mention it, the *Faríd-búti* (or herb on which Shaik Faríd-ud-dín Shakar Ganj sustained life while he acquired the everlasting treasure of knowledge) is supposed to have been this plant. It was first brought to the notice of Europeans by **Rheede** who wrote, "The mucilaginous water

MEDICINE.
Leaves.
364
Herb.
365
Fruit.
366

PEGANUM **Harmala.**	The Farid-buti.

MEDICINE.

derived from the leaves is useful in unnatural heat of the kidneys and in ardor urinæ. It relieves strangury, and dissolves calculi." Ainslie states that the *Vytians* employed the mucilage prepared both from the leaves and fruit in cases of dysuria and gonorrhœa, and adds, "It would seem to be a valuable medicine in all such cases as require mucilaginous mixtures." The plant obtained a place in the secondary list of the Pharmacopœia of India, in which it is remarked on as follows, "Facts communicated to the editor leave little doubt that in cases of gonorrhœa and dysuria it is a remedy of considerable value, and that as a diuretic its action is speedy and marked. Dr. Ives (*Voyage to India, 466*) speaks very favourably of the virtues of this plant, and he adds, to his own testimony, that of Dr. Thomas, as to the power of the mucilage to cure gonorrhœa, without the aid of any other medicine. Water thus rendered mucilaginous soon returns to its original fluidity and it therefore requires to be freshly prepared each time before its exhibition. Its virtues are well deserving of further investigation." Of late years it has been introduced into European medicine as a remedy for spermatorrhœa, incontinence of urine and impotence (*Practitioner, XVII., 381*).

Juice.
367

According to Dr. Emerson the JUICE is used as a local application to aphthæ. Dymock states that in the Konkan a *paushtik*, taken with milk, is made of the powdered fruit with *ghí*, sugar, and spices. Dymock states that the value of the dried fruit in Bombay, to which it is imported from Guzerat and Kathiawár, is R5 per Surat maund of 37½℔.

Seeds.
368

SPECIAL OPINIONS.—§ "Leaves and SEEDS demulcent diuretic; cold infusion freely used as a drink" (*Thomas Ward, Apothecary, Madanapalle*). "Much used as demulcent in gonorrhœa" (*Honorary Surgeon E. A. Morris, Tranquebar*). "The juice of the fruit is used in aphthæ. It is also an emmenagogue; it is employed in puerperal diseases, and to promote the lochial discharge. The leaves are used as a curry in splenic enlargements. Decoction of the root is antibilious" (*Surgeon-Major D. R. Thomas, M.D., C I.E., Madras*). "Demulcent, diuretic, doses ℥ss. to ℥ii used in gonorrhœa, spermatorrhœa, irritability of the bladder and kidney" (*Choonna Lall, Hospital Assistant, Jubbulpore*).

FOOD.
369

Food.—In certain parts of the country butter milk diluted with water is said to be thickened with the LEAVES to make it look rich.

370

PEDICULARIS, *Linn.; Gen. Pl., II., 978.*

A genus of annual or perennial herbs, which belongs to the Natural Order SCROPHULARINEÆ, and comprises about 120 species, natives of northern and alpine regions, with a few South American and South Indian. Of these 37 are regarded by Sir J. D. Hooker (in the *Flora of British India, IV., 306-317*) as Indian. Dr. D. Prain in a recent monograph on the genus has, however, described several new species, making in all 70 (*Jour. As. Soc. Bengal, LVIII. (1889), 255-278*); also *Annals of the Royal Botanic Gardens, Calcutta.*

MEDICINE.
Leaves.
371

Few members of the genus are of economic interest. P. pectinata, *Wall.,* and P. siphonantha, *Don,* are, however, described by Stewart (*Pb. Pl., 162*) as of medicinal value. The pounded LEAVES of the former are employed as an internal remedy for hæmoptysis in Kanáwar, where the plant is known as *michren,* and the species is probably that described by Honigberger as officinal at his time, in Kashmír.

PEGANUM, *Linn.; Gen. Pl., I., 287.*

372

Peganum Harmala, *Linn.; Fl. Br. Ind., I., 486;* RUTACEÆ.

Vern.— *Hurmul, harmal, isband-lahouri, lahouri-hurmul,* HIND.; *Isband* BENG.; *Hurmul, isbund-lahouri, lahouri-hurmul, spelane,* PB.; *Spail*

P. 372

| The Harmal. | (*J. Murray.*) | PEGANUM Harmala. |

anai, PUSHTU; *Spand, spanj, ispanthan*, N. BALUCHISTAN; *Hurmul, isbund-lahouri, lahouri-hurmul*, SIND; *Viláyati-mhéndi, viláyati-isband*, DEC.; *Hurmal, hurmaro, ispand*, BOMB.; *Harmala*, MAR.; *Ispun*, GUZ.; *Hurmaro, viláyati-isband*, DEC.; *Shimai-aravandi-virati, shimai-asha-vanai-virai*, TAM.; *Sima-goronti-vittulu*, TEL.; *Hurmul or harmal*, ARAB.; *Isband** or *ispand,** PERS.

* These names are, in Southern India, incorrectly applied to the seeds of **Lawsonia alba.**

References.—*Dals. & Gibs., Bomb. Fl., 45; Stewart, Pb. Pl., 38; Boiss., Fl. Orient., I., 917; Pharm. Ind., 49; O'Shaughnessy, Beng. Dispens., 260; Moodeen Sheriff, Supp. Pharm. Ind., 195; Mat. Med. of Southern Ind., 84; Dymock, Mat. Med. W. Ind., 2nd Ed., 124; Dymock, Warden & Hooper, Pharmacog. Indica, I., 252; Murray, Pl. & Drugs, Sind, 91; Year-Book Pharm., 1878, p. 288; Irvine, Mat. Med., Patna, 60; Baden Powell, Pb. Pr., 425, 452, 466; Drury, U. Pl., 335; Birdwood, Bomb. Pr., 17; Gasetteers:—N.-W. P., IV., lxix.; Montgomery, 20; Peshawar, 26; Gujrát, 12; Musaffargarh, 27; Rajputána, 30; Settle. Rep., Montgomery, 20; Ind. Forester, IV., 233; XII., App., 9; Agri.-Horti. Soc., India, Journals (New Series), I., 84.*

Habitat.—A bushy herb, 1 to 3 feet in height, much branched, and profusely clothed with leaves; met with in North-Western India from Sind, the Panjáb, and Kashmír, to Delhi, Agra, and the Western Deccan; distributed to Arabia, North Africa, Hungary, and Spain.

Dye.—The SEEDS yield a red dye, employed for browns and blacks, which was, at one time, imported into England from the Crimea. According to Stewart, endeavours were made, some years ago, to send the seeds to Europe from the Panjáb, but it was found that they would have no chance against the cheaper aniline dyes.

DYE.
Seeds.
373

Oil.—Baden Powell mentions the seeds in his List of Oils and Oilseeds of the Panjáb, but, beyond stating that it is employed in medicine, he gives no accounts of its properties or uses.

OIL.
374

Medicine.—In native works on Materia Medica, *harmal* is described as alterative, purifying, aphrodisiac, and lactagogue. The author of the *Makhsan-el-Adwiya* describes it as useful in diseases due to atrabilis, or in those which arise from cold humours, such as palsy, lumbago, &c., and he recommends its administration as a concentrated decoction, mixed with sweet oil and honey, or as an alcoholic preparation formed by boiling the crushed SEEDS in wine (*Dymock*). Stewart writes of the Panjáb: "The burned seeds are in some places applied to the cut navel-string, and other wounds. The seeds are officinal and considered narcotic, being given in fevers and colic. A decoction of the LEAVES is given for rheumatism, and the powdered ROOT with mustard oil is applied to the hair to destroy vermin." Dymock states that the drug has recently been experimented with by Dr. P. Gopal, a medical graduate of the Bombay University. He found that an infusion or tincture acted as a mild emmenagogue, and produced slight intoxication like that of **Cannabis sativa.** He gave the tincture in half-drachm doses to a female suffering from amenorrhœa, with the effect of producing a free menstrual discharge, and found that it was sometimes employed by native midwives to procure abortion. He believed the drug to have properties similar to those of ergot, savine, and rue (*Mat. Med. W. Ind. 125*). *Harmal* occupies a place in the secondary list of the Pharmacopœia of India, in which it is said to be considered anthelmintic, in addition to possessing the properties above detailed.

MEDICINE.

Seeds.
375

Leaves.
376
Root.
377

Moodeen Sheriff (*Mat. Med. of Southern India*) selects the seeds as the most powerful part of the plant, and states that they are narcotic, antispasmodic, hypnotic, anodyne, nauseant, emetic and emmenagogue. He recommends their employment in cases of asthma, hiccough, hysteria,

PENNISETUM.   **Pegs and Poles.**

MEDICINE.

rheumatism, impaction of calculus in the ureter, and of gallstone in the gall-duct, colic, jaundice, dysmenorrhœa and neuralgia; in all of which they relieve pain and procure sleep. The best preparation, he says, is a simple powder prepared in the usual way and kept in a closed vessel. The dose as an anodyne is from thirty grains to two drachms, increased gradually, but in the latter dose it is apt to act as a nauseant and depressant emetic. He sums up his account of the drug in these words, " no hospital should be, in my humble opinion, without a drug so cheap and with so many good qualities as *harmal*." In the Panjáb and Northern India generally, the seeds are frequently burned near the sick with the object of keeping off evil spirits, and averting the evil eye. It has been surmised that the smoke may in reality be antiseptic, thus exercising a beneficial action.

Chemistry.
378

CHEMICAL COMPOSITION.—The seeds contain two alkaloids, *Harmaline*, $C_{13}H_{14}N_2O$, and *Harmine*, $C_{13}H_{12}N_2O$. According to Fritszche, the discoverer of the latter, the yield of the two alkaloids is 4 per cent., of which one-third is *Harmine*, and two-thirds *Harmaline*. The seeds also contain a soft resin of a deep carmine-lake colour having a heavy narcotic odour, like that of the resin of **Cannabis sativa** (*Pharmacog. Indica*).

SPECIAL OPINIONS.—§ " The infusion is administered with benefit in asthma" (*Surgeon-Major A. S. G. Jayakar, Muskat*). " In Peshawar the shrub is collected in great quantity and burnt by the Pathans as a remedy to drive off cholera" (*Surgeon-Major J. E. T. Aitchison, Simla*). " The smoke arising from the burning seeds is used in painful sores, especially venereal, as a sedative" (*Assistant Surgeon Nehal Singh, Saharunpore*). " The seeds are disinfectant and, cooked in oil, are applied to sores, the latter being used while the seeds are thrown away. Wounds are also fumigated by burning the seeds" (*Civil Surgeon R. Gray, Lahore*). " The seeds are generally burnt by the Natives to remove foul smells from sick rooms, especially those of patients suffering from wounds" (*Assistant Surgeon Bhagwan Das, Rawal Pindi*).

PEGS & POLES, Timbers used for—

Acer pictum (shoulder-poles).
Anogeissus latifolia (shoulder-poles).
Bamboo, various species.
Borassus flabelliformis (posts).
Bruguiera gymnorhiza
Butea frondosa (piles).
Cassia Fistula (posts).
Castanopsis rufescens (posts).
Cocos nucifera (ridge poles).
Cupressus torulosa (poles).
Diospyros melanoxylon (shoulder poles).
Excæcaria Agallocha (posts).
Ficus bengalensis (tent & banghy poles).
Fraxinus floribunda (jampan poles).
Grewia tiliæfolia (shoulder poles).
G. vestita (shoulder poles).

Gynocardia odorata (posts)
Heritiera littoralis (posts).
H. Papilio (cart-poles).
Parrottia Jacquemontiana. (tent pegs).
Prosopis glandulosa.
Quercus dilatata.
Q. semecarpifolia.
Rhododendron arboreum (posts).
Salix tetrasperma.
Sesbania ægpytiaca.
Shorea robusta.
S. Tumbuggaia (posts).
Styrax serrulatum (Bhutias' prayer poles).
Taxus baccata (carrying poles.)
Xylia dolabriformis(telegraph posts).
　　&c.,　　　　&c.

(*J. F. Duthie.*)

379

PENNISETUM, *Pers.; Gen. Pl., III., 1105.*

A genus which contains between forty and fifty species, chiefly African. The flowers are arranged in cylindrical spikes, as in SETARIA, but the bristly involucre falls off together with the pedicel as in CENCHRUS. Of the Indian species, about

| The Cumboo Millet. (*J. F. Duthie.*) | PENNISETUM typhoideum. |

five, including the cultivated *bájra* (**P. typhoideum**), occur in the plains, and about the same number are Himaláyan. **P. triflorum** is abundant on the outer ranges, and **P. flaccidum** is characteristic of the inner and drier Himálayan regions. Nearly all the species afford fairly good fodder.

[GRAMINEÆ.

Pennisetum Alopecuros, *Steud. ; Duthie, Fodder Grasses, 17 ;*   380

Syn.—GYMNOTHRIX ALOPECURUS, *Nees ;* CENCHRUS HORDEIFORMIS, *Rottl.*

Vern.—*Mo,* BUNDEL. ; *Moiyar,* RAJ. ; *Mowa, morthan,* C. P.

Habitat.—A coarse perennial grass, plentiful near water in Rájputana and on the black soil of Central India.

Fibre.—On Mount Abu the strong tough LEAVES are collected as fibre for making ropes.

FIBRE. Leaves. 381 382

P. cenchroides, *Rich. ; Duthie, Fodder Grasses, 17 ; Pl. XII., XIII.*

Syn.—CENCHRUS CILIARIS, *Linn.*

Vern.—*Baiba, kusa,* BUNDEL. ; *Kusa, charwa,* N.-W. P. ; *Anjan, dhá-man, kurkán,* PB. ; *Taura,* PUSHTU ; *Andho, dhoula, bharbhunt,* RAJ. ; *Kusa, gulni,* TEL.

References.—*Stewart, Pb. Pl., 259 ; Murray, Pl. & Drugs, Sind, 10 ; Buden Powell, Pb. Pr., 245 ; Atkinson, Him. Dist., 320 ; Duthie & Fuller, Field & Garden Crops, 10 ; Coldstream, Fodder Grasses of S. Punj., No. 11 ; Mueller, Select Extra-Trop. Pl., 284 ; Gazetteers :—C. P., 505 ; N.-W. P., IV., lxxix. ; Ind Forester, X., 220 ; XII., App., 23.*

Habitat.—A perennial tufted grass, usually with decumbent stems, common all over the plains of North-Western India, where the soil is sandy. Not distinguished by natives from **Cenchrus montanus**, a grass occurring in similar localities, and known also under the names of *anjan* and *dháman.* It is probably often confused also with **C. pennisetiformis,** a common grass which becomes subscandent when growing amongst bushes.

Food & Fodder.—Edgeworth mentions that near Multán the GRAIN is swept up from the ground to be used as human food. In the Panjáb this is generally considered to be one of the best of fodder grasses for both horses and cattle. But in the Montgomery district it is stated that horses will not eat it because of its bitterness ; this, however, may prove to be due to some local peculiarity of the soil. In many districts, where *dub* (**Cynodon Dactylon**) is scarce, this is reported to be the best of all grasses for increasing the milk of cows. In the Jhang Settlement Report it is stated that zamindars believe that it is liable to give a semi-intoxicating property to the milk of buffaloes.

FOOD & FODDER. Grain. 383

P. typhoideum, *Rich ; Duthie, Fodder Grasses, 18.*   384

THE BULRUSH, CUMBOO or SPIKED MILLET ; DEKKELÉ, *Fr.*

Syn.—HOLCUS SPICATUS, *Linn.* ; PANICUM SPICATUM, *Roxb.* ; PENICIL-LARIA SPICATA, *Lindl.*

Vern.—*Bájra bájri, bajera, lahra, kasa-jonar,* HIND. ; *Jondhariya, gahuma,* BIHAR ; *Lendha,* SANTAL ; *Bajra-mula,* URIYA ; *Bájra, bájri, lahra, bájra tangunanwa,* N.-W. P. ; *Bájra,* KUMAON ; *Bájra, bajza,* PB. ; *Bajaro,* SIND ; *Bájri, bajera, bájra,* BOMB. ; *Kambu,* TAM. ; *Gan-telú, sajjalú, pedda-ganti, sazza,* TEL. ; *Sajje,* KAN. ; *Mattari,* MALAY.

References.—*Boiss., Fl. Orient., V., 447 ; Roxb., Fl. Ind., Ed. C.B.C., 95 ; Trimen, Sys. Cat. Cey. Pl., 108 ; Dalz. & Gibs., Bomb. Fl., Suppl., 99 ; Stewart, Pb. Pl., 259 ; Aitchison, Cat. Pb. & Sind Pl., 163 ; Sir W. Elliot, Fl. Andhr., 58, 165 ; Bidie, Cat. Raw Pr., Paris Exh., 69 ; Bird-wood, Bomb. Prod., 112 ; Baden Powell Pb. Pr., 238, 383 ; Drury, U. Pl. Ind., 335 ; Atkinson, Him. Dist. (X., N.-W. P. Gaz.), 690 ; Useful Pl. Bomb. (XXV., Bomb. Gaz.), 185 ; Econ. Prod. N.-W. Prov., Pt. IV. (Cultivated Food-Grains), 18 ; Royle, Ill. Him. Bot., 421 ; Mueller, Select*

P. 384

PENNISETUM
typhoideum.   Cultivation of the

Extra Trop. Pl., 285; *Nicholson, Man. Coїmbatore, 218, 219, 229; Moore, Man. Trichinopoly, 71; Gribble, Man. Cuddapah, 43; Grierson, Bihar Peasant Life, 225; Settlement Report:—Panjab, Gujrat, App. xxxviii.; Hazára, 88; Jhang, 85, 92, 93; Montgomery, 107; Lahore, 10; Kángra, 25; N.-W. P., Allahabad, 25; Bareilly, 82; Central Provinces, Chanda, 88; Upper Godavery, 35; Land Rev. Settle., Nursingpore, 52; Bomb. Man., Rev. Accts., 101; Rep. Internal Trade, Punjab, 1884-85, 29; Rep. Agr., Station Cawnpore, 1884, 7; Gazetteers:—Bombay, II., 63, 66, 269, 273, 277, 280, 284, 287, 390, 536, 541, 544, 547; IV., 54; V., 29; VIII,, 182; XII., 149; XVI., 91, 98; XVII., 264; Panjáb, Karnal, 172, 178; Hoshiárpur, 94; Kángra 69; Bannu, 137, 142; Déra Ismail Khán, 126; Montgomery, 110; Rohtak, 94; Sháhpur, 65; Rawal Pindi, 81; Kohát, 105; N.-W. P., I., 57, 85, 152, 198; III., 225; IV., lxxix.; Mysore & Coorg, II., 11; Balfour, Cyclop. Ind., III., 178.*

Habitat.—A tall erect grass, 5 to 6 feet high, most probably of African origin. The spikelets of flowers are crowded into a compact, cylindrical spike, sometimes over a foot long, each spikelet being girt at the base with an involucre of bristles. Mr. Bentham, in his *Notes on Gramineæ* (*Linn. Soc. Journ., XIX., 49*), says:—" The involucres sometimes remain persistent after the spikelet have fallen away, and the filiform styles are remarkably long; but many cultivated specimens, and some East African ones, possibly wild, offer so much variety in those respects, that it seems probable that the peculiarities of habit have arisen from long cultivation." It is largely cultivated in India as a rainy season crop, especially in the Bombay Presidency. It is also grown in Egypt, but the heads are much smaller.

Cultivation.
385

CULTIVATION.—Though grown to a less extent than *juár*, its cultivation is geographically very similar, comprising extensive tracts in North-West, Central, and Southern India. In the Bombay Presidency the area occupied by *bájra* is considerably over four millions of acres, and in many parts it forms the staple food of the people. It is extensively cultivated in Khándésh, Guzarát, and the Deccan. In the Nasik district it is grown in red or *mál* lands mixed with *tur* (Cajanus indicus). The seed is sown in June and the crop ripens towards the end of September. It requires very little artificial irrigation, but if grown on garden land it is sometimes watered from a well. It is the most important *kharif* crop in Sind. Some interesting cultural experiments have been conducted within the last few years at the Hyderabad Farm of that province, one result of which tends to show the advantage of selecting seed for sowing from the middle portion of the spike only.

In the Madras Presidency, according to the statistical returns for 1888-89, the area under *kambu* was nearly 2⅓ millions of acres. · In the Godávari District it is grown on *regada*, *lanka*, and sandy soils. The ground is manured in March by means of cattle or sheep picketed upon it; and in May it is ploughed twice every six days for three weeks. The ground then, as well as at the time of sowing, should be moist. The seed, which should be of the previous year's produce and carefully preserved for the purpose, is sown in June at the rate of four seers to the acre. The heads are cut off when ripe, *i.e.*, about three months after sowing, they are left to dry for three or four days, and are then threshed in the usual manner (*Settlement Rep., p. 137*). This crop is grown pretty generally in the Coimbatore District in garden lands, the garden variety being *arisi* or *aruvathan* (60 days) *kambu*. As a rule, it is sown in April and harvested in June-July. The cultivation is similar to that of *cholam* (Sorghum vulgare). It is cultivated throughout the district on dry lands, except black. It is not sown, as a rule, after the middle of September, for the north-east monsoon would be prejudicial, but in parts of Coimbatore taluk and elsewhere it is occasionally sown in November. Being the staple crop of the

Cumboo Millet.	(*J. F. Duthie.*)	**PENNISETUM typhoideum.**

CULTIVATION

district, upon which both *rayats* and labourers depend for food, the outturn of this crop is all important. It is, however, a delicate crop, deficient or excessive rain being equally prejudicial, while failure at particular crises, or rain coming when the plant is in flower, is very fatal. A *kambu* crop is seldom good on the low-class dry lands, both stalk and ears being poor. It is often grown as a mixed crop with cotton, castor-oil, pulses, &c. In the sub-division, especially in the Erode taluk, lines of *mochai*, castor, &c., are marked features in a *kambu* field, and it is to these that the *rayat* looks to pay his *kist*, while he uses the *kambu* for his family and labourers.

The plagues of *kambu* are *navei-puchi* and *asugani*, both of insect origin, and they occasionally cause immense loss. When properly cultivated, the land is ploughed in the April-heavy rains, after having been manured in the usual way according to the *rayat's* means. In July-August it is again ploughed, and the seed, mixed with various pulses, is sown broadcast, with *dál* or beans in rows of a few feet apart. After six weeks the crop is interploughed and occasionally weeded, partly for the sake of the crop, and partly for the value of the resulting fodder. In November-December the *kambu* is reaped by cutting off the ears as they ripen, the harvest thus extending over several weeks, since the crop matures in stages according to the nature and condition of the soil, seed, &c. The pulses are gathered gradually up to February, when the stalks of *kambu*, beans, &c., are all pulled up together, leaving the cotton and castor only. The practice is the same in all ordinary lands, but manure seldom reaches the dry lands of the poorer *rayats*, except that of the casual droppings of cattle in search of food, and occasional village refuse (*Extract from Nicholson's Manual of the Coimbatore District, pp. 218-220*).

Bájra is largely grown in the districts of the Panjáb, and in the dry elevated tracts south of Rawal Pindi, it constitutes the principal *kharif* crop. In Dera Ghází Khán it is the most important crop after wheat. In favourable years the yield is enormous, but it often suffers from drought or excessive rain; it is also occasionally injured by blight and locusts, and late sowings are sometimes spoilt by the north-west wind, which dries up the ears (*Gazetteer, p. 126*). It is the staple crop of the autumn harvest in the Rawal Pindi District. It thrives everywhere in the plains at the base of the Salt Range, and in unfavourable seasons is replaced by *til* (Sesamum indicum) or some of the pulses. In the Karnál District it thrives best on a sandy soil, and is sparsely sown immediately after the first rainfall, mixed with the seed of some small pulse. Rain falling on the pollen spoils the crop, but nothing else affects it (*Gazetteer, p. 178*). In Kohát it is sown between April and August, and is the principal *kharif* crop on the *barani* lands of the district. The crop is generally cut in October, at whatever time sown. Flocks of small birds do much injury in August and September (*Gazetteer, p. 105*). In the Bannu District it is largely grown on the stiffer *thal* soils. It is sown at the rate of 2 to 5 seers per acre, from the middle of March till the end of July; that which is earliest sown affording the best hope of heavy returns. The first sown crops ripen about the middle of August, and ears are plucked as they ripen till early in December (*Gazetteer, p. 142*).

In the North-West Provinces *bájra* is extensively cultivated, especially in the western districts. It is sown a little later and reaped a little earlier than *juár* (Sorghum vulgare); and is occasionally grown on land intended for *juár*, should the sowing time be delayed in consequence of floods or failure of rain. It is generally mixed with some minor crop, such as *moth* (Phaseolus aconitifolius). If *juár* be taken as the *kharif* counterpart of wheat, *bájra* may be still more aptly compared with barley. Like barley

PENNISETUM
typhoideum.   Cumboo Millet.

CULTIVATION | it often occupies very good as well as very bad soil, but, as a general rule, it is the crop of poor light-soiled outlying land, and requires perhaps rather less rainfall than *juár*. It is never manured, and but rarely irrigated. The land is ploughed once to four times, and the seed, mixed with that of the subordinate crops, is sown broadcast and ploughed in at the rate of 2½ to 3 seers per acre. There should be at least one weeding, if possible, but the place of this is often taken by ploughing up the ground between the plants, exactly as is done to *juár*. The crop should be watched if possible to keep off birds and squirrels for about twenty days before it is cut. The grain ripens towards the beginning of November, when the heads are cut off and carried to the threshing floor, the stalks being frequently left standing on the ground for some weeks. Next to an absolute failure of rain *bájra* suffers most from damp or rainy weather when in flower. It is sometimes attacked by a kind of bunt, and is reported to be often infected with a species of mildew called *bagulia* (**Puccinia penniseti**, *Barclay*). But it owes its liability to these diseases in great measure to the poverty of the soil on which it is cultivated, and the mildew alluded to above is said to be most destructive in cases where *bájra* has been grown too frequently on the same land. The cost of cultivation has been estimated at R9-8 per acre, including R3 for rent of land. The outturn varies from 5½ maunds in the damper localities to 7 maunds in places where the climate is drier. The area under *bájra* and *bájra-arhar* (*bájra* mixed with **Cajanus indicus**) in the thirty temporarily-settled districts of the North-West Provinces has been given as not far short of two million acres (*Field & Garden Crops, North-Western Provinces & Oudh, Part I., p. 30*).

In Rájputana, *bájra* is the principal crop west of the Aravalli range. It thrives well in the Bikanír district, yielding on an average five maunds to the acre.

DYE. Ashes. 386

Dye.—The ASHES of this plant are used as an alkali in dyeing.

FOOD & FODDER. Grain. 387

Food & Fodder.—The GRAIN is used chiefly by the lower classes of Natives, and in many parts of India it is their principal food. It is supposed to be heating, and is, therefore, in Northern India, consumed mostly in the cold weather. The flour, made into cakes or bread with butter-milk, is the staple food of many, and is more nutritious than rice. In Khándesh it is often eaten with butter and various condiments by the well-to-do. The following is the composition of a sample of the unhusked grain according to an analysis made by **Professor Church**:—

									In 100 parts.
Water	.	.	.	.	.	.	.	.	11·3
Albumenoids	.	.	.	.	.	.	.	.	10·4
Starch	.	.	.	.	.	.	.	.	71·5
Oil	.	.	.	.	.	.	.	.	3·3
Fibre	.	.	.	.	.	.	.	.	1·5
Ash	.	.	.	.	.	.	.	.	2·0

The nutrient ratio being 1 : 7·6, and the nutrient value 89½. The grain contains ·42 per cent. of potash and ·68 per cent. of phosphoric acid.

Stalks. 388

In the Panjáb *bájra* is occasionally grown for fodder; and in unfavourable seasons, although sown as a grain crop, it is sometimes given to cattle as green fodder. The STALKS, after the grain has ripened, are more or less utilised as fodder in various parts of India. They are seldom used, however, where *chari* (*juár* straw) is available, whilst in some districts they are thrown away as useless. In the Gazetteer of the Karnál District it is said that in some of the higher villages, the stalks, called *dándar*, are carefully stacked and preserved, and when required for use are chopped up and given to the cattle with green fodder, or even with the *áta* of

grain. In parts of the Bombay Presidency, the stalks (*sarmad*) are given to cattle, but unless trodden into chaff they are said to be very inferior as fodder. They are much more generally used for fodder in the Madras Presidency, where very probably they contain more nutriment than in Northern India; so also with rice-straw, which appears to be much more valued as fodder in Southern India than in the North.	FOOD & FODDER.
Trade.—This grain is to some extent exported to other countries by sea, but the trade returns include it under *juár*.	TRADE. 389

(*J. Murray.*)

PENTACE, *Hassk.; Gen. Pl., I., 231.*

Pentace burmannica, *Kurz; Fl. Br. Ind., I., 381;* TILIACEÆ.

390

Vern.—*Thitka, kathitka,* BURM.
References.—*Kurz, For. Fl. Burm., I., 154; Gamble, Man. Timb., 52; Ind. Forester, I., 207; XII., 73; XIII., 134.*
Habitat.—A tall evergreen tree of Burma, frequent in the tropical forests of the eastern and southern slopes of the Pegu Yomah, and from Martaban down to Upper Tenasserim.

Resin.—Kurz states that the tree exudes a red resin; nothing is, however, known of its properties.

RESIN. 391

Structure of the Wood.—Yellowish-red, shining, soft, even-grained, takes a good polish; average weight 42℔ per cubic foot. It is largely used in Burma for boats, boxes, and other purposes for which a light wood is required. Considerable quantities are annually exported both to Assam and to Europe. In the former country it is employed for making tea boxes.

TIMBER. 392

PENTAPETES, *Linn.; Gen. Pl., I., 222.*

Pentapetes phœnicea, *Linn.; Fl. Br. Ind., I., 371;* STERCULIACEÆ.

393

Syn.—DOMBEYA PHŒNICEA, *Cav.*
Vern.—*Dopahariya, dopohoriá,* HIND.; *Kát-lálá, bándhuli,* BENG.; *Bare baha,* SANTAL; *Gul dupaharia,* PB.; *Tambri-dupári,* MAR.; *Nága-pu,* TAM.; *Bandhuka, bandhujiva, bandúkamu, arka-vallabha, pushparakta,* SANS.
References.—*Roxb., Fl. Ind., Ed. C.B.C., 511; Rheede, Hort. Mal., X., t. 56; Stewart, Pb. Pl., 25; Campbell, Ec. Prod. Chutia Nagpur, No. 9882; Elliot, Fl. Andhr., 22; U. C. Dutt, Mat. Med. Hindus, 293; Pharmacog. Indica, I., 236.*
Habitat.—Found in rice-fields and other wet places, during the monsoon, throughout the hotter parts of India from the Panjáb to Burma and the Western Peninsula.

Medicine.—The CAPSULES are used medicinally on account of their mucilaginous properties, and according to Stewart were at one time officinal in the Panjáb. Campbell states that the ROOT is employed as a medicine by the Santals.

MEDICINE. Capsules. 394
Root. 395

PENTATROPIS, *Br.; Gen. Pl., II., 764.*

[*Wight, Ic., t. 352;* ASCLEPIADEÆ.

Pentatropis microphylla, *Wight & Arn.; Fl. Br. Ind., IV., 20;*

396

Syn.—ASCLEPIAS MICROPHYLLA, *Roxb.;* CYNANCHUM ACUMINATUM, *Thunb.;* EUTROPIS (GEN. NOV.), *Falconer.*
Vern.—*Parparam,* TAM.; *Chékurti tivva, púla pála,* TEL.
References.—*Roxb., Fl. Ind., Ed. C.B.C., 253; Dalz. & Gibs., Bomb. Fl., 149; Grah., Cat. Bomb. Pl., 120; Elliot, Fl. Andhr., 36, 158; Gazetteers:—Bombay, IV., lxxiv.; V., 27.*
Habitat.—A small, delicate, twining perennial, met with in the Sundarbans, the Deccan Peninsula from Bombay southwards, and Burma. It is similar to the following in habit and characters.

PEORI **Dye.**	The Peori or

397

Pentatropis spiralis, *Dcne.; Fl. Br. Ind., IV., 19.*

Syn.—P. MICROPHYLLA, *Wall.;* OXYSTELMA CAUDATUM, *Ham.;* ASCLEPIAS
  SPIRALIS, *Forsk.*

Vern.—*Ambarvel, van veri,* (bazar flower=*ark pushpi*), PB.; *Singarota,*
  BOMB.; *Shigaroti,* GUZ.

Reference.—*Stewart, Pb. Pl., 146.*

Habitat.—A slender climber, found in the Panjáb, Sind, and the country
extending eastwards to the Jumna; distributed to Afghánistan, and westwards to the Red Sea and Nubia.

MEDICINE.
Flowers.
398
Roots.
399
FOOD.
Tubers.
400

Medicine.—Stewart writes, " The FLOWERS of this are apparently
officinal in the Panjáb under the above name." According to a note
received from Sakharam Arjun Ravat the dry ROOTS given in decoction
are considered astringent.

Food.—The small TUBERS which grow on the roots in spring are
peeled and eaten in the Panjáb. Stewart states that they are " sweet and
filling."

PEORI DYE.

401

Peori Dye.

References.—*McCann, Dyes & Tans of Bengal, 92; T. N. Mukharji,
  Rept. on Peori dye, Aug. 27th, 1883, republished in Jour. Soc. of Arts,
  XXXII., 16 (1883); M. C. Græbe, Report on Indian Yellow; Archives des
  Sciences Physiques et Naturelles, Series III., xxii., 497.*

DYE.
402

Dye.—This curious yellow dye-stuff has long attracted the attention
of chemists and physiologists in Europe. In India it has always been
reputed to be made from the urine of cows fed entirely on mango leaves,
a fact which was, however, doubted in England till quite recently. The
ideas previously held regarding its origin were very numerous. Some
maintained, like the Natives of India, that it was of animal origin, and
probably a deposit formed in the urine of the camel, the elephant, or the
buffalo, or an intestinal concretion. In the early part of the last decade
enquiries were made at the Royal Botanic Gardens, Kew, to have the
question cleared up, and as a result **Mr. Mukharji** was deputed to investigate the matter. Since the publication of his report by the India Office
in the *Journal of the Society of Arts,* the origin of *peorí* from the urine
of cows has become an established fact.

It is stated that the dye is rarely employed in India to dye fabrics on
account of the offensive odour of cow's urine which the cloth retains, but
is chiefly used as a pigment for colouring doors, railings, walls, &c. It is
made entirely by inhabitants of a village called Mirzapur near Monghyr,
and is prepared as follows :—

The cows, from the urine of which the dye is to be manufactured, are
fed exclusively on mango leaves and water, with the result that the colouring matter of the urine becomes greatly increased. It is said that the
cows thus treated die after two years, but according to Mr. Mukharji, the
cow-keepers informed him that this was false, and he himself has seen
cows six or seven years old which had furnished *peorí* for at least four
years. These cows, however, appeared very ill and poorly nourished, and
the villagers stated that it was necessary to give them other food from
time to time, but that this admixture at once reduced the quantity of
peorí in the urine. Owing to this the keepers who trade in the dye are
despised as cow-destroyers.

The cows are trained to urinate at certain times only, and are said to
become so habituated to this, that they are incapable of doing so without
the presence of their masters. The urine is collected, allowed to cool for
some time, and then heated. By this means the yellow colouring matter

Yellow Dye.	(*J. Murray.*)

is precipitated. It is then collected and partly dried in cloth. The sediment is then rolled into small balls, which are first of all dried by means of a charcoal fire, afterwards by being placed in the rays of the sun. It is sold in this form to merchants, who also advance money to the cow-keepers. It sells at from R4 to R2 per seer, and is taken by the merchants to Patna and Calcutta, where it is said to fetch R8 per seer. The high price of the dye is owing to the early death of the cows, and to the comparatively expensive mango leaves, of which the quantity produced by a tree of moderate size (about 20 feet) costs about R2. One cow is said to furnish daily three quarts of urine, and 2 ozs. of *peori*. It has been said that the annual production of this substance is as much as from 10,000 to 15,000℔; but **Mr. Mukharji** states that this estimate is too high, when one takes into consideration the small number of cows employed to produce it. The trade seems to be profitable, for **McCann** informs us, one or two dealers in *peori* have paid income tax on the profits.

METHOD OF DYEING.—"The *peori* is dissolved in hot water and the cloth dipped in it. One and a half *chittacks* of *péorí* will dye one yard of cloth a good yellow." As already stated, however, the great objection exists, that the dye imparts an exceedingly offensive odour to cloth, which is retained even after repeated washings. In Europe it is chiefly used as a pigment for water colour drawing under the name of "Indian Yellow."

CHARACTERS AND CHEMICAL COMPOSITION.—*Peorí* consists of small roundish balls weighing from 2 to 4 ozs., of which the interior has a fine yellow colour, while the outer coat is brown or greenish. Its odour is very characteristic and recalls that of castor. The interior yellow substance has been carefully analysed by **M. Græbe**, who found that it contained:—*Euxanthinic acid*, 51; *magnesium*, 4·2; *calcium*, 3·4; *silica* and *albumina*, 1·5; *water* and *volatile matter*, 39 per cent. Commenting on this result he remarks that in preparing the different qualities of pure Indian yellow for use in water colour drawing, the crude *peori* is submitted to a series of washings. The best qualities are those which are most rich in euxanthinic acid and magnesium, and which contain only a very small quantity of calcium. The more ordinary kinds contain, in addition to the acid above mentioned, one of its products of decomposition, namely, *euxanthone*, in large quantities. Thus **M. Græbe**, in analysing seven distinct commercial forms of Indian yellow, found euxanthinic acid to vary from 72·3 per cent. in the best, to 33·34 per cent. in the worst, while euxanthone varied from 0 per cent. in the former to 33 per cent. in the latter. Of magnesium 5·35 per cent. was found in the best quality, and, like euxanthinic acid, gradually diminished in the inferior qualities, being found only to the extent of 3·7 in the worst.

The euxanthinic acid is always found in the condition of a salt, and the composition of the best quality of Indian yellow approaches to that of the euxanthonate of magnesium,—$C_{19}H_{16}O_{11}Mg + 5\ H_2O$; or euxanthinic acid 78·16, magnesium 4·57, and water 16·67 per cent. Thus purified Indian Yellow contains a little less organic, and more inorganic, matter than the pure salt.

In the inferior qualities euxanthone exists partly free, partly as a salt of magnesium or calcium. Thus in the worst quality **M. Græbe** found 20 per cent. in the free state, 13 per cent. as a salt. The first scientific researches on Indian Yellow were published in 1844 by **Stenhouse**, and by **Erdmann**. These chemists proved that the composition of euxanthinic acid was as above described. The free acid has a pale yellow colour, but all its salts are of a deep yellow tinge, and of these the salt of magnesium

CHEMISTRY.

is remarkable for its beauty. By heating the acid with other acids or with water it becomes decomposed into two substances, of which one is colourless, the other yellow. The first of these is chiefly of physiological interest, being an acid which is formed in the body of man and other animals. This substance, *glycuronic acid*, is one of a series of compounds which appears in the urine of man and of the dog, after the administration of several organic drugs such as camphor or chloral. These latter, like euxanthinic acid, are all complex bodies, which decompose easily, and in so doing set the glycuronic acid at liberty, thus:—

$$C_{19}H_{18}O_{11} = C_6H_{10}O_7 + C_{13}H_8O_4.$$

Euxanthinic acid = glycuronic acid + euxanthone.

Glycuronic acid may be considered a derivative of saccharine matter, a product of oxidation intermediate between glucose and saccharic acid. It has not been possible, however, up to the present, to produce it artificially. The coloured product obtained by the decomposition above mentioned is *euxanthone*, an aromatic compound. This substance, when given to a dog or rabbit, can be recovered in the urine as euxanthinic acid, having become combined in the body with glycuronic acid. From this fact it would appear that mango leaves contain either euxanthone, or a body capable of being transformed into that substance in the organs of the cow, and which again combining with glycuronic acid produces the colouring matter. It is difficult to ascertain from whence the magnesium, which plays such an important part in the composition of Indian Yellow, is derived. It may exist in considerable quantity in the water, or in the mango leaves which form the entire nourishment of the *peori*-yielding cows. But **M. Græbe** considers that it may probably by mixed artificially with the food, though the circumstance has not been recorded.

Euxanthone is a beautiful yellow compound which sublimes and crystallines in long needle-like crystals. Until recently it had never been prepared synthetically, but **M. Græbe** has succeeded in obtaining euxanthone "in every way identical with that of Indian Yellow" by combining hydroquinone and resorcine. He states that euxanthone can, up to a certain point, be compared to alizarine. Both have basic properties, and the salts of both are distinguished by their fine colouration (*M. Græbe, Arch. des Sc. Phys. et Nat.*). It would thus appear that the only thing now wanting in chemical knowledge to admit of Indian Yellow being artificially made is a method of synthetically preparing glycuronic acid.

Pepper, Bell, see **Capsicum grossum,** *Willd.;* Vol. II., 139.

Pepper, Black, see **Piper nigrum,** *Linn.;* p. 260.

[II., 137.

Pepper, Cayenne, Goat or Spur, see **Capsicum frutescens,** *Linn.;* Vol.

Pepper Red, see **Capsicum annuum,** *Linn.;* Vol II., 134.

Peppermint, see **Mentha piperita,** *Linn.;* Vol. V., 229.

405

PERFUMES & PERFUMERY.

References.—*Ain-i-Akbari, Blochmann's Trans.,*73; *J. H. van Linschoten, Voyage to the East Indies, Ed. Burnell, Tiele & Yule, II.,* 95, 96; *Forbes Watson, Industrial Survey, II.,* 139-143; *Stewart, Pb. Pl., Index,* 99; *Baden Powell, Pb. Prod.,* 926; *Aitchison, Fl. Kuram Valley,* 25; *U. C. Dutt, Mat. Med. Hindus,* 311; *Indian Agriculturist: April 27th,* 1889; *August 24th,* 1889; *November 2nd,* 1889; *Bomb. Admn. Rep.,* 1871-72, 375, 376; *Bomb. Gazetteer, VIII.,* 26.

Perhaps no branch of industry purely devoted to the supply of luxuries is of such importance in the East as that of Perfumery. Per-

fumed oils for dressing the hair and beard and for anointing the
body, perfumed waters for the bath, and, in the case of the rich, perfumed vapours for scenting the atmosphere of rooms, are all largely
employed by all classes who can afford them, and have been so from the
earliest times. Thus we find **Abul Fazl** in the *Aín-i-Akbári* devoting
many pages to a consideration of the subject, and giving receipts for the
perfumes held in greatest estimation during the reign of Akbar. He commences the chapter as follows : "His Majesty is very fond of perfumes,
and encourages this department from religious motives. The court-hall
is continually scented with ambergris, aloewood, and compositions according to ancient recipes, or mixtures invented by His Majesty ; and incense is
daily burnt in gold and silver censers of various shapes, whilst sweet
smelling flowers are used in large quantities. Oils are also extracted from
flowers and used for the skin and hair " (*Aín-i-Akbari, Blochmann's Trans.*,
73-82). He then proceeds to give a list of sweet-scented flowers, and other
substances used in perfumery, most of which will be found in the list
below.

Perfumes are manufactured and sold in India in the form of scented
oil—called *atar*—not scented spirit or essence as in Europe. The sweet oil
generally employed as the basis of the perfume is *til*, but that of sandal
wood is also occasionally used. The method of extraction, which it will be
seen is similar to that familiar to European perfumers under the name of
enfleurage, is as follows :—

A layer of the flowers from which the perfume is to be derived is
placed on a clean masonry floor, to a thickness of half an inch ; over this
is spread a layer of *til* seed a quarter of an inch thick, then another
layer of flowers and another of seed, and so on till eight or ten altogether have been laid down. This heap is left all night ; next morning
the flowers are removed, and the seeds are left all day to dry in the sun-
shine. In the evening the process is repeated, layers of fresh flowers are
laid down alternating with layers of the now partly perfumed seeds, and
this goes on for ten days. The scented *til* is then put in bags or jars,
till sufficient seed has been collected to press conveniently. It can be
kept thus stored for a long time without losing the scent. As a rule, it is
laid by for about a year, since the pressing can be done at any time
under cover, while the perfuming can be carried on only during the dry
season in the open air.

The scented seeds are ground in a *kolhu* or native press, a rough
wooden mill consisting of the lower part of the trunk of a tree, fixed in the
ground. This has the top hollowed into a large round cup, like a large
mortar, in which rests the *lat*, a huge wooden pestle to which a bullock is
harnessed. As he goes round the *lat* leans over the side of the hollow in
the *kolhu*, pressing the seed. The oil thus extracted flows out through a
small channel into a vessel set to catch the liquid. The oil after being
purified is decanted into *kúpís,* or leather bottles, of various shapes and
sizes (*Ind. Agri., April 27th, 1889*).

Attar of Roses and other essential oils are made by a rude process of
distillation, for an account of which the reader is referred to **Rosa**. Besides
these perfumed oils, and pure essential oils, many substances of a solid
nature are largely employed in Indian perfumery, such as certain gums,
resins, woods, musk, and other animal products.

The following list, classified according as they are of animal, vegetable,
or mineral origin, will be found to contain the chief perfumes, and substances from which perfume is obtained in India. For an account of the
special properties, and methods of preparation of these, the reader is referred to the article on each in its respective position in this work.

VEGETABLE.
406

LIST OF PERFUMES AND PERFUME-YIELDING PRODUCTS.
I. Vegetable.

Abir. A perfumed powder which is mixed with the red *Gulál* and used at the *Holi* festival; see Vol. I., 6-7.

Acacia Farnesiana, *Willd.*; THE CASSIE FLOWER. The round yellow heads are much used in perfumery; see Vol. I., 49.

Acorus Calumus, *Linn.*; THE SWEET FLAG. This yields an essential oil employed by perfumers in Europe; see Vol. I., 99.

Andropogon citratus, *DC.*; THE LEMON GRASS; see Vol. I., 242.

A. muricatus, *Retz.*; THE KHUS-KHUS GRASS; see Vol. I., 245.

A. Nardus, *Linn.*; THE CITRONELLA. A scented grass, largely exported from Ceylon; see Vol. I., 247.

A. Schœnanthus, *Linn.*; THE GERANIUM GRASS; see Vol. I., 250.

Aquilaria Agallocha, *Roxb.*; CALAMBAC, AGALLOCHUM or ALOE WOOD; see Vol. I., 279-281.

Arnebia Griffithii, *Boiss.*; *Fl. Br. Ind.,* IV., *176.* This species is said by Dr. Forbes Watson to be used in perfumery in the Panjáb, where it is known as *paijhambári-phul, sparhi-gul,* or *mumannie.*

Artemisia vulgaris, *Linn.*; INDIAN WORM-WOOD. Used as a perfume for clothes, &c., also in the preparation of an *atar*; see Vol. I., 328.

Balsamodendron; various species, which yield the MYRRHS, BDELLIUMS, and GUM GUGUL of commerce. The two first are very largely employed in the East in the preparation of perfumes; see Vol. I., 365-370.

Blumea balsamifera, *DC.,* and **B. densiflora,** *DC.,* yield camphor; see Vol. I., 158, Vol. II., 85.

Boswellia; various species, which yield the true FRANKINCENSE or OLIBANUM of commerce. Many forms, obtained from different species, are highly prized in Eastern perfumery; see Vol. I., 511-517.

Cananga odorata, *H. f. & T.*; THE ILANG-ILANG of European perfumers. One of the most valuable and highly prized of Indian perfumes; see Vol. II., 94.

Carum Carui, *Linn.*; THE CARAWAY. The volatile oil is employed in the manufacture of cheap essences; see Vol. II., 197.

C. copticum, *Benth.*; THE BISHOP'S WEED, or AJWAIN. The fruit affords an aromatic volatile oil which, on evaporation, yields a stereoptene, sold as a perfume in Western India, under the name of *ajwain-ka-phul;* see Vol. II., 199.

Caryophyllus aromaticus, *Linn.*; THE CLOVE. The oil is extensively utilised in perfumery; see Vol. II., 203.

Chrysanthemum coronarium, *Linn.* A perfume called *Sounti Atar* is prepared from the fragrant flowers; see Vol. II., 272.

Cinnamomum Camphora, *Nees.*; THE JAPAN CAMPHOR of commerce is obtained from this tree; see Vol. II., 84-93, 317.

C. glanduliferum, *Meissn.*; THE NEPAL CAMPHOR WOOD has a highly scented wood; see Vol. II., 17.

C. zeylanicum, *Breyn.*; THE CINNAMON; see Vol. II., 324.

Citrus Aurantium, *Linn.,* vars. **Aurantium** proper, and **Bigaradia.** These yield the *Neroli* oil of European commerce; see Vol. II., 345-347; var. **Bergamia;** THE BERGAMOT ORANGE; yields the oil of Bergamot; see Vol. II., 347-348.

Citrus Medica, *Linn.* All the varieties of this species are used in perfumery, but the most important are **var. Medica** proper, the Citron, and **var. Limonum,** the Lemon, the essential oils of which are both extensively utilised; see Vol. II., 351, 352.

Clerodendron inerme, *Gærtn.* An exquisite perfume is said to be derived from the flowers; see Vol. II., 372.

P. 406

Costus speciosus, *Sm.;* The Kust root; may be used as a substitute for Orris root, but it is possible that the economic literature on the subject confuses this with two species of **Saussurea** ; see Vol. II., 579.

Crocus sativus, *Linn.;* The Saffron; is largely used in India in the preparation of *Jufrán atar ;* see Vol. II., 592.

Cucumis sativus, *Linn.;* The Cucumber. A perfume called *Sasa atar,* made from some part of this plant, was exhibited by Dr. Kanny Lall Dey at the Colonial & Indian Exhibition ; see Vol. II., 632.

Cupressus torulosa, *Don.* The wood yields a resin often burned as incense ; see Vol. II., 646.

Curcuma aromatica, *Salisb.;* The Wild Turmeric. The powdered rhizome is employed by Hindu women as a fragrant cosmetic ; see Vol. II., 657.

Cyperus rotundus, *Linn.* An essential oil obtained from the rhizomes is used in Upper India to perfume clothes ; in Bengal the dried and powdered root is used as a perfume, and in dyeing to impart a scent to the fabric ; see Vol. II., 686.

C. scariosus, *R. Br.,* is similarly employed ; see Vol. II., 687.

C. stoloniferus, *Retz.* The tubers of this species are also used in perfumery ; see Vol. II., 688.

Didymocarpus aromatica, *Wall.* The whole plant is said to be used as a perfume ; see Vol. III., 111.

Dryobalanops Camphora, *Coleb.,* see Vol. II., 84-93.

Elettaria Cardamomum, *Maton.;* The Lesser Cardamom. An essential oil in perfumery is distilled from the seeds ; see Vol. III., 236.

Ferula suaveolens, *Aitch. et Hansl.,* and F. **Sumbul,** *Hook. f.,* yield the musk-scented medicinal root *Sumbul ;* see Vol. III , 339.

Fœniculum vulgare, *Gærtn.;* The Fennel; contains an essential oil used in the preparation of *Mouri atar ;* see Vol. III., 406.

Hedychium spicatum, *Ham.* The aromatic root-stocks are used as an auxiliary in dyeing to impart a pleasant odour to fabrics, and when powdered, form an important ingredient of *Abir ;* see Vol. IV., 207.

Hibiscus Abelmoschus, *Linn.;* The Musk Mallow ; has musk-scented seeds, employed as a substitute for that perfume ; see Vol. IV., 229.

H. ficulneus, *Linn.* The seeds are employed in Arabia to impart a fragrance to coffee ; see Vol. IV., 241.

Illicum verum; *Hook f.;* The Star Anise. An oil very much resembling that of Anise, is occasionally used in Indian perfumery ; see Vol. IV., 333.

Iris florentina, *Linn.;* The Iris or Orris Root; is occasionally cultivated in India ; see Vol. IV., 497.

Jasminum, various species. The essential oil largely used in Indian perfumery ; see Vol. IV., 541-544.

Juniperus, several species. The roots of these are largely collected in the North-West Himálaya for utilisation as incense ; see Vol. IV., 552-556.

Jurinea macrocephala, *Benth.* The roots are used as incense ; see Vol. IV., 556.

Lavandula. Several Indian species are fragrant and probably might be ultilised in the manufacture of Oil of Lavender ; see Vol. IV., 595.

Lawsonia alba, *Lam.;* The Henna plant. The flowers are used in perfumery ; see Vol. IV., 597.

Mangifera indica, *Linn.;* The Mango Tree. One of theper fumes exhibited at the Colonial & Indian Exhibition, and called *Amb atar,* was prepared from some part of this tree ; see Vol. V., p. 147.

Melaleuca Leucadendron, *Linn.* Cajput oil is occasionally employed in perfumery ; see Vol. V., 204.

VEGETABLE.

P. 406

PERFUMES & Perfumery.	Perfume-yielding Products.

VEGETABLE.

Mentha, various species; see Vol. V., 229.

Mesua ferrea, *Linn.;* said by Forbes Watson to be employed as a perfume; see Vol.. V., 236.

Michelia Champaca, *Linn.* The flowers yield a highly fragrant *atar*; see Vol. V., 241.

Mimusops Elengi, *Linn.* This plant is the source of a perfume called *Bakul Atar;* see Vol. V., 249.

Morina Coulteriana, *Royle;* used as incense; see Vol. V., 260.

Myristica fragrans, *Houtt.;* THE NUTMEG. The nut and mace both yield essential oils largely used; see Vol. V., 311.

Myrtus communis, *Linn.;* THE COMMON MYRTLE. Is occasionally met with in cultivation in India, but the essential oil, obtainable from the leaves, does not appear to be prepared; see Vol. V., 316.

Nardostachys Jatamansi, *DC.;* THE SPIKENARD. The root enters largely into the composition of native perfumery, but chiefly in combination with valerian, forming a mixture anything but a favourite with Europeans; see Vol. V., 338.

Nerium odorum, *Soland.* An essential oil may be obtained from the flowers of this plant—the SWEETLY-SCENTED OLEANDER; see Vol. V., 348.

Nyctanthes Arbor-tristis, *Linn.* The flowers contain an essential oil employed in making *Sewli atar;* see Vol. V., 434.

Ocimum Basilicum, *Linn.* It is believed that the Natives of various parts of India distil *atars* from the different species of BASIL or *túlsi;* see Vol., 440.

Pandanus odoratissimus. *Willd.;* employed in the manufacture of a scent called *Keora atar;* see Vol. VI., 5.

Pergularia minor, *Andr.;* used in the preparation of an *atar;* see Vol. VI., 139.

Peucedanum graveolens, *Benth.* The crushed fruit yields an essential oil on being submitted to aqueous distillation, which may be used in mixtures for perfuming soap. The distillate—DILL-water—is chiefly used medicinally; see Vol. VI., 181.

Phœnix dactylifera, *Linn.;* THE DATE PALM; used to prepare an *atar;* see Vol. VI., 199.

Pimpinella Anisum, *Linn.;* THE TRUE ANISE-SEED.; introduced from Europe, and occasionally cultivated in gardens in India; see Vol. VI., 236.

Pinus longifolia, *Roxb.;* is known in Oudh as *Dhub,* and employed for incense; see Vol. VI., 242.

Pistacia vera, *Linn.* The aromatic oil obtained from the Kernels of the PISTACHIO NUT is occasionally used as a perfume; see Vol. VI., 273.

Pogostemon Patchouli, *Pelle.* The essential oil of this plant is a very important perfume called by the Natives *Pachapal atar;* see Vol. VI., 307.

Polyanthes tuberosa, *Linn.;* yields *atar* of Tuberose; see Vol. VI., 312.

Prunus amygdalus, *Baillon.;* yields essential oil of Bitter almonds; see Vol. VI., 342.

Psidium Guyava, *Linn.;* THE GUAVA.; is used to make a perfume called *Payara atar;* see Vol. VI., 351.

Rosa, various species. ATAR OF ROSES is perhaps the most favourite of all perfumes with the rich in India; see Vol. VI.

Salix sp. The flowers of certain species of this genus are used to make a perfume called *atar-bed-muskh;* see Vol. VI.

Santalum album, *Linn.* The aromatic wood, and the essential oil distilled from it, are highly esteemed perfumes; see Vol. VI.

Saussurea Lappa, *Clarke.;* THE SACRED COSTUS ROOT.; is largely exported from Kashmír and used in the preparation of perfumery and expensive forms of incense; see Vol. VI.

Styrax benzoin, *Dryand.* A volatile oil is distilled from Gum Benzoin by the Natives of the Eastern Archipelago, and the balsam itself is also used as a perfume; see Vol. VI.	**VEGETABLE.**

Valeriana Wallichii, *DC.* Aitchison states that the root-stalks of this VALERIAN are collected in the Kuram Valley for export to India where they are used in perfumery; see Vol. VI.

2.—ANIMAL.

Ambergris; an animal concretion used in the preparation of *atar-i-ambar*; see Vol. I., 217.

ANIMAL.
407

Civet; the pouch of the civet cat (**Viverra zibetha,** *Linn.*) contains an unctuous odorous secretion, which is extracted and used in perfumery; see **Tigers & Cats,** Vol. VI.

Musk, see Deer, Vol. III., 58; also **Musk,** Vol. V., 307.

3.—MINERAL.

MINERAL.
408

Earth atar. A perfume called *atar-gil* is prepared by distillation from earth mixed with a little sandal wood oil. It is said to be used chiefly by Muhammadans, who consider that it exudes the odour of earth just after a shower of rain.

According to the recent article in the *Indian Agriculturist*, above quoted, the prices of the most important perfumes at Jaunpore (one of the chief centres of the Indian perfumery trade) are as follows:—

PRICES.
409

Perfume.	Per seer.	Perfume.	Per seer.
	R R		R
Rose atar . . .	40 to 50	Oil of Marigold (*Gehnd*a) .	40
Rose water . . .	as. 8	Oil of Khas-khas (**Andro-**	
Sandal wood oil . .	28	**pogon muricatus**) .	40
Oil of Spanish Jasmine		Oil of Cajeput (*elláchi*) .	30
(*Chambéli*) . .	2 to 4	Lemon-peel oil . . .	20
Oil of Arabian Jasmine		Oil of clove . . .	20
(*Bela*) . . .	2 to 4	Geranium oil . . .	10
Keora oil (**Pandanus**		Earth atar . . .	40
odoratissmus**)	2	Atar of Henna flowers . .	50
Keora water . . .	as. 8		

PERGULARIA, *Linn.; Gen. Pl., II., 773.*

410

A genus of twining sub-glabrous under-shrubs, which belongs to the Natural Order ASCLEPIADEÆ, and comprises about ten species, natives of Asia and Africa. Of these, P. minor, *Andr.* (*Fl. Br. Ind., IV., 38*), *Kanja lúta, kunjalt,* BENG., *Sítá manóharam,* TEL., a native of the Himálaya and the Tregai hills in Burma, is cultivated throughout India as an ornamental and fragrant plant. The flowers are also employed in the preparation of an *atar*.

PERICAMPYLUS, *Miers; Gen. Pl., I., 37, 961.*

[CEÆ.

Pericampylus incanus, *Miers; Fl. Br. Ind., I., 102;* MENISPERMA-
The authors of the *Pharmacographia Indica* state that slender Menispermaceous stems, apparently belonging to this species, are sold in the Bengal bazárs under the name of *Bárak-kánta* (*Pharmacog. Ind., I., 64*).

411

412   **PERIDERMIUM,** *Chev. ; Cooke, British Fungi, 534.*

A genus of parasitic fungi which attacks several Indian forest trees, especially CONIFERÆ, see **Fungoid Pests**, Vol. III., 456.

PERILLA, *Linn.; Gen. Pl., II., 1182.*

413   Perilla ocimoides, *Linn. ; Fl. Br. Ind., IV., 646 ;* LABIATÆ.

Syn.—P. MACROSTACHYA, *Benth. ;* OCIMUM FRUTESCENS, *Linn. ;* MELISSA MAXIMA, *Arduin ;* MENTHA PERILLOIDES, *Willd.*

Vern.—*Bhanjira,* HIND.; *Kenia,* NAGA; *Bhangara, jhutela,* KUMAON.

References.—*Campbell, Ec. Prod., Chutia Nagpur, No. 563 ; Atkinson, Him. Dist., 708 ; Ec. Prod., N.-W. P., Pt. V., 42 ; Settle. Rept., Kumáon, App. 34 ; Madden, Off. Rept., Kumáon, 279 ; Darrah, Note on Cotton in Assam, 33.*

Habitat.—A coarse aromatic herb, found in the Tropical and Temperate Himálaya, from Kashmír to Bhotan, at altitudes of 1,000 to 10,000 feet ; also in the Khásia Mountains from 3,000 to 6,000 feet. It is frequently cultivated on the Himálaya from 4,000 to 5,000 feet for the sake of its small aromatic seeds.

DYE.
Seed.
414   **Dye.**—The SEED is used as a dye auxiliary ; see **Rubia sikkimensis**, *Kurz ;* Vol. VI.

OIL.
Seed.
415   **Oil.**—The SEED yields an aromatic oil by expression, which is used by the hill-men of the North-West Himálaya for culinary purposes and for burning.

FOOD.
Seed.
416   **Food.**—The SEED is used for food, the LEAVES are employed as a vegetable, and the OIL, above mentioned, is used as an adjunct to other substances in the diet of the hill-men of the North-West Himálaya. Dr. Watt notes that the seed and leaves are similarly utilised in Manipur.

Leaves.
417

Oil.
418   **PERIPLOCA,** *Linn. ; Gen. Pl., II., 746.*

419   Periploca aphylla, *Dcne. ; Fl. Br. Ind., IV., 12 ;* ASCLEPIADEÆ.

Syn.—CAMPELEPIS VIMINEA, *Falc.*

Vern.—*Báta, barri, barrarra,* PB.; *Shabbi, barrarra,* PUSHTU ; *Hum, huma,* AFG.; *Um, uma,* BALUCHISTAN ; *Buraye,* SIND.

References.—*Brandis, For. Fl., 330 ; Gamble, Man. Timb., 265 ; Stewart, Pb. Pl., 145 ; Bot. Tour in Hazára, &c., in Jour. Agri.-Horti. Soc. of Ind., XIV. (Old Series), 15 ; Aitchison, Bot. Afgh. Del. Com., 87 ; Murray, Pl. & Drugs, Sind, 159 ; Royle, Fib. Pl., 306 ; Gazetteers :— Bomb., V., 26 ; N.-W. P., IV., lxxiv. ; Peshámar, 27 ; Ind. Forester, IV., 233 ; V., 180 ; XII., App., 16 ; XIV., 361, 371.*

Habitat.—A shrub of the arid, dry zones of the Western Panjáb and Sind ; distributed to Balúchistán, Afghánistán, Persia, Arabia, and Nubia.

FIBRE.
420   **Fibre.**—According to Dr. Stocks the fibre of this plant is employed in Sind with that of **Leptadenia Spartium**, for making into ropes and bands used for wells. Both these fibres are particularly prized for this purpose, as they are supposed to be impervious to the action of water.

MEDICINE.
Milky juice.
421   **Medicine.**—The MILKY JUICE is used in Sind as an external application to tumours and swellings.

FOOD &
FODDER.
Flower-buds.
422   **Food & Fodder.**—The FLOWER-BUDS are sweet, and are eaten, raw or cooked, as a vegetable. The PLANT is cropped by goats and camels, and is largely cut and collected for fodder.

Plant.
423

DOMESTIC.
Dry Shrub.
424   **Domestic & Sacred.**—The DRY SHRUB is used for fuel. It appears probable that this species is closely connected with the *Hóma* of the Parsís (see Vol. III., 246-251).

PERISTROPHE, *Nees; Gen. Pl., II., 1121.*

[THACEÆ.

Peristrophe bicalyculata, *Nees; Fl. Br. Ind., IV., 554;* ACAN-   425

Syn.—P. KOTSCHYANA, *Nees;* JUSTICIA BICALYCULATA, *Vahl.;* J. LIGU-
LATA, *Lamk.;* DIANTHERA MALABARICA, *Linn.;* D. BICALYCULATA,
Rets.; RUELLIA PANICULATA, *Linn.*

Var. rivinoides = β. RIVINOIDES; *Wall.;* P. BICALYCULATA, VAR. β, *Nees.*

Vern.—*Atrilal,* HIND.; *Nasa bhaga,* BENG.; *Barge khode baha,* SANTAL;
Kali andi jahria, MERWARA; *Naspat,* SIND; *Pitpápra,* BOMB.; *Ghati-
pitta papada,* MAR.; *Chébíra,* TEL.; *Maha-nelu,* SING.

References.—*Roxb., Fl. Ind., Ed. C.B.C., 42; Elliot, Flora Andhr., 35;
Ainslie, Mat. Ind., II., 65; Dymock, Mat. Med., W. Ind., 594; S.
Arjun, Bomb. Drugs, 108; Atkinson, Him. Dist., 315; Gazetteers:—
Bomb., V., 27; N.-W. P., I., 83; IV., lxxvi.*

Habitat.—An erect annual, very common in waste places throughout
Tropical and Sub-tropical India, from the Panjáb and Sind, to Assam,
Pegu, and Madras.

Medicine.—According to Rheede the whole PLANT, macerated in   **MEDICINE.**
an infusion of rice, is said, on the Malabar coast, to be a useful remedy in   **Plant.**
cases of poisonous snake-bites. Mr. Sakharam Arjun (*List of Bombay*   426
Drugs) says that it is supposed to have the properties of **Fumaria parviflora,**
and is used in its stead, but has not the bitterness of that plant.

P. tinctoria, *Nees; Fl. Br. Ind., IV., 556.*   427

Syn.—JUSTICIA TINCTORIA, *Roxb.;* J. ROXBURGHIANA, *Roem. & Sch.*

Vern.—*Ráng, bet* or *batia-rang,* BENG.

References.—*Roxb., Fl. Ind., Ed. C.B.C., 41; McCann, Dyes & Tans,
Beng., 66, 191; Wardle, Rep. on Indian Dyes, 2.*

Habitat.—An erect spreading herb, doubtfully wild, but frequently
cultivated in India from Assam to Ceylon, especially near Midnapur,
Bengal.

Dye.—The TWIGS cut into short pieces are used in dyeing, to colour   **DYE.**
the *masland* mats of Midnapur. Roxburgh, though he gives a full de-   **Twigs.**
scription of this plant, and names it **tinctoria,** makes no mention of the   428
dye obtained from it. A similar silence prevails in the works of other
writers, with the exception of **McCann,** from whom the following
passage may be quoted in entirety, giving as it does the only information
available on the subject:—

"Detailed information regarding the cultivation, growth, and use in
dyeing of this plant, has been received from Midnapur, in which district it
seems to give rise to an industry of some importance in connection with
the manufacture of *masland* mats. No mention of it is, however, made
in the returns from any other district. A specimen of the growing plant
was forwarded from Midnapur, and was identified by **Dr. King** as Peris-
trophe tinctoria.

"From the account given by the Collector of Midnapur it appears that
this plant is cultivated in the jurisdictions of thanas Sabang, Patáspur,
Náráyangarh, Pánchkurá, and Raghunáthpur by the classes of people
called Vaishnavs, Báitís, Suktís, and Kaibarttas. The plant is propagated either by cuttings or from seed, but the former method is preferred, as the growth is quicker. The ground is dug up with a *kodáli* in
Kártik (October-November), and again in the month of *Jaishtha* (May-
June) (?) next following, when it is also manured with mud. In the
month of *Ashár* (June-July), when the ground has been sufficiently
moistened by the rain, the cuttings are planted; they begin to grow in
about ten or twelve days. The ground is afterwards dug up with a *kodáli*
once a month, and weeds are carefully removed. The plants grow to a

| PESTS. | Pests which affect |

DYE.

height of about two cubits, and flower in *Kártik* (October-November), in which month their twigs and the matted extremities of the branches are lopped off, to be used in dyeing the sticks for *masland* mats. The twigs are cut into small chips, and are then broken into thin pieces by means of a *dhenki*; they are then dried in the shade and stored in gunny-bags till required. Sometimes a portior of the roots is taken and pounded with the twigs. The same plant serves for two or three years. The area under cultivation of *ráng* is estimated at 1,000 *bighas* (in a previous letter it is given as 2,000 acres), and about 10,000 maunds of the dye-stuff are prepared annually. It sells at R15 per maund.

"*Ráng* is used exclusively in dyeing the sticks from which *masland* mats are manufactured. The sticks are made into convenient bundles, and pounded *ráng* is placed upon the parts to be dyed, which are se-parated from the rest by palm-leaf knots at both ends. One seer of pounded *ráng* is then mixed with 20 seers of water, and the whole boiled. The part of the bundle to be dyed is dipped in this and boiled in it for three hours; then removed and dried. It has acquired a *red* colour. Although exclusively used for this purpose, a piece of cloth dyed with *ráng* by way of experiment has been forwarded from Midnapur, the process adopted being the following:—"For dyeing a yard of cloth a pound of *ráng* is cut or broken into very small chips, and is boiled in about 5℔ of water until about 3℔ of water remain. The solution is then allowed to cool. A pice-weight of alum (about ¼oz.) is then pounded and mixed with the solution. The cloth to be dyed is washed in pure water, and the moisture well wrung out of it. It is then steeped in the above solution and is afterwards put to dry in the shade. The steeping and dyeing is repeated two or three times." The result of this process was a dull red (*McCann, Dyes & Tans, Bengal, p. 66*).

Specimens experimented with by **Mr Wardle** were found to produce colours varying from drab, to red, and brownish red, with tussur, mul-berry silk, and wool. **Mr. Wardle**, writes in his report: "This dye-stuff produces good colours and contains a fair amount of colouring matter. It would be a valuable acquisition to the dye-house if it could be procured at a price which would enable it to compete with other red colouring matters." There would appear to be no reason why it should not be obtained cheaply and plentifully, since the plant is frequent, culti-vated, and spontaneous in the neighbourhood of Calcutta.

PEROWSKIA, *Karel.; Gen. Pl., II., 1193.*

429

Perowskia abrotanoides, *Kiril.; Fl. Br. Ind., IV., 652;* LABIATÆ.
Vern.—*Shanshobai,* PUSHTU.
References.—*Boiss. Fl. Orient., IV., 589; Aitchison, Bot. Afgh. Del. Com., 95; Note by J. H. Lace, Esq.*

Habitat.—A much branched twiggy shrub or undershrub, from 2 to 4 feet high, with the habit of a **Lavandula**, met with in Western Tibet, from 8,000 to 12,000 feet, and distributed to Afghánistán, Balúchistán, Persia, and Turcomania.

MEDICINE.
430
Medicine.—Mr. Lace, in a note in his herbarium, states that, at Ziarat, the plant is used as a cooling medicine.

FODDER.
431
Fodder.—"Eaten by camels, sheep, and goats, but not by cows" (*Lace*).

PESTS.

432
Pests.
References.—*E. T. Atkinson, E. C. Cotes, and L. de Niceville, in Indian Museum Notes, Vol. I., Pts. I., II., IV.; Extracts from above in Indian Agriculturist, Aug. 10, Aug. 24, Aug. 31, Sept. 14, Sept. 21, Sept. 28,*

Oct. 5, Oct. 12, Dec. 7, 1889 ; J. Wood-Mason, Rice Pest of British Burma ; Indian Forester, I., 299 ; III., 25, 27, 88 ; IV., 90, 197, 243 ; VIII., 204 ; IX., 91 ; X., 230, 280 ; XII., 502 ; XIII., 87, 383, 387, 437 ; XIV., 167, 192, 232 ; Thompson, Rep. on Insects injurious to Woods and Forests (1867) ; Tropical Agriculturist, Jan., May, July, 1889 ; Kew Bulletin, 1889, 13 ; Balfour, Agricul. Pests of India ; Cyclop., II., 356-365, 694 ; Agricultural Dept. Reports, in many passages.

The pests which affect our crops, trees, and live stock may conveniently be divided into two great classes, *viz.*—1, Fungoid ; 2, Insect. The former have already been dealt with in Vol. III., 455-458, and space does not admit of a further consideration of that subject. It may be mentioned, however, that a valuable paper has recently been written by **Dr. Barclay** on the subject of RUST in India, which it is hoped will shortly be published, and which contains much useful and interesting information to all interested in the question.

The subject of Insect Pests has lately attracted considerable attention from the able entomologists of the Indian Museum, and it is to be hoped that the labours of these naturalists in economic entomology may be productive of very valuable results. It is of recent years only that any serious attempt has been made to study insect pests as they ought to be studied, a fact which is the more remarkable when one considers the enormous harm which must annually result to all our most valuable crops from their ravages. Within recent times, however, the public in Europe and America have become alive to the great importance of the subject, and interest has been further aroused by the ravages of insects on such crops as the vine, and coffee crops, which, within a short space of time, suffered very severe loss, and in some cases were almost exterminated. Public attention was, perhaps, most fully aroused in England at the time of the threatened invasion of the "Colorado beetle" some few years ago.

The Government of the United States has done much to advance a knowledge of the subject through the labours of the State entomologist, **Dr. Riley ;** but the work of **Miss Ormerod** in England has also been productive of very valuable results.

Notwithstanding the fact that in an agricultural country like India the subject must be one of paramount importance, it has, as already stated, received very little attention till about the beginning of 1889. Up to that time scattered articles in the *Indian Forester*, and in the publications of the Agri.-Horticultural Society of India, constituted almost the entire literature on the subject.

This information was carefully and laboriously compiled by **Surgeon-General Balfour** in his *Cyclopædia*, and later in his small book on "The Agricultural Pests of India." But the first necessary step—the correct identification of the insect and the study of its life-history—does not appear to have been seriously attempted.

Of recent years, however, the Government of India has bestirred itself in the matter, and by inviting contributions and communications from all parts of the country, has succeeded in placing, in the hands of the Indian Museum authorities, sufficient material to have enabled them, in the end of 1889, to publish a series of most valuable papers in the Indian Museum Notes. A study of these publications shows still further the enormous extent to which insects carry on depredations on our most valuable crops, and the large amount of thorough work already accomplished leads to the hope that much more may still be done, and that the subject may emerge from the obscurity in which it has been heretofore involved.

It is impossible with the space at the disposal of the writer, to enter into a complete *resumé* of the papers which have already appeared, but a short account of perhaps the most important pests already determined,

PESTS.	Pests which affect

may be given. For further information the reader is referred to the papers of Mr. Atkinson, Mr. Cotes, and Mr. de Niceville, in the Indian Museum Notes.

LIST OF IMPORTANT INSECT PESTS.
(Arranged according to the plants they infest.)

1. CACAO.

Helopeltis antonii, *Signoret*, is found on the Cacao in Ceylon, and, according to Thwaites, is the only formidable enemy of that tree.

2. CARDAMOM.

Lampides elpis, *Godart.* Mr. L. deNiceville reports that this species of butterfly, the larva of which is a serious enemy to the cardamom plantations of Ceylon, bores a circular hole in the capsule and cleans out the interior of the fruit. It has been estimated by Mr. Owen (*Note on Cardamom Cultivation, Colombo, 1883*) that the damage sometimes done by this pest is as much as 80 to 90 per cent. to young plantations; but another observer states, that 5 to 10 per cent. of the capsules are ordinarily perforated. Mr. deNiceville writes, "The only remedy I can suggest is to catch and kill all the butterflies that can be seen. . . . The butterflies have a slow flapping flight and are very conspicuous, so their capture is very easy. Once the eggs are laid no further remedy is possible, I think."

3. CASTOR OIL.

Achœa melicerte, *Drury.* The larva of this NOCTUES moth has been reported as destructive to the castor oil plants in Lower Bengal, Assam, and Madras. The insect is a common one and occurs in India, Ceylon, the Celebes, and Australia. The report from Assam states that in July 1889 the caterpillar appeared in "millions," rapidly overspread every plant in a large plantation kept for silk-worms, and by the third or fourth day, notwithstanding every exertion, "left literally not a leaf throughout the entire area."

4. CINCHONA.

Helopeltis bradyi, *Waterh.* Does much damage to cinchona in Java.

Helopeltis febriculosa, *Bergroth.* Has been collected in the cinchona plantations at Mungphu in Sikkim. Mr. Atkinson writes, "It has not occurred yet in sufficient numbers to do much damage, but as it belongs to the same genus as the destructive 'mosquito pest' of the tea, its operations should be carefully watched."

Pseudopulvinaria sikkimensis, *Atkinson*, see under Pests of "Forest Trees," p. 149.

Cerataphis sp.—This recently discovered aphis has been found infesting the leaves of cinchona in Sikkim.

Pemphigus cinchonæ, *Buckton.*—An aphis, found infesting cinchona in the same locality.

Disphinctus humeralis, *Walker.*—This is another pest belonging to the Rhynchota, family CAPSIDÆ, which, according to Mr. Atkinson, has also been lately discovered attacking the cinchona at Mungphu in Sikkim. It does not appear to have done much damage.

Lecanium viride. See Pests of "Forest Trees," p. 149, also "Coffee Pests."

5. COFFEE.

Lecanium coffeæ, L. nigrum, and L. viride—THE BROWN, BLACK, AND GREEN COFFEE SCALE INSECTS, also Dactylopius adonidum, *Linn.*—occur in the coffee districts of India, and have done incalculable harm to the plant as well as to several forest trees. The attack is always accompanied by a curious fungoid growth, which covers the twigs, and effectually rots, and kills them. That which accompanies L. nigrum in Ceylon has been found to be due to the fungus **Trisporium Gardneri,** *Berkely.* Mr. **Nietner,** who carefully described and named the pests, believes the fungoid growth to be dependent on the glutinous saccharine secretion ("honey-dew") secreted by the insects. Besides forming a nidus for the fungus this sugary secretion does much harm by stopping up the stomata of the leaves. The injury done by these pests to the coffee industry, especially in Ceylon, has been enormous, and has in many cases caused the planters to direct their attention exclusively to the cultivation of tea instead of coffee. It would appear, however, from the articles in the Indian Museum Notes, that they may be entirely got rid of by applying an emulsion of kerosine oil and milk, or kerosine oil and soap solution to the plants by means of a powerful spray. For a full description of the pests, their ravages, and the methods which have been employed to destroy them, the reader is referred to an interesting account by Mr. **Green,** republished in the Indian Museum Notes, I., 116-122.

Xylotrechus quadrupes, *Chevr.* The "COFFEE BORER."—The larva of this boring beetle attacks the stems of the coffee plant in South India.

6. COTTON.

A pest, commonly known as the "BOLL WORM," causes much damage to the cotton crop in many parts of India. It is apparently not described in the Indian Museum Notes, but is possibly identical with **Heliothis armigera,** the American "Boll worm," which has been frequently reported to be a very harmful opium pest in India, see p. 151.

Oxycarenus lugubris, *Motschulsky,* is a small Lygæid, which infests the ripe pod of the cotton in Ceylon, discolouring and caking the floss.

Lohita grandis, *Gray,* the *kapási-poka* of Chuadanga, occurs very commonly throughout Bengal and Assam, and is supposed to do much harm to the crop.

7. FOOD-GRAINS:

(a) CEREALS AND MILLETS.

Perhaps the most important of this group is the weevil, to which Mr. **Cotes** has devoted much attention.

Calandra oryzæ.—THE WHEAT, BARLEY, MAIZE, AND RICE WEEVIL — It appears that this small beetle may not only attack the crops above enumerated, but that it also commits ravages on *juári.* Numerous careful experiments have proved that—(1) The weevil is a purely granary pest and that grain can therefore be preserved by isolation and other precautions against infection after it leaves the fields; harder grains being easily protected, the softer with considerable difficulty. (2) The weevil has great wandering propensities, which makes isolation very difficult to procure in the neighbourhood of infected localities. (3) That rice is free from the attacks of the insect so long as it lies in husk, which seems to afford complete protection. (4) That the weevil does not develop in grain to any extent after it reaches England. (5) That bisulphide of carbon, *ním* leaves, sulphur fumes (generally believed to exert a preservative action) are but doubtfully efficacious in keeping grain stored in an infected spot free from the insect. (6) That though in wheat and notably in rice the ear is a protection against weevils, barley, on the other hand, is as much subject to attack when in the ear as when husked.

Chætocnema basalis, *Stephen.*—A beetle determined as belonging to this species by Mr. A. E. Shipley, F.L.S., is said to do much damage to the young rice in the Taungyas of the northern part of the Tharrawaddy division of Burma. The insects appear in June, when the paddy is about six inches high, and first attack the leaves, afterwards passing on to the stem and roots. It appears to be well known to the Karens, who state that it has appeared in the same locality before, but the Burmans are doubtful

PESTS.	**Pests which affect**
INSECT PESTS.	

if this is so. It is known to the natives as *wetpo*. As the life-history of the insect has not been properly worked out, it is difficult to suggest any means of combating it. **Mr. Shipley,** however, suggests that some of the following means, which have been found effectual with other members of the HALTICIDÆ, might be found effectual :—(1) Sprinkling the plants with any finely divided matter such as lime, soot, road-dust, ashes, &c. (2) Syringing or sprinkling with whale-oil, soap solution, or an extract of wormwood. (3) Taking great care to keep the land clean. (4) Deeper cultivation. (5) Collecting the insects in kerosine or cloths soaked in kerosine.

CUT-WORMS.—Injuries to rice from insects which are probably "Cut Worms" (**Agrotis suffusa,** *Hübn.*) have been reported from Balasore and Chittagong, while, according to **Mr. Cotes,** "it is not improbable that much of the damage, reported from other districts as due to obscure Lepidopterous larvæ, may also have been done by insects belonging to this group." These larvæ attack the young crop just as it shows above ground, and not only feed on the young leaves, but cut through the stems, dragging the tender plant into their underground burrows where they feed on them at night. The following method of destroying the larvæ has been found effectual by **Dr. Riley** in America :—" Bundles of cabbage, turnip or clover are sprinkled with Paris-green water, and laid at intervals between the rows of the crop to be protected. Before the plants come up these poison the cut-worm, which are thus got rid of before the appearance of the crop which they would otherwise attack."

Leptocorisa acuta, *Thunb.*—THE RICE SAPPER; *Bhoma,* BENG.; *Ghandi,* CH. NAGPUR, N.-W. P., and ASSAM; *Mohua,* ASSAM; *Munju vandu,* TINNEVELLY; *Vandu,* TAM.; *Goyan messa,* SING—is an insect which appears to do considerable damage to the rice crops in Bengal, the North-West Provinces, and Tinnevelly. The Deputy Commissioner of Hazaribagh reports that this pest attacks the *gora* and *badhi* rice while in the ear, destroys up to three-fourths of the crop, and generally appears when the rain sets in early (May). The insect is most destructive in the larval state, sucking out the juices from the halm, which withers and turns yellow. Nothing is known of its life-history, and the only cure suggested for the pest is that of smoking the fields by burning vegetable refuse to windward.

Hispa ænescens, *Baly,* THE BENGAL RICE HISPA.—This rice pest is widely distributed in India, a number of reports having been received by the Indian Museum, of damage done by it in various parts of Bengal. It is a beetle belonging to the family **Chrysomelidæ,** almost all the species of which feed on leaves, both in the larval and mature stage, by far the grearter part of the damage, however, being done by the larvæ. **Mr. Cotes** writes, " From the reports that have been received, it seems that the pest appears often in vast numbers during the rains, when the rice has just been planted out, and is still young and tender, the insect feeding on the parenchyma of the leaves and stalks, leaving the fibre exposed so as to give the plants a white and withered appearance. The effect of the pest would seem to be to stunt and weaken the plants and cause them to yield but a small crop." The rice is apparently in no case completely destroyed by the insect, but the outturn may be reduced by from twelve to fifty per cent.

Only two remedies are mentioned as adopted by cultivators, *viz.,* smoking the insects out of the field by means of fires of paddy-straw, covered over with green leaves to increase the smoke ; and by letting out the water from the fields. The latter is not always practicable, but is recommended by **Mr. Cotes** as likely to be most efficacious. This is the

P. 433

more likely, since it appears that rice which is almost completely submerged is attacked only, the attack lasting so long as the water remains in the field. The insect is known locally as *burma chandali*, *pámari*, and *shanki poka*.

Hieroglyphus furcifer, *Serville*, H., nov. sp., and **Euprepocnemis bramina**, *de Saussure.*—These orthopterous insects, known in the Central Provinces as *papha*, are reported to have done much damage to young rice and *kodo* (small millet) in that locality, during the rains of 1889; and are also said to have been prevalent in Raipur in 1886.

Leucania loreyi.—Specimens of this Noctuid moth were received at the Museum, in 1888, from Sambalpur, Central Provinces, with the information that the larva had done great damage to the rice crop in that district. The insect is allied to the "cut-worms." Several other species of Lepidopterous pests on rice and other food-grain crops are mentioned in various parts of the Indian Museum Notes, but these are too numerous and as yet too incompletely known to render their quotation useful in this work. **Mr. Wood-Mason** described one in 1885 as **Paraponyx oryzalis**, a pest which attacks the rice crop in Burma, where it is known as *tenidoung bo* and *palan byú.* Another, **Tinea granella**, the "wolf moth," has on several occasions done great damage to cargoes of rice shipped from Calcutta and London, while lying in Kingston, Jamaica.

Lachnosterna impressa.—The larva either of this species, or of a beetle very closely related to it, attacks the roots of rice, *kachú*, and Indian corn, in Chittagong, emerging from the ground in July and August. Paddy, kept covered with water in July and August, is said to be unaffected.

Cantharis sp.—A species of this beetle is reported to have eaten up the leaves of yellow *cholum* (millet) in Karnúl, and thus destroyed the crop in one village of that district.

Trogoderma sp.—The larvæ of a Dormestes beetle of this genus were observed in wheat godowns in the Delhi market in 1888 by **Mr. Cotes.** This pest, known as *kapra*, is said to sometimes destroy as much as six or seven per cent. of the wheat.

Cecidomyia oryzæ, *Wood-Mason.*—This insect, which is of special interest as belonging to the same genus as the destructive "Hessian fly" of Europe and America, was described and named by **Mr. Wood-Mason** about ten years ago. It was sent from Monghyr in 1880, where it was said to have played great havoc amongst the rice of the Kurruckpore thana. Nothing appears to have been heard of it since.

Suastus gremius, *Fabr.* (*pattanai*, BENG.), is said to be extremely destructive, while in the larval stage, to rice, consuming the young and tender leaves. **Mr. deNiceville** considers, however, that the damage it effects must only be slight, but recommends its destruction by raising the earthen walls, or *bands*, round the affected fields, and submerging the crop under water for a time.

THE SORGHUM-BORER,—is the larva of a moth which has not as yet been identified. It attacks the stalks of the *juár* or *jowari* (great millet, **Sorghum vulgare**), in much the same way as the sugarcane is tunnelled by the "sugar-cane borer." The larva bears a strong resemblance to that found in sugar-cane, and **Mr. Cotes** suggests that it may, like the latter, set up decomposition and so produce the poisonous symptoms frequently observed to occur in cattle fed on diseased *juar.* The prevailing belief amongst Natives is, that the cause of this poisonous property *is* an insect, and it is worthy of notice that the disease occurs at similar times as it does in sugar-cane, *viz.*, during an exceptionally dry season. The insect, supposed to produce the poisonous property in **Sorghum**, is called in the North-West Provinces *bhaunri* (*Duthie & Fuller, Field & Garden Crops*, 27).

PESTS.	Pests which affect
INSECT PESTS.	

(*b*) **PULSES.**

Leucania extranea.—A Noctues moth said to attack the pea, Pisum sativum, *Linn.*

Agrotis suffusa, *Hübn.*—The cut-worm is reported to have attacked Lathyrus sativus, *Linn.*, in Patna.

Plusia nigrisigma.—Another Noctues moth, which is said to have infested the gram plant, Cicer arietinum, *Linn.*, in the same locality.

In addition to these certain insects, mentioned under "Miscellaneous Pests," attack leguminus along with other crops. Several were lately received from Patna at the Indian Museum, amongst which was **Heliothis armigera** (see Opium Pests), one of the larvæ of which was found with its body half inserted in a full-grown *khesári* pod (Lathyrus sativus), and is also said to largely attack other pulses, especially **Dolichos Lablab.** Others, undetermined, were found infesting **Ervum Lens,** and *arhor,* **Cajanus indicus.**

8. FOREST TREES.

Cælosterna scabrata (?), *Fabr.,* THE SÁL GIRDER BEETLE.—During the rains of 1888 coppice *sál* saplings in Oudh suffered from this insect, which ringed the bark generally within a foot or two of the root. The part above consequently died, and the coppice shoot became crooked or bifurcated (*Wood, Indian Forester* (1888), *503*). Mr. Cotes remarks that in the case of the allied American HICKORY TWIG GIRDER, the eggs are deposited above the groove, and the larvæ on emerging feed on the dead wood. If this also be the case with the *sál* beetle, similar damage in future years might be prevented by collecting and burning the dead tops of the shoots.

Neo-cerambyx holosericeus, *Fabr.,* a species of Cerambycid beetle, attacks the wood of *sál* (Shorea robusta) and *saj* (Terminalia tomentosa). Specimens have been received in the Museum from Calcutta, Múrshidabad, Assam, the Nága Hills, Bangalore, Ellore, Kullu, Rangoon, Perák, the Andamans, and the Nicobar Islands. Its range is therefore considerable. The larvæ appear to live in the first instance upon the sapwood, and afterwards to bore into the heart, which finally becomes riddled with burrows. It attacks the timber only after the tree has been killed, and when the bark has been allowed to remain on, and its ravages can be prevented by removing the bark from felled trees as soon as possible.

Plocederus pedestris, *White,* a beetle of the same family, has been found in Dehra Dun in *sál* and in *jingham* (Odina Wodier.) Mr. Thompson gives an account of apparently the same insect as attacking the latter timber, also **Butea frondosa** (the *dhák*), and **Bombax malabaricum** (the *simul*). Specimens of the insect have been recorded from Calcutta, Maldah, Sikkim, Sibságar, the Nága Hills, Dehra Dun, Ceylon, and the Andaman Islands. Mr. Cotes writes:—"Though no definite information has yet been received of the extent of the injury that the insect does by boring into timber, it is representative of a group of insects of which very little has yet been recorded in India, though they are probably amongst the most destructive with which the forester has to deal." Like the species above described, this pest appears to attack felled trees, laying its eggs between the bark and the wood.

Mr. Cotes describes several other Coleopterous insects, one of which, **Tomicus sp.,** is said to be injurious to the *makai* tree (Shorea assamica, *Dyer*), in Assam; another, the *Kúti,* a species of **Bostrychidæ** beetles (**Apatides**), is very destructive to bamboos, wattles, many jungle woods, solah pith hats, basket work, mats, &c. It is also said to attack horsegram (**Dolichos uniflorus**). The bamboo, however, appears to suffer most, but, according to Mr. Cotes, it can be completely protected by coating i with kerosine.

P. 433

INSECT
PESTS.

Magiria robusta, *Moore*, the "CEDRELA TOONA MOTH."—The larva of this insect attacks the tree from which it is named, devouring the succulent tops, the pith of the young branches, and the leaf-stalks. It is also said to infest young mahogany trees.

Zeuzera sp.—The red caterpillars of a moth, probably belonging to this genus, have been sent to the Museum from Travancore, where they are reported to do very serious injury to teak by boring into the stems. The damage caused is reported to be entirely due to lopping the trees, as manure for paddy-fields. By this process large open surfaces of soft rotting wood are gradually formed on the trees, by means of which the insect is supposed to gain access to the wood.

Pseudopulvinaria sikkimensis, *Atkinson.*—A species of Coccio which attacks the oak, chestnut, and cinchona in Sikkim. It appears in the cold weather in the form of a flour-like substance on the underside of the leaves, and matures about April. Mr. Atkinson states that it has not yet spread sufficiently to do any considerable damage.

Lecanium viride, THE GREEN SCALE BUG, does not confine its attacks to coffee, but is frequently found on several species of Cinchona, also on orange, lime, and guava trees and a large number of forest trees.

L. nigrum, THE BLACK SCALE BUG, according to Mr. Green, is sometimes found in large numbers on the croton-oil plant, and the Ceara Rubber tree, on both of which it produces the usual effects, *viz.*, a heavy fall of leaf and black fungus.

9. FRUIT TREES.

Cryptorhynchus mangifera, *Fabr.*—The MANGO WEEVIL—is a pest which does not as yet affect all the mango-producing districts, but its march is progressive. Restricted at one time to Dacca and the South-Eastern Districts, Backergunge, Faridpur, &c., it is said now to be working its way northward and westward, throughout Bengal and the neighbouring districts. The larval and pupal stages are passed, and the imago form attained in the fruit, and in infected localities nearly every fruit is frequently found attacked.

Idiocerus niveosparsus, atkinsoni and **clypealis,** have been found attacking the Mango at Saharanpur. Unlike the common Bengal mango pest it does not attack the fruit, but appears to subsist on the juices of the flowers, young leaves, and young shoots. According to Mr. Gollan, "they are found upon the mango all summer, but do most harm when the trees are in flower by damaging the productive organs. . . . I have tried mixtures of soap, tobacco, sulphur, kerosine diluted with milk, &c., upon them, but without noticeable effect." Mr. Cotes suggests that a further trial should be given to kerosine emulsion.

Papilio erithonius, *Cramer:*—Caterpillars of this butterfly were sent to the Museum in 1888 from Bangalore, with the information that they attacked Lemon trees; and models of the same were sent from Saharanpur in 1889 with the report that they had done much damage to young budded oranges, "not a plant of which could be raised if boys were not kept to pick off the caterpillars." This pest is said to occur in Lower India most frequently on the *Bél* (Ægle Marmelos), but also to feed on the Orange, Pomelo, Lime, *bér* (Zizyphus jujuba), and on the low-growing wild **Glycosmis pentaphylla.** Mr. Gollan found London Purple to be a complete success in destroying the pest at Saharanpur.

Virachola isocrastes, *Fabr.*, a very abundant butterfly throughout the plains of India (except in desert tracts), and in Ceylon, but not found in Assam or Burma. It attacks the Loquat, the Guava, the Pomegranate, and **Randia dumetorum,** *Lamk.* Mr. de Niceville, who describes it in the Indian

P. 433

PESTS.	Pests which affect

INSECT
PESTS.

Museum Notes, states that no reports have as yet been received regarding damage to fruit caused by this insect, but that, under certain circumstances, it might cause immense loss. The larva bores into the heart of the young fruit, and appears to live more on the seeds than on the pulpy portions. Catching the butterflies and burning all infected fruits have been recommended as likely to exterminate the pest.

Gangara thyrsis, *Fabr.*—The caterpillar of this butterfly has been reported from North Malabar as very destructive to young Cocoa-nut palms.

Rivellia persicæ.—This is a Dipterous insect, which is reported as most destructive to Peaches in Chutia Nágpur, to a great extent preventing their ripening in that district. The larvæ, besides being found in vast numbers in that fruit, occur sparingly in Mangoes and Guavas. An observer writes to **Mr. Cotes**, "I think it safe to assert that in Ranchi from ⅓ to ⅔ of the peach crop is annually destroyed by this pest, according to whether we have early or late rain." It has been suggested, as a remedy, to remove at least six inches of soil from under the trees, during the cold weather, and to kill the contained pupæ by thoroughly baking it.

Diptera (undetermined).—Larvæ, pupæ, and two imagos of a "small dipterous insect not unlike that found on decaying fruit in Bengal" were lately forwarded to the Museum from Simla, but have not yet been identified. These larvæ were found inside diseased grapes in the Bashahr District, and were supposed to be the cause of the disease which has ruined the grape industry of that locality.

Lecanium viride, THE GREEN SCALE INSECT. See Pests of "Forest Trees," p. 149; also "Coffee Pests," p. 144.

L. acuminatum, *Sign.*, has been found on the mango in Ceylon, and is reported to do much damage. It causes the leaves to wither and fall off.

Schizoneura lanigera.—Specimens of a scale insect, supposed to belong to this species, have been recently received at the Museum from Kúnúr, with the information that it has done great damage to nearly every Apple orchard in that district. The insect is said to attack the roots and to be extremely difficult of detection. Tomato planted round the trees is said to preserve them from attack, but this requires confirmation. **Mr. Cotes** recommends kerosine emulsion spray at an early stage.

10. INDIGO.

Chrotogonus sp.—A species of this genus, together with another of Acridid Orthoptera, known as *gadhao* in the North-West Provinces, are reported to have done considerable damage to the young indigo plant in the middle and lower Doab in 1889. They are also said to feed largely on the leaves of Carrots and Cabbages.

11. JUTE.

Spilarctia suffusa.—A moth, probably of this species, has been reported as injurious, while in the larval stage, to jute in Malwa.

12. OPIUM.

THE OPIUM CUT-WORM.—This pest, probably the **Agrotis suffusa** elsewhere mentioned, often does great harm to the young opium poppy in the North-West Provinces and other parts of India. Scott, in his Opium Reports for 1874, 1877, and 1878, describes it as specially destructive to young crops in dry seasons, often stripping *bígah* after *bígah* of plants in a few nights. He estimates that young caterpillars will cut down at least from fifty to one hundred plants in a single night, while another observer estimates that four caterpillars will clear a bed six feet square in a week. **Mr. Cockburn** (Assistant Sub-Deputy Opium Agent, Auchin) wrote in 1889:—"Here, in the Fatehgarh District, their ravages are in full swing among the smaller plants. I have this year seen scores of acres as effectually swept as they could have been by any swarm of locusts, and the larva must do damage to the value of many thousand pounds annually, both to Government and the rayat." The habits of the larva are similar to those of **Agrostis suffusa** already described, whether the species be identical or not, and the same method of destroying the pest might doubtless be equally effectual in both.

Heliothis armigera.—This is another opium pest, which has been fully dealt with by Mr. John Scott in his reports for 1874, 1876, 1877, and 1878. It becomes developed from the egg while the plant is young, and subsists at first on the tender leaves. As the plant matures it eats its way up the stem, and finally bores into and eats the interior of the older capsules, from which much of the drug has been extracted, or which are naturally deficient in narcotic juice. The moth is cosmopolitan and infests many crops. In America the larva is commonly known as the "boll worm," and is perhaps one of the chief agricultural pests of that country. It does the greatest amount of damage to the cotton crop of the Southern States, and Maize also suffers more from it than from any other pest. Tomatoes, Peas, Beans, and other leguminous crops suffer to a less extent, though much damage is occasionally done locally. Mr. Cotes does not state whether this is the same pest which is known in India as the "boll worm" and which does so much harm to our cotton crops.

Several remedies have been suggested and tried in America. Topping the cotton has been much advocated, but Dr. Riley is of opinion that this measure is of little value. Good results have been obtained by catching the moths in lantern traps, and also by basins containing a mixture of molasses and vinegar. Dr. Riley thinks that much may also be done by hand-picking the earlier broods of the larvæ. Spraying the plants with arsenical poisons, such as London Purple and Paris Green, destroys the worms when feeding in the open, but does not reach them if they are in the bolls. Dr. Riley considers that spraying with **Pyrethrum** is a promising remedy, for it appears to destroy the worms both outside and inside the bolls, but this insecticide has only been tried on an experimental scale.

Lasioderma testaceum.—This beetle infests the leaf covering of opium balls, and is also frequently found in Manilla and Indian cheroots. Mr. Cotes writes :—"The writer learns that in England the sale of Indian cheroots (especially Trichinopoly cheroots) is seriously interfered with on account of a boring insect, which is probably this species."

13. SILK.

Dermestes vulpinus, *Fabr.*—THE LEATHER BEETLE—has recently been sent to the Museum with the information that the larvæ attack the eggs, worms, chrysalids, and moths of the mulberry silk-worm. Fully developed insects have been obtained from many parts of India, amongst others from Hazaribagh, where the larvæ are said to attack stored tusser cocoons (*Wood-Mason*). In Rajshahye the insect is reported to be most abundant during the rains, and frequently to destroy cocoons to the extent of R12 per maund (*Cleghorn*).

Trycolyga bombycis—THE BENGAL SILK-WORM FLY—causes considerable loss to silk-rearers in Bengal, but somewhat contradictory statements have been made as to the extent of the evil. Thus Mr. Cotes writes, "Louis estimates this loss in Bengal at between £200,000 and £300,000 annually; Cleghorn notices a loss of five lakhs of rupees in a single crop as indirectly due to the fly; Mukharji recounts how the fly destroyed 90 per cent. of a lot of silk-worms he attempted to rear in Berhampore, while his two village nurseries, which might have been ex-

PESTS.	Pests which affect

pected to yield 40 *khaons* each, if the fly had been kept off, produced only 8½ and 3½ *khaons* respectively." The flies are noted by the last mentioned writer to be specially injurious in the August-September *bund*.

The larvæ are said to be subject to the attacks of a 'midge' (a Dipterous insect, belonging to the family MUSCIDÆ), and it has been suggested that the latter might be bred for the purpose of keeping down the silkworm fly. Too little is, however, known of the life-history of this midge, to admit of this proposal being put forward in a practical form.

Masicera grandis, *Walker,* is a Tachinid fly, said to be parasitic on the Tusser-worm in Singhbúm.

Flacherie—is a fungoid disease which attacks the Mulberry silkworm, and has recently been found also to affect the Eri worm in Assam. See **Silk.**

14. SUGARCANE.

Dæatræa saccharalis, *Fabr.*—THE "SUGARCANE BORER MOTH "—is fully described by Mr. Cotes. The larva of this insect commits great depredations in sugarcane fields, boring into the stalks, often thereby setting up putrefaction, so that the stalks become worthless. **Mr. Cotes** writes, "Sugarcane, in different parts of the world, has, for at least the last hundred years, been known to be subject to the attack, either of this pest or of others so closely allied, as to be scarcely distinguishable from it, and during the last year information has been sent to the Museum of damage done to sugarcane in several parts of India, where the pest would seem to have long been known, though but little has heen recorded concerning it." Amongst these reports is one written in 1857 which describes the total destruction by the pest (*dhosah*) of an imported form of sugarcane (known as Bombay or Red sugarcane) in the districts of Rungpore, Hooghly, and a portion of Burdwan. Another describes the insect as having done great damage in Dhulia in 1885; while in 1888, reports were received from the North-West Provinces that the pest, which appears in dry seasons, had destroyed as much as one-fourth of the crop in the neighbourhood of the Cawnpore Experimental Farm. In the same year the manager of the Dhankora Wards estate wrote that the pest (known locally as *mandranah*) had done considerable injury to the crop; and the Collector of Ganjam also noted injury done by the same insect, known in that district as *monjíkila purugu*. A similar report to that received from Cawnpore came also from the Agricultural Officer of Burdwan and Seebpore, where the pest is known as *majera*. This insect is almost universally supposed to make its appearance only when moisture is deficient.

Mr. Cotes writes, "A large number of remedies have been proposed for the pest, and it seems to be pretty well established that it can be to a great extent controlled by burning or burying all the discarded tops, and clearing the fields of all waste sugarcane stalks after the crop has been taken; for, as the insect passes the winter as a larva inside the sugarcane, if these are destroyed there are no moths in the spring to lay the eggs which produce the next year's 'borers.' The waste tops, however, should be carefully gathered together and removed from the field before being burnt, for if they are burnt carelessly on the field itself, many predaceous insects will be liable to be destroyed, which take shelter in the ground and assist in reducing the number of pests."

Dragana pansalis.—This small moth has been reported to be injurious to sugarcane. Nothing is known of its life-history, nor of the injuries it causes.

15. TEA.

Helopeltis theivora, *Wa'erhouse,* THE MUSQUITO TEA-BLIGHT, *Wúhonce,* ASSAM,—has been fully described by Mr. Atkinson (*Indian Museum*

Notes, 180-187). This insect is the most serious of all tea pests in Assam, where it is known by the above name, or as "black blight or smut." It appears to attack the crop under all conditions of soil and climate. At first the buds and young leaves and shoots alone are attacked, and the more tender and succulent the shoots, the more they suffer. These shew at first only a few small brown spots, but after a time they become completely brown and withered. The dead leaves fall off, the young shoots freshly formed are in their turn attacked, and the whole plant finally loses vitality to such an extent that it ceases to make any endeavour to shoot out. If the insect attack the plants two years in succession they die altogether, and become covered with moss, lichens, &c. On one tea estate in Sikkim alone, the loss caused by this blight in 1889 was estimated at 300 maunds of tea valued at R20,000.

Many remedies have been recommended, amongst others picking off the insects, cutting down the surrounding jungle, anointing the bushes with tar, spraying with kerosine emulsion, &c. Mr. Atkinson considers that the last mentioned remedy appears to promise good results. Cutting down the jungle can at least do no harm, and according to several reports does much good, for the pest lives on and amongst the jungle grasses, creepers, and trees. One planter highly recommends thorough pruning, burning the leaves and twigs in the neighbourhood of the tea bushes, and carefully hoeing round the plants, removing the stale soil, and filling in fresh earth.

Ceroplastes sp., is described by **Mr. Atkinson** as attacking tea in the Kángra Valley. Its attack is evidenced by small ceroid nodules about the size of a pea, attached to the twigs. **Mr. Atkinson** writes, "I do not think that there is any danger of this insect doing much damage to tea. If it does become troublesome, the application of kerosine emulsion by spraying the leaves containing the larvæ will quickly destroy them, and prevent their spreading."

Tetranychus bioculatus, THE "RED SPIDER," is a most destructive pest on tea in Sikkim, where it is reported to do even greater damage than the mosquito blight. It attacks the crop in spring and early summer. Sprinkling with flour of sulphur and washing with a solution of whale oil soap (1℔ of soap to 5 gallons) have been found useful against the allied rust mite of orange trees in Florida, and would probably be efficacious against the Red Spider.

Diapromorpha melanopus.—A Phytophagous beetle, which is said to attack the tea plant in Assam, eating the shoots so that they wither and droop.

Dasychira thwaitesii, *Moore.*—The larva of this moth has been reported to be very destructive to the tea plant and the *sál.* In 1878, it is said to have denuded the *sál* trees in a forest extending over a tract of two hundred square miles in the Eastern Doars and Goalpára; and in 1888, tea gardens in the same district sustained serious damage from the pest.

Eumeta cramerii.—A larva of a Psychid moth; probably this species is said to have attacked a tea garden in Assam, in the autumn of 1889, and to have spread over half the garden.

Oscinis sp.—The larva of a new species of this genus was received at the Museum in 1889, from Ceylon, with the information that it attacked the old leaves, boring below the cuticle. It appears to be of little importance.

Lecanium coffeæ.—THE BROWN COFFEE SCALE INSECT—occasionally attacks tea. See "Coffee Pests," p. 144.

L. theæ has been reported from the Kángra Valley. It is similar to

PESTS.	Pests which affect

the coffee pests of the genus, and can no doubt be destroyed in the same way by kerosine and soap emulsion.

Psychidæ (undetermined).—Larvæ of a Psychid moth have been sent from Ranchi, with the information that they attack tea, *sál*, and other plants. In the case of tea they are removed by hand-picking.

Eumeta cramerii, an allied species to the above, has long been known as injurious to tea in Darjíling while in the larval stage.

16. MISCELLANEOUS.

LOCUSTS—Have from time to time committed extensive depredations in many parts of India. The term is employed in a very loose manner, but in India ought to be restricted to various species of ACRIDIDÆ which have at different times invaded the country. At the present time the literature of the subject is very scanty, being confined to accounts of their depredations at various times, with little or no attempt at scientific classification. **Mr. Cotes**, however, states that he has under preparation a detailed account of the various species of the ACRIDIDÆ which have at different times invaded sections of India. This will doubtless prove a valuable addition to the literature of the subject.

The year 1889 was marked by a general invasion of locusts, which spread over Sind, Rájputana, the Panjáb, North-Western Provinces and Oudh, besides penetrating sporadically into Guzerát, Ahmedabad, Baroda, Khándesh, and other parts of Central India, and even extending so far as the Kistna District of Madras. Specimens forwarded to the Museum all proved, according to **Mr. Cotes**, to be **Acridium peregrinum,** "which is said to range throughout the dry country from Algeria on the west, to North-Western India, on the east." "This species has often proved most destructive in Algeria, and is supposed to be the locust of the Bible, but it must not be confounded either with the locust which has appeared in Algeria during the past two years, or with the locust which invaded the Deccan in 1882-83, though the latter insect was often referred to in reports under the name of **Acridium peregrinum.** The locust which has proved destructive in Algeria during the past two years is **Stauronotus maroccanus,** while that which invaded the Deccan in 1882-83 probably belonged to the species **Acridium succintum.** It is particularly necessary to distinguish carefully between the latter and the Rájputana species, for there are important differences in their habits which make it probable that measures applicable for the destruction of the one may not always be successful with the other."

WHITE ANTS.—The ravages of these pests in many parts of India are too well known to require recapitulation in this place. They appear to be particularly injurious to tea and sugar cane, of which they are said to eat up the roots and thus cause complete destruction of the crops. They can be more or less effectually checked by heavy watering, and by constant deep-hoeing.

CLOTHES MOTHS.—Several larvæ are very destructive to clothing, among which one may be mentioned, probably that of **Tineola baseliella,** *Hum.*, which was received at the Indian Museum after having destroyed a bale of country blanketing. Specimens were also received of **Tinea tapetzella.** The most efficacious method of guarding against their attacks appears to be by pouring a little naphthaline into the centre of a bale.

LEATHER-BEETLES, see SILK PESTS, p. 151.

Tinea lucidella, *Walker.*—The larva of this Micro-lepidopterous insect is said to be "very destructive to the horns of hollow-horned ruminants." No further information is available regarding it.

Oestridæ. The insects of this family, generally known as "Bot Flies," appear to do much harm in India. **Miss Eleanor Ormerod** has noticed that hides shipped from Calcutta, Madras, and Bombay, are occasionally damaged to the extent of 50 per cent. on 10 per cent. of the skins, while certain consignments from Karáchi had an average of about a quarter damaged to the extent of 60 or 70 per cent. From a recent report by a Calcutta hide merchant, it would appear that damaged or "warbled" hides are chiefly found in consignments which come down from the North-West Provinces and Pánjab, fully 50 to 75 per cent. of which are affected, while in Bengal proper there are no "warbles" at all. The affected hides are chiefly found amongst those which come into the market between November and January. The damage is sometimes so great as to reduce the price to but an eighth of the value the goods would otherwise have. Hides are most affected, but deer, horse, and goat skins also suffer.

Mr. **Cotes** writes, "Beyond the fact of the very serious loss which is annually occasioned by Bot Flies in India, little seems to be known about them, though it is probable that their habits are very similar to those of Bot Flies in Europe, where, however, it is likely that the insects belong to a distinct species. Miss Ormerod has shown that much can be done to prevent injury from these insects in England, and it would appear most desirable to study them in India with a view to recording their life-histories, and ascertaining to what extent it may be possible to combat them here."

Aulacophora abdominalis, *Gemminger & Harold.*—" Has been sent to the Indian Museum as destructive to various crops in different parts of India. In Saharanpur it was found to be destructive to all CUCURBI-TACEÆ; London Purple insecticide was tried upon it, but this appeared only to have the effect of making the beetle fly off the plant. Elsewhere in the North-West Provinces it attacks Water Caltrop (**Trapa bispinosa**). In Nuddea it was reported to be injurious to plants and vegetables. In Ganjam it was destructive to cotton, red gram, and cucumber " (*Cotes*). Its habits do not appear to have yet been studied, and no reliable information is available as to its life-history, the amount of injury it occasions, nor as to the most promising methods of dealing with it.

Bruchus chinensis.—This beetle, known in Ganjam as *pesalu purugu,* and in Nuddea as *ghora poka,* is reported to infest Green Gram. A large grey species of the same genus attacks stored Peas in Calcutta, eating out the interior and leaving little more than an empty skin.

Aulacophora abdominalis and **Palæopeda sexmaculata**. are reported to destroy cotton, red gram, and cucumber in the same district, where they are known as *kunkudiya purugu.* The former is probably the true pest.

Shizodactylus monstrosus.—Specimens of this insect were forwarded to the Museum from Darbhunga (where they are known as *bherua*) with the information that it cuts the roots of young plants of *mokai,* tobacco, *morwa,* and other crops growing on high lands, and also injures the leaves of tobacco and cauliflower.

Nezara viridula, *Linn.,* THE GREEN BUG.—Mr. Atkinson writes, " This cosmopolitan insect, found almost in every country in the world, has been reported as occurring on potato halms in Bangalore."

Diatræa saccharalis, *Fabr.,* THE SUGARCANE BORER.—A caterpillar identical with or very nearly allied to this pest has been found doing injury to *brinjál,* paddy, and sugarcane by boring into the stem. It is known in Ganjam as *monjikila purugu.*

Many other caterpillars of less importance injure these and other crops throughout the country.

INSECTICIDES.

With the increase in the knowledge of the habits and life-histories of insect pests, numerous means of destroying them by chemical solutions

PETROLEUM.   Petroleum.

INSECTICI-
DES.

have been tried. Of these the following, issued by the Entomological
Bureau of the Washington Department of Agriculture, are perhaps the
most useful and generally employed :—

LONDON PURPLE.—To 20℔ of flour from ¼ to ½℔ of the dye is added,
and mixed well. This powder is applied with a sifter or blower, or after
mixing the same amount with 40 gallons of water, it may be used as a
spray.

PARIS GREEN.—From ¾ to 1℔ is mixed with 20℔ of flour, or with 40
gallons of water and applied in the same way.

BISULPHITE OF CARBON.—For use in the ground, a quantity is poured
or injected among the roots that are being infected. Against insects
damaging stored grains or museum material, a small quantity is used in an
air-tight vessel.

CARBOLIC ACID.—A solution of one part in one hundred of water is
used against parasites on domestic animals, and insects infesting barns
and sheds ; also on the surface of plants and among roots in the ground.

HELLEBORE.—The powder is sifted on the affected plants alone, or
mixed (one part to twenty) with flour. With one gallon of water ¼℔ is
used for spraying.

KEROSINE MILK EMULSION.—To ¼ of milk add two parts kerosine,
and churn by means of the force-pump or other agitator. The butter
milk emulsion is diluted *ad libitum* with water. An easier method is
simply to mix one part of kerosine with eight of milk.

SOAP EMULSION.—In one gallon of hot water, 1½℔ whale oil soap is
dissolved. This may be mixed with kerosine instead of the milk above
described in the same manner and proportion.

PYRETHRUM (PERSIAN INSECT POWDER)—May be blown or sifted
on dry, or applied with water, in the proportion of one tea-spoonful to a
gallon, well stirred and then sprayed.

TOBACCO DECOCTION.—This is made as strong as possible, and is
used as a wash or spray to kill insect pests both on animals and plants.
Of these insecticides the kerosine soap emulsion is highly recommended in
the Indian Museum Notes as a cure for all " scale insect " blights. It is
directed that the liquids should be at blood heat ; that the emulsion should
be diluted with from nine to fifty parts of water, according to the insect
to be dealt with, as well as to the nature of the plant ; and that it should be
applied by means of a force pump through a spray nozzle, the best of which
is the " eddy " or " cyclone " nozzle.

London purple has also been used in India with considerable success
against several pests, notably the mango Cicadid, and the larva of **Papilio
erithonius.**

Dr. G. Watt has recommended **Adhatoda Vasica,** a common wild plant
all over India, as likely to prove a valuable insecticide in at least some
cases. The smoke of *Mauha* oil cake, made from the seeds of **Bassia
latifolia,** has been reported to prove efficacious in blight on rice and other
plants.

435

PETROLEUM.

A variety of liquids known by the names of petroleum, earth-oil, naphtha,
maltha, mineral tar or oil, Erdöl, Steinöl, bitume liquids, and corresponding
with vegetable and animal oils in the characters of inflammability and insolubility in water, occur in many localities. They all agree in possessing a marked
bituminous odour, but differ considerably in other physical characters. Thus
some are thin, transparent, and clear coloured, others are thick or viscid,
opaque and almost black. The name NAPHTHA or ROCK OIL has generally
been applied to the former class, the latter have been termed MINERAL TAR,
while intermediate forms are known as PETROLEUM.

The countries in which these mineral-oils are produced in greatest abun-

Occurrence of Petroleum in India. (*J. Murray.*) PETROLEUM.

dance are, Persia, the Caucasus and Georgia, Burma, the West Indies, and North America, and by far the largest amount is obtained from the Palæozoic rocks of the United States. They also occur in less abundance in the country to the north of the Danube, in Italy, Bavaria, Hanover, Zante, Switzerland, China, India, and to some extent in England and France.

Petroleum has been known by civilized man from the earliest times. Herodotus described the springs of Zacynthus (Zante) and the fountains of Hit are frequently mentioned by Persian and Arabian writers. Pliny and Dioscorides describe the oil of Agrigentum which was used in lamps under the name of "Sicilian oil." Mention is made of petroleum springs in China in the earliest records of that country. The abundance of petroleum in the neighbourhood of the Caspian and the fire-temple at Baku, have formed sub-jects of description at the hands of many travellers from the time of Marco Polo onwards.

The occurrence of petroleum in North America (the country from which the greater part of the world's supply has practically been derived for many years) was first noticed by a Franciscan Missionary in 1629. In extra-peninsular India petroleum has long been known and its value recognised by the Natives. The most important source (the wells of Burma) are said to have been known to a certain extent for over 2,000 years.

Petroleum, *Ball, in Man. Geol. Ind., III. (Economic), 124-154.*　　**436**

Vern.—*Mitti-ká-tél,* HIND.; *Matti-ká-tailam, matti-ká-tél,* DEC.; *Mátiyá-tail,* BENG.; *Kala salajit* (=bitumen), NEPAL; *Salájit* (=bitumen), KUMAON; *Matti-cha-téla, minak tanah,* MAR.; *Matti-nu-tél,* GUZ.; *Man-yenney, man-tayilam,* TAM.; *Manti-tayilam, bhúmi-tayilam, manti-núne,* TEL.; *Mannu-yanne,* KAN.; *Man-tailam,* MALAY.; *Yé-ná, yená, yenan,* BURM.; *Pruthvi-tailam,* SANS.; *Nift, qafral-yahúd,* ARAB.; *Kafral-yahúd,* PERS.

References.—*Baden Powell, Pb. Prod.,* 20, 58, 115; *Mason, Burma & its People,* 576, 735; *Hunter, Statistical Acct. of Assam, I.,* 281, 290, 299, 379; *II.,* 340, 427; *Yule & Oldham, Mission to Court of Ava,* 18, 316; *Balfour, Cyclop., III.,* 191; *Govt. Rep. on Burma Oil-wells, Rev. & Agri. Dept., Nov. 23rd, 1888; Noetling, Report on Oil-fields of Twing-oung and Beme, 1889; Townsend, Reports on Baluchistan Oil Explora-tions; Report on the Assam Petroleum Deposits, Rev. & Agri. Dept. Pros., No. 6, 1889; Medlicott, Note on the Occurrence of Petroleum in India, Recs. Geol. Survey, Ind., XIX., Pt. 4, 1886; Oldham, Rep. on the Geology &c., of the country adjoining the Sind-Pishin Railway, Recs. Geol. Surv., Ind., XXIII., Pt. 3, 1890; Rev. & Agri. Dept. Pros., Nos. 3 to 9, and 12, 1890 (regarding Shiráni hills); Warden, Rep. on Petroleum from Shirani Hills, (Sept. 30th, 1890); Oldham, Prelim. Rep. on Oil near Moghal Kot (Sheráni hills), Nov. 29th, 1890; Redwood, Petroleum Deposits, India, in Jour Soc. of Chem. Indust., April 30th, 1890; Watts, Chemistry, IV., 383; VIII., pt. ii., 1508; Govt. of India, Rev. & Agri. Dept., Statement of Quantites & Values of Indian Minerals, 1889; In-dian Agriculturist, March 17th June 9th, and 16th, July 7th, 1888; Feby. 2nd and 23rd, June 1st, July 6th and 27th, August 3rd, Oct. 5th, 12th and 26th, Nov. 23rd, Dec. 14th, and 21st, 1889; Settlement Reports: —Bannu,* 91, 92; *Kohat,* 32; *Gazetteers, Bannu,* 21; *Also many papers in the Jour. As. Soc: Beng., and the Rec. Geol. Survey Ind., for an enumeration of which the reader is referred to the Man. Geol. Ind., Vol. III.,* 605.

Occurrence.—The following account of the occurrence of petroleum in OCCURRENCE. India is mainly compiled from Ball's exhaustive article, with the addition **437** of more recent information on the oil wells of Burma, Assam, and Balúchis-tán, obtained from several sources.

According to Ball, petroleum, so far as our knowledge goes, is wholly absent from Peninsular India; supposed discoveries, which have been made from time to time, having proved to be fallacious. Perhaps one of the most notorious of these cases occurred at Khona Oopalapad in Madras; for there the supposed petroleum proved to be merely a substance derived from the accumulated droppings of bats. "Passing from Peninsular

OCCURRENCE

India," writes Ball, "to the extra-peninsular countries where the rocks belong to formations closely related to those of Europe, we meet with abundant supplies of petroleum, some of which are of considerable economic importance. The nature of this product varies at different localities; in some it consists largely of the lighter hydrocarbons, such as naphtha; in others the heavier and less readily combustible varieties prevail, and these, on exposure to the atmosphere, become oxidised and change into the condition of asphalt. There can be little doubt that the formation of petroleum is intimately, though obscurely, associated with the presence of salt, otherwise it would be difficult to account for the simultaneous occurrence of petroleum and lime springs which have been observed in India as well as in Pennsylvania and Virginia." Of recent years the oil-industry in Assam, Burma, the Punjab, and Baluchistan has received much attention and undergone considerable expansion. In Assam the practical results are as yet small, and in the Punjab little success has attended European enterprise. But in Burma and Baluchistan the outturn has greatly increased, and in 1889 the production of all India is returned at 3,298,737 gallons, valued at R2,51,114.

The following are the more important localities from which petroleum has been reported :—

Alwar.
438

ALWAR.—A bituminous deposit was discovered in this State, in 1874, near Tijarah. Surgeon Ffrench Mullen reported in that year that the deposit had been found in two fields, in patches from 3 to 4 inches thick, and at an average depth of 3 inches from the surface. The substance was turned up by the plough and was found to be combustible. The Chemical Examiner of the North-Western Provinces and Oudh found that it yielded 25·56 per cent. of bitumen, and 3·72 of fixed carbon, while other samples forwarded to the Geological Survey Office yielded from 30 to 60 per cent. of combustible matter. "The bitumen had probably been formed by the alteration of the cellulose of an accumulated mass of vegetable matter, in contact with a saliferous soil " (Ball). The small quantity obtained rendered the discovery of no real value. This deposit was again investigated in 1887 by Mr. R. D. Oldham, with the result that the description in the Manual of Geology of India was fully confirmed, and the material was again stated to be of no value.

Cutch.
439

CUTCH.—Small resinous and bituminous masses have been found in the friable brown shales contained in the sub-nummulitic and next succeeding beds in this district, near Mohurr, Julerai, and Lukput. They are called bhut khana, or " spirit-food," in the first mentioned locality and are burned as incense in the native temples.

Baluchistan.
440

BALUCHISTAN.—The existence of petroleum at Khátan, in the Mari Hills, has long been known to the Natives, Khátan being indeed primarily the name of the substance, and only secondarily the name of the place at which it occurs. It is within only the last few years, however, that any attempt to explore the region has been made. With the opening up of the locality by the Sind-Pishin Railway, Government recognised the great value of a fuel, such as petroleum, within easy access, and organised a petroleum exploration under the superintendence of Mr. Townsend. Oil was found flowing in small quantities from the surface, and issuing from fissures in rocks along with an abundance of hot sulphurous matter. Besides this, abundant traces were discovered of former flows in the river gravels at Khátan, and at intervals up the valley for at least 10 miles.

The petroleum at Khátan exudes close to sulphur springs. Three trial borings were at first made, one of which was 524 feet deep and 4¾ inches in diameter. Oil was obtained at 28, 62, 92, 115, 125, 133 and

has been reported to be found.   (*J. Murray.*) PETROLEUM.

374 feet, but owing to breaks and fractures in the rock, the process of | OCCURRENCE
boring was found to be extremely difficult. | Baluchistan.

According to the official report of the Director General of Railways for 1887-88, experiments made by pumping four of the wells in the preceding year showed that the yield of each was from 400 to 600 barrels in the 24 hours. " Thus any one of the existing wells is more than competent to deliver the entire supply of 50,000 barrels of oil a year, which is estimated to be the amount required for the Sind-Pishin Section of the North-Western Railway." In the eleven months ending December 1889, 218,490 gallons of oil were actually despatched from Khátan, the highest monthly total having been 39,070 gallons in June. The works suffered somewhat from the unprecedentedly heavy rains in the following months, with the result that much oil was lost and the outturn became consequently greatly decreased. The total outturn in Baluchistan during the year 1889 was 309,990 gallons, valued at R17,714. Mr. Oldham, in a *Note on the Country adjoining the Sind-Pishin Railway*, written in 1890, states that at the time of his visit four wells were being pumped, of which one was yielding 10 barrels of oil a day; in the other three a good deal of water was being raised, mixed with oil to the extent of 10, 6, and 4 barrels a day respectively, or an average of 7½ barrels a day. "At this rate," he remarks, " it would require at least 25 wells to supply the 50,000 barrels per annum, which is the amount that would be required by the railway alone. It is possible that future borings may yield a higher average than this, as the rocks at the site of the present borings are very much shattered and intersected by fissures, along which the oil finds an outlet to the surface, while to the south of Khátan, there are some three or four square miles over which the shales lie horizontal or nearly so. The cover here is much better than at Khátan itself, and it may be that oil would be found under greater pressure than in the present borings. The experiment will probably be tried in the present working season." In 1889, Mr. Townsend made an experimental boring near Kirta on the Bolan, and at 360 feet found a very good show below a thin stratum of limestone. He writes, " I had great hopes of the Kirta locality furnishing a supply of oil in paying quantity."

In Mr. Oldham's paper above quoted, it is stated that shows of petroleum have also recently been found between Spintangi and Harnai in the Dunghan limestone of the Harnai Valley, but none of the oil actually flows on the surface. This Mr. Oldham thought to be probably due to the fact that the conditions under which it exists do not allow of its rising to the ground. He further arrived at the conclusion that there is probably an abundant underground supply of oil in this district, and that borings judiciously placed would yield it in paying quantities; at any rate, that the experiment was worth trying. It is believed that orders have recently been issued to make experimental borings near Spintangi as soon as possible. The locality, in its nearness to the railway and superior climate, possesses many great advantages over Khátan, and the experiments, if successful, would doubtless yield most valuable results.

Unfortunately the Khátan oil is of remarkably high specific gravity and viscosity, which greatly increases the difficulty of working it. Mr. Boverton Redwood, reports on it as follows:—The oil is black or extremely dark-brown in colour by transmitted light, with comparatively little fluorescence, and it possesses very little odour. Its flashing point is 280° F. (Abel's test) and it contains no hydrocarbons available for use as ordinary burning oil. If 50 per cent. be distilled off, the first third of the product will be found to have a specific gravity of about ·910, and the remaining two-thirds of about ·930, but the viscosity of

PETROLEUM.   Localities in which Petroleum

OCCURRENCE

Baluchistan.

these distillates is extraordinarily low in relation to the specific gravity. By careful distillation with superheated steam about one-fourth of the product (one-eighth of the crude oil) may, however, be obtained of sp. gr. ·958 and viscosity 168 at 70° F. (rape oil at 60° F.=100). This oil would be available for use as a lubricant. The residue in the still, where 50 per cent of the volatile matters have been removed, forms, on cooling, a moderately hard pitch of jet black colour and lustrous fracture. The material might be employed in the manufacture of "patent fuel" and for other purposes. When distilled to dryness the crude petroleum yields rather more than 70 per cent. by volume of a heavy oil containing very little paraffin, and the residue then consists of about 26 per cent. by weight of hard coke. The last portions of the distillate when exposed to a temperature of 32° F. assume the consistence and appearance of vaseline.

Khátan petroleum is by far the most refractory oil to which **Professor Dewar** and I have applied our new process of distillation, but we have succeeded in obtaining from it about 40 per cent. of kerosine, with a sp. gr. of ·810" (*Redwood, Oil Fields of India, Jour. Soc. of Chem. Indust., IX., No. 4*).

The petroleum in question is, however, likely to be of value chiefly as a fuel, and from this point of view is of the utmost importance. It has been proposed to get over much of the present difficulty of imperfect means of transport, by removing the oil from the field to Sibi by a pipe-line, a project which is, however, likely to be rendered difficult by the great viscosity of the oil.

It is to be hoped that the efforts still being continued may be productive of even more marked success than in the past, for not only must an excellent fuel of this description be of the greatest value to the Government for railway purposes, &c., but in a country such as Balúchistán, where fuel is scarcely obtainable at all, it would prove of the greatest value to the people. In addition to other advantages, an abundance of petroleum fuel might also be expected to put a stop to the destruction of forests, which until recently has been going on at an alarming rate.

The attention of the Government of India has been drawn this year (1890) to a mineral oil said to be obtained from Terai in the Shiráni Hills near the Chin Kheyl village of Mogul Kot, in the vicinity of Dera Ismail Khán, and brought to notice by the Superintendent of Police of that district. It is said to issue from small springs in a hill in the pass of the district occupied by the Tsor Kheyl tribe of Shiránis, to be collected by the natives in small holes dug in the ground, and is used for outward application to sores on horses, &c. A sample collected by a Shiráni Police Sergeant, acting under the orders of the Superintendent, was sent to **Dr. Warden**, Chemical Examiner to the Government of Bengal, who reported that it had a sp. gr. at 15·5°C. of 8209, a flashing point by the Abel instrument of 91° F., and that it was of lemon-yellow colour, aromatic odour, and marked fluorescence. He believed the sample not to be a crude oil, but a commercial kerosine oil of Russian origin. Another sample, guaranteed genuine, was again sent from the Panjáb Government and was reported on by **Dr. Warden** as equal to American or Russian commercial kerosine.

He states that it is of a rich straw colour, is perfectly clear, has no sediment, and possesses a well marked blue fluorescence. It is not quite so aromatic as American kerosine, has specific gravity of ·8154 at 15·5°C, and on cooling to 0°C becomes opalescent and of the consistence of *ghí*, from crystallization of paraffin or other solid hydrocarbon. Its flashing point is 84·29° F. (Abel's test). On being subjected to fractional distilla-

P. 440

Products of India. 161

has been reported to be found. ((*J. Murray.*) PETROLEUM.

tion, it yielded 90 per cent. of kerosine with sp. gr. between ·7557 and ·8596. The lowest fractions were practically colourless, the intermediate had faint fluorecesnce, while the last 30 per cent. had a marked yellowish fluorescence. The fraction distilled after 90 per cent. amounted to 7·6 per cent., and was a thickish yellow unpleasant smelling oil, which became of the consistence of *ghi* at 15°C. The small residue left in the retort was solid at 85° F., and had a dark reddish-brown colour. Dr. Warden continues : " Equal volumes of the first nine fractions (90° per cent. of the whole) were mixed and treated with sulphuric acid and caustic soda, the purified oil had a sp. gr. of ·8008 at 15·5C., and a flashing point of 85·29° F. (Abel's test). In colour it was nearly ' water white,' with a slight fluorescence.

"It will be noted that the amount of burning oil, afforded by distilling the sample under report, exceeds that yielded by any of the specimens examined by **Mr. Redwood**" (see table, p. 175). "The highest yield **Mr. Redwood** obtained was from the Persian oil, which amounted to 87·5 per cent. But the sample under report might be placed on the market without any previous distillation, its specific gravity and flash point being those of commercial kerosine. The colour is rather higher than oils usually now met with, but certainly not darker than the grades formerly imported, while by distillation, on the other hand, at least 90 per cent. of a very superior nearly ' water white ' oil would be obtained. Assuming the sample of oil to be what it is stated to be—crude petroleum—its value commercially, if it can be procured in sufficient amount, can hardly be over-estimated. A large supply of a natural oil of this quality would simply drive out all foreign oil from the market ; there could be no competition." Since the above was written a report has appeared on the Shiráni oil locality by **Mr. Oldham, F.G.S.**, who has personally visited and inspected the springs. He found that the petroleum issues in several spots from a band of hard, unfossiliferous sandstone, probably of cretaceous age and about 600 feet in thickness. The most abundant springs are close to the base of this band of sandstone, and the actual points of issue, determined by the profile of the outcrop of the oil-bearing strata, are in the river-bed. Traces of old flows are found for a height of 60 feet above the water, and thus, according to **Mr. Oldham**, demonstrate "that the flow of oil has been continuous for at least the period required to deepen the gorge by 60 feet." The oil, as it issues from the rock, is clear, limpid, of a pale yellow colour, and perfectly free from water, and leaves no reason for doubt as to the genuine character of the samples previously submitted. The actual outflow is probably not more than 10 gallons a day, but this could doubtless be increased to some extent by pits or borings. **Mr. Oldham** concludes as follows :—"The general conclusions to be drawn from what I have seen are, that there can be no doubt of the existence of oil of excellent quality and of great value, in the district, but that it would be premature to undertake any expensive operations at present. It is, however, important that, as soon as the country is sufficiently settled to allow of it, a thorough and systematic exploration should be undertaken, with a view to determining whether there are any localities where it is probable that oil occurs in sufficient quantity, and at a depth which would render its profitable extraction possible."

AFGHÁNISTÁN.—Bituminous products are believed to occur in several localities, and are commonly sold in the bazárs under the name of *mumiai* or *momiai*. A mineral pitch bearing this name is said by **Captain Hutton** to be found in the Shah Makhsud range, but certain samples collected elsewhere have been found to consist of the excreta of birds or bats, mixed with salts of lime, and to contain no trace of bitumen (see Vol. II., pp. 115-116).

OCCURRENCE
Baluchistan.

Afghanistan.
441

11

OCCURRENCE

Panjab.

442

I. PANJÁB.—Petroleum exists in several localities in this province, and is utilised to a considerable extent for local Native consumption. In 1870 a Mining Engineer, Mr. Lyman, brought out by Government to report on the petroleum-yielding districts of India, made an exhaustive enquiry into the oil of the Panjáb. His general conclusion was, that by properly opening up the known localities about 100 gallons a day for eight years might probably be obtained. The accumulated asphalt, he thought, might further yield 100 gallons a day for three or four years, so that in all twelve years' supply might be forthcoming for the gas works at Rawal Pindi at this rate of consumption, after which the railway might make other gas material available. The Superintendent of the Gas Works submitted a report in 1881, from which it appears that the total quantity of oil collected during the preceding year, 1880, was 2,850 gallons, at a total cost when brought to Rawal Pindi of R1,317-0-3, or 7 annas 4¾ pie per gallon. The average production of gas per gallon of oil during the year was 320 cubic feet, with an illuminating power equal to from 14 to 15 standard candles.

The districts from which oil has been reported are the following :—

Shahpur.—Deposits of a tarry or asphalt-like exudation occur on a sandy bed at Duma in this district. Mr. Lyman reported unfavourably on the substance, and discouraged the idea that boring would increase the yield to a profitable extent. Similar reports were made regarding a tarry exudation at Chinnur. A thick bituminous rock oil has also been obtained from a brownish-grey sandstone bed at Hanguch, near the village of Dhuddow.

Jhelum.—Small and unimportant exudations are said to occur at Sadiali and Sulgi.

Bannu.—Three or four springs exist within a distance of 60 yards along the western bank of the Barra Kutta, and at about the water-level, 1¾ miles south of Jaba, 10½ miles south-east of Kalabagh, and 95 miles south-west by west of Rawal Pindi. The oil comes from fissures in a gray limestone rock through a space of about 100 feet in thickness. The main spring is some 6 feet long, a foot or two wide, and quite shallow. The oil is at first dark green, but quickly turns dark brown or black, and tarry. Mr. Lyman examined the spot and estimated the natural yield at three pints daily, but considered the locality of sufficient promise to render boring advisable. The sandstone bed is continuous with that at Chota Kutta, and may possibly be oil-bearing throughout.

Since Mr. Lyman's report, these wells have been utilised to a certain extent. Five springs were found in which the oil rises with the water and is skimmed off the top, and from a small bore hole 4 inches in diameter and 12 feet deep about 1 gallon a day has been obtained. In 1880, the total yield from this locality was 1,400 gallons. At Chota Kutta similar springs occur, three of which are situated on the eastern side of the stream near the water's edge. The oil rises with, and floats upon, water. The estimated natural yield is said to be 3 gallons a day.

Considerable deposits of asphalt occur at Basti Algad, a locality 10½ miles south of Isa-khel, and 124 miles south-west by west of Rawal Pindi. They are said to extend in spots for about a quarter of a mile along the east side of a brook, and to amount in all to about 350 cubic yards.

Kohát.—Springs exist near Panoba, some 87 miles distant from Rawal Pindi, which have been reported to yield, naturally, from half a gallon to 5 gallons a day. The oil is used for burning by Natives.

Rawal Pindi.—Petroleum has been found in many localities in this district, the chief of which are Dulla, Jafir, Boari, Churhut, Gunda or Sudkal, Lundigar, Basala, Chirpar, and Rata Otur. At Dulla (a locality

has been reported to be found. (*J. Murray.*) PETROLEUM.

38½ miles due west of Rawal Pindi and 16 miles west-north-west of Futtehjang), there is no liquid oil exposed, a small amount of accumulated asphalt only having been found. Borings do not promise success. At Jafir boring has revealed a trace of oil, but none is now to be seen at the surface. The quantity of petroleum apparently obtainable at Boari is also small, and the rock from which it has been drawn appears to be only very locally bituminous.

An oil well occurs at Churhut, in rocks of nummulitic limestone. The yield, however, is small; operations carried on in 1869 resulted in only half a gallon daily, and it is not considered likely that borings would be successful. Wells are also found at Gunda or Sudkal about 23 miles slightly. south-west of Rawal Pindi and 2½ north-west of Futtehjang, in nummulitic rocks. Oil was first dug for in this locality in 1866, at which time one well, known as the "main well," yielded about 5 gallons a day. When deepened to 35 feet it yielded at first about 25 gallons, and when a boring was made to a depth of 75 feet from the surface, the first day's yield was about 50 gallons. Between April and October 1870 1,963 gallons were collected from this source. Mr. Lyman estimated at that time, that about 1,000 gallons might be expected, after which the yield would be insignificant. The amount of asphalt on the surface was estimated at about 19 tons, which would give about 11,000 cubic feet of gas. Various borings were recommended by Mr. Lyman to further develop the yield. Ball states that about 1880, the annual yield exceeded 2,000 gallons.

Unimportant traces of petroleum occur at Lundigar near Murat, at Basala, and at Chirpar, and a deposit of asphalt at Rata Otur, at which place Mr. Lyman recommended a small experimental boring.

In March 1888 a concession was granted by Government to Mr. Noble (an American oil refiner of great experience, who represented a syndicate of Canadian capitalists) for the development of the earth oil deposits of the Panjáb. This agreement granted to the concessionaire the exclusive right of boring for oil in the northern part of the Panjáb, with the power (if the boring should prove successful) of selecting five square blocks of land at different points, each block containing ten thousand acres. Up to date it may be gathered that prospects have not been very successful. The total outturn in the province, during 1889, is returned as 2,830 gallons, valued at R1,073.

Hazára.—According to Captain Abbott three springs exist in the Serra mountains, one yielding bitumen, another sulphate of iron, and the third a "mucilaginous substance, resembling the pulp of an orange, and having a pleasant odour." This is probably paraffin. Torches are made of the bitumen, but the use of the other substance is not known.

Kumaon.—Bitumen is said by Captain Herbert to exude from the crevices in a limestone rock on the summit of the range between the Sarju and the Ramaganga. It is used by Natives for medicinal purposes, and from its high price it is concluded that it is not abundant.

II. ASSAM.—The petroleum of Assam has attracted notice for many years. Ball writes, "Hitherto attempts to work it have not been very successful as commercial speculations; but the failure seems to be due to causes other than those which could be attributed to any defects in the quality or quantity of the substance. The consensus of opinion by geologists, and others competent to express judgment on the subject, is favourable to the prospect of a profitable industry being possible when the means of communication shall have been improved, and when other difficulties which have hitherto operated prejudicially shall have been removed." Petroleum in Assam has been shown to be most intimately connected with the presence

II A

OCCURRENCE

Assam.

of coal-bearing rocks. Indeed, according to Mr. Mallet, there is no record of its having been found in Upper Assam except within the limits or in the immediate vicinity of coal-fields. Ball states that marsh gas commonly occurs with the petroleum, and that there appears to be more than chance in the fact of the contiguity of saline and petroleum springs.

The oil fields of Assam were first thoroughly reported on by Mr. Mallet (*Mem. G. S I., XII., 356*), and have been again recently examined by Mr. R. A. Townsend, Superintendent of the Petroleum Works, Balúchistán. The following information has been summarised from the reports of these gentlemen :—

According to Mr. Mallet the oil springs or *pungs* may be conveniently classified for commercial or leasing purposes into the following districts :—

(1) those of the Tipam Hill north of the Dihing ;

(2) those of the range between the Dihing and Disang ;

(3) those of the Makum coal-field, south of the Dihing between the Dirak and Tirap rivers ; and

(4) those to the east of the Tirap.

Petroleum springs have been recorded from Supkong in the Bari Dihing, near an outcrop of coal ; at Namrup Pathar on the Namrup River, in which the yield is considerable, and at Namchick Pathar near the mouth of the Namchick River, where white mud mixed with petroleum is said to be thrown up in certain basins or pools intermittently and with considerable violence. The chief locality, however, is the Makum coal-field. In 1865, the springs in this region were visited by Mr. H. B. Medlicott, who stated that, though their discharge was small, they were the most promising which he had seen. In 1867, boring was attempted by a Calcutta firm with the result that oil was struck, and a large quantity was obtained, but owing to difficulties of transport the enterprise had to be discontinued. A sample of the oil obtained at that period was black, liquid, of rather strong odour, and had a specific gravity of ·971. One thousand parts submitted to distillation yielded eight distinct distillates at different temperatures, and a small residue of coke. The first six fractions were fit for lamp oil, though of rather higher specific gravity than American petroleum. The seventh and eighth distillate contained solid paraffin, which might easily be separated and employed for making candles, while the liquid residue would be a useful lubricant.

The Makum coal-field area was recently carefully prospected by Mr. R. A. Townsend, Superintendent of Government Petroleum Works, Balúchistán, and he has furnished a most favourable report. He writes that the failure above described resulted " not from want of demonstration of the existence of oil in large quantities, but chiefly from want of facilities of cheap and easy transport of the article to consuming markets, and because of general inexperience in making and securing against damage artesian borings, so far back as the time referred to." " All such difficulties," he continues, " have now disappeared. A splendid steam flotilla traverses the Brahmapútra River from Calcutta to Debrughur ; a railway cuts through the very centre of that portion of the field under consideration in its course from Debrughur to Margherita ; the present methods of sinking artesian borings are reduced almost to a science ; and the introduction of cheap foreign kerosine into India and the East has created such a demand for that class of illuminant as to have established it as almost a necessity of life to the millions who require it." " The Government of India many years ago gave various leases conveying the right to search for and mine petroleum in these parts. None of the holders of such leases have done much in the way of testing or developing the lands covered by them. Without giving the history of these transactions it may be said that the Assam Railway

has been reported to be found.   (*J. Murray.*)   PETROLEUM.

and Trading Company hold the only lease from Government now existing, which conveys the right to mine for petroleum in the Makum district. This lease covers an area of 30 square miles, south of the Dihing River, and is, I believe, conterminous with the Company's coal lease over the same area. The Company has lately started a boring near the site of some old ones about 2½ miles south of the Dihing Railway bridge."

Regarding the actual oil tracts and the old borings he writes, " Evidences of the oil yielded by these old borings still remain in the form of rude tanks dug into the soil, the sides of which are secured by wood poles still saturated with oil, and which are in a good state of preservation after twenty years of exposure in a climate very destructive to fallen timber. The soil surrounding is also saturated with petroleum, which exudes with pressure from the foot, especially if the foot be that of an elephant. In a hill close by and to the north, there is, I am told, a seam of coal about three feet thick cropping up, which we were unable to find upon searching ; coal was also, as stated, penetrated by the boring now in progress."

" Sulphuretted hydrogen gas is escaping in considerable quantities from the old borings, especially from No. 5, which is cased with wrought-iron casing to an unknown depth; the hole is filled with water to the surface, within the casing, and is kept in a state of violent ebullition by a constant and copious discharge of gas, having the characteristic odour of oil-gas, and which is highly inflammable. I caused it to be lighted and a splendid flame mounted several feet upwards and burned fiercely until extinguished by smothering at the casing head.

" The gas has been steadily flowing in this way for twenty years past ; and an idea of the waste of valuable fuel can be formed when it is considered that the present supply is sufficient to fire an 8-H. P. steam-boiler, doing its full work, say, at a pressure of 50 to 60℔ of steam. Gas may be seen escaping from some of the shallower borings, but in less quantity. It is evident that the soil and rock near the surface in this locality over a considerable area is saturated with petroleum. A hundred yards up the stream globules of oil rose to the water surface on the removal of each stone from the bottom by coolies who were fishing them out for work-purposes."

Mr. Townsend afterwards explored the Tippam Hills, regarding which his conclusions are as follow :—"Throughout these hills, so far as examined, the oil shows are abundant and of the most surprising character. The sandstones bordering the chief *nallas* are dripping with oil ; the loose sand in the stream beds is so charged with it in places, that a hole made with a walking stick is immediately filled, and a handful of sand, if squeezed, parts with its oil as a sponge with water. Numerous holes on the hill sides are half filled with oil, mud, and leaves ; and there are places so charged with oil vapours or gases as to be unpleasant to one unaccustomed to the fumes of an oil refinery. Nowhere have I seen such shows of oil in a state of nature ; they are not confined to one or two isolated places but extend for miles continuously, every hill-side or *nalla* being more or less marked by its presence. Associated with the oil-springs, there are several places where salt or brackish water comes to the surface, and these are much frequented by wild elephants, deer, and other animals." . . . . "A few miles east of the place under consideration these Tippam Hills sink very gradually to the level of, and below, the closing alluvial plain to the west and south ; they continue increasing in width and height until they join the Naga range a little to the south of Jaipore, the whole forming an ideal anticlinal fold for the storage of petroleum beneath it."

In the hills to the south of Jaipur Mr. Townsend found oil shows similar to those above described, except that they were not so numerous, and the oil was darker, and had a stronger odour. He states that the failure of

P. 443

OCCURRENCE

Assam.

boring in this region is probably due to the fact that they were much too shallow. He concludes his remarks on the Makum field by stating that wells sunk to, say, 400 or 600 feet will penetrate the oil measures in most parts of the field. He was unable, from want of time, to examine the vast tract of country south-east of Jaipore, which is "known to contain shows of petroleum along the northern border of the Nagas and far up the valleys within them." Though unable to speak decisively of their value, he remarks, "They, together with other shows found in water wells on many tea plantations, such as at Bazalona and Talup, far out in the plains, stamp the country in which they appear as being in connection with, and a continuation of, the Makum area, and collectively form an index to an oil deposit of vast extent, and probably of surprising richness." Mr. Townsend's suggestions for the development of these very valuable oil-fields may be found useful. "The time having arrived," he writes, "when the oil deposits of Assam are likely to be worked in a practical way, it will be well for the Government of India to devise some satisfactory method of dividing up the lands into plots for leasing on terms of mutual advantage to all concerned, and it will be found necessary also to establish rules governing their treatment when made, and especially when abandoned. The rainfall of Assam is a phenomenal one, and the bulk of the country in which oil is likely to be found is a flat plain,—conditions which require great care in preventing surface water from flooding the oil measures by means of carelessly or ignorantly constructed boring. It may also be found necessary to relax somewhat the rules governing the reserved forests covering the tracts, at any rate in the pioneering stages of development, if for no other than sanitary reasons affecting the skilled labourers who will be forced to live near their works when once started."

Though the above quoted interesting report has done much to add to our knowledge of the Assam oil-fields, it may be interesting and perhaps of value to give in continuation a short *resumé* of the exact localities mentioned by Ball as especially oil-yielding.

Jaipur.—Petroleum has been reported from the bands of a stream close to the Hukanjuri path, about two miles from this town, where it exudes from below a nine feet seam of coal; and from the Sub-Himaláyan sandstones, at Nahor Pung, quarter of a mile distant. Boring has been attempted, as already stated, in this region but without success, probably as the attempt (195 feet) was too shallow.

Disáng River.—At least two springs are situated in this stream half a mile north of Disáng.

Teok River.—The Sub-Himálayan sandstones on the bank of this river are at a certain spot (Lat. 27°; Long. 95°15') impregnated with petroleum. Petroleum has also been reported as issuing from sandstone, coal, and pyritous and carbonaceous shales, on the banks of the *Saffrai River*, the *Tiru River*, at *Tel-pung* on the *Dikhu River*, at *Tirugaon* on the *Hil Jan stream*, and in the *Disai Valley*.

A good oil spring has recently been struck near Digboi in the Lakhimpur District. The oil from this source has been recently reported on by Dr. Warden, whose analysis, more fully treated of below, tends to show that it contains more illuminating oil constituents than most others from Assam.

Chemistry.

444

CHEMICAL CHARACTER.—Samples of Assam crude petroleum examined on more than one occasion by Mr. Redwood, have been reported on by him as follows:—"The oil is dark brown in colour, of rather high viscosity (the viscosity of a sample of sp. gr. ·940 was 14·2 at 90° F., rape oil at 60° F.=100) and has a slight and not unpleasant odour. Its specific gravity appears usually to range from ·933 to ·940, and its flashing

has been reported to be found. (*J. Murray.*) PETROLEUM.

point is sometimes as high as 212° F. (Abel's test). It begins to distil freely at 280° F., but considerably less than 20 per cent. volatilises within the range of the mercurial thermometer. The oil contains none of the kerosine hydrocarbons, but it yields by the ordinary process of distillation, 89 per cent. by weight of lubricating oil distillates. The proportion of solid hydrocarbons is not large. The carbonaceous residue varies according to my experiments from between 3 and 4 per cent. to over 8 per cent." The Digboi oil, above referred to, is said by Dr. Warden to be semi-solid at 85° F., and to have a dark brown colour and aromatic odour. Fractional distillation yielded 83·8 per cent. of volatile constituents, varying from ·7309 to ·8510 sp. gr. The residue in the retort was solid and of a dark colour. The fractions (of 10 per cent. each) over the 5th could not be blended with the first 50 per cent., owing to the presence of solid paraffins, but by redistillation very slowly they afforded 10 per cent. of their bulk, which could be utilised for this purpose. Though a little difficulty was experienced in obtaining a large percentage of oil suitable for illuminating purposes, owing to the large proportion of light hydrocarbons in the early fractions, and consequent low flashing point, such an oil was eventually prepared. Dr. Warden sums up, "My examination demonstrates conclusively that the crude oil under report is capable of yielding at least over 45 per cent. of a burning oil of practically as good a colour and high flash as the usual grades of American and Russian petroleum now being imported into the market, and I consider the results eminently satisfactory, both as regards yield and quality." The large proportion of solid hydrocarbons, and the amount of naphtha and lubricating oil, present, would also be of commercial value.

Cachar.—Reports have at various times been made of the occurrence of petroleum at Sialtekh on the Barák River, in hills to the north of Cachar (locality not definitely stated), on the banks of the Barak and Sarang rivers, and in other localities not specified. No further information is at present available as to Cachar petroleum.

III. BURMA.—Whether it be true or not that the exploitation of the mineral oils of Upper Burma has been going on for 2,000 years, it is certain that there has been an unfailing supply from this source for some considerable time. As compared with this region, the discovery and working of oil springs in Lower Burma, except on the smallest scale, is still in its infancy, though it is probable that the Natives have collected oil to a limited extent for their own use in the islands off the coast of Arakan for a long period.

Though oil has been reported from many parts of the country, and may exist in quantity in certain of these, only three localities have been regularly worked, *viz.*, Kyaukpyu and Akyab, in the Arakan Division; Thayetmyo and Henzada in the Irrawaddy Division, and Pakokku and Magwè in the Southern Division of Upper Burma. For a description of the other localities of minor importance the reader is referred to Ball's account above cited.

Arakan.—The oil industry of the islands south of Akyab has been carried on in a primitive fashion for many years, shallow wells being dug and a small supply of crude oil obtained morning and evening. An excellent account of the Boronga, Ramri, and Cheduba islands was given by Mr. Mallet, F.G.S., in 1878, from which it appears that the oil season, with the Natives, lasts from December or the beginning of January to the rains. Mr. Mallet states that the wells are of two classes—those which appear to be in communication with a natural reservoir, from which the oil, generally accompanied by large quantities of gas, rises with considerable rapidity, and those sunk in rock more or less soaked with petroleum, from which the oil slowly percolates into the well.

OCCURRENCE

Burma.

The wells of Leedaung and Eastern Boronga are said to belong to the former class, those of Minbyin, about 20 miles from Kyaukpyu in Ramri, are regarded as the most important of the latter. In the year 1877 the attention of capitalists was attracted to this region, and scientific methods of obtaining the oil were adopted. The "Boronga Oil Works Company" was granted a lease for thirty years of the right of extracting oil over 2 square miles, and erected a refinery on a large scale. The efforts were not attended with financial success, and the property was disposed of to a **Mr. Senior** who continues to work the business. In 1884 a Company styled the "Ramree Oil Prospecting Company" was granted a prospecting lease over 1 square mile of Leikmaw on the west coast of Ramri island, but the experiment again proved a failure and the Company broke up. An attempt to work oil was made on the west side of the old cantonment of Kyaukpyu by **Messrs. Bell**, but after sinking 18 wells they also abandoned the enterprise. In a note by the Burma Government in 1888 we find the following :—

"The only Oil Company which is still in existence in Arakan is the Arakan Petroleum Company, which commenced work in August 1883. This Company hold 75 wells, of which 20 are now worked with the assistance of steam power under the supervision of Canadian experts. The remainder of the wells in Kyaukpyu are held by Arakanese merchants. The wells have been worked for many years and the outturn is decreasing yearly. Attempts have been made by clearing and deepening old wells to penetrate to lower oil deposits, but with little or no success. The following statement shows the number of wells worked and abandoned in the Kyaukpyu district with the output in gallons :—

Locality.	Holders.	Wells.		Yearly output in Imperial gallons.
		Working.	Abandoned.	
1. Minbyin, near the western coast of Ramri island	1. Mr. Senior .	.18	2	36,648
	2. Arakan Petroleum Company .	20	55	50,080
	3. Five Arakanese merchants. .	45	53	37,560
2. Leedaung, south-west edge of Ramri.	Arakanese . .	22	...	3,756
3 Leikmaw, west coast of Ramri.	Arakanese . .	15	...	500
4. Kyaukpyu, west side of cantonment.	Messrs. Bell .	...	18	...
5. Cheduba island .	Arakanese . .	124	...	20,000

The thirty years' lease granted to the Boronga Oil Company and now held by **Mr. Senior** is at present rent-free, but from the 1st April 1893 a royalty of 5 per cent. on the gross outturn will become payable. The Arakan Petroleum Company and the Arakanese merchants pay a tax of R2 per annum for each well.

"In the Akyab District, the only place where oil is found is the Eastern Boronga island. Attention was first attracted to the Boronga oil supply by the Akyab Exhibition of 1875. In 1877 prospecting was undertaken by three persons, and two of these, **Mr. W. Savage** and **Mr. William Gillam**, persevered. **Mr. Savage** obtained a lease of 1 square mile for thirty years, rent-free for three years, and subject at the end of that period to a royalty of 5 per cent. on the value of the crude oil. He is

has been reported to be found.   (*J. Murray.*) PETROLEUM.

working two wells and is boring five others. Eight have been abandoned. One well yields about 180 gallons a day, the other, being unfinished, is not fully worked. **Mr. Gillam** obtained, on the same terms as **Mr. Savage,** a lease of 4 square miles. He transferred his concession to the Boronga Company, and the property is now in the hands of **Mr. Senior.** Four wells are in use, five have been abandoned, and one is being bored. The yield of the four wells which are worked is small, being about 43 gallons a day."

Though commercial enterprise in this region would thus appear to have been as yet productive of little success, it is worthy of notice that in the opinion of **Mr. Medlicott** (1886), "There are no doubt very large supplies of high class petroleum to be got from this region, but it must be won by suitable methods."

CHARACTER OF OIL.—Much of the oil obtained in the Arakan District is of very high quality; it has the appearance of brandy or sherry, and can be burned in lamps in its crude state. Specimens from Eastern and Western Boronga submitted to **Mr. Boverton Redwood,** for examination in 1878, are reported on by that gentleman as follow :— "The Eastern Boronga oil was dark brown in colour, and had a pleasant odour. Its sp. gr. was ·835, and I obtained from it 66 per cent. of kerosine of sp. gr. ·810, besides from 2 to 3 per cent. of more volatile products. Owing to the small quantity at my disposal I was not able to ascertain with any approach to precision the percentage of the solid hydrocarbons present, but I formed the opinion that about 3 to 4 per cent. of paraffine, together with lubricating oils, might be obtained from it.

"The Western Boronga oil was of similar colour and odour, but of higher specific gravity, *viz.,* ·888, and it yielded only 7 per cent. of kerosine of sp. gr. ·815. The residue, however, yielded an excellent lubricating oil. About the same time I examined samples of oil from the mud volcanoes near Kyaukpyu, and from the wells (to to 20 feet deep) and Minbyin. The former was an oil of pale colour, and of sp. gr. ·818. It yielded 56 per cent. of kerosine of remarkably high quality, but almost the whole of the material was available for use as burning oil. The Minbyin oil had a specific gravity of 866, and yielded only 15 per cent. of kerosine of sp. gr. ·810."

A sample of Arakan petroleum of unknown origin examined by **Dr. Romanis,** Chemical Examiner, Burma, was found to have a flashing point of 40° F. by Abel's test, to begin to give off gas at 30°C, and to boil at 70°C. Distilled in a current of steam it yielded, up to 130°C., 85·6 per cent. of light oil, and gave a residue of 13·3 per cent. of heavy oil.

IRRAWADDY DIVISION. — Only one well-known well exists in the Henzada District, which is not now worked. The wells in the Thayetmyo District are only two in number, another, that was formerly used, being now filled with water. The works are conducted by a Native contractor. The yearly outturn is about 2,400 *viss* (768 gallons), the oil sells at from R15 to R20 per hundred *viss* or from 5 to 7 annas per gallon, and the lessee is believed to make a profit out of the business. The oil is similar to but thinner than that of Upper Burma, burns rather fast, and is said to be used by Natives for varnishing boats (*Note by Burma Government, 1888*).

Upper Burma.—The most important oil-fields of Burma are those generally but wrongly called the fields of Yenangyaung which means the "creek of oil," or literally "creek of stinking water." These fields are situated 1½ miles to the east of Yenangyaung near the villages of Twingaung (Hill of wells) and Beme. They have always attracted much attention from European travellers, and have been worked by Natives from remote times. This oil district has recently (June 1889) been officially reported on by Dr. Fritz Noetling, Palæontologist, Geological Survey of India. He points out that the field lies on the eastern side of the river Irrawaddy, and is distant about 300 miles from Rangoon, or about 80 miles from Allanmyo. The country forms a tolerably level and flat plateau, rising to 260 feet above the low level of the Irrawaddy at Yenangaung, and consists of strata which, according to Dr. Noetling, unquestionably belong to the tertiary formation, and probably are of miocene age. The strata in descending order may be divided into four groups. The upper consists of sandstones varying in colour from yellowish white to brown; the second of bluish-grey sandstones and clays alternating, and in its lower part is fairly soaked, with oil, though it is not to be considered as the oil-bearing formation proper; the third is of stiff blue clunch, is usually found at a depth of 200 feet, when the well is on the top of a hill, or 100 feet, when in the ravine, and immediately overlies the true oil-bearing strata; the fourth, or oil-bearing group, consists of more or less soft, coarse, or fine micaceous sandstones of bluish-grey colour, changed into more or less yellowish green according to the amount of oil present, and interstratified with hard sandstone and blue clunch.

The superficial area of the Twingaung oil-field, which lies between the villages of Twingaung and Mansu, is about 90 acres, in which space all the productive oil-wells of the district are situated. The total number amounts to 375 wells, of which 166 (44·3 per cent.) are unproductive. Of the remainder 209 (55·7 per cent.) Dr. Noetling writes, "We may call them productive wells, but we have again to distinguish between wells from which oil is regularly drawn, and those from which it is only occasionally drawn." "Of the first class, 'productive wells,' there are 120 (32 per cent. of the total number); of the second class, 'scarcely productive wells,' there are 89 (23·7 per cent. of the total number)." With the exception of two, one of which reaches 310, and the other 305 feet, all the wells are less than 300 feet in depth; indeed, only 33 exceed 250 feet. The greater number of 'productive' and 'scarcely productive' wells are between 200 and 250 feet in depth, but of the 30 between 250 and 300 feet, 20 are 'productive' and 10 'scarcely productive.' Dr. Noetling remarks on this point, "There could hardly be a better illustration to prove the unreasonable style of Burmese working. Beyond 250 feet the difficulties of digging increase so greatly that they could hardly be vanquished. Only 33 wells out of a total of 209 wells could be sunk beyond that limit." "The oil-bearing sandstone is known to have a thickness of at least 200 feet; therefore these wells which we have to consider as being the main oil-producers do not drain more than 30 feet of the upper and worst part of the oil-bearing sandstone, and only 33 wells drain the oil-bearing sandstone up to 80 feet. To work to a greater depth is impossible for the Burmese style of working. Therefore, all over the place where the Burmese wells are situated, nearly 100 feet of oil-bearing sandstone remain untouched, and its wealth of oil, measuring millions of *viss*, unraised." The production of the wells varies greatly, but Dr. Noetling takes as an average for the 'scarcely productive' wells 10 *viss* per diem, and for the 'productive' wells from 80 to 90 *viss* a day. The yield of the latter is probably greater than this estimate, and may even on the average be as much as 120 to 130 *viss*. The daily total yield for all the wells would thus be from 12,000 to 17,000 *viss*. Some of the wells yield from 100 to 300 *viss* a day, and one is stated to produce 500 *viss*. Redwood states that Messrs. Finlay, Fleming & Co. estimate the present total production of the Yenangyaung fields at 200,000 to 250,000 gallons a month. From Dr. Noetling's data it is evident, as might be expected, that the deeper wells are the more productive.

has been reported to be found.   (*J. Murray*.)   PETROLEUM.

The Burman wells are shafts from 4 to 4½ feet square, and are dug by means of a tool called *tayuwen*. This implement consists of a wooden handle, flattened on the top and deeply notched 6 inches below that part, provided with an iron shoe. The man in working grasps it with his two hands, and, leaning his shoulder against the notch, forces the tool into the ground with all his strength. Such an instrument can obviously be used with effect only in soft strata. Over the well a cross-beam supported on stanchions at each side is placed. In the centre of this is a small wooden drum and axis, made out of a single piece of wood, and running in naturally grown fork-shaped supports. The leather-rope used in hauling up oil, and to raise and lower the workmen is passed over the drum. The wooden drum is known as *gyin*, the oil-pot as *yenanoie*. If possible, the well is so placed that the men or women who are drawing up either the pot filled with oil, or the workmen, walk down an inclined plain along the slope of the hill.

The shaft is roughly lined throughout with timber, which requires permanent supervision and frequent repair. The workman is lowered in a rope sling, and as no artificial light can be used in the well, owing to the presence of inflammable gas or vapour, his eyes are bandaged previous to his descent, so that time may not be lost in becoming accustomed to the darkness below. The petroleum vapour renders breathing very difficult, the difficulty increasing with the depth of the well, and **Dr. Noetling** found that the maximum time during which a young and strong man was able to stay down, was 290 seconds.

Dr. Noetling estimates the average cost of making a well, 250 feet deep, to be about R1,500, and states that the time occupied in the operation is from eight to twelve months. From a calculation made in the case of one of the most expensive wells, which cost R3,000 to make, he arrives at the conclusion that, after paying interest on the original outlay, 15 per cent. for incidental loss, and the same for wages and necessary expenses, a net monthly profit of R34, or a net interest of 13·6 per cent., is obtained. From general calculations he arrives at the conclusion that the wells, as a whole, yield a high interest, ranging from 23·6 per cent. upward with amortization of the invested capital after ten years. As a rule four men are actually employed in digging a well, with others for draught purposes. The oil is drawn up in the earthenware pot already alluded to, and is poured into another earthenware vessel of the same shape, which holds from 10 to 16 *viss*. Twelve of these are packed in each country cart for conveyance to the river-side.

The Beme oil-field is much smaller than that above described. It occupies an area of about 35 acres, and the total number of wells does not exceed 151. Of these 72 (47·6 per cent.) are productive. Fifty of the productive wells yield more than 20 *viss* per day, and 22 less than 20 *viss* per day, the daily average amounting to 60 to 70 *viss*. The depth of the wells is not so great as that of the Twingaung wells, and their yield is smaller, none producing more than 165 *viss* per day, while those giving more than 100 *viss* are scarce. According to **Dr. Noetling** the total daily production amounts to about 3,658 *viss*.

He sums up by stating that the Twingaung oil-field is at present in the zenith of its production, though a small increase may be expected for the next few years; that the Beme field, on the contrary, is in its decadence, and that a decrease of production is to be expected; that the industry is not capable of increase by Native methods; but that if worked in the European style by bores, it is susceptible of considerable development. The total outturn of petroleum in the whole of Burma during 1889 is returned as 2,985 gallons, valued at R2,32,327.

PROSPECTS. Burma Oil. **447**	**Prospects of the Oil Industry.**—The only wells which have as yet been drilled in the European way are those of Messrs. **Finlay, Fleming & Co.** at Kodoung. Operations were commenced in 1887, and up to this time nine wells have been bored by Canadian and Pennsylvanian drillers. Of these five have proved productive. The yield varies from 3 to 15 barrels per well each 24 hours. These results do not appear very encouraging, but it must be remembered that in the short time during which the Company has been at work many difficulties have had to be overcome. The transport of machinery has been laborious and expensive, owing to the broken nature of the ground; in every new oil region there is always much to be learnt, and, apart from the difficulties attending all such enterprises, the work has been much interfered with by dacoits. Arrangements are being made to bore new wells, and to seek a greater depth in hopes of obtaining an increased yield, and though the cost of working is great, the Company appears to have no doubt whatever of future success. Transported by country cart to the river-side, the oil is easily conveyed to Rangoon, in native boats or in specially-made flats. A small proportion is also conveyed up and down the river for local sale to Natives, by whom it can be bought wholesale at the river-side for about $1\frac{1}{2}d$. per gallon.
CHEMISTRY. **448**	**Physical and Chemical Characters.**—Mr. Redwood has published reports of analyses of various samples of the crude oil, obtained from Messrs. Finlay, Fleming & Co. All were opaque in bulk, and of dark greenish colour by reflected light; they had little odour, and many samples, owing to the presence of an unusually large proportion of solid hydrocarbon, were solid at ordinary temperatures. The physical characters of five samples are well shown in the following table :—

LOCALITY.	Specific gravity.	Setting point.	Flashing point (Abel's test).	Viscosity by Redwood's viscometer, rape oil at 60° F.=100.
		° F.	° F.	At 90° F.
Yenangoung (from Twinza's wells) .	·887	82	110	10·21
Ditto do. do. . .	·937	Remains fluid at 0° F.	150	25·86
Ditto (from **F. F. & Co.**'s old No. 1 bore).	·869	80	62	...
Ditto (from **F. F. & Co.**'s American bore No. 2, at 260 ft.).	·870	78	80	...
Ditto (from **F. F. & Co.**'s-American bore No. 4, at 272 to 330 ft.).	·875	82	83	10·07

Mr. Redwood writes, " There is a remarkable absence of uniformity in the character of the oil obtained from native wells, even where the wells are contiguous and of the same depth." The wells on the Beme field "show that there are at least two descriptions of oil, *viz.*, one which has a high specific gravity and viscosity, but containing practically no solid

CHEMISTRY.

hydrocarbons, and another of much lower specific gravity but containing a very large proportion of paraffin." The oil containing no paraffin is of higher specific gravity than that which contains a fair proportion. The heavier oils containing but little solid hydrocarbons are to some extent separately collected and employed in the manufacture of lubricating oil of high viscosity.

The sample from **Messrs. Finlay, Fleming & Co.'s** well No. 2 bore began to distil at 260° F. and yielded 27½ per cent., by volume, of kerosine of sp. gr. ·823, and flashing point 73° F. This product was easily refined and of good quality. By the adoption of a new process devised by **Mr. Redwood** and **Professor Dewar**, it was found possible to increase the yield of kerosine to from 70 to 80 per cent.

Up to thirty-five years ago Rangoon oil, which had been well known from remote times, was always used in the crude state. In 1854 **Mr. Warren de la Rue** obtained a patent for improvements in treating products from petroleum, the crude oil with which he experimented being imported from Rangoon. For some time afterwards the material was much used and refined by Price's Patent Candle Company, and by Messrs. Charles Price & Co. From that time onwards the oil has frequently been made the subject of chemical examination and experiment, and " in 1865-66 **Messrs. O. M. Warren** and **F. H. Storer** obtained from Rangoon petroleum the olefines $C_{10} H_{20}$ to $C_{13} H_{26}$; also probably $C_9 H_{18}$, and the paraffins $C_7 H_{16}$ to $C_9 H_{20}$. They also found naphthalene, and their experiments indicated the probable presence of xylene and cumene. In this investigation the more volatile portions of the crude oil only were dealt with, and the results obtained appear to indicate a preponderance of olefines over paraffins." The revival of the old Rangoon oil industry under the altered conditions of improved methods of obtaining the crude oil, and the adoption of the principle of manufacturing the commercial products at the port of shipment, dates from 1870, when the Rangoon Oil Company, Limited, erected a refinery at Rangoon. Owing to political difficulties with Upper Burma oil could not be obtained in sufficient quantity to ensure profit, and in 1876 the concern was wound up. Shortly after the annexation of Upper Burma in 1886, the work having been carried on under great difficulties in the interval by **Messrs. Finlay, Fleming & Co.**, the Burma Oil Company, Limited, was formed, and at once proceeded with the extension of the refinery on the bank of the river at Dunniedaw, close to Rangoon. This factory is capable of refining about 500,000 gallons per month, but at present the total yield of the oil-fields is less than half that quantity. " The crude oil is found to yield on the average about 27 per cent. of fair standard kerosine, but the Company manufacture principally a burning oil of higher specific gravity for use by Natives and Eurasians, and of such oil a much larger percentage is obtained. A sample of this product has a sp. gr. of ·840 and a flashing point of 121° F. (Abel test). Much of the crude petroleum from the Yenangoung field contains, as already stated, from 10 to 12 per cent. of solid hydrocarbons, but it is important to note that, in consequence of the unfavourable conditions under which the work is necessarily conducted, **Messrs. Finlay, Fleming & Co.** do not practically obtain from the average raw material more than 4½ per cent. of paraffin. The 'melting point' (English test) of the crude article is 125½° F., and of the refined is no less than 132° F. The Company has a candle-making department, but finds it impossible to compete against the Dutch stearin candles, which are sold at an extremely low price, and the paraffin is accordingly exported to London. The other products include ' naphtha ' of sp. gr. ·813 and flashing point 67° F. (Abel test) as well as intermediate and lubricating oils. I find

P. 448

PETROLEUM.   Uses of Petroleum.

CHEMISTRY.

that samples of the latter furnish the following results on examination :—

	Specific gravity at 60° F.	FLASHING POINT.		VISCOSITY* RAPE OIL AT 60° F. = 100.		Cold Test.
		Closed.	Open.	At 70° F.	At 140° F.	
		° F.	° F.			
Lubricating oil (3)	·920	238	266	34·23	9·24	Ceases to flow at 30° F.
,, ,, (4)	·930	330	354	73·76	12·47	,, ,, at 30° F.
,, ,, (5)	·931	336	360	124·02	15·71	,, ,, at 42° F.
"Valvoline" (6)	·949	400	422	... (at 200° F. ·02)	25·95	,, ,, at 45° F.

* By Redwood's viscometer.

The colour of the first sample (3) by Lovibond's tintometer, in a 2-inch cell, series 500, is 130; and of the second sample (4) is 224. The specific gravity of these oils is in each case very high in relation to the viscosity, and the viscosity is affected to somewhat more than the average extent by alterations in temperature.

"More than once during the past few years I have had occasion to report upon the commercial products obtained by the Burma Oil Company, and I am in a position to state that the difficulties which at first attended the manufacture have to a large extent been overcome. The extreme heat of the climate, however, is a great drawback, and necessarily renders costly the operations connected with the separation and purification of the paraffin" (*Redwood, Pet. Deposits of Ind.*).

USES.

449

USES.

The uses of this valuable mineral are too well known to require a detailed account in this place; the following passage from **Watts'** *Chemistry* may, however, be useful as indicating the chief products of commercial value which may be obtained from crude petroleums by fractional distillation :—

"By fractional distillation, sometimes with the aid of steam, ordinary or superheated, and purification of the distillates with caustic potash, a number of commercial products are obtained which are applied to various purposes, the lighter oils chiefly as solvents for resins, &c., the heavier for burning in lamps, and as fuel for steam boilers. The following account of them is taken mainly from **Dammer's** '*Kurzes Handwörterbuch der Chemie :*'

"*Rhigolene*, boiling point 30°, used as an anæsthetic.

"*Petroleum-ether* (Kerosolene, Rhigolene, Sherwood-oil) : distils at 45°—60° : clear, colourless oil, having only a faint odour of petroleum ; boiling point 50°—60° ; sp. gr. 0·665 ; absorbs oxygen from the air and becomes heavier (0·670—0·675) ; extremely inflammable ; used as a remedy for rheumatic pains and as a local anæsthetic.

"*Petroleum-ether*, II. (Gasoline, Canadol) : distillate between 60° and 70° ; has a sp. gr. of 0·665 ; second distillate between 70° and 90°.

"*Petroleum benzin :* distillate between 70° and 120° ; has a sp. gr. of 0·680 to 0·700 ; dissolves in alcohol and ether ; boils at 60°—80° ; absorbs oxygen and becomes heavier ; dissolves fats, caoutchouc, asphalt, and turpentine ; less easily colophony, mastic, and dammar resin ; kills all small animal organisms, and is used externally as a remedy for itch and other skin diseases, internally for gastric pains; also for the extraction of oil, the preservation of anatomical preparations, the carbonation of illuminating gas, and for the preparation of lacquers and varnishes.

Ligroin is a similar distillate used for feeding the so-called 'Ligroïn or Wonder-lamps.'

Artificial Turpentine-oil, Petroleum-spirit, Polishing Oil: distillate between 120°—170°; obtained also with superheated steam; sp. gr. 0·740—0·745; does not dissolve resins; used for diluting linseed-oil varnish and for cleaning printers' type.

Illuminating Oil, Petroleum, Kerosine, Paraffin Oil, Refined Petroleum, have a large and increasing consumption for lamps, also for warming greenhouses, &c., sp. gr. 0·78 to 0·81; flashing point (in the open vessel) 90°—110° F.; igniting point 110°—130° F.; includes all the intermediate distillates from crude petroleum, ranging from sp. gr. 0·76 to 0·83. The finest quality of illuminating oil is produced from distillates ranging only from 0·775 to 0·790; this, when treated by æration, is the water-white high-test oil used largely by the higher classes. It has a very high flashing point (120°—140° F. in the open vessel), as it contains none of the lighter constituents of the crude oil.

Lubricating Oil.—Sp. gr. 0·850 to 0·915; of a pale amber colour; coming largely into use as an efficient lubricant in combination with rape, olive, or lard oil. The best qualities have a sp. gr. of about 0·895, the heavier oils from 0·900 to 0·915 containing solid paraffine.

The residues of the distillation are utilised for gas-making.

The present production of petroleum, in the United States only, exceeds 50,000 barrels or 2,500,000 American gallons daily. The quantity exported thence during the year 1879 exceeded 400,000,000 gallons" (*Henry Watts, Dictionary of Chemistry, VIII., ii., 1508*).

Of these products the most important commercially is the kerosine or burning oil, so largely employed for illuminating purposes. The following results published by Mr. Redwood indicate the yield of this product from the crude petroleums from various localities, and are of interest as shewing the relative proportions obtained from Indian and other sources :—

LOCALITY.	Specific gravity.	YIELD OF COMMERCIAL PRODUCTS. Burning oil.		CHARACTERS.
		Per cent.	Specific gravity.	
1. Persia . . .	·777	87·5	...	Straw colour; strong and disagreeable odour.
2. East India . .	·821	62·5	·800	Amber; pleasant odour.
3. Burma, mud volcanoes Kyouk Phyau .	·818	55·7	·800	Reddish-brown colour; slight and pleasant odour.
4. Burma, native pits, Minbyin . .	·866	15·1	·810	Dark-brown colour; slight and pleasant odour.
5. Burma, Western Boronga . . .	·888	7·2	·815	Chestnut brown colour; slight and pleasant odour.
6. Burma, Eastern Boronga . . .	·835	66·1	·810	Dark-brown colour; slight and pleasant odour.
7. Assam . . .	·933	None.	...	Dark-brown colour; viscid; slight and pleasant odour.

PETROLEUM.   Flashing Point of Petroleum.

USES.

LOCALITY.	Specific gravity.	YIELD OF COMMERCIAL PRODUCTS. Burning oil.		CHARACTERS.
		Per cent.	Specific gravity.	
8. India . . .	·935	20·0	·805	Very dark-brown colour; viscid.
9. Russia . . .	·836	40·0	...	Dark reddish-brown colour; mobile, slight and pleasant odour.
10. ,, . . .	·942	None.	...	Brownish-black colour; viscid; very little odour.
11. Hanover . . .	·843	60·0	·812	Dark-brown colour; slight and pleasant odour.
12. South America . .	·852	50·0	·808	Dark reddish-brown colour; little odour.
13. ,, ,, . .	·900	None.	...	Dark-brown colour; somewhat viscid; scarcely any odour.
14. New Zealand . .	·828	60·0	·808	Brownish-colour; little odour.
15. Italy, near Milan .	·787	45·0	·806	Straw colour; pleasant odour.
16. United States, Wyoming. . . .	·910	27·5	...	Black colour; disagreeable odour.
17. ,, ,, ,,	·945	10·0	...	Black colour; disagreeable odour.

(*Journal Soc. of Arts*, XXXIV., *823 to 878.*)

FLASHING
POINT.
450

TESTS AND LEGISLATIVE MEASURES REGARDING FLASHING POINT, &c.—The term 'Flashing-point' is technically employed to designate the temperature at which any sample of petroleum or its products begins to give off sensible quantities of inflammable vapour. The more volatile kinds give off such vapours even at ordinary temperatures, and these vapours, mingling with the air, form explosive mixtures; hence the use of such oil in lamps is attended with great danger. To prevent accidents from this source, it was enacted by the American Petroleum Act, and by the British Petroleum Act of 1871, that no petroleum oil should be used for burning in lamps which should be found to give off inflammable vapours of any temperature below 100° F. (38°C) when tested in an open cup, described in the schedule of the Act. The great increase in the consumption of the article, however, between 1871 and 1877, necessitated the adoption of a system of testing less liable to vary in the hands of different operators. This question having been referred to **Professor Abel**, he proposed the use of a closed vessel to be heated in hot water, with a standard flashing point of 73° F., which was found to be equivalent to 100° on the open cup system. This proposal, having been favourably reported on by the Petroleum Trade Association, was eventually adopted as the basis of the English Petroleum Act of 1879. Abel's testing apparatus consists, in its original form, of a small closed oil-cup fitting into a water bath. On the top of the cup is a brass slide and a minute oil-lamp with a jet. A thermometer is also fixed with its bulb in the oil-cup. The operator watches the thermometer, and at intervals draws out the brass slide, thus causing a jet of flame to dip into the oil-cup, the rate of opening and closing the slide being regulated by observing the oscillations of a pendulum. The 'flashing point' is thereby ascertained (*Watts' Chemistry*). The law under which petroleum is now tested in India is Act XII. of 1886. In this Act the testing apparatus prescribed is an improved form of Abel's, with an automatic form of slide, worked by clock-work. Several other modifications, based on the special conditions attendant on the application of the test in a tropical climate, have also been incorporated in the schedule of the Act. As changes of pressure influence the flash-point of an oil, a table showing the corrections to be applied for variations in barometric pressure is also appended. "Dangerous petroleum" is defined by the Act as "petroleum having its flashing point below 76 degrees of Fahrenheit's thermometer," but a consignment guaranteed to be uniform is not to be considered dangerous if it have an average 'flash-point' of 73° and no sample flashes below 70° F. Dr. Warden states that since the introduction of Act XII. of 1886 the quality of oil imported into Calcutta—based on the flash-point—has much improved, an oil having a flash-point below 76°F. being now rarely met with. In Germany the modified Abel apparatus is used, the minimum flash point being 69·8°F., and an allowance made for variation of barometric pressure. In Russia the legal flash-point for sale or exportation is 82·4°F. In France and Switzerland the limit is 95° F., and in Austria and some of the States of the American Union, 99·5°F (*Allen, Commercial Organic Chemistry, II., 397*). Dr. Warden informs the writer that the 'fire test' or 'burning point,' *viz.*, the temperature at which an oil must be heated before it will permanently inflame, though most unreliable, is still much used in America. The writer has to thank Dr. Warden, Chemical Examiner to Government, Calcutta, for supplying much of the information regarding the above tests and for kindly revising the proofs.

Petroleum Companies.—Mr. O'Conor (*Statistical Tables, 65*) returns six oil-mills and well-boring concerns as existing in India. Of these, two oil refineries exist in Akyab, one in Kyaukpyu, and one in Magwè. The other two, *viz.*, the Arakan Petroleum Oil Company, Limited, and Messrs. Finlay, Fleming & Co.'s Oil Works, have already been mentioned as actually working oil.

In addition to these Mr. Noble's concession is engaged in prospecting in the Panjáb, and the Assam Railways & Trading Company in Assam. The Burma Oil Works Company, the largest refinery in India, is said to have had, in 1888-89, an outturn of 2,81,367 maunds, valued at R6,33,075, and to have employed an average daily number of 650 persons.

TRADE.

INTERNAL.—Kerosine is now thoroughly recognised as a most valuable illuminant by the Native population, and is preferred, whenever it can be cheaply procured, to vegetable oils. Its extended consumption depends to a very great extent on questions of cost and difficulty of transit. Where railways and water transport are available, the oil is very largely employed; but since conveyance by carts on rough roads necessarily entails considerable loss and increased expense, it is but little used in villages far from towns.

In 1888-89 the total recorded internal trade transactions by *road, rail, and river* amounted to 18,15,032 maunds, valued at R78,27,432. Of this Calcutta exported 11,03,959 maunds, Bombay town 4,13,307, Karáchi 1,34,154, and Madras seaports 84,712, while Bengal, Bombay, and Sind

12

PETROLEUM.   Trade in Kerosine.

also exported small quantities to other provinces. The chief importers were Bengal with 8,13,020 maunds, the North-West Provinces and Oudh with 1,59,206 maunds, the Panjáb with 1,50,112, Bombay with 1,48,087, Assam with 1,30,751, and the Central Provinces with 1,08,401 maunds. Madras and Sind are relatively much smaller consumers.

The total *coastwise* imports of foreign kerosine in 1888-89 amounted to 4,016,107 gallons, value R20,61,792. The largest importer was Bombay, which received in all 2,071,952 gallons, nearly all from other ports within the Presidency. Madras imported 1,042,137 gallons, most of which came from Bombay. Bengal received 376,765 gallons, chiefly from British ports within the Presidency. Sind and Burma imported about 250,000 gallons each, the former from Bombay, the latter from ports within the Presidency. The transactions in foreign mineral oil, not kerosine, were unimportant.

The *coastwise* trade in mineral oil of Indian produce (not kerosine) was, however, large. The imports amounted to 965,294 gallons, value R4,83,382, of which 700,196 gallons were imported by Bengal from Burma, and 247,114 gallons by Burma from ports within the Presidency.

EXTERNAL.—The *trans-frontier*, like other branches of trade in petroleum, shows a very marked and steady increase, if the large exports which took place to Upper Burma before its annexation be left out of account.

Thus, during the past five years, the exports have been as follows :—

1885-86.		1886-87.		1887-88.		1888-89.		1889-90.	
Cwt.	R	Cwt.	R	Cwt.	R	Cwt.	R	Cwt.	R
118	1,323	392	4,419	6,223	42,961	22,443	1,48,040	28,270	1,84,096

The development has principally been in the exports to Nepál, which in 1889-90 received 153,89 cwt., in comparison with 1,918 cwt. in 1886-87 (the first recorded). The only other important items in the exports during 1888-89 were trans-frontier by the Sind-Pishin Railway, 6,964 cwt. and Kashmír with 3,458 cwt. Lus Bela, Sewestan, Tirah, Kabul, Ladakh, Thibet, Sikkim, Manipur, Hill Tipperah, Siam, the Shan States, and Karennee, all import small quantities for the use of the better class of population.

Excluding Upper Burma the imports are very small, having amounted in 1889-90 to only 13 cwt., value R180, from Kashmír. Up to 1888-89 inclusive, the imports from Upper Burma were returned in trans-frontier trade statistics. During four years there were ;—1885-86, 122,992 cwt., value R6,68,974; 1886-87, 138,396 cwt., value R7,26,649; 1887-88, 139,002 cwt., value R6,36, 853; 1888-89, 201,676 cwt., value R9,90,255.

The *trade by sea* with foreign countries has increased extremely rapidly, owing to the rapidly extending consumption of mineral oil as an illuminant by the Native population of India. Thus in 1875-76 (the first year in which the quantity of mineral oil imported is separately returned), the imports amounted to 621,530 gallons, in 1879-80 they increased to 7,888,247 gallons, in 1884-85 they were 26,299,091 gallons, while in 1889-90 they attained the large figure of 53,390,000 gallons. The value has similarly increased from R85,18,000 in 1885-86 to R2,47,88,000 in 1889-90.

The bulk of the oil thus imported during the past year was kerosine, only about 1½ million gallons being other mineral oils, used for batching

P. 453

jute and lubricating machinery.   The share taken by each port in the trade during the past year was as follows :—

Port.	Kerosine.		Other kinds.	
	Gallons.	R	Gallons.	R
Bengal	26,788,881	1,14,43,660	1,374,928	7,01,175
Bombay	13,704,072	64,18,850	166,130	2,03,727
Sind	2,391,404	11,48,788	3,227	3,394
Madras	2,509,893	12,98,329	82	92
Burma	6,445,150	33,57,112	6,391	13,038
TOTAL .	51,839,400	2,38,66,839	1,550,758	9,21,426

All mineral oil, except petroleum, with a flashing point at or above 200° F., and which is proved to be used exclusively for batching jute or other fibre, or for lubricating purposes, is subject to a duty of six pies per Imperial gallon.

The amount of "other kinds" free of duty, was 1,504,860 gallons.

The imports were derived from—

Country.	Kerosine.		Other kinds.	
	Gallons.	R	Gallons.	R
United Kingdom . . .	8,796	5,701	1,507,621	8,99,021
United States . . .	35,840,059	1,64,23,845	42,061	20,885
Russia in Asia . . .	15,961,875	74,17,163	...	...
Straits Settlements . .	27,667	19,495	...	...
Other countries . . .	1,003	671	1,076	1,520
TOTAL .	51,839,400	2,38,66,839	1,550,758	9,21,426

Regarding Russian kerosine, Mr. O'Conor writes :—" Russian oil, of which not a gallon was imported before 1886, has taken a firm hold of the market, as will be seen from the subjoined figures showing the imports in each of the last five years from the United States and Russia in Asia, the two countries whence most of the imported mineral oil is received :—

	United States.	Russia in Asia.
	Gallons.	Gallons.
1885-86	20,229,000	...
1886-87	29,145,000	1,577,000
1887-88	25,040,000	5,036,000
1888-89	20,654,000	17,516,000
1889-90	35,882,000	1,596,000

" The enormous increase in the imports from the United States during the year would seem to have been mainly speculative, and to have been designed to keep Russian oil out of the market.   Russian oil has, however, come into use far more largely at Bombay than at Calcutta, the imports into Bombay last year having been about 10¼ million gallons, while Calcutta imported only about 1,600,000 gallons, this being half the quantity imported in 1888-89.   This decline in the imports into Calcutta

TRADE:
External.
Imports.

has been attributed to a prejudice created against the oil in the beginning not being satisfactory. But the oil is freely used in the Western Presidency and in Northern India, and no prejudice against it has ever existed there. It seems more likely that it is not a question of lack of demand in Calcutta, but of convenience of a market. For the steamers which bring oil to India, Bombay is a more convenient port than Calcutta, in which to obtain return cargoes of the merchandise which Russia takes from us in greatest bulk, namely, cotton and seeds" (*Review of Trade of India, 1889-90, 20*).

In the previous trade review (*1888-89, 26*), the following interesting passage occurs :—" These large importations of mineral oil have affected the domestic economy of the Natives in two ways; the oil has displaced vegetable oils, and the tinned iron cans in which the import is made have in many places displaced the earthen jar and the brass pot to a large extent. An importation of 30 or 40 million gallons of oil means also an importation of from 6 to 8 million cans of 5 gallons each. These millions of cans were for some considerable time cut up by tinplate workers for the manufacture of other articles, but now they are also commonly used to hold and convey grain and other stores, water and other liquids, and they are amongst the most common objects in an Indian bazár. They are also used, ingeniously, if unornamentally, as building materials in the repair of roofs."

The re-export trade has increased during the past twenty years with increased imports, but has rather decreased than otherwise during the past five. The figures for these years (1885-86 to 1889-90) are as follows :—

Year.					Kerosine.		Other mineral oils.	
					Gallons.	℞	Gallons.	℞
1885-86	.	.	.	.	361,305	1,57,103	740	404
1886-87	.	.	.	.	272,805	1,22,209	250	392
1887-88	.	.	.	.	358,365	1,66,847	80	240
1888-89	.	.	.	.	535,016	2,69,129	1,491	2,970
1889-90	.	.	.	.	258,634	1,23,044	455	870

The largest exporters are Bombay and Karáchi, the largest importers Persia, Turkey in Asia, the Eastern Coast of Africa, and Ceylon.

Exports.
454

EXPORTS —The exports of mineral oil of Indian production are small and unimportant and show little tendency to increase. During the past five years the figures have been—

Year.									Gallons.	℞
1885-86	.	.	.	.	.	.	.	.	19,830	11,028
1886-87	.	.	.	.	.	.	.	.	33,824	15,701
1887-88	.	.	.	.	.	.	.	.	29,249	16,583
1888-89	.	.	.	.	.	.	.	.	26,417	15,343
1889-90	.	.	.	.	.	.	.	.	24,339	16,120

These exports are " Oils other than kerosine," and are all shipped from Burma, to the United Kingdom, Ceylon, the Straits Settlements, and other countries.

PETROSELINUM, *Hoff.; Gen. Pl., I., 891.*

455

A genus of herbs which belongs to the Natural Order UMBELLIFERÆ, and comprises only one species of economic importance, *viz.,* **Petroselinum sativum,** *Hoff. & Koch (DC. Prodr., IV., 102),* THE PARSLEY. This biennial herb is wild in the south of Europe from Spain to Turkey, and in Sardinia, and is cultivated in many parts of the world for the sake of its finely-cut leaves, which are largely used for flavouring dishes. In India it is frequently met with in market gardens, but is probably cultivated only for European residents. In the *Genera Plantarium* this genus has been reduced to **Carum.**

PEUCEDANUM, *Linn.; Gen. Pl., I., 918.*

A genus of perennial glabrous or pubescent herbs, which comprises about 100 species, of which 9 or 10 are natives of India.

[UMBELLIFERÆ.

456

Peucedanum grande, *Clarke; Fl. Br. Ind., II., 710;*
    Syn.—PASTINACA GRANDIS, *Dalz. & Gibs.*
    Vern.—*Báphali,* BOMB.; *Dúkú,* HIND. & PERS.
    References.—*Dalz. & Gibs., Bomb. Fl., 107; Dymock, Mat. Med. W. Ind., 2nd Ed., 379; Dymock, Warden & Hooper, Pharmacog. Indica, II., 126.*
    Habitat.—A tall, glabrous herb, found on the Bombay Ghâts.

MEDICINE.
457

    Medicine.—Dymock writes, "From the description of *dúkú* in Arabic and Persian works we gather that it was a kind of wild carrot (*dúkú-ágria* of the Greeks), with small fruit and a finely-divided leaf. In India this term has been transferred to quite a different plant. In Royle's *Materia Medica,* Falconer is quoted as describing *dúkú* as a fruit resembling that of asafœtida, and as probably derived from some species of **Ferula.** The *dúkú* of the Muhammadan druggists of Bombay answers to this description, but the same fruit is called by the Hindus, *báphali.* Three years ago I planted some of it, and this year (1880) it has produced a flowering stem and proves to be the **Pastinaca grandis** of Dalzell & Gibson, a native of the Ghâts near Bombay." *Dúkú* was considered by the ancients as carminative, stimulant, and diuretic.

    The authors of the *Pharmacographia Indica* state that other umbelliferous fruits are not unfrequently substituted for this drug, and that they have received those of **Dorema ammoniacum** from Bengal, and of an asafœtida plant from Northern India, under the name of *dúkú.*

Chemistry.
458

    CHEMICAL COMPOSITION.—The authors of the *Pharmacographia Indica* publish an exhaustive analysis of the fruit, from which it appears that twenty-five pounds distilled with water yielded 6-fluid ounces of a light yellow essential oil having the odour of the fruit. It was dextogyre, a column of 100 m.m. rotating 36 degrees. At 15·5 C. its sp. gr. was ·9008, cooled to—14°C. it was still liquid and no crystals separated. After dehydration the oil commenced to boil at 76°C., and up to 185°C. 2 per cent. distilled over. More and more passed off as the temperature was raised, till from 226° to 228° C. a residue (26 per cent.), viscid and of a deep yellow tint, was left in the flask. The fractions up to 210°C. were colourless, the others were of a deep yellow colour.

TRADE.
459
460

    Trade.—The fruit, collected from the hills of the Konkan, is sold in Bombay at R6 per parrah (about 25℔).

P. graveolens, *Benth.; Fl. Br. Ind., II., 709.*
    THE DILL, OR SOWA.
    Syn.—P. SOWA, *Kurz;* ANETHUM SOWA, *Roxb.;* A. GRAVEOLENS, *Linn.*

P. 460

PHASEOLUS aconitifolius.	The Dil or Sowa.

Vern.— *Sowá, soya, sutopsha,* HIND. ; *Súlpha, sowa, shulúpa, súlpa,* BENG. ;
Sowa, soya, sáwa, N.-W. P.; *Soya,* KUMAON; *Soi,* KASHMIR; *Soya,*
PB.; *Baluntshep,* BOMB. ; *Surva, súah,* GUZ.; *Sata kuppi,* TAM. ; *Sa*
myeit, BURM. ; *Misreyá, satapushpi,* SANS. ; *Shubit,* ARAB.

References.—*Roxb., Fl. Ind., Ed. C.B.C., 272 ; Kurz, in Journ. As. Soc.,*
1877, pt., ii., 116 ; O'Shaughnessy, Beng. Dispens., 366 ; U. C. Dutt, Mat.
Med. Hind., 309, 317 ; Dymock, Mat. Med. W. Ind., 2nd Ed., 374;
Fleming, Med. Pl. & Drugs, in As. Res., XI., 309, 313; Flück &
Hanb., Pharmacog., 327 ; Bent. & Trim., Med. Pl., 132; S. Arjun,
Bomb. Drugs, 62 ; Murray, Pl. & Drugs, Sind, 199 ; Year-Book Pharm.,
1874, 144 ; 1879, 214, 466 ; Irvine, Mat. Med. Patna, 97 ; Butler, Med.
Top. & Statis. Oudh & Sultanpur, 31 ; Med. Top. Ajm., 150 ; Trans.
of Med. & Phy. Soc. Bomb. (New Series), IV., 156 ; VI., 330 ; Atkinson,
Him. Dist.,311, 705, 745 ; Lisboa, U. Pl. Bomb., I., 161, 245, 351 ; Birdwood,
Bomb. Pr., 38, 220 ; Smith, Dic., 386 ; Kew Off. Guide to the Mus. of Ec.
Bot., 74 ; Gazetteer, N.-W. P., IV., lxxii.; Settle. Rep. Chanda, 83 ;
Gribble, Man. Cuddapah, 199 ; Nellore Manual, 135 ; Agri.-Horti. Soc.,
Ind.:—Transactions, IV., 103 ; VII., 50, 51 ; Journals (Old Series),
IV., 200 ; X., 9.

Habitat.—A glabrous herb, found throughout tropical and sub-tropical
India; often cultivated in the cold season for use as a vegetable, and for
its carminative seed, the latter of which is employed both for culinary and
medicinal purposes, and is met with in every market.

OIL.
Fruit.
461

Oil.—Dill FRUIT yields by distillation with water from 3 to 4 per cent.
of an essential oil (known as " Bishop's weed oil "). It has a pale yellow
colour, a pungent odour, and a hot sweetish taste. Its principal constituent is a fluid hydrocarbon, $C_{10}H_{16}$, to which the name of *anethene*
has been given. Another hydrocarbon is said by Nietzki to be probably
present ; while the oil contains a third constituent, an oxygenated oil,
probably identical with *carvol*, obtained from carraway fruit. Dill oil
possesses similar properties to the fruit, and is employed medicinally, and
in mixtures for perfuming soap.

MEDICINE.
462

Medicine.—Dill is commonly regarded as the ἄνηθον of Diosecrides,
and the ANETHUM of Palladius and other ancient writers, as well as of the
New Testament. In Greek the name ἄνηθον is at present applied to a
plant of very similar appearance, *viz.,* **Carum Ridolfia,** *Benth. et Hook.*
(*Flück. & Hanbury*).

The properties of dill-oil, dill-water, and the other preparations in which
the fruit is administered, are too well known to require detailed description.
Much less employed in Europe than they were at one time, they are still
very popular remedies with the people of India. It is considered a warm

Leaves.
463
Fruit.
464
FOOD.
Fruit.
465
Leaves.
466

remedy, is much used as a cure for colic, also for all forms of indigestion
and abdominal pains in women, and is considered lactagogue. An infusion is given as a cordial drink to women after confinement. The LEAVES
moistened with oil are used as a stimulating poultice or suppurative.
Muhammadan writers describe the FRUIT as resolvent and deobstruent,
carminative, diuretic, and emmenagogue.

Food.—The FRUIT and the LEAVES are both employed to impart a
flavour to curries.

467

PHASEOLUS, *Linn.; Gen. Pl., I., 538.*

A genus of twiners, usually herbaceous, which comprises about sixty species,
mostly tropical, many of which are widely cultivtated, especially in America. Of
these some fifteen are natives of, or occur spontaneously in, India.

468

Phaseolus aconitifolius, *Jacq.; Fl. Br. Ind., II., 202 ;* LEGUMINOSÆ.

THE ACONITE-LEAVED KIDNEY BEAN.

Syn.—P. TRILOBUS, *Wall. ;* DOLICHOS DISSECTUS, *Lam.*

Vern.— *Moth, mothi, mat, bhringga, urad, meth-kalai,* HIND.; *Banmudga,*

| The Kidney Bean. | (*J. Murray.*) | **PHASEOLUS aconitifolius.** |

kheri, Beng.; *Bir mung, moch, bir móch* Santal; *Matti-kalaie*, Assam; *Moth*, N.-W. P.; *Moth*, Pb.; *Mohar, muhri*, Sind; *Mot*, C. P.; *Math, matha, matki*, Mar.; *Mút, math*, Guz.; *Mote, mut*, Dec.; *Tulka-pyre*, Tam.; *Kúncúma-pesalú, minumulu*, Tel.; *Madki*, Kan.; *Bassunta, vasunta, makushtaka*, Sans.; *Adas*, Pers.

References.—*Roxb., Fl. Ind., Ed. C.B.C., 558; Voigt, Hort. Sub. Cal., 231; Stewart, Pb. Pl., 73; DC., Orig. Cult. Pl., 345; Rev. A. Campbell, Rep. Econ. Pl., Chutia Nagpur, Nos. 7857, 8107, 9492; U. C. Dutt, Mat. Med. Hindus, 308; S. Arjun, Cat. Bomb. Drugs, 209; Murray, Pl. & Drugs, Sind, 126; Birdwood, Bomb. Prod., 120; Baden Powell, Pb. Pr., 240, 342; Atkinson, Him. Dist. (X., N.-W. P. Gaz.), 309, 694; Duthie & Fuller, Field and Garden Crops, 41, Pl. xi.; Useful Pl. Bomb. (XXV., Bomb. Gaz.), 152, 277; Church, Food-Grains, Ind., 152, f. 29; Guide to the Mus. of Ec. Bot. Kew, 44; Stocks, Rep. on Sind; Bomb. Man., Rev. Accts., 101; Settlement Report: Panjáb, Hasára, 88; Montgomery, 107; Kángra, 25; Lahore, 9; Jhang, 85, 93; N.-W. P., Bareilly, 82; Central Provinces, Baitool, 77; Chanda, 81; Gazetteers:—Bombay, IV., 53; V., 29; VIII., 182, 188; XVIII., Pt. II., 43; Panjáb, Hoshiárpur, 93; Ludhiána, 140; Karnal, 172; Gurgaon, 84; N.-W. P., I., 80; III., 225, 453; IV., lxxi.; Rájputana, 254, 269; Agri-Horti. Soc., Panjáb, Select Papers up to 1862, Index, 42; Rept. of Agri. Dept. Bombay, App. viii.; Cent. Prov., Statement D.*

Habitat.—A perennial or annual herb, with stems trailing to a length of 1 to 2 feet; found throughout India from the Himálaya to Ceylon, and extending from the tropical region up to 4,000 feet in the North-West. It is closely related to **P. trilobus**, *Ait.*, with which it agrees in flowers and general habit. It is cultivated in many parts of India, and in certain localities is quite as important a crop as *úrd*. DeCandolle remarks that the absence of a Sanskrit name (by which he probably means an ancient Sanskrit name), and the variety of names in modern Indian languages, point to a recent cultivation. U. C. Dutt gives it the Sanskrit name of *Makushtaka*, a word which, according to **Sir Monier Williams**, is applied to a "species of kidney bean, or of rice."

Cultivation.—This bean is generally cultivated as a hot-weather or *kharíf* crop in the plains, in dry, light, sandy soil, being sown in June-July and reaped in October-November. It is impossible to give the total area annually under the crop in all India, but that it is by no means small is indicated by the figures given for areas when available. The following account of the method of cultivation in different districts and provinces, may be taken as typical :—   **CULTIVATION 469**

1st. Panjáb.—No details exist as to the extent of area under the crop in the whole province. The crop is, however, a fairly important one, and the method of cultivating it may be gathered from the following :—" *Moth* is an extremely hardy plant, and the zamindars say that if it once puts forth sufficient leaves to cover its root, no amount of dry weather affects it. It is supposed to be a capital 'grain,' and the green plant first class fodder for horses. The *bhúsa* is also highly prized. The *bhúsa* of this pulse is of two kinds—' *phaliát*,' the broken shreds of the pods and stalks, and ' *pattari*,' the leaves. Two ploughings are deemed sufficient. The seed is sown broadcast and ploughed in, about the same quantity being used as of *múng* (**Phaseolus Mungo**). The sowings are made rather earlier than those of *mást mung* in the Hathar, as the cultivator has not the fear of floods before his eyes, and the harvest is consequently also earlier " (*Settlement Report, Jhang District*). In the Gazetteer of the Hoshiárpur District it is stated that the crop thrives best in years when the rainfall is scanty, that it is sown on the poorest lands, and on dry sloping hill-sides, and that the experimental returns show an average yield of about two maunds an acre.   **Panjab. 470**

In Ludhiána it is generally grown mixed with *múng* or *másh* (**P. Mungo** and **P. Mungo**, *var.* **radiatus**), in light sandy soils. In richer soils

CULTIVATION in the Panjab.

these pulses are grown mixed with millets. The method of this mixed cultivation is as follows :—" The great millet is either sown wide, when the object is to develop the heads for grain (*jowár*), or thick, with a view to the fodder (*charri*). The times of sowing and reaping are the same as for the pulses. Where, as in the eastern portion of the district, there is a great deal of irrigation, and the well cattle are dependent on the fodder raised in the unirrigated land, the crop is always the mixture of *moth*, &c., with *charri*, except where the soil is sandy, and only a pulse can be grown. The crop grows up very dense, the millet having a very small head, and never reaching more than a height of about four feet. The people begin cutting the whole as green fodder in August, and go on using it for about two months till the crop has ripened. The heads of the *charri* are occasionally picked for grain; but generally the mixed crop is cut down, and given without any attempt to get the grain of the pulses. It is intended that the cattle should get the grain as well as the straw, for it would be a short-sighted policy to keep out the former, as the cultivator well knows. In Jagraon *tahsíl* there is not the same necessity for a strengthening fodder, and very fine *jowár* is grown. There is the same mixture of pulses; but the millet seed is in very small amount. The stalks come up at intervals, and grow to a height often of 8 or 10 feet, and have very fine heads, which almost weight them down. The pulses also have a fair yield of grain; and only the straw and *jowár* stalks are used for fodder. In the Jangal villages the spiked millet (*bájra*) sometimes takes the place of *jowár*. In the Bet, *charri* or fodder alone is grown, the soil not suiting the pulses of the Dháia. There is no yield of grain " (*Gazetteer, 140*).

In Karnál it is cultivated in the same way ; the soil receives two ploughings, and 5 seers of seed are sown per acre.

In Gurgaon 124,347 acres are estimated to be under the crop. It is generally grown mixed with *bájra*, but if sown separately the land is ploughed three or four times, about 10 seers per acre is sown, the crop is weeded once and harvested in October-November. A more spreading and more climbing form, called *gora moth*, is sown under cotton. The estimated yield of *moth* averaged about 4 maunds 6 seers per acre (*Gazetteer, 84, 85*).

Sind.
471

2nd. *Sind.*—Stocks states that *mohar* is a *rabi* crop, but in the Gazetteer the remark is made that the other principal kidney beans are *kharíf* crops. In 1886-87, 8,041 acres were under *mohar* in the Karáchi district, in 1887-88, 8,539 acres, and in 1888-89, 1,153, in the Karáchi, and 23 acres in the Hyderabad, district.

N.-W. Provinces.
472

3rd. *North-West Provinces.*—Messrs. Duthie & Fuller write, " *Moth* in the *kharíf* answers to the coarse pea known as *kesári*" (**Lathyrus sativus**), "in the *rabi*, both being grown on the worst land which can be made to bear a crop. It is in consequence grown much more commonly as a sole crop than either *múng* or *lobia*, and the area which it occupies on its own account is very nearly equal to that under *urd*. It also forms a very common mixture in millet fields, especially in the case of the spiked or bulrush millet (*bájra*), which it resembles in its preference for light sandy soils, and also in its liability to damage from ill-timed rainfall.

"Its cultivation when grown alone is of the roughest possible description. A couple or at most three ploughings are held sufficient, and the seed is sown broadcast at the rate of 4 seers to the acre.

" In a favourable season its produce is often very heavy, but taking into consideration the poverty of the land on which a great portion of it is grown, and its liability to damage from rain while in flower, the highest outturn of grain which can be taken as the average is 8 maunds to the acre, with rather less than double this amount of fodder. The grain is an article of human food, but there are many prejudices against it, the most notable being that it is liable to produce worms in the bowel. On the other hand, it is considered a useful remedy for flatulency. But it is principally used as cattle food, and is said to be a fattening diet, as are also the leaves and stalks " (*Field & Garden Crops*).

During the three years ending 1882, the average total area under *moth* as a sole crop in the thirty temporarily-settled districts, was 211,906 acres, of which 5,879 only were irrigated. The largest area was in the Meerut Division, which had 84,835 acres; then came Rohilkhand with 67,977; Agra, with 31,988; Benares, with 15,662; Allahabad, with 9,853; Kumáon, with 1,012, and Jhansi, with 579.

These figures do not include the large quantity grown as a mixed crop with millets.

4th. Bombay.—*Moth* is an important crop in this Presidency. It is chiefly grown mixed with *bajra*, in shallow, black, or light stony soils. Sowing takes place in June-July, and the crop is harvested in November. The method of cultivation is similar to that already described under the Panjáb and North-West Provinces.

The average area under the crop in Bombay during the past three years (1886-87 to 1888-89) was 263,424 acres. Cultivation appears to have been decreasing during that period, thus in 1886-87 the area was 309,847 acres, in 1887-88 it was 279,193 acres, while in 1888-89, it diminished to 201,234 acres. The chief *moth*-growing districts are, Sátára, with 53,566 acres in 1888-89, Ahmadabad with 39,176, Poona with 23,905, Kaira with 22,735, Sholápur with 21,910, and Dhárwár with 10,491. All the other districts in Gujarát and the Karnátak are returned as having small areas (less than 10,000 acres) under the crop. None appears to be grown in the Konkan. Of the total area, 162,665 acres were *rayatwári*, 21,081 belonged to other Governments, and 17,488 were alienated.

5th. Central Provinces.—In these provinces *moth* is a non-irrigated *kharíf* crop, grown under the same conditions and in the same sort of soil as described elsewhere. Figures are not available as to the area under *moth* alone, that crop being returned together with *mung* and *urad* (see P. Mungo, p. 189-191).

6th. Central India and Rájputana.—*Moth* is cultivated to some considerable extent in these provinces, both alone, and mixed with *bájra*. In the State of Dholpur 40,000 acres are said to be under the crop, of which the largest proportion is in Bári. The average outturn per acre is 10 maunds of pulse and 8 maunds of husk. The latter fetches R2, the former R10, and since the total cost of cultivation is given as R9-12-9, the profit to the cultivator is about R2-3-3.

7th. Bengal.—*Moth* is cultivated to a small extent in most of the districts of this province, particularly in Furruckábád, Patna, Púrneah, and Gorakhpur. It is sown in August and reaped in December or January, and is said to yield " 15 maunds (11 cwt.) of good pulse per acre " (*Hunter, Stat. Acct., Bengal, XIV., 121*).

8th. Assam.—The crop is cultivated in most of the districts in a similar manner to that in Bengal. No statistics are available regarding the area under it.

Medicine.—The PULSE is said to be a useful medicinal diet in cases of flatulence, and is also used as a food in cases of fever; the ROOT, as well as that of P. radiatus, is stated by Royle to be narcotic.

Food & Fodder.—The BEANS are considerably employed in certain parts of India, especially in Bombay, as a vegetable. The PULSE is split and eaten as *dál*, cooked in various ways. Thus it is ground to flour and

CULTIVATION in the N.-W. Provinces.

Bombay. 473

Central Provinces. 474

Central India and Rajputana. 475

Bengal. 476
Assam. 477
MEDICINE. Pulse. 478
Root. 479
FOOD & FODDER. Beans. 480
Pulse. 481

FOOD & FODDER.	used with other grains in making cakes; and is eaten parched or boiled whole with condiments. In the Panjáb and North-West Provinces it is little esteemed, being considered heating and apt to give rise to worms in the bowel.
Plant. 482	The whole PLANT is a valuable fodder, and, as already stated, is frequently grown, either alone or mixed with some millet, for this purpose. The grain is employed as a food for fattening oxen and horses, but is said not to be given to milch cattle, for it is supposed to prevent the flow of milk. When given to oxen it is bruised and steeped in water for a few hours to soften; when for horses it is generally mixed with *bájra*. The stalks are fairly nutritious and are also used as fodder. According to Church the composition of the pulse is as follows:—
Chemistry. 483	Water, 11·2; albumenoids; 23·8; starch, 56·6; fat, ·6; fibre, 4·2; ash 3·6 (0·8 of phosphoric acid) per cent.

Ic., t. 43.

484   Phaseolus adenanthus, *G. F. Meyer ; Fl. Br. Ind., II., 200 ; Wight,*

> **Syn.**—PHASEOLUS ROSTRATUS, *Wall.*; P. ALATUS, *Roxb.*; P. AMARUS, *Roxb. mss.*; P. CIRRHOSUS and TRUXILLENSIS, *H. B. K.*; P. SENEGALENSIS, *Guill. & Per.*
>
> **Vern.**—*Banbarbati*, BENG.; *Kullounda, hullowla*, BOMB.; *Káralsona*, TAM.; *Káralasana, káru alachanda*, TEL.
>
> **References.**—*Elliot, Flora Andhr., 85; Drury, U. Pl., 338; Lisboa, U. Pl. Bomb., 152; Agri.-Horti. Soc., Ind., Journal (Old Series), IX., 417; Ind. Forester, III., 237; XIII., 56.*
>
> **Habitat.**—A sub-glabrous perennial with firmer leaves than the cultivated species, found wild all over the plains from the Himálaya to Ceylon; not cultivated.

FOOD. Root. 485	**Food.**—The tuberous ROOT is cooked and eaten, especially in time of famine.

486   P. calcaratus, *Roxb.; Fl. Br. Ind., II., 203.*

> **Syn.**—PHASEOLUS SUBLOBATUS, *Wall.*; P. HIRTUS, *Wall.*; P. MUNGO, *Wall Cat.*, 5589 C, *ex-parte*; P. PUBESCENS, *Blume.*
>
> **P. torosus**, *Roxb*, is probably a cultivated form of this species, with short suberect stems, and subsessile congested racemes.
>
> **Vern.**—*Sutri, ghurúsh*, HIND.; *Sutri*, SANTALI; *Síta más, pau maia*, NEPAL; *Ghurúsh*, PB.; *Gúrúsh, gúrounsh*, KUMAON.
>
> **References.**—*Roxb., Fl. Ind., Ed.C.B.C., 555, 558; Campbell, Ec. Prod., Chutia Nagpur, No. 8157; Baden Powell, Pb. Pr., 240; Atkinson, Him. Dist., 695; Settlement Rep. Kumáon, App., 32; Madden, Off. Rept. Kumáon, 280; Agri.-Horti. Soc. Ind., Trans., IV., 102.*
>
> **Habitat.**—Cultivated and wild throughout the tropical zone from the Himálaya to Ceylon. It may be grown to a higher elevation than most other pulses, reaching an altitude of 6,500 feet in the North-West Provinces and to 5,000 feet in the Khásia Hills.

CULTIVATION 487	**CULTIVATION.**—Two forms are said to exist—one with red, and the other with cream-coloured, seed. The cultivation of neither is extensive. Like *moth* it is a *kharíf* crop, being sown in June-July and reaped in October-November.
FOOD & FODDER. 488	**Food & Fodder.**—It is used, like the other pulses of this genus, for food and fodder.

489   P. lunatus, *Linn.; Fl. Br. Ind., II., 200.*

> THE LIMA OR DUFFIN BEAN.
>
> **Syn.**—PHASEOLUS VULGARIS, *Wall. non Linn.*; P. INAMŒNUS, *Linn.*; P. XUARESII, *Zucc.*; P. PUBERULUS, *H. B. K.*, and P. TUNKINENSIS, *Lam.*, are cultivated varieties of this species.
>
> **Vern.**—*Kursumbulle-pullie*, HIND.; *Bunbur-butti*, BENG.; *Lobiya*, PB.

| The Green Gram (Mung). | (*J. Murray.*) | **PHASEOLUS Mungo.** |

References.—*Roxb., Fl. Ind., Ed. C. B. C.,* 554; *DC., Origin Cult. Pl.,* 344; *Lisboa, U. Pl. Bomb.,* 153; *Darrah, Note on the People of Assam,* 1888; *Church, Food-Grains of India,* 155; *Firminger, Man. Gard. for India,* 153.

Habitat.—A tall, biennial plant, the legume of which is 2 to 3 inches long, scimitar-shaped, with large seeds, and variable in colour. It resembles the "French Bean" in general aspect, but has smaller and more numerous flowers. It is very commonly cultivated throughout India, for use as a vegetable; indeed, it is frequently termed the "French Bean" by Anglo-Indians. According to DeCandolle it is a native of Brazil, and is believed to have come to India originally from the Mauritius.

CULTIVATION.—The seeds should be sown in October when the rains are over, about 4 inches apart, in rows. The plants are of extensive growth, and require strong sticks for their support. A moderately-shaded situation suits them best (*Firminger*).

CULTIVATION 490

Food.—The young BEANS may be cooked in the same way as the French bean. When old they are rather coarse, and are shelled, the large, flat, ivory-like SEEDS being alone eaten. These may be boiled, or fried in oil, and have in either case a very agreeable roast-chestnut-like flavour. The best variety for cultivation is the large, oval, white-seeded kind, with at the most a brown or black mark close to the hilum. Seeds of this form are much preferable to those which are flattened, and rather reniform with blotches of red or black veinings.

FOOD. Beans. 491 Seeds. 492

A white-seeded form from Mysore was analysed by Church with the following result:—Water, 13·3; albumenoids, 19·7; starch, 57·8; oil, 1·2; fibre, 4·3; ash, 3·7 per cent. One of these seeds weighed 16½ grains.

It is well to remember that this species sometimes exhibits markedly poisonous properties.

[*II., 392.*

Phaseolus multiflorus, *Willd.; Fl. Br. Ind., II.,* 200 *; DC., Prodr.,* THE SCARLET-RUNNER.

493

Syn.—P. COCCINEUS, *Lam.*

References.—*Firminger, Man. Gard. for Ind.,* 151; *Smith, Ec. Dic.,* 371; *Bomb. Gazetteer, VIII.,* 184; *Jour. Agri.-Horti. Soc. (New Series), IV.,* 7.

Habitat.—A climbing bean which differs from **P. vulgaris,** *Linn.,* in its bright scarlet, casually white, flowers, arranged in long racemes which often overtop the leaves. It is a native of Mexico, but is commonly cultivated in India for the sake of its tender green pods. The plant is remarkable as being one of the few that twine in a contrary direction to the sun, that is, from right to left.

CULTIVATION.—The seed should be sown in October when the rains are over, in rows, at a distance of three inches apart. Firminger states that he has sown them earlier, and had plants in blossom in the rains, but the flowers dropped off without forming a single pod. When about 3 inches high they should have sticks for their support. This species does not bear heat well, and, therefore, though perennial, must be raised from seed each season afresh. In the warmer parts of the country its cultivation is attended with but little success (*Firminger*).

CULTIVATION 494

Food.—The tender young PODS are a most excellent vegetable.

FOOD. Pods. 495

P. Mungo, *Linn.; Fl. Br. Ind., II.,* 203. GREEN GRAM.

496

Syn.—P. MAX, *Roxb.;* P. AUREUS, *Ham.;* P. HIRTUS, *Retz.*

Var. glaber, *Roxb.;* P. MUNGO, *Wall. Cat.,* 5589, *ex parte;* P. GLABRESCENS, *Steud.*

Var. Wightianus, *Grah.;* P. WIGHTII, *W. & A.;* P. SUBVOLUBILIS, *Ham.*

Cultivation of Green Gram

Var. radiatus, *Linn.* (sp.) ; P. ROXBURGHII, *W. & A.* ; P. MUNGO, *Wall,* *Cat. 5589, ex parte;* P. SETULOSUS, *Dalz.;* VIGNA OPISOTRICHA, *A. Rich.* It is convenient for the purpose of this work to consider **Phaseolus Mungo,** type form, and varieties **glaber** and **Wightianus** together, separating VAR. **radiatus** as a distinct product, since it is cultivated and recognised as a separate crop

Vern.—*Múng, harri múng, walli múng, múg, munj, pessara,* HIND. ; *Múng, múg, hari-múng, kheruya, bulat, ghora muga, sona múga, krishna múga,* BENG.; *Kalá múg, sauli múg, dhala múg,* URIYA; *Múng,* SANTAL; *Mú, múng,* NEPAL; *Múng, chhimi, chikan,* N.-W. P.; *Múng,* KUMAON; *Múng, múngi, múji,* PB.; *Mai,* PUSHTU; *Múng,* RAJ.; *Múng, mah,* SIND; *Múng, múngu,* C. P.; *Múng,* BOMB.; *Múg,* MAR.; *Mag,* GUZ.; *Pucha-payarú, siru-payarú, patche paira,* TAM.; *Wúthúlú, patcha-pessara* (grain=*patcha-pessalú*), TEL.; *Hesaru hesarnbele,* KAN.; *Painouk, pai,* BURM.; *Munmæ,* SING.; *Mudga,* SANS.; *Mung, múng,* PERS.

References.—*Roxb., Fl. Ind., Ed. C.B.C., 556; Voigt, Hort. Sub. Cal., 280; Stewart, Pb. Pl., 73; DC., Orig. Cult. Pl., 346; Mason, Burma & its People, 467, 768; Sir W. Elliot, Fl. Andhr., 126, 140, 151; U. C. Dutt, Mat. Med. Hindus, 149, 309; S. Arjun, Cat. Bomb. Drugs, 42; Murray, Pl. & Drugs, Sind, 126; Birdwood, Bomb. Prod., 121; Baden Powell, Pb. Pr., 239; Drury, U. Pl. Ind., 337; Atkinson, Him. Dist. (X., N.-W. P. Gaz.), 695; Duthie & Fuller, Field and Garden Crops, 37, pl. ix.; Useful Pl. Bomb. (XXV., Bomb. Gaz.), 152, '77; Church, Food-Grains, Ind., 148, f. 28; Guide to the Mus. of Ec. Bot., Kew, 44; Rep. Agri. Station Cawnpore, 1884, 7; Man. Madras Adm., I., 289; Madden, Off. Rept., Kumáon, 279; Bomb. Man. Rev. Accts., 101; Settlement Report:—Panjáb, Montgomery, 107; Lahore, 10; Gujrát, App., xxxviii.; Hasára, 88; Peshawar, cxvii., App., ix.; N.-W. P., Bareilly, 82; Kumáon, App., 32; Allahabad, 27; Central Provinces, Chanda, 81; Baitool, 77; Upper Godavery, 35; Land Rev. Settle., Nursingpore, 52; Papers relating to Settle. of Chellumbrum and Manargúdi, South Arcot, 110; Gazetteers:—Bombay, II., 63; IV., 53; V., 25; VII., 30, 79, 89; VIII., 182; XII., 151; XIII., 289; XVI., 100; XVII., 269; Panjáb, Hoshiárpur, 93; Ludhiána, 140; Gurgaon, 84; Karnal, 172; N.-W. P., III., 225, 463; IV., lxxi.; Orissa, II., 15, 45, 133, 180; Mysore & Coorg, I, 59; II., 11; Rájputana, 269; Agri.-Horti. Soc., Ind.:—Trans., I., 41; II., 101, 128; VIII., 259; Journals (Old Series), IX. (Sel.), 59; XIII., Sel., 51, 52.*

Habitat.—This pulse is a native of India, and has been cultivated for at least 3,000 years. It is met with, both wild and cultivated, throughout the plains, ascending to 6,000 feet in the outer ranges of the North-West Himálaya.

CULTIVATION.—*Múng* is one of four pulses which resemble each other very closely in appearance and habit of growth, the other three being *móth* (**P. aconitifolius),** *lobia, rausa* or *rawas* (**Vigna Catiang,** *Endl.*), and *urd* or *másh* (**P. Mungo** *var.* **radiatus**). It can be easily distinguished from the two first, but belonging to the same species as the last, the resemblance between the two is very great. The most popular distinction between the two, as met with in the field, lies in the fact that *múng* has dark green, and *úrd* yellowish green, leaves, but the chief difference is in the shape of the grain, that of the latter being much larger and longer than that of *múng.* "Exclusive of *úrd,*" Messrs. Duthie & Fuller write, "there are three well-marked varieties, having respectively green, yellow, and black seeds." The green-seeded is the typical and commonest variety (*harri-múng*), that with yellow seed (*sona-mung*) has been described as **Phaseolus aureus,** while that with black seeds (*krishnamung*) has also been described by Roxburgh as distinct under the name of **Phaseolus Max.**

Múng is cultivated all over the Peninsula, and an idea of its importance may be gathered from provincial figures, which are cited when available, though no estimate can be given of the total area under the crop.

CULTIVATION
in the
Panjab.
498

1st. The Panjáb.—The cultivation of P. **Mungo** in this province is very extensive, but no idea of the area can be given from the fact that it is almost always sown mixed with either *moth* (**P. aconitifolius**), or with *másh* (*var.* **radiatus**), and that all are generally cultivated as mixed crops with certain of the millets. In the Karnál Gazetteer, it is stated that this is more particularly so in the case of *múng*, which is nearly *always* grown with *juár* or *bájra*. It is also occasionally sown broadcast with cotton, the tall plants of the latter affording the shade so beneficial to its growth.

The method of cultivation differs in no essential point from that of P. **aconitifolius** already described. The plant grows well in loams and light soils, is not suited to clays, and succeeds best with a fair amount of moisture in the soil, though a heavy rainfall is deleterious. The seed is sown broadcast in July or August after one or two ploughings and is then ploughed in. It ripens at the same time as *móth* (in October-November) and is cut in certain districts, in others it is pulled. In the Settlement Report of the Jhang District it is stated that fields that have been cropped with *mung* are usually covered with a strong after-crop of *talla* grass (**Cynodon Dactylon**, *Pers.*). The plants when young are said to suffer from the attacks of grasshoppers (*tidda*), and later from caterpillars which attack the pods.

2nd. Sind.—The crop is grown in a similar manner to *móth*, in the *kharíf* season. It is apparently increasing in importance at the expense of P. **aconitifolius**, the area of which, as already stated, is steadily decreasing. The only area from which it is returned is the Karáchi district. During the past three years, the average area has been 10,822 acres; in 1886-87 it was only 5,624 acres, while in 1887-88 it increased to 13,553, and in 1888-89, to 13,288 acres.

Sind.
499

*3rd. North-West Provinces —*Messrs. Duthie & Fuller write, " *Mung* is grown in every district of the Provinces, but almost invariably as a subordinate crop in fields of millet or cotton, and very seldom by itself. It is therefore a *kharíf* crop, being sown at the commencement of the rains, and reaped in October. It is in some respects remarkable that it is not more frequently grown alone, since its grain commands a far higher price than that of millet, but this is no doubt partly explained by the precariousness of its growth; heavy and continuous rain, especially in September (when it is in flower), often causing absolute ruin. But, as a counterpoise to this, it bears, and justly, the reputation of being able to withstand a great deal of drought, and in a season of scanty rainfall, when millets have utterly failed, it, with *urd, lobia,* and *moth,* forms a most valuable food resource, the so-called ' subordinate' crop, becoming, in this case, of firstrate importance. Another advantage which these pulses share with *arhar* is that of not impoverishing the soil, or, at all events, not to the extent of gramineous crops, such as the millets. Not only does the depth to which the roots penetrate, enable them to gain moisture from land on which their shallow-rooted companions wither of drought, but it also leads to the æration of the ground, and whether it be true or not that they actually add to the fertility of the soil by fixing atmospheric nitrogen, they at all events increase the fertility of its surface by accumulating from below foodsubstances which were beyond the reach of shallow-rooted plants.

N.-W.
Provinces.
500

Its cultivation is exactly the same as that of cotton or millet. When grown alone it is sown at the rate of about 12 seers to the acre. When associated with millet or cotton it shares the benefit of the weeding which these crops receive, and only receives irrigation when they require it. It is reaped about a fortnight before the millets and is threshed out by bul-

**CULTIVATION
in the
N.-W.
Provinces.**

locks in the usual manner. The crushed stalks and leaves are much prized as fodder, and are used to give a tempting flavour to trash that even Indian cattle might otherwise reject as uneatable. Its cost of cultivation may be assumed to be the same as that of *juár* or *bájra*" (from R12-13 to R13-13 per acre). "When grown alone, the average outturn per acre is reported from most districts as about 5 maunds of grain and three times the weight of fodder."

In 1881, 29,684 acres were returned as being under *múng* in the thirty temporarily-settled districts of the North-West Provinces. Of this amount the Rohilkhand Division was returned as having 15,235 acres; Meerut had 8,256; Jhansi, 2,799; Agra, 1,988; Allahabad, 1,357; and Benares 49 acres. These figures, however, give no real idea of the importance of the crop in the agriculture of the Provinces. Mr. Fuller states that it is grown in greater or less amount on fully one-fourth of the total area under *kharíf* crops, and represents, so to speak, the cultivator's *insurance* against a shorter allowance of rainfall than his millets can make shift with.

Bombay.

501

4th. Bombay.—The crop is grown to a greater or less extent in all districts of the Province, but is cultivated on much the largest scale in the districts of the Karnatak, a locality in which *móth* is but little grown. It is cultivated, either by itself or mixed with millets, in shallow, light, stony soils, and often as a first crop on rich lands. Sown in June it is harvested in September. A variety called *múgi* is said to be most frequently grown as a mixed crop with *bájra*.

The average area under the crop during the past three years has been 158,868 acres, *viz.*, 178,973 in 1886-87, 156,839 in 1887-88, and 140,793 acres in 1888-89. Last year this area was distributed as follows:—Dhárwár, 43,035 acres, Bijápur, 37,667 acres, Sátára, 15,383 acres, Belgaum, 8,367 acres, and the other districts with smaller areas.

**Central
Provinces.**

502

5th. Central Provinces.—*Múng* is cultivated as an unirrigated *kharíf* crop in the same way as **P. aconitifolius**, and, together with that pulse and *urád* (*var.* **radiatus**), covers a large area. The areas under these three crops returned during the past three years have been :—1886-87, 207,157 acres; 1887-88, 181,247 acres; 1888-89, 169,138 acres. The falling off is due to a diminution of the figures from Ráipur and Bilaspur. The districts which had the largest areas under the crop in 1888-89 were :—Bhandárá with 37,032 acres, Ráipur with 32,709, Bálághát with 27,250, Nimár with 22,380, and Jabalpur with 13,550. All the other districts, except Bilaspur and Sambalpur, were returned as having small areas (less than 10,000 acres) under the crop. The remarks made above regarding figures of area in the North-West Provinces apply equally to these.

In several Settlement Reports it is stated that though chiefly grown with *jowári* as a *kharíf* crop it is also cultivated in the *rabi* season. In the report on the Nagpur Experimental Farm for 1887-88, it is said that the average outturn per acre, obtained at the farm in 1887, was 450℔ to the acre.

Rajputana.

503

6th. Rájputana and Central India.—No statistics are available regarding the area, but it may be gathered that the crop is cultivated to a considerable extent in the same way as elsewhere. In the Rájputana Gazetteer it is stated that the estimated outturn is 8 seers of *múng*, and 3½ seers of the millet with which it is grown, per acre. The cost of cultivation is estimated at R10-0-9, the value of the crop at R13-6-3, and the profit at R3-5-6 per acre.

Madras.

504

7th. Madras.—It is grown in this province in calcareous loamy soils, and is considered a good rotation after cereal crops. It is sown both as a *kharíf* and *rabi* crop, either alone or mixed with *rági* (**Eleusine Coracana**) and the great and the spiked millets. **Drury** states that it forms a very large portion of the pulse crops, and that large quantities are annually shipped from Madras, to Pegu, Bengal, Bombay, Mauritius, and other places. In the Madras Experimental Farms Report for 1871 it is stated, that sown in October and reaped in February, 24℔ of seed yielded on an average 480℔ of pulse. Statistics of the area under the crop are not available.

Bengal.—*Múng* is grown to a large extent as a rotation after *aus* paddy, and is followed by the same crop, *arahar, kachu,* or *begun.* In the *Agri. Dept. Report for 1886,* it is stated that three varieties, *sona múg, krishna múg,* and *ghora múg* are grown. All require a light friable soil, high and well drained, and cannot be grown in clay. Four or five ploughings suffice, but the soil should be fairly pulverised. The seed is sown in September-October, about 1¾ seers being thrown broadcast over a *bigha.* The after cultivation consists in hoeing the field once with the *kodali.* Harvest time is the end of December and beginning of January, and a yield of 2 to 2½ maunds per *bigha* is obtained.

The pulse if heated in any way is liable to the attacks of weevils, and should consequently be kept in a cool place. Some mix the grain with sand, some rub it with oil, and others mix with tamarind seeds to preserve it. The last is said to be the most efficacious remedy (*Agri. Dept. Report, 1886*). No figures can be given of the area under the crop.

Medicine.—The PULSE is employed as a diet in cases of fever, and though difficult to digest is considered cool, light, and astringent. Thus U. C. Dutt writes, "*Mudga,* especially its green variety, is considered most wholesome and suited to sick persons. A soup made of this pulse is often the first article of diet prescribed after recovery from acute illness." It is also said to be "used to strengthen the eyes."

Food & Fodder.—The green PODS are eaten as a vegetable. The ripe PULSE is used boiled whole, or split hke *dál.* Parched, and ground to flour, it is made into balls with spice, or is employed for making a sort of porridge, and in times of scarcity for bread. It is also eaten parched, or boiled whole with condiments. The grain has a pleasant taste, is wholesome and nutritious, and commands a comparatively high price. By some authorities it is stated to cause flatulence unless eaten with asafœtida, but it must be remembered that all pulse has this tendency, if it form too large a proportion of the day's ration, or be insufficiently cooked.

The pulse is a valued food for cattle and horses, being considered very fattening. The crushed STALKS and LEAVES, though not so valuable as those of *P. aconitifolius,* are useful as fodder, and are said to be much prized, since they give a tempting flavour to "trash that even Indian cattle might otherwise reject as uneatable."

CHEMICAL COMPOSITION.—According to Church a large number of analyses have been made of the different forms of *múng* without disclosing any decided difference in chemical composition. One hundred parts contain :—water, 10·8 to 11·4; albumenoids, 22·2 to 23·8; starch, 54·1 to 54 8; oil, 2·7 to 2·0; fibre, 5·8 to 4 2; ash, 4·4 to 3·8. The former figures are for typical green-seeded *múng,* the latter for typical yellow-seeded. In unhusked beans the fibre is reduced to 1·1 per cent., all the other constituents being proportionately increased.

Domestic.—The FLOUR of the green form is said to be an excellent substitute for soap. It leaves the skin soft and smooth, and is an invariable concomitant of the Hindu bath (*Elliot*).

Phaseolus Mungo, *Linn.* ; var. radiatus, *Linn.* ; *Fl. Br. Ind., II., 203.*
　　Syn.—P. ROXBURGHII, *W. & A.*; P. SETULOSUS, *Dals.*
　　Vern.—*Urud, urid, dord, thikiri,* HIND.; *Tircorai-kalai, másh-kulái,* BENG.; *Ramra, bir sang,* SANTAL; *Úrd,* KUMAON; *Másh, máh, úrad,*

Margin notes
CULTIVATION in Madras.

Bengal.
505

MEDICINE.
Pulse.
506

FOOD & FODDER.
Pods.
507
Pulse.
508

Stalks.
509
Leaves.
510

CHEMISTRY.
511

DOMESTIC.
Flour.
512

513

P. 513

PB.; *Máh, maga, urad,* SIND.; *Udid, maga,* MAR.; *Adad, arad,* GUZ.; *Urud, kala úrd, udid,* DEC.; *Patchay-pyre, panny-pyre,* TAM.; *Minumulu, karu-minumulu, patsa-pesalu,* TEL.; *Hasaru, uddu,* KAN.; *Cheru-poiaar,* MALAY.; *Ulundu-mae,* SING.; *Hurita, másha, danie-másha,* SANS.; *Másh,* ARAB.; *Benú mash,* PERS.

References.—*Roxb., Fl. Ind., Ed. C.B.O.,* 557; *Voigt, Hort. Sub. Cal.,* 231; *Stewart, Pb. Pl.,* 73; *Rev. A. Campbell, Rept. Econ. Pl. Chutia Nagpur, Nos.* 7512, 8145; *Sir W. Elliot, Fl. Andhr.,* 87, 115; *O'Shaughnessy, Beng. Dispens.,* 317; *U. C. Dutt, Mat. Med. Hindus,* I, 49, 309; *Murray, Pl. & Drugs, Sind.,* 126; *Baden Powell, Pb. Pr.,* 239; *Drury, U. Pl. Ind.,* 338; *Man. Madras Adm.,* I., 288; *Nicholson, Man. Coimbatore,* 223; *Moore, Man., Trichinopoly,* 71, 72; *Madden, Off. Rept. Kumáon,* 280; *Bomb. Man. Rev. Accts., App.,* 32:—*Settlement Report; Panjáb, Lahore,* 10; *Montgomery,* 107; *Hasára,* 88; *Kángra,* 25, 27; *Simla, App. ii., H;—Jhang,* 85, 93; *N.-W. P., Bareilly,* 82; *Kumaon, App.,* 32; *Allahabad,* 27; *Asamgarh,* 116; *Central Provinces, Chanda,* 81; *Upper Godavery,* 36; *Baitool,* 77; *Land Rev. Settle., Nursingpore,* 52; *Gazetteers:—Bombay, II.,* 63, 269, 273, 277, 280, 284, 287, 291, 295, 390, 536, 538, 541, 544, 547; *IV.,* 53; *VII.,* 30; *VIII.,* 182; *XIII.,* 289; *XVI.,* 91, 100; *Panjáb, Ludhiána,* 140; *Hoshiárpur,* 93; *Gurgáon,* 84; *Karnal,* 172; *Simla,* 58; *N.-W. P., I.,* 80; *III.,* 2, 25, 463; *Agri.-Horti. Soc. Ind.:—Trans., IV.,* 128; *Journals (Old Series), IV.,* 185-189, *Sel.* 56; *XIII.,* 387. *Sel.* 51, 52; *(New Series), II., Sel.* 16.

Habitat.—This variety, which has a similar distribution to the typical P. Mungo, differs in having a longer and more trailing habit, in the greater hairiness of the plant, and in the seeds being fewer, larger, longer, and usually of a dark brown colour.

Urd has two distinct cultivated forms, one with large black seeds, which ripens in August and September, the other with smaller green seeds, ripening in October and November. Both are, however, sown at the commencement of the rains, but may be also occasionally grown as a spring crop in low wet fields, in which case it is sown in February and reaped in May.

CULTIVATION
514

CULTIVATION.—*Urd* is the most highly prized of all the pulses of the genus Phaseolus, and is largely cultivated in all parts of India. The method followed is exactly the same as that with *móth* and *mung,* with the exception that it thrives best on the heavier classes of soil, and it is therefore unnecessary to repeat the information already given, province by province. Like the others it is generally grown as a subsidiary crop to certain of the millets, rice, and Indian corn, and to cotton, but is sometimes cultivated alone. Messrs. Duthie & Fuller state that the average area under this crop in the thirty temporarily-settled districts of the North-West Provinces, was, during the three years ending 1881, 258, 495 acres. Of the total area under *kharíf* crops it constituted a percentage of 3·6 in the Meerut Division, 4·7 in Rohilkhand, ·2 in Agra, ·2 in Allahabad, 1·4 in Benares, 1·2 in Jhansi, and 3·1 in Kumaon. This, however, as in the case of P. Mungo, represents but a very small fraction of the total area on which the crop is grown. Mr. Fuller indeed states that, if the area were included on which it is cultivated as a subordinate crop, the total would be at least twelve times increased.

The average amount of seed sown is from 4 to 6 seers per acre, the average outturn (when grown as a sole crop) is 5 maunds of grain, with three times its weight of straw. The plants suffer much from mildew if there be a long continuance of damp winds, and are also much damaged by heavy rain.

It is the most important leguminous crop in the valley of Nepál, and constitutes a considerable portion of the area under kidney beans in the Central Provinces (see P. Mungo, p. 190).

In certain districts of Madras it is grown as a separate crop, either

in gardens or on dry lands. It is sown in August or September, and | CULTIVATION
reaped from December to January. The crop is said to obtain greater
attention in this Presidency than it does elsewhere ; sheep manure is used,
the soil is well ploughed, and it is weeded by hand after about six weeks.
As much as 700℔ per acre is said to be sometimes obtained (*Coimbatore
Dist. Man.*)

Medicine.—*Másh* is a highly valued article of Hindu medicine. It is | MEDICINE.
prescribed both internally and externally in paralysis, rheumatism, and
affections of the nervous system, and enters into the composition of
several decoctions recommended by Chakradatta for these diseases.
Several oils for external application also contain this PULSE as their basis | Pulse.
or principal ingredient, for example, the *svalpa másha taila,* a prepara- | 515
tion said to be useful in rheumatism, contracted knee, stiff shoulder,
&c. (*U. C. Dutt*).

In modern Indian medicine it is prescribed as laid down by Sanskrit
writers, and is also considered hot and tonic and useful in piles, paralysis,
and affections of the liver. The ROOT is supposed to be narcotic, and | Root.
according to Campbell is prescribed by the Santals as a remedy for | 516
" aching bones."

SPECIAL OPINIONS.—§ " Used as a poultice for abscess " (*Assistant
Surgeon Nehal Sing, Saharunpore*). " Sometimes used as a lactagogue in
the form of ordinarily cooked *dál* (*Assistant Surgeon S. C. Bhattacharji,
Chanda*). " A poultice of the seeds is used in inflammation " (*Assistant
Surgeon T. N. Ghose, Meerut*). " *Mash kalie* is the most commonly used
pulse in Bengal. In seasons when cholera prevails this *dál* is often
preferred. It is considered an easily digestible and cooling article of
diet. My personal experience accords with this popular impression "
(*Surgeon R. L. Dutt, M.D., Pabná*).

Food & Fodder.—The RIPE GRAIN is the most esteemed of all pulses | FOOD &
in India, and fetches the highest price. It is eaten in the same way as | FODDER.
múng, namely, in the form of *dál,* bread, boiled whole, parched, or as | Ripe Grain.
spice balls, &c., and is the chief constituent of the wafer biscuits known | 517
in Bombay as *pápad.* In the Panjáb it is also used in the form of two
preparations known as *bári* and *sepa.* Both of these are similarly prepared,
by soaking the seed for a couple of days in slightly warmed water. It is
then crushed to a pulp by means of a stone, dried, and eaten mixed with
ghí, or with butter-milk. According to Stewart it is not so highly esteemed in this province as elsewhere, being considered the most heating,
and apt to give colic, of all pulses. The green PODS are used as a | Pods.
vegetable. Like all other pulses it is well adapted, from its chemical | 518
composition, to supply the food elements wanting in the ordinary diet of
a rice-eating people. It is of interest to note in this connection that the
ASH of the straw, or stems and leaves, is sometimes eaten, *e.g.,* in Dinajpur, | Ash.
in place of salt, probably because it satisfies the craving for mineral matter | 519
brought about by the marked deficiency of rice in ash constituents. The
grain is a valuable fattening cattle and horse food, and the STRAW is a | Straw.
highly esteemed fodder. | 520

CHEMICAL COMPOSITION.—One hundred parts of the unhusked grain | Chemistry.
contain :—water, 10·1 ; albumenoids, 22·7 ; starch, 55·8 ; oil, 2·2 ; fibre, | 521
4·8 ; and ash, 4·4 (of which 1·1 consists of phosphoric acid) (*Church*). It
therefore resembles the typical species very closely in all its constituents,
but has a larger proportion of starch, oil, and ash than the yellow-seeded
form of P. Mungo.

Domestic & Sacred.—The SEED is the reputed origin of the weight | DOMESTIC.
known as *másha,* twelve of which go to the *tola,* and 960 to the *seer* (2℔). | Seed.
| 522

Bread made from the finely ground pulse is said to be used in many of the religious ceremonies of the Hindus (*Roxburgh*).

523

Phaseolus trilobus, *Ait. ; Fl. Br. Ind., II., 201 ; Wight, Ic., t. 94.*
THE THREE-LOBED KIDNEY BEAN.

Syn.—DOLICHOS TRILOBATUS, *Linn.*
Vern.—*Mugáni, trianguli,* HIND.; *Mugáni,* BENG.; *Arkmut, mukuya,* BOMB.; *Pani-pyre, nari-payir,* TAM.; *Pilli persara,* TEL.; *Mudga-parni,* SANS.
References.—*Roxb., Fl. Ind., Ed. C.B.C., 558 ; Burm., Fl. Ind., t. 50, fig. I. ; Voigt, Hort. Sub. Cal., 231 ; Dals. & Gibs., Bomb. Fl., 71 ; DC., Origin Cult. Pl., 345 ; Elliot, Flora Andhr., 152 ; Mason, Burma & Its People, 468,768 ; Drury, U. Pl., 338 ; Lisboa, U.Pl.Bomb., 152, 198, 277 ; Balfour, Cyclop., III., 196 ; Gasetteers :—Mysore & Coorg, I., 59 ; Bomb., XV., 432 ; N.-W. P., I., 80 ; IV, lxxi. ; Madras Man., Administration, I., 289 ; Nicholson, Man., Coimbatore, 223.*
Habitat.—A pulse which is distributed throughout India, wild and cultivated, from the Himálaya, where it attains an altitude of 7,000 feet, to Ceylon and Burma. It is chiefly cultivated as a mixed crop for fodder purposes, and as food for the poorer classes.

MEDICINE.
Leaves.
524
FOOD &
FODDER.
Pulse.
525
Root.
526
Straw.
527
528

Medicine.—The LEAVES are alleged to be tonic and sedative, and are used in the form of a cataplasm for weak eyes. According to Ainslie, they are given by the *Vytians* in Behar, in decoction, as a remedy for "irregular fever."

Food & Fodder.—The PULSE, though highly nutritious and much esteemed in certain localities, is generally eaten only by the poorer classes. The ROOT is said to be eaten in China, and a sort of arrowroot is obtained from it in the same country. It is also employed as a cattle food. The STRAW is said to afford good fodder.

P. trinervius, *Heyne ; Fl. Br. Ind., II., 203.*

Syn.—PHASEOLUS FARINOSUS, *Linn.*
Vern.—*Múkani, mataki,* BOMB.
References.—*Dals. & Gibs., Bomb. Fl., 71 ; Lisboa, U. Pl. Bomb., 198 ; Gasetteer, Bombay, XV. (Kanara), 432 ; Agri.-Horti. Soc. Ind., Jour. (Old Series), VI., 43.*
Habitat.—A pulse, closely allied to **P. Mungo,** which is found on the plains of the Western Peninsula and Ceylon, ascending to 4,000 feet.

FOOD.
Seed.
529

Food.—The SEED, which is said to be rich in nitrogenous principles, was largely used as food in the Deccan famines.

530

P. vulgaris, *Linn. ; Fl. Br. Ind., II., 200.*
THE KIDNEY, FRENCH, or HARICOT BEAN.

Syn—P. NANUS, *Linn. ;* P. COMPRESSUS, OBLONGUS, SAPONACEUS, TUMIDUS, HÆMATOCARPUS, SPHÆRICUS, and GONOSPERMUS, *DC.*
Vern.—*Bakla, loba,* HIND.; *Sheto seru,* NEPAL ; *Shiuchana, bákula,* N.-W. P.; *Bábri,* PB.; *Barigalú,* TEL.; *Dambala,* SING.
References.—*Roxb., Fl. Ind., Ed. C.B.C., 554 ; DC., Orig. Cult. Pl., 338 ; Atkinson, Him. Dist , 694 ; Lisboa, U. Pl. Bomb., 152 ; Birdwood, Bomb. Prod., 122 ; Church, Food Grains of India, 147 ; Firminger, Man. Gard. for India, 152 ; Ind. Gard., 163-164 ; Agri.-Horti. Soc. of Ind.:—Trans., I., App., 248 ; III., Pro., 226 ; IV., 145, 146, 215 ; V., 64 ; VII., Pro., 108, 184 ; Journals (New Series), IV., 24 ; Gasetteers :— Mysore & Coorg, I., 59 ; N.-W. P., I., 80 ; Moore, Man. Trichinopoly, 16, 67 ; Gribble, Man. Cuddapah, 43 ; Settlement Repts. :—Simla, App., ii., H, xli. ; Kumáon, App. 33.*
Habitat.—The common French Bean is universally cultivated in tropical, sub-tropical, and temperate regions, but is nowhere clearly known to be a wild plant. Botanists held for a long time that it was of Indian or Kashmír origin, but DeCandolle (*Orig. of Cul. Pl.*) has proved this

P. **530**

not to be the case. According to that learned writer, the concensus of evidence supports the theory that the plant was originally American. In favour of this view, he sums up as follows :—" (1) **Phaseolus vulgaris** has not been long cultivated in India, the South-West of Asia, and Egypt ; (2) it is not certain that it was known in Europe before the discovery of America ; (3) at this epoch the number of varieties suddenly increased in European gardens, and all authors commenced to mention them ; (4) the majority of the species of the genus exist in South America ; (5) seeds apparently belonging to the species have been discovered in Peruvian tombs of an uncertain date, intermixed with many species, all American."

CULTIVATION.—Many cultivated races of this species are cultivated in gardens all over India for the sake of their young green pods. The following extract from **Firminger** may be quoted as describing the method of cultivation most likely to prove successful :— Cultivation. 531

" (1) *Runners.*—Runner French Beans I have found to be far less prolific in this country than the Dwarf kinds ; and as they in no way compensate by their flavour for the scanty produce, it is perhaps advisable to make the principal sowings consist of the Dwarf kinds. The Dutch, bearing small ivory-like seeds, next to the Dwarf kinds, has proved to be the most productive. The plant is of slender habit, does not grow high, and bears long, narrow, very delicate pods. The seed should be sown in October, in a row, about three inches apart. I have had the plants in blossom in the rains, but found them utterly unproductive at that season. Runners. 532

" *Dwarfs.*—Of the Dwarf kinds of French Bean, as those are called which require no sticks for their support, there are a great many named varieties. The principal, or only, difference, however, between them seems to consist in the form and colour of the ripened seeds. In flavour, at least as regards those cultivated in this country, the several varieties are as much alike as possible. The first sowing may be made about the beginning of October in a good soil. The seeds should be put in two inches apart, about an inch deep, in rows, two feet between each row. The seed, if sound, will germinate in three or four days ; and the plants will come into full bearing in about six weeks from the time of sowing. As the crops are of short continuance, sowings should be made in succession, at intervals of about ten days, to keep up a constant supply. Dwarfs. 533

" I have not found Dwarf French Beans thrive well except in a situation considerably shaded. Where much exposed to the sun the plants not only make slow growth, but are apt to have their leaves preyed upon and much injured by insects ; they then become entirely unproductive" (*Firminger, Manual of Gardening for India*).

Food.—The young PODS are a well known green vegetable, and are cooked in India in the same way as elsewhere, for the use of Europeans. They are scarcely, if ever, used by the Natives, who prefer the cheaper and better known pulses. FOOD. Pods. 534

The dried SEED is highly nutritious, is largely employed as a vegetable, and has much to recommend it as a portable diet for camp and field use. According to Church, one hundred parts of the bean contains :—water, 14 ; albumenoids, 23 ; starch, 52·3 ; oil, 2·3 ; fibre, 5·5 ; and ash, 2·9 parts. Seed. 535

PHEASANTS, JUNGLE-FOWL, PARTRIDGES, &c.

Pheasants, &c., *Hume & Marsh., Game Birds, 183 et seq.; Murray, Avifauna of Br. Ind., II., 529 et seq.* 536

The pheasant families of birds PHASIANIDÆ, MEGAPODIDÆ, and GALLINÆ, comprise the pea-fowl, pheasant, jungle-fowl, and spur-fowl, while the

13 A

partridge family, TETRAONIDÆ, include the partridge, snow cock, and certain forms of quail. It would probably be of little value to describe at length each species in a work such as the present; the reader may, therefore, be referred to the two works above quoted for zoological descriptions of each species with an account of its habitat, &c. The forthcoming Second volume of BIRDS, in the series of the *Fauna of British India,* will also soon be available for reference.

The Indian birds included in the above enumerated families are as follow :—

FAM. PHASIANIDÆ.

Pavo cristatus, *Linn.;* The Peacock.
P. muticus, *Linn.;* The Burmese Pea-fowl.
Argusianus argus, *Linn.;* The Argus Pheasant.
Polyplectron thibetanum, *Gm.;* The Grey Peacock Pheasant.

FAM. MEGAPODIDÆ.

Megapodius nicobariensis, *Blyth;* The Nicobar Mound Bird, or Megapode.
Crossoptilon thibetanum, *Hodgson;* Hodgson's Eared Pheasant.
Lophophorus Impeyanus, *Lath.;* The Monaul Pheasant.
L. Sclateri, *Jerd.;* Sclater's Crestless Monaul
Ceriornis satyra, *Linn.;* The Sikkim Horned Pheasant, or Indian Crimson Tragopan.
C. melanocephalus, *Gray;* The Simla Horned Pheasant, or Western Tragopan.
Ithaginis cruentus, *Hardw.;* The Green Blood Pheasant.
Pucrasia macrolopha, *Lesson;* The Koklás Pheasant.
Phasianus Wallichii, *Haraw.;* The Chír Pheasant.
Euplocamus albocristatus, *Vigors;* The White Crested Kalij Pheasant.
E. leucomelanus, *Lath.;* The Nepál Kalij.
E. melanonotus, *Blyth;* The Black-backed Kalij.
E. horsfieldi, *G. R. Gray.;* The Black-breasted Kalij.
E. lineatus, *Elliot;* The Lineated Silver Pheasant.
E. Cuvieri, *Temm.;* The Arracan Silver Pheasant.
E. Andersoni, *Elliot;* Anderson's Silver Pheasant.
E. Vieilloti, *G. R. Gray;* Vieillot's Fire-backed Pheasant.

FAM. GALLINÆ.

Gallus ferrugineus, *Gm.;* The Common Jungle Fowl.
G. Sonneratti, *Tem.;* The Grey Jungle Fowl.
Galloperdix spadiceus, *Gmel.;* The Red Spur Fowl.
G. lunulatus, *Valenc.;* The Painted Spur Fowl.

FAM. TETRAONIDÆ.

Tetraogallus himalayensis, *G. R. Gray;* The Himálayan Snow Cock.
T. thibetanus, *Gould;* The Thibetan Snow Cock.
Lerwa nivicola, *Hodgs.;* The Snow Partridge.
Francolinus vulgaris, *Steph.;* The Black Partridge.
F. pictus, *Jerd. & Selby.;* The Painted Partridge.
F. chinensis, *Osb.;* The Chinese Francolin.
Caccabis chukor, *Gray;* The Chukor Partridge.
Ammoperdix bonhami, *Gray.;* The Seesee Partridge.
Ortygornis ponticeriana, *Gmel.;* The Common Grey Partridge.
O. gularis, *Temm.;* The Kyah Partridge.
Arboricola torqueola, *Valenc.;* The Common Hill Partridge.
A. atrogularis, *Blyth;* The Black-throated Hill Partridge.

P. 536

Meadow Cat's Tail Grass.	(*J. Murray.*)	PHLEUM pratense.

Arboricola brunneipectus, *Tickell ;* The Brown-breasted Hill Partridge.
A. chloropus, *Tickell ;* The Green-legged Hill Partridge.
A. intermedia, *Blyth ;* The Arrakan Hill Partridge.
A. rufogularis, *Blyth ;* The Red-throated Hill Partridge.
A. mandelli, *Hume ;* The Bhotan Hill Partridge.
Bambusicola fytchii, *Anderson ;* The Western Bamboo Partridge.
Caloperdix oculea, *Temm. ;* The Ferruginous Wood Partridge.
Rollulus roulroul, *Scop. ;* The Red-crested Wood Partridge.
Perdicula asiatica, *Latham ;* The Jungle Bush Quail.
P. argoondah, *Sykes ;* The Rock Bush Quail.
Ophrysia superciliosa, *J. E. Gray ;* The Mountain Quail.
Microperdix erythrorhyncha, *Sykes ;* The Painted or Red-billed Bush Quail.
M. blewitti, *Hume ;* The Eastern Painted Bush Quail.
Coturnix communis, *Linn. ;* The Common or Large Grey Quail.
C. coromandelica, *Gmel. ;* The Black-breasted Rain Quail.
Excalfactoria chinensis, *Linn. ;* The Blue-breasted Quail.

**FOOD.
537**

Food.—Many of the above enumerated game birds are prized as articles of food. Several are obtainable during the season at most Indian hill stations, where they are used at the tables of Europeans. By Sanskrit writers they are included in the class of *vishkíra,* or birds which scatter their food, and, with the class *jángalá,* or animals which live in the wilds, are considered superior to all others as articles of diet (*U. C. Dutt, Hindu Mat. Med.,* 286, 287). They are also prized by Muhammadans ; indeed, as far back as the time of Akbar, we find them mentioned as prized for the table. Abul Fazl states in a list of " Living Animals and Meats " that the *durráj* or black partridge cost 3 *dam,* a partridge called the *kabg,* 20 *dam,* and the *búdanah* and *lawah,* 1 *dam* each. In another passage he describes the method of catching partridges by means of a call-bird, a practice common with-native fowlers to the present day.

**DOMESTIC.
538**

Domestic.—The feathers of some of the handsomer species of pheasant are highly valued for ornamental purposes. See **Feathers.** Partridges and quail are kept by Natives in Northern India for fighting purposes. They are confined in small cages, and carefully trained for this purpose.

PHILADELPHUS, *Linn. ;* Gen. Pl., I., 642.
[*II.,* 407 ; SAXIFRAGACEÆ.

539

Philadelphus coronarius, *Linn.,* var. **tomentosus ;** *Fl. Br. Ind.,*
Syn.—P. TOMENTOSUS, *Wall. ;* P. TRIFLORUS, *Wall ;* P. NEPALENSIS, *Loud. ;* P. CORONARIUS, *Brand.*
Vern.—*Dalunchi, bhoj,* SIMLA.
References.—*Brandis, For. Fl.,* 212; *Gamble, Man. Timb.,* 173; *Stewart, Pb. Pl.,* 93; *Royle, Ill., t. 46, f. 1.*
Habitat.—A shrub met with in the Temperate Himálaya, from Kashmír to Bhután, at altitudes from 5,000 to 9,000 feet ; often planted for ornamental purposes.

**FIBRE.
540**

Fibre.—Stewart describes an undetermined species of **Philadelphus** (probably **P. coronarius,** var **tomentosus,** the only member of the genus described in the *Flora of British India*) as occurring on the Sutlej, where it bears the names of *búsrú, múdnú,* and *shbang.* It is stated to be used for ropes.

PHLEUM, *Linn. ;* Gen. Pl., III., 1146.

541

Phleum pratense, *Linn. ;* GRAMINEÆ.
TIMOTHY, or MEADOW CAT'S-TAIL GRASS.
Syn.—P. NODOSUM, *Linn. ;* Boiss., *Fl. Or., V.,* 484.

P. 541

References.—*Hook., Students' Flora Brit. Islands, 427; Sutton, Permanent & Temporary Pastures, 58; Stebler & Schröter, Best Forage Plants, 52 (Eng. Ed.).*

Habitat.—A medium-sized perennial grass, growing in tufts, and thriving on moist retentive soils. It is indigenous in Europe and in North America, and is extensively cultivated as a pasture grass. Royle found it on the Chor near Simla, and it doubtless occurs in other localities on the Himálaya. The popular name of "Timothy Grass" was named after Timothy Hansen, who first cultivated it in Carolina about the middle of the last century, soon after which time it came into general cultivation in Europe.

Cultivation.
542

CULTIVATION.—Usually grown mixed with clovers and other grasses. It gives a large yield, since the hay is very heavy, heavier, in fact, than that of any other grass, a cubic yard weighing from 600 to 850℔. For hay it should be cut before flowering.

FODDER.
543

Fodder.—" Timothy," whether by itself, or forming a large proportion of a clover mixture, makes an excellent green fodder, much esteemed for horses. The hay is excellent, containing a large amount of nutritive matter in small bulk.

PHLOGACANTHUS, *Nees.; Gen. Pl., II., 1100.*
[ACANTHACEÆ.

544

Phlogacanthus thyrsiflorus, *Nees ; Fl. Br. Ind., IV., 512 ;*

Syn.—JUSTICIA THYRSIFLORA, *Roxb.*

Vern.—*Tita phul, bakha tita,* ASSAM; *Sua, shechin,* NEPAL; *Sumcher,* LEPCHA; *Bashkah,* MICHI.

References.—*Kurz, For. Fl. Burm , II., 246; Gamble, Man. Timb., 281; Atkinson, Him. Dist., 315 ; Journ. Agri.-Horti. Soc., Ind. (New Series), IV., 119; Gazetteer, N.-W. P., I., 83; IV., lxxvi.*

Habitat.—A large evergreen shrub, found in the sub-Himálayan tract from Kumáon to Assam, the Khásia Hills, and Burma. Often cultivated, for it is a very handsome shrub with long spikes of flame-coloured flowers.

TIMBER.
545

Structure of the Wood.—White, moderately hard, close-grained. Weight 37℔ per cubic foot.

PHŒBE, *Nees ; Gen. Pl., III., 157.*
[LAURINEÆ.

546

Phœbe lanceolata, *Nees ; Fl. Br. Ind., V., 141 ; Wight, Ic., t. 1821 ;*

Syn.—OCOTEA LANCEOLATA, *Nees ;* LAURUS LANCEOLARIA, *Roxb.;* L. LANCEOLATA, *Wall.* ; L. SALICIFOLIA and L.? CAMPHORATA, *Hamilt.*

Vern.—*Haulia, dandorla, káwal, kaula, sún káwal, bilphari,* HIND.; *Kaula, lali,* DARJILING; *Dupatti,* MECHI; *Nuni ojhar,* GARO; *Sun kanwál,* KUMAON; *Chan, chandra, badror, shalanghi,* PB.

References.—*Brandis, For. Fl., 377 ; Kurz, For. Fl. Burm., II., 290; Gamble, Man. Timb., 308 ; Beddome, For. Man., 184; Ind. Forester, I., 95, 99 ; V., 195 ; VIII., 106, 404; Gazetteer, N.-W. P., X., 316.*

Habitat.—A small evergreen tree, found in the sub-Tropical Himálaya from Simla to Bhután, ascending to 6,000 feet ; also met with in the Khásia Hills, Sylhet, Burma, and the mountains of South India.

TIMBER.
547

Structure of the Wood.—White, hard, close-grained; turns brown on exposure, and varies in colour in different localities ; weight 46 to 55℔ per cubic foot. It is used for planking, especially in native houses.

548

P. paniculata, *Nees ; Fl. Br. Ind., V., 142 ; Wight, Ic., t. 1820.*

Syn.—PHŒBE PUBESCENS, *Nees ;* P. WIGHTII, *Meissn. ;* OCOTEA PUBESCENS and PANICULATA, *Nees ;* LAURUS PANICULATA, *Wall.;* L. PUBESCENS, *Wall. ;* CINNAMOMUM TOMENTOSUM, and C. CATHIA, *Don.*

	PHŒNIX
The Dwarf Date Palm. (*J. Murray.*)	dactylifera.

Vern.—*Boltigachu,* GARO; *Kapua kanwál,* KUMAON; *Kumara,* NIL-
GHIRIS; *Chekio,* MAGH.
References.—*Brandis, For. Fl., 377; Kurz, For. Fl. Burm., II., 290;
Beddome, For. Fl., t. 292.*
Habitat.—An evergreen tree found in the Central Himálaya, Burma,
and the Nilghiri Hills, where it attains an altitude of 5,000 to 6,000 feet.

TIMBER.
549

Structure of the Wood.—Dense, heavy, close-grained, soft, of a light
red colour; used by the Natives of Southern India for many purposes. In
Burma it is soon attacked by xylophages (*Beddome, Kurz*).

PHŒNIX, *Linn.; Gen. Pl., III., 921.*

550

A genus of Palms which comprises some five or six species, all of which yield
more or less edible fruits, and are of very great interest from an economic
point of view. Considerable confusion, however, still exists as to the names
of the various palms of India. The following descriptions of the species of
Phœnix and their synonomy have been adopted from **Brandis.**

Phœnix acaulis, *Roxb.; Brandis, For. Fl., 555; PALMÆ.*

551

THE DWARF DATE PALM.

Syn.—P. OUSELEYANA, and P. PEDUNCULATA, *Griffith;* a variety with black
fruit is described by **Griffith** as P. ACAULIS, *var.* MELANOCARPA.
Vern.—*Khajuri, pind khajúr, jangli khajúr,* HIND.; *Schap,* LEPCHA;
Chindi, hindi, jhari sindi, GOND.; *Juno,* KURKU; *Pind khajúr,* PB.;
Boichind, MAR.; *Yíta,* TEL.; *Thin boung,* BURM.
References.—*Roxb., Fl. Ind., Ed. C.B.C., 722; Voigt, Hort. Sub. Cal.,
642; Griffith, Palms of Br. East India, 137, 139, pl. ccxxviii.; Kurz,
For. Fl. Burm., II., 535; Gamble, Man. Timb., 419; Stewart, Pb. Pl.,
243; Atkinson, Him. Dist., 318; Simmonds, Trop. Agri., 418; For.
Adm. Rept., Chutia-Nagpur, 1885, 6, 34; Jour. As. Soc., Pl. II., No.
II., 1867, 82; Gazetteers, Mysore and Coorg, I., 66; Central Provinces,
286; Settlement Report, C. P., Upper Godavery, 39; Ind. Forester,
VIII., 301; IX., 399; X., 325; XIII., 55; XIV., 393.*

Habitat.—A low, often almost stemless, palm; commonly met with in dry
ground in the sub-Himálayan tract from the Jumna eastward to Behar,
and southwards to Central India; found also in Burma. It flowers in
the cold season, and ripens its fruit in April and May.

The tree described by **Griffith** as P. **pedunculata** is referred by
Brandis to this species. It is common and gregarious on open ground of
the hilly country about Courtallum and Kunur on the Nilghiris, at 6,000
feet elevation.

Fibre.—Stewart states that in certain localities rope is made from the
bruised LEAVES. The fibrous leaves are also used, in the Central Provinces, to thatch houses.

FIBRE.
Leaves.
552
FOOD.
Fruit.

Food.—The FRUIT is small but eatable. · The Natives of Chutia Nagpur make a sort of sago from the PITH, but the stem is apparently nowhere tapped for its juice.

553
Pith.
554
555

P. dactylifera, *Linn.; Brandis, For. Fl., 552.*

THE EDIBLE DATE.

Vern.—Tree = *Khajúr, khhaji,* fruit = *khúrma, chúhára, kukyán, khujiyán,
kujran, pindakhejúr,* HIND.; tree=*Khájúr,* fruit=*khurmá, pindakhejúr,*
BENG.; *Kasser,* BHOT.; tree = *Khajúr, khaji,* fruit = *pind, chirwi, bag-
ri, khajúr, kukyán, khugiyan,* cabbage of leaves = *gadda, galli,* gum =
hokmchil, gond, sher-i-darakht-i-khurma, PB.; tree = *Mach,* fruit =
khurma, N. BALUCH.; *Kajura,* PUSHTU; *Karmah,* TURKI; *Pind chir-
di, kurma, tár, khaji,* fruit = *jarkha, clanuko,* SIND; tree = *Khajúr,*
fruit = *tamara, rájib, nakel, kurma, chuara,* BOMB.; *Kharjúr,* MAR.;
Khajúr, kárek, GUZ.; fruit=*Périch-chankay,* TAM.; *Kharjúrapu, périta;
mudda kharjúrapu,* fruit = *karjúru-káya,* TEL.; *Kharjúra,* KAN.; fruit
= *Ténich-chan-káya,* MALAY.; tree = *Swonpalwon,* fruit = *somblón-si,*
BURM.; *Indi,* SING.; *Pindakharjura,* fruit = *kharjjúraha,* SANS.;
fruit = *Khurmáe-yábis,* ARAB.; fruit = *Khurmae-khushk, nakhl,* PERS.

P. 555

References.—*Roxb., Fl. Ind., Ed. C.B.C., 723; Voigt, Hort. Sub. Cal., 642; Griffith, Palms of Br. East Ind., 142; Gamble, Man. Timb., 419; Stewart, Pb. Pl., 243; Aitchison, Rept. Pl. Coll. Afgh. Del. Com., 120; DC., Orig. Cult. Pl., 301; Mason, Burma & Its People, 425; Sir W. Elliot, Fl. Andhr., 90, 117, 151; O'Shaughnessy, Beng. Dispens., 641; Irvine, Mat. Med. Patna, 21, 101; U. C. Dutt, Mat. Med. Hindus, 313; Murray, Pl. & Drugs, Sind, 16; Med. Top., Ajm., 130; Birdwood, Bomb. Prod., 184; Baden Powell, Pb. Pr., 379; Useful Pl. Bomb. (XXV., Bomb. Gaz.), 181; Econ. Prod. N.-W. Prov., Pt. V. (Vegetables, Spices, & Fruits), 89; Royle, Fibrous Pl., 96; Shortt, Man. of Agriculture, 217; E. Bonavia, The Future of the Date Palm in India, 1885; Kew Bulletin, 1889, 28; Stocks, Rep. on Sind; Guide to the Mus. of Ec. Bot., 35; Ayeen Akbary, Gladwin's Trans., I., 83; Ain-i-Akbari, Blochmann's Trans., I., 66, 71; Linschoten, Voyage to East Indies (Ed. Burnell, Tiele, & Yule), I., 48; II., 119; Westland, Rept. on Jessore, 161-176; Settlement Reports:—Panjáb, Peshádwar, 13; Reports of Agri. Depts., Bombay, N.-W. P., Bengal, & Madras; Proceedings of Agri., Rev. & Commerce Dept., 1871, November, Nos. 9 to 11; 1782, March, No. 8, October, Nos. 18 & 19; Proceedings of Rev. & Agri. Dept., 1886, May, Nos. 1 to 6; Sept., Nos. 19 & 20; Gazetteers:—Bombay, VIII., 95, 96; XVII., 25; Panjáb, Rohták, 14; Orissa, II., 5; Mysore & Coorg, I., 66; Trans. of Med. & Phys. Soc., Bombay (New Series), VI., 76; Agri.-Horti. Soc. Ind.:—Journals (Old Series), II., 198; VII., pro. 28; VIII., 143; X., 351; XI., 263; (New Series), VI., 8; Agri.-Horti. Soc., Panjáb, Select Papers up to 1862, 208, 238; Tropical Agriculturist, March 1889, 645; July 1889, 40; Ind. Forester, XIV., 116, 321; Indian Agriculturist, Sept. 18th, 1886; January 1st, and Sept. 3rd, 1887; March 17th, 1888; Smith, Ec. Dict., 151.*

Habitat.—A tall tree, attaining 100 to 120 feet, but which differs from **P. sylvestris,** *Roxb.* (The Wild Date Palm of India), in having the foot of the stem often surrounded by root-suckers and in the leaflets, forming a very acute angle with the common petiole. It is cultivated and self-sown in Sind and the Southern Panjáb, particularly near Multan, Muzaffargarh, the Sind Sagar Doab, and in Trans-Indus territory. Near Dera Gházi Khán the trees are very numerous on a strip 10 to 12 miles long from north to south. A few are found planted at many places in the Eastern Panjáb, at Saharanpur, here and there in the Ganges Doab, and in Bandelkhand. It is also grown in the Dekhan and Guzerát, but does not thrive in Bengal.

The date-palm, according to DeCandolle, has existed from prehistoric times in the warm dry zone, which extends from Senegal to the basin of the Indus, principally between parallels 15 and 20. From Egyptian and Assyrian remains, it is evident that the tree was known, and its value appreciated, by the ancient peoples of these nations. Herodotus, in a more recent age (fifth century before Christ), mentions the date-palms of Babylonia, and still later. **Strabo** refers to those of Arabia. A certain amount of doubt exists as to whether the palm of the early books of the Old Testament was the true date-palm, but it appears certain that if so, it was then a much less valuable fruit tree than it is at the present time.

David, about 1,000 years before Chirst, and 700 after Moses, does not mention the date-palm in the list of trees to be planted in his gardens, and many years later **Herodotus** and **Strabo** principally remark on the wood of the tree, and indeed the former says of the date-palms of Babylonia that only the majority produced fruit sufficiently good to be used for food. On this point DeCandolle remarks, "This seems to indicate the beginning of a cultivation, perfected by the selection of varieties and of the transport of male flowers into the middle of the branches of female trees, but it perhaps signifies also that **Herodotus** was ignorant of the existence of the male plant."

To the west of Egypt it had probably existed for centuries before **Herodotus** mentioned them. Philological evidence points to its great antiquity

Products of India. 201

The True Date Palm. (*J. Murray.*) PHŒNIX dactylifera.

HISTORY.

in Africa and Western Asia; indeed, the number of Persian, Arabian, and Berber names is incredible. Some of these are said to be derived from the Hebrew, others are of unknown origin. The Greek name 'Phœnix,' refers to Phœnicia and the Phœnicians, who possessed the date-palm; the names "Dactylus," and "date" are, according to Henn, derived from *dachel* in a Hebrew dialect (*Origin Cult. Pl.*). The origin of the modern Sanskrit and vernacular names of India is evidently from the Persian or Arabic; a fact which, along with other evidence, tends to the inference that the plantations in North-Western India are not very ancient. In Sind, two legends exist regarding the introduction of this palm—one that the trees originated from date seeds thrown away by the army of Alexander the Great, the other that the seeds were introduced in a similar way by the western conquerors of Multán and Sind, in the seventh century A.D. (*Bonavia, Date-palm in India, 15*).

According to certain writers the plant is derived from the Indian wild species—**P. sylvestris.** Roxburgh, and following him Brandis and most other Indian botanists, have, however, agreed in separating the two species, though Griffith remarks that the materials at his disposal did not enable him to establish any specific distinction. In any case, DeCandolle has arrived at the conclusion, from geographical, historical, and philological evidence, that if the two do in reality belong to one species, the so-called wild **P. sylvestris** must have been derived from the true date-palm by the results "of recent naturalization in an unfavourable soil," and would thus become reduced to a variety of **P. dactylifera.** The universal distribution of P sylvestris in India, the very large extent to which it grows in many localities, and the absence of any evidence of intermediate forms, tend, however, to controvert this view, and these facts, together with the constant characters in which the two trees are said to differ, render it more probable that they are, as Roxburgh considered them to be, distinct, though nearly allied, species.

From Persia the date-palm spread farther east to China in the third century A.D., where its cultivation was spasmodically resumed at different times, but has now been abandoned. DeCandolle, remarking on this fact, sums up his interesting account of the history of the tree as follows:— "As a rule, beyond the arid region which lies between the Euphrates and the south of Atlas and the Canaries, the date-palm has not succeeded in similar latitudes, or at least it has not become an important culture. It might be grown with success in Australia and at the Cape, but the Europeans who have colonised these regions are not satisfied, like the Arabs, with figs and dates for their staple food. I think, in fine, that in times anterior to the earliest Egyptian dynasties, the date-palm already existed, wild or sown here and there by wandering tribes, in a narrow zone extending from the Euphrates to the Canaries, and that its cultivation began later as far as the North-West of India on the one hand, and the Cape de Verd Islands on the other, so that the natural area has remained very nearly the same for about five thousand years. What it was previously palæontological discoveries may one day reveal."

CULTIVATION 556

CULTIVATION.—The true date-palm, as already stated, occurs in large numbers over the dry arid tracts of the Western Panjáb and Sind, both in a state of cultivation and self-sown. Some idea of the extent to which it is grown may be gathered from the following remarks by Mr. O'Brien, Deputy Commissioner of Múltan, quoted by Brigade-Surgeon Bonavia:— "The date palm grows literally in hundreds of thousands in the Múltán, Muzaffargarh, Dera Gházi Khan, and Dera Ismail Khán Districts, and perhaps in Bannu; also in Jhang, Bhawalpur, and Sindh. I can form no idea of their number, but the Government revenue, at a nominal rate of

P. 556

CULTIVATION
in
Multan.

one anna per female tree, comes, in this district (Múltán), to R12,084. There are, however, many exemptions.

"From experiments made in the Muzaffargarh District, I made out the average produce of a date-palm to be 20 seers; but I have seen a tree which produced three maunds of dates. The Settlement Officer here esti. mated the average outturn, per tree, at from 1½ to 2½ maunds, and the price at two maunds per rupee, thus making the value of the gross produce from 12 annas to R1-4 per tree." With regard to fertilisation and cultivation he writes:—

"I have heard of owners of date-palms placing a cluster of male flowers among those of a female tree, but I have never seen this done. The date tree grows everywhere, so I should say that artificial fertilisation was not necessary, and that insects would do all that was wanted. The date-palm grows spontaneously everywhere from seed; occasionally one sees an offset planted by watercourses. With this exception, the date-palm here is never cultivated."

Later information received from the Panjáb Government shows that the number of assessed (female) trees in the Province were, in 1886;—Muzaffargarh, 414,509, Múltán (excluding two tahsils in which the number is not known), 86,041; Dera Ismail Khán, 95,659; Jhang, 4,819; and Dera Gházi Khán, 239,869

Though the tree is thus a very important one in these districts of the Panjáb, and is admirably suited to the soil and climate, it is allowed to grow altogether spontaneously from seed, and no attempt appears to be made to improve it by cultivation, nor to propagate the best kinds by means of offsets. Nor is artificial propagation apparently practised anywhere. As a consequence, though many eatable and fairly palatable fruits are produced, they are much inferior to those imported from Arabia, and the surplus production over local consumption is exported only to the adjoining districts of the Panjáb and North-West Provinces. The reason of this is not far to seek. The Arabs devote the greatest attention to the cultivation of the date, and, as a consequence, by careful selection and by rearing from offshoots, are constantly improving the cultivated races. More than a hundred of these are known by distinct names, races which differ in their adaptability to surroundings and in the quality of their fruits. Vigorous offsets, from three to four years old, are planted in groves, a locality being selected where abundance of water is available, or where it is possible to irrigate during the rainy season. The grove, which consists of 80 to 200 trees, is surrounded by a high *band* for the purpose of retaining water. For the first two or three months the offsets are watered separately, and daily, by hand. After they have struck root they are watered weekly, fortnightly, or at longer intervals as necessary. This interval should not exceed one month, and even after they have grown and attained some age, watering should be continued monthly during the hot dry weather. When water is abundant the grove is irrigated by means of deep trenches between the rows, where less plentiful by small channels leading to a depression round the foot of each tree. Occasionally vegetables are grown between the trees, and in every case the soil is said to be ploughed and turned once a year. In localities near the sea a fish manure is employed, in other cases the dung of cows, sheep, goats, &c., is used.

Excessive rain and consequent flooding, and a damp atmosphere when the fruit is forming, are prejudicial.

As a rule, only one or two male trees are planted in each grove, and the female trees are artificially fertilised by detaching the male flower, splitting the spathe, and when the female flowers are fit for impregnation, inserting one or two sprigs of the former into each branch of the latter. Each tree

CULTIVATION
in
Multan.

is capable of yielding only a certain number of fruits, and more than twelve
bunches are seldom left, the weakest and least promising being carefully removed. The pollen of the male flower is said to remain active for
one to two months after its removal from the tree, so the flower is carefully kept and used as occasion demands. After fertilisation of the female
flower, water is considered injurious, and the supply is cut off for one and-
a-half to two months.

Offsets for planting are removed from the parent tree, when of sufficient size and vigour. Under favourable conditions they are generally
ready for removal at the age of three or four years. The average weight
of a shoot suitable for removal is said to be 6℔, though heavier offsets are
more hardy and vigorous, and require less after-care. The young shoots
if not wanted for propagation, and any buds or offsets which may appear
on the crown or stem, should be removed in the spring of each year, since
they derive their nourishment at the expense of the main stem (*Bonavia,
54-60*).

It is not surprising then, that with such a careful and advanced
method of cultivation the dates of Arabia should be infinitely superior
to the product of the neglected palm of Sind and the Panjáb. Dr.
Bonavia recommends that offsets or carefully collected seed from the
better varieties should be obtained on a large scale and cultivated with
care, with the view of obtaining a stock from which finer races might be
distributed. He recommends that the offsets should be planted in September or October on the east side of other trees, so as to obtain shade
during the hotter part of the day, that they should be transplanted next
rains to high ground within reach of water, and that otherwise they
should be carefully cultivated in the manner above indicated. If propagated from seeds, these should be selected from good fruit, and sown
in September and October in richly manured beds; the seedlings should be
very carefully tended and watered, and finally transplanted when sufficiently vigorous.

In 1871-72 attempts were made by Government to introduce the true
date-palm into Oudh, through the Lucknow gardens. The experiments
were attended with a fair amount of success, and scattered specimens of
the tree now occur here and there in that Province and in the North-West
Provinces. Of recent years the attention of the Government of India
has been attracted by Brigade-Surgeon Bonavia to the great value of
the date as a famine food, in the districts in which it already grows, and it
has been suggested by that gentleman that further endeavours might be
made to increase the cultivation of such a valuable tree. He also pointed out
that the tree grows in Arabia in ground strongly impregnated with salt,
which seems to suit it; that the young palms could be planted in soil
the water of which is so strongly impregnated with salt as to be undrink-
able even by the Natives, and that on the whole it might be said to thrive
best in sandy, granitic, schistic and calcareous soils. He therefore recommended that experiments should be carried out with seedlings in localities in which the subsoil water was salt, on *úsar* or *réh* soils, and in other
lands, using a manure of mixed salt and lime. This scheme was warmly
supported by the Kew authorities, and arrangements were made in 1886 to
obtain seed and shoots from Her Majesty's Consul General in Algiers.

There would appear to be no doubt from the large array of facts
brought forward by Dr. Bonavia in support of his proposal, that with a
little preliminary care and expenditure, the cultivation of the date-palm
might be very largely extended in such localities as those indicated by him,
and the security of a good crop of valuable food-stuff, in dry seasons
when the ordinary food staples of the people inhabiting the more arid

P. 556

PHŒNIX
dactylifera.   Uses of the Edible Date Palm.

CULTIVATION

tracts of India are apt to fail, cannot but be a question worthy of much consideration.

Dr. Bonavia, in concluding his description of the tree. sums up as follows :—(1) It is eminently suited to tracts of country in India which normally have a rainfall. (2) It is also well suited to tracts of country with sometimes good, and sometimes scanty, rains; for these reasons (a) that it stands rain well, and that even if the moisture be too great for it to fruit well, the other crops will be good, and that the unripe fruit, which is extremely nourishing, will always be obtained ; (b) that it fruits best when the rain fails, and when other crops may be deficient to the verge of famine. (3) In Sind and adjoining localities, where it already grows, there is great room for the cultivation of the finer races of Persian Gulf dates. (4) The date tree, when once fairly established, will grow and fruit with comparatively less labour and attention than any other fruit tree. (5) The tree will grow in tracts and in classes of soil in which no other timber tree will thrive, and that, consequently, the annual decay of the lower whorl of woody leaves might be made to yield a valuable supply of fuel in such localities.

GUM.
557
FIBRE.
Leaves.
558
Petioles.
559
Sheathing
Base.
560
Fruit Stalks
561
Spathes.
562
PERFUMERY.
Spathes.
563

Gum.—The tree yields a gum, called *hukm chil*, which is used medicinally in the Panjáb.

Fibre.—In the Panjáb mats, fans, baskets, and ropes are made from the LEAVES, which are known as *bhútrá, pattra,* and *khúshab*. The PETIOLES (*chhari*) make excellent light walking sticks, and, when split up, furnish material for making crates and baskets. The fibrous net-work which forms the SHEATHING BASE of the petioles, called *kabál, khajúr ka bokla,* or *khajúr múnj*, is used for making pack saddles for oxen, and the fibre separated from it for cordage. The bunch of FRUIT STALKS, *buhárá*, is said to make a good broom, and is employed for that purpose in the Panjáb. In Arabia the green flower SPATHES are also used for rope-making, yielding a quantity of fibre on being beaten with a wooden mallet.

Perfumery.—The Assistant of the Political Resident, Persian Gulf, writes, " From the fresh SPATHES (called *tara*), is obtained, by distillation, *tara*-water. This has a strong but agreeable scent, and is mostly prepared at Busra and Bahrain. It is sold in carboys at 1½ to 2 rupees each, and is used for making *sherbet*. It is greatly prized by Arabs and Persians." This " water " does not appear to be known or prepared in India.

MEDICINE.
Dates.
564
Gum.
565
Fruit.
566
Juice.
567
Seeds.
568

Medicine.—DATES are considered demulcent, expectorant, laxative, nutrient, and aphrodisiac. They are prescribed in cases of cough, asthma, and other chest complaints, also in fever, gonorrhœa, &c. The GUM is esteemed as a useful remedy in diarrhœa and diseases of the genito-urinary system. Long continued use of the FRUIT is said to produce soreness of the gums (*Emerson*). Honigberger states that the inspissated JUICE was, in his time, officinal in Lahore.

SPECIAL OPINIONS.—§ " The Natives of South India make a paste of the SEEDS by trituration with water, and apply it over the eyelids for opacity of the cornea" (*Surgeon W. F. Thomas, 33rd M. N. I., Mangalore*). " The paste of the seeds is used in opthalmia and keratitis. The fruit is used in fœtid breath" (*Surgeon-Major D. R. Thomson, M.D., C.I.E., Madras*). " The fresh juice is cooling and laxative. In the cold season, when the juice does not soon undergo fermentation, it is an excellent medicine. " DATE SUGAR " is more nutritious and agreeable than cane sugar. It can be used as a substitute for maltine and its various preparations" (*Surgeon R. L. Dutt, M.D., Pubna*).

Date Sugar.
569

FOOD &
FODDER.
Date.
570

Food & Fodder.—The DATE forms a very large part of the food-supply of the countries in which it grows. Even in the Panjáb and Sind, where it is, comparatively speaking, neglected, and is only a secondary source of

FOOD &
FODDER.

food, it is very largely utilised. Thus Mr. O'Brien, when Commissioner of Múltán, wrote, "The fruit is ripe all July and in the first fifteen days of August. The people eat it universally in large quantities." All the great pilgrimages take place in these months, because date-picking is going on, and the pickers are allowed to give a handful of the fruit to each passer-by. The date remains in good condition after picking for six months. It may also be preserved by boiling in water, and then frying in oil, in which state it will last for a year. It is largely exported into the Northern Panjáb from Múltán.

In the Persian Gulf the fruit is eaten in all stages of its growth. It begins to form about May, and does not ripen thoroughly till September. The young green dates which fall from the tree are called *khamál*. In June or July it becomes red or yellow according to the variety, and is then known as *kharak*. A considerable portion of the total yield is cut down while in this state and sold. In August it nearly reaches maturity, becoming sweet, soft, and juicy, and is then known as *rutub*. In this condition it is sold in the bazárs, but can only be kept for two or three days, as it readily ferments. In September the date ripens thoroughly, and at this stage is called *khúrma*. It is now removed from the tree, gathered in troughs and exposed to the air and sun, when it throws out its surplus juice, *doshab*, and becomes drier and harder. It is then packed and prepared for exportation. The juice or *doshab* is used for preserving dates with other fruits and spices, in jars, and is also employed as a substitute for sugar by the poorer classes.

Certain races do not ripen beyond the hard or *kharak* stage. These are preserved by first boiling them in water, and then exposing them till they harden. They are then called *kharak pokhta*, and if carefully packed will keep a long time. Many preparations are made of this *kharak*, all of which, though not so palatable as those of the ripe fruit, are valuable articles of food. In India a similar preparation of unripe dates, known in Múltán as *bhúgrian*, is made, and it is probable, according to Dr. Bonavia, that the *chohara* brought down by Afghán merchants are similarly prepared.

Other kinds of date, called *sáhidí*, do not ripen beyond the *rutub* stage. The fruit of these is allowed to remain on the trees till it hardens, and is then collected and packed in baskets for export to India, where it is in great request by distillers of *arak*.

Fully ripe or *khurma* dates are made into preparations of various kinds, besides being packed for export. The best, intended for sale, are very carefully packed one by one in small card-boxes, or boxes lined with paper. The crushed fruit sold in foreign markets consists of the inferior and damaged sorts.

In Múltán the hard and unripe date is called *gandorá*, when it just turns yellow *doká*, when one side becomes soft *dang*, and when quite ripe *pind*. *Pind* dates may be *van di pind*, ripened on the tree, or *pind lúni*, ripened after gathering. Dates which have shrivelled on the tree are known as *kuk* or *kukan*. In Sind, the fruit is called *khurma* when ripe, and *chuwarar* when plucked before it is fully ripe, in which case it is boiled, and dried in the sun. In Muzaffargarh, according to Coldstream, the most esteemed kind is called *chirni*. This is the date from the best palms split in the middle and dried in the sun. The second best is called *pind*; it is eaten as it comes from the tree, without further preparation. The least esteemed kind, *búgri*, is taken from inferior trees, and boiled in oil and water.

The terminal bunch or heart of YOUNG LEAVES (*gáchi*, within which is | Young leaves.
the tender cabbage-like edible part, *gari*) is preserved when a tree is cut | 571

P. 571

FOOD &
FODDER.

down and eaten as a vegetable, both raw and cooked, by Natives. It is said to be excellent, and Coldstream vouches for its making a good curry.

Sugar.
572
Drink.
573

The true Date-Palm, like **P. sylvestris**, yields a saccharine juice, from which SUGAR and a FERMENTED DRINK may be made. It is, however, but little used for this purpose, since the fruit is more valuable. **Mr. Edgeworth**, when Commissioner of Múltán, employed Natives of Jessore for the purpose of trying to make sugar from trees in the district, but found the trial unsuccessful, only a small quantity of sugar, unequal in value to the loss of fruit entailed, being obtained. The method followed is similar to that with the toddy-yielding date-palm, **P. sylvestris** (see p. 208).

Kernel.
574

The hard KERNEL of the fruit is ground for food for camels, goats, sheep, horses, and other domestic animals. The green date or *khamál* is also given as food to sheep and cattle.

Of recent years the roasted kernels have been considerably used as a substitute for coffee; a Company has been found for its preparation and sale under the name of " Date Coffee."

TIMBER.
575

Structure of the Wood.—Light, soft internally, but fairly durable; used in Múltán and Sind as beams for supporting roofs in native architecture, also for water channels, bridges, and other purposes. In the countries adjoining the Persian Gulf, the midribs or leaf-stalks called *gors* are made into *jhowlies* for covering sheds, roofs of houses, and for various other economic purposes. The lower and thicker parts of the midribs called *tapúl* are generally used for fuel, but when broad and light they are made into floats for fishing-nets. In the Panjáb the midribs are known as *chhari* and are employed to make light strong walking sticks, and enclosures, called *khori*, within which the dates are dried ; split up they furnish excellent material for crates and baskets.

576

Phœnix farinifera, *Willd. ; Roxb., Fl. Ind., Ed. C.B.C., 722.*

Vern.—*Palawat,* HIND.; *Chilta-eita, chittita, ita-koyya,* TEL.; *Itcham thattu, kasangu,* TAM.; *Ichal,* KAN.

References.—*Brandis, For. Fl., 556; Griffith, Palms of Br. East India, 140; Elliot, Fl. Andhr., 44; Gamble, Man. Timb., 419; Drury, U. Pl., 339; Lisboa, U. Pl. Bomb., 207 ; Ind. Forester, III., 237 ; Madras, Man. Admin., I., 361 ; Gasetteers :—Mysore & Coorg, I., 49 ; III., 24 ; Bombay, XV., 445 ; Settlement Rept., Upper Godavary Dist., C. P., 38.*

Habitat.—A considerable amount of confusion regarding this species appears to exist amongst botanists. Since **Roxburgh's** description no writer appears to have been able to identify the plant meant by him. **Brandis** writes that were it not for the description of the stem and leaves he would be disposed to identify **P. farinifera** with a slender-stemmed form, found in South and Central India, and doubtfully referred by **Dalzell** to **P. paludosa.** It is, however, desirable in the present state of our knowledge, to retain **Roxburgh's** species, for the purpose of recording the various uses to which it is said to be put.

According to **Roxburgh** it is a dwarf species with scarcely any stem, a native of dry barren ground, chiefly of the sandy lands at a small distance from the sea near Coringa. It flowers in January and February, and the fruit ripens in May.

FIBRE
Leaflets
577
Petioles.
578
FOOD.
Fruit.
579

Fibre.—The LEAFLETS are, by the poorer classes wrought into coarse mats, for sleeping on, &c. The common PETIOLES are split into three or four, and used to make baskets of various kinds, but are less valuable for this purpose than the bamboo, which is more elastic, much more durable, and splits more easily.

Food.—The FRUIT has a small quantity of pulp, but it is sweet and mealy, is said to have the flavour of chestnut, and is eaten by the Natives.

The Indigenous Date Palms of India. (*J. Murray*.)	**PHŒNIX rupicola.**

The TRUNK of the tree yields a farinaceous substance, eaten by the Natives, the account of which, by Dr. Roxburgh, is of much interest and may be quoted in full:—

FOOD.
Trunk.
580

"The small trunk, when divested of its leaves, and the strong, brown fibrous web that surrounds it at their insertions, is generally about fifteen or eighteen inches long, and six in diameter at the thickest part; its exterior or woody part consists of white fibres matted together; these envelope a large quantity of farinaceous substance which the Natives use for food in times of scarcity. To procure this meal, the small trunk is split into six or eight pieces, dried, and beat in wooden mortars, till the farinaceous part is detached from the fibres; it is then sifted to separate them; the meal is then fit for use. The only further preparation it undergoes is the boiling it into a thick gruel, or as it is called in India, *kangi*; it seems to possess less nourishment than the common sago, and is less palatable, being considerably bitter when boiled; probably a little care in the preparation and varying the mode might improve it; however, it certainly deserves attention, for during the end of the last, and beginning of this, year, and even again at this present time, May 1792, it has saved many lives. Rice was too dear, and at times not to be had, which forced many of the poor to have recourse to these sorts of food. Fortunately it is one of the most common plants on this part of the coast, particularly near the sea." The identification of Roxburgh's P. farinifera is thus a matter of some considerable importance.

According to a writer in the *Indian Forester*, the CABBAGE or young top of leaf-buds, is eaten similarly to that of P. dactylifera.

Cabbage.
581

Phœnix paludosa, *Roxb.; Fl. Ind., Ed. C.B.C.*, 724.

582

> Vern.—*Hintál, hital, golpatta*, BENG.; *Thinboung*, BURM.; *Giruka táti, hintalamu*, TEL.; *Hintála*, SANS.
>
> References.—*Griffith, Palms Br. East Ind., 144, pl. ccxxix., A and B; Voigt, Hort. Sub. Cal., 643; Brandis, For. Fl., 556; Kurz, For. Fl. Burm., II., 536; Gamble, Man. Timb., 419; Elliot, Fl. Andhr., 60, 69; Mason, Burma & Its People, 426, 812; U. C. Dutt, Mat. Med. Hind., 300; Gazetteer, Mysore & Coorg, II., 8; Ind. Forester, I., 7, 9; III., 46; VIII., 401; IX., 324.*

Habitat.—A gregarious palm, which forms impenetrable thorny thickets throughout the Sundarbans, and in the deltas of rivers in Burma and the Andaman Islands.

Fibre.—Its LEAVES are used in the Sundarbans to make rough ropes for tying boats and logs, and also for thatching.

FIBRE.
Leaves.
583
TIMBER.
584

Structure of the Wood.—"The trunks of the smaller trees serve for walking sticks, and the Natives have an idea that snakes get out of the way of any person having such a staff. The longer ones serve for rafters to their houses" (*Roxburgh*).

Domestic, &c.—Griffith states that this species is well worth cultivating on account of its elegance, and owing to its being so well adapted for "back scenery."

DOMESTIC.
585

P. rupicola, *T. And.; Jour. Linn. Soc., XI., 49, p. 13, 1869.*

586

> Vern.—*Schiap*, LEPCHA.
>
> References.—*Gamble, Man. Timb., 419; List, Pl. of Darjiling, 86.*

Habitat.—"A beautiful palm of the lower hills of Darjiling and Bhután, generally growing on rocks, often to a height of 20 feet" (*Gamble*).

Food.—The interior of the STEM is, according to Gamble, often eaten by the Lepchas.

FOOD.
Stem.
587

PHŒNIX sylvestris.	The Wild Date, or Date-Sugar Palm of India.

588   **Phœnix sylvestris,** *Roxb.; Brandis, For. Fl., 554.*
     THE WILD DATE, OR DATE-SUGAR PALM.

    Syn.—PHŒNIX HUMILIS, *Rcyle.*

    Vern.—*Sendhi, kejur, khajúr, khaji, salma, thalma, thakil,* HIND.; *Kajar, kejur,* BENG.; *Khejuri,* URIYA; *Khajur,* KÓL.; *Khijur,* SANTAL; *Sindi,* GOND; *Khajúr, khaji, juice = sendhi, tári,* PB.; *Seindi,* BERAR; *Sendi, khajura, khajuri,* BOMB.; *Boichand, sendri, shindi,* MAR.; *Kharak,* GUZ.; *Sandole-ka-nar,* DEC.; *Itchumpannay, peria-itcham, itcham-nar, itham pannay,* TAM.; *Ita, pedda íta, íta-nara, ishan-chedi,* TEL.; *Ichal, kullu, ichalu mara,* KAN; *Khurjjúri, kharjura, madhukshir,* SANS.

    References.—*Roxb., Fl. Ind., Ed. C.B.C., 723; Voigt, Hort. Sub. Cal., 642; Griffith, Palms Br. East Ind., 141, t. ccxxviii.; Brandis, For. Fl., 554; Kurz, For. Fl. Burm., II., 535; Gamble, Man. Timb., 419; Dals. & Gibs., Bomb. Fl., 278; Stewart, Pb. Pl., 245; Rev. A. Campbell, Rept. Econ. Pl., Chutia Nagpur, No. 9246; Mason, Burma and Its People, 425, 812; Sir W. Elliot, Fl. Andhr., 71, 148; Irvine, Mat. Med. Patna, 101; U. C. Dutt, Mat. Med. Hind., 250, 305; Dymock, Mat. Med. W. Ind., 2nd Ed., 801; Birdwood, Bomb. Prod., 214, 250, 338; Baden Powell, Pb. Pr., 587; Drury, U. Pl. Ind., 340; Atkinson, Him. Dist. (X., N.-W. P. Gas.), 318; Useful Pl. Bomb. (XXV., Bomb. Gas.), 134, 181, 206, 212, 237, 398; Royle, Fibrous Pl., 91, 96; Liotard, Mem. Paper-making Mat., 11, 19; Cooke, Gums and Resins, 24, 28; Indian Fibres and Fibrous Substances, Cross, Bevan, King & Watt, 67; Kew Off. Guide to Bot. Gardens and Arboretum, 35; Buchanan, Statistics, Dinajpur, 150; Man. Madras, Adm., II., 27, 59; Boswell, Man. Nellore, 97, 117; Westland, Rept. on Jessore, 161-176; Bomb. Man. Rev. Accts., 102; For. Adm. Rept. Chutia Nagpur, 1885, 6, 34; Dollard, Med. Top., Kulu, Kumáon and Shore Valley, 32; Settlement Reports:— N.-W. P., Shajehánpur, ix.; Central Provinces, Upper Godavery, 38, 39; Mandla, 89; Chanda, App. vi.; Gasetteers:—Bombay, II., 39, 355; V., 23, 28, 107; VI., 13, 38; VII., 38, 39; VIII., 21, 95, 96; XV., 445; XVII., 26; XVIII., 48; Panjáb, Karnál, 16; Musaffargarh, 23; Shahpur, 69; Sialkote, 11; N.-W. P., I., 84; III., 33; IV., lxxviii.; Orissa, II., 180; Mysore & Coorg, I., 49, 66; II., 7, 8; Rev. & Agri. Dept., Note on Sugar Plants and Sugar, 1888; Agri.-Horti. Soc. Ind.:—Journals (Old Series), V., Sel., 75; X., 243; XII., 148; (New Series), I., 105; IV., Sel., 49; V., 73; Ind. Forester, III., 237; X., 308; XI., 6; Smith, Dic., 801.*

    Habitat.—A tree which attains a height of 30 to 40 feet, and which may be distinguished from **P. dactylifera,** by the fact that it has no root-suckers, and that the leaflets make half a right angle with the common petiole. It is indigenous in many parts of India, being most abundant in Bengal, Behar, on the Coromandel Coast, and in Guzerát. It also forms extensive forests in Rohilkhand, on the low ground along the Ram-ganga river, and on the plateau of Mysore between Shimoga and Tum-kúr. It is not uncommon in the Siwalik tract and outer Himálaya, attaining an altitude of 5,000 feet in Kumáon. It is also commonly culti-vated and self-sown in most parts of India and Ceylon, except in Sind, and the South-Western Panjáb, where **P. dactylifera** takes its place. The wild date flowers in March and ripens its fruit in September and October.

    Gum.—This species, like **P. dactylifera,** yields a gum of which little is known.

    Fibre.—The LEAVES are employed in Bengal for making mats, baskets, and bags, and in Bombay, for brooms and fans. The PETIOLES are beaten and twisted into ropes, which are employed for drawing water from wells. According to Lisboa, the fibre obtained from the leaves is plentiful, soft, bleaches well, and is admirably adapted for paper-making.

    Medicine.—The FRUIT, pounded and mixed with almonds, quince seeds, pistachio nuts, spices and sugar forms a *paushtik,* or restorative remedy, much in vogue. A paste formed of the KERNELS, and the root of

GUM.
589
FIBRE.
Leaves.
590
Petioles.
591
MEDICINE.
Fruit.
592
Kernels,
593

The Wild Date or Date-Sugar Palm of India. (*J. Murray*)	**PHŒNIX sylvestris.**

Achyranthes aspera, is eaten with betel leaves as a remedy for ague (*Dymock*). The JUICE obtained from the tree (*toddy* or *tári*) is considered a cooling beverage.

SPECIAL OPINIONS.—§ "The CENTRAL TENDER PART is used in gonorrhœa and gleet. The ROOT is used in toothache" (*Surgeon-Major D. R. Thomson, M.D., C.I.E., Madras*). "Good in nervous debility" (*Surgeon-Major C. R. G. Parker, Pallaveram, Madras*).

Food.—The trees flower at the beginning of the hot weather, and produce an inferior yellowish or reddish FRUIT, eaten by Native boys and the poorer classes.

In many localities, however, especially in Jessore and other districts of Bengal, it is of the greatest importance as a source of food-supply, owing to the extensive utilisation of its SAP in making sugar. The wild date tree is met with in almost every part of Bengal proper, but flourishes most congenially, and is found plentifully, only in the alluvial soils which characterise the south-eastern portion of the province.

Mr. Robinson, in a prize essay communicated to the *Agri.-Horticultural Society of India* in 1859, states that the extent of country most suited for its growth may be taken to be within an area stretching east and west about 200 miles, and north and south about 100 miles, thus comprising within an irregularly triangular space, an extent of about 9,000 square miles. He remarks, "The practice of extracting its JUICE, however, for the production of sugar, extends at present over a much smaller area, probably not more than two-thirds of the above described space; and if we consider, further, how small a portion of even these favourite date districts are as yet occupied by date-tree cultivation, the room for its future extension, even if confined to these tracts alone, appears a wide one indeed. If we trace an irregular parallelogram, stretching eastward from Kishenganj in the Nadiyá District to Bákarganj, and from Mádehpur in Farídpur District southward to the borders of the Sundarbans, we shall find a space of about 100 miles long by 80 broad, and comprehending the district of Jessor, with portions of Farídpur, Nadiyá, and Burrisaul, to which the production of date-sugar is mainly confined." *Gúr*, or the fresh raw product made by boiling down the juice, is, however, commonly manufactured for local consumption in most localities in which the tree is found at all abundantly.

In the most important sugar-producing district, *viz.*, Jessor, the industry is an old one. It is mentioned as a source of income to the inhabitants as early as 1788, and in a statistical table prepared in 1791, 20,000 maunds is stated to have been the annual produce, of which half was exported to Calcutta. This figure must, however, be remembered to include cane sugar, of which there was a considerable production in the district at that time, while in later years the date-sugar has driven sugar-cane almost entirely out of the fields of the district, as well as from the market. The trade did not acquire very large dimensions till about forty or fifty years ago, at which time English factories set up in the district. These factories, according to Mr. Westland, though they were not themselves pecuniarily successful, did a great deal towards increasing the trade and extent of cultivation. "The truth was," he writes, "that when they gave a great impulse to the sugar cultivation, Native merchants stepped in and appropriated all the trade which the factories had given birth to. English refining is good only for one market— the European market. The demand for sugar among Natives is very great, but they do not care to have it so thoroughly refined. Thus, as the Native market is now, and has been for very many years, the chief market for sugar, it follows that expensive methods of refinement are thrown away,

MEDICINE.
Juice.
594
Central Part.
595
Root.
596
FOOD.
Fruit.
597

Sap.
598

Juice.
599

**PHŒNIX
sylvestris.**   The Wild Date or Date-Sugar Palm of India.

CULTIVATION
in
Jessor.

and the methods used by Native merchants impart to the sugar all the purity which is required by the consumers."

The method of cultivation and manufacture of date-sugar cannot be better described than in the following excellent account extracted from Mr. Westland's *Report on the District of Jessore, 1874* :—

Planting.
600

PLANTING.—"The ground chosen for date cultivation is the higher ground, that which is too high for rice to grow well, and the rent paid for such ground is at least three times that paid for rice land. One often finds date trees ranged round the borders of fields cultivated with cold weather crops, and, indeed, in the west and north of the district almost every village is thickly studded with these trees ; but a very large amount of cultivation is upon land especially set apart for it. The trees are planted in regular rows, each tree being about twelve feet from its neighbour. If so planted, and left for seven years before being touched, good healthy trees may be expected. Those who cultivate dates, keep the land, especially in the cold season, perfectly bare of any vegetation, ploughing up the turf, so that the whole strength of the ground may expend itself in the trees. Of course, there are people who cultivate other crops upon the land where the date trees grow, and there are very many who have not patience enough to wait for the expiration of the full seven years; such people, however, lose in the end by their trees failing to give the same richness in juice that is obtained from trees more carefully tended.

Tapping.
601

TAPPING.—" When the tree is ripe the process of tapping begins, and it is continued each year thereafter. There are in the date-palm two series, or stories as it were, of leaves ; the crown leaves, which rise straight out from the top of the trunk, being, so to speak, a continuation of it ; and the lateral leaves, which spring out of the side of the top part of the trunk. When the rainy season has completely passed, and there is no more fear of rain, the cultivator cuts off the lateral leaves for one-half of the circumference, and thus leaves bare a surface measuring about ten or twelve inches each way. This surface is at first a brilliant white, but becomes by exposure quite brown, and puts on the appearance of coarse matting. The surface thus laid bare is not the woody fibre of the tree, but is a bark formed of many thin layers, and it is these layers which thus change their colour and texture.

"After the tree has remained for a few days thus exposed, the tapping is performed by making a cut into this exposed surface, in the shape of a very broad V, about three inches across and a quarter or half inch deep. Then the surface inside the angle of the V is cut down, so that a triangular surface is cut into the tree. From this surface exudation of the sap takes place, and caught by the sides of the V, it runs down to the angle, where a bamboo of the size of a lead pencil is inserted in the tree to catch the dropping sap and carry it out as by a spout.

"The tapping is arranged, throughout the season, by periods of six days each. On the first evening a cut is made as just described, and the juice is allowed to run during the night. The juice so flowing is the strongest and best, and is called *jiran* juice. In the morning the juice collected in a pot hanging beneath the bamboo spout is removed, and the heat of the sun causes the exuding juice to ferment over and shut up the pores in the tree. So in the evening the new cut is made, not nearly so deep as the last, but rather a mere paring, and for the second night the juice is allowed to run. This juice is termed *do-kat* and is not quite so abundant or so good as the *jiran*. The third night no new cutting is made, but the exuding surface is merely made quite clean, and the juice which then runs is called *jarra*.

P. 601

The Toddy Palm of India. (*J. Murray.*)	**PHŒNIX sylvestris.**

"It is still less abundant and less rich than the *do-kat*, and towards the end of the season, when it is getting hot, it is unfit even for sugar manufacture, the *gúr* made from it being sold simply as 'droppings.'

"These three nights are the periods of activity in the tree, and after these three it is allowed to remain for three nights at rest, when the same process again begins. Of course, every tree in the same grove does not run in the same cycle. Some are at their first, some at their second night, and so on; and thus the owner is always busy.

"Since every sixth day a new cut is made over the previous one, it follows that the tree gets more and more hewed into as the season progresses, and towards the end of the season the exuding surface may be, and often is, as much as four inches below the surface above and below. The cuts are during the whole of one season made about the same place, but in alternate seasons alternate sides of the tree are used for the tapping; and as each season's cutting is thus above the previous season's and on the opposite side, the stem of the tree has a curious zigzag appearance. The age of a tree can of course be at once counted up by enumerating the notches and adding six or seven, the number of years passed before the first year's notch. I have counted more than forty notches on a tree, but one rarely sees them so old as that, and when they are forty-six years old they are worth little as produce-bearing trees. I have said that at first the size of the bared surface, previous to the notching, is about ten inches square; but it gets less and less as the notches come to the higher and narrower part of the trunk, and I have seen old trees where not more than four inches square could be found.

"It is somewhat remarkable that the notches are almost always on the east and west sides of the tree and very rarely on the north and south sides; also, the first notch appears to be made in by far the majority of instances on the east side.

PRODUCE OF ONE TREE.—"As to the produce of one tree, one may expect from a good tree a regular average of five seers per night (excluding the quiescent nights). The colder and clearer the weather, the more copious and rich the produce. In the beginning of November tapping has begun. In December and January the juice flows best, beginning sometimes as early as 3 P.M., and dwindles away as the warm days of March come. If the cultivator begins too early, or carries on too late, he will lose in quality and quantity as much as he will gain by extending the tappinng season. But high prices begin in October, and I am afraid there are not many who can resist the temptation of running into market with their premature produce.

"During the whole of the tapping season a good cultivator will keep his grove perfectly clean and free from jungle or even grass.

BOILING.—"So much then for the tapping : the next process is the boiling, and this every *rayat* does for himself, and usually within the limits of the grove. Without boiling, the juice speedily ferments and becomes useless; but once boiled down into *gúr*, it may be kept for very long periods. The juice is, therefore, boiled at once in large pots placed on a perforated dome, beneath which a strong wood-fire is kept burning, the pared leaves of the trees being used among other fuel. The juice, which was at first brilliant and limpid, becomes now a dark brown, half-viscid, half-solid mass, which is called *gúr* (molasses), and when it is still warm, it is easily poured from the boiling pan into the earthenware pots (small *gharras*) in which it is ordinarily kept.

PRODUCE OF GÚR.—"As it takes from seven to ten seers of juice to produce one seer of *gúr* or molasses, we can calculate the amount of *goor* which one ordinarily good tree can produce in a season. We may

PHŒNIX
sylvestris   The Wild Date or Date-Sugar Palm of India.

CULTIVATION
in
Jessor.

count four and a half months for the tapping season, or about sixty-seven tapping nights. These, at five seers each, produce 335 seers of juice, which will give about forty seers, or one maund of *gúr*, the value of which, at present rates, is from R2 to R2·4 A *bigah* containing one hundred trees will therefore produce from R200 to R225 worth of *gúr* if all the trees are in good bearing.

Pottery.
605

POTTERY.—" It is not all sorts of pottery which will bear the continuous hard firing required for boiling down the juice, and some potters have obtained a special reputation for the excellence of their wares in this respect. The whole of the region about Chaugachha and Kotchandpur is supplied principally from a village, Bagdanga, a little west of Jessor, where the clay seems to be of an unusually good quality. The southern part of the district, again, is supplied chiefly from Alaipur, a bázar near Khulna.

Refining.
606

REFINING.—" A *rayat*, after boiling down his juice into *gúr*, does not ordinarily do more; it is then sold to the refiners, and by them manufactured into sugar. Near Keshabpur, however, a large number of *rayats* manufacture their own sugar, and sell it to the exporters only after manufacture. There are also in almost all parts of the district a class of refiners, different from those who are refiners and only refiners by profession. These are the larger *rayats* in the villages, many of whom combine commercial dealing with agriculture. They receive the *gúr* from the *rayats* in their vicinity, and sometimes also purchase it in the adjacent *háts*, and after manufacturing what they thus purchase, they take their sugar to some exporting mart and sell it there to the larger merchants.

" These, however, are the outsiders in the sugar trade, for by far the greater quantity of the sugar is manufactured by regular refiners, and it becomes necessary to describe how the *gúr* finds its way from the *rayats'* hands into theirs.

" Few of the sugar refiners purchase direct from the *rayats*, for the small quantities which each man brings would render this inconvenient; there are consequently a number of middlemen established, called *byapáris* or *dalláls* (the latter name prevails principally near Chaugachha), who collect the produce from the *rayats* and sell it at a small profit to the refiners. They do it sometimes by giving advances to the *rayats* to aid them in their cultivation, getting the advances repaid in produce; but the *rayats* are not, as a rule, dependent on such advances, and the greater number of *byapáris* simply make excursions round the country, buying up the *gúr* from *rayats* and bringing it in to the merchants.

" On *hát* days also another class of *byapáris* will be seen (some of whom have a very large business) lining the roads by which the *rayats* bring their produce to the *hát*. They pick up the pots of *gúr* by ones and twos from the smaller class of cultivators and profit by selling them in bulk to the refiner. *Rayats* who have extensive cultivation sometimes bring in quantities large enough to be sold direct to refiners, but by far the bulk of the *gúr* comes through the hands of intermediaries, in the various methods just described. Of course, the earthen pot is transferred along with the *gúr* that is in it; separation is in fact impossible, and the refiners always smash the pots to get out the *gúr*. Hence there is a great trade in pottery during the whole of the sugar season, for every *rayat* must buy for himself as many new pots as he sells pots of *gúr*. Those *rayats* who bring their own produce to the *hát* always buy and take away with them the new pots they require.

MANUFAC-
TURE
OF SUGAR.
Dhulua.
607

" MANUFACTURE OF *dhulua* SUGAR.—" There are several methods of refining, and two or three sorts of sugar produced. We shall take them in order, and describe first the method of manufacturing "*dhulua*" sugar—that soft, moist, non-granular, powdery sugar used chiefly by Natives, and specially in the manufacture of Native sweetmeats.

"The pots of *gúr* received by the refiner are broken up and the *gúr* tumbled out into baskets, which hold about a maund each and are about fifteen inches deep; the surface is beaten down so as to be pretty level, and the baskets are placed over open pans. Left thus for eight days, the molasses passes through the basket, dropping into the open pan beneath and leaving the more solid part, namely, the sugar, in the basket. *Gúr* is, in fact, a mixture of sugar and molasses, and the object of the refining is to drive off the molasses which gives the dark colour to the *gúr*.

"This eight days' standing allows a great deal of the molasses to drop out, but not nearly enough; and to carry the process further, a certain river weed, called *syála* " (Hydrilla **verticillata**) " which grows freely in the Kabadak especially, is placed on the baskets so as to rest on the top of the sugar. The effect of this weed is to keep up a continual moisture, and this moisture, descending through the sugar, carries the molasses with it, leaving the sugar comparatively white and free from molasses. After eight days' exposure with *syála* leaves, about four inches on the surface of the mass will be found purified, and these four inches are cut off and *syála* applied on the newly-exposed surface. This and one other application will be sufficient to purify the whole mass.

"The sugar thus collected is moist, and it is, therefore, put out to dry in the sun, being first chopped up so as to prevent caking. When dry it is a fair, lumpy, raw sugar, and weighs about thirty per cent. of the original mass, the rest of the *gúr* having passed off in molasses. Dishonest refiners can get more weight out of it by diminishing the exposure under *syála* weed, so as to leave it only five or six days, instead of eight. The molasses is less perfectly driven out, and the sugar, therefore, weighs more. Of course it has also a deeper colour, but that is in a measure remedied by pounding under a *dhenki*. There are also other dishonest means of increasing the weight: for example, the floors of the refineries are sometimes a foot or more beneath the level of the ground outside, the difference representing the amount of dust which has been carefully swept up with the sugar when it is gathered up after drying. Also, it is very easy so to break the pots that fragments of them remain among the sugar.

THE DROPPINGS.—"The 'first droppings,' gathered in the open pans in the manner described above, are rich in sugar, and are used, especially in the North-West, for mixing with food. It entirely depends, therefore, upon the price offered for them for this purpose, whether they are sold at once or reserved for a second process of sugar manufacture. In this second process the first droppings are first boiled and then placed under ground in large earthenware pots to cool. Unless thus boiled they would ferment, but after being boiled in this fashion they, on cooling, form into a mass somewhat like *gúr*, but not nearly so rich. After this the previous process is again gone through, and about ten per cent. more weight in sugar is obtained. This sugar is, however, coarser and darker in colour than the first.

"If the refiner is not very honest, and if he is sure of finding immediate sale, he will use a much more speedy process. Taking the cooled *gúr*, he will squeeze out the molasses by compressing the mass in a sack, and then, drying and breaking up the remainder, will sell it as sugar. It does not look much different from that prepared in the more elaborate way, but it will likely soon ferment, and hence the necessity of finding an immediate purchaser.

"The remainder, after all this sugar has been squeezed out, is molasses,

MANUFAC-
URE
of Sugar
in Jessor.

Pucka.
609

'*chitiya gúr*' as it is called. It forms a separate article of commerce, being exported to various places.

MANUFACTURE OF '*pucka*' SUGAR.—"The sugar produced by the method just prescribed is called *dhulua* sugar—a soft yellowish sugar. It can never be clean, because it is clear, from the process used, that whatever impurity there may originally be in the *gúr*, or whatever impurity may creep into the sugar during its somewhat rough process of manufacture, must always appear in the finished article. Another objection to it is that it tends slightly to liquefaction, and cannot, therefore, be kept for any considerable time.

"The '*pucka*' sugar, the manufacture of which I am now about to describe, is a much cleaner and more permanent article. It has also a granular structure, which the *dhulua* has not. The manufacture of it is more expensive than the other, and the price when finished is about R10, whereas *dhulua* costs only about R6 per maund.

"In this process the *gúr* is first cast upon flat platforms, and as much of the molasses as then flows off is collected as first droppings. The rest is collected, put into sacks and squeezed, and a great deal of molasses is thus separated out. The sugar which remains behind is then boiled with water in large open pans, and as it boils all scum is taken off. It is then strained and boiled a second time and left to cool in flat basins. When cooled it is already sugar of a rough sort, and now *syála* leaves are put over it and it is left to drop. The result is good white sugar, and should any remain at the bottom of the vessels still unrefined, it is again treated with *syála*.

"The first droppings, and the droppings under the *syála* leaves, are collected, squeezed again in the sacks, and from the sugar left behind, a second small quantity of refined sugar is prepared in exactly the same way by twice boiling. The droppings from the sacks are *chitya gúr*, and are not used for further sugar manufacture. About thirty per cent. of the original weight of the *gúr* is turned out in the form of pure *pucka* sugar."

Another method peculiar to Keshabpur is also described in which the *gúr* is first boiled in large open pots, into each of which a handful of *fichh* (the last droppings under the *syála* leaf, burnt) is put. It is then left to cool and in doing so coagulates, after which it is treated with *syála* leaf and thus refined.

Besides these methods the English system of refining is used in the few existing factories. In this the raw material is mixed with a certain amount of water, and boiled in cisterns by means of steam. The lighter impurities float and are skimmed off, while the boiling solution is made to flow away through blanket strainers into another cistern. The sugar solution thus formed is then evaporated slowly in vacuum pans, and the resulting viscid mass is run off into sugar loaf moulds. These have a hole at the vertex, and being placed upside down, allow the small amount of molasses remaining to percolate away by its own weight, aided by the addition of a little moistened sugar on the top.

If the raw material used be *gúr* the result is a yellowish sparkling loaf-sugar, while if native refined *dhulua* sugar be employed the loaf is brilliantly white.

The total area in Jessor under date trees is given in 1875 as 17,500 acres. The produce of one acre at 9 tons of *gúr* fetches from R500 to R600, and the rent of one acre is R9 to R15.

AREA.
610

The area under the crop in all India is given as follows in the Revenue and Agricultural Department's *Statistics of Sugar-plants and Sugar in 1888*:—

		Area. Acres.	Outturn. Maunds.	AREA.
Madras	. . . . .	1,600	1,21,000	
Bengal {24-Pergunnahs	. . .	15,000	3,53,000	
{Jessore	. . .	15,000	3,90,000	
Mysore	. . . . .	29,000 (say)	6,00,000	

More complete figures are not available, though date-sugar is extracted to a minor extent in other localities.

A small proportion of the juice, instead of being used in the two ways above described, is consumed as a drink either unfermented or fermented, under the name of *tari*. The popularity of this liquor is said to be increasing. For a description of the revenue derived from these trees, the amount of *tari* consumed, and the method of manufacture, the reader is referred to **Spirits**, under the article **Narcotics**, Vol. V., 332; also the article SUGAR under **Saccharum officinarum**.

Trade.—The proportion of the total trade in sugar, which consists of the produce of the date, cannot be accurately ascertained. According to Westland, exports from Jessor are made chiefly to Calcutta and Nalchiti: nearly all the *dhulua* sugar produced in the district is said to find its way to the latter market, while a small proportion of *dhulua* and nearly all the *pucka* sugar goes to Calcutta for export to Europe.

The *chitiya gúr* or refuse of the sugar-refining process is employed to a small extent locally for mixing with tobacco, but the bulk is exported to Calcutta, Nalchiti, and Serajganj.

Mr. Westland estimates the whole trading profit in Jessor, distributed among *rayats* and professional traders, to amount to at least six or seven lakhs of rupees. For an account of the trade in SUGAR as a whole, the reader is referred to the article on that subject in Vol. VI., where further particulars regarding the date palm cultivation and sugar manufacture will also be found.

TRADE.
611

PHORMIUM, *Forst.; Gen. Pl., III., 773.*

Phormium tenax, *Linn. f.; Linn. Soc. Jour., XI., 357;* LILIACEÆ.
NEW ZEALAND FLAX.

612

Habitat.—Originally a native of New Zealand, now largely cultivated in waste lands bordering on the sea in tropical or warm temperate countries, such as St. Helena, Algiers, and South France. It has been introduced into Southern India of late years.

Fibre.—The plant yields a soft white fibre, with a silky lustre, which is now largely used for making sail cloth, cordage, and paper. Many recommendations have been made that the plant should be introduced into India, and it is said to succeed well in the South Wynaad (*Cameron*). But there does not seem much chance of its ever becoming introduced to any large extent. India already teems with valuable fibre-yielding plants, many of which are almost entirely neglected, while several are probably at least as valuable as New Zealand Flax.

FIBRE.
613

PHRAGMITES, *Trin.; Gen. Pl., III., 1179.*

614

Phragmites communis, *Trin.; Duthie, Fodder Grasses, 60;*
COMMON REED.   [GRAMINEÆ.

Syn.—ARUNDO PHRAGMITES, *Linn.*   **Vern.**—*Dila*, PB.

References.—*Stewart, Pb. Pl., 250; Aitchison, Cat. Pb. & Sind Pl., 166; Boiss., Fl. Or., V., 564; Treasury of Bot., II., 879; Ind. Forester, XIV., 372.*

Habitat.—A coarse perennial grass with creeping root-stock, found in the plains of North-West India, and up to 14,000 feet on the Panjáb

PHRYNIUM dichotomum.	The Sital-pati Mats.

FODDER.
Grass.
615
DOMESTIC.
Stalk.
616
Grass.
617
618

Himálaya. It is an excellent plant for binding loose soils, and on this account is valued in many parts of India.

Fodder.—Cattle eat the young GRASS, and, according to Aitchison, it is collected largely for fodder in Northern Afghánistán.

Domestic.—In the Panjáb the STALKS are employed for making sandals, and in Lahoul the GRASS is much used for roofing (*Stewart*).

Phragmites Roxburghii, *Kunth. ; Duthie, Fodder Grasses,* 60.

Syn.—ARUNDO KARKA, *Roxb.*

Vern.—*Narkul,* HIND.; *Nal,* BENG.; *Karka, nal, khaila, khailuwa,* KUMÁON; *Bichhra,* GARHWÁL; *Nar, nara, naria, nal, bag-narrí, nái, nalu,* PB.; *Drúmbi, dwárena, ghwarga,* PUSHTU; *Naga-sara maitantos,* TEL.; *Kaing* (name applied also to other similar reed-like grasses), BURM.; *Nala,* SANS.

References.—*Roxb., Fl. Ind., Ed. C.B.C., 117; Stewart, Pb. Pl., 250; Aitchison, Cat. Pb. & Sind Pl., 166; U. C. Dutt, Mat. Med. Hind., 311; Taylor, Topograph. Dacca, 59; Atkinson, Him. Dist., 806.*

Habitat.—A tall reed with stems 8 to 12 feet high, common in the plains along the banks of streams and in marshy places; ascending to 3,500 feet in the Himálayan valleys.

FIBRE.
Stems.
619
Flower Stalks
620
FODDER.
621
DOMESTIC.
Stems.
622
Stalks.
623
624

Fibre.—Stewart says that a fibre is extracted from the upper part of the STEMS; and according to Atkinson (*Him. Dist.*) the fibre of the FLOWER STALKS is manufactured into rope in the Kumáon Bhábar.

Fodder.—Watson mentions that this grass has proved poisonous to cattle in Kumáon; it is in any case much too coarse a grase for fodder purposes.

Domestic Uses.—The STEMS are used for making baskets, chairs, hurdles, *chiks* (screens), and the tubes of *hukahs.* According to Roxburgh the common *durma* mats are made of the STALKS split open; he also adds:—" Vessels from the port of Calcutta are generally dunnaged with them " (*Conf.* with **Mats and Matting,** Vol. V., 196.)

PHRYNIUM, *Willd. ; Gen. Pl., III.,* 652.

Two of the species described by Roxburgh as belonging to this genus have been isolated in the *Genera Plantarum* under the generic name **Clinogyne,** *Salisb.* Since the *Fl. Br. Ind.* containing these plants has not as yet appeared, we have preferred to retain **P. dichotomum,** *Roxb.,*—the species of economic interest—in this place. This at least would appear a more correct position, than to assign the Indian species (as was done by Wallich) to the American genus **Maranta,** *Linn.*

625

Phrynium dichotomum, *Roxb., Fl. Ind., Ed. C.B.C.;* SCITAMINEÆ.

Syn.—MARANTA DICHOTOMA, *Wall.;* DONAX ARUNDASTRUM, *Lour.*

Vern.—*Muktá-páta, pati-patá, sitalpátir gách, madar-pati,* the mat= *sital-pati,* BENG.; *Thin pin, pin pwa, then,* BURM.; *Sitalapa-triká,* SANS. Roxburgh gives in addition the following Sanskrit names:— *Godúnika, bhúdra, bhúdroudúni, shúrákasthika, kúlianini, bhúdrá-búla, ghatá, pati, vulahwa.*

References.—*Voigt, Hort. Sub. Cal., 575; Mason, Burma & Its People, 521, 806; O'Shaughnessy, Beng. Dispens., 647; U. C. Dutt, Mat. Med. Hind., 318; Treasury of Bot., 720.*

Habitat.—Found in Eastern Bengal, Assam, the Coromandel Coast, and Burma.

FIBRE.
626

Fibre.—This plant yields a sufficient quantity of fibre to be useful for paper-making; while it is highly probable that other and entirely new uses may be found for it, such as the manufacture of hats, table mats, straw braid, &c.

The following interesting note is extracted from Royle's *Fibrous Plants :*—"The stems, which are straight and tapering, about as thick as

P. 626

The Emblic Myrobalan. (*J. Murray.*) **PHYLLANTHUS Emblica.**

FIBRE.

a man's thumb, and from three to five or six feet high, of a beautiful highly polished green colour, are employed in making some of the mats for which Calcutta is famous. **Mr. Colebrooke** says :—' Mats made of the split stems of this plant, being smooth and particularly cool and refreshing, are termed in Hindu *sital-pati*, which signifies a cool mat, whence the plant itself is said to bear the name. Suspecting, however, this to be a misappropriation of the term, I have enquired of Natives of the eastern parts of Bengal, who assure me that the plant is named *mukta-pata* or *patti-pata*, and the mat only is called *sital-pati*.' The split stems, as prepared for making mats, are about four feet in length, one-twentieth of an inch in breadth, thin as paper, greyish coloured, compact and shining almost like cane on the outside; finely striated on the in side, and apparently made up by the agglutination of very fine fibres.''

These mats are much employed by Europeans as well as by Natives in the hot weather owing to their coolness, being placed for this reason beneath the sheet of the bedding. Some of the finest are said to cost about R50 each, while those in general use average from R2 to R3. (*Conf.* with **Mats and Matting**, Vol. V., 196.)

PHYLLANTHUS, *Linn. ; Gen. Pl., III., 272.*

627

Phyllanthus distichus, *Muell. Arg.; Fl. Br. Ind., V., 304;*
[EUPHORBIACEÆ.

Syn.—P. CHEREMILA, *Roxb.;* P. LONGIFOLIUS, *Jacq.;* P. CICCA, *Muell. Arg.;* P. TETRANDRUS, *Wall.;* CICCA DISTICHA, *Linn.*

Vern.—*Harfarauri, chalmeri,* HIND.; *Noári, loda.* fruit = *hariphul,* BENG.; *Narkuli,* URIYA; *Cherámbola,* GOA; *Arunelli,* TAM.; *Rácha usirike,* TEL.; *Kirnelli,* KAN.; *Nelli,* MALAY.; *Thinbozihpyú, thinbawnibyu,* BURM; *Rata-nelli,* SING.; *Lavani,* SANS.

References.—*Roxb., Fl. Ind., Ed. C.B.C., 684; Kurz, For. Fl. Burm., II., 353; Beddome, For. Man., 191; Gamble, Man. Timb., 351; Grah., Cat. Bomb. Pl., 180; Dals. & Gibs., Bomb. Fl., Suppl., 78; Elliot, Fl. Andhr., 162; U. C. Dutt, Mat. Med. Hind., 307; Lisboa, U. Pl. Bomb., 116, 171; Birdwood, Bomb. Pr., 76, 175, 229; Bidie, Cat. Raw Pr., Paris Exh., 57; Hunter, Orissa, II., 180; Ind. Forester, VI.*

Habitat.—A small elegant tree, common in gardens in South India, Burma, and the Andaman Islands.

Medicine.—The FRUIT is acid and astringent, the ROOT is an active purgative, and the SEED is also cathartic.

Food.—It produces numerous small reddish flowers at the beginning of the hot season, which are succeeded by fleshy FRUIT not unlike gooseberries. These are much used as an article of food, either raw or dressed in various fashions, pickled, or made into preserves.

MEDICINE.
Fruit.
628
Root.
629
Seed.
630
FOOD.
Fruit.
631
632

P. Emblica, *Linn.; Fl. Br. Ind., V., 289; Wight, Ic., t. 1896.*
THE EMBLIC MYROBALAN.

Syn.—EMBLICA OFFICINALIS, *Gærtn.;* CICCA EMBLICA, *Kurz;* ?PHYLLANTHUS TAXIFOLIUS, *Don;* P. GLOMERATUS, *Herb. Roxb.*

Vern.—*Aonlá, ánúli, ámlaki, dúngra, ánwerd, ánvulá, ánvurah, ámalaci, ánolá, dunra, dáula, ámla, áura,* HIND.; *Amlá, ámlaki, dunlah, yeonlah, ambolati, amulati,* BENG.; *Gondhona, amlaki, ohalu,* URIYA; *Alá thanda,* CUTTACK; *Aura,* KOL.; *Meral,* SANTAL; *Ambari,* GÁRO; *Amluki, sohmyrlain,* ASSAM; *Amla,* NEPÁL; *Suom,* LEPCHA; *Nelle,* MAL. (S.P.); *Nilli, milli, nalli, dunri, usir, lalla,* GOND; *Aunre,* KURKU; *Amla, áoula,* N.-W. P.; *Ambal, ámbli, ámla, áonla,* PB.; *Amla, ánla,* C. P.; *Avalkati, ávala, ávla, ámla,* BOMB.; *Aonli, bhui dwali, dvalá,* MAR.; *Bhoza ámali, ámla, ámbala,* GUZ.; *Owla, ownla, ámla,* DEC.; *Nelli-kái, tóppi, nelli,* TAM.; *Usereki, wúsheriko, osirka, usri, usirika, nelli, vusirika,* TEL.; *Nelli, nilika,* KAN.; *Boa-malacca,*

P. 632

**PHYLLANTHUS
Emblica.**   The Emblic Myrobalan.

MALAY.; *Shabju, tasha, hsi-phyú, siphiyu-sí,* BURM.; *Nelli, nellika, awusada-nelli,* SING.; *Amulki, umrita, ámalaki, dhátri, ámala, kamu,* SANS.; *Amlaj,* ARAB ; *Amuleh, ámelah,* PERS.

References.—*Roxb., Fl. Ind., Ed. C.B.C., 684; Voigt, Hort. Sub. Cal., 153; Brandis, For. Fl., 454, t. 52; Kurz, Fl. Burm., II., 352; Beddome, Fl. Sylv., t. 258; Gamble, Man. Timb., 351; Thwaites, En. Ceyl. Pl., 282; Dals. & Gibs., Bomb. Fl., 235; Stewart, Pb., Pl., 193; Rev. A. Campbell, Rept. Econ. Pl., Chutia Nagpur, No. 9228; Graham, Cat. Bomb. Pl., 180; Mason, Burma and Its People, 458, 761; Sir W. Elliot, Fl. Andhr., 14, 187; Sir W. Jones, Treat. Pl. Ind., V., 151; Pharm. Ind., 204; Ainslie, Mat. Ind., I., 240; II., 244; O'Shaughnessy, Beng. Dispens., 551; Irvine, Mat. Med. Patna, 7, 106; U. C. Dutt, Mat. Med. Hindus, 225, 290; Murray, Pl. & Drugs, Sind, 35; Dymock, Mat. Med. W. Ind., 2nd Ed., 699; Cat. Baroda Durbar, Col. & Ind. Exhib., Nos. 149, 150; Year-Book Pharm., 1880, 249; Med. Top., Ajm., 124; Birdwood, Bomb. Prod., 78, 175, 229; Baden Powell, Pb. Pr., 374; Drury, U. Pl. Ind., 194; Atkinson, Him. Dist. (X., N.-W. P. Gas.), 317, 745 777; Useful Pl. Bomb. (XXV., Bomb. Gas.), 115, 171, 249, 253, 259, 279, 284, 388; Econ. Prod., N.-W. P., Pt. III. (Dyes & Tans), 36, 84; Pt. V. (Vegetables, Spices, & Fruits), 87; Liotard, Dyes, App. ii., v., vii.; McCann, Dyes & Tans, Beng., 137, 138, 140, 143-44, 146, 160, 165, 167; Wardle, Dye Report, 9, 20, 24, 44, 45; Darrah, Note on Cotton in Assam, 33; Kew Bulletin, 1889, 26; Cat. Col. & Ind. Exhib., Raw Products, No. 2481; Selections, Records Govt. India (R. & A. Dept.), 1888-89, 86, 91, 92; Kew Off. Guide to the Mus. of Ec. Bot., 116; Linschoten, Voyage to East Indies (Ed. Burnell, Tiele & Yule), II., 124; Buchanan, Statistics Dinajpur, 162; Man. Madras Adm., I., 313; Boswell, Man., Nellore, 101, 143; Moore, Man., Trichinopoly, 79; Gribble, Man., Cuddapah, 263; Aplin, Rep. on Shan States, 1887-88; Settlement Reports:—Panjab, Guzrát, 134; N.-W. P., Shahjehanpore, ix ; Central Provinces, Chandwara, 110; Baitool, 127; Bhundara, 20; Chanda. App. vi.; Seonee, 10; Raipore, 76; Nimar, 307; Gazetteers:—Bombay, V., 285; VII., 37, 39, 41; VIII., 11; XIII., 23; XV., 74; XVII., 23; XVIII., 42; Panjáb, Peshávar, 27; Sialkot, 11; Hoshiárpur, 10; Gujrát, 11; N.-W. P., I., 84; IV., lxxvii.; Central Provinces, I., 138; Burma, I., 138; Orissa, II., 27, 180; Mysore & Coorg, I., 48, 53, 56, 65; II., 7; III., 23; Agri.-Horti. Soc. Ind., Journals:— (Old Series), X., 34; XIII., 303; (New Series), II., 234; VII., 135; Ind. Forester, II., 19, 171; III., 23, 204; IV., 321; VI., 104; VIII., 29, 30, 270, 439; IX., 438; X., 325; XI., 485; XII., App., 21; XIII., 121; XIV., 147; Trans. Med. and Phys. Soc., Bombay (New Series), XII., 174.*

Habitat.—A moderate-sized deciduous tree, met with, wild or planted, throughout the forests of Tropical India and Burma.

Gum. — It yields a gum, of which little is known.

**Gum.
633**

Dye & Tan.—The FRUIT, known as the Emblic Myrobalan, is used as a medicine, and in dyeing and tanning. The dye from the fruit is said by McCann to be a blackish-grey, but it is very rarely used alone, being generally employed like other myrobalans to produce a black with salts of iron, or the barks of other trees. According to Liotard it plays the part more of a colour concentrator than of a dye. Wardle obtained several beautiful light brown colours on silk, both tussur and mulberry, and remarks that the Emblic Myrobalan appears to be valuable when used alone as a dye-stuff. The results obtained with cotton were poor. The BARK and LEAVES are also similarly employed, and produce the same colours. Wardle found that the leaves contained a small amount of colouring matter, and produced, by various processes, light drab and brownish-yellow colours. He writes of a sample received from the Central Provinces, "these leaves, of which I have received a very small sample for examination, contain a brownish-yellow colouring matter, soluble in water, which gives very beautiful, although somewhat faint, shades of colour to tussur silk, mulberry silk, and wool."

**DYE & TAN.
Fruit.
634**

**Bark.
635
Leaves.
636**

| The Emblic Myrobalan. (*J. Murray.*) | PHYLLANTHUS Emblica. |

DYE & TAN.

The fruit, leaves, and bark all contain tannin, and are all used for tanning purposes in various parts of the country, generally in combination with some stronger tanning material, such as the true myrobalan. **Professor Hummel** examined samples of the fruit and leaves, sent to the Colonial & Indian Exhibition, and reported as follows:—" The concentration of tannic acid in the leaves of **Phyllanthus Emblica**, while the fruit contains mere traces, is remarkable. Unless, therefore, the sample of fruit examined has been damaged, or that it is wrongly labelled, the Hindus make a great mistake in employing it as a tannin matter." **Dr. Watt** commenting on this report states that the fruits were correctly labelled, but conjectures that they may have been too ripe (*Sel. Rec. Govt. of Ind., R. & A. Dept., 1888-89, 92*). It is a well known fact that the percentage of tannic acid diminishes in many fruits as they ripen, and in the case of this fruit it is certain that when young it is intensely bitter, while when ripe it becomes edible and even pleasant to eat. **Professor Hummel's** analysis proved the leaves to contain 18 per cent. of tannic acid, to yield a decoction of a pale yellow colour (slightly turbid), and to have a commercial value relatively to Divi-divi of 4/10½*d.*, to Valonia cups of 9/1½*d.*, to Ground Sumuch of 12/2½*d.*, and to Ground Myrobalans of 5/10½*d.* The leaves appeared to contain some essential oil or fat. **Professor Hummel** was of opinion that the leaves, carefully dried and ground, ought to find a ready sale in England.

It is to be hoped that the *young* fruits may be subjected to a similar analysis and compared with standard tanning materials, to enable the question of their possible success as a commercial tan to be settled. The remark of **Dr. Æ. Ross** (see " Medicine ") regarding the product obtainable from the ROOT also appears well worthy of investigation.

Root. 637

McCann states that the price given for the fruit in Bengal varies from ₨2 to ₨5 per maund, while **Sir E. C. Buck** writes that the leaves (which appear to be more largely employed in the North-West Provinces than the fruit) are valued at about ₨4, and that in 1875, 128 cwt. was exported from the Gorakhpur forests at that average price. According to **Atkinson** the annual export from the Kumáon forest division is about four tons, and **Daveis'** Trade Report gives the annual export *viâ* Peshawar to Afghánistán at 50 maunds.

MEDICINE.

Medicine.—The Emblic Myrobalan is an important article in Hindu Materia Medica. It is mentioned by **Chakradatta**, and in the *Bhávaprakása*, many preparations in which it plays an important part being described at length. Generally used in combination with the two other myrobalans, it is supposed to possess much the same properties. The JUICE of the fresh fruit is considered cooling, refrigerant, diuretic and laxative, and enters into many prescriptions for dyspepsia. It is also ordered to be given with honey as a diuretic. The dried FRUIT is said to be astringent and useful in hæmorrhage, and together with iron is considered a valuable remedy in anæmia, jaundice, and dyspepsia. A paste of the fruit is also described as a useful external application over the public region in irritability of the bladder. The exudation from incisions made in the fruit, while on the tree, is believed to be a useful external application in recent inflammation of the conjunctiva. A fermented LIQUOR prepared from the fruit is used in jaundice, dyspepsia, cough, &c. (*U. C. Dutt, Mat. Med. Hind.*).

Juice. 638

Fruit. 639

Liquor. 640

Dymock states that Muhammadan physicians esteem it equally with the Hindus. They describe it as astringent, refrigerant, cardiacal, and a purifier of the humours of the body. It is much prescribed by them in " fluxes," and is also applied externally on account of its cooling and astringent properties.

P. 640

**PHYLLANTHUS
Emblica.**
                              The Emblic Myrobalan.

MEDICINE.

By the people of India generally, the fresh fruit is used as a purgative, while a *sherbet* prepared from it, sweetened with honey or sugar, is a favourite cooling drink, and is believed to be diuretic. Campbell states that in Chutia Nagpur, it is, when boiled till it becomes like oil, applied as a remedy for *khasra*, a skin disease. The dried fruit is much prescribed as an astringent in diarrhœa and dysentery. Bontius (*Diseases of India*, 200) testifies to its value in the treatment of these diseases. Antiscorbutic virtues have also been attributed to it by Dr. McNab (*Cal. Med Phys., Trans., VIII.*), and by Dr. Irvine (*Mat. Med. Patna*, 7). Ainslie writes, " The um-

Flowers.
641

belled yellow FLOWERS of this species have an odour much resembling that of lemon-peel, and are supposed, by the *Vitians*, to have virtues of a cooling and aperient nature; they are prescribed, in conjunction with other articles in the form of an electuary, in the quantity of about a tea-spoonful twice daily. The BARK partakes of the astringency of the fruit, and is similarly

Bark.
642
Juice.
643

employed. Dymock states that the JUICE of the fresh bark is given with honey and turmeric in the Konkan, as a remedy for gonorrhœa. The editor of the *Pharmacopœia of India* (in the secondary list of which the plant occupies a place) states, upon the authority of Dr. Æ. Ross, that the root, by decoction and evaporation, yields an astringent extract equal to catechu, both for medicinal purposes and in the arts. The LEAVES are, in Baroda,

Leaves.
644

used as an infusion with fenugreek seeds in cases of chronic dysentery, and are also considered a bitter tonic. In the same locality the MILKY JUICE

Milky Juice
645

is considered a good application to offensive sores (*Cat. Baroda Durbar, Collections at Col. & Ind. Exhib.*, No. *150*).

SPECIAL OPINIONS.—§ " An excellent astringent for genito-urinary discharges when used with beleric myrobalans in decoction; laxative when combined with chebulic myrobalans " (*Surgeon-Major H. D. Cook, Calicut*). " The fresh fruit preserved as a confection is used for bilious affections " (*Surgeon-Major Robb, Ahmedabad*). " Astringent, demulcent, tonic, dose ʒi to ʒiii, used in diarrhœa, dysentery, hæmoptysis and hæmatemesis " (*Hospital Assistant Choonna Lal, Jubbulpore*). " In some cases of disturbed cerebral circulation, attended by a burning sensation at the crown of the head, a tendency to headache, confusion of thoughts, and probably loss of hair, I have used, with great success, mustard oil in which *amloki* fruit had been allowed to remain for some time. The oil so prepared was applied on the head; in a few days the burning sensation diminished, the mental faculties became relieved of confusion, and in two cases the hair grew very rapidly on the part " (*Civil Surgeon D. Basu, Farídpore, Bengal*). " Infusion of seeds used in eye diseases" (*Assistant Surgeon Nehal Sing, Saharunpore*). " Infusion of *amloki* is cooling and astringent and a useful adjunct to other medicines in diarrhœa and dysentery; also found efficacious in hæmaturea. It is an ingredient of *triphala*, that is, the infusion of three myrobalans— emblic, chebulic, and beleric. This infusion has been found to be a cooling, stomachic, and general tonic to the digestive system " (*Assistant Surgeon S. C. Bhattacharji, Chanda*). " The pulp is used for the same purposes as liquorice, *viz.*, for cough, &c." (*Surgeon-Major Lionel Beech, Cocanada*). " It is one of the ingredients of *triphala* (three fruits) commonly used by natives as stomachic and laxative primarily; and astringent finally. The other two are *harar* and *baherá* " (*Assistant Surgeon Bhagwan Dass, Rawal Pindi*). " A decoction of the fruit is useful in chronic dysentery and diarrhœa " (*Civil Surgeon S. M. Shircore, Murshedabad*). " It is a stimulating expectorant in chronic bronchitis. It is also antibilious. An infusion of the roots is given to allay thirst" (*Surgeon-Major D. R. Thomson, M.D., C.I.E., Madras*). " The preserved, as well as the dried, fruit, is exported as a medicinal agent to

P. 645

| The Emblic Myrobalan. (*J. Murray.*) | PHYLLANTHUS maderaspatensis. |

Yarkhand from the Panjáb" (*Surgeon-Major J. E. T. Aitchison, Simla*). "Fresh fruit useful in chronic dysentery when eaten rather freely" (*Apothecary Thomas Ward, Madanapallee*). "The dried fruits, immersed in water in a new earthen vessel a whole night, yield a decoction which is used as a collyrium in ophthalmia. It may be applied cold or warmed" (*Surgeon R. Gray, Lahore*).

Food & Fodder.—The acid FRUIT, which is of the size of a small gooseberry, with a fleshy outer covering, and a hard three-celled nut containing six seeds, is used, among other purposes, for food and preserves by the Natives. It is made into a sweet-meat with sugar, or eaten raw as a condiment and is also prepared as a pickle. The fruit is said to be eaten greedily by cattle and the LEAVES are considered good fodder.

Structure of the Wood.—Red, hard, close-grained, warps and splits in seasoning; no heartwood; weight from 45 to 56℔ per cubic foot. Though apt to warp, it is flexible, tough, and tolerably straight-grained, and when well seasoned is much used for building purposes, furniture, agricultural implements, and gun-stocks, and is adapted for turning. It is durable under water, and partly for this reason, partly because it is supposed to clear muddy water, it is considerably employed for making well rings.

Domestic & Sacred.—Chips of the WOOD and small branches are said to clear muddy water effectually, and are considerably used for this purpose. The dried FRUIT is detergent and is employed for washing the head instead of soap. It is also used for making ink, and enters into the composition of a number of hair dyes. The tree is included by Lisboa in his list of sacred plants, where he states that Chapter I. of *Kartik Máhátma* orders the worship of this tree, and that a Brahmin couple should feed under it, whereby all their sins are washed off. At page 73 of the *Vrat Kaumudi* also, the *vrat* and worship of the tree are ordered. Buchanan records an interesting superstition regarding the fruit, as follows:—"The Natives have an idea that the fruit of Chebula Myrobalan (Terminalia Chebula) have never been found ripe except by some very holy persons, and it is supposed that such as have been favoured with eating such a rarity have been ever afterwards exempted from hunger" (*Statistics of Dinájpur, 162*).

Phyllanthus indicus, *Muell. Arg.; Fl. Br. Ind., V., 305.*

Syn.—PROSORUS INDICA, *Dals.*; PHYLLANTHUS STOCKSII, *Muell. Arg.*; P. ANOMALUS, *Muell. Arg.*; P. HYSTERANTHUS, *Muell. Arg.*; CICCA ANOMALA, *Baill.*; BRIDELIA BERRYANA, *Wall.*

References.—*Beddome, For. Man., 191; Dals. & Gibs., Bomb. Fl., 236; Thwaites, En Ceylon Pl., 281; Lisboa, U. Pl. Bomb., 116.*

Habitat.—A deciduous tree found on the Gháts of the Konkan and Kanara, and in the Central and Southern forests of Ceylon.

Structure of the Wood.—White and tough, used for building purposes
[*1895, f. 3.*]

P. maderaspatensis, *Linn.; Fl. Br. Ind., V., 292; Wight, Ic. i.,*

Syn.—PHYLLANTHUS ANDRACHNOIDES, *Willd.*; P. OBCORDATUS, *Willd.*; P. JAVANICUS, *Poir.*; P. ANCEPS, *Herb. Heyne.*; P. MALABARICUS, *Herb. Wight*; P. NIRURI, *Wall. not of Linn.*

Vern.—*Kanocha,* HIND.; *Nala usereki,* TEL.

References.—*Roxb., Fl. Ind., Ed. C.B.C., 678, 679; Grah., Cat. Bomb. Pl., 180; Dals. & Gibs., Bomb. Fl., 233; Ainslie, Mat. Ind., II., 245.*

Habitat.—A herb or small shrub, met with throughout the drier parts of India, from Banda through the Deccan Peninsula to Ceylon.

Medicine.—Ainslie informs us that the wedge-shaped LEAVES are used in infusion by the *Vytians,* as a remedy for headache.

Fodder.—Cattle eat the HERB (*Roxb.*).

P. 656

Marginal notes (right column):

MEDICINE.

FOOD & FODDER.
Fruit.;
646
Leaves.
647

TIMBER.
648

DOMESTIC.
Wood.
649
Fruit.
650
SACRED.
651

652

TIMBER.
653
654

MEDICINE.
Leaves.
655
FODDER.
Herb.
656

**PHYLLANTHUS
Niruri.**

The Bhui-amla.

657

Phyllanthus Niruri, *Linn.; Fl. Br. Ind., V., 298; Wight, Ic., t. 1894.*

Syn.— Phyllanthus urinaria, *Herb. Russ., not of Linn.;* P. Lonphali *Herb. Madr.;* P. polyphyllus, *Herb. Wight;* Nymphanthus Niruri *Lour.*

Vern.—*Jarámlá, bhúin-án-valah, sada-hazur-mani,* Hind.; *Bhui-ámlá, sada-hazur-mani,* Beng.; *Bhui-dólá,* Uriya; *Niruri,* Sind; *Bhui-áva-lá,* Bomb.; *Bhuin-ánvaláh,* Dec.; *Kishkáy-nelli, kishá-nelli,* Tam.; *Néla-usirika, néla-vusari,* Tel.; *Kiranelli-gidá,* Kan.; *Kirganelli, kis-há-nelli, kishkkáyi-nelli,* Malay.; *Mi-siphiyu,* Burm.; *Pittawáka,* Sing.; *Bhumyámalaki,* Sans.

References.—*Roxb., Fl. Ind., Ed. C.B.C., 680; Voigt, Hort. Sub Cal., 154; Thwaites, En. Ceylon Pl., 282; Elliot, Fl. Andhr., 133; Pharm. Ind., 205; Ainslie, Mat. Ind., II., 150; O'Shaughnessy. Beng. Dispens., 551; Moodeen Sheriff, Supp. Pharm. Ind., 198; U. C. Dutt, Mat. Med. Hind., 294; Dymock, Mat. Med. W. Ind., 2nd Ed., 701; Arjun, Bomb. Drugs, 125; Murray, Pl. & Drugs, Sind., 35; Bidie, Cat. Raw Pr., Paris Exh., 42; Year-Book of Pharm., 1880, 249; Atkinson, Him. Dist., 317; Drury, U. Pl., 342; Gazetteers:—Mysore & Coorg, I., 65; Bomb., V., 28; VI., 15; XV., 442; Orissa, II., 159; N.-W. P., I., 84; IV., lxxvii.; Boswell, Manual, Nellore, 132; Trans. Med. & Phys. Soc., Bomb. (New Series), IV., 155.*

Habitat.—A small herb met with throughout the hotter parts of India from the Panjáb eastward to Assam, and southward to Travancore, Malacca, and Ceylon, ascending the hills to 3,000 feet.

MEDICINE.
Juice.
658
Leaves.
659
Root.
660

Medicine.—According to Muhammadan writers the milky JUICE is a good application to offensive sores; a poultice of the LEAVES with salt cures scabby affections, and without salt may be applied to bruises, &c. (*Dymock*). Roxburgh noticed the medicinal value of the ROOT, writing, " The Rev. Dr. John informs me that he has known the fresh root prove an excellent remedy for jaundice. About half an ounce, while fresh, was given, rubbed up in a cup of milk, night and morning; the cure was completed in a few days, without any sensible operation of the medicine." According to Ainslie, the white root, small bitter leaves, and tender

Shoots.
661

SHOOTS are all used in medicine by Indian practitioners, who consider them to be deobstruent, diuretic, and healing. The two first are commonly prescribed in powder or decoction, in cases of over-secreted acrid bile, and in jaundice; an infusion of the latter, together with fenugreek seeds, is supposed to be a valuable medicine in chronic dysentery. The dose of the powdered leaves is about a tea-spoonful. The whole PLANT is

Plant.
662

also considered a useful diuretic, and is said to be much employed in native practice for dropsy, gonorrhœa, and other genito-urinary affections. It has obtained a place in the secondary list of the Pharmacopœia of India, along with P. urinaria, *Linn.,* both of which are said to have been very favourably reported on by Horsfield and others. Dymock states that in the Konkan the root rubbed down with rice water is given as a remedy for menorrhagia.

Special Opinions.—§ " The tender leaves are ground into a paste with a few corns of pepper, and a piece of this paste, about the size of a nutmeg, is taken internally with good effect in the milder forms of intermittent fever. The bruised leaves are applied in scabies and aphthæ " (*Surgeon-Major D. R. Thomson, M.D., C.I.E., Madras*). " The leaves bruised with milk are used in urinary and dropsical disorders" (*Surgeon-Major W. D. Stewart, Cuttack*). " The leaves are used locally in decoction, to the scalp, as a refrigerant" (*Native Surgeon T. Ruthnum Moodelliar, Chingleput*). " A useful remedy for jaundice " (*V. Ummegudien, Mettapolliam, Madras*).

P. 662

Phyllanthus reticulatus, *Poir.; Fl. Br. Ind., V., 288; Wight, Ic.,* 663
[*t. 1899.*

Syn.—PHYLLANTHUS MICROCARPUS, *Muell. Arg.;* P. MULTIFLORUS, *Willd.;* P. KIRGANELIA, *Herb. Ham., & Roxb.;* R. DALBERGIOIDES, MYRTIFOLIUS, WIGHTIANUS, & GRISEUS, *Wall.;* P. PRIEURIANUS, *Muell. Arg.;* P. SINENSIS, *Muell. Arg.;* P. PUBERULUS, *Miquel;* P. PENTANDRUS, *Herb. Roxb.;* P. VIROSUS, & SPINESCENS, *Wall.;* ANISONEMA RETICULATUM, *A. Juss.;* A. MULTIFLORUM, *Wight;* A. ZOLLINGERI, *Miquel;* A. DUBIUM, *Blume;* KIRGANELIA MULTIFLORA, *Thwaites;* CICCA RETICULATA, *Kurz;* C. MICROCARPA, *Benth.;* RHAMNUS ZEYLANICUS, *Burm.*

Vern.—*Panjoli, mákhi, buin-owla, kálé-madhká-pér,* HIND.; *Panjúli,* BENG.; *Panjúli,* PB.; *Kabonan,* RAJ.; *Kámohi,* fruit=*pikapirú,* leaves=*kámohi-jo-pun,* bark=*kámohi-jo-chodo,* SIND; *Pavana,* BOMB.; *Datwan,* GUZ.; *Buin-owla, kálé-madh-ká-jhár,* DEC.; *Púlavayr-puttay, pillánji, karappu-pillánji,* TAM.; *Nalla-purugudu, purugudu, nella-purúdúdú, phuls r,* TEL.; *Welkyla,* SING.; *Krishna-kámboji,* SANS.

References.—*Roxb., Fl. Ind., Ed. C.B.C., 681; Voigt, Hort. Sub. Cal., 154; Brandis, For. Fl., 453; Kurz, For. Fl. Burm., II., 354, 355; Beddome, For. Man., 190; Gamble, Man. Timb., 353; List, Flowering Pl., &c., Darjiling, 68; Thwaites, En. Ceylon Pl., 282; Grah., Cat. Bomb. Pl., 180; Dals. & Gibs., Bomb. Fl., 234; Ainslie, Mat. Ind., II., 323; Moodeen Sheriff, Supp. Pharm. Ind., 198; Dymock, Mat. Med. W. Ind., 2nd Ed., 703; Murray, Pl. and Drugs, Sind, 35; Bidie, Cat. Raw Pr., Paris Exh., 58, 115; Agri.-Horti. Soc., Ind., Jour. (New Series), VII., Sel., 205; Gazetteers:—Mysore & Coorg, I., 56; Bomb., XV., 442; Ind. Forester, III., 204; XII., App. 21.*

Habitat.—A large, often scandent, shrub, common throughout Tropical India, Burma, and Ceylon, especially on low moist ground, along river banks.

Dye.—Dr. Bidie states that the ROOT is used in Madras as a dye for producing a red colour.

Medicine.—The LEAVES are employed as a diuretic and cooling medicine in Sind (*Stocks*). Ainslie writes, "The BARK, as it appears in the Indian bazárs, is commonly in pieces about a foot long and as thick as the wrist, of a dark colour outside, and of a faint sweetish taste; it is considered as alterative and attenuant, and is prescribed in decoction in the quantity of four ounces or more twice daily." According to Bidie and Murray it is similarly employed in Madras and Sind at the present time. Dymock remarks that the JUICE of the leaves is used medicinally in the Konkan. It is made into a pill with camphor and cubebs, which is allowed to dissolve in the mouth as a remedy for bleeding from the gums; it is also reduced to a thin extract along with the juice of other alterative plants and made into a pill with aromatics. This pill is given twice a day rubbed down in milk as an alterative in "heat of the blood."

DYE.
Root.
664
MEDICINE.
Leaves.
665
Bark.
666

Juice.
667

[*1, 3 & 4.*

P. simplex, *Retz.; Fl. Br. Ind., V., 295; Wight, Ic., t. 1902, ff.* 668

Syn.—P. SIMPLEX PROPER = P. SIMPLEX, *Roxb.;* P. ANCEPS, *Heyne;* P. SIMPLEX, VAR. GENUINUS, *Muell.*

Var. oblongifolia, *Muell. Arg.*

Var. Gardneriana, *Muell. Arg.,*=P. GARDNERI, *Thw.,* P. GARDNERIANUS, *Baill.;* P. MIQUELIANUS, *Muell. Arg.;* P. FRUTICOSUS, *Herb. Heyne;* P. DEBILIS, *Wight;* P. MARGINATUS, *Herb. Heyne;* MACRAEA GARDNERIANA, & OVALIFOLIA, *Wight.*

Var. pubescens.

Vern.—*Tandi meral,* SANTAL; *Bhuiavali,* MAR.; *Uchchi usirika,* TEL.

References.—*Roxb., Fl. Ind., Ed. C.B.C., 678; Dals. & Gibs., Bomb. Fl., 234; Thw., En. Ceylon Pl., 282; Campbell, Ec. Pl., Chutia Nagpur, No 7593; Elliot, Fl. Andhr., 185; O'Shaughnessy, Beng. Dispens., 551; Gazetteers:—Bombay, XV., 442; N.-W. P., I., 84; IV., lxxvii.*

P. 668

PHYSALIS minima.	**The Physalis Fruit.**

Habitat.—A most variable shrub, native of the plains and low hills throughout India, from Kumáon to Assam and southward to Travancore, Malacca, and Ceylon, ascending to 6,000 feet.

MEDICINE.
Leaves.
669
Flowers.
670
Fruit.
671

Medicine.—Roxburgh writes, "The Natives use the fresh LEAVES, FLOWERS, and FRUIT, with common " (? cummin) " seeds and sugar, of each equal parts, made into an electuary, for the cure of gonorrhœa ; a tea spoonful is given twice a day. The fresh leaves bruised and mixed with buttermilk make a wash to cure the itch in children." With the exception of O'Shaughnessy, who quotes Roxburgh's observation, no writer on Materia Medica appears to have noticed these uses. Campbell states

Root.
672

that the ROOT is used in Chutia Nagpur as an external application for mammary abscess.

673

[*1895, f. 4.*
Phyllanthus urinaria, *Linn ; Fl. Br. Ind.,* V., *293 ; Wight, Ic., t.*

Syn.—PHYLLANTHUS LEPROCARPUS, *Wight ;* P. ALATUS, *Blume ;* P. ECHINATUS, *Herb. Ham.*
Var. Hookeri=P. HOOKERI, *Muell. Arg.*
Vern.— *Lal-bhuin-ánvalah, hazar mani,* HIND.; *Hazar mani,* BENG.; *Badar shapni,* SANTAL ; *Shivappu-nelli,* TAM.; *Erra-usirika,* TEL.; *Chiru-kizhuká-nelli, chukanna-kizhánelli,* MALAY.; *Kempu-kiranelli,* CAN.; *Rat-pittawáka,* SING.; *Mi-ziphiyu-ani,* BURM.
References.—*Roxb., Fl. Ind., Ed. C.B.C., 680 ; Grah., Cat. Bomb. Pl., 180 ; Thwaites, En. Cey. Pl., 282 ; Campbell, Ec. Pl., Chutia Nagpur, Nos. 9417, 9814 ; Elliot, Fl. Andhr., 54 ; Pharm. Ind., 205 ; Moodeen Sheriff, Suppl. Pharm. Ind., 198 ; Ainslie, Mat. Ind., II., 151, 437 ; Dymock, Mat. Med. W. Ind. 2nd. Ed., 701.*

Habitat.—A tall branched erect or decumbent herb, met with throughout India, from the Panjáb to Assam, Burma, Malacca, Penang, and Ceylon.

MEDICINE.
Root.
674

Medicine.—It would appear from the descriptions of writers on Indian Materia Medica that the properties of this species are identical with those of P. Niruri (see p. 222). Campbell informs us also that in Chutia Nagpur the ROOT is believed to be sudorific, being given to sleepless children along with *tandi shapni* (Zornia diphylla, *Pers.*).

FODDER.
Leaves.
675

Fodder.—Cattle eat the LEAVES (*Roxb.*).

PHYLLOSTACHYS, *Sieb. & Zucc.; Gen. Pl., III., 1208.*

676

[*427 ;* GRAMINEÆ.
Phyllostachys bambusoides, *Sieb. & Zucc. ; Gamble, Man. Timb.,*
Habitat.—A bamboo, met with in the Mishmi Hills.

TIMBER.
677

Structure of the Wood —Walking sticks are made of the stems, similar to the Chinese *Whangee* cane.

PHYSALIS, *Linn.; Gen. Pl., II., 890.*

Physalis flexuosa, *Link.,* see **Withania somnifera,** *Dun. ; Vol. VI.*

678

[SOLANACEÆ.
P. minima, *Linn.; Fl. Br. Ind.,* IV., *238 ; Wight, Ill., 166 b., f. 6 ;*

Syn.—P. PARVIFLORA, *Br. ;* P. DIVARICATA, *Don ;* P. VILLOSA, *Roth ;* P. ROTHIANA, *Roem. and Sch.;* P. PUBESCENS, *Wight, not of Linn.*
Var. indica=P. PSEUDO-ANGULATA, *Blume ;* P. ANGULATA, *Wall. mss. ;* NICANDRA INDICA, *Roem. & Sch.*
Vern.—*Túlati-pati,* HIND.; *Ban-tepariya,* BENG. ; *Handi khandi,* SANTAL.
Var. INDICA=*Túlati-pati,* HIND.; *Ban tepariya,* BENG.; Fruit-*habbi káknaj, káknaj,* PB.; *Thánmori, nanvachi-wel,* BOMB.; *Phopéti, chirboti, kappárphodi, tahan popti,* MAR.; *Kupanti,* TEL.
References—*Roxb., Fl. Ind., Ed. C.B.C., 189 ; Stewart, Pb. Pl., 158 ; Rev. A. Campbell, Rept. Econ. Pl., Chutia Nagpur, No. 9818, Ainslie,*

Mat. Ind., II., 15; Dymock, Mat. Med. W. Ind., 2nd Ed., 643; Baden Powell, Pb. Prod., 364; Atkinson, Him. Dist., 314; Gazetteers:—N.-W. P., I., 83; IV., lxxv.; Pb., Gujrát, 12; Ind. Forester, XII., App., 18; Agri.-Horti. Soc., Ind., Journals (Old Series), VI., 50; X., 27.

Habitat.—An herbaceous, pubescent annual, found commonly throughout Tropical India.

Medicine.—The FRUIT of variety **indica** is used medicinally in the Panjáb, where it is considered tonic, diuretic, and purgative, and a useful remedy in ulceration of the bladder (*Stewart, Baden Powell*). Royle made the remark in 1839 that the PLANT was at that time successfully grown (cultivated) and converted into extract for medical depôts, in several stations in the plains of Northern India and the Himálaya. Cultivation for this purpose, if it ever existed, must have been very short-lived, since Stewart writes only a few years later, "So far as I am aware, this is not cultivated in the Panjáb." Ainslie states that the plant has been noticed by Dr. Heyne as medicinal amongst the Hindus, but adds, "He says nothing about its supposed virtues." Dymock informs us that in the Konkan the plant is made into a paste with rice-water and applied to restore flaccid breasts, in accordance with the doctrine of signatures.

MEDICINE.
Fruit.
679
Plant.
680

Food.—The FRUIT of the cultivated P. **minima** is eaten by Natives.

FOOD.
Fruit.
681
682

Physalis peruviana, *Linn.; Fl. Br. Ind., IV., 238.*

THE CAPE GOOSEBERRY.

Syn.—P. EDULIS, *Sims.*; P. PUBESCENS, *Don.*

Vern.—*Macao, tipári, tepáriyo,* HIND.; *Tepuriá, tepíriya, tophli, tekári,* BENG.; *Tipári, tipáriya,* N.-W. P.; *Chirput,* GOA & KONKAN; *Phopti,* MAR.; *Budda-búsara, pámbudda, búsara káya,* TEL.; *Pung-ben, pong-pin,* BURM.; *Tankári,* SANS.

References.—*Roxb., Fl. Ind., Ed. C.B.C., 189; Voigt, Hort. Sub. Cal., 514; Dals. & Gibs., Bomb. Fl., App., 61; Mason, Burma & Its People, 455, 798; Elliot, Flora Andhr., 32, 143; U. C. Dutt, Mat. Med. Hind., 320; Atkinson, Him. Dist., 314; Econ. Prod., N.-W. P., Pt. V., 19; Lisboa, U. Pl. Bomb., 167; Birdwood, Bomb. Pr., 170; Firminger, Man. Garden, India; Gazetteers:—Mysore & Coorg, I., 63; N.-W. P., I., 83; IV., lxxv.; Agri.-Horti. Soc., Ind.:—Transactions, III. (Proc.), 229; VI., 235; Journals (Old Series), II. (Sel.), 382, 386; X., 26; (New Series), V. (Pro. 1875), 21; VII. (Proc.), 81.*

Habitat.—A weak sub-erect plant, native of tropical America; cultivated to a limited extent throughout India for its fine-flavoured, luscious fruit; occasionally spontaneous from cultivation, especially in certain parts of Ceylon, Burma, and here and there in India.

Food.—The FRUIT, which closely resembles that of the Winter Cherry of English gardens (**P. Alkekengi**), is enclosed in a large inflated calyx. It is of the size and shape of a cherry and has a bright amber colour. Being very palatable it is eaten raw by all classes, and is largely used by Europeans to make jams, preserves, tarts, &c.

FOOD.
Fruit.
683

According to Firminger the seeds should be sown in May or June, and the seedlings planted out two feet distant from each other in the open ground in rows four feet apart. They thrive in common garden soil, but do better if manured; when about eight inches high they should be earthed up to half their height. As soon as they come into blossom the ends of the shoots should be nipped off. The fruit ripens in January and February. Although naturally perennial, the plants must be raised annually from seed, since they are not fruitful after the first year. They thrive best in a climate in which they are not subject to extreme cold.

In certain localities, especially of Burma and Ceylon, it has run wild freely from cultivation, and it may be said to form a not unimportant part of the food-supply of these and the warmer regions of India generally.

| PHYTOLACCA acinosa. | The Lac Plant. |

Physalis somnifera, *Link.,* see **Withania somnifera,** *Dunal.;* S OLA-
[NACEÆ Vol. VI.

Physic-nut plant, s ee **Jatropha Curcas,** *Linn. ;* EUPHORBIACEÆ Vol.
[IV., 545.

684

PHYSOCHLAINA, *G. Don. ; Gen. Pl., II., 902.*

Physochlaina præalta, *Hook. f.; Fl. Br. Ind., IV., 244 ;* SOLANACEÆ.
Syn.—P. GRANDIFLORA, *Hook. f. ;* BELENIA PRÆALTA, *Dcne. ;* HYOSCYA-
MUS PRÆALTUS, *Walp. ;* SCOPOLIA PRÆALTA, *Dunal.*
Vern.—*Sholar, bajar-bang, nandrú, dandarwa, lang thang, khardag,* PB.
Reference.—*Stewart, Pb. Pl., 159.*
Habitat.—An'erect, nearly glabrous, herb, met with in North Kashmír
and Western Tibet, at altitudes of 12,000 to 15,000 feet ; distributed to
Kashgar and Yarkand.

MEDICINE.
Leaves.
685

Medicine.—Stewart states that this plant occurs in the Chenáb basin,
Zanskar, and Spiti, occasionally Trans-Indus, and perhaps in one place
near Lahore. Regarding its properties he writes, " In the hills the LEAVES
are applied to boils, and are also said to be poisonous, the mouth swelling
from their touch, and the head and throat being affected when they are
eaten. A man was poisonously affected by eating the plant gathered in
the Lahore habitat, and the Negí of Lahoul, when at Lé in 1867, suffered
from its narcotic effects for two or three days, some of its leaves having
been gathered by mistake with his *ság.* At the same time they can hardly be very poisonous to all animals, for in Lahoul they are browsed by
cattle. In a recent communication to the Agri.-Horticultural Society of
India, Dr. Christison of Edinburgh states that this has the same property
of dilating the pupils as Belladonna."

FODDER
Leaves.
686

Fodder.—As remarked above the LEAVES are reported to be browsed
by cattle in Lahoul (*Stewart).*

687

PHYTOLACCA, *Linn.; Gen. Pl , III., 84.*

A genus of herbaceous or arborescent plants, which comprises some ten
species ;. mostly American and tropical or sub-tropical. The generic name is
derived from " phyton "= a plant, and "lacca "= lac, in allusion to the crimson
fruits of certain of the species. Several American forms such as P. decandra,
Linn., P. dioica, *Linn.,* and P. octandra, *Moq.,* are of considerable value
in medicine and in the arts.

688

Phytolacca acinosa, *Roxb.; Fl. Br. Ind., V., 21 ;* PHYTOLACCACEÆ.
Syn.—P. DECANDRA, VAR. β. ACINOSA, *Moq.;* P. KÆMPFERI, *A. Gray ;* PIR-
CUNIA LATBENIA, *Moq.*
Vern.—*Jirrag,* KUMAON ; *Lúbar, búrgú, dentúrú, rinság, jirka, matazor,
sarunga,* PB.
References.—*Roxb., Fl. Ind., Ed. C.B.C., 389 ; Brandis, For. Fl., 371 ;
Stewart, Pb. Pl., 176 ; O'Shaughnessy, Beng. Dispens., 527 ; U. S. Dis-
pens., 1100 ; Royle, Ill. Him. Bot., 320 ; Atkinson, Him. Dist., 708 ;
Ec. Prod., N.-W. P., Pt. V. (Foods), 42 ; Balfour, Cyclop., III., 209.*
Habitat.—An herbaceous erect plant from 3 to 5 feet in height ; met
with in the Temperate Himálaya, from Kashmír and Hazára to Bhután, at
altitudes of from 5,000 to 9,000 feet.

MEDICINE
689

Medicine.—The Natives do not appear to use any part of the plant as
a medicine, but in every district in which it is cultivated they seem
to be fully aware of its power of producing delirium. It is commonly
stated that the poisonous property is only destroyed by complete boiling.
The narcotic virtues of certain American species are well known, and it is
possible that the Indian plant may be equally valuable.

FOOD.
Fruit.
690
Leaves.
691

Food.—The FRUIT is said to be occasionally eaten, and in the Simla
district it is certainly used to flavour curries. Stewart states that it is
reported, in some localities, to produce cerebral symptoms. The LEAVES

P. 691

FOOD.

are eaten as a pot-herb or as a vegetable in curries. **Royle** states as the result of his own experience that they are very palatable. **Stewart** remarks that a "variety" occurs in Kanawar with dark green petioles, which is said to produce delirium.

PICEA, *Link.; Gen. Pl., III., 439.*

692

Picea Morinda, *Link.; Fl. Br. Ind., V., 653;* CONIFERÆ.

Syn.—ABIES SMITHIANA, *Forbes;* A. SPINULOSA, *Griff.;* PINUS SMITH-IANA, *Lamb.;* P. KUTROW, *Royle.*
This tree has been already treated of under the name of **Abies Smithiana**, *Forbes* (Vol. I., 3), to which the reader is referred. Since that article was written, however, the fifth volume of the *Flora of British India* has appeared, in which this species has been reduced to **Picea Morinda,** *Link.,* with the synonymy above detailed.

PICRASMA, *Blume; Gen. Pl., I., 311.*

Picrasma quassioides, *Benn.; Fl. Br. Ind., I., 520;* SIMARUBEÆ.

693

Syn.—SIMABA QUASSIOIDES, *Ham.;* NIMA QUASSIOIDES, *Ham.*
Vern.—*Charangi, kashshing, ? bharangi,* HIND.; *? Bhurungi,* BENG.; *Shama baringi,* NEPAL; *Puthorín, bera, máthú, mont, bering, pesho, khashbar, birgo, tuthai, tithu, hala,* PB.; *Thityúben,* BURM.
References.—*Gamble, Man. Timb., 64; Stewart, Pb. Pl., 39; Pharm. Ind., 50; O'Shaughnessy, Beng. Dispens., 269; Irvine, Mat. Med., Patna, 16; Dymock, Warden & Hooper, Pharmacog. Ind., I., 287; Pharmaceutical Journal, July 20, 1889; Atkinson, Him. Dist., 307; Royle, Ill. Him. Bot., 158.*
Habitat.—A large scrambling shrub, with stout, often spotted branches, and very bitter bark; common in the sub-Tropical Himálaya, from Jamu to Nepál, and met with from 3,000 to 8,000 feet in altitude.

Medicine.—**Royle** first drew attention to the BARK, WOOD, and ROOT of this plant which he described to be quite as bitter as the true Quassia of the West Indies, and suggested that any of these parts would doubtless prove an excellent substitute for that drug. Some years later **Stewart** remarked that the leaves were used in Chamba to cure itch. The bark has been recommended by **Macardieu** as a febrifuge, and is noticed in the secondary list of the *Indian Pharmacopæia* as a likely substitute for quassia. In that work it is stated (probably quoting from **Irvine**, who made the same remark) that the bark is brought to Bengal from the hills, and is sold under the name of *bharangi* as a heating and stimulating medicine (*Mat. Med. Patna, 16*). The authors of the *Pharmacographia Indica* are unable to confirm this, and state that all specimens of *bharangi* which they have been able to secure have proved to be the roots and-stems of **Clerodendron serratum** (see Vol. II., 374).

MEDICINE.
Bark.
694
Wood.
695
Root.
696

CHEMISTRY.
697

CHEMICAL COMPOSITION.—The wood of this plant has recently been subjected to careful chemical examination by **Drs. Dymock & Warden**, the full details of which were published in the *Pharmaceutical Journal, July 20, 1889,* while a condensed account appears in the *Pharmacographia Indica, I., 287,* from which the following extracts may be taken:—"Our experiments indicate that the wood contains a crystallisable principle, probably *quassiin*, a fluorescing, bitter, resin-like principle, and at least one other non-crystallisable, bitter, resinous body, probably the uncrystallisable *quassiin* of **Adrin & Mordeaux.** There are several points of interest connected with the examination of **P. quassioides** to which we would refer. Firstly, the wood is not so bitter to the taste as ordinary quassia wood. Secondly, the authors of the *Pharmacographia* state that they obtained 7·8 per cent. of ash from quassia wood dried at 100°C.; the

P. 697

15 A

PICRORHIZA
Kurrooa.   The Kutki Root.

CHEMISTRY.

ash of **P. quassioides** obtained by us amounted to only 1·7 per cent. Thirdly, a watery solution of ordinary quassia wood is stated to display a slight fluorescence, especially if a little caustic lime has been added. According to **Fluckiger & Hanbury** this is apparently due to quassiin. We have repeated the experiment with a sample of ordinary quassia wood with negative results. The **P. quassioides** wood, when treated with water or alcohol, affords solutions which display a very marked greenish fluorescence. Regarding the content of quassiin it appears to vary considerably." The yield recorded by various chemists varies from 0·01 to 0·15 per cent. **Drs. Dymock & Warden** write, " The amount of crystallisable principle present in the wood we examined we are unable to accurately give; as a rough approximation, we do not consider that it would amount to more than ·02 to ·03 per cent. as an outside limit."

" Regarding the methods of analysis, extraction of the wood by alcohol, and subsequent boiling of the dry alcoholic extract with water, concentrating and precipitating with tannin, appears to give the best results, as far as a crystalline product is concerned."

Further experiments were made later in order to ascertain whether any of the Jaborandi alkaloids were present in the wood. As a result, the presence of a distinctly alkaloidal principle, in addition to other matters, was distinctly indicated, but the analysts write, " As far as our experiments have gone there is no evidence to show that the alkaloidal principle is related to the Jaborandi alkaloids. We have also examined **P. nepalensis** for these alkaloids with a negative result. It, however, contains an alkaloid, which does not appear to be identical with that found in **P. quassioides**."

These experiments, though not quite conclusive, point to the probability that quassiin does exist in **P. quassioides**, and that possibly to as great an extent as in average specimens of quassia wood, thus confirming **Royle's** supposition that it might be utilised as an efficient substitute for the officinal drug.

FOOD &
FODDER.
Fruit.
698
Shrub.
699
700

Food & Fodder.—The shrub produces green flowers, and a small red FRUIT or drupe which is eaten by Natives in certain localities The SHRUB is browsed by goats and sheep.

PICRORHIZA, *Royle; Gen. Pl., II., 962.*

Picrorhiza Kurrooa, *Benth.; Fl. Br. Ind., IV., 290;* SCROPHULA-
Syn.—VALERIANA? LINDLEYANA, *Wall.*   [RINEÆ.

Vern.—*Kutki, katki, kurú,* HIND, & BENG.; *Kutki,* NEPAL; *Kuruwa,* KUMAON; *Kaur, karrú,* ROOT=*karrú, káli kútki,* PB.; *Káli-kutki, bálkadu,* BOMB.; *Kutaki,* MAR.; *Kadu,* GUZ.; *Kali-kutki,* DEC.; *Katuku-rogani,* TAM.; *Katuku-roni, katuka-rogani,* TEL.; *Katuká, katurohini,* SANS.; *Kharbaqe-hindi,* ARAB. & PERS.

References.— *Stewart, Pb. Pl., 163 ; Pharm. Ind., 160 ; O'Shaughnessy, Beng. Dispens., 478 ; Irvine, Mat. Med. Patna, 58 ; U. C. Dutt, Mat. Med. Hindus, 212, 304; Dymock, Mat. Med. W. Ind., 2nd Ed., 577 ; Cat. Baroda Durbar, Col. & Ind. Exhib., No. 151 ; Official Corresp. on Proposed New Pharm. Ind., 234, 239, 291, 320 ; Year-Book Pharm., 1874, 135 ; 1878, 289; Atkinson, Him. Dist. (X., N.-W. P. Gas.) 314, 746 ; Royle, Ill. Him. Bot., 291, t. 71 ; Bidie, Cat. Raw Pr., Paris Exh., 36 ; Cat. Col. & Ind. Exhib. Raw Products, No. 90 ; Kew Off. Guide to the Mus. of Ec. Bot., 102 ; Gazetteer, Panjáb, Simla, 12 ; Agri.-Horti. Soc. Ind., Journals (Old Series), XIII., 389.*

Habitat.—A low, more or less hairy, herb, with a perennial, woody, bitter root-stock, common on the Alpine Himálaya from Kashmír to Sikkim, at altitudes of 9,000 to 15,000 feet.

P. 700

Medicine.—The ROOT is described in Sanskrit medicine as bitter, acrid, and stomachic, and in large doses as a moderate cathartic. It is directed to be used in fever and dyspepsia, and as an ingredient of various purgative combinations. **Chakradatta** recommends two drachms of the powdered root in sugar and warm water as a mild purgative, and a compound decoction with aromatics and *nim* bark, as a remedy for "bilious fever" (*U. C. Dutt*). Muhammadan writers have caused great confusion by assigning the name *kutki* to Black Hellebore, at the same time describing a totally different drug, probably, in certain cases, **P. Kurrooa.** This subject has been already dealt with under **Actæa** (*Vol. I., 103*), **Cimicifuga** (*Vol. II., 288*), **Coptis Teta** (*Vol. II., 522*), **Gentiana Kurroo** (*Vol. III., 486*), and at length under **Helleborus niger** (*Vol. IV., 216-17*). It, therefore, need not be again entered into here.

Although some doubt exists regarding the exact plant meant by Muhammadan writers, the *kutki* of the bazárs is certainly in some cases the root of this plant.

Royle notices it as bitter, and employed medicinally by the Natives. Irvine (*Mat. Med. of Patna*) mentions its use as a "tonic in purges." Antiperiodic virtues have been assigned to it by **Dr. Tripe** (*Madras Med. Reports, 1855, 422*), and later by **Moodeen Sheriff, Dymock,** and others. Thus **Dymock** writes, "I can state from personal observation that it is used successfully as an antiperiodic in native practice in Bombay; its slight laxative action is rather beneficial than otherwise." The dose as a tonic is from 10 to 20 grains, as an antiperiodic, from 40 to 50 grains; it is best administered in combination with aromatics. **Surgeon Gimlette,** Residency Surgeon in Nepál, states that it is brought from the Thibetan border, and that the root is used in ague and enlargement of the spleen, in doses of from four to eight *mashas*. The plant occupies a place in the non-officinal list of the *Indian Pharmacopœia*.

SPECIAL OPINIONS.—§ "From my experience of the root of **P. Kurrooa** I can say that it is a good stomachic and very useful in almost all forms of dyspepsia and in nervous pain of the stomach and bowels. Doses, as an antiperiodic, from 20 to 40 grains, and as a stomachic and tonic, from 10 to 15 grains, three or four times a day" (*Honorary Surgeon Moodeen Sheriff, Khan Bahadur, G.M.M.C., Triplicane, Madras*). "If a strong decoction of this drug be given three or four times a day and continued for three or four days in cases of dropsy, copious watery evacuations are discharged, and the dropsical effusion is relieved. In some cases the medicine must be continued for about a week to bring about the desired result" (*Surgeon-Major D. R. Thomson, M.D., C.I.E., Madras*).

PIERIS, *Don. ; Gen. Pl., II., 588.*

[ERICACEÆ.

Pieris ovalifolia, *Don. ; Fl. Br. Ind., III., 460 ; Wight, Ic., t. 1199;*
   Syn.—ANDROMEDA OVALIFOLIA, *Wall.*; A. ELLIPTICA, *Sieb. & Zucc.*;
      THIBAUDIA, *Griff.*
   Var. lanceolata=P. LANCEOLATA, *D. Don*; ANDROMEDA LANCEOLATA,
      Wall.; A. SQUAMULOSA, *D. Don.*
   Vern.—*Ayár,* HIND. ; *Anjir, angiar, aigiri, jagguchal,* NEPAL; *Piasay,*
      BHUTIA ; *Kangshior,* LEPCHA ; *Rattankát, arur, eilan, eilaur, eran, ellal,
      bhel, erana, ayatta, sarlakhtei, arwán, aira, yarta, piru,* PB.
   References.—*Griffith, Ic. Pl. Asiat., t. 514 ; Brandis, For. Fl., 280 ; Kurs,
      For. Fl. Burm., II., 92 ; Gamble, Man. Timb., 235 ; Cat., Trees, Shrubs,
      &c., Darjiling, 50 ; Stewart, Pb. Pl., 133 ; Baden Powell, Pb. Pr., 567 ;
      Gazetteers, Panjáb, Simla, 10 ; Agri.-Horti. Soc. Ind.:—Journals (Old
      Series), VI., 167 ; XIII., 384 ; XIV., 261 ; (New Series), II., Pro. 1870,
      37 ; Ind. Forester, XI., 4, 284 ; XIII., 59 ; Smith, Ec. Dict., 16.*

PIGEONS & Doves.	The Pigeons & Doves of India.

Habitat.—A small deciduous tree found in the outer Himálaya from the Indus to Assam, usually between 4,000 and 8,000 feet, also met with on the Khásia Hills and Hills of Martaban from 5,000 to 7,000 feet.

MEDICINE.
Young
leaves.
703
Buds.
704
TIMBER.
705
DOMESTIC.
Leaves.
706
Buds.
707

Medicine.—An infusion of the YOUNG LEAVES and BUDS is used as an external application in certain cutaneous diseases (*Gamble*).

Structure of the Wood.—Light reddish-brown, moderately hard, weight 41℔ per cubic foot. It is not durable, warps and shrinks badly in seasoning, and is consequently only employed for fuel and for making charcoal. The smoke of the burning wood is said to cause inflammation of the eyes and face.

Domestic, &c.—The young LEAVES and BUDS, according to **Stewart**, are poisonous to goats in the spring months only. They are said to produce cerebral symptoms, which can be cured by administering *lassí* (sour milk) to the affected animals. This poisonous property is said to render the young leaves a useful insecticide. According to **Madden** honey collected from the flowers is poisonous.

PIGEONS & DOVES.

708

Pigeons & Doves. *Murray, Avifauna of Br. Ind., II., 497-519.*

These birds are comprised in the families TRERONIDÆ, COLUMBIDÆ, and GOURIDÆ of the Natural Order GEMITORES, and are nearly all of economic value. The following occur in India:—

TRERONIDÆ.

Treron nipalensis, *Hodgs.;* The Thick-billed Green Pigeon.
Crocopus phœnicopterus, *Lath.;* The Bengal Green Pigeon.
C. chlorigaster, *Blyth;* The Southern Green Pigeon.
C. viridifrons, *Blyth;* The Yellow-fronted Green Pigeon of Burma.
Osmotreron bicincta, *Jerd.;* The Orange-breasted Green Pigeon.
O. vernans, *Linn.;* The Pink-necked Green Pigeon.
O. malabarica, *Jerd.;* The Grey-fronted or Malabar Green Pigeon.
O. Phayrii, *Blyth* Phayre's Green Pigeon.
O. fulvicollis, *Wagler;* The Ruddy Green Pigeon.
O pompadoura, *Gmel.;* The Yellow-fronted Green Pigeon of South India.
Sphenocercus sphenurus, *Vigors;* The Wedge-tailed Green Pigeon or the Kokla.
S. apicaudus, *Hodgs.;* The Pin-tailed Green Pigeon.
Carpophaga ænea, *Linn.;* The Imperial Green Pigeon.
C. insignis, *Hodgs.;* The Bronze-backed Imperial Pigeon.
C. griseicapilla, *Wald.;* The Grey-headed Imperial Pigeon.
C. bicolor, *Scop.;* The Red Imperial Pigeon.

COLUMBIDÆ.

Palumbus pulchricollis, *Hodgs.;* The Darjíling Wood Pigeon.
P. casiotis, *Bonap.;* The Himalayan Cushat.
P. Elphinstonei, *Sykes.;* The Nilghiri Wood Pigeon.
Palumbæna eversmanni, *Bp.;* The Indian Stock Pigeon.
Columba intermedia, *Strickl.;* The Indian Blue Rock Pigeon.
C. livia, *Bp.;* The Rock Dove.
C. rupestris, *Pall.;* The Blue Hill Pigeon.
C. leuconota, *Vigors;* The White-bellied, or Imperial Rock Pigeon.
Alsocomus puniceus, *Tickell;* The Purple Wood Pigeon.
A. Hodgsonii, *Vigors;* The Speckled Wood Pigeon.
Macropygia leptogrammica, *Tremm;* The Bar-tailed Cuckoo Dove.
M. assimilis, *Hume;* The Tenasserim Cuckoo Dove.
Turtur pulchratus, *Hodgs.;* The Ashy Turtle Dove.

Turtur meena, *Sykes ;* The Rufous Turtle Dove.
T. senegalensis, *Linn. ;* The Little Brown Dove.
T. suratensis, *Gm. ;* The Spotted Dove.
T. tigrinus, *Tem. ;* The Malayan Spotted Dove.
T. risorius, *Linn. ;* The Indian Ring Dove.
T. humilis, *Temm. ;* The Red Turtle Dove.

GOURIDÆ.

Chalcophaps indica, *Linn. ;* The Emerald Ground Dove.
Geopelia striata, *Linn. ;* The Barred Ground Dove.
Calœnas nicobarica, *Linn. ;* The Hackled Ground Pigeon.

Medicine.—The Arabs and Persians place pigeons and pigeons' eggs amongst their aphrodisiacs (*Ainslie, I., 313*).

Food.—The pigeons are all used as articles of food, and are prized by most of the Natives of India, who consider them specially stimulating and nourishing. The most highly appreciated by Europeans are the green pigeons.

Domestic & Sacred.—The feathers like those of most other birds are very little utilised in India. Siva and his spouse are said by Hindu Pandits to have dwelt at Mecca in the form of pigeons. The pigeon is closely connected with certain traditions of the Muhammadan faith, and is carefully protected in certain localities such as Mecca.

MEDICINE.
709

FOOD.
710

DOMESTIC.
711

PIGMENTS.

712

Pigments are coloured powders which, when mixed with some fluid such as oil, water, varnishes, &c., in which they are insoluble, form paints. They differ from dyes and washes in their entire insolubility in the media in which they are mixed. Such insoluble colours, when used in printing, or colouring textile fabrics, are distinguished as pigment colours.

It will be seen from the foregoing definition that the class of pigments must be very large. Any powder possessed of colour, whether of animal, vegetable, or mineral origin, and at the same time insoluble in the medium with which it is mixed for use, may be employed as a pigment. But the colour must, to be of value, be more or less unaffected by atmospheric influence, must possess a certain amount of depth, covering power or "body," must be unaffected in appearance by the medium with which it is made into paint, and must dry in an even, uniform, and adhesive coat. From an artistic point of view they must also, as satisfactory decorative or pictorial materials, possess purity or lightness of tone, with intensity, and a capacity for mixing with other colours so as to produce an agreeable compound tint.

The following note by **Mr. J. L. Kipling**, Principal, School of Art, Lahore, on the methods of house decoration in India, is of interest, and as it has an important bearing on the pigments of this country, may be quoted in full :—"House painting in the European sense of painting in oil is entirely unknown. The oxidising or drying properties of linseed oil may be known, but practically the country is in the state of Europe before the invention of oil-painting, popularly attributed to Hubert Van Eyck, but probably of earlier date.

"Large surfaces are seldom, indeed never, covered with coats of colour with a brush, even in distemper or water-colour. Wood-work, when it is painted at all, is done in water-colour, which is afterwards covered with a resinous varnish. The decoration of large mosque fronts in colour was accomplished by men whose largest paint-pot was no bigger than a tea-saucer. The supply of colour, thus sparingly used, has always been irregular and fortuitous, and it is only in recent years, when oil-painting has been introduced by the railway workshops, and colour washing by the Public Works Department, that a regular trade has arisen in colours, with which in former times the painter supplied himself after much research and a considerable expenditure of labour in their preparation.

"All the indigenous notions of painting involve the use of water as a medium. In fresco-painting, now seldom practised, no other vehicle is used, and the colours being laid on fresh, wet chunam, with which they are mechanically and chemically incorporated as the lime sets. In ordinary painting on wood or dry plaster, gum, starch, or rice water is used to bind the colours. It is obvious, however, that most of the colours used can be worked in oil as well as water."

713

Pigments.

References.—*Baden Powell, Pb. Pr., 22-24: Forbes Watson, Ind. Survey, I., 418; Mason, Burma & Its People, 567, 730; Reports from various Governments on Pigments, in answer to Government of India's Circular No. 5—54-1G., 2nd February 1887.*

In enumerating the chief pigments produced, or largely used, throughout India, it is convenient to follow the usual custom of grouping them according to their colours.

COLOURS.
White.
714

(1) White.

WHITE-LEAD (see Vol. IV., 602, 603).—This, one of the most important colours in wall-painting, and largely used as a ground colour and white paint, is the chief white pigment. It is extensively employed by Native painters and washers all over India, and is chiefly imported, though manufactured to a small extent.

ZINC-WHITE.—The above remarks apply also to this pigment.

WHITE EARTH.—Mr. Kipling states that a white earth, *karya mitti*, which is not true whiting or chalk, serves in the Panjáb as white in water-colour. It is much employed also by the poorer classes as a whitewash, and is sold at about six annas a maund (80℔). No good indigenous white appears to exist. A white earth called *soonka kul*, or *mauve sonkan* (TAM.), is obtained in the Tiruvallúr taluk in Madras, where it is employed as a primary coat and for making caste marks. Another called *baitur padi* is obtained in the Nizam's dominions near Beder, and is used as a substitute for whitening.

LIME—Is extensively used for whitewashing all over the country. Calcined chalk from shells was formerly much utilised in Madras under the names of *vellai*, TAM., and *thelupu*, TEL., as a high class white paint. It is now almost entirely superseded by white lead.

Blue.
715

(2) Blue.

SULPHATE OF IRON (see Iron Salts, Vol. IV., 523).—This salt is used as a pigment either in its natural state or artificially manufactured. It is sold for about R1 for 16℔.

SULPHATE OF COPPER (see Vol. II., 649)—Also occurs naturally and as a manufactured salt. Its rupee price varies from 2½ to 4℔.

ARSENIC.—Arsenic, *sankhia*, is used with sulphate of copper and potash to produce various shades of blue and green; it costs about eight annas per pound.

LAPIS LAZULI, or ULTRAMARINE (see Vol. IV., 587)—Was formerly extensively used in paintings, illuminated manuscripts, and fine decorative wall-painting. It is now almost entirely superseded by imported English or German artificial blues; indeed, its name *lajward* has been applied to these colours.

INDIGO (see Vol. IV., 389-469).—This colour is used not only as a dye but also as a pigment. Alone, it is employed for blue tints, mixed with yellows it forms green, with reds it produces purple, and with salts of iron, black.

(3) Yellow.

YELLOW ORPIMENT.—A sulphuret of arsenic of a heavy liver yellow colour, known as *hartal*, is imported chiefly from China, and sells for 2 to 4℔ per rupee. According to Mr. Kipling it was probably formerly employed in painted decoration. It is a favourite colour with the water-colour painting of wood-work, and the lac-turner's yellow is generally orpiment. It is used in the Panjáb by liners, and for the lines in horoscopes, round the margins of manuscripts and other paper-work. See Vol. I. 321-22.

YELLOW OCHRE (see Iron, Vol. IV., 520, 521)—Is, like red ochre, much used in painting, and as a pigment colour for cloth. It sells for about 27℔ per rupee. Mr. Kipling states that though cheaper than anything imported, and when well ground quite satisfactory as to colour, the trouble of grinding and its comparative dulness, when compared with English chromes, are causing it to become less and less used in the Panjáb.

MÚLTÁNÍ MATI.—A soft saponaceous earth, varying in colour from yellow, to red, sometimes greenish or lavender-coloured, is much employed as a pigment colour by the poorer classes near Múltán and in the Dérajat (see Clays, Vol. II., 362). It is exported to other parts of India, and sells in Bombay for 6 to 12℔ per rupee.

YELLOW OXIDE OF LEAD (see Lead, Vol. IV., 602).

YELLOW CHROMATE OF LEAD (see Lead, Vol. IV., 603).—This pigment is rapidly increasing in popularity and is taking the place of the indigenous yellows formerly used.

GAMBOGE (see Garcinia, Vol. III., 472.)

PEORI (see p. 132).

PEVADI.—A yellow pigment, said to be obtained by soaking pounded flowers of **Butea frondosa**, in cow's urine, is used for decorative house painting in Bombay, where it costs R20 per cwt.

NAKLI PEORI.—An artificial yellow pigment, prepared from the yellow colour extracted from the flowers of the *dhak* (**Butea frondosa**). This colour, called *tessú*, is mixed with lime to form the pigment. It is manufactured at Agra and Hattras, and costs about three annas a seer. It gives a dull yellow colour, not so brilliant as that of true *peori*, but is cheaper, and is said to be used in large quantities in Delhi.

(4) Green.

VERDIGRIS.—*Jangal, sangar*, a subacetate of copper, of a very beautiful bluish-green colour, is much used for painting on paper and in decorative house-painting, and in the Panjáb and North-Western Provinces is made by treatment to produce a turquoise as well as greener tints. It fetches R1 for ½ to 2℔.

GREEN CHROMATE OF LEAD—Is imported to a small extent as a pigment.

EMERALD GREEN.—A cupric arsenite and acetate, is largely imported from Europe, and from its brilliancy and crudity is a favourite colour.

SILICATE OF PROTOXIDE OF IRON.—An earth composed of this mineral is described by Mr. Baden Powell as formerly much used, under the name of *sangsabz*. Mr. Kipling states that it is a bluish green like the *tena verte* of European colour boxes, is now less frequently met with in the bazárs than it used to be, and costs about R5 per seer. Surgeon-Major Lyon found what appears to be the same colour in the Ajanta caves, Bombay.

COLOURS.

Green.

Many greens are also formed by mixing yellow pigments, such as chromate of lead, sulphuret of arsenic, &c., with indigo, or some imported colour, such as Victoria blue.

Red.
718

(5) Red.

Red pigments embrace two distinct series of colours, *viz.*, the reds of inorganic origin, and the red lakes obtained from animal and vegetable colours. The most important of these in India belongs to the former class, of which the following are chiefly used.

RED OCHRE (see Iron Oxides, Vol. IV., 520).—Red ochre is largely used in all parts of India for colouring cloth a red or brownish-red colour, and as a paint for walls and woodwork. It sells in Bombay for 12 to 20℔ per rupee, but a specially good kind, distinguished as *kapil*, fetches as much as R1 for 4℔. Of recent years the oxide has been extensively mined in Jabalpur for the purpose of manufacturing Olpherts' patent mineral paint. Mr. Kipling distinguishes two forms in the Panjáb, one called *hurmachi*, a fine, rich, madder-like, deep Indian red, costing about 1 anna a pound, and a cheaper light brick-red ochre, known as *géru*, which is useful for light tones. The former is described as varying in quality, but the best samples are said to be remarkably solid and rich in colour. It works equally well in water and oil, and when finely ground its covering qualities are equal to those of any other pigment. Mr. Kipling writes, "In old plaster work *hurmachi* was sometimes mixed with *chunam* for coloured dados and afterwards polished to resemble red or pink marble. It is the predominant colour in much good mosque painted decoration. Nothing imported is half so good or cheap."

VERMILLION (see Mercury, Vol. V. 232)—Is manufactured to a small extent in India, but is chiefly imported from China. It is much employed in India as a pigment for all purposes for which a brilliant red is required, and in Bombay sells for about ½℔ per rupee.

RED LEAD (see Lead, Vol. IV., 602, 603).—This important pigment is very largely used in India for ordinary purposes of painting, washing, &c., also for making caste marks, and for adorning sacred objects. It is manufactured in the country and also imported, and sells at about 8℔ per rupee in Bombay, and 5℔ per rupee in the Panjáb.

REALGAR (see Vol. VI.,)—Is much used in the preparation of saffron-coloured pigments, and sells at about 2℔ per rupee. It is ichiefly imported.

CINNABAR (see Vol. II., 316; Vol. V., 233).—Crude cinnabar, under the name of *shingarf*, was formerly much employed as a brilliant metallic red, but has been almost entirely superseded by the small packets of Chinese vermillion now universally obtainable. It is imported into Northern India from Calcutta and Central India, and sells at about R2-8 a seer.

LAKES.—A lake prepared from lac-dye (see Vol. IV., 516) is used to some extent in India as a paint for wood. Lac-dye has greatly diminished in commercial value of late years and now sells for about R24 the cwt. or 5℔ for the rupee. A lake is also prepared for use as a pigment from cochineal, with tartrate of potash and alum. The cochineal from which it is made costs R2-8 per pound in the North-Western Provinces.

Purple.
719

(6) Purple.

Dark purple is a favourite colour in high class house decoration, and is produced by various mixtures of vermillion and indigo, vermillion and Victoria blue, red-lead and Victoria blue, and lake with indigo.

(7) Grey.

At Bijapur a pearl-grey pigment is said by Mr. Ozanne to be largely used for painting walls. It consists of mica mixed with the mucilage of *bhendi* (Hibiscus esculentus). The chief ingredient, mica, sells at Bijapur for 4℔ per rupee.

Other greys are generally prepared by mixing some white pigment with charcoal, or black earth.

KHÁKÍ.—The term *kháki* was originally the name of a sect of Vaishnava Hindus founded by Kil, a disciple of Krishna Das. They apply ashes of cow-dung to their dress and persons, hence the name has come to be applied to a grey-coloured dye. In many parts of the country this *kháki* colour is given to cloth by means of a natural grey-coloured earth, but the use of such colouring matter, which from its nature is always fleeting and requires renewal, is being gradually superseded by that of *kháki*-coloured dyes, generally prepared with myrobalans, galls, and sulphate of iron. One of the most important *kháki* clays is the *leingang* of Manipur. Major Trotter in his report on that country describes it as occurring nearly everywhere in the valley. Two *tolas* of the clay is mixed with 1½ quarts of water, coloured with a *chittak* of *haldi* (wild turmeric), ¼ pint of milk is added, the whole strained, and the cloth steeped in the fluid. It is then wrung out, dried in the sun, again steeped, wrung out in *heibúng* water and dried in the shade. Major Trotter states that the colour may be made permanent by boiling in alum solution. If so, the earth must not act as a simple pigment, but also as a true dye.

(8) Black.

Most black pigments owe their colour to carbon, obtained by charring animal or vegetable substances.

LAMP BLACK, or *Kajal*, is the most important pigment of this class in India. It is generally collected from earthen shades or reeds placed over burning oil lamps, and sells from R4 to R8 per rupee. Of recent years, however, this pure lamp black has been largely ousted from the market by the much cheaper, but equally inferior, imported chimney soot.

COCOANUT BLACK and ALMOND BLACK are obtained by burning these nuts. They are chiefly used as a colour for tinting whitewash. By variety in the amount added several shades of grey may be obtained.

BLACK EARTH.—A black earth is used for colour washing in Rájputana, and is said to be procurable in the bazár at Abu for 3 pie a basketful.

The foregoing list includes the indigenous pigments chiefly employed in the bazárs of India, and though it would be possible to mention a few others, the enumeration includes all that can be reckoned of practical value. Commenting on the short list of pigments available in the Panjáb, Mr. Kipling remarks :—"I am not certain that it indicates the mineral poverty of the country, but it shows, at least, with how limited a colour scale, skilful ornamentists could produce artistic effects." This remark applies to India generally, and though European importations have placed many new and brilliant colours at the disposal of the Indian workman, the result has been by no means an improvement in the class of work turned out.

The question of pigments in India may be fittingly summed up by the remark that most if not all chemically prepared colours are more cheaply procured and of better quality when imported, than when of native make; and that, with the exception of the ochres, the supply of naturally coloured mineral and vegetable colours of indigenous origin is small, and not likely to exceed local demands. In the case of the red and yellow ferruginous earths, however, this is not so. Already they have been utilised in the manufacture of mineral paints for European purposes, and it appears not

PIMPINELLA
Anisum.   The Pilea Fibre; Anise.

COLOURS.

unlikely that the large deposits existing in India may be still further taken advantage of for this purpose.

TRADE.
722

Trade.—A considerable foreign trade exists in paints and colours, which, however, consists chiefly of imports. This import trade has increased considerably during the past ten years, the average for the five years ending 1883-84 having been 115,196 cwt., value R17,13,699; while that of a similar period ending 1888-89 was 139,670 cwt., value R19,15,445. In 1888-89 it reached the amount of 184,383 cwt., value R24,16,086; and in 1889-90 154,124, cwt., value R20,40,443.

The chief source of supply is the United Kingdom (with, in 1889-90, 123,475cwt.), followed by Persia, Belgium, France, the Straits Settlements, and Arabia, named in order of importance. Bombay takes the greatest share (73,383 cwt. in 1889-90), followed by Bengal (60,131 cwt.), while Madras, Sind, and Burma import insignificant quantities.

The re-export trade is small. During the five years ending 1883-84 it averaged 1,351 cwt., value R25,838; during a similar period ending 1888-89, 2,615 cwt., value R37,352. Like the imports, however, the re-exports shewed a sudden rise in 1888-89, *viz.*, to 3,364 cwt., value R41,199, and to 3,180 cwt. valued at R40,781 in 1889-90. Nearly the whole re-export trade is from Bombay, which supplies small quantities of foreign colours to the Eastern Coast of Africa, Aden, Arabia, Ceylon, Hong-Kong, Persia, Siam, Turkey in Asia, and other countries.

The exports of Indian pigments are very unimportant and shew no tendency to increase. Thus in 1883-84 they were 205 cwt., valued at R15,790, in 1888-89 227 cwt., value R14,286. The exports are made from Madras, Bengal, and Bombay, and received by Arabia, Aden, and other countries.

Pig, see Hog, Vol. IV., 253.

PILEA, *Lindl.; Gen. Pl., III., 384.*

723

Pilea scripta, *Wedd.; Fl. Br. Ind., V., 556;* URTICACEÆ.
Syn —P. GOGLADO, *Wedd.;* URTICA SCRIPTA, *Ham.;* U. HAMILTONIANA, *Wall.;* U. IRIPLINERVIS, *Ham.*
Reference.—*Atkinson, Him. Dist., N.-W. P., 317.*
Habitat.—A tall-branched, glabrous, large-leaved herb, native of the Temperate Himálaya from Dalhousie to Bhután, from 3,500 to 6,000 feet; also met with in the Khásia Mountains and Manipur.

FIBRE.
724
725

Fibre.—This is referred to by **Royle** as one of the fibrous plants of the North-West Himálaya.

P. smilacifolia, *Wedd.; Fl. Br. Ind., V., 553.*
Syn.—P. GOGLADO and GLABERRIMA, *Blume;* P. MICONIÆFOLIA, *Miq.;* URTICA GLABERRIMA, *Blume;* U. SMILACIFOLIA, *Wall.;* U. GOGLADO, *Ham.*
Habitat.—A herbaceous or suffruticose glabrous plant, found in the Eastern Tropical Himálaya from Sikkim to Mishmi, between the altitudes of 2,000 and 4,000 feet, also in Assam, the Khásia Mountains, Sylhet, and Chittagong.

FIBRE.
726

Fibre.—It is stated to yield a fibre, regarding which no definite information can be found.

PIMPINELLA, *Linn.; Gen. Pl., I., 893.*

727

Pimpinella Anisum, *Linn.; Bent. & Trim., Med. Pl., t. 122;*
THE ANISE.   [UMBELLIFERÆ.
Vern.—*Saurif, saónf, somp,*? *anisún,* HIND.; *Muhúri, souf-ka-jur, mithá-jirá,* BENG.; *Pán-mahúri,* URIYA; *Sop,* NEPAL; **Anisun, mahori,*

| The Anise-seed. | (*J. Murray.*) | PIMPINELLA Anisum. |

sauraf, N.-W. P.; **Anisún, sounf*, Pb.; *Badián*, West Afg.; *Ervados, echra, sataphushpha, sonf*, Bomb.; *Sómp, badi-shep*, Mar.; *Sewa, anisa, variali*, Guz.; *Burri-shep, sónf*, Dec.; *Sombú, shómbu, perunshiragam*, Tam.; *Kuppi, pedda-jilakara, sópu, pedda sadapa*, Tel.; *Dodda-jirage, sómpú, subbasije*, Kan.; *Jeramanis, perin-chirakam*, Malay.; *Sa-mung-sa-ba, samusaba*, Burm.; *Mahá-duru, déva-duru, lokka-duru, sinhala-asamodagan*, Sing.; *Setapushpa*, Sans.; *Anesún,* anisún,* rásiyánaj, shamár*, Arab.; *Razaneh-rúmi, razani, rúmi, rásiyanah, bádiyán, váláne-buzarg*, Pers.

* According to **Moodeen Sheriff**, these names (all of Arabic origin) ought to be properly applied to the fruit of **Carum Roxburghianum.**

References.—*Stewart, Pb. Pl., 107; Mason, Burmal & Its People, 496, 740; Sir W. Elliot, Fl. Andhr., 104; Pharm. Ind., 100; British Pharm., 112, 219; Fluck. & Hanb., Pharmacog., 310; U. S. Dispens., 15th Ed., 194, 1000; O'Shaughnessy, Beng. Dispens., 358; Irvine, Mat. Med., Patna, 4, 96; Moodeen Sheriff, Supp. Pharm. Ind., 199; Sakharam Arjun, Cat. Bomb. Drugs, 64; Murray, Pl. & Drugs, Sind, 197; Dymock, Mat. Med. W. Ind., 2nd Ed., 377; Dymock, Warden & Hooper, Pharmacog. Ind., II., 131; Year-Book Pharm., 1873, 348; 1879, 465; Med. Top., Ajm., 125; Butler, Med. Top. & States., Oudh & Sultanpore, 46; Vety. Aide Memo., 13; Birdwood, Bomb. Prod., 42, 222; Baden Powell, Pb. Pr.; 301; Econ. Prod., N.-W. P., Pt. V. (Vegetables, Spices, & Fruits), 30; Cat. Cal.\& Ind. Exhib. Raw Prods., No. 85; Piesse, Perfumery, 90: Mys. Cat. (C. Ex.), 22; Bomb. Man. Rev. Accts., 103; Settlement Report: —Panjáb, Kángra, 25, 28; Agri.-Horti. Soc., Ind.:—Transactions, I., 41; VI., 241; Spons, Encycl., I., 334; Balfour, Cyclop. Ind., 109; Smith, Ec. Dict., 17.*

Habitat.—An annual herb, with an erect stem, native of Egypt, Crete, Cyprus, and many islands of the Greek Archipelago, introduced from Persia into Northern India, where it is cultivated, by the Muhammadans. The chief part of the supply, however, still comes from Persia. It is also said to be cultivated to a small extent in other parts of India, *e.g.*, Orissa (*Hunter, Orissa, II., 134*).

Oil & Perfumery.—An odorous principle is obtained by distilling the FRUIT, the product being the "oil of aniseed" of commerce. The perfume is very powerful, and is well adapted for toilet soap and for scenting pomatums, but is not a favourite for mixing with handkerchief perfumes. *Arak badián* or water of anise is a favourite perfume in India (see **Perfumes**, p. 138).

OIL & PERFUMERY. Fruit. 728

Oil of anise is colourless, of a pale yellow tint, and has an aromatic spicy taste. Its specific gravity varies from 0·977 to 0·983. At 10° to 15°C. it solidifies to a hard crystalline mass which does not again become fluid under 17°C. It agrees very closely in character with the oil of Star Anise, and like oil of fennel consists almost entirely of ANETHOL (see Vol. IV., 331).

Anethol. 729

Medicine.—Anise is amongst the oldest of medicines and spices, and is mentioned in many classical and early European writings. It does not, however, appear to have been known to the ancient Hindus, and is not mentioned in Sanskrit works. After its introduction into Northern India by the Muhammadans, the knowledge of its virtues appears to have spread rapidly, and now it is used by Native practitioners in the same way as by Europeans. Thus Irvine states that in Patna it is imported for use as a carminative and deobstruent, and to promote digestion. The OIL and an ESSENCE prepared from that product are officinal in the *Pharmacopœia of India*. The properties and actions of the drug are too well-known to warrant any description in a work such as the present. Irvine states that the ROOT is supposed in Patna to have similar properties to the fruit.

MEDICINE. 730

Oil. 731 Essence. 732 Root. 733

Special Opinions.—§ "Is much used as a carminative in the form of bruised and cleaned fruit, and also as a water of Anise, *Araq Badian*" (*Assistant Surgeon Bhagwan Das, Rawal Pindi*). "Semi-parched seeds

PINUS **excelsa.**	**The Chir.**

mixed with sugar are useful in mild cases of dysentery" (*Surgeon R. Gray, Lahore*). " Useful in cardialgia, pyrosis, bilious giddiness and indigestion " (*Surgeon-Major D. R. Thomson, M.D., C.I E., Madras*).

FOOD.
Leaves.
734
Fruit.
735

Food.—Stewart states on the authority of **Bellew** that the LEAVES are used as a vegetable in the Peshawar Valley. The FRUIT is largely employed in Europe as an ingredient in cordial liqueurs, and in India is employed to a considerable extent as a flavouring agent in native confectionery.

Pimpinella crinitam, *Boiss.*, see **Psammogeton biternatum,** *Edgw.* ; p. [351 FERÆ; Vol II , 201.

P. involucrata, *W. & A.*, see **Carum Roxburghianum,** *Benth.*; UMBELLI.

Pine-apple, see **Ananas sativa,** *Linn.*; BROMELIACEÆ; Vol. I., 236.

736

PINUS, *Linn.*; *Gen. Pl., III., 438.*

A genus of coniferous forest trees of great economic value. In addition to the species, described below, several are cultivated to a greater or less extent in India, but not sufficiently to warrant their being included amongst the economic plants of the country. Three at least of the Indian forms promise well as sources of turpentine, and it is possible that the wood of certain species may be found suitable for the manufacture of wood wool, pine wood paper, pine fibre cloth, &c., now extensively made in Europe from that of the "Silver Fir," Pinus sylvestris.

Pinus Deodara, *Roxb.*, see **Cedrus Deodara,** *Loudon;* Vol. II., 235, now reduced to **C. Libani,** *Barrel.*, var. **Deodara,** *Fl. Br. Ind., V., 653.*

737

P. excelsa, *Wall. ; Fl. Br. Ind., V., 651 ;* CONIFERÆ.

THE INDIAN BLUE OR FIVE-LEAVED PINE.

Syn.—P. PENDULA, *Griff.;* P. GRIFFITHII, *M'Clelland.*

Vern.—*Tongschi, lamshing,* BHUTAN ; *Chila, karchilla, chil, chir, chilu, kail, lim,* N.-W. P.; *Raisalla, lamshing, byans, dolchilla,* KUMAON; *Yári, yara, yúr, yiro, kaiar, tser, chil,* KASHMIR ; *Biár, yero, yári, káiar, tser, chil, chir, kachir, dárchir, keiiri, lhim, chiti, biár, shim, partal, andal, lhim tser, sam, pálsam, dárchil, som shing, káil, lim,* PB.; *Limansa, piuni,* PUSHTU.

References.—*Lambert, Pin., Ed. 2, 40, t. 26 ; Brandis, For. Fl., 510 ; Griffith, Notul., IV., 18 ; Ic. Pl. Asiat., 365, 366 ; Gamble, Man. Timb., 398 ; Stewart, Pb. Pl., 225 ; Aitchison, Lahoul, Journ. Linn. Soc., X., 79 ; Cleghorn, Pines of N.-W. Him., t 2 ; Baden Powell, Pb. Pr. 588 ; Drury, U. Pl. Ind., 343 ; Atkinson, Him. Dist. (X., N.-W. P. Gaz.), 318, 829 ; Gums & Resinous Prod. (P. W. Dept. Rept.), 53, 56, 58, 61 ; Cooke, Gums & Resins, 124 ; Wardle, Dye Rept., 26 ; Brown, Forester, 322 ; Kew Off. Guide to the Mus. Ec. Bot., 134 ; Kew Off. Guide to Bot. Gardens & Arboretum, 20, 111. 150, 154 ; Settlement Reports :— Panjáb, Hazára, 10 ; Kángra, 140 ; Simla, App. II. H., xliii.; Gazetteers:—Panjáb, Rawalpindi, 15 ; Bannu, 23 ; Gurdáspur, 54; Simla, 10 ; Hazára, 14 ; Mysore & Coorg, I., 66 ; Agri.-Horti. Soc., Ind. :— Journals (Old Series), IV., Sel. 223, 226-228, 243, 264; VI., 167 ; VII., 77, 80-83 ; VIII., Sel. 25, 192, 193, 197 ; XIII., 363 ; XIV., 12, 265, 268, 270, 272 ; Agri.-Horti. Soc., Panjáb, Select papers to 1862, p. 48 ; Ind. Forester, I., 55, 306 ; II., 182, 183, 185, 186, 188 ; VII., 208, 211 ; VIII., 37, 106, 119, 121, 123, 269, 301, 375, 399, 404, 406, 408, 409 ; IX., 74, 180. 211, 303 ; X., 2, 12, 50, 159 ; XI., 5, 44, 270 ; XII., 262 ; XIII., 61, 62, 63, 66, 70, 317 ; XIV., 21, 63, 69, 247, 255, 256 ; Balfour, Cyclop. Ind., III., 220.*

Habitat.—A lofty symmetrical gregarious tree, found on the Temperate Himálaya between 6,000 and 12,500 feet, but most common up to 8,000 feet. It is readily recognised by the blue tint of its tufted foliage. the long penduous cone, the compact greenish blue bark of the young stems and leaves remaining on the tree for three years.

OLEO-RESIN.
Wood.
738

Oleo-resin.—The WOOD is highly resinous and affords turpentine and tar. These are, however, rarely extracted, and though **Stewart** men-

The Chir. (*J. Murray.*)

OLEO-RESIN.

tions that the resin is a Kángra export, little is known of the extent of its preparation, or the uses to which it is put. The more highly resinous parts of the wood are much employed for torches, which are known as *mashál* in Hindustani and Panjábi, *jagni* in Pushtu. The resinous cones are valuable for lighting fires. In certain dry winter seasons the leaves and twigs, together with those of the deodar (**C. Libani**, var. **Deodara**), **Pinus longifolia**, and a few other trees not conifers, become covered with a copious sweet exudation. This occurred to a large extent during the winter of 1889-90 in the Simla District, and has already been noticed under **Fraxinus ornus**, Vol. III., 443, and **Manna**, Vol. V., p. 165. The " Manna " thus found is collected and eaten by Natives, and is said to have been used in Bashahr for adulterating honey. A similar shedding of manna was noticed by Madden as having occurred in Kumáon.

Dye.—The BARK contains a fair amount of colouring matter. Samples experimented with by **Mr. Wardle** were found to produce tints which varied much with different mordants. With one process it produced good yellows on bleached Indian tussur and corah silk, and a deep orange on wool. **Mr. Wardle** writes, " The effect of this process on wool is remarkable and promises a great utility for this dye-stuff. The yellow on corah silk is a very good colour which might be brought to use with little pains." " As for the other colours " (varying from blackish-brown to drab) " on the card they are not very remarkable, and do not vary much. A cotton sample has not been able to be dyed on account of the smallness of the sample of bark." The bark possesses a small quantity of tannin which is shown by the dark colour it produces with an iron salt. Should the dyeing property of *chíl* bark become of commercial value, as **Mr. Wardle** predicts, large quantities of the material might be easily and cheaply obtained from the North-West Himálaya.

DYE.
Bark.
739

Structure of the Wood.—Sapwood whitish; heartwood, distinct red, compact, even-grained, soft and easy to work; weight from 26 to 33℔ per cubic foot, average 30℔. Of all the North-West Himálayan Conifers it ranks next to the *Deodar* as regards durability, and is preferred to P. longifolia and the species of *Abies*. Brandis states that it is said to last 7-8 years as shingles, 8-10 years as beams in walls, and 15 years as ordinary inside planking. It is largely used for house-building, shingles, water-channels, water-troughs, wooden spades, and other implements. According to **Stewart**, quoting **Mr. Watson**, it is the best of all PANJÁB woods for pattern-making, since it works easily. **Stewart**, however, remarks that this can be the case only with carefully selected wood which contains little resin. It yields an excellent charcoal for iron-smelting. In short, wherever *Deodar* is not available or is very expensive, the wood of this tree, though only fairly durable and liable to warp and rot is employed for all purposes of construction. **Brandis** states that a variety of the wood, valued less than the ordinary kind, is called *dar chíl* at the Rávi and Chenáb timber depôts. Trees grown on the southern aspect of a hill are said to yield heavier and more highly resinous wood than those on the Northern slope.

TIMBER.
740

Domestic.—The more resinous parts of the WOOD are largely used for torches; chips and small pieces of the wood form an article of traffic in the treeless Inner Himálaya, under the name of *láshi*, or *clanshing;* and the CONES are employed for lighting fires. The BARK is used to roof huts, and the trees are often from this cause denuded of their bark to a considerable height. The LEAVES and TWIGS form admirable litter, and in certain localities the leaves are mixed with mortar and plaster in building (*Brandis*).

DOMESTIC.
Wood.
741
Cones.
742
Bark.
743
Leaves.
744
Twigs.
745

P. 745

PINUS Gerardiana.	The Neosia or Edible Pine.

746 | **Pinus Gerardiana,** *Wall.; Fl. Br. Ind., V., 652.*

THE NEOSIA or EDIBLE PINE.

Vern.—*Gunober, rhi, newr,* SEEDS=*chilgoza, neosa,* HIND.; *Kannuchi, koniŭnchi, kaninchi, shangti,* W. TIBET; *Ronecha, rolecha,* SEEDS= *neosa,* KUMAON; *Rhi, ri, shangti,* KUNAWAR; *Chiri, prita, mirri, galboja, galgoja, kashti, ri, rhi, neosa, shangti, newr, ruminche, roniunchi,* PB.; *Chir chil, kashti, san-ghŏsa,* SEEDS=*chilghŏsa, jalghŏza, san-ghŏs;* AFG.; *Chilgosá,* PERS.

References.—*Brandis, For. Fl.,* 508, *t. 67 ; Gamble, Man. Timb.,* 398 ; *Stewart, Pb. Pl.,* 225 ; *Cleghorn, Pines of N.-W. Himal., t. 4; Sakharam Arjun, Cat. Bomb. Drugs,* 133 ; *Murray, Pl. & Drugs, Sind,* 25 ; *Baden Powell, Pb. Pr.,* 268, 378, 588 ; *Atkinson, Him. Dist. (X., N.-W. P., Gaz.),* 318, 828 ; *Royle, Ill. Him. Bot.,* 353, t. 85 ; *Gums & Resinous Prod (P. W. Dept. Rept.),* 56, 61 ; *Cooke, Gums & Resins,* 126 ; *Church, Food-Grains, Ind.,* 177 ; *Brown, Forester,* 305 ; *Kew Off. Guide to the Museum of Ec. Bot.,* 149 ; *Gazetteer :—Panjáb, Simla,* 10 ; *Agri.-Horti. Soc. Ind. :—Journals (Old Series), IV., Sel.* 223, 228, 265 ; *VI., Pro.* 6 ; *VII.,* 77, 83, 84 ; *VIII., Sel.* 25, 193 ; *XIII.,* 383, 386 ; *XIV.,* 49, 76, 266, 267, 270, 272 ; *Panjáb, Select. papers to 1862, Index,* 43 ; *Ind. Forester, V,* 183, 184, 186 ; *VIII.,* 408 ; *IX.,* 15 ; *XIII.,* 70 ; *Balfour, Cyclop. Ind., III.,* 221.

Habitat.—A moderate-sized tree of the inner dry and arid valleys of the North-West Himálaya from Kunawar westwards, and in Garhwal, found in isolated areas of no great extent, and generally at altitudes between 6,000 and 12,000 feet.

OLEO-RESIN. Wood. 747 Cones. 748 | **Oleo-resin.**—The WOOD is very resinous and though generally considered the best for torches and fuel, on account of the value of the fruit, is seldom cut for these purposes. Stewart states, on the authority of Major Longden, that the Kanáwarís do not use the resin as it "gets too hard," but that excellent tar may be extracted from the wood by destructive distillation. According to **Gordon** it affords abundance of fine turpentine, and the CONES exude a copious white resin.

OIL. Oleo-resin. 749 Seeds. 750 MEDICINE. Seeds. 751 Oil. 752 | **Oil.**—The OLEO-RESIN yields turpentine, and an oil used medicinally is said to be extracted from the seed.

Medicine.—The SEEDS are "idly supposed by the Natives to have many good qualities" (*Irwin*). They are considered anodyne and stimulant. The OIL extracted from them is highly esteemed for its stimulating and healing powers when applied as a dressing to wounds, ulcers, &c. It is also said by Stewart to be employed as an external application in diseases of the head.

FOOD. Seed. 753 | **Food.**—The chief product of this tree is the almond-like SEED, contained in the cones. The cones ripen in October, are plucked before they open, and heated to make the scales expand. The seed is then easily taken out. They are largely eaten by the Natives, and are stored for winter use. According to **Brandis** they form a staple food of the inhabitants of Kunáwár, by whom they are often eaten ground and mixed with flour. Amongst these people the proverb is current, "one tree, a man's life in winter." They are also exported to the plains, from the hills of the Panjáb, and large quantities are imported annually into India from Afghánistán by the Khyber and Bolan passes.

No statistics are available of the probable annual production, but a full-sized cone yields more than 100 seeds, and each tree produces from 15 to 25 cones. The seed is sold in the Simla bazár at from 6 to 8 seers per rupee, and is considerably eaten as a dessert fruit by Europeans, in the same way as pistachio nuts. It is somewhat oily and difficult of digestion, but has a very delicate terebinthinous flavour.

According to **Church** the chemical composition of the seed is as

| The Khasya Turpentine. | (*J. Murray.*) | PINUS khasya. |

follows :—Water 8·7 per cent.; albumenoids, 13.6; starch, &c., 22·5; oil, 51·3; fibre, 0·9, and ash, 3 per cent.

Structure of the Wood.—Heartwood yellowish brown, hard, durable, very resinous, weight 44 to 47℔ per cubic foot. It is rarely utilised, since the tree is so much valued for its seeds that it is rarely felled; it is, however, sometimes hollowed out for water-courses, and, according to **Brandis**, is used for the hook which supports the seat on the single rope swing bridge, so common on the Himálayan streams. It is, therefore, probably tough and durable.

Domestic.—The wood is occasionally employed for torches, the RESIN for patching cracked wooden vessels, and the BARK for making baskets and rough water buckets.

TIMBER. 754

DOMESTIC. Resin. 755 Bark. 756 757

Pinus khasya, *Royle; Fl. Br. Ind., V., 652.*

Syn.—P. KASYA, *Kurz*; P. KHASYANA, *Griff.*; P. KESEYA, *Royle.*

Vern.—*Dingsa,* KHASIA; *Tinyu,* BURM.

References.—*Griffith, Notul., IV., 18; Ic. Pl. Asiat., tt., 367, 368; Hook. f., Himal. Journ., II., 282; Brandis, For. Fl., 508; Kurz, For. Fl. Burm., II., 499; Gamble, Man. Timb., 397; Aplin, Report on the Shan States, 5; Home Dept., Procgs., Forests, August 1884, Nos. 16 to 21; December, 1885, Nos. 84 to 101; Gazetteer:—Mysore & Coorg, I., 66; Ind. Forester, I., 363; VI., 128, 308; VII., 125, 245; VIII., 399, 404, 405; IX., 181; XIII., 56; XIV., 63, 64, 69.*

Habitat.—This species is a native of the Khásia Mountains, Chittagong, and the hills of Burma, at altitudes between 3,000 and 7,000 feet. In Assam it is a small tree, but in Burma attains a height of 100 to 200 feet, with a trunk 10 feet in girth. The forests of this tree are estimated to cover about 270 square miles, of which 230 are in Assam, the remainder in Burma. It is believed that the smaller area in the latter country can be largely extended and improved, nothing but fire conservancy being required to transform the hill-sides into pure pine forests.

Oleo-resin.—The resin of this pine is perhaps the most valuable obtainable from any Indian species of the genus. In Assam it is collected, and crude oil of turpentine is made from the resinous WOOD. The method followed has been fully described by Mr. Mann (*Indian Forester, VII., 125*), who states that highly resinous chips of the wood are sold in the bazárs of the Khásia Hills, and are used by Natives and Europeans for kindling fires. This wood is produced artificially, by cutting a hole into the lower part of the trunk of the tree, about one foot above ground. A blaze about 4 feet long and 12 inches wide is then made above this notch. A good deal of resin exudes from the blaze, and is collected, and the constant outflow in one direction excites an unusually copious secretion of resin in the trunk. The wood, at least in the vicinity of the blaze, becomes soaked with oleo-resin. Twelve months after the tree has been thus treated, it is felled. Mr. Mann calculates that a full grown tree will yield 68℔ of crude resin from one of these wounds, and that the resin-encrusted wood contains in addition 16 per cent. of its total weight of crude resin. Liquid turpentine is extracted from this wood as follows :—The wood is cut into chips, and placed crosswise in layers in an ordinary earthenware cooking pot, the mouth of which is closed by large smooth leaves perforated with small holes. The vessel is then inverted on a support over another receptacle, and fire is made on or above it, the vessel being surrounded with hot ashes and burning charcoal. As a consequence the turpentine exudes from the wood and drops through the holes in the leaf-covering into the pot below.

The resinous wood used in this process is said by Mr. Mann to have sold in 1882 in Shillong at the rate of about R1 for 80℔.

OLEO-RESIN. Wood. 758

16

P. 758

PINUS. longifolia	The Long-Leaved Pine

OLEO-RESIN.

In Burma no resin is collected from this tree. In 1884-85 the subject of the turpentine yield from this species, along with that from **P. longifolia** and **P. Merkusii,** came before the Government of India, the possibility of opening up an extensive trade in the oil between India, and England being under consideration. Papers received from the Forest Department and other sources showed that the cost of crude turpentine amounted to R32 per 100℔ delivered in Calcutta, and that the supply was limited though no doubt capable of increase by careful conservancy of the forests. The very high price of the crude product, compared with that of the oil obtainable from **P. longifolia,** and with that of American or French turpentine in Europe, render prospects of any development in the trade very small.

A specimen of the turpentine from **Pinus khasya,** however, sent from the Indian Museum to **Professor Armstrong, F.R.S.,** appears to show that the product is purer, and probably of higher commercial value than other turpentines. It was reported on as follows :—" Of all the substances which you sent me from the India Collection some time ago, the only one which has hitherto furnished any result of interest is that labelled Crude Turpentine (**Pinus khasyana,** Burma). It consists of a solid resin similar to colophony and of a liquid 'turpentine oil.' The latter is remarkably pure and free from smell, and ought, I should say, to be very valuable for purposes for which the French and American oils are used. Scientifically it is interesting in several ways; it has the greatest amount of action on polarized light of any *coniferous* oil of turpentine with which I am acquainted, the rotatory power of a 200 mm. column being about 76° for light of the refrangibility of the D line, whereas that of French oil of turpentine is about 60°, and that of American oil about 30° on the average. Neither French nor American oil, especially the latter, are homogeneous ; but this **P. khasyana** oil, so far as I can judge from the examination of the small quantity at my command, is almost a pure substance. It may, therefore, be worth while to call attention to it."

TIMBER.
759

Structure of the Wood.—Very resinous, heartwood red, weight 38℔ per cubic foot. It is extensively used in the Khásia Hills for building purposes, and in Burma for torches.

760

Pinus longifolia, *Roxb. ; Fl. Br. Ind., V., 652.*
THE LONG-LEAVED, OR THREE-LEAVED PINE.

Vern.—*Salla, saral, chir, chil,* OLEO-RESIN=*ganda-birosa, chir-ka-gond,* HIND. ; *Dhúp, sala, dhúp, sula,* OLEO-RESIN=*dhúp, koto,* NEPAL ; *Gniet,* LEPCHA ; *Teadong,* BHUTIA ; *Kolan, chir, salla, sapin, kolon, kolain,* SEED=*kalghosa, chalhatti,* N.-W. P. ; *Dhúp,* OUDH ; *Chir, salla, sapin, kolon, kolan, kolain,* KUMAON ; *Salla, sarl,* KASHMIR ; *Chir, chil, dráb chir, nashtar, nakhtar, ransuru, gula, thansa, anander, saral,* OLEO-RESIN=*ganda-birosa,* PURIFIED OLEO-RESIN=*birosa, sat-birosa,* PB.; *Nashtar, nakhtar.* PUSHTU ; OLEO-RESIN=*Gandah-birosah,* BOMB. ; *Sarala,* OLEO-RESIN=*sarala drava, srivása, kshira,* SANS. ; OLEO-RESIN =*Biroseh,* PERS.

References.—*Lambert, Pinatum, Ed. 2, 32, t. 22 ; Roxb., Fl. Ind., Ed. C.B.C., 677 ; Griffith, Notul., IV., 18 ; Ic. Pl. Asiat., tt. 369, 370; Brandis, For. Fl., 506 ; Gamble, Man. Timb., 396 ; Cat., Trees, Shrubs, &c., Darjiling, 81 ; Stewart, Pb. Pl., 226; Cleghorn, Pines of N.-W. Himál., t. 3 ; Pharm. Ind., 225 ; U. C. Dutt, Mat. Med. Hind., 316, 318 ; Sakharam Arjun, Cat. Bomb. Drugs, 133 ; K. L. De, Indig. Drugs, Ind., 88 ; Murray, Pl. & Drugs, Sind, 26 ; Dymock, Mat. Med. W. Ind., 2nd Ed., 756 ; Official Corresp. on Proposed New Pharm. Ind., 234, 283 ; Year-Book Pharm., 1880, 251 ; Baden Powell, Pb. Pr., 378, 410, 588 ; Drury, U. Pl. Ind., 343 ; Atkinson, Him. Dist. (X., N.-W. P. Gaz.), 746, 824 ; Royle, Ill. Him. Bot., 353, t. 85, f. 1 ; Prod. Res., 233 ; Gums & Resinous Prod. (P. W. Dept. Rept.), 37, 52, 53, 54,*

or Ganda biroza Tree. (*J. Murray.*)

56, 57, 58, 60 ; Cooke, Gums & Resins, 125 ; Wardle, Dye Rept., 27, 28 ; Cat., Raw Prods., Col. & Ind. Exhib., No. 97 ; Hamilton, Acct., Nepal, 67, 197 ; Kew Off. Guide to the Mus. of Ec. Bot., 134 ; Home Dept. Procgs., Forests, 1884, Nos. 16 to 21 ; 1885, Nos. 84 to 101 ; Gazetteers :—Panjáb, Rawalpindi, 15 ; N.-W. P., IV., lxxviii. ; Mysore & Coorg, I., 66 ; Agri.-Horti. Soc. Ind.—Trans., VII., Pro. 90 ; Journals (Old Series), IV., Sel., 223—226, 253 ; VII., 75-80, 162, 163 ; VIII., Sel., 25, 150, 152, 154, 191, 192 ; XIII., 313, 363, 383 ; XIV., 12, 41, 265, 268, 270, 272 ; Ind. Forester, I., 82, 306 ; II., 172, 182, 390, 392 ; III., 88, 92, 358, 366, 370, 371 ; V., 183, 190 ; VI., 15, 148, 308, 309 ; VII., 245 ; VIII., 37, 102, 106, 113, 118, 120, 123, 124, 125, 128, 130, 185, 271, 301, 371, 374, 375, 388, 399, 404, 405; 406, 408, 412 ; IX., 13, 125, 126, 211, 301, 321, 361, 402 ; X., 2, 56, 125, 176, 183 ; XI., 270, 387, 388 ; XII., 63, 262, 418 ; XIII., 2, 38, 56, 57, 59, 60, 61, 63, 67, 70, 280, 314 ; Balfour, Cyclop. Ind., III., 221.

Habitat.—A large gregarious tree of the outer and drier Himálayan slopes, from the Indus to Bhután, met with as low down as 1,500 feet and ascending to 7,000 feet; distributed to Afghánistán. In a recent report called for on the subject of turpentine, the area under this species in the North-Western Provinces (Kumáon, British Garhwál, Naini Tal, and Ranikhet) is estimated at 615,914 acres, and the total area in North-Western India, including Kashmír and other Native States at from 2,000 to 4,000 square miles. The tree is easily recognised by its pale green tint, rounded, large, cone, brown corky bark, breaking off in slabs, and by the leaves (at least in Simla) falling every year, about the beginning of the rains (June).

Oleo-resin.—The oleo-resin of this species is more largely collected and used than that of any other Himálayan conifer. A certain amount naturally exudes from the bark; it is known as *berja* in the Panjáb and is occasionally collected. The oleo-resin is, however, more generally procured by incision, conducted so carelessly as to be detrimental and finally destructive to the tree. The product is valued by the Natives chiefly for its resin, which is obtained by exposing the oleo-resin to heat, by which means the oil or more valuable product is dissipated. Oil of turpentine is, however, extracted to a small extent by the Natives, but the practice was probably introduced by the Europeans, with that of preparing tar by destructive distillation of the wood. This latter process, which is much more common than the former, is conducted as follows :—Dry chips of wood are put into a large earthen pot with a narrow neck, in the bottom of which four or five small holes are drilled. The pot so filled is luted over with stiff wet mud on the sides and top; a hole is dug in the ground, in which a smaller pot holding about 3 seers is placed, and the larger one is put on the top. The joint is then luted, the sorrounding space is filled up with earth, and a fire of cow-dung fuel is lit over the whole. The fire is kept up from eight to nine hours, at the end of which time 5 chittaks of tar and 1 seer of charcoal are obtained from a pot which originally held about 10 seers of chips. Four men easily make 2¼ maunds, or nine large pots full of tar in a month, and the cost is about R21, or 3 annas 8 pie per seer. The product appears to be equal to tar imported from Europe. Several improvements on the native method have been used from time to time with considerable success. Thus it was found at the Madhapore workshops that by conducting the process in several pots close together at once, and by silghtly modifying the native apparatus, 1 seer of tar could be produced from 6 seers 4 chittaks of fresh chips as a charge, and 2 maunds 2 seers of chips for fuel. The estimated cost was only 1 anna per seer.

The tar thus obtained is said to be of a superior description. It is a mixture of resin and oil of turpentine, blackened by the presence of empyreumatic products; it thickens after exposure to the atmosphere, and may be used in the same way as ordinary coal tar. It is also

OLEO-RESIN.
761

PINUS longifolia.	The Long-Leaved Pine

OLEO-RESIN.

employed in the preparation of the large skin floats, on which rivers are crossed.

The method of obtaining the oleo-resin followed by Natives has formed the subject of an exhaustive note by Mr. Fernandez. From this it appears that the tapping season commences in February and ends in June. A notch is cut in the tree about 3 feet from the ground; the bottom of this notch is hollowed out to receive the resin, which is collected as the cup, thus formed, fills, sometimes as often as every second or third day, but usually between the fourth and eighth days. The notch is deepened and lengthened from time to time, so as to freshen the wound and favour the flow of resin; and unless this is done the old resin hardens over the walls, and thus effectually prevents the exudation of fresh material. The same cut is, as a rule, used for two years, and sometimes even for three, when no fire occurs in the time to scorch or burn the resin incrusted wood. The yield of resin per tree, obtained by this process, is very variable. Brandis, quoting Mr. Thompson, estimates it at from 10 to 12lb in Garhwál. The Conservator of Forests, School Circle, North-Western Provinces, states that in Naini Tal it is said to be from 4 to 6lb. the first year, and about half that amount, the second. The largest flow is said, by the same authority, to take place when the niche is freshly made, as much as 1lb being obtained in the first collection from an average tree. In 1883 experiments in tapping according to the European method were started in the Déra Dún School Circle, and it was found as a result, that the average yield by the system of *gemmage à vie* (that is, tapping the tree so as not to injure it), was an average of 3 to 3·5 seers in trees from 7 to 9 feet in girth. Mr. Greig states that trees from 4 to 6 feet in girth may be made to yield an average of 5 to 6lb each, for a period of eight to ten years, without causing death.

No estimate can be given of the present production, but it may be worth notice that Mr. Thompson, in 1867, estimated the quantity annually exported from Garhwál to the markets at the foot of the hills, to be between 1,000 to 1,200 maunds. Reports called for by the Government of India, in 1885, shew that the cost of collection and delivery is very variable. One maund collected in Hazára was delivered in Rawalpindi for about R2-8, and sold for R7. Oleo-resin collected in the Beás valley cost, at the Railway, R9-8 per maund; in Chamba the cost of collection and carriage was estimated at R6-12 per maund, and in Kalel and Ranikhét at R5-12. It was found possible to collect the resin in the School Circle, North-Western Provinces, and convey it to Deoban at R4 per maund, but this price was said to be susceptible of great reduction, if large quantities were collected. The crude product is said to sell in Bombay for R6 to R6½ per maund of 41lb. From these figures it would appear that the average cost of 1 maund (82lb) of oleo-resin collected and delivered at one of the stations in the Panjáb or North-West Provinces might amount to about R6, though with increased care in collection under intelligent supervision, it might be considerably reduced.

It must, however, be observed that increased tapping for pure resin would necessarily entail a greatly augmented risk of forest fires, and that, therefore, strict fire protection on a large scale would be indispensable to the success of the operations. This would involve increased outlay, which would only be warranted if the extraction promised to be a financial success, a question which depends entirely on the price the crude product, or turpentine distilled from it, would be likely to fetch in the English market.

It may be mentioned that the Government of India have taken steps to settle this question, and have it at present under consideration.

P. 761

Products of India. 245

or Ganda-beroza Tree. (*J. Murray.*) PINUS longifolia.

CHEMICAL COMPOSITION.—Mr. Fernandez states that he obtained
from the crude oleo-resin about 30 per cent. of oil of turpentine, equal
to the best imported oil, except that its odour was much less pungent.
Dymock, on the other hand, writes that 56℔ of the crude drug, distilled with
water, yielded only 8℔ (or about 1·43 per cent.) of a colourless limpid oil
having the peculiar odour of *gandah-birozah*. The resin remaining in the
still was of a dull brown colour; after straining, to remove impurities and
stirred with a small quantity of boiling water until hard, it was found to
weigh 43℔, and afforded a very fair substitute for Burgundy Pitch.

The oil, according to Lyon, has a specific gravity of ·875 at 82° Fh.;
it commences to boil at about ·310° Fh., and is dextro-rotatory.

Oil of turpentine is extracted from the crude product in the Panjáb,
Bijnaur, and elsewhere in North-West India; in the former locality the
oleo-resin is first mixed with water and carbonate of soda. The residue
(pale resin, colophony) is called *súndras* in Bijnaur. The value of colophony and of turpentine in the arts is too well known to necessitate re-
petition in this work. In India the crude product is chiefly used medicinally
and is also said to be a common adulterant of lac; the tar is employed in
the same way as elsewhere, and the purified resin is used by the tinsmith
in soldering metals.

Dye & Tan.—The BARK is used for tanning; the charcoal of the LEAVES
mixed with rice-water is used for the purposes of ink. Samples of the
bark examined by Mr. Wardle gave results which would seem, like those
with **P. excelsa**, to promise a new and valuable economic use. He writes:—
"This bark seems to be rather remarkable for its yellowish dye on bleached
Indian tussur and eria silk, and for its rich orange shade on wool, this
bark and that of **Pinus excelsa** forming remarkable exceptions to the
majority of Indian barks, yielding, with the aid of the new mordants,
very rich colours. On unbleached Indian tussur, when an iron salt is
used as a mordant, a darkish drab is produced owing to the small quantity
of tannin the bark contains; the unbleached Indian tussur dyed in a bath
of the extracted colour alone produces a nice pale-brown shade."

Oil.—Oil of turpentine is obtained from the oleo-resin, see remarks
under "Resin."

Medicine.—The aromatic WOOD and the OLEO-RESIN are mentioned
in Sanskirt works as medicinal. Oil of turpentine, however, does not
appear to have been known,—at all events it is not mentioned by Sanskrit
writers. The wood is considered stimulant, diaphoretic, and useful in
"burning of the body," cough, fainting, and ulcerations, and is generally
prescribed in combination with other drugs. The crude oleo-resin is described as useful in the preparation of plasters, ointments, and pastilles for
fumigations, and is also directed to be applied to buboes and abscesses
for the purpose of promoting absorption. Chakradatta describes a pas-
tille made of turpentine, bdellium, wood of **Aquilaria Agallocha**, and resin
of **Shorea robusta**, as useful for the fumigation of unhealthy and painful
ulcers (*U. C. Dutt*). The purified and crude oleo-resins are found in all
Indian bazárs and are employed both as external and internal remedies.
They appear to have the properties of ordinary turpentine, and are chiefly
used for making a pectoral plaster, like the pitch-plaster of Europe. They
have also a reputation in veterinary practice as a remedy for mange.
Dymock states that the *Vaids* obtain from the oleo-resin, by distillation
without water, a limpid, sherry-coloured oil, having the peculiar odour of the
drug, which they call *Khanno* oil in the Deccan; it is in much repute as
a remedy for gleet or long-standing gonorrhœa. According to Baden
Powell the crude oleo-resin is used in the Panjáb as a stimulant diuretic
in diseases of the urinary organs, chronic bronchitis, and hæmorrhages,

CHEMISTRY.
762

DYE AND TAN.
Bark.
763
Leaves.
764

OIL.
765
MEDICINE.
Wood.
766
Oleo-resin.
767

PINUS longifolia.	The Long-Leaved Pine.

MEDICINE.
Tar.
768

also in rheumatism and fevers. The TAR (*zift rúmí*) is employed in chronic bronchitis, phthisis, and skin diseases, and as an application to the skin of camels for "itching" in the cold season. In Nepál, turpentine is imported from Bhután, and given in doses of four to eight *mashas* in cases of ague and splenic enlargement.

SPECIAL OPINIONS.—§" The turpentine, *ganda-biroza*, is distilled at the Bombay Medical Store Depôt for the sake of the oil which is issued to veterinary charges in the Central Provinces and Berars as a remedy for mange in horses, under the name of Chir Pine oil. The native *vaids* also distil it and call it Khanno oil; they prescribe it in gleet. It resembles oil of turpentine, but has a peculiar agreeable aromatic odour like that of the Chir turpentine. One cwt. of *Ganda-biroza* yields 16℔ of oil. The resin which remains has a similar agreeable odour, is of a light brown colour, and is used by us as a substitute for Pix Bargundica" (*Dr. Dymock, Bombay*). "The oleo-resin, mixed with cocoa oil or glycerine, is a useful application to buboes" (*W. Forsyth, F.R.C.S. Edin., Civil Medical Officer, Dinajpore*). "The application of *gand-biroza* has been found to dissipate buboes and other glandular swellings, by preventing suppuration" (*S. C. Bhattacharji, Chanda*). "The resin of **Pinus Longifolia** is soft, opaque, ductile, and greenish white. It is stimulant both internally and externally. Internally it acts chiefly on the mucus membrane of the genito-urinary organs, and is therefore a very good remedy for gonorrhœa. I have used it with success in many cases of this disease, and in a few, with decided benefit, after the failure of copaiba, cubebs, gurjan balsam, and turpentine. Dose, from one to three drachms in emulsion with mucilage, four times in the 24 hours. As it is very thick it requires to be mixed well and gradually with the mucilage" (*Honorary Surgeon Moodeen Sheriff, Khan Bahadur, G.M.M.C., Triplicane, Madras*). "A good stimulant application especially for boils" (*Assistant Surgeon Nehal Singh, Saharunpore*). "I have found *gandha baroza* to be an excellent application for the ulcers known as Frontier sores in the Panjáb; it is a powerful external stimulant." (*Civil Surgeon J. Mullane, Gauhati, Assam*). "Have used it externally to ripen boils, abscesses, and buboes with good effect" (*Surgeon D. Picachy, Purneah*). "*Gondh-biroza* certainly promotes suppuration when externally applied, and is specially useful in indolent abscesses and buboes" (*Civil Surgeon S. M. Shircore, Murshedabad*). "A decoction of the chips is given for the purpose of checking suppurative swellings" (*V. Ummegudien, Mettapolliam, Madras*).

FOOD.
Seed.
769

Food.—Dr. Stewart writes: "In parts of the Jhelum basin, the turpentiny SEED is at times eaten when food is scarce, but it cannot be a pleasant, and is probably not a nutritious, food."

TIMBER.
770

Structure of the Wood.—Yellowish, reddish white or brown, no distinct heart-wood, weight from 37 to 45℔ per cubic foot when unseasoned, about 27℔ when seasoned. It is easy to work and is extensively employed in the hills for building purposes, rafters, shingles, tea-boxes, and the bottoms of boats. Though durable when used for interior house-work—roof-trees are said to last two generations in Kumáon—it decays rapidly when exposed to wet, and is not valued for exterior work, except where *deodar* is not available. Brandis states that in Kumáon, about Piura, and in several places on the Wardwan branch of the Chenáb, a large proportion of the trees have the bark and the fibres of the wood spirally twisted, in the same way as, though to a much smaller degree than, is often seen in the horse-chestnut in Europe. The wood of the twisted trees is useless for any purpose save fuel. The wood is much used for making charcoal. The TIMBER of stumps and trees which have been notched and mutilated is often so full of the resin as to be translucent, and is much employed for

The Pan Leaf.	(*J. Murray.*)

torches in houses and mines (*Brandis*). Stewart states that in certain localities in the Panjáb baskets are made of thin slips of the wood. The BARK is used as a fuel in iron-smelting.

Pinus Merkusii, *Jungh. & De Vriese; Fl. Br. Ind., V., 652.*   **771**

Syn.—P. LATTERI, *Mason;* P. SUMATRANA, *Jungh.;* P. FINLAYSONIANA, *Wall.*

Vern.—*Tinyú, htenrú,* BURM.

References.—*Kurz, For. Fl. Burm., II., 499; Gamble, Man. Timb., 308; Mason, Burma & Its People, 545, 803; Aplin, Report on the Shan States, 5; Home Dept. Progs., Forests, 1885, Nos. 84 to 101; Ind. Forester, VI., 128, 308; VIII., 399, 403; XIII., 56; XIV., 63, 69.*

Habitat.—A tree, of 50 to 60 feet in height, found in the Shan States, Martaban, and Upper Tenasserim, between 500 and 2,800 feet, and in Lower Tenasserim between 3,000 and 4,000 feet; distributed to Sumatra and Borneo, where it sometimes attains the height of 100 feet. In Burma it almost always grows mixed with **Dipterocarpus tuberculatus.** In the Tenasserim Circle alone it occupies an area of about 50 square miles.

Oleo-resin.—Perhaps the first Indian writer to notice the abundance of resinous matter in the wood of this tree was Mason, who wrote that it appeared to contain more than any other species of conifer he had ever seen, and that large quantities of both pitch and tar might be manufactured in the forests. Reports called for by the Government of India in connection with the enquiry on turpentine, shew that the trees have been much cut up by *taungya* cultivation; but it is stated that successful fire protection would suffice to restock the area now denuded, and that forests could be extended without difficulty by propagating the tree. Experiments have shown that the oleo-resin can be extracted without killing the tree, and that a tree of six feet girth can yield 12℔ at the first tapping. The quantity decreases in the second and third years, after which the tree requires a lengthy period of rest. The cost of delivering 100℔ of crude turpentine at Moulmein amounts at present to R12, and the quantity available is small, though capable of considerable increase. With augmented facilities for transport and more careful conservancy of the forests, the supply could be largely increased and the price diminished.

Structure of the Wood.—Heart-wood yellowish brown with dark streaks, moderately hard, exceedingly resinous, weight from 51 to 54℔ per cubic foot. The wood is sometimes brought to Moulmein for mast pieces, but the difficulties of land and water transport are very great, almost preventing its being exported at a profit. Splinters are extensively employed for torches (*Gamble*).

PIPER, *Linn.; Gen. Pl., III., 129.*   **774**

A genus of shrubs, rarely herbs or trees, which comprises about five hundred named species (probably greatly exaggerated), all tropical or subtropical. Of these forty-five are described in the *Flora of British India* as natives of, or long established in, this country. Several are of great economic interest and value.

[PIPERACEÆ.

Piper Betle, *Linn.; Fl. Br. Ind., V., 85; Wight, Ic., t. 2926;*   **775**

Syn.—P. SIRIBOA, *Linn.;* P. PEEPULOIDES, *Wall.;* P. CHAVYA, *Ham.;* CHAVICA BETLE, *Miq.;* C. CHUVYA, *Miq.;* C. SIRIBOA, *Miq.*

Vern.—*Pán, támbulí,* HIND.; *Pán,* BENG.; *Pán, vilyadele,* BOMB.; *Videcha-pána,* MAR.; *Pán, nágur-vel,* GUZ.; *Pán,* DEC.; *Vettilai. Tám.; Tamalapáku, nágavalli,* TEL.; *Vile-dele,* KAN.; *Vetta, vetrila, vettila,* MALAY.; *Kúnyoe, kwán, kwanynet, kwon-rwet,* BURM.; *Balát,* SING.; *Támbula, nágavalli,* SANS.; *Tanból,* ARAB.; *Barge-tanból, tamból,* PERS.

Betel Leaf Cultivation

References.—*Roxb., Fl. Ind., Ed. C.B.C.,* 53, 569 *; Voigt, Hort. Sub. Cal.,*
299 *; Thwaites, En. Ceyl. Pl.,* 292 *; Dalz. & Gibs., Bomb. Fl., Suppl.,*
89 *; Mason, Burma & Its People,* 495, 778 *; Burmann, Fl. Ind.,* 14 *;
Thesaurus, Zey., t.* 88, *f.* 2 *; Pharm. Ind.,* 208 *; Flück. & Hanb., Pharmacog.,* 583, 669 *; U. S. Dispens.,* 15th *Ed.,* 380 *; Ainslie, Mat. Ind.,*
II., 465 *; O'Shaughnessy, Beng. Dispens.,* 575 *; Beng. Pharm.,* 89 *;
Moodeen Sheriff, Supp. Pharm. Ind.,* 97 *; U. C. Dutt, Mat. Med.
Hind.,* 92, 244, 310, 320 *; Dymock, Mat. Med. W. Ind.,* 2nd *Ed.,* 727 *;
Drury, U. Pl. Ind.,* 129 *; Useful Pl. Bomb. (XXV., Bomb. Gas.),*
173 *; Econ. Prod. N.-W. Prov., Pt.* IV. (*Vegetables, Spices, & Fruits*),
32 *; Butler, Top., & Statis., Oudh and Sultanpur,* 40 *; Ayeen Akbary,
Gladwin's Trans.,* II., 27, 28, 39, 44 *; Ain-i-Akbari, Blochmann's Trans.,*
I., 72, 73 *; Linschoten, Voyage to East Indies (Ed. Burnell, Tiele &
Yule),* I., 213, 214, 215 *; II.,* 53, 62, 66, 130 *;Man. Madras Adm., I.,* 289 *;
Nicholson, Man. Coimbatore,* 231, 232 *; Morris, Account Godavery,* 10 *;
Boswell, Man. Nellore,* 137 *; Moore, Man. Trichinopoly,* 74 *; Adm. Rep.
Beng.,* 1882-83, 13 *; Bombay, Man. Rev. Accts.,* 103 *; Darrah, Note on
the Condition of the People of Assam, App.* D *; Settlement Reports :—
N.-W. P., Allahabad,* 37 *; Central Provinces, Bhandara, para.* 177 *;
Nâgpore, Suppl.,* 274 *; Wardah,* 60, 61 *; Nimar,* 199-200 *; Madras, Goda-
very,* 148 *; Gazetteers :—Bombay, XI.,* 97 *; XIII.,* 293 *; XV., Pt. ii.,* 11 *;
XVI.,* 104 *; XVII.,* 276 *; N.-W. P., I.,* 84 *; Central Provinces,* 56 *; Orissa,*
II., 136, *App. iv.,* 160 *; Mysore & Coorg, I.,* 66, 136 *; II.,* 11 *; Agri.-
Horti. Soc., Ind. :—Trans.,* III., 68, 200 *; VI.,* 239 *; Journals (Old
Series), VII.,* 68 *; IX., Sel.,* 50, 59 *; X.,* 43 *; XIII., Sel.,* 64 *; Indian Agri-
culturist,* 17-7-86 *;* 8-4-88 *; Spons, Encycl.,* 1305 *; Balfour, Cyclop. Ind.,*
III., 222.

Habitat.—A perennial diœcious creeper, probably a native of Java,
cultivated, for the sake of its leaves, in the hotter and damper regions of
India and Ceylon.

CULTIVATION
776

Cultivation.—The cultivation of betel leaf is attended with many difficulties, and requires a constant temperature, a fairly steady degree of moisture,
and much attention on the part of the husbandman. The methods pursued
vary considerably in different localities, and may, perhaps, be best considered under the heads of the Presidencies and Provinces in which betel
leaf is chiefly grown.

Madras.
777

MADRAS.—The plant is very widely distributed under cultivation in
this Presidency, being chiefly found in the moister regions. It is specially described in the District Manuals of Coimbatore, Godavery, Nellore,
and Trichinopoly, in all of which districts the method followed appears
to be much the same. It is described as follows in Nicholson's *Coimbatore Manual* :—

"The land, which has been well prepared, is divided by water channels
into beds of about two feet wide, and in August, *agathi* (Sesbania grandiflora) seeds are sown and occasionally watered till October. Betel cuttings from a two-year old garden are then planted between the *agathi*
trees, at two per tree, and during three months are watered on alternate
days for the first fortnight, after which they are watered once a week.
In January the dung of cattle, asses, and pigs is applied to the roots, and
covered with silt from the water-channels, and the creepers are then tied
to the *agathi* trees. As they grow they are repeatedly tied on with plantain fibre, till at the end of the first year they twine naturally. In July
manure is again applied. From the end of the first year leaves are
picked daily for about sixteen months, if the soil be good and well
manured.

The best monthly yield of a good plantation, per acre, is said to be 15
konies=240 *palageis*, each *palagei* containing 25 *kattus* (bundles) of 100
leaves each. The *palagei* is ordinarily sold at two annas, so that the
monthly yield is worth R30 per acre, being for the sixteen months of crop,
a gross yield of R480 during the three years of cultivation. The labour is

Products of India.   **249**

in the Central Provinces.   (*J. Murray.*)   **PIPER
                                                            Betle.**

great and the profits long deferred, so that it is not much engaged in, though so profitable."

CENTRAL PROVINCES.—In these provinces the crop is apparently an important one, and is described at length in the Settlement Reports of the districts of Bhandárá, Nágpur, Wardhá, and Nimár. The method of cultivation is described as follows in the account of Wardha :—" The *pán* leaf is cultivated by a class of people called *Burehs* and the *pán* garden is *Bureja*, or sometimes *Pan-ka-tanda*. The plant is very delicate and susceptible in a high degree of the influence of heat, light, disease, &c.; great care and watchfulness is, therefore, necessary in rearing it. If, however, a good crop is obtained, the profit is fair and sufficient to compensate for the anxiety of two years of watching. The *pán* gardens are generally held by a cultivating brotherhood of old standing, forming quite an independent section in the village, who pay their rent to the landholder through the head of their body. A portion of the village lands is recognised as belonging to this caste; in this they dig wells or tanks, and make their gardens, and, being careful and steady cultivators, contribute much to the general prosperity of the neighbourhood. The *pán* garden is enclosed on all sides with a bamboo and mat covering to shield the delicate plant from the weather, and cool plantain leaves and the graceful wide-spreading leaves of the *Arun*, which shelter and support the young plant, are massed within the walls. The interior of these gardens is strikingly pretty and inviting, the *pán* leaf carefully trellised in all directions, the broad leaves of the plants grouped beside it, affording a grateful shade, whilst the constant supply of water renders the garden agreeably cool, even in the hottest weather. These spots are fully appreciated by tigers and panthers, which often seek shelter here during the hot weather, and the cultivators on going to their work not unfrequently find one of these animals entangled in the maze of the trellised plants.

" The leaf is planted in ridges, varying of course in length with the area of the enclosure. After the leaf is planted the ridges or drills are measured, and the garden is found to contain a certain number of units of length called *lani*, which are portioned off among the brotherhood. The betel leaf requires constant care and much water. Manure too is essential. When young and delicate, the plant is even fed with milk, which is found to be an excellent manure. It ceases bearing leaves of any marketable value in two years. New gardens are then made. The first year of cultivation is called *Wotuk*, and the second *Korwa*, the produce of the latter being much more esteemed and sought after, and fetching a higher price than the leaves yielded by the first year's crop. At the expiration of the second year the ground is allowed to remain fallow for periods varying, according to the nature of the soil, from two to three years.

" The betel-growing brotherhood are so careful, and are so much respected, that landholders allow them many privileges that are not granted to the ordinary race of agriculturists. Thus custom has prescribed that no rent is to be paid or demanded during the time the land remains uncultivated. Rent, too, is only paid on the exact quantity of land sown, *i.e.*, the rent is paid per *lani*, the rate on which varies in the first and second year of cultivation. Although all the members of the brotherhood or caste give their labour towards the construction of the garden, share in the expenses of watering, weeding, watching, &c., yet at the same time each individual is the *owner* of a fraction of either one or more *lanis*, the produce of which he himself gathers, and disposes of on his own account, and each has an independent interest in the undertaking."

In Nimár the method differs from that just described, and presents several features of interest. Instead of only lasting two or three years a

**CULTIVATION
in the
Central
Provinces.**

pán tanda once established is said to yield steadily for ten to twelve years. The ground is similarly prepared by repeated ploughing and manuring, but, as in Madras, trees are sown to afford shade. In this case they consist of *saora* (**Sesbania ægyptiaca**), a tree which shoots up ,rapidly and affords good shade for two or three years. The plantation is also surrounded by poles of *pángrá* (**Erythrina indica**) to which bamboo mats are attached. After the *saora* trees die down they are replaced by poles of the *salei* (**Boswellia serrata**), of which the bark is believed to give a specially favourable hold to the climber. These poles give no shade, but in the meantime plantains have been planted all over the garden at intervals of 10 or 12 yards, and thus supply complete protection. No matting is used overhead or for partitions as in Wardhá. Every year after the leaves have been gathered the creeper is coiled down at the root until only some 3 feet of it are left above ground; a fresh root is thus struck and the old coil is next year cut away altogether. The crop is irrigated steadily at all seasons except during the rains, and at the commencement of each hot season (March) the plants are pruned, manure is applied to the roots, and fresh earth, brought from another field for the purpose, is piled up round them. This supplies the new soil necessary, and which, in other systems of cultivation, renders a short-lived crop and long fallow essential. Owing to this superposition of fresh earth the garden soil increases in depth by about 2 feet, before the ten or twelve years are over. After that time the garden is removed, hemp is sown to clean the land, and alternate fallows and irrigated spring crops, with liberal manuring, follow for some years to render the soil again fit for another crop of *pán* (*Settlement Rep., Nimár, 199-200*).

This system appears to be preferable in many ways to that ordinarily followed; it entails less labour, less retardation of profit, and with the annual supply of fresh earth ought to produce as good a yield. It will be seen that much the same method, though combined with artificial shade, is pursued in Bengal and other localities in which betel cultivation is extensively carried on.

**Bengal.
779**

BENGAL.—*Pán* is largely cultivated in Bengal, but the system adopted would appear to be fairly constant, and the following report on the method pursued in Burdwán may be taken as typical :—

"Betel is extensively grown in some parts of the division, especially along the banks of the Hooghly. The following is a description of the mode of cultivation followed in the village of Bantul, in the Uluberia sub-division, which has the reputation of producing the finest betel leaves in Bengal.

**Varieties.
780**

"*Varieties.*—Three different varieties of betel are grown here, *viz.*, the *bangla* or country, the *khas*, generally known by the name of *sanchi pan*, and the *karpurkath*, or the sweet. The cultivation of this last variety, which has a very fine flavour and an odour like camphor, is very limited. It grows only in a few fields, and even in these few plants out of many planted succeed.

**Soil.
781**

"*Soil.*—Betel grows best in the friable black clay resembling pond mud and containing a large amount of organic matter. It is, however, worthy of notice that the betel gardens at Bantul, which used to grow the most highly flavoured *karpurkath*, consist of a rather light loam, very slightly reddish. Soils of this description are no longer available in the village, and new gardens are mostly made of the ordinary clay of the description mentioned above. The land must be high and well drained, stagnant water being most injurious to betel.

**Preparation.
782**

"*Preparation of garden.*—To make a new betel garden, or *boros*, as it is called by the cultivator, a plot of land situated on high ground, well-drained, and close to a pond or *khal*, is selected. The soil generally is a blackish clay, containing a large percentage of decomposed vegetable matter. Underbushes, if there be any, are cut down, and their roots dug out with the spade. The whole land is then thoroughly turned over by the same implement to a depth of 18 inches. Trenches are dug around the field, and the earth thereby raised is spread over it. In some places, when making a new *boroz*, the whole field is raised by 5 or 6 inches with pond mud, but this is not the practice at Bantul. By the repeated use of the *kodali* the ground is thoroughly stirred to more than a foot, and pulverised as fine as possible. The land is levelled, and over it a roof is placed made of bamboo slips and jute stalks supported on bamboo posts to a height of 4¾ cubits from the ground. The four sides of the *boroz* are enclosed by *tattis* made of the same materials as the roof. Great care is taken in making the bamboo supports. Good, sound, and well-ripe bamboos are cut into pieces, each about 10 feet long. These are slit in the middle, dried in the sun, and then kept under water for about a month. The uprights are placed in parallel lines alternately 18 and 27 inches apart. The opposite uprights of lines 18 inches apart are made to intersect one another a little over half the way up. The betel cuttings are planted between the other two lines of uprights at intervals of 6 inches.

"*Planting.*—Cuttings are made from plants taken from old *borozes*. These cuttings are about 12 to 18 inches long, and contain five or six joints each, of which two are buried in the earth, and the portions left above ground are made to recline on the surface. These are then covered with date leaves, and watered every morning and evening till they strike root and put forth buds. The planting time extends from *Jaistya* to *Kartik*" (May to November). "In some places the cuttings are first planted in nurseries, and when they have struck root and sent forth new leaves and buds, they are taken out for transplantation.

"*After-treatment.*—As the plants go on increasing in length, one or two jute stalks are struck into the ground close to each, the upper ends of the stalks reaching the roof. The betel plants are tied to these supports by the straw of *ulu* cut into lengths of one foot; when the plants reach the roof they are pulled down, and a portion is made to rest on the ground in a coil which is covered with a little earth. This process is repeated as often as necessary, and every time this is done two or three leaves are plucked from below.

"Pond mud and other kinds of earth well-dried and pulverised are placed between the lines, and are used as occasion requires in earthing up the plants. In this way the whole ground is slowly raised; and to keep the roof at the proper height from the ground, it has to be changed every second year. Some of the old *borozes* at Bantul have in this way been raised above the level of the first floor of surrounding buildings.

"*Manures.*—The manures generally used are—(1) cowdung dried and well ground into a fine powder, (2) pond mud dried and powdered, and (3) oil-cake. The mode of applying these manures is to put them round the roots and cover them with a little earth. Mustard cake only is used in *borozes*. Castor cake destroys the plants. Brick-dust is likewise sometimes used. All through the summer the *borozes* require to be constantly irrigated, but excessive watering is very injurious, and stagnant water should on no account be allowed to accumulate.

"*Diseases.*—Betel is liable to the attack of many diseases, most of which seem to be of a fungoid nature. Some attack the leaves only, others the stalk and whole plant." The following are described :—

"*Tuto.*—Black stains, which increase in size and destroy the entire leaves; *Bont angáre*—from which the leaf stalks rot; *Noná*—from which the leaf stalks wither away; *Tasare*—from which the edges of

CULTIVATION
in
Bengal.
Preparation.

Planting.
783

After-treatment.
784

Manures.
785

Diseases.
786

P. 786

PIPER **Betle.**	Betel Leaf Cultivation

CULTIVATION
in
Bengal.
Diseases.

leaves turn red; *Chittigábri.*—from which the edges of leaves rot. The diseases above enumerated will be seen to affect the leaves only; those mentioned below attack the whole plant:—*Angáre* is an infectious disease from which the joints of the plant turn black and then rot. Any plant on which water from the affected parts of the diseased plant falls is attacked. Diseased plants are therefore carefully removed and thrown away, together with a portion of the earth on which the plant was growing; *Gándi*—the part of the plant close to the ground is affected. The plant turns red and dies. Onion juice and cowdung mixed together and sprinkled over the plant are said to arrest the attack to a certain extent (*Agri. Dept. Rept., Beng., 1886, lvi.*).

Orissa.
787.

ORISSA.—Cultivation is carried on in this province in the same way as in Bengal. Shoots are planted in July, the plant reaches maturity in twelve or fifteen months, and yields leaves for fifty or sixty years. The crop is constantly irrigated, and protected from the sun by a reed roofing, "so that a *pán* garden is simply a vast mat green-house, very steamy inside, but of a uniform temperature all the year round." An acre is said to yield from £88 to £100, of which, from £50 to £75 represent the capital laid out by the cultivator, including rent; a fair average profit to the husbandman is thus from £25 to £35 per acre (*W. W. Hunter, Orissa, II., 136*).

Bombay.
788.

BOMBAY.—*Pán* is an important crop in Dhárwár, the Deccan, and the Konkan. The system of cultivation differs in no essential particular from the method followed in other localities, but the accounts in several District Gazetteers present minor points of interest which may be briefly noticed.

In *Ahmednagar*, it is grown in open fields. As much as 75 cart-loads of manure is said to be applied per acre, and the plantation is surrounded by a thorn fence. Shoots of *shevri* (**Sesbania ægyptiaca**) are sown for the young plant to climb on. These die down in three to six years and are replaced by *pángára* (**Erythrina indica**) or *sheoga* (**Moringa pterygosperma**), which in their turn are allowed to grow to 8 or 9 feet high and are then pollarded. The plants are watered at least once in ten days, and do not yield leaves fit for consumption till the third year, after which time a crop is plucked every eight to ten days. The vine is cut to the ground once a year, and remanured, after which new shoots spring up and are trained on the trees.

In *Khandesh* the method is very similar to that in vogue in the districts of Ahmednagar and Nasik. When the vine, which is reared on the same trees, has grown to the proper height, it is turned back and trained down till it reaches the ground, where it is covered for a short distance with earth and again turned up. This is repeated till the tree stem is fully clothed with vines, when the whole is firmly tied with *laváli* grass. The leaves are gathered after from 1½ to 2 years, by pickers who have the thumbs sheathed in sharp-edged thimble-like plates which nip the leaves clean off without wrenching the plant. A wicker-work fence of split bamboo is made outside the thorn hedge, and castor oil plants are frequently planted inside. The leaves retail at 1 to 2 annas per hundred.

In *Poona* it is said to be an important garden crop, especially in the Haveli villages of Kondvi Budruk, Kondvi Kurd, Undri, Muhammadvádi, and Phúrsangi. It is grown in light red soils, requires much manure and constant watering, and if well cared for, lasts from fifteen to twenty years. The *pán mala* or betel-garden generally covers about an acre, the vines are trained on slender *hadga, pángára, shevri,* and *bakán* trees planted in rows and pollarded, and the whole garden is sheltered by high hedges, or screens of grass or mats. Every year in March, April, and May, the upper half of the vine is cut, while the lower half is coiled away and buried under fresh red earth and manure. The other details of cultivation are similar to those practised elsewhere. Irrigation is always by well-water.

In *Násik* the method is very similar, but an outer fence of coarse grass matting is added to that of thorn. The garden is divided into beds three feet long by two wide, each with its own water channel and trench in which the shoots are planted. At the end of the first year the top shoots are drawn down, covered with earth, and are thus made to shoot afresh. New earth is added as described in Bengal, and owing to the consequent variability of the height of the soil, the plots are always watered from wells, never from rivers. The plants furnish eatable leaves after one year, and the acre yield is said to value R150 to R700 annually. From R400 to R600 is often spent, however, before the crop yields any return. Stunted shoots (*nakhi*) are said to give the best leaves, soft, smooth, and full of flavour, while those on a more luxuriant shoot are coarse. Two kinds of vines are distinguished in the district, *viz.*, *bakshi* and *támbdya*, of which the latter yields the quicker return, the former the better and larger crop.

In *Dhárwár pán* is the most important garden crop, but the method of cultivation is apparently less elaborate than in many other districts. The shoots are planted in the open, trained on quick-growing trees, an acre of land containing upwards of two thousand plants, and the crop lasts only from four to seven years. The vines are then dug up, the leaves of the trees on which they were trained are used for vegetable manure, and the wood for fuel. The garden is deeply dug all over, allowed to lie fallow for one year and afterwards planted with sugar-cane. After the sugar-cane it enjoys another year of fallow, and betel vines plantains are again laid out.

In *Kánara* the plant is described as cultivated in gardens on mango trees. The leaves are picked three years after planting the young shoots, and the yield is estimated at 100 to 200 leaves from a full-grown vine every fortnight. An acre of spice-garden containing 500 plants is said to produce some 40,000 leaves annually, worth R20, and costing R8 to grow.

In *Thána* the crop is grown on any soil, if not salt, stony or too damp and stiff. Such land is generally selected after a crop of rice, and a booth is built to cover the young plants. Pits are dug one foot and a half across and one foot deep, at a distance of $1\frac{1}{2}$ feet apart; in December and January they are filled with water, and when the ground is still moist four shoots, 18 inches long, are set in each. These shoots are liberally watered, till they begin to grow freely, after which they are trained on reeds, and 5 ozs. of oil-cake is given to each pit. A month after this first manuring, $\frac{1}{2}$lb of oil-cake is again put into each pit, and the young plants are watered every second day till the rains. About the end of June the plants have attained a considerable height, they are unfastened from their supports, allowed to droop to within a foot of the ground, and the side shoots are gathered into the pits, and covered with earth so as to produce new roots. Half pound oil-cake is again applied to each pit, and the main stem is again bound to the reeds and trained. A second thatched booth is reared over the first in September, and the creeper is trained up its posts and allowed to climb over the roof. The leaves are ready for cutting after 12 months, and yield a steady crop for another year, after which time the vine is cut down and young shoots planted in another place.

In *Kolába* the cuttings are planted in ridges, trained on *túr* sticks, shaded with palm leaves and manured with fish. They yield leaves in the first year.

Lisboa states that in Bombay the best *pán* leaves are grown in Poona, Sátára, and above the Gháts, and that a large quantity is exported from these localities to the town of Bombay.

P. 788

PIPER **Betle.**	Use of Betel Leaf

CULTIVATION
in
Bombay.

The total area under the crop in 1888-89 is returned by the Bombay Agricultural Dept. as 4,492 acres, of which 1,204 were in Dhárwár, 904 in Poona, 412 in Ahmednagar, 380 in Khándesh, 365 in Sátára, 264 in Kánara, 205 in Thána, 194 in Belgaum, and 179 in Násik.

N.-W.
Provinces.
789

NORTH-WEST PROVINCES.—Atkinson, in his account of *pán*, quotes Roxburgh's description as applicable to these Provinces. The plants are raised from slips and cuttings, carefully planted in a rich moist soil, well inclosed and shaded, so that they are in a great measure protected from both sun and wind. In certain localities small plantations of **Sesbania grandiflora** are laid out, for support and to keep off the sun, in others poles are employed for the former, and a thin shed of mats for the latter purpose; the plants require to be frequently watered during the dry weather. " *Pán* cultivation," writes **Atkinson**, " is extensively pursued in these Provinces, and the ' variety' grown in Bundelkhand is celebrated over Upper India. Betel cultivation requires considerable care and labour, and is conducted by *Barehs* or *Tamolis*, the betel garden being known as *bareh-ja* in Bundelkhand. The garden is enclosed on all sides by matting attached to poles which are also thickly studded within to support the plants."

Burma.
790

BURMA.—No statistics are available as to the extent to which *pán* is cultivated in Burma. The only information the writer is able to procure is the short note by **Mason**, who writes, " Karens plant the vines on their uplands, where there are tall forest trees. The branches of the trunks are lopped off, leaving only the topmost boughs, and the vines readily climb up and weave their dark, glossy leaves all over the summits, making a betel-vine form a most beautiful object. Karen boys and maidens engage in these leaf harvests with great zest, and it is not uncommon for young men, in seeking companions, to enquire who are the most agile climbers of *poo-lah* or betel-leaf trees." He further states that the Karen forests produce a wild species of pepper, the leaf of which is used as a substitute for the common betel-leaf.

From the above, it will be seen that the cultivation of *pán* is one of the most widely spread and highly developed forms of agriculture in India. Unfortunately, no statistics are available of the total area under the crop, nor of the outturn and consumption in the country.

MEDICINE.
Leaves.
791

Medicine.—The following account of the medicinal use of *pán* in India is extracted from U. C. Dutt's *Materia Medica of the Hindus:*—" The LEAVES of this creeper are, as is well known, masticated by the Natives of India. The poorer classes make their packet of betle with the addition of lime, catechu, and betle-nuts. The rich add cardamoms, nutmegs, cloves, camphor, and other aromatics. Betle-leaf thus chewed acts as a gentle stimulant and exhilarant. Those accustomed to its use feel a sense of languor when deprived of it. Ancient Hindu writers recommend that betle-leaf should be taken early in the morning, after meals, and at bedtime. According to Susruta, it is aromatic, carminative, stimulant, and astringent. It sweetens the breath, improves the voice, and removes all foulness from the mouth. According to other writers it acts as an aphrodisiac.

Juice.
792

Medicinally it is said to be useful in diseases supposed to be caused by deranged phlegm and its JUICE is much used as an adjunct to pills administered in these diseases ; that is, the pills are rubbed into an emulsion with the juice of the betel-leaf and licked up. Being always at hand *pán* leaves are used as a domestic remedy in various ways. The STALK of the leaf smeared with oil is introduced into the rectum in the constipation and tympanitis of children, with the object of inducing the bowels to act. The leaves are applied to the temples in headache for relieving pain, to painful and swollen glands for promoting absorption, and to the mammary glands

Stalk.
793

MEDICINE.

with the object of checking the secretion of milk. *Pán* leaves are used as a ready dressing for foul ulcers which seem to improve under them." Excessive use produces effects somewhat similar to those of alcoholic intoxication.

In the Konkan the FRUIT is employed with honey as a remedy for cough, and in Orissa the ROOT is said to be used to prevent child-bearing. Ainslie states that the warm juice of the leaf is prescribed by the Vytians as a febrifuge, in the quantity of a small tea-spoonful twice daily ; it is also given in the indigestion of children, and in conjunction with milk in cases of hysteria. A paste composed of the root of a " cotton plant," beaten up with juice of betle leaves, is employed by Sanskrit chemists in the process of reducing a diamond to dust for medicinal purposes (*see Vol. III., 101*).

Fruit.
794
Root.
795

It is somewhat astonishing that a narcotic stimulant so much used by all Natives of India, and which has been known for so long by European physicians, should have attracted so little attention in writings on medicine. The plant is included in the secondary list of the Indian Pharmacopœia, but nothing is said as to its therapeutic value. The editor, however, notices an interesting form of utilisation of the leaves : "amongst the Indo-Britons of Southern India," he writes, " a use is made of the leaves which merits notice. In catarrhal and pulmonary affections generally, especially of children, the leaves warmed and smeared with oil are applied in layers over the chest ; and the editor from personal observation in many instances can testify to the relief afforded to the cough and dyspnœa, far more than can be accounted for by the warmth and exclusion of air, or by any rubefacient effect it produces, which indeed is very slight in most cases. Dr. Gilson, who corroborates this statement, states that he has often seen the application afford marked relief in congestion and other affections of the liver. Mr. J. Wood reports that the leaves warmed by the fire and applied in layers over the mammæ are used effectually for arresting the secretion of milk. Their use in this manner is also noticed by Dr. J. Shortt, who adds that the leaves are similarly employed as a resolvent to granular swellings."

But no European physician in India appears to have experimented on the value of the drug as a tonic, stomachic, and slight stimulant. In Java, on the contrary, it has attracted considerable attention. Acting on the great reputation enjoyed by *pán* all over the East, and on the remarks made on the drug by such early travellers as **Marco Polo**, Dutch botanists and physicians have used it experimentally, and have come to the conclusion that the chewing of betel-leaves *does* promote health in the damp and miasmatic climate of that country. The Netherlands Indian Government has enjoined that the leaves be served out to invalids and even convicts, under the belief that sickness is thereby reduced. The favourable effect of the drug on catarrhal and pulmonary disease has also been noticed. Probably the reason that the drug has not attracted a greater share of attention in Europe is, that the fresh leaves alone are active; when dry they wholly lose their aromatic volatile oil, and with it all their valuable qualities. It appears to be possible, however, that the volatile oil which is separable by distillation may be a fairly stable therapeutic agent of some value. It has already been exported to Germany from Java, and appears to have been favourably reported on. Dymock states that **D. S. Kemp** in 1885 obtained two pale yellow oils by distilling the fresh leaves with water, one heavy and the other light. Both had the peculiar odour of the leaf, but the light was the more aromatic of the two. From an ethereal solution of the leaves an alkaloid, called *arakene*, has also been recently obtained, from which salts similar to those of *cocain* have been produced. The

PIPER Chaba.	Medicinal Uses of Betel Leaf.

MEDICINE.

taste of the alkaloid and its salts is slightly acrid. They increase the flow of saliva, slow the action of the heart, and have a purgative effect.

Experiments on the physiological and therapeutic actions of this oil and the alkaloid would be of great interest, and might lead to valuable results.

SPECIAL OPINIONS.—§ "The juice of the leaves is dropped into the eye in painful affections of that organ; it is also used to relieve cerebral congestions and satyriasis, and to allay thirst" (*Surgeon-Major D. R. Thomson, M.D., C.I.E., Madras*). "The juice of the leaves is prescribed as a stimulant" (*Surgeon-Major Robb, Civil Surgeon, Ahmedabad*).

Leaf-stalks.
796

"The LEAF-STALKS dipped in *til* oil are put in the anus of young children and infants to promote the action of the bowels" (*Assistant Surgeon N. R. Banerjee, Etawah*). "Fresh juice is applied to the eyes as a collyrium in conjunctivitis, and is a reputed remedy for hemeralopia" (*Assistant Surgeon T. N. Ghose, Meerut*). "I have used the leaves warmed over a fire and mixed with mustard oil in cases of sore-throat with good effect" (*Surgeon D. Picachy, Purneah*). "The slender roots taken with black pepper are used to produce sterility in women. It is said that they produce paralysis and subsequent atrophy of the ovaries. The juice of the leaves is stomachic, carminative, astringent, and expectorant" (*Civil Surgeon J. H. Thornton, B.A., M.B., Monghyr*). "*Bangla pán* is especially valued for allaying bronchial irritation. The practice of introducing the mid-rib of the leaves into the anus of infants, to relieve constipation, is very common among Natives" (*Surgeon A. C. Mukerji, Noakhally*). "The powdered root in combination with honey and liquorice is useful in catarrhal affections. The root is commonly chewed by public singers and criers to improve their voices" (*Narain Misr, Hoshangabad*). "Juice of leaves is stimulant, carminative, and stomachic. I have frequently used it with honey in coughs of children with decided benefit" (*Assistant Surgeon S. C. Bhattacharji, Chanda*). "The juice of the leaves is a good expectorant for children" (*Surgeon W. F. Thomas, 33rd M. N. I., Mangalore*). "The leaves administered in the form of syrup with spices are useful in general debility in doses of an ounce three times a day" (*Lal Mahomed, Hospital Assistant, Central Provinces*).

797

Piper Chaba, *Hunter; Fl. Br. Ind., V., 83; Wight, Ic., t. 1927.*

Syn.—PIPER MARITIMUM, *Blume;* P. LONGUM, *Blume* (*excl. various synonyms*); P. CALLOSUM, *Opis;* P. OFFICINARUM, *Cas. DC.;* P. ARNOTTIANUM, *Cas. DC.;* P. GLABRUM, *Roxb.;* P. PEEPULOIDES, *Wall.;* CHAVICA OFFICINARUM, *Miq.;* C. MARITIMA, *Miq.;* C. PEEPULOIDES, *Wight* (*not of Roxb.*).

Vern.—*Cháb, chavi,* HIND.; *Chai, choi,* wood and roots=*chaikath,* BENG.; *Kankala,* BOMB.; *Chaviká, chuve,* SANS.

References.—*Roxb., Fl. Ind., Ed. C.B.C , 52, 53 ; Voigt, Hort. Sub. Cal., 299 ; U. C. Dutt, Mat. Med. Hind., 244, 295 ; McCann, Dyes & Tans, Beng., 136 ; Wardle, Dye Rept., 13, 32.*

Habitat.—A native of the Moluccas, cultivated in India for the sake of its fruit, the *Chaba* of Indian medicine.

DYE.
Wood.
798
Root.
799

Dye.—The WOOD and ROOT are used in Bengal for dyeing; by itself *chaikath* gives a pale brown on cotton, mixed with *bakam* (Cæsalpinia Sappan) it communicates a brownish-red colour to the same fabric. Specimens examined by Mr. Wardle were reported on as yielding a small amount of colouring matter readily given up to boiling water. The colours produced were, on corah silk, light drab to light-brown; on tussur, eria silk and wool, a fawny-drab; and on cotton a light-grey. Mr. Wardle states that the amount of colouring matter is not sufficient to bring the wood into use as a dye in Europe.

		PIPER
Cubebs.	*(J. Murray.)*	**Cubeba.**

Medicine.—The FRUIT is considered stimulant, anti-catarrhal, and carminative, and is much employed as an adjunct to medicines for coughs, cold, and hoarseness (*U. C. Dutt*). **Taleef Shereef** mentions its use in hæmorrhoids.

MEDICINE.
Fruit.
800

Piper Cubeba, *Linn. f. ; Kew Bulletin, Dec. 1887, 2 ; Baillon,*
    CUBEBS.   [*Histoire des Plantes, III., 508.*

801

    Syn.— CUBEBA OFFICINALIS, *Miq.*
    Vern.—*Kabáb-chini, val-milaku,* HIND.; *Kábáb-chini,* BENG. ; *Timmue,*
    NEPAL ; *Luit-mars,* KASHMIR ; *Kabáb-chini,* BOMB. ; *Kababa-chini,*
    himsi-mire, chinnkabale, kankola, MAR.; *Kabáb-chini, tada-miri, chinn-*
    kabale, GUZ. ; *Kabáb-chini, dumki mirchi,* DEC. ; *Val-mellaghu, vál-*
    milaku, TAM. ; *Chalavamiriyálu, tóka-miriyálu, sinban-karawa,* TEL. ;
    Bála-menasu, KAN. ; *Komunkus, vál-mulaka, lada barekor,* MALAY. ;
    Sinban-karawa, BURM. ; *Walgumdris, válmolagu, val-molavu, wal-*
    gummeris, SING. ; *Sugandha-muricha,* SANS. ; *Kabábah,* ARAB. ; *Kiba-*
    beh, kabáb-chini, PERS.
    References.—*Roxb., Fl. Ind., Ed. C.B.C.,* 53 ; *Pharm. Ind.,* 206 ; *Fluck.*
    & Hanb., Pharmacog., 584 ; *Irvine, Mat. Med. Patna,* 20 ; *Moodeen*
    Sheriff, Supp. Pharm. Ind., 121 ; *Sakharam Arjun, Cat. Bomb. Drugs,*
    129 ; Murray, Pl. & Drugs, Sind, 106 ; *Dymock, Mat. Med. W. Ind.,*
    2nd Ed., 724 ; *Year-Book Pharm.,* 1878, 362 ; 1879, 466 ; *Med. Top.,*
    Ajm., 130 ; *Trans., Med. & Phys. Soc., Bomb., No.* 12, 173 ; *Birdwood,*
    Bomb. Prod., 80 ; *Kew Bulletin,* 1887, *Dec. 1st ; Kew Off. Guide to the*
    Mus. of Ec. Bot., 109 ; *Butler, Top. & Statis., Oudh & Sultanpur,* 38 ;
    Linschoten, Voyage to East Indies (Ed. Burnell, Tiele & Yule), II.,
    130 ; Tropical Agriculturist, May 1888, 781 ; *Indian Agriculturist, 4th*
    Aug. 1888 ; *Balfour, Cyclop. Ind., III.,* 223.

    Habitat.—A native of Java and the Moluccas, cultivated to a small extent in India and the fruit imported.

    Gum-resin.—It yields an unimportant gum-resin.

GUM-RESIN.
802
OIL.
Fruit.
803

    Oil.—Four to thirteen per cent. of a thick, colourless essential oil, with a faint aromatic odour and a warm flavour of camphor and peppermint, is obtained from the FRUIT. It is a mixture of an oil $C_{10} H_{16}$ with two others of the formula $C_{15} H_{24}$. On cooling it deposits camphor of cubebs, $C_{30} H_{48} + 2OH_2$ (*Pharmacographia*).

    Medicine.—The FRUIT, commonly known as CUBEBS, has been known and used in European medicine from the middle ages. It is mentioned as a production of Java by **Marco Polo** and later by **Linschoten**, the latter of whom writes, "The Javers hold it in so great estimation, that they sell it not before it is sodden, because the strangers that buy it should not plant it." The action of cubebs on the mucous membrane of the genito-urinary tract was known to the old Arabian physicians, but the drug appears to have been used chiefly as a carminative spice up to the present century. **Linschoten** describes it as a powerful aphrodisiac, but makes no mention of its value in gonorrhœa, &c., dwelling at length on its power as a carminative and dispeller of "phlegm."

MEDICINE.
Fruit.
804

    Cubebs was a favourite condiment in Europe in ancient times, but was displaced by black pepper, and, according to **Flückiger & Hanbury**, quoting **Crawford**, its importation into Europe, which had long been dis-continued, recommenced in 1815, in consequence of its medicinal virtues having been brought to the knowledge of English Medical Officers, serving in Java, by their native servants. The majority of Indian *hokíms* regard it as stimulant, resolvent, and diuretic, and prescribe it in the same manner as black pepper, but its special use in affections of the genito-urinary sys-tem is now becoming well known throughout the country. Thus **Dymock** states that in Bombay, which is supplied by Singapore, a good demand for the drug exists, and that its consumption in native practice appears to be increasing. The virtues of the drug are well known, and since the

17   **P. 804**

| PIPER longum. | Long Pepper. |

MEDICINE

product is not, strictly speaking, an Indian one, they need not be further discussed in this work.

SPECIAL OPINIONS.—§ " Is used as an expectorant and is believed to have the power of producing tension of the vocal cords and of clearing the throat of tenacious mucus. It is much used by singers in this country" (*Civil Surgeon J. H. Thornton, B.A., M.B., Monghyr*). "The fruit is kept in the mouth, for stomatitis" (*Surgeon-Major Robb, Ahmedabad*). "Used by singers to remove hoarseness of voice" (*Surgeon R. Gray, Lahore*). " Procurable of good quality in all bazárs. The powder is best taken in milk, the oil in mucilage " (*Civil Surgeon C. M. Russell, Sarun*). " *Kabáb-chíní* and *cubebs* sold in the bazárs differ very materially from each other in their physical properties. The former looks quite like the heads of cloves with a small thin appendix as a tail. It is a very effectual remedy in dysentery if given combined with opium in pills, but Native Doctors invariably prescribe along with this treatment a curry composed entirely of green gram and raw plantains with rice, cooked according to native taste. The medicine proves efficacious only when this diet is prescribed " (*Surgeon-Major D. R. Thomson, M.D., C.I.E., Madras*).

805

Piper longum, *Linn. ; Fl. Br. Ind., V., 83 ; Wight, Ic., t. 1928.*

LONG PEPPER.

Syn.—CHAVICA ROXBURGHII, *Miq.* ; C. SARMENTOSA, *Miq.* ; PIPER SARMENTOSUM, *Wall* ; P. LATIFOLIUM, *Hunter.*

Vern.—*Pipulmúl, pipli, pipal, gas pipal,* ROOT=*pipla-múl, pipal-ki-jer,* HIND. ; *Pipul, pipli,* ROOT=*pipla-múl, pipla-mor,* BENG. ; *Ralli,* SANTAL ; *Pipla mol, popal-pipal, pipal, maghs-pipal, filfil darás, darfilfil, pipla múl,* PB. ; *Filfildray,* SIND ; *Pipli, bangáli-pim-pali,* ROOT= *pipla-múl,* BOMB. ; *Pimpli,* MAR. ; *Pipli, pipér,* GUZ. ; *Pipulmul, pipplie,* DEC. ; *Tippili, pipili, pippallu,* ROOT=*tippilimúlam,* TAM. ; *Pippali katte, pipili,* TEL. ; *Yippali,* KAN. ; *Lada, mulagu, cutta terpali, chabai, jawa, tippili,* MALAY. ; *Peik-khyen, pesining-oun, perk chin,* BURM. ; *Tippili,* SING. ; *Pippali, kaná, krishná, pippalu, úpukúlya, videhi, magudhi, chupula, kuna, ushuna, pippuli, shoundi, kola,* SANS. ; *Dár-filfil,* ARAB. ; *Filfildray, pipal, maghs-pipal, pilpil, filfil-i-darás,* PERS.

References.—*Roxb., Fl. Ind., Ed. C.B.C., 52, 53 ; Voigt, Hort. Sub. Cal., 299 ; Dals. & Gibs., Bomb. Fl. Suppl., 84 ; Rev. A. Campbell, Rept. Econ. Pl., Chutia Nagpur, No. 8191 ; Graham, Cat. Bomb. Pl., 199 ; Mason, Burma & Its People, 494, 778 ; Sir W. Elliot, Fl. Andhr., 153 ; Pharm. Ind., 208 ; Fluck. & Hanb., Pharmacog., 582 ; U. S. Dispens., 15th Ed., 1120 ; Fleming, Med. Pl. & Drugs (Asiatic Reser., XI.), 174 ; Ainslie, Mat. Ind., I., 309 ; O'Shaughnessy, Beng. Dispens., 574 ; Beng. Pharm., 89 ; Irvine, Mat. Med. Patna, 82 ; Moodeen Sheriff, Supp. Pharm. Ind., 98 ; U. C. Dutt, Mat. Med. Hind., 243, 313 ; K. L. De, Indig. Drugs, Ind., 90 ; Murray, Pl. & Drugs, Sind, 105 ; Bent. & Trim., Med. Pl., III., 244 ; Dymock, Mat. Med. W. Ind., 2nd Ed., 721 ; Official Corresp. on Proposed New Pharm. Ind., 238 ; Year-Book Pharm., 1874, 301, 324 ; 1880, 250 ; Macleod, Med. Top., Bishnath, 16 ; Baden Powell, Pb. Pr., 376 ; Drury, U. Pl. Ind., 130; Cat., Raw Prods., Col. & Ind. Exhib., Nos. 94, 154 ; Kew Off. Guide to the Mus. of Ec. Bot., 109 ; Linschoten, Voyage to East Indies (Ed. Burnell, Tiele & Yule), II., 73 ; Hove, Tour in Bombay, 119 ; Settlement Report :—N.-W. P., Kumáon, App., 34 ; Agri.-Horti. Soc., Ind.:—Trans., II., 200 ; III., 60, 67 ; Journals (Old Series), VI., 46 ; X., 43 ; Balfour, Cyclop. Ind., III., 223.*

Habitat.—A perennial shrub, native of the hotter parts of India from East Nepál, eastwards to Assam, the Khásia Mountains, and Bengal, westwards to Bombay, and southwards to Travancore, Ceylon, and Malacca. It is cultivated extensively in Bengal and the Southern Presidency on account of its fruit. The plant flowers in August and September, and the fruit matures in January.

| Long Pepper. (*J. Murray*) | PIPER longum. |

CULTIVATION.—No recent account of the cultivation of long pepper appears to exist, the following description of the method pursued in Bengal is, therefore, compiled from the descriptions of Roxburgh and of Major Bruce (*Agri.-Horti. Soc. of Ind. Trans, III., 60*).

The plant is propagated by suckers, and requires a rich, high and dry soil. The suckers are transplanted soon after the periodical rains set in, and are placed at a distance of five feet between each, so that a *bígah* contains 196 plants. Each *bígah* is said to produce two maunds of pepper the first year, four maunds the second, and six maunds the third, after which the plant becomes annually less and less productive; the roots are therefore grubbed up dried, and sold, and fresh roots or young shoots are set in their place, the soil requiring no treatment, save a slight covering of manure. The plants require no irrigation, but at the commencement of the hot season the roots are carefully covered with straw to preserve them against the heat of the sun. Radishes, barley or *brinjals* are usually cultivated in the space between the plants. The fruit ripens and is picked annually in January. It is gathered when green or unripe, and preserved by drying in the sun. Roxburgh states that in his time large quantities of the fruit and roots were exported to Bombay, the former for culinary, the latter for medicinal, purposes. Dymock writes that Bengal is still the chief source of true long pepper in Bombay, and that the fruit fetches R9, the root R7 to 7½ per maund of 41℔. The root is, however, also obtained from Mirzapore and Malwa, and appears to be more valuable when brought from these localities, that from the former fetching R10 to R40, that from the latter R50 per maund of 41℔.

Medicine.—The dried unripe FRUIT and the ROOT have long been used in medicine. "They are considered heating, stimulant, carminative, alterative, laxative, and useful in cough, hoarseness, asthma, dyspepsia, paralysis, &c. Old long pepper is said to be more efficacious than the fresh article. In the form of *trikatu*, or the three acrids, it is much used as an aromatic adjunct in compound prescriptions. Powdered long pepper, administered with honey, is said to relieve cough, asthma, hoarseness, hiccough, and sleeplessness. A mixture of long pepper, long pepper root, black pepper and ginger in equal parts, is prescribed by several writers as a useful combination for catarrh and hoarseness.

"As an alterative tonic long pepper is recommended for use in a peculiar manner. An infusion of three long peppers is to be taken with honey on the first day, then for ten successive days the dose is to be increased by three peppers every day, so that on the tenth day the patient will take thirty at one dose. Then the dose is to be gradually reduced by three daily, and finally the medicine is to be omitted. Thus administered it is said to act as a valuable alterative tonic in paraplegia, chronic cough, enlargements of the spleen and other abdominal viscera, &c." The foregoing passages are translations by U. C. Dutt from the Bhava prakasha and Chakradatta. The same authority states that long pepper and black pepper enter into the composition of several irritating snuffs, for administration in coma and drowsiness, and also describes a rubefacient oil containing long pepper and ginger, which is prescribed by Chakradatta for application in sciatica and paraplegia.

Dymock states that Muhammadan writers describe the drug as a resolvent of cold humours; they say that it removes obstructions of the liver and spleen, and promotes digestion by its tonic properties, and is also aphrodisiac, diuretic, and emmenagogue. Both the fruit and the root are much prescribed in palsy, gout, lumbago, and other diseases of a similar nature. A collyrium of long pepper is recommended for night blindness; made into a liniment it is applied to the bites of venomous

PIPER nigrum.	Cultivation of Pepper

MEDICINE.

reptiles (*Mat. Med. W. Ind.*). Ainslie writes, "The *Vytians* on the Coromandel coast prescribe it in infusion, mixed with a little honey, in catarrhal affections, when the chest is loaded with phlegm." " The root is a favourite medicine with the Hindus; it possesses the virtues of the berry but in a weaker degree; and is prescribed by them in cases of palsy, tetanus, and apoplexy." The plant has obtained a place in the secondary list of the *Pharmacopœia of India*, where it is stated that the fruit possesses stimulant and carminative properties similar to, but more powerful than, those of black pepper. Dr. Herklots is said to report favourably on a combination of this drug with black pepper, ginger, and arak in *Beri-beri*. The root, in infusion, is prescribed in Travancore after parturition, with the view of causing the expulsion of the placenta.

In the Panjáb the fruit is considered vermifuge, in addition to the other properties generally ascribed to it by Muhammadans. Campbell states that in Chutia Nagpur a preparation of the root is given to women who have "cough complicated by some menstrual complaint," and that it is also eaten raw to allay thrist in fever and employed as an application to reduce swelling. According to Dymock the "roasted aments " are, in the Konkan, beaten up with honey and given in rheumatism.

Chemistry.
809

CHEMICAL COMPOSITION.—The active constituents of long pepper appear to be the same as those of black pepper, and consist essentially of a *volatile oil*, a *resin*, and *piperin*. The two former are contained entirely in the pericarp.

SPECIAL OPINIONS.—§ "Prescribed for children in lung affection as a stimulating expectorant " (*Civil Surgeon John McConaghey, M.D., Shaja-hanpore*). "Used in charitable hospitals as a stimulating expectorant " (*Civil Surgeon S. M. Shircore, Moorshedabad*). "Have given it in asthma and spleen without any marked effect " (*Surgeon D. Picachy, Purneah*). " The pepper when triturated for a few days is considered a very powerful tonic and stimulant, and the longer it is rubbed in a mortar the more powerful is considered its effect. Long pepper is a very common in-gredient of most native prescriptions " (*Surgeon-Major Robb, Civil Sur-geon, Ahmedabad*). "Is stomachic and carminative, and said to be one of the best medicines for colic. Is largely used by Native physicians in cases of fevers, colic, indigestion and bronchitis, and is given to women after the time of delivery to check hœmorrhage and ward off fever " (*Civil Surgeon J. H. Thornton, B.A, M.B., Monghyr*). "Used as a confection with honey in catarrhal affections of children " (*Surgeon E. Borill, Motihari, Champarun*). "Found useful in coughs and colds. Invariably used by native midwives in the puerperal state, along with other durgs, to restore the womb to its normal state " (*Assistant Surgeon S. C. Bhattacherji, Chanda*).

FOOD.
Root.
810

Food.—LONG PEPPER, though considered inferior to black, is consi-derably used as a spice. The ROOT is said by Campbell to be used in Chutia Nagpur to ferment rice-beer.

811

Piper nigrum, *Linn.; Fl. Br. Ind., V., 90 ; Wight, Ic., t. 1935.*

Syn.—PIPER TRIOICUM, *Roxb.;* P NIGRUM, *var.* TRIOICUM, *Cas. DC.;* P MALABARENSE & BACCATUM (the Indian synonyms only), *Cas. DC.;* MULDERA MULTINERVIS, & WIGHTIANA, *Miq.*

Vern.—*Gúlmirch, filfilgird, mirch, káli-mirch, habush, choca mirch,* white form = *saféd mirch,* HIND.; *Vellajung, murichung, kolukung, muricha, kálá-morich, gól-morich,* BENG.; *Spót,* BHOTE; *Marts,* KASHMIR; *Gol-mirich,* PB.; *Dáru-garm, daur-garm, march,* AFG.; *Gúlmirien,* SIND; *Miri, kala-miri,* white form = *saféd-miri, pándhári-miri,* BOMB.; *Kali mirch, miré,* MAR.; *Kálámari, kálo-mirich, miri,* GUZ.; *Choca, kali mirchingay, káli-mirchi,* DEC.; *Milágu,* TAM.; *Miryála tige, miriyálu* TEL.; *Menasu, kare menasu, molúvukodi, mirialu,* KAN.; *Lada, kuru mulaka,* MALAY.; *Sa yo mai, nya-yoke koung, náyukon,* BURM.; *Gam-mirris-wil, gam-miris, kalu-miris,* SING.; *Maricha, ushana, hapushá,* SANS.; *Filfiluswud,* ARAB.; *Filfile-siyáh, pilpil, filfile-asvad, filfile-gird,* PERS.

References.—*Roxb., Fl. Ind., Ed. C.B.C.,* 51, 53; *Thwaites, En. Ceyl. Pl., Suppl.,* 84; *Mason, Burma & Its People,* 494, 777; *Sir W. Elliot, Fl. Andhr.,* 116; *Rheede, Hort. Mal., VII.,* 23, t. 12; *Burmann, Fl. Ind.,* 13; *Pharm. Ind.,* 205; *British Pharm.,* 241; *Fluck. & Hanb., Pharma-cog.,* 576; *U. S. Dispens., 15th Ed.,* 1119; *Fleming, Med. Pl. & Drugs (Asiatic Reser., XI.),* 174; *Ainslie, Mat. Ind., I.,* 302, 621; *II.,* 385; *O'Shaughnessy, Beng. Dispens.,* 571; *Moodeen Sheriff, Supp. Pharm. Ind.,* 200; *U. C. Dutt, Mat. Med. Hind.,* 241, 299, 309; *Sakharam Arjun, Cat. Bomb. Drugs,* 129; *K. L. De, Indig. Drugs, Ind.,* 90; *Murray, Pl. & Drugs, Sind.,* 105; *Bent. & Trim., Med. Pl., t.* 245; *Dy-mock, Mat. Med. W. Ind., 2nd Ed.,* 718; *Official Corresp. on Proposed New Pharm. Ind,* 234, 292; *Year-Book Pharm.,* 1880, 506; *Med. Top., Ajm.,* 140; *Trans. Med. and Phys. Soc., Bombay (New Series), No.* 4, 153. *No.* 12, 173; *Birdwood, Bomb. Prod.,* 81, 230; *Baden Powell, Pb. Pr.,* 376; *Drury, U. Pl. Ind.,* 131; *Useful Pl. Bomb. (XXV., Bomb. Gas.),* 173; *Royle, Prod. Res.,* 53; *Christy, New Com. Pl., VII.,* 18; *Lin-schoten, Voyage to East Indies (Ed. Burnell, Tiele & Yule), II.,* 73; *Milburn, Oriental Commerce (1825),* 154-156; *Man. Madras Adm., II.,* 69; *Bomb. Man., Rev. Accts.,* 103; *Rept. Director, Land Rec. & Agri., 1889, App. x., xi.; Gazetteer:—Bombay, XV., Pt. ii.,* 10; *Agri.-Horti. Soc. Ind.:—Trans., II.,* 60, 67; *VII.,* 88; *VIII.,* 247-248, *Pro.,* 363, 377; *Journals (Old Series), XIII., sel.,* 58; *Spons, Encycl.,* 1425; *Balfour, Cyclop. Ind., III,* 223; *Smith, Ec. Dic.,* 317.

Habitat.—A climber, which is usually diœcious though the female often bears two anthers, and the male a pistillode. It is found wild in the forests of the Circars, doubtfully indigenous in those of Assam and Malabar, and is cultivated in the hot damp localities of Southern India.

CULTIVATION —Pepper was one of the earliest articles of Indo-Euro-pean trade, and has been extensively cultivated on the Western Coast of Southern India for many centuries. Linschoten in 1598 gives perhaps the first detailed account of the method of growing and gathering the spice, stating that "much pepper" was there to be found along the whole coast of Malabar. Though cultivated from remote times in Sumatra, the Straits, Siam, and the Malay Peninsula generally, Malabar has always been considered to produce the best pepper. | CULTIVATION 812

MADRAS.—The method of cultivation followed is very simple. Cuttings are put down before the commencement of the rains in June, in rich soil, not subject to excessive accumulation of moisture. Places at the foot of trees which have rough or prickly bark are chosen, such as the jack tree, **Erythrina indica**, the cashew nut, mango tree, and others of a similar description. The vines shoot up, and if allowed, will grow to a height of 20 to 30 feet, but are generally kept down by cutting and pruning. All suckers should be removed, and the ground around should be kept clear of weeds. In three years the vine begins to bear, producing long clusters of 20 to 50 berries abundantly from all its branches. From the third to the seventh year the plant improves, after that time it remains in good condition for three to four years, and then deteriorates for about the same period, until it is no longer worth keeping. It is then cut down and new shoots are planted. The fruit is gathered as soon as the berries at the base of the spike begin to change colour from green to red, since if allowed to ripen fully it becomes less pungent, and easily falls off. The day after they are gathered the berries are separated from the stalk by hand-rub-bing, and picked clean. They are then dried by exposure to the sun, or more frequently by the heat of a gentle fire. During this process hand-rubbing is occasionally performed to separate any remaining portions of stalk. "White pepper" is prepared by divesting the ripe berry of its | Madras. 813

PIPER nigrum.	Pepper Cultivation

skin by maceration in water, after which it is rubbed and finally bleached in the sun. It is occasionally bleached still further by means of chlorine. It is twice as expensive as black pepper, but is in little demand.

Towards the end of last century **Dr. Roxburgh** discovered the plant growing wild in the hills north of Samulcottah, and commenced a large plantation in that neighbourhood. In 1789 this garden contained about 40,000 or 50,000 vines occupying about 50 acres of land. Considerable success was attained, about 1,000 vines yielding from 500 to 1,000℔ of berries. Many of the plants, however, were found to bear only male or female flowers, and the latter if not associated with a hermaphrodite flower on the same ament, were found to produce pepper less pungent than the ordinary spice. The attempt was abandoned in a few years. No recent statistics are available regarding the total yield of the Malabar Coast. Milburn in 1825 estimated it at 30,000 *piculs* (1 *picul*=133℔). Drury estimates the average annual production of the hilly districts of Travancore to be 5,000 candies.

BOMBAY.—The only district in which pepper is grown to any extent in this Presidency is Kánara, in which 1,096 acres were under the crop in 1888-89. The pepper-vine cuttings are planted in August in betle-palm gardens when the trees are thirteen years old. Four cuttings, about 2 feet 3 inches long, are made for every betle-palm. One end of each is set four to six inches deep and the other is trained on to the palm. The vine requires no further care except tying its branches once a year in May. It bears in six or seven years and lives about twenty-five, so that one betle-palm outlives three or four sets of vines. The pepper is picked from ladders in March and April as soon as the berries are full grown but before they are ripe. One man gathers and cures at the most about three pounds a day. The bunches are piled in a heap under shelter and kept for three days. They are then rubbed with the foot, to separate the stalks and all foreign matter, after which the pepper is fit for sale.

Three different forms of vine are recognised, called respectively, *kari malisaru, sambar,* and *arsina murtiga,* which differ, not in quality but in yield. Of these the first is the best bearer, each vine yielding as much as 3 pounds a year, but it is not so easy to grow as the others, since it thrives only on strong red soil. The other two grow well in the ordinary light-coloured soil known as *arsina munnu,* but *sambar* yields only about ⅔ of a pound, and *arsina murtiga* 1⅛℔ annually. The average yearly yield of each pepper vine, taking all forms into consideration, is said to be about 1 1/10℔. The acreage outturn averages 280℔ in a first class 140℔ in a second class, and 56℔ in a third class garden. The selling price is about 3½d. a pound (R4 the *man* of 28℔).

In Kánara the vine is not only cultivated, but also grows spontaneously in "pepper forests," or *menasu káns.* These are occasionally tended for the sake of the fruit. The branches of the vines should be stripped of all other climbing plants, all the bushes in the forest should be cut down every third year, and every fifth year the side branches of the trees ought to be lopped, as the vine clings best round straight slender stems. Where the supporting trees are too far apart, a branch or cutting should be planted, and if no pepper vine be near, a shoot or two should be set in the earth near the young tree. When thus cared for a vine is said to live about ten years. When an old vine dies a young shoot must be trained to take its place. Nothing is done besides attending to the training and distribution of the vines. When a tree withers its branches are lopped off, and a circle round the foot is stripped of bark. Under this treatment the dead tree gradually rots, and does not fall, which is necessary since the fall might produce great havoc amongst surrounding vines. The rotten wood also forms a valuable manure for the pepper.

Owing to want of tillage the yield is inferior in quantity and quality to that of pepper grown in gardens, one vine seldom yielding more than half the quantity produced by a garden vine. A man in one day can gather the produce of about twenty trees, or rather more than 12℔, at the same time training and tying up the plants where necessary (*Kánara Gazetteer, pt. ii., 10-11*).

Medicine.—Sanskrit authors describe BLACK PEPPER as acrid, pungent, hot, dry, carminative, and useful in intermittent fever, hæmorrhoids, and dyspepsia. It is generally prescribed in combination with long pepper and ginger, under the name of *trikatu*, or the three acrids. Dutt states that very few Sanskrit prescriptions are free from these three, which, however, appear frequently to be added without reason, and sometimes only for the sake of rhyme. Externally it is used as a rubefacient in alopecia and skin-diseases (*Mat. Med. Hind.*). Dymock informs us that Muhammadan writers describe the spice as deobstruent, resolvent, and alexipharmic. It is prescribed by them internally as a nervine tonic, and applied externally in paralytic affections, while in toothache it is used as a mouth wash. It is esteemed as a tonic and digestive, and is believed to be diuretic, emmenagogue, and a good stimulant in cases of bites from venomous reptiles.

In modern Indian medicine the properties assigned to the drug by Sanskrit, Persian, and Arabic writers are still believed in. It is much employed as an aromatic stimulant in cholera, weakness following fevers, vertigo, coma, &c, &c.; as a stomachic in dyspepsia and flatulence; as an antiperiodic in malarial fever, and as an alterative in paraplegia and arthritic diseases. Externally it is valued for its rubefacient properties, and as a local application for relaxed sore-throat, piles, and some skin diseases. According to Dymock abortive pepper-corns, known as *poklimiri*, have long been known and used by the Hindus. Garcia d'Orta notices them under the name of "Canarese Pepper," and observes that they are valued by the Natives as a medicine to purge the brain of phlegm, to relieve toothache, and as a remedy for cholera.

Pepper was one of the earliest spices used in Europe, both as a condiment and medicine. In the fourth century B C., Theophrastus noticed two kinds of pepper (πέπερι), probably the Black Pepper and Long Pepper of modern times. Dioscorides knew pepper to be a production of India and was acquainted with White Pepper (λευχὸν πέπερι). Pliny, Arrian, and others also mention it (*Pharmacographia*). Many writers of the early Middle Ages refer to the spice, all describing it as coming from Male or Malabar. Perhaps the earliest to accurately describe the extent of its cultivation and uses was Linschoten, whose friend and commentator Paludanus enters into a long dissertation on its medicinal virtues: "It warmeth the mawe," he writes, "and consumeth the cold slymenes thereof, to ease the payne in the mawe which proceedeth of rawnesse and winde, it is good to eat fyve pepper cornes everie morning. He that hath a bad or thick sight, let him use pepper cornes, with annis, fennel seed and cloves, for thereby the mystinesse of the eyes which darken the sight is cleared and driven away."

But in modern European medicine it is very little used, being rarely prescribed except indirectly as an ingredient of some compound preparation.

CHEMICAL COMPOSITION.—Pepper contains a resin, a small quantity of fatty oil and an essential oil, to the first of which its sharp pungent taste is due. The essential oil has the odour, but little of the taste, of the

PIPER nigrum.	History of the Trade

CHEMISTRY.

spice. The drug yields from 1·6 to 2·2 per cent. of this volatile oil, which agrees with oil of turpentine in composition as well as in specific gravity and boiling point. The most interesting constituent of pepper is *piperin*, of which the spice contains from 2 to 8 per cent. Like morphine it agrees with the formula $C_{17} H_{19} N O_3$, though without alkaline reaction it combines with hydrochloric acid in the presence of mercuric and other metallic chlorides. **Anderson**, in 1850, resolved *piperin* into *piperic acid*, $C_{12} H_{10} O_4$, and *piperidine*, $C_5 H_{11} N$. The latter is a liquid colourless alkaloid, boiling at 106°C, having the odour of pepper and ammonia, and directly yielding crystallizable salts (*Flückiger & Hanbury, Pharmacographia*).

SPECIAL OPINIONS.—§ " Made into infusion and given in colic and cholera" (*Surgeon H. D. Masani, 30th Bo. N. I., Karáchi*). "Stimulant, carminative, antiperiodic; dose 5 to 20 grains; used in intermittent fever, debility, hæmorrhoids, prolapsus ani, gonorrhœa, and cholera" (*Chúnna Lall, Hospital Assistant, Jubbulpore*). " It is used extensively as an external application to inflammation" (*Civil Surgeon J. McConaghey, M.D., Shajahanpore*). " A very strong decoction of pepper given in that stage of cholera in which all vomiting and purging cease and the abdomen becomes tympanitic affords great relief" (*Surgeon-Major D. R. Thompson, M.D., C.I.E., Madras*). " Valuable carminative and stomachic. Highly useful in indigestion and dyspepsia. Forms a useful ingredient in tooth powder" (*Assistant Surgeon S. C. Bhattacharji, Chanda*). " Black pepper powder mixed with *ghi* is said to be a useful application in cases of urticaria" (*Surgeon Joseph Parker, M.D., Poona*). " Used as antiperiodic, also in hæmorrhoids and cholera" (*Assistant Surgeon Nehal Sing, Saharanpore*). " Is a useful application to boils and pimples. It is carminative, antacid, and stomachic, and is used in dyspepsia, diarrhœa, and indigestion" (*Brigade Surgeon J. H. Thornton, B.A., M.B., Monghyr*). " Aromatic, stimulant, and stomachic. Combined with calumba and bismuth, it is useful in dyspepsia; also with asafœtida and camphor in the flatulency of dyspepsia. Largely used by me in preparation of cholera pills for distribution during epidemic outbreaks of the disease" (*Civil Surgeon S. M. Shircore, Murshedabad*). " Locally it is applied to boils, &c., in the form of a thin paste as a resolvent" (*Assistant Surgeon T. N. Ghose, Meerut*).

FOOD.
817

Food.—The use of black and white pepper as condiments is too well known to require any remark in this work.

TRADE.
818

TRADE.—HISTORY.—The trade in Pepper is perhaps the oldest, and during the Middle Ages was certainly much the most important branch of commerce between Europe and the East. The history of its development from earliest times is well brought out in the interesting and exhaustive account given in the *Pharmacographia*. According to **Fluckiger & Hanbury**, the learned authors of that work, the spice was well known at least as early as the fourth century B C. **Arrian**, the author of *Periplus of the Erythrean Sea*, written about A.D. 618, states that pepper was then imported from Baraké the shipping place of Nelkunda, localities which have been identified with points on the Malabar Coast. During the Middle Ages it was the most esteemed and important of all spices, indeed it formed "the very symbol of the spice trade, to which Venice, Genoa, and the commercial cities of Central Europe were indebted for a large part of their wealth "

Tribute was levied on pepper, the ransom demanded from Rome in 408 A.D., by Alaric, King of the Goths, included 3,000 pounds of pepper, and after the conquest of Cæsarea in 1101 A.D., by the Genoese each soldier received two pounds of pepper as part of his booty. Pepper rents were also not uncommon, by means of which the wealthier classes secured from

P. 818

TRADE;

their tenants a supply of their favourite condiment, at a time when the market was by no means certain.

The earliest reference to a trade in pepper in England is in the statutes of Ethelred, A.D. 978—1016, where it is enacted that traders bringing their ships to Billingsgate should pay at Christmas and Easter, with other tribute, 10 pounds of pepper. The "Pepperers" formed a fraternity mentioned as existing in London in the reign of Henry II., and subsequently became incorporated as the Grocers' Company. The price of the spice during the Middle Ages was always very high, owing to the large tax imposed on it in its passage through Egypt. Thus in England in the thirteenth and fourteenth centuries it averaged 1s per pound, equivalent to about 8s. of our present money. This high price, amongst other reasons, incited the Portuguese to seek for a sea-passage to India, after the discovery of which (1498 A.D.) the price fell considerably, and the cultivation was extended to the Western Islands of the Malay Archipelago. The trade then departed from the ports of Central Europe, and became practically a monopoly of Portugal till the eighteenth century (*Flückiger & Hanbury, Pharmacographia*) During this period of Portuguese supremacy in the trade we find Linschoten writing that the "King of Portingal" contracted with middlemen in each of his "Forts" on the Coast of Malabar for an annual supply of thirty thousand "quintales" of pepper, and bound himself to send five ships every year to export that amount. All risk was held by the middlemen or "farmers," who obtained the price of twelve ducats a quintal for all pepper landed in Portugal. As a compensation for this risk the middlemen had "great and strong priviledges: first that no man of what estate or condition so ever he bee, either Portingall, or of any place in India, may deale or trade in pepper, but they, upon paine of death, which is very sharply looked unto." And "although the pepper were for the King's own person, yet must the farmer's pepper be first laden, to whome the Viceroy and other officers and Captaines of India, must give all assistance, helpe and favour with watching the same, and all other things whatsoever that shall by the said farmers be required for the safetie and benefite of the saide pepper."

In more recent times the Malabar Coast monopoly in the trade became gradually destroyed. More and more pepper was cultivated in, and exported from, the Malay Archipelago, and localities further east, till, as early as the beginning of this century, we find the Indian trade much smaller in proportion than that of other places. Thus Milburn in 1829 states that the produce of Sumatra was estimated at 168,000 peculs, that of the islands at the mouth of the Straits (Burtang, Linga, &c.) 12,000 peculs, the Malay Peninsula produced 28,000 peculs, the east coast of the Gulf of Siam 60,000 peculs, Borneo 20,000 peculs, and the west coast of India, only 30,000 peculs. But he remarks "the pepper of Malabar is esteemed the best."

A review of the present percentage imports into Great Britain shews a still further decrease in the relative importance of India as a source of supply. Thus, in 1872, the imports amounted to 27,576,710℔, of which the Straits Settlements supplied 25,000,000℔, and British India only 256,000℔. In 1889 the total imports from the British East Indies (including the Straits Settlements, Ceylon, &c., along with India) amounted to 28,555,394℔, of which 28,041,096℔ came from the Straits Settlements.

Internal.—It is impossible, in considering internal trade, in pepper, to separate that in the Indian and foreign articles, since they are dealt with conjointly in trade returns.

The total registered transactions by road, rail, and river are small, nearly all the Indian pepper being shipped from Madras coastwise to the large centres of demand. Thus, in 1888-89, they amounted to only 80,114 maunds, valued at R27,13,393. Of this amount Madras seaports exported

Internal.
819

PIPER nigrum.	History of the Trade

TRADE :
Internal.

5,373 maunds; Madras Presidency, 6,629; Bombay Port, 19,973 maunds; Bombay Presidency, 2,868; Calcutta, 32,149 maunds; Bengal, 3,263; and Karáchi, 8,769 maunds. The largest importers were Bengal with 36,152 maunds; the North-West Provinces and Oudh with 9,700; the Panjáb with 9,489; and Madras with 7,173. It will be seen from these figures that the trade results almost entirely from the distribution of European and Madras pepper from the large seaport towns, to the various inland provinces.

The coastwise trade in Indian pepper, which represents the first stage in this distribution, is fairly large and important. In 1889-90 the total imports into all ports amounted to 10,539,676℔, value R 39,24,141, of which 5,063,211 came from Madras District Seaports, and 1,373,699 from Travancore. The largest importing provinces were Bombay with 5,154,036℔, and Bengal with 2,402,562℔ Madras, Sind, and Burma received smaller amounts. The coastwise trade in foreign pepper is included in the general class of "spices," and cannot be given separately.

External.
820

External.—Notwithstanding the fact brought out in the previous *resumé* of the history of Indian trade in pepper, namely, that this country, once the centre of supply for the whole world, has been to a very large extent supplanted by the great development of Malayan trade, it is encouraging to note a considerable and steady increase in the exports from India during the past twenty years; and not only has the quantity exported increased, but the proportionate value has risen to a marked extent. The averages during the past fifteen years have been as follows :—

Quinquennial period.	℔	R
1875-76 to 1879-80	5,420,963	9,77,924
1880-81 to 1884-85	5,961,098	12,20,083
1885-86 to 1889-90	7,652,334	28,91,805

In 1889-90, the total amounted to 8,249,100℔, of which Madras exported 6,154,380, Bombay, 2,078,131, Sind, 9,244, and Bengal, 7.345℔., probably all received from Malabar by coastwise traffic. The distribution of these exports during the past three years may be shown by the following table :—

Importing Countries.	1887-88.	1888-89.	1889-90.
	℔	℔	℔
United Kingdom	830,361	898,309	869,827
Austria	...	11,200	33,600
Belgium	39,200	78,400	44,800
France	3,763,914	4,692,830	4,049,238
Germany	75,824	229,376	377,948
Holland	...	...	33,600
Greece	5,488	4,480	...
Italy	509,600	305,172	450,900
Malta	...	11,592	...
Turkey in Europe . . .	6,944	...	...
Russia	10,528	...	67,200
Abyssinia	5,320	3,024	12,655
East Coast of Africa {Mozambique .	7,112	6,188	11,151
{Zanzibar .	73,885	59,521	99,365
Egypt	313,292	223,083	271,426
Mauritius	19,712	41,552	32,504
Natal	5,656	8,204	18,788

| in Pepper. | | | (*J. Murray.*) | | PIPER
sylvaticum. |

Importing Countries.	1887-88.	1888-89.	1889-90.
	℔	℔	℔
United States	63,000	14,860	...
Aden	321,972	223,649	287,196
Arabia	352,526	280,601	269,476
Ceylon	36,774	19,030	76,348
Mekran and Sonmiáni . . .	3,572	3,716	5,026
Persia	275,470	227,139	717,713
Turkey in Asia	397,018	320,969	518,129
Australia	23,800	...	...
Other countries	5,268	3,112	2,210
TOTAL .	7,146,236	7,666,007	8,249,100

The imports from foreign countries are, on the contrary, steadily decreasing. During the past fifteen years the averages have been as follows :—

Quinquennial period.	℔	R
1875-76 to 1879-80	7,322.176	13,81,545
1880-81 to 1884-85	4,705,934	11,03,452
1885-86 to 1889-90	2,816,798	10,43,673

In 1889-90 the imports were, however, considerably above the average, amounting to 5,707,147℔, value R20,04,850. Of this 5,649,629℔ were received from the Straits Settlements, and unimportant quantities from Ceylon, Zanzibar, Java, and other countries. The largest importer was Bengal with 3,828,400℔, followed by Bombay with 1½ million ℔, Burma and Madras with insignificant quantities.

The re-export trade is small and unimportant, but, like the export trade, has increased to a slight extent during the past ten years. In the five years ending 1884-85 it amounted to 79,699℔, valued at R19,569, while in a similar period ending 1889-90, it attained the quantity of 119,123℔, value R47,763. In 1889-90 the re exports were 141,817℔, valued at R56.057. Of this amount 52.416℔ went to China, 20,081 to Mauritius, 17,920 to Persia, and insignificant quantities to Cape Colony, Australia, Natal, Egypt, and other countries. The largest re-exporting province is Bengal, which shipped 103,278℔ out of the total.

Piper sylvaticum, *Roxb. ; Fl. Br. Ind., V., 84 ; Wight, Ic. t. 1930.*

821

Syn.—? PIPER BETLE, *Wall ;* P. MALAMIRI, *Roxb. ;* CHAVICA SYLVATICA, *Miq.*

Vern.—*Pahari pípal,* BENG.

References.—*Roxb., Fl. Ind., Ed. C.B.C., 52 ; O'Shaughnessy, Beng. Dispens , 575 ; Irvine, Mat. Med., Patna, 85.*

Habitat.—A low creeping species, met with in the *jhíls* and low hills of Bengal, in Upper and Lower Assam, and in Burma.

Medicine.—The FRUIT is used by the Natives of Bengal as a carminative similarly to long pepper. Atkinson confuses this species with P. longum, in his *Himálayan Districts, 705,* and in the *Economic Products of the N - W. P., V., 32.*

MEDICINE.
Bruit.
822

Food.—Roxburgh states that the Natives use this PEPPER both green and ripe in their dishes.

FOOD.
Pepper.
823

Pipal Tree, see Ficus religiosa, *Linn. ;* Vol. III., 357.

PISONIA, *Linn. ; Gen. Pl., III., 9.*

[*1763-4 ;* NYCTAGINEÆ.

824

Pisonia aculeata, *Linn.; Fl. Br. Ind., IV., 711 ; Wight, Ic., t.*

Syn.—P. GEORGINA, *Wall ;* P. VILLOSA, *Poir. ;* TRAGULARIA HORRIDA, *Koen.*

Vern.—*Baghachura,* BENG. ; *Háti-ánkusá,* URIYA ; *Embúdi chettu, konki, kanki putri,* TEL.

References.—*Roxb., Fl. Ind., Ed., C.B.C.,* 312 ; *Gamble, Man. Timb.,* 302 ; *Kurz, For. Fl. Burm., II.,* 279 ; *Beddome, For. Man., clxxv., t. xxii., f. 3 ; Elliot, Fl. Andhr.,* 50, 82, 98 ; *Graham, Cat. Bomb. Pl.,* 167 ; *Gazetteers :—Mysore & Coorg, I.,* 65 ; *Orissa, II.,* 181.

Habitat.—A very common, strong, large, straggling shrub, met with in South India and the coast forests of Burma, the Andaman Islands, and Ceylon.

MEDICINE.
Bark.
825
Leaves.
826
Juice.
827
DOMESTIC.
Shrub.
828

Medicine.—The BARK and the LEAVES are used as a counter-irritant in cases of inflammation and rheumatism.

SPECIAL OPINION.—§ " The JUICE mixed with pepper and other ingredients is given to children suffering from pulmonary complaints " (*V. Ummegudien, Mettapoliam, Madras*).

Domestic.—The SHRUB makes most excellent impenetrable fences (*Roxb.*).

829

P. alba, *Spanoghe ; Fl. Br. Ind., IV., 711 ; Wight, Ic., t. 1765.*

Syn.—? P. INERMIS, *Forst. ;* ? P. MORINDÆFOLIA, *Br. ;* ? P MALABARICA, *Poir. ;* ? P. MITIS, *Linn.*

Vern.—*Chinai sálit,* BOMB.

References.—*Kurz, For. Fl., Burm., II.,* 279 ; *Dalz. & Gibs. Bomb. Fl., Suppl.,* 72 ; *Dymock, Mat. Med W. Ind.,* 2nd Ed., 656 ; *S. Arjun, Bomb. Drugs,* 220 ; *Lisboa, U. Pl. Bomb.,* 109 ; *Gazetteer, Mysore & Coorg, I.,* 65.

Habitat.—A tree from 30 to 40 feet high, found sparsely wild in the beach forests of the Andaman Islands ; cultivated to a small extent in India and Ceylon.

MEDICINE.
Leaves.
830

Medicine.—The fresh LEAVES, moistened with Eau de Cologne, are used to subdue " inflammation of an elephantoid nature " in the legs and other parts (*S. Arjun*).

TIMBER,
831

Structure of the Wood.—Lisboa states that this tree is occasionally, though rarely, cut down for the sake of its wood.

832

PISTACIA, *Linn. ; Gen. Pl., I., 419.*

A genus of trees or shrubs which comprises some six species, natives of Western Asia and the Mediterranean region, one Mexican. Several of these are found in Afghánistán and Balúchistán, but do not cross the frontier, only one being a native of India. Nearly all are of economic importance, however, and yield valuable articles of trade with India, besides being occasionally cultivated in this country. They are, therefore, worthy of brief consideration.

[DIACEÆ.

833

Pistacia integerrima, *Stewart ; Fl. Br. Ind., II., 13 ;* ANACAR-

Syn.—RHUS INTEGERRIMA, *Wall ;* R. KAKRASINGEE, *Royle.*

Vern—*Kákra,* galls=*kákrasingí,* HIND., MAR., GUZ. ; galls=*Kakra-sringi,* BENG. ; *Kakkar, drek, gurgú,* KASHMIR ; *Kangar, khangar, kakar, kakkar, khakkar, kakkrei, kákrá, kakkeran, kakrain, kakkrangche, kakla, drek, gurgú, tánhári, túngú, shné, sarawán, masna,* galls=*kakra singí,* fruit=*súmak,* PB. ; *Sarawán, shné, masna,* PUSHTU ; galls=*Kákka-tashingí,* TAM. ; galls=*Kákara-shingí,* TEL ; galls=*Dushtapuchattu,* KAN. ; galls=*Karkata sringí,* SANS.

References.—*Brandis, For. Fl., 122; Gamble, Man. Timb., 106; Stewart, Pb. Pl., 47; Aitchison, Kuram Valley Rep., Pt. I., 42; Pt. II., 156; O'Shaughnessy, Beng. Dispens., 282; Sakharam Arjun, Cat. Bomb. Drugs, 33; Dymock, Mat. Med. W. Ind., 2nd Ed., 191; Dymock, Warden & Hooper, Pharmacog. Ind., I., 374; Birdwood, Bomb. Prod., 19; Baden Powell, Pb. Pr., 589; Atkinson, Him. Dist. (X., N.-W. P. Gas.), 746; Royle, Ill. Him. Bot., 175; Settlement Report: Panjáb, Hazára, 11; Gazetteers:— Panjáb, Bannu, 23; Dera Ismail Khán, 19; Rawalpindi, 15; Hoshiárpur, 11; Hazára, 14; Gujrát, 12; Gurdáspur, 55; Sháhpur, 69; Agri.-Horti. Soc. Ind.:— Journals (Old Series), VII., 160; XIV., 14, 15; (New Series), I., 85; Agri.-Horti. Soc., Panjáb, Select. papers to 1862, Ind. 43; Ind. Forester, V., 180; VIII., 35; IX., 15, 68, 291; XI., 320; XIII., 58, 543; Balfour, Cyclop. Ind., III., 226.*

Habitat.—A tall, nearly glabrous tree, from 40 feet upwards in height, met with on the North-Western Frontier, the Peshawar valley and Salt range, and on the hot slopes of the Western Himálaya from the Indus to Kumáon, at altitudes of 1,200 to 8,000 feet.

Dye & Tan.—The hard, rugose, hollow, irregular GALLS, which form in October on the leaves and petioles, are used to a small extent for dyeing and tanning. These galls sometimes attain a very large size, and from their peculiar shape well warrant the name of *singhí* or "horns." The writer has recently seen them growing, near Simla, to a length of more than a foot, though Dymock states the average to be 1½ inches.

DYE & TAN.
Galls.
834

Medicine.—"The GALLS have long held a place in the Materia Medica of the Hindus. They are considered tonic, expectorant, and useful in cough, phthisis, asthma, fever, want of appetite, and irritability of the stomach. The usual dose is about 20 grains combined with demulcents and aromatics. Muhammadan writers describe them as hot and dry, useful in chronic pulmonary affections, especially those of children, also in dyspeptic vomiting and diarrhœa; they notice their use in fever and want of appetite, and say that they are a good external application in cases of psoriasis" (*Dymock, Mat. Med. W. Ind.*).

MEDICINE.
Galls.
835

Stewart states that the drug known in the Panjáb bazárs as *sumák* is the FRUIT of this tree; it is administered to strengthen digestion.

Fruit.
836

CHEMICAL COMPOSITION.—The authors of the *Pharmacographia Indica* state that, according to Prebble, the galls contain 75 per cent. of tannin.

Chemistry.
437

TRADE.—The drug is exported from Northern India to bazárs in other localities. Atkinson states that the average annual export from the Kumáon forest division is about 70 maunds. Dymock writes that in Bombay the galls fetch from R2-8 to R3 per maund of 37½℔.

Trade.
838

SPECIAL OPINIONS.—§ "Twenty grains of the powdered gall may be given in cream, twice daily, in the treatment of dysentery" (*Surgeon J. Mc-Cloghey, Jakobabad, Sind*). "The galls, also called *kadú kazípú* in Tamil, powdered, fried with *ghí*, and a little sugar added, may be given internally with good effect in cases of dysentery" (*Surgeon-Major D. R. Thompson, M.D, C.I E., Madras*).

Fodder.—The LEAVES are lopped for fodder for buffaloes and camels.
Structure of the Wood.—Hard, close, and even-grained, brown, beautifully mottled with yellow and dark veins, capable of a fine polish, weight 54℔ per cubic foot. The sapwood is liable to be attacked by insects, but the heartwood is durable, and highly prized. The tree is consequently often cut recklessly. In Hazára the TIMBER is used for roofing, and for making spinning wheels, furniture, &c. Elsewhere it is chiefly employed for furniture, for which its beautiful grain renders it specially suitable. Gamble states that it is generally obtainable in the bazárs of the North-West Himálaya, and particularly at Simla, in the form of thick, short planks.

FODDER.
Leaves.
839
TIMBER.
840

P. 840

| PISTACIA Lentiscus. | The Mastic Tree. |

841

Pistacia Lentiscus, *Linn.; Boiss., Fl. Orient., II., 8.*

THE MASTIC TREE, or MASTICHE.

Vern.—Resin=*rúmi mastiki, kúndur-rúmi,* HIND.; *Rúmi mastungi, kúndur-rúmi,* BENG.; ? seeds=*tantarík,* PB.; resin=*arah, auluk-bagdadi, uluk baghdani, mustoka,* ARAB.; *Kandar-i-rumi, kundar-i-rumi, mastaká-i-rumi, kinneh, kinnoli,* PERS.

References.—*Voigt, Hort. Sub. Cal., 274; Stewart, Pb. Pl., 46; Aitchison, Notes on Products of W. Afghánistán and N. E. Persia,* 154; *Pharm. Ind.,* 58; *Ainslie, Mat. Ind., I.,* 214; *O'Shaughnessy, Beng. Dispens,* 278; *Fluck. & Hanb., Pharmacog.,* 161, 598; *U. S. Dispens.,* 15th Ed., 931; *S. Arjun, Bomb. Drugs,* 32; *Murray, Pl. & Drugs, Sind,* 86; *Med. Top., Ajmir,* 149; *Year Book Pharm.,* 1874, 624; 1875, 260; *Irvine, Mat. Med., Patna,* 54 90; *Birdwood, Bomb Pr.,* 19; *P. W. D. Rep. on Gums & Gum-resins,* 15, 22, 60; *Cooke, Gums & Gum-resins,* 104; *Smith, Dic.,* 31, 269; *Kew Off. Guide to the Mus. of Ec. Bot.,* 36; *Kew Off. Guide to Bot. Gardens and Arboretum,* 72, 127; *Ind. Forester, XI.,* 55; *XIII.,* 58; *Agri.-Horti. Soc. Ind., Trans., III.,* 41.

Habitat.—A diœcious evergreen shrub of the Mediterranean region, which yields the mastic of Chios, imported into India.

RESIN. 842

Resin.—The resin, MASTIC, occurs in small irregular, yellowish tears, brittle and of a vitreous fracture, but soft and ductile when chewed. It has a faint agreeable odour, which is increased by the application of heat or friction. It has been known and much valued in medicine and the arts from the earliest period, and has until comparatively late years been obtained almost entirely from the island of Scio. It is collected in that island as follows :—"About the middle of June incisions are made in the bark of the stem and principal branches. From these incisions, which are vertical and very close together, the resin speedily flows, and soon hardens and dries. After fifteen to twenty days it is collected with much care in little baskets lined with white paper, or clean cotton wool. The ground below the trees is kept hard and clean, and flat pieces of stone are often laid on it, so that the droppings of the resin may be saved uninjured by dirt. There is also some spontaneous exudation from the branches, which is of very fine quality. The operations are carried on by women and children, and last for a couple of months. A fine tree may yield as much as 8 to 10 pounds of mastich. The dealers in Scio distinguish three or four qualities of the drug, of which the two finer are called κυλιστό and φλισκάρι, that collected from the ground πῆττα, and the worst of all φλõνδα " (*Flückiger & Hanbury*).

Chemistry. 843

CHEMICAL COMPOSITION.—The authors of the *Pharmacographia* state that mastic consists of two resins called, respectively, α & β resin of mastich. The former constitutes about 90 per cent. of the whole, possesses acid properties, has the formula $C_{20} H_{32} O_3$,—and is soluble in cold alcohol. The latter, left as an insoluble residue, is translucent, colourless, tough and soluble in ether or oil of turpentine. It is somewhat less rich in oxygen than the α. resin. Mastic also contains a little volatile oil.

TRADE. 844

TRADE.—According to Aitchison mastic is imported from Turkey to Western Afghánistán and North-Eastern Persia, for re-export to India.

MEDICINE Mastic. 845

Medicine.—MASTIC is considered by Hindu physicians to be corroborant and balsamic, and is generally ordered by them along with *salep.* Muhammadan women of high rank employ it as a masticatory to preserve the teeth and sweeten the breath, and Muhammadan physicians consider it aphrodisiac, stimulant, and diuretic.

In the middle ages it held a high reputation in Europe, and appears to have entered into a large number of the prescriptions of that period. It is now, however, regarded as devoid of any important therapeutic property,

and as a medicine is becoming obsolete. Dentists employ it to a considerable extent for stopping teeth.

Domestic, &c.—In India MASTIC is used in the preparation of a
perfume. The Turks employ it in the manufacture of a liqueur called *raki*.
A varnish made from mastic is still commonly used in the arts for pictures,
maps, &c., on paper and canvas, but its value for this purpose is rapidly
decreasing, with the great number of other less expensive resins which
are daily becoming more and more utilised.

[*Fl. Orient, II.,* 6.

Pistacia Terebinthus, *Linn.*, var. mutica, *Aitch. et Hemsley ; Boiss.*,
THE TEREBINTH TREE.

847

Syn.—PISTACIA MUTICA, *Fisch. et Mey.* ; P. KHINJUK, *Stocks* ; P. CABU-
LICA, *Stocks.*

Vern.—Resin=*mastáki, kábulí mastaki,* galls=*gul-i-pista, buzghanj,*
HIND., BOMB. ; ? *Khinjak, shne,* PB. ; *Ban, wan, wana, gwa, gwana,*
BALUCH. ; *Kanjak, kinjak, kunjad,* resin=*kunjad, kunjada, khunjad,*
khunjada, kinjad, kinjada, wanjad, wanisad, kandur, kundar, kund-
arud, kunderu, shilm, leaves=*gosh-wára, barg-a-bana,* oil=*réghan-*
i-kanjak, W. AFGHAN. & N.-E. PERSIA.

References.—*Voigt, Hort. Sub. Cal, 273 ; Stewart. Pb. Pl., 46 ; Aitchison,*
Kuram Valley Rept., Pt. I., 42 ; Pt. II., 156 ; Rept. Pl. Coll. Afgh. Del.
Com., 47 ; DC., Orig. Cult. Pl., 316 ; Pharm. Ind., 59 ; Fluck. & Hanb.,
Pharmacog., 165, 598 ; U. S. Dispens., 15th Ed., 1428, 1430 ; Ainslie,
Mat. Ind., I., 458; O'Shaughnessy, Beng. Dispens., 277; Moodeen Sheriff,
Supp. Pharm. Ind., 201 ; Murray, Pl. & Drugs, Sind., 86 ; Bent. &
Trim., Med. Pl., 69 ; Dymock, Mat. Med. W. Ind., 2nd Ed., 194 ; Dy-
mock, Warden & Hooper, Pharmacog. Ind., I., 377 ; Year-Book Pharm.,
1879, 466 ; 1880, 234 ; 1881, 142 ; Trans. Med. & Phys. Soc., Bom.
(New Series), III., 146 ; Baden Powell, Pb. Pr., 589 ; Aitchison, Notes on
Products of W. Afghán. and N.-E. Persia, 155 ; *Gums & Resinous*
Prod. (P. W. Dept. Rept.) 22, 55; Cooke, Gums & Resins, 104, 105 ;
Christy, New Com. Pl. IV., 40 ; V., 58 ; VI., 91, 99 ; Kew Off. Guide to
the Mus. of Ec. Bot., 36 ; Kew Off. Guide to Bot. Gardens and Arbo-
retum, 127 ; Gazetteers :— Panjáb, Bannu, 23 ; Ind. Forester, X.,
516 ; XI., 55 ; XIII., 58 ; XIV., 361, 367 ; Balfour, Cyclop. Ind., III.,
226 ; Smith, Ec. Dict., 409.

Habitat.—The plant defined as P. Terebinthus, *Linn.*, is generally a
tree, 20 to 40 feet or more in height, in some countries a shrub, common
on the islands and shores of the Mediterranean, as well as throughout
Asia Minor. It has many varieties extending as **P. palæstina** to Syria and
Palestine, and eastward as **P. cabulica, P. mutica** or **P. Khinjuk** to Baluchistán and Afghánistán. It is found and has been described as **P. at-
lantica** in Northern Africa, where it grows to a large size, and in the Canary
Islands. Aitchison describes var. **mutica,** to which the four last mentioned have been reduced, as follows :—" This is *the* tree of Baluchistán, and
hence its name *ban, wan, gwan ;* on Persian territory and near Herat its
name is altered to *kinjad, kunjad.* It is usually a small tree about 18
feet in height and with a bole of from 3 to 5 feet in circumference, occurring occasionally in clusters, but usually scattered singly at long distances
on limestone formation."

Resin.—The typical species, and the Asian var. **mutica,** appear to produce resinous products varying in character and composition. That of
the former is the terebinth of the ancients, the τὲρμινθος of Theophrastus,
τερεβινθος of other authors, and the *alah* of the Old Testament. It is an
oleo-resin and was the true primitive turpentine, ῥητίνη τερμινθίνη, celebrated as the finest of all analogous products and preferred both to mastic and the pinic resins. Like mastic it is produced chiefly in the island
of Scio where it is collected from incisions made in the stem and branches

PISTACIA Terebinthus.

The Terebinth Tree.

RESIN.

in spring (*Pharmacographia*). Of late years it has again, under the name of Chian-turpentine, become of importance in European medicine.

The resin of var. **mutica** resembles that of **P. Lentiscus** and is used in the East as a substitute for that substance. It also bears the same vernacular names and is in every way considered identical with the resin of **P. vera.** Aitchison states that throughout the region where the tree grows, the resin is not usually to be found for sale, but is to be met with in all households, since it is looked upon as an every-day remedy for cuts and bruises. A turpentine is also occasionally obtained from the resin, but only by any one specially making it for himself (*Notes on Prods. of W. Afghánistán and N.-E. Persia*). In Christy's *New Commercial Plants*, V., 58, the resin of var. **mutica** is recommended for the preparation of varnishes for articles requiring washing with soap, such as oil cloth, painted surfaces, floor-cloth, &c., since it is not affected by soda.

Chemistry.
849

CHEMICAL COMPOSITION.—Chian turpentine consists of a resin, probably identical with the *a*. resin of mastic (*see p*. 270), and an essential oil. The former dissolves entirely in alcohol. Flückiger & Hanbury obtained 14½ per cent. of an essential oil from the drug, which had the odour of Chian turpentine, a specific gravity of 0·869, a boiling point of 161°C. and deviated the ray of polarised light 12·1° to the right. By combination with sodium it lost a small proportion of oxygenated oil and changed its characters, becoming more agreeable in odour, and resembling a mixture of cajeput, mace, and camphor. After this treatment and rectification the oil was found to have the same composition as oil of turpentine (*Pharmacographia*).

The resin of var. **mutica** is also entirely soluble in alcohol, but differs from true mastic in being partly insoluble in turpentine, and from the resin of European Chian turpentine in being quite soluble in ether. In solution with acetone or benzol it has the same optical properties as true mastic.

DYE & TAN
Leaves.
850

Dye & Tan.—Aitchison writes, " The LEAVES, almost without exception, are affected by a flat horse-shoe shaped gall, that extends round the margin of the leaf; the gall is so very distinct in form, much resembling the lobe of the ear, that the leaves get their name *góshwára*, meaning ear-like, owing to this resemblance; by these galls alone the leaves of this species may be identified from those of **P. vera.** These galls, the Natives say, are of no use, but the leaves are valued for dyeing and tanning. May not the presence of the galls on the leaves be the reason why these leaves are employed, and the galls really be the active agent?" The galls are small and succular, are of a pink colour, have a terebinthinate and astringent taste, and appear to be caused by the presence of an **Aphis.** The authors of the *Pharmacographia Indica* state that they observed the galls in the Bombay market for the first time in 1889, where they were offered for sale as Pistachio galls.

OIL.
Kernels.
851

Oil.—From the KERNELS is obtained a mixture of an essential and a fatty oil, which is eaten as a relish with *karut* (dried oxygal) and bread (*Aitchison*). This substance is the original oil of turpentine, τερεβἰνθινον-έλαιον.

MEDICINE.
Fruit.
852
Galls.
853
Resin.
854

Medicine.—In Afghánistán the FRUIT is considered warm, stimulating, and stomachic, and is prescribed in colic and dyspepsia (*Stewart*). The GALLS are employed as an astringent. The RESIN is used in every way similarly to that of **P. Lentiscus**, the true mastic, and is almost universally employed as a substitute for that drug in India, Persia, Baluchistán, and Afghánistán. It is also said to be used in Sind and Afghánistán, like true mastic in Europe, for stopping decayed teeth. The vernacular names borne by it imply the belief in its virtues as an external styptic application to wounds, signifying " the remedy (or resin) for stopping blood," or " for dressing wounds," or " *the* tree-resin " (*Aitchison*). The greater part of the mastic imported into India consists of this drug ; only a small quantity of the true mastic being obtained direct, or through Persia and Afghánistán, from Turkey. The price of the former, known commercially as Bombay mastic, ranges in Bombay from 8 to 12 annas a pound (*Dymock*).

Chian turpentine has stimulant and diuretic properties, and was formerly used for these purposes in European medicine. Of late years it dropped almost completely out of use, until in 1880 **Professor Clay** of Birmingham recommended its employment in cases of uterine cancer.

Food & Fodder.—Mr. **Lace** informs the writer that the LEAVES are eaten by camels, goats, and sheep in Southern Afghánistán, and that the FRUIT, called *shiné*, is eaten by Natives in the same locality. **Aitchison** makes a similar statement regarding Northern Baluchistán.

Structure of the Wood.—Aitchison writes, " In the districts where it is to be met with, trees are so scarce that to cut one down would be almost sa acrilege, hence I can say nothing regarding its value as timber, but its dry branches make excellent fuel."

FOOD & FODDER.
Leaves.
855
Fruit.
856
TIMBER.
857

Pistacia vera, *Linn. ; Boiss., Fl. Orient., II., 5.*

THE PISTACHIO NUT.

Syn.—PISTACIA NARBONENSIS, *Linn. ;* P. RETICULATA, *Willd.*

Vern.—Tree and nut=*Pista, pistá,* HIND., BENG., BOMB., AFGH., PERS. ; country where the tree abounds=*pistalik,* galls=*bóz-ghánj ;* gum-resin=the same as the names of that of P. **Terebinthus, var. mutica,** PERS. and W. AFGHAN.

References.—*Stewart, Pb. Pl., 47 ; Aitchison, Rept. Pl. Coll. Afgh. Del. Com., 47 ; DC., Orig. Cult. Pl., 316 ; O'Shaughnessy, Beng. Dispens , 276 ; Irvine, Mat. Med., Patna, 34, 86 ; Sakharam Arjun, Cat. Bomb. Drugs, 33 ; Murray, Pl. & Drugs, Sind. 85 ; Dymock, Warden & Hooper, Pharmacog. Ind., I., 379 ; Birdwood, Bomb. Prod., 19 ; Aitchison, Notes on Products of W. Afghánistán & N.-E. Persia, 156 ; Liotard, Dyes, 11 ; Wardle, Dye Rept., 8 ; Ayeen Akbary, Gladwin's Trans., I., 81, 84 ; Gazetteer, Bombay, XIII., 384 ; Agri.-Horti. Soc. Ind., Trans., III., 41 ; Agri.-Horti Soc., Panjáb, Select papers to 1862, 197 ; Ind. Forester, XIV., 365, 367 ; Smith, Ec. Dict , 325.*

Habitat.—A small tree, forming forests at altitudes of 3,000 feet and upwards, usually on sandstone formations, in Syria, Damascus, Mesopotamia, Terek, Orfa, the Badghis, and Khorasan ; extensively cultivated in Syria, Palestine, and Persia (*Boissier, Aitchison*). The tree was introduced into Italy by **Vitellius** in the end of the reign of **Tiberius,** and thence into Spain by **Flavius Pompeius.** De**Candolle** writes, " There is no reason to believe that the cultivation of the pistachio was ancient even in its primitive country, but it is practised in our own day in the East, as well as in Sicily and Tunis. In the south of France and Sicily it is of little importance."

Brigade-Surgeon Aitchison, in his recent interesting and valuable *Notes on the Products of Western Afghánistán and North-Eastern Persia,* gives a full account of the occurrence of the tree in the tracts over which he has travelled,—the country from which India draws its large supply. He states that in suitable localities it forms large forests, the most celebrated of which occur in Badghis, near Kala-nao, and at Zulfikar. Others exist in the hills of Khorasan, and small clumps to the south of Bezd. In Persia it is cultivated in orchards, but it is not grown artificially in Afghánistán. " It is, however, " he writes, " a common thing to see trees of the indigenous pistachio growing round shrines where they are carefully protected. The cultivated tree of orchards has usually a

PISTACIA vera.	The Pistachio Nut.

Occurrence.

good stem, showing a fair amount of wood, and growing altogether more luxuriantly and more like a tree than the wild form.

The value of the forests of indigenous pistachio lies in their yield of nuts, but the harvest is a precarious one, generally due to the tree being diœcious, and to fertilisation being frequently unaccomplished. The appearance of staminate flowers on these trees are the first signs of spring, and as they appear long before there is any sign of the leaves, they are unprotected and easily injured by frost, and I have no doubt but that a late recurrence of frost is one of the most frequent causes of a bad nut harvest. The Natives say that there is only a good nut harvest every second year, and that when the nuts fail, the galls on the leaves are more numerous.

" The nuts on some of the trees are partially dehiscent, whereas in others they are quite indehiscent. So well is this known to the people of the country that in collecting nuts for eating, should they chance to come on a tree of which the nuts are indehiscent, they just move on until they come to a tree bearing dehiscing nuts. In the latter case a slight crushing of the nut with the fingers gives exit to the kernels, whereas in the former each nut has to be broken up, as we would a hazel, before the kernel can be got at. On many trees the female flowers are found not to have been fertilised ; these develop into a nut-like form, and when these unfertilised ovaries are examined they are found to be quite hollow, the walls being apparently analogous to the covering of the fertilised nut. " " From the great trade-value of the nuts and of the galls, there is much jealousy as to the forest rights, as to whom they belong, and in what proportion to each tribe. Half the blood-feuds of the nomads originate in their quarrels over the rights of produce in their forests.

" All persons connected in the rights to the forest and produce unitedly collect the nuts, and the general harvest is subsequently divided in the allotted proportions to those to whom they may belong. In the meanwhile the Amír's tax collectors are at hand ready to carry off the usual tax imposed on produce before it is permitted to leave the ground. "

A few trees have been cultivated here and there in North-Western India with success. Thus Aitchison mentions some large bushes in Rawalpindi, and a very fine fruit-bearing tree in Srinagar, Kashmír. There appears to be no reason why, with a little trouble, the cultivation of this valuable tree should not be carried on with success, at least along the whole North-Western Frontier.

RESIN.
860

Resin.—A resin, similar to mastic, and employed in every way like that of **P. Terebinthus**, var. **mutica**, is obtained from this species. It is chiefly consumed locally, and is known by the same names as the resin of the preceding species. Aitchison states that when fresh it has a most pleasant fruity odour ; at first it is very liquid, but gradually hardens on exposure, till it becomes very brittle, and almost transparent.

DYE, TAN, & MORDANT
Galls.
861
Pericarp.
862.
Ovaries.
863

Dye, Tan, & Mordant.—The leaves are very frequently affected by GALLS, which are irregularly-shaped spheroids, from the size of a cherry to that of a large gooseberry, borne on a short stalk and usually growing from the surface of the leaf. These, with the PERICARP of the fruit, and the unfertilised fruit-like OVARIES above described are used locally for dyeing silk, and are also largely exported to Persia and Turkistan, and a small proportion to India (*Aitchison*). In this country the galls are known as *gul-i-pista*, *baz-ganj* or *boza-ganj*, the fruit husks as *post-i-pista*, and are imported into Bombay from the Persian Gulf, and into Northern India transfrontier. Davies' *Trade Report* gave the quantity annually imported from Afghánistán into Peshawar as 50 maunds, and about 100 maunds in addition are said to be brought down by the Bolan Pass yearly. They are used as a dye, or as a mordant for silks, and as a tan.

The Pistachio Nut.	(*J. Murray.*)	PISTIA Stratiotes.

The galls contain 45 per cent. of tannin allied to gallo-tannic acid, also gallic acid, and 7 per cent. of an oleo-resin to which the odour is due (*Pharmacog. Ind.*).

Oil.—The NUTS contain about 60 per cent. of a fatty oil which is occasionally extracted for use in medicine. It is of a greenish colour, sweet-flavoured, and aromatic.

Medicine.—Eastern writers consider the NUT warm, digestive, restorative, tonic and, aphrodisiac, and prescribe it also as a sedative in nausea and vomiting. Arabian physicians prepare a *logh* with them known in French Pharmacy as ' Looch vert des pistaches.' The GALLS are used as an astringent, and the RESIN for the same purposes as mastic. The OIL above described is employed medicinally as a demulcent and restorative. The BARK is believed to be tonic and useful in indigestion.

Food & Fodder.—The FRUIT, well known as the pistachio nut, is oval-shaped and varies in size with the amount of cultivation which the tree has received. It has a brittle shell, enclosing a kernel, generally about half an inch in length, of a greenish colour and agreeable flavour. These nuts are exported in immense quantities from the localities above enumerated to Afghánistán proper, and to India, as well as to Persia and Turkistan. No exact estimate of the quantity annually imported into India can be formed, since they are included in trade returns under the heading of Fruits, Nuts, and Vegetables. The wild fruit is smaller and more terebinthinate in flavour than the cultivated, but is preferred by many. In India the nut is much appreciated by all classes, is a very common article of food with the more well-to-do classes, and a frequent ingredient of confectionery. It is also considerably used in dessert at the tables of Europeans. By Natives, the nuts are generally roasted in their shells in hot sand, and then thrown into a hot paste of salt-water, and stirred so as to make the salt adhere to the shell, much as sugar does to a burnt almond. They are hawked about the streets of large towns under the name of *Khára pistá* (salted Pistachio nuts). The OIL is said to be also occasionally used as food, but rapidly becomes rancid. Church gives the chemical composition of the nut as follows:—Water, 5·9; albumenoids, 24·4; starch, 3·5; oil, 62·5; fibre, 1·3; ash, 2·4 per cent., and states that this (an Afghán) sample contained 11 per cent. more oil than European kernels. Sheep, camels, and goats feed greedily on the FOLIAGE, hence the name applied to the galls, " the goats' store " (*Aitchison*).

Structure of the Wood.—Aitchison informs us that the wood is highly valued in Afghánistán and Persia for the manufacture of agricultural implements, especially ploughs, and for making spoons, also that it certainly affords the best fire-wood of any tree in these countries.

OIL.
Nuts.
864

MEDICINE.
Nuts.
865
Galls.
866
Resin.
867
Oil.
868
Bark.
869
FOOD &
FODDER.
Fruit.
870

Oil.
871

Foliage.
872
TIMBER.
873

PISTIA, *Linn.; Gen. Pl., III., 964.*

Pistia Stratiotes, *Linn.; DC., Monogr. Phaner., II., 634;* AROIDEÆ.

874

> Var. α CUNEATA, *Engl.;* P. STRATIOTES, *Linn., Roxb.,* in *Fl. Ind.,* and following him other Indian writers; KIAMBAM KITSII, *Rumph.;* P. CRISPATA, *Blume.*
> Three other varieties are described, and each has many synonyms, but as they are not Indian they need not be enumerated in this work.
> **Vern.**—*Jal-kunbhi, jalkhumbi, tákápáná,* HIND.; *Tákápáná,* BENG.; *Baujhánjhe,* URIYA; *Prashni, gondála,* BOMB.; *Prásni, gondála,* MAR.; *Anter-ghunga,* DEC.; *Agasatamaré,* TAM.; *Akása támara, autara támara, níru budiki,* TEL.; *Kodda-pail,* MALAY.; *Deya-parandella,* SING.; *Kumbhiká,* SANS.
> **References.**—*Roxb., Fl. Ind.,* Ed. C.B.C., 502; Mason, *Burma & Its People,* 504, 815; Sir W. Elliot, *Fl. Andhr.,*13, 15, 135; Rheede, *Hort. Mal.,* XI., 63, t. 32; Rumphius, *Amb.,* VI., 177; Ainslie, *Mat. Ind.,* II., 7;

18 A

The Grey or Field Pea.

U. C. Dutt, Mat. Med. Hind., 306 ; Sakharam Arjun, Cat. Bomb. Drugs, 209; Dymock, Mat. Med W. Ind., 2nd Ed., 811; Drury, U. Pl. Ind., 346 ; Boswell, Man. Nellore, 118 ; Gazetteers :—Orissa, II., 182 ; Mysore & Coorg, I., 66 ; Agri.-Horti. Soc. Ind.:—Journals (Old Series), IV., 233 ; X., 340 ; Ind. Forester, XIV., 390, 392 ; Smith, Ec. Dict., 435.

Habitat.—An aquatic herb, found widely dispersed in the tropical and subtropical regions of both hemispheres. The Indian variety a CUNEATA has been recorded from Bengal, the Coromandel Coast, Madras, the Deccan, Burma, and Ceylon.

MEDICINE.
Plant.
875

Medicine.—Ainslie writes, "The Hindu Doctors consider a decoction or infusion of this PLANT as cooling and demulcent, and prescribe it in cases of dysuria, in the quantity of about ten pagodas weight twice daily; the LEAVES are made into poultice for piles." Mixed with rice and cocoanut milk they are also given in dysentery, and with rose water and sugar in cough and asthma. The ROOT is laxative and emollient. The ASHES are applied to ringworm of the scalp, and in some parts of India are known as *páná* salt.

Leaves.
876
Root.
877
Ashes.
878
Chemistry.
879

CHEMICAL COMPOSITION.—Dr. Warden has kindly furnished the following note, the result of his analysis of the plant, which has been fully described in the *Chemical News* :—

" The incinerated plant yields a saline substance, which is known as *páná* salt, and which is almost entirely used for medicinal purposes. A specimen of the plant, obtained from the Dinajpur district, yielded 31·4 per cent. of mineral matter, calculated upon the plant dried at 130°C., of which 6·1 per cent. was soluble in water. *Páná* salt consists of the soluble portions of the ash. A specimen from the same district contained 75 per cent of chloride, and 22·6 of sulphate of potassium."

FOOD.
Herb.
880
DOMESTIC.
Herb.
881

Food.—The HERB was eaten in the Poona district during the Deccan famine of 1877-78.

Domestic.—Mason notes that the HERB is frequently transferred to tubs of water near public buildings to keep the water fresh, but remarks that it is well to remember that in Jamaica the same plant, in hot dry weather, has been observed " to impregnate the water in the tanks with its particles to such a degree as to give rise to the bloody flux."

SPECIAL OPINION.—§ " The plant purifies water in which it grows, and Natives use it largely in their drinking tanks" (*Surgeon-General W. R. Cornish, F.R.C.S., C.I.E., Madras*).

882

PISUM, *Linn. ; Gen. Pl., I., 527.*

Pisum arvense, *Linn. ; Fl. Br. Ind., II., 181 ;* LEGUMINOSÆ.
THE GREY or FIELD PEA.

Vern.—*Mattar, mattar rewari, kulon, desi-mattar, chota mattar,* HIND. ; *Keiao, desi-mattar, chota mattar,* BENG. ; *Kalon, kulai batana,* N.-W.P. ; *Kulon,* KUMAON ; *Korani, karain,* KASHMIR ; *Kuláwan, kála muttar, matar rewari, karain, ghále, kalao, kulah,* PB. ; *Karain,* GUZ.

References.—*Roxb., Fl. Ind., Ed. C.B.C., 566 ; DC , Orig. Cult. Pl., 327 ; Birdwood, Bomb. Prod., 122; Baden Powell, Pb. Pr., 242; Atkinson, Him. Dist. (X., N.-W. P. Gaz.), 694 ; Duthie & Fuller, Field & Garden Crops, II., 17, Pl. xxxii., b. ; Royle, Ill. Him. Bot., 200 ; Church, Food-Grains, Ind., 120, 135 ; Madden, Note on Kumáon, 279; Settlement Report :—Panjáb, Simla, App. II., H., xxxix. ; Kangra, 24 ; N.-W. P., Bareilly, 82 ; Kumáon, App., 132 ; Gazetteers :—Panjáb, Simla, 56 ; Balfour, Cyclop. Ind., III., 227.*

Habitat.—This pea is, according to DeCandolle, undoubtedly wild in Italy, where it grows, not only in hedges and near cultivated ground, but also in forests and wild mountainous districts. Evidence regarding its being truly wild in Spain, Algeria, Greece, the South of Russia, and the East affords no positive proof. It has, however, for long been extensively

| The Garden Pea. | (*J. Murray.*) | PISUM sativum. |

cultivated, and consequently specimens occur in abundance spontaneously, in many parts of North-Western Asia, which renders it very difficult to say when it is truly wild. **Royle** appears to be the only Anglo-Indian botanist who has considered the plant to be indigenous to India, and from philological and other evidences, there would seem to be little doubt, not only that it is always cultivated or spontaneous in this country, but that it is of comparatively recent introduction. It has recently been found wild on the slopes of the Zard Kuh in Bakhtriana by Major Sawyer. It is considered by many botanists to be the wild parent of the cultivated **Pisum sativum**, and by Roxburgh is treated as a variety of that species. It is extensively cultivated in many parts of India during the cold weather, in the same way as **Pisum sativum**; indeed, in agricultural reports, the two crops are generally treated of as one, under the general name of "peas." For an account of the methods employed, the reader may, therefore, be referred to the description of the next, and more important, species.

Food & Fodder.—This pea produces a small, round, compressed, greenish, and marbled SEED, which is generally eaten as *dal* by Natives. It is rightly regarded as more indigestible than the garden pea, but, according to Church, there is no constant difference of chemical composition between the two. It must be carefully distinguished from *kesári dal* (**Lathyrus sativus**), which it considerably resembles both in the appearance of the grain and in the mode of cultivation. The STRAW is highly esteemed as a fodder.

FOOD &
FODDER.
Seed.
883

Straw.
884

885

Pisum sativum, *Linn.; Fl. Br. Ind., II., 181.*

THE GARDEN PEA

Vern.—*Bara-mattar, bahtahna, wattahna, matar, gol-mattar, battani-chola, buttani,* HIND.; *Burra-mattar, matar, kuda,* BENG.; *Shánmá, ahandíl,* LADAK; *Kalon, kulai, batana, mattar, gol-mattar,* N.-W. P.; *Ahsa, matra,* OUDH; *Sen, mattar, khandú, bára mattar, sén,* PB.; *Lárkána,* SIND.; *Butana,* C. P.; *Vátána, watana,* BOMB.; *Vátáne, watana,* MAR.; *Vatána, patana,* GUZ.; *Watana,* DEC.; *Pattanie, vella patani,* TAM.; *Gúndú-sani-ghelú, patanlu,* TEL.; *Batgadle,* KAN.; *Kachang,* MALAY.; *Pai,* BURM.; *Ratagoradiya,* SING.; *Harenso, satilá, kaláya,* SANS.; *Hummus,* ARAB.

References.—*Boiss., Fl. Orient., II., 622; Roxb., Fl. Ind., Ed. C.B.C., 566; Stewart, Pb. Pl, 73; DC., Orig. Cult. Pl., 328; Mason, Burma & Its People, 466, 768; Ainslie, Mat. Ind, I., 297; U. C. Dutt, Mat Med, Hindus, 302, 317; Murray, Pl. & Drugs, Sind, 120; Watts, Dict. Chemistry, IV., 358; Birdwood, Bomb. Prod., 123; Baden Powell, Pb. Pr., 242; Atkinson, Him. Dist. (X., N.-W. P Gas.), 309, 694; Duthie & Fuller, Field & Garden Crops, II., 17; Useful Pl. Bomb. (XXV, Bomb. Gas.), 152; Royle, Ill. Him. Bot.. 200; Church, Food-Grains, Ind., 120, 135; Man. Madras Adm., I., 288; Bombay, Man. Rev. Accts., 101; Settlement Reports:—Panjáb, Dera Ismail Khan, 345; N.-W. P., Allahabad, 33; Azamgarh, 120; Central Provinces, Baitool, 77; Nimár, 197; Chanda, 81; Gazetteers—Bombay, II., 63, 67, 269, 273, 277, 280, 284, 287, 291, 295, 390, 536, 538, 541, 544, 547; III., 233; VII., 30; XII, 151; XIII., 289; XVI., 91, 100; XVII., 269; XVIII., Pt. ii, 44; XIX., 164; Panjáb, Dera Ismail Khan, 127; Musaffargarh, 94; Mooltan, 96; Jhang, 111; N.-W. P, I., 80; II., 159, 376; IV., lxxi.; Mysore & Coorg, I., 59; Annual Reports of Directors of Land Rev. & Agri.:—Bombay, App., viii.; Cent. Prov., D (2); Bengal, 1886, App. lxxi.; Agri.-Horti. Soc., Ind.—Trans., I., 47; (App.), 237-240, 256; II., 35, (App.), 305, 307, 310; III., 70, 198; Pro., 227, 239; IV., 145, 147; V, 64; Journals (Old Series). III., 180, 228, IV., 214; VII., 63; VIII., Pro., 48; IX., 416; X., 90; XI., Pro., 20; XIII., Pro. (1863), 54, (1864), 76; (New Series), III., Sel., 11-12; IV., 33, 34; V., 34, 42, Sel. 5; Balfour, Cyclop. Ind., III., 227.*

Habitat.—The garden pea had probably a more southern and more restricted original wild area than **P. arvense.** According to DeCandolle it has not hitherto been found wild, either in Europe or in the west of Asia, whence it is supposed to have come Certain botanists have regarded

PISUM sativum.	Cultivation in India

it as probably but a cultivated race, variety or at most sub-species of **P. arvense**, but further observation and experiment is required before this view can be entertained. Evidence, at present available, points to the following facts, summarised by the learned author of the *Origin of Cultivated Plants*, "The species seems to have existed in Western Asia, perhaps from the south of the Caucasus to Persia, before it was cultivated. The Aryans introduced it into Europe, but it perhaps existed in Northern India before the arrival of the Eastern Aryans. It no longer exists in a wild state, and when it occurs in fields, half-wild, it is said not to have a modified form, so as to approach some other species.

CULTIVATION
886

CULTIVATION.—The field-pea and garden-pea are, as already stated, generally considered together in Agricultural Reports, and by writers on Indian crops. The true garden-pea, **P. sativum**, is, however, much the more valuable and prolific of the two.

In all parts of the country it is a *rabi* crop, and, as will be seen from the following provincial accounts, the methods employed differ but little.

Panjab.
887

Panjáb.—In this Province, peas are cultivated, as a field crop, almost entirely for fodder. The seed is sown on *sailáb* lands in October and November, by means of a drill, after a couple of ploughings at the most. It is sometimes sown broadcast on the same description of soil when too moist to plough at all, and even under these conditions frequently yields a good crop. The average amount of seed per acre is from 20 to 30 seers. The crop is pulled, not reaped, in April, an outturn in grain of from 3 (Muzaffargarh) to 8 maunds (Dera Ismail Khan) being obtained. In the Gazetteer of the latter district it is stated that good *matar* crops sell at from R1 to R2 a *kanal* (⅛ acre) even in remote villages, and are bought up by graziers, especially such as own buffaloes. The finer kinds of garden-pea are also extensively grown here, as elsewhere, for the use of Europeans by native gardeners, both in the plains and in the hills near large stations. Stewart states that **P. sativum** may be found under cultivation in Kanáwar, Spiti, Lahoul, and Ladák, up to 13,000 or 14,000 feet. At the latter height it does not ripen its seeds, but is chiefly used for fodder. No statistics are available of the total area under the crop.

N.-W.
Provinces.
888

North-West Provinces.—Messrs. Duthie & Fuller write that both species of pea are largely grown in certain districts of these Provinces. Two distinct kinds of **P. sativum** are recognised, called, respectively, *kabli* and *patnai*, and distinguished by their difference in size. As in the Panjáb, peas are a *rabi* crop, being sown from the end of September to the middle of October, and reaped in March. In the western and central districts they are most commonly grown as a second crop after indigo or rice in the preceding *kharif*, and since they are rarely irrigated the outturn is very small. As a rule, they are sown on very heavy soil, which receives very little, if any, previous preparation, and is rarely, if at all, manured. The seed is sown broadcast at the rate of ⅓ to 1 maund per acre, and ploughed in. In the Oudh and Benares Division the crop is once watered, in other localities it is not always irrigated. It is harvested in the same manner as other *rabi* crops, but the green pods are regularly picked for home consumption from the time when they first reach their full size.

The whole crop, but especially the finer kinds of garden-pea, suffer from frost, and from the ravages of a caterpillar called the "*bahadura.*"

The average outturn in the Meerut, Rohilkhand, Agra, Allahabad, and Jhansi Divisions is said to be about 10 maunds per irrigated, and 7 maunds per unirrigated, acre. In Oudh, it rises to 16 maunds, and in Benares it is said to be 8 maunds. The average outturn of chaff (*bhúsa*) may be taken as equal to that of the grain. The cost of production per acre may be put at R12-13 for the coarse kinds when no irrigation is used, and R17-13 for

of the true Garden Pea. (*J. Murray.*)

the fine kinds, assuming that two waterings are given and that the land is of rather higher rent.

Regarding the area under the crop, Messrs. Duthie & Fuller state that, in the years 1879-80 and 1881, the average total acreage in the 30 temporarily-settled districts was 379,852, of which 262,034 was irrigated and 117,818 unirrigated. The division which grew the crop to the largest extent was Benares with 271,907 acres, followed by Meerut with 47.350 acres, Allahabad with 42,783, Rohilkhand with 10,323, and Agra, Jhansi, and Kumaon with insignificant areas. It will be seen from these figures that, with the exception of the Meerut district, in which the average area amounts to nearly 4 per cent of the *rabi*-cropped area, the cultivation of peas is carried on only on a very trifling scale in the localities west of Allahabad. In the Allahabad District the area rises to 5'9 per cent. on the *rabi*-cropped area, while in the districts of the Benares Division and of the east of Oudh, it forms an important part of agriculture, being largely grown, with careful tillage and irrigation. In Azamgarh, Gorakhpore, and Basti the area amounts to 13'4 per cent. of the total *rabi* crop, and in wet seasons, when the ground is too damp to admit of wheat being sown in time, peas frequently take it place.

The figures above given are as accurate as can be obtained with present statistics, but are believed to include a certain amount of land under *kesári*, a crop which, as already stated, is very similar to the coarse field-pea, and is largely cultivated in the Allahabad Division (*Field & Garden Crops*).

Central Provinces.—Pea cultivation in these Provinces very much resembles that already described in the North-West Provinces and need not be again detailed. The crop is, however, never irrigated. The area in 1888-89 amounted to 94,569 acres, of which 50,007 were in the Raipur District.

Bombay.—Peas are grown to a considerable extent in this Province. They are sown in moist land in October and November, receive, as a rule, no manure nor irrigation, and little care during the time of their growth. The crop takes four-and-a-half months to ripen, and is reaped in the end of February and March. The green pod s picked before ripening to be used as a vegetable, or is allowed to mature, the seed being used as an article of food either whole or split. In Ahmadnagar, however, it is said never to be split for use as *dál*. The leaves and stalks are much valued as fodder, under the name of *haleem*, and appear to be considered the most valuable part of the crop.

In 1888-89, 17,307 acres were returned as under peas, of which 5,263 were in Belgaum, 2,256 in Násik, 1,967 in Khándesh, 1,802 in Poona, 1,394 in Broach, 1,359 in Ahmadnagar, and 1,342 in Sátára. All the districts of the Deccan thus cultivate the crop to some considerable extent; with the exception of Belgaum it is unimportant in the Kárnátak, while in Gujarát and the Konkan it is scarcely grown at all.

Bengal.—In the *Report of the Director of Land Records and Agriculture for 1886*, a full account of the method of cultivation of peas in the Burdwán Division is given. From this it appears that the field pea, known as *deshi* and *bhurro*, and three distinct kinds of P. sativum, namely, the white or *kabulli*, the *pahari*, and the *piyara* are grown. In addition to these, peas from an English stock, "*olondo*," are grown for the Calcutta market near the railway line between Howrah and Húghli. These are sold green and fetch a high price.

The pea crop comes after *áus* paddy and is followed by the same crop or by sugar-cane or *kachú* (Colocasia antiquorum). It is sown by preference on a naturally rich soil, especially loam, and the land is carefully

PISUM sativum.	The true Garden Pea.

CULTIVATION in Bengal.

ploughed unless it receives a yearly deposit of river silt. A good crop is not usually obtained with fewer than eight or nine ploughings. The crop is sown either alone or mixed with mustard, in both cases 20 seers per *bigha* (⅓ acre) being sown, in the end of October and beginning of November. No after-cultivation, weeding, or irrigation is required. The produce is harvested in the end of February and beginning of March by pulling the entire plants up by the hand. They are then spread out and exposed to the sun, for if kept in heaps the colour becomes rapidly spoiled. The seed is threshed out in the usual manner, by subjecting the whole to the treading action of bullocks. In lands which receive a yearly deposit of silt, the seed is often sown broadcast immediately after the inundation water recedes, without any previous cultivation.

The outturn is very variable, but 4 maunds per *bigha* (about 12 maunds per acre) is considered an average crop. This outturn is, however, liable to be considerably affected by various diseases to which the crop is subject. In cloudy weather it is said to be sometimes attacked by a yellowish-green fly which lays its eggs on the immature pods, and these, developing into larvæ, may do immense damage. Stagnant water exercises a very injurious influence, and too much organic matter in the soil causes the plants to grow very rankly and produce little fruit. A kind of weevil is said to attack the seed when stored in barns, especially, in the opinion of the *rayat*, when the grain heated in the sun has not been allowed to cool properly before being stored.

Madras. 892

Madras.—Peas are mentioned in the *Madras Manual of Administration* as one of the pulse crops of that Presidency, but no information is available as to the extent of its cultivation, from which it may be assumed to be unimportant.

FOOD & FODDER. Green Pods. 893 Ripe Seed. 894

Food & Fodder.—The GREEN PODS are collected in many localities while the plant is growing, and are either cooked and eaten whole, or the young seed is extracted and eaten in the same way as by Europeans. The roasted green pods are known in the Panjáb under the name of *dadhrián* and *ámián*. The RIPE SEED is used, whole, split as *dál*, or ground and made into bread. In some districts it is little esteemed in comparison with other pulses. It undoubtedly has a tendency to produce flatulence and is unwholesome when imperfectly cooked, or carelessly freed from the tough coat of the seed. Indeed, in the south of the Purneah District, where it is eaten uncooked to some extent, it is believed to cause dysentery and diarrhœa, and certainly aggravates these diseases. In certain districts of the Panjáb sleeping in a pea field is believed to produce a kind of paralysis called *manda*, and a diet of peas is said to cause the disease known as *wá* (*Musaffargarh Gazetteer*).

CHEMISTRY. 895

The composition of peas is given as follows by Church, in 100 parts :—

	Water.	Albumenoids.	Starch.	Oil.	Fibre.	Ash.
Husked .	11.8	28.2	55.0	1.5	1.0	2.5
Unhusked .	12.5	23.6	54.5	1.3	5.7	2.4

The ash of the husked grain contains 1·0, that of the unhusked 0 8, of phosphoric acid. As already stated, the whole plant and the "straw" (stems, leaves, and empty pods) are both extensively used and highly valued as fodder for cattle, sheep, and goats. It is generally considered to be as nourishing as hay, and is in many localities, especially of the Panjáb and Bombay, the most highly valued part of the crop. Analyses of pea straw made in Europe by **Mayer** and others, shew it to contain 12 per cent. nitrogenous matter, 21·9 matter soluble in potash, 47·5 non-nitrogenous matter insoluble in potash, 6·0 ash, and 12·0 water. The ash of pea straw from Duderstat was found by **Hertwig** to consist of Carbonate of Calcium 47·8 per cent., Sulphate of Potassium, 10·7 per cent., Carbonates of Sodium and Potassium, 12·4 per cent., Phosphates of Calcium and Magnesium, 9·5 per cent., and smaller quantities of other salts (*Watts, Dic. Chem.*).

CHEMISTRY.

Pita Fibre, see **Agave americana,** *Linn.*; AMARYLLIDEÆ; Vol. I., 134.

Pitch, see Pinus, pp. 238—247.

PITHECOLOBIUM, *Mart.*; *Gen. Pl., I., 597.*
[LEGUMINOSÆ.

896

Pithecolobium bigeminum, *Benth.*; *Fl. Br. Ind., II., 303;*

> **Syn.**—MIMOSA BIGEMINA, *Linn.*; M. LUCIDA, *Roxb.*; M. MONADEIPHA, *Roxb.*; INGA BIGEMINA, *Willd.*; I. LUCIDA, *Wall.*; I. ANNULARIS, *Grah.*; I. WIGHTIANA, *Grah.*
>
> **Vern.**—*Kachlora,* HIND.; *Kachlora,* BOMB.; *Ta-nyen,* BURM.; *Calateya,* SING.
>
> **References.**—DC., *Prodr., II., 439; Roxb., Fl. Ind., Ed. C.B.C., 417; Brandis, For. Fl., 173; Beddome, For. Man., 96; Gamble, Man. Timb., 145; Thwaites, En. Ceylon Pl., 100; Dalz. & Gibs., Bomb. Fl., 89; Mason, Burma & Its People, 459, 772; Atkinson, Him. Dist., 746; Lisboa, U. Pl. Bomb., 71; Gazetteer, Bombay, XV., 433.*

Habitat.—A large tree, met with in the forests of the outer Himálaya from the Ganges eastward, ascending to 3,000 feet, also in South India and Ceylon.

Medicine.—A decoction of the LEAVES is used externally in the North-West Provinces as a remedy for leprosy and, as a stimulant, to promote the growth of hair (*Atkinson*). There is no record of the plant being recognised as medicinal in other parts of India.

MEDICINE. Leaves. 897

Food.—**Mason** states that, though the SEEDS are poisonous and sometimes produce disastrous consequences, the Burmese and Karens are extravagantly fond of them as a condiment with preserved fish. They are said to fetch a high price in the bazárs of Burma.

FOOD. Seeds. 898

Structure of the Wood.—Dark-coloured, heavy, sometimes called "Iron-wood."

TIMBER. 899 900

P. dulce, *Benth.*; *Fl. Br. Ind., II., 302; Wight, Ic., t. 198.*
MANILLA TAMARIND.

> **Syn.**—MIMOSA DULCIS, *Roxb.*; INGA DULCIS, *Willd.*
>
> **Vern.**—*Vilaití imlí, dakhani babúl,* HIND.; *Dakhini babúl,* N.-W. P.; *Vilayti ambi, chinch, deccani babul, vilaiti ámli,* BOMB.; *Hatichinch,* MAR.; *Karkapilli, korukápuli,* TAM.; *Sima chinduga, sima chinta,* TEL.; *Sime hunase,* KAN.; *Kwaytanyeng, kywétanyin,* BURM.
>
> **References.**—*Roxb., Fl. Ind., Ed. C.B.C., 421; Brandis, For. Fl., 173; Kurz, For. Fl. Burm., I., 431; Beddome, Fl. Sylv., t. 188; For. Man., 96; Gamble, Man. Timb., 145; Dalz. & Gibs., Bomb. Fl. Suppl., 25; Sir W. Elliot, Fl. Andhr., 167; Murray, Pl. & Drugs, Sind, 135; Dymock, Warden & Hooper, Pharmacog. Ind., I., 553; Drury, U. Pl. Ind., 257; Useful Pl. Bomb. (XXV., Bomb. Gaz.), 71, 154, 217; Cooke, Oils & Oilseeds, 65; Church, Food-Grains, Ind., 173; Kew Bulletin, 1889, 24; Man. Madras Adm., II., 138; Moore, Man. Trichinopoly, 78; Bombay Agri. Dept. Rept., 1884-85, 38; 1885-86, 37; Madras Experimental Farms Rept., 1879, 131; 1885-86, 44; Gazetteers:—N.-W. P., I., 80; IV., lxxi.; Mysore & Coorg, I., 59; Ind. Forester, III., 237; IV., 96, 200; VI., 240; VII., 180, 367; IX., 200; Spons, Encycl., II., 1415; Balfour, Cyclop. Ind., III., 229; Smith, Ec. Dict., 111.*

P. 900

PITHECOLOBIUM lobatum.	Manilla Tamarind.

Habitat.—A large tree, introduced from Mexico and now cultivated throughout India, especially along railway lines in the Madras Presidency.

In this country it flowers during the cold season, and annually produces many pods from 4 to 5 inches in length and ½ inch in breadth with six to eight seeds.

GUM.
901

Gum.—The authors of the *Pharmacographia Indica* state that this tree yields a gum, usually in spheroidal tears, about half an inch in diameter, of a deep reddish-brown colour, transparent, and with a polished surface. It is said to be freely soluble in water, forming a thick brown mucilage. The solution is unaffected by neutral acetate of lead, but is gelatinized by the basic acetate, ferric chloride, and borax. It freely reduces Fehling's solution.

OIL.
Seeds.
902

Oil.—The SEEDS yield a fatty oil, light coloured, and as thick as castor oil. It was first brought to notice, in 1857, by Lieutenant Hawkes, who exhibited a sample at Madras, but its qualities and uses do not appear to have been determined. Church states that the bean contains 17·1 per cent. of fatty matter.

FOOD.
Fruit.
903

Food.—The FRUIT, which ripens from April to June, consists of a number of large seeds, each of which is enveloped in a sweet, wholesome, whitish pulp. These are contained in a cylindrical, irregularly swollen pod curled at the end. The pulpy aril is eaten by the poorer classes. Church states that 100 parts of the bean contain: Water, 13.5 parts; albumenoids, 17·6; starch, 41·4; fat, 17·1; fibre, 7·8, and ash, 2·6.

TIMBER.
904

Structure of the Wood.—Sapwood small, heartwood reddish brown; weight when seasoned, 40℔ per cubic foot; smells unpleasantly when fresh sawn; used for making country carts, packing boxes, and the panelling of doors. It coppices well in South India, where it is largely grown for fuel.

DOMESTIC.
905

Domestic, &c.—It is a good avenue tree, and is also extensively cultivated as a hedge plant. The following account of the best method of cultivating it for this purpose is of interest :—" The ground was dug about a foot deep and then heavily manured. The seed was sown in small beds in lines 9 inches apart; after sowing, the beds were watered at first every day, and after the seed had germinated every second or third day, until the plants were about 6 inches high, when gradually the supply of water was diminished, and, when about one foot high, watering was discontinued. This stopped the growth of the plants, but they still remained green; they were then transplanted out to form a hedge. In a few small beds enough plants were raised to form a hedge 500 yards long. Nearly all the plants grew after being transplanted; water was applied occasionally. By this means a cheap hedge can be raised very conveniently in a short time; the planting may be performed at any time during the dry season when there is little circulation of sap in the plants " (*Madras Exp. Farm Rep.*, *1879*, *131*). This same fact has been urged by many writers: it is a useful hedge, grows rapidly, and affords food for men and fodder for horses.

906

Pithecolobium lobatum, *Benth.*; *Fl. Br. Ind., II., 305.*

Syn.—MIMOSA KOERINGA, *Roxb.*; M. JIRINGA, *Jack.*; INGA LOBATA, *Grah.*; I. ATTENUATA, *Grah.*

Vern.—*Danyin, tanyeng*, BURM.; *Kœringa*, MALAYS.

References.—*Roxb., Fl. Ind., Ed. C.B.C., 416; Kurz, For. Fl. Burm., I., 429.*

Habitat.—A large unarmed tree, met with in the tropical forests of Pegu, Tenasserim, and Malacca; distributed to the Malaya Islands and the Phillipines.

P. 906

Resin.—The WOOD exudes a blackish resin (*Kurz*).

Food.—Roxburgh describes the FRUIT as very similar to that of **P. dulce**, and states that the seeds are covered with a quantity of edible, fleshy pulp.

RESIN.
Wood.
907
FOOD.
Fruit.
908
909

Pithecolobium Saman, *Benth.; Lond., Jour. Bot., II., 423.*

THE UANGO or RAIN-TREE.

Syn.—INGA SAMAN, *Willd.*

References.—*DC., Prodr., II., 441 ; Drury, U. Pl., 257 ; Dymock, Warden
& Hooper, Pharmacog. Ind., I., 553 ; Balfour, Cyclop., III., 229 ;
Man. Madras Adm., II., 138 ; Agri.-Horti. Soc., Ind., Journals
(New Series), VI.,'xxxiv. xxxvi. ; Ind. Forester, III., 314, 315 ; IV., 96,
151, 152, 174, 199, 200, 288, 347 ; V., 461 ; VI., 149, 326 ; VII., 181, 185,
232 ; VIII., 38, 187 ; IX., 605, 606 ; X., 405 ; XIV., 378.*

Habitat.—A small, low-branching tree, native of America, introduced into the neighbourhood of Calcutta, the plantations of Kadápah and Kadur, the horticultural gardens of Lucknow and Saharanpur, and other parts of India, as an ornamental tree of rapid growth.

Gum.—It yields a clear yellow gum which, according to the authors of the *Pharmacographia Indica*, is of very inferior quality. It occurs in irregular tears and vermicular pieces with waved transverse ridges, is soft and tough, swells up in water into tough cartilage-like masses, and on keeping turns a deep reddish brown, or black colour.

GUM.
910

Structure of the Wood.—"Sapwood white, heartwood brown, soft, perishable, and to judge by our specimen by no means so good as it has been reported to be; weight 26ḷ per cubic foot" (*Gamble*). Thwaites recommended it as likely to be a tree of great value for railway fuel.

TIMBER.
911

PITTOSPORUM, *Banks; Gen. Pl., I., 131.*

[PITTOSPOREÆ.

Pittosporum floribundum, *W. & A.; Fl. Br. Ind., I., 199 ;*

912

Syn.—SENACIA NEPALENSIS, *DC.* ; CALASTRUS VERTICILLATA, *Roxb.*

Vern.—*Tibiliti,* NEPAL; *Prongsam,* LEPCHA; *Yekdi, yekaddi,* BOMB.;
Vehkali, vikhari, vehyenti, yekaddi, MAR.

References.—*Roxb., Fl. Ind., Ed. C.B.C., 209 ; Brandis, For. Fl., 19 ;
Gamble, Man. Timb., 19 ; Dals. & Gibs., Bomb. Fl., 44 ; Dymock, War-
den & Hooper, Pharmacog. Ind., I., 153 ; Atkinson, Him. Dist., 305 ;
Lisboa, U. Pl. Bomb., 8 ; Aplin, Rept. on the Shan States.*

Habitat.—A small tree, found in the sub-Tropical Himálaya, from Sikkim to Garhwál, ascending to 5,000 feet on the hills, and at Mishmi, also in the Western Peninsula from the Konkan to the Nilghiris. Mr. Aplin, while travelling with the Southern Shan Expeditionary Column in 1887-88, found it in Burma also.

Oleo-resin.—See paragraph "CHEMICAL COMPOSITION."

Medicine.—The medicinal virtues and utilisation of this plant have been brought to light by the authors of the *Pharmacographia Indica*, who state that the BARK is bitter and aromatic, and is said by Natives of the Western Ghâts to possess narcotic properties. It is used in doses of 5 to 10 grains as a febrifuge, and, in doses of 50 grains, is believed to be a specific for snake poisoning. The Maratha names signify, "an antidote for poison." A Brahmin practitioner of Poona is cited as having given 5 to 10-grain doses of the dried bark with benefit in chronic bronchitis. He found it to be a good expectorant and never found any objectionable symptoms arise from its use. In one or two cases, however, in which it was tried in Bombay, it is said to have given rise to dysenteric diarrhœa.

MEDICINE.
Bark.
913

CHEMICAL COMPOSITION.—The plant yields a similar principle to the bitter glucoside *Pittosporin*, obtained by Baron F. von Mueller from **P. undulatum**, *Vent.*, a species found in New South Wales. It also contains

Chemistry.
914

P. 914

PLANTAGO lanceolata.	Plantago Herbs.

TIMBER.
915

an aromatic yellow resin or oleo-resin having very tenacious properties (*Pharmacog. Ind*).

Structure of the Wood.—Light-coloured, strong and tough, but of small size.

PLANCHONIA, *Blume ; Gen. Pl., I., 721.* [CEÆ.

916

Planchonia littoralis, *Van Houtte ; Fl. Br. Ind., II., 511 ;* MYRTA-

Syn.—PIRIGARDA VALIDA, *Blume ;* GUSTAVIA VALIDA, *DC.*

Vern.—*Bamwbay nee*, BURM. ; *Baila dá*, AND.

References.—*Kurz, For. Fl. Burm., I., 500 ; Gamble, Man. Timb., 198.*

Habitat.—A moderate-sized evergreen tree, frequent in the coast forests of the Andamar Islands.

TIMBER.
917

Structure of the Wood.—Yellowish or reddish brown, with yellow specks, very hard and close-grained ; weight from 61 to 64℔ per cubic foot ; seasons well and takes a fine polish. Gamble remarks that it is a valuable wood which should be better known.

918

PLANTAGO, *Linn. ; Gen. Pl., II., 1224.*

A genus of herbs which comprises about 50 species, of which ten are natives of India. Besides the properties noticed under the various species, it is worthy of notice that, according to Gmelin, the leaves of several members of the genus yield, on fermentation, a volatile oil, which may be obtained by distilling the fermented mass. These oils do not exist ready formed in the fresh plant, and differ from ordinary volatile oils in their more ready solubility in water.

[GINEÆ.

919

Plantago amplexicaulis, *Cav. ; Fl. Br., Ind., IV., 706 ;* PLANTA-

Syn.—P. BAUPHULA, *Edgew.* ; P. SALINA, *Dcne.* ; P. LAGOPOIDES, *Desf.*

Vern.—*Spighwol, isafghol, gajpipali*, PB.

References.—*Boiss., Fl. Orient., IV., 883 ; Stewart, Pb. Pl., 173 ; DC.,* *Prodr., XIII., i., 719 ; S. Arjun, Bomb. Drugs, 110 ; Baden Powell, Pb.* *Pr., 369 ; Agri.-Horti. Soc., Ind., Journals (Old Series), XIV., 6, 7, 40.*

Habitat.—A stemless or sub-caulescent herb, found in the plains of the Panjáb from the Sutlej westwards, also in Malwa, and on the Búgta Hills in Sind.

MEDICINE.
920

Medicine.—Baden Powell states that it is said to be astringent, useful in intermittent fever, and as an application to the eyes in ophthalmia ; and is also used as a remedy for snake-bite and pulmonic diseases. Dr. Dymock, in a note to the editor, throws some doubt on these remarks referring to a **Plantago**, and states that the *gajpipal* of Bombay appears to be a **Balanophora.**

921

P. brachyphylla, *Edgew., Fl. Br. Ind., IV., 706.*

Syn.—P. REMOTIFLORA, *Stocks.*

Vern.—*Parhar pángi*, PUSHTU.

Habitat.—A herb found in the Western Himálaya from Kumáon to Kashmír, at 9,000 to 13,000 feet, in Western Tibet, from 11,000 to 14,000 feet ; distributed to Afghánistán.

MEDICINE.
Leaves.
922

Medicine.—Mr. Lace informs the writer that the LEAVES, slightly bruised, are, in Ziarat, used as an application to wounds.

923

P. lanceolata, *Linn. ; Fl. Br. Ind., IV., 706.*

Syn.—P. ATTENUATA, *Wall.*

Vern.—*Baltanga*, HIND. ; *Baltung, bartung*, BENG. ; *Parhar pangi,* *parbar pangi, bartang*, PUSHTU.

References.—*Thwaites, Enum. Cey. Pl., 245 ; Aitchison, Bot. Afgh. Del.* *Com., 100 ; Lace, MSS. on Quetta Pl. ; MSS. notes by collector, Trans-* *Indus, in Mr. Duthie's Herb. ; Irvine, Mat. Med., Patna, 13 ; Ind. For-* *ester, VI., 238 ; Agri.-Horti. Soc., Ind., Jour. (Old Series), XIV., 15.*

P. 923

Habitat.—A variable herb, met with in the Western Himálaya from Kashmír to Simla, the Salt Range, and Waziristan ; distributed to Europe and Northern Asia.

Medicine.—The LEAVES are similarly employed to those of the last species, as an application to wounds, inflamed surfaces and sores. Mr. Duthie's collector states that the SEEDS are used with sugar as a drastic purgative, while Irvine believed certain angular glassy-looking seeds imported into Patna under the name of *bartung* and employed as an astringent and cooling remedy for diarrhœa, to be the produce of this species. Irvine's seeds may possibly be those of P. **major,** *Linn.*

MEDICINE.
Leaves.
924
Seeds.
925

Plantago major, *Linn. ; Fl. Br. Ind., IV., 705.*

    926

   Syn.—P. EROSA, *Wall. ;* P. ASIATICA, *Linn. ;* P. LONGISCAPA, *Jacquem.*

   Vern.—*Luhuriya,* HIND. ; *Luhuriya,* KUMAON ; *Gúl, isafghol,* KASHMIR ; *Gúl, isafgol, karet, ghuzbe,* FRUIT=*gaz pípal,* PB. ; *Ghuzbhe,* PUSHTU ; *Bártang, bárhang,* BOMB. ; *Bártang, bárhang,* PERS.

   References.—*Boiss., Fl. Orient., IV., 878 ; Thwaites, En. Ceylon Pl., 245 ; Stewart, Pb. Pl., 174 ; Bot. Afgh. Del. Com., 100 ; O'Shaughnessy, Beng. Dispens., 511 ; Dymock, Mat. Med. W. Ind., 2nd Ed., 650 ; Year-Book of Pharmacy, 1874, 628 ; Atkinson, Him. Dist., 316, 746 ; Gaz., Mysore & Coorg, I., 65 ; Ind. Forester, VI., 238 ; Agri.-Horti. Soc., Ind., Journal, XIV., 28.*

Habitat.—A large herb, found on the Temperate and Alpine Himálaya, from Pesháwar and Kashmír to Bhotán, from 2,000 to 8,000 feet ; and in Western Tibet, at 10,000 to 12,000 feet. It has also been reported from Assam, the Khásia Mountains, Burma, Malacca, Singapore, Bombay, the Nilghiri Hills, and the higher parts of Ceylon.

Dye.—The ROOT and LEAVES contain a red colouring matter, which, when these parts are chewed, tinges the saliva red.

DYE.
Root.
927
Leaves.
928

Medicine.—This plant, known as "Plantago" to the Romans, and as αρνογ λωσσον to the Greeks, had medicinal properties ascribed to it in ancient times. Dioscorides recommends the LEAVES and ROOT in a number of diseases as a valuable astringent and febrifuge, and the SEEDS in wine as a cure for "bloody fluxes." Dymock informs us that Arabian and Persian writers describe it under the name of *lisan-el-hamal,* and state that it is the *kasrat-el-azlaa,* and *sabaat azlaa* of Dioscorides (Arabian translations of επτάπλευρον and πολύνευρον) meaning "seven-ribbed" and "many-ribbed." These writers repeat with a few trifling additions the observations of the Greeks. The seeds at the present time are the part most utilised in Indian medicine. They have the same properties ascribed to them as those of P. **ovata,** being considered an efficient remedy in dysentery, stimulant, warm, and tonic. Dymock states that they are largely imported from Persia. The root and leaves possess slightly bitter and astringent qualities, and, according to O'Shaughnessy, were formerly much used as a febrifuge, though now justly neglected. They are, however, still employed as a domestic remedy in England, and in Tuscany a decoction of the leaves is believed to form an excellent eye-wash, and to have styptic properties.

MEDICINE.
Leaves.
929
Root.
930
Seeds.
931

P. ovata, *Forsk. ; Fl. Br. Ind., IV., 707.*

    932

   Syn.—P. DECUMBENS, *Forsk. ;* P. ISPAGHUL, *Roxb. ;* P. LANATA, *Wall.*

   Vern.—*Ispaghúl, isbaghól, isabghul, issufgúl,* HIND. ; *Eshopgól, ispaghúl, isabgul,* BENG. ; *Isabgul,* URIYA ; *Ispaghul,* N.-W. P. ; *Is-mó-gul,* KASHMIR ; *Bartang, isafghol, ispaghól, isabghol,* PB. ; *Ishpagul, isafgól,* PUSHTU ; *Ispungur,* SIND ; *Isapghol,* BOMB. ; *Isabagóla,* MAR. ; *Isapghól, urthamujirú, isafghol,* GUZ. ; *Isapghól, ispoghul,* DEC. ; *Ishappukól-virai, iskól-virai, ispoghol verai, ispoghol,* TAM. ; *Isapag ála-vittulu, isphagula,* TEL. ; *Isabakólu, isabgólu,* KAN. ; *Bazre-quatúná, bazre-*

PLANTAGO
ovata.

The Ispaghul.

katima, ARAB.; *Ispoghul, ispaghol, isbaghól, isparsah, shikam-daridah,* PERS.

References.—*Boiss., Fl. Orient., IV., 885; Roxb., Fl. Ind., Ed. C.B.C., 135; Voigt, Hort. Sub. Cal., 437; Stewart, Pb. Pl., 174; Sir W. Elliot, Fl. Andhr., 71; Pharm. Ind., 182; Fluck. & Hanb., Pharmacog., 490; Fleming, Med. Pl. & Drugs (Asiatic Reser., XI.), 174; Ainslie, Mat. Ind., II., 116; O'Shaughnessy, Beng. Dispens., 510; Irvine, Mat. Med. Patna, 40; Moodeen Sheriff, Supp. Pharm. Ind., 201; Sakharam Arjun, Cat. Bomb. Drugs, 110; Murray, Pl. & Drugs, Sind., 168; Waring, Basar Med., 118; Bent. & Trim., Med. Pl., t. 211; Dymock, Mat. Med. W. Ind., 2nd Ed., 648; Official Corresp. on Proposed New Pharm. Ind., 223; Med. Top., Ajmere, 148; Birdwood, Bomb. Prod., 68; Baden Powell, Pb. Pr., 368; Drury, U. Pl. Ind., 347; Atkinson, Him. Dist. (X., N.-W. P. Gas.), 316; Royle, Ill. Him. Bot., 312; Stocks, Rep. on Sind; Christy, Com. Pl. & Drugs, IV., 45; VI., 99; Rept. on the Govt. Bot. Gardens & Park on the Neilgherries, 1881-82; Gazetteers:—Bombay, VI., 15; Panjáb, Gujrát, 12; N.-W. P., II., 506; IV., lxxvi.; Orissa, II., 159; Agri.-Horti. Soc., Ind.:—Journal X., 28; Balfour, Cyclop. Ind., III., 231.*

Habitat.—A herb found in the Panjáb plains and low hills, from the Sutlej westwards; distributed westwards to Spain and the Canaries. Edgeworth states that it was, in his time, cultivated at Múltán, but according to Stewart it is never cultivated int he Panjáb, except, perhaps, sparingly at Lahore. It is also said to be cultivated to a small extent in Bengal, Mysore, and the Coromandel coast.

MEDICINE.
Seeds.
933

Medicine.—*Isbaghúl* or "spogel" SEEDS are not mentioned by Sanskrit writers on medicine, and were apparently unknown to them. They (and probably the seeds of several other species of Plantago) are, however, very frequently referred to by Arabian and Persian writers, who appear to have long held the medicine in high esteem. The Persian physician Alhervi mentions them in the tenth century, a little later Avicenna refers to the drug, and nearly all subsequent writers on Muhammadan medicine ascribe valuable properties to *Ispaghul* or *basre-katúna.* Flückiger & Hanbury state that its valuable qualities were first brought to the notice of European science in 1719 by Luick, and, towards the end of the century, Fleming, Ainslie, and Roxburgh have all something to say in its favour. Ainslie writes, "These seeds are of a very cooling nature, and form a rich mucilage with boiling water, which is much used by Native practitioners, and indeed of late years by the European medical officers of India, in cases of catarrh, gonorrhœa, and nephritic affections." Fleming makes a similar statement regarding the utilisation of the drug in Bengal, "in all diseases in which acrimony is to be obviated or palliated." Later, the valuable demulcent properties which they possess in cases of dysentery and certain forms of diarrhœa, were forcibly brought to notice by Twining (*Diseases of Bengal, I., 212*), and in 1868 the seeds were admitted as officinal to the *Indian Pharmacopœia.* Some authors recommend that the seeds should be administered dry in doses of two and a half drachms; others prefer its exhibition as a mucilaginous decoction, four drachms being boiled in two pints of water, till the quantity is reduced to a pint, and the whole given, in divided doses, in the course of a day. The latter course appears to be best adapted for cases of dysentery, while the former is most efficacious in diarrhœa, especially the chronic diarrhœa of children, and "hill" diarrhœa. A slight degree of astringency may, it is said, be imparted to the seeds by slightly heating them. By Natives they are considered cooling and demulcent, and are chiefly employed in diarrhœa, dysentery, other inflammatory and functional derangements of the digestive organs, and fever. The crushed seeds made into a poultice are applied to rheumatic and gouty swellings. A cooling lotion for the head is prepared from the mucilage; and a decoction is prescribed in cases of cough and colds. There is little doubt that the mucilaginous decoction is a valuable adjunct to other curative agents in the treatment of dysentery and diarrhœa, important points in its favour being that it has no unpleasant taste, and may be prescribed, while any other drug is being, at the same time, administered. *Ispaghúl* has attracted little attention in European medicine outside India, though it is probably worthy of a fair trial in cases of ordinary and summer diarrhœa. **Mr. Christy** recommends the seeds as a valuable remedy in fowl-diarrhœa.

MEDICINE.

CHEMICAL COMPOSITION.—"*Mucilage* is so abundantly yielded by these seeds that one part of them, with 20 parts of water, forms a thick tasteless jelly. On addition of a larger quantity of water and filtering, but little mucilage passes, the majority adhering to the seeds. The mucilage separated by straining with pressure does not redden litmus, is not affected by iodine, nor precipitated by borax, alcohol or ferric chloride" (*Pharmacographia*). A *fat oil* and *albumenous matter* also exist in the seed, but neither of these has been carefully examined.

Large quantities of the seed are said to be imported into Bombay from Persia; value R4 per maund of 37½ḷ̣ (*Dymock*).

Chemistry.
934

SPECIAL OPINIONS —§"Is most useful as a drink in dysentery and diarrhœa when giving other remedies" (*Civil Surgeon G. C. Ross, Delhi, Panjáb*). "A very valuable remedy in the treatment of chronic dysentery. I have used it largely with good results and can bear personal testimony of its efficacy. I use it in spoonful doses of the whole seed, steeped for 15 or 20 minutes in water, the resulting mucilaginous mass being swallowed. Many of the swollen seeds pass out whole with the motions, and I believe their actions to be mechanical as well as astringent to the intestinal ulcers. I have used it largely for dysentery and chronic diarrhœa in both Europeans and Natives, and consider it a very valuable medicine" (*Surgeon C. H. Joubert, Darjíling*). "Useful in gonorrhœa, as well as in diarrhœa and dysentery" (*Assistant Surgeon T. N. Ghose, Meerut*). "Extremely useful in inflammatory affections of the mucus membrane of the alimentary canal. The seeds alone may be eaten, powdered and mixed with sugar, or may be soaked in a small quantity of water. Dose one to two drachms of the seeds frequently during the day" (*Surgeon G. Price, Shahabad*). "The mucilage prepared from this seed is useful in cases of chronic diarrhœa and dysentery" (*Officiating Civil Surgeon E. Borill, Motihari, Champarun*). "Useful in chronic dysentery. The seeds should be given intact mixed with sugar" (*Surgeon J. Maitland, M.B., Madras*). "A good demulcent in dysentery, diarrhœa, acute catarrh" (*Assistant Surgeon Nehal Sing, Saharunpore*). "Is a useful demulcent in gonorrhœa" (*W. Forsyth, F.R.C.S., Edin., Civil Medical Officer, Dinajpore*). "The seeds when steeped in water yield a mucilaginous matter which is demulcent and soothing in gonorrhœa. The seeds should be strained off and the mucilage beaten up with the water and taken *ad libitum*. The seeds, *warmed* and reduced to powder, are of signal use in diarrhœa and dysentery. Dose ʒi to ʒii" (*Civil Surgeon C. M. Russell, Sarun*). "The mucilage is very useful in cases of heat and scalding during micturition, taken in the form of a *sherbet* with sugarcandy" (*Assistant Surgeon S. C. Bhattacharji, Chanda, Central Provinces*). "It is very useful in cases of suppression of urine, ardor urine, and irritability of bladder, &c., either alone or combined with alkalies and diuretics, &c. It is much used by the Natives here in similar cases" (*Apothecary J. G. Ashworth, Kumbakonam, Madras*). "Is emollient, demulcent, and laxative; also diuretic and is useful as an internal remedy for piles. Poultices made with the seeds are used for unhealthy ulcers, sores, and sinuses" (*Civil Surgeon J. H. Thornton, B.A., M.B., Monghyr*). "I have seen this medicine do good

PLATINUM.   The Plane Tree; Platinum.

MEDICINE.

in mucous-dysentery when other remedies failed to influence the disease " (*V. Richards, Civil Medical Officer, Goalundo, Bengal*). " Demulcent and mild astringent combined with Ipecacuanha, useful in dysentery, but not in diarrhœa. With cubebs and nitrate of potash most useful in gonorrhœa " (*Civil Surgeon S. M., Shircore, Murshedabad*). " Is also used for piles " (*Surgeon-Major P. N. Mukerji, Cuttack, Orissa*).

Plantain, see **Musa sapientum,** *Linn.;* Vol. V., 290.

Plasma.—A bright to leek-green form of Chalcedony, used in the Panjáb as a substitute for Jade—see Vol. II., 172; also Vol. IV., 541.

Plaster, see **Cements;** Vol. II., 245.

Plaster of Paris, see **Gypsum;** Vol. IV., 195.

PLATANUS, *Linn.; Gen. Pl., III., 396.*

935

Platanus orientalis, *Linn.; Fl. Br. Ind., V., 594.* PLATANACEÆ.

THE PLANE TREE.

Syn.—P. VULGARIS, *Spach.*

Vern —*Búin, búná, bóin, bonin,* KASHMIR; *Búin, búna, chinár, chanár,* PB.; *Chintar, chinár,* PUSHTU; *Chinár,* PERS.

References.—*Boiss., Fl. Orient., IV., 1161; Brandis, For. Fl., 434; Gamble, Man. Timb , 345; Stewart, Pb. Pl., 202; Bot. Afgh. Del. Com., 110; Notes on Prods.. W.Afghánistan & N.-E. Persia, 160; Baden Powell, Pb. Pr., 589; Smith, Dic., 328; Gasetteers:—Panjáb, Ráwalpindi, 13; Dera Ismail Khan, 19; Bannu, 23; Ind. Forester, II., 181, 408; IV., 252; V., 180, 181, 478; IX., 14; XI., 55; XIV., 371; Settle. Rept., Panjáb, Kohat, 30; Agri.-Horti. Soc., India:—Journals (Old Series), XIV., 15; Agri.-Horti. Soc., Panjáb, Select Papers to 1862, Index, 43.*

Habitat.—A large, deciduous tree, cultivated in the North-West Himálaya, from the Sutlej westwards between the altitudes of 5,000 and 8,000 feet. Indigenous from Northern Persia (west of Meshad, according to Aitchison), westwards to Southern Italy. It is chiefly valued for its wood, but it is also much admired as an ornamental tree, and attains a great size and immense age.

MEDICINE.
Leaves.
936
Bark.
937
TIMBER.
938

Medicine.—Honigberger states that the fresh LEAVES were in his time bruised and applied to the eyes in cases of ophthalmia, and that the BARK, boiled in vinegar, was given in diarrhœa.

Structure of the Wood.—Yellowish white, somewhat resembling Beech-wood; no distinct heart-wood; weight from 38 to 41℔ per cubic foot. It is compact, fine-grained, but not strong; and in Kashmir, where other timber is plentiful, is valued only because it is easily worked, being made into small boxes, trays, pen-cases, and similar articles which are painted and lacquered. Aitchison, however, informs us that it is largely employed in Northern Persia for making much larger articles, such as the large gateways of villages, doors, lintels, and sometimes roofing. In Afghán-istán, it is said to be used for making gun-carriages. Brandis states that it takes a beautiful polish and recommends it for cabinet-work.

939

PLATINUM, *Ball, Man. Geol. Ind., III. 167, 607.*

This metal rarely if ever occurs pure in nature. It is generally found com-bined with one or oftener more of the rare metals, palladium, rhodium, iridium, ruthenium, and osmium, in an ore called *polyxene.* It is also found in associa-tion with stream gold.

940

Platinum, *Mallet, Man. Geol. Ind., IV., 3.*

Vern.—*Saféd soná,* HIND.; *Shintan, shwéphyu,* BURM.

P. 940

References.—*Baden Powell, Pb. Prod., 14; Forbes Watson, Industrial Survey, II., 406; Balfour, Cyclop. Ind., III., 234; also many papers in the publications of the Geological Survey of India, and of the Asiatic Society of Bengal, for an enumeration of which the reader is referred to the Manual of the Geology of India, l.c.*

Occurrence.—According to Ball, platinum, where it has been found in India, is originally derived from metamorphic rocks. It has been found only in association with gold in auriferous sands. There is no trade in Indian platinum, and, since the gold-washers are ignorant of its value, our knowledge of the possible productiveness of Indian platiniferous sands is far from accurate. The following are the localities cited by Ball and Medlicott, from which platinum has been recorded :— | OCCURRENCE 941

Madras.—According to Mr. Rice platinum in small quantities has been obtained in the gold-washings in the Kolar District. | Madras. 942

Bombay.—The metal has been stated to occur as fine particles in the gold dust obtained in the Kappatgode region. Ball throws doubt on these particles being really platinum, and draws attention to the fact that native silver has been obtained in the same locality. | Bombay. 943

Panjab.—Baden Powell states that grains of platinum are occasionally found in small quantities, along with gold, in the Indus valley in the Tavi river, and in the Kabul river at Naushíra. The gold-seekers called it *saféd soná* and reject it as useless. | Panjab. 944

Bengal.—Medlicott informs us that minute grains of platinum were, in 1882, noticed in stream gold from the Guram river, near Dhadka in Mánbhum, from Lándu in Chaibassa, and from the Bráhmini river in the Tributary Mahals of Orissa. In all cases, however, the amount was extremely small, being not more than a trace, in comparison with the accompanying gold. | Bengal. 945

Tibet.—Dr. Saunders (*Turner's Embassy to Tibet* (1800), 405) records that he extracted 12 per cent. of refuse from some gold dust obtained between Eastern Bhután and the Sangpo river, and on examination found it to be sand and filings of iron "which last was not likely to have been with it in its native state, but probably employed for the purpose of adulteration." Medlicott commenting on this statement writes, "Was the supposed iron platinum? The former metal would be a very clumsy adulterant of gold on account of its colour." | Tibet. 946

Assam.—Gold sand, obtained from the Noa Dihing river in Upper Assam, has been found to contain platinum. | Assam. 947

Burma.—As far as our present knowledge extends, platinum appears to be more plentiful in Burma than in any other part of India. In 1831 a button, obtained by melting some metallic grains "having every appearance of iron," which had been found mixed with gold from the Chindwin river in Upper Burma, was subjected to analysis by Mr. J. Prinsep. He found it to be a mixture of platinum, gold iridium and osmium, arsenic and lead, with doubtful traces of rhodium and palladium. Twenty-five per cent. consisted of platinum, forty per cent. of iridium and osmium. A sample of the original ore, afterwards examined by the same chemist, gave about twenty per cent. of platinum and twice at much iridium. Medlicott considers it likely that the substance in question was a mixture of platinum and iridosmine, not a definite ore. The gold dust, with which this ore was obtained, was collected by means of wild cow horns, placed in the stream (see **Gold**, Vol. III., 531). It is stated that considerable quantities are collected, and that the Burmans are supposed to be able to work the metal. | Burma. 948

Platinum has also been obtained from the streams which fall into the Irrawadi in the direction of Bhámo, and from certain parts of the Shan

19

States. A sample of gold from the Meya valley, recently analysed by Dr. **Romanis**, contained 2·53 per cent. of platinum, mixed with a little iridium.

USES.
949

Uses.—The qualities of platinum are too well known to require notice in such a work as the present. Unsuited by its unattractive appearance to the jewellers' purposes,—though one of the " noble metals" chemically, —it has never been utilised in India. Almost all the metal now made is employed for making chemical utensils, for which it is admirably suited owing to its infusibility and chemical inertness.

PLECOSPERMUM, *Tréc.; Gen. Pl., III., 361.*
[*t. 1963;* URTICACEÆ.

950

Plecospermum spinosum, *Tréc.; Fl. Br. Ind., V., 491; Wight, Ic.,*
    Syn.—BATIS SPINOSA, *Roxb.;* B. AURANTIACA, *Wall.;* TROPHIS SPINOSA, *Heyne;* T. ACULEATA, ·*Roth.*
    Vern.—*Banabana,* URIYA; *Mainakat-lara, maidal-lara,* NEPAL; *Gumbengfong,* MICHI; *Alaṣále, koriti, sali, goriti, goriti donka,* TEL.; *Kattotimbúl,* SING.
    References.—*Roxb., Fl. Ind., Ed. C.B.C., 715; Brandis, For. Fl., 401; Beddome, For. Man., 220; Gamble, Man. Timb., 327; Elliot, Fl. Andhr., 13, 62, 98, 166; McCann, Dyes & Tans, Beng., 90; Wardle, Dye Report, 48.*

Habitat.—A large thorny shrub, or small tree, met with from the Salt Range, altitude 3,000 feet, eastwards along the foot of the Himálaya, and southwards to Travancore and Ceylon.

DYE.
Wood.
951
Bark.
952

Dye.—**Gamble** states that the WOOD is used in the Darjíling Terai in colouring silks yellow. According to **McCann** this BARK is generally employed along with turmeric and the bark of **Symplocos racemosa.** The latter author further states that some of the specimens he had received as *gumbengfong* belonged to a species of **Morinda.** This remark makes the following report by **Wardle** doubtfully referable to **Plecospermum.** " This specimen " (*Gumbengfong*) "of **Morinda** root appears to contain less colouring matter than any I have examined." The colours he obtained with unbleached tussur varied from brownish yellow to light brown; those with bleached tussur were light brownish yellow; those with corah silk and wool a brown gold. But the authenticity of his sample may be doubted; he himself certainly considered it to be a root of a species of **Morinda.**

TIMBER.
953

Structure of the Wood.—Greyish-white, with a small bright orange-yellow heartwood, which is very hard and is filled with a yellow resinous matter; growth slow; weight 50℔ per cubic foot. It makes an excellent fuel.

DOMESTIC.
954

Domestic.—Roxburgh and Beddome state that the shrub is well adapted for planting as a hedge.

PLECTOCOMIA, *Mart.; Gen. Pl., I., 934.*

A genus of canes which belongs to the Natural Order PALMÆ, and which, according to Indian writers, comprises four species, natives of this country. They are probably all utilised in the same way as other canes. The following are the species described up to the present.

955

Plectocomia assamica, *Griff., Palms, Br. East India, 107.*
    Habitat.—A cane met with in Upper Assam. It resembles **P. khasyana,** and is united to that species by **T. Anderson.**

956

P. himalayana, *Griff., Palms, Br. East India.*
    Vern.—*Taki bet,* NEPAL; *Runúl,* LEPCHA.
    References.—*Gamble, Man. Timb., 424; Trees, Shrubs, &c., Darjíling, 88.*

Habitat.—A cane met with in the Sikkim hills at 4,000 to 7,000 feet altitude.

Plectocomia khasyana, *Griff., Palms, Br. East India, 106.*   **957**
Habitat.—Found in the Khasia hills.

P. macrostachya, *Kurz, For. Fl. Burm., II., 514.*   **958**
Habitat.—An evergreen lofty climber, met with on the Bithoko range in Tenasserim at an altitude of 3,000 feet.

PLECTRANTHUS, *Lher.; Gen. Pl., II., 1175.*

Plectranthus Patchouli, *Clarke; Fl. Br. Ind., IV., 624;* see **Pogostemon Patchouli,** *Pelle.;* LABIATÆ; p.

P. rugosus, *Wall.; Fl. Br. Ind., IV., 620.*   **959**
Vern.—*Solei,* KASHMIR; *Búi, piumár* (=FLEA-KILLER), *chúgú, sola, solei, kot, siringrí, pek, rosbang, chichri, tsarbs, kwangere, itsit,* PB.
References.—*Gamble, Man. Timb., 300; Stewart, Pb. Pl., 171; Agri-Horti. Soc. Ind., Journals (Old Series), XIV., 26.*
Habitat.—A small shrub with slender branches, found on the dry hills of the Western Himálaya from Kashmír to Garhwál, at altitudes from 3,000 to 8,000 feet; also met with in Bhután and Merwar; distributed to Afghánistán.
Domestic.—The PLANT is, in certain places, used as bedding to keep off fleas (*Stewart*).

DOMESTIC.
Plant.
960

PLUCHEA, *Cass.; Gen. Pl., II., 290.*

[(*colour of flower wrong*); COMPOSITÆ.   **961**

Pluchea indica, *Less.; Fl. Br. Ind., III., 272; Wight, Ill.; t. 131*
Syn.—P. FOLIOLOSA, *DC.;* CONYZA CORYMBOSA, *Roxb.;* C. INDICA, *Miq.;* BACCHARIS INDICA, *Linn.*
Vern.—*Munjhú rúkha, kukronda,* BENG.; *Kayu,* BURM.
References.—*Roxb., Fl. Ind., Ed. C.B.C., 601; Kurz, For. Fl. Burm., II., 83; Gamble, Man. Timb., 232; Irvine, Mat. Med. Patna, 73; Gazetteers:—Mysore & Coorg, I., 62; Bombay, V., 26.*
Habitat.—A low evergreen shrub, met with from the Sundarbans to Malacca and Penang; distributed to the Malay Islands and China.
Medicine.—Irvine states that the ROOT and LEAVES are employed medicinally in Patna as astringents and in cases of fever.

MEDICINE.
Root.
962
Leaves.
963
964

P. lanceolata, *Oliv.; Fl. Br. Ind., III., 272.*
Syn.—BERTHELOTIA LANCEOLATA, *DC.;* CONYZA LANCEOLATA, *Wall.*
Vern.—*Reshæ, marmandai, reshami, resham-búti, sármei,* leaves=*rásanna,* PB.; *Marwande,* PUSHTU; *Chota kalia,* RAJ.; *Kúra-sanna,* SIND.
References.—*Stewart, Pb. Pl., 122; Lace, Quetta Pl., MSS.; Dymock, Mat. Med. W. Ind., 2nd Ed., 449; Murray, Pl. & Drugs, Sind, 183; Dymock, Warden & Hooper, Pharmacog. Ind., II., 256; Baden Powell, Pb. Pr., 358; Ind. Forester, XII., App., 15.*
Habitat.—An annual shrub which grows abundantly in Upper Bengal, Oudh, and westwards to the Panjáb and Sind; distributed to Afghánistán, Baluchistán, and North Africa. It has long-spreading roots which extend to several yards and send up shoots as they go. Though, according to Stewart, not considered particularly troublesome to the Panjáb cultivator, it is said to be one of the chief agricultural evils against which cultivators near Agra, Jaunpur, &c., have to contend.
Medicine.—The LEAVES are aperient and are used as a substitute or adulterant for senna.

MEDICINE.
Leaves.
965

19 A

P. 965

Plum-tree, see **Prunus communis,** *Huds.*, **var.domestica,** Rosaceæ; p. 347.

966   **PLUMBAGO,** *Ball, in Man. Geol. Ind., III., 50, 597.*

Plumbago or Graphite, a mineral form of Carbon, occurs in many parts of the world, always in rocks belonging to the earliest formations. It exists native in two varieties, one of which, like the Bonorodale graphite, is fine-grained and amorphous; the other, of which Ceylon plumbago is an example, is composed of small flat plates. The crystalline structure of the mineral is that of a flat hexagon, it has a metallic lustre, a black or dark-grey colour and streak, it feels greasy, and conducts electricity. Chemically native graphite contains from 95 to nearly 100 per cent. of carbon, the impurity generally consisting of small quantities of silicates. At the present day nearly all the plumbago of commerce comes from Ceylon.

967   **Plumbago,** *Mallet, in Man. Geol. Ind., IV., 9.*

Graphite, Black Lead.

References.—*Mason, Burma & Its People, 576; Theobald, Natural Productions, Burma, 2nd Ed., 10; Baden Powell, Pb. Prod., 17, 57; Atkinson, Him. Dist., 292; Ec. Mineralogy Hill States, N.-W. P. (1877); Madras Man. of Admin., II, 32; Balfour, Cyclop. Ind., I., 1245; also many pamphlets in the publications of the Geological Survey of India and of the Asiatic Society, for a list of which the reader is referred to the Man. Geol. Ind., l.c.*

OCCURRENCE
968

Occurrence.—The localities where graphite has been found most abundantly in India are in the Madras Presidency, especially in Travancore. The evidence of experts on the mineral from this locality, however, where it is of better quality than elsewhere, was unfavourable, and it appears to be doubtful whether the Indian deposits could ever be worked with financial success. It may, however, be noted that **Ball** takes a rather more favourable view. "It is conceivable," he writes, "that an increased demand, and improved and more economical methods of purification, might render its exploitation a profitable undertaking. At the same time the possibility of further search proving the existence of a quality equal to the better kinds found in Ceylon cannot be denied, as the probability is that the including rocks are of identical age." With these prefatory remarks a short account of the localities enumerated by **Ball** and **Medlicott** may be given.

Madras.
969

Madras.—In 1845 General **Cullen** discovered graphite in Travancore and traced it in the gneiss from the south of Trivandram northwards as far as Cochin. Samples sent to the Asiatic Society were reported on as too soft and scaly for the manufacture of pencils by the old method, and **Ball** states that from the specimens in the Geological Museum he considers that the Travancore graphite could not be made available even for inferior purposes, without much grinding and washing.

Specimens of apparently better quality were described by **Dr. Royle** in 1855, and though these also were reported on as too gritty and impure to be of much value, a crucible manufacturer estimated the value of the mineral for his purposes at 8 shillings a cwt. About ten years ago a still purer-looking graphite was found by **Dr. King** close to Vellurnad near Arinand. Graphite has also been found in Tinnevelly, in the red hills near Madras, in the Kistna district, and near Kasipuram, and one or two other localities in Vizagapatam, but all apparently of little value. That from the last-mentioned locality can, according to **Ball**, be obtained in any quantity at Vizianagram at a rupee for 24℔. It is used locally for giving a polish to pottery.

P. 969

Black Lead is found. (*J. Murray.*) PLUMBAGO.

Bengal.—A loose fragment of gneiss containing lamellæ of graphite was found in the bed of the Koel river near the village of Hutar in the Loharaaga District. **Medlicott** states that a peculiar carbonaceous mineral was recently discovered by **Dr. Emil Stœhr** in the Jamjura Coppermine, Singhbhum, 37 feet below the surface. On analysis it was found to contain 94 per cent. of carbon, and was considered by **Breithaupt** as intermediate between anthracite and graphite.

Central Provinces.—Graphite, probably commercially valueless, has been found at Daramgarh and Domaipalı ın the Sambalpur District, as a constituent of gneissose schist. **Medlicott** writes that in 1882 a sample of fairly good quality was sent to the Museum from the Chhattisgarh Division with the remark that it was "said to occur in large quantities near the villages of Lanjigan and Dınsargi in the feudatory state of Kalahandi." No report is given as to its possible commercial value.

Rájputana.—Worthless deposits occur near Sohna in the Gurgaon District.

Afghánistán.—**Captain Drummond** states that he obtained a specimen of graphite said to have been found in the vicinity of Koh-i-daman, and includes the mineral as one of the regular productions of North Afghánistán.

Panjáb.—**Baden Powell** states that impure graphite has been found in Haripúr, Kangra, where it is known as *Kalimitti,* and in Sonah and Bhúndi, in Gurgaon. None of these appears to be of value.

North-West Provinces.—Graphite was discovered in the neighbourhood of Almora by **Captain Herbert,** which, when roughly analysed, was found to contain about 72 per cent. of carbon. In 1850 the Government of India took measures to find the extent and value of the deposit, and though the amount was found to be large the opinions of several experts were decidedly against the mineral having any commercial value. Samples, carefully assayed by **Mr. Medlicott,** were found to contain only 52 per cent. of carbon, together with much gritty, insoluble impurity. The graphite also was much mixed with schist, occurred in unworkably thin layers, and though great hopes were at the time held out of its being possible to work it with financial success, all the conditions were in reality such that **Ball** remarks, "There are no grounds for believing that the graphite can ever be worked so as to become a profitable commercial commodity." The probable presence of iron in considerable quantity renders its utility doubtful, even for the coarse purposes of the crucible maker.

Darjíling.—Very worthless plumbago, which contains from 83 to 93 per cent. of siliceous impurity, has been found at the foot of the Darjiling Hills. Even the little carbon these shale-like deposits contain has only been partially changed from the amorphous to the graphitic condition. Needless to say they are utterly valueless.

Burma.—According to **Mr. Theobald,** "**Colonel Bogle** forwarded specimens of graphite of fair quality from the Tenasserim Provinces, and **Dr. Mason** records having seen fine specimens from the Kanee valley, 20 miles north-east of Toung-ngoo, where the Karens report the substance abundant." **Major Strover,** some years ago, reported that the mineral was to be found in Upper Burma to the east of Nat-taik on a low range of hills near the village of Nzoketoke. A carbonaceous mineral, obtained from Tenasserim and of doubtful affinity, has been called Tremenheerite. **Dr. Mason** states that it is abundant in several localities in the vicinity of both Tavoy and Moulmein, but **Medlicott** considers it not altogether certain that the mineral he refers to is identical with the substance originally called Tremenheerite by **Mr. Piddington.** Analysis of that substance showed it to contain nearly 86 per cent. of carbon, mixed

OCCURRENCE

Bengal.
970

Central
Provinces.
971

Rajputana.
972
Afghanistan.
973

Panjab.
974

N.-W.
Provinces.
975

Darjiling.
976

Burma.
977

PLUMBAGO rosea	A Stimulating Medicine.

with water, sulphur and ferric peroxide (iron pyrites), and silica. Its physical properties do not appear to be such as to render it of probable commercial value, but this point cannot be determined without further investigation. It differs from pure graphite and from anthracite. Dr. Oldham states that it is not coal, **Professor Dana** suggests that it may be "impure graphite or between coal and graphite," and **Medlicott** seems to suggest that it may be graphite, the physical characters of which are altered by a large admixture of pyrites.

USES.
978

Uses.—Graphite is used in the manufacture of pencils, crucibles, grate and iron work polish, as a lubricator for machinery and wood work, for electro-typing, as facings for moulds, for refractory mixtures, and for giving a protecting surface to the interior of blowing cylinders in blast furnaces (*Ball*). The native Indian mineral appears to be only occasionally utilised, and then merely for polishing pottery.

PLUMBAGO, Linn. ; Gen. Pl., II., 627.

979

Plumbago rosea, *Linn. ; Fl. Br. Ind., III., 481 ;* PLUMBAGINEÆ.

Syn.—P. COCCINEA, *Boiss. ;* THELA COCCINEA, *Lour.*

Vern.—*Lál-chitarak, chitra, lál-chitrá, lál-chitá, rakta-chitrá,* HIND.; *Lálchitá, rakto-chitra, chitra, rakto chitá,* BENG.; *Lál-chitá,* URIYA; *Shitray, shitranj,* KASHMIR; *Chitrak,* C. P.; *Lál-chitra,* BOMB.; *Támbada-chitramúla,* MAR.; *Lál-chitrak,* GUZ.; *Shivappu-chittira, chitturmol, kodimúli, cittra-molum, yerra-kódivaylie,* TAM.; *Lál-chitarmúl, lálchitarmul, lál-chitarmulam,* DEC.; *Yerra-chitra, erra-chitramúlam,* TEL.; *Kempu-chitramúlá,* KAN.; *Schettiecodvalie, chukanna-kotupéli, chenti-kotuvéli.* MALAY.; *Ken-khyoke-ni, kan-chop-ni, chuvondacoduvallie, kin-khen-ni,* BURM.; *Rat-nitúl, rat-nettol,* SING.; *Raktachitraka, arúna-chitraca, chitraca, rakta-shikha,* SANS.; *Shitturridge, chittermúl, shitaraje-ahmar,* ARAB.; *Shitarake-surkh,* PERS.

References.—*Roxb., Fl. Ind., Ed. C.B.C., 155 ; Kurz, in Jour. As. Soc., 1877, Pt. ii, 218 ; Dals. & Gibs., Bomb. Fl., Suppl., 71 ; Mason, Burma & Its People, 432, 789 ; Pharm. Ind., 169 ; Fleming, Med. Pl. & Drugs (Asiatic Resear., XI.), 175 ; Ainslie, Mat. Ind., II., 78, 379 ; O'Shaughnessy, Beng. Dispens., 142, 508 ; Irvine, Mat. Med., Patna, 22 ; Moodeen Sheriff, Supp. Pharm. Ind., 202 ; U. C. Dutt, Mat. Med. Hindus, 186, 315 ; Sakharam Arjun, Cat. Bomb. Drugs, 109 ; K. L. De, Indig. Drugs, Ind., 92 ; Murray, Pl. & Drugs, Sind., 167 ; Dymock, Mat. Med. W. Ind., 2nd Ed., 623 ; Official Corresp. on Proposed New Pharm. Ind., 228 ; Birdwood, Bomb. Prod., 68 ; Drury, U. Pl. Ind., 347 ; Useful Pl. Bomb. (XXV., Bomb. Gaz.), 266 ; Boswell, Man., Nellore, 143 ; Settlement Report:—Central Provinces, Chanda, App. VII. ; Gazetteers:—Bombay, VI., 15 ; XV., 437 ; Orissa, II., 178 ; Mysore & Coorg, I., 65 ; Balfour, Cyclop. Ind., III., 237.*

Habitat.—A shrubby perennial, frequently cultivated in gardens in India. It is doubtfully wild in valleys in Sikkim and Khásia, and may be only a cultivated variety of **P. zeylanica** (*C. B. Clarke*).

MEDICINE.
Root.
980

Medicine.—The ROOT is mentioned by ancient Sanskrit and Muhammadan writers as an abortifacient and vesicant. It was noticed by **Rumphius, Rheede,** and **Ainslie,** all of whom remarked on its vesicant qualities. The last mentioned author wrote, "The bruised root of this plant is, in its natural state, acrid and stimulating, but when tempered with a little bland oil, it is used as an external application in rheumatic and paralytic affections; it is also prescribed internally in small doses for the same complaints in combination with other simple powders." He then remarks that, according to **Dr. Horsfield,** "the root is used by the Javanese for the purpose of blistering, and causes more inflammation than cantharides, but produces less effusion." More recent writers appear to have directed their attention to the latter quality, hoping to prove the drug a cheap and efficient substitute for cantharides. O'Shaughnessy instituted a

| | A Vesicant Medicine. | (*J. Murray.*) | **PLUMBAGO zeylanica.** |

very complete series of trials of the root, and arrived at the conclusion that it had proved itself a cheap and excellent substitute. He used the ROOT-BARK rubbed into a paste with water and a little flour, and applied to the skin for about half an hour. It was then removed, and within twelve or eighteen hours afterwards a large uniform blister was found to result. **Dr. Waring**, commenting on these opinions in the non-officinal list of the *Pharmacopœia of India*, expresses a much less favourable verdict. The results which appeared to be shewn by his experience were that it caused more pain than an ordinary blister, that the resulting vesication was less uniform and not always easily healed. These remarks are supported by **Dr. Dymock**. There is no doubt, however, that though perhaps inferior to cantharides, it is nevertheless a powerful vesicant, which may be resorted to with benefit when the officinal drug is not available.

Taken internally it is an acrid stimulant, and in large doses acts as an acro-narcotic poison, in which character it is said to be not infrequently employed in Bengal. It is also taken internally for the purpose of procuring abortion, and still more commonly the scraped root is introduced into the *os uteri* for that purpose. Death not infrequently results from the introduction of this highly acrid agent. In Southern India the dried, comparatively inert root is in high repute as a remedy for secondary syphilis and leprosy.

CHEMICAL COMPOSITION.—The active principle of the root, called *plumbagin*, occurs in delicate yellowish tufts of needles, slightly soluble in cold water. An aqueous solution throws down a carmine precipitate with basic acetate of lead. Plumbagin is freely soluble in ammonia or potassic hydrate solutions (*Dr. Wrden*).

SPECIAL OPINIONS.—§ "It is not only used to procure abortion but also as a remedial agent for post partum and other hæmorrhages from the uterus, hence its importance in medico-legal cases" (*Assistant Surgeon S. C. Bhuttacharji, Chanda, Central Provinces*). "The MILKY JUICE is useful in ophthalmia. The STALK is introduced into the uterus to produce criminal abortion" (*Assistant Surgeon T. N. Ghose, Meerat*). "The juice and the tender TWIGS are used as an application to ulcers. The roots are used (per vaginum) to produce abortion" (*Civil Surgeon J. H. Thornton, B.A., M.B., Monghyr*). "The milky juice is useful as an external application for scabies" (*Surgeon A. C. Mukerji, Noakhally*). "Its vesicating properties have been successfully utilised in curing certain cases of leucoderma." (*Surgeon R. Gray, Lahore*).

[*t. 179.*

Plumbago zeylanica, *Linn.; Fl. Br. Ind., III., 480; Wight, Ill.,*

Syn.—P. AURICULATA, *Blume Bijd.* ; THELA ALBA, *Lour.*

Vern.—*Chitrá, chitá, chitarak, chitawár, chiti,* HIND.; *Chitá, chitruk sufaid,* BENG.; *Chitá, chitámúl-nil,* URIYA; *Chitrak,* PB.; *Chitra, chitrack,* BOMB. ; *Chitra-múla, chitraka,* MAR. ; *Chitaro,* GUZ. ; *Chitarmúl, chitarmul, chitar-mulam,* DEC.; *Ven-chittira-múlam, chittiramúlam, chittira,* TAM.; *Chitra-múlam, agnimáta, tella-chitra-múlam,* TEL.; *Chitra-múlá,* KAN.; *Kotu-véli, tumpa-kotu-véli,* MALAY.; *Kin-khen-phiú, kan-chop-phijú,* BURM.; *Sudu-nitul, ella-nitul,* SING.; *Agnishikha, chitraka-vrikshaha, chitraka,* SANS. ; *Shitaraj,* ARAB.; *Shitarak, shitirah,* PERS.

References.—*Roxb., Fl. Ind., Ed. C.B.C., 155 ; Voigt, Hort. Sub. Cal., 438 ; Kurz, in Jour. As. Soc., 1877, II., 217 ; Dals. & Gibs., Bomb. Fl., 220 ; Stewart, Pb. Pl., 173 ; Mason, Burma & Its People, 432, 789 ; Sir W. Elliot, Fl. Andhr., 12 ; Sir W. Jones, Treat. Pl. Ind., V., 86 ; Rheede, Hort. Mal., X., t. 8 ; Pharm. Ind., 170 ; Fleming, Med. Pl. & Drugs (Asiatic Reser., XI.), 175 ; Ainslie, Mat. Ind., II., 77 ; Irvine,*

MEDICINE.
Root-bark.
981

Chemistry.
982

Milky juice.
983
Stalk.
984
Twigs.
985

986

PLUMBAGO
zeylanica.   A vesicant Drug.

Mat. Med. Patna, 22; Medical Topog., Ajmir, 130; Taylor, Topog., Dacca, 54; Moodeen Sheriff, Supp. Pharm. Ind., 202; U. C. Dutt, Mat. Med. Hindus, 185, 295; Sakharam Arjun, Cat. Bomb. Drugs, 110; K. L. De, Indig. Drugs, Ind., 93; Dymock, Mat. Med. W. Ind., 2nd Ed., 620; Cat. Baroda Durbar, Col. & Ind. Exhb., No. 154; Official Corresp. on Proposed New Pharm. Ind., 240; Year-Book Pharm., 1880, 250; Trans., Med. & Phys. Soc., Bomb. (New Series), No. 4, 155; Drury, U. Pl. Ind., 348; Atkinson, Him. Dist. (X., N.-W. P. Gaz.), 746; Useful Pl. Bomb. (XXV., Bomb. Gaz.), 266; Butler, Top. & Statis., Oudh & Sultanpur, 35; Boswell, Man. Nellore, 121; Gazetteers:—Bombay, XV., 436; N.-W. P., I., 83; IV., lxxiii.; Orissa, II., App., iv., 159; Mysore & Coorg, I., 65; Agri.-Horti. Soc.:—Ind., Trans., VI., 242; Journals (Old Series), VII., Pro., 10.; X., 28; Ind. Forester, VIII., 418; XII., App., 15; Balfour, Cyclop. Ind., III., 237.

Habitat.—A diffuse rambling herb or undershrub, much cultivated, and readily spreading throughout India, probably wild in the Southern Peninsula and Bengal.

MEDICINE.
Root.
987

Medicine.—The ROOT appears to possess similar properties to those of **P. rosea** though to a much milder degree. It was known to Sanskrit writers, who state that it increases the digestive powers, promotes the appetite, and is useful in dyspepsia, piles, anasarca, diarrhœa, skin diseases, and other complaints. It is much used as a stimulant adjunct to other medicines, several examples of such compounds being given by **Dutt** from **Chakradatta** and other writers. The root, like that of the former species, was also highly esteemed as a local application. Reduced to a paste it was applied to abscesses with the object of opening them, and it also entered into the composition of several caustic preparations (*Mat. Med., Hindus*). **Dymock** writes regarding its utilisation by Muhammadan physicians, "Muhammadan writers treat of the drug under the name of *shitaraj*, a corruption of the Hindustani *chitrak;* they describe it as caustic and vesicant, and expellant of phlegmatic humours, useful in rheumatism and spleen, digestive; it also causes abortion. For external application it is made into a paste with milk, vinegar, or salt and water. Such a paste may be applied externally in leprosy and other skin diseases of an obstinate character, and be allowed to remain until a blister has formed. In rheumatism it should be removed after 15 or 20 minutes. When administered internally the dose is one dirhem."

This drug attracted the notice of several early European writers on Indian drugs, among others **Sir W. Jones, Fleming, Ainslie, Taylor,** and **Irvine.** They describe it as a vesicatory and useful tonic in dyspepsia, and **Taylor** comments on its sialogogue properties. It is also mentioned in the secondary list of the *Pharmacopœia of India*, where it is said to be well worthy of further trial. The supposed antiperiodic effects of a tincture of the root-bark is specially drawn attention to, and **Dr. Oswald** is said to have employed this preparation with good effect in the treatment of intermittents. It acts as a powerful sudorific.

The drug is a common one in bazárs all over India to this day, the shops being supplied by herbalists who cultivate or collect the roots in the neighbourhood. The virtues ascribed to it, and the uses to which it is put, are the same as those described by ancient writers. Like the root of the former species it is occasionally employed criminally either as a local application to procure abortion, or as an acrid poison.

Milky Juice.
988

SPECIAL OPINIONS.—§ "Parts of this plant have been brought to me as having been used in causing abortion" (*E. Borill, Offg. Civil Surgeon, Motihari, Champaram*). "The MILKY JUICE is used as an application to unhealthy ulcers and in cases of scabies" (*J. H. Thornton, B.A., M.B., Civil Surgeon, Monghyr*).

| Caoutchouc from Plumeria. | (*J. F. Duthie.*) | **POA.** |

PLUMERIA, *Linn.; Gen. Pl., II., 704.*
[*471*; APOCYNACEÆ.

Plumeria acutifolia, *Poiret; Fl. Br. Ind., III., 641; Wight, Ic., t.*   **989**
Syn.—P. ACUMINATA, *Roxb.*
Vern.—*Gálachin, goburchamp, golainchi, chameli,* HIND., *Gorur-champa,*
BENG.; *Kátchámpá,* URIYA; *Gulanj baha,* SANTAL; *Champa pungár,*
GOND; *Khair-champa, dolochápá, khadchampo, gutáchin, chameli,*
BOMB.; *Rhuruchápá, khairchampa,* MAR.; *Rhadachampo,* GUZ.; *Váda
gannéru,* TEL.; *Kanagala, ganagalu,* KAN.; *Tayopsagah, china cham-
pac, taroksoga,* BURM.
References.—*Roxb., Fl. Ind., Ed. C.B.C., 248; Voigt, Hort. Sub. Cal.,
528; Brandis, For. Fl., 323; Kurz, For. Fl. Burm., II., 178; Gamble,
Man. Timb., 261; Dals. & Gibs., Bomb. Fl., Suppl., 52; Mason, Bur-
ma & Its People, 406, 799; Elliot, Fl. Andhr., 187; Rev. A. Campbell,
Rept. Econ. Pl., Chutia Nagpur, No. 7591; O'Shaughnessy, Beng.
Dispens., 449; Dymock, Mat. Med. W. Ind., 2nd Ed., 508; S. Arjun,
Bomb. Drugs, 210; Cat. Baroda Durbar, Col. & Ind. Exhb., No. 148;
Lisboa, U. Pl. Bomb., 99, 391; Birdwood, Bomb. Pr., 270; Gums &
Resinous Prod. (P. W. Dept. Rept.), 28; Aplin, Rept. on the Shan States;
Gazetteers:—Mysore & Coorg, I., 62; N.-W. P., I., 82; IV., lxxiv.;
Orissa, II., 178; Ind. Forester, III., 203; Agri.-Horti. Soc., Ind.,
Journal, X., 16.*
Habitat.—A small tree, which is a native of Tropical America, but is
cultivated and naturalised in many parts of India.

Caoutchouc.—Attempts have been made, with no success, to manu-   **CAOUTCH-
OUC.**
facture caoutchouc from the abundant, tenacious, MILKY JUICE which   **Milky Juice.**
exudes plentifully when any part of the tree is wounded.   **900**

Medicine.—By the Persians the BARK, known as *aachín,* is recom-   **MEDICINE.**
mended as a cure for gonorrhœa (*Dymock*). Dr. Hove, in 1787, found   **Bark.**
the tree growing abundantly on Malabar Hill, and mentions that the   **991**
inhabitants used it for intermittents as we do cinchona. S. Arjun (*Bombay   **Leaves.**
Drugs*) writes that the LEAVES, made into a poultice, are used to dispel   **992**
swellings; the milky JUICE is employed as a rubefacient in rheumatism,   **Juice.**
and the blunt-ended BRANCHES are introduced into the uterus to procure   **993**
abortion. According to Dymock, the bark is given, in the Konkan, with   **Branches.**
cocoanut, *ghí,* and rice as a remedy for diarrhœa; the FLOWER-BUDS are   **994**
eaten with betel leaves in scurvy, and the juice with sandal wood oil and   **Flower-buds.**
camphor is employed as a cure for itch. Campbell states that in Chutia   **995**
Nágpur the leaves and root are used medicinally, but that the part best
known to the forest tribes of Manbhum is the core of the young WOOD,   **Wood.**
which is given to lying-in women to allay thirst and for cough. In the   **996**
Baroda Durbar Catalogue of Medicinal Plants at the Col. & Ind. Exhb.,
it is stated that the bark is purgative and used in cases of leprosy.

SPECIAL OPINION.—§ "This plant is known as *Dalána phula* in Northern Bengal, where its milky juice has been tried and found to be an
effectual purgative. The dose is as much as a grain of parched rice
(*khaí*) will absorb, the grain being administered as a pill" (*Surgeon-Major
C. T. Peters, M.B.*).

(*J. F. Duthie.*)

POA, *Linn.; Gen. Pl., III., 1196.*   **997**
A genus containing nearly 100 species, inhabiting chiefly temperate or
mountainous regions. Several of them are well known as very nutritious pasture grasses, and are cultivated as such. In addition to those specially mentioned below, about twenty others have been discovered at various elevations on
the Himálayas.

Poa annua, *Linn.; Duthie, Fodder Grasses, 67;* GRAMINEÆ.   **998**
Syn.—P. SUPINA, *Schrad.*

PODOCARPUS
   neriifolia.   Meadow Grasses.

Vern.—*Chirua*, N.-W. P.
References.—*Hook., Stud. Fl. Brit. Islands, 442; Boiss., Fl. Or., V., 601; Agri.-Horti. Soc., India, Trans., VIII., 105.*
Habitat.—An annual with stems less than a foot high. It is common on the Himálayas up to 9,000 feet, descending as far as the plains in the Panjáb, where, in some of the submontane districts, it forms in the early spring a green sward.

FODDER.
999
Fodder —A very nutritious grass as long as it lasts, but the yield of foliage is too small to be of much account.

1000
Poa pratensis, *Linn.*
SMOOTH-STALKED MEADOW GRASS.
Syn.—P. ANGUSTIFOLIA, *Linn.*
References.—*Hook., Stud. Fl. Brit. Islands, 442; Boiss., Fl. Or., V., 601; Sutton, Perm. & Temp. Pastures, 60; Stebler & Schröter, Best Forage Pl., 72 (Eng. Ed.).*
Habitat.—A perennial stoloniferous grass, found within the temperate and alpine zones of the Himálayan range, and in the temperate regions of Europe and America.

FODDER.
1001
Fodder.—Much esteemed both in Europe and America as one of the best kinds of pasture grasses. In the latter country it is known as Kentucky Blue Grass.

1002
P. trivialis, *Linn.*
ROUGH MEADOW GRASS.
References.—*Hook., Stud. Fl. Brit. Islands, 443; Boiss., Fl. Or., V., 602; Sutton, Perm. & Temp. Pastures, 62; Stebler & Schröter, Best Forage Pl., 77 (Eng. Ed.).*
Habitat.—Confined to the Himálayan ranges in India. A common European grass, extending through Siberia to Japan, and to North Africa. Introduced into America. It differs from the preceding by its rough stems, and by its stolons rooting above ground.

FODDER.
1003
Fodder.—In stiff moist soils this grass yields an abundance of excellent fodder, but it is soon affected by drought.

(*J Murray.*)
PODOCARPUS, *L'Herit.; Gen. Pl., III., 434.*

1004
Podocarpus latifolia, *Wall; Fl. Br. Ind., V., 649;* CONIFERÆ.
Syn —NAGEIA LATIFOLIA, *Gord.*
Vern.—*Soplong,* KHASIA; *Nirambali,* TINEVELLY; *Thitmin,* BURM.
References.—*Gordon, Pinetum, 138; Gamble, Man. Timb., 414; Beddome, Fl. Sylv., t. 257; Kurz, For. Fl. Burm., II., 500; Madras, Man. Admin., II., 27; Ind. Forester, III., 21; XI., 98, 486; XIV., 317; Agri.-Horti. Soc. Ind.:—IV., Sel. 251; Smith, Ec. Dict., 446.*
Habitat.—A large, evergreen, glabrous tree, found in the Khásia mountains, at 3,000 feet altitude, in the hill forests of Burma and the Malay Peninsula, and in the mountains of South Tinnevelly, at altitudes of 3,000 to 5,000 feet.

TIMBER.
1005
Structure of the Wood.—Grey, aromatic, similar in structure to that of P. neriifolia; weight 33℔ per cubic foot. It is used for the same purposes as the timber of the commoner species.

1006
P. neriifolia, *Don.; Fl. Br. Ind., V., 649.*
Syn.—P. BRACTEATA, *Bl.;* P. MACROPHYLLA, *Wall;* NAGEIA BRACTEATA, *Kurz.*
Vern.—*Dingsableh,* KHASIA; *Jinari,* KACHAR; *Gúnsi,* NEPAL; *Thitmin* (=prince of woods), BURM.; *Welimáda,* AND.
References.—*Brandis, For. Fl., 541; Kurz, For. Fl. Burm., II., 500; Gamble, List of Trees, Shrubs, &c., of Darjiling, 83; Man. Timb., 414;*

P. 1006

Rumphius, Amb., III., 214, t., 172, f. 1; Ind. Forester, III., 181; IV., 241; XII., 74; Agri.-Horti. Soc. Ind.:—Transactions, V., 110; Journals (Old Series), IV., Sel., 251; VII., 159.

Habitat.—A large, glabrous, evergreen tree, met with on the Tropical Himálaya in Nepal and Sikkim, also in the Khásia hills, Burma, the Malay Peninsula, and the Andamans; distributed to Java, Sumatra, and Borneo.

Structure of the Wood.—Grey, moderately hard, of very uniform grain and texture; weight, according to **Brandis**, 50℔, **Gamble**, 39℔, and **Bennett** 34℔ per cubic foot. It is highly esteemed by the Burmese, as is shewn by their name for it, and is used for making oars, masts, and planking.

Mason states that the Burmans have a superstition that the beams of balances should be made of it, and, according to **Major Berdmore**, a peg driven into a house-post or boat is supposed to avert evil (*Gamble*).

TIMBER. 1007

(*G. Watt.*)

PODOPHYLLUM, *Linn., Gen. Pl., I., 45.*

1008

A herbaceous genus of BERBERIDIÆ, which contains in all some three or four species, one met with in North America, a second in the Himálaya, and two in China. The American species, **P. peltatum**, affords the officinal root of commerce, the Indian **P. emodi**, *Wall*, has hitherto been entirely overlooked as a possible source of the drug, and the Chinese forms, **P. pleianthum**, *Hance*, and **P. versipelle**, *Hance*, appear to be only known botanically. The generic name is derived from ποῦς, a foot, and φύλλον, a leaf, in allusion to the resemblance of the leaf of the American species to the feet of certain aquatic birds.

Podophyllum emodi, *Wall; Fl. Br. Ind., I., 112;* BERBERIDEÆ.

1009

The MAY APPLE or DUCK'S FOOT, sometimes also called the AMERICAN MANDRAKE; RIZONE DE PODOPHYLLUM, PODOPHYLLÉE, *Fr.;* FUSSBLATTWARZEL, ENTENFUSS, *Germ.*

Syn.—P. HEXANDRUM, *Royle.*

Vern.—*Pápra, pápri, bhavan-bakra, bakra chimyaka,* HIND.; *Papri, bankakri, banbakri, kákra, bankákra, chimyáka, chijákri, gúl-kákrú, wanwángan,* PB.

References.—*Stewart, Pb. Pl., 8; Report of Tour in Hasára (Jour-Agri.-Horti. Soc. Ind., XIV., 55); Aitchison, Flora of Lahoul, in Jour. Lin. Soc., X., 71; Kuram Valley Report, Pt. I., 32; Pharm. Ind., 11-12; British Phram., 245; Flück & Hanb., Pharmacog., 36; Flückiger in Sonder-Abdruck aus der Apotheker-Zeitung, 1890, Nos. 102, 103; U. S. Dispens., 15th Ed., 637, 1133, 1138; Martindale, Extra Pharm., 288; Garrod, Mat. Med., 163; Per. Mat. Med. (Ed. B. & R.), 1007; O'Shaughnessy, Beng. Dispens., 170; Bentley, in Pharm. Jour., III., 2nd Series, 457; Hooper, in Pharm. Jour. Jan. 26th, 1889; Bent. & Trim., Med. Pl., 17; Dymock, Warden & Hooper, Pharmacog. Ind., I., 69; Watt, Cal. Exh. Cat., 1883, V., No. 929; VI, No. 604; Atkinson, Him. Dist. (X., N.-W. P. Gas.), 304, 329; Forbes Watson, Ind. Survey, 192; Royle, Ill. Him. Bot., 64 and note to 379; Henderson & Hume, Lahore to Yarkand, 309; Settlement Report:—Panjab, Simla, App. II., H, also p. xliii.—Kakra, said to be a small tree, which yields good wood, cannot therefore be Podophyllum, though the vernacular name given to it is that generally applied to this plant; Indian Agriculturist, July 27th, 1889; March 16th, 1890; Treasury of Botany, II., 909; Spons, Encycl., I., 820; Balfour, Cyclop. Ind., III., 238; Smith, Dict. Econ. Prod., I., 820; Balfour, Cyclop. Ind., III., 238; Smith, Dict. Econ. Prod., 270.*

Note.—Some of the above references are to **P. peltatum**, but as they have a direct bearing on **P. emodi**, its medicinal and chemical properties, they have been here quoted as worthy of consultation.

Habitat.—A small, erect, herbaceous plant, met with in the higher, rich, and shady temperate forests from Sikkim to Simla, Kashmír, Hazára,

Tibet, the Kuram Valley, and Afghánistan. On the main line of the Himálaya it rarely occurs much below 9,000 feet, and ascends to 14,000 feet; but in Kashmír it is said to occur as low down as 6,000 feet. It chiefly abounds on the northern slopes of the outer ranges; but in the more interior tracts, becomes prolific on any exposure. As far as the writer's personal experience goes, it is less abundant on the Eastern than on the Central and Western Himálaya. It approaches nearest to the plains of India on the Chor mountain—a high ridge south-south-east of Simla—and is fairly plentiful on the northern forest-clad slopes of the Shalai, north-east of Simla, and on the Nagkanda hill. It becomes abundant farther to the north, namely, on the Murale hill and the Jalauri pass. It is also plentiful in many parts of Kumáon, but to the west of Simla seems to attain its greatest degree of prevalence. In Kulu and in the Chumba State there are many mixed forests with their glades almost exclusively covered with this pæony-rose-like herb; and the lower part of the valley leading from the Sach Pass to Kilar in Pangi may be specially mentioned.

DESCRIPTION.
TION.
1010

DESCRIPTION.—**Podophyllum emodi** may popularly be described as a herbaceous plant with a perennial rhizome, the part above ground being annual. It appears about the middle of April, its reddish succulent *stem* bearing a pair of curiously reflexed leaves, which droop umbrella-like from the top of the petiole. The stem rises to a height of 6 to 12 inches before the *leaves* (which attain a diameter of 6 to 10 inches) are fully expanded. The leaves are often quaintly spotted, and give origin in May to a large subsessile pinkish-white *flower*, which appears to arise from the leaf-stalk of one of the leaves, and to have passed maturity before the leaves have been fully developed. The *fruit* ripens about August or September, when the leaves have quite faded; it becomes about the size of a small lemon, is of a bright orange colour, and hangs suspended from the stout, naked, and often mottled stem.

MEDICINE.
Rhizome.
1011
Roots
1012

Medicine.—The official parts of the American plant are the RHIZOME (or underground stem) with its attached cord-like ROOTS. From these rhizomes and roots is prepared the so-called "Resin of Podophyllum." Preparations of the rhizomes are now seldom, if ever, used direct in the practice of medicine: it is the resin or podophyllin that is employed. The British and also the Indian Pharmacopœias direct the officinal resin to be prepared as follows:—"Take of Podophyllum root in coarse powder, one pound; rectified spirits, three parts, or a sufficiency; water, a sufficiency; hydrochloric acid, a sufficiency. Exhaust the Podophyllum with the spirit by percolation: place the tincture in a still, and draw off the greater part of the spirit. Acidulate the water with $\frac{1}{24}$ of its bulk of hydrochloric acid, and slowly pour the liquid which remains after the distillation of the tincture into three times its volume of the acidulated water, constantly stirring. Allow the mixture to stand for 24 hours to deposit the resin, wash the resin on a filter with distilled water, and dry it in a stove." The resin is further described as "a small greenish-brown amorphous powder, soluble in rectified spirit and in ammonia; precipitated from the former by water, from the latter by acids. Almost entirely soluble in pure ether." In the United States, in addition to the "Resina" there is also an "Extractum Podophylli" authorised to be used officinally. The process of preparation recommended for the resin differs but little from that given above, and the resin is described as of "a bright brown colour, an acrid bitter taste, and a slight odour of the root." Last year a correspondence took place in the *Chemist and Druggist* as to the advisability of the British Pharmacopœia method of preparing the resin. A correspondent pointed out that, as prepared by him, following the process prescribed the resin was more often of a much paler colour than that described above

| The Resin of Podophyllum. (*G. Watt.*) | PODOPHYLLUM emodi. |

The employment of acid to precipitate the resin was condemned by subsequent correspondents, and alum water recommended instead. The fact that the resin as sold to the dispensing chemist is found to be less soluble in rectified spirit than stated in the Pharmacopœia was contended to be due to a fraudulent or careless high percentage of ash being present in the resin. A sample of resin prepared by **Dr. Hooper** from the Indian root assumed a pale straw colour; but doubtless this may have been due to the process of manufacture which had been pursued, and not to any special peculiarity of the Indian root.

Martindale (*Extra Pharmacopœia*) gives the following brief account of the resin, which, for all practical purposes, may be accepted as expressing the results of the chemical investigations of that substance up to date: " The crude resin may be divided by treatment with ether, which dissolves a portion and leaves another which is soluble in alcohol but not in ether. The former has a bright yellow colour, an herby odour and acrid taste; the latter has a pale brown colour, is odourless, and has a less acrid taste than the other. The author found little difference in their purgative action. The brown resin was more prompt. The crude resin is a slow and rather uncertain purgative requiring from 12 to 20 hours to act."

" In a later research by **Podwissotzki**, he obtains from a chloroformic extract of the root an amorphous principle which is free from the fatty and colouring matter of the officinal resin. This he names *Podophyllotoxin;* it is more certain in its action than Podophyllin and is given in doses of $\frac{1}{10}$ to $\frac{1}{8}$ grain; to children $\frac{1}{60}$ to $\frac{1}{30}$ grain. It is best administered by dissolving 1 grain in 2 drachms of rectified spirit. Dose, 2 to 12 drops in a teaspoonful of syrup. *Podophyllotoxin* is in its turn capable of being separated into a bitter crystalline acid (*Picropodophyllic acid*), a bitter crystalline neutral body (*Picropodophyllin*), the latter of which is the more medicinally active, and an amorphous substance (*Podophyllic acid*) which is inert."

Cadbury, Power, and others affirm that the medicinal virtue is due to the resin soluble in ether, and that the other resin left behind in the alcoholic medium is comparatively inert. The experiments of **Power, Maisch, Quereschi, Podwissotzki,** and others prove that Podophyllum does not contain berberine (as was announced by **Mayer**), or any alkaloid, and that its activity is due to the resin or resins.

Care must be taken in handling the resin in quantity, since it is a powerful irritant and may produce painful conjunctivitis.

Mr. Martindale describes six preparations of Podophyllin as follows :—

1. Pilula Podophyllin.
2. Pilula Podophyllin Composita, U. C. H.
3. Pilula Podophyllin et Quininæ.
4. Tinctura Podophyllin (*Dobell*).
5. Tinctura Podophyllin (off.).
6. Tinctura Podophyllin Ammoniata.

Medicinal Properties and Uses.—It is commonly stated that Podophyllin may be given in all cases where mercury is indicated. On this account it is often designated " vegetable calomel." It may be described as hydragogue carthartic in large doses, and alterative in small. It has its more specific action on the liver, and differs from jalap or scammony in causing an increased secretion or flow of bile. It is liable, however, to cause severe griping, and is accordingly rarely prescribed by itself but is combined with other purgatives, and aloes or colocynth are those generally preferred. The *Pilula Podophyllin et Quininæ* is a favourite " dinner pill :" it contains sulphate of quinine 1 grain, podophyllin $\frac{1}{12}$ grain, sugar

Uses.
1013

MEDICINE.
Uses.

of milk $\frac{1}{12}$ grain, extract of belladonna $\frac{1}{8}$ grain, and extract of socotrine aloes 1 grain. The belladonna is given to check the tendency to griping, and it seems desirable to adopt some such combination. The *Tincture of Podophyllin Ammoniata* is described as a powerful hepatic stimulant, and in large doses a violent purgative. In dropsies, podophyllin is sometimes given in combination with calomel and bitartrate of potash.

Externally applied, the resin of podophyllum acts as a powerful irritant, and, dissolved in alcohol, it is sometimes employed as a counter-irritant.

TRADE.
1014

Trade in Podophyllum Root.—No trustworthy statistics can be discovered on this subject either as to the British or the Indian trade. It has often been recommended that England should commence to grow her own root. But to what extent this suggestion has been acted on the writer cannot discover. At present she would seem to draw her entire supplies from America, in which country a very considerable trade may be said to be done (especially at Cincinnati) in the manufacture of podophyllin. But recently a similar enterprise has been started in England. India apparently imports podophyllin—the root in small quantities only coming to this country. It is thus highly desirable that the Indian root should come to be recognised. Were this done, the importation of American podophyllin would at once be checked, and a large export trade of the native drug most probably established. Lloyd's podophyllin in Trade Reports is generally put down at 7*s*. 6*d*. to 10*s*.[*] a pound in wholesale quantities, (*Chemist and Druggist*). Since the Indian root affords 12 per cent. of the resin as compared with the American 4 per cent., some idea of the likely profits from an Indian manufacture of podophyllin may be arrived at. But should the technical knowledge be wanting in India, there still remains what appears a possible lucrative trade, *viz.*, the exportation of the dried root. It is somewhat surprising that the natives of India, who have discovered so many drugs, should have failed to detect the properties of the podophyllum root. According to **Flückiger & Hanbury**, the properties of the rhizome as an anthelminthic and emetic have been long known to the natives of North America. Only one Indian writer makes the slightest allusion to the medicinal properties of P. emodi, *viz.*, Dr. **Stewart.** All he says, however, amounts to little: "In Lahoul it seems to be used medicinally." **Royle,** in his *Himalayan Botany*, remarks: "I know not if the Himalayan Podophyllums possess any of the properties ascribed to the American species, but this could be easily grown." **Sir W. O'Shaughnessy** only briefly alludes to the drug, remarking that the sample furnished by **Dr. Falconer** "was too small for the requisite experiments being instituted upon its virtues." The *Pharmacopœia of India* says of P. emodi, "its properties may be deserving of investigation." Thus, although the fact that India possesses a species of Podophyllum has been known to Europeans for nearly a century, the root, until within the past few months, has neither been chemically examined nor medicinally experimented with.

────────

The above account of **Podophyllum emodi** was drawn up by the Editor in connection with a correspondence instituted by him which ultimately appeared in the Selections from the Records of the Government of India in 1889-90. Since considerable extra information was brought to light by the correspondence, and the more so since that correspondence has been conducted to the present date (Feb. 1891), a *précis* of the chief letters may be of public interest.

────────

[*] The root is sold at 5 pence a pound.

From GEORGE WATT, M.B., C.M., C.I.E., Reporter on Economic Products, to the Secretary to the Government of India, REVENUE AND AGRICULTURAL DEPARTMENT,—dated 4th March 1889.

During a rapid botanical trip into Kúllú in November 1888, I collected some of the roots of a wild plant—**Podophyllum emodi**—with the object of having them chemically analysed. I knew that the resin obtained from the North American plant—**Podophyllum peltatum**—had recently been giving indications of becoming daily a more and more important drug. It, therefore, occurred to me that the Indian species might be found to possess, in a profitable degree, the active principle. **Dr. Dymock,** Medical Store-keeper, Bombay, was good enough to offer to co-operate with me in the matter; and thus encouraged I took some little trouble to discover the plant, though during the late season at which I visited Kúllú, only portions of the withered leaves could be found. The result has proved highly satisfactory. From the letter below, it will be observed that **Dr. Hooper,** Quinologist to the Madras Government, has discovered, on analysing the roots furnished by me, that whereas the American root contains only 4 per cent. of the resin, the Indian yields as much as 12 per cent.; and that the Podophyllin thus obtained has been found to possess the same medicinal properties as that presently being used in Europe.

Under these circumstances, I venture to place before the Government of India the results of the enquiry up to date, in the hope that they may be deemed of such importance as to justify their publication. It would seem highly probable, as suggested by **Dr. Dymock,** that, with a little encouragement vouchsafed by Government during the initial stage of the proposed new industry, a lucrative trade of the utmost value to the hill tribes of the higher Himálayas, might be organised. With this object in view I beg to submit a brief note, which will be found to contain the main facts regarding Indian and American Podophyllum, together with the correspondence conducted on this subject.

BOMBAY, *December 26th, 1888.*

You will be pleased to learn the **Podophyllum emodi** is a great success. **Dr. Hooper** got 12 per cent. of resin (podophyllin) from it, whereas the yield from **P. peltatum** is about 4 per cent. only. Medicinally the resins act just in the same way as the officinal drug.

As Podophyllin is by far the most popular purgative at the present time, your Department should take steps to collect some and put it in the market. A good supply would drive the American drug out of the European market. I enclose a sample of the resin for your inspection.

Believe me, yours sincerely,

W. DYMOCK.

From MESSRS. KEMP & Co., Limited, Pharmaceutical Chemists, Bombay, to the Reporter on Economic Products, dated 21st February 1889.

Mr. Hooper, Government Quinologist of Ootacamund, has referred us to you with the view of obtaining a supply of the rhizomes of **Podophyllum emodi** which he, in conjunction with **Dr. Dymock,** have found to contain a large percentage of Podophyllin. We should be glad of obtaining a preliminary supply of about 20℔. As it is stated to grow very plentifully, we presume that it could be obtained at a low rate.

Aconitum ferox and **Napellus, Nardostachys Jatamansi,** and **Ophelia chirata,** we obtain in the Bombay bazaar, but if they could be obtained at a lower rate from Simla, we should be glad to purchase direct. Any formation regarding these and other drugs in ordinary use would greatly oblige.

Extract from the Reporter on Economic Products' reply to the above, dated March 4th, 1889.

I am laying the correspondence before **Sir E. O. Buck**, Secretary to the Government of India, in the Revenue and Agricultural Department, and shall, I trust, be able to report shortly. I shall also draw up a detailed note on the subject which very probably the Government may consider of such importance as to permit me to publish it in the Selections from the Records of the Government of India which I edit. Such a note would set forth the localities where the plant occurs, and might thus be expected to bring to light many persons who would be willing to collect or sell the root. It is probable that private enterprise would prove commercially more satisfactory than official. In bringing a new drug into the market, I am well aware that high prices are not likely to be paid at first, but as the root occurs abundantly at some distance from Simla only and would thus have to bear heavy coolie charges, it would be of great value to know the price you would be disposed to offer and the amount approximately you would be prepared to purchase annually. It might be thought worth while to cultivate the plant in some of the forests where now it exists but sparingly, so as to have at hand a larger supply than could be procured except from the more distant forests.

Extract from a letter received from Sir B. Simpson, *Surgeon-General, dated 20th March 1889.*

If the Indian plant yield so much as 12 per cent. of Podophyllin resin, and if the resin be of as good quality as the American drug, the initiation of a trade in the Indian drug will require very little help from Government.

In order to discover whether the Indian root, as is the case with the American, possessed the greatest amount of Podophyllin when in fruit, the writer visited, during March 1889, one of the Simla habitats of the plant, and collected some of the roots. These were forwarded to **Dr. Dymock**, and the following passage from the reply he was good enough to furnish would appear to settle that point :—

Extract from letter by Dr. Dymock, Bombay, May 17th, 1889.

The root of **Podophyllum emodi** last sent contains 6 per cent. of resin. It gives just as much spirit extract as that collected when in fruit, so it must contain a larger proportion of sugar and other nutritious materials stored up for the use of the growing plant. The root must therefore be collected when the plant is in fruit.

Captain C. W. Losack at Gulmury in Kashmír forwarded specimens of the plant in August 1889. In his letter accompanying these he says :—

The specimen sent was collected here at an elevation of 8,000 feet. It is not sufficiently abundant to make its collection remunerative. I learn, however, that it is very plentiful in other localities of Kashmír which I hope to visit shortly.

P. 1014

Resin of Podophyllum.	(*G. Watt.*)	PODOPHYLLUM emodi.

Several Mercantile firms having asked the Reporter on Economic Products whether he could procure for them a supply of the roots the following reply was issued :—

In my office I have no agency by which I could undertake to collect the root. My connection with this or any other product must be confined to merely placing in the hands of the public the facts I am able to collect regarding the products. But I shall be very pleased to be of any service in that direction; and should you establish an agency of your own for collecting the root, I will further your views by furnishing your collector with a dried specimen of the plant, and by identifying his collections until he has been trained to recognise the plant. There are many localities where it grows—the Black Mountains are literally covered with the plant. It is fairly plentiful in Garhwal and Kumaon ; but in Pangi on the Chenab (a tract of country belonging to the Raja of Chamba) it is particularly plentiful, especially in the valley leading from the Sach Pass to Kilar. There is a Forest officer in Kilar and Pangi ; and perhaps he might be willing to procure the root for you. Although the plant occurs in the Simla district, it would most probably not pay to collect it here unless the people could be induced to cultivate the plant. I have a few plants of it growing in pots in Simla now, and have no fear as to the possibility of its succeeding as a regular crop if the price paid for the root would prove remunerative.

In October 1889, the Reporter on Economic Products received, through the Conservator of Forests, Panjáb, two sets of specimens accompanied with letters of explanation. It is perhaps unnecessary to print these letters, since the following extract from the report on the samples analyses the main facts brought to light by the Deputy Conservators in Kúllú and Chamba.

Extract from a report by the Reporter on Economic Products, dated 13th December 1889.

" It is remarkable that both from Kúllú and from Chamba the vernacular names given for the plants forwarded should be almost identical. Unless these names were contained in the official correspondence and were thus adopted, they may prove of considerable use hereafter in identifying the root. In the new *Pharmacographia Indica* **Podophyllum emodi** is said to be known as *Pápra* or *Pápri*, also *Bhavan-bakra* or *bakra* : hence the authors of that work think it was probably one of the bile-expelling plants, described by Sanskrit writers under the names of *Parpata* and its synonym *Vakra*. The samples marked B. and C. forwarded by the Deputy Conservator of Forests, Kúllú, are **Podophyllum emodi**, and he reports that on the Rupi Range the plant is called *Bankukri, gulkúkri*, while in the Seoraj Range it is *Gulákra*. Both the Deputy Conservator of Forests, Kúllú, and the Deputy Conservator, Chamba, however, forward fragmentary specimens of the leaves and fruits of a Cucurbitaceous creeper (**Zehneria umbellata** *var.* **nepalensis**), with red fruits, which of course is not **Podophyllum emodi**. It would, therefore, appear that the roots of that creeper possess a distinct medicinal value (since recognised by the inhabitants of two regions so widely distinct) and probably that, like **Podophyllum**, it is a purgative. As opposed to this idea, however, on microscopic examination, the roots are seen to contain a large quantity of starch, and, according to the Rev. A. Campbell, the Santals are said to eat the roots of **Zehneria umbellata**, These roots require, therefore, more careful study, and they will be examined hereafter and reported on, but it is worthy of note that the creeper should bear in Kúllú the names *Gulkukru, gulale-kukri*, and *gulákri*, while in Chamba the same plant is called *Bankákra, bankákrú.* The singular uniformity in these names, if not the outcome of a suggestion that the **Podophyllum** plant required, bore some such names, would point to both plants having in native minds the same properties.

" Most writers allude to the fact of the fruits of **Podophyllum** being eaten, but the Deputy Conservator of Forests in Kúlú is the first person to have recorded that the ground root is used as jalap. Hitherto it has always been supposed that the natives of India were ignorant of the medicinal properties of **Podophyllum**. Now we know that they use the plant as a purgative, the development of trade may be simplified. It seems, how-ever, probable that the word *kákru* or *kákri* which occurs in all the verna-cular names mentioned above, means ' purgative ' or ' bile-expelling,' and in that case the danger would have to be guarded against of sending into trade the roots of the Cucurbitaceous plant here noticed."

FOOD.
Fruits.
1015

Food.—**Podophyllum** emodi produces handsome red FRUITS, which ripen in September and October and are eaten by the Natives in most parts of the Himálaya. Europeans consider the fruit insipid.

(*J. Murray.*)

PŒCILONEURON, *Beddome ; Gen. Pl., I., 981.*

1016   **Pœciloneuron indicum,** *Bedd. ; Fl. Br. Ind., 1., 278 ;* GUTTIFERÆ.

Vern.—*Kirballi, ballangi,* KAN.

References.—*Beddome, Fl. Sylv., t. 3; Gamble, Man., Timb., 21; Gazet-teer, Mysore & Coorg, I. 51 ; Ind. Forester, II., 20.*

Habitat.—A large tree, found on the western slopes of the Gháts,'from South Kanara to Malabar, at altitudes between 3,000 and 4,000 feet.

TIMBER.
1017

Structure of the Wood.—According to Beddome the timber, though little known, is likely to be of value. It is very compact and hard, and is employed, by the Natives of South Kanara, to make rice-pounders.

1018   **P. pauciflorum,** *Bedd. ; Fl. Br. Ind., I., 278.*

Vern.—*Podungali,*'TINNEVELLY.

References.—*Beddome, Fl. Sylv., t. 93; Gamble, Man. Timb., 21 ; Ind. Forester, III., 22.*

Habitat.—A large tree, abundant on banks of rivers on the hills of South Tinnevelly and Travancore, up to nearly 4,000 feet.

TIMBER.
1019

Structure of the Wood.—The timber is hard, reddish, and, according to Beddome, valuable. It is used by the Natives for building, for making walking-sticks, and for other purposes.

POGOSTEMON, *Desf. ; Gen. II., Pl., 1179.*

1020   **Pogostemon parviflorus,** *Benth. ; Fl. Br. Ind., IV., 632 ;* LABIATÆ.

Syn.—P. PUBESCENS, *Benth.* ; P. FRUTESCENS, *Grah.* ; P. PURPURICAULIS, *Dals.* ; P. INTERMEDIUS, *Wall.*

Vern.—*Phángla, pángla, pháng,* BOMB.

References.—*Graham, Cat. Bomb. Pl., 149; Dals. & Gibs., Bomb. Fl., 207 ; Dymock, Mat. Med. W. Ind., 2nd Ed., 609, 888.*

Habitat.—A small bush found on the sub-Tropical Himálaya from Kumaon to Bhotan, between the altitudes of 500 and 4,000 feet; also in Assam, the Khásia mountains, Sylhet, Chittagong, Tenasserim, and Western India from the Konkan to the Anamallays.

MEDICINE.
Leaves.
1021
Roots.
1022

Médicine.—Dymock writes, " The whole plant has a strong black-currant odour, . ., the fresh LEAVES have a slightly pungent taste, and when bruised are applied as a cataplasm in order to clean wounds and promote healthy granulation. The ROOTS are reputed to be a remedy for the bite of the *Phúrsa* snake (**Echis carinata**) ; they are directed to be

P. 1022

| The Patchouli Oil. | (*J. Murray.*) | POGOSTEMON Patchouli. |

MEDICINE.

chewed, but as this plant, like other labiates, has simply stimulant properties, it is probable that peppermint would be just as efficient. The bite of
the *Phúrsa*, though very dangerous, is by no means always fatal" (*Mat.
Med. W. Ind.*).

Food.—The LEAVES were eaten in Khándesh District during the famine
of 1877-78.

FOOD.
Leaves.
1023
1024

Pogostemon Patchouli, *Pellet.; Fl. Br. Ind., IV., 633; Wight,*

PATCHOULI. [*lc., t. 1440.*

Syn.—P. HEYNEANUS, *Benth.* ; ? ORIGANUM INDICUM, *Roth.*
Var. suavis=P. SUAVIS, *Tenore*; P. PATCHOULI, *Hook., Kew Journ.
Bot., I., 328, t. 11, exel. syn. putcha pat.*
Vern.—*Peholi, pachóli, HIND.; Pachápát, BENG.; Panel, SIND; Patcha,
patchpan, máli, MAR.; Pachpanadi, GUZ.; Poko nilam, búr kalif,
MALAY.; Gang-kolang-kola, SING.*
References.—*Thwaites, En. Ceyl. Pl., 239; Dals. & Gibs, Bomb. Fl., 207,
App., 66; Graham, Cat. Bomb. Pl., 149; Rheede, Hort. Mal., X., 77;
Hooker, Him. Journals, II., 314; Pharm. Ind., 169; Moodeen Sheriff,
Supp. Pharm. Ind., 201; Sakharam Arjun, Cat. Bomb. Drugs, 103;
Year Book Pharm., 1879, 467; Trans., Med. & Phys. Soc., Bomb. (New
Series), III., 149; Drury, U. Pl. Ind., 349; Useful Pl. Bomb. (XXV.,
Bomb. Gaz.), 224; Piesse, Perfumery, 57, 174–175; Kew Bulletin, 1888,
71–74, 133; Kew Off. Guide to the Mus. of Ec. Bot., 104; Kew Off. Guide
to Bot. Gardens & Arboretum, 73; Gazetteer:—Mysore & Coorg, I.,
69; Agri.-Horti. Soc., Ind., Jour. (New Series), VIII., 283; Tropical
Agriculturist, I., 983; II., 161; Ind. Agriculturist., May 5th, 1888.*
Habitat.—P. Patchouli, *var. proper*, is said to be found wild and cultivated in Western India from Bombay southwards and in Ceylon. The
Patchouli plant of commerce is, however, *var.* suavis, which, according to
Professor Oliver, is probably not indigenous in any part of India. The
same authority is of opinion that this variety may ultimately prove to have
originated in China. This opinion is supported by Mr. Wray, Curator of
the Government Museum, who, in a recent paper on the subject (*Agri.-
Horti. Soc., Ind., Jour. (NewSeries), VIII., 283*), states that the plant is
doubtfully wild in the Malay Peninsula. "There appears to be no evidence," he writes, "that it has been met with in the jungle, except in
places where it could be clearly traced to some old cultivation. It is
grown and much esteemed by the aboriginal tribes of Perak and Pallang,
and this should be borne in mind when cases of its being found in out-of-
the-way places are brought forward in support of its being a native of
the Peninsula." It is abundantly cultivated in the Straits Settlements,
Perak, and other parts of the Malay Peninsula, and is grown as a garden
plant in several localities in India.

Oil & Perfume.—The PLANT is grown entirely for its value as a per-
fume-yielder. A full description of the method of cultivation, curing, and
preparation is given by Mr. Wray in the paper above quoted, to which the
reader may be referred for particulars. Shortly, it may be said that in the
Straits Settlements the cultivation is carried on almost exclusively by Chinamen, each man cultivating a patch of only from a half to one acre. The
land is trenched, the plants are put in rows, two feet apart, during the wet
season, and are very carefully shaded until thoroughly rooted. After six
months the first crop is obtained, two others are secured at intervals of six
months, after which the plants are dug up, and the land re-trenched and
manured.

The plants are gathered by cutting down all except one stalk, and are
cured by being placed in the sun during the day and put under cover at
night. When thoroughly dry it is done up into bales, and sold, either to

OIL &
PERFUME.
Plant.
1025

20 A

| POGOSTEMON Patchouli. | The Patchouli Oil. |

OIL & PERFUME.

the dealers or to distillers. Adulteration is said to be freely practised by the former, either with the leaves of *ruku* (Ocimum Basilicum, Linn., *var.* pilosum), or with those of *perpulut* (Urena lobata).

The volatile oil of Patchouli is prepared by passing steam through the leaves in a large copper cylinder and condensing the distillate. One *pikul* (133½℔) of the raw material yields from 24 to 30 ounces of the oil, and if free from the heavier stalks, about double that amount. **Mr. Wray** states that the oil may be of two kinds, one sage-green, the other golden-brown, each of which possesses certain peculiar chemical and physical properties, but both are of equal value commercially.

Patchouli oil is very largely employed in European perfumery. Its odour is said to be the most powerful of any volatile oil ; accordingly, a very small amount suffices to impart a powerful and permanent odour to a large quantity of essence. Its introduction into the European market is said to have been due to the Lyons' shawl manufacturers, who, finding that the Indian shawls were always strongly scented with the perfume, import-ed the oil from the East to scent goods of their own manufacture. The

Leaves 1026

dried LEAVES are also extensively employed for scenting linen and other clothes in the same way as lavender, and are said to be a very efficacious preventative against the attacks of insects, hence, probably, the use of the material to scent Indian shawls.

Flowering Tops 1027 Oil 1028

In 1888, selected leaves were valued in the London market at from 9*d.* to 1s. 3*d.* the pound, the FLOWERING TOPS at 4*d.* to 1s. per ℔, and the pure OIL at 2s. 6*d.* to 3s. 9*d.* per oz. A letter from a Mincing Lane importer pub-lished in the *Kew Bulletin, 1888*, states that at that time the demand was good, and that 50 to 100 bales would sell if of good quality. There would appear to be no doubt that the demand certainly did exceed the supply at that time, and possibly does so still, chiefly owing to the fact that the steamers plying between Europe and the Straits refused to accept consign-ments of patchouli on account of the danger of the oil communicating its powerful odour to other goods.

The above rather lengthy remarks on this oil, not of strictly Indian origin, have been called for by the fact that it has been frequently suggest-ed that it should be cultivated on an extensive scale in this country. Commercial demand in Europe, the large profit derivable from its culti-vation, and the fairly large market existing in India, all point to the possi-bility of success. There appears to be no reason why it should not do well in many localities, though the commercial variety does not occur wild in any part of the country. The species *var.* **proper,** as already stated, is tru-ly wild, but little of a definite nature is known regarding its perfume-yield-ing powers. Lisboa states that the **P. Patchouli of Dalzell & Gibs.,** (*Bomb. Fl.,* addend. 66), which is *var.* **proper,** is employed for similar pur-poses to the patchouli of commerce in Bombay, and in the *Transactions of the Med. and Phys. Soc. Bombay,* it is remarked that it is similarly used in Sind. Whether it or imported leaves furnished the material for scenting Kashmír shawls, is doubtful, and the subject seems to be one worthy of attention. The ROOTS of a plant with a distinct odour of the oil are occa-

Roots 1029

sionally shipped to Europe from Bombay, and may be derived from the same plant.

A doubtful species of **Plectranthus (Plectranthus Patchouli,** *Clarke*), which occurs in Assam and Khásia, is said to have the characteristic odour of the volatile oil, and may prove to be of commercial importance. The material hitherto collected, however, has not been sufficient to render it possible to determine accurately its genus even, and the point is one which calls for further enquiry.

P. 1029

| Articles used as Poisons. | (*J. Murray.*) | POISONS. |

Medicine.—The PLANT is included in the secondary list of the *Pharma-copœia of India*, where, however, it is only stated that it is not highly valued as a medicine. The writer can find no mention of its therapeutic properties.

MEDICINE.
Plant.
1030

SPECIAL OPINION.—§ "The DRIED TOPS are mixed with the ingredients from which country-spirit is distilled in order to give to it an agreeable flavour" (*Surgeon A. C. Mukerji, Noakhally*).

Dried Tops.
1031

POINCIANA, *Linn.; Gen. Pl., I., 569.*

Poinciana elata, *Linn.; Fl. Br. Ind., II., 260,* LEGUMINOSÆ.

1032

Syn.—CÆSALPINIA ELATA, *Swarts.*

Vern.—*Váyni,* BOMB.; *Sankásúra,* MAR.; *Padenarayan,* TAM.; *Chiti-késwarum, chilikeswarapu, sunkéswaram, sunkeswar,* TEL.; *Nirangi, sunkanthe mara,* KAN.

References.—*DC., Prodr., II., 484; Roxb., Fl. Ind., Ed. C.B.C., 356; Brandi . For. Fl., 157; Beddome, Fl. Sylv., t. 178; Gamble, Man. Timb., 124; Elliot, Fl. Andhr., 43, 170; Lisboa, U. Pl. Bomb., 62, 392; Gazet-teer — Mysore & Coorg, I., 59; Ind. Forester, IX., 357; XII., App., 27; Boswell, Man. Nellore, 97, 100, 121.*

Habitat.—An erect tree found truly wild in the forests of South India, cultivated elsewhere.

Structure of the Wood.—Yellow, close, and even-grained, easily worked, gives a smooth surface, warps slightly but does not crack; weight, when seasoned, from 45 to 47℔ per cubic foot. It is well suited for cabinet work (*Beddome ; Gamble*).

TIMBER.
1033

Domestic. The LEAVES and TWIGS are said to be employed in Madras as a manure.

DOMESTIC.
Leaves.
1034
Twigs.

P. pulcherrima, *Linn.;* see **Cæsalpinia pulcherrima,** *Swarts*; Vol. II., 10.

1035

P. regia, *Bojer.; Fl. Br. Ind., II., 260.*

1036

Vern.—*Gulmohr,* MAR.

References.—*Kurz, For. Fl. Burm., I., 404; Beddome, For. Man., 91; Gamble, Man. Timb., 124; Mason, Burma & Its People, 412; Dymock, Warden & Hooper, Pharmacog. Ind., I., 549; Lisboa, U. Pl. Bomb., 62; Birdwood, Bomb. Prod., 266; Baden Powell, Pb. Prod., 590; Mysore Gazetteer, I., 59; Ind. Forester, I., 154; X., 304.*

Habitat.—A native of Madagascar, which was introduced into India within the last 70 years, and is now to be found planted almost all over the country.

Gum.—This tree yields a gum plentifully, which has been examined and described by the authors of the *Pharmacographia Indica* as follows :— "Irregular, granular or warty tears of a yellowish or reddish-brown colour, soluble in water, forming a thick opalescent mucilage. The solution is gelatinized by basic acetate of lead and ferric chloride, but not by the neutral acetate, nor by borax. Fehling's solution is slightly reduced. The gum contains a large quantity of oxalate of lime. The surface of some of the tears is of an opaque yellow colour; this portion consists large-ly of beautiful sphæro-crystals of oxalate of lime, closely resembling in formation the sphæro-crystals of inulin. On moistening this gum with water a cloud of small crystals often separates, and the sphæro-crystals attempt to arrange themselves into bundles of acicular crystals."

GUM
1037

Structure of the Wood.—White, light, soft, and loose-grained, takes a fine polish (*Kurz*).

TIMBER.
1038

POISONS.

Poisons, ARTICLES USED AS,—

1039

Many articles are frequently employed in India as poisons, either cri-minally, with the object of destroying the lives of human beings and cattle,

POISONS.	Articles used as Poisons.

or for the purpose of intoxicating or killing fish, to secure them as articles of food. The utilisation of the various poisonous products for these objects will be found detailed in the article on each in its respective position in this work. It is, therefore, only necessary, here, to give a list of the articles which fall under this designation.

For this purpose, poisons may be conveniently classified into I., those which may be inadvertently eaten by, or criminally administered to, human beings and cattle; and II., poisons, employed to intoxicate or kill fish. Each of these classes may be again sub-divided into A. Vegetable, B. Mineral.

References.—*Stewart, Pb. Pl., 96; Lisboa, U. Pl. Bomb., 264; O'Shaughnessy, Bengal Dispens., 718.*

POISON TO MEN & CATTLE.

I.—Poison to Men and Cattle.

A.—Vegetable.

Vegetable.
1040

Aconitum ferox, *Wall;* RANUNCULACEÆ.
A. Napellus, *Linn.*
A. palmatum, *Don.*
Actæa spicata, *Linn.;* RANUNCULACEÆ.
Ammania baccifera, *Linn.;* LYTHRACEÆ.
Anamirta Cocculus, *W. & A.;* MENISPERMACEÆ.
Andrachne cordifolia, *Muell. Arg;* EUPHORBIACEÆ.
Andropogon sp, *Stewart;* possibly A. scandens, *Roxb.;* GRAMINEÆ.
Arisæma curvatum, *Kunth.;* AROIDEÆ.
A. speciosum, *Mart.*
Atropa Belladonna, *Linn.,* SOLANACEÆ.
Avena fatua, *Linn.;* GRAMINEÆ.
Baliospermum axillare, *Blume;* EUPHORBIACEÆ.
B. montanum, *Muell. Arg.*
Buxus sempervirens, *Linn.;* EUPHORBIACEÆ.
Calotropis procera, *R. Br.;* ASCLEPIADEÆ.
Caltha palustris, *Linn.;* RANUNCULACEÆ.
Cannabis sativa, *Linn.;* URTICACEÆ.
Casearia graveolens, *Dalz.;* SAMYDACEÆ.
Cerbera Odollam, *Gærtn.;* APOCYNACEÆ.
Chrozophora plicata, *A. Juss.; Fl. Br. Ind., V., 409;* EUPHORBIACEÆ.
　　The whole plant has acrid poisonous properties.
Coriaria nepalensis, *Wall;* CORIAREÆ.
Crinum asiaticum, *Linn.;* var. toxicarum, *Herbert;* AMARYLLIDEÆ.
Crocus sativus, *Linn.;* IRIDEÆ.
Croton Tiglium, *Linn.;* EUPHORBIACEÆ.
Daphne oleoides, *Schreb.;* THYMELÆACEÆ.
Datura fastuosa, *Linn.;* SOLANACEÆ.
D. Stramonium, *Linn.*
Dioscorea ? deltoidea, *Wall.;* DIOSCOREACEÆ.
Diospyros montana, *Roxb.;* EBENACEÆ.
Elæodendron glaucum, *Pers.;* CELASTRINEÆ.
Euphorbia Tirucalli, *Linn.;* EUPHORBIACEÆ.
Excæcaria Agallocha, *Linn.;* EUPHORBIACEÆ.
Flueggea microcarpa, *Blume (Fl. Br. Ind., V., 328);* EUPHORBIACEÆ.
　　The bark is said to be poisonous.
Gloriosa superba, *Linn.;* LILIACEÆ (doubtfully poisonous).
Hura crepitans, *Linn.;* EUPHORBIACEÆ.
Hyoscyamus niger, *Linn;* SOLANACEÆ.
Jatropha Curas, *Linn.;* EUPHORBIACEÆ.

Articles used as Poisons.	(*J. Murray.*)	POISONS.

Jatropha multifida, *Linn.* (*Fl. Br. Ind., V., 383*), is cultivated, and natu- POISON TO
ralised in India. The seeds are a powerful acrid poison, similar to MEN &
those of the preceding. CATTLE.

Kalanchœ spathulata, *DC.*; CRASSULACEÆ. Vegetable.
Lactuca tatarica, *C. A. Meyer.*; COMPOSITÆ.
Lagenaria vulgaris, *Seringe*; CUCARBITACEÆ.
Lasiosiphon eriocephalus, *Dcne.*; THYMELÆACEÆ.
Lathyrus sativus, *Linn.*; LEGUMINOSÆ.
Lobelia nicotianæfolia, *Heyne.*; CAMPANULACEÆ.
Manihot utilissima, *Pohl.*; EUPHORBIACEÆ.
Meconopsis aculeata, *Royle*; PAPAVERACEÆ.
Melilotus alba, *Lamk.* (*Fl. Br. Ind., II., 89*); LEGUMINOSÆ. Said to
produce poisonous symptoms in cattle, which browse it.
Nerium odorum, *Soland.*; APOCYNACEÆ.
Papaver somniferum, *Linn.*; PAPAVERACEÆ.
Paspalum scrobiculatum, *Linn.*; GRAMINEÆ.
Phœnix sylvestris, *Roxb.*; PALMÆ (fruit said to be occasionally poisonous).
Physochlaina præalta, *Hook. f.*; SOLANACEÆ.
Phytolacca acinosa, *Roxb.*; PHYTOLACCACEÆ.
Plumbago rosea, *Linn.*; PLUMBAGINEÆ.
P. zeylanica, *Linn.*
Ranunculus arvensis, *Linn.*; RANUNCULACEÆ.
Rauwolfia serpentina, *Benth.*; APOCYNACEÆ.
Rhododendron arboreum, *Sm.*; ERICACEÆ.
R. campanulatum, *Don.*
Ricinus communis, *Linn.*; EUPHORBIACEÆ.
Sapium insigne, *Benth.*; EUPHORBIACEÆ.
Semecarpus Anacardium, *Linn.*; ANACARDIACEÆ.
Sophora mollis, *Grah.*; var. hydaspis; LEGUMINOSÆ.
Sorghum halepense, *Pers.*; GRAMINEÆ.
Strychnos colubrina, *Linn.*; LOGANIACEÆ.
S. Nux-vomica, *Linn.*
Taxus baccata, *Linn.*; CONIFERÆ.
Thevetia neriifolia, *DC.*; APOCYNACEÆ.
Trianthema pentandra, *Linn.*; FICOIDEÆ.
Trigonella Fœnum-græcum, *Linn.*; LEGUMINOSÆ.
T. polycerata, *Linn.*
Tylophora fasciculata, *Ham.*; ASCLEPIADEÆ.
Withania somnifera, *Dunal.*; SOLANACEÆ.

The above list comprises most of the plants, parts of which are poisonous when eaten, or most commonly cause deleterious effects when browsed
for fodder by cattle, sheep, goats, or camels.

B.—MINERAL.

Mineral.
1041

ACIDS.—The strong mineral acids are not infrequently used criminally
for poisoning.

ANTIMONY.—The tartrate of antimony and potash, or tartar emetic,
is occasionally employed criminally.

ARSENIC.—The white arsenic of the bazárs is perhaps the drug most
commonly resorted to in India for criminal poisoning.

COPPER. - The salts are occasionally employed.

LEAD.—White lead (carbonate) being much used in the arts, and
readily obtainable in the bazárs, is occasionally used for poisoning.

MERCURY.—The crude mixture of chlorides of this metal, which contains both mercuric and mercurous salts, and is easily obtainable in the
bazárs, is considerably used for criminal purposes.

POLIANTHES
tuberosa.   Articles used as Foisons.

POISONS TO FISH.	**II.—Poisons used for intoxicating or killing fish.**

A short list of the more important fish poisons is given in Volume III., 366; but as several others are worthy of notice, and as more than one printer's error unfortunately occurs in the former list, a more complete enumeration may be given in this place.

Vegetable.
1042

A.—VEGETABLE.

Albizzia procera, *Benth.;* LEGUMINOSÆ. The bark pounded and thrown into water stupifies fish.
Anagalis arvensis, *Linn.,* var. cœrulea; PRIMULACEÆ.
Anamirta Cocculus, *W. & A.;* MENISPERMACEÆ.
Balanites Roxburghii, *Planch.;* SIMURABEÆ.
Barringtunia acutangula, *Gærtn.;* MYRTACEÆ.
Bassia latifolia, *Roxb.;* SAPOTACEÆ.
Casearia graveolens, *Dalz.;* SAMYDACEÆ.
C. tomentosa, *Roxb.*
Eremostachys Vicaryi, *Benth.;* LABIATÆ.
Euphorbia Tirucalli, *Linn.;* EUPHORBIACFÆ.
Flueggea microcarpa, *Blume;* EUPHORBIACEÆ
Gnetum scandens, *Roxb.;* GNETACEÆ. Leaves used to poison fish in the Konkan
Gynocardia odorata, *R. Br.;* BIXINEÆ.
Hydnocarpus venenata, *Gærtn.;* BIXINEÆ.
H. Wightiana, *Bl.*
Lasiosiphon eriocephalus, *Dcne.;* THYMELACEÆ.
Mæsa indica, *Wall.;* MYRSINEÆ.
Mundulea suberosa, *Benth;* LEGUMINOSÆ.
Ougeinia dalbergioides, *Benth.;* LEGUMINOSÆ.
Randia dumetorum, *Lamk.;* RUBIACEÆ.
Sapindus trifoliatus, *Linn.;* SAPINDACEÆ.
Strychnos Nux-vomica, *Linn.;* LOGANIACEÆ.
Walsura piscidia, *Roxb.;* MELIACEÆ
Zanthoxylum alatum, *Roxb.;* RUTACEÆ

[RIDEÆ; Vol. II., 370.
Polanisia icosandra, *W. & A.,* see Cleom· viscosa, *Linn.;* CAPPA-

1043   ### POLIANTHES, *Linn; Gen. Pl.,* III., 757.

The name of this genus is sometimes written by mistake **Polyanthus** or **Polyanthes**. It is, however, evidently derived from πολιός, white or glistening, and ἄνθος, a flower, not from πόλις, a city, nor from πολύς, many, as some authors try to make out. (*Gen. Pl.*)

1044   ## Polianthes tuberosa, *Linn.;* AMARYLLIDEÆ.

Vern.—*Gulshabbo, gulchéri, gulshabbá,* HIND.; *Rajanigandha, runjuni,* BENG.; *Gul shabbo,* PB; *Gulchéri,* BOMB.; *Néla sampenga, véru sampenga,* TEL.; *Hnen-ben,* BURM.; *Rajanigandha,* SANS.
References.—*Roxb., Fl. Ind.,* Ed C.B.C., 295; *Stewart, Pb. Pl.,* 235; *Elliot, Fl. Andhr.,* 132, 192; *Mason, Burma & Its People,* 429, 814; *Ainslie, Mat. Med.,* II., 481; *U. C. Dutt, Mat. Med. Hind.,* 315; *Dymock, Mat. Med. W. Ind.,* 2nd Ed., 832; *Murray, Pl. & Drugs, Sind,* 23; *Piesse, Perfumery,* 213; *Settlement Report, C. P., Chanda,* 83; *Gazetteers:—Mysore & Coorg,* I., 67; *Bomb., VII.,* 40; *N.-W. P., I.,* 85.

Habitat.—A native of Mexico or South America, cultivated in gardens in India, Ceylon, and Java. It is much prized for the fragrance of its flowers.

Oil & Perfume.—Attar of Tuberose is obtained from the FLOWERS of this plant. One of the most exquisite of perfumes is obtained from the flowers by *enfleurage*, that is, by extracting the odorous principle with oily matter. From the pomatum thus manufactured, an extract is prepared with rectified spirit. The essence, like that of jasmine, is exceedingly volatile, and, consequently, is not used alone. The best fixing ingredients with which to mix it are tincture of storax or extract of vanilla (*Piesse*).

OIL & PERFUME.
Flowers.
1045

Medicine.—Ainslie mentions that the PLANT was shown to him, by Natives, as medicinal, but he had reason to doubt the assertion. Dymock states that it is considered to be hot, dry, diuretic, and emetic. The BULBS, after being dried and powdered, are used as a remedy for gonorrhœa. In the Konkan, when rubbed up with turmeric and butter, they are applied to remove *watiya*, small red pimples which often trouble new-born children. They are also rubbed into a paste with the juice of *Durva* grass (**Cynodon dactylon**), and applied to buboes (*Mat. Med. W. Ind.*).

MEDICINE.
Plant.
1046
Bulbs.
1047

Pollard, see **Coppice**, Vol. II., 520; also **Hedges & Fences**; Vol. IV, 206.

POLYALTHIA, *Blume.*; *Gen. Pl., I., 25.*

[ANONACEÆ.

Polyalthia cerasoides, *Benth. & Hook. f.; Fl. Br. Ind., I., 63;* 1048

Syn.—UVARIA CERASOIDES, *Roxb.*; GUATTERIA CERASOIDES, *Dun.*

Vern.—*Kudumi,* HIND.; *Panjon,* SANTAL; *Húm, vubbina,* BOMB.; *Húm,* MAR.; *Nakulsi, mútili,* TAM.; *Chilka dúdúgú, chilta dudaga, dudduka,* TEL.; *San heesare, vubbina,* KAN.

References.—*Roxb., Fl. Ind., Ed. C.B.C., 456; Brandis, For. Fl., 5; Kurz, For. Fl. Burm., I., 38; Beddome, Fl. Sylv., t. 1; Gamble, Man. Timb., 9; Dalz. & Gibs., Bomb Fl., 3; Elliot, Fl. Andhr., 43, 47; Rev. A. Campbell, Rept Econ. Pl., Chutia Nagpur, No. 9461; Drury, U. Pl., 341; Lisboa, U. Pl. Bomb, 3; Birdwood, Bomb. Pr., 323; Gazetteers:— Mysore & Coorg, I., 52, 57; Bombay, XV., Pt. I., 74; Settle. Rept., Upper Godavery, 37; Ind. Forester, III., 200; Agri. Horti. Soc. India, Trans., VII., 49.*

Habitat.—A large evergreen tree, found in dry forests on eastern exposures from Behar to Travancore.

Structure of the Wood—Olive-grey, moderately hard, close-grained, weight 52℔ per cubic foot (*Gamble*). According to Beddome it is, in the Central Provinces and Bombay, much used in carpentery and for making masts and small spars for boats. He remarks that, notwithstanding the abundance of the tree in dry forests near the foot of all the mountains on the western side of the Madras Presidency, in the Salem forests, the Nullay Mullays, Mysore, Orissa, and the Godavery forests,—the timber is apparently but little known in Madras.

TIMBER.
1049

1050

P. coffeoides, *Benth. & Hook. f.; Fl. Br. Ind., I., 62.*

Syn.—GUATTERIA COFFEOIDES, *Thwaites.*

References.—*Beddome, Ic. Pl. Ind, t. 53; Thwaites, Enum Pl Cey., 10; Gamble, Man Timb., 8.*

Habitat.—A spreading tree found in the forests of the Wynaad and Ceylon.

Fibre.—The BARK is said to be made into ropes on the Western Gháts.

FIBRE.
Bark.
1051

P. longifolia, *Bth. & Hook. f.; Fl. Br. Ind., I., 62; Wight, Ic., t. 1.* 1052

THE INDIAN FIR, OR MAST TREE.

Syn—GUATTERIA LONGIFOLIA, *Wall*; UVARIA LONGIFOLIA, *Lam.*

Vern.—*Asok, debdari, dévadaru, dévadár,* HIND.; *Dévadáru, dévadár,* BENG.; *Deb-dáru,* URIYA; *Asók, asoka, asúpála, devadaru, ásopálav, ásupál,* BOMB.; *Ashopalo,* GUZ.; *Assothi,* TAM.; *Asoká, devadaru, asókam,* TEL.; *Asoka, putrajiva,* KAN.; *Maraillepe,* SING.

P. 1052

POLYCARPÆA
corymbosa.   The Indian Fir or Mast Tree.

References.—*Roxb., Fl. Ind., Ed. C.B.C.,* 455 ; *Brandis, For. Fl.,* 4 ; *Beddome, Fl. Sylv., t.* 38 ; *Gamble, Man. Timb.,* 8 ; *Thwaites, En. Ceylon Pl.,* 10 ; *Dals. & Gibs., Bomb. Fl. Suppl.,* 2 ; *Elliot, Fl. Andhr.,* 17 ; *Sir W. Jones, Treat., Pl. Ind. V.,* 130 ; *Lisboa, U. Pl. Bomb.,* 2, 399 ; *Gazetteers :—Mysore & Coorg, I.,* 57 ; *Bombay, V.,* 23, 360 ; *VII.,* 42, 43 ; *Orissa, II.,* 68, 179 ; *Ind. Forester, IV.,* 389 ; *X.,* 31 ; *Agri.-Horti. Soc. India, Trans., VII.,* 49 ; *Journals (Old Series). IV., Sel.,* 246, 268 ; *VII.,* 118 ; *IX.,* 401.

Habitat.—A large, erect, evergreen glabrous tree, wild in the drier parts of Ceylon and Tanjore ; cultivated throughout the hotter parts of India. It is commonly planted in avenues along roads in Bengal and South India.

FIBRE.
Bark.
1053

Fibre.—Dr. G. Watt writes, " A good bast fibre was shown me by Babu T. N. Mukharji, which was said to have been prepared from the inner BARK of this tree, and was sent to the Amsterdam Exhibition."

MEDICINE.
1054

Medicine.—Hunter states that " it " (he does not mention what part of the tree) " is used as a febrifuge in the Balasor District of Orissa."

FOOD.
Fruit.
1055

Food.—The tree flowers in February, and during the rains ripens its FRUIT, which is ovoid or oblong, one-seeded, and purple. It is a favourite food of birds, and is said to be eaten during times of scarcity.

TIMBER.
1056

Structure of the Wood.—White or whitish-yellow, light and very flexible, tolerably close and even-grained ; weight when seasoned 37℔, when un-seasoned 4¼ to 48℔, per cubic foot (*Beddome*). It is used for making drum cylinders in Southern India, and, according to the Rev. **J. Long,** is employed in Bengal for making pencils and boxes, and, in China, for matches. Dr. Watt remarks, " It might be found suitable for tea-boxes, as also the wood of **P. simiarum,** *Bth. & Hook. f.,* a large tree of Eastern Bengal and Burma, and **P. Jenkinsii,** *Bth. & Hook. f.,* which is found in Assam and Sylhet.

DOMESTIC

Leaves.
1057

Domestic & Sacred.—The tree, owing to its handsome appearance and shade-giving powers, is much esteemed for avenue-making It bears a semi-sacred character, as certain of the vernacular names imply. The LEAVES, strung into wreaths, are in Baroda hung up to adorn doors, on Hindu festive occasions, and in Southern India generally, according to Lisboa, they are employed in marriage ceremonies for covering *mándvás.*

1058

Polyalthia suberosa, *Bth. & Hook. f.* ; *Fl. Br. Ind., I.,* 65.

Syn.—UVARIA SUBEROSA, *Roxb.* ; GUATTERIA SUBEROSA, *DC.*
Vern.—*Bara chali,* BENG. ; *Sandi omé,* KOL. ; *Bandorkola,* ASSAM ; *Chilaka dúdúga,* TEL.
References.—*Roxb., Fl. Ind., Ed. C.B.C.,* 456 ; *Brandis, For. Fl.,* 5 ; *Kurs, For. Fl. Burm , I.,* 38 ; *Beddome, Ic. Pl., Ind. Or., t.* 56 ; *Gamble, Man. Timb.,* 9 ; *Elliot, Fl. Andhr.,* 38 ; *For. Ad. Rep., Ch. Nagpur,* 1885, 28 ; *Ind. Forester, III.,* 200 ; *Agri.-Horti. Soc. Ind., Trans., VII.,* 49.

Habitat.—A small tree or shrub, with corky bark, met with in the forests of Bengal, South-Western India, Tenasserim, and Ceylon.

TIMBER.
1059

Structure of the Wood.—Hard, close-grained, tough, and durable ; weight, according to Brandis, 40℔, and Kyd, 45℔, per cubic foot ; structure similar to that of **P. cerasoides.** It is employed for similar purposes to the wood of that species.

POLYCARPÆA, *Lam.* ; *Gen. Pl., I.,* 154.

[*A.,* 712 ; CARYOPHYLLEÆ.

1060

Polycarpæa corymbosa, *Lamk.* ; *Fl. Br. Ind, I.,* 245 ; *Wight, Ic.*

Syn.—P. SPADICEA, *Lamk.* ; P. DENSIFLORA, *Wall.* ; P. INDICA, *Lamk.* ; PARONYCHIA SUBULATA, *Lamk.* ; ACHYRANTHES CORYMBOSA, *Linn.* CELOSIA CORYMBOSA, *Willd.* ; MOLLIA SPADICEA & CORYMBOSA, *Willd.* ; LAHAYA SPADICEA & CORYMBOSA, *Schult.*

P. 1060

| The Milkworts. | (*J. Murray.*) | POLYGALA
crotalarioides. |

Vern.—*Janhe nanjour,* SANTAL; *Nilai sedachi,* TAM.; *Bomma sári, rajuma,* TEL.

References.—*Roxb., Fl. Ind., Ed. C.B.C.,* 229; *Burm., Zeyl., t.* 65, *f.* 2; *Rev. A. Campbell, Econ. Pl., Chutia Nagpur, No.* 8736; *Sir W. Elliot, Fl. Andhr.,* 30; *Murray, Pl. & Drugs, Sind,* 95; *Dymock, Warden & Hooper, Pharmacog. Ind., I.,* 158; *Atkinson, Him. Dist.,* 306; *Gazetteers:—Bombay, V.,* 23; *N.-W. P., I.,* 79; *IV., lxviii.*

Habitat.—An annual or perennial herb, found in many parts of India, from the Western Himálaya, where it ascends to 7,000 feet, and Burma to Central and North-Western India, Sind, Southern India, and Ceylon.

Medicine.—The authors of the *Pharmacographia Indica* state that the HERB is "administered in Pudukota, both externally and internally, as a remedy for the bites of venomous reptiles." They believe that it may possibly contain a little *saponin,* but did not consider it of sufficient importance to be worthy of examination.

MEDICINE.
Herb.
1061

POLYGALA, *Linn.; Gen. Pl., I., 136.*

Polygala chinensis, *Linn.; Fl. Br. Ind., I., 204;* POLYGALEÆ.
THE COMMON INDIAN MILKWORT.

1062

Syn.—P. ARVENSIS, *Willd.;* ? P. PROSTRATA, *Willd.;* P. ROTHIANA, *W. & A.;* P. TRANQUEBARICA, *Mart.;* P. GLAUCOIDES, *Wight;* P. GRANDIFLORA, *Hb. Wight.;* P. BRACHYSTACHYA, *DC. (not of Blume).*

Var.—α. triflora, *Linn. (sp.);* P. MULTIBRACTEATA, *Wall.;* P. RAMOSA, *Hb. Wight;* P. KLEINII, *Hassk.;* β. brachystachya, *Blume Bijd.* (*sp., not of DC.).* The latter variety, according to the *Fl. Br. Ind.,* is, perhaps, a distinct species.

Vern.—*Merádú,* HIND.; *Gaighura,* SANTAL; *Négli,* MAR.

References.—*Roxb., Fl. Ind., Ed. C.B.C.,* 531; *Thwaites, En. Ceylon Pl.,* 400; *Dalz. & Gibs., Bomb. Fl.,* 12; *Rev. A. Campbell, Rept., Ec. Pl., Chutia Nagpur, No.* 7592; *Dymock, Warden & Hooper, Pharmacog. Ind., I.,* 155; *Atkinson, Him. Dist.,* 305; *Gazetteers:—Mysore & Coorg. I.,* 57; *Bomb., XV.,* 427; *N.-W. P. I.,* 79; *IV., lxviii.; Agri.-Horti. Soc., Ind., Trans., VII.,* 79; *Journals (Old Series), IV.,* 205.

Habitat.—A very variable, usually procumbent, leafy, stout herb, found throughout India, from the Panjáb to Burma, South India, and Ceylon.

Medicine.—The authors of the *Pharmacographia Indica* state that this species is not used medicinally, but the Rev. A. Campbell writes that in Chutia Nagpur the ROOT is given medicinally in cases of fever and dizziness.

MEDICINE.
Root.
1063

Food.—In the Kaladgi District and other localities in Bombay the young LEAVES are always more or less used as an article of food, and are freely employed in times of scarcity. They are eaten with salt, chillies, and other condiments, and are said to be pleasant in flavour and perfectly wholesome (*Lisboa*).

FOOD.
Leaves.
1064

P. crotalarioides, *Ham.; Fl. Br. Ind., I., 201.*
Vern.—*Lil kathi,* SANTALI.

References.—*Rev. A. Campbell, Rept. Econ. Pl. Chutia Nagpur, No.* 8464; *Pharm. Ind.,* 29; *O'Shaughnessy, Beng. Dispens.,* 211; *Dymock, Warden & Hooper, Pharmacog. Ind., I.,* 154; *Drury, U. Pl. Ind.,* 352; *Atkinson, Him. Dist. (X. N.-W. P. Gaz.),* 305; *Royle, Ill. Him. Bot.,* 19.

Habitat.—An erect herb, found on the Temperate Himálaya from Chamba to Sikkim, between 4,000 and 7,000 feet, also on the Khasia Hills.

Medicine.—The entire PLANT and ROOT have a reputation as remedies for cough and pulmonary catarrh. Royle states that the plant was sent to him with the information that the root was employed as a cure for snake-bite by the hill-people of the Himálaya. This fact, though not corroborated by other writers, is of interest, since P. senega is similarly used in South America.

MEDICINE.
Plant.
1065
Root.
1066

P. 1066

| POLYGONATUM verticillatum. | Senega or Seneka ; Solomon's Seal. |

1067

Polygala Senega, *Linn. ; Bentl. & Trim., Med. Pl., I., t. 29.*
SENEGA, or SENEKA.
This species, which is the one most commonly used medicinally, is officinal in all Pharmacopœia. It is not a native of, nor is it cultivated in. India, and as its properties are well known, it need not be discussed in this work. The remark may, however, be made that certain of the Indian species may prove efficien substitutes for the imported drug. This would appear to be the moret probable from the fact that analyses of **P. chinensis, P. crotalarioides,** and **P. telphoides,** by the authors of the *Pharmacographia Indica,* have proved them, like Senega, to owe their medicinal properties to the presence of a substance closely related to, if not identical with, *Saponin.*

MEDICINE.
1068

1069

P. telephioides, *Willd. ; Fl. Br. Ind., I., 205.*
Syn.—P. SERPYLLIFOLIA, *Poir.* ; P. BUXIFORMIS, *Hassk.*
References.—*Roxb., Fl. Ind., Ed. C.B.C., 531 ; Drury, U. Pl. Ind., 352 ; Pharm. Ind., 29 ; Dymock, Warden & Hooper, Phamacog. Ind., I., 155 ; Gazetteer, N.-W. P., I., 79 ; Agri.-Horti. Soc. Ind., Trans., VII., 79.*
Habitat—Met with in the Karnatic, Travancore, and Ceylon.
Medicine —This species, like **P. crotalarioides,** is used medicinally in catarrhal affections. Both species occupy a place in the secondary list of the *Indian Pharmacopœia,* where they are recommended as worthy of further attention.

MEDICINE.
1070

POLYGONATUM, *Adans. ; Gen. Pl., III., 768.*

1071

Polygonatum multiflorum, *All. ; Baker, Linn. Soc. Jour., XIV.*
SOLOMON'S SEAL.   [*555 ;* LILIACEÆ.
Syn.—CONVALLARIA MULTIFLORA, *Linn.* ; C. GOVANIANA, *Wall.*
References. —*Boiss., Fl. Orient., V., 331 ; Aitchison, Fl. Lahoul, Linn. Soc. Jour. X., 178 Fl. Kuram Valley, Linn. Soc. Jour., XVIII., 189 ; Atkinson, Him. Dist., X., 319 ; Royle, Ill. Him. Bot., 380, 382 ; Smith, Ec. Dict., 384.*
Habitat.—A perennial herb, found over the whole of Europe, also in Siberia, Japan, Central Asia, Afghánistan, and the Western Temperate Himálaya.
Medicine.—The RHIZOMES are used in Europe as a popular remedy for removing bruises, and discolouration of the skin resulting from blows. The writer can find no record of its being utilised medicinally in India.
Domestic.—Aitchison informs us that the ROOT is employed in Lahoul instead of soap.

MEDICINE.
Rhizomes.
1072
DOMESTIC.
Root.
1073

1074

P. verticillatum, *All. ; Baker in Linn., Soc. Jour., XIV., 560.*
Syn.—CONVALLARIA VERTICILLATA, *Linn.* ; EVALLARIA VERTICILLATA, *Neck.* ; CAMPYDORUM VERTICILLATUM, *Salisb.* ; POLYGONATUM LEPTOPHYLLUM, *Royle* ; CONVALLARIA LEPTOPHYLLA, *D. Don.* ; P. JACQUE-MONTIANUM, *Kunth.*
Vern. — *Mitha dádiq,* rhizome=*shakákalmisrí,* HIND.; rhizome=*shakákal,* AFGH.
References.—*Boiss., Fl. Orient., V., 333 ; Aitchison, Fl. Lahoul, in Jour. Linn. Soc., X., 100 ; Fl. Kuram Valley, in Linn. Soc. Jour., XVIII., 103; Royle, Ill. Him. Bot., 380, 382 ; Aitchison, Notes on Products of W. Afghán, and N.-E. Persia, 161.*
Habitat.—Met with over the whole of Europe, widely spread in Northern and Central Asia, also found in the Kuram Valley (Afghanistan), and the Temperate Himálaya,
Food.—Aitchison s ates that the ROOT is much sought after and valued as an article of food in the Kuram Valley. They are also collected and

FOOD.
Root.
1075

| The Knot-Grass. | (*J. Murray*) | POLYGONUM barbatum. |

sent to Kabul, and are thence exported, *viá* Peshawar, to India, where they are valued in the same way as orchis tubers, as a strength-giving food. Royle states that the plant (he does not mention which part) is considered poisonous by the hill-people on the Himálaya.

FOOD.

POLYGONUM, *Linn.*; *Gen. Pl., III., 97.*
[POLYGONACEÆ

Polygonum alatum, *Ham.,* var. **nepalense**; *Fl. Br. Ind, V., 42;*

1076

Syn.—P. NEPALENSE, *Meissn.*; P. GUTTULIFERUM, *Miq.*
Vern.—*Sat balon,* PB.
References.—*Stewart, Pb. Pl., 185 ; Atkinson, Him. Dist., 316.*
Habitat.—An annual herb, abundant throughout the Himálaya from Sikkim to Kashmir, between elevations of 4,000 and 10,000 feet.
Medicine.—In Kángra the LEAVES are employed as a local application to swellings (*Stewart*).

EDICINE.
Leaves.
1077
1078

P. aviculare, *Linn.*; *Fl. Br. Ind., V., 26.*

THE KNOT-GRASS.

Syn.—P. AVICULARE, γ diffusum, *Meissn.*

Vern.—*Macháti, nisomali, ban-natia, endrani, bigbund, hunráj,* HIND.; *Machútie,* BENG.; *Kesrú, bandúke,* PB.; *Tandulai,* PUSHIU; *Miromati, nisomali,* SANS.

References.—*Stewart, Pb. Pl., 184; Bot. Afgh. Del. Com., 105; O'Shaughnessy, Beng. Dispens., 522; Dymock, Mat. Med. W. Ind., 2nd Ed., 659; Murray, Pl. & Drugs, Sind, 96; Irvine, Mat. Med., Patna, 16; Year-Book Pharm., 1874, 33 ; Atkinson, Him. Dist., 316; Butler, Top. & Statis., Oudh & Sultanpur, 34; Balfour Cyclop., III., 254; Agri.-Horti. Soc., Panjab, Select papers to 1862, 214.*

Habitat. — Found in the North-West Himálaya from Kashmír to Kumaon, between 6,000 and 10,000 feet ; also in Western Tibet, between 10,000 and 12,000 feet
Dye.—According to Thunberg a blue dye, not much inferior to indigo, is prepared from the PLANT in Japan.
Medicine.—In Chamba the HERB is applied externally as an anodyne (*Stewart*). Honigberger states that in his time it was official in Kashmír. The SEEDS are said to be powerfully emetic and cathartic, and in Europe the whole plant is considered vulnerary and astringent. In the *Pharm. Jour.* above cited, an interesting account is given of the reputed value of the decoction of the herb in cases of vesical calculus. A case is described in which a dose of two tumblerfuls of the decoction is said to have been followed by almost immediate relief.
Food & Fodder.—Mr. Duthie's collector states that it is occasionally employed in the Trans-Indus country, to make curry, but is generally given as fodder to sheep and goats, for which purpose it is highly prized.
Domestic.—A writer in the Publication of the Panjáb *Agricultural and Horticultural Society* draws attention to the fact that knot-grass has been used with success for feeding silk-worms in France, instead of mulberry, and suggests that it be tried for the same purpose in the Panjáb during the cold weather.

DYE.
Plant.
1079
MEDICINE.
Herb.
1080
Seeds.
1081

FOOD &
FODDER.
1082
DOMESTIC.
1083

P. barbatum, *Linn.*; *Fl. Br. Ind, V., 37 ; Wight, Ic., t. 1798.*

1084

Syn.—P. RIVULARE, *Kæn.*; P. HORNEMANNI, *Meissn.*; P. MARMORAMÆ and FLUVIATILE, *Herb Ham.*

Vern.—*Bekh-unjubaz,* BENG. ; *Narri,* PB.; *Atalari,* TAM.; *Konda malle, niru gannéru,* TEL. ; *Velluta modela mukku,* MALAY.

P. 1084

POLYGONUM Hydropiper.   The Water-Pepper.

References.—*Roxb., Fl. Ind., Ed. C.B.C., 335 ; Voigt, Hort. Sub. Cal., 325 ; Graham, Cat., Bomb. Pl., 172 ; Dalz. & Gibs., Bomb. Fl., 214 ; Stewart, Pb. Pl., 185 ; Elliot, Fl. Andhr., 96, 135 ; Ainslie, Mat. Ind., II., 1 ; O'Shaughnessy, Beng. Dispens., 522, 523 ; Murray, Pl. & Drugs, Sind, 96 ; Irvine, Mat. Med., Patna, 15 ; Atkinson, Him. Dist., 316 ; Drury, U. Pl., 352 ; Gazetteers:—Mysore & Coorg, I., 65 ; Bombay, XV., 441 ; N.-W. P., IV., lxxvii.*

DYE.
Herb.
1085

Habitat.—Common throughout the hotter parts of India, from Assam to the Indus, and southwards to Malacca, Penang, and Ceylon.

MEDICINE.
Root.
1086

Dye.—A blue dye, like indigo, is said to be obtained from this HERB in China and Japan.

Seeds.
1087

Medicine.— The ROOT, according to Irvine, is imported into Patna from the North-West Provinces, and is used medicinally as an astringent and cooling remedy. The SEEDS are said by Ainslie to be employed to relieve colic in Southern India. In China, a decoction of the LEAVES and STALKS is said to be used as a stimulating wash for ulcers.

Leaves.
1088

Stalks.
1089

Fodder.—Balfour remarks, "Cattle eat it greedily."

FODDER.
1090

Polygonum Fagopyrum, *Linn. ;* see **Fagopyrum esculentum,**
[*Moench* ; Vol. III., 310.

1091

P. glabrum, *Willd. ; Fl. Br. Ind., V., 34 ; Wight, Ic., t. 1799.*

Syn.—P. PERSICARIA, *Wall.* ; P. POIRETII, *Meissn.*

Var.—α. SCABRINERVIS, P. SCABRINERVE, *Royle ;* P. QUADRIFIDUM, *Ham. ;*
  β. MACRANTHA.

Vern.—*Sauri arak', jióti,* SANTAL ; *Larborna, bih agni, bih langani, patharua,* ASSAM ; *Rakta rohida,* BOMB. ; *Aatlaria,* TAM. ; *Schovanna mudela muccu,* MALAY.

References.—*DC., Prodr., XIV., i., 115 ; Roxb., Fl. Ind., Ed. C.B.C., 334 ; Dalz. & Gibs., Bomb. Fl., 214 ; Rev. A. Campbell, Rep. Econ. Pl., Chutia Nagpur, Nos. 8227, 8423, 9139 ; Graham, Cat. Bomb. Pl., 172 ; Rheede, Hort. Mal., XII., t. 77 ; Dymock, Mat. Med. W. Ind., 2nd Ed., 659 ; Note on the Condition of the People of Assam, App. D ; Gazetteer, N.-W. P., IV., lxxvii. ; Agri.-Horti. Soc., Panjáb, Select papers to 1862, 214.*

Habitat.—Found in ditches and wet places from Assam, Sylhet, and Bengal, westwards to the Indus and Sind, and southwards to Burma, ascending the Himálaya to 6,400 feet in Garhwal ; also common in Ceylon.

MEDICINE.
Leaves.
1092

Medicine.—An infusion of the LEAVES is used in Bombay to relieve pain in cases of colic ; in Chutia Nagpur it is employed as a cure for "stitch in the side," and in Assam as a remedy for fever.

FOOD.
Leaves.
1093

Food.—Campbell states that it is eaten as a pot-herb in Chutia Nagpur, a custom which, according to Darrah, also obtains in Assam. The latter author states that the LEAVES and YOUNG SHOOTS are always, when eaten, cooked with other vegetables. They are pungent in flavour, and are used only in small quantities.

Young shoots
1094

1095

P. Hydropiper, *Linn. ; Fl. Br. Ind., V., 39.*

THE WATER-PEPPER.

Syn.—P. MITE, *Wall.*

Vern.—*Packur-múl,* BENG.

References.—*O'Shaughnessy, Beng. Dispens., 522 ; Irvine, Mat. Med., Patna, 82 ; Balfour, Cyclop., III., 255.*

DYE.
Plant.
1096

Habitat.—Met with in wet places throughout India, from Assam, Sylhet, Chittagong, and Bengal to North-West India, ascending the Khasia mountains to 5,000 feet, and the Himálaya to 7,000 feet ; also reported from Madras.

MEDICINE.
Root.
1097

Dye.—The PLANT can be made to dye wool yellow (*O'Shaughnessy*).

Medicine.—Irvine states that the ROOT is stimulating, bitter, and tonic, and is used for these properties in Patna. In other localities the JUICE is

Juice.
1098

P. 1098

Polygonum Indigo.	(*J. Murray.*)	POLYGONUM tortuosum.

said to be considered diuretic, carminative, and anthelmintic, and is also employed externally as a wash in itching affections of the skin. O'Shaughnessy states that the whole PLANT is reputed to be a powerful diuretic, but to lose its activity on drying.

    Domestic, &c.—The plant is said by Balfour to be "used as a flux in operating on metals."

Polygonum microcephalum, *Don.; Fl. Br. Ind., V., 42.*

    Syn.—P. STATICIFLORUM, *Wall.*; P. STRIGOSUM, *Herb. Ham.*; ? P. CILIA-TUM, *Ham.*

    Vern.—*Madhu fulong, madu suleng,* ASSAM.

    References.—*Darrah, Note on the Condition of the People of Assam.*

    Habitat.—A large herb of Nepál, Sikkim, the Khásia mountains, and Sylhet.

    Food.—The young TOPS are sparingly used in Assam, as a flavouring agent to other vegetables, but the plant is never cooked nor eaten alone.

P. molle, *Don.; Fl. Br. Ind., V., 50.*

    Syn.—COCCOLOBA TOTNEA, *Ham.*

    Vern.—*Totnye, tuknu, patú-swa,* NEPAL.

    Reference.—*Gamble, Man. Timb., 303.*

    Habitat.—A robust shrub of the Central and Eastern Himálaya in Nepál and Sikkim, between 5,000 and 7,000 feet, and of the Mishmi Hills.

    Food.—"The young SHOOTS are pleasantly acid and are eaten like rhubarb" *Gamble*).

[*t. 1808.*

P. plebejum, *Br.;* var. indica, *Fl. Br. Ind., V., 28; Wight, Ic.,*

    Syn.—P. INDICUM, *Heyne.*; P. ROXBURGHII, *var.* α. and γ., *Meissn.*; P. DRYANDRI, *Wall.*

    Vern.—*Raniphul,* SANTAL.

    References.—*Dalz. & Gibs., Bomb. Fl., 214; Rev. A. Campbell, Rept. Ec. Pl., Chutia Nagpur, No. 7859.*

    Habitat.—Common in Mysore and the Karnatic; occurs also in Chutia Nagpur and Bombay.

    Medicine.—"The ROOT is given for bowel complaint" (*Campbell*).

P. polystachyum, *Wall.; Fl. Br. Ind., V., 50; Wight, Ic., t. 1807.*

    Syn.—P. MOLLE, *Wight.* Four varieties are enumerated in the *Fl. Br. Ind.,* but they have no important economic distinctions and need not be detailed in this work.

    Vern.—*Amldandi, chúchi, tror,* PB.

    Reference.—*Stewart, Pb. Pl., 185.*

    Habitat.—A shrub met with on the Temperate Himálaya from Mishmi to Kashmír, generally between the altitudes of 7,000 feet and 12,000 feet, but reaching 14,000 feet in Sikkim; distributed to Afghánistán.

    Food.—Stewart informs us that in the Panjáb Himálaya the young LEAVES are eaten as a pot herb, while the STALKS are consumed either raw after peeling, or stewed in imitation of rhubarb.

P. tinctorium, *Lour.;* see Indigo substitutes, Vol. IV., 451.

P. tortuosum, *Don.; Fl. Br. Ind., V., 52.*

    Syn.—P. TATARICUM, *Wall.*

    Vern.—*Niála, niálo,* PB.

    References.—*Stewart, Pb. Pl., 185; Aitchison, Fl. Lahoul, in Jour. Linn. Soc., X., 78.*

    Habitat.—Found in Garhwal, Kunawar, and Lahoul in the Western Himálaya, between 9,000 and 13,000 feet, also in Western Tibet at 15,500 feet, and Ladak at 16,500 feet.

Margin notes
MEDICINE.
Plant.
1099
DOMESTIC.
1100

1101

FOOD.
Tops.
1102

FOOD.
Shoots.
1103

1104

MEDICINE.
Root.
1105
1106

FOOD.
Leaves.
1107
Stalks.
1108

1109

POONYET, Pwenyet.	Poonyet Wax.

DYE

Plant.
1110

Dye.—The PLANT is used in Lahoul for dyeing yellow (*Stewart; Aitchison*).

FODDER.
1111

Fodder.—It is browsed by goats and yaks in Ladák (*Stewart*).

1112

Polygonum viviparum, *Linn.; Fl. Br. Ind., V., 31.*

Syn.—P. BULBIFERUM, *Royle;* P. AFFINE, *Wall.;* P. BISTORTA, *Gærcke;* P. ANGUSTIFOLIUM, *Don.;* P. BRACTEATUM, *Spreng.*

Vern.—*Maslún,* KASHMIR; *Maslún, mamech, dori, bajír, bílauri,* ROOT= *anjabár,* PB. "Not the *anjubár-i-rumi* of the Bombay bazárs, which is a drug imported from Persia." (*Dymock*).

References.—*Stewart, Pb. Pl., 183; in Journ. of the Agri.-Horti. Soc. Ind. (Old Series), XIV., 42; O'Shaughnessy, Beng. Dispens., 522; Murray, Pl. & Drugs, Sind, 96; Baden Powell, Pb. Prod., 372; Ind. Forester, VIII., 273; XI., 276.*

Habitat.—Found in the Alpine and Sub-Alpine Himálaya, from Kashmír to Sikkim, also in Western Tibet between 9,000 and 15,000 feet.

MEDICINE.
Root.
1113

Medicine.—The ROOT is powerfully astringent, containing a large quantity of tannic and gallic acids, and is used as an injection in the treatment of gleet and leucorrhœa, and internally in diarrhœa, dysentery, hæmoptysis, passive hæmorrhage generally, and intermittent fever. It is also employed to make a gargle for sore-throat, and a wash for ulcers with much discharge.

FOOD.
Leaves.
1114

Food.—The young LEAVES are edible (*Baden Powell*).

1115

[*Ed., 304.*

POLYPODIUM, *Linn.; Hooker & Baker, Synopsis Filicium, 2nd*

A large genus of FERNS, which comprises some 400 species, of which many are natives of India. A common European species, **P. vulgare,** *Linn.,* has already been described in the article on "Ferns" Vol. III., 323, as possibly forming part of *bisfáij,* a bazár drug in common use in India. Dymock states that an Indian species, **P. quercifolium,** *Linn.,* (*Hooker & Baker, Syn. Fel., 367;* (*Kadikapana, Kadickpan,* BOMB.; *Básingh, báshíng, wándar-báshíng,* MAR., *Sae-mway-pa* BURM.) is also employed medicinally in phthisis, hectic fever, dyspepsia, and cough (*Mat. Med. W. Ind., 2nd Ed., 863*). The leaves of plants growing on **Strychnos Nux-vomica** are said to be preferred.

[456.

Polyporus Fungi, see **Agaricus,** Vol. I., 129; also **Fungi,** Vol. III., 453,

Pomegranate tree, see **Punica Granatum,** *Linn.;* LYTHRACEÆ; p. 368.

POMETIA, *Forst.; Gen. Pl., I., 407.*

1116

Pometia tomentosa, *Kurz; Fl. Br., Ind., I., 691;* SAPINDACEÆ.

Syn.—P. EXIMIA, *Bedd.;* ECCREMANTHUS EXIMIUS, and NEPHELIUM EXIMIUM, *Thw.*

Vern.—*Thabyay,* BURM.; *Badoh,* AND.

References.—*Kurz, For. Fl., Burma. I., 294; Beddome, Fl. Sylv., t. 157; Thwaites, En. Cey. Pl., 57; Gamble, Man. Timb., 98.*

Habitat.—A large tree, met with in the South Andamans, Malacca, and the central province of Ceylon; distributed over the Malay Archipelago.

TIMBER.
1117

Structure of the Wood.—Whitish, very light and coarsely fibrous, heartwood red; weight 48℔ per cubic foot. Said by the authors of the *Flora of British India,* to be useful.

POONYET or PWENYET.

1118

Poonyet, Pwenyet or Pwai-nget.

References.—*Mason, Burma & its People, 350; Madras Reports of Exh., 1857; Rev. C. S. Parish, Science Gossip, 1866, 198; Cooke, Gums & Resins, 95.*

The Black Resin of Burma.	(*J. Murray.*) POONY-ET.

For many years the origin of the peculiar honey-combed black resin of Burma, employed by the natives of that country to caulk their boats, has formed a subject for speculation. It is generally supposed that at least a portion of the substance is derived from the Black Dammar Tree, **Cananium strictum.** Thus Mr. **Brown,** in forwarding specimens of the exudation of that tree to the Madras Exhibition of 1855, wrote, "It seems to be a great favourite of several species of insect, especially of one resembling a bee, called by the hillmen, *kulliada,* which lives in holes in the ground." **Mason** described the substance as the "wax" of a bee which built its nest in trees, and his opinion was upheld by **Helfer,** who, however, recorded it as "a kind of balsamic gum-resin." **McClelland,** writing three years later, stated that it was obtained from **Shorea robusta,** a remark which, according to **Cooke,** is manifestly an error."

RESIN.
1119

Three years later the subject was taken up by the *Agri.-Horticultural Society of India,* the Secretary to which wrote to the Rev. **O. S. Parish** requesting him to supply the necessary information. That gentleman consequently enquired closely into the matter, and in 1866 wrote an interesting account of his researches in *Science Gossip.* According to this article (which the reader desiring further information will find extracted in **Cooke's** *Report on Gums & Resins, &c.,*), true *poon-yet* is "a combination of various gums or resins, and probably also of oils, gathered from various sources while in a soft state by a bee" (identified as **Trigona læviceps),** and built up, and moulded very much as wax is moulded, with this difference, that whereas wax is formed by the honey-bee into cells of perfect and uniform symmetry, the cells in *pyai-ngyet* assume no regular form at all." **Mr. Parish** inclined to the belief, that the mass was built up chiefly from the resin of **Hopea odorata,** *Roxb.,* and the oils of the various species of **Dipterocarpus.** He states that these opinions are strengthened by the fact that the texture, colour, and smell of the substance are all such as would apparently result from a combination of the two, and that the trees are among the commonest in the forests in which *poon-yet* is generally found.

The bee generally builds its nest in the hollow of a tree, entering by a small aperture. The cavity is lined with the resin, and if left undisturbed, the insects build a large, projecting, wide-mouthed, trumpet-shaped entrance outside the tree. The mouth of this peculiar structure has a vertical diameter of a foot or more, and a horizontal width of three or four inches. The nest may be built in any tree, or even in the hollow parts of an old hut, the ground, or in a hollow amongst rocks,—facts which render it certain that the substances from which it is built are collected by the bees from other sources. From 5 to 10 *viss* (about 18 to 38℔) are said to be generally obtained from one nest, and the substance sells for about 4 annas a *viss.* For use it is boiled in water, thus becoming soft, and is then mixed, and thoroughly kneaded with sufficient petroleum to render it of the consistence of putty. In this state it is extremely viscid, and tenacious, and is admirably fitted for the use to which it is put, *viz.,* caulking boats.

Mr. Parish remarks that though true *pwai-ngyet* is produced in the manner described, the same name is frequently used in the bazárs to denote any kind of resin. **Cooke,** commenting on the available information on the subject up to the time at which he wrote, sums up by stating that it may reasonably be assumed that the name is applied in Burma to at least three substances, *viz.,* 1. True *Pwai-ngyet,* formed as above described; 2. Honey-combed or perforated black resin, with a strong elemi odour, found in amorphous masses, often very large, and which *may* be the internal structure above alluded to, although this seems doubtful; and

Chemistry.
1120

3. Black dammar with a vitreous fracture and without odour. No further information is available on the subject, and it will thus be seen that, notwithstanding **Mr. Parish's** careful investigation, much doubt still exists regarding the source of the second and third kinds of *poon-yet*.

CHARACTERS AND CHEMICAL COMPOSITION.—Cooke describes the specimens examined by him as follows:—" The samples in the Museum collection from Burma differ from the ordinary black dammar of Southern India, in being less brittle, and making the fingers sticky after handling them, in having a more or less earthy and less shining fracture; and in a more decided, and often very distinct and strong, balsamic, elemi-like odour when rubbed or broken." "Samples of the honey-combed resin from Burma are found to be partly soluble in spirits of wine and partly in turpentine, but not wholly in either. It appears to consist of a dark-coloured resin which is soluble in cold spirits of wine, and a pale resin which is disintegrated, but not dissolved, by spirits of wine, but soluble in turpentine. In this respect it differs from the black dammar of Travancore, and a similar black dammar from Burma, which are not at all soluble in cold alcohol, but dissolve at once and entirely in turpentine. This kind of *pwai-ngyet* is very similar to the Canara resin and the Peruvian resin of Venezuela, partaking more of the character of elemi than of copal."

PONGAMIA, *Vent.; Gen. Pl., I., 549.*

[LEGUMINOSÆ.

Pongamia glabra, *Vent.; Fl. Br. Ind., II., 240; Wight, Ic., t. 59;*
Syn.—GALEDUPA INDICA, *Lam.;* G. ARBOREA, *Roxb.;* ROBINIA MITIS, *Linn.;* DALBERGIA ARBOREA, *Willd.*

Vern.—*Karanj, karanjh, papar, kiramál, karanjaca,* HIND.; *Karanja, karan̄-gáchh, dalkaramcha, darkaranja, dahar karanja, khawári, karmuj,* BENG.; *Koranjú, karansá,* URIYA; *Kuruinj,* SANTAL; *Ga-ranji,* GOND; *Pápar, sukh-chain,* KUMAON; *Súkhchein, paphri, rárá-pod-bará karanj,* PB.; *Charr,* RAJ.; *Kurunji,* C. P.; *Karanj, kiramál,* BOMB.; *Karanj, karanjicha-jháda, karanja,* MAR.; *Karanj, karanj-nu, jháda, kanaji,* GUZ.; *Karanj, karanjh,* DEC; *Pungam-maram, pongá,* TAM.; *Kánuga, kaggera, kránuga, kanu-gamanu, kanga, pungu, kaniga, ganuga, ranagu,* TEL.; *Hónge, kanaga, húngé, pong,* KAN.; *Unna-maram, punnam, pongam,* MALAY.; *Simisu or timisu, thinwin, tha-wen,* BURM; *Magúl-karanda,* SING.; *Tamála-vrikshaha, karanja, naktamála,* SANS.

References.—*Roxb., Fl. Ind., Ed. C.B.C.,* 538; *Voigt, Hort. Sub. Cal.,* 239; *Brandis, For. Fl.,* 153; *Kurz, For. Fl. Burm., I.,* 335; *Beddome, Fl. Sylv., t.* 177; *Gamble, Man. Timb.,* 133; *Dals. & Gibs., Bomb. Fl.,* 77; *Stewart, Pb. Pl.,* 73; *Rev. A. Campbell, Rept. Econ. Pl., Chutia Nagpur, No.* 7545; *Mason, Burma & Its People,* 504, 525, 769; *Sir W. Elliot, Fl. Andhr.,* 76, 82, 100; *Sir W. Jones, Treat. Pl. Ind., V.,* 146; *Rheede, Hort. Mal., VI., t.* 3; *Ainslie, Mat. Ind., II.,* 332; *Moodeen Sheriff, Supp. Pharm. Ind.,* 205; *U. C. Dutt, Mat. Med. Hindus,* 153, 303, 310; *Sakharam Arjun, Cat. Bomb. Drugs,* 210; *Murray, Pl. & Drugs, Sind,* 134; *Dymock, Mat. Med. W. Ind.,* 2nd *Ed.,* 243; *Dymock, Warden & Hooper, Pharmacog. Ind., I.,* 468; *Cat. Baroda Durbar, Col. & Ind. Exhb., Nos.* 152, 153; *Official Corresp. on Proposed New Pharm. Ind.,* 234, 338; *Trans., Med. & Phys. Soc., Bomb. (New Series), No.* 4, 155; *Birdwood, Bomb. Prod.,* 281; *Baden Powell, Pb. Pr.,* 590; *Drury, U. Pl. Ind.,* 353; *Atkinson, Him. Dist. (X., N.-W. P. Gas.),* 309, 746; *Useful Pl. Bomb. (XXV., Bomb. Gas.),* 62, 217, 260, 399; *Cooke, Oils & Oilseeds,* 65; *Christy, New Com. Pl., VI.,* 100, 102; *VII.,* 79; *Nicholson, Man., Coimbatore,* 41, 192; *Boswell, Man., Nellore,* 98, 125; *Moore, Man., Trichinopoly,* 79; *Gribble, Man., Cuddapah,* 14, 84, 211, 227; *For. Adm. Rept., Ch. Nagpur,* 1885, 30; *Settlement Reports:—Central Provinces, Chánda, App. VI.; Burma, Hathawaddy & Pegu Dists.,* 1882-83, 4; *Gazetteers:—Bombay, V.,* 285; *VI.,* 13; *VII.,* 39, 41

VIII.,11 ; XIII., 25 ; XV., 74; XVI., 16 ; XVII., 18 ; XVIII., 47 ; Panjáb, Gurgaon, 16 ; N.-W. P., I., 80 ; IV., lxxi. ; Orissa, II., 176, 179 ; Mysore & Coorg, I., 50, 59; II., 7, 8, 11 ; Agri.-Horti. Soc., Ind., Journals(Old Series), VIII., Sel., 141,1 76 ; IX., 418 ; Sel., 53, 54; X., 223, 224; Ind. Forester, III., 201 ; IV., 322, 411 ; VI., 124; VIII., 117, 401, 410 ; IX., 357 ; XII., 188, App. I. (xxii.) ; XIII., 120 ; Spons, Encycl., 1393, 1694.

Habitat.—A tall erect tree or climber, met with all over India from the Central and Eastern Himálaya to Ceylon and Malacca.

Gum.—It yields a thick, black, opaque gum (*Spons' Ency.*). **Dr.** Dymock informs the editor that he has never seen any gum on this tree, and that gum does not exude when the tree is wounded.

GUM.
1122

Dye.—According to **Mr. Campbell** the ASH of the wood is employed, in Chutia Nagpur, for dyeing.

DYE.
Ash.
1123

Fibre.—A coarse BARK fibre of a brown colour, extracted from this tree, was obtained by **Dr. Watt** from the Salem district. In the *Poona Gazetteer* it is stated that the "rind or *pend* of the bark is pressed and rolled by *Pinjaris* or cotton-teazers into a felt."

FIBRE.
Bark.
1124

Oil.—The SEEDS yield a red-brown thick oil, used for illuminating and medicinal purposes. In 1858, **Mr. Grose** made some observations on this product in the *Journal, Agri.-Horti. Soc. Ind.*, from which it appears that a maund of seed will yield 8 seers of oil, or allowing forty trees to a Bengal *bigah*, 10 maunds of oil would be the produce of one *bigah*; at the then bazár rate of R10 a maund, this amount would fetch R160, a sum which would much more than cover the expenses of cultivation and preparation.

OIL.
Seeds.
1125

CHEMICAL COMPOSITION.—According to Lepine (*Pharm. Jour.* (*3*) XL., 16,) the seeds yield 27 per cent. of a yellow oil, having a sp. gr. of 0·945 and solidifying at 8°C. It has been examined by the authors of the *Pharmacographia Indica*, who write : "The oil which we have examined (called *Houge* oil in Mysore), and expressed purposely from fresh seeds, was thick, of a light orange-brown colour, and bitter taste The sp. gr. at 18°C was 09·458. It yielded 93·3 per cent. of fatty acids melting at about 30°. With sulphuric acid it became yellow with orange streaks, and when stirred formed an orange-red mixture, which, after standing, became yellow. With nitric acid it formed an orange emulsion. With the elaïdin test it remained liquid for several hours, and was of the colour and consistence of honey after two days. The fresh oil deposits solid white fats if kept at the temperature of 16° for a few weeks, and the clear oil then has the specific gravity of 0·935. The bitter principle of the oil appears to reside in a resin, and not in an alkaloid as is the case with Margosa oil."

Chemistry.
1126

Medicine.—In Sanskrit medicine the SEEDS are described as a useful application in skin diseases ; the expressed OIL is said to be valuable in these diseases as well as in rheumatism, and a poultice of the LEAVES is mentioned as a good dressing for ulcers infested with worms (*U. C. Dutt, quoting Chakradatta*). The value of the oil was early recognised by European writers in India ; thus, **Ainslie** notices its use in itch and rheumatism as well as the employment of the juice of the ROOT as an application for cleansing foul ulcers and closing fistulous sores. The plant obtained a place on the secondary list of the *Pharmacopœia of India* in 1868, where it is stated that, according to **Dr. Gibson,** "No article of the vegetable kingdom is possessed of more marked properties" as a remedial agent in cases of scabies, herpes, and other cutaneous diseases than **Pongamia** oil. **Dr. Dymock** confirms this statement to a certain extent, mentioning that he has employed a liniment of the oil and lime or lemon juice with success in prurigo, pityriasis, and psoriasis. From these

MEDICINE.
Seeds.
1127
Oil
1128
Leaves.
1129
Root.
1130

21 A

POPPY.	Pongamia Oil.

MEDICINE.
Stem.
1131

facts it would appear that the juice of the STEM and root and the oil both possess marked antiseptic properties.

The leaves are said to be given internally in leprosy along with those of Plumbago, pepper, salt, and curds, and, according to Dymock, also enter into the composition of many complicated prescriptions for "epilepsy and abdominal enlargements." The most recent writings on the subject are given as follows in the *Pharmacographia Indica* :—"Dr. P. S. Mootooswamy mentions the use of the juice of the root with cocoanut milk and lime water as a remedy for gonorrhœa in Tanjore, and of the leaves in flatulency, dyspepsia, and diarrhœa. He also informs us that broken rice is boiled with the leaves and those of Morinda citrifolia, dried in the shade, cleaned and crushed, and from this preparation a thin salt gruel is made, to feed young children with instead of cow's milk, which is supposed to cause glandular enlargements of the abdomen. He has noticed

Flowers.
1132
Pods.
1133

the use of the FLOWERS as a remedy for diabetes, and of the PODS worn round the neck as a protection against whooping cough (*Ind. Med. Gaz., 1888*). Dr. B. Evers has seen the seeds administered internally for the last-named affection."

SPECIAL OPINIONS.—§ "The pulp of the seed said to be good in leprosy" (*Surgeon H. W. Hill, Manbhum*). "The young leaves of this plant have been recommended as an application for bleeding piles. I have not used it myself, but I have known a case cured by its use" (*Civil Surgeon R. Macleod, Gya*). "The expressed oil is used in rheumatism with good effect" (*Honorary Surgeon E. A. Morris, Tranquebar*). "Powder of rind of legumes useful in whooping cough, vide *Ind. Med. Gazette for March 1875, page 66*" (*Civil Surgeon B. Evers, M.D., Wardha*). "The milky juice of the root-bark is injected into fistulous tracts to hasten the healing process. Equal quantities of this milky juice, and gingelly oil, together with sulphate of copper (3 grs. to ℨi) is a good preparation for sinuses" (*Surgeon W. F. Thomas, 33rd Regiment, M. N. I., Mangalore*). "I have used the oil successfully in pityriasis versicolor, and prurigo capitis" (*W. Dymock, Bombay*).

FOOD &
FODDER.
Fruit.
1134
Leaves.
1135
TIMBER.
1136

Food & Fodder.—The FRUIT is eaten in Chutia Nagpur (*Campbell*). The LEAVES form a good fodder, and are said to act as a lactagogue on cows.

Structure of the Wood.—Moderately hard, white, turning yellow on exposure; tough, fibrous, coarse, and even-grained, not easily worked; not durable and easily attacked by insects, but improved by seasoning in water; weight 40℔ per cubic foot, when seasoned. In Lower Bengal it is employed for making oil-mills and for firewood, while in Southern India it is chiefly used for making solid cart-wheels (*Beddome*).

DOMESTIC.
Leaves.
1137

Domestic.—The LEAVES are extensively employed as manure for wet cultivation; indeed, so much are they valued for this purpose that Mr. Gribble states (*Manual of the Cuddapah District, Madras,*) that the question of rights to the tree causes many of the disputes amongst the villagers. Mr. Hardinge (*Rep. Sett. Oper. Hanthawaddy and Pegu District, 1882-83, 4*), states that the leaves are very effective in destroying blight if taken in time, and that this fact is well known to the Burmans. It would be interesting to find out whether the use of the leaves as manure, which is expressly stated to be specially employed in *wet* cultivation, be not similar in its cause to that of Adhatoda vasica, to which attention has been drawn by Dr. G. Watt. All information on the qualities of the leaves seems to point to their being more or less antiseptic, and scientific research into their chemical composition might produce results of much interest.

[VERACEÆ, p. 17

Poppy, see **Papaver Rhœas,** *Linn.,* and **P. somniferum,** *Linn.;* **PAPA**

POPULUS, *Linn. ; Gen. Pl., III., 412.*

Populus alba, *Linn. ; Fl. Br. Ind., V., 638 ;* SALICINEÆ.   1138

THE ABLE or WHITE POPLAR.

Vern.— *Fras,* KASHMIR ; *Saféda, chita bagnú, fras, jangli-frast, rikkan, prasti, prist, sannan, chanúni, mál, baid,* PB. ; *Sperdor, spelda, speda,* AFG.

References.— *Boiss., Fl. Orient., IV., 1193 ; Brandis, For. Fl., 473 ; Gamble, Man. Timb., 378 ; Stewart, Pb. Pl., 203 ; Baden Powell, Pb. Pr., 385, 590 ; Gazetteers :—Simla, 10 ; Bannu, 23 ; Ind. Forester, V., 181 ; XI., 52 ; XIII., 70 ; Agri.-Horti. Soc., Ind., Journals (Old Series), XIII., 384 ; Panjáb, Select Papers to 1862, Index, 44.*

Habitat.—A large tree, wild and cultivated in the North-West Himálaya and Western Tibet from 4,000 to 10,000 feet ; distributed to Afghánistán, Baluchistán, and Europe.

Medicine.—Baden Powell writes that " it " (? the bark) " contains some *salicin,* acts as a tonic, and is used for purifying the blood, and in skin diseases. The BARK is also said to be useful in strangury." Dr. Warden makes the following comment on these remarks : —   MEDICINE. Bark. 1139

" The P. tumala, or aspen, according to Braconnet, contains a white, crystalline glucoside—*Populin* or *Benzoylsalicin*—which is present both in the bark and leaves. Nitric acid oxidises it into Benzohelicin, and by the action of either sulphuric or hydrochloric acid, it is converted into saligenol, benzoic acid, and glucose ; by prolonged action of these acids saligenol is converted into salictin, and this principle, by the action of other reagents, yields salicin and benzoic acid. It would be interesting to determine whether any of the Indian species contain populin, in addition to salicin. It is well known that, like the willows, many species of poplar contain salicin."   Chemistry. 1140

Structure of the Wood.—Like that of all poplars it is soft, white, easily worked and suited for carving. It is, however, neither strong nor durable, and is little valued, except for making the small round boxes in which grapes are exported from Afghánistán to India.   TIMBER. 1141

P. balsamifera, *Linn. ; Fl. Br. Ind., V., 638.*   1142

Syn.—P. SUAVEOLENS, *Loud. ;* P. LAURIFOLIA, *Ledeb.*

Vern.— *Berfa, changma, yarpa, maghal, máhal,* W. TIBET ; *Pkalsh, makkal, pakhshu, pakh, bút, kramal, máal, changma, mághal, yárpa,* PB.

References.— *Brandis, For. Fl., 476 ; Gamble, Man. Timb., 379 ; Stewart, Pb. Pl., 204 ; Baden Powell, Pb. Pr., 590.*

Habitat.—A tall tree, from 60 to 70 feet in height, remarkable for its handsome foliage and the pleasant balsamic odour of the leaves and buds. It is met with, wild and planted, in the inner ranges of the North-Western Himálaya, from Kunawar, altitude 8,000 to 13,000 feet, westwards. In Western Tibet it is found up to 14,000 feet.

Gum.—The LEAVES and BRANCHES are full of balsamic juice, which also exudes on a fresh cut from between the bark and the wood (*Gamble*). The writer can find no account of this exudation being utilised.   GUM. Leaves. 1143 Branches. 1144

Fodder.—The BRANCHES are lopped for cattle fodder.   FODDER. Branches. 1145

Structure of the Wood.—Light grey, soft to moderately hard, weight 32℔ per cubic foot. It is much used for fuel in Nimar and the Himálaya (*Gamble*).   TIMBER. 1146

Sacred.—Aitchison states that it is never cut in Lahore, as it is supposed to be the abode of a *deva* or god, and that festivals are celebrated under some of the finer trees.   SACRED. 1147

P. 1147

1148

Populus ciliata, *Wall.; Fl. Br. Ind., V., 638.*

Syn.—P. ROTUNDIFOLIA, *Griff.;* P. PYRIFORMIS, *Royle.*

Vern.—*Bangikat,* NEPAL; *Sungribong,* LEPCHA; *Garpípal, chalníya, chauniya, chan,* KUMAON; *Fálsh, palách,* KASHMIR; *Sufédar, sháwa, bagnú, phálja, suláli, krambal, saki, krammal, paluch, asán, dúd- fras, supida, rikkan, pabe, chanún, flassu, ban frastú, fálsh, tallon, tálúng, chalon, palách, chalonwa, kramali, falis, suáli, pahári pípal,* PB.; *Shodar,* PUSHTU.

References.—*DC., Prodr., XVI., ii., 329; Brandis, For. Fl., 475; Gamble, Man. Timb., 379; Stewart, Pb. Pl., 204; Griffith, Notulæ, I., 82; Ic. Pl. As., t. 546; Baden Powell, Pb. Pr., 590; Atkinson, Him. Dist., 317, 747; Royle, Ill. Him. Bot., 345, t. 84, and Notes on the Ec. Pl., Balu- chistán; Gazetteers:—Panjab, Hazara, 14; Gurdaspur, 55; Simla, 10; Rawalpindi, 15; Ind. Forester, IV., 90; XIII., 67; Settle. Repts., Hazara, 11; Simla, App. II., (H.), p. xlii.; Agri.-Horti., Soc., Ind., Journals (Old Series), VI., 167; XII., Pro. (1862), 3; XIII., 384.*

FIBRE.
Seeds
1149
MEDICINE
Bark
1150
FODDER.
Leaves
1151
TIMBER.
1152
1153

Habitat.—A large deciduous tree of the Himálaya, from the Indus to Bhotan, between 4,000 and 10,000 feet.

Fibre.—Baden Powell states that the coma of the SEEDS, which covers the ground in many places, like snow, is a good paper-material.

Medicine.—Some part of the tree (? the BARK) is, according to Atkinson, occasionally used as a tonic stimulant and purifier of the blood.

Fodder.—The LEAVES are employed as fodder for goats.

Structure of the Wood.—Grey, or brownish-grey, shining, soft; weight on an average 29·5℔ per cubic foot. It is of little value, but is used for making water-troughs, and for fuel.

P. euphratica, *Oliv.; Fl. Br. Ind., V., 638.*

Syn.—P. DIVERSIFOLIA, *Schrank;* BALSAMIFLORA DELTOIDES, *Griff.*

Vern.—*Hodung, hotung,* LADAK; *Bahan, bhan, bhani, labhán, junglí, bentí, sperawan, saféda,* PB.; *Patki,* BRAHUI (BALUCH); *Bahán,* PASH- TU; *Pada, padak,* W. AFGHANISTAN; *Saféda, bhan, bahan,* SIND.

References.—*DC., Prodr., XVI., ii., 326; Brandis, For. Fl. 474, t. 63; Gamble, Man. Timb., 378; Stewart, Pb. Pl., 204; Rept. Pl. Coll. Afgh. Del. Com., 111; Notes on Prod. W. Afgh. & N.-E. Persia, 161; Boiss., Fl. Orient., IV., 1194; Murray, Pl. & Drugs, Sind, 29; Baden Powell, Pb. Pr., 590; Gazetteers:—Panjab, Jhang, 17; Mozaffergarh, 22; Bannu, 23; Dera Ismail Khan, 12, 19; Dera Ghazi Khan, 10; Settle. Repts., Jhang; 22; Dera Ghazi Khan, 4; Dera Ismail Khan, 23; Ind. Forester, V., 478; IX., 174; XIV., 362, 371.*

Habitat.—A large, deciduous tree, met with on the banks of the Indus, in Sind, in the Upper Valley of the Indus, and its tributaries in Western Tibet, where it ascends to 13,500 feet.

MEDICINE.
Bark.
1154
FODDER.
Leaves.
1155
TIMBER.
1156

Medicine.—In Sind and the Panjáb the BARK is employed as a vermi- fuge.

Fodder.—The LEAVES form good fodder for sheep, goats, and camels.

Structure of the Wood.—Sapwood white, heartwood red, often nearly black in the centre, moderately hard, compact, even-grained, fairly tough, but not durable, and when unseasoned very subject to the attacks of white ants; weight 32 to 37℔ per cubic foot (*Gamble*). In the Southern Pan- jáb it is, for the most part, used only for lining wells, in Ladak and Tibet it is grown chiefly for firewood, and in Sind, the coppice shoots are em- ployed for rafters, and the cut timber for other building purposes, and for lacquered and turned work. Aitchison informs us that though little valued in Afghánistán except for fuel, it makes excellent rafts, owing to its extreme lightness, and Stewart states that it is said to be largely employed for boat-building on the Euphrates.

DOMESTIC
Twigs.
1157

Domestic & Sacred.—The tree coppices well, and bears pollarding long. The smaller trees are much cut in this way in Sind. The TWIGS are used as tooth-brushes by Natives in the Panjáb and Sind and are said

Products of India. 327

The Black or Lombardy Poplar. (*J. Murray*.) | PORANA racemosa.

to be imported into Lahore and other large towns in considerable quantities. The INNER BARK is made into gun matches in Sind. Aitchison states that in the desert country of Baluchistán the tree is cultivated only in the neighbourhood of shrines. [*638.*

DOMESTIC.
Inner Bark.
1158

Populus nigra, *Linn.*, VAR. **pyramidalis**, *Spach ; Fl. Br. Ind., V.,*
THE BLACK or LOMBARDY POPLAR.

1159

Syn.—P. FASTIGIATA, *Desf.*

Vern.—*Changma, yarpa, yúlatt, kabúl, kaúll,* LAD. ; *Frast,* KASHMIR ; *Frast, prost, farsh, suféda, makkal, pakhshu bút, biúns, kramali, do, sufedár,* PB. ; *Safedár, saféda, kabuda,* AFG.

References.—*Boiss., Fl. Orient., IV., 1194; Brandis, For. Fl., 472; Gamble, Man. Timb., 378; Stewart, Pb. Pl., 205; Aitchison. Bot. Afgh. Del. Com., 111; Notes on Prod., W. Afghán., & N.-E. Persia, 162; Year-Book of Pharmacy, 1874, 629; Royle, Ill. Him. Bot., 344; Panjáb Gazetteer, Gurdaspur, 55; Ind. Forester, XI., 55; Agri.-Horti. Soc., Ind., Journal (Old Series), IV., Sel. 249.*

Habitat.—This, the cupressiform or pyramidal variety of the Black Poplar, is not uncommonly planted in the North-Western Himálaya, especially in Kashmír, and in the basins of the Jhelam, Chenab, and Sutlej between 3,000 and 11,500 feet, and in Ladak as high as 12,500 feet. It is also occasionally met with cultivated in the plains at Peshádwar, Lahore, Hushiárpur, and elsewhere. It is commonly cultivated in certain parts of Afghánistán, and, according to Griffith, occurs truly wild at Shekkabad, near Kabul, at 7,500 feet.

Medicine.—Stewart states that the BARK is employed medicinally in the plains of the Panjáb, an *arak*, which is considered depurative, being extracted from it. Mr. Groves informs us (*Year-Book Pharm.*) that in Tuscany, an ointment prepared from the BUDS is used for hœmorrhoids, and the BALSAM obtained from the same source is a popular remedy for colds.

MEDICINE.
Bark.
1160
Buds.
1161
Balsam.
1162

Fodder.—The tree is lopped, in certain localities, for fodder.

FODDER.
1163

Structure of the Wood.—Whitish-brown, very soft and even-grained near the centre; structure similar to that of P. alba; weight 24 to 27.5℔ per cubic foot. It is little used except in Afghánistán, where, like that of P. alba, it is employed for making the small round boxes in which grapes are packed for exportation.

TIMBER.
1164

PORANA, *Burm. ; Gen. Pl., II., 876.*

Porana racemosa, *Roxb. ; Fl. Br. Ind., IV., 222 ; Wight, Ic., t.*
THE SNOW-CREEPER. [*1376.*

1165

Vern.—*Bhauri, gariya,* DECCAN. These names are said by Dymock to be used indifferently for this plant and Ipomæa muricata.

References.—*DC., Prodr., IX., 436; Roxb., Fl. Ind., Ed. C.B.C., 156; Dals. & Gibs., Bomb. Fl., 162 (?); Atkinson, Him. Dist., 314; Dymock, Mat. Med. W. Ind., 2nd Ed., 890; Ind. Forester, II., 26; Agri.-Horti. Soc., Ind., Journal (Old Series), VI., 47.*

Habitat.—This, one of the most beautiful of Indian plants, occurs in dense, not lofty, masses, climbing over other plants in the jungle, the dazzling white flowers closely massed together, resembling patches of snow. It is common in the sub-Tropical Himálaya from the north-west to Bhotan, at altitudes of 2,000 to 6,000 feet, and from Khásia to Martaban between the same altitudes. Mr. C. B. Clarke (*Fl. Br. Ind.*) states that the plants named P. racemosa by Dalzell and collected in the Deccan, all belong in reality to P. malabarica.

Food.—The PEDUNCLES are said by Dymock to be eaten in the Deccan, during famine seasons, but his remarks probably refer to P. malabarica, no true specimen of P. racemosa having been recorded from that region.

FOOD.
Peduncles.
1166

P. 1166

The vernacular names above cited would, in this case, also belong to P. malabarica.

1167 Porana malabarica, *Clarke; Fl. Br. Ind., IV., 223.*
Habitat.—A species, nearly allied to P. racemosa, found in Bombay, Malabar, and the Konkan, Kanara, and Mysore—see remarks under P. racemosa.

1168   ## PORCUPINES.

A well-known group of animals, which derive their popular name (signifying "spiny-pig") from their large size, grunting voice, and dense armour of quills. The true porcupines constitute the genus HYSTRIX of the family HYSTRICIDÆ.

1169 Porcupines, *Jerdon, Mammals of India, 218.*
The species of economic value are the following :—

1170 1. Hystrix bengalensis, *Blyth.* ; RODENTIA.
Vern.—*Serr*, BENG.
Habitat.—The Bengal porcupine is found in Lower Bengal, extending into Assam and Arracan, and also, if Blyth's identification be correct, to South Malabar (*Jerdon*).

1171 2. H. leucura, *Blyth.*
Vern.—*Sajru*, BENG.; *Dumsi*, NEPAL; *Ho-igu* GOND.; *Saori*, GUZ.; *Salendra*, MAR.; *Yed*, KAN.; *Yeddu pandi*, TEL.
Habitat.—This, the commonest species, generally known as the "Indian porcupine," is found over the greater part of India, from the lower ranges of the Himálaya to the extreme south, but does not occur in Lower Bengal, where it is replaced by the preceding.

1172 3. H. longicauda, *Marsden.*
Vern.—*Anchotia dumsi*, NEPAL; *Sathung*, LEPCHA; *O'-e'*, LIMBUS.
Habitat.—The "crestless porcupine" is met with in the central region of Nepal and Sikkim, and extends through Burma into the Malayan peninsula and islands (*Jerdon*).

FOOD.
1173 Food.—All three species are eaten by Natives, the flesh of H. bengalensis being more highly esteemed than that of the common Indian porcupine. Even the latter is described by Jerdon as "not bad eating, the meat, which is white, tasting something between pork and veal." The crestless porcupine, however, appears to afford the best eating of all, and is described as follows by Hodgson : "The flesh is delicious, like pork, but much more delicate flavoured, and they are easily tamed, so as to breed in confinement. All tribes and classes, even high-caste Hindus, eat them, and it is deemed lucky to keep one or two alive in stables, where they are encouraged to breed."

DOMESTIC.
1174 Domestic.—In certain parts of the country porcupines are very destructive to various crops, potatoes, carrots, and other vegetables. They may generally be tracked and killed by means of dogs which take up their scent very keenly. They are also trapped or killed by spring guns. The quills of all the Indian species are smaller and less handsomely marked than those of the porcupine of Africa and Southern Europe, and do not appear to be utilised in any way.

Porpoises, see Whales.

[*togam., 299.*
1175 PORPHYRA, *Ag. ; Baill., Traité de Botanique Médicale Cryp-*

A genus which belongs to the PORPHYRACEÆ, one of the lowest families of FLORIDEÆ or coloured Algæ, and which is characterised by a thallus formed in its vegetation portion of a single layer of cells, and shaped like thin transparent plate with an irregular outline. All the members of the genus are marine, and are annual, dying down in winter and being reproduced from spores in spring.

P. 1175

| Sea-weed; Indian Purslane. | (*J. Murray*.) | PORTULACA oleracea. |

Porphyra vulgaris, *Linn.; FLORIDEÆ.*   1176

THE PURPLE LAVER.

Vern.—*Las, luch,* SIND.

References.—*Murray, Pl. & Drugs, Sind, 2; Dymock, Mat. Med. W. Ind., 2nd Ed., 874.*

Habitat.—A ribbon-like sea-weed, met with on rocky coasts from the United Kingdom to the Mediterranean and the Indian Ocean. It has been recorded in India only from the Manora Rocks, Sind.

Medicine.—The SEA-WEED is gathered just before the monsoon, and is used as a demulcent and as an alterative in cases of scrofula. It owes the latter property probably to a small quantity of iodine, the former to the large amount of gelatinous matter which it contains.

MEDICINE.
Sea-Weed.
1177

Food.—The gelatinous matter it contains renders it nutritious, and it is frequently eaten in the south of England and the Western Isles. In Sind it is used as food by the poorer classes near the coast, in times of scarcity, and when dressed with lemon-juice is said to sometimes find a place at the tables of the rich.

FOOD.
1178

PORTULACA, *Linn.; Gen. Pl., I., 156.*

Portulaca oleracea, *Linn.; Fl. Br. Ind., I., 246;* PORTULACEÆ.   1179

THE COMMON INDIAN PURSLANE.

Syn.—P. LÆVIS, *Ham.;* P. SUFFRUTICOSA, *Thw. (not. of Wight).*

Var. erecta,=P. OLERACEA, *var.* SYLVESTRIS, *Hb. Royle (no*ᵗ*. of DC.).*

Vern.—*Khursa, khurfah, khurfé-ká-ság, lonia, múncha, lúnia, kurfa, mánya, kúlfa, lúnuk, nonkha, chhota-luniá, baraluniá, nonkha lunuk, lúniya-kúlfah;* SEEDS=*khurfé-ké-binj,* HIND.; *Bara-lóniya, mánya, chhotaluniá, kulfi;* SEEDS=*tukhm-kúlpha, bará-loniyá-bij,* BENG.; *Puruni-ság,* URIYA; *Mota uric' alang,* SANTAL; *Lúniya, núniya, lúnak, desi-kulfah,* N.-W. P.; *Lúniya-kúlfah, lúnak,* KUMAON; *Lonak, kulfa, lúniya, kundar;* SEEDS=*dhamni,* PB.; *Murlai, tursbuk, warkhárai,* PUSHTU; *Lónk,* SIND; *Ghol, gholú,* C.P.; *Kurfáh, gól, moti ghol,* BOMB.; *Bhuigholi,* MAR.; *Loni,* GUZ.; *Khulfé-ki-bháji;* SEEDS= *khulfé-ké-binj,* DEC.; *Parpu-kire, passelie kiray, caril-kiray, parúpú, puropú-kiray, carie-kiray;* SEEDS=*parpu-kire-viral, pedda-pail-kuru, boddu-pavili-kúra, ganja-pávili-kúra, batchali aku;* SEEDS=*pappu-kúra- vittulu, pedda-pávila-kurá-vittulu, boddu-pávili-kurá-vittulu,* 'I AM.; *Pappu-kura, pedda-pávili-kura, boddu-pávilikúra, ganga-pávili-kúra* TEL.; *Dúda gorai,* KAN.; *Korie chira,* MALAY.; *Mya byit,* BURM.; *Gendakola,* SING.; *Lonika, lúnia, loni,* SANS.; *Kourfa kara-or, baglatul humqú, buklut-ul-kukema, khurfa;* SEEDS=*basrul-baglatul-humqá-* ARAB.; *Cholæa, khurfáh, turuk, kurfáh, kherefeh, túrk;* SEEDS= *tukhme-khurfah,* PERS.

References.—*Roxb., Fl. Ind., Ed. C.B.C., 391; Thwaites, En. Ceyl. Pl., 24; Stewart, Pb. Pl., 99; DC., Orig. Cult. Pl., 87; Rev. A. Campbell, Rept. Econ. Pl., Chutia Nagpur, No. 8785; Mason, Burma & Its People, 473, 763; Sir W. Elliot, Fl. Andhr., 29, 57; Rheede, Hort. Mal., X., t. 36; Ainslie, Mat. Ind., II., 287; O'Shaughnessy, Beng. Dispens., 353; Irvine, Mat. Med., Patna, 48, 114; Medical Topog., Ajmir, 142; Mat. Med. S. Ind. (in MSS.), 57; U. C. Dutt, Mat. Med. Hind., 308; Sakharam Arjun, Cat. Bomb. Drugs, 61; Murray, Pl. & Drugs, Sind., 96; Dymock, Mat. Med. W. Ind., 2nd Ed., 75; Dymock, Warden & Hooper, Pharmacog. Ind., I., 158; Birdwood, Bomb. Prod., 38,160; Baden Powell, Pb. Pr., 245, 336; Drury, U. Pl. Ind., 364; Atkinson, Him. Dist. (X., N.-W. P. Gas.), 306, 708, 747; Useful Pl. Bomb. (XXV., Bomb. Gas.), 146; Pt. V. (Vegetables, Spices, & Fruits), 39; Stocks, Rep. on Sind.; Boswell, Man. Nellore, 116; Settlement Report:— Central Provinces, Chánda, 82; App. vi.; Gazetteers:—Bombay, V., 23; Panjáb, Peshawar, 26; N.-W. P., I., 81; IV., lxviii.; Orissa, II., 180; Mysore & Coorg, I., 68; Agri.-Horti. Soc.:—Ind., Journals (Old Series), X., 7; XIII., Sel., 63; Ind. Forester, III., 237; XII., 329.*

P. 1179

| PORTULACA quadrifida. | Indian Purslane. |

Habitat.—An annual, usually prostrate, herb, found throughout India, ascending to 5,000 feet in the Himálaya. According to DeCandolle the evidence of philology, and botany alike, shows that the species is indigenous in the whole of the region which extends from the Western Himálaya to the south of Russia and Greece.

MEDICINE.
Plant.
1180

Leaves.
1181

Medicine.—The PLANT has long been used as a domestic remedy by the Hindus, and was early noticed by European writers. Thus Ainslie writes of **P. quadrifida** which possesses the same properties: "The bruised fresh LEAVES of this acid and pleasant tasted purslane are prescribed by the Tamool practitioners as an external application in *akki*, erysipelas; an infusion of them is also ordered as a diuretic in dysuria to the extent of half a teacupful twice daily." He further mentions that in Jamaica **P. oleracea** is employed as a cooling and moistening herb in "burning fevers," bruised it is applied to the temples to allay "excessive heat" and pain, and that the JUICE is "of use in spitting of blood." Dymock informs us that both species are supposed by Arabian and Persian writers to be cold and moist, and to have detergent and astringent properties. The plant and SEEDS are recommended by them in a great many diseases of the kidneys, bladder, and lungs which are supposed to be caused by hot or bilious humours. They are also praised as an external application in burns, scalds, and various forms of skin disease (*Mat. Med. W. Ind.*). Moodeen Sheriff, in his forthcoming *Mat. Med. of Southern India*, describes the seeds as demulcent, slightly astringent, and diuretic; the leaves as refrigerant, astringent, diuretic, and emollient. He believes both to be "very useful" in some cases of strangury, dysuria, irritation of the bladder, hæmaturia, hæmatemesis, hæmoptysis, and gonorrhœa. "In addition to this," he writes, "the seeds seem to have some beneficial influences over the mucous membrane of the intestinal canal, and therefore relieve tormina, tenesmus, and other distressing symptoms in many cases of dysentery and mucous diarrhœa. This is particularly the case when they are combined with some other drugs of similar nature." He recommends the fresh succulent leaves as a cooling external application in the place of ice or cold lotion. The seeds and juice of the fresh leaves may be best administered in the form of a draught, from thirty grains to one drachm of the former, and from one to two fluid ounces of the latter (obtained by pressing the leaves) being the dose. He recommends either of these as substitutes for spirit of nitrous ether, Pareira-brava, tragacanth, elm-bark, rhatany, copaiba, and ice.

Juice.
1182
Seeds.
1183

By Natives, generally at the present day, the herb is chiefly valued as a refrigerant and alterative pot-herb, particularly useful as an article of diet in scurvy and liver disease. In addition to the properties above detailed, the seeds are believed in the Panjáb to be vermifuge.

The seeds cost about R4 per maund wholesale, retail about 3 annas per pound. The leaves are very cheap, sufficient for one or two doses may be obtained for one pie (*Moodeen Sheriff*).

Chemistry.
1184

CHEMICAL COMPOSITION.—The authors of the *Pharmacographia Indica* state that the leaves contain oxalate of potash and mucilage.

FOOD.
Herb.
1185
Shoots.
1186

Food.—The HERB is largely eaten as a vegetable, either as a simple pot-herb or as a semi-medicinal, antiscorbutic article of diet. The young SHOOTS make an excellent salad. Three varieties are to be met with in gardens, *viz.*, the Green, the Golden, and the Large-leaved Golden.

1187

Portulaca quadrifida, *Linn.; Fl. Br. Ind., I.,* 247.

Syn.—P. MERIDIANA, *Linn.;* P. GENICULATA, *Royle;* ILLECEBRUM VERTICILLATUM, *Burm.*

Vern.—*Chounláyi, loniyá, khate chawal,* HIND.; *Núniya, chhota luniya,* BENG.; *Lúnak, haksha, lún-ki-búti,* PB.; *Kota, chaval-ke-bhaji, bárika*

Indian Purslane   (*J. Murray.*)   **POTAMOGETON lucens.**

ghola, Bomb.; *Luni*, Guz.; *Chounláyi-ki-bháji, ghól-ki-bháji, chowli,*
Dec.; *Soin-parpu-kirai, pasarai-kirai, siru-pasarai-kirai, passeli-kirai,*
Tam.; *Sanna-pappu, sanna-pávili, goddu pavili, pedda pávili, sunpail
kura, pávili, kura, payalaku, sanna payala,* Tel.; *Hali bachcheli,* Kan.;
Hin-genda-kola, Sing.; *Upadyki,* Sans.; *Baqlatul-yamániyah, baqlatul-
aarabbiyah, budelut-ul-mobarik,* Arab.

References.—*Roxb., Fl. Ind., Ed. C.B.C., 391; Stewart, Pb. Pl., 100;
Elliot, Fl. Andhr., 61, 147, 149, 167; Rev. A. Campbell, Rept. Ec. Pl.,
Chutia Nagpur, No. 8786; Ainslie, Mat. Ind., II., 286; O'Shaughnessy,
Beng. Dispens , 353; Moodeen Sheriff, Mat. Med. S. Ind. (in MSS.),
38; Dymock, Mat. Med. W. Ind., 2nd. Ed., 75; Dymock, Warden &
Hooper, Pharmacog. Ind., I., 158; S. Arjun, Bomb. Drugs, 210; Year-
Book Pharm., 1878, 290; Lisboa, U. Pl. Bomb., 146; Royle, Ill. Him.
Bot., 221; Gazetteers:—Mysore & Coorg, I., 55, 68; N.-W. P., I., 81;
IV., lxviii.; Boswell, Man., Nellore, 133; Agri.-Horti. Soc., Ind.,
Journals (New Series), I., 33.*

Habitat.—A succulent herb, common throughout the warmer parts of India, occasionally cultivated.

Medicine.—The LEAVES are similar in every way to those of P. olera-cea and need not be further described. The SEEDS also, according to the authors of the *Pharmacographia Indica*, possess identical qualities to those of the former species; but they are rarely, if ever, obtainable in the bazárs, those of P. oleracea being alone sold.

Food.—The HERB is much cheaper than the preceding species, and is accordingly much used as a pot-herb by the poorer classes.

MEDICINE.
Leaves.
1188
Seeds.
1189

FOOD.
Herb.
1190
1191

Portulaca tuberosa, *Roxb.; Fl. Br. Ind., I., 247.*

Syn.—*P.* CRISTATA, *Ham.; P.* PILOSA, *Hb. Madr. in Wall. cat. (not of Linn.).*

Vern.—*Lúnuk,* SEED=*dhamni,* Sind. *Bodda kúra,* Tel.

References.—*Roxb., Fl. Ind., Ed., C.B.C., 391; Rev. A. Campbell, Rept.
Ec. Pl., Chutia Nagpur, No. 8787; Sir W. Elliot, Fl. Andhr., 29;
Murray, Pl. & Drugs, Sind, 96; Gazetteer, N.-W. P., IV., lxviii.*

Habitat.—Met with in Behar, Sind, the Western Peninsula, and Ceylon.

Medicine.—The fresh LEAVES are employed similarly to those of the preceding species.

Food.—Eaten as a pot-herb. Murray states that in Sind it is consi-dered superior as a vegetable to P. oleracea.

MEDICINE.
Leaves.
1192
FOOD.
1193

POTAMOGETON, *Linn.; Gen. Pl., III., 1014.*

Potamogeton crispus, *Linn.; Stewart, Pb. Pl., 241;* NAIADACEÆ.

Vern.—*Sawál,* Pb.; *Chúsbal,* Ladak.

References.—*Gazetteer, N.-W. P., IV., lxxviii.; Ind Forester, XIV., 39;
Agri.-Horti. Soc., Ind., Journals (Old Series), XIII., 313; XIV., 393.*

Habitat.—"Not uncommon in the Panjáb plains, and (apparently) abundant at 9,000 to 11,000 feet in Ladak" (*Stewart*).

Fodder.—It is said to be used for fodder in Ladak.

Domestic.—The plant "is probably one of those used in refining sugar here as elsewhere" (*Stewart*).

1194

FODDER.
1195
DOMESTIC.
1196

P. gramineus, *Linn.; Stewart, Pb. Pl., 242.*

Vern.—*Jála, sawáli,* Ladak; *Zimbil, chupein, phús,* Pb.; *Jhala, phas,* Sind.

1197

P. lucens, *Linn.; Stewart, Pb. Pl., 242;* and **P. natans,** *Linn.; Stewart, Pb. Pl., loc. cit.; Phas,* Sind; are all similarly employed in the parts of the Panjáb plains, Sind, and North-West Himálaya in which they occur.

1198
1199

P. 1199

| POTENTILLA Salessovii. | Potassium ; Ratanjot. |

POTASSIUM.

1200

Potassium.—The only important salts of this metal in India are the carbonate and the nitrate. For a description of the former the reader is referred to the articles **Alkaline ashes,** Vol. I., 167, and **Carbonate of Potash,** Vol. II., 152; and for an account of the latter to **Reh,** Vol. VI., also Saltpetre Vol , VI. The other salts are of too little importance to demand notice.

Potato, see **Solanum tuberosum,** *Linn.*; SOLANACEÆ; Vol. VI.

POTENTILLA, *Linn.; Gen. Pl., I., 620.*

1201

Potentilla fruticosa. *Linn. ; Fl. Br. Ind., II., 347 ;* ROSACEÆ.
　Syn.—P. RIGIDA, *Wall. ;* P. ARBUSCULA, *Don.*
　Var. α. glabrata.
　　　β. ochreata, *Lehm. ;* P. OCHREATA, *Lindl.*
　　　γ. pumila,=P. LINDENBERGII, *Lehm.*
　　　δ. Inglisii,=P. INGLISII, *Royle.*
　　　ε. armerioides.
　Vern —*Pinjúng, penma,* LADAK. ; *Spang jhá, merino,* PB.
　References.—*Stewart, Pb. Pl., 81 ; Gamble, Man. Timb., 161 ; Royle, Ill. Him. Bot., 207, t. 41 ; Atkinson, Him. Dist., 309.*
　Habitat.—A much branched, rigid, robust, variable shrub, met with on the Temperate and Sub-alpine Himálaya, from Kashmír to Sikkim, between the altitudes of 8,000 and 16,000 feet.

FOOD & FODDER. Leaves. **1202**

　Food & Fodder.—Stewart states, on the authority of **Aitchison** and **Longden,** that the fragrant LEAVES are used in the upper parts of the Chenáb basin as a substitute for tea, and that the plant is browsed by sheep. Dr. **Aitchison** confirms the former remark in the case of Lahoul, where, he informs the editor, the leaves of var. δ **Inglisii** are known as *shang-cha.*

1203

P. nepalensis, *Hook. ; Fl. Br. Ind., II., 355.*
　Syn.—P. BIFURCATA, *Wall.';*.P. FORMOSA, *Don.;* P. COLORATA, *Lehm·; *P. COCCINEA, *Hoffm.*
　Vern.—*Rattanjot,* PB.
　References.—*Stewart, Pb. Pl., 81 ; Atkinson, Him. Dist., 309.*
　Habitat.—Met with on the Western Temperate Himálaya between 5,000 and 9,000 feet, from Marri to Kumáon.

DYE. Root. **1204**

　Dye—The reddish ROOT is said by **Stewart** to form part of the dye-stuff and medicinal root exported under the name of *rattanjot* from the hills to the plains. It is employed to impart a red colour to wool.

MEDICINE. Ashes. **1205**

　Medicine.—*Rattanjot* is considered depurative and is "used externally in the Yunani system, the ASHES being applied with oil to burns" (*Stewart*).

1206

P. reptans, *Linn.*
　Habitat.—Found ın Kashmir, and distributed to Afghánistán, Abyssinia, Europe, North China, and Japan.

MEDICINE. **1207**

　Medicine.—This species does not appear to be employed medicinally in India, but is said to be used as a hæmostatic in Tuscany (*Year-Book Pharm., 1874, 624*).

1208

P. Salessovii, *Steph. ; Fl. Br. Ind., II., 348.*
　Syn.—P. DISCOLOR, *Camb. ;* COMARUM SALESSOVII, *Bunge.*
　Vern.—*Shour,* LADAK.
　Reference.—*Stewart, Pb. Pl., 80, 81.*

P. 1208

Products of India. 333

Timbers used for Pounders and Presses. (*J. Murray*) POUNDERS.

Habitat.—Found in Western Tibet, Lahoul, and the northern border of Kashmír, between 11,000 and 14,000 feet. The under-surface of the leaves is covered with fine dust, which, when the plant is shaken, causes violent sneezing.

Fodder.—It is browsed by sheep (*Stewart*).

FODDER.
1209

Potentilla supina, *Linn.; Fl. Br. Ind., II. 359.*

1210

Syn.—P. DENTICULATA, *Ser.;* P. HEYNII, *Roth.;* P. CANA, *Wall.;* P. RUTHENIA, ? *Ham.;* COMARUM FLAVUM, *Roxb.*

References.—*Roxb., Fl. Ind., Ed. C B.C., 409; Dymock, Mat. Med W. Ind., 2nd Ed., 304; Murray, Pl. & Drugs, Sind, 143; Atkinson, Him. Dist., 309; Gazetteers:—N.-W. P., I., 81; IV., lxxi.; Ind. Forester, II., 24; Agri.-Horti. Soc., Ind., Jour. XIV., 11.*

Habitat—Met with throughout the warmer parts of India from Kashmír to Malacca and the Nilghiri Hills, ascending the Himálaya to 8,500 feet in the North-West and in Tibet.

Medicine.—The ROOT is employed in Sind as a febrifuge (*Murray*). Like most other members of the genus it is rich in tannin, and is probably astringent and tonic.

MEDICINE.
Root
1211

Pothos officinalis, *Roxb.,* see **Scindapsus officinalis,** *Schott;* AROIDEÆ.

Pottery clays, see Clay, Vol. II., 364-367.

Pounce.—A resinous substance obtained chiefly from Sandarac, and used by paper-makers to prepare the surface of parchment and paper for writing. See **Callitris quadrivalvis,** *Vent.*; Volume II., 28.

1212

POUNDERS, PRESSES, &c.

Pounders, Presses, Rollers, Timbers used for—

Acacia arabica, *Willd.;* sugar, oil-presses, rice-pounders.
A. Catechu, *Willd.;* rice pestles, oil and sugarcane-crushers.
A. modesta, *Wall.;* sugarcane-crushers.
Ægle Marmelos, *Corr.;* pestles of oil and sugar-mills.
Alangium Lamarckii, *Thwaites;* pestles for oil-mills.
Albizzia Lebbek, *Benth.;* sugarcane-crushers, oil-mills.
A. odoratissima, *Benth.;* oil-mills.
A. procera, *Benth.;* sugarcane-crushers, rice-pounders.
Avicennia officinalis, *Linn.;* mills for husking paddy, rice-pounders, and oil-mills.
Barringtonia acutangula, *Gærtn.;* rice-pounders.
Capparis aphylla, *Roth.;* oil-mills.
Carallia integerrima, *DC.;* rice-pounders.
Cassia Fistula, *Linn.;* rice-pounders.
Cedrela Toona, *Roxb.;* rice-pounders.
Chloroxylon Swietenia, *DC.;* oil-mills.
Dillenia pentagyna, *Roxb.;* rice-mills.
Eriolæna Candollei, *Wall;* rice-pounders.
Feronia Elephantum, *Correa;* oil-crushers.
Mimusops hexandra, *Roxb.;* sugar-mill beams, oil-presses.
Odina Wodier, *Roxb.;* oil-pressers and rice-pounders.
Parrotia Jacquemontiana, *Dcne.;* rice pestles.
Pongamia glabra, *Vent.;* oil-mills.
Schleichera trijuga, *Willd.;* oil, rice, and sugar-mills.
Soymida febrifuga, *Juss.;* oil-mills.
Tamarindus indica, *Linn.;* rice, oil, and sugar-mills.
Terminalia tomentosa, *Bedd.;* rice-pounders.

1213

P. 1213

PRANGOS pabularia.	**The Pouzolia Fibre.**

Vitex alata, *Heyne.*; sugarcane-crushers.
Zizyphus Jujuba, *Lamk.*

POUZOLZIA, *Gaudich.*; *Gen. Pl., III., 387.*

[*696, 2096* ; Urticaceæ.

1214   **Pouzolzia pentandra**, *Benn.*; *Fl. Br. Ind., V., 583*; *Wight, Ic., t.*
     Syn.—Memorialis, pentandra, *Wedd.*; M. ciliaris, *Ham.*; Urtica
     pentandra, *Roxb.*; U hippurioides, *Griff.*; Gonostegia oppositifo-
     lia, *Turcs.*; Hyrtanandra pentandra & javanica, *Miq.*; Bœhmeria
     melastomoides, *Griff.*

     Var. α. proper,=P. pentandra, *Roxb.*

       β. Stocksii,=P. integrifolia, β. *Dals.*; P. Stocksii, *Wight.*

       γ. ramosissima, *Wedd.*=P. ramosissima, *Wight, Ic., t. 2095, f.*
         17, and P. Dalzellii, *t. 2096, f. 21.*

       δ. integrifolia,=P. integrifolia, *Dals.*

       ε. Walkeriana,=P. Walkeriana, *Wight, Ic., t. 2095, f. 16, and*
         glabra, *f. 15.* Memorialis aquatica, *Wedd.*; Hyrtanandra
         Walkeriana, *Thwaites.*

     Vern.—*Pippira-sari*, Hind.; *Jaiphal-jari*, N.-W. P.
     References.—*Roxb., Fl. Ind., Ed. C.B.C.*, 654; *Dals. & Gibs., Bomb. Fl ,*
       240 ; *Thwaites, En. Cey. Pl., 261* ; *Giff., Itin. Notes, 362* ; *No:ul., 386* ;
       Ic. Pl. Asiat., IV., 563, f. 2. ; *Atkinson, Him. Dist , 798.*

     Habitat.—A stout variable annual, met with in the Tropical Himálaya
     from Kangra, eastwards to Assam, the Khásia mountains, and Bengal, and

FIBRE   southwards to Orissa and Kanara.
1215   Fibre.—According to Atkinson it yields a useful cordage fibre.
1216

   P. tuberosa, *Wight*; *Fl. Br. Ind., V., 582*; *Wight, Ic., t. 697.*
     Syn.—Urtica tuberosa, *Roxb.*
     Vern.—*Pilli-dumpa*, Tam.; *Eddu mukka-dumpa, eddu-mutte dumpa,*
       Tel.
     References.—*Roxb., Fl. Ind., Ed. C.B.C.*, 654; *Sir W. Elliot, Fl. Andhr.,*
       49 ; *Lisboa, U. Pl., Bomb., 204.*

     Habitat.—A native of banks of water-courses, hedges, &c., in the
     Circars.

FOOD.   Food. —" The roots are esculent and nutritious; the Natives eat them
Roots.   raw, boiled, or roasted " (*Roxburgh*).
1217

[*f. 44.*

1218   **P. viminea**, *Wedd.*; *Fl Br. Ind., V., 581*; *Wight, Ic., t. 2100,*
     Syn.—P. borbonica, *Wight*; P. ovalis, *Miq.*; Bœhmeria viminea,
       Wall.; Urtica viminea, *Wall.*; U. sanguinea, *Blume Bijd.*; U.
       chiple, punctata & parvifolia, *Ham.*; Margarocarpus vimi-
       neus, *Wedd.*; Leptocnide borbonica, *Blume.*
     Vern.—*Chhota kúail*, Nepal; *Kyingbi*, Lepcha.
     References.—*Brandis, For. Fl.*, 405 ; *Kurz, For. Fl. Burm., II.*, 425 ;
       Gamble, Man. Timb., 325 ; *Atkinson, Him, Dist.*, 317.

     Habitat.—A small shrub, found on the Tropical and Sub-tropical
     Himálaya, from the Sutlej eastwards, ascending to 7,000 feet in Sikkim,
     also from Assam to Tenasserim and Perak.

FIBRE.   Fibre.—A cordage fibre is prepared from the bark in the Eastern
Bark.   Himálaya (*Gamble*).
1219
FOOD.   Food.—The leaves are eaten in Sikkim (*Brandis*).
Leave.
1220

PRANGOS, *Lindl.*; *Gen. Pl., I., 904.*

1221   **Prangos pabularia**, *Lindl.*; *Fl. Br. Ind., II., 695*; Umbelliferæ.
     Syn.—Laserpitium sp., *Wall.*

| A Substitute for Rock Parsley. (*J. Murray.*) | PRANGOS
pabularia. |

Vern.—*Prangos,* ? Ladak ; fruit=*Fitrasúliu n, fitrásálium,* Pb., Sind ; *Komál, komal, bádián-kóhi,* Afgh. ; fruit=*Fituras-aliyán, phattars-dlam,* Bomb.

References.—*Stewart, Pb. Pl., 108 ; Aitchison, Bot. Afgh. Del. Comm., 66 ; Notes on Products of N.-W. Afghánistán. & N.-E. Persia, 164 ; O'Shaughnessy, Beng. Dispens., 360 ; Dymock, Mat. Med. W. Ind., 2nd Ed., 378 ; Dymock, Warden & Hooper, Pharmacog. Ind., II., 138 ; S. Arjun, Bomb. Drugs, 64 ; Murray, Pl. & Drugs, Sind, 201 ; Year-Book, Pharm., 1878, 287 ; Baden Powell, Pb. Prod., 352 ; Birdwood, Bomb. Prod., 43 ; Royle, Prod. Res., 179, 225 ; Smith, Dic., 338 ; Agri.-Horti. Soc., Ind. :—Transactions, I., 74-82 ; II., 191-195 ; III., App., 214 ; V. (Pro.), 66, 85 ; VI., 32, (Pro.) 51 ; VII. (Pro.) 51, 52, 90 ; Journals (New Series), VII., 359.*

Habitat.—A tall perennial herb, met with in Kashmír and Baluchisthán between the altitudes of 6,000 and 11,000 feet ; distributed to Afghánistán and Persia.

Medicine.—In ancient times this plant appears to have been held in considerable repute by the physicians of Northern India. Thus, in a letter from the Maharaja Ranjít Singh to Lord Bentinck in 1833, it is stated that the learned men of Lahore had extracted a treatise from their medical books on the plant, a treatise which was forwarded to the Secretary to Government. The writer can, however, find no further reference to this, nor any translation of the passages extracted. Even at the present day it commands a certain amount of interest The root is considered diuretic and emmenagogue, and a valuable external application to itch. Murray states that the plant is poisonous to lower animal life, a decoction destroying snails. It might in all probability act as an efficient antiseptic wash. The fruit is generally obtainable in the bazárs under some of the above cited vernacular names, names which are corruptions of the Greek πετροσέλινον, or Rock Parsley, a plant for which it is employed as a substitute. It is believed to be stomachic, stimulant, carminative, aphrodisiac, and diuretic, and, Dymock informs us, it is also said to promote the explusion of the fœtus. In Kashmír a decoction of the fruit is employed as a wash to cure "rot" (the 'liver-fluke," *Fasciola hepatica*) in sheep.

Chemical Composition.—An analysis of the air-dried fruit has recently been published by the authors of the *Pharmacographia Indica,* from which it appears that the following constituents occur in its composition;—an essential oil, traces of a fixed oil, resins, traces of an alkaloid, quercitrin in large amount, and an etherial salt of valeric acid. The oily constituents amounted to only about half an ounce from sixty pounds of fruit, and were almost entirely soluble in the water of the distillate. The etherial oil recalled the odour of menthol and **Xanthoxylon** oil, with an after-odour of caraways. It was a mixture of more than one oil, but the amount obtained was too small to admit of fractional distillation. The alkaloid afforded marked precipitates with alkaloidal reagents : Concentrated nitric acid, yellow ; sulphuric acid, brown ; no reaction with ferric chloride. "An alcoholic extract was agitated with ether, and after driving off the ether, the etherial extract was heated with caustic soda, when an odour was developed very similar to that of otto of roses."

Fodder.—Much has been written at different times regarding the properties of this plant as a fodder. The authors of the *Pharmacographia Indica* are of opinion that the *komal* and *avi-priya* or "dear to sheep" of Sanskrit writers may have been **P. pabularia.** Royle believes that it was, in all probability, the **Sílphium** mentioned by Arrian, the historian of the campaigns of Alexander the Great, a plant which he describes as growing with pines on Paropanessus, and attracting the sheep from afar by its fragrance. In 1822, shortly after the formation of the *Agricultural*

MEDICINE.

Root.
1222
Plant.
1223
Fruit.
1224

Chemistry.
1225

FODDER.
1226

PREMNA herbacea.	The Premna Shrub.

FODDER.

and *Horticultural Society of India,* the attention of that Body was directed by **Mr. Moorcroft,** a veterinary surgeon who was then on deputation in Upper Asia, to the value of the plant as a fodder. It was described by him as possessing highly nutritious qualities, giving a very large yield, easy to cultivate, of great vitality, and capable of flourishing on lands of the poorest quality. "When once in possession of the ground," he writes, "for which the preparation is easy, it requires no subsequent ploughing, weeding, manuring, nor other operation, save that of cutting and of converting the foliage into hay." He also commented on the facts that it appeared to be heating, produced fatness, "in a space of time singularly short," and when eaten proved destructive to the liver-fluke in sheep. He sent large consignments of the fruit, hay, and root to the Society; but though the first arrived apparently in good condition and every precaution was taken for its preservation, it entirely failed to germinate. An effort was subsequently made by the Society in 1834, through the intervention of Lord William Bentinck, then Governor-General, with Maharaja Ranjít Singh, to obtain more seed, but without success. Subsequently seed was obtained in good condition and attempts were made to introduce it into cultivation in England, the Cape, and other localities, but in no case does it appear to have succeeded.

Stewart considers that **Moorcroft** unintentionally much exaggerated the virtues of the plant, and states that it is nowhere highly prized, not even in Afghánistán, where almost every plant of any size is cut for fodder. In opposition to this statement, it may, however, be remarked that Lieut. **Burnes,** an early traveller in Afghánistán, found it "greedily cropped by sheep," and even eaten by his fellow-travellers on account of its supposed fattening powers; and that **Aitchison,** in his recent *Notes on the Products of Western Afghanistan and North-East Persia,* states that it is a common plant at Badghis, where it is considered an excellent fodder for goats and sheep.

It may be remarked, in conclusion, that horses fed on the fruit are said to have suffered severely from inflammation of the eyes, and even from temporary blindness.

Precious Stones, see Carnelian; Diamond; Pearl; Ruby &c., &c.

1227

PREMNA, *Linn.; Gen. Pl., II., 1152.*

A genus of trees or shrubs, sometimes climbing, which comprises some 40 species, of which 34 or 35 are natives of India.

1228

Premna esculenta, *Roxb.; Fl. Br. Ind., IV., 580;* VERBENACEÆ.

References.—*Roxb., Fl. Ind., Ed. C.B.C., 485; Voigt, Hort. Sub. Cal., 467; Kurz, For. Fl. Burm., II., 261; Sir W. Elliot, Fl. Andhr., 104, 133; Drury, U. Pl., 355; Boswell, Man. Nellore, 132.*

Habitat.—A branching shrub, recorded in India from Assam and Chittagong. Mr. C. B. Clarke states, in the *Flora of British India,* that all the specimens seen by him appear to have been cultivated, and that the native habitat is uncertain.

MEDICINE. Leaves. 1229
Medicine.—The Natives of Chittagong employ the LEAVES medicinally (*Roxb.*).

FOOD. Leaves. 1230
Food.—The LEAVES, in the same locality, are used as an article of diet (*Roxb.*).

1231

P. herbacea, *Roxb.; Fl. Br. Ind., IV., 581.*

Syn.—P. PYGMÆA, *Wall.*

Vern.—*Bhuijám,* BENG.; *Kada met,* SANTAL; *Néla nirédu,* TEL.; *Bhumijambu, bhúmi-jambúka,* SANS.

P. 1231

The Premna Shrub. (*J Murray.*)

References.—*Roxb., Fl. Ind., Ed. C.B.C.,* 485 ; *Brandis, For. Fl.,* 368 ;
Elliot, Fl. Andhr., 131 ; *Rev. A. Campbell, Rept. Econ. Pl., Chutia
Nagpur, Nos.* 7554, 8759 ; *U. C. Dutt, Mat. Med. Hind.,* 294 ; *Atkinson,
Him. Dist.,* 315 ; *Gazetteer, N.-W.P., IV., lxxvi.* ; *Ind. Forester, IX.,* 255.
Habitat.—A small undershrub, met with in the sub-Tropical Himálaya,
from Kumáon to Bhotán, between the altitudes of 500 and 3,000 feet ; also
found in Southern India.

Medicine.—The Rev. Mr. Campbell remarks, "a preparation of the
ROOT is given internally for rheumatism."

MEDICINE.
Root
1232
1233

Premna integrifolia, *Linn.; Fl. Br. Ind., IV.,* 574; *Wight, Ic., t.* 1469.

Syn.—P. SERRATIFOLIA, *Linn.* ; P. SPINOSA, *Roxb.*; P. SAMBUCINA *and*
VIBURNOIDES, *Wall.*; P. OVALIFOLIA, *Wall.*; GUMIRA LITTOREA,
Rumph.

Vern.—*Agetha, arni, ustabunda,* HIND. ; *Bhút-bhiravi, ganiári,* BENG. ;
Aguyábát, URIYA ; *Gineri,* NEPAL ; *Bakorcha,* GARHWAL ; *Ganniari,*
OUDH ; *Arni, nárvel,* BOMB. ; *Chámári,* MAR. ; *Arni,* GUZ. ; *Munnay,
munni-vayr, múney kiray,* TAM. ; *Ghebu-nelli, pinna-nelli,* TEL. ; *Appel,*
MALAY. ; *Toung-than-gyee,* BURM. ; *Karnika, middi-gass,* SING. ; *Agni-
mantha, ganikáriká,* SANS.

References.—*Roxb., Fl. Ind., Ed. C.B.C.,* 484; *Brandis, For. Fl.,* 366 ;
Kurz, For. Fl. Burm., II., 262 ; *Beddome, For. Man.,* 172 ; *Gamble, Man.
Timb.,* 295 ; *Thwaites, En. Ceylon Pl.,* 242 ; *Mason, Burma & Its People,*
792 ; *Elliot, Fl. Andhr.,* 84, 153; *Ainslie, Mat. Ind., II.,* 210 ; *O'Shaugh-
nessy, Beng. Dispens.,* 486 ; *U. C. Dutt, Mat. Med. Hind.,* 219, 289, 298 ;
Dymock, Mat. Med. W. Ind., 2nd Ed., 604 ; *Trans., Med. & Phys. Soc.,
Bomb. (New Series), No.* 4, 154 ; *Atkinson, Him. Dist.,* 747 ; *Drury, U.
Pl.,* 365 ; *Lisboa, U. Pl. Bomb.,* 203 ; *Gazetteers :—Mysore & Coorg, I.,*
64 ; *Bombay, XIII.,* 24; *XV.,* 440 ; *Orissa, II.,* 158 ; *Ind. Forester, III.,*
237.

Habitat.—A small deciduous tree, or shrub, met with near the sea from
Bombay to Malacca ; also in Sylhet, Ceylon, the Andamans, and the Nicobar Islands.

Medicine.—The ROOT is described in Sanskrit works as bitter,
stomachic, and useful in fevers, anasarca, &c. Chakradatta recommends
the root rubbed into a paste with water to be taken with clarified butter,
for a week, in cases of urticaria and roseola. It also forms an ingredient of
dasamula, a favourite compound decoction of ten plants, and is thus largely
used in a variety of diseases. The LEAVES are said to be bitter and carminative, and a soup prepared from them is recommended as stomachic
(*U. C. Dutt*). Ainslie states that the root is prescribed in decoction as a
gentle cordial and stomachic in fevers, to the quantity of half a tea-cupful
twice daily, and quotes Rheede to the effect that the leaves are employed
on the Coromandel Coast, for the purposes indicated above. Atkinson informs us that the leaves, rubbed up with pepper, are administered in colds
and fever, and that the whole PLANT is used in the form of decoction, as a
remedy for rheumatism and neuralgia.

Food & Fodder.—The LEAVES are said by Rheede to be eaten on the
Coromandel Coast, probably as a semi-medicinal, carminative article of
diet. In certain localities they are used for feeding cattle.

Structure of the Wood.—White, with purple streaks, moderately hard,
close-grained ; weight 35℔ per cubic foot ; used for firewood (*Gamble*).

MEDICINE.
Root.
1234

Leaves.
1235

Plant.
1236
FOOD &
FODDER.
Leaves.
1237
TIMBER.
1238

P. latifolia, *Roxb. ; Fl. Br. Ind., IV.,* 577.

1239

Var. cuneata,—P. VIBURNOIDES, *Kurz.*
mollissima,—P. MOLLISSIMA, *Roth*; P. VIBURNOIDES, *Wall., Cat.,*
2646, a.
mucronata,—P. MUCRONATA, *Roxb.* considered separately, see below.
viburnoides, *Wall., Cat.* 2 6, b. ;—P. LATIFOLIA, *Wight, Ic., t.* 869.

22

P. 1239

PREMNA **tomentosa.**	The Premna Shrub.

Vern.—*Gondhona,* URIYA; **var.** cuneata=*Dangra seya,* SANTAL; *Dauli,* RAJBANSHI; *Gineri,* NEPAL; *Michapgong,* LEPCHA; *Chambadi, chambari,* BOMB; *Nella,* TAM.; *Pedda-nella-kúra, nelli kúra, pedda nelli kúra, nelli,* TEL.; *Middi,* SING.

References.—*Roxb., Fl. Ind., Ed. C.B.C.,* 483; *Brandis, For. Fl.,* 366; *Kurz, For. Fl. Burm., II.,* 261; *Beddome, Fl. Sylv.,* 172; *Gamble, Man. Timb.,* 294; *Dals. & Gibs., Bomb. Fl.,* 200, 203; *Elliot, Fl. Andhr.,* 133, 149; *Rev. A. Campbell, Report, Ec. Pl.,* Chutia Nagpur, *No.* 9471; *Atkinson, Him. Dist.,* 315; *Drury, U. Pl.,* 354; *Lisboa, U. Pl. Bomb.,* 107, 202; *Gazetteers:*—Mysore & Coorg, *I.,* 64; *Bombay, V.,* 28; *XV.,* 440; *Ind. Forester, III.,* 237.

Habitat.—A small deciduous tree of the Sub-Himálayan tract from Kumáon eastwards, and of South India.

FOOD &
FODDER.
Leaves.
1240

Food & Fodder.—The LEAVES, which have a strong, but not unpleasant, odour, are eaten in curries by Natives, especially in South India, and are also sometimes given as fodder to cattle.

TIMBER.
1241

Structure of the Wood.—Grey, with yellow, green, and purple streaks; softer than that of *P. tomentosa*; weight 38℔ to 43℔ per cubic foot; used by the hill-tribes in Sikkim to obtain fire, by friction, and in the Darjíling Terai for firewood and charcoal.

1242

Premna latifolia, *Roxb.; var.* mucronata, *Fl. Br. Ind., IV.,* 578.

Syn.—P. MUCRONATA, *Roxb.*

Vern.—*Bakar, bakarcha, basóta, ágniún, tumari, jhatela,* HIND.; *Agniú,* KUMAON; *Ganhila, gian, ganhin, bankár,* PB.

References.—*Roxb., Fl. Ind., Ed. C.B.C.,* 485; *Brandis, For. Fl.,* 366; *Gamble, Man. Timb.,* 295; *Stewart, Pb. Pl.,* 166; *Atkinson, Him. Dist.,* 315, 747; *Gazetteer, Hoshiarpur,* 11; *Ind. Forester, VIII.,* 410; *Agri.-Horti. Soc., Ind., Journals (Old Series), XIII.,* 313.

Habitat.—A small tree of the sub-Himálayan tract, from the Chenab eastwards to Bhotán and the Khásia mountains, ascending to 5,000 feet.

MEDICINE.
Milk.
1243
Juice.
1244

Medicine.—Atkinson states that in Kumáon the MILK of the bark is applied to boils, and the JUICE given to cattle suffering from colic. Stewart also writes that the juice is employed medicinally in the Panjáb.

TIMBER.
1245

Structure of the Wood.—Moderately hard, light purple, structure similar to that of *P. integrifolia.* It is used, like that of the other varieties of *P. latifolia,* for obtaining fire by friction, and as fuel.

1246

P. longifolia, *Roxb. ; Fl. Br. Ind., IV.,* 575.

Vern.—*Gohora,* ASSAM; *Gwyheli,* NEPAL; *Sungna,* LEPCHA; *Dhaoli,* MICHI; *Gabbu nelli,* TEL.

References.—*Roxb., Fl. Ind., Ed. C.B.C.,* 484; *Gamble, Man. Timb.,* 294; *Elliot, Fl. Andhr.,* 55.

Habitat.—An evergreen tree, found in the Eastern sub-Himálayan tract of Bengal and Assam.

TIMBER.
1247

Structure of the Wood.—Greyish-brown, hard, close-grained; weight 47℔ to 50℔ per cubic foot. It is used in Assam for making house-posts (*Gamble*).

1248

P. tomentosa, *Willd. ; Fl. Br. Ind., IV,* 576; *Wight, Ic., t.* 1468.

Syn.—P. FLAVESCENS, *Juss.* ; CORNUTIA CORYMBOSA, *Lamk.*

Var. detergibilis.—P. LATIFOLIA, *Thwaites, not of Roxb.*

Vern.—*Katokoi,* KOL.; *Chambara,* MAR.; *Nagal, naoru, naura, nagaru, navuru,* TEL.; *Ije,* KAN.; *Kyunnalin,* BURM.; *Búsairu, gulseru,* SING.

References.—*Roxb., Fl. Ind., Ed. C.B.C.,* 483; *Brandis, For. Fl.,* 367; *Kurz, Fl. Burm., II.,* 260; *Beddome, Fl. Sylv., t.* 251; *Thwaites, En. Ceylon Pl.,* 242; *Elliot, Fl. Andhr.,* 122, 130; *Drury, U. Pl.,* 355; *Madras Man. Admin., II.,* 123; *Ind. Forester, III.,* 204; *X.,* 31.

Habitat.—A moderate-sized deciduous tree, common in South India, Burma, and Ceylon.

P. **1248**

Structure of the Wood.—Smooth, light-brown, somewhat resembling teak in colour, moderately hard, close and even-grained, seasons and polishes well. It is used in Burma for making weaver's shuttles, and has been recommended as suitable for turning and fancy work.

TIMBER.
1249

PRIMULA, *Linn.; Gen. Pl., II., 631.*

Primula reticulata, *Wall; Fl. Br. Ind., III., 483;* PRIMULACEÆ.

Syn.—P. ALTISSIMA and SPECIOSA, *Don.*
Vern.—*Bish-kopra, jal-kutra,* KUMAON.
References.—*Atkinson, Him. Dist., 747; Gazetteer, Gujrat, 11.*
Habitat.—Found in the Central and Eastern Himálaya, Nepal, and Sikkim, between the altitudes of 11,000 and 15,000 feet.
Medicine.—Atkinson writes that this PLANT is poisonous to cattle, and is used externally as an anodyne.

1250

MEDICINE.
Plant.
1251

PRINSEPIA, *Royle; Gen. Pl., I., 611.*

Prinsepia utilis, *Royle; Fl. Br. Ind., II., 323;* ROSACEÆ.·

Vern.—*Bhekal, bekkra, karanga, cherara, dhatela, jhatela,* HIND.; *Chirara, jhatela, dhatela, phaláwa, bhekla, dintüi, bhekara, bhekala,* KUMAON; *Gurinda, chamba, arund, tatúa, phálwára, rári, jinti, bekklí, behkul, bhékhr, karngúra, bhekal, bekhar, behhúl, bekhwa, bekling, bhekal, bhekar, garandu, bekrúl,* OIL=*baikar,* PB.
References.—*Voigt, Hort. Sub. Cal., 265; Brandis, For. Fl., 196; Gamble, Man. Timb., 164; Stewart, Pb. Pl., 81; O'Shaughnessy, Beng. Dispens., 28; Baden Powell, Pb. Pr., 346, 422, 591; Atkinson, Him. Dist., 309, 747; Econ. Prod., N.-W. P., Pt. V., 44, 65; Royle, Ill. Him. Bot., 206, t. 38, f. 1; Cooke, Oils & Oilseeds, 66; Gazetteers:—Ráwalpindi, 83; Simla, 12; Settle. Rep., Simla, App. II. (H.), xliii.; Ind. Forester, XI., 3; Agri.-Horti. Soc., Ind., Trans. III. (Pro.), 267; Journals (Old Series); II., Sel., 298; XIII., 390; XIV., 53.*
Habitat.—A deciduous, thorny shrub, found on the dry rocky hills of the Temperate Himálaya, from Hazára to Sikkim and Bhotán, between 4,000 and 9,000 feet, also on the Khásia hills.
Oil.—The SEED yields an oil by expression which is much used in the North-West Himálaya, for food, illuminating, and occasionally in medicine. Atkinson states that a small quantity is exported from the forests of Garhwál and Kumáon.
Medicine.—The OIL is rubefacient and is applied externally for rheumatism and pains resulting from over-fatigue.
Food.—The OIL is used as an article of food.
Structure of the Wood.—Sapwood white, heartwood red, very hard and compact, close and even-grained but very liable to split; growth slow; weight 69℔ per cubic foot. It is employed for fuel, and occasionally for making walking-sticks.
Domestic.—The OIL is burned for illuminating purposes.

1252

OIL.
Seed.
1253

MEDICINE.
Oil.
1254
FOOD.
Oil.
1255
TIMBER.
1256
DOMESTIC.
Oil.

PROSOPIS, *Linn.; Gen. Pl., I., 591.*

A genus of erect trees or shrubs, which comprises about eighteen species, dispersed over the tropical and sub-tropical regions of both hemispheres. Of these two are natives of India. Several are of considerable economic value, and since a few have been introduced into India, they are worthy of a short notice. **P. dulcis,** the "Alagaroba," or "Paray," a native of America, produces sweetish succulent pods, from 20 to 24 inches long, which are largely used for feeding cattle. It has been introduced into Madras, where it is planted along railway lines, and is known locally as "Tamarind." **P. glandulosa.** *Torr.,* the "Mesquit Bean," is a native of the mountainous regions of Western Texas. It has been successfully introduced into the North-West Provinces. The tree yields a large quantity of gum, resembling gum-arabic, and often used

1257
1258

The Prosopis Bean.

as a mucilage in making jujubes and other sweetmeats. The pods are filled with a sweet pulp, which, by fermentation and boiling, makes a not unpleasant drink. The seeds or beans, powdered and mixed with water, form a paste, which, when dried, is used for tood. The beans are also useful for fodder, and have apparently been employed for this purpose in India (*Atkinson*). The wood is very hard and durable, has a beautiful grain, and is much used for furniture, and in the manufacture of charcoal (*Gamble*).

P. pallida, *Kunth.,* a native of South America, has been successfully introduced into Ceylon. Its pods contain as much as 90 per cent. of tannic acid, and are highly valued in tanning. They are imported into Europe under the name of " Algarobilla " or " Balsamocarpon " (*Gamble*).

P. pubescens, *Bth.,* the " Screw Bean,'' or " Screw Mesquit," also a native of New Mexico and Texas, is being experimentally cultivated in the Royal Botanic Gardens, Calcutta. The bean of this species is of much economic value. It is screw-shaped, is borne in abundance, and ripens at all times of the year. It contains a large quantity of saccharine nutritious matter, from which molasses may be made by boiling. The pods form an important article of food to the Natives, and are a valuable fodder for cattle, but great caution is said to be necessary in using them as fodder for horses (*Gamble*).

                           [*Sylv., t. 56 ;* LEGUMINOSÆ.

1459

Prosopis spicigera, *Linn ; Fl. Br. Ind., II., 288 ; Bedd., Fl.*

Syn.—P. SPICATA, *Burm.* ; ADENANTHERA ACULEATA, *Roxb.*

Vern. - *Jhand, jand,* HIND. ; *Shami, somi,* BENG. ; *Shami, savandal,* URIYA ; *Chaunkra,* N.-W. P. ; *Jhand, khár, seh, jandi, jand, kanda, kandi, khunda, jánt,* POD=*Sangar, sankhri, shángar,* GALLS=*kharnúb hindí,* PB. ; *Aghsakai, aghsakár, kandi,* PUSHIU ; *Khejra, sangri, kajra, sé,* RAJ. ; *Kandi, kundi, kemdo,* SIND. ; *Kandi, shumi,* C. P. ; *Shemi, sumri, sounder, shemú, saundad, shami, shemri, shamri,* BOMB. ; *Shemi, saunder, savandal,* MAR. ; *Semru, hamia, khijro, khijado,* GUZ. ; *Soundar,* DEC. ; *Perumbe, vunne, jambu, vanni,* TAM. ; *Chani, shumi, priyadarsini, jammi, chami,* TEL. ; *Perumbe, perumbai, vunne,* KAN.

References.—*Boiss., Fl. Orient., II., 634 ; Roxb., Fl. Ind., Ed. C.B.C., 361 ; Voigt, Hort. Sub. Col., 259 ; Brandis, For. Fl., 169, t. 25 ; Gamble, Man. Timb., 147 ; Dalz. & Gibs, Bomb. Fl., 84; Stewart, Pb. Pl., 74 ; Sir W. Elliot, Fl. Andhr., 72, 158 ; Murray, Pl. & Drugs, Sind, 136 ; Dymock, Warden & Hooper, Pharmacog. Ind., I., 550 ; Birdwood, Bomb. Prod., 344 ; Baden Powell, Pb. Pr., 263, 288, 397 ; Drury, U. Pl. Ind., 355 ; Useful Pl. Bomb. (XXV., Bomb. Gas.), 66, 154, 278, 279, 280, 290, 397 ; Liotard, Dyes, 33 ; Cooke, Gums & Resins, 22 ; Watson, Rep., 36, 51 ; Rept. Exper. Farms, N.-W. P., 1883, App. ii. ; Boswell, Man., Nellore, 97 ; Moore, Man., Trichinopoly, 79 ; Stocks, Report on Sind. ; Settlement Report :—Panjáb, Dera Ismail Khán, 25 ; Rohták, 78 ; Lahore, 14 ; Dera Ghási Khán, 4 ; Montgomery, 17 ; Central Provinces, Chanda, App. VI ; Raipore, 10 ; Gazetteers :—Bombay, II., 355 ; V., 25, 285 ; VI., 13 ; VII., 38, 39, 40 ; XV., 74 ; XVII., 18, 26 ; XVIII., 52 ; Panjáb, Montgomery, 18 ; Bannu, 23 ; Musoffargarh, 21 ; Ludhiána, 9 ; Lahore, 13 ; Dera Ismail Khán, 19 ; Múltan, 102 ; Karnal, 16 ; Rohták, 14 ; Amritsar, 4 ; Dera Ghási Khán, 10 ; N.-W. P., I., 80 ; IV , lxxi. ; Mysore & Coorg, I., 52, 59 ; Agri.-Horti. Soc., Ind. :—Journals (Old Series), VIII. Sel., 176 ; XIII., 172 (New Series), VI., 6 ; Panjáb, Select Papers to 1862, 18 ; Index, 44 ; Ind. Agriculturist, Oct., 1886 ; Ind. Forester, II., 171 ; III., 23, 202, 237 ; IV., 153, 227 ; V., 330 ; VIII., 121, 131, 439 ; IX., 42, 174, 327 ; XI., 388 ; XII., 32 ; App., 1, 2, 12 ; XIII., 120.*

Habitat.—A moderate-sized, deciduous, thorny tree, found in the arid dry zones of the Panjáb, Sind, Rájputana, Guzerát, Bundelkhand, and Deccan. It is easily raised from seed and coppices well. It has an enormously long tap root ; one specimen, of which pieces were sent to the Paris Exhibition of 1878, had a root 86 feet long, penetrating vertically to a depth of 64 feet. The tree is said to specially favour a heavy soil.

GUM.
1260

Gum.—Stewart writes, " From the stumps of pruned branches and other scars, a gum exudes similar to gum arabic, but I do not know that

The Prosopis Bean.	(*J. Murray.*)	PROSOPIS spicigera.

GUM.

it is collected or used." It has recently been examined by the authors of the *Pharmacographia Indica*, who state that it is "unusually friable, occurs in small angular fragments of a yellowish colour, more or less deep, sometimes in large ovoid tears about two inches long, of an amber colour internally, but having a frosted or candied appearance externally from the presence of numerous minute cracks which cause the tears to crumble under pressure. With water it forms rather dark-coloured tasteless mucilage of about the same viscosity as gum arabic. The solution is precipitated by the normal acetate of lead, and gelatinized by the basic acetate, also by ferric chloride, borax, and alkaline silicates. It rather freely reduces Fehling's solution. This is a valuable gum, and appears to resemble, except in its behaviour to reagents, the Mozquite gum of Mexico and Texas, which is now coming into use in America." It would appear from these remarks, that the gum is well worthy of further investigation as to its probable utility in the arts, and the degree to which it could be obtained.

Tan.—The BARK is used as a tan.

TAN.
Bark.
1261

Medicine.—The POD, according to **Stewart**, is probably sometimes used in the Panjáb for medicinal purposes. On many trees, large spongy wooden GALLS are produced, which are known in the bazárs of the Panjáb under the name of *Kharnúb Hindí*, and are believed to be astringent. In the *Settlement Report of Chanda District, Central Provinces*, the BARK is said to be locally employed as a remedy for rheumatism.

MEDICINE.
Pod.
1262
Galls.
1263

Bark.
1264

Food & Fodder.- The PODS which ripen before and during the rains contain, when scarcely ripe, a considerable quantity of a sweetish farinaceous substance, which has the flavour of the fruit of the carob tree. It is largely consumed as food in the Panjáb, Guzerát, and the Deccan; in some localities by all classes, in others by only the poor, and in times of scarcity. It is eaten in various ways, green or dry, raw and alone, boiled with salt, onions, and *ghí*, with bread, or mixed with *dahí*. **Stewart** remarks that the quantity which may be consumed at once by a man, has been variously stated to be from one *chitták* to a *sér*. During the great Rájputana famine of 1868-69 the BARK was largely eaten, ground into flour and made into cakes. A writer in the *Indian Agriculturist* states that it was the means of saving thousands of lives. Grain was so scarce as to be beyond the reach of the poor, who had consequently to depend to a very large extent on this bark as almost their only article of food. It is sweetish to the taste, and is said to be not at all unpleasant in flavour.

FOOD
& FODDER.
Pods.
1265

Bark
1266

The pods are good food for camels, cattle and goats, and the LOPPINGS, called *lanji* in Dera Ismail Khán, are said to be highly valued in the Trans-Indus region as fodder for sheep and goats. The trees are carefully preserved for this purpose, and are pollarded during the cold weather when the grass and other fodder supply is short. The *lanji* is made to last if possible for three months, from December to February. In other localities of the Panjáb the lopped leaves seem to be equally valued.

Loppings.
1267

Structure of the Wood.—Sapwood large, perishable; heartwood purplish brown, extremely hard; weight about 58lb per cubic foot. It is tough, but not durable, liable to dry, rot, and very easily destroyed by insects. It is used for building, carts, well-curbs, furniture, and agricultural implements, but is chiefly valued for fuel, as its heating power is great (*Gamble*). According to **Brandis**, 1,374lb were consumed in evaporating 11·8 cubic feet of water per hour during seven hours, the pressure of steam being kept at 27lb per square inch, while 1,388lb of **Acacia arabica** wood, and 1,627lb of **Tamarix gallica**, were required to produce the same result. Owing to the value of the wood for this purpose it has been extensively consumed on loccmotives and steam boats in Sind and the Panjáb, a fact

TIMBER
1268

PRUNUS	
Amygdalus.	**The Almond.**

FOOD
& FODDER.

that is said to have brought about a steady decrease in the area under the tree in certain districts (*Settlement Report, Montgomery*). Good charcoal is also made from the wood, though it is said to have the disadvantage of emitting a great many sparks when burned.

DOMESTIC
& SACRED.
1269

Domestic & Sacred.—In the Panjáb the tree is found not uncommonly planted near the grave of some saint, and when in such situations, it is said not to be pollarded, and to be generally protected. **Mr. Duthie** informs us that in Rájputana it is worshipped at certain times of the year, when large crowds of Natives form themselves into a procession, headed by the Maharája, Rajá or head Thákur of the place, and make their way to some particular tree set apart for worship. With the Hindus generally, all over India, it is sacred. Its worship is enjoined in the *Vratráj* to be performed on the tenth of *Ashwin Sudhapaksha*—the *Dasera* festival—because on this tree the five *Pándav* princes hung up their arms when they entered *Hirát Nagri* in disguise. On the tree the arms turned to snakes and remained untouched till the owners returned. It is also believed that the tree is a transformation of the goddess that pleased Rám. It is worshipped to obtain pardon for sins, success over enemies, and the realisation of the devotee's wishes. The dry TWIGS are employed as *samidhas* in feeding the sacred fire, and the LEAVES as *patri* (leaf-offerings in the worship of Ganpatti – (*Lisboa*).

Twigs
1270
Leaves.
1271

1272

Prosopis Stephaniana, *Kunth. ; Fl. Br. Ind., II., 288.*

Syn.—LAGONYCHIUM STEPHANIANUM, *M. Bieb.* ; MIMOSA AGRESTIS, *Sieb.*

Vern.—*Chiggak, chogak, khar-i-jinghak,* AFGH.

References.—*Boiss, Fl. Orient., II., 633 ; Brandis, For. Fl., 170 ; Aitchison, Rept. Bot. Afgh. Del. Com., 60 ; Ind. Forester, IV., 153.*

Habitat.—A low bushy tree, found in the Panjáb near Pesháwar, distributed to the Caucasus and Afghánistan.

TAN.
Pods.
1273

Tan.—Aitchison writes of the Perso-Afghán border, "The PODS are usually infested with insects, and become immensely large and irregular in form. They are called *Hæ-chi*, and are employed in tanning, for which purpose they are collected and exported."

1274

PRUNUS, *Linn. ; Gen. Pl., I., 609.*

Prunus Amygdalus, *Baill. ; Fl. Br. Ind., II., 313 ;* ROSACEÆ.

THE ALMOND.

Syn.—AMYGDALUS COMMUNIS, *Linn.*

Vern.—*Bádám,* HIND. & PB. ; *Biláti-badám,* BENG. ; *Bádam,* MAR. ; *Badám,* GUJ. ; *Bádam,* DEC. ; *Vádam-kottai,* TAM. ; *Bádam-vittulu,* TEL. ; *Bádámi,* KAN. ; *Bádam, vátam-kotta,* MALAY. ; *Bádan,* BURM. ; *Rata-kotambá,* SING. ; *Bádámitte,* SANS. ; *Louz,* ARAB. ; *Bádám,* PERS.

Sweet and bitter almonds are distinguished by adding the necessary adjective to the above names in all the languages.

References.—*Roxb., Fl. Ind., Ed. C.B.C., 403 ; Brandis, For. Fl., 190 ; Stewart, Pb. Pl., 77 ; Rept., Bot. Afgh. Del. Com., 62 ; DC., Origin Cult. Pl., 218 ; Pharm. Ind., 83 ; Ainslie, Mat. Ind., I., 6, 582 ; O'Shaughnessy, Beng. Dispens., 319 ; Dymock, Mat. Med. W. Ind., 2nd Ed., 292 ; Dymock, Warden & Hooper, Pharmacog. Ind., I., 563 ; Fluck. & Hanb., Pharmacog., 244, 247 ; Bent. & Trim., Med. Pl., t. 99 ; S. Arjun, Bomb. Drugs, 51 ; Murray, Pl. & Drugs, Sind, 140 ; Year-Book Pharm., 1879 465 ; Irvine, Mat. Med., Patna, II., 38 ; Med. Top., Ajm., 127 ; Trans. Med. & Phys. Soc., Bomb. (New Series), No. XII., 172 ; Baden Powell, Pb. Pr., I., 346, 421, 463 ; Atkinson, Him. Dist., 711 ; Econ. Prod., N.-W. P., Pt. V., 44, 62 ; Lisboa, U. Pl. Bomb., 154, 390 ; Birdwood, Bomb. Pr., 32 ; Butler, Top. & States Oudh & Sultanpore, 33 , Smith, Ec. Dic., 11 ; Kew Off. Guide to*

| The Almond. | (*J. Murray*.) | **PRUNUS** **Amygdalus**. |

the Mus. of Ec. Bot., 59, 60 ; Ayeen Akbary, Gladwin's Trans., I., 83 ; Rep. Horti. Gar. Lucknow, March 1884, 3 ; Gazetteer, Bomb., XVIII., 43 ; Ind. Forester, II., 291 ; Agri.-Horti. Soc., Ind., Trans., I., 66 ; II. (App.), 307 ; Journals (Old Series), II., Sel., 539.

Habitat.—The original indigenous area of the almond cannot be ascertained with precision. DeCandolle, however, after considering the statements of various authors, arrives at the conclusion that its native country possibly originally extended from Persia westward to Asia Minor, Syria, and even Algeria. It is cultivated in Kashmír, the Panjáb, Afghánistán, and Persia.

Two varieties are recognised, *viz*, α. **amara** and β. **dulcis**, the bitter and sweet almond, but these cannot be distinguished by any permanent botanical character, and appear to occupy the same area of growth.

Gum.—This tree yields the *Badam* or "Hog-tragacanth" exported from Persia into Bombay and re-exported to Europe. It is used as a substitute for true Tragacanth.

GUM. **1275**

Oil.—The almond yields two distinct oils—an essential oil and a fixed or fatty oil. The latter oil is obtained by expression from either bitter or sweet almonds, the average produce being from 48 to 52℔ from 1 cwt. of fruit. The yield is greater by hot expression 2℔ 2oz. being obtained from 5½℔. The oil is clear, yellow, with an agreeable flavour, but without odour. It is much used by perfumers, but is frequently largely adulterated with gingelly, poppy, or mustard, oil.

OIL. **1276**

The essential oil is obtained from *var.* **amara**, and is well known under the name of "oil of bitter almonds." It is prepared by submitting bitter almond cake, left after expression of the fixed oil, to distillation with water either alone, or more commonly with salt. The yield is liable to great variation, the amount obtained by Unmey having been, in some cases, as much as 0·95 per cent., while in others it was only 0·42 per cent. This oil is a mixture of various substances (see paragraph "CHEMICAL COMPOSITION"), and is highly poisonous owing to the presence of prussic acid. An essence prepared from it is largely used in perfumery and confectionery, but great caution is necessary in employing it. The oil itself has medicinal properties.

Medicine.—An interesting history of the almond may be found in the *Pharmacographia*, from which it appears that it has attracted much attention, and has been familiarly known in Europe for many centuries. In India, on the other hand, it appears to have been but little esteemed. It is but rarely mentioned in Sanskrit works and is not described as of medicinal value by older writers on Hindu medicine. **Dymock** states, however, that Arabian and Persian writers on materia medica discuss its properties at considerable length. Two kinds of sweet almond are described—a thick-shelled, and a thin-shelled or *Kaghazi* (*amendes des dames*, or *amandes sultanes*, Fr.), and the uses of these appear to have been essentially the same as in Europe. The author of the *Makhzan-el-adwiya*, however, describes almonds as having been employed to extract perfumes of flowers by a sort of *enfleurage*, similar to that for which sesamum seeds are now more commonly used in India (see Perfumery, p. 135). He also notices the fact that the burnt SHELLS are made into toothpowder, and the unripe FRUIT (*chugala*) given as an astringent application to the gums and mouth.

MEDICINE. **1277**

Shells. **1278** Fruit. **1279** Bitter Almonds. **1280**

"BITTER ALMONDS (*louz-el-murr*) are described by Muhammadan writers as attenuant and detergent; they are recommended both externally and internally for a variety of purposes. As a plaster made with vinegar they are used to relieve neuralgic pains; as a collyrium, to strengthen the sight; in emulsion with starch and peppermint, to allay cough. They are

P. 1280

| PRUNUS armeniaca. | Bitter Almonds ; Apricot. |

MEDICINE.

Root.
1281
Juice.
1282

Chemistry.
1283

also considered to be lithontriptic and diuretic, and efficient in removing obstructions of the liver and the spleen. Applied to the head they kill lice; as a suppository they relieve pain in difficult menstruation; as a poultice they are a valuable application to irritable sores and skin eruptions. The ROOT of the tree is described as discutient and alterative; it is used both internally and externally" (*Mat. Med. W. Ind.*). Emerson states that the JUICE of the fruit with sugar is used in coughs, and that almonds mixed with figs are considered to be laxative and to relieve intestinal pain.

The therapeutic properties and uses of both bitter and sweet almonds, as recognised in Europe, are too well known to require recapitulation.

CHEMICAL COMPOSITION.—The writer is indebted to Dr. Warden, Chemical Examiner, Calcutta, for the following note :—

"Both the sweet and bitter almond contain about 50 per cent. of fixed oil, which is slightly yellow and fluid, and does not solidify below 7·6° C. Almond oil, like castor oil, is soluble in alcohol. Both varieties of almond also contain a white, albumenous substance—emulsin,—but in the bitter almond there is, in addition, a white, crystalline glucoside—amygdalin. Dry emulsin and amygdalin do not react, but when moistened with water a species of fermentative action ensues, by which the amygdalin is decomposed into prussic acid, hydride of benzole or essential oil of almonds, grape sugar, and formic acid. A similar decomposition takes place when bitter almonds are moistened with water. Amygdalin also yields prussic acid when treated with acid. The volatile oil of bitter almonds, when freed from prussic acid, is not poisonous, but commercial samples usually contain a considerable amount of the acid, and several fatal accidents have followed the incautious use of this popular flavouring essence. On exposure to air, volatile oil of bitter almonds absorbs oxygen, and a crystalline deposit of benzoic acid occurs. Artificial essence of bitter almonds is a solution of nitrobenzol in alcohol, and is prepared by the action of nitric acid on benzol, a volatile liquid obtained during the distillation of coal-tar. In addition to bitter almonds, amygdalin is contained in cherry, plum, and peach kernels, apple seeds, cherry laurel leaves, and the bark of the **Rhamus frangala**. The amount varies from 2·5 to 2·3 in bitter almonds and peach kernels. Apple seeds, cherry, and plum kernels and the bark of R. frangala contain under 1 per cent., while cherry laurel leaves contain 1·38 per cent. In bitter almonds, apple seeds, &c., no prussic acid exists ready formed, but in the bark of R. frangala and in cherry laurel leaves free prussic acid has been detected."

FOOD.
Kernels.
1284

1285

Food.—The seed lobes or KERNELS of the sweet almonds are largely used as a dessert and in confectionery. Considerable quantities are imported into India from Afghánstán.

Prunus armeniaca, *Linn. ; Fl. Br. Ind., II., 313.*

THE APRICOT, MISHMUS, OR "MOON OF THE FAITHFUL."

Var DASYCARPA, the Black-fruited Apricot.

Vern.—*Chúari, sardalu, khúbani,* HIND. ; *Pating, chuli,* BHOTE ; *Chúdru, chola, sard-áru, jald-áru, kushm-áru,* KUMAON ; *Chúli,* LAD. ; *Galdam,* TIBET ; *Iser, gurdálú, cherkúsh,* KASHMIR ; *Hári, gurdalu, jaldárú, shiran, cheroli, shári, cherkúsh, serkuji, harian, chúli, chir, sári, chiran, kush, jaldárú chúli, mandata, arti, sardalu, kishta, alú kashmiri,* PB. ; *Nakhter, mandata,* DRIED = *khubáni,* MOIST = *khista,* PUSHTU ; *Zard-alu,* AFG. ; *Bin-kúk, tuffa-urmena,* ARAB. ; *Mishmish,* PERS.

References.—*Boiss., Fl. Orient., III., 641 ; Roxb., Fl. Ind., Ed. C.B.C., 403 ; Brandis, For. Fl., 191 ; Gamble, Man. Timb., 162 ; Stewart, Pb. Pl., 81 ; Aitchison, Cat. Pb. and Sind Pl., 56 ; Rept. Pl. Coll. Afgh. Del. Com., 61 ; DC., Orig. Cult. Pl., 215 ; Baden Powell, Pb., Pr. 268, 269, 346,*

397, 422, 567; Atkinson, Him. Dist. (X., N.-W. P. Gas.), 309, 711; Econ. Prod. N.-W. Prov., Pt. V. (Vegetables, Spices, & Fruits), 44, 63; Royle, Ill. Him. Bot., 204; Cooke, Oils & Oilseeds, 67; Ayeen Akbáry, Gladwin's Trans., I., 81, 83; II., 135, 169, 170; Settlement Report :—Panjáb, Simla, App. II. (H.), xliii.; Kohat, 30; Peshawar, 13; Agri.-Horti. Soc., Ind. Trans., I., 64-65 II., 96; III., 95, (Pro.), 254; IV., 106, 141, 146, 148; VI. (Pro.), 41; VII. (Pro.) 94, 152; Journals (Old series), XIII., 371, 384, 390 (New Series), VII., (Pro.) 7; Agri.-Horti. Soc., Panjáb, Select Papers to 1862, 46; Ind. Forester, VIII., 36; Smith, Ec. Dict., 20.

Habitat.—A moderate-sized deciduous tree commonly cultivated between the Indus and Sarda, in the North-West Himálaya, in the plains of the Panjáb, in Afghánistán, Western and Central Asia, Europe and China. In the Himálaya the fruit ripens well up to 10,000 feet, but is best between 6,000 and 9,000 feet. The tree is cultivated in West Tibet as high as 12,000 feet, but at that elevation, the fruit seldom ripens properly. It is very frequently spontaneous, and occurs far away from human habitations, so much so that **Stewart** writes, " I believe I have seen the apricot wild in many places from 4,000 to 6,000 feet in the Panjáb Himálaya." In this opinion, however, he is opposed to the general opinion of other Indian botanists. Older writers were of opinion that the tree was, as its name implies, of Armenian or Caucasian origin, but **DeCandolle** follows **Roxburgh** in considering it to have been originally a native of China. " The Chinese," he writes, " knew it two or three thousand years before the Christian era. **Changkien** went as far as Bactriana, a century before our era, and was the first to make the West known to his fellow-countrymen. It was then, perhaps, that the apricot was introduced in Western Asia, and that it was cultivated and became naturalised here and there in the north-west of India, and at the foot of the Caucasus, beyond the limit of population by the scattering of the stones."

Gum.—The tree yields a gum similar to Tragacanth, which, as with most of the gums from other members of this genus, is known commercially as Cherry gum. Of this series the gum from the true cherry is the most valuable, being more soluble than the others, and is used commercially in France.

Oil.—A clear oil, of a pale yellow colour, which smells strongly of hydrocyanic acid, and which, indeed, contains a fairly large proportion of that substance, is extracted from the SEED. This oil is used in burning and cooking, and for the hair. It is said to be one of the most largely utilised oils in Kashmír, and has a pleasant flavour, with an odour of bitter almonds.

Medicine.—In Tibet the APRICOT is employed after mastication as an external application to the eye in cases of ophthalmia, and in Afghánistán it is considered laxative, and a refrigerant in cases of fever (*Stewart*).

Food.—In the North-West Himálaya, dried APRICOT forms a considerable portion of the food of the people, particularly during autumn and winter. It ripens from May to September according to elevation, and in the apricot-districts, *e.g.*, on the Sutlej, Tonse, and Jumna, the roofs of all the houses are covered during that season with the fruit spread out to dry. It is eaten both fresh and dried; the KERNELS sometimes separately, or more frequently pounded with the pulp to mitigate the acidity of the latter. By Europeans it is largely employed in the fresh state for making jam, and when dried for cooking. The dried FRUIT is an important article of trade, and is brought to the plains from the hills, and from Afghánistán. The black variety is not often met with, but is occasionally cultivated in Kashmír. The OIL, as already stated, is largely used as food in Kashmír.

GUM.
1286

OIL.
Seed.
1287

MEDICINE.
1288

FOOD.
1289

Kernels.
1290

Fruit.
1291

Oil.
1292

P. 1292

PRUNUS Cerasus.	**The Sweet and Sour Cherries.**

TIMBER.
1293

DOMESTIC.
Juice.
1294
1295

FOOD.
Fruit.
1296
1297

GUM.
1298
MEDICINE.
Bark.
1299
Kernel.
1300
FOOD.
Kernel.
1301
Fruit.
1302
TIMBER.
1303

Structure of the Wood.—Sapwood white, heartwood greyish-brown, mottled with dark-brown streaks; weight 49 ℔ per cubic foot, handsome, and used for various purposes in the Panjáb Himálaya (*Gamble*). The drinking-cups of Tibet, described by Cunningham are said to be made from it (*Stewart*). In Lahoul and Upper Kanáwar it is the chief firewood.

Domestic.—The JUICE of the fruit is employed to clean brass-work.

Prunus Avium, *Linn.; Fl. Br. Ind., II., 313.*

THE SWEET CHERRY, GEAN OR WILD CHERRY.

Syn.— P. CERASUS, var. α. AVIUM, *Brandis*.

Vern.— *Gilás, krusbál,* KUMAON.

References.—*Boiss., Fl. Orient., II., 649 ; Brandis, For. Fl., 193 ; DC., Origin Cult. Pl., 205 ; Atkinson, Him. Dist., 712 ; Econ. Prod., N.-W. P., Pt. V., 44, 63.*

Habitat.—The Sweet cherry, from which the white and black cherries are developed, is indigenous in Europe, from the South of Sweden to the mountainous parts of Greece, Italy, and Spain, in Algeria, in the Russian provinces to the south of the Caucasus, in Armenia, and in the north of Persia. It is cultivated in the North-West Himálaya up to 8,000 feet. It is said to be most frequently met with in Kashmír, where it flowers in April-May, and the fruit ripens in June. Many of the European varieties introduced have not succeeded in these hills owing to the effect of the heavy rains on the young fruit (*Atkinson*). The true cherry is a scarce fruit in India.

Food.—The FRUIT is well known and much valued for dessert purposes. The cherry of Himálayan writers is mostly **P. Puddum.**

P. Cerasus, *Linn.; Fl. Br. Ind., II., 313.*

THE SOUR CHERRY OR DWARF CHERRY.

Syn.—CERASUS CAPRONIANA, *DC.;* C. VULGARIS, *Miller*.

Vern.—*Alu-bálu,* N.-W. P. ; *Gilás, olchi,* PB. ; *Jerasayna, kerásya,* ARAB.; *Alu-ba-lú, alu-bu-ali,* PERS.

References.—*Roxb., Fl. Ind., Ed. C.B.C., 403 ; Stewart, Pb. Pl., 78 ; DC., Origin Cult. Pl., 207 ; Atkinson, Him. Dist., 712 ; Econ. Prod., N.-W. P., Pt. V., 44, 63 ; Birdwood, Bomb. Prod., 32 ; Smith, Dict., 107, 336 ; Ayeen Akbary, Gladwin's Trans , I., 82 ; II., 135 ; Ain-i-Akbari, Blochmann's Trans , 228, 616 ; Agri.-Horti. Soc., Ind. ;—Trans., I., 65 ; III., 95 ; IV., 106, 141, 151 ; VI. (Pro.), 60 ; VII. (Pro.), 94, 150, 151, 184 ; Journ., VIII., 241 ; XIII., 385.*

Habitat.—Like the preceding species, the natural habitat of this, according to DeCandolle, extends from the Caspian Sea to Western Anatolia. Available evidence points to its having been introduced at a later period into India, where it is now cultivated in the Himálaya of the Panjáb and North-West Provinces up to 8,000 feet. It is the parent of the " Montmorency," " Griotte," and several other cultivated races known to horticulturists (*Origin of Cultivated Plants*). Atkinson states that several kinds have been recently introduced from European stock, and thrive where the rain is not excessive. In Kashmír it is commonly cultivated, and several kinds are said to bear good fruit (*Stewart.*) (*Conf.* with remark Vol. III., 449.)

Gum.—The tree yields a gum, similar in character to tragacanth.

Medicine.—The BARK, which is bitter and astringent, is said to possess febrifugal properties. The KERNEL is supposed to be a nervine tonic, and is used for the same purposes as hydrocyanic acid, of which it contains a considerable proportion.

Food.—The KERNEL is used for flavouring in Europe several liqueurs. The FRUIT is employed for making preserves, &c.

Structure of the Wood.—The timber of this and the preceding species is highly valued in Europe by cabinet, musical instrument and pipe makers, but is not utilised to any extent in India.

P. 1303

The Plum and Damson	(*J. Murray.*)	PRUNUS communis.

Prunus communis, *Huds.,* **var. insititia ;** *Fl. Br. Ind., II., 315.*
1304

THE PLUM, ALUCHA.

Syn.—P. INSITITIA, *Linn.* (the Bullace) *;* P. BOKHARIENSIS & P. ALOOCHA, *Royle ;* & var. DOMESTICA (the Damson or Sloe or Blackthorn.)

Vern. The following vernacular names are used for the fruit :—*Alú, álú bo ; khárá,* HIND. & BENG.; *Alu-bukhára* (blue), *alecha, alúcha* (yellow), *chhota álu* (small kind), *bhotiya badám* (blue damson). *ladákhí badam* (large red plum), KUMAON ; *Alúcha, olchi, er, aor, gurdalu, álú bokhárá, aru bukhára, zardalu, luni, shaft álu,* PB.; *Allocha,* PUSHTU ; *Alú-bokhárá* GUZ.; *Alpogádá-pasham,* TAM.; *Alpogádá-pandlu,* TEL.; *Aálu-paká, álú-pakárá,* BURM.; *Ijás,* ARAB.; *Alu,* PERS.

References.—*Boiss., Fl. Orient., II., 651, 652 ; Brandis, For. Fl., 192, Gamble, Man. Timb., 162; Stewart, Pb. Pl., 82; DC., Orig. Cult. Pl., 211, 214; Pharm. Ind., 86; Flück. & Hanb., Pharmacog., 251; O'Shaughnessy, Beng. Dispens., 324; Sakharam Arjun, Cat. Bomb., Drugs, 52; Bent. & Trim., Med. Pl., t. 96; Dymock, Mat. Med. W. Ind., 2nd Ed., 294; Dymock, Warden & Hooper, Phamacog. Ind., I., 568 ; Official Corresp. on Proposed New Pharm. Ind., 234; Year-Book Pharm., 1879, 213 ; Trans., Med. & Phys. Soc., Bomb. (New Series), No. 12, 174; Baden Powell, Pb. Pr., 265, 270, 315, 346, 396, 591 ; Atkinson, Him. Dist. (X., N.-W. P. Gaz.), 309, 712, 747; Econ. Prod., N.-W. Prov., Pt. V. (Vegetables, Spices, & Fruits), 44, 64; Royle, Ill. Him. Bot., 205 ; Settlement Report:—Panjáb, Peshâwar, 13; Delhi, 27 ; App. XXV., ccii.; Hasára, 12; Lahore, 12; Simla, App. II. (H), x'ii. ; Gazetteers:—Panjáb, Siálkote, 11 ; Delhi, 19 ; Amritsar, 4 ; N.-W. P., III., 238 ; Mysore & Coorg, I., 68 ; Agri-Horti. Soc., Ind.:—Trans., I., 65 ; II., 254, 256 (App.), 310 ; III., 103 (Pro.), 228 ; IV., 106, 145, 148, 149 ; VII. (Pro.), 150, 151, 184; Journ. (Old Series), II., 253, 506 ; Smith, Ec. Dict., 68, 330.*

Habitat.—**Var** insititia, the common yellow bullace, is wild in the south of Europe, Cilicia, Armenia, and to the south of the Caucasus, and is believed by Thomson to be also truly wild in the Western Temperate Himálaya, where it is very commonly met with from Garhwál to Kashmír between 5,000 and 7,000 feet. A dark blue damson, probably **var. domestica,** and an orange red larger one, are said by Madden to be cultivated about Almorah ; they are fairly common throughout the North-West Himálaya. The fruit succeeds admirably in all these localities. It is also cultivated in gardens and orchards in the plains near large stations as far south at least as Saharanpur, but is much less successful than in the hills. (*Conf.* with remark Vol. III., 449.)

Gum.—The plum tree yields a yellow gum of little value, which somewhat resembles gum-arabic.

GUM.
1305
OIL.
Kernels.
1306

Oil.—An OIL prepared from the KERNELS, the "Plum Oil" of Europe, is used for illuminating purposes. It is of little value, and rapidly becomes rancid.

Medicine.—The Bokhara PLUM, which Sir J. D. Hooker believes to be referable to the same variety as the Yellow Bullace, is commonly met with in Indian bazárs, and in Native Materia Medica supplies the place of the Prune in Europe. According to Dymock it "may be made use of in the preparation of the confection of Senna and for any other purpose to which prunes are applicable." The same author informs us that it is considered the best plum for medicinal purposes by Persian writers, who describe it as sub-acid, cold, and moist, digestive and aperient, especially when taken on an empty stomach, and useful in bilious states of the system and heat of the body. The ROOT is considered astringent, and the GUM a substitute for gum-arabic. As met with in commerce the Bokhara plum is about the size and shape of the dry prune of Europe, but is of lighter colour, the skin having been removed ; it is very acid, but on the addition of a little sugar, the taste is agreeable and refreshing (*Mat. Med. W. Ind.*).

MEDICINE.
Plum.
1307

Root.
1308
Gum.
1309

P. 1309

PRUNUS Padus.	The Plum and Alucha.

MEDICINE.
Chemistry.
1310

CHEMICAL COMPOSITION.—The dried fruit has recently been subjected to analytical examination by the authors of the *Pharmacographia Indica,* who find it to have the following percentage composition:—Moisture in vacuo over sulphuric acid, 6·24; ash, 3·39; extractive matter soluble in boiling water, 74·10; ash in extractive matter, 4·58; principles precipitated by absolute alcohol from aqueous extract, 12·68; ash in absolute alcohol precipitate, ·226; saccharine matter possessing a reducing action on alkaline copper solution, without previous ebullition with acids, 44·63; total free and combined citric acid, 3·05; total free and combined malic acid, 1·98. The total alkalinity of the ash expressed as KHO was equal to 61·75 per cent., calculated on the ash, while the total free acidity of the fruit expressed as NaHO was equivalent to 3·8 per cent.

SPECIAL OPINIONS.—§ "Digestive and aperient. Infusion of fruit useful in fever" (*Assistant Surgeon Nehal Sing, Saharanpore*). "Retained in mouth to allay thirst or parched condition" (*Surgeon-Major Robb, Civil Surgeon, Ahmedabad*). "Used to allay thirst in fever" (*Surgeon A. C. Mookerji, Noakhally*). "A cooling laxative" (*Civil Surgeon C.M. Russell, Sarun*).

FOOD.
Fruit.
1311

Food.—The FRUIT, when ripe, is large, yellow, sweet, and juicy. It is eaten by all classes and much esteemed. The dried fruit of the Bokhara plum is imported from Afghánistán in large quantities, and is much used as an article of diet, being stewed or otherwise prepared by Europeans, and eaten in various ways by Natives. It forms an ingredient of a common *chatni.*

TIMBER.
1312

Structure of the Wood.—Reddish-brown, hard, very close-grained, but apt to warp and split; weight 52℔ per cubic foot. It is smooth to work, and is used in Kashmír for the skeleton of the so-called papier maché boxes (*Gamble*).

1313

Prunus Mahaleb, *Linn.; Fl. Br. Ind., II., 312.*

Vern.—*Mahalib,* SIND; *Gávala* or *gahula,* BOMB.; *Priyunger,* SANS.; *Mahalib,* ARAB.; *Paiwand-i-miryam,* PERS.

References.—*Sakharam Arjun, Cat. Bomb. Drugs, 32, 218; Dymock, Mat. Med. W. Ind., 2nd Ed., 295; Dymock, Warden & Hooper, Pharmacog. Ind., I., 567; Trans., Med. & Phys. Soc. Bomb. (New Series), No. iii., 148.*

Habitat.—A small tree, native of Central Asia and Europe, cultivated in Baluchistán, and probably in the North-West Frontier.

MEDICINE
Kernels.
1314
DOMESTIC.
Kernels.
1315

Medicine.—The scented KERNELS, sold in the bazárs of North-West India, serve in Native practice as a substitute for prussic acid.

Domestic.—"The KERNELS, strung as necklace, are much valued by the women of Sind" (*Trans., Med. & Phys. Soc., Bomb.*).

1316

P. Padus, *Linn.; Fl. Br. Ind., II., 315.*

THE BIRD CHERRY.

Syn.—CERASUS CORNUTA, *Wall.*

Vern.—*Jamana,* HIND.; *Likh-aru, arupatti,* NEPAL; *Hlo sa hlot-kúng,* LEPCHA; *Jámana, bombaksing, bombali,* KUMAON; *Jáman, samb-chule,* KASHMIR; *Páras, kálakát, gidar-dák, bart, zúm, zam, sambu, jamú, chúle, krún, dúdla, jamún, jammú, jamna,* PB.

References.—*Brandis, For. Fl., 194; Gamble, Man. Timb., 163; Stewart, Pb. Pl., 83; Baden Powell, Pb. Pr., 346, 574, 591, 594; Atkinson, Him. Dist., 309, 712; Ec. Prod., N.-W. P., Pt. V., 44, 65; Gazetteers:— Hasára, 14; Gurdáspur, 55; Rawalpindi, 15; Simla, 11; Ind Forester, IV., 91; V., 184, 186; XI., 3; XIII., 66; Agri.-Horti. Soc., Ind., Journ., XIII., 385; XIV., 18.*

Habitat.—A small deciduous tree, native of the Temperate Himálaya from Murrí, at altitudes of 6,000 to 9,500 feet, to Sikkim between 8,000 and 12,000 feet, and Bhotán; distributed westward to Great Britain, northward to Siberia. The specific name **cornuata**, given to it by Wallich, arose

P. 1316

Products of India. 349

The Peach and Nectarine. (*J. Murray.*) | PRUNUS persica.

from the fact that the fruit is frequently punctured by an insect, and consequently developes into a horn-like irregular mass.

Gum.—The tree yields an inferior gum in small quantities.

Oil.—Like most others of the genus, this species yields, by expression from the KERNELS, a poisonous oil, like oil of bitter almonds, which may be used medicinally.

Medicine.—See above.

Food.—The FRUIT, though mawkish, astringent, and not attractive to European taste, is eaten by Natives, and, according to Atkinson, may be employed for making liqueurs. The LEAVES form an excellent fodder for cattle.

Structure of the Wood.—Sapwood large, whitish; heartwood reddish-brown, with an unpleasant smell, beautifully mottled, and moderately hard; average weight 41℔ per cubic foot. It is little valued by Natives, but is occasionally used for making railings, agricultural implements, and spoons. Gamble remarks, however, that it has often a very handsome grain and deserves to be better known.

Prunus persica, *Benth. & Hook. f.; Fl. Br. Ind., II., 313.*

THE PEACH AND NECTARINE.

Syn.—AMYGDALUS PERSICA, *Linn.;* A. COLLINUS, *Wall.;* PERSICA SALIGNA, *Royle;* P. VULGARIS, *Miller.*

Vern.—*Arú,* HIND.; *Takpo,* LEPCHA; *Aru, rek,* N.-W. P.; *Arú, súnnú, tsúnú, arúi, chimnánú, aor, bem beimi, bembi, katherti,* NECTARINE=*shaftálú, múndla, árú,* PB.; *Mandata, mandala, sháftálú, ghwareshtai, ghargashtái,* PUSHTU; *Aru,* RAJ.; *Arú, shúftálú,* PERS.

References.—*Roxb., Fl. Ind., Ed. C.B.C., 403; Brandis, For. Fl., 191; Kurz, For. Fl. Burm., I., 433; Dalz. & Gibs., Bomb. Fl., Suppl., 32; Stewart, Pb. Pl., 78; Rept. Pl. Coll. Afgh. Del. Com., 62; DC., Origin Cult. Pl., 221; Boiss., Fl. Orient., II., 640; Ainslie, Mat. Ind., I., 299; O'Shaughnessy, Beng. Dispens., 28; Baden Powell, Pb. Pr., 567; Atkinson, Him. Dist., 711; Ec. Prod., N.-W. P., Pt V., 44, 62; Lisboa, U. Pl. Bomb., 155; Royle, Ill. Him. Bot., 204; Wardle, Dye Rept., 29, 44; Strettell, Report on Ficus elastica in Burma, 120; Gazetteers:—N.-W. P., III., 238; IV., lxxi.; Panjáb, Siálkot, 11; Déra Ismail Khán, 19; Settle. Reps., Delhi, 27; Hazára, 94; Kohát, 30; Simla, App. II. (H.), xlii.; Lahore, 12; Peshawar, 13; Agri.-Horti. Soc., Ind.:—Trans., I., 64, 65; II., 13-16, 96, 254-256 (App.), 208, 307; III., 58, 66-68, 70, 95, 103; IV., 106, 141, 145, 150, 232-235; VI. (Pro.), 60; VII. (Pro.), 94, 95, 184; VIII. (Pro.), 389, 390; Journals (Old Series), IV., 212; VI., 241, 242, Sel., 112 (Pro.), 121; VIII., 241; X. (Pro.), 40; XI., Sel., 31; XIII., 385; (New Series), V. (Pro. 1876), 25, 26; Agri.-Horti. Soc., Panjáb, Select Papers to 1862, 46; Ind. Forester, VI., 240; XII., App. 27; XIV., 391.*

Habitat.—The peach, according to DeCandolle (who devotes many pages to an interesting discussion of the arguments for and against his theory), had its original habitat in China, where it was cultivated two thousand years before its introduction into the Graeco-Roman world, and a thousand years before it appears to have been known in the lands of the Sanskrit-speaking race. It is, however, now widely spread all over the North-West Himálaya, and, according to Dr. Watt, has a greater claim to be considered indigenous than any of the fruit trees belonging to the genus, except perhaps the cherry. It occurs near almost every village up to 10,000 feet, though frequently entirely neglected, and not utilised in any way. It is found in the same naturalised state, in the cooler parts of the plains of Northern India, on the North-West Frontier, in Afghánistán, North Persia, and in Trans-Caucasia. It was introduced into Europe from Persia about the commencement of the Christian era, and now grows well as far north as Great Britain.

Margin notes:
GUM. 1317
OIL. Kernels. 1318
FOOD. Fruit. 1319
Leaves. 1320
TIMBER. 1321
1322

The Peach and Nectarine.

In India it is extensively cultivated in the neighbourhood of European centres throughout the plains, and, even in the neighbourhood of Calcutta, produces excellent fruit. It is also cultivated in Southern India, Ceylon, and Burma, though Strettell informs us that in the latter country it is valued only for its flower, the fruit never being eaten by the Burmans. The tree flowers, according to elevation, from January to May, and the fruit ripens between May and October. (*Conf.* with remark Vol. III., 445, also 449.)

GUM.
1323

Gum.—The peach, like other species of **Prunus**, yields small quantities of an unimportant gum.

DYE.
Root-bark.
1324

Dye.—Samples of the ROOT-BARK were sent to **Mr. Wardle** as a dye-stuff employed in India; the writer can find no record of where, or how, it is utilised. **Mr. Wardle** reports as follows, " A likely-looking, colour-producing material, does not possess any more colouring matter than most of the Indian barks sent to be examined; in fact they all, with a few exceptions, give a peculiar reddish fawn tint. This bark does not possess much tannin which may be seen by its not producing much colour with an iron salt." The colours obtained with silk varied from fawn to brown, while with wool a rich golden fawn was produced.

OIL.
Kernels.
1325

Oil.—An oil is obtained, from the KERNELS, resembling that of bitter almonds, for which it may be substituted. It is used by the Natives of the North-West Himálaya and Kashmír for cookery, for illuminating purposes, and as a dressing for the hair.

MEDICINE.
Flowers.
1326
Leaves.
1327
Kernels.
1328
Fruit.
1329

Medicine.—The FLOWERS are purgative, the LEAVES and KERNELS yield prussic acid on distillation, the FRUIT is demulcent, antiscorbutic, stomachic, and is believed by the Natives to have the property of killing the intestinal worm *kenchwa* (**Ascaris lumbricoides**). **Vigne** states that in Ladak " foot-rot " in sheep is treated by a decoction of the leaves.

Food. – See Fruits, Vol. III., 449.

TIMBER.
1330

Structure of the Wood.—Heartwood brown, compact, even-grained and smooth to work. The timber of trees past bearing is used for building and other work (*Brandis*).

Prunus prostrata, *Labill.; Fl. Br. Ind., II., 313.*

1331

Vern.—*Paltú, tára, ter, talle,* Pb.

References.—*Brandis, For. Fl., 193; Stewart, Pb. Pl., 33; Aitchison, Rept. Pl. Coll. Afgh. Del. Comm., 60, 61.*

Habitat.—A scraggy shrub from 5 to 6 feet in height, met with on the Temperate Himálaya, from the Sutlej westwards, at altitudes from 5,000 to 10,000 feet; distributed westwards to Spain, ascending to 12,000 feet in Afghánistán.

FOOD.
Fruit.
1332

Food.—Produces a small ovoid FRUIT, reddish purple when ripe, with scanty juicy pulp. It is by no means palatable, but is eaten by Natives.

P. Puddum, *Roxb.; Fl. Br. Ind., II., 314.*

1333

THE WILD CHERRY OF THE HIMÁLAYA.

Syn.—P. CERASOIDES, *Don;* P. SYLVATICA, *Roxb.;* CERASUS PUDDUM, *Wall.;* C. PHOSHIA, *Ham.*

Vern.—*Paddam, páya, padma-káshtha,* HIND.; *Kongki,* LEPCHA; *Púya, páiya, padam, paddam,* KUMAON; *Chamiári, amalgúch, pájá, pajia, pajja, paddam,* PB.; *Padma-kasta,* BOMB.; *Padma kastha, padmaka,* MAR.; *Padma kathi, padmak,* GUZ.; *Padmaka, padmaksh,* SANS.

References.—*Roxb., Fl. Ind., Ed. C.B.C., 403; Brandis, For. Fl., 194; Kurz, For. Fl. Burm., I., 434; Gamble, Man. Timb., 163; Stewart, Pb. Pl., 83; Dymock, Mat. Med. W. Ind., 2nd Ed., 295; S. Arjun, Bomb. Drugs, 52; Irvine, Mat. Med., Patna, 83; Dymock, Warden & Hooper, Pharmacog. Ind., I., 567; Baden Powell, Pb. Pr., 574; Atkinson, Him. Dist., 309, 712; Ec. Prod., N.-W. P., Pt. I., 3, 11; V., 44, 64; P. W. D, Rept. Gums & Gum-resins, 15; Cooke, Gums & Gum-resins, 23; Aplin, Report on the Shan States; Ind. Forester VIII., 404; XI., 355; Agri.-Horti. Soc., Ind., Journ., VI. (Pro.), 128; XIV., 31.*

P. 1333

Himalayan Wild Cherry: The Guava. (*J. Murray.*)	PSIDIUM Guyava.

Habitat.—A moderate-sized (in Sikkim, a large) deciduous tree, wild on the Temperate Himálaya, from Garhwál, at altitudes of 3,000 to 6,000 feet, to Sikkim and Bhotán, from 5,000 to 8,000 feet; also met with in Burma. (*Conf.* with remark Vol. III., 449.)

Gum.—The tree yields a gum similar to that obtained from other members of the genus, and of little value.

GUM.
1334

Oil.—The KERNELS contain an oil similar to that of bitter almonds, and with a strong flavour of prussic acid.

OIL.
Kernels.
1335

Medicine.—The BARK contains amygdalin, and the smaller branches are sold in the bazars as substitutes for hydrocyanic acid in native practice The KERNEL is said to be employed as a remedy for stone and gravel.

MEDICINE.
Bark.
1336
Kernel.
1337

Food.—The tree flowers in October-November, and produces an abundance of small oblong FRUIT, with a scanty, yellow, orange, or reddish pulp in spring. It is little, if at all, eaten by the Natives, but is largely collected and sold to Europeans for making the well-known hill cherry-brandy.

FOOD.
Fruit.
1338

Structure of the Wood—Sapwood large, greenish-white; heartwood reddish, beautifully mottled; weight 44℔ per cubic foot. In the Panjáb Himálaya it is employed for making walking-sticks and pipe-stalks, also occasionally for building and for making furniture. Gamble remarks, "It deserves to be better known and to be more extensively used, as, at any rate in Sikkim, it is common and reaches a large size"

TIMBER.
1339

PSAMMOGETON, *Edgw.; Gen. Pl., I., 929.*

[UMBELLIFERÆ·

Psammogeton biternatum, *Edgw. ; Fl. Br. Ind., II., 719 ;*

1340

Syn.—P. CRINITUM, *Boiss.* ; PIMPINELLA CRINITUM, *Boiss.*
Vern.—*Gar gira,* PUSHTU.
References.—*Boiss., Fl. Orient., II., 1078 ; Aitchison, Bot. Afgh. Del. Com., 72; Notes, Ec. Pl., Baluchistán, No. 53; Murray, Pl. & Drugs, Sind., 201 ; Gazetteer, N.-W. P., IV., lxxii.*

Habitat.—A small pubescent annual, frequent in the Panjáb and Sind, ascending the Himálaya to 3,000 feet.

Medicine.—Reputed stomachic in Sind (*Murray*). The FRUIT of the allied species, P. setifolium, *Boiss., Fl. Or., II., 1079 (Kara-bia, N.-E. AFGHÁNISTÁN,)* is largely collected and employed in native medicine in the Hari-rud valley (*Aitchison*).

MEDICINE.
Fruit.
1341

PSEUDOSTACHYUM, *Munro ; Gen. Pl., III., 1212.*

1342

A genus of large, nearly scandent, tufted bamboos, which belongs to the Natural Order GRAMINEÆ, and comprises three Indian species which are ever-green, except after they have flowered when they die off. Kurz describes two (*For. Fl. Burm., II., 567*) ;—P. compactiflorum, *Kurz*, a bamboo which grows on the drier hill forests of Martaban, east of Tounghoo, at elevations between 4,000 and 6,000 feet; and P. helferi, *Kurz*, a gregarious species met with in the tropical and moister upper mixe l forests of the Pegu Yomah and Martaban, up to 3,000 feet. The other, P. polymorphum, *Munro* (*Gamble, Man. Timb., 429*) ; (*Purphiok, paphok,* LEPCHA ; *Filing,* NEP.), is a small bamboo found in Sikkim from 4,000 to 6,000 feet, and in Assam. The soft green stems are used for baskets, mats, &c., and for tying the rafters, &c., of Native houses and huts (*Conf.* Bamboo, *Vol. I.,* 370-389).

PSIDIUM, *Linn.; Gen. Pl., I., 713.*

1343

Psidium Guyava, *Linn.; Fl. Br. Ind., II., 468 ;* MYRTACEÆ.
THE GUAVA.
Var. pyriferum, *Linn.,* sp.
Var. pomiferum, *Linn.,* sp.

PSIDIUM Guyava.	The Guava.

Vern.—*Amrút, amrúd, ám,* HIND. ; *Peyara, piyárá, góaáchhi-phal,* BENG.;
Madhuriam, múhuriam, ASSAM ; *Amuk,* NEPAL ; *Gaya,* MAGH. ; *Amrúd,
piyára,* N.-W. P.; *Amrút, amrúd, anjír sard,* PB.; *Amrút,* RAJ.;
Zaitun,* SIND ; *Perala, peru,* BOMB.; *Jámba, túp-kél,* MAR.; *Piyára,
péru, jamrúd, jamrukh,* GUZ.; *Guava,†* jám, DEC.; *Segápu, koaya,
koyya, goyyá-pasham,* TAM.; *Jama, cova, jám-pandu, goyyá-pandu,*
TEL.; *Sebe, shibé-hannu, shépe,* KAN.; *Pela, péra, pérakka, malák-kop-
péra,* MALAY.; *Málaká beng, málaká,* BURM.; *Péra, péra-gadi,* SING.;
Amruta-phalam, bahu-bíja-phalam, SANS.; *Amrúd,†* ARAB.; *Amrúd,†*
PERS.

* **Stocks** informs us that this name, though applied to the Guava, strictly signifies the olive.

† **Moodeen Sheriff** remarks *amrúd* is a proper Arabic and Persian name for
the Pear, but in India is often used for the Guava. *Jám* is applied to Guava in
Dukhni, but in Bengali to the fruit of **Eugenia jambolana.**

References.—*Roxb., Fl. Ind., Ed. C.B.C., 396; Brandis, For. Fl., 232;
Kurz, For. Fl. Burm., I., 476; Gamble, Man. Timb., 190; Dals. & Gibs.,
Bomb. Fl., Suppl., 34; Stewart, Pb. Pl., 94; DC., Orig. Cult. Pl., 241;
Mason, Burma & Its People, 448, 744; Sir W. Elliot, Fl. Andhr., 72;
Rheede, Hort. Mal., III., t. 34; Rumphius, Amb., I., 121, t. 37; Pharm.
Ind., 92; O'Shaughnessy, Beng. Dispens., 337; Moodeen Sheriff, Supp.
Pharm. Ind., 206; Sakharam Arjun, Cat. Bomb. Drugs, 211; Murray,
Pl. & Drugs, Sind, 192; Bidie, Cat. Raw Pr., Paris Exhib., 29, 30, 80;
Dymock, Mat. Med. W. Ind., 2nd Ed., 334; Official Corresp. on Proposed
New Pharm. Ind., 238; Birdwood, Bomb. Prod., 154; Baden Powell,
Pb. Pr., 349, 591; Drury, U. Pl. Ind., 355, 356; Useful Pl. Bomb.
(XXV., Bomb. Gaz.), 79, 156; Econ. Prod., N.-W. Prov., Pt. III. (Dyes
& Tans), 84; Pt. V. (Vegetables, Spices, & Fruits), 44, 73, 74; Royle,
Prod. Res., 40; Liotard, Dyes, 116; Darrah, Note on Cotton in Assam,
30; Stocks, Rep. on Sind; Note on the Condition of the People of Assam,
App. D; Boswell, Man., Nellore, 97; Moore, Man., Trichinopoly, 79;
Bombay, Man. Rev. Accts., 102; Settlement Reports :—Panjáb, Lahore,
12; Simla, App. xlii.; Delhi, 27; Gujerát, 135; Amritsar, 4; N.-W. P.,
Allahabad, 38; Central Provinces, Chánda, 82; Nimár, 200; Port Blair,
33; Gazetteers :—Bombay, V., 25; VII., 39, 40; VIII., 184; XV., 74;
XVI., 103; XVIII., 50; N.-W. P., I., 81; III., 33, 238; IV., lxxii.;
Orissa, II., 180, 188; Mysore & Coorg, I., 53, 61; III., 18, 48; Agri.-
Horti. Soc., Ind.:—Trans., II., 208, App., 307; III., 39, 68, 69; IV.,
105, 149, 234, Pro., 229; VI., 235; VII., 60, 63, Pro., 95, 160; Journals
(Old Series), IV., 202; VI., 37; IX., 423; Ind. Forester, VI., 301;
XII., App., 28; Smith, Ec. Dict., 201.*

Habitat.—A small tree, originally a native of the country, extending
from Mexico to Columbia and Peru, possibly including Brazil, before the
discovery of America (*DeCandolle*), now naturalised and largely cultivated throughout India. It was probably introduced into this country by
the Portuguese (*Royle*). Linnæus and some later botanists admitted two
species, the one with elliptical or spherical fruit, **P. pomiferum**; the other
with a pyriform fruit, and white or pink flesh, **P. pyriferum.** These are
now considered to be simply varieties of one species. The trees are raised
from seed, and planted out when three or four years old. They bear fruit
from the second or third year after planting, and continue to yield for
about six or seven years, after which the tree is destroyed. The fruit requires to be protected whilst on the tree from rats and squirrels by which
it is greedily eaten.

DYE & TAN.
Leaves.
1344
Bark.
1345

Dye & Tan.—In Assam the LEAVES and BARK are used for dyeing.
The latter, boiled with the bark of **Terminalia citrina, Mangifera indica,**
and a species of **Eugenia,** produces a black colour in yarn or cloth
(*Darrah*). In the North-West Provinces and Bengal the leaves are
occasionally employed in tanning by the poorer classes, either alone or
along with mango or *mahua* leaves.

P. 1345

| The Guava. | (*J. Murray.*) | SORALEA corylifolia. |

Medicine.—The FRUIT is astringent, and is employed by Natives as a cure for diarrhœa. The BARK, especially that of the ROOT, is noticed in the *Pharmacopœia of India* as an astringent worthy of notice. Dr. Waitz (*Dis. of Children in Hot Climates*, 225) states that he employed it with much success in chronic infantile diarrhœa. He administered it as a decoction (½ oz. of the root-bark, with 6 ozs. water, boiled down to 3 ozs.) in doses of one or more teaspoonfuls three times a day. He also recommends this preparation as a local application in the prolapsus ani of children. Many other writers have noticed the astringent properties of the bark, which appears to be worthy of more extended utilisation. The LEAVES possess similar qualities, though perhaps to a less marked extent, and are also said, when pounded down, to make an excellent poultice. In cases of scurvy, a decoction is employed as a mouth-wash for swollen gums, and has been used in cholera with some success, for arresting the vomiting and diarrhœaic symptoms.

SPECIAL OPINIONS.—§ " The unripe fruit has been found serviceable in diarrhœa" (*Surgeon R. Gray, Lahore.*). " As a poultice, leaves are useful in unhealthy ulcers" (*Surgeon A. C. Mukerji, Noakhally*). " While the bark and leaves act as an astringent, the ripe fruit is generally used as a good aperient" (*Assistant Surgeon N. L. Ghose, Bankipore*). "A decoction of the root-bark, and of the young leaves, is a useful astringent in chronic diarrhœa complicated with dyspepsia. It may be specially recommended for charitable dispensary practice, because it can be obtained without cost in any part of Bengal" (*Civil Surgeon S. M. Shircore, Murshedabad*). " The young leaves with the bud of the pomegranate and *bábul* leaves, given in the form of a cold infusion, are useful in the diarrhœa of children. A decoction of the unripe fruit is used with benefit in the diarrhœa of adults" '*Hospital Assistant Lal Mahomed, Hoshangabad*).

Food.—The FRIUT is eaten by all classes, and is a special favourite with Natives who like its strong aromatic flavour. By Europeans it is generally preferred stewed, and in the form of jelly, or of the well-known " Guava cheese." The fruit of more carefully cultivated trees has a thin, bright yellow rind, and a pulpy cream-like or reddish coloured flesh with a pleasantly acid sweet taste.

Structure of the Wood.—Whitish, moderately hard, even-grained ; weight about 42℔ per cubic foot. It is said to work well and smoothly and to be used for wood-engraving, spear handles, and instruments of various kinds (*Gamble*).

PSORALEA, *Linn.* ; *Gen. Pl., I.*, 491.

Psoralea corylifolia, *Linn.* ; *Fl. Br. Ind., II.*, 103 ; LEGUMINOSÆ.

Syn.—TRIFOLIUM UNIFOLIUM, *Forsk.*

Vern.—*Bhávanj, bukchi, babchi, bábachi, bávanchiyán, bávchiyán, bávanchi*, HIND. ; *Latakasturi, hakúch, bávachi*, BENG. ; *Bákuchi*, URIYA ; *Bábchi*, PB. ; *Bawachi*, BOMB. ; *Bávachya, bavachi*, MAR. ; *Babchi, bawachi, bávachá*, GUZ. ; *Kárpó-karishi, kárpuvá-arishi*, TAM. ; *Kala ginja, bhávanji, korjáshtam, karú-bógi, kálu-gechcha*, TEL. ; *Sugandha kantak, avalguja, vákuchi, sóma rája*, SANS.

References.—*Roxb., Fl. Ind., Ed. C.B.C.*, 588 ; *Dals. & Gibs., Bomb. Fl.*, 60 ; *Stewart, Pb. Pl.* 75 ; *Sir W. Elliot, Fl. Andhr.*, 25, 79, 99 ; *Burmann, Fl. Ind.*, 172, *t.* 49, *fig .2* ; *Ainslie, Mat. Ind., II.*, 141 ; *O'Shaughnessy, Beng. Dispens.*,, 316 ; *Moodeen Sheriff, Supp. Pharm. Ind.*, 208 ; *Dymock, Mat. Med. W. Ind.*, 2nd Ed., 216 ; *K. L. Dey, Indig. Drugs, Ind.*, 97 ; *Dymock, Warden & Hooper, Pharmacog. Ind.*, *I.*, 412 ; *Cat. Baroda Durbar, Col. & Ind. Exhib., No.* 155 ; *Baden Powell, Pb. Pr.*, 339 ; *Drury, U. Pl.*, 357 ; *Smith, Dic.*, 41 ; *Gazetteers :—Bombay, VI.*, 15 ; *XV.*, 431 ; *N.-W. P., I.*, 80 ; *Orissa, II.*, 159 ; *Agri.-*

MEDICINE.
Fruit.
1346
Bark.
1347
Root.
1348
Leaves.
1349

FOOD.
Fruit.
1350

TIMBER.
1351

| **1352** |

PSYCHOTRIA. A Tonic Nervine.

OIL.
Seeds.
1353
MEDICINE.
Seeds.
1354

Horti. Soc., Ind., Journals (Old Series), IX., 414; Ind. Forester, VI., 240.

Habitat.—A common herbaceous weed found in the plains, from the Himálaya through India to Ceylon.

Oil.—Dymock states that in the Konkan the SEEDS are used for making a perfumed oil which is applied to the skin.

Medicine.—The SEEDS are described by certain authors as hot and dry, by others as cold and dry, laxative, stimulant, and aphrodisiac. "They are recommended in leprosy and other skin diseases which depend upon a vitiated state of the blood, and are given internally and applied externally as a plaster, whence the synonym *Kushtanásini*; they are also said to be useful in febrile bilious affections and as an anthelmintic and diuretic. The Hindus class them with their *rasáyán*, or alchemic drugs" (*Pharmacog. Ind.*) Ainslie states that they are used in South India as stomachic and deobstruent, and are prescribed in cases of lepra and other inveterate cutaneous affections. Of recent years they have been extensively tried by various practitioners as a remedy for leprosy with a certain amount of success (*Dymock*).

Dr. Kanai Lall De in particular strongly recommends an oleo-resinous extract of the seeds, diluted with simple ointment, as an application in cases of leucoderma. "After applications for some days," he writes, "the white patches appear to become red or vascular; sometimes a slightly painful sensation is felt. Occasionally some small vesicles or pimples appear, and if these be allowed to remain undisturbed, they dry up, leaving a dark spot of pigmentary matter, which forms, as it were, a nucleus. From this point, as well as from the margin of the patch, pigmentary matters gradually develope, which ultimately coalesce with each other, and thus the whole patch disappears. It is also remarkable that the appearance of fresh patches is arrested by its application" (*Pharm. Jour., Sept. 24th, 1881*). The authors of the *Pharmacographia Indica*, commenting on this passage, state that, in the hands of later investigators, only negative results have been obtained by this mode of treatment.

"Several species of **Psoralea** have been used medicinally in America, and have been found to act as gentle, stimulating, and tonic nervines" (*Dymock*).

CHEMICAL COMPOSITION.—A long account is given in the *Pharmacographia Indica*, of a recent analysis of the seeds. It is worthy of notice that one of the constituents separated was a colourless oil lighter than water, and possessing in a marked degree the odour of the seeds, which was obtained on distillation with water.

SPECIAL OPINION.—§ "In cases of leucoderma it does good temporarily, but the effect is not lasting" (*Assistant Surgeon Nehal Sing, Saharanpur*).

Psoralea plicata, *Delile; Fl. Br. Ind., II., 103.*

Vern.—*Bakhtmal,* PB.

Reference.—*Stewart, Pb. Pl., 75.*

Habitat.—A low, much-branched shrub found in the arid plains of the Panjáb; distributed to Arabia, Egypt, and Tropical Africa.

Fodder.—Edgeworth states that camels are very fond of the PLANT.

PSYCHOTRIA, *Linn.; Gen. Pl., II., 123.*

A genus of shrubs or small trees, rarely herbs, which belongs to the Natural Order RUBIACEÆ, and comprises about 500 species, all tropical or sub-tropical. Of these the only one of economic importance is **P. emetica**, *Mutis*, a native of New Granada, the root of which is used as a substitute for ipecacuanha. Though none of our fifty to sixty Indian species are known to possess roots with emetic properties, it appears probable that some of them might do so. The subject is, at all events, worthy of investigation.

PTERIS, *Linn.; Syn. Fil., Hook. & Baker, 153.*
[FILICES.   **1359**

Pteris aquilina, *Linn.; Beddome, Ferns of British India 115;*
BRAKE or BRACKEN.

Vern.—*Kakhash, kakei, lungar, dio,* PB.

References.—*Stewàrt, Pb. Pl., 266 ; Mason, Burma & Its People, 441 ; 821, 827 ; Atkinson, Him. Dist., 321 ; Aplin, Report on the Shan States ; Agri.-Horti. Soc., Ind., Journals (New Series), VI., Sel., 64 ; VII., Pro. (1885), 148 ; Smith, Ec. Dict., 58.*

Habitat.—A fern, abundant in the Panjáb Himálaya.

FOOD. Stems. 1360

Food.—The underground STEMS contain a quantity of mucilage and starch, which, on being prepared by washing and pounding, and mixed with meal, make bread in times of scarcity. Even in England attempts have been made to use it as food. **Dr. Clark** considered it a wholesome table vegetable when young and blanched like asparagus, but it is rather astringent. The FRONDS, in quite a young state, are eaten, cooked as a pot-herb, and are juicy, though rather insipid.

Fronds. 1361 DOMESTIC. 1362

Domestic.—Bracken is extensively used by coffee planters in the Nilghiris, Wynaad, and Coorg, as a litter for cattle, horses, &c. The "box manure," thus formed, is said to be very valuable for coffee trees. **Professor Wolff** found the ash of the fern to contain 42·8 per cent. of potash, 14 of lime, 10·2 of chlorine, 9·7 of phosphoric acid, 7·7 of magnesia, 6 of silica, 5·1 of sulphuric acid, and 4·5 of soda. The total percentage of ash was 7·01. A consideration of this analysis shows that more than one half of the ash consists of the highly fertilizing minerals, phosphoric acid and potash, both very essential constituents of soil on which coffee trees are growing.

PTEROCARPUS, *Linn.; Gen. Pl., I., 547.*

Pterocarpus indicus, *Willd.; Fl. Br. Ind., II., 238;* LEGUMINOSÆ.
THE ANDAMAN REDWOOD, or PADAUK.

1363

Syn.—P. DALBERGIOIDES, *Roxb.* ; P. WALLICHII, *W. & A.*
Vern.—*Erra végisa,* TEL.; *Padauk, toung-kha-yai,* BURM.; *Chalangada,* AND.

References.—*Roxb., Fl. Ind., Ed. C.B.C., 537 ; Brandis, For. Fl., 153 ; Kurz, For. Fl. Burm., I., 349 ; Beddome, Fl. Sylv., t. 23 ; Gamble, Man. Timb., 130 ; Mason, Burma & Its People, 405, 484, 485, 531, 769 ; Elliot, Fl. Andhr., 54 ; Cooke, Gums & Gum-resins, 37 ; P. W. D. Rept. on Gums & Resinous Prods., 15 ; Aplin, Report on the Shan States ; Smith, Dic., 356 ; Ind. Forester, IV., 292, 411, 425 ; V., 497, VIII., 29, 403, 414 ; X., 465, 532 ; XI., 490 ; XII., 85 ; XIV., 373 ; Agri.-Horti. Soc., Ind., Journals (Old Series), IV., 55 ; V., Sel., 145 ; VII., 73 ; IX., Sel., 43, 51 ; XI., 447 ; XII., Pro., 77 ; XIV., Pro., 69.*

Habitat.—A large and lofty tree found in Burma and the Andaman Islands.

GUM. 1364

Gum.—According to **Mason** this tree produces a gum almost identical in characters with kino, which has long been known in the trade, since he states that, "long before **Dr. Royle** had written his valuable article on gum-kino, more than one consignment had been made by parties in Moulmein, to houses in London, of gum-kino to the amount of a thousand pounds." He then carefully discusses the identification of the *padauk*, and arrives at the conclusion that it is **Pterocarpus Wallichii**, *Wight,* while a slightly different form is **P. dalbergioides**, *Roxb.,* two forms now united into one, the species under consideration. **Mason** then goes on to

23 A

PTEROCARPUS indicus.   The Andaman Redwood—Padauk.

GUM.

state that "these provinces can furnish the commercial world with a large quantity of gum-kino. The exudation of our *padauk*, one of the most abundant forest trees, has been proved by experiment to possess all the qualities of gum-kino, while the product of the neighbouring provinces, whose only avenue to market is through our territories, has been bought by London druggists for the gum-kino of the Pharmacopœia." Subsequent to the date of the above, samples were submitted to Mr. Daniel Hanbury for examination. He reported somewhat unfavourably, stating that the liquid drug sent was intensely black and opaque, and did not give a solution like good kino. The fault in colour may, however, have been due to the sample having been sent in a vessel of tinned iron. Thus, while in the fluid condition, it is quite unsaleable and valueless, there is, nevertheless, little doubt that if carefully dried by solar heat in vessels of earthenware or wood, it is as good a drug as the true Indian kino derived from **Pterocarpus Marsupium.**

MEDICINE.
Gum.
1365
TIMBER.
1366

Medicine.—The GUM might be largely utilised as a substitute for true kino.

Structure of the Wood.—Sapwood small; heartwood dark-red, close-grained, moderately hard to hard, with a slight aromatic scent. It is durable and not attacked by white-ants. When thoroughly seasoned it is almost unaffected by alternate dryness and moisture of the atmosphere. It seasons well, works well, and takes a very fine polish. The weights of specimens from Burma differ considerably from those from the Andamans, the latter being much lighter, the wood softer, and the colour rather paler. The Burmese specimens give from 56 to 81℔ per cubic foot, while those from the Andamans weigh only about 49℔ per cubic foot.

It is used for furniture, carts, gun-carriages, and other purposes, and is said to be the most useful wood in the Andamans, where it grows to an enormous size. Major Protheroe describes a tree felled in 1876 with a clear stem of 65 feet and a girth of 17 feet, and says that the wood of the root is closer-grained, darker-coloured, and more beautifully marked than that of the stem. A plank sent to the Paris Exhibition of 1878 measured nearly 4 feet across. In London, a portion of the same log from which the plank was cut fetched a price of £17-10s. per ton, or nearly R4 per cubic foot, while three logs lately sold in Calcutta fetched R60 per ton. Furniture made from *Padauk* wood and exhibited at Paris in 1878 by Messrs. Jackson and Graham was much admired. They reported on it as follows :—

"This is a straight-grown wood, with rather a coarse open grain, but without any strong figure or markings. When first cut it is of a reddish-brown colour, but it fades to much the same colour as teak—a wood it resembles very much, and it is about as hard, but much heavier. From the six specimen trees sent us, we imagine that it does not grow to any great size. The largest sent to us measured 16 feet long × 1 foot 7 inches × 1 foot 6 inches. We consider it suitable for all kinds of furniture. We manufactured it into a suite of morning-room furniture, which was exhibited at the Paris Exhibition, and which stood the test of a very hot summer in a most satisfactory manner" (*Gamble*). It may be mentioned that the timber attracted much attention at the Colonial and Indian Exhibition in 1886, where it was considered to be specially suited for cabinet-work and carriage-building. At a conference on the Timbers of the Indian Section Dr. Brandis stated that teak could be profitably sold at from £10 to £15 per load, and that *Padauk* could probably be obtained more cheaply.

Padauk is extremely abundant in the mixed forests from Martaban down to Tenasserim in Burma, while in the Andamans Home's valuation survey gives an average of seven mature trees per acre, so a large quantity

The Indian Kino Tree. (*J. Murray.*)	**PTEROCARPUS Marsupium.**

of the timber might be made available for export, should the demand increase.

Pterocarpus macrocarpus, *Kurz ; Fl. Br. Ind., II., 239.*

1367

Vern.—*Patauk,* BURM.

References.—*Kurz, For. Fl. Burm., I., 349; Gamble, Man. Timb., 130.*

Habitat.—A deciduous tree, frequent in the *In* and upper mixed forests of Martaban and Tenasserim.

Resin.—Yields a red resin, a sort of gum-kino (*Kurz*). This resin may possibly, like that of the preceding species, be occasionally used or even exported as true gum-kino, but no information is available on the subject.

RESIN. 1368

Structure of the Wood.—Sapwood pale brown, streaked, rather light, close-grained (*Kurz*).

TIMBER. 1369

P. Marsupium, *Roxb. ; Fl. Br. Ind., II., 239.*

1370

THE INDIAN KINO TREE.

Syn.—P. BILOBUS, *Roxb.*

Vern.—*Bija, bijasár, bijasál, piasál, pit-shola, banda,* HIND.; *Pít-shál, pit-sál,* BENG.; *Byasa, piasál, bijasa,* URIYA; *Hitun, hid, hed, hilum, paisar,* KOL.; *Murga, banda,* SANTAL; *Peddei, bijo,* GOND; *Radat bera,* BHIL; *Bijasal, biah, bija sah, dhorbenla,* C. P.; *Bijaira,* BIJERAGHO-GARH; *Bibla, bibla, húni, hvni, honne, bibala, bija, piasal, dorbenla, asan,* BOMB.; *Dhorbeulá, ásan, húni, asana,* MAR.; *Bibla, bia,* GUZ.; *Bibla, lewba, bia,* DEC.; *Vengai, yeanga, vengai-maram,* TAM.; *Peddagi, yeanga, yeggi, yegisa, pedéga, pedéi, végisa, égisa, végi, pedyegi,* TEL.; *Benga, honné, bibla, honnemaradubanke,* KAN.; *Karinthagara, vénna, vénna-maram,* MALAY.; *Gummálú, gañmalu,* SING.; *Kati mukki,* ARAB. The GUM=*Hirá-dókhi, khólar-mandá, rang-barat,* HIND.; *Nát-ká-dam-mul-akhvain,* DEC.; *Kándámiruga-mirattam,* TAM.; *Gándámrugam-netturu,* TEL.; *Vénnap-paska,* MALAY.; *Dammul-akhvaine-hindi,* ARAB.; *Khúne-siyávusháne-hindi,* PERS.

References.—*Roxb., Fl. Ind., Ed. C.B.C., 536; Voigt, Hort. Sub. Cal., 242; Brandis, For. Fl., 152; Beddome, Fl. Sylv., t. 21; Gamble, Man, Timb., 132; Dalz. & Gibs., Bomb. Fl., 76; Rev. A. Campbell, Rept. Econ. Pl., Chutia Nagpur, Nos. 9465 9466; Sir W. Elliot, Fl. Andhr., 49, 190; Pharm. Ind., 70; Flück. & Hanb., Pharmacog., 194; Ainslie, Mat. Ind., II., 264; O'Shaughnessy, Beng. Dispens., 299; Medical Topog., Ajm., 131; Moodeen Sheriff, Supp. Pharm. Ind., 208; Murray, Pl. & Drugs, Sind., 128; Dymock, Mat. Med. W. Ind., 2nd Ed., 239, 888; Dymock, Warden & Hooper, Pharmacog. Ind., I., 464; Trans., Med. and Phys. Soc., Bomb. (New Series), No. 12, 173; Birdwood, Bomb. Prod., 30, 266, 311, 329; Drury, U. Pl. Ind., 357; Useful Pl. Bomb. (XXV., Bomb. Gaz.), 62, 251, 394; Econ. Prod., N.-W. Prov., Pt. I. (Gums & Resins), 3; Gums & Resinous Prod. (P. W. Dept. Rept.), 2, 3, 5, 7, 9, 25, 32, 65, 67; Cooke, Gums & Resins, 35; Wardle, Dye Rep., 26; Christy, New Com. Pl., V., 39, 40, 44; Man. Madras Adm., I., 313; II., 64; Nicholson, Man. Coimbatore, 401; Boswell, Man., Nellore, 101; Moore, Man., Trichinopoly, 79; Gribble, Man., Cuddapah, 199, 262; Settlement Report:—Central Provinces, Nimár, 305; Raepore, 75; Chánda, App. VI.; Godávery, 37; Seonee, 10; Chindwára, 110, 111; Bhundara, 18; Gazetteers:—Bombay, III., 199; VII., 32, 35, 39; XIII., 23; XV., 33; XVI., 19; XVII., 24; Central Provinces, 58, 503; Mysore & Coorg, I., 48, 71; III., 21; Agri.-Horti. Soc., Ind.:—Journals (Old Series), IV., Pro., 21; V., 110 113; VIII., Sel., 140; IX., 294, Sel., 51, Pro., 149; XIV., Sel., 166; (New Series), IV., Pro. (1873), 38; VII., 136; Ind. Forester, I., 274; II., 18, 19; III., 23, 189, 201; IV., 292, 322, 366, 411; VI., 104, 304; VIII., 105, 118, 125, 126, 378, 411, 414, 415, 416, 417, 438, 439, 440; IX., 294, 356, 427; X., 33, 222, 325, 326, 547, 550, 552; XI., 230; XII., 85, 188, 313; XIII., 120; XIV., 151.*

Habitat.—A large deciduous tree of Central and South India and Ceylon, extending northwards as far as the Banda District of the North-West Provinces.

PTEROCARPUS Marsupium.	The Indian Kino Tree.

GUM.
1371

Gum.—This species yields the gum kino of European Materia Medica. The gum as it exudes has the appearance of red currant jelly, but hardens in a few hours after exposure to the air. It is obtained as follows, " A perpendicular incision with lateral ones leading into it is made in the trunk, at the foot of which is placed a vessel to receive the outflowing juice. This juice soon thickens, and, when sufficiently dried by exposure to the sun and air, is packed into wooden boxes for exportation " (*Hanbury*, quoting *Cleghorn*). Dymock states that in the Kanara District of Bombay it is collected in little cups made with leaves, and consequently assumes the form of concavo-convex cakes 3 to 4 inches in diameter. These are said to be always broken up and garbled by the wholesale dealers. The supplies are chiefly obtained from the Government Forests in Malabar, permission to extract the drug being granted on payment of a small fee and on the understanding that the tapping be performed skilfully and without damage to the timber. Fluckiger & Hanbury inform us that the introduction of kino into European medicine is due to Fothergill, an eminent physician of the last century. The drug which he examined was brought from the river Gambia in West Africa, as a rare sort of "Dragon's Blood," and was described by him in 1757 under the name of *Gummi rubrum astringens Gambiense.* Twenty years before it had been noticed by Moore, factor to the Royal African Company, who says that the tree yielding it is called in the Mandingo language *Kano.* In 1805 specimens of the tree were sent to England by Mungo Park, and were recognised some years later as belonging to the species Pterocarpus erinaceus, *Poivet.* African kino continued to be the only kino to reach Europe for some years, and is found to be regularly valued in the stock of London druggists up to 1792. In the beginning of the present century substitutes from Jamaica, from Malabar, and from New South Wales, all similar in properties, though obtained from distinct trees, gradually began to oust the African drug, till in 1811 it is recorded that the African kino is no longer to be met with. The Indian gum gradually gained in favour, till in a few years it became the only recognised legitimate kino in all the principal pharmacopœias of Europe. The first supply for the European market is said to have been prepared on a plantation of the East India Company, called Anjarakandy, a few miles from Tellicherry on the Malabar Coast, the exudation being brought from the Ghâts a few miles inland (*Pharmacographia*).

Chemistry
1372

CHEMICAL COMPOSITION.—By the action of dilute mineral acids on aqueous solutions of kino, *kino-tannic* acid is precipitated; by boiling a solution of this acid for some time a red precipitate of *kino-red* is obtained. By destructive distillation kino affords *pyro-catechin*, and, when melted with caustic alkalies, *protocatechuic* acid, and *phloroglucin*; thus yielding the same products as catechu. By the action of ether a small quantity of a crystalline astringent principle can be separated. According to earlier investigators, this was assumed to be catechin. Fluckiger & Hanbury are, however, of opinion that it possesses rather the properties of *pyrocatechin* than catechin. Pyrocatechin also occurs in the fresh leaves of the Virginian creeper. Broughton, late of the Ootacamund Cinchona Plantations, failed to obtain any indication of pyrocatechin in the fresh bark, almost saturated with liquid kino, while in kino itself the reaction was distinct (*Pharmacographia*). " The tannin of kino, sometimes called *mimotannic* acid, or *catechu tannic acid*, is, according to Stenhouse, distinct from *gallo-tannic acid*, and the experiments described by the authors of the *Pharmacographia* appear to indicate that kino tannic does not agree in all reactions with catechu tannic acid. A red gummy matter is also a constituent of kino " (*Warden*).

OIL.
1373

Oil.—It is said to yield an oil of which little is known.

P. 1373

The Red Sandal Wood. (*J. Murray.*)	**PTEROCARPUS** santalinus.

Dye & Tan.—The GUM, if cheap enough, might be used in dyeing and tanning. The BARK—occasionally employed for dyeing – was found by Mr. Wardle to contain a brownish-red colouring substance, which produced reddish fawn colours with tussur silk.

Medicine.—Rumphius is perhaps the first writer in India to notice the GUM, neither Hindu nor Muhammadan authors making any mention of it. He states that it cures diarrhœa, and that the bruised LEAVES are useful as an external application to boils, sores, and skin diseases. Ainslie remarks that the Natives of the Coromandel Coast suppose the exudation as well as the BARK to have virtues in cases of toothache, but makes no mention of its valuable astringent properties which he would appear not to have been acquainted with. Even to the present day Malabar kino is chiefly reserved for the European market, and there is little demand for it in Native practice, in which Dragon's Blood and Butea kino take its place (*Dymock*). It is well known as a valuable astringent, with properties similar to those of catechu, but milder in its action. It is, therefore, better adapted for children and delicate patients.

Fodder.—The LEAVES are an excellent fodder for cattle and goats, and are much in demand; indeed, cattle-keepers are said to often do great damage to the trees.

Structure of the Wood.—Sapwood small; heartwood brown, with darker streaks, very hard, durable, seasons well, and takes a fine polish; it is full of red gum resin and stains yellow when damp; weight 47℔ to 52℔ per cubic foot. It is much used for door and window frames, posts and beams, furniture, agricultural implements, cart and boat building; and has also been employed for sleepers. Twenty-five sleepers, which had been down seven to eight years on the Mysore State Railway, were found, when taken up, to comprise nine good, eleven still serviceable, and five bad; sleepers of this timber have also been used to a certain extent on the Holkar and Neemuch and other lines (*Gamble*).

DYE & TAN.
Gum.
1374
Bark.
1375
MEDICINE.
Gum.
1376
Leaves.
1377
Bark.
1378

FODDER.
Leaves.
1379
TIMBER.
1380

Pterocarpus santalinus, *Linn. f.; Fl. Br. Ind., II., 239.*

1381

THE SANDERS RED or RED SANDERS TREE, sometimes also called RED SANDAL WOOD, *Eng.;* SANTALE ROUGE, *Fr.;* ROTHES SANDELHOLZ, *Ger.;* SANDALO ROSE, *It.;* SANDEL-HOUT, *Dan.*

Vern.—*Rukhto-chandan, undum, lálchandan, rarat chandan,* HIND.; *Kúchunduna, tilaparni, ranjana, rakta-chandana, raktachondon, lál chandan, rukt chundun,* BENG.; *Raktachandan,* URIYA; *Chandan lál,* PB.; *Ratánjli, rakta-chandan, lálachandana,* BOMB.; *Rakta-chandan, támbada chandana, tambada-gand-hácha-chekká,* MAR.; *Ratánjli,* GUZ.; *Lalchandan, undum,* DEC.; *Seyapu chandanum, shen-shandanam, lal chandan, rakta chandan,* TAM.; *Kuchandanam, erragandhapu-chekka, rakta chandan, lal chandan, seyapu chandanum, chandam, erra chandanum, rakta gandham, rakta-chandanam, gerra chandan,* TEL.; *Kem-pugandha-cheke, honné, rakta-candana, agaru,* KAN.; *Uruttah-chundanum, rakta-channanam,* MALAY.; *Sandaku, nasa-ni,* BURM.; *Ruct-handún, rat-handun,* SING.; *Raktachandana, agaru-gandhakáshtaha, rajana kuchandana, tilapari,* SANS.; *Sandaleahmar, undum,* ARAB.; *Buckum, sandale-surkh, sun, undum, dul-surkh,* PERS.

References.—*Roxb., Fl. Ind., Ed. C.B.C., 536; Voigt, Hort. Sub. Cal., 242; Brandis, For. Fl., 153; Beddome, Fl. Sylv., t. 22; Gamble, Man. Timb., 131; Sir W. Elliot, Fl. Andhr., 34, 51, 101, 162; Pharm. Ind., 71; Flück. & Hanb., Pharmacog., 199; Fleming, Med. Pl. & Drugs (Asiatic Reser., XI.), 175; Ainslie, Mat. Ind., I., 385; O'Shaughnessy, Beng. Dispens., 298; Irvine, Mat. Med. Patna, 20; Moodeen Sheriff, Supp. Pharm. Ind., 209; U. C. Dutt, Mat. Med. Hindus, 154, 315; Sakharam Arjun, Cat. Bomb. Drugs, 45; Bent. & Trim., Med. Pl., t. 82; Dymock, Mat. Med. W. Ind., 2nd Ed., 237; Dymock, Warden & Hooper, Pharmacog. Ind., I.,*

P. 1381

462; Birdwood, Bomb. Prod., 297; Baden Powell, Pb. Pr., 342; Drury U. Pl. Ind., 358; Useful Pl. Bomb. (XXV., Bomb. Gaz.), 243; Econ. Prod. N.-W. Prov., Pt. III. (Dyes & Tans), 44; Liotard, Dyes, 60; McCann, Dyes & Tans, Beng., 67; Boswell, Man., Nellore, 96, 100; Gribble, Man., Cuddapah, 7, 199, 267; Gazetteers:—Orissa, II., 160, 179; Mysore, & Coorg, I., 436, 437; II., 7; III., 28; Agri.-Horti. Soc.:—Ind., Trans., II., App., 314; VI., 127; Journals (Old Series), V., Sel., 112; Ind. Forester, III., 359; IX., 137-140, 356; X., 547, 548; XI., 186; Smith, Ec. Dict., 367.

Habitat.—A small tree of South India, chiefly found in Cuddapah, North Arcot, and the southern portion of the Karnul District. Considerable plantations are now cultivated in these districts of Madras, and to a smaller extent in Bombay and Bengal. It favours a dry, rather rocky soil, and a hot fairly dry climate.

Cultivation.
1382

CULTIVATION.—The seeds are gathered in May or June, and sown in July in small beds about 8 feet square, within easy reach of water. They are thrust into the light soil perpendicularly or at an angle, and about an inch deep. From seven hundred to eight hundred may be put into nursery beds of the above-mentioned dimensions. The seeds should be watered by hand every second evening. They germinate, if previously soaked for a night in cold water, in twenty to twenty-five days, if not so treated, in thirty to thirty-five. After germination, the beds should be moderately watered, particu ar care in this respect being very necessary during the first six months. Too much water proves equally destructive to giving none at all. At six months the shoot has attained some height and requires the support of a forked stick. The nursery must be kept free of weeds.

When six months' old the plants should be transferred to wicker and bamboo baskets the removal being carefully performed so as not to injure the long tap-roots. These baskets must be placed in a shady place, and watered every second or third day. When the roots have taken firm hold, the baskets should be buried in pits and watered till the rains set in. The shoots may then be put down in the plantation (*Mr. Yarde, at For. Conf. Meeting in 1875*).

In Bombay it is reported to be cultivated in the Barsi taluk to the extent of about 300 acres. It is there sown in September, and the plants are allowed to grow for only three years, after which they are pulled up, the small roots are cut off and dried in the sun, and yield the dye. The cost of cultivation is said to be about R1¼ per acre, and the profit R3. About 600 maunds are thus annually produced, of which about 30 are absorbed locally, and the rest exported to Sholapur, Poona, and Ahmednugger. In the Ratnagiri District the tree is not cultivated but grows in the woods, whence the dye-stuff is collected for export to Bombay (*Liotard*).

DYE.
Wood
1383

Dye.—The WOOD contains a red colouring matter called *santalin*, which is easily dissolved out by means of any alkaline solution, and is used as a dye. In India it appears to be chiefly employed as a stain or pigment for marking idols and the forehead in ceremonies. It is, however, occasionally employed as a red dye in the same way as sappan. The cloth to be coloured is boiled with the wood, not merely soaked in it, and the tint produced is said to be lasting. In Europe, to which considerable quantities are exported, it is employed as a colouring agent in pharmacy, for dyeing leather, and for staining wood. The authors of the *Pharmacographia* inform us that red-sandal-wood was noticed by **Marco Polo**, and later by **Garcia d'Orta** who clearly distinguished the fragrant sandal of Timor, from the red inodorous wood of the Coromandel. During the middle ages it was well known and was used as a medicine, as well as for an alkanet for culinary purposes, such as colouring sauces and other articles of food. The price in England, between 1326 and 1399, was very variable, but on an average exceeded 3*s.* per ℔ (*Flückiger & Hanbury*).

DYE.

TRADE.—The felling of trees to obtain this dye-stuff is under strict Government control, and yearly yields a considerable revenue. The price varies considerably with the quantity in the market. In the North-Western Provinces it is said to fetch R9-8 per cwt. (*Buck*); in Bombay the average price is said by Dymock to be R15 to R28 per kandy of 7⅓ cwt.

TRADE. 1384

Medicine.—According to U. C. Dutt, Sanskrit writers describe several varieties of sandal or *chandana*. Of these, *srikhanda*, white; *pitachandana*, yellow; and *raktachandana*, red; are best known. The first two are simply the wood of the true sandal, **Santalum album**, of different shades. The last is that at present under consideration. It has long been a matter of question, how woods differing so entirely in character as sandal-wood and red sanders-wood should have come to bear the same Sanskrit name of *chandana*, and the same English appellation. On this subject Dutt remarks, " I am inclined to think that the name is owing to the similarity in the uses to which Hindus put both these articles. Both sandal-wood and red sandal-wood are rubbed on a piece of stone with water, and the emulsions are used for painting the body after bathing and in religious services."

MEDICINE. 1385

The same author states that Hindu physicians describe the WOOD as an astringent tonic. It enters into numerous prescriptions of an astringent character, and of cooling external applications for inflammation, headache, &c., some of which are given by Chakradatta ; it is, however, rarely directed to be given alone. It is also much employed as a colouring agent for the preparation of medica'ed oils (*Mat. Med. Hindus*).

Wood. 1386

Dymock informs us that Muhammadan writers follow the Hindus in describing the three kinds of sandal-wood and their uses. " The author of the *Shafa-el-askam*," he writes, " says that in bilious fluxes white sandal is used, when blood is being passed, red sandal, and when the stools contain both bile and blood, the two woods are combined. This treatment must be based upon the doctrine of signatures." Ainslie states that in Southern India it is little used medicinally, though the Indian medical practitioners sometimes recommend it in powder, in conjunction with certain herbs, and mixed with gingelly oil, as an external application and purifier of the skin after bathing. In the North, on the other hand, we find it supposed to possess many virtues,—" Considered by Natives a hot remedy ; useful in bilious affections and skin diseases ; also in fever, boils, and to strengthen the sight. It also acts as a diaphoretic, and is applied to the forehead in headache" (*Baden Powell*).

Though officinal in most modern pharmacopœias, it is used only for pharmaceutical purposes, as a colouring agent, and is believed to be inert.

CHEMICAL COMPOSITION.—The wood contains a resinoid substance— *Santalic acid*, or *Santalin*, which is readily soluble in ether, alcohol, alkaline solutions or concentrated acetic acid, and is said to form ruby red crystals, devoid of odour and taste. Weidel, in 1870, obtained from the wood colourless crystals of *Santal* $C_8 H_6 O_3$. By the action of ether on the wood a red powder is obtained, having a green hue in reflected light, and which, when melted with potash, yields *Resorcin* (*Pharmacographia*).

Chemistry. 1387

SPECIAL OPINIONS.—§ " The wood is rubbed on a stone with water and the resulting paste applied over boils and other inflammatory affections of the skin. It has a cooling effect " (*Surgeon-Major A. S. G. Jayakar, Muskat*). " The wood rubbed up with water is advantageously

---

MEDICINE.	employed as a wash in superficial excoriations of the genital organs" (*Surgeon R. Gray, Lahore*). " A decoction of the legume is useful as an astringent tonic in chronic dysentery after separation of the slough " (*Civil Surgeon S. M. Shircore, Murshedabad*). " Useful application in headache" (*Assistant Surgeon S. C. Bhattacharji, Chanda, Central Provinces*).
TIMBER. 1388	**Structure of the Wood.** — Sapwood white ; heartwood purplish-black, dark orange-red when fresh cut, extremely hard, the shavings giving a blood-red orange colour ; weight about 76℔ per cubic foot (*Gamble*). It is used for building and for turning. and is said to be much prized because it is not subject to the attacks of white ants.

PTEROSPERMUM, *Schreb. ; Gen. Pl., I., 220.*

[*Ic. t. 63 ;* STERCULIACEÆ.

1389   **Pterospermum acerifolium,** *Willd. ; Fl. Br. Ind., I., 368 ; Wight,*
Syn. — P. ACEROIDES, *Wall.*
Vern. — *Kanak-champa, kaniár, katha-champa,* HIND. ; *Kanak-champá, mús,* BENG. ; *Machkunda,* SANTAL ; *Hattipaila,* NEPAL ; *Gaik,* MAGH. ; *Laider,* MICHI ; *Karni-karo, kanak-champa,* BOMB. ; *Matsa kanda,* TEL. ; *Toungpetwún, tha-ma jam wai-soke,* BURM. ; *Karnikára,* SANS.
References. — *Roxb., Fl. Ind., Ed. C.B.C., 511 ; Brandis, For. Fl., 35 ; Kurz, For. Fl. Burm., I., 145 ; Beddome, Fl. Sylv., t. 35 ; Gamble, Man. Timb., 49 ; Elliot, Fl. Andhr., 113 ; Mason, Burma & Its People, 536, 754 ; Rev. A. Campbell, Rept. Econ. Pl., Chutia Nagpur, Nos. 9276, 9482 ; U. C. Dutt, Mat. Med. Hind., 303 ; Dymock. Warden & Hooper, I., 233 ; Baden Powell, Pb. Pr., 591 ; Atkinson, Him. Dist., 306 ; Lisboa, U. Pl. Bomb., 23 ; Gazetteers :—Mysore & Coorg, I., 68 ; Bombay, XV., 429 ; Ind. Forester, I., 86 ; XIV., 208 ; Agri.-Horti.; Soc., Ind.:—Trans., VII., 81 ; Journals (Old Series), VIII., Sel., 177 IX., 408, Sel., 44 (New Series), VII., 150.*

	Habitat. — A tall tree found in the sub-Himálayan tract from the Jumna eastwards to Bengal, Chittagong, and Burma ; also met with in the Konkan. It is often planted for ornament.
MEDICINE. Down. 1390 Leaves. 1391 Flowers. 1392 Bark. 1393	**Medicine.** — The DOWN on the LEAVES is used by the hill-people in Sikkim to stop bleeding in wounds (*Gamble*). In the Konkan the FLOWERS and BARK, charred and mixed with *kamala,* are applied in suppurating small-pox (*Pharmacog. Ind.*).
TIMBER. 1394	**Structure of the Wood.** — Sapwood white ; heartwood soft to moderately hard, red ; weight on the average 47·5℔ per cubic foot. In structure it somewhat resembles *Thitka* (Pentace burmanica, *Kurz*), and is worthy of notice ; it is occasionally used for planking in Bengal, is said to take a fine polish, and to be suitable for making furniture.
DOMESTIC. Leaves. 1395 Flowers. 1396	**Domestic.** — In Bombay and Bengal the LEAVES are employed as plates, and for packing tobacco. The FLOWERS are used by Bengalis as a disinfectant and to keep insects away from bed clothes, &c. They are also said to render water gelatinous.

1397   **P. suberifolium,** *Lam. ; Fl. Br. Ind., I., 367.*
Syn. — P. CANESCENS, *Roxb. ;* P. OBLONGUM, *Wall. ;* PENTAPETES SUBERIFOLIA, *Linn.*
Vern. — *Muchkand,* HIND. & BENG. ; *Baelo, giringa,* URIYA ; *Muchunda, muchukunda,* BOMB. ; *Taddo,* TAM. ; *Lolagu,* TEL. ; *Naji,* BURM. ; *Velenge, venangu,* SING. ; *Muchukunda,* SANS.
References. — *Roxb., Fl. Ind., Ed. C.B.C., 512 ; Beddome, Fl. Sylv., t. 34 ; Gamble, Man. Timb., 56 ; Thwaites, En. Ceylon Pl., 30 ; Dals. & Gibs., Bomb. Fl., 24 ; Mason, Burma & Its People, 536, 754 ; Rev. A. Campbell, Rept. Econ. Pl., Chutia Nagpur, No., 9895 ; U. C. Dutt, Mat. Med. Hind., 123, 309 ; Dymock, Mat. Med. W. Ind., 2nd Ed., 114 ; Dymock, Warden & Hooper, Pharmacog. Ind., I., 233 ; Lisboa, U. Pl. Bomb., 22 ; Ind. Forester, Jan. 1884, X., 26, 31, 547.*

Habitat.—A moderate-sized tree of the Northern Circars, Carnatic.

Medicine.—"The FLOWER rubbed into paste with *kángika* (rice vinegar) is an ancient and well known application for hemicrania. It is mentioned by many writers, and is used even to the present day, as a domestic remedy (*U. C. Dutt*).

MEDICINE. Flower. **1398**

Structure of the Wood.—Light red, moderately hard, tough, structure the same as that of **P. acerifolium**; weight 36 to 40℔. It is very tough and is used for building, carts, gunstocks, and other purposes (*Gamble*).

TIMBER. **1399**

Domestic.—The FLOWERS, like those of the preceding species, render water gelatinous.

DOMESTIC. Flowers. **1400**

Ptychotis Ajowan, *DC.* ; also **P. coptica**, *DC.*, see **Carum copticum**, *Benth.;* UMBELLIFERÆ ; *Vol. II., 198.*

P. Roxburghiana, *DC.*, see **Carum Roxburghianum**, *Benth. ; Vol. II.,*
[*201.*

PUERARIA, *DC.; Gen. Pl., I., 537.*

[*412 ;* LEGUMINOSÆ.

Pueraria tuberosa, *DC. ; Fl. Br. Ind., II., 197; Wight, Ic., t.*
   Syn.—HEDYSARUM TUBEROSUM, *Roxb.*

1401

   Vern.—*Tirra, patal kohnda, bedari kand, siali, badár, billi, pona, bilaikand, surál,* HIND. ; *Shimia batraji,* BENG. ; *Tirra, patal kohnda, jang tirra,* SANTAL ; *Birali kund,* NEPAL ; *Debrelara,* PAHARIA ; *Surál, sarár, sarwála, surált, siáli, badár,* N.-W. P.; *Bildi-kand, bili, birálipanwa, biráli-púna,* KUMAON; *Sidli, badár, saloha, salor, surál,* (TUBER=) *bilai kand, bidári kand,* PB.; *Gorahel,* RAJ. ; *Dári,* BOMB. ; *Karwí-naí,* GUZ. ; *Dári, gúmodi, dári gummadi,* TEL.

   References.—*Roxb., Fl. Ind., Ed. C.B.C., 580; Brandis, For. Fl., 141; Dalz. & Gibs., Bomb. Fl., 67; Stewart, Pb. Pl., 75; Rev. A. Campbell, Rept. on Ec. Pl., Chutia Nagpur, Nos. 8444, 9251; Sir W. Elliot, Fl. Andhr., 46; O'Shaughnessy, Beng. Dispens., 316; Dymock, Mat. Med. W. Ind., 2nd Ed., 234; Dymock, Warden & Hooper, Pharmacog. Ind., I., 422; Baden Powell, Pb. Pr., 260; Atkinson, Him. Dist., 309, 748; Econ. Prod., N.-W. P., Pt. V., 91, 94; Drury, U. Pl., 360; Lisboa, U. Pl. Bomb., 151; Cat. Col. Ind. Exhib., Raw Products, No. 145; Gazetteers:— N.-W. P., IV., lxxi.; Hoshiarpur, 11; Ind. Forester, I., 84; XII., App., 11; Agri.-Horti. Soc., Ind., Journals (Old Series), I., 396; XIII., 313.*

Habitat.—A twining shrub found in the Western Tropical Himálaya ascending to 4,000 feet in Kumáon; also met with in the hills of Southern India and in Orissa.

Gum.—The TUBER, when wounded, exudes a bitter acrid gum of an opalescent colour (*Dymock*).

GUM. Tuber. **1402**

Medicine.—The pounded TUBER, made into a poultice, is applied to reduce inflammatory swellings of joints (*Roxburgh*). **Stewart** states that it is "officinal" in the plains of the Panjáb as a cooling medicine, while Dr. Gimlette, Residency Surgeon, Nepál, informs us in the *Col. & Ind. Exhib. Cat.*, that it is employed in that country as an emetic and tonic, and is also believed to be lactagogue. It is given as a tonic in doses up to two *mashas* with milk; a larger quantity has an emetic effect. According to Cleghorn it is exported from the North-West Himálaya to the plains. Dymock conjectures that it may be the *shurava* of Sanskrit writers.

MEDICINE Tuber. **1403**

CHEMICAL COMPOSITION.—A recent analysis by the authors of the *Pharmacographia Indica* shows that the tuber contains saccharine matter, a principle probably allied to *inulin,* a bitter principle, an easily oxidizable resin, and a resin acid.

Chemistry. **1404**

Food.—The yam-like TUBER is said to be employed in the Panjáb as an article of food (*Stewart*).

FOOD. Tuber. **1405**

PULSES.	The Pulses of India.

1406

PULICARIA, *Gærtn.; Gen. Pl., II., 335.*

Pulicaria crispa, *Benth.; Fl. Br. Ind., III., 299 ;* COMPOSITÆ.

Syn.—FRANCŒURIA CRISPA, *Cass. ;* INULA QUADRIFIDA, *Ham. ;* DUCHES-NIA CRISPA, *Cass.*

Vern.—*Burhna,* HIND. ; *Búí, gidí, sutei, phatmer,* PB.

References.—*Stewart, Pb. Pl., 126 ; Gazetteer :—N.-W. P., I., 82 : IV., lxxiii. ; Ind. Forester, XII., App., 15.*

Habitat.—A stout, shrubby plant, found in the Panjáb, the Upper Gangetic plain, and eastwards to Behar ; distributed to Arabia and Africa.

MEDICINE.
Plant.
1407

Medicine.—" In the Salt Range, the dried PLANT is bruised and applied as a vulnerary to bruises, &c., of bullocks " (*Stewart*).

1408

P. foliolosa, *DC.; Fl. Br. Ind., III., 298.*

Syn.—BLUMEA SENECIONIDEA, *Edgew. ;* CONYZA FOLIOLOSA, *Ham.*

Vern. - *Arambu,* PUSHTU.

References.—*Aitchison, Bot. Afgh. Del. Com., 75 ; Notes by Mr. Duthie's Trans-Indus Collector, No. 31 ; Gazetteer ·—N.-W. P., I., 82 ; IV., lxxiii.*

Habitat.—Found in the Upper and Lower Gangetic plains, Central India, Banda, and the Konkan ; distributed to the North-West Frontier, Afghánistán, and Persia.

FODDER.
Plant.
1409

Fodder. – The PLANT is used in the Trans-Indus territory as fodder for camels (*Mr. Duthie's Collector*).

PULSES.

1410

Pulses. **Dr. Watt** gives the following account of these edible seeds in the Colonial & Indian Exhibition Catalogue (the figures of trade only being altered):—

They may be briefly defined as leguminous seeds which are boiled and eaten as an article of diet. Amongst the most curious and interesting features of an Indian bazár, are the baskets of whole and split pulses, peas, beans, and lentils of all colours—white, black, brown, green, grey, red, &c.—shown by the dealers.

There are probably not more than two or three kinds of pulses exported from India, but as the quotations of sea-borne trade are given collectively, the relative amounts of each kind cannot be accurately expressed. In the trade returns also gram is treated separately instead of being included with the other pulses. It may be stated of the pulses as a whole, that they form an exceedingly important article of internal trade, and including gram the exports are by no means inconsiderable. During the five years, 1884-85 to 1889-90, the exports averaged ;—gram, 315,229 cwt., valued at R10,20,101 ; other pulses, 631,489 cwt., valued at R17,90,098. The exports of gram are very regular and fluctuate little; those of other pulses exhibit marked irregularities. Thus the average would have been only about 400,000 cwt., but for a remarkably large export, in 1887-88, of 1,426,058 cwt. The imports are small, and the total foreign trade in these pulses may approximately therefore be put down at £270,000, but this sum must bear but a small relation to the internal trade. The total area under pulses is about 48,000,000 of acres, but as the crop is chiefly grown in the cold season the same land yields in addition some other product such as wheat or rice.

1411

Cajanus indicus, *Spreng.;* LEGUMINOSÆ.

THE PIGEON-PEA; *Arhar dál,* HIND. & BENG.

This is apparently a native of equatorial Africa. Cultivated in most parts of India, constituting an important article of food.

There are two chief varieties : **C. flavus,** *DC.,* the plain yellow pea known in the vernacular as *thor*; and **C. bicolor,** *DC.,* the pea veined with purple, known as *arhar*. The latter is the one most commonly cultivated in the North-West Provinces and Oudh, while in the Central Provinces and the Deccan *thor* takes the place of *arhar*. It is grown mostly as a subordinate crop with *juar, bajra,* and cotton, *sunn,* jute, &c., and to a smaller extent by itself.

The leaves are considered an excellent fodder; the stalks are used for roofing, basket-making, and the tubular wicker-work fascines (*bira* or *ajar*) for lining wells to prevent the earth from falling in.

Cicer arietinum, *Linn.*; LEGUMINOSÆ.

    THE COMMON GRAM, or CHICKEN-PEA; *Chena* or *Chola*, HIND. & BENG.

This plant appears to be a native of the countries to the south of the Caucasus and to the north of Persia. It was probably carried into India by the Western Aryans at an early date, and prior to the time of its cultivation in Europe. There seems some probability that the plant may have also been indigenous in Greece.

Gram is cultivated throughout India in any soil, giving, however, the largest return in heavy soils. There are two sorts—one, a large reddish grain; the other, a small light-brown one.

1412

Dolichos biflorus, *Linn.*; LEGUMINOSÆ.

    THE HORSE-GRAM; *Kurti-kalai,* BENG.; *Kulthi gahat,* HIND.

There are two very distinct varieties of this pulse—the one an erect annual (*var.* **uniflora**), the other a twining herb (*var.* **biflora**), met with chiefly cultivated as a pulse crop on the Tropical and sub-Tropical Himálaya, to Burma and Ceylon. It is sown either singly or along with other grains; it seems to succeed best near the coast. The sowing is made in October and November, generally in dry, light, rich soils; and the crop is reaped in February.

The grain is eaten by the poorer classes of Natives, and is also given to horses and cattle. The pods are flat and curved like a sickle; they are given to cattle. The straw is a useful fodder.

1413

D. Lablab, *Linn.*

    THE INDIAN BEAN; *Sim, makham-sim,* BENG.; *Sim,* HIND.

Wild and cultivated throughout India; ascends to 6,000 to 7,000 feet on the Himálaya.

There are several varieties of this bean. Roxburgh describes thirteen cultivated. Most of the forms are eaten cooked in curry by the Natives as a vegetable or bean, not as a pulse. When young and tender, the pods constitute a good substitute for the common French beans.

1414

Glycine hispida, *Maxim.*; LEGUMINOSÆ.

    THE SOYA BEAN; *Gari-kulay,* BENG.; *Bhat, bhatwan,* HIND.

This plant, densely clothed with fine ferruginous hairs, is sub-erect. It is met with in the tropical regions and the outer Himálaya, from Kumáon to Sikkim, and the Khásia and the Naga Hills to Upper Burma. Dr. Stewart mentions a field of *bhat* having been observed in Bisahir in the Panjáb, at an altitude of 6,000 feet. Dr. Roxburgh first saw it grown from seed received from the Moluccas in 1798.

DeCandolle considers Soya, and apparently correctly, as a native of Cochin China, Japan, and Java. But he remarks that "it is of modern introduction into India." "There are no common Indian names" for it. This seems to be a mistake; the plant is well known in India under the names given above. In Manipur and the Naga Hills it is one of the most abundant of pulses. In the Naga Hills it is known as *Tsu dza,* a name not unlike *Soja,* but at the same time it may be related to the old Chinese

1415

P. **1415**

|

name *Shu*. The Soya most likely reached India from China, passing by way of the mountain tracts of Assam. The importance of these hills in settling questions of the nativity of cultivated Indian and Chinese plants has not been fully appreciated, and we might fairly anticipate that many statements at present accepted as facts will be considerably modified with an extended knowledge of the wild and cultivated plants of the Assam and Chinese frontier. Soya is an important article of food in Tibet. It is made in India into a sauce called "Soy." The advisability of extending its cultivation on the Himálayan tracts was pressed on the Government of India in 1882 by Professor Kinch, and the attention of local Governments also was called to it, but it does not appear to have made much progress.

1416

Lathyrus sativus, *Linn.*; LEGUMINOSÆ.
THE JAROSSE or GESSE; *Khésari*, BENG.
This pulse is common in the Northern Provinces of India, from the plains of Bengal to Kumáon, where it reaches 4,000 feet in altitude. It is generally cultivated, but in some places grows wild. When cultivated, it is sown about the close of the rains (October) in heavy clay soils and on land hardened through submersion, and occasionally in rice-fields before the rice is cut. Its cultivation in the North-West Provinces and Oudh is more common in the eastern districts and in parts of Allahabad and Azimgarh. It is also extensively grown in Bengal and other parts of India.

It is chiefly used as a green fodder for cattle, and seems to spring up as a weed among other crops. The seeds are very irregular in form, generally wedge-shaped, gray, and minutely spotted with a dark line. They have in Europe the reputation of causing paralysis of the lower extremities. In the Proceedings of the Government of the North-West Provinces for 1866, pp. 265 to 295, Dr. Irving gives an interesting account of the prevalence of a form of palsy in the Barrah and also in the Khyragarh divisions of the Allahabad district. He attributes the disease to the habit of eating, as an article of daily food, the *khesari* vetch. He bases his opinion on the fact that the peculiar disease is met with only in districts where this pulse is eaten as a regular article of food. The disease appears suddenly, having none of the premonitory symptoms of paralysis. The sufferer having shown no previous disease which could be supposed to give origin to palsy, suddenly becomes paralysed. There is no pain, and the affected part, instead of becoming deformed, continues to grow. The disease is confined to the lower limbs, and is much more prevalent amongst men than women, but boys are often found quite lame. The symptoms and history are entirely in favour of the disease being of a paralytic nature and not rheumatic, such as might result from exposure to wet and cold.

The split pea is largely used to adulterate *dál*, from which it can scarcely be distinguished when sold in the split form. It is used, in fact, by the poorer classes as an occasional substitute for other pulses, but is hard and indigestible. It is, however, only injurious when eaten continuously as a regular article of food. The troops under General Elphinstone in the first expedition to Kabul suffered much from having to mix this pulse with their food. Pigs fed on the *khesari* are said to lose the use of their limbs, but to fatten well.

1417

Lens esculenta, *Mœnch ;* LEGUMINOSÆ.
THE LENTIL; *Masúri,* HIND.
This plant appears to be a native of Western Asia, Greece, and Italy. The lentil was introduced into Egypt as a cultivated plant at an early date, and from this centre it spread east and west, reaching India about

2,000 years ago. The meal from this pulse is sold in Europe as a food for invalids under the name of EVALENTA or REVALENTA.

In India it is largely grown as a winter crop, and it is universally eaten, both by Natives and Europeans. In the Panjáb, excluding perhaps the more arid tracts, it is grown everywhere in the plains and hills, up to an altitude of 11,000 feet in the Himálaya. It is also common in the North-West Provinces, the Central Provinces, and in Bengal.

Phaseolus aconitifolius, *Jacq.;* LEGUMINOSÆ.   1418
    Moth or *Mothi,* HIND.
    'Found from the North-West Himálaya to Ceylon, ascending to 4,000 feet in altitude; cultivated as a hot-weather or *kharif* crop in the plains, in dry, light, sandy soil. In many parts of India it is quite as much cultivated as *úrd.* In the North-West Provinces and Oudh it is grown both alone and among millets.

    The grain is used as food by the Natives, but is not considered wholesome. It is also used as cattle-food, and is regarded as fattening. The leaves and stalks are also given to cattle.

P. Mungo, *Linn.*   1419
    and Var. **glaber,** *Roxb.*
    Mung or *Mug,* HIND.
    This is sometimes spoken of as 'green gram.' It is a native of India, but has been cultivated for at least 3,000 years. It is met with throughout the plains, and ascends to 6,000 feet in the outer ranges of the North-West Himálaya. It requires a strong, rich dry soil and is generally grown as a subordinate crop in fields of millet or cotton.

    The ripe grain is wholesome and nutritious, much esteemed, and commands a comparatively high price. The crushed stalks and leaves are prized as fodder for cattle.

    Var. **radiatus,** *Linn.*   1420
    Mash-kolai, BENG.; *Urd mash,* HIND.
    This variety differs from **P. Mungo** proper in having longer and more trailing stems, in the plant being much more hairy, in the seeds being fewer, larger and longer, and usually of a dark colour.

    It has two distinct sub-varieties—one with large black seeds, ripening in August and September, and the other with smaller green seeds, ripening in October and November. Both are, however, sown at the commencement of the rains. Heavy soils suit the crop best. It is cultivated in most parts of the plains as a subordinate crop with millet or cotton.

P. trilobus, *Ait.*   1421
    Mugani, BENG.
    Common throughout India, both wild and cultivated; it ascends in the North-West to 7,000 feet in altitude. Seeds are gathered and eaten by the poor. The plant is regarded as good fodder.

P. vulgaris, *Linn.*   1422
    THE KIDNEY or FRENCH BEAN, or HARICOT.
    Cultivated, for the sake of its young pods, in all parts of India, chiefly in gardens. The green pods are cut up into slices, boiled, and eaten by Europeans; they are scarcely, if ever, used by the Natives.

Pisum arvense, *Linn.*; LEGUMINOSÆ.   1423
    THE GREY or FIELD PEA; *Desi Mattar,* HIND. & BENG.
    This is supposed to be originally a native of Greece and of the Levant, and is probably the parent of **P. sativum.** It is cultivated in many parts of India during the cold weather.

PUNICA Granatum.	Properties and Uses of

It produces small, round, compressed, greenish, and marbled peas. It bears some resemblance, in appearance of grain and in mode of cultivation, to **Lathvrus sativus**, the latter being frequently mixed with it.

1424

Pisum sativum, *Linn.*

THE COMMON PEA; *Mattar, Gol Mattar,* N.-W. P.

An annual plant, climbing by means of tendrils. A native of southern Europe. Cultivated in many parts of India during the cold weather, and extensively so by Europeans. It includes the white pea, known as *Kabuli* and *Patnai*, according as they are large or small. It is one of the oldest and most valuable of cultivated legumes. **P. sativum** is more valuable and prolific than **P. arvense**.

1425

Vigna Catiang, *Endl.*; LEGUMINOSÆ.

Chowli, HIND.; *Barbati,* BENG.; *Urohi Mahorpat,* Ass.

Universally cultivated in the tropical zone of India on account of the grain, which forms one of the summer crops raised along with the millets; it ripens in October and November.

The pod is sometimes as much as two feet in length, and contains a number of seeds. These constitute a considerable article of food, but the crop is not much valued, the seed being difficult of digestion. In Bengal, the young green entire pods are cooked in curry. The leaves and stems are used as cattle-fodder.

This plant is said to afford a green dye.

Pumpkin, see Cucurbita Pepo, *DC.,* CUCURBITACEÆ; Vol. II., 641.

Puneeria coagulans, *Stocks,* see Withania coagulans, *Dunal;* SOLANACEÆ.

PUNICA, *Linn.; Gen. Pl., I., 784.*

[*t. 97;* LYTHRACEÆ.

1426

Punica Granatum, *Linn.; Fl. Br. Ind., II., 581; Wight, Ill.,*

THE POMEGRANATE; GRANADES, *Fr.;* GRANATS, *Germ.*

Syn.—P. NANA, *Linn.*

Vern.—*Anár-ká-pér, dhalim, dharimb,* FLOWERS=*gulnár, julnár, dárim pushp,* FRUIT=*anár, dáram, damú,* HIND.; *Dálim-gáchh,* FLOWERS=*gúl-anár, unnum,* FRUIT=*anár, dalim, dárim, darmi,* RIND=*anár-kachilka,* SEED=*habul-kilkils,* BENG.; *Dalim, dalimba,* URIYA; *Dalim,* ASSAM; *Anár-ká-j ar,* FLOWERS,=*gule anár,* FRUIT=*anár,* DEC.; *Madala,* MICHI; TREE & FRUIT=*anár, dárim,* RIND=*náspál, kushiála, post-anár,* N.-W. P.; TREE & FRUIT=*dárú, darúni, dariún, danú, dáán, jaman, dáran, ánár,* FLOWERS=*gul-anár, darim pashk,* RIND=*náspál, chál-anár,* SEEDS=*ánár-dána,* PB.; TREE & FRUIT=*anor, anár, nargosh, ghar nangoi,* PUSHTU; TREE & FRUIT=*anar, dhalim, dharimb, darhú,* BARK=*darú-jo-kul,* SEEDS=*darú-bij,* RIND,=*khashiala-chodi,* SIND; *Anara, dalimba,* BOMB.; *Dalimba-jháda,* FRUIT=*dálimba,* MAR.; *Dádam-nu-jháda,* FLOWERS=*gál-anar,* FRUIT=*dáram, dadúr, dádam,* GUZ.; *Mádalai, madalam, magilan,* FLOWERS=*pú-madalai,* FRUIT=*mádalaip-pasham,* TAM; *Danimma, dadima, dálimba,* FLOWERS=*puerri-danimma,* FRUIT=*dádima-pandu, dalimbapandu, danimmapandu,* TEL.; *Dálimbe-gidá,* FLOWERS=*hushi-dálimbe,* FRUIT=*dalimbekáyi,* KAN.; *Mátalam,* FLOWERS=*pú-mádalam,* FRUIT=*mátalam-pasham,* MALAY.; *Salé-bin, tali-bin, thalé,* FRUIT=*salé-si, tali-si,* BURM.; *Delun-gahá,* FLOWERS=*mal-delun,* FRUIT=*delun, dellun,* SING.; *Dadima-vrikshaha,* FRUIT=*dádima-phalam,* SANS.; *Shajratur rummán,* FRUIT=*rumman, ráná,* ARAB.; *Darakhte-nár,* FLOWERS=*gulnár,* FRUIT=*anar, nar,* PERS.

References.—*Roxb., Fl. Ind., Ed. C.B.C.,* 402; *Voigt, Hort. Sub. Cal.,* 50; *Brandis, For. Fl.,* 241; *Kurz, For. Fl. Burm., I.,* 528; *Beddome, Fl. Sylv., t. 119; Gamble, Man. Timb.,* 205; *Dals. & Gibs., Bomb. Fl., Suppl.,* 34; *Stewart, Pb. Pl.,* 94; *DC., Orig. Cult. Pl.,* 237; *Mason,*

| The Pomegranate. | (*J. Murray.*) | PUNICA Granatum. |

Burma & Its People, 453 ; *Sir W. Elliot, Fl. Andhr.*, 45, 159, 161 ; *Lace, Quetta Plants ; Pharm. Ind.*, 93, 447 ; *Flück. & Hanb., Pharmacog.*, 289, 290 ; *Fleming, Med. Pl. & Drugs (Asiatic Resear., XI.)*, 175 ; *Ainslie, Mat. Ind., I.*, 322 ; *II.*, 175 ; *O'Shaughnessy, Beng. Dispens.*, 338 ; *Irvine, Mat. Med. Patna*, 38, 76, 93, 121 ; *Medical Topog., Ajmir*, 125 ; *Moodeen Sheriff, Supp. Pharm. Ind.*, 209 ; *U. C. Dutt, Mat. Med. Hind.*, 166 ; *Sakharam Arjun, Cat. Bomb. Drugs*, 56 ; *Murray, Pl. & Drugs, Sind*, 191 ; *Bent. & Trim., Med. Pl., t.* 113 ; *Dymock, Mat. Med. W. Ind., 2nd Ed.*, 309 ; *Cat. Baroda Durbar, Col. & Ind. Exhb.*, *No.* 157 ; *Trans., Med. & Phys. Soc., Bomb. (New Series), No.* 4, 152 ; 12, 174 ; *Birdwood, Bomb. Prod.*, 36, 154, 193, 312 ; *Baden Powell, Pb. Pr.*, 349, 452, 591 ; *Drury, U. Pl. Ind.*, 360 ; *Atkinson, Him. Dist.* [*L., N.-W. P. Gaz.*), 310, 715, 748, 777 ; *Useful Pl. Bomb. (XXV., Bomb. Gaz.)*, 80, 156, 246, 257, 259, 389 ; *Econ. Prod., N.-W. Prov., Pt. III. (Dyes & Tans)*, 34 ; *Pt. V. (Vegetables, Spices, & Fruits)*, 44, 74 ; *Ec. Pl., Baluchistan, No.* 17 ; *Royle, Ill. Him. Bot.*, 208 ; *Crookes, Hand-book, Dyeing, &c.*, 512 ; *Liotard, Dyes*, 11 ; *McCann, Dyes & Tans., Beng.*, 56, 85, 144-145, 151 ; *H. Z. Darrah, Note on Cotton in Assam*, 31 ; *Modden, Note on Kumáon*, 280 ; *Wardle, Dye Rep.*, 6, 11, 54 ; *Stocks, Rep. on Sind ; Christy, New Com. Pl., V.*, 24, 44, 71 ; *VI.*, 100 ; *Ayeen Akbary, Gladwin's Trans., I.*, 81-82 ; *II.*, 180 ; *Note on the Condition of the People of Assam ; Man., Madras Adm., II.*, 52 ; *Nicholson, Man. Coimbatore*, 241 ; *Boswell, Man., Nellore*, 122 ; *Gribble, Man., Cuddapah*, 199 ; *Man., Rev. Accts., Bomb.*, 102 ; *Settlement Reports :—Panjáb, Delhi*, 27 ; *Kohát*, 29, 30 ; *Lahore*, 12 ; *Hazára*, 94 ; *Gujrát*, 135 ; *Simla, App. ii. (H), xliii. ; Peshawar*, 13 ; *N.-W. P., Allahabad*. 38 ; *Central Provinces, Chánda*, 82 ; *Nimár*, 200 ; *Port Blair*, *1870-71*, 33 ; *Gazetteers :—Bombay, V.*, 25 ; *VII.*, 39 ; *VIII.*, 184 ; *XV., ii.*, 21 ; *XVIII.*, 45 ; *Panjáb, Shahpur*, 69 ; *Bannu*, 23 ; *Rawalpindi*, 11, 15 ; *Peshawar*, 27 ; *Dera Ismail Khan*, 19 ; *N.-W. P., I.*, 81 ; *II.*, 506 ; *III.*, 33, 238 ; *IV., lxxii. ; Orissa, II.*, 27, 159, 182 ; *Mysore & Coorg, I.*, 53, 61 ; *Agri.-Horti. Soc., Ind. :—Trans., I.*, 66 ; *II., App.*, 307 ; *III.*, 67 ; *IV.*, 105, 141, 150 ; *VI., Pro.*, 55 ; *VII.*, 60, 62, *Pro.*, 95 ; *VIII., Pro.*, 379 ; *Journal (Old Series), IV.*, 202 ; *VI.*, 37 ; *IX.*, 424 ; *XI., Pro,* *(1860)*, 25 ; *XII.*, 348 ; *XIII.*, 389, *Sel.*, 61 ; *XIV.*, 15 ; *(New Series), V., Pro.*, 2 ; *Agri.-Horti. Soc., Panjáb, Select papers to 1862*, 46 ; *Ind. Forester, V.*, 184 ; *XII., App.*, 28 ; *Smith, Ec. Dict.*, 332.

Habitat.—A small tree or large shrub, wild in stony ground in Persia, Kurdistán, Afghanistán, and Baluchistan, cultivated, or spontaneous, throughout India.

Dye & Tan.—The FLOWERS are used in various parts of India to impart a light red colour, said to be fleeting, to cloth. The astringent RIND of the fruit is a valuable tan, and is also frequently employed as an auxiliary to other colouring agents (generally turmeric or indigo) in dyeing. Alone, it imparts to cloth the greenish colour known in the North-West Provinces as *kakrezi*. When used for this purpose the rind is boiled in water till three-fourths of the latter has evaporated and the cloth then dipped in the concentrated infusion.

DYE & TAN. Flowers. 1419 Rind. 1420

The BARK is similarly used as a dyeing auxiliary, but its chief value is as a tan and dye for leather. It is largely employed in preparing the Morocco leather of Tangiers. Large quantities are said to be exported from the forest divisions of the North-West Provinces. Its price is said to vary from 3℔ per rupee in Meerut to 82℔ per rupee in Garhwál (*Sir E. C. Buck*).

Bark. 1421

Samples of the rind examined by **Mr. Wardle** were found to contain a moderate amount of yellow colouring matter, readily given up to boiling water, which gave colours varying from a dull-yellowish green to a bright reddish-drab with tussur and corah silk, and cotton. **Mr. Wardle** appears to think little of its value for this purpose, but remarks, " It is remarkable, however, for its large quantity of tannic acid (25·8 per cent.), and with salts of iron may be made to produce an almost black dye on wool."

Medicine.—The pomegranate has been highly prized both as a fruit

MEDICINE. 1422

24

**PUNICA
Granatum.**   Medicinal uses of

MEDICINE.

Juice.
1423
Rind.
1424
Root-bark.
1425
Flowers.
1426
Seeds.
1427
Pulp.
1428
Buds.
1429

and for its medicinal qualities from the remotest antiquity, as is shown by the reference to it in the Old Testament, and by the numerous representations of the fruit in the sculptures of Persepolis and Assyria, and on the ancient monuments of Egypt (*Pharmacographia*). In India it is frequently referred to by Sanskrit writers. The fresh JUICE is used as an ingredient of cooling and refrigerant mixtures, and of some medicines for dyspepsia. The RIND is valued as an astringent in cases of diarrhœa and dysentery. The ROOT-BARK, though known to **Dioscorides** and **Pliny**, appears not to have attracted attention; according to **U. C. Dutt**, it is not noticed in any Sanskrit work. **Ainslie**, however, informs us that Muhammadan physicians, besides using the FLOWERS and rind in a variety of ways on account of their astringency, recommended the root-bark as the most astringent part of the plant, and, moreover, considered it a perfect specific in cases of tape-worm. The SEEDS were believed to be stomachic; the PULP cardiacal and stomachic; the flowers and BUDS, styptic and cicatrizing. At the present time all the parts above enumerated are largely employed by Natives. The seeds also are considered refrigerant and astringent. **Dymock** remarks that the Arabs apparently derived their knowledge of the medicinal qualities of the plant from the Greeks, with whom similar beliefs obtained.

The value of the root-bark as an anthelmintic was first brought to the notice of European medicine by **Dr. Buchanan** of Calcutta about 1805, in the *Edinburgh Medical & Chururgical Journal, IX., 22.* It was afterwards highly recommended by **Ainslie, Fleming,** and others, and is now officinal in the Pharmacopœias of India, Great Britain, and many other European countries. It is recommended to be taken in decoction, with the addition of cloves or other aromatics, preceded and followed by a purgative. **Ainslie** prescribed as much as two ounces of the bark in this form, a wine-glassful being taken every half hour till the whole was finished. "This quantity," he remarks, " occasionally sickens the stomach a little, but seldom fails to destroy the worm, which is soon after passed."

The root-bark is also employed as an astringent application for relaxed sore-throat, and as a wash in cases of uterine disease. **Dr. Kirkpatric** is quoted in the *Pharmacopœia of India* as stating that a decoction of the rind with cloves was the best remedy he knew for the chronic forms of dysentery from which the native poor of India so often suffer. **Waring** adds his testimony in its favour, stating that he had long used it with the best effect. Both the rind and root-bark are now almost obsolete in European medicine, though retained in the Pharmacopœias. The former is easily replaced by other more easily obtainable and active astringents, while the latter appears to lose a great part of its virtue when not quite fresh, and has given way in favour of Male Fern. Both are, however, admittedly efficacious while in the fresh state, and are daily employed with benefit.

Chemistry.
1430

CHEMICAL COMPOSITION.—The chief constituent of the rind is tannin. It also contains sugar and a little gum, and yields at 100°C. 5·9 per cent. of ash. The root-bark yields over 22 per cent. of tannin, of which the greater proportion consists of a peculiar variety called *punico-tannic acid* ($C_{20}H_{16}O_{13}$). When boiled with dilute sulphuric acid this is resolved into *ellagic acid* ($C_{14}H_8O_9$), and sugar. A considerable amount of *mannite*, probably the *punicin* or *granatin* of earlier observers, is also present. The tænicide power, according to **Tanret**, is due to *pelletierine* ($C_8H_{13}NO$), a colourless dextogyre alkaloid, boiling at 180° to 185°C. This substance has a somewhat aromatic odour, and is readily soluble in water, alcohol, or chloroform. It may be extracted in the proportion of ⅛ per cent., or 2 per cent. of crystallized sulphate. (*Flückiger & Hanbury*). **Dr. Warder**

MEDICINE.

states that this alkaloid has been successfully employed in the expulsion of tape-worm, and has sometimes succeeded when extract of Male Fern had failed.

"SPECIAL OPINIONS.—§" The root-bark an undoubted anthelmintic. The powdered rind is an excellent vegetable astringent in chronic dysentery and diarrhœa" (*Civil Surgeon S. M. Shircore, Murshedabad*). " A decoction of the root-bark is also given in cases of uterine inertia" (*Surgeon W. F. Thomas, 33rd M. N. I., Mangalore*). " The decoction of the root-bark is a common remedy for round-worm in Bengal" (*Assistant Surgeon N. L. Ghose, Bankipur*). "No other remedy so certainly cures tape-worm as the root-bark of pomegranate, but it should not be given as usually directed. My plan of giving it is this :—Two or three ounces of the bark are macerated in *cold* water for 24 hours, then strained, and the *strained* liquor reduced by evaporation to a pint. Of this one-third is taken every two hours, beginning in the morning on an empty stomach. Thus given it acts safely, quickly, and pleasantly, without sickening the stomach or causing any other disturbance. I have never failed to cure tape-worm with the remedy thus given in many cases when other remedies had been fruitlessly tried for years. The worm is generally expelled head and all, so that there is no relapse" (*Brigade-Surgeon C. Joynt, M.D., Poona*). "The alkaloid of this drug has been lately introduced and is said to be effectual in the treatment of Tænia and Bothricephalus-latus" (*Brigade-Surgeon, G. A Watson, Allahabad*). "The decoction of the root-bark is the most valuable anthelmintic for tape-worm. It should be prepared as directed, taken on an empty stomach, and followed by castor oil" (*Assistant Surgeon Nehal Sing, Saharunpore*). "The expressed juice of the leaves and the young fruit is used in dysentery, also the decoction of the bark. The latter is useful as an anthelmintic in tape-worm" (*Civil Surgeon J. H. Thornton, B.A., M B., Monghyr*). "In a few cases of epistaxis, I have successfully used the juice of the red flower, dropped or snuffed into the nostril" (*Civil Surgeon D. Basu, Faridpur, Bengal*). "For dysentery boil 2 ounces of the rind of the fruit in milk (2 pints) down to one-half and strain" (*Surgeon-Major G. Y. Hunter, Civil Surgeon, Karachi*). "*Anar-dana* is the technical trade term for the small fruits of the wild shrub exported from Kashmir for dyeing purposes" (*Surgeon-Major J. E. T. Aitchison, Simla*). "Seeds cooling and useful in fevers. The rind of the fruit forms a valuable astringent decoction in diarrhœa. The root-bark is a specific in tape-worm" (*Assistant Surgeon S. C. Bhuttacharji, Chanda*). "Dried wild fruit rind is used as a mordant in wool-dyeing. The flowers are useful in decoction for dysentery" (*Civil Surgeon G. C. Ross, Delhi, Punjab*). "First-rate anthelmintic for tape-worm" (*Surgeon-Major Beech, Coconada*). "Snuffing the juice expressed from the leaves and flowers stops bleeding from the nose" (*Surgeon A. C. Mookerji, Noakhally*). "Have used the decoction of the root-bark in tape-warm, and the decoction of the rind in chronic dysentery and diarrhœa, with good effect" (*Surgeon D. Picachy, Purneah*). "The flower-buds powdered in doses of 4 to 5 grains are useful in bronchitis. They are also much used by Natives in dysentery and diarrhœa" (*Narain Misser, Hoshangabad*). "I have known eating the fruit to arrest traumatic hœmorrhage of the tongue in a hæmaphilic child. The bleeding had gone for some weeks and had resisted every styptic, mechanical and chemical, that I could think of" (*Brigade Surgeon Joynt, M. D., Poona City*).

Food.—The FRUIT, usually about as large as a full-sized apple, has a hard rind of a brownish red colour, composed of two whorls of carpels, one placed above the other ; the lower whorl consists of three or four, the upper of from five to ten segments. This double calyx contains nu-

FOOD
Fruit.
1431

24 A

PUTRANJIVA
Roxburghii.   The Putranjiva Fruit.

FOOD.

merous yellowish seeds, imbedded in a pellucid reddish pulp. It is universally eaten and much esteemed. The quality varies in different localities;
in the Lower Provinces and Bengal it is inferior to the pomegranate of the
North-West and hilly regions. The best kinds, indeed, are produced still
farther west, in Afghanistán and Persia, whence large quantities are imported into India. It keeps very well, remaining moist, palatable, and
fresh for a long time.

A pleasant cooling *sherbet*, made from the pulp, is highly esteemed by
Natives, and is justly appreciated by all who have tasted it.

TIMBER.
1432

Structure of the Wood.—Light yellow, with a small dark-coloured, irregularly-shaped heartwood, compact and close-grained; weight 52 to 63℔
per cubic foot. It is not used for any particular purpose, but might be
tried as a substitute for box-wood (*Gamble,*.

[*217.*

Purging Cassia, see Cassia Fistula, *Linn.;* Leguminosæ; Vol. II.,

PUTRANJIVA, *Wall.; Gen. Pl., III., 277.*

[*t. 1876;* Euphorbiaceæ.

1433

Putranjiva Roxburghii. *Wall.; Fl. Br. Ind., V., 336 ; Wight, Ic.,*
Syn.—P. sphærocarpa & amblyocarpa, *Muell. Arg.;* Nageia Putran-
jiva, *Roxb.*
Vern.—*Jiaputa, joti, júti, pútr-jiva, patji, jivputrak, patigin,* Hind.;
Putranjiva, jiáputa. Beng.; *Pitoj,* Santal; *Júti, joti, jin-púta, pútra.
jiva,* N.-W. P.; *Patji,* Oudh; *Patáian, jiyaputra,* seeds=*jiapota;*
leaves=*pútrajivak,* Pb.; *Puta-jan, putra-jivu, jiv-putrak, jewan-putr,*
Bomb.; *Jewan-putr,* Mar.; *Karupale,* Tam.; *Kadrajuvi, kudrajinie,
maháputra jivi putrajivi, yárala, kuduru juvir,* Tel.; *Pongalam,*
Malay.; *Toukyap, toukvat,* Burm.; *Putranjiva,* Sans.
References. Roxb., Fl. Ind., Ed. C.B.C., 716 ; Voigt., Hort. Sub. Cal.,
296 ; Brandis, For. Fl., 451, t. 53 ; Kura, For. Fl. Burm., II., 366 ;
Beddome, Fl. Sylv., t. 275 ; Gamble, Man. Timb., 353 ; Dals. & Gibs.,
Bomb. Fl., 236 ; Stewart, Pb. Pl., 196 ; Rev. A. Campbell, Rept. Econ.
Pl., Chutia Nagpur, No. 8202 ; Sir W. Elliot, Fl. Andhr., 100, 101,
110, 160, 161, 194 ; O'Shaughnessy, Beng. Dispens., 611 ; U. C. Dutt,
Mat. Med. Hind., 314 ; Irvine, Mat. Med. Patna, 89 ; Birdwood, Bomb.
Pr., 336 ; Baden Powell, Pb. Pr., 376, 591 ; Atkinson, Him. Dist., 317,
748 ; Drury, U. Pl., 361 ; Lisboa, U. Pl. Bomb., 117, 286 ; Royle, Ill. Him.
Bot., 347, t 83 ; Cooke, Oils & Oilseeds, 67 ; Gazetteers:—Bombay.
XV., 75 ; N.-W. P., IV., lxxvii.; Gurdaspur, 53 ; Hoshiarpur, 11 ; Ind.
Forester, IV.. 389 ; VIII., 402 ; X., 325 ; Agri.-Horti. Soc., Ind., Journals
(Old Series), IV., 217 ; VI., 46.*

Habitat.—A moderate-sized evergreen tree with pendant branches, met
with, wild and cultivated, throughout Tropical India from the Lower Himálaya in Kumáon, eastwards and southwards, to Pegu and Ceylon.

OIL.
Seeds.
1434

Oil.—The seeds yield a rather turbid oil of an olive-brown colour, which
on standing soon deposits the more solid portion. It is used for burning.

MEDICINE.
Leaves.
1435
Stones.
1436

Medicine.—The leaves and the stones of the fruit are officinal in
some parts of the Panjáb *Dr. Stewart).* They are given in decoction
in colds and fevers.

FODDER.
Leaves.
1437

Fodder.—The leaves are lopped for fodder.

TIMBER.
1438

Structure of the Wood.—Grey, shining, moderately hard, close-grained ; weight 36·6 to 49℔ per cubic foot It is sometimes used for making
tools, for turning, and in South India for house-building and making
agricultural implements.

DOMESTIC.
Stones.
1439

Domestic.—The stones of the fruit are strung together to form rosaries by Hindu *fakirs.* Brahmans, and by parents to put round the necks of
their children. These are supposed to preserve the wearer from harm,
hence the vernacular names, which signify "the life of the child."

Putu Manga.—The Tamil name of a fungus, which see under its scientific name **Sclerotium stipitatum.**

PYGEUM, *Gærtn.; Gen. Pl., I., 610.*

[*I., 203;* Rosaceæ. **1440**

Pygeum ceylanicum, *Gærtn.; Fl. Br. Ind., II., 321 ; Wight, Ill.,*
Syn.—P. Walkerii, *Blume;* Polyodontia Walkerii, *Wight.*
Vern.—*Galmorre, galmora,* Beng.; *Gal-mora-gass,* Sing.
Reference.—*Thwaites, En. Cey. Pl., 102.*
Habitat.—A large tree of South India and Ceylon.
Structure of the Wood.—Close-grained, yellow, resembles **Eriobotrya**
in structure; weight, 65℔ per cubic foot. It makes a good fuel for burning
bricks or lime.

TIMBER.
1441

P. Gardneri, *Hook. f. ; Fl. Br. Ind., II., 321 ; Wight, Ic., t. 993.*
Syn.—P. ? acuminatum, *Wight;* P. zeylanicum, *Dalz. & Gibs., Bomb.
Fl., excl. syn., not of Gærtn.*
Vern. *Dacca, kaula,* Bomb.
References.—*Dalz. & Gibs., Bomb. Fl., 89; Lisboa, U. Pl. Bomb., 72;
Ind. Forester, X., 33.*
Habitat.—A large tree on the Nilghiri Hills and Bombay Gháts.
Structure of the Wood.—"Sapwood dark red; heartwood whitish,
coarse-grained; used, I am informed, for making boxes, planks, rafters,
and beams" (*Lisboa*).

1442

TIMBER.
1443

Pyrethrum indicum, *DC.;* see **Chrysanthemum indicum,** *Linn.;* Compositæ, Vol. II., 272.

PYRULARIA, *Mich.; Gen. Pl., III., 223.*

[*255;* Santalaceæ. **1444**

Pyrularia edulis, *A. DC ; Fl. Br. Ind., V., 230 ; Wight, Ic., t.*
Syn.—Sphærocarya edulis, *Wall.;* S. vestita, *Wall.*
Vern. — *Amphi,* Nepal; *Safhyi,* Lepcha.
Reference — *Gamble, Man. Timb., 320.*
Habitat.—A small or moderate-sized thorny tree of the Central and
Eastern Tropical Himálaya from 4,000 to 5,000 feet; also met with in the
Khásia Hills at the same altitude.
Food.—The fruit is eaten by Natives.
Structure of the Wood.—White, moderately hard, close-grained;
weight 47 to 50℔ per cubic foot. It is used by the Bhutias for making
dairy utensils.

FOOD.
Fruit.
1445
TIMBER.
1446

PYRUS, *Linn.; Gen. Pl.. I., 626.*

Pyrus Aucuparia, *Gærtn. ; Fl. Br. Ind., II., 375;* Rosaceæ.
The Rowan Tree, or Mountain Ash.
Syn.—P. ursina, *Herb., Strach. & Wint. ;* Sorbus Aucuparia, *Linn.*
Vern —*Battal, ránthúl, wámpú litsí, rangrek,* Pb.
References.—*Stewart, Pb. Pl., 84; Smith, Dict., 27; Ind. Forester, X.,
516; XI., 3.*
Habitat.—A small tree occurring in India in the Western Temperate
Himálaya, from Kashmír to Kumáon, at altitudes from 11,500 to 13,000
feet; distributed widely, on the west through Turkistan and the Caucasus
to the Atlantic, and north-eastward to N. China, Japan, and Siberia.
Food.—The red fruit, which resembles in size and flavour that of the
same tree in Europe, is not eaten in India. In certain parts of Europe it is
employed to make a jelly, and in the preparation of a liqueur or cordial.

1447

FOOD.
Fruit.
1448

P. 1448

1449

Pyrus baccata, *Linn.; Fl. Br. Ind., II., 373.*

THE SIBERIAN CRAB.

Syn.—P. BACCATA, *var.* HIMALAICA, *Maxim.* ; MALUS BACCATA, *Desf.*

Vern.—*Ban mehal, gwálam,* HIND.; *Ban-mehal, gwála -mehal, rutripuli,* KUMAON ; *Liú, liwár, lhijo, litsi, baror, choda,* PB.

References.—*Brandis, For. Fl., 205 ; Gamble, Man. Timb., 168 ; Stewart, Pb. Pl., 84 ; Aitchison, Lahoul, Jour. Linn. Soc., X., 75 ; Irvine, Mat. Med. Patna, 89 ; Baden Powell, Pb. Pr., 592 ; Atkinson, Him. Dist., 309, 713 ; Econ. Prod., N.-W. P., Pt. V., 44, 71 ; Gazetteer, Rawalpindi, 15 ; Settle. Rep., Simla, App. II. (H), xliii. ; Ind. Forester, XI., 3.*

Habitat.—A small tree of the temperate Himálaya, from the Indus to Bhután, between 6,000 and 11,000 feet, and of the Khásia Hills at an altitude of 6,000 feet.

FOOD.
Fruit.
1450

Food.—It produces a small, very sour FRUIT, of a red or scarlet colour, with the true apple flavour, which ripens towards the end of the rains and is eaten by the Natives of the tracts where it grows.

TIMBER.
1451

Structure of the Wood.—White, with pale-brown heartwood; structure similar to that of **P. Pashia,** but medullary rays slightly broader; weight 53℔ per cubic foot. It is hard, tough, and useful for a variety of purposes, though it is said to warp easily.

1452

P. communis, *Linn.; Fl. Br. Ind., II., 374.*

THE COMMON PEAR.

Vern.—*Náshpáti,* HIND. ; *Naspati, nak,* N.-W. P. ; *Kishta bahira, naspáti, tang, batang, batank, amrúd,* KASHMIR ; *Tang, tangi, batang, batank, nák, náshpáti, naspáti, charkeint, sunkeint, li,* PB. ; *Amrúd, nak, amrucha,* AFGHAN. ; *Amritaphala,* SANS. ; *Amrúd,* ARAB. ; *Náshpáti,* PERS.

References.—*Roxb., Fl. Ind., Ed. C.B.C., 406 ; Brandis, For. Fl., 203 ; Gamble, Man. Timb., 168 ; Stewart, Pb. Pl., 84 ; DC., Origin Cult. Pl., 229 ; Aitchison, Bot. Afgh. Del. Com., 63 ; U. C. Dutt, Mat. Med. Hind., 291 ; Baden Powell, Pb. Pr., 347, 592 ; Atkinson, Him. Dist., 309, 713 ; Econ. Prod., N.-W. P., Pt. V., 44, 70 ; Dollard, Med. Top. Kalee Kumaon & Shore Valley, 32 ; Kew Bulletin, 1889, 25 ; Smith, Dic., 316 ; Kew Off. Guide to the Mus. of Ec. Bot., 61 ; Rep. Horti. Gar., Lucknow, March, 1884, 2 ; 1885, 1 ; Ind. Gar., 271 72 ; Ayeen Akbary, Gladwin's Trans., I., 82, 83 ; Gazetteers :—Mysore & Coorg, I., 68 ; Sialkot, 11 ; Ind. Forester, XI., 51 ; XIV., 420 ; Settle. Repts. :—Hazára, 4 ; Simla, App. II. (H), xlii. ; Agri.-Horti. Soc. Ind., Transactions, I., 64 ; II., App., 298 ; III., 103 ; IV., 106, 149 ; VII., Pro., 44, 101, 150, 184 ; Journals (Old Series), VIII., 241 ; XIII., Sel., 61.*

Habitat.—A small, thorny tree, believed to be wild in Kashmír, cultivated in the North-West Himálaya, altitude 2,000 to 8,000 feet, in Western Tibet to an altitude of 10,000 feet, and occasionally in the plains. It is also found wild and cultivated over the whole of temperate Europe and Western Asia. Of recent years most of the European cultivated races have been introduced with success in Kulu, Kashmir, and in the neighbourhood of the larger Himálayan hill stations. The tree flowers in spring and ripens its fruit during the rains.

FOOD.
Fruit.
1453

Food.—The FRUIT of the ordinary neglected hill-pear tree is generally hard and flavourless, and, though eaten to a considerable extent by the Natives, is little used by Europeans except for cooking purposes. According to Irvine a SPIRIT is manufactured from the fruit in Kashmír. In Afghánistán the fruit of the wild tree is dried, ground into flour, and mixed with ordinary wheaten flour to increase its bulk (*Aitchison*). Of late years the quality of fruit obtained from introduced European races

Spirit.
1454

in Kulu has very greatly improved; indeed, large yellow, soft, lucious pears are now to be obtained which compare favourably with anything produced in Europe.

Structure of the Wood.—Hard, close and compact-grained, valued in Europe for engraving purposes, turning, and for making mathematical instruments. According to Brandis it is not esteemed in India, being used only for ordinary purposes, but Baden Powell mentions it as valued for carving.

TIMBER. 1455

Pyrus Cydonia, *Linn.*, see Cydonia vulgaris, *Pers.*; Vol. II., 676.

P. foliolosa, *Wall.; Fl. Br. Ind., II., 376.*

1456

Syn.—P. URSINA, *Wall.*; SORBUS URSINA, and FOLIOLOSA, *Dcne.*; S. URSINA, *Wensig.*

Vern.—*Kharsani*, NEPAL; *Wampu litsi*, LEPCHA; *Húliya, súliya, martili*, KUMAON; *Súlia, hulia*, PB.

References.—*Brandis, For. Fl., 206, 207; Gamble, Man. Timb., 169; Atkinson, Him. Dist., 309; Ind. Forester, XI., 3.*

Habitat.—A small tree of the Himalaya, from the Indus to Sikkim between 6,000 and 12,000 feet.

Structure of the Wood —White, with small, darker-coloured heartwood; weight 45 ℔ per cubic foot. It is hard, and close-grained, but like the timber of other trees of the section **Pyrus** proper, its utility is lessened by the readiness with which it warps.

TIMBER. 1457

P. kumaoni, *Dcne.; Fl. Br. Ind., II., 374.*

1458

Vern.—*Doda, chitana, mahaul, bun-pala, gun-palos*, PB.

References.—*Atkinson, Econ. Prod., N.-W. P., Pt. V., 71; Gazetteer, Simla, 9.*

Habitat.—Confined to the western regions of the Himálaya, from Kashmír to Kumáon.

Food.—This species produces a small FRUIT of very indifferent taste, which is generally eaten half-rotten by the poorer Natives.

FOOD. Fruit. 1459 1460

P. lanata, *Don.; Fl. Br. Ind., II., 375.*

Syn.—P. KUMAONENSIS, *Wall.*; P. ARIA, *Hook. f. & Thoms.*; P. ARIA, *var.* KUMAONENSIS, *Maxim.*; SORBUS LANATA, *Wensig.*

Vern.—*Galion, máuli, paltu, ban palti*, HIND.; *Galion, mehali, paltu, ban-palti*, KUMAON; *Doda, chola, chilana, maila, paltu, ban-pála, maul, kanghi, thánki, morphal, marpol, chota, chitana, máil tang, máil, mahaul, litsi, chótta, kandlu, polú, bisir*, PB.; *Gún palos*, PUSHTU.

References.—*Brandis, For. Fl., 206; Gamble, Man. Timb., 169; Stewart, Pb. Pl., 84; Baden Powell, Pb. Pr., 592; Atkinson, Him. Dist., 714; Econ. Prod., N.-W. P., Pt. V., 44, 71, 72; Gazetteer, Gurdáspur, 55; Settle. Rep., Simla, App. II. (H) xlii.; Agri.-Horti. Soc. Ind., Journals (Old Series), IV., Sel., 239.*

Habitat.—A moderate-sized deciduous tree, met with on the Himálaya from Kashmír to Kumáon, between altitudes of 5,000 and 10,000 feet.

Food.—It produces a large FRUIT which ripens about October, and is eaten by the Natives when half rotten. Like the fruit of the preceding species and that of **P. Pashia**, it is quite uneatable till it softens and loses its acidity by over-ripeness.

FOOD. Fruit. 1461

Structure of the Wood.—White, moderately hard, close and even-grained, seasons well; weight 40 to 47 ℔ per cubic foot. It might be useful for making boxes and other similar purposes for which a close and even-grained wood is required (*Gamble*).

TIMBER. 1462

P. 1462

PYRUS Malus.	The Apple.

1463

Pyrus Malus, *Linn. ; Fl. Br. Ind., II., 373.*

THE APPLE.

Syn.—MALUS COMMUNIS, *Desf.*

Vern.—*Seb, seo, sev,* HIND. ; *Seb,* BENG. ; *Kúshú,* LAD. ; *Seo, seb, sheo, sco,* N.-W. P. ; *Sher, tsúnt (amrú), sún, chung, senu, kashú, khajú, choli, chunt, pálu, chúi,* PB. ; *Manra, mána,* PUSHTU ; *Seb, shewa,* N.-E. AFGH. ; *Súf,* SIND ; *Sévu,* KAN. ; *Seba, seva,* SANS. ; *Tuffah,* ARAB. ; *Sib, séb, séf,* PERS.

References.—*Roxb., Fl. Ind., Ed. C.B.C., 406 ; Brandis, For. Fl., 205 ; Stewart, Pb. Pl., 85 ; DC., Origin Cult. Pl., 233 ; Aitchison, Bot. Afgn-Del. Com., 63 ; Dymock, Mat. Med. W. Ind., 317 ; Baden Powell, Pb. Pr., 592 ; Atkinson, Him. Dist., 309, 713 ; Econ. Prod., N.-W. P., Pt. V., 44, 71 ; Lisboa, U. Pl. Bomb., 155 ; Birdwood, Bomb. Pr., 150 ; Stocks, Rept. on Sind ; Smith, Dic., 18 ; Ind. Gar., 267-270 ; Rep. Horti. Gar., Lucknow, March 1884, 4 ; Ayeen Akbary, Gladwin's Trans., I., 82, 83, 88 ; II., 135 ; Gazetteers :—Mysore & Coorg, I., 53, 68 ; Bannu, 23 ; Dera Ismail Khán, 19 ; Simla, 11 ; Settle. Repts. :—Kohát, 30 ; Simla, App. II. (H); xlii. ; Agri.-Horti. Soc., Ind., Transactions, I., 63, 64 ; II., 96, App., 298, 310 ; IV., 105, 106, 141, 148, 149 ; V., 20, 21, 69, Pro., 69 ; VI., 38-40, 234, 247, Pro., 40, 50, 55, 60, 112 ; VII., Pro., 12, 101, 150, 160, 184 ; Journals (Old Series), VI., Sel., 112 ; VIII., Sel., 109 ; XIII., 385, Sel., 61 (New Series), VI., Pro., 2.*

Habitat.—The apple grows wild throughout Europe, except in the northern part, in Anatolia, the south of the Caucasus, and in the Persian province of Ghilan. In the mountains of the north-west of India, according to Sir Joseph Hooker, Stewart, Brandis, and others, it is apparently wild. DeCandolle sums up a long and interesting review of the evidence obtainable as to its earliest home as follows :—"From all these facts, I consider the apple to have existed in Europe, both wild and cultivated, from prehistoric times. The lack of communication with Asia, before the Aryan invasion, makes it probable that the tree was indigenous in Europe as in Anatolia, the south of the Caucasus and Northern Russia, and that its cultivation began early everywhere."

In India, at the present day, it is described as "apparently wild" in the North-West Himálaya, ascending to 9,000 feet, and to 11,400 feet in Tibet. Dr. Watt is, however, disinclined to admit any of the apples as indigenous ; the tree nowhere occurs in the wild profusion of, for example, the wild pear, and may certainly in most cases be no more than spontaneous from early cultivation. It is largely cultivated, however, in the Himálaya, the Panjáb, Sind, North-Western Provinces, Central India, and the Dekkan. The fruit from Indian stock is very indifferent in the plains, but improves in the Himálaya, and with a certain amount of cultivation is by no means unpalatable. In Tibet and Afghánistán really good apples are produced, and large quantities are imported from the latter country into India.

Of recent years finer races have been largely introduced from England into Kulu, Kumaon, &c., and the vicinity of the larger Himálayan hill stations. The fruit now obtainable from the large orchards in Kulu and also near Simla, is almost equal to the best produced in Europe. The Kumaon fruit is finding ready sale in Calcutta. The credit of having developed this new fruit-growing industry is largely, if not entirely, due to Sir E. C. Buck.

FOOD.
1464

Food.—See **Fruits,** Vol III., 450. In addition to the remarks there made on the different qualities of Indian apples, it may be noticed that some years ago the Mahárája of Kashmír made an attempt to start the manufacture of cider in his territory. He obtained the services of a European for the purpose, but the experiment appears to have been unsuccessful.

P. 1464

	PYRUS vestita.
The Pear. (*J. Murray.*)	

Structure of the Wood.—Hard and close-grained, good for making cog-wheels and gun-stocks, though inferior to pear-wood.

TIMBER.
1465

Pyrus Pashia, *Ham.; Fl. Br. Ind., II., 374.*

1466

> **Syn.**—P. variolosa, *Wall.*; P. verruculosa, *Bertol.*; P. nepalensis, *Ham.*
>
> **Vern.**—*Mehal, mol,* Hind.; *Passi,* Nepal; *Li,* Lepcha; *Mehal, mol,* Kumaon; *Tang, batangi, shindar, shegúl, shegal, tángi, ku, katári, kithú, kiat, keint, kent, kainth, sanjad, s ogul, gádkúji, keitha, mehal,* Pb.
>
> **References.**—*Brandis, For. Fl., 204, 575; Kurz, For. Fl. Burm., I., 441; Gamble, Man. Timb., 168; Stewart, Pb. Pl., 85; Baden Powell, Pb. Pr., 592; Atkinson, Him. Dist., 309, 713; Econ. Prod., N.-W. P., Pt. V., 44, 70, 72; Gazetteers:—Hazára, 13; Gurdáspur, 55; Rawalpindi, 15; Simla, 9, 11; Kángra, 31; Ind. Forester, VIII., 406, 412; XIV., 420; Settle. Repts:—Kangra, 22; Simla, App. II., (H), xlii.; Agri.-Horti, Soc., Ind., Transactions. IV., 105; Journals (Old Series), IV., 105, Sel., 229; XIII., 385; XIV., 21.*

Habitat—A moderate-sized deciduous tree of the temperate Himálaya, from the Indus to Bhotán, between 2,500 and 8,000 feet, also met with in the Khasia mountains, at an altitude of 5,000 feet, and in Burma. According to Brandis it is occasionally cultivated. The flowers appear in spring and the fruit ripens in September-December.

Food.—The FRUIT, which is dark yellowish-brown, scurfy, and covered with raised white spots, is eatable when over-ripe. Natives use it as an article of food in this state and when half rotten. Sir E. C. Buck has succeeded in Simla to graft on to this hardy tree the better class forms of the pear.

FOOD.
Fruit.
1467

Structure of the Wood.—Heartwood light reddish-brown, hard, close, and even-grained; weight 47℔ per cubic foot. It cracks and warps readily but is tolerably strong, and is used for making walking-sticks, combs, tobacco pipes, and for various other purposes.

TIMBER.
1468

P. vestita, *Wall.; Fl. Br. Ind., II., 375.*

1469

> **yn.**—P. crenata, *Lindl.*
>
> **Vern.**—*Mayhell, gúhor,* Nepal; *Singka,* Bhutia.
>
> **Reference.**—*Atkinson, Econ. Prod., N.-W. P., Pt. V., 72.*

Habitat.—A deciduous tree of the Eastern Himálaya, from Garhwál to Sikkim, between 9,000 and 10,000 feet.

Food.—The FRUIT is edible (*Gamble*).

FOOD.
Fruit.
1470

P. 1470

QUARTZ.	Quails ; Quartz.

(*W. R. Clark.*)

QUAILS.

1

Quails, *Hume & Marshall, Game Birds of India, II., 105-202.*

The Quails of British India are commonly arranged under the genera —**Perdicula** (or Bush (Quails, **Ophrysia** (or Mountain Quails), **Microperdix** (or Painted Quails), **Coturnix** (or True Quails), **Excalfactoria** (or Blue-breasted Quails), and **Turnix** (or Bustard Quails).

Mr. Gould, in his *Birds of Asia*, describes the common quail of Asia as distinct from the European species; but both **Jerdon** and **Hume & Marshall** and other Indian authors are agreed in stating that both are of one species—**Coturnix communis,** *Linn.*

FOOD.
2

Food.—Numbers of the different species are netted or shot throughout India in the various regions where they occur but the Common or Grey Quail may be said to be not only the most plentiful but the one which alone is sold in the chief markets. They are much valued as articles of food.

DOMESTIC.
3

Domestic.—The male birds of the common grey quail are very pugnacious and are kept for fighting purposes by Muhammadans in most parts of India.

Quamoclit, see Ipomæa Quamoclit ; Convolvulaceæ; Vol. IV., 491.

Quarries, see **Stones**, Vol. VI., Pt. II.

QUARTZ.

4

Quartz, *Ball, in Man. Geol. of India, III., 502.*

References.—*Mallet in Man. Geol. of India, IV., 62, 78 ; Encyclop. Brit., XVI., 389 ; XX., 160 ; Balfour, Cyclop. Ind., III., 325.*

This name is applied to one of the most prevalent of minerals; it includes nearly all the native forms of silica. It thus embraces a great number of different minerals, several of which are cut as ornamental stones or used in the arts. In its purest form, as rock crystals which are found crystallised in six-sided prisms with pyramidal ends, (belonging to the hexagonal series), quartz consists entirely of silicic anhydride (S_iO_2). Slightly impure forms, which owe their value to the tints produced by the presence of small quantities of foreign minerals, are " amethyst, rose quartz, false topaz or citrine, smoky quartz, milky quartz, prase and aventurine quartz " (*Ball, Economic Geology*).

Common quartz, crystallised or massive and of a great variety of colours, is a frequent constituent of many rocks, and some impure varieties are properly rocks. Such are ferruginous quartz, often associated with iron ores, jasper, lydian stone or flinty slate, hornstone or chut.

Other silicious minerals, such as flint, chalcedony. and chrysoprase seem to be intimate mixtures of quartz and opal.

Quartz crystals are found most abundantly among the older acidic crystalline rocks, but they also occur in basic traps and basalts, and to some extent in sedimentary formations (*Ball, Economic Geology*).

For a description of the quartz crystals and other inferior gems of the quartzose group, the reader is referred to Vol. II., 167-176, of this work.

Quartz rocks have been found the richest in minerals of all the Indian rocks. Thus lead and silver ores were formerly mined in the quartz rocks of the Kadapah and Karnúl districts of Madras, in the Raipur district of the Central Provinces, and in Hazara in the Panjáb. None of these lead mines, however, are now worked in India. Copper ores have been found in a quartz vein at Somadapilly and Nellore in Madras, at Hazaribagh in Bengal, and in various parts of Nepal. Micaceous iron ore is a very common product of this rock, and gold occurs in it in many parts of India. For an account of the use of quartz in the manufacture of Glass, see Vol. III., 503.

Flint.—True flints are of rare occurrence in India, but the allied hornstones and agates have been largely employed from prehistoric times for the manufacture of knives and arrowheads.

In the Trichinopoli district (Madras) flints almost indistinguishable from English chalk flints exist, it is believed, in a continuous bed and not in detached nodules as in the cretaceous formations. In the Dharwar district (Bombay) flint pits exist, from which it is said Haider and Tippu obtained the gun flints for their troops. In Afghanistan also, near the Kurram river, there are large flint beds, whence the supply used by the Afghan tribes for their flintlocks was formerly obtained.

Flint.
5

Quassia, see Picrasma quassioides, *Benn;* SIMARUEÆ ; Vol. VI., 227.

QUERCUS, *Linn.; Gen. Pl., III.,* 407.

Quercus acuminata, *Roxb.; Fl. Br. Ind.,* V., 607 ; *Wight, Ic., t.*
[*221, f. 6-9;* CUPULIFERÆ.

6

Syn.—Q. FENESTRATA, *Roxb., var.* ACUMINATA, *Wensig.*
Vern.—*Sanu arkaula,* NEPAL ; *Kanta gola batana,* CHITTAGONG.
References.—*Roxb., Fl. Ind., Ed. C.B.C., 672 ; DC. Prod., XVI., ii., 90 ; Kurz, For. Fl. Br. Burm., II., 484 ; Gamble, Man. Timb., 386 ; King, Annals of Bot. Gard. Calc., II. (1889), 41, Pl. 32B. ; Ind. Forester, VIII., 404.*

Habitat.—An evergreen tree, met with in the Eastern Himálaya, Khasia Hills, and down to Chittagong, ascending to 6,000 feet.

In a note by Dr. King in the Annals of the Calcutta Botanical Gardens (*l.c.*), it is stated that this specimen was first collected by Roxburgh, but after him was never seen till 1885, when specimens were sent to the Calcutta Herbarium by a private collector. "The species," Dr. King continues, "is a very distinct one, and does not much resemble Q. fenestrata, *Roxb.*"

Structure of the Wood.—Light red and very hard, weight about 55℔ per cubic foot.

Domestic.—"The tree coppices well and is very good for firewood" (*Kurz*).

TIMBER.
7
DOMESTIC.
8

[*Gard. Calc., II. (1889),* 40, *Pl. 32A.*

Q. Amherstiana, *Wall ; Fl. Br. Ind.,* V., 607 ; *King, Ann. Bot.*

9

References.—*A. DC., XVI., ii., 83 ; Kurz, For. Fl. Burm., II., 484.*
Habitat.—A large evergreen tree, which seems to be confined to the Upper Tenasserim Province of Burma.
Structure of the Wood.—Seems of inferior quality, but is said by Kurz to be used for boat-building.

TIMBER.
10

[*Gard. Calc., II. (1889),* 31, *Pl. 25A.*

Q. Brandisiana, *Kurz ; Fl. Br. Ind.,* V., 604 ; *King, Ann. Bot.*

11

References.—*Kurz, in Journ. As. Soc. Bengal (1873), II., 108 ; For. Fl. Br. Burm., II., 488 ; Ind. Forester, VIII., 416.*
Habitat.—An evergreen tree, 35-40 feet high, found in the dry forests of Martaban at altitudes up to 4,000 feet, and distributed to the Shan Hills in Burma.
Structure of the Wood.—Kurz remarks that the sapwood is whitish.

12
TIMBER.

Q. 12

[*Gard. Calc. II.* (1889), 23, *Pl.* 15B.

13 Quercus dilatata, *Lindl.; Fl. Br. Ind., V., 602; King, Ann. Bot.*
THE GREEN OAK OF THE HIMÁLAYA.

Vern.—*Moru, tilanga, kilonj, tilonj, timsha,* N.-W. P.; *Tilonj, kilonj, ramshing,* KUMAON; *Káli ring, chora, parúngi, bán, banji, banchar, barachar, baráin, banni, márú, maur, morú, marghang, karsh,* PB.; *Záih,* KAFIRISTAN.

References.—*Brandis, For. Fl., 482; Gamble, Man. Timb., 383; Stewart, Pb. Pl., 199; Atkin on, Him. Dist., 317; Royle, Ill. Him. Bot., 346, t. 84, f. 2; Agri.-Hort. Soc. Ind. Journ., XIII., 384; Gazetteers:—Rawalpindi, 82; Hazara, 14; Simla, 10; Gurdaspur, 55; Ind. Forester, VIII., 114, 123, 127, 406, 407, 408, 409; IX., 126, 197; XI., 5, 284; XIII., 60, 66, 124, 319; XIV., 320; Settle. Repts., Simla, App. II. H., xliii.; Hazara, 11.*

Habitat.—A gregarious, evergreen tree, sometimes attaining a height of 80 feet. It is met with in Afghanistan, the Sulaiman range, and the North-West Himálaya, at altitudes between 7,000 and 9,000 feet. Near Simla, at Mahasu, and on the east side of the ridge between Theog and Matiyana, almost pure forests of this tree occur.

TIMBER. **Structure of the Wood.**—The heartwood is reddish grey with darker
14 streaks, very hard, and seasons well. It is very elastic, easily worked, and warps and splits less than any other species of oak in the North-West Himálaya. Weight 61℔ per cubic foot.

FODDER. **Fodder.**—The LEAVES and SHOOTS of this tree are extensively lopped
Leaves. as fodder for sheep and goats. Unlopped forests are indeed rare. The
15 Himálayan cultivator's idea of rich pasture lands for cattle might be defined as a profus on of stunted and distorted oak trees.

Shoots. **Domestic Uses.**—The wood is used in building, and for agricultural
16 implements and jampan poles. On the Sutlej the timber of the species
DOMESTIC. is prized more than that of the other Oaks. It has been recommended for
17 use in cask-making, and experiments are said to have been made in Jaunsár to test its suitability for this purpose, but the results do not yet appear to have been made known (*Ind. For.*). The tree coppices well. In the Simla district the wood is extensively used in the manufacture of charcoal.

18 **Q. fenestrata,** *Roxb.; Fl. Br. Ind., V., 608; Wight, Ic., t. 219.*

Vern.—*Kala chakma,* BENG.; *Kuhi,* ASSAM *and* MANIPUR; *Patlé, katús,* NEPAL; *Kashiendúng,* LEPCHA; *Dingjing,* KHASIA; *Thitkya,* BURM.

References.—*Roxb., Fl. Ind., Ed. C.B.C., 671; Kurz, For. Fl. Burma, II., 483; Gamble, Man. Timb., 385; Mason, Burma & Its People, 397, 775; King, Ann. Bot. Gard Calc., II. (1889), 45; Pl. 39; Kew Off. Guide to the Mus. of Ec. Bot., 128; Agri.-Hort. Soc., Ind. Journal, V., 186; Ind. Forester, I., 112; VIII., 405; Trotter, Report on Manipur dyes (1883); Darrah, Report on Cotton in Assam, 32.*

Habitat.—A large evergreen tree of the Eastern Himálaya, met with at altitudes between 5,000 and 8,000 feet. It occurs also on the Khásia Hills, in Eastern Bengal, Martaban, and Upper Tenasserim.

DYE. **Dye.**—A decoction of the BARK is used by the Manipuris for changing
Bark. the blue dye of the *rum* (Strobilanthes flaccidifolius) into a black. The
19 Manipuri process has been thus described by the late Major Trotter in his report on Manipur dyes:—"Add 8 or 10 tolas of *kumbang* to the pot containing the blue dye; stir up thoroughly, and then mix 4 tolas of shell lime with it. Take a cloth that has recently been dyed blue, and steep it in two pints of *kuhi* water (prepared by filling a pot with the bark of the *kuhi* tree, and adding enough water to submerge it, letting it stand for 24 hours and then pouring off the water), press, and squeze the cloth in it for about 8 or 10 minutes, then wring out and dry in the sun. When dry, steep thoroughly in the pot containing the blue dye, wring out and

Q. 19

dry in the sun. When dry, again steep it in two pints of *kuhi* water, dry in the sun, and again steep it in the blue dye and dry in the sun, keep on repeating this process until the cloth has taken the proper colour, then wash with clear fresh water and finally dry in the sun." The ACORNS as well as the bark are said by **Major Trotter** to be used in Manipur for dyeing.	DYE. Acorns. 20
Structure of the Wood.—Heartwood red, very hard. Weight about 56℔ per cubic foot.	TIMBER. 21
Domestic Uses.—In the Khásia Hills the wood is used for building and farm purposes.	DOMESTIC. 22

Quercus glauca, *Thunb.; Fl. Br. Ind., V., 604; King, Ann. Bot.* [*Gard. Calc., II., (1889), 29, Pl. 23.*]

23

THE GREEN OAK OF LOWER ALTITUDES OF THE HIMÁLAYA.

Syn.—Q. ANNULATA, *Smith;* Q. PHULLATA, *Ham.;* Q. DENTOSA, *Lindl.;* Q. LAXIFLORA, *Lindl.*

Vern.—*Phalat,* NEPAL; *Siri,* LEPCHA; *Pharonj, phanát, phaliant, inai,* N.-W. P.; *Brán, brén, barin, banni, imbri, indri, bankau,* PB.

References.—*Brandis, For. Fl., 488, t. 65; Gamble, Man. Timb., 387; Stewart, Pb. Pl., 198; Atkinson, Him. Dist., 317; Kew Off. Guide to the Mus. of Ec. Bot., 128; Ind. Forester, V., 196; VIII., 404, 406; XI., 5; Agri.-Hort. Soc., Ind., Trans.. VIII., 359; Jour. XIV., 36; Gazetteers:—Hazara, 13; Rawalpindi, 15; Settle. Rept., Hazara, 15.*

Habitat.—A large evergreen tree, found in the valleys of the Outer Himálaya, from Kashmír to Bhutan, between altitudes of 2,000 and 5,000 feet. It is met with also on the Khásia Hills, and is distributed to Japan.

Structure of the Wood.—Tough, close-grained, very hard, grey or greyish brown in colour. Handsomely mottled and polishes well, but is apt to warp and crack. It is, however, fairly durable if not much exposed to wet. Average weight, 60℔ per cubic foot. | TIMBER.
24 |

Domestic Uses—It is used in the construction of bridges, and of houses, for door-posts, window-frames, rafters, &c., but the wood is not so much esteemed for these purposes as that of **Q. lamellosa.** | DOMESTIC.
25 |

Q. Griffithii, *Hook. f. & Thoms.; Fl. Br. Ind., V., 602; King, Ann.* [*Bot. Gard. Calc., II. (1889), 24, Pl. 18.*]

26

Vern.—*Dingim,* KHASIA.

Reference.—*Gamble, Man. Timb., 321.*

Habitat.—A large deciduous tree, found in the Eastern Himálaya, in Sikkim and Bhutan, at an altitude of 3,500 feet. It also occurs on the Khásia Hills, between 3,500 and 5,000 feet in Manipur, and, according to Brandis, in Burma.

Structure of the Wood.—Brown, very hard, and closely resembling in appearance that of the English oak. | TIMBER.
27 |

Domestic Uses.—It is much used in the Khásia Hills for building and agricultural purposes. | DOMESTIC.
28 |

Q. Ilex, *Linn.; Fl. Br. Ind., V., 602; King, Ann. Bot. Gard. Calc.,* THE HOLM OAK. [*II. (1889), 24, Pl. 17.*]

29

Syn.—Q. BALLOOT, *Griff.;* Q. BALLOOTA, *Desf.*

Vern.—*Spercheref, pargái, kharanja, chúr, keharsu, khareu, irri, yuru, bán, h ru, bré, brezche, yirú, khareo, kathun bán,* PB.; *Khárpata, cherai,* PUSHTU; *Charrei, serei, balút, sháh balút,* AFG.

References.—*Brandis, For. Fl., 480; Gamble, Man. Timb., 383; Stewart, Pb. Pl., 199; Atkinson, Him. Dist., 317; Christy, Com. Pl. & Drugs, V., 48; Gazetteers:—Gurdaspur, 55; Bannu, 23; Dera Ismail Khan, 19; Simla, 10; Ind. Forester, V., 182, 183, 184, 185; VI., 113; VIII., 406; X., 303; XIII., 70, 246, 251, 542; Agri.-Hort. Soc., Ind., Journals (Old Series), XIII., 383; XIV., 79.*

Q. 29

QUERCUS incana.	**The Grey Oak,**

Habitat.—An evergreen shrub or small tree, found on the Western Temperate Himálaya. It occurs usually on the drier ranges from Kumaon westward, at altitudes between 3,000 and 8,500 feet, but was discovered by **Dr. Watt** as far east as Manipur. It is distributed to Afghanistan and from Syria westwards to the Atlantic.

TAN. Bark. 30

Tan.—It is probable that some of the galls of the Panjáb are got from this species. The BARK is used in many parts of France for tanning purposes.

FOOD & FODDER. Leaves. 31 Acorns. 32

Food & Fodder.—The LEAVES when not spinose are used as winter fodder, for which purpose they are stored. In France, the ACORNS are eaten.

TIMBER. 33

Structure of the Wood.—Heartwood red or reddish-brown, very hard and durable. It warps and twists, but when well-seasoned, works admirably and takes a fine polish. Weight generally between 60 and 70℔ per cubic foot.

DOMESTIC. Branches. 34

Domestic Uses.—It is largely used for tool handles, and pieces are brought from the Sulaiman range for that purpose. Agricultural implements are made of it. It yields good fuel and charcoal. The BRANCHES, with strongly spinose leaves, are used for fencing.

35

Quercus incana, *Roxb.; Fl. Br. Ind., V., 603; King, Ann. Bot.*
THE GREY OAK. 　　　　　　　*[Gard. Calc. II. (1889), Pl. 20.*

Syn.—Q. DEALBATA, *Wall.;* Q. LANATA, *Smith ;* Q. LANATA, *Don,* var. INCANA, *Wensig.*

Vern.—*Banj, ban* or *bán,* KUMAON ; *Sila supári,* KASHMIR ; *Bán, ban, rin, rinj, vari, banj, márú, hkarshu, shindar, kharpata serei, daghún-bán,* PB.

References.—*Roxb., Fl. Ind., Ed. C.B.C., 674; Brandis, For. Fl., 482¹ ; Gamble, Man. Timb., 384 ; Stewart, Pb. Pl., 199 ; Baden Powell, Pb. Pl., 385, 593 ; Atkinson, Him. Dist., 317, 748 ; Wardle Rep. on Dyes, Sect. 17 ; Agri-Hort. Soc., Ind. Journals (Old Series), V., 186, VI., 167 (Pro.) 104 ; XIII, 383 ; XIV., 35.; Ind. Forester :—I., 19, 306 ; III., 46, 81 ; VIII., 101, 114, 117, 122, 124, 301, 406, 407, 408, 438 ; IX., 54, 197 ; X., 56, 61 ; XIII, 59, 60, 69, 70 ; XIV., 21, 320 ; Gazetteers :—Rawalpindi, 15 ; Hazara, 13 ; Bannu, 23 ; Gurdaspur, 55 ; Dera Ismail Khan, 19 ; Simla, 10 ; Settle. Repts. :—Hazara, 12 ; Simla, App. II. H. ; Kangra, 21.*

Habitat.—A large, gregarious, evergreen tree, found on the temperate Himálaya from the Indus to Nepal, generally between altitudes of 3,000 and 8,000 feet. It is associated with the Deodar, the tree Rhododendron, and the long-leaved pine. It can be grown also on the Panjáb plains and is distributed to the Shan States of Upper Burma. In spring it becomes of a purplisn colour, owing to the appearance of a fresh burst of softly tomentose leaves.

DYE & TAN. Bark. 36 Galls. 37

Dyes & Tans.—The BARK yields a small quantity of a reddish fawn colouring matter, which can be used in dyeing silk and cotton (*Wardle*). The GALLS of this and other species, chiefly imported from the west, are used in the Panjáb for dyeing the hair (*Stewart*). The bark is extensively employed for tanning purposes.

MEDICINE. Acorns. 38

Medicine.—The ACORNS form the medicine known in the Panjáb bazars as *balút.* They are given as a diuretic and in gonorrhœa, and also as an astringent in indigestion, diarrhœa, especially of children, and in asthma. Before being administered, they are usually buried in the earth to remove their bitter principle, then washed and lastly ground ; dose 3 *mashas.*

FOOD & FODDER. Acorns. 39 Leaves. 40

Food & Fodder.—The ACORNS are greedily eaten by monkeys and bears, which may to some extent account for the sparse natural reproduction of the tree in spite of profuse seeding (*Gamble).* The LEAVES are extensively lopped for fodder.

Q. 40

	QUERCUS infectoria.

The Gall or Dyers' Oak.   (*W. R. Clark.*)

Structure of the Wood.—Heartwood very hard, reddish brown, very liable to warp and split. It is a very difficult wood to season. Weight 64℔ per cubic foot.

TIMBER.
41

Domestic Uses.—The wood is employed for building and for agricultural implements such as plough-handles and beams. Much of the fuel used at Himálayan stations consists of this wood ; it is also made into charcoal (*Stewart*).

DOMESTIC.
42

Quercus infectoria, *Oliver.*

43

THE GALL or DYERS' OAK.

Vern.—*Majuphul, másúphal, masu, múphal,* HIND. ; *Májuphal,* BENG. ; *Máyá, mawa,* SIND. ; *Maithal, mája,* BOMB. ; *Maapul, masu,* DEC. ; *Machakai, mashik-kai,* TAM. : *Majakani,* MALAV ; *Pyintogar-ne-thi,* BURM. ; *Majuphul,* SANS. ; *Uffes,* ARAB. ; *Masú,* PERS.

References.—*Dymock, Mat. Med. W. Ind., 2nd Ed., 729 ; S. Arjun, Bomb. Drugs, 132 ; Murray, Pl. & Drugs, Sind. 36 ; Ajm. Med. Top., 147 ; Irvine, Mat. Med. Patna, 68 ; Birdwood, Bomb. Pr., 83 ; Christy, Com. Pl. & Drugs, V., 35, 40, 43 ; Buck, Dyes & Tans,, N.-W. P., Sect. 16 ; Liotard. Dyes, 112 (App. II) Smith. Dic., 295 ; Kew Off. Guide to Bot. Gardens & Arboretum, 128 ; Ind. Forester, XI, 55.*

Habitat.—A middle-sized tree or shrub, native of Greece, Bosnia, Asia Minor, and Syria, extending east to the confines of Persia. It produces cylindrical acorns, and has the leaves grey underneath. It yields the galls (used in medicine and dyeing), which are imported into Europe from the Levant and the Persian Gulf *viá* Bombay.

Dye.—"The real gall or dyer's oak is not indigenous to India, but is found in Greece, Asia Minor, and Syria. It is believed, however, that some of the GALL-NUTS employed in India are obtained from Oak-trees found in the Kumaon, Garhwál, and Bijnor forests. The best gall-nuts, used by the dyers of India, are mainly obtained from Calcutta.

DYE.
Gall-nuts.
44

"In dyeing, the galls are boiled in water in the proportion of 2 oz. of the former to a quart of the latter, till three-fourths of the water is evaporated. The cloth is then dipped in the decoction" (*Buck, Dyes & Tans, N.-W. Provinces*).

CHEMISTRY.—"The rough taste of galls is due to their chief constituent, tannic or gallo-tannic acid. Tannic matter was long supposed to be of one kind, namely, that found in the oak gall, but in 1843 it was pointed out by Stenhouse that the tannin of oak-galls differs from that contained in oak-bark, hence the name of gallo-tannic, and to distinguish the former kind which is principally derived from galls. It has, however, been since shown that the tannic acid found in the leaves of **Rhus coriaria** is identical with that of oak-galls. Gallic acid to about 3 per cent. exists in galls. Free sugar, resin and protein substances have also been found" (*Flückiger & Hanbury, Pharmacographia*).

Chemistry.
45

Medicine.—It seems probable that a large proportion of the GALLS imported into India are obtained from this Oak. Galls are the most powerful of vegetable astringents, and are recommended as an antidote in poisoning with vegetable alkaloids. Tincture of galls, diluted with water, is a very useful astringent gargle and wash. The ointment of galls, either alone or in combination with opium, forms a valuable application to hæmorrhoids (*Bentley & Trimen*). In the Pharmacopœia of India, galls are described as astringent, tonic, and antiperiodic. They are prescribed by the native practitioners in dysentery and diarrhœa and are also given as tonics in intermittent fever ; the powder, moistened with a little water, is applied to chapped nipples, and, made into a soft ointment, it is a useful application to blind piles (*Ainslie*). "Two varieties of galls are sold in the bazars of India, the black and the white.

MEDICINE.
Galls.
46

Q. 46

They are used indiscriminately as astringents in the bowel complaints of children (*Arjun*).

SPECIAL OPINIONS.—§ "A powerful astringent useful in passive hæmorrhages, diarrhœa, and dysentery. The best antidote in poisoning by tartar emtic" (*Civil· Surgeon C. M. Russel, Saron*). "Valuable astringent and styptic" (*Assistant Surgeon S. C. Bhattacherji, Chanda, Central Provinces*). "Astringent locally, in leucorrhœa, gonorrhœa, and hæmorrhoids " (*Assistant Surgeon Nehal Sing, Saharunpore*). " Powerful astringents. Decoction used as gargle. Powder used with opium as gall-ointment" (*Civil Surgeon S. M. Shircore, Moorshidabad*).

47

Quercus lamellosa, *Smith ; Fl. Br. Ind., V., 606 ; King, Ann. Bot·*
[*Gard. Calc., II. (1889), 36, Pl. 30.*

Syn.—Q. IMBRICATA, *Ham ;* Q.ˊ LAMELLATA, *Roxb. ;* Q. PAUCILAMELLOSA, *DC.*

Vern.—*Shalshi, pharat-singhali, budgrat,* NEPAL ; *Búk,* LEPCHA.

References.—*Roxb., Fl. Ind., Ed. C.B.C., 673 ; Brandis, For. Fl., 488 ; Gamble, Man. Timb., 387 ; Kew Off. Guide to the Mus. of Ec. Bot., 128 ; Agri.-Hort. Soc., Ind. Trans., V., 126 ; Journals, V, 186, 192 ; VI., (Pro.), 120 ; Ind. Forester, I., 94 ; III., 46 ; V., 196, VIII., 404 ; XI., 315, 355.*

Habitat.—A very large evergreen tree of the Eastern Himálaya, from Nepal to Bhutan, which occurs at altitudes between 5,000 and 8,000 feet. It is also found in the Naga and Duphla Hills and in Manipur. It is by far the commonest of the oaks in the Darjiling Forests, where it often attains a height of 100 to 120 feet and a girth of from 20 to 30 feet. but old trees are very frequently hollow.

TAN.
Bark.

Tan.—In Darjiling, the BARK is used for tanning.

Medicine.—The BARK and ACORNS are employed in medicine.

48
MEDICINE.
Bark
and Acorns.
49
TIMBER.
50
DOMESTIC.
51

Structure of the Wood.—Heartwood greyish brown, showing a beautiful silver grain on a radial section. " In appearance the wood is very much like that of the English oak, but has the medullary rays exceedingly developed. When well worked it takes on a good polish" (*Ind. For.*). It is durable if not much exposed to wet, and does not warp to the same extent as do the timbers of **Q. incana** and **Q. annulata**. Weight 59℔ per cubic foot.

Domestic Uses.—"The wood is used for beams and posts in the construction of houses and bridges " (*Gamble*).

52

[*Gard. Calc., II. (1889), 79, Pl. 74.*
Q. lanceæfolia, *Roxb. ; Fl. Br. Ind., V., 616 ; King, Ann. Bot.*

Syn.—CASTANOPSIS LANCEÆFOLIA, *Kurz ;* CASTANEA TRIBULOIDES *Wall. ;* Q. GLOMERATA, *Wall.* (non *Roxb.*).

Vern.—*Shingra, chauko,* GARO ; *Bucklai,* ASSAM. ; *Dingsning,* KHASIA ; *Hingori,* CACHAR ; *Patle katús,* NEPAL ; *Siri,* LEPCHA.

References.—*Roxb., Fl. Ind., Ed. C.B.C., 671 ; Gamble, Man. Timb.; 388 ; Agri.-Hort. Soc., Ind. Jour. V., 186 ; Ind. Forester, III., 47 ; IV., 286.*

Habitat.—A small evergreen tree, met with in the Sub Himálayan tracts of Sikkim and Bhutan, up to an altitude of 2,000 feet. It occurs also in Assam, Manipur, Chittagong, and Upper Burma.

TIMBER.
53
DOMESTIC.
54

Structure of the Wood.—Greyish white, hard ; weight 42℔ per cubic foot.

Domestic Uses.—The wood is used in Assam for building.

55

[*Gard. Calc., II. (1889), 25, Pl. 19.*
Q. lanuginosa, *Don ; Fl. Br. Ind., V., 603 ; King, Ann. Bot.*

Syn.—Q. LANATA, *Smith ; also Wall. ;* Q. BANGA, *Ham. mss.*

Vern.—*Banga,* NEPAL ; *Ranj, rionj, rai banj,* KUMAON.

Q. 55

The Brown Oak. (*W. R. Clark.*) **QUERCUS semecarpifolia.**

References.—*Brandis, For. Fl., 481 ; Gamble, Man. Timb., 384 ; Atkinson, Him. Dist.. 317 ; Agri.-Hort. Soc. Ind. Jour. (Old Series), IV., 239 ; Ind. Forester, III., 47 ; VIII., 406 ; XI., 5.*

Habitat.—A large evergreen tree of the Temperate Himálaya, from Kumáon to Bhutan; found at altitudes between 6,000 and 7,500 feet. It is gregarious or associated with **Q. incana.**

Fodder.—The LEAVES and young TWIGS are lopped for fodder.

Structure of the Wood.—Greyish brown in colour, very hard, but apt to warp and split. "It is very liable to the attacks of a small black hymenopterous insect which often riddles it completely in a few years" (*Atkinson*). Weight 55lb per cubic foot.

Domestic Uses.—It is used as fuel.

[*Bot. Gard. Calc., II.* (*1889*) *41, Pl. 33.*

Quercus lappacea, *Roxb. ; Fl. Br. Ind., V., 607 ; King, Ann.*

Syn.—Q. HIRSUTA, *Lindl.* ; Q. MACKIANA, *Hook.*

Vern.—*Ulu chakma,* BENG. ; *Thitcha,* BURM.

References.—*Roxb., Fl. Ind., Ed. C.B.C., 672 ; Brandis, For. Fl., 489 ; Kurz, For. Fl. Burm., II., 484 ; Gamble, Man. Timb., 386 ; Dals. & Gibs., Bomb., Fl., 484 ; Agri.-Hort. Soc., Ind. Jour., V. 186.*

Habitat—An evergreen tree of Eastern Bengal, the Khásia Hills, and Tennasserim, ascending to altitudes of 4,000 feet. It is distributed to the Malayan Peninsula.

Structure of the Wood. Sapwood light brown or yellow. Heartwood very hard, reddish in colour. Weight about 56lb per cubic foot.

Domestic Uses.—It has been suggested that this might be a valuable wood for cabinet work. (*Jour. Agri.-Hort Soc. Ind.*)

[*Gard. Calc., II.* (*1889*), *Pl. 38.*

Q. pachyphylla, *Kurz ; Fl. Br. Ind., V., 608 ; King, Ann. Bot.*

Vern.—*Bara katás,* NEPAL ; *Hlosiri,* LEPCHA.

References —*Gamble, Man. Timb., 386 ; Ind. Forester, V., 196 ; VI., 146 ; VIII., 405. IX., 322 ; XI., 355.*

Habitat.—An evergreen tree or shrub found in the Sikkim-Himálaya, between altitudes of 6,000 and 10,000 feet, and in Manipur between 7,000 and 9,000. Dr. King, in his *Annals of the Calcutta Botanical Gardens,* remarks, "In Sikkim this forms a magnificent tall tree, but on the Eastern Manipur frontier, Dr. Watt found it only as a bush, a singular variation in habit, within limits so narrow and under climatic conditions so similar."

Dyes & Tans.—The BARK and ACORNS are said to be used in Manipur for dyeing and tanning. "If this be so," says Dr. Watt, "the Naga Hills might afford an unlimited supply, since the higher forests of Manipur are covered with miles upon miles of this species of oak, mixed with **Q. dealbata, Q lamellosa,** and **Q. Ilex.**"

Medicine.—In Sikkim the BARK and ACORNS are used medicinally as astringents.

Structure of the Wood.—Grey in colour, seasons well. does not warp or split, and is much more durable under exposure to damp than that of **Q. lamellosa** or of **Q. annulata.** Weight 50lb per cubic foot.

Domestic Uses.—It is largely employed in Darjiling for planks, palings, shingles, and other purposes (*Gamble*).

[*Bot. Gard. Calc., II.* (*1889*), *21, Pl. 154.*

Q. semecarpifolia, *Smith ; Fl. Br. Ind., V., 601 ; King, Ann.*

THE COMMON BROWN OAK OF THE HIMÁLAYA.

Syn.—Q. OBTUSIFOLIA & CASSURA, *Don.*

Vern.—*Ghesi, kasru,* NEPAL ; *Kharen, karshu,* KUMAON ; *Banchar, jangal ka parúngi, kreu, khareu, krúi, karshú, karsúi, sañj, karsu, khareo, kharshú, khuriu,* PB.

FODDER. Leaves. 56 Twigs. 57 TIMBER. 58 DOMESTIC. 59 60

TIMBER. 61 DOMESTIC. 62

63

DYE & TAN. Bark. 64 Acorns. 65 MEDICINE. Bark. 66 Acorns. 67 TIMBER. 68 DOMESTIC. 69 70

| QUERCUS serrata. | The Oak. |

References.—*Brandis, For. Fl., 479 ; Gamble, Man. Timb., 382 ; Stewart, Pb. Pl., 200 ; Atkinson, Him. Dist., 317 ; Agri.-Hort. Soc., Ind. Jour., VII., 89, 103 ; XIII., 384 ; Gazetteers :—Gurdaspur, 55 ; Simla, 10 ; Ind. Forester:—III., 47, 82 ; V., 183, 185 ; VIII., 38, 101, 114, 123, 124, 127, 400, 404, 406, 407, 408, 409, 412 ; IX., 50, 126, 196, 198, 302 ; X., 316 ; XI., 284; XIII., 65, 68, 69, 318, 320 ; Settle. Repts :—Kangra, 21 ; Simla, App. II. H., xliii.*

Habitat.—A small or large, sub-evergreen gregarious tree, which occurs on the temperate Himálaya from Afghánistán to Bhutan, at elevations between 6,000 and 12,000 feet. Specimens of it were collected also on the Naga-Burma frontier by **Dr. Watt** at an elevation of 10,000 feet. "It is usually rather a small tree with gnarled tortuous branches, but occasionally attains a height of 80 to 100 feet, and then has a fine clear stem" (*King, l.c.*). It coppices well, reproduces readily from seed, and often forms considerable extents of almost pure forests.

FODDER.

Leaves.
71

TIMBER.
72

DOMESTIC.
73

Fodder.—The LEAVES are used as fodder and are stored in winter for that purpose.

Structure of the Wood.—Heartwood grey, often with a reddish tinge. It is said to warp on exposure and to be liable to the attacks of insects. Weight 53 and 54℔ per cubic foot.

Domestic Uses.—Though probably better than that of the other Himálayan oaks, the wood, on account of its great weight and the remote localities where it grows, is little exported, but on the spot is used in house-building and for door-frames, bed-steads, carrying poles, and ploughs. It is a good firewood and an excellent source of charcoal.

74

[Bot. Gard. Calc., II. (1889, 28), Pl. 22.
Quercus semi-serrata, *Roxb.; Fl. Br. Ind., V. 604; King, Ann.*

Syn.—Q. HORSFIELDII, *Miq.*

Vern.—*Thitkya,* BURM.

References.—*Brandis, For. Fl., 488 ; Kurz, For. Fl. Br. Burm., II.,488 ; Agri.-Hort. Soc., Ind. Jour., V.,186 ; Ind. Forester, VIII.,405.*

Habitat.—An evergreen tree, 40 to 50 feet high, found on the plains of Assam and Cachar and on the Garo and Khásia Hills, up to an elevation of 3,000 feet. It occurs also in Burma from Pegu to Tennasserim, and is distributed to Sumatra and Banka.

TIMBER.
75

DOMESTIC
76

Structure of the Wood.—Weight 48℔ per cubic foot.

Domestic Uses.—The wood of this tree is used for the plugs or pins to join together the three pieces that compose a Burmese cart-wheel (*Kurz*).

77

[Calc., II. (1889), 22, Pl. 16.
Q. serrata, *Thunb.; Fl. Br. Ind,, V., 601 ; King, Ann., Bot. Gard.*

Syn.—Q. POLYANTHA, *Ldl;* Q. ROXBURGHII, *Endl.*

Vern.—*Dingrittiang,* ASSAM.

References.—*Brandis, For. Fl., 486 ; Gamble, Man. Timb., 384; Ind. Forester, VI., 25 ; XIII., 52, 74.*

Habitat.—A moderate-sized, deciduous tree of the Eastern temperate Himálaya, extending from Nepal to Bhutan, at altitudes between 5,000 and 6,000 feet. It is found in Manipur and on the Khásia Hills at altitudes between 3,000 and 5,500 feet, and is distributed to the Shan Hills, China, and Japan.

DYE.
78

TIMBER.
79

OMESTIC.
80

Dye.—With the addition of sulphate of iron, the Japanese produce a black dye from the galls of this species.

Structure of the Wood.—Brown, very hard ; much resembles that of Q. Griffithii in structure (*Conf.* with **Silk**).

Domestic Uses.—The timber is used in Assam for building. In Japan the "Yamamai" silkworm is raised on its leaves.

Q. 80

The Cork Oak.	(*W. R. Clark.*)	**QUINOA** **Seed.**

[*Gard. Calc., II. (1889), Pl. 41, 42, 43.*

Quercus spicata, *Smith ; Fl. Br. Ind., V., 609 ; King, Ann. Bot.*   81
    Syn.—Q. ARCAULA, *Ham.* ; Q. ELEGANS, *Blume* ; Q. GRANDIFOLIA, *Don* ;
      Q. PYRIFOLIA, *Blume* ; Q. RACEMOSA, *Jack* ; Q. SQUAMATA, *Roxb.*
    Vern.—*Bara chakma,* BENG. ; *Sahu hingori,* ASSAM ; *Dingjing,* KHASIA ;
      Danwa singali, phaco singali, arkaula, NEPAL ; *Kaching,* LEPCHA ;
      Thitcha, BURM.
    References.—*Roxb., Fl. Ind., Ed. C.B.C., 673 ; Brandis, For. Fl., 489 ;*
      Kurz, For. Fl. Burm., II., 486 ; Gamble, Man. Timb., 385 ; Kew Off.
      Guide to the Mus. of Ec. Bot., 128 ; Ind. Forester, III., 46, 47 ; VIII.,
      404, 405.
    Habitat.—A large or small evergreen tree, found on the Tropical
Himálaya from Nepal eastwards, between altitudes of 2,000 and 4,000 feet.
It occurs also from Assam and Manipur to Tennasserim and the Malay
Peninsula. It is distributed to the Malay Archipelago. It is often almost
gregarious, or mixed with chestnut. Being very widely distributed, it
presents a number of different forms.
    Structure of the Wood.—Red, very hard. It is very durable and does   **TIMBER.**
not warp. Weight 58℔ per cubic foot.   82
    Domestic Uses.—The timber is used for building in Assam and for   **DOMESTIC.**
charcoal in Darjiling.   83

84

Q. Suber, *Linn. ; Brandis, For. Fl., 485.*
    THE CORK OAK ; CHÊNE LIÈGE, *Fr.* ; SOVERO, SUGHERO, *It.*
    References.—*Atkinson, Him. Dist., 884 ; Agri-Hort. Soc., Ind. Trans.,*
      III., 40, 41 ; Journ., VII. (Pro.), 87 ; X. (Pro.), 90 ; Ind. Forester, X.,
      303 ; XIII., 253 ; XIV., 58, 430 ; Smith, Econ. Dict., 133.
    Habitat.—A stout, middle-sized tree of the Oak family, native of
South Europe and North Africa. " At various times, seeds of the Italian
and Spanish cork oak have been planted in Dehra Dun, and have germinated freely. The young trees thrive well, and may, hereafter, prove of
value, but the success of the experiment has yet to be seen " (*Atkinson*).

[*Calc., II. (1889), 73, Pl. 69A.*   85
Q. Thomsoni, *Miq ; Fl. Br. Ind., V., 615 ; King, Ann. Bot. Gard.*
    Syn.—Q. LEUCO-CARPA, *Hook. f. & Thoms.* ; Q. TURBINATA, *Roxb.*
    Vern.—*Bansua batana,* CHITTAGONG.
    References.—*Roxb., Fl. Ind., Ed. C.B.C., 672 ; Kurz, For. Fl. Br. Burma,*
      II., 486.
    Habitat.—A large tree, often attaining a height of 80 to 100 feet, found
in Sylhet and the Khásia Hills up to altitudes of 5,000 feet. It occurs also
in Chittagong and Burma.
    Domestic Uses.—The timber is used for fuel only.   **DOMESTIC.**

Quicklime, see **Carbonate of Lime,** Vol. II., 142.   86

Quicksilver, see **Mercury,** Vol. V., 232.

QUILLS.   87

    Quills.—The strong outer-wing feathers of the various species of geese
and swans used for writing, and, in India, for various forms of ornamental
work (see **Ducks, Teals, Geese and Swans,** Vol. III., 196). For porcupine
quills the reader is referred to **Porcupines,** p 328.

Quince, see **Cydonia vulgaris,** *Pers.* ; ROSACEÆ ; Vol. II., 676.

Quinia or **Quinine,** see **Cinchona,** *Linn.* ; RUBIACEÆ ; Vol. II., 289.

Quinoa seed, see **Chenopodium Quinoa** ; CHENOPODIACEÆ ; Vol. II., 268.

An Anthelmintic Medicine.

QUISQUALIS, *Linn.; Gen. Pl., I., 689.*

[*357;* COMBRETACEÆ.

88   Quisqualis indica, *Linn.; Fl. Br. Ind., II., 459; Wight, Ill., t.*

Vern.—*Rangún-ki-bel,* HIND.; *Vilayati chambeli,* BOMB.; *Rangún-ki-bíl,* DECCAN; *Irangún-malli,* TAM.; *Dawéhmaing,* BURM.

References.—*Roxb., Fl. Ind., Ed. C.B.C., 397 ; Brandis, For. Fl., 220 ; Kurz, For. Fl. Burm., I., 467 ; Gamble, Man. Timb., 178 ; Mason, Burma & Its People, 421, 743 ; Pharm. Ind., 90 ; Dymo k, Mat. Med. W. Ind., 2nd Ed., 324; S. Arjun, Bomb. Drugs, 211 ; Gasetteers:—Mysore & Coorg, I., 60 ; Bombay, XV., 434.*

Habitat.—A scandent shrub, indigenous to the Malay Archipelago. Cultivated in gardens throughout India up to altitudes of 1,000 feet.

MEDICINE.
Seeds.
89
Leaves.
90

Medicine.—In the Moluccas, the SEEDS have for long been considered anthelmintic. Rumphius in his *Amboyan Flora* states, that in Amboya the seeds were, even in his time, used as anthelmintics for children and the LEAVES given in a compound decoction for flatulent distension of the abdomen, and that in China the ripe seeds were roasted and given in diarrhœa and fever. In 1883, Dr. Oxley and Mr. Gordon of Singapore recommended the use, in cases of lumbrici of children, of four or five of these seeds, bruised and given in an electuary with honey or jam. Bouton, in his *Plants of Mauritius,* states that, if beyond four or five of these seeds are given, they are apt in some cases to cause colic (*Waring, Pharm. Ind.*).

(*W. R. Clark.*)
RACONDA.

Raconda russelliana, *Day ;* see Fish, Vol. III., 392.

Radish, see Raphanus sativus, *Linn. ;* CRUCIFERÆ; p. 393·

Rafters, Timbers used for, see Vol. IV., 300.

Rags, see Paper Materials, Vol. VI., 106-109.

Raisins, see Vitis vinifera, *Linn. ;* AMPELIDEÆ ; Vol. VI., Pt. II.

Raji or rágí, see Eleusine Coracana, *Gærtn.*; GRAMINEÆ ; Vol. III., 237·

RANDIA, *Linn.; Gen. Pl., II., 88.*
[*t. 580-583 ;* RUBIACEÆ.

Randia dumetorum, *Lamk. ; Fl. Br. Ind., III., 110 ; Wight, Ic ,*   **I**

> **Syn.**—RANDIA NUTANS, LONGISPINA, and FLORIBUNDA, *DC. ;* R. ROTT-LERI, *W. & A.;* R. MALABARICA, *Wall. ;* GARDENIA FLORIBUNDA, LONGISPINA, NUTANS, *Roxb. ;* G. DUMETORUM, *Retz. ;* POSOQUERIA DUMETORUM, *Willd.*
>
> **Vern.**—*Mainphal, manyúl, karhar, main, arar, mainhúri, manneal,* HIND. ; *Menphal, madan,* BENG. ; *Guról,* RAJBANSHI ; *Pativa,* URIYA; *Pato, portoho,* KOL.; *Loto, boi bindi,* SAN'AL ; *Guról,* ASSAM ; *Maidal, amuki, maida phul,* NEPAL ; *Panji,* LEPCHA ; *Gundrow,* MICHI ; *Kuay, katúl,* GOND ; *Mainphal, mányúl, karhár, main, maini,* N.-W. P. ; *Mainphal, manyúl, karhar,* KUMAON ; *Mindhal, mindla, mand-kolla, arara,* PB. ; *Ghétu* (MELGAT), MERWARA ; *Jous-ul-maindal,* SIND.; *Bhita,* KURKU ; *Ghela, gelaphal, gelaphala, gehela,* BOMB. ; *Gera, galay, gelaphala,* MAR. ; *Medhola, mindhal, mirdhala,* GUZ. ; *Gehela, piar-alú,* DEC.; *Maruk-kallán-kay, madu-karray, makarung-kai,* TAM. ; *Mandó, manga,* TEL. ; *Karé,* KAN. ; *Sethanbaya, hsay-than-paya,* BURM. ; *Wali-kukuru-man,* SING.; *Madana,* SANS. ; *Jous-ul-kosul,* ARAB. ; *Jús-ul-kueh,* PERS.
>
> **References.**—*Roxb., Fl. Ind., Ed. C.B.C., 239-242 ; Voigt, Hort. Sub. Cal., 380 ; Brandis, For. Fl., 273 ; Kurz, For. Fl. Burm., II., 45 ; Beddome, Fl. Sylv., t. 132 ; Gamble, Man. Timb., 227 ; Dalz. & Gibs., Bomb. Fl., 119 ; Stewart, Pb. Pl., 116 ; Rev. A. Campbell, Rept. Econ. Pl., Chutia Nagpur, Nos. 7801, 9279 ; Sir W. Elliot, Fl. Andhr., 111 ; Pharm. Ind., 118 ; O'Shaughnessy, Beng. Dispens., 399 ; U. C. Dutt, Mat. Med. Hind., 177, 308 ; S. Arjun, Cat. Bomb. Drugs, 71 ; Murray, Pl. & Drugs, Sind, 194 ; Dymock, Mat. Med. W. Ind., 2nd Ed., 406-408 ; Baden Powell, Pb. Pr., 354, 593 ; Drury, U. Pl. Ind., 363 ; Atkinson, Him. Dist. (Vol. X., N.-W. P. Gaz.), 311, 748 ; Useful Pl. Bomb. (Vol. XXV., Bomb. Gaz.), 85, 162, 252, 256, 272, 278 ; Liotard, Dyes & Tans, App., iv ; For. Ad. Rep., Ch. Nagpur, 1855, 32 ; Home Dept. Cor., 226, 239, 283 ; Gazetteers :—Bombay. V., 285 ; XIII., 24 ; XV., 35, 75 ; XVIII., 45 ; N.-W. P., I., 81 ; IV., lxxii. ; X., 311. ; Mysore and Coorg. I., 70 ; Agri.-Horti. Soc. :—Ind., Journ. (Old Series), X., 10 ; XI., 143-146 ; XIII., 313 ; Ind. Forester, VIII., 101, 417 ; X., 325 ; XII., App. 15 ; XIII., 121 ; XIV., 112 ; Balfour, Cyclop. Ind., III., 263.*

    Habitat.—A deciduous thorny shrub or small tree, found throughout India, or which may be said to extend in the North-West, outer Himálaya, as far as the Beas, and thence southwards to Chittagong, Pegu, Martaban, the Western Peninsula, and Ceylon. In Sikkim it occurs up to altitudes of 4,000 feet. It is distributed to Java, Sumatra, South China, and Eastern Tropical Africa.

    Dye.—The FRUITS are used as a colour intensifier, in calico-printing (*Buck, Liotard*). They are said to be used in China to produce a yellow dye (*Agri.-Horti. Soc Journ. l. c.*).

    Medicine.—The BARK, RIND, and FRUIT are all used medicinally. The first is given internally and is also applied externally when tne bones ache during fever (*Rev. A. Campbell*). Mixed with cowdung it is applied to

DYE.
Fruit.
2
MEDICINE.
Bark.
3
Rind.
4
Fruit.
5

MEDICINE.

Pulp.
6

bruises to relieve pain (*Stewart*). It is also said to be astringent, and is given by native practitioners in diarrhœa and dysentery. The rind and PULP of the fruit are used as an emetic; but with regard to the former of these the late **Dr. Moodeen Sheriff**, in his Supplement to the *Pharmacopœia of India*, says that the shell or rind of the fruit has no emetic properties, but, if anything, is slightly irritant. He, therefore, considers that the pulp only, separated from the hard rind and the seeds, ought to be administered. He continues his remarks as follows : " The contents of two or three nuts are generally a sufficient dose; they should be bruised, macerated for ten or fifteen minutes in three or four ounces of water, rubbed and strained through cloth. The draught is now ready for use and produces nausea and vomiting in about ten minutes; emesis should be promoted by the administration of warm water."

"The pulp of **R. dumetorum** is certainly a very valuable emetic, and also the cheapest one in India, one pie worth of it being sufficient for a dose. Its action is very safe, certain, and regular, and it is equal, if not superior, to Ipecacuanha in every respect except its dose, which is larger than that of the latter. In smaller doses the pulp is nauseant, expectorant, and diaphoretic; in fact it possesses all the medicinal properties of Ipecacuanha." "The pulp is very useful in all the diseases in which Ipecacuanha is indicated, and it should be used in the same manner and form. It possesses a great beneficial influence over the mucous membrane of the alimentary canal, through its combined action as an emetic, depressant, nauseant, and diaphoretic. Like Ipecacuanha, of course, the pulp of **R. dumetorum** alone is not always sufficient to effect or complete a cure in dysentery, but often requires the assistance of other medicines, particularly the preparations of opium."

"PREPARATIONS.—Simple powder is the best and most convenient form for keeping the pulp ready for use. This powder is prepared as follows : —After removing the rind the interior should be well bruised and passed through an ordinary sieve or thin cloth, by which means all the seeds will be separated. The coarse powder thus obtained should be powdered well and passed again through a fine sieve or cloth. It is now fit for use and may be kept in corked or stoppered bottles."

"DOSES.—Of the powder 15-40 grains as an emetic, 15-30 grains or more in dysentery, 5-10 grains as a nauseant, expectorant, and diaphoretic."

The pulp of the fruit is believed by many native practitioners to have also anthelmintic properties, and is said to be used sometimes as an abortefacient. Ground into a coarse powder and applied thus to the tongue and palate it is highly esteemed as a domestic remedy for the fevers and incidental ailments which children are subject to while teething (*Murray*). Externally it is applied to rheumatic swellings, abcesses, acne, and other cutaneous affections. In colic, the fruit is rubbed into a paste with rice water, and applied over the navel to relieve the pain. In a recent article in the *Lancet* **Sir James Sawyer** of Dublin says that his attention having been drawn to this drug by **Mr. D. Hooper**, he has used it as " a nervine calmative and antispasmodic in cases in which the vegetable antispasmodics, such as Valerian and Asafœtida, appear to be indicated. I now desire to invite my brethren to examine this drug for themselves." **Sir James** adds that he has had four tinctures prepared of the drug, *viz.*, with ether, rectified spirits of wine, aromatic spirits of ammonia, and proof spirits, respectively of the strength of one part of the drug in five parts of the tincture. When any one of these is diluted with water and acidulated with acetic acid, the odour of Valerian is very apparent, especially in the first one, which is the tincture **Sir James** has selected for therapeutic

R. 6

A Useful Astringent. (*W. R. Clark.*) RANDIA uliginosa.

use. The dose of the Ethereal Tincture given by **Sir James** is xv. to lx. *M.* with water.

Food.—The fresh, ripe FRUIT is roasted and eaten by the natives in many parts of the country. FOOD. Fruit. 7

Structure of the Wood.—White or light-brown, compact, hard. Weight 55℔ per cubic foot. TIMBER. 8

Domestic and Sacred.—The WOOD is used for agricultural implements, fences, and fuel. It is not much valued. The bruised ROOT, as also the FRUIT, before it is ripe, are thrown into ponds to poison fish. By the hill people in many parts of the Himálaya, the fruit is used instead of SOAP, but is said to destroy the clothes. In the Konkan it is mixed with grain to preserve it from the attacks of insects (*Dymock*). DOMESTIC. Wood. 9 Root. 10 Fruit. 11 Soap. 12

Among the Hindus this plant is considered sacred to the god Siva, and at the marriage ceremonies of the Vaisya caste the fruit is, together with that of **Helicteres Isora,** tied upon the wrists of both bride and bridegroom. SACRED. 13

Randia tetrasperma, *Bth. & Hook. f.; Fl. Br. Ind., III., 109.* 14

Syn.—GARDENIA DENSA, *Wall.;* G. TETRASPERMA and DENSA, *DC.;* G. LONGISPINA, *Wall.*

Vern.—*Bara garri, batya gingaru,* KUMAON.

References.—*DC., Prod., IV., 381, 382; Roxb., Fl. Ind., Ed. C.B.C., 238; Brandis, For. Fl., 272; Gamble, Man. Timb., 227; For. Ad. Rep., Ch. Nagpur, 1885, 32; Gazetteers, N.-W. P., IV., lxxiii.; X., 311; Journ. Agri.-Horti. Soc., Ind. (Old Series), XIV., 37.*

Habitat.—A small shrub of the Salt Range and sub-tropical Himálaya, from Kashmír eastward, found up to altitudes of 6,000 feet in Kumáon, and of 7,000 feet in Sikkim and Bhután. It occurs also in Assam.

Structure of the Wood.—White, very hard. Weight 56℔ per cubic foot. In structure it resembles that of **R. uliginosa.** TIMBER. 15

R. uliginosa, *DC.; Fl. Br. Ind., III., 110; Wight, Ic., t. 397.* 16

Syn.—GARDENIA ULIGINOSA, *Retz.* (?); G. POMIFERA, *Wall.;* POSOQUERIA ULIGINOSA, *Roxb.*

Vern.—*Pindálu, panar, paniah, katul, pindar, bharani, katúl,* HIND.; *Piralo,* BENG.; *Pendra,* URIYA; *Kúmkúm, pindar,* KOL; *Pinde,* SANTAL; *Maidal,* NEPAL; *Katil, pender,* GOND; *Banpindalu,* (BIJNOR), *Panár,* (OUDH), *Paniya,* (GORAKPUR), N.-W. P.; *Pindaru,* KUMAON; *Púrputá,* MELGHAT; *Gangru, gangáru,* KURKU; *Kaurio,* PANCH MEHALS; *Pendari, telphetru, pindra, panar, katul,* BOMB.; *Tapkél,* BHIL; *Panelva, cindra, telphetru, phetra,* MAR.; *Wagata,* TAM.; *Nalaika, devátá-malle, nalla kákasi, gúdu,* TEL.; *Karé, pendri, pandri,* KAN.; *Mhaniben mhan-hpyoo, ny-an-gyee,* BURM.; *Hmanbyu,* SHAN.

References.—*Roxb., Fl. Ind., Ed. C.B.C., 239; Voigt, Hort. Sub. Cal., 381; Brandis, For. Fl., 273; Kurz, For. Fl. Burm., II., 44; Dalz. & Gibs., Bomb. Fl., 119; Sir W. Elliot, Fl. Andhr., 46, 63, 125, 149; Dymock, Mat. Med. W. Ind., 2nd Ed., 406; Atkinson, Him. Dist. (Vol X., N.-W. P. Gaz.), 311; Useful Pl. Bomb. (Vol. XXV., Bomb. Gaz.), 85, 162; For. Ad. Rep., Ch. Nagpur, 1885, 32; Gazetteers:—Bombay, XV., 35; N.-W. P., IV., lxxiii.; Agri.-Horti. Soc., Ind.:—Trans., IV., 103; Journals, IX., Sel., 57; XIII., 314; Ind. Forester, III., 203; VIII., 402; IX., 238; X., 325; XIII., 121; Balfour, Cyclop. Ind., III., 363.*

Habitat.—A small, deciduous tree of Eastern, Central, and Southern India, but not commonly found in the more northern parts of the Peninsula. It is distributed to Burma, where it occurs very commonly in the low forests of Martaban and Tenasserim.

Dye.—The FRUIT, which closely resembles that of **R. dumetorum,** is similarly used in dyeing as a colour intensifier (*Agri.-Horti. Soc. Journ.*). DYE. Fruit. 17

Medicine.—The FRUIT is astingent, and is used as a remedy in diarrhœa and dysentery, especially in the case of pregnant women. For this purpose, MEDICINE. Fruit. 18

RANUNCULUS
sceleratus   The Water Celery.

MEDICINE. Root. 19 FOOD & FODDER. Fruit. 20 Leaves. 21 TIMBER. 22 DOMESTIC. Fruit. 23	it is cooked in wood-ashes, but the pulp alone is employed. The ROOT, boiled in *ghi*, is, however, sometimes given in similar cases. Food & Fodder.—The FRUIT, boiled or roasted, is often eaten by the Natives as a vegetable either alone or more frequently in curries. "The LEAVES are boiled and eaten as greens" (*Atkinson*). They also serve as fodder for cattle. Structure of the Wood.—Whitish-grey, close-grained, hard, no heart-wood. Weight 48℔ per cubic foot. It is not used for any special purpose. Domestic.—The unripe FRUIT of this species, like that of **R. dumetorum**, is employed for poisoning fish.

RANUNCULUS, *Linn.; Gen. Pl., I., 5.*

24   **Ranunculus aquatilis,** *Linn.; Fl. Br. Ind., I., 16;* RANUNCULACEÆ.
THE WATER CROWFOOT.
References.—*Gazetteers, N.-W. P., IV., lxxii.; X., 304; Agri.-Horti. Soc., Ind., Journ. (Old Series), XIV., 29, 39; Aitchison, Rep. of Afgh. Del. Com., 29; Smith, Ec. Dict., 434.*
Habitat.—An aquatic plant found on the West Himálaya from Ku-máon to the Indus. It is also met with in the plains of the Panjáb as far south as Delhi and Saharunpore. It is distributed to the temperate regions of the North and South hemispheres

FOOD & FODDER.
25

Food & Fodder —This is one of the few wholesome members of the family, but in India no domestic use appears to be made of the plant. In some parts of England cows are fed on it during the winter, and in the Indian hills, where it occurs abundantly, it might with advantage be used for that purpose.

26   **R. arvensis,** *Linn.; Fl. Br. Ind., I., 20.*
Vern.—*Chambul,* PB.
References.—*Royle, Ill. Him. Bot., 53; Stewart, Pb. Pl., 5; Aitchison, Rep. Afgh. Del. Com., 30; Gazetteers, N.-W. P., IV., lxvii.; X., 304.*
Habitat.—An erect herb of the Western Himálaya from Kashmír to Kumáon, which is found also in the plains of the North-West Panjáb. It is distributed to Afghánistán, Western Sibeia, Asia Minor, and Europe.

FODDER.
27

Fodder.—Stewart states that it is eaten greedily by sheep and goats, but frequently produces in them symptoms of irritant poisoning.

28   **R. sceleratus,** *Linn.; Fl. Br. Ind., I., 19.*
THE WATER CELERY.
Syn.—R. INDICUS, *Roxb.*
Vern.—*Kaf-es-saba,* ARAB.; *Kabikaj,* PERS.
References.—*Roxb., Fl. Ind., Ed. C.B.C., 458; Voigt, Hort. Sub. Cal., 3; Royle, Ill. Him. Bot., 9, 43, 46, 53, 113; Murray, Pl. & Drugs, Sind, 73; Dymock, Warden & Hooper, Pharmacog. Ind., I., 38; Watt, E on. Prod., V., 236; VI., 153; Smith, Dic., 434; Don, Prod., 195; Gazetteers, N.-W. P., I., 78; IV., lxvii.; X., 304; Agri.-Horti. Soc. Ind., Trans., V., 119; VII., 44; Journ. (Old Series), VIII., Sel, 154.*
Habitat.—An erect annual, 1-3 feet high, met with on the river banks in Bengal and Northern India, in the marshes of Peshawar and in the warm valleys of the Himálaya. It appears in India during the cold weather and remains up till the rains. In distribution it extends through-out the whole of the Northern Temperate Zone.

MEDICINE.
Plant.
29

Medicine.—The fresh PLANT is poisonous, and produces violent effects if taken internally. Draughts of tepid water constitute the best remedy in cases of poisoning by these plants by diluting the acrid juices on which their effects chiefly depend. The bruised leaves form an application to raise blisters, and may also be used to keep open, sores caused by vesication, or

| The Radish. | (*W. R. Clark.*) | RAPHANUS sativus. |

by other means (*Murray*). It was formerly used in Europe by professional beggars to produce or maintain blisters or open sores intended to excite sympathy. Roxburgh remarks that it has no native name, and that its properties are apparently unknown. It certainly possesses a very powerful principle, and one would expect to find it taking a place in the practice of herbalism. Water distilled from a decoction retains its acrid character, and if this be allowed to slowly evaporate it leaves behind a quantity of highly insoluble crystals of a very inflammable character.

MEDICINE.

Food.—The inhabitants of Wallachia use it as a vegetable when boiled, a remarkable fact when it is remembered that it is poisonous, and a powerful vesicant when uncooked.

FOOD.
30

Rape, see Brassica campestris, *Linn.*; CRUCIFERÆ; Vol. I., 522.

RAPHANUS, *Linn.; Gen. Pl., I., 101.*

Raphanus sativus, *Linn.; Fl. Br. Ind., I., 166;* CRUCIFERÆ.
THE RADISH.

31

Vern.—*Múli, muro,* HIND.; *Múla,* BENG.; *Mula sinki,* NEPAL; *Múli,* N.-W. P.; *Tara mira, múli, mŭngra,* PB.; *Muri,* SIND; *Mula, muro,* BOMB.; *Muri, mula,* MAR.; *Mala, mura,* GUZ.; *Mulli,* DEC.; *Magunigadde,* MYSORE; *Mullangi, mulŭnghie,* TAM.; *Mullangi,* TEL.; *Bili, mullangiyanne, mullangi,* KAN.; *Lobak,* MALAY.; *Móulá, mone-lah-mung-la,* BURM.; *Rabu,* SING.; *Múlaka,* SANS.; *Fugil, fioyl, bokel,* ARAB.; *Turb,* PERS.; *Fidjel,* EGYPT.

References.—*Roxb., Fl. Ind., Ed. C.B.C.,* 500; *Voigt, Hort. Sub. Cal.,* 72; *Stewart, Pb. Pl.,* 15; *DC., Orig. Cult. Pl.,* 29; *Mason, Burma and Its People,* 469, 749; *Sir W. Elliot, Fl. Andhr.,* 118; *O'Shaughnessy, Beng. Dispens.,* 191; *Irvine, Mat. Med. Patna,* 70; *Moodeen Sheriff, Supp. Pharm. Ind.,* 212; *Mat. Med. S. Ind. (in MSS.),* 25; *U. C. Dutt, Mat. Med. Hind.,* 310; *S. Arjun, Cat. Bomb. Drugs,* 211; *Dymock, Mat. Med. W. Ind.,* 60; *Dymock, Warden & Hooper, Pharmacog. Ind., I.,* 129; *Year-Book Pharm.,* 1875, 262; *Med. Top. Ajm.,* 147; *Birdwood, Bomb. Prod.,* 138; *Baden Powell, Pb. Pr.,* 327; *Atkinson, Him Dist. (Vol. X., N.-W. P. Gas.),* 702, 748; *Useful Pl. Bomb. (Vol. XXV., Bomb. Gaz.),* 145; *Linschoten, see Note under Cucumis sativus; Bombay Man. Rev. Accts.,* 101; *Settlement Reports:— Panjáb (Hazara Dist.),* 93; *(Kangra Dist.),* 25, 28; *(Montgomery Dist.),* 109; *(Simla Dist.), xli, App. ii., H.; Central Provinces,* 53, 82; *Ann. Rep. Settlement, Port Blair (1870-71),* 41; *Gazetteers:—Bombay, VII.,* 40; *VIII.,* 183; *XII.,* 171; *XVII.,* 276; *N.-W. P., I.,* 79; *IV., lxvii; X.,* 305, 702; *Mysore and Coorg, I.,* 55, 57; *W. W. Hunter, Orissa, II.,* 180; *App. VI.; Agri-Horti. Soc:—Ind., Trans., I. (App.),* 244; *II.,* 12 (*App.*), 306; *III.,* 10, 69, 198; *IV.,* 103, 133, 145; *V.,* 64; *VI.,* 234; *VII.,* 69; *Journ. (Old Series), IV.,* 203, *Sel.,* 211, 213; *IX.,* 404, *Sel.,* 58; *X.,* 91; *XI., Pro.,* 46; *XIII., Sel.,* 54; *(New Series), IV.,* 36; *V.,* 35, 44, *Sel.,* 5; *Reports of the Agri. Dept. and the Experimental Farms of Bengal,* 1886, *App. II., XXVII., lxxx.; Smith, Econ. Dict.,* 344.

Habitat.—An annual herb of the cabbage family, unknown in its wild state, but cultivated throughout the plains of India and in the Himálaya up to 16,000 feet. It is a cold-weather crop in the plains, and grows nearly all the year round in the hills. There are several varieties cultivated in India—the large long pale-pink, the small longish pale-pink, and the small round bright red. The last is generally raised in gardens from selected seeds intended for consumption by the Europeans in India.

History.—There seems to be some difference of opinion as to the origin of the radish. Bentham thinks it may possibly have come from the British wild plant, **R. Raphanistrum,** *Linn*, and DeCandolle is of the same opinion. The latter author, in support of this, adduces a series of experiments made by **Carriese,** the head gardener of the Nurseries of the Natural History Museum in Paris, who sowed the seeds of the **Raphanistrum** in rich well-manured ground, with the result that in the fourth gene-

HISTORY.
32

| RAPHANUS | The Radish. |
| sativus. | |

HISTORY.

ration he obtained fleshy radishes of varied colour and form, and having the pungent taste of the ordinary European garden radish. **R. Raphanistrum**, however, is a European plant, which does not exist in Asia, and is not likely, therefore, to be the species that has furnished the inhabitants of India, China, and Japan with the radishes which they have cultivated for centuries. It would seem the most natural explanation that the European radish may have been derived from **R. Raphanistrum**, but that the Indian form, from an Indian and Chinese indigenous wild plant, now apparently lost. In part support of this view it may be pointed out that the Indian radish is almost tropical in its habit instead of temperate. It is often transplanted from one field to another, yielding its seed in the second year. The root grows to an enormous size, sometimes as large as a man's leg, and rises partly above the ground like a stem.

OIL.
Root.
33
Seed.
34

Oil.--The ROOT and SEED yield oils apparently similar to those obtained from other cruciferous plants. They have a most disagreeable odour, but they are said to be sometimes used for burning and for culinary purposes Radish seed oil is colourless, heavier than water in which it dissolves pretty freely, and contains a considerable quantity of sulphur. With perchloride of mercury it forms a white precipitate, with bichloride of platinum a yellow one (*Pleiss in Gmelin's Hand-Book, 10, 56*).

MEDICINE.
Seeds.
35
Roots.
36
Juice.
37
Leaves.
38

Medicine.—In India the SEEDS are considered diuretic, laxative, and lithontriptic. In the Panjáb, they are also believed to have emmenagogic properties. The ROOTS are used in native medicine for urinary and syphilitic diseases. In Bombay, the JUICE of the fresh LEAVES is used medicinally for the same purposes as the seeds.

SPECIAL OPINIONS.—§ " The root of **Raphanus sativus** (the radish) is stimulant, diuretic, stomachic, and antilithic, and the seeds demulcent and diuretic. In full and repeated doses, the seeds produce vomiting sometimes, but this is so rare that they cannot be considered as an emetic. The juice of the radish is useful in dysuria and strangury, and also in some slight cases of ischuria and calculus in the bladder. Eaten before a meal the radish improves appetite, and increases the digestive power. The dry seeds of the radish are also useful in some slight cases of dysuria and strangury, but their action is rather uncertain and irregular. The juice of the radish is to be pressed out through a cloth by bruising it without water. The seeds are used in the form of a draught by bruising and rubbing them with water and straining the liquid through a cloth. The dose of the juice is from one ounce and a half to three ounces repeated frequently till the desired effect is produced ; of the seeds from one to two drachms" (*Honorary Surgeon Moodeen Sheriff, Khan Bahadur, G.M.M.C., Triplicane, Madras*). "The seeds in doses of one drachm are useful in gonorrhœa" (*Narain Misser, Kathe Bazar Dispensary, Hoshangabad, Central Provinces*). " The root is a reputed medicine for piles and gastrodynic pain " (*Assistant Surgeon T. N. Ghose, Meerut*).

FOOD.
Root.
39
Seed-pods.
40
Leaves.
41
Pickle.
42

Food.—The radish is cultivated in India not only for the sake of the ROOT, but also for the young SEED-PODS and the LEAVES, all of which are eaten. The root is eaten both raw and boiled as an ordinary vegetable ; it is also occasionally PICKLED and consumed in that form. The unripe seed-pods are boiled in water and eaten with *ghi* or cooked in curries. They also are used for pickling alone or with other vegetables, and are regarded as a fair substitute for capers. The leaves are occasionaly employed as a pot-herb.

43

Raphanus sativus, *Linn.. var.* caudatus, *Linn. ; Fl. Br. Ind., I., 166.*

THE RAT-TAIL RADISH.

Vern.—*Mugra*, HIND.; *Mungra*, PB.; *Mogri*, BOMB.

R. 43

| The Rat-tail Radish. | (*W. R. Clark.*) | RATS. |

References.—*Stewart, Pb. Pl., 15; Baden Powell, Pb. Pr., 260; Birdwood, Bomb. Pr., 138; Smith, Econ. Dict., 344; Gazetteer, Bomb., VIII., 184.*

Habitat.—This variety of R. **sativus**, said to be a native of India and China, is commonly cultivated in Western India and the Panjáb. It is highly prized on account of its succulent pods which under cultivation attain a length of two or three feet. **Mr. Baden Powell**, in his *Panjáb Products*, states that "the natives have an idea that this plant is only the ordinary radish, subjected to a peculiar treatment, *viz.*, by being taken up and having all its roots cut close round and then replanted." There seems little doubt of the origin of this plant from the same stock as the ordinary Indian radish, and the habit of removing the tap-root as a vegetable and replanting the stock for the production of seed is quite common with the poorer classes. The Rat-tail radish is, however, so peculiar as a cultivated plant and possesses even certain botanical characteristics as would almost justify the suggestion made above that it and the ordinary field radish of India may have been derived from an independent source from that of the European radish. If this idea be confirmed by future investigators, it would then be desirable to assign them a distinctive name. The information on this subject is at present too imperfect to justify more than the suggestion that the Indian plants may not be forms of R. sativus.

Food.—The elongated PODS are eaten by the natives of the Panjáb either cooked as a pot-herb or in the form of a salad. They are also PICKLED.

FOOD.
Pods.
44
Pickle.
45
46

RASPBERRY.

The Raspberry of England (**Rubus Idœus**) is not a native of India, but it is now being experimentally cultivated at several hill stations, particularly Simla. Many other species, however, of the same genus are found abundantly, some of which are even used for the same purposes as the ordinary Raspberry, although none of them are cultivated. For further information on the subject see **Rubus.**

RATS, MICE, MARMOTS, SQUIRRELS, & MOLES.

47

Under this heading, the writer proposes to deal not only with the various species of **Rodentia** not treated of under **Hares & Porcupines**, but with the insectivorous animals popularly known as Rats, but which should be designated SHREWS, and also with other allied species of insectivorous animals, the economic uses of which are so slight that it has not been considered necessary to treat of them separately. The brief enumeration here given is, however, simply an abstract of the information contained in **Jerdon's** work on the Mammalia of India, together with a small amount of information obtained from other sources as to the economic uses of the different members of the series.

RODENTIA.

The Indian rodents are sub-divided into four families—**Sciuridæ** or Squirrels; **Muridæ** or Rats; **Hystricidæ** or Porcupines, and **Leporidæ** or Hares.

48

I.—SCIURINÆ.

Characters.—The Squirrels form a well-marked group of animals, which have mostly arboreal habits. They are widely distributed both in the Old and New Worlds. In India there are some 10 or 12 well-marked species of Sciurus. Of these the following may be specially mentioned: S. **malabaricus**—The Malabar Squirrel; S. **maximus.**—The Central Indian Red Squirrel; S. **elphinstonei**—The Bombay Red Squirrel; S. **macrouroides**—The Black Hill Squirrel; S. **macrourus**—The Grizzled Hill Squirrel; S. **lokrioides**—The Hoary Grey Squirrel; S. **palmarum**—The Common

49

RATS.	Indian Rodentia.

SCIURINÆ.

Striped Squirrel; **S. tristriatus**—The Jungle Striped Squirrel; **S. layardi**—The Travancore Striped Squirrel; **S. sublineatus**—The Nilghiri Stripped Squirrel; and **S. mcclellandi**—The Small Himalayan Squirrel. Two smaller genera embrace the Flying Squirrels, *viz.*, **Pteromys** and **Sciuropterus.** Under the former there are three species: **P. petaurista**—The Brown Flying Squirrel; **P. inornatus**—The White Bellied Flying Squirrel; and **P. magnificus**—The Red Bellied Flying Squirrel. Under the latter there are five species. *viz.*, **S. caniceps**—The Grey Headed Flying Squirrel; **S. fimbriatus**—The Grey Flying Squirrel; **S. alboniger**—The Black and White Flying Squirrel; **S. villosus**—The Hairy Footed Flying Squirrel; and **S. fusco-capillus**—The Small Travancore Flying Squirrel. One group only is quite terrestrial, *viz.*, the marmots (**Arctomydinæ**).

USES.
Fur.
50

Uses.—The FURS of many of the animals of this family especially are much valued as articles of commerce. They are chiefly exported to Europe where they are used by furriers. For information as to the species and vernacular names of those chiefly employed for that purpose, see **Furs,** Vol III., 458. The flesh of some species of squirrels and marmots is eaten by many of the aboriginal tribes of India.

51

II.—MURINÆ.

This group comprises all the remaining rodents which have tolerably perfect clavicles and subquadrate lower jaws. They are divided into five sub-families, only two of which have representatives in India, *viz*, the **Murinæ** and **Arvicolinæ.** The MURINÆ comprises the true rats and mice. The flesh of several of these is eaten by the rural inhabitants of India. The **Arvicolinæ** or VOLES are mostly Palæacrtic, but one or two species occur on the Himálaya. They are of no economic value. The following may be specially mentioned as the better known species of MURINÆ :—

1. GERBILLUS INDICUS.
    THE GERBILLE-R T or JERBOA of some writers.
    Vern.—*Hurna mús,* HIND.; *Jhenku indúr,* BENG.; *Billa ilei,* KAN.
    Habitat.—A common field rat met with in Africa and Asia. It is plentiful all over India except on the western side and in Sind where its place is taken by an allied species—**G hurrianæ.** This latter frequents the more desert sandy tracts of the country, and in that respect resembles the Jerboa of Egypt.
    Food.—In most parts of India the rural inhabitants dig one or other species of Jerboa out of their burrows and eat them.

2. NESOKIA INDICA.
    THE INDIAN MOLE-RAT.
    Vern.—*Kok,* KAN.; *Golattu koku,* TEL.
    Habitat.—Found throughout India up to altitudes of 7,000 feet and also in Ceylon.
    Food.—" The race of people known by the name of Wuddurs. or tank-diggers, capture this animal in great numbers as an article of food, and during the harvest plunder their earths (*sic*) of the grain stored up for their winter consumption, which in favourable localities they find in such quantities as to subsist almost entirely upon it during that season of the year. A single burrow will sometimes yield as much as half a seer of grain, containing even whole ears of jowaree (**Sorghum vulgare**)" (*Jerdon*).

3. MUS BANDICOOTA.
    THE BANDICOOT-RAT.
    Vern.—*Ghus,* HIND.; *Ikria, ikara,* BENG.; *Heggin,* KAN.; *Pandkoku,* TEL.
    Habitat.—Found throughout India, also in Ceyon and in Malaya.
    Food.—In many parts of India its flesh is eaten by the natives.

MURINÆ.

Besides the species here mentioned there are some 20 otherspecies of Rats and Mice met with in India. These are, however, more destructive than economic and need scarcely be here separately dealt with.

4. LEGGADA LEPIDA.
   THE SMALL SPINY MOUSE.
   Vern.—*Chitta burkani, chityelka, chitta ganda,* TEL.
   Habitat. – Most parts of Southern India.
   Domestic Uses.—The natives of Southern India very generally use this animal as a bait in their attempts to catch the Indian Jay with bird-lime.

5. RHIZOMYS BADIUS.
   THE BAY BAMBOO-RAT.
   Habitat.—General throughout the Eastern Himálaya, Assam, Burma, and the Malayan Peninsula.
   Food.—Some varieties of this species attain the size of a small rabbit. They feed exclusively on vegetable matter and mainly on the shoots of the bamboo. They live in burrows under the roots of these plants whence they are dug out by the hill people who esteem their flesh as a great delicacy.
   Domestic Uses.—Their fur is much valued as an article of commerce, see Furs, Vol. III., 458.

INSECTIVORA.

52

Few animals of the insectivorous order are of much economic value. They are usually of small size and some of them superficially resemble certain of the Rodents above described. The members of the order are chiefly confined to the Old World. They may be divided into Talpidæ or Moles ; Sorecidæ or Shrews ; Erinaceidæ or Hedge-hogs ; and Tupaiadæ or Tree Shrews, all which have representatives in India.

I.—TALPINÆ.

53

The animals of this family are better known in Europe than in India, where they are mainly represented on the Eastern Himálaya and the Khasia Hills. The only one of any economic value is the short-tailed mole.

1. TALPA MICRURA.
   Vern.—*Pariam,* LEPCHA ; *Biyu-kantyem,* BHOr.
   Habitat —The Eastern Himálaya.
   Domestic Uses.—The skins are used as Furs, Vol. III., 461.

II.—SORECINÆ.

54

Many of the animals of this family are from their size, shape and nocturnal habits confounded with rats and mice, as is the case with the common Indian shrew which is frequently known as the Musk-rat.

2. SOREX CÆRULESCENS.
   THE MUSK-RAT.
   Vern. – *Chachundar,* HIND. ; *Sondeli,* KAN.
   Habitat —A common animal throughout India frequenting houses at night and hunting round rooms for cockroaches or any other insects It occasionally utters a shrill sharp cry. When alarmed it emits a musky odour, which is supposed by natives to be nauseous to snakes. As in England the common shrew, so in India the musk-rat is credited by popular superstition with most injurious qualities, and it is believed that if a musk-rat runs over any part of an animal, it will cause a sore to break out in the place over which it has passed.
   Perfumery.—The ducts of the musk-rat are not used in perfumery (*Piesse*).

| RAUWOLFIA serpentina. | A Valuable Antidote for Snake-poison. |

55

III.—ERINACEINÆ.

Several species of hedge-hog exist in India, but they are of no economic value. The FLESH of some of them are said in the Panjáb to be pounded up with *ghi* into an ointment for application to long-standing ulcers. Their QUILLS are burned in an open fire, and persons suffering from hœmorrhoids are locally fumigated in the smoke. The FAT of hedge-hogs is made into an oil which is externally applied to rheumatic swellings.

56

IV.—TUPAIANÆ.

None of the members of this family are of economic value.

Rattans, see Calamus Rotang, *Linn.;* PALMÆ ; Vol. II., 22; also **Canes,**
[Vol. II., 98.

RAUWOLFIA, *Linn.; Gen. Pl., II., 697.*
[*t. 849;* APOCYNACEÆ.

57

Rauwolfia serpentina, *Benth.; Fl. Br. Ind., III., 632; Wight, Ic.,*
Syn.—OPHIOXYLON SERPENTINUM, *Linn.;* O. TRIFOLIATUM, *Gœrtn.;* TABERNÆMONTANA CYLINDRACEA, *Wall.;* RAUWOLFIA DENSIFLORA, *Benth.*
Vern.—*Chota-chánd,* HIND.; *Chandrá, chota-chárd,* BENG.; *Chandra, chota chand, karavi, harkai, tsjovanna amelpodi* (Malabar Coast), BOMB.; *Harkaya,* MAR.; *Pátala gandhi, pátála garuda,* TEL.; *Chu-vanna-avilpori,* MALAY.; *Bongmaiza,* BURM.; *Aika-waireya,* SING.; *Sarpagandhá, chundrika,* SANS.
References.—*Roxb., Fl. Ind., Ed. C.B.C., 233; Voigt, Hort. Sub. Cal., 532; Kurz, For. Fl. Burm., II., 171; Gamble, Man. Timb., 262; Dals. & Gibs., Bomb. Fl., 143; Mason, Burma and Its People, 800; Sir W. Elliot, Fl. Andhr., 146; Sir W. Jones, Treat. Pl. Ind., V., 156; Rheede, Hort. Mal., VI., 47; Rumphius, Amb., VII., 27, t. 16; Ainslie, Mat. Ind., II., 441; O'Shaughnessy, Beng. Dispens., 447; U. C. Dutt, Mat. Med. Hind., 317; Dymock, Mat. Med. W. Ind., 2nd Ed., 505; Useful Pl. Bomb. (Vol. XXV., Bomb. Gaz.), 274; Gazetteers:—Bombay, XV., 438; N.-W. P., IV., lxxiv; X., 313; Journal, Agri.-Horti. Soc., Ind. (Old Series), VI., 14; Ind. Forester, IX., 255.*
Habitat.—A large, climbing or twining shrub, found in the tropical Himálaya and plains near the foot of the hills from Sirhind and Morad-abad to Sikkim. It occurs also in Assam, Pegu, Tennasserim (at altitudes up to 4,000 feet), and in the Deccan Peninsula along the Gháts to Travan-core and Ceylon. It is distributed to the Malay Peninsula and Java.

MEDICINE.
Root.
58

Medicine.—In India and the Malayan Peninsula the ROOT of this plant has been, from ancient times, much valued as an antidote for the bites of poisonous reptiles and the stings of insects, also as a febrifuge, and as a remedy for dysentery and other painful affections of the intestinal canal. In Sanskrit works it is mentioned under the name of *Sarpaganda.*

Rumphius speaks of it under the name of **Radix mustelœ,** and says that in his time it was widely used in India and Java as an antidote against every sort of poison. It was administered both internally in the form of a decoction of the root and externally by making a plaster of the roots and fresh LEAVES and applying them to the soles of the feet. For snake poisons, he continues, it was esteemed as specially valuable and the poisonous effects of even the Cobra's bite were viewed as rendered harmless by the administration of this wonderful root. It is said by him to have been universally employed as an internal remedy against fevers, cholera, and dysentery, and the JUICE of the leaves was instilled into the eyes as a remedy for the removal of opacities of the cornea. He states also that this is the plant to which the mongoose is believed to have recourse when bitten by poisonous snakes. (*Conf.* with **Ophiorrhiza Mungos,** Vol. V., 488.) **Sir W. Jones** gives a similar account of the supposed medicinal virtues of

Leaves.
59

Juice.
60

the plant, but expresses a doubt as to whether it really is the so-called Ich-neumon plant. Roxburgh states that it is used by the "Telinga physicians, *first*, in substance, inwardly, as a febrifuge; *secondly*, in the same manner, after the bite of poisonous animals; and *thirdly*, it is administered, in substance, to promote delivery in tedious cases." Horsfield remarks that the root yields a strong bitter infusion and that its sensible properties indicate considerable activity.

Altogether the popular beliefs with regard to this plant and the testi-mony of medical men in India who have practically tried it as a remedy for fevers, seem to indicate that it possesses strong and well-marked proper-ties; it might, therefore, be advantageous to have a more complete ana-lysis of its composition and more careful determinations of its actions.

SPECIAL OPINION.--§ " Used in dysentery and diarrhœa by the labour-ers in the Koncan" (*W. Dymock, Bombay*).

REALGAR, *Mallet, Manual of Geology of Ind., IV., 12.* 61

REALGAR, NATIVE SULPHURET OF ARSENIC, DISULPHIDE OF AR-SENIC, RED ARSENIC, RED ORPIMENT, $As_2 S_2$.

Vern.—*Mansil, naushádar káni,* HIND. & PB.; *Manahsila,* SANS.

References.—*Pharm. Ind., 346; U. C. Dutt, Mat. Med. Hind., 45; Ir-vine, Mat. Med. Patna, 37; ball, in Man. Geol. Ind., III., 162; Watt, Cal. Exh. Cat., V., 237; Baden-Powell, Pb. Pr., 102; Balfour, Cyclop., III., 375.*

Occurrence and Characters.—Realgar is said to occur native in Yunan, 62
Kwei-chaw, and Kansut. Specimens of orpiment containing some realgar and also a lump of pure massive realgar weighing over a pound are in-cluded in the collections of the Indian Museum, Calcutta, and are said to have been obtained from Munsiari in Kumáon. Orpiment mixed with some realgar was found in Chitral, north-west of Kashmír, by Dr. Giles, naturalist to the Gilgit Mission (*Mallet*). When native it occurs in the form of ruby-red prismatic cystals.

It may be prepared artificially by heating together 198 parts of arseni-ous anhydride with 112 parts of sulphur. When heated in a closed vessel, realgar melts and at a higher temperature it sublimes without decomposi-tion. The sublimed mass is hard, brittle, transparent and of a beautiful red colpur (*Miller*).

Medicine.—" Realgar is purified for medicinal purposes by being rub- MEDICINE,
bed with the juice of lemons or of ginger. It is used internally in fever, 63
skin diseases, cough, asthma, &c., and externally in skin diseases. In fever it is generally used in combination with mercury, orpiment, &c." (*U. C. Dutt*).

Domestic Uses.—It is employed in soldering gold and for the manu- DOMESTIC.
facture of ornamental vessels and medicine cups (*Balfour*). 64

Reana luxurians, see Euchlæna luxurians, *Ascheron;* GRAMINEÆ;
[Vol. III., 282.

REAUMURIA, *Linn.; Gen. Pl., I., 161.*

Reaumuria hypericoides, *Willd.; Boiss., Fl. Orient., I., 761;* 65

HYPERICUM-LIKE REAUMURIA. [TAMARISCINEÆ.

Vern.—*Lanisah,* SIND.

References.—*Murray, Pl. and Drugs, Sind, 70; Aitch., Afgh. Del. Com., 42; Balfour, Cyclop., III., 376.*

Habitat.—A small shrubby plant, found in Sind and distributed to Afghánistán, Balúchistán, Persia, Arabia, Syria, the Mediterranean coast and the milder parts of North Asia. MEDICINE.

Medicine.—It is used in Sind in the treatment of porrigo and itch, the 66

R. 66

bruised leaves being applied externally, and a decoction of $2\frac{1}{2}$ to 3 of the juice of the leaves diluted and internally administered *(Murray).*

Rectified Spirit, see **Spirits,** Vol. V., 332-334.

Red Lead, see **Pigments,** Vol. VI., 232 ; also **Lead,** Vol. IV., 603.

Red wood, see **Pterocarpus santalinus,** *Linn. f. ;* LEGUMINOSÆ ; Vol. VI., 359-362.   [Pt. II.

Red wood, see **Soymida febrifuga,** *Adr. Juss. ;* MELIACEÆ ; Vol. VI.,
(*G Watt.*)

REH.

67

Reh is the vernacular name in India for efflorescent salts which accumulate in the soil and sub-soil waters of large tracts of India. When this takes place to a serious extent, the land is rendered sterile. In the Upper Panjáb efflorescence is known as *kallar,* and in Oudh and other parts of India the affected lands are called *usar.*

Numerous reports and papers on this subject have appeared both in the official proceedings of Committees held to consider the subject and in the Records of the Geological Survey. (*Conf.* with *Man. Geol., Ind., Vol. I., 413-415.*) Space cannot be afforded in this work to deal with the subject in the detail which its importance to the people of India demands. The reader should, therefore, consult the works alluded to in this brief review of the arguments advanced and the main facts which have been brought to light regarding the origin of the efflorescence. Ball (*Economic Geology, III., 696*) gives a brief sketch of the subject. and his article may, therefore, be quoted here as an introduction. He writes : " As affecting the general prosperity and revenues of the country, the subject of how to mitigate or diminish the evil has naturally attracted a good deal of attention, and numerous reports and papers have been published which contain more or less trustworthy information, and more or less of practical suggestion." " Primarily the saline matters are derived from the decomposition of rocks, and, taking the case of Northern India, the rivers descending from the Himálayas carry down in solution proportions of salt which vary with the character of the strata traversed. The salts so carried in solution consist principally of calcium and magnesium carbonates, and sodium sulphate and chloride. In addition, of course, the alluvium or silt which is brought down, consisting of finely comminuted minerals, includes materials which, on decomposition, are capable of supplying bases for the ultimate formation of the same salts under suitable conditions In a region of intense evaporation, and where there is not a free drainage outlet of water, these salts, by long continued concentration, accumulate in

Chemical
Formation.

the soil or in the sub-soil waters, and over and above this rain water charged with carbonic acid, falling on a porous soil, has the effect of decomposing its mineral constituents and of carrying down the salts so formed in solution either to the region of sub soil water, or else for only a few inches or feet below the surface. When the surface of the ground again becomes dry, this saline water rises by capillary attraction and evaporates, and a salt efflorescence remains, which at length so permeates the superficial layer of soil that cultivation becomes impossible. With free underground drainage, which would admit of the rain passing through and washing the soil, this would not occur, especially where the surface was well protected from evaporation by vegetation."

" Irrigation by canal water, when not accompanied by deep drainage, has had the effect of increasing the amount of reh deposit, and large tracts have been in consequence thrown out of cultivation. The indirect action which has produced this result has been fully explained by Mr. Medlicott

(*Selections from the Records of the Government of India, No. XLII., p. 32, of 1864*). In this case, the direct increase in the amount of saline matter is inconsiderable owing to the comparative purity of canal water; but the so-called table of sub-soil saline water has, by the addition of irrigation water, without an increase in the drainage, had its level raised to an extent which has rendered capillary attraction operative, and so these saline waters, which were previously to a great extent innocuous, have been brought into injurious contiguity with the superficial layers of soil. Thus is explained the apparently paradoxical fact that irrigation, by comparatively pure canal water, has been followed by an increase of salts in the superficial soils.

"A very exhaustive paper on the subject of *reh* by Dr. Center (*Records of the Geological Survey, Vol. XIII., 253*) gives an interesting account of the methods which are adopted by farmers in the salt-lake regions of America to cure lands which are similarly affected as are the sterile tracts of India. Among these, thorough washing of the superficial soil, and the removal of the salt by solution, heavy manuring, and protection of the surface from evaporation, have been tried with good results, and sterile land has been brought under cultivation.

Properties & Uses.—"The uses to which *reh* may be put have been enumerated by Dr. Center. SODIUM SULPHATE for medicinal purposes can be easily obtained from it, and when it is abundant it might be used to manufacture into CARBONATE for glass or soap manufacture. The natural average mixture of sodium sulphate and sodium chloride is similar to that produced by the manufacturers of sodium carbonate, who add sulphuric acid to common salt. By evaporation a salt cake of sodium sulphate might be obtained free from sodium chloride, and, with the aid of charcoal and *kankar*, the rest of the process might be performed. Sodium carbonate itself, as already stated, occurs in some abundance in certain of these soils. In former times an impure salt for comestible purposes was largely manufactured from *reh*, but it was naturally very much mixed.

"It has been proposed by Dr. Brown to cure the soil by the application of lime nitrate, which may easily be obtained by mixing pounded *kankar* with manure; by double decomposition, it would produce alkaline nitrate and calcium sulphate, the sodium carbonate would also be neutralised, but the sodium chloride would remain unaltered."

While the above brief statement of the perplexing subject of *Reh* touches on its chief features, *viz.*, the origin of the salts as derived from the decomposition of the rocks, washed in the higher courses of the rivers; the long continued concentration of these salts in the soil and sub-soil of the plains; the action on these and on the soil, of the heavy periodic falls of rain charged with carbonic acid; and the effect of the severe drought which follows, in sucking again to the surface these salts and thus forming an efflorescence over tracts where the drainage is either naturally imperfect or has been interfered with by any operation, such as clearing of forests, levelling of lands for cultivation, formation of canal and railway embankments, &c., it by no means conveys a clear conception of the various arguments that have been advanced or refuted by the writers who have dealt with the subject. Indeed, it is impossible to convey a definite idea of the extensive controversy that for some years past has occupied the attention of the public and of Government without reprinting a voluminous correspondence. Mr. Baden-Powell devotes many pages of his admirable *Panjáb Products* to the views held prior to the date of publication of that work (1868) in which he reviews in some detail the opinions which had then been advanced by Mr. Medlicott and Dr. Brown. The reader should consult Mr. Baden-Powell's work, but for the purpose of the present article it seems desirable to give here the subsequent facts brought

Method of curing Reh.

PROPERTIES. USES. Sodium Sulphate. 68 Carbonate. 69

26

REH.	Accumulation of Efflorescent Salts

PROPERTIES.
USES.

out by the " Reh Committee " which was convened to consider the matter in 1878 and also the able statement furnished by Dr. W. Center, Chemical Examiner for the Panjáb, furnished to the Government of India in 1880.

The following were the members of the " Reh Committee,"—President, Mr. Stewart Reid, Senior Member of the Board of Revenue ; Members, Messrs. Medlicott, Forrest, Ibbetson, and Michel with Mr. (now Sir) E. O. Buck as Secretary. The Review, furnished to the Government by the President, conveys the chief opinions arrived at by the members of the Committee, after the consideration of all the facts which had been placed before them, and it may, therefore, be here given as a *résumé* of the more detailed information which will be found in the final Report of the Committee. Mr. Reid wrote :—

" 1 shall attempt in the following paragraphs to show, *firstly*, the conditions under which *reh* is developed in the North-West Provinces ; and, *secondly*, how these conditions are affected by the introduction of canal irrigation. When dissent is not expressed, it may be assumed that I accept the facts and theories of my colleagues as agreeing with the results of my own enquiries and personal observation. I shall travel as little as possible outside their notes, and refer no more than is absolutely necessary to former discussions on the subject.

" 2. Mr. Medlicott holds that *reh*, which appears on the surface of the ground in the form of a white flocculent efflorescence, is formed of ' highly soluble sodium salts, the result of the decomposition by air and water of the particles of rock-minerals to be found in almost every soil, being the waste products of soil formation—the elements unassimilated by vegetation—the gradual removal of which is ordinarily effected by the rain-water draining through the soil and carrying with it any excess of these soluble salts.' That it is hardly ever developed in sandy soil both Mr. Forrest and Mr. Ibbetson agree. The former, however, has found it in

Influence of Soil.

clayey (*matiyár*) soil, while Mr. Ibbetson states that clay being too close in texture for capillary attraction never shows *reh* other than the scum left by the evaporation of the surface drainage. It may be assumed on Mr. Medlicott's authority that the salts which form *reh* are found in most soils, even where *reh* is not developed. As the gradual removal of these soluble salts is effected by rain-water drainage through the soil, sub-soil drainage water must generally contain *reh*, while mere surface drainage water contains little or none.

" 3. *Reh*, where it shows as a white efflorescence on the surface, has been brought up from below the soil by the force of capillary attraction, drawing up and evaporating the sub-soil water which holds it in solution. The effect of the evaporating force depends on the distance of the water-table from the surface of the ground and the nature of the intervening soil (*Ibbetson*). Under the influence of long and continuous evaporating action, fresh accretions of salt are thus brought continually to the surface ; and the fact that the most extensive outcrops of *reh* occur chiefly on the scorched plains of Northern India is due probably to the circumstance that evaporation acts more freely in a dry atmosphere, through which radiation can pass unimpeded by vapour (*Buck*). Mr. Medlicott points out (*paragraphs 11 and 5) that the immediate active cause of the mischief is the *extreme climatal* condition of the area affected, the sun's scorching heat and the desiccated hot winds from which the ground has no protection. Within the continuous area of the Indo-Gangetic plains, it is only in the scorching climate of the North-West that *reh* has become a scourge. It is quite certain, Mr. Medlicott adds, that evaporation is the immediate cause of

* The paragraphs referred to are those of the detailed report of which the present is, as explained, but an abstract.

PROPERTIES.
USES.

the appearance of *reh* as a white crust on the surface, and that its concentration there is the efficient cause of ruined vegetation. Evaporation being thus the immediate developer of *reh*, and water in the soil the necessary vehicle for its operation, the conditions of the water circulation on, in, and under the soil became a main feature of the whole question.

"4. It has been proved by facts, Mr. Buck observes (*paragraph 5), that a rise in the level of a main body of water underlying the surface is frequently accompanied by a greater development of *reh* at the surface, and often in new places. These results may possibly be ascribed as well to an increase in the evaporating process due to the approach of the main body of water to the surface, as to a change in the geographical distribution of under-ground accumulations, which with every change in level may find new centres of rest. Mr. Medlicott points out that, though the greatest amount of evaporation takes place at a surface supplied from a shallow water-table, similar results may occur at any elevation wherever the water soaked during the rains is withdrawn again by evaporation. The apparently capricious distribution of *reh* a great obstruction to a right comprehension of its mode of production) over a uniform surface does not imply a greater amount of *reh* in the ground under that surface originally, but only that more evaporation has taken place there owing to a freer supply of soaking,—this condition being determined by original difference of texture in these old alluvial deposits: a very little more or less of clay quite inappreciable to the eye in different layers, or in different parts of the same bed, would quite account for the very irregular appearances of the efflorescent *reh* and the peculiar aspect of a *úsar* plain. On the same even surface, level, or sloping with no visible difference in the soil, *reh* often shows itself most irregularly—bald patches in the midst of cultivation, or cultivated patches surrounded by *reh*; sometimes the efflorescence is the strongest in the slight depressions, sometimes in the slightly raised area. But, however great the peculiarities above described which shake all preconceived theories, there can be no doubt that when *reh* exists in the soil, it is more largely developed on a surface where the water-level is nearer at the surface, and that *reh* will develope in places in which it was formerly unknown if the water-level is raised by whatever cause.

Influence of
evaporation.

"5. Besides the primary conditions at work which have been considered in the preceding paragraphs, there are, what Mr. Medlicott styles, 'secondary influences,' which, though insignificant when compared with the primary conditions, must yet be taken into account. These secondary influences are—

(*a*) Local obstruction of surface drainage.
(*b*) The possible spread of *reh* by rain-wash and wind-drift.

"6. Local obstruction of surface drainage must result in raising the water-level within the area the natural drainage channels of which have been obstructed. The effect of the rise in the water-level in developing *reh* has been explained fully above. On this point there can hardly be two opinions. But Mr. Buck and I differ with Mr. Medlicott and Mr. Ibbetson in our estimate of the influence of rain-wash and wind-drift on the spread of *reh*. Mr. Ibbetson writes—'I do not believe that *reh* is spread over uncultivated land by wind or surface washing.' Mr. Medlicott states that it has been shown experimentally that mere surface drainage is ineffectual for the removal of the salts of which *reh* is formed, although they are accumulated at the very surface as the first drops of rains dissolve, and carry them into the top soil, where they are untouched by the water draining off the surface. Mr. Buck, on the other hand, contends that some part of the early rainfall, which has the opportunity of taking up *reh*

Drainage.

PROPERTIES.
USES.

Drainage.

in solution, is not absorbed into the soil, but spreads over surrounding land or finds its way along drainage lines; and also that where efflorescent *reh* is collected in large quantities, and is placed at the disposal as it were of the strong winds of the hot weather, *reh* must be largely dispersed. I have seen frequent instances of the effect of water carried across a *úsar* plain impregnated with *reh*, injuring that portion of the field into which it flows. Last February I was riding back to camp after inspecting a *reh*-infested tract. A strong west wind was blowing and the whole atmosphere was filled with *reh* particles. This was immediately after I had seen an accumulation of *reh* in the western border of fields lying to the east of *úsar* land white with *reh*, the villagers informing me that the *reh* had been blown thither by the west wind. I can also bear **Mr. Buck** out when he speaks of land which formerly grew stunted grass, and which is now white with *reh*. Zamindars have often complained to me that land of this description was formerly of some value on account of the pasturage it afforded, and is now good for nothing. This is a very serious matter, for extension of cultivation has very largely diminished the area of the pasture lands, especially of canal-irrigated estates. **Mr. Wright** in his report on his inspection of the *reh*-affected villages of Sikandra Rao (letter * No. K., dated 22nd March 1878, states (paragraph 11) that the severest loss is that of grazing land, which it is almost impossible to estimate.

" 7. I now proceed to consider how far the conditions under which *reh* is developed are influenced by the introduction of canal irrigation, taking up *seriatim* the questions noted in paragraph 17 * of the *Reh* Committee's Preliminary Report (page iii).

" 8. First with regard to the extension of *reh*.

Canal
irrigation.

" *A.*—Is the introduction of canal irrigation believed to be a principal cause of *reh* extension ? **Mr. Medlicott** points out that where great canals have been led over the face of the country, they have not only altered in a very marked way the level of water in the upper strata, but have also unquestionably introduced an independent and an inexhaustible source of *reh*. The more immediate and startling effects have been indirectly raising the water-level over an immense area, and so bringing the under-ground *reh* water more within reach of rapid evaporation; but it is also evident that where the conditions of *reh* formation are set up, *i.e.*, where the removal of soakage water takes place almost exclusively by evaporation, irrigation by canal must end in a destructive crop of *reh*, where the indigenous resources of the ground in that way were comparatively harmless. The process, Mr. Medlicott adds, may be slow according to circumstances, but that it is certain no man in his senses will deny. **Mr. Ibbetson** states that the Western Jumna Canal was opened in 1820. An irrigation far more extensive than had previously existed was established in 1838, and swamp and *reh* were the immediate results.

" Mr. Buck assuming with **Mr. Medlicott** that an almost permanent condition of *reh* distribution had been reached in the north of India, observes that on the sudden introduction of a canal system over the face of the country, the underground level is disturbed, the relative positions of *reh* and water are changed, equilibrium is upset, and a fresh activity sets in. The supply of a large body of water brought into close proximity of the surface adds fuel to the evaporating machinery, which is thus enabled to raise larger as well as more exclusive crops of the devastating substance. **Mr. Forrest** points out that the physical side of the question—that introducing a large body of water into a country, and not drawing on the subterraneous resources as before, will make a change—has been long recognized. He considers that canal water carries *reh* in solution, and also carries it along the surface of the ground, washing it out of its place of formation into good land. (This statement hardly fits in with his assertion in another part of his note, that the canal has never produced *reh* where it did not exist before, or to his reply to question A*.) Captain Howard is of opinion that the rise of spring-level ' in recent years has been caused by obstruction of drainage and by percolation of land water partly from the fields over which it is spread and partly direct from the carrying channels. Any cause which accelerates the rise of the spring-level tends to increase the area of *reh*.' There are, I would observe in parts of the North-Western Provinces, Oudh, and Panjáb, tracts of country in which *reh* abounds in immense quantities. But there the *reh* lands do not increase year by year. We do not hear of their encroaching largely on the cultivated area. The complaints which reach us of *reh* extension do not come from countries irrigated by wells, but chiefly, if not solely, from canal-irrigated districts. Does not this very fact justify the belief that canal irrigation is a principal cause of the spread of *reh* ?

" 9. *B*.—How far is the extension supposed to be due to sub-soil percolation of the water ? Mr. Ibbetson holds that percolation from the canal does to a considerable extent take place. He reads the word percolation as I do, *viz.*, ' percolation under the surface from the canal and its distributaries, as opposed to saturation caused by actual irrigation, *i.e.*, through water on the surface of the soil.' Mr. Buck treats of percolation of canal water in connection with evaporation rather than with raising the water-level. Mr. Ibbetson doubts whether the percolation can be great when the canal is in soil. None of the members of the committee lay any stress on the effect of percolation under pressure, which condemned in the eyes of Colonels R. Strachey, Baird-Smith and Turnbull, canals in embankment which carried water above the level of the country. Captain Howard attributes the rise of spring-level to the percolation of canal water partly from the fields over which it is spread and partly direct from carrying channels. Mr. Forrest does not attribute the extension of *reh* to sub-soil percolation. The sub-soil movements must be slow. That percolation from canal affects the land in the neighbourhood of the canal, in the form of raising the water-level, is a fact that can hardly be gainsaid. How far it affects the land by bringing with the water particles of *reh* is a more doubtful question.

" 10. *C*.—How far is the extension of *reh* due to irrigating with canal water ? Mr. Ibbetson considers that canal irrigation as practised does more injury than percolation, and that it is the chief cause of the high water-level in canal-irrigated tracts. This fact Mr. Medlicott appears to doubt. He contends that if the water did penetrate so easily from the surface, the *reh* scourge would hardly be what it is, as the non-penetration of water is a general accepted condition of *reh* production. Mr. Buck thinks that *reh* is frequently introduced by canal irrigation, as the canal water is brought over soil in which *reh* is deposited, and that canal water adds to *reh* deposit by bringing with it a fresh supply of *reh* from the canal itself. Analysis, he says, may be able to give actual proof of the facts, which it indicates by a scientific examination of the soil of adjacent fields irrigated for a series of years, in one case by well water and in the other by canal water. But analysis will not show whether the presence of *reh* is due to super-saturation or to percolation, or to *reh* having been brought to the land by canal water, unless care is taken to see that the canal-irrigated land is not over-irrigated. Mr. Forrest thinks that canal water may help to distribute *reh* by super-saturation, the water remaining stagnant in the soil and drawing up the *reh* from below the surface by capillary attraction. I doubt whether *reh* is largely conveyed in canal water, but have no doubt whatever that irrigation, especially flush, is so carried on as to inundate the land rather than irrigate it. There are no means available for draining off the superfluous water, which sinks into the soil, and thus adds to

R. 69

REH.	Accumulation of Efflorescent Salts

PROPERTIES.
USES.

Canal
Irrigation.

the sub-soil water, which is the vehicle for the operation of the evaporative forces which bring the *reh* to the surface.

"11. *D.*—How far is the extension of *reh* supposed to be due to the interruption of drainage by canals or distribution channels? **Mr. Medlicott** regards local obstruction of surface drainage as a secondary influence, and insignificant when compared with the primary conditions at work. **Mr. Ibbetson** attributes the high water-level in canal-irrigated tracts to the interference with the natural drainage. In Karnal every drainage line in the country is crossed at intervals by high banks. The stoppage of the natural channels by which the surface water should be carried off aids most materially in water-logging the country. **Mr. Buck** is of opinion that much secondary mischief is caused by water-logging and interference with drainage, which prevents the escape of *reh*-infected water. **Mr. Forrest** thinks that such obstructions would have an injurious effect if it led to super-saturation. I have frequently seen that even where canals have been carefully aligned, the *rajbahas* (distributaries) have been taken across the natural drainage of the country. The obstruction of channels, by which the country is drained off its superfluous water, *must* result in water-logging the soil and raising the water-level, and thus bringing the sub-soil water within the influence of capillary attraction.

"12. *E.*—What relation (if any) does the extension of *reh* bear to the proximity of a canal or a canal watercourse? This subject has not been taken up by the members in detail. It was decided in committee as to the question—'Has the increase of *reh* been greater near the canal or at a distance from it,' that it was difficult to give a precise answer to this question without discussing the conditions of tracts other than that inspected, and that at present it was sufficient to say that no instance is known to any of the committee of serious extension of *reh* at a distance from a canal system. After the committee had broken up, I marched along the Cawnpore Branch of the Ganges Canal in its course through the Mainpuri, Farukhabad and Cawnpore districts, and rode over much of the adjacent country. The development of *reh* was far greater in the vicinity of the canal and of its distributaries than at a distance. Such was the rule to which, indeed, I saw exceptions. But even in the latter cases it might happen that the land nearer the main canal was less irrigated than that lying alongside a distant distributary, from which water was largely dispensed.

"13. *F.*—Is there any relation between the extension of *reh* and the surface level of land affected? I would refer for an answer to this question to the 7th and 8th paragraphs of **Mr. Medlicott's** first note * (pages ix-x).

"14. The questions relating to the prevention of *reh* extension will now be considered.

Improved
drainage.

"Firstly, *G.*—Is any improvement of drainage desirable? **Mr. Michel** considers that the one alternative to remodelling the entire canal system is deep drainage; to take and keep down the water to some fixed level; to straighten the natural water drainage channels of the country; to provide embankments with the necessary culverts, and thus relieve water-logging. **Mr. Medlicott** holds that the one measure that is of obvious necessity under any aspect of the case, and which may be put in hand at once, is *deep drainage,* the original cause of the accumulation of *reh,* whether at or below the surface, being the *stoppage* or the obstruction of the percolation of the atmospheric water from the surface. If the irrigation water is only disposed of by evaporation, it is certain that a crop of *reh* will be the ultimate result of irrigation by canal water, whether by *lift* or *flush* is only a matter of time. *Reh*-charged water being simply the accumulated drainings of *reh* water from the surface, which under proper (natural) surface conditions, for the distribution or the atmospheric water from the

R. 69

Products of India. 407

in large tracts of India. (*G. Watt.*) REH.

ground, would not have become so lodged : this present most abundant source of *reh* might be completely removed by the proper management of drainage cuts. Mr. Ibbetson attaches great importance to keeping the natural drainage of the country open. No dams or cultivation should be allowed in them. The more water we can get to flow away from canal tracts the better, while any *reh* it may carry away in solution is so far a gain. But this evil can always be met by proper alignment of the channels, and by scrupulously keeping the drainage lines open under the Canal Act. The drainages should of course be cleared *from the bottom*, and not from the upper end, as has been done in several instances to the great injury of the people.

" Mr. Buck is in favour of *immediate* action for draining such areas as are manifestly swamped by interference with surface drainage, and in which injury is due to this immediately local cause. He looks on a system of very deep drainage as almost impossible. The drainage system must be accessible to the very shallow surface drainage of the earliest rainfall, wherever there is an accumulation of *reh* in the surface. Mr. Forrest thinks that improvement of drainage is always desirable. It is very much to be regretted that the subject of drainage did not receive in the earlier days of canal construction the attention which it now commands. It was the wish and intention of Colonel Cautley, the able projector of the Ganges Canal, that measures for drainage and for irrigation by canal should proceed *pari passu*. Had this been done, the evils which the Government must now cure at any cost would have been prevented. It is sufficient to say that the Canal Department is now fully alive to the evils resulting from the obstruction of natural drainage channels, and that canals and distributaries are now aligned with great care. In Cawnpore the *rajbahas* have been re-aligned. It seems to be quite superfluous to insist on the absolute necessity of taking immediate measures for lowering the water-level in canal-irrigated tracts, in which it has been raised to such a height as to bring the sub-soil water within the grasp of the evaporating forces the action of which is graphically described by Mr. Buck in the 2nd paragraph of this note.

" 15. *H.*—Is any alteration of the canal system desirable ?

" Mr. Michel considers the remodelling of the whole canal system is the only thing that would ensure its being made remunerative beyond any comparison with existing arrangements, while the country might benefit to the extent of double, if not treble. Mr. Medlicott holds that the project of lowering the canal, so as to carry the water everywhere deep in soil, is a move in the wrong direction for a radical cure, inasmuch as the very fact complained of, *viz*, that the high level canal forces the *reh*-charged water to the surface, causes the extreme production of *reh* by this very means [see remarks on drainage (G*)]. Mr. Ibbetson would distribute water from the canal as at present, *i.e.*, at a high level, and would give enhanced powers to canal officers. He would charge different rates for fields irrigated as a whole and by *kiáris*, and would charge progressive rates for consecutive seasons of canal irrigation. (Mr. Ibbetson is speaking rather of the 'system of canal management' than 'the canal system.') Mr. Buck doubts whether lowering the canal a few feet would sufficiently curtail the evaporating process. But if he is wrong, he would certainly advocate construction of canals below, and not above the surface. (This is with especial reference to flush irrigation noticed further on.) Mr. Forrest says the main line cannot be changed. The alignment of distributaries is being improved. I do not consider any alteration (that is, remodelling) of the canal system desirable, for I hold it to be virtually impossible. Even if the idea that swamping and *reh* are mainly caused

REH.	Accumulation of Efflorescent Salts

by percolation under pressure, *i.e.*, that such evils are greater where the canal is in embankment than where it is in soil, were correct, it would be far easier to remedy those evils by efficient drainage than re-align a main line of canal. The last measure appears to me to be simply an impossibility, except at a most enormous and prohibitive cost.

" 16. *J.*—Is the substitution of lift for flush irrigation, or any other measure of controlling the surface supply of canal water, recommended ?

" Mr. Medlicott treats this subject with the preceding one —' Is any alteration of the canal system desirable.' He apparently disapproves of the proposal (see paragraph 15, p xii*) on the score of the mechanical waste of efficiency. Mr. Ibbetson, on the other hand, lays very great stress on the advantages which would result from the substitution of lift for flush irrigation. He would make the people lift every drop of water they use, water being distributed as before at a high level, but into a small reservoir, from which it would be raised for the irrigation of the field. He observes that such apparently wanton waste of power is of course revolting to the soul of an engineer, but contends that under the flush system economy in the use of canal water is out of question, and that it is the fact that water is lifted up by scoop, the fields are terraced channels kept in good order, and the severest economy is practised. He further contends by placing a fixed charge per scoop, bucket, or wheel for crop rates per acre (the amount of water that would be irrigated from a single scoop, etc., being practically a constant quantity) : that necessity for half-yearly canal measurements with their attendant annoyances would be avoided. Mr. Buck considers it absolutely necessary to do anything possible to put an end to the vicious system of swamping fields for irrigation purposes, which is the result of the accessibility of flush water. No check is so certain as that of making every cultivator lift his own irrigating water. Mr. Forrest thinks it would be a great economic error ; it would throw away a great part of the advantage of the canals. The true remedy is the greater economy in the distribution of the water. The rates of flush irrigation should be raised. Mr. Forrest says very truly that the true remedy is greater economy in the distribution of the water, but this remedy the Canal Department have not yet provided, and, I believe, admit that they are unable to provide. In the absence of the best remedy we must be content with an inferior one. By substituting lift for flush irrigation you substitute perhaps waste of labour for waste of water, and at the same time you diminish the canal revenue, as the flush water rates are much higher than the lift. But of the two evils waste of water is by far the more serious. It means swamping and inundation in place of irrigation, resulting in deterioration both of the soil and of the health of the people, and eventually in diminution of the income of the people and the reduction of the land revenue, resulting in a loss immeasurably greater than the loss in canal revenue which the substitution of lift for flush irrigation might entail. Waste of power is a far lesser evil. The additional power (that is, labour) which would be required would be supplied from power (labour) now lying idle. The sum representing the difference between the dearer flow rate and the cheaper water rate would purchase that additional power. It cannot be denied that the lavish expenditure (amounting to ' wicked waste') of canal water which the flow system encourages causes excessive saturation of the soil, which is one of the most common and most powerful causes of *reh* extension. Until the Canal Department succeeds in securing economy of distribution by giving out their water by measure, the uneducated agriculturist will lay as much water as he can on his field, whenever he can do so without trouble. I think with Mr. Forrest that flow rates should be raised, on the ground that the Government are not getting the proper price for the water. As a remedy against the evils resulting from lavish waste of water under the

R. 69

PROPERTIES. USES.

flow system, enhancing the flow rates will be of no avail. The choice will not be (as far as the people are concerned) between taking water by lift at a cheaper rate cr water by flow at a dearer rate, but between taking water by flush or getting no water at all. In short, until canal water is dispensed by quantity, lift irrigation should be substituted wherever it is possible for flow, even if this can be effected only in the mode suggested by Mr. Ibbetson.

Result of Experiments.

"17. Before proceeding to discuss the question (lettered K*)—'What experiments or enquiries (if any) should be undertaken with view of ascertaining the possibility of eradicating *reh* from the *reh*-affected land.' I would invite the attention of the Lieutenant-Governor to Mr. Michel's interesting account of the experiments made by him with very great care and under favourable conditions for cleansing and reclamation of certain *reh*-infected land. on his estate. Mr. Michel brought his large farming experiences to bear on the experiment, which he carried out at no small cost. For his *modus operandi*, I must refer to his letter of the 14th February last (page iv,* etc., of the printed papers). It will be sufficient to state in this place that the result of his experiments was the conviction that surface drainage is not a guarantee against *reh* infection, and that *reh* once fairly developed can never be cured under existing conditions of water-level. Mr. Medlicott takes the same view. He writes: 'Mere surface drainage is ineffectual for the removal of the highly soluble salts of which *reh* is formed, though they are accumulated at the very surface the first few drops of rain dissolving and carrying the top soil, where they are untouched by the water draining off the surface. The attempt to make any permanent impression on *reh* ground by any system of manuring or of special crops seems to be hopeless.'

"18. In regard to the proposals made by the sub-committee, Messrs. Medlicott, Buck, and Forrest, to whom the question of the experiments or enquiries which should be undertaken with the view of ascertaining the possibility of eradicating *reh* was referred, I have only to recommend their adoption. The attention of His Honour is particularly requested to Mr. Medlicott's letter No. 121, dated 6th May (page xlii* of the printed papers).

"Mr. Wright's letter of the 22nd March (page xli*) is of interest as describing the actual condition of *reh* infected tracts. Mr. Wright's local and personal enquiries have proved beyond the possibility of doubt that the water level has been raised very considerably in the *reh*-infected country he inspected.

"In conclusion, I must express my regret that the committee did not meet and inspect the *reh* tract in the Mainpuri and Farukhabad districts, through which the Ganges Canal runs, under the able guidance of that most excellent officer, Captain Harrison, R.E., to whom my best thanks are due for the assistance he gave me. Mr. Forrest would not then have had reason to notice the small extent of the area affected by *reh*. I saw more *reh* in one day from Bhawant and Dharos in the Mainpuri and Baghosi and Aima in the Farukhabad district than I should have seen in ten days in that part of the Aligarh district which was visited by the committee."

It need scarcely be here stated that the "Reh Committee," embracing as it did members with an intimate acquaintance with the conditions of the country and the people, with the geological problems of India and with engineering enterprise, were highly qualified for the accomplishment of the task

---

* The references to letters and other documents parenthetically alluded to above are to papers which the reader will find in the detailed report. These references have been left since they exhibit the detailed nature of the information which was collected by the Reh Committee.

REH.	Dr. Center's Note on Alkaline Soils

**PROPERTIES.
USES.**

**Result of
Experiments.**

entrusted to them, and consequently their detailed report embraces all that has been definitely learned on the subject. Subsequent to its appearance, the Director of Land Records and Agriculture in the North-West Provinces has continued to devote a considerable portion of his time to the consideration of the problem of the reclamation of *úsar* lands by the cultivation of trees and shrubs found suitable for the agriculturally sterile tracts and to the utilisation of the salts found in the efflorescence in glass-making, etc. The following passages in the Annual Reports of his Department may be specially consulted :—Report of 1881, pp. 6-7, App. D, pp. 29-32; 1882, pp. 13-15; 1884, App. I., pp. 1a to 8a; 1887, p. 9; 1888, 6; 1889, pp. 3-4, App. C.

The report furnished by **Dr. Center** approaches the subject, however, from a distinct and very instructive point of view, *viz.*, the chemical. Having analysed the waters of springs, wells, tanks, and rivers, over an extensive area of the Panjáb, **Dr. Center** arrived at certain very definite conclusions. His remarks regarding the fact that wells in the vicinity of canals manifested no greater percentage of alkaline salts than those remote from any source of artificial or local percolation are peculiarly interesting in the light of the contention that the evil is largely due to canals. But, indeed, **Dr. Center's** statement of the *Reh* question is throughout so able that it would be unwise to single out any feature of it as more valuable than another; an abridgement of his report would at the same time mar its utility. It may, therefore, be here given in its entirety, the more so since it is not easily accessible :—

70

NOTE ON REH OR ALKALI SOILS AND SALINE WELL WATERS.

"A reference was made to this Office (*Chemical Examiner to the Panjáb*) by Government regarding the treatment of *reh* or saline soils by chemical manures. My predecessor, **Dr. Brown**, had written a report regarding the use of nitrate of lime as a remedy, and a copy of this was asked for. It could not be found in the records of the office, but I afterwards found that it had been published in the Selections from the Records of the office of the Financial Commissioner, and the gist of it was embodied in **Baden Powell's** book on Panjáb Products. As I had made numerous analyses of such efflorescences, and studied their connection with saline well waters, samples of which I had analysed from all parts of the Panjáb, and as I had an opportunity of observing and learning something of similar soils known as alkali soils in the Utah Basin and other parts of America, and of the methods used to reclaim them, I beg to submit a few notes on my observations. I am indebted to **Captain Ottley**, of the Irrigation Department, and **Mr. Miller**, Secretary to the Financial Commissioner, for access to the literature on the subject in the form of reports to Government. In the valuable report of the Aligarh Committee on the action of canals and irrigation in producing or extending reh, **Medlicott** gives a masterly exposition of the subject from a geological point of view, and most valuable information is contributed by **Messrs. Buck, Ibbetson,** and others who had studied the matter more from a practical point of view. I propose considering more especially the chemistry of the production of those salts and the conditions of their accumulation in soils and in the underground water,—points intimately connected with each other, and equally important in the agricultural and sanitary aspects. The efflorescences consist chiefly of sodium chloride and sulphate in varying proportions. In addition there is sometimes carbonate of soda, and I have usually found some magnesian sulphate. In certain localities the last-named salt is in very considerable proportion. In other cases nitrate of lime or alkali is present.

R. 70

" Various theories have been started regarding the origin of these efflor-
Marine theory. escences, the oldest being probably the marine theory.
According to this the Indo-Gangetic depression
was considered to be an old sea bed, the soil of which became impregnated with salts from the existence of shallow 'rans' and lagoons in a
former geological age. In favour of this it might be mentioned that there
is certain geological evidence that an Eocene sea covered the Panjáb
plain, its shore coinciding with some part of the outer slope of the Himálaya, with a gulf or gulfs penetrating the mountains as far as the valley of the
Upper Indus. On the other hand, to the east of Kumáon and to the
north of the Gangetic valley the situation of this shore line is obscured till
the Assam region is reached. The theory of recent marine impregnation
is now entirely to be abandoned. It is proved beyond doubt that the
whole of the materials of the Indo-Gangetic basin are fresh water alluvia
to an unknown depth, and consist in fact of the debris of the Himálayas
carried down by its drainage and deposited in this immense depression. There are no deep natural sections in which to observe the structure,
but in the Umballa boring of 701 feet, the Calcutta boring of 481 feet,
and that near Rajanpur of 464 feet, nothing but fresh water alluvia
were met. We do not speak here of the Salt Range region, in which are
accumulations of salt as old as the Silurian period.

" The true origin of reh or alkali efflorescence is the decomposition of
True origin of reh. the elements of rocks and soils which is continually
going on under the action of air and water. The
accumulation of the resulting salts in superficial soils or in subsoil waters
depends on various conditions of chemical constitutions and permeability
of soils, and on the nature of the surface and subsoil drainage, which will
be considered in detail.

" If the rain water that runs off the surface of the hills be examined, it
Decomposition in hill re- is found to have washed out appreciable amounts
gions. of soluble salts, chiefly carbonate of lime and alkaline chloride and sulphate. If such water runs
off crystalline or schistose rocks the amount of salts washed out may be
extremely small,—even 2 grains per gallon as at Dalhousie. If it runs off
a loose decomposing rock the quantity may be considerable,—for example,
8 grains near Murree. The rain-fall that percolates the debris of the decomposed rock which covers the surface of the hill-sides and fills up the
channels of ravines issues in springs at lower levels, and is found to contain much greater proportions of the same salts. This water not only
comes in contact with a larger quantity of degraded rock, and washes out
its soluble salts, but it takes up more carbonic acid from the air in the
pores of the ground, which is rich in this gas, and this dissolves more lime
and magnesian carbonate. From 10 to 25 grains per gallon are found in
springs in clean soils in various hill stations. In the hill stations themselves, where the porous subsoil becomes loaded with sewage impurity
from human habitation, the dissolved salts and organic impurity may be
very great. For example, in the bazar well at Murree I found 35 grains
per gallon in which were 12 grains of common salt. This last is, however,
a sanitary fact, and I wish at present to speak generally of the saline ingredients washed out of such oils not contaminated by human occupation.

" The soluble substances produced by rock decomposition and dissolved
by water are remarkably uniform in their nature, though varying in
amounts, both relative and total, according to the nature of the decomposing rock or soil. It may be generally stated that the earth water shews
a fugitive acidity from the presence of free carbonic acid and a slight permanent alkalinity from the presence of alkaline carbonate, but that the

REH.	Dr Center's Note on Alkaline Soils

main ingredients are carbonates of alkaline earths, chiefly of lime, and alkaline chlorides and sulphates, chiefly of soda. Other ingredients are generally in smaller amount, such as lime and magnesian chlorides or sulphates forming the permanent hardness, also silica, traces of iron, etc. Of course in special formations it may be highly charged with peculiar salts, and may even form what are called mineral springs; but we are speaking generally of the body of water that filters from the hill-sides and either sinks into the underground strata of the plains or finds its way into the streams and rivers, and thence into the sea, the great natural reservoir of the soluble salts washed out of the earth. The waters of the Panjáb rivers which I have examined, the Ravi, Jhelum and Indus, contain from 8 to 15 grains per gallon, varying according to the floods. The amount of soluble salt capable of efflorescence varies from about 2 to $4\frac{1}{2}$ grains. The river waters are most concentrated when they are at the lowest. At that time they are supplied by the water that has filtered through the soil and sub-soil of the higher regions, and has thus taken up more salts. In the hot weather, when the glacial water comes down, and in the rain floods at the end of the hot season, the dilution is at its highest. Other glacial rivers and those subject to annual floods shew the same thing. For example, the total solids in Nile water vary from $9\frac{1}{2}$ to $14\frac{1}{2}$ grains per gallon.

" To explain the ultimate origin of these salts we have to consider the action of the oxygen and carbonic acid in rain water on the rock elements. With the exception of the limestone strata which consist of carbonate of lime, often with carbonate of magnesia, all great rock formations are composed of silica and silicates, chiefly of alumina, lime, magnesia, soda and potash, with smaller amounts of iron and other metals. Such is the constitution of the granites, gneisses, slates, traps, etc. The old sedimentary rocks are similar in composition, being formed by the disintegration of these. The recent alluvia of the plains consist of finely-divided debris of the limestone and silicious groups, and in them the chemical decomposition going on under the influence of air and water is much intensified, owing to the state of fine division which favours chemical action, and because the constituents of the soil are further advanced in the path of degradation.

Ultimate origin of reh salts.

" In order to understand the slow chemistry going on in the ground, we have to conceive the outer shell of the earth generally covered with more or less vegetable mould, and permeated to its greatest known depth by meteoric water. There is no rock, however compact, and there is no depth to which man has penetrated, in which water is not found to have permeated by pores, cracks or fissures. The great agent of change is the carbonic acid of the air. This is dissolved in rain water, which also dissolves more from the decaying vegetable mould and from the air in the pores of the ground, which is rich in this gas. It has two great functions. It attacks the silicates of the alkalis and lime, forming carbonates. It further dissolves the carbonate of lime and enables it to be transported by water, and on its evaporation it deposits it. From the limestone rocks the water takes up carbonate of lime and magnesia which dissolve in its free corbonic acid, and in such formations it becomes very hard. The amount of carbonate of magnesia dissolved is always much less than that of lime. In the silicious rocks the felspathic family of minerals decomposes most readily. These consist of silicates of alumina and alkali, with generally small quantities of lime and magnesia. The white or soda felspar, which contains more soda than potash, is a common ingredient of the Himálayan rocks, and the decomposition of this in soils may possibly to some extent

Origin of carbonate of lime, alkaline carbonate.

R. 70

ORIGIN OF
CARBONATE
OF LIME.

account for the very great excess of soda over potash salts. The chief reason, however, depends on the fact which has been experimentally verified, that in a silicate containing both potash and soda the latter is dissolved out with greater facility and in much larger quantity than the former. The process of decomposition consists in removal of the alkali by the action of carbonic acid, while water is taken up, leaving hydrous silicate of alumina or clay. The presence of alkaline water also assists in promoting the breaking up by dissolving some silica. Another group, the lime silicates, is also readily decomposed by the action of carbonic acid or alkaline carbonate, and forms an additional source of carbonate of lime. On the other hand, the talcose rocks, which contain magnesian silicate, are hardly attacked at all. This magnesian metamorphosis of rock, which is very extensive and very ancient, is also the most permanent, and apparently a final one. We have thus accounted for the alkaline carbonate and carbonate of lime. The earth water is almost always slightly alkaline, and this plays a most important part in the decomposition of the silicious rocks and their metamorphosis. The alkaline carbonate rarely, however, appears in large amount, because it partly expends itself in decomposing silicate of lime, thus forming carbonate of lime, and if free carbonic acid is present this will be dissolved and carried away by the water. If magnesian or lime sulphate be present, the carbonate of soda with these will produce lime or magnesian carbonate; while sulphate of soda will be found in the solution. It thus happens that the waters of the rivers contain apparently no alkaline carbonate, but shew a permanent neutral reaction. All the river waters, in addition to carbonate of lime, which is their chief ingredient, contain also lime and magnesian sulphates which there has not been enough alkaline carbonate to decompose

"The circulation of the sulphur that occurs in the earth is very interesting. That which forms the sulphates in the earth water appears to be derived from the sulphurets, especially of iron, which are so universally diffused in rocks, and from the gypsum rocks, which, however, form an insignificant portion of the strata. The sulphate of lime being moderately soluble may be readily taken up by water. The sulphurets become oxidised by the oxygen in air or water leaving red iron oxide, which gives the yellow or red colour to soils and clays; while the sulphuric acid attacks the silicates and unites with soda or lime. In the strata of the earth are found deposits of sulphates of lime, but these appear to have been deposited from solution by infiltration or by evaporation, as in the Salt Range. Their ultimate origin is probably the same as that just indicated. The presence of sulphate of lime in soils leads to the production of sulphate of soda. The former salt is slightly soluble, and as the earth water contains alkaline carbonate, mutual decomposition leads to the formation of carbonate of lime and sulphate of soda. This partly accounts for the excessive proportions of sulphate of soda often found in reh. The sulphates may be again reduced to sulphides by organic matter from the vegetable mould or other sources, which accounts for the presence of sulphuretted hydrogen in dirty well waters rich in sulphates.

Origin of sulphates.

"As regards the chlorine of the alkaline chlorides, there is more difficulty. Chlorine is not an important chemical constituent of any common minerals forming rocks, but there is no rock that on being powdered and washed with distilled water does not shew its presence. The only explanation known of its appearance lies in the fact that, though generally in minute quantities, it is the most universally diffused substance we know. Even in air a chemically clean platinum wire cannot be exposed for some time without shewing the

Origin of chlorides.

REH.	Dr. Center's Note on Alkaline Soils

ORIGIN OF
CHLORIDES.

sodium lime in the spectroscope due to sodium chloride which can be extracted from the air dust.

"There is next to be considered the chemistry going on in the decomposition of the debris of the rocks forming the plain. It is in a more finely-divided state, and is therefore in a condition more favourable for chemical action, and besides the constituents are in a further advanced state of decomposition than in the fresh rocks. The action that has been described is therefore intens.fied. It has been proved by experiment that it is from the most finely-divided clay (felspathic) particles of soils that most of the soluble substances can be extracted. These particles are so extremely fine that under the microscope they are seen only as minute dots. The other small particles which are of measureable dimensions are silicious, and yield to acids only a slight amount of soluble matter.

Formation of salts in the plains.

"There are three points to be considered,—the action that takes place on the surface, that which takes place in the strata permeated by the underground water, and also the relations between the two.

"On the surface undoubtedly the greatest amount of decomposition goes on from the united action of air, moisture, heat and light. This produces the perennial supply of soluble salt necessary for the growth of plants, and in cultivation it is assisted by turning up and pulverising the soil and acting on it by water. In countries with good surface and underground drainage there is a constant escape of these salts, and the difficulty may be how to get enough of them. In many parts of our plains circumstances favour their accumulation, and the question is how to get rid of the excess. I have frequently taken samples of soil and sub-soil from places where there were efflorescences and where there were none, and on washing out the soluble substances with boiled distilled water found that they were similar, but different in amounts. They always consisted chiefly of alkaline chlorides and sulphate, with often small quantities of alkaline carbonate, and frequently larger amounts of soluble magnesian salt, sulphate or chloride. Another experiment was to take a sample of reh soil and wash it repeatedly till no trace of soluble salts could be found. It was then dried and thoroughly mixed and a portion tested again to see that no soluble salt was present. It was then placed on a filter and covered with porous filter paper so as to exclude dust but allow evaporation, and the bottom of the glass filter was corked. It was frequently watered with distilled water charged with carbonic acid and exposed to the heat and light of the sun in the hot weather for nearly three months. At the end of that time it showed no efflorescence, but on being washed with distilled water the solution shewed the presence of considerable quantities of alkaline chlorides and sulphates. This experiment proved that in that species of soil a sensible production of reh salt may take place in a few months. A similar sample irrigated with ordinary well water rapidly developed an efflorescence owing to the presence of salts in the water. This is nothing particularly novel in these results. Experiments have often been made of grinding down the solid rock from the debris of which the adjoining country was formed. On washing out the powdered rock the solutions were found to contain the salts of the water of the district; indeed it is always possible to account for the composition and proportions of the ingredients in a water draining any area if the structure and composition of the rocks are known.

Surface production of salts.

"To estimate approximately the decomposibility of a soil, the simplest method is to dry and weigh a sample and wash out from it the soluble salt already present. On drying

Approximate estimate of decomposition in soils.

and weighing the residue and deducting its weight from the original there will be found the soluble salt (along with some organic matter). This is the result of decomposition already accomplished. The solution may be tested in the usual way by evaporation to find the total dissolved matter and by ignition to find organic matter. The washed residue of earth is then ignited to expel all remaining organic matter and treated with hydrochloric acid, which will decompose and dissolve the materials, which are in an easily decomposable state. The solution will contain lime, magnesia, alumina and iron, and also the alkaline basis capable of efflorescing. On deducting the undissolved residue from the former, a figure will be found which will be an approximate index of the facility of decompositon of the soil. The hydrochloric acid solution can be examined in the usual way if required by first precipitating the iron, alumina, and phosphates of the alkaline earths, then the lime, and afterwards separating the magnesian and alkaline bases. The last will shew the salt capable of efflorescing. A more correct way is to perform an experiment similar to what I have described before with the glass funnel. A zinc box is made open at the top and closed at the bottom, with a false bottom of perforated zinc half way down. The section is usually 1 square foot. Earth is placed above the perforated zinc and the whole is exposed to the varying conditions of the season and climate, as rainfall, heat, moisture, etc. All water that falls sinks through the perforated bottom and is collected or evaporates. After some months or a season the solution in the bottom of the box and the earth are examined in the usual way to find the results of decomposition. Such an instrument is called a Lysimeter, and has the advantage of demonstrating the changes that take place, not by the action of acids, but by the ordinary operations of nature.

"Another source of generation and accumulation of these salts takes place in the strata moistened by the underground water. This is partly derived from percolation of rainfall from the surface where it is sufficiently porous. In its passage downwards it washes out any soluble salts it meets and carries them down till it reaches the impermeable stratum. In the second place the air contained in the vegetable mould and porous ground is rich in carbonic acid, and this is absorbed by the water and enables it to dissolve more lime and magnesian carbonate, which accounts for the much greater hardness of sub-soil waters. In the third place the alkaline water charged with carbonic acid not only promotes the decomposition of the strata through which it filters, but by a constant soakage action on that which it moistens produces still more. The amount produced would be in a great measure proportional to the time the water remains in contact with the stratum. In stagnant underground waters in the middle of the plains, as at Chunga Manga and Wanradaram, the dissolved salts amount to 400 grains per gallon. Another feeder of the underground water is the percolation of hill water that sinks into the porous fringe at the base of the hills. This, however, affects particularly the plain near the base of the hills. The solution formed from the debris on the hillside is much less saline than that from the finely-divided and more degraded materials of the plain. The hill percolation, therefore, affects the underground water near the hills in two ways It raises its level by hydrostatic pressure and it makes it less saline by dilution. There is still another source of underground waters in the percolation from rivers, streams, and canals. The neighbourhood of rivers affects the water level, and very sensibly influences the quality of the sub-soil water. Analyses of waters taken from wells near them shew that they closely approximate to the river waters, being little more than those filtered. For example, the well

Side notes:

ESTIMATE OF DECOMPOSITION OF SOILS.

Underground production of salts.

UNDER-GROUND PRODUCTION OF SALTS.

water near the Ravi was found to contain from 8 to 15 grains per gallon, that near the Jumna 9·8 to 14 grains. Advantage is now being taken of this in supplying water from such wells to some large cities in the Panjáb. The influence on the quality of the sub-soil water, however, only exists in the khadar land or low river valley. In the bhangar or bar land, the upland that lies between neighbouring rivers, even at short distances from the valley, the water may be highly saline. In the case of canals, as far as my observation goes, there is very little percolation in the districts I have seen irrigated by the Bari Doab Canal on account both of the impermeability of the soil and the disposition of the strata. If, however, a canal were made on a natural line of drainage, as I have heard the Western Jumna Canal is, it might influence the adjoining ground in the way I have mentioned in the case of rivers, both as to the water level and quality of percolating water.

"When rain water sinks into a soil containing soluble salts it dissolves them and carries them down till it reaches an impermeable stratum. Medlicott has pointed out the action of the first rain drops in carrying efflorescent salts down, so as to be in a great measure out of the reach of the surface scour of the succeeding rainfall. If the soil is porous it may gravitate down to the water stratum, which then becomes a reservoir of the surface salts. If it is only slightly porous, as in alluvial soils containing much clay, the soakage is only superficial to one or more feet in depth, and generally in such cases the surface soil is more or less porous from atmospheric exposure, and below it lies a more compact clay sub-soil. As rain water contains free carbonic acid, it dissolves also carbonate of lime and magnesia if these be present in the soil. When evaporation succeeds it draws up the moisture in the more porous surface soil by capillary action. As the water and carbonic acid pass off, the solution becomes more concentrated and carbonate of lime is re-deposited. This last action takes place first, and as the concentrated solution is drawn up to the surface, it finally deposits its most soluble salts on drying as an efflorescence on the surface. An essential condition is the dryness of the climate. In more temperate but dry regions, as in the Utah Basin and the elevated parks or plateaus of the rocky mountains, efflorescences appear as well as in the scorching plains of India. The action, however, is intensified by heat, which increases evaporation. By similar capillary action the moisture will creep up the sides of objects lying on the ground, such as pieces of brick, and deposit a copious efflorescence. At first it appears in glittering crystals, but as the sodium sulphate gradually loses its water of crystallization it breaks up into a copious white powder of anhydrous salt, and it is then that it is most apparent. The carbonate of soda behaves similarly, but the sodium chloride does not, having no water of crystallization. Nitrate of soda and lime deliquesce in damp air. During the hot months the salts, if brought up by rain, melt in their water of crystallization. By the word efflorescence we do not mean here what is known as such in chemical language, that is, the breaking up of a crystallized salt into a powder from loss of water of crystallization. What is meant is efflorescence in the physical sense, or the appearance on the surface of the ground of soluble salts brought up by capillary evaporation. It is true that sulphate and carbonate of soda efflorescence in the chemical sense, but chlorides and nitrates do not. From what has been explained regarding the origin of the salts dissolved out of the earth, it can be understood how the solutions can naturally be divided into two groups, whether they be river and canal waters, or well waters, or solutions formed when rain water soaks a saline soil. There are first the neutral solutions from which carbonate of soda

Nature of reh and its varieties.

COMPOSITION OF SALINE WELLS.

has almost or entirely disappeared, having been used up in decomposing any soluble lime or magnesian sulphate or chloride and precipitating their carbonates. To this belong the river and canal waters, the chief ingredient of which is carbonate of lime with less amounts of magnesian carbonate held in solution by free carbonic acid. There is present probably next in amount soluble salt of lime and magnesia, sulphate or chloride—the magnesia in smaller amount. The alkaline chloride, though the most constant ingredient in all waters, is in small amount, from $\frac{1}{2}$ grain to 2 grains, and the alkaline sulphate in about equal or larger quantity. In the majority of well waters in the plains in my experience there is high permanent hardness, indicating lime or magnesian sulphate or chloride, and sodium carbonate is deficient. The total dissolved salts is in fresh well waters about double that in rivers and canals, and may rise in saline wells from 10 to 40 times the amounts, the increase being chiefly in carbonate of lime and alkaline chlorides and sulphates. Though we speak usually of individual salts existing in a solution, this is not, strictly speaking, scientifically correct. If, for example, sodium chloride and lime sulphate be made into a solution, it will really contain quantities also of sodium sulphate and lime chloride, and the amounts of the four salts will depend on the masses of the first two, temperature, concentration, &c. Properly speaking, in recording an analysis, the total amounts of acids and bases should be separately recorded. By a conventional rule, however, it is customary to arrange the salts hypothetically. The second group of waters or solutions is that containing carbonate of soda. In these there is generally little permanent hardness, or soluble lime or magnesian salt. If these two groups on evaporating produce efflorescence, in the first we may have sodium chloride and sulphate, and any magnesian sulphate, if present; in the second we may have sodium carbonate with sodium sulphate and chloride, but no lime or magnesian salt. During the process of drying, which leads to the efflorescence, the first thing that occurs is the deposition of lime and magnesian carbonate, as the free carbonic acid disappears. Subsequently, sulphate of lime being only little soluble would deposit and the highly soluble salts including sodium carbonate, chloride and sulphate, magnesium and calcium chloride and nitrate and magnesium sulphate would be capable of efflorescence. These salts, however, are not deposited, as they exist in solution as new laws come into play. The chief of these is that during evaporation the least soluble salt that can be formed is first deposited; but this is modified by two other laws, the tendency of certain compounds to form double salts, and the tendency of substances with the same crystalline form to crystallize out together. The efflorescences thus produced consist of three groups: *1st*, the neutral, which contain no carbonate of soda (these consist chiefly of sodium chloride and sulphate, and frequently magnesium sulphate); *2nd*, the alkaline, which contain carbonate of soda, and alkaline chlorides and sulphates, but no lime or magnesian salt; *3rd*, the nitrous efflorescences. These generally contain no alkaline carbonate and consist chiefly of nitrate of lime and alkaline chlorides. Others contain alkaline nitrate, chloride and sulphate. They are developed where the soil has become loaded with organic nitrogenous matter. In several places about Lahore there is a good deal of magnesian sulphate, and I have observed on twigs of *farash* trees a saline coating of this salt. Reh is thus not a special salt or mixture of salts, but a very variable compound. It is really the most easily soluble salt in the earth-water, remaining in solution after the deposition of carbonate of lime, &c., on evaporation. The ingredients and their relative proportions are found to vary in different places, exactly as the well waters at different spots differ in saline

REH.	**Dr. Center's Note on Alkaline Soils**

contents, and in the same area there is a close relation between the two. The relative proportion of common salt to sodium sulphate was found by Medlicott to vary from 4 to 24 per cent.

"The re-deposit of carbonate of lime gives rise to those nodules known as *kankar*. It takes place at the upper margin of the impermeable sub-soil. They are not formed by
Formation of kankar.
the lime depositing round a nucleus and pushing the other elements of the soil aside. A portion of rather porous soil, consisting of a mixture of lime, sand, and clay, is infiltrated with water retained in it by an impermeable bottom. The carbonate of lime is deposited throughout this porous mass and cements its particles together till it becomes of a stony hardness. Deposit no doubt also takes place along the outer surface, as each former minute crystal deposited acts as a nucleus for further deposit. The formation is often seen in an incomplete state, nodules of soil having become only partially hardened. The process is essentially one of segregation from the soil itself. Such nodular formations, which are very common with other minerals, as iron oxide, silica, &c., are an example of the simplest kind of metamorphosis going on in rocks and soils. It is not necessarily connected with efflorescences on the surface. The essential condition of its existence is the presence of carbonate of lime or its ready production by ordinary decomposition in the soil. In soils and sub-soils which supply little lime there may be efflorescences without formation of kankar as in those consisting of clay and silicious sand. On the other hand, in marly soils, in which there may be little production of alkaline salt, kankar may form without any efflorescence. The analysis of kankar very well illustrates their mode of formation. They show from 20 to 50 per cent. of carbonate of lime, the rest consisting of the mixture of clay and sand of which the soil is composed.

"To estimate practically the amount of injurious reh in any soil, it should be washed with boiled distilled water and the
Estimation of reh in soils and water.
solution evaporated, then burned to expel organic matter, and finally weighed. In the case of the waters of rivers, canals, and wells, they should be evaporated, ignited, re-carbonated, and weighed. The easily soluble salts should then be washed out with a little distilled water, and the residue weighed. The portion undissolved consists of lime and magnesian carbonates and some sulphate of lime with small amounts of silica, &c. The difference between the two weights is the amount of salt capable of efflorescing. If one have a record of the analysis of any water, a rough approximation is got by deducting from the total solids the volatile matter (almost all organic), also the removable hardness consisting of carbonate of lime. In addition two grains per gallon of carbonate of lime should be further deducted as in boiling, in order to remove carbonate of lime ; two grains per gallon still remain dissolved. A still further deduction would require to be made for silica, iron, &c., but these are in small amount. I mention these methods of approximate estimation because they are readily applied and are useful for all practical purposes.

"In considering the conditions that lead to accumulation of salts on the surface or in the underground water, it is to be
Causes of accumulation of salts.
borne in mind that all soils exposed to moisture, air and heat are continually generating them, and that in some in which the felspathic elements are undergoing rapid decay the production may be profuse. Also all water, river, canal or underground, that has washed over or filtered through the ground, contains similar salts and promotes their further production.

and Saline Well Waters. (*G. Watt.*) | REH.

"The simplest case of accumulation is that of a closed basin like the Utah Basin. The surface water washing the salts off the ground has no escape to the sea and forms an inland salt lake. The soil in such cases is very saline, except in places where there is slope to allow thorough surface washing by rainfall, or permeability to allow the surface salt to be washed down to a deep ground water. In the centre of the depression both the surface and sub-soil and the sub-soil water are loaded with salt. The efflorescences in Utah closely resemble those in the Panjáb, the main common ingredients being sulphate of soda, common salt, and often sulphate of magnesia. In some places there is a large amount of carbonate of soda, in others borax is present. In the Caspian Basin the main ingredients are sulphate of soda and common salt. The very opposite case is a hilly or undulating country with sufficient rainfall and good natural surface drainage, the strata of which are also inclined, thus allowing of natural sub-soil drainage till the underground water finds an outlet at the outcrops of the strata or where they are laid open by natural sections of the country. Here the salts continually formed are either washed off the surface or are carried down to the sub-soil water which drains them off.

Accumulation in basins.

"In examining the state of things in the Indo-Gangetic plain, it is necessary to consider the structure of the country. The Himálayan axes stretching along the north of the plain are elevated cores of granitic gneiss flanked by metamorphic and limestone rocks. To the south of this is the Siwalik fringe with its dúns consisting of clays, sandstones and conglomerates. These are fresh water deposits formed by river and torrent action in the tertiary period, and having suffered displacement by the Himálayan elevation, they are seen to pass with great undulations and numerous fractures under the strata of the plain. This formation conducts water under the plain. There succeeds to this the recent gravel deposits from the ·outer hills, brought down by river and torrent action, similar to that which caused the Upper Siwaliks, and known as the Bhábar. This is extremely porous, and a great part of the water of the streams passing over it sinks into the ground and issues in springs at a lower level in the adjoining part of the plain, which is known as the Terai. Part also sinks beneath the plain and raises the ground-water level. The great alluvial plain itself is composed of horizontal strata. Near the hills are gravel deposits, but further off the soil and sub-soil to an unknown depth are composed of deposits of clay, sand and mixtures of the two in various proportions, according to the stream or lake action that deposited them. Diffused through these are found mica and small quantities of carbonate of lime, which makes soils more or less marly, and iron oxide which gives them a yellow or red colour, and minuter amounts of sulphate of lime and other salts. From numerous well sections it is seen that these alternating permeable and un-permeable beds of sand and clay are not continuous, but that they thin out and are replaced horizontally by others. This is observed even at short distances. Possibly many of the sheets of clay may have more or less of a basin form. The important points for us to remark in consider-ing the surface and sub-soil drainage are that this immense plain has an average breadth of about 200 miles, that practically the Gangetic and Panjáb plains are one, the watershed between the two being only percep-tible by accurate scientific measurements, and that its length is about 1,200 miles. There are also no deep natural sections exposing outcrops of the deep strata so as to allow of escape of underground water to the sea. In consequence of the very small surface slope, and on account of the horizontal disposition of the strata over such an enormous area, the

Accumulation in plains.

**ACCUMULA-
TION IN
PLAINS.**

conditions as regards drainage approach to those of a basin. The surface
drainage is weak, but ultimately finds its way by the rivers to the sea, but
the underground drainage is usually imperceptible. As regards the production of efflorescences, we have further to consider that in the Panjáb
there are three belts of plain. That adjoining the hills, the sub-montane
tract has a plentiful rainfall and moister air; south of this is a sub-desert
tract with small rainfall, and still further south is the desert country with
deficient rainfall. In the sub-montane belt the rainfall is sufficient to scour
the surface, and as it is more permeable from the presence of gravel and
sand, and has greater slope, the surface and sub-soil drainage are more
efficient. In the other two tracts the working of these agencies is defective. In the hills themselves the annual rainfall of a series of years is as
follows: Murree, 56·8 inches; Dharmsala, 123·2 inches; Simla, 68·6
inches. This does not include snowfall however. In the sub-montane
belt we would have—Rawalpindi, 32 inches; Sialkot, 39·3 inches;
Gurdaspur, 33·1 inches; Hoshiarpur, 36·5. Of the less watered region
there is Lahore with 19·3 inches; Shahpur, 14·5 inches; Sirsa, 14·5 inches;
while about Mooltan the rainfall is 6·9 and at Dera Ismail Khan 8·2.
The simplest case to consider is that which occurs in the more desert
country in which the rainfall is only enough to moisten the surface and
promote decomposition. If the soil is sandy the dissolved salt is carried
down to the underground water and the accumulation takes place there.
If the ground is not porous, as where clay predominates, only the upper
portion is soaked, and on drying the soluble salts are brought to the
surface. Instances of both these cases are found everywhere along the
southern portion of the Panjáb plain. In the middle portion of the plain,
where the rainfall may go up to 20 inches, similar actions take place. The
first drops of rainfall dissolve any efflorescence and sink into the ground,
carrying it out of the reach of surface scour, which, on account of the flat-
ness of the plains and small rainfall, is slight. In the more porous portions
the salt is carried down to the underground water; in the more impermeable it is brought to the surface by evaporation. It thus happens that in
certain places there is a scum of efflorescence on the surface while generally the ground water is saline. These remarks apply to the Doab or
Bhangar land, the more elevated part of the plain lying between adjacent
rivers. In this the water lies at a considerable depth from 30 to 100 or
more feet, and is more or less saline; in many places on digging deeper to
another stratum, fresher water is found. In the other great plains of the
earth where the climate is dry and like conditions of soil prevail, similar
efflorescences are developed. In the dry pampas of South America they
consist chiefly of sodium sulphate with some common salt; in the Siberian
steppes, of sulphate of magnesia along with sulphate of soda and common
salt. They are likewise found in the Russian steppes and the Thibetan
plateaus. The Khadar or low-lying river valley, cut out by recent erosion
from the old alluvial plain, usually shews little or no saline accumulation on
the surface and none in the underground water. Here the circumstances
are all different. In fact the river occupies the line of natural drainage
of the country and its deposits are parallel to the line of slope. Accordingly the water percolating from the river forms a subterranean stream,
gravitating down the river course and accompanying the main stream.
Its extent depends on the permeability and arrangement of the strata
and the resistance of the porous beds along which it moves. In the beds
of dry nullahs this gravitating water may be met on digging in the dry
channels. If the underground water were stagnant, remaining long in
soakage contact with the water bed, it would become more or less saline,
whereas it is found to resemble the river water filtered, though of course

it has taken up some ingredients from the earth, chiefly more carbonate of lime. In two cases in which I examined the water in beds of dry nullahs, I found it much less saline than that of the surrounding plain. In the Khadar land the water lies near the surface, and may be within the reach of capillary evaporation, which would produce efflorescences, as it often does to some extent. But in consequence of the occasional washing by floods, and of the underground circulation I have described, there is no permanent accumulation either on the surface or in the ground water.

" One of the most interesting and important cases is that in which the Accumulation by evapora- ground water lies close to the surface within the tion from a shallow water reach of capillary evaporation, thus furnishing an table. unlimited supply of efflorescence. The enquiry made by the Aligarh Committee chiefly referred to this instance. It was considered that the ground-water level had been raised by percolation from the canal assisted by hydrostatic pressure in consequence of the canal being above the level of the country. Other causes assigned for the rise were the obstruction to surface drainage by canal and railway embankments acting as bunds, and the practice of profuse irrigation in flooding. All these would lead to an increased body of water sinking into the ground, carrying earth salts in solution to be again brought up by capillary evaporation from the shallow water table. It is very important to be able to estimate how much is due to each of these agencies, as on the decision of this point would depend the remedial measures to be applied, such as the lowering the level of the canals, their realignment on the high Bhangar land instead of on the lower ground, the restriction of profuse irrigation, the relieving of the surface drainage, and the establishment of artificial sub-soil drainage. I am unable to enter into the merits of these most interesting points, because I have never had an opportunity of making observations on an area where this mode of generation of reh was going on to a serious extent. The only portions of country I have seen in which the ground water lies very near the surface are the plains adjoining the hills and the Khadar lands or river valleys. In the former the rainfall is more plentiful, the slope of the surface and deep strata are better, there is more moisture in the air, and therefore less evaporation. All these tend to prevent accumulation of salt below and efflorescence above. In the latter the washing of the surface by the floods and better subterranean drainage may account for the want of accumulation. In the parts of the Bari Doab Canal which I have seen, the ground water lies at a depth that is totally out of the range of capillary action, and the strata consisting of alternating clays and sands are so impenetrable that percolation can have little effect on the water level. Captain Ottley informs me that on the Bari Doab and Upper Sutlej Inundation Canals the curves of the rise and fall of the well waters markedly follow those of the rainfall and do not appear to be affected by irrigation. I did not find any marked difference in the water levels of the wells near and at a distance from the canal about Lahore. A still better proof was that the salinity of the wells was not altered by proximity to the canal. If percolation to any extent existed, the wells close to the canal ought to be fresher than those at a distance. In the part of Lahore occupied by the railway station and barracks the ground water is salt. At the end of the hot weather I found that a well a few yards from the canal contained as much salt as others far off. After the rains the same well waters were found to be so diluted as to contain less than one-half of the former amounts. The depth from which capillary evaporation can take place is also a question that ought to be investigated by observation and experiment. Much of course depends on the porosity

REH.	Dr. Center's Note on Alkaline Soils

ACCUMULA-TION BY EVA-PORATION.

of the soil, but in the most favourable cases one would fancy, from the known laws of capillary force, that the action would only be through a few feet, unless assisted by hydrostatic pressure. At the village of Baoli, on the Western Jumna Canal, where the reh action is very pronounced, the depth from the surface of the ground to the water table (as shewn by measurements of an unused well) is 8 feet. It is said that, before the Western Jumna Canal was re-opened in 1819, the water in wells about the part lay at a depth of 60 to 70 cubits, and this tradition appears to be confirmed by inspection of the records of other wells which had been sunk to as much as 116 feet, and in which now there are 62 feet of water. On the banks of water-courses and canals about Lahore in salt soils one often observes two lines of efflorescence, one a few feet above the water level at the upper limit of capillary soakage, and another some distance from the surface, at the base of the surface percolation. As regards the rise in the well water levels said to be caused by canals, it would be necessary to have accurate information as to what those levels were before the canals were made. Probably no accurate record was made before the earlier canals were started, as attention was not directed to the point.

"There are last to be noticed some other modes of distribution and accumulation of alkali salts. Irrigation by flooding and allowing the water to dry on the soil, unless
Other modes of accumulation.
it is very permeable, of necessity leads to production of salt. Not only does the irrigating water contain salt which it deposits as an efflorescence, but it also promotes further decomposition in the soil. The amount of reh in ordinary canal water might be from 2 to 6 grains per gallon. If well water is used the accumulation is much greater, because it contains much more salt. In places where the water is sweet, the reh may be about 6 to 15 grains per gallon; where it is salt, it may amount to more than 200 grains per gallon, as at various places on the Railway Line between Lahore and Mooltan. An extraordinary instance is mentioned in the Aligarh Report of a reh soil tried by the most energetic measures without effect. An analysis of this soil would probably have proved that the elements of the soil itself were in such a state of decomposition that most of the measures employed assisted the process. Again, water running off a saline field must necessarily dissolve a portion of its salt, and if it be allowed to run into another and dry, that salt will be deposited. The agency of wind appears to be a slight and very variable one. There is no doubt that wind blowing over a saline country and raising dust transports saline particles. Travellers over the alkali plateaus of the Rocky Mountains are familiar with the irritation caused to the eyes by this mode of transport. All these, however, are of secondary importance. The main points to bear in mind are that there are several factors causing production and accumulation, and others leading to the removal of earth salts. Of the former there is first the soil itself. This is always generating them, and in certain cases its materials so readily undergo decomposition that perhaps even artificial means may fail to cure the evil. The next chief factor is the water used in irrigation. This always contains reh salts,—the river and canal water in small amount, but the well water often in enormous quantities. In addition, the irrigation water may not only deposit its salt in the soil, but it causes further production in the soil itself. Another cause is the special condition in which the sub-soil water lies within the reach of capillary action from the surface, which may give rise to an inexhaustible supply. The factors concerned in the removal are, first, permeability of the soil, which may allow the salts to be washed down to the underground water. If this have a ready outlet, they are removed; if not, there will be a saline ground water; but the surface may shew no accumulation if the

water table is deep. If, however, the ground water is a very short distance from the surface, there may be a profuse efflorescence. The second great cause of removal is surface scour. If the rainfall is copious it may thoroughly wash off excess of salts, and for this reason in rainy regions alkali is rare. If it is slight and only moistens the soil without scouring it, there will be a continuous production and accumulation on the surface, except when the soil is porous and allows it to be carried down to the ground water. The third means of removal is by vegetation, which annually takes up its necessary portion of salts and assimilates them. It is frequently observed that in cultivated spots the reh is kept under; while the uncultivated ground around may be covered with it. In connection with this it is to be remarked that for land plants potash salts are necessary, but it is doubted whether soda salts are essential, except in the case of Salsolæ, &c., which grow in soda soils. This may have something to say to the barrenness of our soda reh soils. Another factor to be noticed is the effect of shade produced by vegetation, which prevents the excessive evaporation which brings the salt to the surface. It thus remains more diffused through the moisture in the soil. Lastly, plants also induce capillary currents towards themselves. The absorbing parts are the rootlets and myriads of hairs surrounding each. These, by the act of absorption, set up capillary currents in the moisture of the soil towards themselves, which compete with capillary evaporation at the surface and tend to the diffusion of the moisture and its salts through the soil as far as the roots extend. It is to be noted that if a soil remain damp, so that the salts are diffused through it, they may do no harm. It is their concentration as a scum on the surface that poisons crops. The moisture round the rootlets forms a solution so saline, that the osmose currents by which the plants are nourished are interfered with and they perish.

"I conclude this paper with some practical remarks regarding the me-
Methods of cure. thods of dealing with saline efflorescence agriculturally; but these I wish to be considered suggestive more than anything else, as I cannot pretend to any experience in that line. When visiting Utah I was very much struck on finding that the saline efflorescences of that basin were similar in nature to those I had seen and studied in India. I made enquiries into the ideas current on the subject and the methods of reclaiming the soils. **Brigham Young's** notions of natural philosophy were both extremely simple and at the same time shrewd, as would be expected from an uneducated but practical and successful man. He said: 'There is salt in everything. Water has salt, plants have salt, and earth has salt; and the Bible tells us that if the earth have lost its salt it is useless. A certain quantity of salt is necessary for vegetation; in our country we have too much of it, and we get rid of part of it.' He referred me to **Mr. Woodruff,** who was Secretary to the Agricultural Society, and to some of the best farmers, to see what was done.

By sluicing and irrigation. The plans adopted were the following: A salt field was ploughed and small runlets of fresh water were sent down the field, at short distances apart, washing the soil and running off into the drainage of the country. Another method was to plough up a field and make a terrace round it and then flood it. The water was allowed to soak for some time till it had dissolved the salt and was then run off. Another plan was to terrace a ploughed field and dig a deep trench round it. The field was flooded, and, the unploughed sub-soil being less permeable, the water holding the salt in solution filtered into the trench. I observed similar processes carried out on the salt marshes round the Bay of San Francisco. This is gradually silting up, and surrounding it are miles of low flats impregnated with sea salt and growing only saline

REH.	Dr. Center's Note on Alkaline Soils

EFFECTS OF MANURE.

plants. Through these pass shallow delta channels, scoured by the rise and fall of the tide. To reclaim this soil, low earth embankments are raised round the farms. These are fitted with floodgates closed by the rise of the tide and opening on its fall. The salt in the soil is washed out by the fresh water of the streams falling into the bay by a process of sluicing such as I have described, and is run off as the tide falls. In the depression between the coast range and the second range of hills artesian wells can be made, and these were used where none of the mountain streams were available. An English Company was working on a salt marsh by the aid of artesian water only; but it was generally considered that it would not be a success, as the amount of artesian water was after all only trifling compared with the area to be reclaimed. The universal opinion in Utah was that if they once succeeded in covering an alkali field with a crop of any kind the victory was won. After the land was half cured, they generally covered it with a hardy grass, the most approved being red-top American grass. Beetroot was also said to grow well as an early crop; after that Indian corn and other crops by degrees. Tuberous crops grow well in the country, and the potatoes are said to be the best in the

By manure and cultivation.

world. The last method I shall mention was that employed by **Brother Fenton**, an energetic Devonshire farmer. It happened to be impossible for him to get fresh water to wash the salt out of his fields, and he tried large quantities of manure,—20 to 50 tons per acre. Barn-yard manure was considered the best, and, as his great object was to keep the surface from the sun, which drew up the salt, he also used litter to cover it. The first crops he covered the ground with were the red-top grass and oats, and he sowed his crops in September, so that the ground should be covered with vegetation when the alkali would be appearing. As soon as by this means he got his first crop of red Timothy grass he found he had succeeded. **Mr. Fenton** complained that after partly curing one field he ruined it by trenching and bringing up a saline sub-soil. His idea was that the salt was a sort of perspiration of the earth, and, therefore, mostly on the surface, and that by turning up the sub-soil he would get a better soil. In India it is certainly the case that a short distance below the surface less reh is found. It may be different in a closed basin like that of Utah, where the sub-soil also may become saturated with salt. Utah city is partly situated on a bench at the base of the Wasatch hills adjoining the plain, and at first the farms surrounding it were made on the ground that was not saline. About one-fourth of the land under cultivation was salt, and three-fourths of this had been cured by sheer cultivation, much in the way I have described in the case of **Mr. Fenton's** farm. For the other fourth sluicing and irrigation had been available. The cultivation of saline soils is also carried out in other settlements. In most old-settled countries, and especially in India, agriculturists are very conservative in following the practices of their forefathers. In America, where the population is composed of emigrants from all countries, every man brings the methods used in his own, and all sorts of trials are made and the fittest survives. These are made in a new country under new circumstances, and people are not bound by traditional customs, but are anxious to try whatever succeeds in the hands of others, and also make experiments according to their own ideas. These may be crude, but still a vast number of experiments are made,—not isolated ones by a Government, but everywhere generally by the people themselves—and anything that is successful is hailed as a discovery. Some of the methods I have described as used in America may not always be practicable in the plains of India. To run off the saline water requires a slope and lines of natural drainage that may not be available. It might be possible to run off the salt-impregnated water into

EFFECTS OF CULTIVATION.

absorption wells, thus returning the salt to its natural destination, the underground water. It is a law that a well will absorb as much water without raising its level as it would give out without sensibly lowering it. This
By arboriculture. means has been used in some cases to get rid of liquid sewage, but was found to poison the wells. The plantation of trees is also proved to be a very efficient means of cure. The kikar is well known as capable of flourishing in such soils. They not only assist in moderating excessive evaporation by shade, but they also absorb and remove a certain amount of salt from the soil. As the alkali exists chiefly in the surface soil and in much less amount at a small depth, trees may grow readily where annual crops could not. The latter have their rootles only in the surface soil, and are poisoned by the excess of salt; while the roots of trees extend deeper into less saline ground; also plants not only consume portion of the salt, but they prevent its concentration on the surface. A most conclusive experiment made near the Western Jumna Canal by the Irrigation Department is reported by Colonel Fulton. A piece of utterly useless reh land, for which revenue was remitted, was taken up by the Department and planted with kikar trees. These flourished and a very fine crop of doab grass, two feet high, came annually up under the trees, and the efflorescence disappeared. The villagers, seeing that the land was improved and fearing it would be alienated by the new settlement, applied for the restoration of both trees and land, and carried their point in the courts of law. A few days after the restoration the wood was sold to a wood merchant and every tree cut down. At present the doab grass is all gone, and the soil is encrusted with salt Such an experiment made among American farmers would have excited the keenest interest and given rise to numerous trials of the same.

"The method of cure by nitrate of lime as a manure, suggested by Dr.
Chemical manure. Brown, would act in two ways. It would partly serve as a manure favouring vegetation, and in addition it would act on the alkaline and magnesian sulphate by double decomposition, producing nitrate of alkali and sulphate of lime, which last is a slightly soluble salt which is not hurtful to vegetation and would not form an efflorescence. Carbonate of soda would be similarly neutralised, but the sodium chloride would remain unaltered. The natives are well acquainted with this use of nitrous efflorescences, which can be distinguished from the sulphate of soda by its moistness due to deliquescence and by the brown colour and by not efflorescing in fine powder. It consists mainly of common salt and nitrates of lime and soda. This production of nitrate is due to the decomposition of nitrogenous animal or vegetable matter, first producing ammonia which is afterwards oxidised to nitric acid. An essential condition of the nitrification process is the presence of alkaline carbonate or carbonate of lime to fix the nitric acid. For example, ordinary dung-heaps may produce plentiful supplies of ammonia, but no nitric acid. Indeed, nitric acid, if present, is changed by the reducing action of the decomposing organic matter to ammonia. If wood-ashes containing carbonate of potash or lime be mixed with the heap, the acid becomes fixed. Artificial nitre beds, called *nitrières* or nitre plantations were first introduced by the chemists of France to supply nitre for gunpowder during the wars of the Revolution, when the ports of France were blockaded by the English and imports prevented. Animal manure is mixed with carbonate of lime and wood-ashes and frequently watered with urine, which produces much ammonia. This is cultivated for two or three years. In tropical countries the production of nitrates is more plentiful and rapid. A manure of a valuable quality could probably be made by municipalities or by the zamindars themselves by mixing pounded kankar, or even marly

R. 70

REH.	Dr. Center's Note on Alkaline Soils

CHEMICAL MANURES.

soil with manure and moistening it frequently during one or two hot seasons. If it were moistened with liquid sewage, which would tend to produce more ammonia, the production would be increased. This artificial production is an exact imitation of what takes place naturally in soils in which nitre is produced. In the Panjáb nitrates effloresce near villages where the soil becomes impregnated with animal sewage, which undergoes nitrification in presence of the carbonate of lime and alkaline carbonate in the soil. The most plentiful supply is in the soil on the mounds that indicate the sites of old villages. This is the main source of the manufacture of saltpetre in the Panjáb. Similarly near buffalo ponds and watering-places for cattle, where dung is trodden into the soil, nitrates effloresce and are swept up by the zamindars as manure. A similar process no doubt takes place when a field is well manured with animal refuse. The conditions of the production of nitrate of lime in the soil are present, and this may account to some extent for the reclamation of alkali soils by manuring alone. For this purpose animal manures would be far superior to vegetable. In plants there is comparatively little nitrogenous matter, which alone can generate nitrates or ammonia. In Utah a favourite manure is the refuse of slaughter-houses, which would be capable of supplying large amounts of ammonia and nitrates.

Uses.

" As regards the uses to which the alkali efflorescence might be put, sulphate of soda can easily be separated by evaporation and forms a useful purgative. It might be possible to utilize those more rich in alkaline sulphate for the manufacture of carbonate of soda for glass or soap work. The average mixture of sodium chloride and sodium sulphate resembles the product of the first step of manufacture of this carbonate, which is done by the addition of sulphuric acid to common salt. By evaporation the sulphate which crystallizes out first can be freed from most of the common salt, and this would resemble the salt cake. The materials for the further reduction, charcoal and lime, would be readily available, the latter from the kankar beds. Certain soils contain carbonate of soda in such quantities that it can readily be separated by the crystallization process. At one time an enquiry was made as to whether the nitre manufacturers defrauded the revenue to any extent by disposing of the alimentary salt left in the refuse saltpetre earth after extracting the nitre. Samples have from time to time been forwarded to this office, and these were found to contain from 35 to 70 per cent of common salt. It would certainly be possible, and not very difficult, to obtain a rather impure alimentary salt by rough crystallization processes not only from the saltpetre earth but also from suitable kinds of reh. "

A writer in the *Englishman* newspaper recently attempted to show that an important discovery had been made by Mr. Cockburn that " *Reh* was essentially produced by the solution of *kankar*, which subsequently combines with salt to form caustic soda." The argument advanced seemed to hinge on the observation that *usar* lands were not confined to the vicinity of canals, and that, therefore, some other explanation would have to be sought, though, it was admitted, canal water, like the rain and carbonic acid of the air, would tend to produce the necessary decomposition of the *kankar*, and perhaps more rapidly so than would be the case in the absence of canal flooding. But even were it possible to establish the accuracy of the statement that *kankar* is, so to speak, a resting stage for the lime of *reh*, the explanation, and still more so the prevention, of the *reh* devastation of soils is as far off as ever. The source of the formation of the *kankar* itself would be the primary source of the evil, acted on by some interruption to the drainage, natural or artificial, which allowed of the salts on percolation through the soil with the rain or canal inundation to return

Products of India. 427

and Saline Well Waters. (*W. R. Clark.*) RENNET.

again to the surface by capillary attraction. Though much has been written for and against the injurious effects of canal embankments as interrupting the drainage and thus tending to cause *reh* efflorescence, all writers seem agreed on the chief features of the explanation that interruption to the natural drainage of the country is an all-important factor in the accumulation in the soil of the salts of disintegration which by the summers drought are brought to the surface of the soil (*Conf.* with **Barilla,** Vol. I., 394-396; also **Beads,** 426).

(*W. R. Clark.*)

REINWARDTIA, *Dum.; Gen. Pl., I., 243.*

[*1100;* LINEÆ. **71**

Reinwardtia trigyna, *Planch., Fl. Br. Ind., I., 412; Bot. Mag., t.*
    Syn.—LINUM TRIGYNUM, *Roxb.;* L. REPENS, *Don.*
    Vern.—*Gulashruf, karkún, kaur, gud batal, basant, bál-básant,* PB.
    References.—*Roxb., Fl. Ind., Ed. C. B. C., 277; Voigt, Hort. Sub. Cal., 100; Stewart, Pb. Pl., 20; Wall. Cat., 1505; W. & A., Prodr., 134; Don, Prodr., 217; Watt, Cal. Ex. Cat., V., 237; Gazetteers:—N.-W. P. (Agra Div.), IV., lxix; (Him. Dist.), X., 306; Mysore and Coorg, I., 58; Agri.-Horti. Soc., Ind., Trans., V., 122; VII., 80; Jour., IV., 206; VI., 40; XIV., 14; Ind. Forester, VI., 240.*
    Habitat.—A tufted, glabrous undershrub, 2-3 feet high, met with in the hilly parts of India from the Panjáb eastward to Sikkim, ascending to 6,000 feet; also in Behar, Assam, and Chittagong, and southward from the Bombay ghâts to the Nilgiri hills.
    Medicine.—It is said to be used as a medicine for "founder" in cattle (*Stewart*).

MEDICINE. **72**

Remija, see **Cinchona,** Vol. II., 315.

73

RENNET and RENNET SUBSTITUTES.

    Vern.—*Panir maya,* HIND., GUZ., PB., PERS.; *Wfeheh, maya shutr,* ARAB.
    References.—*Rennet in Ainslie, Mat. Ind., I., 334; Baden-Powell, Pb. Prod., 113; Forbes Watson, Indust. Surv. Ind., II., 330; Rennet Substitutes in Stewart, Pb. Pl., 64, 164, 168; Murray, Pl. & Drugs of Sind, 112. 157; Kew Report (1881), 36; Dymock, Mat. Med. W. Ind., 645; Drury, U. Pl., 445.*
    Chemical Composition.—By rennet, in ordinary English, is meant an aqueous infusion of the dried stomach of a calf which is used to coagulate the casein of milk, preparatory to the manufacture of cheese. It appears to contain a soluble ferment which acts directly on the milk. In India, the stomach of the calf is not used for making rennet, but that of the kid is substituted, while medicinally the dried stomachs of the camel, horse, hare, male kid, and ewe, imported from Arabia, are much valued for various purposes, among the Muhammadans of India.

HEMISTRY. **74**

    Medicinal Uses.—In Arabian works of medicine rennet is described as possessing "deobstruent attenuant and aphrodisiac" qualities, and the Muhammadans of India have long been in the habit of regarding it as of value for such purposes.

MEDICINE. **75**

    Domestic Uses.—A knowledge of the method of preparing cheese by means of rennet seems to be entirely confined to the higher classes of Muhammadans in India and they employ as a source of their rennet the dried and preserved stomachs of kids, not those of calves.

DOMESTIC. **76**

    Rennet substitutes.—Among the Hindus cheese-making is a branch of agricultural industry unknown, since the use of ordinary rennet in cheese-making is altogether contrary to the tenets of their religion. Cheese to be saleable among the Hindus would have to be made with some vegetable rennet. Many such substitutes have been suggested, but as yet no progress has been made in the making or sale of cheese among the Hindus of India.

SUBSTITUTES. **77**

R. 77

SUBSTI-
TUTES.

Perhaps the best known and most valued substitute for rennet is a plant, known in the Panjáb and Sind as *panír-bund* (cheese-maker, **Withania coagulans**). This has been used from very early times by the Beluchis and Afghans as a substitute for the more expensive animal rennet. The value of this plant as a coagulant for milk was tested at Kew in 1881, and it was reported on, as follows :—

"A quantity of the dried capsules of this plant was obtained and part of it tried here and found to be most suitable for the purpose. Being a member of the poisonous nightshade family, its safety was in the first p'ace carefully and gradually tested. It has been ascertained that an ounce of the pounded capsules in a quart of water is a very suitable strength for use ; a tablespoonful of this decoction coagulates a gallon of warm milk in about half an hour." On analysis of the capsules of **Withania coagulans** they were found to contain a ferment closely resembling animal rennet. Attempts have been made, but as yet with only partial success, to separate and preserve this ferment by means of sugar and so render available to the people of India a purely vegetable rennet. For further information on the subject, the reader is referred to **Withania coagulans**, *Dunal ;* Vol. VI.

Stewart (*Panjáb Plants*) also mentions **Crotolaria Burhia,** *Hamilt.* (Vol. II., 595), and **Leucas cephalotes,** *Spreng.* (Vol. IV. 633), as possessing the power of coagulating milk, and **Dr. Stocks** in the *Journal of the Asiatic Society of Bombay* speaks of **Rhazya stricta** as being in universal use for this purpose throughout Sind, but no further information with regard to this property seems to be available concerning these plants. The milk-coagulating properties of the berries of **Streblus asper,** *Lour.*, have been made the subject of an interesting series of experiments at the Madras Experimental Farm (see **Streblus asper,** Vol. VI., Part II.)

(G. Watt.)

78

REPTILES; *Boulenger, Reptilia, in Fauna of British India.*

Reptiles may be briefly defined as vertebrate animals that breathe throughout their existence by means of lungs, and have the body covered with horny or bony plates (scales or scutes). Their respiratory movements are slow and irregular, hence Reptiles are cold-blooded animals. A basi-occipital bone is present in the skull, which articulates with the vertebral col‌umn almost invariably by means of a single convex condyle. All Reptiles are oviparous or oviviviparous.

The class—Reptilia—is divided into numerous orders, most of which however, embrace only extinct species. Indian recent reptiles belong to three orders, *viz.* :—

     I.—**Emydosauria,** or Crocodiidæ, Crocodiles.
     II.—**Chelonia,** Tortoises and Turtles.
     III.—**Squamata,** Lizards, and Snakes.

The last is more commonly broken up by naturalists into two distinct orders—the **Lacertilia** and **Ophidia** ; but the arrangement followed in the present article will be that of **Boulenger** in the text-book above quoted, who treats these as sub-orders of **Squamata.**

References.—*Fayrer, Thanatoph. Ind. ; Günther, Reptiles of British India ; Murray, Vertebrate Zool. of Sind., 333 ; Ewart, Poisonous Snakes of India ; Aitchison, Zoology of Afgh. Del. Com., 94 ; Tennent, Nat. Hist. Ceylon, 271 ; Ainslie, Mat. Ind., II., 263, 276, 291 ; Irvine, Mat. Med., Patna, 99 ; U. C. Dutt, Mat. Med. Hindus, 278 ; Baden Powell, Pb. Pr., 153 ; Useful Pl. Bomb. (Vol. XXV., Bomb. Gas.), 276, 277 ; Forbes Watson, Indust. Survey Ind., 345, 392 ; Home Dept. off. correspondence and statements of destruction to human life caused by reptiles ; Gazetteers, Bombay, V., 362 ; VIII., 105 ; Encycl. Brit., XX., 432 ; Balfour, Cyclop. Ind., III., 384.*

The Reptiles of India.   (*G. Watt*)   REPTILES.

Distribution in India. —The chief characteristic of the Reptilian Fauna of the Indian region is the great variety of the generic types and number of species of its Snakes. The latter amounts to no fewer than 450, which is nearly one-third of the total number of species known in the world. They are referable to about 100 genera, of which the majority do not range beyond the limits of India.

DISTRIBU-TION.
79

The lizards are numerous and highly specialised, but unlike the snakes not one of the families is peculiar to the Indian continent. The crocodiles are represented by two, and the gavials by one, species. Aligators are altogether absent. The Chelonians comprise altogether 44 species, of which 43 belong to the sub-order **Thecophora.** The Athecæ are represented by a single species (**Dermochelys coriacea,** the leathery turtle} which is only an occasional visitor on the coasts of India. (*Jour. As. Beng., 1862, p. 367.*)

As might be expected, in so immense an area as that of the Indian continent, the Reptilian fauna is immensely diversified ; it varies according to the physical characters of the different districts, or their proximity to neighbouring regions. With reference to the distribution of Reptiles in India, Günther divides the country into seven provinces, each of which is characterised by certain leading types which are seldom found out of these areas. The first of these includes Mysore, the Carnatic, Malabar, and Travancore, together with the Island of Ceylon. The leading type in that area is the **Uropeltidæ** or "Earth snakes" which are found nowhere else, and of which only one or two species extend into the Deccan. The second province is that of the Deccan, which contains a large number of forms peculiar to itself. The Reptilian fauna of the plains formed by the Ganges and the Indus, present features entirely different from one another, although both are inhabited by a small number of such species as are characteristic of the Indian region generally. An Indo-African character is assumed by the fauna of Sind, which extends for some distance southward along the coast of the Koncan. In Bengal, the Reptilian fauna is chiefly composed of the common species which range over a greater or lesser part of the other provinces. The Himálaya is characterised by the appearance of a large number of new or peculiar species, and by the disappearance of such forms as are abundant in Bengal. This change commences to be very marked from an altitude of 4,000 feet above the sea. The last province described by Günther is that which extends from the east coast of the Bay of Bengal round to the Malayan Peninsula and Siam. This belt of land is distinguished by a Reptilian fauna which assumes more and more the archipelagic character, the nearer we approach to the Malay Peninsula where more than half the reptiles are species found also in the Islands of the Archipelago (*Günther, Rept. Br. Ind.*).

Food.—Although the members of the class Reptilia are by no means so important a source of food to the natives of India, as are the Fishes, still many of them are largely used as such chiefly by men of the lower castes and by aboriginal tribes. Thus the Nagas and Kukis capture and eat the snakes and lizards found in their territory. The low caste men and aboriginal tribes of Southern India and Ceylon regularly hunt for and kill as food many of the large lizards which are there numerous. Several species of turtle, besides the true edible one (**Chelone mydas,** *Schweigg* are regularly exposed for sale in the markets of many coast-towns, while the eggs of a still greater number of species are much sought after for culinary purposes by the inhabitants of the coasts which these animals frequent for breeding purposes.

FOOD.
80

The market for these articles of food is, however, almost purely a local one, although the flesh of the Iguana of Southern India and Burma is sometimes eaten by Europeans, and that of the edible turtle is much relished by all classes. India is not, however, the source of the edible turtle used in Europe: that is chiefly imported from the West Indies, but it is the same or a closely allied species to the edible turtle of the Indian seas.

MEDICINE.
Flesh.
81

Oil.
82
Poison.
83

Medicine.—The FLESH of several species of lizards is recommended by Native practitioners as a medicine, and is in common use as a domestic remedy among the people. It is credited with tonic, stimulant and alterant properties, and is thought particularly useful in cases of syphilis. An OIL, extracted from several species of lizard, is supposed to be a powerful aphrodisiac. SERPENT POISON is occasionally used in Hindu medicine. It is collected by making the snake bite on a piece of wood, the poison being received on a plantain-leaf. The liquid poison is then allowed to dry in a cup, or is rubbed up with a fourth part of mustard oil and spread out to dry on a piece of plantain-leaf. It is used in fever in combination with other remedies.

Snake-bite.
84

SNAKE BITE.—The annual loss of life in India occasioned by the bites of poisonous snakes is, as might be expected, very considerable. In 1889, 22,480 persons lost their lives by snakes, and 3,793 herd of cattle were similarly destroyed. The average number of persons killed by snakes, during the past ten years, may be given at 20,000. The deaths occur almost entirely among Natives; comparatively few instances are recorded of Europeans having been killed by the bites of poisonous reptiles. The accidents seem to occur chiefly at night when the animal, having been surprised or trodden on, inflicts the wound in self-defence. The fact that the dwellings of Natives are not usually raised above the surface of the ground, together with the universal custom of the people in taking off the shoes when in their houses, renders them particularly liable to the attacks of these animals. The death-rate is in fact higher or lower more in relation to the habits of the people than to the prevalence of poisonous snakes.

Symptoms.
85

SYMPTOMS.—The local symptoms of snake-poisoning are a stinging sensation of the part penetrated which is soon followed by pain, at first dull and aching, subsequently lancinating and piercing in character. The ultimate effect is numbness, terminating in local paralysis of sensation. There may, or may not, be slight swelling of the part. The marks or points where the two fangs have entered are usually to be found in bites from terrestrial snakes. In the bites of salt-water snakes these are more difficult to distinguish, since the fangs are not much larger than the fish-like teeth, immediately behind them. Very soon after an effective bite, general symptoms make their appearance. The patient becomes restless and excited. Nervous depression, languor and muscular exhaustion then occur. The face becomes pallid and bathed in perspiration, the pupils are dilated, the pulse is quickened, there is loss of appetite, with nausea or vomiting. General muscular paralysis supervenes, there is lethargy and drowsiness, ending in unconsciousness. Involuntary evacuations usually occur, which are sometimes tinged with blood. The breathing becomes slow, laboured, shallow and eventually ceases. The pulse remains full, quick and compressible, and beats for a few minutes (from three to four), after all breathing has ceased. The pupils are widely dilated. Death is ushered in by convulsions or convulsive twitchings of the muscles of the extremities and face. Recovery from the general symptoms may take place where a person has been bitten by an exhausted snake or by one whose aggregate supply of poison is small or by a vigorous cobra, from which only a minute quantity of poison has been injected, or where a person has been greatly protected against the absorption of the poison by the early application of the ligature. In such cases the above symptoms are present, but are modified in severity.

Treatment.
86

TREATMENT.—Immediately on infliction of the bite, a ligature should be applied very tightly about the limb at a short distance above the wound

and several other ligatures at suitable distances further up. The flesh
around the marks should be excised freely, the surface of the wound should
be scarified, washed and squeezed and bleeding should be encouraged.
The amount to be excised around the fang-marks must depend upon the
looseness or denseness of the surrounding tissues, and their consequent tendency to allow the poison to diffuse freely or not. The actual cautery may be
applied to the deeper portions of the wound, where the poison may not have
been removed by the excision. Stimulants, such as alcohol or liquor ammonia in 15-drop doses in an ounce of water, may be given, but over-stimu-
lation should be avoided, and during the administration of stimulants, liquid
nourishment should also be exhibited. If depression be marked, mustard
plasters may be applied over the heart, on the pit of the stomach, or on the
nape of the neck. The patient should be allowed to rest in a well-ventilated
room, protected from the sun. In cases where the prostration is extreme,
methods of artificial respiration and galvanism may have to be employed.
Amputation should at once be had recourse to in cases when the snake is
known to have been one of a markedly poisonous character, and the bite
is in one of the fingers or toes. In wounds of the larger members or of the
trunk where free excision is possible, amputation may not be necessary.
The intravenous injection of ammonia or of liquor potassæ do not apparently do much good, but they may be injected freely into the poisoned part.
The local application of permanganate of potash is strongly recommended
by some authorities.

Various methods of treatment, all more or less founded on superstition, are
employed by the natives of India, and every now and then alleged specifics
for the poison of snake-bite have been vaunted, but as yet no antidote is
known which is capable of neutralising the action of the poison. The so-
called "snake-stones" can have no other effect than at best to act as local
absorbents. They are for the most part merely pieces of charred bone which
by absorbing from the wound the poison-charged blood may be of use in
slight cases of snake-bite.

MEDICINE.

Domestic and Industrial.—The SKINS of some species of crocodiles and
lizards have been for long utilised by the Natives of many parts of India
in the manufacture of leather, and of late years these have been exported for
a similar purpose to Europe where they are used for hand-bags, cigar-cases,
and other small articles.

DOMESTIC.
Skins.
87

The imbricated epidermic plates chiefly from the hawk-bill turtle
(**Chelone imbricata**) form the TORTOISE-SHELL of commerce. The plates
consist of horny matter, but are harder, more brittle, and less fibrous than
ordinary horn. Their value in European commerce depends on the rich mottled colours they display—a warm translucent yellow dashed and spotted
with rich brown tints and on the high polish they take and retain. In
China, on the other hand, shells which have white and black spots that
touch each other and are as much as possible similar on both sides of the
blade are most valued. The finest tortoise shell is obtained from the Eastern Archipelago, but it is also exported from the southern coasts of the
Indian continent, from Ceylon, and from the West Indian Islands and
Brazil. The epidermic scales are detached from the turtle either by actual
force after the animal is killed or by suspending the living animal over a
fire till heat makes the scales start or by immersing the dead animal in boiling water. The last is the most usual method. If taken from the animal
after death and decomposition, the shell becomes clouded and milky.

**Tortoise-
shell.**
88

From very ancient times tortoise-shell has been a prized ornamental
material. It was brought from the East to ancient Rome by way of Egypt,
and was used by the wealthy Romans as a veneer for their furniture.
In modern times it has been employed in Europe for the characteristic

inlaying work known as buhl furniture. It is also used as a veneer for small boxes and frames, it is moulded into snuff-boxes and cigar-cases, formed into knife and razor handles or cut into combs. This last use is its most important method of employment in the East, where in Ceylon, especially, tortoise-shell combs are largely worn by both men and women.

SACRED.
89

Sacred.—Throughout India the serpent, in some form or other, is worshipped by many Hindus as an incarnation of the deity, and ordinarily no Hindu will kill a snake, but will rather turn aside from its path on seeing it. At the time of Asoka, B C. 255, Buddhism appears as a system of pure abstract morality, but about the beginning of the Christian Era, a Naga or Turanian revelation became incorporated with it, and still later, about 700 A.D., Buddha was represented as an object of worship with the serpent as his co-equal. At the present day, the naga or cobra-di-capello is revered by all Hindus, and Hindu women resort for worship to the white-ants' nests where the cobra generally takes up his home. If a cobra be killed by the Hindus of rural districts, they give it a funeral as if it were a human being. Their gods and deified warriors are figured, as shadowed with the outspread hoods of 3, 5, 7, 9, and 11 cobras. In their mythological language, the snake is the emblem of immortality, its endless figure with the tail inserted into the mouth, and the repeated renewal of its skin affords symbols of continued youth and eternity. A large number of temples in India are dedicated to snakes, and Hindus resort thither to worship both the sculptured form and the living animal. Many legends relating to snakes are told by the people. Hidden treasures are believed always to have a serpent guardian. Leprosy, ophthalmia, and barrenness in women are supposed by Hindus to be a punishment to those who in a former or in the present life may have killed a snake, and the curse is thought to be removed only by serpent worship.

Various species of snakes are tamed by professional charmers, but the cobra (**Naia tripudians**) and the Hamadeyad (**Naia bungarus**) are perhaps the favourites for this purpose, since, in addition to their size, formidable appearance and well known venomous properties, they are the most docile and easily trained of this class of reptiles.

For the following list of Reptiles the writer is chiefly indebted to Boulenger's *Reptilia in the Fauna of British India* and to Günther's *Reptilia of British India.* The remarks on their economic uses are for the most part derived from the latter work :—

90

I.—EMYDOSAURIA, *Boulenger, Reptilia Fauna, Br. Ind., 1.*
                              [*Günther, Rept. Br. Ind.,* 61.

91

1. **Crocodilas palustris,** *Lesson ; Boulenger, Rept., in Fauna Br. Ind.,* 5 :
THE CROCODILE.

Habitat.—The common Crocodile of India is found in rivers, marshes, and ponds throughout the country, its area being extended westward to Baluchistán. It is distributed to Ceylon, Burma, the Malay Peninsula and Archipelago.

FOOD.
Flesh.
92

Food.—In Siam the FLESH of this and other species of crocodiles is sold for food in the markets and bazárs.

DOMESTIC.
93

Domestic.—The large TEETH of this species are sometimes mounted by the natives of Southern India and Ceylon with silver lids and used as boxes to carry the powdered *chunam,* which they chew with the betel-leaf. For further information see the account CROCODILE, Vol. II., 591.

94

II.—CHELONIA ; *Boulenger l. c.,* 6.
                              [*Günther, Rept. Br. Ind.,* 37.

95

2. **Batagur baska,** *Gray ; Boulenger, Rept., in Fauna Br. Ind.,* 38;
THE BATAGUR.

Habitat.—A water tortoise, the shell of which attains a length of 20 inches; found in the rivers of Bengal, Burma, and the Malay Peninsula.

Food.—Mr. Blyth says it abounds at the mouth of the Hughli, and that great numbers are brought up to Calcutta, where they are eaten by particular castes of Hindus, and are even kept for sale in tanks.

[*ther, Rept. Br. Ind., 54.*

3. **Chelone imbricata,** *Boulenger, Rept., in Fauna Br. Ind., 49; Gün-*
THE HAWK-BILL TURTLE OR INDIAN CARET.

Vern.—(Tortoise-shell=) *Kachakra,* HIND., GUZ.; *Alúngú-thadú,* TRAVAN-CORE; *Ammah,* MALYAL.; *Sisik-kurakura, sisik-panu,* MAL.

Habitat.—A marine turtle, found in tropical and sub-tropical seas. Although it occurs throughout the East Indian Archipelago, it is plentiful only in certain localities, for instance, on parts of the coasts of Ceylon, of the Maldives and the Celebes. As, however, turtles always resort to the locality where they were born or where they have been used to propagate their kind, and as their capture is very profitable, they become scarcer and scarcer at places where they are known formerly to have been abundant (*Günther*).

Industrial.—This species is the most prevalent source of the TORTOISE-SHELL of commerce, and Kelaart (*Rept. Ceyl., p. 181*) says that some specimens sell for as much as £4, the price depending on the quality of the shell.

Food.—The natives eat the FLESH of this turtle, but it is unpalatable to Europeans; the eggs, however, are regarded as equal to those of the other turtles.

[*Br. Ind., 52.*

4. **C. mydas,** *Boulenger, Rept., in Fauna Br. Ind., 48; Günther, Rept.*
THE GREEN OR EDIBLE TURTLE.

Vern.—*Leik-pyen-won, leik-kyæ,* BURM.

Habitat.—The Green Turtle is found in tropical and sub-tropical seas, It occurs plentifully on all the coasts of the East Indies, but is rather rare in the Bay of Bengal.

Food.—In habit it is herbivorous, and its FLESH forms an important article of food. The turtle of Indian seas rivals in size the Atlantic turtle, and in flavour it is not inferior. The EGGS are very rich and have a taste somewhat like that of marrow; they will keep for weeks although exposed to the air. At certain seasons, the FLESH of this species becomes poisonous, and deaths have been ascribed to its use. [*Fauna Br. Ind., 17.*

5. **Emyda granosa,** *Günther, Rept. Br. Ind., 45; Boulenger, Rept., in*
THE BUNGOMA.

Habitat.—A river turtle, found in the Indus and Ganges, and in rivers and canals throughout India.

Food.—It grows to a length of ten inches, and its FLESH is relished as food by certain castes of Hindus. [*Rept. Br. Ind., 29.*

6. **Nicoria trijuga,** *Boulenger, Rept., in Fauna Br. Ind., 27; Günther,*
THE COMMON CINGHALESE POND TORTOISE.

Habitat.—A common pond tortoise in the Peninsula of India and in Ceylon.

Domestic.—It is thoroughly aquatic and carnivorous in its habits, and is put by the natives of Southern India and Ceylon into wells and tanks to keep them clear of impurities.

[*Günther, Rept. Br. Ind.*

7. **Testudo elegans,** *Schœpff.: Boulenger, Rept., in Fauna Br. Ind., 21;*

Habitat.—A land tortoise, found all over India except in Lower Bengal. This, or a closely allied species (**T. platynota**), is distributed also to Ceylon and Burma.

Food.—Its FLESH is eaten by the natives of some parts of the country, and in Burma especially it is esteemed a great delicacy.

Marginal notes
FOOD. Flesh. 96 97
INDUSTRIAL. Tortoise shell. 98
FOOD. Flesh. 99 100
FOOD, Flesh. 101 Eggs. 102 103
FOOD. Flesh. 104 105
DOMESTIC. 106 107
FOOD. Flesh. 108

109	**III.—SQUAMATA,** *Boulenger, l. c. 52.*
	Sub-order, LACERTILIA.
110	8. **Gecko verticillatus,** *Laur. ; Boulenger, Rept., in Fauna Br. Ind.,* 102 ; THE COMMON GECKO. [*Günther, Rept. Br. Ind.,* 102.
	Vern.—*Chipkuli,* HIND.; *Paillie,* TAM.; *Bullie,* TEL.; *Chapkali,* DEC.; *Touktai, eing-hmyoung,* BURM.; *Mus-ali, sarata,* SANS.; *Chilpásah,* PERS.
	Habitat.—Found on trees as well as in houses, throughout the greater part of British India, Burma, Siam, Cochin China, and Southern China.
MEDICINE. Body. 111	**Medicine.**—"The bruised BODY of this animal, made into electuaries in conjunction with certain aromatics, the Hindu doctors think, possesses virtues in leprous affections" (*Ainslie*).
	[*Ind.,* 86 ; *Günther, Rept. Br. Ind ,* 107.
112	9. **Hemidactylus gleadovii,** *Murray ; Boulenger, Rept., in Fauna Br.* THE HOUSE GECKO.
	Habitat.—The most common house gecko in India, Ceylon, Burma, and Southern China.
MEDICINE. Body. 113	**Medicine.**—Its bruised BODY is thought to have the same properties as that of the preceding.
114	10. **Mabuia carinata,** *Boulenger, Rept., in Fauna Br. Ind.,* 188 ; *Günther,* THE COMMON INDIAN SKINK. [*Rept. Br. Ind.,* 79.
	Vern —*Reg-mahi,* PB.
	Habitat.—This is one of the most common and widely-spread lizards of the East Indies. It is found in almost every part of the continent as well as of the Archipelago from Afghanistán to China and to the Philippine Islands. It does not occur above an altitude of 8,000 feet. It is found both on trees and in houses, but more particularly in loose rocky soils or tumbled-down walls; in its higher areas it chooses sunny exposed spots.
MEDICINE. Oil. 115 116	**Medicine.**—An OIL made from this and other allied species is supposed to have restorative, stimulant, aphrodisiac, and anti-syphilitic properties.
	11. **Varanus bengalensis,** *Boulenger, Rept., in Fauna Br. Ind. ; Günther,* THE IGUANA. [*Rept. Br. Ind.,* 65.
	Vern.—*Gho-samp,* HIND., BENG. & DEC.; *Gho-samp,* PB.; *Udumu,* TAM.; *Udumbu,* TEL.; *Biyawak, manawak,* MAL.; *Talla-goyá,* SING.; *Ghoda-sala, gandhera,* SANS.; *Zib,* ARAB. The young of this and other pecies of **Varanus** are often known in Northern India by the name of *bis-cobra* and are regarded as venomous, probably on account of their long forked tongues.
	Habitat.—This is a most common species of **Varanus**; occurs over the whole of India, Ceylon, and Burma. It is terrestrial in its habits and lives in holes in dry places.
MEDICINE. Body. 117 Tongue. 118 FOOD. Flesh. 119 DOMESTIC. 120	**Medicine.**—The dried BODY of the Iguana, made into an electuary with *ghi,* is recommended by the Vytians as a strengthening medicine in consumptive complaints. The head, tail, and feet are not, however, employed in medicine. The TONGUE, if plucked from the living animal and swallowed whole, is believed by the Singhalese to be a specific for consumption.
	Food.—Its FLESH is relished by the Muhammadan inhabitants of Southern India, and is supposed to be very strengthening. The skin of the Iguana is made by the Singhalese into shoes. It is often badly infested with a tick or parasitic insect which assumes the exact shape and colour of the scales.
	[*Günther, Rept. Br. Ind.,* 67.
121	12. **V. salvator,** *Cantor ; Boulenger, Rept., Fauna Br. Ind.,* 166 ;
	Vern.—*Kabara-goyá,* SING.
	Habitat.—A still larger species than the preceding; is found in marshy localities and on trees overhanging rivers in Bengal, Ceylon, and Burma; it is distributed to the Malay Peninsula and Archipelago, and to South China.

R. 121

The Reptiles of India. (*G. Watt.*)	**REPTONIA boxifolia.**

Medicine.—The Singhalese believe that the FAT of this species if externally applied is a cure for cutaneous disorders, but that taken internally it is poisonous. The froth from the lips of this iguana is supposed to be one of the ingredients of the celebrated Singhalese poison, *Kabara-tel.*

Sub-order, OPHIDIA, *Boulenger l. c. 232.*

13. Naia tripudians, *Merr.; Boulenger, Rept., Fauna Br. Ind., 391 ;*
THE COBRA or NAGA.   [*Günther, Rept. Br. Ind., 339.*

Vern.—*Nag, nag samp, kala samp, gokurrah* (the spectacled) *keautiah.* (with only one ocellus), HIND.; *Keute sap,* BENG.

Habitat.—The Cobra or Naga is the best known and most deadly snake of India; it is found all over that country and in Ceylon, Burma, the Andamans, Southern China, Indo-China, the Malay Peninsula and Archipelago. On the Himálaya it occurs up to altitudes of 8,000 feet, and to the west it is distributed to Afghánistán, North-Eastern Persia, and Southern Turkistan, as far as the eastern coast of the Caspian Sea.

Medicine.—As before mentioned, the POISON is used as a medicine.

Sacred.—The cobra is perhaps the one most frequently trained by snake charmers, and both the sculptured form and the living animal are revered by Hindu serpent worshippers throughout India.

[*Ophiophagus elaps, in Günther, Rept. Br. Ind., 341.*
14. N. bungarus, *Schleg. ; Boulenger, Rept., in Fauna Br. Ind., 392;*

THE HAMADRYAD, often called the *Black Cobra.*

Vern.—*Sunkerchor,* BENG.; *Airaj,* URIYA; *Gnan-pok,* BURM.

Habitat.—This snake, from its larger size and fiercer habits, is still more dangerous than the Cobra. Fortunately, although it has a rather large geographical range, it is comparatively rare. It is a native of Southern India, Orissa, Bengal, Assam, Burma, the Andamans, Siam, the Malay Peninsula, Java, Borneo, Sumatra, and the Philippines. It is found in hollow trees and is sometimes found resting on the branches, and lives on other snakes.

Sacred.—It is a favourite snake with charmers and is particularly docile, and easily trained when in captivity.

[*Günther, Rept. Br. Ind., 330.*
15. Python reticulatus, *Gray ; Boulenger, Rept., Fauna Br. Ind., 246;*

Vern.—*Saba-gyee,* BURM.; *Ular, sawa,* MAL.

Habitat.—The Rock snake of Burma, the Nicobar Islands, the Malay Peninsula and Archipelago.

Medicine.—The GALL-BLADDER of the Python is much sought after by the natives of Burma on account of its supposed medicinal virtues.

Food.—Its FLESH is eaten by the Karens who capture them with a slip noose at the end of a long bamboo.

MEDICINE.
Gall-Bladder.
130
FOOD.
lesh.
131

REPTONIA, *A. DC.; Gen. Pl., II., 648.*

Reptonia buxifolia, *A. DC.; Fl. Br. Ind., III., 534 ;* MYRSINEÆ.

Syn.—EDGEWARTHIA BUXIFOLIA, *Falc.;* MONOTHECA MUSCATENSIS, *A. DC.*

Vern.—*Gúrgúra, garar,* PB.; *Garar,* AFG.; *Gurgura,* PUSHTU.

References.—DC., *Prod., VIII., 153 ; Boiss., Fl. Orient., IV., 32 ; Brandis, For. Fl., 287, t. 34; Gamble, Man. Timb., 241 ; Stewart, Pb. Pl., 135; Murray, Pl. & Drugs, Sind., 169 ; Watt, Cal. Ex. Cat., VI., 154; VII., 212 ; Baden-Powell, Pb. Pr., 594; Settlement Report :—Panjáb, Peshawar Dist., 13 ; Kohat Dist., 29 ; Gazetteers :—Panjáb, Peshawar (1883-84), 27 ; Dera Ismail Khan (1883-84), 19 ; Bannu, 23 ; Rawalpindi, 83 ; Indian Forester, V., 179; Balfour, Cyclop. Ind., III, 398.*

RESINS	The Resins and Gums of India.

Habitat.—A large evergreen shrub or small tree, common in the Trans-Indus Hills, at altitudes between 2,000 and 3,000 feet, and occasionally met with in the western part of the Salt Range. It is distributed to Afghánistán and Muscat.

FOOD.
Fruit.
133

Food.—The FRUIT which is globose and fleshy is collected in April and eaten by the Natives. By European taste, it is regarded as insipid, but it is much esteemed by the Afghans.

TIMBER.
134

Structure of the Wood.—Light brown, with irregular purplish-brown heart-wood, very hard, heavy, close and even grained. Weight 71℔ per cubic foot.

DOMESTIC.
Wood.
135
Seeds.
136
137

Domestic and Sacred.—The WOOD is used in the parts of the Panjáb where it occurs, as fuel and for other domestic purposes. The SEEDS are strung together and used as rosaries.

RESEDA, *Linn.; Gen. Pl., I., 112.*

Only two members of this genus are indigenous to the Indian Peninsula, *viz.,* R. Aucheri and R. pruinosa, neither of which are of economic value but several others are found cultivated, principally for ornamental purposes.

138

Reseda luteola, *Linn.; Fl. Br. Ind., I., 181;* RESEDACEÆ.
THE WELD.

References.—*Boiss., Fl. Orient., I., 434; Aitch., Afg. Del. Com., 38; Voigt, Hor. Sub. Calc., 75; Gasetteer, Mysore and Coorg, I., 69; Crookes, Dyeing, &c., 408; Agri.-Hort. Soc., Ind. Trans., V., 121; VII., 71; Smith, Dict., 437.*

Habitat.—A native of Western Europe, but distributed thence eastward to Afghánistán, where it is a common field weed. In India it has been found, as an escape from cultivation, at many localities such as Simla, Calcutta, &c., and it is mentioned as cultivated in the Bangalore Gardens. It is said to occur wild in China.

DYE.
139

Dye.—It was at one time cultivated in England, and in several parts of France, Germany, Austria and Hungary it is so at the present day, on account of the colouring matter it yields, which, according to the different mordants employed, gives green, yellow or blue results. The colour known as Dutch pink is also chiefly obtained from this material.

140

R. odorata, *Linn.; Fl. Br. Ind., I., 181.*
MIGNONETTE.

References.—*Gasetteer, Mysore and Coorg, I., 69; Agri.-Hort. Soc., Ind. Trans., VII., 71; Smith, Dict., 272.*

Habitat.—A plant of doubtful habitat; but commonly cultivated during the cold weather in most European and many Native gardens in India; it is often seen especially in the temperate tracts as a perennial.

DOMESTIC.
141

Domestic.—Among the Romans, this plant was applied as a charm to allay the irritation of wounds. It is now universally cultivated on account of its sweet odour.

142

RESINS AND GUMS.

A detailed list of the RESINS and RESIN-YIELDING plants as also those that afford the GUMS of India will be found in the writer's *Calcutta International Exhibition Catalogue*, Vol. I.

It may suffice here to explain that by a "Resin" is meant a hard friable natural plant substance, externally resembling a Gum, but insoluble in water, although soluble in ether and alcohol. Resins are rich in carbon, poor in oxygen, devoid of nitrogen, and burn with a smoky flame. No resin is a definite chemical compound, but is rather a complicated mixture. The essential ingredients of every resin are the resin acids, volatile

R. 142

Products of India. 437

The Resins and Gums of India. (*G. Walt.*) | RESINS.

RESINS.

oils, gums and often cinnamic and benzoic acid as well as the ordinary components of plant tissues, *vis.*, cellulose, tannin, &c.

The classification of Resins, which has been adopted in this work, is that given by Dr. M. C. Cooke in his Report on Indian Gums and Resins.

Omitting the gums, which have been already dealt with (Vol. IV., 188), the resins may be divided into three groups as follow : —

I. **Gum Resins**, or those which contain a gum soluable in water, associated with a resin and a volatile oil, soluble in alcohol.

II. **Oleo-Resins**, or natural compounds composed of a resin and an essential oil.

III. **True Resins** which may be hard or soft.

I.—GUM RESINS.

143

(A.)—EMOLLIENT GUM RESINS.

1. Garcinia Cambogia, *Desr.;* GUTTIFERÆ, Vol. III., 464.
2. G. Cowa, *Roxb.;* Vol. III., 465.
3. G. Mangostana, *Linn.;* Vol. III., 470.
4. G. Morella, *Desr.;* Vol. III., 472.
5. G. pedunculata, *Roxb.;* Voi. III., 476.
6. G. travancorica, *Bedd.;* Vol. III., 478.
7. G. Xanthochymus, *Hook.;* Vol. III , 478.

(B.)—FETID GUM RESINS.

8. Dorema Ammoniacum, *Don.;* UMBELLIFERÆ, Vol. III , 191.
9. Ferula alliacea, *Bois.;* UMBELEIFERÆ, Vol. III., 333.
10. F. fœtida, *Regel.;* Vol. III., 335.
11. F. galbaniflua, *Buhse.;* Vol. III., 338.
12. Gardenia gummifera, *Linn.;* RUBIACEÆ, Vol. III., 480.
13. G. lucida, *Roxb.;* Vol. III., 482.

(C.)—FRAGRANT GUM RESINS

(α) *Bdellium kind.*

14. Balsamodendron Mukul, *Hook.;* BURSERACEÆ, Vol. I., 366.
15. B. Myrrha, *Nees.;* Vol. I., 367.
16. B. Roxburghii, *Arn.;* Vol. I., 369.
17. Mangifera indica, *Linn.;* ANACARDIACEÆ, Vol. V., 146.

(β) *Benzoin kind.*

18. Styrax Benzoin, *Dryand;* STYRACEÆ, Vol. VI., Pt. II.

(γ) *Olibanum.*

19. Bosmellia Bhan-Dajiana, *Bird.;* BUR ERACEÆ, Vol. I, 512.
20. B. Carteri, *Bird.;* Vol. I , 512.
21. B. serrata, *Roxb.;* Vol. I., 515.

(δ) *Dragon's Blood.*

22. Calamus Draco, *Willd.;* PALMÆ, Vol. II., 17.
23. Dracœna Draco, *Linn.;* LILIACEÆ, Vol. III., 193.
24. Pterocarpus indicus, *Willd.;* LEGUMINOSÆ, Vol. VI., 355.
25. P. Marsupium, *Roxb.;* Vol. VI., 357.
26. P. santalinus, *Linn.;* Vol. VI., 359.

II.—OLEO RESINS.

144

(A.)—BALSAMS.

27. Altingia excelsa, *Noronha;* HAMAMELIDEÆ, Vol. I., 202.
28. Balsamodendron Berryi, *Arn.;* BURSERACEÆ, Vol. I., 366.

R. 144

RESINS.	The Resins and Gums of India.

RESINS.

29. Chlorokylon Swietenia, *DC.;* MELIACEÆ, Vol. II, 270.
30. Dipterocarpus alatus, *Roxb.;* DIPTEROCAEPEÆ, Vol. III., 157.
31. D. incanus, *Roxb.;* Vol. III., 158.
32. D. tuberculatus, *Roxb.;* Vol. III, 160.
33. D. turbinatus, *Gærtn;* Vol III., 161.
34. Dryabalanops Camphora, *Coleb.;* DIPTEROCARPEÆ, Vol. III., 196.
35. Hardwickia piunata, *Roxb.;* LEGUMINOSÆ, Vol. IV., 201.
36. Melaleuca Leucadendron, *Linn.;* MYRTACEÆ, Vol. V., 204.
37. Mesua ferrea, *Linn.;* GUTTIFERÆ, Vol. V., 236.

(B.)—NATURAL VARNISHES.

38. Buchanania latifolia, *Roxb.;* LEGUMINOSÆ, Vol., I., 544.
39. Holigarna Arnottiana, *Hook.f;* ANACARDIACEÆ, Vol. IV., 259.
40. Melanorrhœa usitata, *Wall.;* ANACAEDIACEÆ, Vol. V., 208.
41. Odina Wodier, *Roxb.;* ANACARDIACEÆ, Vol. V., 445.
42. Rhus succedanea, *Linn.;* ANACARDIACEÆ, Vol. VI., Pt. I.
43. Semecarpus Anacardium, *Linn.;* ANACARDIACEÆ, Vol. VI., Pt. II.
44. S. travancorica, *Bedd.;*   Ditto.

(C.)—TURPENTINE AND TAR.

45. Cedrus Libani, *Barrel;* CONIFERÆ, Vol. II., 235.
46. Erythroxylon monogynum, *Roxb.;* LINEÆ, Vol. III., 270.
47. Picea Morinda, *Link;* CONIFERÆ, Vol. I., 3; VI., 227.
48 Pinus excelsa, *Wall.;* CONIFERÆ, Vol. VI., 238.
49. P. Gerardiana, *Wall.;* Vol. VI., 240.
50. P. khasya, *Royle;* Vol. VI., 241.
51. P. longifolia, *Roxb.;* Vol. VI., 243.
52. Tectona grandis, *Linn.;* VERBENACEÆ, Vol. VI., Pt. II.

145

III.—TRUE RESINS.

(A.)—COPALLINE RESINS.

53. Canarium bengalense, *Roxb.;* BURSERACEÆ, VOL. II, 94.
54. C. strictum, *Roxb.;* Vol. II., 96.
55. Dammara orientalis, *Lamb.;* CONIFERÆ, Vol. III, 18.
56. Hopea micrantha, *Hook.;* DIPTEROCAEPEÆ, Vol. IV., 271.
57. H. odorata, *Roxb.;*   Ditto.
58. H. parviflora, *Bedd.;*   Ditto.
59. Shorea robusta, *Gærtn.;* DIPTEROCARPEÆ, Vol. VI, Pt. II.
60. S. sericea, *Dyer.;*   Ditto.
61. S. Tumbuggaia, *Roxb.;*   Ditto.
62. Vateria indica, *Linn.;*   Ditto.
63. V. acuminata, *Heyne;*   Ditto.
64. Vatica lanceæfolia, *Blume;*   Ditto.
65. V. Roxburghiana, *Blume;*   Ditto.

(B.)—ELEMI 'OR SOFT RESINS.

66. Ailanthus malabarica, *DC.;* SIMARUBEÆ, Vol. I., 148.
67. Calophyllum inophyllum, *Linn.;* GUTTIFERÆ, Vol. II., 29.
68. Canarium commune, *Linn.;* BURSERACEÆ, Vol. II., 94.
69. Cassia auriculata, *Linn.;* LEOUMINOSÆ, Vol. II., 215.

TRADE.
146

Trade.—It has been found impossible to furnish an exact statement of the traffic in Resins as distinct from Gums. The aboove numeration gives the chief resin-yielding plants of India, and the gums will be found separately dealt with in Vol. IV., 188. Taking the trade returns conjointly, it may be said that a considerable EXPORT takes place from India in Gums and Resins which may be divided into Exports of Indian Produce

The Resins and Gums of India. (*G. Watt.*)

and Re-exports of Foreign Produce. The extent of the trade under these
sections may be shown by the following tables :—

I.—Exports Indian Produce for the year 1889-90.

	Quantity.	Value.
	Cwt.	R
Gum, Arabic	5,259	2,06,291
Benjamin	1,125	37,477
Olibanum	20,516	4,39,508
Rosin	1,067	5,946
Other sorts	1,456	46,884
TOTAL	29,423	7,36,106

The United Kingdom was the principal importing country, having
taken 22,634 cwt. valued at R6,38,381; the rest went in small amounts to
China, Arabia, Persia, Austria, and other European, Asiatic, and African
countries.

II.—Foreign Produce re-exported from India in 1889-90.

	Quantity.	Value.
	Cwt.	R
Gum, Arabic	37,246	12,99,146
Benjamin	228	2,521
Rosin	10	100
Other sorts	1,421	13,856
TOTAL	38,905	13,15,623

The United Kingdom took 31,138 cwt. valued at R10,64,570, and other
countries received only very small shares in these re-exports. The gums
and resins registered under this section usually come from the United
Kingdom, Arabia, and the Straits Settlements. Thus, in 1889-90 India
Imported from the countries shown in the following table the supply from
which its re-exports were mainly made :—

III.—Imports of Foreign Produce.

COUNTRIES.	Cwt.	R
United Kingdom	36,519	1,85,313
France	156	10,670
East Coast Africa—Zanzibar	1,053	48,050
Aden	9,148	1 61,749
Arabia	23,027	3,15,078
Persia	4,564	54,171
Straits Settlements	13,795	4,45,395
Turkey in Asia	243	4,598
Other Countries	90	1,326
TOTAL	88,595	12,26,350

Of the total imports of Foreign Gums and Resins Bombay took 56,368
cwt. and Bengal 24,688 cwt.

It will be seen from the comparison of the tables given of exports and
re-exports that India is itself a large consuming country for Foreign gums
and resins. The imports from Arabia, Persia, Aden, Zanzibar, and the
Straits, afford the materials for the re-export traffic. The imports from the
United Kingdom are mainly used up in this country.

RESINS.	**The Resins and Gums of India.**

TRADE.

The internal trade in Gums and Resins is not very extensive, so far as the registration of transactions by coastwise are concerned. The most important exporting province appears to be Sind which annually consigns large quantities to Bombay. The total quantity of gums and resins exported from province to province was in 1889-90, 31,354 cwt., of which Sind consigned nearly half to Bombay. That amount was largely drawn from countries across the land frontier.

The internal trade in Gums and Resins shown by trade returns is, however, but a very small portion of what is known to take place. The dyers and calico-printers of India use up large quantities and many gums such as that of **Acacia arabica** are extensively eaten. Many features of the local transactions, therefore, escape registration, and little or nothing can be said of the actual total consumption in India. Certain of the resinous substances, included in the above classification (such as asafœtida, camphor, myrrh, &c.), appear in the trade returns of India under the heading of "*Medicines*" and are not, therefore, embraced by the figures of the trade given above.

Excluding from consideration the trade under asafœtida, the following were the chief imports into and exports from British India, of Gums and Resins during the last three years, by trans-frontier routes :—

Trans-frontier Imports.
150

IV.—Trans-frontier Imports into British India.

COUNTRIES FROM WHICH IMPORTED.	1887-88.		1888-89.		1889-90.	
	Quantity.	Value.	Quantity.	Value.	Quantity.	Value.
	Cwt.	R	Cwt.	R	Cwt.	R
Lus Bela	270	1,748	2,773	22,421	4,366	28,107
Sewestan	146	1,999	84	1,283	100	2,816
Kabul	8	297	32	756	17	519
Kashmir	2	75	...	...	55	2,660
Nepal	4,465	90,982	5,839	1,13,128	7,110	1,21,128
Bhutan	4	60	1	5	...	...
N. Shan States . . .	...	...	...	...	365	1,498
S. Shan States . .	...	...	...	...	41	276
Karennee . . .	34	827	...	...	28	540
Trans-frontier by S. P. Ry. .	157	4,047	...	...	...	...
TOTAL .	5,086	1,00,035	8,729	1,37,593	12,082	1,57,544

Trans-frontier Exports.
151

V.—Trans-frontier Exports from British India.

COUNTRIES TO WHICH EXPORTED.	1887-88.		1888-89.		1889-90.	
	Quantity.	Value.	Quantity.	Value.	Quantity.	Value.
	Cwt.	R	Cwt.	R	Cwt.	R
Sewestan	9	336	...	...	...	...
Kashmir	13	403	79	1,224	17	535
Nepal	59	780	...	...	6	96
Western China . .	...	...	...	...	7	200
N. Shan States . .	...	...	...	...	12	327
S. Shan States . .	...	...	...	...	2	24
Trans-frontier by S. P. Ry. .	1	19	1	38	...	...
TOTAL .	82	1,538	80	1,262	44	1,182

It will thus be seen that the chief imports are obtained from Nepal and the major portion of the exports go to Kashmir.

R. 151

| A pigment and dye. | (*W. R. Clark.*) | RHAMNUS davuricus. |

Revalenta arabica, see **Lens esculenta,** *Mænch ;* Vol. IV., 621 ;
[LEGUMINOSÆ.

RHAMNUS, *Linn. ; Gen. Pl., I., 377.*

Rhamnus davuricus, *Lawson in Fl. Br. Ind., I., 639 ;* RHAMNEÆ.   **152**

Syn.—With regard to the synonymy of this species, Messrs. **Forbes &**
Hemsley in their Enumeration of the plants of China (*Journal Linn. Soc.,*
XXIII., 129), say Authentically named R. DAVURICUS is quite like UTILIS,
but that it is preferable to retain DAVURICUS as a species than to unite it
with CATHARTICUS. Under the name R. DAHURICUS. Lawson (in *Hook.*
f., Fl. Brit. Ind., I., p. 639) unites R. GLOBOSUS, *Bunge. ;* R. VIRGATUS,
Roxb. ; and several other forms, which certainly differ from each other as
much, or even more than some of the species retained. On the other hand,
it is difficult to define satisfactorily the limits of the Eastern Asiatic species
of RHAMNUS.

Vern.—*Chato, chedwala, chadua,* HIND. ; *Tsápo, mail,* TIBET ; *Spiti,*
KUMAON ; *Phipai, dádúr, tadru, seta, pajja, kanji, mamral, shomfol,*
reteon, gogsa, sindrol, mútni, nior, chatr, romúsk, thalot, chetain, PB. ;
Wurak, PUSHTU.

References:—*DC., Prod., II., 25 ; Roxb., Fl. Ind., Ed. C.B.C., 203 ;*
Brandis, For. Fl., 92 ; Gamble, Man. Timb., 91 ; Stewart, Pb. Pl., 42 ;
Baden-Powell, Pb. Pr., 594 ; Gazetteers :—Bombay, VII., 40 ; Panjáb,
Bannu District, 23 ; Rawalpindi, 15 ; Dera Ismail Khan, 19 ; N.-W.
P., IV., lxx ; X., 307 ; Agri.-Horti. Soc., Ind., XIV., 31 ; Indian
Forester, XIII., 68 ; XIV. 267, 391.

Habitat.—A deciduous tree or shrub found in the Panjáb and the Temperate Himálaya at altitudes between 4,000 and 9,000 feet, from the
Jammu to Bhutan. It occurs also on the Gháts of the Western Himálaya,
and is distributed to China and Japan.

Dye.—Several species of **Rhamnus** are, in Europe, and more especially   **DYE.**
in China, much valued for the dyes and pigments obtained from them.   **153**
Thus, in Europe, the juice of the ripe berries of R. **catharticus** is used in the
preparation of the pigment called " sap green," while the bark of this, and
of R. **infectorius** and R. **Frangula** is the source of a yellow dye, much used
by calico printers. The most important dye, however, yielded by the
members of this genus is the "*Lokao*" of China which imparts beautiful
shades of green to silk fabrics, and is extensively imported into England
and France in the form of cakes called " Chinese Green Indigo."

According to Messrs. **Forbes & Hemsley** the sources of "*Lokao*"
are three species of **Rhamnus,** *vis,* R. **davuricus** (dahuricus, *Lawson* in
part) R. **tinctorius** and R. **sp.**

No information seems to be obtainable as to whether any of the Indian
species of **Rhamnus** yield dyes, but **Dr. Prain,** in an interesting paper
on the Chinese green dye, published in the *Journal Agri.-Hort Soc.* (*New*
Series), *Vol. VIII., Part II., p. 278,* suggests that it would be interesting to
ascertain whether any such dye could be extracted. The process by which
the Green Indigo of China is obtained is described both by **Dr. Prain** (*loc*
cit) and by **Fortune** (*Journal Agri.-Hort. Soc. Ind. (Old Series),*
IX., 274). A *résume* of the information contained in these articles may be
given here. The barks of two kinds of **Rhamnus,** one wild and one
cultivated (both apparently varieties of R. **davuricus**), are used to obtain the
dye. Both are stripped of their leaves, chopped up into convenient
lengths and boiled thoroughly. The residuum is left undisturbed for
several days, after which it is placed in large earthen-ware vessels, and
cotton cloth prepared with lime is immersed several times in it, till the
colouring matter is taken up. After five or six emersions, the colouring
matter is washed from the cloth with water and placed in iron pans to be
again boiled. Lastly, it is taken up in cotton yarn, washed off and

R. 153

| RHAMNUS Wightii. | The Buckthorn. |

sprinkled on thin paper, which is pasted on light screens and exposed to the sun till it is thoroughly dry. The reader will find a detailed review of the subject of Green Indigo in Vol. IV., 451 to 455.

MEDICINE.
Fruit.
154

Medicine.—The FRUIT, which is bitter even when ripe, has emetic and purgative properties and is given (Trans-Indus) in affections of the spleen. SPECIAL OPINION.—§ "Fruit employed as a purgative by the Afghans" (*Surgeon-Major J. E. T. Aitchison, Simla*).

TIMBER.
155

Structure of the Wood.—Sapwood whitish, heartwood reddish-brown, very hard and close-grained. Weight 55 to 57℔ per cubic foot.

DOMESTIC.
156

Domestic Uses.—The wood is used as fuel. Baden-Powell remarks that, although small, it is valued in the Panjáb for ornamental purposes.

157

Rhamnus persicus, *Boiss.; Fl. Br. Ind., I., 638.*

BUCKTHORN.

Vern.—*Kukai, jalidar, kúchni, wúrák, nikki, kander,* PB.; *Sherawane, wurak,* AFG.

References.—*Boiss., Fl. Orient., II., 17; Brandis, For. Fl., 93; Stewart, Pb. Pl., 41; Aitch., Afgh. Del. Com., 46; Agri.-Horti. Soc., Ind., XIV., 5; (New Series), I., 84; Indian Forester, XIV., 267.*

Habitat.—A shrub or small tree, common in the Sulaiman and Salt Ranges and found also in the Temperate Himálaya from the Jhelum eastward to Garhwal, at altitudes between 2,000 and 5,000 feet. It occurs in Western Tibet between 9,000 and 14,000 feet, and is distributed to Afghánistán and Baluchistán.

FOOD.
Fruit.
158

Food.—The small black FRUIT is said to be sweet, but to affect the head when eaten in excess (*Stewart*).

159

R. purpureus, *Edgar.; Fl. Br. Ind., I., 639.*

Vern.—*Bat sinjal, tunani sanani, tandra, tundhi, mimarari, kunji, chaterni, kárn, kart, rári, tadru,* PB.

References —*Brandis, For. Fl., 91; Gamble, Man. Timb., 92; Stewart, Pb. Pl., 42; Baden-Powell, Pb. Pr., 594; Gasetteers, N.-W. P., X., 307; Indian Forester, XIII., 68; XIV., 267.*

Habitat.—A deciduous, unarmed, large shrub or small tree, found on the Western Himálaya, fro.n Marri to Kumaon, at altitudes between 4,500 and 1,000 feet.

MEDICINE.
Fruit.
160

Medicine.—In Hazara, the FRUIT is used as a purgative (*Stewart*).

TIMBER.
161

Structure of the Wood.—Brownish grey, close-grained, weight 41℔ per cubic foot.

162

R. triqueter, *Lawson, Fl. Br. Ind., I., 639.*

Syn.—Under this name, R. TRIQUETRUS, *Wall.*, appears in the *Flora of British India.*

Vern.—*Gogs ', ghant,* N.-W. P.; *Gudlei, fagora, gardhan, phulla,* PB.

References.—*Brandis, For. Fl., 92; Gamble, Man. Timb., 92; Gasetteers, N.-W. P., X., 307; Indian Forester, XIV., 267.*

Habitat.—A deciduous shrub or small tree, of the North-West Himálaya from the Jhelum to Nepal, between 3,000 and 6,000 feet. It occurs also in the Salt Range.

TIMBER.
163

Structure of the Wood.—Yellowish white, moderately hard, close grained.

164

R. Wightii, *W. & A.; Fl. Br. Ind., I., 639; Wight, Ic., t. 159.*

Syn.—CEANOTHUS WIGHTIANA, *Wall.*

Vern.—*Rakta-rohida, rugt-rorar,* BOMB.

References.—*Beddome, Fl. Sylv., lxx.; Dals. & Gibs., Bomb., Fl., 50; Dymock, Mat. Med. W. Ind., 2nd Ed., 181; S. Arjun, Bomb. Drugs, 31; Birdwood, Bomb. Pr., 18.*

Habitat.—A large glabrous shrub, found in the Western Peninsula; from the highest hills of the Konkan southward to the Nilgiris; also met with in Ceylon, in the elevate 1 parts of the Central Provinces.

R. 164

Rhea and China Grass. (*G. Watt.*)	**RHEA.**

Medicine.—The BARK is highly esteemed in the Western Peninsula for its "tonic, astringent, and deobstruent properties" (*Dymock*).

MEDICINE.
Bark.
165

Rhatany, see **Krameria triandra,** *Ruis. et Pad.;* Vol. IV., 568; POLY-
[GONACEÆ.

RHAZYA, *Dcne.; Gen. Pl., II., 703.*

Rhazya stricta, *Dcne.; Fl. Br. Ind., III., 640;* APOCYNACEÆ.

166

Vern.—*Sunwar,* HIND.; *Vena, gandera,* PB.; *Wargalion, vargalum,* PUSHTU; *Ishwarg,* MUSREE; *Sewar, sihar, ishwarg,* SIND; *Iswarg,* PERS.

References.—*DC., Prod., VIII., 386; Boiss., Fl. Orient., IV., 46; Brandis, For. Fl., 322; Gamble, Man. Timb., 261; Stewart, Pb. Pl., 143; Aitchison, Bot. Afgh. Del. Com., 87; Pharm. Ind., 139; Murray, Pl. & Drugs, Sind , 149; Dymock, Mat. Med. W. Ind., 2nd Ed., 496, 497; Baden-Powell, Pb. Pr., 361; Drury, U. Pl. Ind., 446; Notes by Mr. J. F. Duthie's Trans-Indus Collector; Gasetteers:—Panjáb, Rawal Pindi, 83; Sind, 589; Agri.-Horti. Soc., Ind., XIV., 4; Journal, Bombay, As. Soc., Jan., 1849; Balfour, Cyclop. Ind., III., 403.*

Habitat.—A small, glabrous, erect shrub, common in the Trans-Indus territory, and found also in the Salt Range and at Peshawar. It is distributed to Afghánistán, Baluchistán, and Arabia.

Medicine.—The JUICE of the leaves is given with milk to children for eruptions, and an infusion of them is said to be useful in the treatment of sore throats, low fevers, and general debility (*Stewart*). The dried leaves and twigs are sold in the bazárs of Sind, where the Natives are in the habit of using them in the preparation of cooling bitter infusions. Dr. Stocks describes the infusion as a good and peculiar bitter tonic, and recommends it for trial (*Pharm. Ind.*).

MEDICINE.
Juice.
167

"The FRUIT and upper LEAVES were sent as officinal to the Panjáb Exhibition in 1863, being considered efficacious in cases of boils and eruptions" (*Stewart*). In Afghánistán, the roots, stem, leaves, and flowers are dried and used in infusion for the treatment of syphilis, in all its stages, and of chronic rheumatism, old joint affections, and pains of every kind (*A note by Mr. J. F. Duthie's Trans-Indus Botanical Collector*).

Fruit.
168
Leaves.
169

Food & Fodder.—The LEAVES are very bitter to the taste; but after being steeped in water for some days, they are given as fodder to goats (*Stewart*). The dried leaves and twigs are employed by the natives of Sind in the preparation of cooling drinks during the hot weather.

FOOD &
FODDER.
Leaves.
170

Domestic.—The wood is used as fuel. The dried FRUIT is in universal use throughout Baluchistán, for coagulating milk in the process of cheese manufacture (*Stocks in Jour. Bomb. As. Soc.*) (see **Rennet,** 427).

DOMESTIC.
Fruit.
171

(*G. Watt.*)

RHEA AND CHINA-GRASS.

172

The fibres obtained from two forms of Bœhmeria are sometimes separately distinguished by the names RHEA (Ramie) and CHINA-GRASS at others; these names are viewed as synonymous and are indifferently applied to both fibres or the rough ribbons of bark (whether obtained from India or China) are commercially designated rhea and the cleaned fibre, China grass. It would, however, appear that whether distinct species, recognisable varieties, or only cultivated races from a common species, the two forms are widely distinct from each other by two features of the greatest importance commercially. The former Rhea is a *tropical* plant which yields a considerably *inferior* fibre to the latter, a *temperate* species.

Nearly two years ago the writer, under instructions from the Government of India, drew up a *précis* of the official correspondence and reports that existed, in the Proceedings of the Revenue and Agricultural Department, on the subject of Rhea. The facts there brought together were printed in the volume of *Selections from the Records of the Government of India* issued in 1888-89. That volume appeared before the part of Hooker's *Flora of British India*, which contains Bœhmeria, had reached India, and prior also to the receipt of the *Kew Bulletin* for 1888 and for 1889. The last mentioned publication contains, in the volumes quoted, a series of most valuable papers which chiefly deal with the appliances and methods experimented with in Europe to extract the fibre. The admission is, however, made both in the *Flora of British India* and in the *Bulletin* that the Rhea of India may be recognised as a geographical variety for which the name **B. tenacissima** may be retained. This admission will thus be seen to uphold the chief contention advanced by the writer in the *précis* (above alluded to), and it only remains to ascertain whether the further statement made by him be also correct, namely, that it is a more tropical loving form than the type of the species—the true **B. nivea.** If this be confirmed then there would seem little room for doubt but that the major portion of the experiments hitherto made in India involved an error which made failure almost a necessity. The temperate plant of China was cultivated in tropical tracts of India and found to yield little or no separable fibre, owing to its growth being arrested with the return of the hot dry season. That success may be possible with the China plant in the temperate tracts of this country, and with the Indian in the moister tropical portions would, from the correspondence and reports quoted below, appear highly probable.

In preparing the material, at the writer's disposal, for the present article, the effort has been made to bring the facts given in the *Selections* up to date and to arrange these in the customary form pursued in this work. The present article may, therefore, be accepted as a more practical and commercial account than the scientific and botanical statement that will be found under Bœhmeria in Vol. I. (461 to 484) of this work.

173   **Rhea and China-grass.**

References.—*Fl. Br. Ind., V., 576; Selections from the Records of the Government of India, Revenue and Agricultural Department, 1888-89, 283-322; Roxb., Fl. Ind., Ed. C.B.C., 656; Benth., Fl. Hong-Kong, 331; DC., Orig. Cult. Pl., 146; Rumphius, Amb., t. 79, f. 1, p. 215; Growth and Preparation of Rhea Fibre by the People of India, by Burows, Thomson, and Milne, 1881; Royle in Journal Society of Arts, June 1882; Dr. Forbes Watson, Journal Society of Arts, December 1883; Sansone, on China Grass or Rhea Fibre in Soc. Chem. Indust., 1886; Kew Bulletin, 1888, 145-149; 273-280; 297-298; 1889, 268-278; Linschoten, Voyage to East Indies (Ed. Burnell, Tiele and Yule), Vol. I., 96 (under the name of Herbe of Bengalen); Marsden, Sumatra, 57; Opuscule sur le Traitement Industriel de la Ramie, et de la Ketmie, par A. V. de Bethounée (1876); La Ramie, Utilisation Industrielle, Culture et Récolte Prix de Revient, par E. Royer (1888); A Propos du Concours de la Ramie du 25th Septembre 1888, par E. Royer et N. de Landtsheer; Ramie, Its Cultivation and Preparation—the Problem solved; Gasetteers, N.-W. P., Saharanpur, 173; Ind. Agr., July 21, 1888; Jan. 15th, 1889; Ind. Forester, X., 83. Note.*—The above references are intended to supplement only those given in Vol. 1., 468, of this work.

REGIONS OF
CULTIVA-
TION.
174

Habitat and Regions of Cultivation.—According to Sir J. D. Hooker, Bœhmeria nivea, *Hook. & Arn.*, is a native of the Malay Islands, China, and Japan, but is cultivated in the warmer parts of India, especially Assam and Bengal.

The peculiar form of the plant met with, however, in Bengal, Assam, Burma, &c., is that more particularly characteristic of the Malay, *viz.*, the variety known as **B. tenacissima**, which chiefly differs from the type condition of the species (a native of China) by having the under-surface of the leaves green with whitish veins, and borne on longer petioles than is the

R. 174

case with the type form where the leaves are silvery below. It would also appear that the so-called Assam plant, though cultivated over a wide area, has not been observed in a truly wild condition in India or Burma. There is still, however, some confusion on this point, as it has by no means been definitely settled what is meant by *ban-rhea* (=wild-*rhea*). It is, indeed, just possible that *ban-rhea* may be a truly wild form of B. tenacissima, and that the cultivated plant of North-Eastern Bengal and Assam may be one of the numerous cultivated races of the true B. nivea. As opposed to this suggestion we have Buchanan-Hamilton's statement that the plant grown at Rungpore was identical with Roxburgh's imported Malayan species which he named B. tenacissima, and the further consideration also that *ban-rhea* appears likely to prove a species of Villebrunea. From the *Rhea* plant not having been seen in a wild condition in India, it has been assumed that it must have been early introduced. But it is probable that B. nivea itself may also have existed in India prior to the efforts which, from the beginning of the century to the present day, have been continuously maintained with the object of establishing Rhea and China-grass as agricultural crops in India. In China it would appear the plant grown in the colder tracts is B. nivea; but that towards the south and south-west the Malay form (B. tenacissima) takes its place. An interesting series of specimens, recently received by the Government of India from Her Majesty's Consul at Wenchow, prove, as was to be expected, the true B. nivea. So far it is satisfactory, however, to have obtained authentic botanical specimens of the China-grass plant of Wenchow. But if the enquiry into the question of the form best suited to the various provinces of India is to be prosecuted with energy, it will be necessary to institute a thorough investigation in India itself as well as in China. Specimens of the Bœhmeria plants, therefore, which are grown in each of the Chinese producing districts, accompanied with details of methods of cultivation, would afford a valuable clue to the likely adaptability of these to certain tracts of India. But we are at present absolutely ignorant of the forms of Bœhmeria grown in India itself. In the experiments hitherto made it would appear effort was entirely confined to attempting the cultivation of B. nivea, without having taken any steps to ascertain whether the Indian indigenous or long-acclimatized stock was or was not better suited for the purpose.

The *précis* of official correspondence (above alluded to, as having appeared in the *Selections from the Records of the Government of India*) was thought of such merit as to render its submission to Her Majesty's Secretary of State for India desirable, with the view of obtaining the opinion of the Director of the Royal Botanic Gardens, Kew, on the issues raised, namely, the probability that the superior price obtained for the Chinese fibre was due to the inherent greater value of that fibre as compared with the Indian. Again that the writer had contended that up to the date of the publication of the *précis*, the opinion (advanced originally by Sir. W. Hooker) seemed to prevail amongst botanists, namely, that the Indian plant was identically the same as the Chinese. Several facts adduced in the *précis* seemed to justify the suspicion that this early opinion was probably a mistaken one; but chiefly the report that European and American cultivators had found the green-leaved plant to succeed better in tropical than in temperate climates. This fact would not appear to have been recognised during the period that might be viewed as that of India's greatest interest in the subject, and hence on a review of the official reports that had been published regarding the experiments made, the conviction matured that the form of Bœhmeria least suited to the regions of experimental cultivation had in all probability been alone tried. It is, therefore, satisfactory to learn from the reply furnished to Her Majesty's

RHEA.	Commercial Forms and place of growth

Secretary of State that this view of the case was apparently also held by Mr. W. T. Thiselton Dyer, Director of the Royal Botanic Gardens. The subject of Rhea may thus be regarded as having assumed a new and, perhaps, more hopeful turn, so that planters interested in the subject are likely to regard Mr. Dyer's remarks as of peculiar interest. He writes :—
" The point raised by Dr. Watt from the Botanical point of view has been dealt with by Sir Joseph Hooker in the *Flora of British India, part XV., 576-577.* The further distinction in adaptability to climatic conditions which chiefly concerns planters has been observed in nearly every part of the world where these plants are grown for fibre purposes. The Ramie or Rhea, properly so-called, may be looked upon as the tropical representative of the China-grass (**Bœhmeria nivea**), and it is on that account probably better adapted for cultivation in hot and moist countries. Under such conditions it is a very robust plant and yields valuable fibre.

Whether this fibre at its best is really as good as the best China-grass is a point that appears not to have been definitely settled. It may turn out to be simply a question of soil or climate. At Kew we find that we cannot successfully grow B. **tenacissima** in the open, whereas B. **nivea** itself remains in the ground all the winter, and furnishes in the summer a large crop of vigorous stems. The China-grass may, therefore, give a larger and better supply of fibre under cool conditions, whereas the Ramie or Rhea may do equally well under essentially tropical conditions. The question as regards India may be settled by cultivating under various conditions of climate and soil authentic specimens of each plant and by instituting, as suggested by Dr. Watt, a careful chemical and microscopic analysis of the fibres yielded by Indian-grown plants of what are known to be the true **Bœhmeria nivea** and the true **Bœhmeria tenacissima.**"

COMMERCIAL FORMS OF BŒHMERIA.

It will thus be seen that a point of great importance, upon the full appreciation of which depends the future possible expansion of Rhea as an agricultural crop in India, is the question of the form of Rhea suited to the region where cultivation is to be attempted. Doubt has been thrown on the identification of the Rhea or Ramie of India being the Grass-cloth of China. This position has been maintained by practical men ; botanists regard the two as at most cultivated forms of one species. From an economic standard it is of little consequence whether the differences amount to specific distinctions, or justify the formation of varieties under one species, or can alone be admitted as establishing cultivated races. The point of importance, from the view it is here desired to uphold, is therefore—whether or not there are superior and inferior qualities of fibre obtained from plants all designated Rhea, China-grass or **Bœhmeria nivea**, or any other collective appellation. Further, whether these plants can or cannot be all cultivated under the same conditions of soil and climate ; in other words, whether a form exists that would give in India better results than have hitherto been obtained with the so-called Rhea of the major portion of past experiments. The writer became aware of the possibility of a mistake in the identification of Rhea while preparing (in 1883) the account of **Bœhmeria nivea** as given in this work, see Vol. I., 461-484. The opening sentence or two of that article may be here reproduced :—

" Kurz regards the *ban-rhea* of Assam as the China-grass-cloth which would thus be quite distinct from the Rhea fibre proper. If this be correct we have in India been trying to produce from the wrong plant a fibre to compete with the Chinese grass-cl th. This might account for the fact that the samples of Indian Rhea fibre exported to Europe have uniformly been pronounced inferior to the China fibre. It seems highly desirable

of the Rhea (Ramie) and China-Grass. (*G. Watt.*) | **RHEA.**

that the grass-cloth of China should be carefully looked into with the object of confirming the opinion which generally prevails that it is obtained from the same species as the Rhea fibre of India." Then again Dr. Roxburgh, without apparently being aware of the existence of Rhea in Assam and parts of Bengal, and of the fact that it was being cultivated and used by the natives there, procured from Sumatra in 1803 four plants of the Caloce (or Caluse) and planted these in the Botanic Gardens, Calcutta. He gave the plant thus obtained the name of **Urtica tenacissima.** His stock grew and multiplied so rapidly that shortly after he had several thousand plants. About this time the discovery was made by Dr. Buchanan-Hamilton (in 1810) that the *konkura* or *kankhura* of Rungpore and Dinagepore was identical with the plants Dr. Roxburgh was cultivating. In the *Journal of the Agri.-Horti. Society of India* [Vol. VI. (Old Series), p. 30] will be found a description by Dr. Campbell of the method of cultivation as practised in Rungpore.

The account given by Roxburgh of his **Urtica tenacissima** was most probably taken, however, from the imported Malay stock and not from the Indian plant. In this connection Roxburgh's report to the Board of Trade (1809), on the introduction of the *Calúe* into India may be read with interest. It is reproduced in the *Jour. Agri.-Hort. Soc. Ind.* [Vol. VI. (Old Series), p. 181]. In that paper will also be found a reference to Major Jenkins' discovery of the plant in Cachar in the year 1833 and to Colonel Burney's discovery of it in the Shan provinces of Pevela and Youkzouk, which are distant from Ava seven or eight days' journey. Major Macfarquhar wrote (1836) that it was cultivated by the Shans, Siamese, and the Chinese, "the latter, with whom I have spoken on the subject, are loud in its praise for its fineness of texture and durability both as cloth and cordage." Major Hannay and Major Jenkins lay stress on the fact that Rhea is cultivated and *ban-rhea* a wild plant, and neither of these early observers seems to have seen Rhea except under cultivation. But to continue the quotations from the Dictionary (Vol. I.) " A few specimens from Assam were sent to the Agri.-Horticultural Society of Calcutta, and from cuttings thus obtained plants were grown in the Society's garden. From this date (1840) the Society received contributions from several writers, from time to time, giving new facts regarding the growth and preparation of the fibre in Northern India. Dr. McGowan furnished information and samples from China, and Dr. Falconer and afterwards Sir William Hooker identified Rhea as the same plant from which the Chinese grass-cloth is prepared."

(Conf. with 464.)

(Conf. also with Ville-brunea.)

It is worth recollecting in this connection that from Roxburgh's Malay stock, the introduced plant was being rapidly distributed all over India, and this being so, doubt may be admitted as justifiable as to whether the introduced or the indigenous plant was sent to Europe for determination. Indeed, it is just possible that the true China-grass plant may have, long anterior even to Roxburgh's day, been introduced into India. At all events, in his account of U. tenacissima, Roxburgh maintains that it is a perfectly distinct plant from U. nivea as described by Louriero. The word " Rhea," though used in Assam, is questionably of Indian origin. One of the dominant races on the eastern frontier of India and in the valley of the Brahmaputra, by successive waves of conquest, came from Siam. May not these people have brought the Rhea and its name to the country of their adoption ? In part support of this suggestion it may be added that although species of Bœhmeria are met with all over India, it is among the invaders of Assam that the fibre is mainly used for textile purposes. That B. nivea and B. tenacissima are both found in certain parts of India, at the present day, goes without saying, and therefore the

RHEA.	Commercial Forms and place of growth

FORMS OF RHEA.

Indian samples identified by the early botanists may have been China, grass and not Rhea. Indeed, **M. A. deCandolle** (*Origin Cultiv. Pl., p. 146*) goes so far as to doubt the existence of **B. tenacissima** as a truly wild plant in India, and the last number of the *Flora of British India* treats it as an introduced plant. Speaking of the writings of authors on Rhea and China-grass **M. deCandolle** says : " We must not trust the vague expressions of most authors, " nor put faith in " the labels attached to the specimens in herbaria, since frequently no distinction has been made between cultivated, naturalized, or truly wild plants, and the two varieties of **Bœhmeria nivea** (**Urtica nivea**, *Linn.*, and **B. tenacissima**, *Gaud.*, or **B. candicans**, *Kussk.*) have been confounded together ; forms which appear to be varieties of the same species, because transitions between them have been observed by botanists. There is also a sub-variety, with leaves green on both sides, cultivated by Americans and by **M. demalartic** in the south of France. " **M. deCandolle** then proceeds to show that, according to **Linnæus**, **B. nivea** is most probably only a cultivated plant in China, but that according to many authors it is an abundant wild plant in Cochin-China, in Hong-Kong, in the Philippine Islands, and in the Malay Archipelago. He then adds : " The other varieties have nowhere been found wild, which supports the theory that they are only the result of cultivation. " The wild rhea of Assam may therefore not be Rhea at all (*Conf.* with **Villebrunea**, *Vol. VI, Pt. II.*)

In the account given in Vol. I. of this work stress has been laid on the fact that Indian Rhea always fetches a lower price in the market than China-grass. " It is remarkable that the Chinese grass-cloth should be much finer than Rhea ; that on being boiled it should lose only 0·89, while Rhea under the same treatment parts with 1·51 of its weight. These and other facts, in addition to the pronounced superior quality of, and therefore higher price paid for, China-grass-cloth as compared with Rhea, would seem to confirm the suspicion that these two fibres may after all be obtained from different plants. This remark is made purely as a suggestion, but it seems highly desirable that we should thoroughly examine all the plants met with in India which afford Rhea-like fibres, as well as re-examine the plant from which the Chinese grass-cloth is obtained, before much more money be spent on experiments with new machinery " (*p. 481*).

There would be nothing remarkable in assuming that in the vast area of China, possessing like India every gradation in climate and soil, tropic and temperate, upland and plain, there should exist many forms of the Asiatic genus, **Bœhmeria.** Forty years ago, **M. Decaisne** (the most distinguished French botanist of his day) cultivated in Paris the China-grass brought from the Celestial Empire by **M. Leclancher**, and having carefully examined the living plants came to the conclusion that they represented two readily distinguishable species. These he designated **Urtica nivea** and **U. utilis**,—the latter was the **U. tenacissima** of Roxburgh, a tropical plant, while **U. nivea** is temperate. The subject of Rhea or China-grass for some time after the date of **M. Decaisne's** report laggest in popular favour, and when it was reawakened the two plants were treaded as one and the same. To this day the utmost that botanists have consented to, is to assign **U.** (**Bœhmeria**) **tenacissima** the rank of a variety under **B. nivea.** Structurally the differences do not appear to be very great, but the temperate plant (**B. nivea**) is generally viewed popularly as recongnisable from the tropical (**B. tenacissima**) by having silvery white under-surfaces to its leaves instead of being green with white veins.*

Tropical form.
177
Temperate form.
178

* **Hooker** says : " **Weddell** distinguishes as **B. candicans** the form under this name, " (**B. nivea**), " and that of **tenacissima**, by the more robust habit, larger, longer-petioled concolorous leaves " (*Fl. Br. Ind., V.,* 577).

of the Rhea (Ramie) and China Grass. (*G. Watt.*) | RHEA.

FORMS OF
RHEA.

If these be the only differences, it must be admitted the botanical position is unassailable, but from a practical point of view it would be fatal to force the cultivation of a tropical form of a species in a temperate region or of a temperate in a tropical. It is true, however, that both these plants have shown a sufficient power of adaptation to at first sight permit of this being disregarded, at least to some extent. But the experience of European cultivators is to assign B. **nivea** to the more northern and colder areas and B. **tenacissima** to the more southern and warmer tracts of the Continent of Europe. According to most writers B. **nivea** has also been found best suited to America. In India, it would appear that we have disregarded these eminently practical considerations, and it is, therefore, probably safe to say that some share of our failure is attributable to our negligence in not experimentally discovering which form or forms of Rhea or China-grass were suitable to each area before extensive cultivation was undertaken. But even had the distinction between the above commercial forms of Bœhmeria been observed we should, in India, have still erred, since there can be little room for doubt that there exist in this country numerous other plants generically treated as Rhea, the claims of which to independent recognition have never been considered.

(Conf. with Debregeasia, Vol. III., 52—54; Villebrunea, Vol. VI.

Different Systems of Cultivation and different Conditions of Climate and Soil necessary for the two Commercial Forms.—Among the records of the Government of India there occurs a correspondence accompanied with a copy of M. Favier's *Textile Nettles.* A passage from that useful publication may be here given, since it records the results of the French efforts to cultivate the two commercial forms of Bœhmeria :—

"**Yield.**—The textile nettles cultivated in France must not be expected to grow with the same strength of vegetation which they possess in tropical countries, under the double influence of heat and moisture. These plants, therefore, only give two crops in our country, the first towards the 15th of July and the second towards the end of October or the beginning of November. In order to obtain fibre of uniform quality from both crops it is necessary that the stems of the two crops should come to exactly the same degree of maturity. It is well therefore to hasten the growth of the first crop by means of liquid manures, especially if the spring has been cold, so that it may be gathered about the middle of July; this will give time to obtain the second crop in satisfactory condition.

Yield per acre.
179
(*Conf. with pp. 464, 465, 468, 469, 473, & 483.*

"According to the calculations of **Mr. Hardy**, ex-Director of the Botanical Gardens near Algiers, a field of textile nettles over a year old, whose stems had reached a height of about six feet, would produce 48,000℔ per acre of green stems with their leaves. In that weight there would be 20,400℔ of leaves, and 27,600℔ of stems, which would be reduced to 4,900℔ in drying, and would give 1,400℔ of fibrous thongs. **Mr. Hardy** admits that two such crops may be had in Algeria, which would raise the yield of an acre to 9,800℔ of dry stems and 2,800℔ of useful fibres. Under these circumstances the cultivation of textile nettles would be extremely remunerative.

"We must acknowledge that in France such great crops have not generally been obtained. We find, in a very interesting pamphlet on the Ramia, published in 1877 by **Monsieur Goncet de Mas**, exact statements of the results which he obtained in his trials of the culture of this plant. **Monsieur Goncet de Mas** grows the Ramia in the neighbourhood of Padua, in similar conditions of climate to those of the south of France, and the figures which he gives seem to correspond with the results to be expected from the cultivation of textile nettles in our southern departments.

"The first year, according to his own report, he obtained 14,400℔ of green stems per acre in two crops, of which weight one half was composed of leaves. The 7,200℔ of green stems without leaves yielded 1,440℔ of dry stems and 320℔ of filament.

"The second year he gathered 52,600℔ of stems including leaves, *viz.*, 26,300℔ of green stems without leaves, 5,260℔ of dry stems, and 944℔ of filament.

"The third year, the field being in its definite state, and the plants three feet apart in both directions, but joined by their shoots and roots, he gathered in two crops 64,720℔ of green stems with their leaves (which correspond to 32,360℔ of stems without leaves), 6,400℔ of dry stems, and 1,280℔ of filament.

29

RHEA.   Chemistry of the

FORMS OF RHEA.

"Other agriculturists have obtained in the south of France as high as 1,600lb of filament per acre. It may therefore be admitted that the Ramia, in conditions which may be considered normal in our country, will yield from 1,280 to 1,600lb of filament in two crops per annum, so that each cutting will yield more filament than the best annual crop of hemp or flax.

"The figures which we have just quoted refer to the **Urtica utilis** or *Ramia*. **Monsieur Goncet de Mas** thinks that the **Urtica nivea** or Chinese White Nettle would yield about a third less. This being the case, the cultivation of this nettle would still be very remunerative in the regions where it is not possible to grow the Ramia.

Price. 180
(*Conf. with pp. 463, 464, 470, 484.*)

"As to the price which may be obtained for the filament of the textile Nettles, this must evidently vary according to its quality and condition, but we think it must at least be equal to that of Harl in the same conditions of preparation and quality. The present value in London of the China-grass filament, imported from India and China is from £45 to £50 per ton.

"A point to which we must call the attention of agriculturists is that the textile Nettles do not bring forth their full product until the third year of their cultivation. It is only when all the subsoil of the field is permeated with their roots that these plants throw out numerous shoots and yield the quantity of stems above stated."

RACES. 181

The best form of B. **nivea**, according to many writers, is that known as **sanguinea**, a name given by Hasskarl, owing to the red colour of the mature stems. There are doubtless, however, many forms both of **nivea** and **tenacissima**, and the intending planter would do well to procure and experiment on a small scale first, with as many forms as can possibly be obtained, in order to discover which yields the maximum of superior fibre under the conditions of soil and climate peculiar to his estate. Indeed, the Saharanpur failures would point to the by no means improbable result of such experiments proving that in that district at least no form of Rhea or China-grass is ever likely to become a commercial success.

CHEMISTRY of the two forms. 182

Relative Merits of China-grass and Rhea.—Dr. Forbes Watson's figures of loss of weight on boiling these two fibres under a high pressure have already been quoted. The markedly different hydrolysis of the two fibres points to their being distinct,[*] but this has been established more conclusively by Muspratt (*Theoretische un Peaktische Chemie, 1880*) in his comparative chemical analysis of a sample of the Chinese and the Indian fibre.

	Chinese Nettle.	Ramia.
Ashes .	2·87	5·63
Water .	9·05	10·15
Matters soluble in water	6·47	10·31
Wax and fatty matter .	0·21	0·59
Cellulose .	78·07	66·22
Intercellular substance .	6·10	12·70

Thus, according to this result, the two fibres must the viewed as entirely different, the Chinese white nettle being ever so much superior to the Indian Rhea or Ramie. In connection with the Colonial and Indian Exhibition a special effort was made to procure as extensive a series of Rhea and allied Rhea fibres as possible. This failed signally owing to few if any of the samples of fibre, contributed by local officers, having been accompanied with dried botanical specimens of the plants from which they were prepared. It was thus impossible to tell which was which and all were stated to be *Rhea* or *Poi*. The Glen Rock Company, Limited, of

* NOTE.—Mr. Montgomery's results in this light are highly instructive. He regards China-grass and Rhea as distinct. (See under Panjab below, pp. 469, 471.)

| Rhea (Ramie) and China Grass. | *(G. Watt.)* | RHEA. |

CHEMISTRY
of the
forms of
Rhea.

Wynaad, exhibited admirable samples of Rhea, as also did Messrs. Reinhold Bros. of Calcutta, and a French firm showed Rhea manufactures, the fibre having been prepared by the Favier process. These special mercantile exhibits were over and above the very extensive collections made by Government agencies. Samples of all the Rhea fibres, in the Exhibition, were given to Messrs. Cross, Bevan & King for the purpose of examination, in connection with their report, but as they publish but one analysis, it is impossible to discover to which it refers. The presumption is, however, that it was a sample of B. nivea, and to contrast with Muspratt's result Messrs. Cross & Bevan's analysis may be here reproduced.

Moisture	9·0
Ash	2·9
Hydrolysis (*a*) (5 minutes' boiling in Na_2O 1 p.c.) . . .	13·0
(*b*) (1 hour ditto ditto). . . .	24·0
Cellulose	80·3
Mercersing (1 hour in solution of Na_2O of 33 p c.) . .	11·0
Nitration	125·0
Acid purification	6·5

Thus in moisture, ash, and cellulose, this agrees very closely with Muspratt's analysis. In several of the preceding descriptions of fibres it has been contended that the percentage of cellulose present in a fibre is the best criterion of its industrial value, and, accepting Muspratt's analysis of Ramie, we are driven to the conclusion that it is a very inferior fibre to the Chinese nettle. A sample of *Poi-rhea* (Maoutia Puya) was examined by Messrs. Cross & Bevan, which gave only 32·7 per cent. of cellulose, and by the (*b*) Hydrolysis lost 62·7 per cent. of its weight. Messrs. Cross & Bevan explain, however, that according to Mr. King's microscopic examination, the particular sample of *poi* given them bore evidence of having been badly prepared. They, therefore, in their report, set aside their analysis, and placed the *poi* fibre near Rhea as one of the fibres belonging to the first class. The possibility, however, of their analysis being even approximately a correct indication of the relative value of *poi* to B. nivea, justifies, in the strongest possible terms, the line of argument the writer has adopted, namely, to urge that all the Rhea and allied Rhea fibres should be thoroughly investigated. The people of Darjiling and perhaps also of Assam prepare a by no means inconsiderable quantity of *poi*, whereas they might just as well grow the Rhea or even the Chinagrass. The injury to a future Indian Rhea trade by a fibre of a distinctly inferior quality finding its way to Europe under the name of Rhea may well be imagined.

(Conf. with Vol. V., 177—180.

Rhea Patents.—Since the writer published the brief account of Rhea fibre given in the first volume of this work very little of a new nature has been brought to light. The development of this fibre has progressed slowly and steadily more than been the subject of any violent evolution. At the same time a comparison of the present position of the Rhea industry with that attained twenty years ago or even half or quarter that period will reveal how wonderfully the public attention has been directed to the subject. The annals of the Patent Offices of Great Britain, France, the United States, and of India have been crowded with the registration of processes and appliances, each improving, or claiming to have improved, the methods formerly in use for separating, cleaning, and utilising the fibre. Appliances and machines have been invented to (*a*) decorticate the twigs into ribbons of bark containing the fibre; (*b*) to separate the fibre from the ribbons of dry bark; (*c*) to clean and de-gum the fibre; and (*d*) to spin and weave it into yarns, ropes, and textiles. By some of these patents it is contended that good results can be obtained only with the green twigs, by

PATENTS.
183

PATENTS.

others that the dry twigs are equally utilisable. The increased freight, however, in carrying to a distant manufacturing centre the contained useless wood of the twigs, has brought into existence special decorticating appliances to be used locally. In some patents, retting has been recommended, in others condemned; and both chemical and mechanical methods of decortication have each been stoutly defended. Dried ribbons of bark have been prepared in the neighbourhood of cultivation and shipped to the manufacturer to undergo the further treatment—separation and purification of the fibre. Perhaps, however, the best modern advance has been the combination of decortication and cleaning in one and the same operation. The justification of this rests on the economy effected, and on the theory that the inner layers of fibre are finer and hence more easily injured by chemical reagents, that they also contain less gum than the outer, and hence the treatment of ribbons or thongs of bark would be accompanied by injurious effects on certain proportions of the fibre. On the other hand, the chemical and miscroscopic examinations accomplished by some of the French experts might be viewed as throwing a degree of doubt on the statement that a heavier deposit of gum exists on the outer than the inner layers of fibre.

Resting satisfied with this brief review as indicating some of the chief features of the keen competition in Rhea patents, it may be said that, by whatever means the fibre is separated, Rhea has now practically attained a recognised commercial position. The supposed insurmountable difficulties, which all early writers dwelt on, have virtually disappeared, and it is customary to have thrust into one's hands, as proof of the merits of this process and that, samples of a beautiful silvery-white rhea fibre, of yarns spun from it, and of table-cloths, curtains, and other such textiles woven of this admirable fibre. The writer has had the pleasure to examine several such samples, in the files of correspondence conducted by the Revenue and Agricultural Department. But in preparing the present article from the Selections from the Records of the Government of India, he has been compelled to issue an abstract only, of a few of the more notable papers, drawn from the voluminous materials placed at his disposal.

Since the date at which the above remarks regarding patents first appeared the *Kew Bulletin* has published **Mr. D. Morris'** Report of the experiments he witnessed at Paris in October 1888, in connection with the separation and cleansing of Ramie fibre. Four machines were exhibited, three of which were worked. These were the Delandtsheer Machine, the Barbier Machine, and a Machine of the American Fibre Company. These appear to have been more or less similar to the Death & Ellwood machine noticed in the article **Bœhmeria nivea,** Vol. I., 481. of this work. The Royer Chemical Process was also shown, but the nature of the chemicals used was not explained. Commenting on these methods of extraction of the fibre **Mr. Morris** says:—

"These are briefly stated, the results of the Paris trials on Ramie. That the results are unsatisfactory and disappointing and fall far short of the estimates of the inventors there can be no matter of doubt. It is probable that a fresh series of trials will be inaugurated next year in connection with the Paris Exhibition of 1889; and if the value of the prizes is increased there will doubtless appear a larger and better representation of machines and processes."

Favier
system.
184

The Favier System.

"It will be noticed that there was no trial this year of the Favier system which is in operation in Spain, and is described in the *Kew Bulletin* for June 1888, pp. 145-149. Nor was there a trial of the Death machine

| Other Processes and Machines. | (*G. Watt*) | RHEA. |

(constructed by Death and Ellwood of Leicester) which has been used experimentally in many parts of the world. The Favier process is being worked privately, and is therefore not available to the public. The fibre hitherto produced has been exclusively used in France, but the quantity so available has not been sufficient to base an opinion as to the permanency of the enterprise. M. Favier, who has long taken a deep interest in the Ramie fibre, was a member of the jury at the Paris trials, and the articles which he has contributed on the subject to the *Journal l'Industrie Progrissive* of October 7 *et seq.*, may be looked upon as embodying the views of one of the best informed of French experts on the present position of the Ramie question."

THE TREATMENT OF DRY AS AGAINST GREEN RAMIE STEMS.

" Amongst the French there is attached an importance beyond their value, to machines for cleaning Ramie in the dry state. This has arisen partly, no doubt, from the fact that, the Favier system, the only one which hitherto has obtained a measure of success, requires the stems to be dried before they are treated. An idea was also prevalent in France that in some parts of the country it might be possible for the farmers to grow one or two crops of Ramie, and cut and harvest the stems in summer; and work them off at their leisure during the winter. If a machine were devised to treat Ramie successfully, it is improbable that France could compete with tropical and subtropical countries, where three or four crops of stems could be reaped in the year. This conclusion is now being gradually adopted in France, and the future exploitation of Ramie is treated as a question which more nearly concerns Algiers and the French tropical Colonies.

" As regards India and our own Colonies, it is essential that Ramie machines should work upon the green stems, not upon the dry. In the rainy season, when the air is impregnated with moisture, to dry Ramie stems in the open air after cutting would be an impossibility. To attempt to dry by artificial means the enormous quantity of stems yielded even by a few acres would entail so much labour in handling, and so much expense for buildings and fuel that it would be altogether a hopeless task.

" The percentage of crude fibre yielded by Ramie stems is estimated at about 10 per cent. If the stems must be first dried before they are treated it would be necessary to handle, to cart in and cart out again from drying sheds 100 tons of stems for every 10 tons of fibre produced. It might be suggested that harvesting the stems should take place in the dry season when the conditions would be most favourable to drying them in the open air.

" This unfortunately would not be practicable. The stems grow best during the rainy season, and when once ripe they must be cut at once. Besides, it is evident the sooner one crop is removed the better will be the prospects of the next. During the dry season the stems grow very slowly, and it has been noticed that such stems have short internodes, are very woody, and offer relatively greater resistance to the process of decortication."

OTHER PROCESSES AND MACHINES.

" Of processes and machines not already mentioned, it is desirable to refer to one or two for the information of persons who may not otherwise become aware of them. In June of last year Mr. O. Maries of Durbhunga, Bengal, forwarded a series of specimens of Ramie fibre in different states of preparation to Kew and asked for an opinion upon them. It appeared that he had invented a machine worked by two men in the field, capable of operating upon two or three hundred stems per hour. This machine simply separated the fibrous bark from the wood. The bark was

RHEA.	Processes and Machines for the

then operated upon by other processes and eventually it was deprived of gum and mucilage and worked into a tolerable fair fibre suitable for manipulation by textile manufacturers. This fibre was reported by Messrs. Ide & Christie as 'long, fairly cleaned Ramie fibre, worth £28 per ton.' The particulars of Mr. Maries' methods have not been made public; but we understand that a well-known firm of merchants in Calcutta has acquired the patent connected with them, and the system is now in course of being practically tested on a large scale. In the columns of the *Times* there recently appeared an account of a machine invented by Mr. John Orr Wallace, and placed on view at the Irish Exhibition. This was termed 'patent scutching machine for cleaning Ramie, flax, hemp, &c.' The apparatus is about 6 feet high by 4 feet long. It consists of an upper feed table 36 inches wide, on which the stems are fed to three pairs of fluted rollers, which deliver the stems downwards between five pairs of pinning tools alternating with six pairs of guide rollers. The pinning tools somewhat resemble hand-hackles, and may be popularly described as very coarse wire brushes. They are attached to two vertical frames to which a horizontal to-and-fro motion is imparted, and the pins interlace as the two sides approach. The fibrous material is drawn downwards by rollers which have an intermittent motion, and at each momentary pause, the pricking pins enter the material and are rapidly withdrawn from it. By degrees this fibrous, descending curtain is delivered on to a sloping receiving table at the bottom of the machine, over which table the woody substance has previously passed, to a receiver in a crushed and semi-pulverized condition, and perfectly free from fibre. This machine, it may be mentioned, was not constructed for the special treatment of Ramie. In spite of this, however, it has cleaned Ramie in a fairly satisfactory manner, and the inventor claims that with a few necessary alterations in detail he will be able to treat the stems either green or dry and produce clean fibre at the rate of 1 cwt. per hour. The machine can be driven by a two-horse power engine, and it requires two persons to feed and tend it.

"Small quantities of Ramie stems grown at Kew have been successfully passed through the machine. It is proposed by the inventor, when he has completed the alterations, to submit this machine to a public test similar to that adopted at the Paris trials. For this purpose he states that a large supply of Ramie stems will be obtained from France.

"There are some special advantages connected with this machine which deserve to be mentioned. In the first place, the feed table is so large that at least 40 stems can be fed to the rollers at once. When the stems have been fully grasped by the rollers, the operator need not retain his hold upon them any longer. They pass on uninterruptedly through the machine, and they can be followed immediately by a fresh lot without the return action, which is an essential part of the treatment by the Death and the Delandtsheer machines. There is here a considerable saving in time, and there is also a complete absence of the rough usage to which the fibre is subjected in nearly all the purely mechanical processes which have hitherto come under my notice.

"Personally, I am unable to express an opinion upon the Wallace machine. To say that it is more promising than any machine exhibited at the Paris trials is merely to affirm that it is not altogether a failure. When the machine is fairly tested on its merits, and it is worked continuously on large quantities of Ramie stems, the results will speak for themselves. Until that is done it is obviously undesirable to do more than draw attention to a machine which possesses merit, and which, with further improvements, may be rendered of service in the production of fair marketable fibre."

R. 186

GENERAL CONCLUSIONS.

" An eminent firm of brokers recently informed me : ' There is no doubt that Ramie is exciting great interest in many parts of the world, and many people are experimenting with various processes for extracting the fibre cheaply and quickly. We cannot say that any results submitted to us up to the present time are quite satisfactory. The fibre is either imperfectly freed from gummy matter, or the process breaks down in the matter of cost or owing to the local conditions under which it must be carried on. We consider that no system of preparation which cannot produce the clean, unbleached fibre under £30 per ton is likely to succeed in establishing this article firmly in the estimation of English textile manufactures.' This opinion expresses very briefly and clearly the conclusion at which I have arrived in connection with the preparation of Ramie fibre. It is quite possible that some machine or process will eventually solve the problem, but at present the exploitation of Ramie, in spite of years of labour and the expenditure of large sums of money upon it, cannot be said to have yet emerged from the experimental stage" (*Kew Bulletin*).

As announced by Mr. Morris an interesting series of experiments were actually held at Paris during the Exhibition of 1889. Mr. Morris was again deputed to attend these on behalf of the India Office and his report appeared in the *Kew Bulletin*. It may in general terms be said that the machines, &c., shown manifested a vast improvement, so that Mr. Morris was induced to take a more hopeful view of the future prospects of the *Ramie* question. The following review of Mr. Morris' report gives the main facts and expresses at the same time the opinion held by the Editor of the *Indian Agriculturist* :—

" *Barbier Machine.*—The first machine dealt with was that of M. Armand, constructed by Barbier ; it is designed to be worked by hand or steam power, and the result of the trial is thus summed up : ' Taking into consideration the cost of this machine, and the power necessary to drive it, the outturn of ribbons is too small to prove remunerative, and the machine in its present form is useless. Better results than these have been obtained by decorticating Ramie by hand.'

" *Favier Machine.*—This machine is driven by ¾-horse power, and consists of a feeding trough, and a somewhat complicated system of rollers and beaters. In the trial the ribbons caught once or twice in the rollers and the machine had to be stopped. The mean of two trials of 4½-minutes and 18 minutes gave a result equivalent to about 360℔ of dry fibre in the day of 10 hours. There appeared to be no waste. Mr. Morris says :— ' These results I regard on the whole as satisfactory.'

" The somewhat intricate character of the various parts of the machine would be against its general use by planters in the Colonies, but there can be but little doubt it is a great advance on most other Ramie machines now available. It might, however, be adapted for use in central factories or *usines*, where skilled labour would be obtainable and for this and similar purposes the Favier machine may be recommended.

" *Michotte Machine.*—Of this machine a description is given, but Mr. Morris says :—' This machine in its present state possesses no merits whatever. It is difficult to realise under what circumstances it could have been entered for trial.'

" *De Landtsheer Machines.*—There were two machines exhibited, but the larger appears to be more likely to succeed than the other ; it is driven by two-horse power, and appears to consist of an arrangement of rollers and beaters. A trial of two and a half minutes was given, and in which the results were held as equivalent to 176℔ of dry ribbons per day of 10 hours.

RHEA.	Other Processes and Machnies for the

PATENTS.

General
Conclusions.

The second trial was for 11½ minutes, and the results are held to have been equal to 575℔ of dry ribbons per day of 10 hours.

"Mr. Morris remarks: 'It is not at all improbable that M. de Landtsheer will be able to effect some further improvement in this machine. In any case the machine is worthy the attention of planters, who, with a single instrument, could work off about 50 tons of the green stems per week. This is an exceptionally good result, and serves to show what progress has now been made in perfecting machines for treating the Ramie plant on a commercial scale.'

Fleury Moriceau Process.
192

"*Fleury-Moriceau process.*—. . . . 'This was singularly simple, and consisted of steeping the fresh (or dry) stems for a short period in boiling water and removing the ribbons by hand. An open galvanised tank, about six feet long, two feet wide, and about four feet deep, filled with water was raised on bricks (or stones) about eighteen inches from the ground over an open fire. When the water had reached boiling point, a crate, containing 50 or 100 fresh stems, was lowered into it (and depending on their age and character) left in it for 5 or 15 minutes. At the end of that time the crate was lifted out, the stems left to drain, while another lot was put in. The stems already steeped were then taken up by a couple of workmen and quickly and effectually cleaned by hand. The action of the boiling water had apparently thoroughly loosened the attachment of the cortex to the wood, and ribbons were produced perfectly clean and regular, and apparently without any loss of fibre.'

"The results obtained were equal to 166℔ of dry fibre per day of 10 hours: the trial lasted 46 minutes. "Summing up the conclusions arrived at of the trials of 1889, Mr.Morris says:— . . . 'Thetrials of 1889 have proved much more favour able than those of 1888, and the subject is evidently ripening for solution in many directions not thought of before It will be noticed that the best results in 1888 were at the rate of 120 pounds of dry ribbons per day of 10 hours. This was the de Landtsheer small machine. In 1889 this machine, with improvements, produced at the rate of 287 pound of dry ribbons (more than double the quantity) for the same period. With the large machine (making due allowance for the pith and wood lightly adhering to the wet ribbons) the return of dry ribbons would be at the rate of over half a ton per day.'

"Mr. Morris concludes his report with a 'review of a few of the machines and processes not represented at Paris, which have recently come into notice in this country and elsewhere.'

"The following remarks are of great importance, and deserve the most careful consideration of all who are interested in the subject of the production of Rhea fibre :—'With regard to what is known in commerce as ' China Grass,' this is hand-cleaned fibre shipped usually from Chinese ports. It arrives in this country in small parcels, the yearly importation being only about 100 tons. It is nearly all taken up by Continental buyers. Rhea is the term applied to machine-cleaned fibre, generally in the form of ribbons or half cleared stuff. The price is much less than China Grass, and in case of large shipments, would probably not exceed 7*l*. or 8*l*. per ton. It is important, therefore, for Ramie planters to aim at the production of ribbons at a cost not exceeding 4*l*. or 5*l*. at the port of shipment. Important elements in such production would be to plant Ramie only in places where the soil and climate will allow of 3 or 4 crops to be reaped per annum, where labour is very cheap and abundant, and where good facilities exist for transport and shipment' (*D. Morris*).

"Should Mr. Morris be right in his forecast (and there is no one in better position than himself to make such a forecast) it would be questionable whether the cultivation of Rhea will pay better than say Jute or other

Extraction of the Rhea Fibre. (*G. Watt.*)	**RHEA.**

crops already grown in India. The price quoted is equivalent to say R6-8 per maund in London, and even if three cuttings can be made from the plant in Bengal—a doubtful matter—the total weight of fibre per acre would hardly exceed (in Bengal), one crop of jute, the habit of the two plants and the amount of fibre on each being taken into consideration. Jute, again, is a one *fasl* crop, and a second crop of another kind can therefore be taken off the land in the same year" (*Ind. Agriculturist, Dec. 14th, 1889.*)

General Conclusions.

Since the appearance of **Mr. Morris'** report on the experiments held at Paris, a further series of similar trials were made at Gennevilliers, a suburb of Paris, where the plant had been grown. The "Concours" took place on the 27th to 30th September 1891, when a number of new apparatus and machinery were experimented with. The opinion seems to have been arrived at that a great advance had been effected by M. **Faure's** Decorticator. This may be described as a simple machine adapted for field use without necessitating skilled labour.

But although progress has been accomplished in the matter of machinery to separate and clean rhea fibre, the writer by no means holds the opinion advanced by the *Indian Agriculturist* as to the prospects of a future large Indian cultivation of rhea. At the same time it is probable that an invention calculated to deal with the dried or partially dried stems would prove more successful in India than any invention or contrivance however simple in which the preparation of ribbons devolved on the cultivator. This view, it will be seen, is at variance with that advanced by **Mr. Morris** who holds that inventors are rightly directing attention to simple appliances or processes to treat the green stems. By the stems, procured over a wide area, finding a market at some not very distant factory, where they could be decorticated, the chief difficulty in inducing the Indian cultivator to embark in this new crop would be overcome. The poverty of the Indian *rayat* is such that the iron tank in which to boil the stems, recommended in the *Fleury-Moriceau* process would be beyond his means. If he cannot profitably separate the bark by hand labour or by a system that costs nothing more than labour, or at least does not involve the purchase of any European-made machinery or chemicals, then decortication must be effected at factories for the purpose. Were it possible to enforce cultivation within a restricted radius around such factories, the machinery might be designed to deal with green stems. But it seems likely that past experience in Indigo manufacture would retard capitalists from embarking in an enterprise that might be confronted with the total loss of supply of material. To be safe, therefore, the stems would have to be drawn from an extensive area, and to lessen freight (were there no other considerations) they would have to be sold in a dry state. Everything points to this Indian view of the case being the natural one, if it be intended to, in time, see rhea classed with the other numerous crops which are open to the cultivators' choice. It might be quite otherwise were planters to take the matter into their own hands and grow rhea on estates of their own, as in the case of tea and coffee. In other words, it will take some centuries before the Indian cultivators see the advantages of purchasing special appliances for a new crop. They have not even at the present day discovered the economy of time, and hence money value, of the commoner appliances of European agriculture. They continue, as their fathers did, to reap their crops with a small prunning hook, to tread out the corn by the feet of the patient bullock and to look to the breezes that blow to winnow in handfuls their grain. With such a backward state of agriculture the profits from rhea cultivation would have to be demonstrated as far greater than they are ever likely to be, before the *rayat* could be induced to spend more than a very few rupees in the pur-

Conf. with pp. 470, 474 475.

USES of
RHEA.
192

chase of necessary apparatus. Were it possible to liberate the fibre by
boiling the stems in an earth-pot, such as the village potter might supply,
the Indian cultivator would probably look at the question more favourably,
but, as matters stand, he is likely only to be induced to cultivate rhea
if his crop can be sold in the field or after it has been cut and stacked for
a short time.

European Uses and Recognised Properties of Rhea—"The inherent
physical properties of the fibre place it in a pre-eminent position. In
strength it is second to no vegetable fibre and in some trials it has proved
to be more than twice as strong as Russian hemp (**Cannabis sativa**). It
also presents unusual resistance to the effects of moisture and other climatic
conditions, to judge by the slight action of high pressure steam upon it.
Samples of the fibre, exposed for two hours to steam at about two atmospheric pressures, boiled in water for three hours, and again steamed for
four hours, lost only 0·89-1·51 per cent.; while flax lost 3·50 per cent.;
Manilla hemp, 6·07; New Zealand flax, 6·14; hemp, 6·18—8·44; and jute,
21·39 per cent. At the same time, the fineness of the fibre places it ordinarily before flax, though, according to the method of cultivation, it varies from
an extreme degree of attenuation, equalled only by the pine-apple fibre.
While in strength, resistance, and fineness it equals or surpasses the best
known fibres, it possesses a silky lustre, shared only by jute, which is by
far its inferior in strength and durability. On the other hand, must be
mentioned the peculiar hairiness of the fibre, which, while enabling it to
combine readily with wool, renders it difficult to spin, on account of its
stiffness and brittleness interfering with the twist, and rendering the yarn
rough, despite the silky smoothness of the individual filaments.

"The combination of qualities exhibited by the fibre endow it with
affinities to other fibres, both animal and vegetable, which favour a wide
range of application. During the cotton famine, it was tried as a substitute or for mixing purposes, being first cut into lengths of two inches,
and treated with alkalies and oil. Fabrics made with equal proportions
of the fibre and Egyptian and Indian cotton gained in strength and gloss,
and offered no difficulty in spinning and weaving; they also took dyes
as well as Egyptian and American cotton, and better than Indian cotton,
a little modification of the mordant, and of the strength of the vat, being
necessary with a few colours."

"As a rival to the finest varieties of flax, it has perhaps a better
prospect. Technical difficulties, however, arise in spinning the fibre on
flax machinery, and owing to the stiffness of the fibre, the yarn produced
is often very rough. A number of processes have been devised and
patented by J. H. Dickson (Godalming), by Marshall (Leeds), Moerman
(Ghent), Bonsor (Wakefield), &c., for working up the fibre on flax
machinery; but the real conditions of success, where it is attained, are
kept jealously secret. In comparing the two fibres, account must be
taken, not only of their relative market values, but also of the fact that
the commercial Bœhmeria fibre still contains much of its natural gum,
involving the cost of labour and chemicals for its removal, and consequent
loss in weight, before it is ready for combing and spinning. The loss in
weight amounts to 23-38 per cent., generally 30-34 per cent., so that
the price of the available fibre is increased by 33-50 per cent., without
including the cost of treatment. With the finest descriptions of flax,
it might compete in price, but the demand for such is limited; that it will
ever supplant ordinary flax, appears doubtful.

"The hairy nature and length of the fibre point it out as likely to
compete successfully with wool, especially long-shaped kinds, the market
values of which are very high in comparison. Several manufacturers,

e.g., Lister, Sangster, Wade & Sons, Whitaker (Bradford), China-grass Co. (Wakefield), have energetically followed up this promising outlet, though not always with success. The fibre is subjected to a chemical treatment, which causes the cells to separate, the longest varying from 4-9 inches. The loss by chemical treatment generally amounts to $\frac{1}{3}$ the weight of the imported fibre; combining leaves about equal proportions of long fibre, and tow, or 'noils.' Thus prepared, the fibre has been spun on worsted machinery, and used like mohair, for glossy goods; as a rule, the warp was cotton, and the weft was Bœhmeria yarn of comparatively little twist. The success of the experiment was foiled by the ease with which the fabric took and retained creases; this evil has since been remedied by using very thick cotton warps or by mixing with wool. A new effect in woollen goods is now obtained by mixing 10-20 per cent. of Bœhmeria fibre with 90-80 per cent. of wool, combining before spinning, say, on the carding-engine or willow, and taking the former a little longer than the latter. The yarn is used for both warp and weft; the wool employed may be either carded or combed; and the cloth can be raised, milled, and woven as usual. In dyeing the fabric, the advantage arises that the two fibres do not take the same dye. Moreover, the noils has been found very suitable for admixture with coarse wools, for blankets, shoddy, and other rough purposes.

"Many experiments have been made in applying the fibre as a substitute for, or in admixture with, silk; but the cost of the fibre, and the difficulties encountered in its preparation, preclude it from competition with jute for this purpose. At the same time, it must be remembered that the study of the industrial applications of this beautiful fibre is yet in its infancy, and the inherent virtues of the fibre must ensure its extended use in textile fabrics, when the cultivation of the plant, and the extraction and preparation of the fibre, have received higher development. Even now, new uses are cropping up: Baker, Hill & Sons (Nottingham) are employing it extensively for ladies' scarves; and the Yorkshire Fibre Co. (Wakefield) are converting it into handkerchiefs, umbrella and parasol covers, &c.

"The combined strength and lightness of the fibre, and its great durability and resistance to water, favour its application to the manufacture of ropes, cordage, and nets. In all respects, save price, it is much superior to ordinary hemp, and even in the matter of price it does not compare so very unfavourably, as the cost and loss in preparing hemp is very considerable. Its competitors on this ground will probably be Manilla hemp (**Musa textilis**), **Phormium tenax**, and the **Agaves.** For canvas and sailcloth its superiority over flax seems undoubted. To the paper-maker its price is prohibitive; but an admixture of a proportion of noils will impart strength and cohesion to very inferior materials. The average weight sustained by slips of sized paper, each weighing 39 gr. made of this fibre, was 60 ℔, as against Bank of England note pulp, 47 ℔, and 'raw' **Agave americana** fibre, 89 ℔.

"The market values and supplies of this fibre have hitherto been subject to the greatest fluctuation. The former will depend upon the degrees of success with which the fibre may be made to replace others as already indicated; and an important condition necessary to the welfare of the industry will be the possibility of obtaining constant supplies, of uniform quality or qualities, and at a figure not exceeding 40*l*. a ton" (*Spons' Encycl.*, *931*).

The attempt made in this paper to establish the fact that there are superior and inferior qualities of fibre regularly sold as Rhea may, therefore, be fitly concluded by the statement that we possess no authentic

Uses.
193

RHEA.	Suitability of India for Cultivation of

chemical and microscopic analysis of the Indian-grown **Bœhmeria nivea** as distinct from **Bœhmeria tenacissima**.

CULTIVATION.
Climate required for Rhea.

CULTIVATION
OF RHEA.
194

For information regarding methods of cultivation and climates suited for Rhea and for China-grass the reader is referred to the account given in Vol. I. of this work. Since the publication of that volume very little of a practical nature has come to light that calls for special modification of the opinions there given. Indeed, the one point of importance has already been indicated, namely, that we have still to discover whether or not the two forms of Rhea have been separately investigated in India ; and if not, whether there are areas in India peculiarly suited to the one or to the other. In general terms, it may, however, be stated that past experience

Important to India. 195

would seem to point to the inference that while many species of **Bœh-meria** are indigenous to India and can be freely enough grown,—some being even cultivated for their fibre,—the climate of the greater part of India is apparently not suited to Rhea as a fibre crop. The plant is a perennial, and is not therefore, like Jute or Sunn-hemp, capable of being restricted in its growth to the months of the year suitable to it. The marked transitions from the hot damp rainy season to the dry hot season are unfavourable to the formation of Rhea fibre. During the former, long succulent stems are formed highly suitable for fibre purposes, but during the latter, growth is slow or suspended and the sappy stems of the formei are dried up. The final result of this is that long and short joints are formed with numerous knots which oppose great difficulties to mechanical appliances for decortication. On the other hand, such chemical means as **M. Favier's** would probably be workable even with freely jointed stems, since the formation of fibre is continuous and not necessarily interrupted by joints. The experiments hitherto conducted in India, and which have been universally admitted to have been failures, were all directed towards the discovery of a machine that would cheaply and conveniently decorticate.

In the face of these failures, however, interested parties have continued to prosecute the enquiry; and numerous communications accompanied with admirable samples of fibre are yearly produced. In Assam, where Rhea is fairly extensively cultivated, it has been contended that land for tea purposes is so valuable and labour so expensive that it is highly probable Rhea would not pay as a European industry. On the other hand, there are large portions of Eastern Bengal where these objections would have less weight and in which it seems probable some of the forms of Rhea or China-grass might be profitably cultivated. This would also apply to a large part of Madras and along the western coast to the high-lands of Bombay. It is being successfully grown in the Wynaad and in Mysore. and apparently also in Tirhut. Perhaps certain parts of Burma, now that vast tracts, particularly in the upper territories, have been thrown open, would. be found highly suitable, though of course the labour question would be serious. It is, however, impossible to hazard an opinion as to a future extended cultivation in India until systematic experiments have been performed with all the better fibre-yielding plants allied to the so-called Chinese grass-cloth.

OFFICIAL CORRE-SPONDENCE REGARDING RHEA. 196

NOTES FROM THE PROCEEDINGS OF THE GOVERNMENT OF INDIA ON
THE SUBJECT OF RHEA.

Having indicated the spirit of some of the numerous communications received by the Government of India, a few extracts from the more import-

R. 196

Rhea and China Grass.	(*G. Watt.*)	RHEA.

ant papers may be here given, arranged as far as possible under the name of the province to which they refer.

For commercial purposes it is perhaps not necessary to trace the history of Rhea further back than to the time (1869) when that enlightened statesman, the late Lord Mayo, took an active personal interest in the subject. His Excellency was induced to offer a reward of £5,000 for the best machine which would decorticate and clean the fibre from the green stems; and do so cheaply and conveniently. The first exhibition of machines and appliances was held in 1872, and the result having proved a failure, the reward was a second time offered. In 1879 a trial of some ten machines was made, but again, while some of the inventors received rewards, none of the machines were pronounced to have fulfilled the conditions desired by Government. At the same time the opinions briefly indicated above regarding the unsuitability of a great part of India for Rhea, became currently accepted, and accordingly the Governor General in Council felt called upon to withdraw the reward which had been twice unsuccessfully competed for. These trials had, however, a world-wide influence. Inventors in every country turned their attention to the subject, and as the result, numerous contrivances and processes have been patented. The outcome of this awakened interest has been the creation for Rhea of a much more promising position than it ever enjoyed before. The demand for the fibre is steadily increasing, and even India can boast of at least one Company—the Glen Rock Fibre Company of Wynaad—that is devoting its attention largely to Rhea.* Thus in the face of the disappointing features that have been recorded above India still bids fair to hold its own among the Rhea-producing countries of the world.

Before the Report of the Commissioners appointed by Government to adjudge the merits of the Rhea machines, shown at Saharanpur, had found its way into the hands of commercial authorities, several applications for waste land had been received by the Government of India. The fact that many species of Bœhmeria were natives of India, suggested that country as a hopeful field for Rhea and China-grass cultivation. It is needless to publish these letters or the replies given by Government. The fact had to be made known that the information possessed by Government would not justify sanguine hopes of a rapidly successful future for Rhea cultivators. This, it is feared, had the effect of diverting capitalists to other tropical countries. The following official note issued by the Revenue and Agricultural Department gives a complete summary of the opinions arrived at by Government:—

"There appears to be a very general impression prevailing in England, and to some extent in India, that this country is on the whole well suited for the extensive cultivation of the Rhea plant. This is not the case, and as many enquiries have been recently received from various sources asking for information, and in some cases for grants of land and other assistance from Government, it seems desirable that the existing evidence on the subject should be made generally known. The fact is that the greater part of India is unsuited for the profitable cultivation of Rhea as a commercial product, although it may be quite true that the plant will grow anywhere in India.

"Rhea was cultivated for some nine or ten years in and near the Saharanpur Botanical Gardens in connection with the prize offered by Lord Mayo's Government for a successful fibre machine. It was then proved that the Rhea stalk of Saharanpur was usually quite unfit for conversion into fibre. The cause was alleged by a Dutch planter of great experience

---

* It is reported that a second Company has been established in Bombay—the Bœhmeria Company.

OFFICIAL
CORRE-
SPONDENCE
REGARDING
RHEA.

from Java to be that the stalk was of uneven quality owing to the alternations of dry and wet heat during the season of its growth. Between one pair of joints the stalk was shown to be short, hard, and dry, and between another pair to be long, soft, and pliant. If the top of the stalk was ripe, the foot of it was green; if the end was green, the base was only half-grown, and so on, whereas, for purposes of fibre, the whole stalk should be a green flexible wand of even quality throughout, such as can only be produced in an equable climate like that of Java, where the temperature varies but little throughout the year, and the atmosphere never becomes desiccated : a gentle shower or two usually falling every day. It is true that occasionally a spell of rainy moist weather lasts sufficiently long even at Saharanpur to produce one serviceable crop in the year, but Rhea can never be grown profitably unless three or four cuttings can be made in the twelve months.

"Now Saharanpur may be taken as typical of the whole of the North of India. Some localities may be a little more favourable, but the majority are less so. In Bengal rain is more abundant in the monsoon (June to September) than in the country of which Saharanpur is typical, while there are, in many parts of it, Calcutta for instance, earlier rains in April and May. But the following is the testimony of **Dr. King**, Superintendent of the Botanical Gardens at Calcutta, who tried practical experiments with Rhea on the banks of the Hooghly :—

" 'The experience I have already had of the growth of Rhea in Bengal, supplemented as it has recently been by the opportunity afforded me of visiting Java and the Malay Archipelago, makes me doubt very much whether Rhea can ever be grown to commercial advantage in Bengal. The soil in this Province is poorer than that of Java and the islands of the Malay Archipelago, and manure is obtainable with difficulty even near large stations, while in the mofussil it is almost an impossibility. Rhea will certainly not yield well in Bengal without plenty of irrigation during the hot weather and copious manuring once a year. It is, moreover, a plant which naturally grows in an equatorial climate. The low temperature (and especially the low soil temperature) of the cold weather and the dry heat of the hot season in Northern India are equally ill-suited for it. I believe that even if Burma had every other condition in perfection, the high price of labour in that Province would be fatal to financial success.'

" The following extract from a note, * called for by the Government of India in 1881, gives a clear summary of the results gained up to that date, and is generally confirmed by later experience :—

" ' The idea of opening a public competition of a machine to clean Rhea fibre I think grew up very naturally from the correspondence regarding the cultivation of the plant in India. No one reading that correspondence could avoid the conclusion "that the fibre can be grown ; the *crux* is, how to prepare it as a marketable commodity when it is grown." The two questions unfortunately " see-saw " one upon the other; it is no use growing the fibre if we cannot prepare it for market ; it is of no use to invent machines if the fibre cannot be grown in suitable quantities.

" ' Among plants which are recommended for acclimatisation, we find three classes: (1) plants that seem to take to almost any climate ; (2) plants that will only grow under certain specific conditions like *Ceara* rubber, *Pithecolobium*, &c.; (3) plants that *will grow*, but under what I may briefly call, " garden cultivation." It is the last class that cause expenditure, raise great hopes, give rise to flaming reports of success and great deal of correspondence, from which it is very likely that a false conclusion will be drawn, false in fact, though seemingly quite in accordance with the evidence. To this class I am afraid Rhea belongs. There is no doubt that you can grow it with care in Lucknow, Bareilly, Saharanpur, Dehra Dun, and even in Lahore, as well as in Calcutta, the Sunderbans, Assam, and other places. But to grow a thing in such a way as to produce a fine garden of ten acres and evoke an "interesting" report is one thing ; to grow it commercially is another.

" ' I think if we look back to the collected evidence, there was much reason to conclude that Rhea would grow, but not that it would grow commercially. It is an

Note.—It is regularly grown in Assam p. 464; in Eastern Bengal, see p. 465 et seq. and perhaps also in Kangra, see pp. 471-476.
198

---

* By Mr. Baden Powell, C.S.

| Rhea aud China Grass. | (*G. Watt.*) | RHEA. |

equatorial* plant and this clearly points to its requirements being a great *equability* of conditions. Dry heat burns it up; drought kills it outright; frost it will not stand (though frost is one of the conditions that is more easily overcome after some time of establishment). Even *cold soil*, such as Calcutta has during the winter, is unfavourable.

" 'What Rhea wants is a moist air, no long dry hot months, a naturally rich soil, which does not need much manure, plenty of rain, and no extremes of temperature. The manure difficulty is not of such importance, except in places where it is difficult to procure or costly. It does not need much horticultural science to perceive that a plant may be coaxed into living where these conditions are not general. A sheltered garden may produce the plant when the fields all round would not. Poverty of soil may be overcome by manure. Artificial irrigation may obviate natural effects of a dry climate. The more care is given, the more these helps remove defects of climate and the better the plant. Their cost, and the difficulty of applying them on a large scale, is not appreciated when the area of trial is small, and if the experimenter is an enthusiast, his report is full of confidence. Even when very good results are not obtained, still partial results are.

" 'I believe then that the just conclusion regarding Rhea is this: it is no use trusting to results obtained under special conditions. Even the Howrah experiments are of no use as a test, except to show that Rhea will not grow in a commercial sense in Calcutta. Saharanpur and all the other trials have been " garden cultivation," and even so have not been successful. Unless Rhea can be grown in some districts as an open field crop, with just so much care as an ordinary well-farmed field for a rather superior crop gets, there is no hope for it. In China even it is evident from Dr. Watson's report that manure is largely given. In India this is always a drawback, but not an insuperable one on a large estate.

" ' The only chance is to look out for places where the natural conditions are as like those of Sumatra, Java, &c., as possible : where there are no extremes, where the soil temperature is not low in winter, and where there is plenty of moisture. Burma is probably out of the question, because of the labour difficulty. The only districts that do not seem to me hopeless are Upper Assam, Dacca, Dinajpur, &c. No flooded land is of the least use—floods kill the plant.

" ' Now as regards Assam, there is a report from the Conservator of Forests, which says that even there the fields require digging carefully, fencing and plenty of manure; and that " if its extensive introduction into the home markets depends on its being supplied at an average price of £30 to £40 per ton of rough fibre, * * this province will not be a source of supply, since it cannot be produced at double that rate at present.'"

(*Conf. with p. 450.*)

"The conclusion drawn in the final paragraph of the above extract is probably correct, though it may be the case that some parts of Southern Madras should be added to the list of likely localities. There is reason, too, to believe that certain tracts in Burma (probably the lower hills of Pegu and Tenasserim) would be found suitable, and as the labour difficulty is decreasing under a system of free emigration, the time may come when that province may afford facilities for Rhea cultivation.

"At present, however, India cannot be looked upon as a favourable field for the Rhea industry." (*Note by Revenue and Agricultural Department.*)

ASSAM.

A reference was made to the Assam Government on the subject of allotting waste land in that province to intending Rhea cultivators, and the following extract from an official letter expresses clearly the prospects that might be expected from such a venture :—

"The Rhea plant is commonly grown in Upper Assam by Doms and other fishermen who weave their nets out of the fibre. It is cultivated in small patches close under their houses, rudely fenced in and manured liberally with cowdung. Under this treatment, which involves no manipulation of the soil, nor any trouble in cultivation, the plant thrives luxuriantly, and supplies a sufficiency of fibre for all their wants. It is much to be doubted, however, whether looking to the cost of living and of labour in

---

* See the contention (pages 444 to 449) that there is both a temperate and tropical form.—*Ed*.

RHEA.	Experimental Cultivation of

CULTIVATION in Assam.

Assam, a European could make Rhea cultivation on a large scale pay. Three years ago, the Director of Agriculture in Assam collected some statistics about Rhea fibre in view of the large prices offered for it in England by manufacturers who combine it with silk and other materials for hangings and similar fabrics. The English prices appeared at first sight to promise a large profit, but on working the calculation out with reference to the actual price of Rhea fibre in its coarsest shape on the spot, it appeared that there would be scarcely any margin left even to pay the cost of carriage."

Report by GUSTAV MANN, Esq., *Conservator of Forests, on the Cultivation of the Rhea plant in Assam.*

(Conf. with p. 462.)

"The Rhea plant (**Bœhmeria nivea**) is cultivated everywhere in the Brahmaputra valley districts, but grows much better in Upper than in Lower Assam on account of the greater moisture and heavier rain. No more congenial climate than Assam exist, probably in India, for the cultivation of this plant, for it grows to the greatest perfection possible where it receives the requisite care and cultivation.

2. It is at present only grown by the fishermen in this province, on account of the superiority of its fibre for the manufacture of nets and not for sale, since, even at the present high price, varying from R1 to R2 per seer, nobody would consider the cultivation of Rhea worth the trouble, as it only grows well where it receives a very great deal of attention in the way of loosening of the surface soil, plenty of manure, and careful fencing.

3. Soil open and friable, slightly loamy, and not too sandy, is best suited for its growth. The locality where it is planted must be high, so that there is no risk of inundations, since, even if the water remained only a short time on the land, the plants would be ruined.

(Conf. with p. 447.)

4. The cultivation has been described by **Major Hannay** and others correctly, so that there is nothing for me to add on this subject. The plant is propagated by parting the roots, and improves in the same degree as the soil is worked or loosened, kept clear of weeds, and manured. It is cut from three to five times annually, and reaches a height of from 4 to 5 feet.

(Conf. with p. 449.)

5. The outturn per acre, according to the statements of the dooms, or fishermen, in the different districts where I made inquiries, is only from 200 to 300℔ of clean fibre an acre per annum; but their statements are very unreliable.

6. The main question at issue is, whether the Rhea plant can be cultivated sufficiently cheaply in this province so as to allow the fibre to be used to a greater extent in the manufacture of cheap articles, produced in large quantity, so that it may become a great staple and develop into a large trade, as is pointed out by **Dr. Watson**, in paragraph 45 of this report. If its extensive introduction into the home markets depends on its being supplied at an average price of £30 to £40 per ton of rough fibre, as stated by **Dr. Watson** in paragraph 46 of his report, this province will not be a source of supply. since it cannot be produced here at even double that rate at present, or in future either as far as can be judged now, for its production requires as much time and labour as tea does, whilst the latter plant produces on an average 280℔ per acre, and fetches on an average one shilling and eight pence per pound in Calcutta.

Price. 200
(Conf. with pp. 450, 463, 470, 484.)

7. In fact, at the above low value of Rhea fibre, as quoted by **Dr. Watson**, it would only give a return about equal to rice, whilst its cultivation requires double and treble the time and attention. For this reason, I do not even see a likelihood of its being grown in the Sylhet district, where there is a greater population, and labour is comparatively cheap.

R. 200

Rhea in Assam.	(*G Watt.*)	**RHEA.**

CULTIVA-TION.

8. From the above remarks it will be seen that Rhea fibre has no chance in this province, since the success of tea cultivation will, as far as can be judged at present, always prevent European capital being employed on Rhea cultivation, and it is far too laborious for natives of this province to take to, as they have done in Bengal to jute cultivation, for the sake of gain, as long as the fibre has to be produced at £40 per ton.

(*Conf. with pp. 450 & 463.*)

While Rhea cultivation is referred to in many official publications subsequent to the above Selections from the Records of the Government of India, very little information of a definite nature has been brought to light regarding the yield of fibre per acre. The following passage from the Agricultural Report of Assam for the year 1885-86 will, however, be read with interest. It confirms in general terms the reports published by the jails regarding their experiments in Rhea cultivation.

"A small quantity of Rhea was grown in the Nowgong Jail during the year under report. The object of the experiment was double. Details as to the cost of producing the fibre were required, and a comparison between the crop as grown in shade and as grown in the sun was wanted. The second object was quickly gained. The plants put down in the shade refused to grow at all and were a total failure. The patch grown in the sun, on the other hand, did well. An area 71' × 74' (= 1 k. 3 l.) was planted in the jail garden in the middle of April.

"The first cutting yielded 3 seers of dried fibre in July. The second cutting yielded 10 seers, 9 chittacks in September. The third cutting yielded 7 seers, 4 chittacks in October. Total yield in six months = 20 seers, 13 chittacks = value (@ R1 per seer) R20-13. The total cost of planting, cutting, and extracting the fibre was R13; consequently, on an expenditure of R13 there was in six months a profit of R7-13. When I saw the crop in the middle of December it was nearly fit to cut, and might safely have been estimated to yield one more crop before the following April. Therefore, five crops might be calculated on from the above data in one period of twelve months. But the produce from three crops was 20 seers, 13 chittacks; therefore the produce from five crops would be 34 seers, 11 chittacks. (I have allowed for slower growth in the cold weather by only taking one crop between December and April.) But the cost of cutting and extracting fibre from one crop was R2. Therefore, the cost of the two additional crops would be R4. Therefore, the total expenditure in twelve months would be R17. Therefore, the net profits in twelve months would be R17-11, or, roughly, 100 per cent. per annum. Working out the figures for the acre we see that the weight of fibre obtainable would be 911℔, and the cost R222 per annum.

"The above calculation of course applies only to a very small area, and it is exceedingly dangerous to argue from such limited data that profits will necessarily accrue if the undertaking were conducted on a commercial scale. Still what evidence there is goes to show that Rhea is a profitable crop, and there is no doubt that the climate of the province suits it admirably."

(This highly favourable opinion is opposed in every point to that given at the various passages quoted by marginal note on p. 449.)

BENGAL.

**CULTIVATION in Bengal.
201**

In the official papers, which the writer has been able to consult, nothing of any importance occurs regarding the rhea of the Lower Provinces. It is necessary, in fact, to turn for information to the accounts given by Buchanan-Hamilton, Campbell, and other such authors who wrote of the rhea cultivation of Rungpore and Dinagepore half a century ago. Dr. Campbell's description of the former district is, perhaps, the best account which has as yet appeared. He writes:—"In the *Eastern Star* of the 26th ultimo, there is a very interesting article on the plant supposed to

30

Dictionary of the Economic

RHEA.	Experimental Cultivation of

CULTIVATION in Bengal.

yield that fine fabric—the China grass-cloth. The authority of Buchanan is quoted in support of the supposition that the plant is also a native of Dinagepore and Rungpore, where it is known and cultivated under the name of *kankhura.* As the *Star* is anxious to obtain further information on this plant, I beg leave to offer the little I know about it, and to tell you where and how you can procure additional and valuable particulars.

"In the month of January last, when I was returning from the Bhután Frontier, through the district of Rungpore, my attention was attracted by patches of a small green crop, cultivated with much care, close to the villages along the banks of the Teesta River. I had never seen the plant before, and as winter crops are rare in that part of the country, it was an object of additional interest.

"It turned out to be the *kankhura,* and is considered by the people to be a species of hemp. It is sown at the end of the rains, and cut and pulled in February and March. It is of a dark green colour, grows to 3 or 4 feet high, and does not particularly resemble any of the nettles I am familiar with (the grass-cloth plant is supposed to be a nettle). The leaf is not unlike that of the black currant. It is cultivated with much care, principally by fishermen and others along the river banks, and exclusively for making fishing nets, for which purpose it is considered unequalled by any other kind of hemp. The fibre is wonderfully strong and stands wetting for a long time without injury. It is not used, I believe, in making any description of cloth or for ropes. The preparation of thread is similar to that of hemp.

"When at the Durwany sugar-works about the same time, I learnt that Mr. Henley of Calcutta was very anxious to procure a large quantity of the *kankhura* for use in the factory, and that efforts were made to purchase a supply, but without success, as the people who had grown it for their own use would not sell it ; and previously there was no extraneous demand for it. The fibre of the *kankhura* is extraordinarily tough, and would be highly valuable in rope works or the purposes proposed by Mr. Horsfall of Leeds, in the manufacture of cloth. I would suggest that samples of the Rungpore *kankhura* be procured through the manager of the Durwany sugar-works, Mr. Ahmuty, and submitted to the valuation of a competent person, and the price it is worth in Calcutta and England made known in the districts in which it is cultivated, with a view of producing a quantity of it for the English market. If the requisite information as to price can be obtained, I shall be glad to assist in communicating it to the people in the portions of Rungpore along the Bhután Frontier.

In addition to the remarks by Dr. Campbell on the Rungpore *kankhura* hemp, the following memorandum contains the results of Mr. T. F. Henley's enquiries on the subject :—" Feeling convinced of the excellence of the description of fibre in question, I endeavoured, when in Rungpore, to collect a quantity, and succeeded in procuring a moderate-sized bale of it, which has been forwarded to a house in London, requesting a report of some of the large hemp and flax spinners of Lancashire. This information may shortly be expected. It is difficult in the present state of the enquiry to ascertain the price at which it might be procured eventually, if the stimulus of advances of funds for its cultivation were made. It is now only procurable at very high and variable rates in small parcels from the fishermen. Under any circumstances it appears to me that it must necessarily be a much more expensive article than either *sunn* or *jute,* inasmuch as a labourer can prepare one-and-a-half to two mounds of *jute* per day's work, whilst of the *kankhura,* he cannot manufacture more than as many seers. The *jute* and *sunn* fibres

CULTIVATION
in
Bengal.

are separated from the woody stems by the process called water-retting, in a similar manner to that employed in the preparation of the true hemp and flax; a process by which the removal of the fibre from the stem is rendered of easy accomplishment. The above nettle, on the contrary, requires a tedious manipulation. The bark or epidermis of each individual stem must be carefully scraped off the fresh-cut plant, a most tedious and delicate operation. The natives declare that the fibre cannot be separated by the water-steeping process, and they are doubtless correct on that point. The scraped stems are then spread out in the sun and dried to a certain point, after which they are beaten carefully one by one in order to facilitate the separation of the fibrous coat which is now removed by being pulled off each plant, still operating one by one on the plants, and not in bundles as is the case when operating on other fibres. The fibres now require to be carefully washed, in order to separate the remaining impurities. This sketch will serve to show whence it arises that a labourer employed in the manufacture of the nettle hemp can only prepare about one-fortieth part of what he could have accomplished with *jute* fibre. The plant itself is perhaps of easier cultivation, and more productive from a given surface than *sunn*, as it will yield several cuttings from one planting. It requires, however, a rich, free soil, and plenty of manure. The second and third cuttings produce a much weaker but a finer fibre, and it is not improbable but that the Chinese in their grass-cloth manufacture select the after-crops for their purpose. Some experiments made by bleaching and heckling a portion of first crop *kankhura* did not produce so fine and silky a material as that of the China grass-cloth, but as my experiment was hastily and imperfectly made, it is by no means conclusive. As I have noticed above also, it is probable that a very different result would arise from the employment of the after-crop fibres, or by the selection of the young and delicate shoots, which actually do produce a much more flax-like material. It is possible the Chinese by adopting these means may produce textile materials of very different qualities. The large stems, near the lower part of the plant, yield a very tough but coarse product, admirably adapted, however, from their immense strength for many purposes.

"Some comparative experiments were made with European hemp compared with the Rungpore *kankhura* by loading small bundles (four-picked fibres loosely twisted) of each kind to the breaking point, and taking the average of several trials. These experiments showed that the nettle fibre possesses about three times the strength of Russia hemp. The latter being of an excellent quality, imported into Calcutta for special purposes.

"I had the *kankhura* also employed for packings of steam engines on which it proved quite efficient, the ordinary country fibres, such as *sunn* and *jute*, being totally useless for that purpose" (*Journ. Agri -Horti. Soc., Ind. (Old Series), Vol. VI., p. 30*).

In Volume I. of this work, page 473, will be found a notice of the comparatively recent experiments which were made by the Rajah of Dinagepore to cultivate rhea, also a brief mention of certain efforts in Shahabad. In a report submitted by Dr. King in January 1878, the results (unfavourable on the whole) of the experiments conducted by him to cultivate "Rhea" at the Rungbee cinchona plantations were made known to Government. It is not stated whether the India, *i.e.*, true rhea plant was grown or the China form. The endeavour to grow it seems, however, to have for some years been abandoned as no mention of any importance occurs of further experiments. At the same time it may be said that frequent allusion is made in the Indian newspapers to private experiments in Tirhut, Dhurbungah, and other localities, but it is not known how far these could at present be regarded as of commercial importance. Indeed, we

RHEA.	Experimental Cultivation of

CULTIVATION
in
Burma.
202

know nothing for certain as to the extent or value of the Bengal cultivation nor the form of the plant grown.

BURMA.

The admission of imperfect knowledge, already made regarding Bengal, has to be repeated in connection with Burma. No further information has come to light than will be found in tne early volumes of the *Journal of the Agri.-Horticultural Society of India,* chiefly from the pens of Colonel Burney and Major Macfarquhar. The following brief note from Royle's *Fibrous Plants of India* may be furnished as containing practically the chief facts :—" It is known in Siam and at Singapore ; the string made of it is called *tali rami,* and the fishing nets manufactured with it are conspicuous for their elegance and strength. Colonel Burney, in 1836, obtained it from Pivela and Youkyouk, in the Shan province of Ava, where it is called *pan,* and where Mr. Landers afterwards found it. Plants sent by the Colonel to Moulmein and to Tavoy succeeded well, but required much water." The remarks in Mason's *Burma and Its People* are to the same effect. It will thus be seen that we know very little indeed of a definite nature as to the extent of rhea cultivation in Assam, Bengal, and Burma—the three most hopeful of all provinces for the enterprise—and absolutely nothing as to the nature of the plant that is actually grown in these provinces. As will be seen from the extensive correspondence below, the bulk of the Indian literature on rhea is connected with the North-West Provinces and Oudh, and that in these Provinces as also in the Panjáb, the freshly-imported Chinese plant had alone been experimented with.

MADRAS.

CULTIVATION
in
Madras.
203

That Rhea can and is being successfully cultivated, however, in India, may be judged of from the following communication received from the Manager of the Glen Rock Company, Limited, Wynaad :—

"Mr. W. Gollam, Superintendent of the Botanical Gardens at Saharanpur, has kindly given me some information as to his past experience in the cultivation of Rhea, but tells me that the cultivation of Rhea has been discontinued since 1880, and that he cannot now recall many facts with which he was then familiar. He advises me to write to you and beg for copies of all the official reports on Rhea, and thinks if you have spare copies you may be able to help me.

"This Company has now 250 acres of land planted with Rhea in 1885, and a further 250 acres will be planted this year. I am most anxious to get information as to past experience in this cultivation. What I have hitherto learnt of the past is from experience obtained on very small areas of land and under exceptional circumstances.

Yield.
204
Conf.
with p. 449.

"I want to know what has been the actual weight of green stems per acre for cutting from any considerable extent, and what the percentage of clean fibre or Filasse obtained from the green stems by any mode of treatment.

"I have the pleasure to send you a sample of the Filasse as treated here. The result and cost of the treatment is considered satisfactory, but the percentage of Filasse to the green stems cut is disappointing in comparison with statements given as to results elsewhere obtained, and the weight of green stems per cutting per acre, which varies here considerably according to the lay and circumstances of the ground, is generally less than was anticipated.

"The information given me by Mr. Gollam of his recollection of average results agrees much more nearly with my experience here than the figures given in French and other books published on the subject, or with the estimates in Dr. Forbes Watson's Report of 1875."

R, 204

| Rhea in Kangra. | (*G. Watt.*) | RHEA. |

The question asked by the Company as to yield of fibre cannot, it is feared, be satisfactorily answered in India. All the reports and publications in the possession of Government deal only with experimental cultivation on a very small scale. In Vol. I., page 470 of this work, the various opinions hitherto published have been briefly reviewed with this one only certain conclusion, that so far as India is concerned, the results do not confirm those published by continental authorities. Either these are overstated, or the plant in India is by no means so productive of fibre as in Europe and China.

PANJAB.

From the reports and correspondence regarding the experiments conducted in this province, it would appear that the greatest degree of success hitherto recorded regarding the true China-grass has been attained. It appears that the authorities in Kangra early recognised that the tropical plant of India was not suited to their climatic conditions. Accordingly seed was imported from China of Bœhmeria nivea and its cultivation prosecuted with remarkable results as will be seen from the following correspondence :—

From J. G. CORDERY, Esq., Deputy Commissioner, Kangra, to the Commissioner and Superintendent, Jullundur Division,—No. 106—228, dated 15th March 1876.

With regard to your No. 2314, dated 25th October last, in forwarding to you a copy of the report which has been furnished to me by **Mr. Montgomery**, the conductor of the experiments which have been made in the cultivation of Rhea-grass in this valley, I confine my own remarks to the points on which the results coincide with, or differ from, those shown in **Dr. Forbes Watson**'s printed report.

2. It will be noted at once that the plant with which **Mr. Montgomery** has been concerned, and which alone has been grown in this valley, is the China variety of the species, and not that indigenous to India. This fact will, perhaps, serve to explain the variation in the results as compared with those detailed in other reports.

(*Conf. with p. 450.*)

3. On the question of the propagation, actual cultivation, and reaping of the plant, which is treated at some length by **Mr. Montgomery**, and in detail, which is rendered of much value by an experience of twelve years, the printed report does not touch, and the paragraphs in which these subjects are dwelt upon may probably be held to have added a good deal to our information as to the best method of introducing the Chinese variety into this country.

4. With regard to the very important point of the outturn per acre, it will be seen that longer experience has induced **Mr. Montgomery** to enhance his former estimates. If an acre, as he argues, can hold 3,000 plants, each of which is allowed to grow to a height of six feet, and each of which yields six stems, and if out of every 1,000 plants the average yield of clean fibre is 18℔, the total result is, as he expresses it,

(*Conf. with p. 449.*)

$$\frac{3,000 \times 6 \text{ stems} \times 3 \text{ crops} \times 18℔}{1,000} = 972℔$$ per acre for a year. About the height to which the plant should be permitted to grow he appears to differ from some of the authorities quoted by **Dr. Watson**. But he is careful to explain that he has been led to his own conclusion by finding on his own account that, when he grew smaller stems, though the return of peel was longer, the return of fibre was smaller. He goes further, and seems to imply that larger stems, from seven to eight feet, will yield more fibre, though not in the same proportion to the loss of the weight in peel. It is remarkable, however, that his whole view as regards the height to which the stems should reach, is opposed to that maintained in the 61st paragraph of **Dr. Watson**'s report, where the plants of from three to four feet

RHEA.	Experimental Cultivation of

CULTIVATION
in the
Panjab.

Kangra a
favourable
locality for
China-grass.
207

are declared to yield the fibre of the most uniform, finest and most valuable quality.

5. This estimate per acre exceeds by 200℔ that which **Mr. Montgomery** gave in 1876, and which is the highest quoted by **Dr. Watson**. But it will of course be borne in mind that the fibre, thus extracted and cleaned by the hand, has to undergo a further chemical process in which it loses a portion of its weight before it can be subjected to machinery. **Mr. Montgomery** gives this loss at nine per cent. in fibre prepared by himself, but **Dr. Watson** estimates it as 25 to 30 per cent. The former ascribes this difference to the steps taken by him in its previous preparation in this country, as compared with those ordinarily taken in the preparation of the purely indigenous species.

*Conf. with
pp. 457, 475.*

6. **Mr. Montgomery** is sanguine concerning the invention of machinery for the separation of the fibre, even from the green stem, in a state fit at once for the operation of machinery. His reasons are given in paragraphs 21-23 of his report, and though stated in this form they do not afford grounds for any confident expectation, yet they make it a matter of regret that he is unable, from want of funds, to make any attempt at experiment in this direction. But it will be observed, in **Dr. Watson's** report, that the general concourse of opinion is in favour of operation on the dried stems, and not upon the green. And the reasons given for this in the 58th paragraph of the report appear to me, with due deference to **Mr. Montgomery's** practical experience, to outweigh those urged by him. The description of the working of **Mr. Greig's** machine, which was tested at Calcutta by **Colonel Hyde**, serves to show that, up to the present time, no successful machinery for the required purpose has been devised.

7. How essential such an invention is to any success in the cultivation of the plant is seen clearly when we turn to the all-important point of the cost of production. The lowest estimate at which **Mr. Montgomery** can rate the cost of turning out one ton of the fibre at **Kángra** is R369. The fibre so turned out has not only to be carried to England, but, before it can be manufactured, has to be subjected to a chemical process, apparently involving a further expenditure of 50 per cent. on the original cost, and certainly causing a loss in the weight of the fibre which is variously calculated at from 9 to 30 per cent. This calculation will make its total cost, before it can be applied to manufacture, £54 or 55 per ton at the *very lowest computation*. And when this figure is compared with the considerations contained in paragraph 46 of **Dr. Watson's** report, the chance of successful competition with flax, hemp, and the China-grown Rhea already in possession of the market, cannot but appear to be somewhat dubious. Nor, as a matter of fact, has there been any return whatever for the capital which has been sunk in this speculation at **Kángra** during the past twelve years.

Cost of
production
(*Conf. with
pp. 450,
463, and 484.*)
£55 a ton.
208

8. On the other hand, out of the R369, which is the present cost of the fibre at **Kángra** per ton, not less than R247, or 67 per cent., are expended on the process of extracting the fibre by hand labour. Unless, therefore, machinery can be substituted for this method, the hopes of any profitable cultivation of Rhea far from being increased, will be diminished every year, for it is not likely that the wages in the valley will remain long stationary at their present rates. But the invention of such machinery is surely possible at any moment.

(*Conf. with
pp. 452, 458.*)

9. The final result of the experiment made at Kangra may be said to show that both the climate and the soil of the valley are admirably adapted to the production of the Chinese variety of the plant reared at a slight expense, and yielding a fibre of a quality inferior to none. But the stems will not bear the plan of retting which can be successfully applied to jute,

Rhea in Kangra. (*G. Watt.*)	**RHEA.**

flax or hemp; and the fibre can therefore only be extracted by a slow, laborious and difficult process, which, however remunerative it may be in China, has been proved to be too expensive here for the foundation of a profitable trade. Only let a more expeditious and cheaper method be discovered for this process, and there would be every reason to hope for success in the introduction of a new and most valuable product and trade.

    10. The specimen of fibre promised in the concluding paragraph of **Mr. Montgomery's** report has now been received and accompanies this letter.

REPORT OF CHINA-GRASS CULTIVATION AND PREPARATION FOR EXPORT.

In submitting, for the information of the Government, the results of my experience in the cultivation of this valuable plant, I wish pointedly to note that my remarks refer solely to that variety of the plant cultivated and known in China under the appellation "Tchow Ma." My stock of plants has been derived from seed procured with great difficulty from that country in 1863.

    2. Whether the variety of the plant known in Assam as "Rhea," or that known as "Rami" in the eastern islands, is identical with the Chinese plant, I do not venture to offer an opinion. The Government of India have apparently adopted the former appellation, "Rhea," in designating the fibre; the American Government have adopted the latter, "Rami." I have not had an opportunity for comparing growing plants of each variety with mine, but I have had many specimens of fibre from each supplied to me, and there appear to me well-marked distinctions between the three, in colour and texture of the fibre.

Seed procured
direct from
China.
210
(*Conf. with
pp. 450, 469.*)

    3. At the time I succeeded in establishing the growth of the plant here (1863-64) the tea plantation at Holta was the property of Government, and several Chinese were then employed there. These men recognised my plants with much surprise, and showed me the Chinese method of separating the fibre.

    4. *By seed.*—This course must be adopted in some cases, when the germ of the plant has to be carried over great distances; but probably much disappointment will attend the result. To obtain the seed great care is requisite, and a favourable atmospheric season. For this purpose young spring shoots should be carefully reserved in a well-sheltered position. These plants should receive special care and be well manured. During the rainy season they must be kept thoroughly drained, and after that has passed, the ground should be carefully loosened round the plants. If the rains cease early in October, a fair amount of seed may be obtained; but as far as I can judge no amount of care can ensure success, so much depending on the season—a dry one being most favourable for the full development of the seed. The only method of sowing which I found successful was on a gentle hot-bed, under glass, in March and April; the seed scattered over the surface, covered very thinly with sifted earth, and carefully shaded from the sun, until the plants were about three inches high, when sunlight may be gradually admitted. When sufficiently strong they should be planted out a foot apart every way.

    5. *By cuttings of the stems.*—The stems should be spring-grown ones allowed to ripen well and not cut until duly ripe. Then divide the ripened portion of the stem where the cuticle has turned fully brown into short lengths, each including three eyes or buds. Cut a quarter of an inch below the bottom bud and as much above the top one, and plant with the centre

CULTIVATION
in the
Panjab.
Report on,
in Kangra.

bud level with the surface. If the weather be damp and cloudy, they will readily strike root, otherwise they will require shading for a week or ten days, the soil being kept moist. As with seedlings, I find a foot apart every way the most advantageous distance, as very few shoots are thrown up the first year.

6. *By division of the roots.*—This is by far the most advantageous and profitable method. The plants for this purpose should be three or four years old. After gathering the spring crop, dig up each plant carefully and remove the earth from the roots. I generally put the mass of roots into running water for a short time; this cleanses them thoroughly, and enables the gardener to see his work clearly. The tuberous portions of the roots will be found to show a large number of eyes similar to those on a potato. From these carefully separate portions, each containing five or six eyes, let the cuts be clean and reject all fibrous and decayed matter. Expose these sets to the sun for a couple of hours to dry the surface of the wounds, and then plant six inches deep, and at the full distance of four feet apart every way. In this way two good crops will be obtained from them the first year.

7. *Soil and situation for plantation.*—A rich loam suits the plants best, but they will grow in any kind of soil, provided a full supply of moisture be available, combined with thorough drainage. The latter is emergently required, particularly during the rainy season, as should the land be retentive and become swampy the plants will wholly decay in a very short period. If the land be poor, a liberal supply of manure is requisite, otherwise the stems will be short and weak, yielding scarcely any fibre. In no part of Upper India can the plant be successfully cultivated unless water for irrigation be available during the dry season. The facilities for obtaining an ample supply of water, combined with the moderate temperature at all seasons, renders this district particularly favourable to the plant.

8. *Cultivation.*—Should the land have been stocked with seedlings or cuttings (paragraphs 4 and 5), then in the following spring, after having reaped the first crop of available shoots, every other plant should be transferred to fresh ground, and put down at two feet apart. The following year the same course should be pursued, taking up each alternate root and replanting at four feet apart. After this the plants may well remain undisturbed for four years, hoeing well between after each crop, clearing away weeds, irrigating moderately during the dry season, and supplying manure where necessary. The only manure I had at command has been vegetable, consisting mainly of the leaves and wood portion of the plant itself, and of tree and vegetable leaves stored up for the purpose, with which I mix a considerable amount of wood-ashes. With the aid of this only I have kept plants growing in the same spot for upwards of six years; but consequent on the then very crowded state of the ground, the stems were short and very weak. I would therefore recommend a thorough removal after four years, the land to be then well ploughed, cleaned and manured.

Three Crops.

Conf. with p. 477.

9. *Gathering the crop.*—The periods of reaping will vary slightly according to difference of season. I find that in this district three good crops can be relied on each year;—the first during the latter half of April, the second about the commencement of August, and the third about the end of November. It will be found of much advantage to postpone reaping the second and particularly the third, as long as the condition of the plants will admit. If the third crop be cut in the middle of November, the weather here during the remainder of that month is not sufficiently cold to keep back the new growth; and should the young shoots appear above ground early in January, the frosts which are usual at that period seriously

Rhea in Kangra. (G. Watt.) RHEA.

injure them and lessen the spring crop. My own experience indicates that the stems should be gathered so soon as the cuticle shows a clear brown colour for about one-third of the length. At this stage, if the soil be good and the plant healthy, the stems will be clean from butt to point, the leaves of a rich dark green above, and pearly white below, and the branch-buds at the axil of each leaf-stalk just showing If gathered earlier than this I find the connection of the fibres very weak, and that a considerable portion separates in the operation of scraping the peel. If allowed a further growth, the axillary branches will have been thrown out, which will cause breakages at every point both in peeling and cleaning.

10. The average height of stems grown here has been six feet, after cutting off the soft portion at the top. In gathering I supply each coolie with a sharp pruning knife. With this they cut the ripe stems close to the butt; these are removed in bundles by boys to the nearest manure pit. Here the boys cut off nine inches of the top and pass one hand with a gentle pressure from top to butt; this removes every leaf. The stems are then placed in clean water from whence the peelers remove them and separate the peel, which is again thrown into water, from which it is withdrawn as wanted by the men who clean it. These lay three or four strips of peel on a flat board, scrape it a few times on the inner side from butt to point, then turn it over and repeat the scraping, which removes the cuticle : it is then hung up or thrown on clean grass to dry.

11. Taking the distance of four feet apart for fully bearing plants, an acre will contain (allowing for paths and water-channels) 3,000 plants : more than this I find to be too crowded and to increase labour while lessening the actual yield during a four years' period. Thus planted the yield will be a steadily increasing one, and the plants will not show any deterioration.

12. From repeated experimental weighings, I have deduced the following average proceeds from 1,000 freshly cut 6-feet stems :—

						℔	
Weight as cut	.	.	.	.	.	286	
Do. when dried	.	.	.	.	.	77·5 =27 per cent.	
Do. { fresh peel	.	.	.	.	.	83 =29	,,
Do. { dry peel	.	.	.	.	.	21·5=7·5	,,
Do. { fresh wood	.	.	.	.	.	203 =71	,,
Do. { dry wood	.	.	.	.	.	56 =19·5	,,
Do. clean dry fibre	.	.	.	.	.	18·7= 6·5	,,
Do. water	.	.	.	.	.	208·5=73	,,

(Conf. with p.449.)

13. If large stems, from seven to eight feet, be taken, the average is less in the weight of peel, but in the outturn of clean fibre it is slightly greater. With small stems from three to four feet, the percentage of peel is markedly greater, but the return of fibre is barely 35 per cent. Moreover, the extra labour in cutting, peeling, and cleaning these small stems is an important consideration. The crop cut during the rainy season will always contain a larger percentage of water, and that of clean fibre be found rather less, the fibre being also softer than at the other periods of cutting. This I consider due to the fact that at this period the resinous matter in the plant is in a more diluted state, and consequently a greater portion of it is removed during the process of washing and scraping the peel.

14. I have already expressed my opinion against the use of either immature or small stems as likely to give a result inferior both in quality and quantity; yet I am fully satisfied as to the advisability of not only sorting the crop, as cut, according to length of stem when necessary, but I would further recommend that the peel from all stems of five feet and upwards should be divided into two, and the fibre from the upper and lower por-

CULTIVATION
in the
Panjab.

Report on,
in Kangra.

tions kept distinct. If cultivated as I suggest, the difference in length of the stems at each cutting will be found very small, the monsoon crop always giving the longest stems.

15. Taking the above as a basis for calculation, and knowing that each plant established as I recommend will give at least an average of six stems during the first year, I assume : $\frac{3,000 \text{ plants} \times 6 \text{ stems} \times 3 \text{ crops} \times 18\text{lb}}{1,000}$ $=972\text{lb}$ per acre per annum.

In earlier estimates, calculating on closely-planted crops and stems four to five feet, I was cautious to restrict my estimate to 750lb per acre, but five years' additional experience has shown me that with proper open cultivation 1,000lb per acre may be fairly assured.

16. I would now allude to the cost of growing and separating the fibre into a state fit for export. After a careful review of actual outlay, I estimate this as under :—

	R	a.	p.
Land rent per acre per annum . . .	10	0	0
Cultivation ¼ man per acre @ R5 per mensem .	15	0	0
Cutting and training stems, two men for three months @ R4 per mensem each . . .	24	0	0
Peeling and scraping, seven men @ R5 per mensem each	105	0	0
Native supervision @ R10 per mensem, for 50 acres, say	2	8	0
Cost of 950lb of fibre	156	8	=R369 per ton,

of which R247'5, or 67 per cent., has accrued in the preparation only of the fibre. This outturn has been obtained under the strictest supervision, and I do not think more could be obtained by native hand labour when doing daily work.

17. The best means of reducing the excessive cost of production have been, and are now being, earnestly sought for, and the result is anxiously awaited. Many anticipate that the separation of the fibre may be effected by mechanical means, others that the object may be obtained by chemical processes. Hitherto I think we have been led astray by our knowledge of the Chinese method of preparing the fibre. But, so far as I am informed, the Chinese do not use the fibre in a *spun* state, but that of divided filaments united into threads by manipulation peculiar to themselves. This process would be equally unsuitable and expensive in Europe as that of the first separation of the fibre has been shown to be. We want the fibre in a state in which it can be at once operated upon by machinery and reduced to yarn, and I am deeply impressed by the conviction that this may be accomplished without the aid of any expensive machinery and of the mechanical power requisite to work it.

(Conf. with p. 457.)

18. The plan of retting as applied to flax, hemp, sunn, jute, &c., is stated to have been in some localities successfully employed with Chinagrass. I have tried it in every manner at my command on the green and dried stem, as well as on the green and dried peel in running water and in stagnant, both cold and heated. The results have been uniformly unsuccessful. From the peel retted in still water, frequently changed, the fibre was cleanly separated and looked well ; but after rinsing and drying was worthless, being weak, dull and discoloured. In all other attempts the fibre itself decomposed equally with the resinous matter. I may add that I have succeeded in growing and retting flax here which has been valued in England at £6-5-0 per ton, so that my management in retting could not have been so very inaccurate as to have solely caused my failures in these attempts.

19. When proceeding to England in 1871, I took with me from the

Rhea in Kangra. *(G. Watt.)*	**RHEA.**

produce of this estate, dried stems, dried peel, and hand-cleaned fibre. All these I succeeded in getting experimented upon by manufacturers who had been accustomed to working the fibre. I may here remark that **Dr. Watson** refers to the fibre having been worked up by the aid of machinery used for the preparation of flax and wool. Mine was prepared by the machinery used for the utilising waste silk, and China-grass, in the state in which it is usually imported, goes through precisely the same process, stage by stage. The result of these operations showed clearly that both dried stem and peel could be operated upon, each giving a good clean fibre. My cleaned fibre suffered a loss of barely 9 per cent. in preparing it for the operation of the machines. **Dr. Watson** estimates the loss at 25 to 30 per cent. I can fully understand this after examining the specimens of Rhea and Rami I obtained in England. These I doubt not were roughly prepared in the manner described in that gentleman's report (page 37, column 2) where a bunch of the peel is tied by one end to a hook and a scrape on each side of each strip is supposed to finish the work. In this procedure a large amount of evaporation must have taken place before each strip of peel had been operated upon. In my procedure there was no opportunity for evaporation until the clean fibre was exposed to the air ; and the repeated scrapings on both sides of the ribband of peel, water being frequently applied during the process, must naturally have removed a much larger portion of the gum and resinous matter than the rude procedure stated.

20. With the knowledge at present attained it is evident that, however cleanly prepared, the fibre of China-grass has to undergo a manipulatory chemical process prior to machinery acting upon it. This process involves the use of heat, cheap chemicals, and appliances of small cost compared with machinery. I have already endeavoured to show that operating on the plant in its fresh state must be most profitable, inasmuch as under the present system the cost of carriage is reduced to far less than it would be by transport of the produce in any other form not yet known.

(*Conf. with p. 470.*)

21. As this chemical process is the first step enforced on the manufacturer, and by it the fibre loses a portion of its weight, it would evidently be most desirable that the process should be carried out by the cultivator, or in his immediate vicinity, who would thus save 10 to 30 per cent. in cost of transport, besides obtaining a better price for his produce. The results of the experiments made for me in England, showing that clean fibre could be extracted from the *dried peel*, without the aid of machinery, naturally forced upon me the conviction that a similar process would be equally effective on the fresh peel ; and as in the latter case the gum and resin would be in a liquid state, they would be far more readily acted upon than after they had been dried and concentrated ; therefore that weaker and consequently less expensive solutions would produce the desired effect. I have not had means at my command to procure appliances properly constructed for the purpose, but I have fully satisfied myself of the feasibility of my idea of procedure, and that it will dispense with all costly machinery in the preparation of the fibre in this country, unless it be desired to convert it into yarn, and then weave it, in which case a factory properly fitted must be established.

22. Many years back I recollect reading an account of an instrument or small machine which had been invented in America for the use of basket-makers, by the aid of which one man could peel as many osiers in a day as would formerly have employed a score. One or more instruments of this kind, according to the size of the plantation, would meet our first want, as peeling the China-grass stems even by an expert hand is a slow process. A properly constructed and fitted boiler in which to

RHEA.	Experimental Cultivation of

subject the peel to the action of the chemicals is the next requirement; and some suitable vessels in which to thoroughly wash the cleaned fibre would complete the necessary plant for the factory. The interest on the outlay for these, added to cost of chemicals used, would, I firmly believe, not amount to one-fourth of that of hand labour as at present, and be a small sum compared with the cost of machinery and engine-power to drive it.

23. I fear the above expression of my ideas will be considered very startling, and I should not have ventured at present to promulgate them had not this report been asked for by Government. I have now given my opinions, and, with due deference to those of the many clever men whose attention has been devoted to this subject, I believe they will be found worthy of consideration. I have spent twelve years and utterly exhausted my means in the persistent effort to firmly establish China-grass as an important product of this district, and I still trust that some other individual will benefit by my losses and succeed where I have failed from want of means to protract the struggle.

24. It was my earnest wish to have forwarded a specimen of my fibre prepared in the manner stated, but I have not been able to obtain the necessary materials. Should I do so shortly, a specimen shall be sent.

THE RAM BAGH,   JAMES MONTGOMERY.
    KANGRA;
The 24th February 1876.

CULTIVATION
in the
North-
Western
Provinces.
211

NORTH-WEST PROVINCES.

The correspondence and reports published in connection with the experiments to test the value of the apparatus and appliances submitted in competition for the Government reward of £5,000 constitute the most valuable papers that have hitherto appeared on the subject of Indian rhea. Space cannot, of course, be afforded to reprint the whole of these papers, but as the experiments were conducted at Saharanpur, the following selection of reports may be here given under the heading of the North-West Provinces. It will be understood, however, that as in the Panjáb so in these provinces the cultivation of the plant is purely an exotic industry and nowhere followed except under the fostering hand of European influence :—

Report by J. F. DUTHIE, Esq., *Superintendent, Botanical Gardens, Saharanpur, on the Cultivation of Rhea from June 1877 to November 1878.*

A short account of the history of Rhea cultivation at Saharanpur, from its commencement in 1870 up to June 1877, has already been given by the Director of Agriculture and Commerce, North-Western Provinces and Oudh, in a note No. 155 T. A., dated 27th June 1877.

The present report will, therefore, relate to the operations which have taken place since then, including certain experiments which I have undertaken in accordance with some suggestions of Dr. Forbes Watson's in his report published in 1875. The total nominal area in June 1877 was 32 acres, but a good deal of this land had got into very bad order, owing partly to irregular demand in previous years for supplies of stems, and mainly by reason of insufficient funds to keep up a large area in a high state of cultivation. In a letter No. 690 L. A., dated 13th May 1878, instructions were received from the Director of Agriculture and Commerce, North-Western Provinces and Oudh, to get ready 20 acres of land for the supply of Rhea stems in the autumn of 1879, to the competitors for the prize of R50,000 offered by the Government of India for the best fibre-cleaning machine.

For this purpose it was considered advisable, in order to secure a good uniform crop, to plough up those portions of land which had got out of condition, and to replant afresh during the succeeding rains. It was proposed, in the meantime, to make use of the ground by raising a crop, the profits from which it was hoped would help to defray the cost of removing and replanting the Rhea. In December last, the ground was, accordingly, sown with various kinds of melons, the fruit of which was sold by auction. Very unseasonable weather, however, occurred just at the time when the fruit was ripening, and the greater portion of the melon crops in this district were destroyed, so that the sum realized fell far short of what was expected. After this (27th May) the land was again well ploughed and prepared for Rhea, and the roots were put in during the following rains.

Planting.—The method of planting was as follows : The land after being well ploughed and thoroughly cleaned was divided into the usual longitudinal beds for irrigation purposes. Holes were dug 2½ feet apart, and these were filled in with manure composed of a mixture of horse-dung or of city manure, and the ashes of burnt Rhea leaves in the proportion of about 7½lb of the former and 1½lb of the latter, and the whole thoroughly mixed with a small quantity of the ordinary soil of the ground. Into this some healthy roots were planted, 6 to 10, according to their size. The beds were then kept constantly irrigated and carefully weeded, until the plants became high enough to prevent any further growth of weeds.

Cutting.—At Saharanpur it has been the practice to take only two crops during the year, one in June and the other in October or November. By this arrangement the stems are allowed to grow to their maximum degree of ripeness, any further delay causing injury, at the former period by the growth of lateral branches, and at the latter by the production of flowers. It is more than probable, however, that a much finer quality of fibre can be obtained from crops cut more frequently than twice during the year, and this I hope to be able to show from the results of experiments undertaken to ascertain this point. The first crop of this year was cut towards the end of June. From the newly-planted portion of ground, no stems were ready to be cut at this time. The second crop was cut during the latter half of October; this included about an acre and a half of the newly-planted ground. A few more cuttings can still be made from the more recently planted portions.

Two Crops. (*Conf. with pp. 472 and 483.*)

Great care has been taken in the preparation of the newly-planted ground, especially with regard to a uniform system of manuring and planting, so as to secure a good even crop at the time required. Experiments have been undertaken during the present year with the object of ascertaining the effects on the quality of the fibre of different methods of cultivation.

These relate chiefly to—(1) Monthly cuttings. (2) Spacing of plants. (3) Manuring.

Samples of fibre have also been prepared from stems from certain pieces of ground where the conditions of soil and former treatment were sufficiently marked to lead me to expect a corresponding variation in the condition of the fibre.

For the preparation of the various samples of fibre, I am entirely indebted to a gentleman at Agra, who has been good enough also to give me his opinion on the quality of each. The fibre was extracted in the majority of cases from dried stems, and all appear to have received the same treatment in the process.

RHEA.   Experimental Cultivation of

Statement shewing some results of experiments on fourteen samples of stems.

Number on samples.	Date of cutting.	WEIGHT. GREEN. Stems. ℔	oz.	Leaves. ℔	oz.	Dry stems. ℔	oz.	Average length of stems when cut. in.	Fibre extracted.	Conditions of soil and cultivation.
9	March 1878	...	...	...	...	...	...	25	28th October 1878	These are monthly cuttings taken from about one-tenth acre forming part of a small rich piece of ground somewhat shaded by trees. The plants have remained undisturbed here for upwards of seven or eight years, so that the roots have taken complete possession of the ground to the exclus on of weeds. The best crops are produced here, and the stems are usually free of knots and of uniform thickness.
1	June 1878	98	...	146	...	22	...	19	Ditto	
2	July 1878	54	...	106	...	17	...	20	Ditto	
3	August 1878	80	...	124	...	12	...	17	Ditto	
4	September 1878	114	...	192	...	14	...	21	Ditto	
8	October 1878	60	...	116	...	10	8	18	Ditto	
15	November 1878	18	...	32	...	2	8	0,	Ditto	
12	16th November 1878	...	...	...	...	...	...	13	23rd November 1878	From stems of 14 days' growth.
5	14th October 1878	...	...	...	...	1	1	10·24	28th October 1878	Stems from plants put in 1 foot apart. *(Planted about the end of February 1878.)*
6	Ditto	...	...	...	...	2	...	14·26	Ditto	Ditto. 2 feet apart.
7	Ditto	...	...	...	...	2	9	15·36	Ditto	Ditto. 3 feet apart.
11	June 1877	...	...	...	...	...	...	33	Ditto	From dried stems taken at random from 1st crop of last year.
10	November 1877	...	...	...	...	...	...	37	Ditto	ditto 2nd crop ditto.
13	16th November 1878	...	...	...	...	...	...	23	21st November 1878	Stems half ripe. *(From a piece of ground replanted in March 1878. The roots were put in 6 inches apart.)*
14	Ditto	...	...	...	...	...	...	29	Ditto	Stems fully ripe.

N.B.—The fibre of Nos. 12, 13, and 14 was worked up from green stems; all the other samples were prepared from dried stems.

J. F. DUTHIE, *Superintendent, Botanical Gardens, Saharanpur.*

CULTIVATION in the N. W. Provinces.

A short account by J. F. DUTHIE, Esq., *Superintendent, Botanical Gardens, Saharanpur, of each sample of fibre extracted from the stems enumerated in the preceding Statement.*

" *No. 1.*—The stems are smooth and free from knots. The boon' is brittle and easily broken. The harl is dry, tough, and tenacious, and does not break much when breaking out the boon. The outer bark separates freely from the fibre and the fibre is fine and soft, and of a uniform colour of whity-brown. It is slightly glossy.

" *No. 2.*—These stems are about the same length and thickness as No. 1. A sort of dry rot appears to have set in among them, and the boon crumbles on the breaking of it. The harl is not so tenacious as that of No. 1, and breaks very readily in passing through the machine. This causes an excess of boon waste. The fibre is not so fine or soft as No. 1, or even so tenacious. The colour is a sort of dirty whity brown. The crumbling of the boon renders its separation from the harl an easy task, but the agent that works the boon into this crumbling state works also on the harl and takes away its toughness, renders it more brittle, and easily broken. As a consequence, the fibre produced is shorter and there is more waste.

" *No. 3.*—These stems are not quite so thick as Nos. 1 and 2. They have something of the appearance of stem No. 1. The boon is brittle and breaks readily. It separates freely from the harl. The harl is not so brittle as No. 2, but resembles it a great deal. The fibre appears wanting in tenacity; it is stiff and rough to the touch. It is much the same colour as No. 2.

" *No. 4.*—These are nice clean-looking stems, and such as take the eye; they are not quite so thick as No. 1, but more uniform. They are lighter in colour than Nos. 1, 2, and 3. The boon is more solid than that of either Nos. 1, 2, and 3, and yet just as brittle, if not more so. It separates easily from the harl. The harl is tougher than either Nos. 1, 2, or 3, and is more tenacious and not so ready to break. There is not much boon waste. The adhesion between the bark and fibre is rather stronger, but separate readily notwithstanding. The fibre obtained is fine and soft, and almost full length. It has a nice soft feel and a glossy appearance, but has a slightly greenish tinge.

" *Nos. 5, 6, and 7.*—These stems are hard and knotty, and not easy to work, are very irregular in thickness, and are of a very dark colour and thick-skinned. The fibre obtained from them is rough and hard, and they produce a good deal of waste on account of their hardness and being full of knots. The harl breaks off at every knot.

" *No. 8.*—These stems are much more slender than any of the previous numbers, are clean and smooth. A little darker in colour than No. 4, but not so dark as Nos. 1, 2, and 3 The boon is brittle and breaks easily, and is easily separated from the harl. The harl is very tender and breaks with much less force than any of the other monthly cuttings. It requires to be very carefully operated upon until the bark is freed from it and it gets a working up. This working up increases the tenacity of the fibre considerably. The fibre is very fine and soft, has not that stiff, rough feel which characterises some of the other samples. Its colour is greenish, more so than No. 4, which may be owing to the freshness of the stems.

" *No. 9.*—These stems are regular in thickness, have a nice clean appearance, are smooth and of a light colour, much lighter than No. 4. The skin or bark is not so thick as on any of the other stems, but in this particular Nos. 1 and 8 approach very near. They are very easy stems to work up. The boon is brittle and the harl tough; it comes from the machine without being much broken, and with scarcely any waste. The

CULTIVATION in the N. W. Provinces.

bark separates very easily from the fibre. The fibre is very smooth and soft, and has not that stiff rough feel referred to as possessed by the other specimen. It is of a lighter colour.

" *No. 10.*—These stems do not appear to be so uniform in thickness as No. 11. They work up better than No. 11. The boon is solid and brittle, and the harl is more tenacious than No. 11. The bark separates readily from the fibre. The fibre is hard and rough, and appears to be stained by the juice of the stem in irregular drying.

" *No. 11.*—The stems are fairly regular, but inclined to be knotty. The boon crumbles in the breaking, and the same agent appears to have been at work here as described in No. 2. The harl is also affected. It is brittle and breaks across very readily, and more waste is produced. It is not difficult to remove the bark. The fibre is much shorter in comparison with the length of the stems than any of the samples, and is hard and rough. It is also of a darker colour. The stems are attacked by some kind of insect which bores a small round hole into the boon, causing a break in the fibres. The whole of the fibres which are cut by this insect tears off into waste in breaking up the stems and working the fibre out.

"*No. 12.*—These stems are very thin and are awkward to work up through being so short. The bark separates freely from the fibre, but the adhesion between the woody portion and the harl is rather strong. The fibre produced is soft and fine, and of a light colour, and separates readily into finer filaments.

" *No. 13.*—The woody portion of these stems was very soft and tough, the core of the woody-portion quite green and full of juice. The harl separates from the woody portion better than in No. 12. The bark separates very readily indeed from the fibre. The fibre obtained is soft and fine, but not so fine as No. 12. It is of a light colour, slightly greenish.

" *No. 14.*—The woody portion of these stems is hard, but not very brittle, yet not so tough as No. 13, and the core is more or less dry. There is more dryness between the harl and boon, and more adhesions of the one to the other, the juice having dried to a great extent and fixed the fibre to the boon or woody portion, which is still in a fresh juicy state. The bark does not separate readily from the fibre. The fibre is not so soft as either No. 12 or 13 ; it is more stiff, hard, and brittle; it is just as fine as No. 13.

" I will now state briefly some conclusions which may be drawn from the above experiments. It is clear from the habit of growth of the Rhea plant that whatever distance apart the roots or plants are originally put in, they will, unless purposely prevented, gradually approach each other and fill up the interspaces. I have observed that in proportion as these conditions are allowed to act the better and more uniform does the crop become, and, what is of more importance, the quality of the fibre is also improved.

" It is again evident from observation that the greater the spaces are between the plants the more room there is for the growth of weeds, which spring up with great rapidity in a soil which has to be so liberally supplied with water and manure. In those pieces of ground where the plants have had time to spread over the whole ground, few weeds can exist. Another advantage in a compact crop, and especially in a dry climate, is the shade afforded by the leaves over the entire ground. During the hot weather a certain number of leaves are continually falling, and these also contribute to moisten and manure the ground—conditions which are particularly favourable to a shade-loving plant like the Rhea.

" I believe, therefore, that in planting out Rhea, good crops will be more quickly obtained and with greater economy by putting in the roots or plants as close as possible in lines, as far apart only as is necessary for weeding and hoeing purposes, until the plants have strength enough to maintain their own ground."

OUDH.

The following extracts from the Proceedings of the Government of India give the main facts which have been learned up to date regarding rhea cultivation in this province:—

Report by the Officiating Inspector General of Prisons, Oudh, on the Cultivation of Rhea (Bœhmeria nivea) in the Jails of Oudh.

In paragraphs 4 and 5 of letter No. 57, the Secretary of State for India calls for information on the following points:—

(A) The measures which seem best adapted for the purpose of conducting experiments on Rhea fibre.

(B) The cultivation of the Rhea plant, precise information being required on—

    (1) The variations in quantity and quality due to difference in—

(*a*) Season.	(*d*) Soil.
(*b*) Growth.	(*e*) System of cultivation.
(*c*) Age.	(*f*) &c., &c.

    (2) The most economical system of cultivation generally.

2. The Rhea plant has been cultivated to a small extent in nearly all the Oudh Jails for some years past,—in one or two for a period of eight or nine years; and, with the view of obtaining such information as might be procurable from the officers in charge of jails, a series of questions was submitted to them by **Dr. Sutherland**, to which categorical replies were required.

These questions embraced the following points bearing on the cultivation of the plant, the outturn, and the methods actually employed in the Jails to separate the fibre:—

    I.—(*a*) The soil, and (*b*) the manure best suited to the plant, and (*c*) the number of years during which the plants will produce freely.

    II.—The amount of irrigation required in hot weather.

    III.—The effect of frost, whether destructive or not.

    IV.—The best interval to be left between the plants to ensure straight growth.

    V.—(*a*) The number of crops 4 feet high to be obtained each year after abundant irrigation and manuring.

    (*b*) The seasons for cutting the crops.

    (*c*) The yield per acre.

    VI.—(*a*) The time, and (*b*) the method of separating the bark from the fibre and the fibre from the wood (1) in the fresh green, and (2) in the dried state.

    VII.—The (*a*) object and (*b*) duration of the retting process if employed; whether it is used with (*c*) green or (*d*) dried stems; (*e*) whether it can be carried out in the rains, and (*f*) whether chemicals are required.

    VIII.—Method of removing gum if not steeped.

    IX.—Specimens and description of fibre as it can be best produced, and the rate at which it can be offered per maund at the jail.

3. These questions by no means embrace the whole points on which information is sought, as they leave out of sight those embodied in it as being beyond the experience of Superintendents of Jails, and do not touch at all on many of the most important in connection with relations of quality and quantity to season, soil, age of plant, &c.

RHEA.	Experimental Cultivation of

CULTIVATION in Oudh.

Exact information could scarcely be looked for from jail establishments, as the cultivation has in no case been in any way extensive, the largest area placed under Rhea cultivation in any one jail probably not exceeding 1½ to 2 acres, and no experiments have been systematically made on the influence of soil, amount of irrigation, &c., on the quality and quantity of the outturn of fibre.

4. The conclusion to be derived from the reports submitted is that attention has not been directed to those points, the plants having been planted and cared for in a way, no very great attention having been paid to the cultivation, at any rate in most instances, as the fibre, when extricated, with the rude methods at command, would never be remunerative even when the day's wage of a prisoner was placed at so miserable a sum as 6 pie.

5. The information obtained from the Superintendents of the different jails, such as it is, is embodied in the following paragraphs.

Replies to the circular have been obtained from all the Jail Superintendents, the last being received on 11th March 1876.

6. (*a*) *Season.*—The information on this point is virtually *nil*, as regards its relation to quantity and quality, **Dr. McReddie** only mentioning that the autumn produce after the cessation of the rains is the most abundant, the stems then growing to a much greater height.

The number of crops that can be obtained in a year with free irrigation and manuring is three after the rains, in early spring and in June; but on this point there is no agreement between the various reporters.

Three crops are obtained—

Crops (Conf. with pp. 449-477.)

In Partabgarh .	.	.	February, June, and October.
„ Fyzabad .	.	.	March, June, and October.

Two crops—

In Lucknow	.	.	June and October.
„ Hardoi	.	.	February and September.
„ Gonda	.	.	April and October.

One crop—

In Bahraich	.	.	November.
„ Sultanpore .	.	.	September.

In the other two jails and even in the two last it is cut irregularly and when required.

The only information submitted on the total outturn relates to the whole year's produce, and is mentioned in but three reports.

The yield per acre per annum in these is given below : —

	Green stem.	Rough uncarded fibre.
Lucknow .	. 160 maunds	8 maunds.
Partabgarh .	. ...	1 maund.
Hardoi .	. 10 maunds	...

The difference in these estimates is certainly remarkable, and I am inclined to place no reliance at all on those of Partabgarh or Hardoi. In the absence of reliable information as to soil, manuring, &c., it is impossible to guess even at the cause of this discrepancy.

7. (*b*) The relation of quantity and quality to the period of growth must remain unanswered. **Dr. McReddie** states that the fibre is much finer and more delicate when obtained from the young stem.

8. (*c*) As regards age, the only information given is that the first plants, planted some eight or nine years ago, still produce freely; but there is no evidence as to whether the older plants are deteriorating or not.

R. 212

CULTIVATION in Oudh.

9. (*d*) Nothing absolute can be learnt as to the nature of the soil most suited to the successful cultivation of the Rhea, as it appears to thrive in all kinds so long as water and manures are freely supplied, whether it be clay as at Lucknow, or sand, with only a small portion of clay, as at Partabgarh, or light and sandy as at Hardoi, or *domatti* (sand and clay mixed) as at Fyzabad.

10. (*e*) System of cultivation.

All are agreed that free manuring is necessary, but not as to the nature of the manure, some preferring vegetable and some animal.

At Lucknow the ground was at first starting very freely manured with the ordinary jail manure, but has subsequently had nothing more than that obtained from the leaves of the plants themselves, which are stripped and left on the spot after the plant has been cut. Certainly, the plants thrive on this, and it is by far the most economical plan.

The plants require to be placed on raised ridges, for though requiring and using a considerable amount of water, they are liable to be injured if water is allowed to lodge at the roots. With reference to the amount of irrigation required opinions differ. In Fyzabad I gather that it is carried on all the year round except during the rains, and this I believe to be necessary. The question, however, was only put on this point as to the requirement during the hot weather, and on this opinions differ in a very marked degree, from two or three times in the month as at Sitapur, Lucknow, and Hardoi, to twice or even three times a week, as at Unao, Partabgarh, Gonda, and Fyzabad. This is a matter of considerable importance in connection with the economical cultivation of the Rhea, and I am induced to think the Lucknow estimate the correct one.

11. Another point for consideration is the best distance apart at which to place the plants. In the Saharanpur garden two feet is the distance found necessary to ensure free and straight growth without wasting space, and this is the standard most generally accepted by Jail Superintendents, though one or more consider three or four feet as necessary. There is nothing to show, however, what influence the greater or less distance has on the outturn either in fineness of fibre or amount.

12. No remarks are made on the influence of shade on the growth, though it is my impression that this is supposed to proceed more satisfactorily when there is a fair amount of shade.

13. The remainder of the enquiries were addressed to the processes in use to obtain the fibre from the stem, the use of chemicals in retting, and the price at which it can be produced.

The method employed for separating the fibres is so primitive, and the instruments so rough, and the process so slow, that it is scarcely needful to enter into consideration of them, especially as the whole question has arisen out of the desire to find some machine that will perform the first process rapidly and economically so as to make the production pay. This hand labour never can do, and hand labour alone is available. Dr. McReddie estimates that cleaned fibre will cost R7-8 per maund at the jail. One prisoner cannot, he considers, prepare more than 4 chittacks a day, calculating from the number employed during the whole process from first to last and the amount of outturn; this, at 6 pie per diem per man as wage, would give R5 per maund. To this must be added R2-8 cost of cultivation, making in all R7-8 per maund at the jail or £28 per ton. Dr. Forbes Watson gives the price to be paid in London at £30 to £40 per ton for the better, and £20 to £25 per ton for the lower qualities, so that the above rate for the uncombed fibre would leave no margin for profit after freight, &c., has been paid, supposing even £30 per ton were obtained in London.

Price £28 a ton.
213
(*Conf. with pp. 450, 463, 470.*)

CULTIVATION
in Oudh.

The total outturn also would be very small, as Dr. McReddie considers that half a ton per annum would be fully as much as he could produce. Other estimates of the cost per maund have been given, but being little more than mere guesses, need not be quoted.

Other questions raised with reference to the preparation of the Rhea fibre is the need of retting for the removal of gum and the use of chemicals. The fullest description of the process comes from the Lucknow jail. Dr. McReddie, who seems to have paid attention to the subject, expresses his opinion that the process should commence immediately after the stems are cut, as any delay increases the amount of gum found in the fibrous layer and the difficulty of separation and clearing the fibre. He considers that it should not be left one night even. If this precaution is attended to, the amount of gum in the fibre is small and easily removed during the subsequent retting process.

The first step is to clean off the outer epidermal layer by scraping with sharp-edged pieces of bamboo and by rubbing with gunny, and then to remove the fibre in strips with the fingers. This process is never undertaken in the dry state.

The next state is that of retting, by which fibres are separated, and any remaining gum removed.

A solution of impure carbonate of soda (*sajjí*) is prepared, two ounces being added to a gallon of water and boiled down till two-thirds remain. The clear portion after standing is decanted off, and about five seers of the raw fibre are placed in it, sufficient water being added to cover the fibre completely.

It is allowed to remain thus from 7 to 20 days according to season, and the fibre is then removed from the water, well washed, and dried. The washing is continued until the fibre is perfectly clean and bleached.

For spinning, the fine fibre is prepared from this by combing or by hand-picking.

In some jails the stem is allowed to partially or wholly dry, and then retted, *sajjí* or alum being used to assist in the removal of impurities.

15. In conclusion, I must express my regret at the absence of clear and definite information on the points in question, which arises wholly from the fact, I presume, that the attention of those superintending the cultivation of Rhea has not been directed to them, at any rate, in a constant and prominent manner. Jail Superintendents as such are naturally liable to treat any employment of prisoners that is not remunerative with callousness. Should a further series of experiments be considered advisable in jails, I would suggest that they should be confined to three or four jails, and be carried out in them on a more extended scale and in a systematic manner."

There would thus appear to be no spontaneous rhea cultivation in Oudh. In one or two works on India, published in Europe, it is stated that rhea is grown in the Tarai and in Nepál. It is believed the plant there alluded to is *Poi-rhea* (Maoutia Puya, which see) and not a form of Bœhmeria.

(*W. R. Clark.*)

214

RHEUM, *Linn.; Gen. Pl., III., 100.*

A genus of stout herbs, with large woody roots, comprising about twenty species, all of them indigenous to Central Asia and the Himálaya. In the *Flora of British India*, Sir J. D. Hooker describes seven species as natives of the Himálayan region, all of which are used medicinally in the districts where they occur, but it is very doubtful if any of those that occur in British India or its tributary states are the source of the true Turkey, Russian or Chinese Rhubarb of commerce. This latter is said to be the dried root stock of **Rheum**

Products of India. 485

The Rhubarb of Commerce. (*G. Watt.*) | RHEUM emodi.

officinale, *Baillon,* R. palmatum, *Linn.,* and probably of several other species indigenous to South Eastern Tibet and Western and North-Western China. With reference to the origin of the Rhubarb of commerce Flückiger & Hanbury write as follows :—" No competent observer, as far as we know, has ever ascertained, as an eye witness, the species of Rheum, which affords the commercial rhubarb. Rheum officinale, from which it appears at least partly derived, is the only species yielding a root stock, which agrees with the drug." "The plant was discovered in South-Eastern Tibet, where it is said to be often cultivated for the sake of its medicinal root, but it is supposed to grow in various parts of Western and North-Western China whence the supplies of rhubarb are derived. It was obtained by the French Missionaries about the year 1867 for Dabry, French Consul at Hankon, who transmitted specimens to Dr. Sonbeiran of Paris. From one of these, which flowered at Montmorency in 1871, a botanical description was drawn up by Baillon. To what extent the rhubarb of commerce is derived from this plant is unknown. But that the latter may be a true source of the drug is supported by the fact that there is at least no important discrepancy between it and the accounts and figures, scanty and imperfect though they are, given by Chinese authors and by the old Jesuit Missionaries, and still more by the agreement in structure which exists between its root and the Asiatic rhubarb of commerce." "R. palmatum, *Linn.,* a species known as long as 1750, has always been supposed to yield rhubarb also and this has again been asserted by the Russian Colonel Przelvalski, who observed, in 1872 and 1873, that plant in the Alpine parts of Tangat, round the lake Kuku-nor, in the Chinese province of Kansu in 36°-38° North Lat. Rheum palmatum has been frequently cultivated in Russian Asia, and in many parts of Europe since the last century, but without producing a root agreeing with Chinese rhubarb. Specimens of the root were largely brought to St. Petersburgh by Przelvalski, but Dragendorff expressly points out in his Jahresbericht for 1877 (p. 78) that it is dissimilar to true rhubarb."

In England, at the village of Bodicott near Banbury, R. Rhaponticum is now extensively cultivated, and is the source of the "Banbury" rhubarb of commerce. When well prepared this is of similar appearance and size to China rhubarb. It is less bitter, but more mucilaginous and astringent, and the root is of a more spongy, soft and brittle texture and is chiefly sold for exportation in the state of powder.

None of the Himálayan rhubarbs are of much commercial importance, the rhubarb sold in Indian bazárs being usually imported from London. It is principally derived from English grown rhubarb, which, on account of its low price, is most in demand in India. "None of the commercial rhubarb known as Chinese or East Indian is imported into Bombay unless specially ordered from China, but it often passes through the port on board the P. and O. Company's steamers " (*Dymock*).

Rheum emodi, *Wall. ; Fl. Br. Ind., V., 56 ;* POLYGONACEÆ.   215

Syn.—R. EMODIUM, *Wall. ;* R. AUSTRALE, *Don. ;* " Don considered R. emodi, *Wall.* (R. australe, *Don.*), a Himálayan and Central Asiatic species to be the chief source of the rhubarb of commerce. It is grown for Rhubarb in Silesia and is a very distinct species " (*Bentley & Trimen*). In India this species, together with R. Moorcroftianum, R. spiciforme, and R. Webbianum are the chief sources of Himálayan rhubarb. All appear to be used medicinally by the natives of the regions where they occur, and the different species seem to be very insufficiently distinguished, both in their vernacular names and in the literature on the subject of Rhubarb.

Vern.—*Hindi-révand chini, dolu,* HIND. ; *Banglá-révan-chini,* BENG. ; *Padam-chal,* NEPAL ; *Archu,* GARHWAL ; *Lachú,* LADAK & SPITI ; *Chutiál, chotiál, chúchi, chúkri, khabiún, kandául, lachú, pambash, átsú, artso, arts ;* (Stalks =) *ribas,* (Root =) *rewand chini,* PB. ; *Rawásh, chúkri,* AFG. ; *Ládaki-révanda-chini,* BOMB. ; *Mulká-cha-révalchinni,* MAR. ; *Gamni-révanchini,* GUZ. ; *Náttu-iréval-chinni, náttu-manjat-chinaá-kishangu,* TAM. ; *Nát-ki-révanchini,* DECCAN ; *Náttu-réval*

RHEUM Moorcroitianum.

The Himalayan Rhubarb.

chinni, náttu-pasupu-china-gadda, TEL.; *Nat-révá-chinni*, KAN.; *Rávande-hindí*, ARAB.; *Révande-hindí*, PERS.

Except in the districts where indigenous plants occur, it will be observed that all the vernacular names for Indian rhubarb are merely modifications of *revand* or *revand-chini*, the name of the imported article.

References —*Stewart, Pb. Pl., 186; Pharm. Ind., 187; O'Shaughnessy, Beng. Dispens., 518; Moodeen Sheriff, Supp. Pharm. Ind., 213; Fleming, Med. Pl. and Drugs, as in As. Res., Vol. XI., 188; Flück. & Hanb., Pharmacog., 502; Bent. & Trim., Med. Pl., III., 215; S. Arjun, Bomb. Drugs, 114; Year-Book Pharm., 1874, 95; 1876, 243; Baden-Powell, Pb. Pr., 371; Atkinson, Him. Dist., 748; Royle, Ill. Him. Bot., 315, 317; Watt, Ec. Prods., II., 52; V., 238; VI., 155; Pereira, Mat. Med., II., 485; Gazetteer:—N.-W. P. (Himálayan Districts), X., 316; Ind. Forester, XIV., 370.*

Habitat.—An herb with very stout stem, 5-6 feet high, found in the Sub-Alpine and Alpine Himálaya of Nepál and Sikkim at altitudes between 11,000 and 12,000 feet. This species was described by **Edgeworth** as occurring on the Chór, a mountain near Simla, although his figure has the larger flowers of **R. Webbianum** but the habit and glabrous panicle of **R. emodi.** It is described by **Aitchison** as indigenous to Lahoul.

Dye.—Moorcroft states that the Bhutias of Garhwal use the root of this species with madder and potash for dyeing red. "The colour would probably be derived from the madder and the rhubarb is most likely only an auxiliary" (*Watt*).

Medicine.—The ROOTS of this species constitute, according to the *Pharmacopeia of India,* the "large" variety of Indian or Himálayan Rhubarb. "It occurs in twisted or cylindrical pieces of various sizes and shapes, furrowed; cut obliquely at the extremities, about 4 inches long and an inch and a half in diameter; of a dark brown colour, feeble rhubarb odour, and bitter astringent taste; texture radiated rather spongy, not presenting on fracture the marbled texture characteristic of ordinary rhubarb; pulverized with difficulty; powder of a dull brownish-yellow colour.

Himálayan rhubarb is used as a purgative and astringent tonic in place of the imported article. It is less active, and has often been pronounced worthless **Dr. Cleghorn** (*Madras, Quart. Med. Journ*, 1862, V., 464) states, however, that only an inferior variety reaches the plains of Hindustan, and that he tested the action of the fresh root and found it to resemble that of Russian rhubarb. He, therefore, suggested that if the plant were cultivated with due care, a good serviceable drug equal to Chinese or Turkish rhubarb might be obtained from the Himálayan species (*Pharm. Ind.*). **Dr. Aitchison** in his *Flora of Lahoul* states that "the Natives of that region recognize no medicinal quality in this plant, but frequently eat the stems raw which are agreeably acrid to the taste and refreshing when one is tired." Although not recognised by the natives as a purgative, the young leaves and stalks, when used by Europeans as a salad, had distinctly purgative effects.

Food.—The STALKS are eaten by the Natives either boiled or in their natural state pounded and mixed with salt and pepper; they are also dried stored, and eaten with other food and sometimes are made into preserves Dr. Watt cooked and ate the stems of the wild plant in Lahoul, but he records that he found them to act as a powerful purgative.

Trade.—The average annual export of **R. emodi** from the Kumáon forest division is said to be about 1,000℔ (*Atkinson*).

Rheum Moorcroftianum, *Royle; Fl. Br. Ind., V., 56*

Syn.—Sir J. D. Hooker, with reference to this species, writes:—"I am uncertain about this plant, which differs from **R. spiciforme** in the very much larger pubescent peduncles and racemes which together are two feet long, and in the form of the fruit. The only specimens ar **Wallich's**

Products of India. 487

The Medicinal Rhubarb. (*W. R. Clark.*) **RHEUM spiciforme.**

are very bad, and have neither locality nor collector's name. They are ticketed ' large broad-leaved, small stalked rhubarb, the root more purgative than the long stalked.' Another sheet has attached to it a ticket in the same handwriting ' narrow round-leaved, long-stalked rhubarb.' R. **Moorcroftianum** is written in pencil on the sheets, I think, by Royle (certainly not by Wallich). Hence they are no doubt the plants mentioned by Royle (*Ill. Him. Bot.*, 315) as brought by Moorcroft from Niti, altitude 12,000 feet in Kumáon, and of which Royle says that.Major **Hearsay**, Moorcroft's companion, has described two kinds to me, one round-leaved and short-stalked, and the other short-stalked, but large and broad-leaved. (R. **Moorcroftianum**, *Nob.*) with the root more purgative than that of the former.' From this it appears that Royle, not Wallich, as hitherto supposed, is the author of R. **Moorcroftianum**, and that **Meissner** is further in error, in describing it as everywhere glabrous " (*Fl. Br. Ind.*)

Vern.—For vernacular names, see R. **emodi**. No distinction appears to be made by the natives between the two species.

References.—*Royle, Ill., 315, 318 ; Stewart, Pb. Pl., 186 ; Baden-Powell, Pb. Pr., 369, 371 ; Liotard, Rep. on Ind. Dyes, 96.*

Habitat.—This species is said to occur on the Western Himálaya, principally about Kumáon.

Dye.—" The ROOTS of R. **Moorcroftianum** known in Sialkote as *rewand*, are powdered and steeped in cold water for two days. The decoction is then boiled and woollen stuff immersed in it, while still boiling. The result is a fleeting yellow, which can be deepened by the addition of turmeric to the dye " (*Liotard*).

DYE.
Roots.
221

Medicine.—Same properties as R. **emodi**.

MEDICINE.
222
223

Rheum nobile, *Hookf. and Thoms. ; Fl. Br. Ind., V., 57.*

Vern.—*Tchuka*, SIKKIM.

References.—*Hooker, Him. Journals, II., 58 ; Hooker f. & Thoms., Ill. Him. Pl., t. 19 ; DC., Prod., XIV., 1, 136.*

Habitat.—A handsome herbaceous plant, with stem 3-4 feet high, and as thick as the wrist below, found in the inner ranges of the Sikkim Himálaya, at altitudes between 13,000 and 15,000 feet.

Medicine.—The ROOT resembles that of medicinal rhubarb, but is spongy and inert (*Hooker*).

MEDICINE.
Root.
224

Food.—The acid STEMS are eaten both raw and boiled.

FOOD.
Stems.
225

Domestic Use.—The dried leaves afford a substitute for tobacco; a smaller kind of rhubarb is, however, more commonly used in Tibet, for this purpose; it is called *chula*.

DOMESTIC.
226
227

R. spiciforme, *Royle ; Fl. Br. Ind., V., 55.*

Syn.—R. MOORCROFTIANUM, *Meissn. not of Royle.*

Vern.—For vernacular names and medicinal properties, *see* Rheum emodi.

References.—*DC., Prod., XIV., 1, 36 ; Stewart, Pb. Pl., 186 ; O'Shaughnessy, Beng. Dispens., 519 ; Royle, Ill. Him. Bot., 318, t. 78.*

Habitat.—This species' is found on the drier ranges of the Western Himálaya from Kumáon (altitude 14,000 to 16,000 feet) to Western Thibet (altitude 9,000 to 14,000 feet) and is distributed to Afghánistán.

Food.—" In Afghánistán, the plant is always wild, and appears to grow abundantly in many parts. When green, the leaf-stalks are *rawash*, and when blanched by heaping up stones and gravel round them, they are called *chukri*; when fresh (in which state they are sometimes brought to Peshawar in spring) they are eaten either raw or cooked. They are also dried for use, to be eaten with other food, and are sometimes made into a preserve " (*Stewart*).

FOOD.
228

| RHINACANTHUS communis. | Tong-pang-chong, a cure for Ringworm. |

229

Rheum Webbianum, *Royle; Fl. Br. Ind., V., 57.*

Vern.—For the vernacular names of this species, *see* those of R. emodi.

References.—*Royle, Ill. Him. Bot., 318, t. 17a; Pharm. Ind., 187; Bent. & Trim., Med. Pl., 215.*

Habitat.—This plant, which is very difficult to distinguish from R. emodi, occurs on the Central and Western Alpine Himálaya from Nepál to Kashmír, at altitudes between 10,000 and 14,000 feet.

MEDICINE.
230

Medicine.—This constitutes, according to the *Pharmacopœia of India,* the small variety of Himálayan Rhubarb which "consists of short transverse segments of the root-branches; of a dark brownish colour, odourless or nearly so with a very bitter astringent taste."

RHINACANTHUS, *Nees.; Gen. Pl., II., 1112.*

231

Rhinacanthus communis, *Nees.; Fl. Br. Ind., IV., 541; Wight,* [*Ill., 164, bis., fig. 9 & Ic., t. 464; ACANTHACEÆ.*

TONG-PANG-CHONG.

Syn.—R. ROTTLERIANUS, *Nees.;* JUSTICIA NASUTA, *Linn.;* J. ROTTLERIANA, *Wall.*

Vern—*Palik-juhia, pálak-juhi, Jui-pani,* HIND.; *Jui-pana,* BENG.; *Pul-colli, Pushuk-kolli, pushpa-kedal, nagamallich-cheti,* MAL. (S.P.); *Gach-karan,* BOMB.; *Gajakarni,* MAR.; *Kabutar-ka-jhár,* DEC.; *Naga-malli,* TAM.; *Nargamollay, nága malle,* TEL.; *Nága-mallige,* KAN.; *Anitia,* BURM.; *Yuthika-purni,* SANS.

References.—*DC., Prod., XI., 442; Roxb., Fl. Ind., Ed. C.B.C., 40; Dals. & Gibs., Bomb. Fl., 194; Grah, Cat. Bomb. Pl., 164; Sir W. Elliot, Fl. Andh., 122; Rheede, Hort. Mal., IX., 135, t. 69; O'Shaugh-nessy, Beng. Dispens., 481; S. Arjun, Cat. Bomb. Drugs, 107; Dymock, Mat. Med. W. Ind., 2nd Ed., 588, 589; Year-Book Pharm., 1881, 197; Drury, U. Pl. Ind., 363; Royle, Ill. Him. Bot., 298; New Com. Pl. & Drugs, IV., 46; V., 75; VI., 102; VII., 73; Gazetteers:—Bombay, VI., 15; XV., 440; Mysore and Coorg, I., 64; Ind. Forester, XI., 231; Lancet (1881), 887; (1882), 78; Balfour, Cyclop. Ind., III., 405; Smith, Ec. Dict., 284.*

Habitat.—A shrub 4-5 feet high, met with in a state of cultivation throughout India, and doubtfully wild in the Deccan Peninsula and Ceylon. It is cultivated in the Straits Settlements, Java, and Madagascar.

MEDICINE.
Root.
232
Leaves.
233

Medicine.—Used for various medicinal purposes by the natives of India. Milk, in which the ROOT has been boiled, is considered aphrodisiac; the fresh root and LEAVES, bruised and mixed with lime juice and pepper, are employed as an external application in eczema and ringworm, especially in cases of the variety of that disease known in India as "Dhobie's Itch" (*Tinea circinata tropica*). The roots are believed in some parts of India to be an antidote to the bites of poisonous snakes, hence the name *nagamulli* (cobra root). Of late, under the name of "Tong-pang-chong," it seems to have attracted considerable attention in Europe, on account of its reputed value in the treatment of ringworm, but the results obtained by its use are very contradictory, some practitioners asserting that it is utterly useless, while others declared that it is most efficacious. It seems, however, to be universally used with good results in cases of *Tinea circinata tropica,* although its utility in ordinary ringworm (*Tinea tonsurans*) seems very doubtful. Dr. Liborius analysed the root in his laboratory at Dorpaat, and found that it contained a substance which he called *rhinacanthin,* and which resembled *chrysophanic* and *frangulic* acids in its antiseptic and antiparasitic properties.

R. 233

(*John Watt.*)

234

RHINOCEROS, *Linn.; Jerdon, Mam. Ind., 232.*

This mammalian belongs to the great order UNGULATA (or hoofed animals), in the sub-order of **Perissodactyla**, Family RHINOCEROTIDÆ. The horse and tapir are its nearest allies.

CHARACTERS.—The most striking peculiarity of this animal is the horn or horns on the nose (ρινοκερως). It is terrestrial in its habits, and feeds on vegetables. Its limbs are adapted for progression and not for prehension. It has no collar-bones, and the digits on each foot are three in number and each terminates in a small hoof-like nail. The third or middle digit is much more developed than the others, and has its two sides similarly formed. The thigh bone bears a third trochanter. The molar and premolar teeth are similar and. form a continuous series. The bones of the foreleg are well developed and remain distinct. The head is large, the eyes small, and the ears moderate. The horns grow throughout the life of the animal, but are reproduced if lost. The skin is thick, and thrown into deep folds in certain places; it is also thinly clad with hair. The Rhinoceros is, as a rule, a quiet inoffensive animal.

The classification adopted here, for the various species, is that given by **Mr. W. T. Blanford** in the recently published volume of *Fauna of British India.* The species of Rhinooeros are grouped into two main sections, *viz.* one-horned, and two-horned. The former is represented in Iadia by two, the latter by one, species. In Geological formations, how ever, a larger number of forms are represented than at the present time.

235

Rhinoceros sondaicus, *Cuv.; Fauna, Brit. Ind., I., 474.*

THE SMALLER ONE-HORNED RHINOCEROS.

Syn.—R. JAVNICUS, *Cuv.;* R. INRRMIS, *Lessow.;* R. NASALIS, *Gray.*
Vern.—*Gondu,* BENG.; *Gainda,* HIND.; *Kunda, kedi, kweda,* NAGA; *Kajeng, kyjantsheng,* BURM.; *Bádák,* MALAY.
References.—*Blyth, Mam. Burm., 50; Sterndale, Mam. Ind., 410.; Mason, Burma and Its People, 166, 669; Balfour, Cyclop. Ind., III., 406; Encycl. Brit., xx., 522.*
Habitat.—The Lesser Rhinoceros is found in the Sunderbands and parts of Eastern Bengal. It occurs also abundantly in Burma and thence throughout the Malay Peninsula to Java and Borneo. **Kinloch** shot it also in Sikkim Tarai. It is distinguished by the fold in front of the shoulder, being continued over the back of the neck; the skin of the sides is also divided into small polygonal scales.

236

R. sumatrensis, *Cuv.; Flora, Brit. Ind., I., 476.*

THE ASIATIC TWO-HORNED RHINOCEROS.

Syn.—RHINOCEROS SUMATRANUS, *Raffles;* R. CROSSEI, *Gray;* R. LASIO TIS, *Sclater.;* CERATORHINUS CROSSEI and C. SUMATCENSIS, *Blyth* C. SUMATRANUS, C. NIGER, C. CROSSEI, and C. BLYTHII, *Gray.*
Vern.—*Kyaw, kyaw-shaw,* BURM.; *Bádák,* MALAY.
References.—*Blyth, Mam. Burma, 52; Sterndale, Mam. Ind., 412; Mason, Burma and Its People, 167, 669; Balfour, Cyclop. Ind., III., 406; Encycl. Brit., XX., 522.*
Habitat.—This species is rare in Assam; but from that province it ranges to Siam, the Malay Peninsula, Sumatra, and Borneo.

237

R. unicornis, *Linn.; Fauna Brit. Ind., I.; 479.*

THE GREAT ONE-HORNED RHINOCEROS.

Syn.—R. INDICUS, *Cuv.;* R. STENOCEPHALUS, *Gray.*
Vern.—*Gainda, gargadan,* HIND.; *Karkadan,* PERS.; *Gonda,* BENG.; *Gor,* ASSAM; *Kyan-hsen,* BURM.; *Khadga, kadgin, gandaka,* SANS.

RHINOCEROS
unicornis.   The Rhinoceros.

References.—*Blyth, Mam. Burma, 51; Sterndale, Mam. Ind., 407; Smith, Chinese Mat. Med., 185; Buchanan-Hamilton, Acc. of Kingdom of Nepal, 63; Ayeen Akbery (Gladwin's transl.), II., 96; Balfour, Cyclop. Ind., III., 406; Encycl., Brit., XX., 522.*

Habitat.—This species at the present day is almost restricted to the Assam Valley. Formerly it was extensively distributed in the Indian Peninsula. It is chiefly characterized by the fold in front of the shoulders not being continued over the back of the neck, and by the skin of the sides bearing tubercles.

From an economic point of view the various species of Rhinoceros scarcely need to be separately considered. The following facts may, there-fore, be given of all three species collectively :—

MEDICINE
Horn.
238

Medicine.—In Indian medicine, the Rhinoceros was formerly much esteemed, and even now we find cups of Rhinoceros HORN valued in many parts of the country as tests for poisons. In **Linschoten's** *Voyage to the East Indies*, published in Holland in 1590, there occurs the following account of the medicinal uses of the Rhinoceros:—"Their hornes are much esteemed, and used against all venime, poyson, and many other diseases; likewise his teeth, clawes, flesh, skin and blood and his very dung, and water and all whatsoever is about him, is much esteemed in India, and used for the curing of many diseases and sicknesses, which is very good and most true, as I my selfe by experience have found ; but it is to be under-stood, that all Rhinocerotes are not alike good, for there are some whose hornes are sold for one, two or three hundred *Pardawes* the peece, and there are others of the same colour and greatnes that are sold but for three or fours *Pardawes*, which the Indians knowe and can discerne. The cause is that some Rhinocerotes which are found in certaine places in the countrie of Bengala have this virtue, by reason of the hearbes which that place only yeeldeth and bringeth forth which in other places is not so and this estima-tion is not onely held of the horne, but of all other thinges in his whole bodey, as I saide before."

Flesh.
239

The same belief in the medicinal value of the various parts of the Rhi-noceros seems to have held good in India even in later days than those of **Linschoten**, for in the *Taleef Shereef* (translated by **Playfair** in 1833) we find the use of Rhinoceros' FLESH recommended in " disorders of the wind " and for the purpose of " decreasing the urine and fæces," the smoke from the burning horn is pronounced excellent for the cure of piles and for producing easy labour, and drinking from a cup made of the horn there stated to be efficacious for the cure of piles. **Dr. Hamilton** also, in his manuscripts written in Berar and quoted by **Ainslie** (*Mat. Ind., II., 480*) states that the Hindu physicians, even of his time, considered the flesh of the Rhino-ceros medicinal, and ordered it boiled and in combination with *ghí* in the last stages of typhus fever.

Hides.
240

The horns of the Rhinoceros imported from India, Siam, Cochin China, and Sumatra, are much sought after in China for their supposed medicinal virtues. They receive the names *sí-koh, sí-niu-koh,* and are credited with tonic, alterative, and many other medicinal properties. The black and pointed horns are thought to be the best. Cups also are made of the horns which are believed to have the valuable property, alluded to above, of detecting poisonous draughts placed in them. Rhinoceros' HIDES also are exported to China where they are made into a kind of jelly which is used medicinally.

DOMESTIC.
241

Domestic Use.—Rhinoceros, hide is, in India, made into shields, sword handles, and ramrods, which are much prized by the native grandees of the regions where the animal occurs.

R. 241

(*W. R. Clark.*)

RHIZOPHORA, *Linn.; Gen. Pl., I., 678.*

Two species of Rhizophora are indigenous to India. The one here described in detail and the other—**R. conjugata,** *Linn.*—a small tree nearly related to it and frequently associated with it, but which, being of no special economic value, need not be further mentioned.

Rhizophora mucronata, *Lamk. ; Fl. Br. Ind., II., 435 ; Wight, Ic.,*

THE MANGROVE.   [*t. 238 ; Ill., I., 209 ;* RHIZOPHOREÆ.

Syn.—R. MACRORRHIZA, *Griff. ;* R. CANDELARIA, *W. & A. ;* R. MANGLE, *Linn.*

Vern.—*Kamo, bhora, bhara,* BENG. ; *Rái,* URIYA ; *Kamo, kunro,* SIND ; *Upu-poma, adavi ponna,* TEL. ; *Byu, pyu, byuma,* BURM. ; *Bairada, jumuda,* ANDAMAN ; *Kadol,* SING.

References.—*DC., Prod., III., 32 ; Roxb., Fl. Ind., Ed. C.B.C., 389 ; Voigt, Hort. Sub. Cal., 41 ; Kurz, For. Fl. Burm., I., 447 ; Beddome, Fl. Sylv., t. XIII, fig. 4 ; Gamble, Man. Tim., 176 ; Dalz. & Gibs., Bomb. Fl., 95 ; Graham, Cat. Bomb. Pl., 68 ; Sir W. Elliot, Fl. Andhr., 11, 186 ; Rheede, Hort. Mal., VI., t. 34 ; Rumphius, Amb., III, t. 71 ; Murray, Pl. & Drugs, Sind, 190 ; Drury, U. Pl. Ind., 272 ; Useful Pl. Bomb. (Vol. XXV., Bomb. Gaz.), 72 ; Watt, Ec. Pr., II., 53 ; VI., 155 ; VII., 213 ; Strettell, Nar. of Journey in search of Ficus elastica, 5 ; Gazetteers :— Bombay, V., 25 ; (Kanara) XV., 434 ; Burma, I., 137 ; Orissa, II., 175, Appendix 6 ; Tropical Agriculturist, IV., 503 ; Ind. Forester, XI., 490 ; Smith, Dict. Econ. Pl., 264 ; Balfour, Cyclop. Ind., III., 407.*

Habitat—A small evergreen tree, found on tidal, muddy shores throughout India, Burma, and the Andaman Islands.

Dye and Tan.—The BARK is said to be used to give a chocolate dye ; it is used also in tanning. This tan **Christy** recommends to be used as a preliminary preparation for cheap leathers. It is recommended that the leather should be about half prepared in India and exported to Europe in that condition, to be redone and have the colour improved by myrabolans or other tanning materials. Mangrove bark has been exported to Europe, but leather prepared with it solely is always inferior in colour and quality. Except, therefore, as a preliminary tan, or in the preparation of cheap leathers, it is not likely to become an article of European trade. Instead of the bark, which is bulky, **Strettell** suggests that an extract should be prepared for the purpose of exportation. He further states that this extract would perform its office in half the time of oak-bark. It would, however, have to be made in an earthen vessel, since any contact with iron would make the leather prepared by the extract, brittle and discoloured.

Medicine.—**Rheede** states that the BARK mixed with dried ginger or long pepper and rose water is said to be a cure for diabetes. No other writer appears to have made this observation ; indeed, **O'Shaughnessy** expressly states that none of the mangroves are reputed to have any medicinal virtue.

Food.—The FRUIT is reported to be sweet and edible, and the JUICE to be made into a kind of light-wine. Salt is extracted from the AERIAL ROOTS.

Structure of the Wood.—Sapwood light red, heartwood dark red, extremely hard, but warps and splits in seasoning ; it is very durable. Weight 70·5℔ per cubic foot.

Domestic Uses.—The WOOD, although good, is rarely used in India. **Rumphius,** speaking of the uses of the wood, states that in Amboyna it was a much valued fuel, and that the Chinese employed it to make charcoal for use in their workshops. He also adds that the larger AERIAL ROOTS were used as anchors for small boats, and remarks that in the Moluccas a curious custom, and one at variance with European ideas, prevails in that,

242

DYE &
TAN.
Bark.
243

MEDICINE.
Bark.
244
FOOD.
Fruit.
245
Juice.
246
Roots.
247
TIMBER.
248
DOMESTIC.
Wood.
249
Aerial Roots.
250

R. 250

RHODODENDRON arboreum.   **The Rhododendron.**

while the anchors are made of wood, the boats are constructed of a light stone thrown up by the volcanoes in these islands.

Rhizophora caseolaris, see **Sonneratia acida**, *Linn. f.;* LYTHRACEÆ ;
[Vol. VI.

251

RHODODENDRON, *Linn. ; Gen. Pl., II., 599.*
[*Royle, Ill., 260, t. 64, fig. 2 ;* ERICACEÆ.

Rhododendron Anthopogon, *D. Don.; Fl. Br. Ind., III., 472 ;*

Syn.—R. AROMATICUM, *Wall. ;* AZALEA LAPPONICA, *Pall. ;* OSMOTHAM-NUS FRAGRANS and PALLIDUS, *DC.*

Vern.—*Palu,* BHUTEA; *Tasak-tsun,* KASHMIR; *Nichni, rattankát, nera, kái zabán, morúa, talisa, talisri,* (Bazár name for leaves) *talisfar,* PB.

References.—*DC., Prod., VII., 715, 725; Brandis, For. Fl., 282; Stewart, Pb. Pl., 133; Watt, Cal. Exhb. Cat., V., 239; Gasetteer:—N.-W. P., X., 312; Ind. Forester, V., 183; Sir J. D. Hooker in Agri.-Hort. Soc., Ind., Journ., (Sel.), VIII., 47, 58, 63, 81, 83.*

Habitat.—A small shrub, with very aromatic, strongly-scented leaves, common at altitudes between 11,000 and 16,000 feet on the Alpine Himá-laya, from Kashmír to Bhután, and distributed to Central and Northern Asia.

MEDICINE.
Leaves.
252

Medicine.—The LEAVES of this plant and of R. **lepidotum,** are aromatic, and their smoke is considered by the Natives to be useful in some diseases. They are supposed to have stimulant properties, and are collected and exported to the plains, where they are officinal (*Watt*). This is one of the species which is thought by the Bhutias to excite the headache and nausea which attends ascents to the high elevations of the Eastern Himálaya (*Sir .J. D. Hooker*).

253

R. arboreum, *Sm. ; Fl. Br. Ind., III., 465.*

Syn.—R. PUNICEUM, *Roxb. ;* R. CINNAMOMEUM, *Wall. ;* R. CAMPBELLIᴎ. *Hook. f.*

Vern —*Bhoráns, gurás, ghonás, taggá, lalgurás,* NEPAL; *Etok,* LEPCHA & BHUTIA; *Brus* or *brás,* KUMAON; *Ardáwal, mandál, chiú, árú, brás, broá, chacheon,* PB.; *Trikhgandere,* TRANS-INDUS; *Billi, pumaram,* NILGHIRIS; *Ma-ratmal,* SING.

References.—*DC., Prod., VII., 720; Roxb., Fl. Ind., Ed. C.B.C., 373; Voigt, Hort. Sub. Cal., 333; Brandis, For. Fl., 281; Kurz, For. Fl. Burm., II., 93; Beddome, Fl. Sylv., t. 228; Gamble, Man. Timb., 236; Cat., Trees, Shrubs, &c., Darjiling, 51; Stewart, Pb. Pl., 133; Mason, Burma and Its People, 403, 781; Med. Top. Ajmir, 140, 151; Baden Powell, Pb. Pr., 594; Drury, U. Pl. Ind., 364; Kew Off. Guide to Bot. Gardens and Arboretum, 122, 124; Settlement Report:—Simla District, Panjáb, App. II., H., xliii; Gasetteers:—Panjáb (Ráwal Pindi), 13, 15; Simla, 10; Gurdaspore, 55; N.-W. P., Himálayan Districts, X., 312; Agri.-Horti. Soc. Ind.:—Trans., VI., 235; Journ. II., Sel , 385; IV., Sel., 264; VI., 167; VIII., Sel., 38, 43, 44, 51, 52, 58, 59, 81, 83; XIII., 384; Ind. Forester, I., 97; II., 25; IV., 198; VIII., 26, 404, 407, 409, 412; Sel., 41, 53, 58; X., 35; XI., 4; XIV., 247.*

Habitat.—A tree, which often attains a height of 25 feet; common on the temperate Himálaya from the Indus to Bhután, at altitudes between 5,000 and 10,000 feet. It is frequent on the Khásia hills, between 4,000 and 6,000 feet, and occurs also on the hills of Southern India and Ceylon, very abundant in Manipur, and on the Kareen hills in Burma.

MEDICINE.
Leaves.
254

Medicine.—"Madden states that the young LEAVES of this species are poisonous. It is also stated that they are medicinal, and on the Bias are applied to the forehead for headache " (*Stewart*). The honey of wild bee is said in Sikkim to be poisonous at the flowering time of this species.

FOOD.
Flowers.
255

Food.—The FLOWERS have a sweet, sour taste, and are said to make a good sub-acid jelly. They are, in some parts of the Himálaya, eaten by the Natives, who, according to Madden, get intoxicated if they consume a large quantity.

R. 255

| The Rhododendron. | (*W. R. Clark.*) | RHODODENDRON cinnabarinum. |

Structure of the Wood.—Soft, reddish, white or reddish-brown, close and even-grained, apt to warp and shrink. Weight 41·4℔ per cubic foot.

Domestic and Sacred.—The WOOD is chiefly used for fuel and charcoal, but is also sometimes employed for building and for making dishes. In Sikkim for *kukri* handles, boxes, and other small articles, and on the Nilgiris for gun-stocks and posts (*Gamble*). Hoffmeister mentions that a SNUFF made from the bark of the tree is excellent. He also notes that in Nepál the FLOWERS are used for offerings at temples.

TIMBER.
256
DOMESTIC.
Wood.
257
SACRED.
Flowers.
258

Rhododendron barbatum, *Wall.; Fl. Br. Ind., III., 468.*

259

Syn.—R. NOBILE, *Wall.* partly; R. LANCIFOLIUM, *Hook. f.*

Vern.—*Gurás, chimal,* NEPAL; *Kému,* BHUTIA.

References.—DC., *Prod., VII., 721; Gamble, Man. Timb.,* 237; *Gazetteer:—N.-W. Prov. (Himálayan Dist.),* X., *312; Ind. Forester, I., 97; VIII., 409; Agri.-Hort. Soc. Ind. Jour.; VIII., Sel.* 40, 41, 43, 52, 53, 55, 81.

Habitat.—A tree, 30 to 40 feet high, met with in the Temperate Himálaya from Kumáon to Bhután, at altitudes between 8,000 and 12,000 feet.

Structure of the Wood.—Pinkish-red, shining, of slow growth. Weight 39℔ per cubic foot.

TIMBER.
260

R. campanulatum, *Don.; Fl. Br. Ind., III., 466.*

261

Syn.—R. ÆRUGINOSUM, *Hook. f.*; R. NOBILE, *Wall.* (chiefly) R. EDGARII, *Gamble.*

Vern.—*Cherailu,* HIND.; *Cheraidhu, cheriala, teotosa,* NEPAL; *Chimul,* N.-W. P.; *Gaggar yurmi,* KASHMIR; *Sarngar, shinwala, shargar, simrung* (Himálayan Hill names), PB.; (Bazár, leaves=*tamákú hulas* or *pattikashmiri*); *Bargí,* TIBET.

References.—DC., *Prod., VII., 721; Brandis, For. Fl.,* 281; *Gamble, Man. Timb.,* 237; *Darjíling List,* 52; *Stewart, Pb. Pl., 134; O'Shaughnessy, Beng. Dispens.,* 48; *Baden Powell, Pb. Pr.,* 359, 594; *Atkinson, Him. Dist.,* 749; *Birdwood, Bomb. Pr.,* 51; *Watt, Cal. Exhb. Cat., V., 239; VII., 214; Gazetteers:—N.-W. P., X, 312; Simla Dist., 12; Ind. Forester, IX., 198; XIII.!, 65, 80; Agri.-Horti. Soc. Ind., Journ., VIII., Sel.* 36, 38, 41, 44, 59, 83.

Habitat.—A large shrub found in the inner Himálaya from Kashmír to Bhután at altitudes between 9,000 and 14,000 feet. It occurs also on the outer ranges of the Chór and Kedarkanta, and is very abundant in Sikkim.

Medicine.—The LEAVES are exported to the plains, where they are ground up with tobacco and used as a SNUFF, which is said to be useful in cold, and hemicrania. They are also said by Baden Powell to be used in chronic rheumatism, syphilis, and sciatica. The dried TWIGS and WOOD are used in Nepál, as a medicine in phthisis and chronic fevers.

Structure of the Wood.—Light pinkish-red, moderately hard. Weight 39℔ per cubic foot.

Domestic Use.—The WOOD is small and crooked, but makes excellent fuel, the smoke of which is, however, very acrid and irritant.

MEDICINE.
Leaves.
262
Snuff.
263
Twigs.
264
Wood.
265
TIMBER.
266
DOMESTIC.
267
268

R. cinnabarinum, *Hook. f.; Fl. Br. Ind., III., 474.*

Syn.—R. ROYLEI, *Hook. f.*; R. BLANDFORDIÆFLORUM, *Hook.*; R. SP., *Griff.*

Vern.—*Búlú,* NEPAL; *Kema, kechung,* LEPCHA.

References.—*Gamble, Man. Timb.,* 51; *Darjíling List,* 51; *Watt, Cal. Exhb. Cat., VII.,* 215; *Agri.-Horti. Soc., Ind. Journ., VIII., Sel.,* 36, 41, 45, 61, 81.

Habitat.—A shrub, 4 to 8 feet high; met with on the Eastern Himálaya, in Sikkim and Bhután, at altitudes between 10,000 and 12,000 feet.

Medicine.—The LEAVES are universally considered poisonous to cattle and goats. If employed as fuel, the smoke causes the eyes to inflame and the face to swell (*Hooker, Him. Journ.*).

MEDICINE.
Leaves.
269

R. 269

RHODODENDRON
lepidotum.   The Rhododendron.

TIMBER.
270

Structure of the Wood.—Grey, moderately hard, even-grained, is apt to warp. Weight 42℔ per cubic foot.

271

Rhododendron Falconeri, *Hook. f.; Fl. Br. Ind., III., 465.*

Syn.—R. EXIMIUM & VENOSUM, *Nutt.*

Vern.—*Kurlinga*, NEPAL; *Kégu, kalma*, BHUTIA.

References.—*Gamble, Man. Timb., 237; Kew Off. Guide to Bot. Gardens and Arboretum, 124-25; Watt, Cal. Exhb. Cat., VII., 215; Ind. Forester, I., 97; VIII., 409; XIII., 146; Agri.-Horti. Soc. Journ., VIII., Sel., 42, 57, 58, 81.*

Habitat.—A moderate-sized tree or frequently a gregarious shrub, abundant on the Eastern Himálaya from East Nepál to Bhután, at altitudes between 9,000 and 13,000 feet.

TIMBER.
272

Structure of the Wood.—Reddish-white, shining with a satiny lustre, takes a beautiful polish, is hard and does not warp. Weight 39℔ per cubic foot.

DOMESTIC.
Leaves.
273

Domestic Uses.—" As it is easily worked, and not apt to split, it is admirably adapted for use in the parched and arid climate of Tibet, and the Bhutias make from it cups, spoons, ladles, and the saddles by means of which loads are slung upon the yak. The LEAVES are employed as platters, and serve for lining the baskets, which contain the mashed pulp of **Arisæma** root (a kind of colocass). The customary present of butter or curd is always enclosed in this glossy foliage " (*Hooker in Journ. Agri-Horti. Soc., Ind.*).

274

R. fulgens, *Hook. f.; Fl. Br. Ind., III., 466.*

Syn.—R. NOBILE, *Wall.*, in part.

Vern —*Chimal*, NEPAL.

References.—*Watt, Cal. Exhb. Cat., VII., 215; Sir J. D. Hooker in Journ. Agri.-Horti. Soc., Ind., VIII., 44, 59.*

Habitat.—A small tree or large shrub of the Nepál and Sikkim Himálaya, found at altitudes of from 10,000 to 14,000 feet.

TIMBER.
275

Structure of the Wood.—Grey, darker in the centre, moderately hard, even-grained. Weight 36℔ per cubic foot.

276

R. grande, *Wight; Fl. Br. Ind., III., 464; Wight, Ic., t. 1202.*

Syn.—R. ARGENTEUM, *Hook. f.*

Vern.—*Kali guras, putlinga*, NEPAL; *Etok-amal*, LEPCHA.

References.—*Gamble, Man. Timb., 236; Watt, Cal. Exhb. Cat., VII., 214; Kew Off. Guide to Bot. Gardens and Arboretum, 124, 125; Journ. Agri.-Horti. Soc., Ind. (Old Series), VIII. (Sel.), 41, 42, 51, 59, 81, 83, 84.*

Habitat.—A tree frequent in the Sikkim and Bhután Himálaya, at altitudes between 7,000 and 11,000 feet.

TIMBER.
277

Structure of the Wood.—Yellowish with darker heartwood, shining, soft, close and even-grained. Warps less than that of **R. arboreum.** Weight 39℔ per cubic foot.

278

R. lepidotum, *Wall.; Fl. Br. Ind., III., 471.*

Syn.—R. SALIGNUM, ELÆAGNOIDES, & OBOVATUM, *Hook. f.*

Vern.—*Tsaluma, tsuma*, BHUTIA; *Talisfur*, NORTHERN INDIA; *Taliori* SIMLA.

The vernacular names of this species and of **R. Anthopogon** seem in many districts of the Panjáb to be in common.

References.—*DC., Prod., VII., 724; Brandis, For. Fl., 282; Gamble, Man. Timb., 236; Darjiling List, 52; Royle, Ill. Him. Bot., 260, t. 64, fig. 1; Dymock, Mat. Med. W. Ind., 759; Birdwood, Bomb. Pr., 51; Watt, Cal. Exhb. Cat., V., 240; Gazetteers:—Simla Dist., 12; N.-W. P., Himálayan Dist., X., 312.; Ind. Forester, V., 184; Agri.-Horti. Soc., Ind. Journ., VIII., Sel., 38, 46, 58, 62, 63.*

R. 278

The Sumach tree of Europe. (*W. R. Clark.*)	RHUS Coriaria.

Habitat.—A shrub found on the Temperate and Alpine Himálaya, from Kashmír to Bhután, at altitudes between 8,000 and 15,000 feet.

Medicine.—The medicinal properties of this species and of **R. Anthopogon** (*q.v.*) are similar.

MEDICINE.
279

Rhododendron setosum, *Don; Fl. Br. Ind., III., 472.*

280

Syn.—R. ANTHOPOGON, *Wall.*, partly.

Vern.—*Tsallu*, BHUTIA & TIBETAN.

References.—*DC., Prod., VII., 724; J. D. Hooker in Journal Agri.-Horti. Soc., Ind. (Old Series), VIII., (Sel.) 46, 64.*

Habitat.—A small and elegant shrub found in Sikkim and Nepál at altitudes between 13,000 and 16,000 feet.

Medicine.—"The Sikkim Bhutias and Tibetans attribute the oppression and headaches attending the crossing of the loftiest passes to the strongly resinous odour of this and **R. Anthopogon** (*q.v.*) A useful volatile OIL of no less marked character than that of the American Gaultheria might probably be obtained from the foliage by distillation" (*Hooker*).

MEDICINE.
Oil.
281

RHODOMYRTUS, *DC.; Gen. Pl., I., 713.*

Rhodomyrtus tomentosa, *Wight; Fl. Br. Ind., II., 469; Wight,* [*Ill., II., 12, t. 97, f. 3, Ic., 522;* MYRTACEÆ.

282

HILL GOOSEBERRY.

Syn.—MYRTUS TOMENTOSA, *Ait.;* M. CANESCENS, *Lour.*

Vern.—*Thaontay* (=The jam), NILGIRIS.

References.—*Roxb., Fl. Ind., Ed. C.B.C., 402; Voigt, Hort. Sub. Cal. 46; DC., Prod., III., 240; Beddome, Fl. Sylv., civ., Anal. Gen., t. xiv.; Gamble, Man. Timb., 187, 188; Drury, U. Pl., 364; Extract from Madras Manual of Administration, I., 314; Extract from Trichinopoly Manua of Administration, 16; Rep. Govt. Bot. Gar., Saharanpur and Mussoori (1884), 5; Ind. Forester:—II., 25; VIII., 26; Agri.-Horti. Soc., Ind. Trans., VII., 60.*

Habitat.—A shrub much resembling the common myrtle; found in the higher mountains of South India, and distributed to Malacca, Singapore, Penang, and Ceylon.

Food.—The FRUIT, which is about the size of a cherry, of a dark purple colour and with a fleshy sweet and aromatic pulp, is very palatable. It is eaten raw, or made into a jam or jelly, which is very similar in flavour to apple jelly.

FOOD.
Fruit.
283

Structure of the Wood.—White with a pink heart, close-grained, but easily worked, apt to split in seasoning.

TIMBER.
284

Domestic Uses.—The WOOD is much employed in turnery, and many small articles are made from it.

DOMESTIC.
285

Rhubarb, see Rheum emodi, *Wall.; p. 486';* POLYGONACEÆ.

286

RHUS, *Linn.; Gen. Pl., I., 418.*

A genus of trees and shrubs, indigenous chiefly to the warmer temperate regions of both hemispheres. They are nearly all poisonous, and have most of them a very acrid juice; all are highly astringent and are used for tanning. Twelve species are natives of the East Indies, most of which are of economic value. The true Sumach Tree of Europe, although not known to be a native of India proper, is found both wild and cultivated in the adjacent countries of Afghánistán and Persia, and as its leaves and fruit are extensively imported thence into India for industrial and medicinal purposes; it has therefore been thought proper to give it a place in this work.

Rhus Coriaria, *Linn.; Boiss., Fl. Orient., II., 4;* ANACARDIACEÆ.

287

THE SUMACH TREE OF EUROPE.

Vern.—*Tatrak, mutchli,* HIND.; *Sumok,* BENG.; *Samáhk, sumahk, samagk*

R. 287

RHUS Cotinus.	**The Elm-leaved Sumach.**

or *sumagh*, AFG. ; *Sumák*, BOMB.; *Sumák, hút, tumtum*, ARAB.; *Sumak, mahi*, PERS.

References. — *Aitchison, Rept. Pl. Coll. Afgh. Del. Com., 41; also Products of Western Afghan & N.-E. Persia, 174; DC., Orig. Cult. Pl., 133; O'Shaughnessy, Beng. Dispens., 280; Irvine, Mat. Med. Patna. 107; S. Arjun, Cat. Bomb. Drugs, 33; Dymock, Mat. Med. W. Ind., 2nd Ed., 192-194; Dymock, Warden & Hooper, Pharmacog. Ind., Vol. I., 372; Journal of the Soc. of Arts (Feb. 24th, 1882), 399; Watt, Cal. Exhb. Cat., V., 240; Birdwood, Bomb. Prod., 19, 310; Baden-Powell, Pb. Pr., 339; Kew Off. Guide to Mus. of Ec. Bot., 20; Christy, Com. Pl. and Drugs, V., 28, 40, 43; Agri.-Horti. Soc. :—Ind., Trans., III., 41; Journ. (New Series), VII., 60; Proc., xci; Ind. Forester:—XI., 55; XIV., 364, 366; Smith, Dic. Econ. Pl., 397; Spons, Encycl., 1988.*

Habitat. — A small tree, extensively cultivated in Spain, Italy, and Sicily, for the young shoots and leaves, which are dried and made into a powder used in tanning. It grows wild in the Canaries and Madeira, in the Mediterranean region, around the Black Sea, in the Trans-Caucasian Provinces, and in Persia and Afghánistán. Aitchison records it as cultivated in orchards in Khorasan and Western Afghánistán.

DYE. Leaves. 288

Dye. — The LEAVES contain colouring matter and are employed by the Afgháns and Persians for dyeing purposes. The dried leaves are sold in the bazárs of Afghánistán under the name of *barg-a-sumaghk.*

TAN. Leaves. 289 Twigs. 290

Tan. — The LEAVES and TWIGS are largely used in Europe for tanning. They are employed locally in Persia and Afghánistán for this purpose, and are imported into India usually from Sicily for use in the large European tanneries. They do not appear to be much employed in native tanning, as effectual and low-priced substitutes are found in several of the other species of this genus.

MEDICINE. Fruit. 291

Medicine. — The FRUIT is a small, flattened, drupe of a red colour, containing one lenticular polished brown seed; it is acid and very astringent. It is imported in considerable quantities into India from Persia for medicinal purposes. It is not used among the Hindus; but Muhammadan medical writers universally recommend it as an astringent and tonic. Thus in the *Makzan-el-Adwiya* it is said by Mir Mahomed Husain to check bilious vomiting and diarrhœa, hæmoptysis, hæmatemsis, diuresis, and leucorrhœa. It is used to strengthen the gums, and as an astringent in ophthalmia. To sores it is applied in the form of a poultice either alone or mixed with charcoal. A liquid extract obtained from the LEAVES and fruit is used as an astringent. The leaves are made into a poultice which is applied to the abdomen in the case of children suffering from diarrhœa or dysentery (*Dymock*). In the Panjáb it is much used among the Muhammadans, as an astringent and tonic, in cases of cholera, dysentery, diarrhœa, and indigestion.

Leaves. 292

293

Rhus Cotinus, *Linn.; Fl. Br. Ind., II., 9.*

THE ELM-LEAVED SUMACH.

Syn. — R. VELUTINA & LÆVIS, *Wall.*

Vern. — *Túnga, tung, chaniát, ámi*, N.-W. P.; *Páán, phán, bhán, bana, bauru, largá, manú, túng, tittri*, PB.; *Darengri*, KASHMIRI; *Erandi*, MAHR.

References. — *Boiss., Fl. Orient., II., 4; Brandis, For. Fl., 118; Gamble, Man. Timb., 104; Stewart, Pb. Pl., 49; Baden Powell, Pb. Pr. 453; Christy, Com. Pl. and Drugs, V., 40; Hummel, Dyeing of textile fabrics, 364; Watt, Cal. Exhb. Cat., II., 53; VII., 215; Kew Off. Guide to the Mus. of Ec. Bot., 36; Gazetteers :—Shahpur Dist., 70; Rawalpindi, 15; N.-W. P., X., 308; Journ., Agri.-Horti. Soc. Ind. :—VII., 160; XIV., 15; Ind. Forester :—I., 185; XIII., 57.*

Habitat. — A shrub or small tree of the Western Sub-Tropical Himá-

| The Elm-leaved Sumach. | (*W. R. Clark.*) | RHUS mysorensis. |

laya, from Marri to Kumáon; found up to altitudes of 5,000 feet. It is distributed westwards through Afghánistán and Persia to France.

Dye.—Throughout the area of its indigenous habitat, the LEAVES, BARK, and WOOD seem all more or less to be locally employed in dyeing. Thus Dr. Aitchison, in the *Flora of the Kuram Valley,* says:—" I was informed that the old wood of **R. Cotinus** is used as a dye for wool-stuffs, chiefly to produce an orange red colour in felts. Baden Powell, speaking of the use of this plant in Kashmír, states that "the astringent leaves are used in dyeing with *kahi* to produce black and grey shades." Under the name of "Young Fustic," the wood is used in Europe, principally for the purpose of producing orange or scarlet tints in woollen fabrics. It was formerly much employed in dyeing brown colours on silk, the yarn being mordanted with alum and afterwards dyed with Young Fustic, Peachwood and Logwood. The colours produced by this dye stuff are, however, of a fugitive character, two or three months' exposure to light being sufficient to bleach them entirely (*Hummel*).

Tan.—The BARK and LEAVES are much used locally by Native tanners in various parts of India; but there seems as yet to be no general demand for the article by European tanners in India, although it is exactly the same plant that is known in Europe as Venetian Sumach and is much valued in the Tyrol for tanning purposes. An unlimited supply exists on the Lower Himálaya and if desired large tracts of useless country might soon be covered with it—the only care necessary being the initial one of planting the bushes. The subject of the utilization of this tanning material seems worthy of the consideration of the Indian importers of Sumach.

Structure of the Wood.—Moderately hard, sapwood small, heartwood mottled, of a rich, dark-yellow colour.

Domestic Uses.—In South Europe the wood is used for inlaid and cabinet work. In the Himálaya the twigs are employed for basket making.

Rhus insignis, *Hook. f.; Fl. Br. Ind., II., 11.*

Vern.—*Kagphulai,* NEPAL; *Serh,* LEPCHA.
References.—*Gamble, Man. Timb., 105; Darjíling List, 24; Watt, Cal. Exhb. Cat., VII., 215.*
Habitat.—A tree of moderate size; found in the interior valleys of the Sikkim Himálaya at altitudes of from 3,000 to 6,000 feet, and in the Khásia hills at an altitude of 4,000 feet.

Oils and Oil-seeds.—The FRUIT is a drupe, composed of a very thin epicarp, which encloses a globose white mass of wax, containing a very small flattened stone. The wax is probably used or may be used for similar purposes to that obtained from **R. semi-alata** and **succedanea.**

Medicine.—The JUICE of this species is a powerful vesicant (*Gamble*).

Structure of the Wood.—Grey, soft; heartwood yellowish-brown. Weight about 26℔ per cubic foot.

R. kakrasingee, see Pistacia integerrima, *Stewart;* Vol. VI., 268.

R. mysorensis, *Heyne; Fl. Br. Ind., II., 9.*

Vern.—*Dasarni,* AJMERE; *Dasan, davan, dasni,* MERWARA.
References.—*W. & A., Prodr., 172; Brandis, For. Fl., 119; Beddome, Fl. Sylv., 78, t. xi., f. 3; Gamble, Man. Timb., 104; Gazetteers:—Mysore and Coorg, I., 71; Bombay, V., 24; Ind. Forester:—VII., 261; VIII., 418; XI., 465, 467; XII., 33, App. 2, 10.*
Habitat.—A small shrub, met with on the Sulaiman range, at altitudes between 2,500 and 5,000 feet, and on the plains of Sind, the Panjáb, Rájputana, and the Deccan.

Side notes:

DYE.
Leaves.
294
Bark.
295
Wood.
296

TAN.
Bark.
297
Leaves.
298

TIMBER.
299
DOMESTIC.
300

301

OIL.
Fruit.
302

MEDICINE.
Juice.
303
TIMBER.
304

305

Indian forms of Sumach.

TAN. Bark. 306 TIMBER. 307 DOMESTIC. Wood. 308 309

Tan.—Mr. E. A. Fraser, Assistant Agent to the Governor General in Rájputana, reports that there "the BARK of this shrub is very largely used by native tanners, and gives a splendid buff or brown colour to leather."

Structure of the Wood.—Hard, pinkish yellow, close-grained, heavy.

Domestic.—The WOOD is employed only for fuel.

Rhus paniculata, *Wall. ; Fl. Br. Ind., II., 10.*

References.—*Kurz, For. Fl., Burm., 319; Report on the Shan States by Mr. Aplin; Mason, Burma and Its People, 540, 774.*

Habitat.—A tree not uncommon in the dry forests of Prome and Ava (*Kurz*), and found also in Bhotán (*Griffith*).

DOMESTIC.
310

Domestic.—Mason speaks of this as a useful timber tree but gives no information with regard to the uses made of it.

311

R. parviflora, *Roxb.; Fl. Br. Ind., II., 9.*

Vern.—*Túng, raitúng, túmra,* HIND. & PB.; *Túnga, túngla, dúngla, rái túng, tumra, raunel,* N.-W. P.; *Túngla,* (Bazar, seed *tantarík, túng*), PB.; *Samák,* KASHMIRI.

References.—*Roxb., Fl. Ind., Ed. C.B.C., 274; DC., Prodr., II., 70; Brandis, For. Fl., 119; Dalz. & Gibs., Bomb. Fl. Suppl., 19; Stewart, Pb. Pl., 49; Baden Powell, Pb. Pr., 339, 595; Gazetteers:—N.-W. P., X., 308; Panjáb, Simla, 11; Ind. Forester, XI., 232; Agri.-Hort. Soc. Ind., XIII., 385.*

MEDICINE.
Fruit.
312
TIMBER.
313
DOMESTIC.
Wood.
314
Leaves.
315
316

Habitat.—An unarmed shrub of the Western Himálaya, found from Kumáon to Nepál, at altitudes between 2,000 and 5,000 feet. It also occurs in Central India on the Pachmarhi hills (*Brandis*).

Medicine.—"The '*tantarík*' of the bazárs appears, in the Panjáb, to be the FRUIT of this plant. It is used in Hindu medicine, and, mixed with salt, is said to act like tamarinds" (*Stewart*).

Structure of the Wood.—Hard and yellow.

Domestic Uses.—The WOOD is small, but excellent for turning (*Baden Powell*). The LEAVES of this shrub are reported, says Mr. J. F. Duthie, to be used in Tirhri Garhwál instead of tobacco.

R. punjabensis, *Stewart; Fl. Br. Ind., II., 10.*

Vern.—*Títari, tetar, arkhar, palai, choklu, kangar, kakkrein, dor, rashtu,* PB.

References.—*Brandis, For. Fl., 120; Gamble, Man. Timb., 105; Gazetteer of the Gurdaspur Dist., 55; Ind. Forester, VIII., 35; IX., 20, 68, 291.*

Habitat.—A fairly sized tree, 35 feet high, which occurs in the mixed forests of the North-West Himálaya, at altitudes between 2,500 and 8,500 feet.

TIMBER.
317

Structure of the Wood.—It consists of alternate layers of soft, porous spring wood and hard autumn wood. The heartwood is yellowish grey, with dark longitudinal streaks, moderately hard. Weight 34℔ per cubic foot.

318

R. semi-alata, *Murray; Fl. Br. Ind., II., 10; Wight, Ic., t. 561.*

Syn.—*R.* BUCKIAMELA, *Roxb.;* R. JAVANICA, *Linn.;* R. AMELA, *Don.*

Vern.—*Arkhar, tatri, titri, tetar, thissa, knitri, arkol, dúdla kakkari, kakkeran, chechar, vrásh, wánsh, káshin, hulashing, wásho, hulúg, rashtu, Pb.; Dakhmila, dáswila,* N.-W. P.; *Bakkiamela, bhagm ili,* NEPAL; *Takhril,* LEPCHA.

References.—*Roxb., Fl. Ind., Ed. C.B.C., 273; Voigt, Hort. Sub. Cal., 274; DC., Prodr., II., 67; Brandis, For. Fl., 119; Kurz, For. Fl., Burm., I., 319; Gamble, Man. Timb., 105; Stewart, Pb. Pl., 48; Baden Powell, Pb. Pr., 595; Gazetteers:—N.-W. P., X., 308; Simla Dist., 11; Gurdaspur Dist., 55; Rawalpindi, 15; Ind. Forester, VI., 24; XIII., 57; Agri.-Hort. Soc. Ind., XIII., 385.*

R. 318

The Japan Varnish & Wax.   (*W. R. Clark*).   **RHUS succedanea.**

Habitat.—A small tree, met with on the outer Himálaya, from the Indus to Assam, up to altitudes of 6,000 feet, and on the Khásia mountains, at altitudes between 3,000 and 5,000 feet.

Oils and Oil-seeds.—The DRUPES are covered with a small quantity of pulp, from which is prepared the vegetable wax known in Nepál as *omlu*. This wax is similar to the "Japanese wax" of commerce, which is sometimes imported into England and used in the manufacture of candles and night lights.

OIL.
Drupe.
319

Medicine.—The FRUIT is given by the hill tribes of the Himálaya as a remedy for colic.

MEDICINE.
Fruit.
320

Food.—The BERRIES have a sharp acid taste, and are sometimes eaten by the Nepalese and Lepchas.

FOOD.
Berries.
321

Structure of the Wood.—Soft, shining, grey in colour with darker streaks. Weight about 27℔ per cubic foot. It is valueless except as fuel (*Stewart*).

TIMBER.
322

Rhus succedanea, *Linn. ; Fl. Br. Ind., II., 12 ; Wight, Ic., t. 560.*

323

Syn.—R. ACUMINATA, *DC.*

Vern.—*Kakra-singi, kakarsing,* HIND. ; *Kákrásringi,* BENG.; *Raniwalai,* NEPAL; *Serhnyok,* LEPCHA; *Dingkain,* KHASIA; *Arkhar, arkhol, choklú, hala, halai, holashi, hulashing, kakkrin, lakhar, rikhúl, tatri, titar, títri,* (Bazar, fruit=) *habat ul khizra,* (the galls in Kangra=) *kakursinghi,* PB.; (The galls=) *kakadashingi,* BOMB.; (The galls=) *kákarasingi,* DEC.; (The galls=) *kakada-shingi,* MAR.; (The galls=) *kakkatashingi,* TAM.; *Karkatasringi,* SANS.

References.—DC., *Prodr., II.,* 68 ; *Roxb., Fl. Ind., Ed. C.B.C.,* 273 ; *Brandis, For. Fl.,* 121 ; *Gamble, Man. Timb.,* 106 ; *Darjíling List,* 24 ; *Stewart, Pb. Pl.,* 48, 49 ; *Pharm. Ind.,* 59 ; *O'Shaughnessy, Beng. Dispens.,* 282 ; *Irvine, Mat. Med. Patna,* 45 ; *Med. Top. 'Ajm.,* 141 ; *U. C. Dutt, Mat. Med. Hind.,* 140, 303, 319 ; *S. Arjun, Cat. Bomb. Drugs,* 33 ; *Murray, Pl. & Drugs, Sind,* 86 ; *Year-Book Pharm.,* 1873, 46 ; *Baden Powell, Pb. Pr.,* 594 ; *Kew Off. Guide to the Mus. of Ec. Bot.,* 20 ; *Watt., Cal. Exhib. Cat.:—I.,* 56 ; *IV.,* 59 ; *V.,* 240 ; *VII.,* 216 ; *Report of Bot. Gar. Gonesh Khind, Poona, 1882-83,* 3 ; *Gazetteers :— Panjáb, Shahpur Dist.,* 70 ; *Simla Dist.,* 11 ; *N.-W. P., X.,* 308 ; *Agri.- Hort. Soc. Ind.:—Journ. VII.,* 172 ; *XIII.,* 314, 384. *Proc.,* 3 ; *Ind. Forester :—II.,* 290 ; *IX.,* 516, 576 ; *XIII.,* 57 ; *Smith, Econ. Dic.,* 437.

Habitat.—A tree, about 30 feet high, found on the Temperate Himálaya, from Kashmír to Sikkim and Bhotán, at altitudes between 3,000 and 8,000 feet. It also occurs on the Khásia mountains between 2,000 and 6,000 feet, and is distributed to Japan.

Resin.—Koempfer calls this the Wild Varnish Tree, and says that in Japan, the STEM of this as well as that of R. vernicifera is incised, and the exudation collected, for the manufacture of the varnish used in Japanese lacquer-work. Stewart states that in Bombay a varnish is said to be yielded by this species ; but as no further information can be obtained with regard to this, it seems probable that, in India at any rate, the use of this varnish is very inconsiderable. Dr. Dymock does not allude to this plant in his *Materia Medica of Western India*, so that the statement regarding the use of the varnish in Bombay is probably incorrect.

RESIN.
Stem.
324

Oils and Oil-seeds.—In Japan, the FRUITS are crushed, boiled, mixed with the fruit of another tree (*Sindan,* **Melia Azedarach** ?), and pressed while hot. A wax, analogous to bees' wax, is thus produced which is made into candles, and is sometimes sent to Europe under the commercial name of "Japan Wax." It is said that in Sikkim and Nepál this plant is utilized as a source of the wax known as *omlu*, but this method of employing it appears to be quite local and unimportant.

OIL.
Fruit.
325

Medicine.—The milky JUICE of the tree is very acrid and is said to possess vesicant properties (*Vigne*). The FRUIT is considered official and

MEDICINE.
Juice.
326
Fruit.
327

32 A

R. **327**

RHUS vernicifera.	**The Japan Varnish Tree.**

MEDICINE.
Excrescences
328

is used in Kashmír in the treatment of phthisis. The horn-like EXCRE-SCENCES caused by insects on the branches are commonly confused with those on **Pistacia integerrima** (the true *karkata-sringí* of Sanskrit medical writers), and are given medicinally in place of the latter. For an account of their action and uses in Hindu and Muhammadan medicine, see **Pistacia integerrima**, *Stewart*, ANACARDIACEÆ; Vol. VI., 268.

FOOD.
Fruit.
329

Food.—The bear is said to eat the FRUIT (*Stewart*).

TIMBER.
330
331

Structure of the Wood.—White, shining, soft, with small darker coloured heartwood. Weight 32℔ per cubic foot. It is not used.

Rhus vernicifera, *DC., Prodr., II., 68.*

References.—*Thunb. Fl. Jap., 121; Linn. Soc. Journ., XX., 411; Report of Agric. Dept. & Experim. Farms, Madras (1884-85), 7, 34; N.-W. P. (1884), 12; Bombay Admn. Rept. (1871-72), 369, 394; Report of Nilghiri Bot. Gardens (1884-85); Report of Govt. Bot. Gardens, Saharanpur (1885), 13; (1886), 12; (1888), 15; (1889), 13; Gazetteers:—Panjáb, Hoshiarpur Dist., 116; Lahore Dist., 101; N.-W. P., V., 727; Ind. Forester:—I, 362; II., 181; VI., 24; IX., 576-579; XIII., 57; Smith, Econ. Dict., 426; Strettell, Ficus Elastica, 22, 23.*

Habitat.—The Japanese Varnish tree grows all over the main Island of Japan, and also in smaller quantities in Kinshin and Shikoku; but it is from Tokio northwards that it principally flourishes, and it grows freely in the colder regions on the mountains as well as on the plains.

CULTI-VATION.
Japan.
332

Cultivation.

Japan.—Attempts to cultivate this species in India are now being made. In consequence of recommendations from **Sir J. D. Hooker** elaborate enquiries were made by the Secretary of State for India with regard to the lacquer industry in Japan, and reports on it were received from the British Consul at Hakodate, one of the principal centres of the trade. These reports gave full details as to the cultivation of the tree, and may therefore be in part reproduced here :—"The tree can be propagated by seed sown at the end of January or the beginning of February. The first year the seedlings reach a height of from 10 inches to 1 foot. The following spring, the young trees are transplanted about 6 feet apart, and in ten years an average tree should be 10 feet high, the diameter of its trunk 2½ to 3 inches, and its yield of lacquer sufficient to fill a 3-ounce bottle. Another and more speedy method of propagating is, however, generally adopted. The roots of a vigorous young tree are taken, and pieces six inches long and the thickness of a finger are planted out in a slanting direction a few inches apart, one inch being left exposed above the ground. This takes place in the end of February and through March, according to the climate of the locality. These cuttings throw a strong shoot of from 18 to 20 inches the first year, and are likewise planted out the following spring. Under equally favourable circumstances, these trees would in ten years be nearly 25 per cent. larger in girth, some 2 or 3 feet higher, and would yield nearly half as much more sap than the trees raised from seed."

In his account of the Japanese lacquer industry **Dr. Dresser** gives the following details as to the method adopted for the extraction of the sap used in the preparation of the Japan varnish :—"The tree is diœcious, and wax is extracted from its seeds as well as from those of **Rhus succedanea**. The lactiferous vessels, unlike the wax, are found in both the stamniferous and pistilliferous trees. The quality of the lacquer depends in some degree on the nature of the soil in which the tree grows. Incisions are made in the stem, the punctures being repeated every fourth day at successively higher parts of the tree. The juice which oozes out is scraped off with a flat iron tool. When the tree has been thus tapped to the topmost branches it is

Japan Lacquer Industry.	(*W. R. Clark.*)	**RHUS vernicifera.**

felled. The log is cut into lengths, which are tied into faggots and steeped in water for ten to twenty days, after which the bark is pierced and the oozing lacquer is collected in the same way as from the stem.

The juice thus collected, is a tenacious fluid of a greyish brown colour. It is allowed to stand and settle when first obtained. A kind of skin forms over the surface; the better quality rises to the top, and the impurities sink to the bottom. It is thus easy to separate the finest from the inferior qualities, and the former are strained through cotton or porous paper. By stirring in the open air the lacquer partially dries, absorbs oxygen, and gains a brilliant dark colour. In the fluid state it is highly corrosive, and if a drop falls upon the skin it will produce a bad sore.

India.—In 1884 some seed of **Rhus vernicifera** was sent out by **Sir Joseph Hooker** for experimental cultivation in Madras and Saharanpur, and the Revenue and Agricultural Department also obtained a supply direct from Her Majesty's Consul at Kanagawa, which was made over to the Superintendents of the Botanical Gardens at Calcutta and Saharanpur. The only details as to the results of the cultivation of the tree, yet available, are from **Mr. J. F. Duthie**, who, in 1885, reported that, "the plants of this valuable tree are in as healthy a condition as could be desired, but they are growing slowly. Unless the rate of growth increases as they become older, it will take many years before they are sufficiently large for tapping purposes. Another supply of seed was received from **Dr. G. King**, Calcutta, in August last, and sown as soon as received. A number of these germinated shortly after being sown, and several others have since appeared and are still appearing above ground. The stock at present numbers 35 plants." In the report for the year 1886 it is stated that "the young seedlings of this species have at last started into growth and are now shooting up fast. There is now no reason to doubt that this useful tree will thrive in this climate. A small plantation will be made next rainy season, and it will then be a question of time as to when the plants will be ready for tapping." Subsequent reports, however, throw some doubt on the success of the experiment, and in 1889 the plants are said to "require more than ordinary care," and it is remarked that the only chance of their doing any good in Saharanpur is under high garden cultivation. It seems, therefore, very doubtful if the high expectations which were at first formed of establishing a lacquer industry in India, similar to the Japanese one, will ever be fulfilled, unless some of the indigenous species are found to serve the same purpose.

Japan Lacquer Industry.

The following account of the industry in Japan is reproduced from **Dr. Dresser's** account :—

"The lacquer-workers kneel on the usual matted floors, and the chief care taken, is to keep the apartments clean and free from dust. The lacquer is spread on the substratum employed, which is almost invariably wood in coats of successively increasing fineness; the first coat being usually mixed with powdered earth. Each coat when dry is rubbed down with a cutting stone. In an object intended to be of excellent quality as many as eleven coats are thus laid on "before the decoction is commenced. After the application of the last coat the surface is ground down with lumps of hard charcoal, which are kept wet, and the final polish is given by the ashes of deer's horns."

"The pattern to be borne by the object is sketched in outline in lacquer upon fibrous elastic paper; the paper is warmed and fitted to the surface to be decorated, and the pressure of the hand is enough to transfer the pattern, after which the paper is removed. If the pattern is to be in gold,

RHYNCHOCARPA
   foetida.   The Japan Varnish Tree.

LACQUER INDUSTRY.

the outline is then followed by a fine hair pencil, dipped in lacquer which is intended to act as a size. When this has so far dried as to be sticky, fine gold dust is shaken on it from a spoon. The gold dust looks grey at first, but its yellow colour is brought out by burnishing."

Thus the lacquer industry of Japan will be seen to be entirely different to that of India, both in the materials used and the workmen employed. In India it is accomplished by the application of a stick of hard shellac, like sealing wax, to a rapidly revolving surface which, by means of the friction, develops sufficient heat to make the shellac adhere irregularly; whereas in Japan the method consists in laying on with a brush successive coats of a purely vegetable varnish. The Indian craftsman has plenty of patience, but does not possess that passion for finish that distinguishes the Japanese, and it is doubtful if anything so perfect, as some of the best Japanese work, could ever be wrought by an Indian hand. even if the Japanese varnish tree were found ultimately to flourish in India (*J. J. Kipling*). It would thus seem desirable that the *loc-work* of India and the *lacquer-work* of Japan be regarded as perfectly distinct.

335

Rhus Wallichii, *Hook. f.; Fl. Br. Ind., II., 11.*

> **Syn.**—For some years this species was regarded as at most the Indian form of **R. vernicifera**, the Japanese Varnish Tree, but **Brandis**, in his *Forest Flora* and latterly **Sir J. D. Hooker**, in the *Flora of British India*, have pointed out that the latter is a distinct species, having sessile flowers and laxes, longer panicles. The Himálayan tree is not known to yield any Varnish.
>
> **Vern.**—*Bhálaio, chosi,* NEPAL; *Akoria, kaunki, bhaliún, kaunui,* N.-W. P.; *Kambal, galámbal, arkhar, arkol, harkú rikhali, lohása, urkúr, hulása, ríkhúl,* PB.
>
> **References.**—*DC., Prodr., II., 68; Brandis, For. Fl., 120; Gamble, Man. Timb., 106; Stewart, Pb. Pl., 49; Watt., Cal. Exhb. Cat., IV., 59; V., 241; VII., 216; Royle, Ill. Him. Bot., 175; Gazetteers:—Panjáb, Gurdaspur, 55; N.-W. P., X., 308; Agri.-Hort. Soc., Ind., XIV., 57; Indian Forester, IX., 576-579; XIII., 57.*

OIL. Seed. 336

    Habitat.—A tree of the Temperate Himálaya from Garhwál to Nepal, occurring at altitudes between 6,000 and 7,000 feet.

    Oil and Oil-seed.—The SEEDS are said to yield a wax, similar to Japan wax.

MEDICINE. Juice. 337

    Medicine.—Vesicating properties are in some places attributed to the JUICE.

TIMBER. 338

    Structure of the Wood.—Sapwood white, soft; heart-wood reddish brown, yellow when dry (*Brandis*).

DOMESTIC. Juice. 339

Leaves. 340

    Domestic Uses.—The WOOD is used in the Sutlej valley for saw frames and axe handles. The JUICE of the LEAVES is in some places rubbed on thread to strengthen it (*Stewart*).

RHYNCHOCARPA, *Schrad.; Gen. Pl., I., 831.*

                   [CUCURBITACEÆ.

341

Rhynchocarpa foetida, *Schrad.; Fl. Br. Ind., II., 627;*

> **Syn.**—R. ROSTRATA, *Kurz;* ÆCHMANDRA ROSTRATA, *Arn.;* BRYONIA PILOSA, *Roxb.;* B. FILICAULIS, *Wall.;* B. ROSTRATA, *Rottler.;* B. PEROT-TIANA, *Seringe;* MELOTHRIA FŒTIDA, *Lamk.*
>
> **Vern.**—*Cucuma-dunda,* TEL.; (The root=) *Appakovay kalung,* TAM.
>
> **References.**—*DC. Prodr., III., 304; Roxb., Fl. Ind., Ed. C.B.C., 703; Dals. & Gibs., Bomb. Fl., 100; Ainslie, Mat. Ind., II., 21; Dymock, Mat. Med. W. Ind., 2nd Ed., 355; Useful Pl. Bomb. (Vol. XXV., Bomb. Gaz.), 200.*

    Habitat.—A climbing herb found in Bombay, Gujarat, the Deccan Peninsula, and in Ava. It is distributed to Tropical Africa and Natal.

R. 341

The Gooseberry rarely cultivated in India. (*W. R. Clark.*)	**RIBES.**

Medicine.—Speaking of this plant **Ainslie** says:—"The ROOT, as it appears in the medicine bazárs, is about the size of a finger and of a light grey colour; it has no particular smell, but a slightly sweetish and mucilaginous taste: it is prescribed internally, in electuary, in cases of piles; in powder it is sometimes ordered as a demulcent in humoral asthma; dose of the electuary two teaspoonfuls thrice daily."

Food.—The FRUIT and LEAVES are said by **Lisboa** to be eaten.

RHYNCHOSIA, *Lour.; Gen. Pl., I., 542.*

A genus of twining or erect herbs or shrubs, belonging to the Natural Order LEGUMINOSÆ, and comprising about eighty species, which are distributed everywhere in the tropics of both hemispheres, and some of which reach the Cape and the United States. Twenty-two species are described in the *Flora of British India*, as found in the Indian Peninsula; but of these only one is mentioned as being of any economic value.

Rhynchosia minima, *DC.; Fl. Br. Ind., II., 223;* LEGUMINOSÆ.

Syn.—R. MEDICAGINEA, *DC.;* R. RHOMBIFOLIA, *DC.;* R. PROSTRATA, *Grah.;* R. MICROPHYLLA, *Wall.;* R. NUDA, *DC.;* GLYCINE RHOMBIFOLIA, *Willd.;* DOLICHOS MINIMUS, *Linn.;* D. MEDICAGINEUS, *Lamk.*

Vern.—*Gadi chikkudu káya, néla alumu,* TEL.; *Ghattavare,* KAN.

References.—*DC., Prodr., II., 385; Roxb., Fl. Ind., Ed. C.B.C., 564; Voigt, Hort. Sub. Cal., 228; Dals. & Gibs., Bomb. Fl., 74; Sir W. Elliot, Fl. Andhr., 56, 130; Thesaurus, Zey., 188, t. 84, f. 2; Gazetteers: —Bombay, V., 25; N.-W. P., Bundelkhand, I., 80; IV., lxxi; X., 309; Mysore and Coorg, I., 55; Ind. Forester, XII., App. 2, 11.*

Habitat.—A twining or wide trailing annual, with very slender stems; found throughout the plains of India, from the Himálaya, where it ascends to an altitude of 3,000 feet, to Ceylon and Burma. It is cosmopolitan in distribution, occurs everywhere in the tropics, and is distributed to the Cape and the United States.

Fodder.—Roxburgh, writing of **D. scarabæoides**, says:—"Cattle eat this sort: I know of no other use it is put to."

RHYNCHOSPERMUM, *Reinw.; Gen. Pl., II., 263.*

[*248;* COMPOSITÆ.

Rhynchospermum verticillatum, *Reinw.; Fl. Br. Ind., III.,*

Syn.—LAVINIA RIGIDA, *Wall.;* CARPESIUM RACEMOSUM, *Wall.*

Vern.—*Hukmandás,* PB.

References.—*DC., Prodr., V., 280; Stewart, Pb. Pl., 123.*

Habitat.—A slender, puberulous herb, with long spreading branches, found in the Temperate Himálaya, from Kashmír (altitude 5,000 feet) to Sikkim (6,000 feet) and Bhután. It occurs also on the Khásia mountains at an altitude of 5,000 feet, and in Burma. It is distributed to the Islands of the Malayan Archipelago and to Japan.

Medicine.—Stewart remarks that **Honigberger** states it is used medicinally in Kashmír, but neither of these writers mention for what diseases it is a reputed cure.

RIBES, *Linn.; Gen. Pl., I., 654.*

A genus of prickly or unarmed shrubs, belonging to the Natural Order SAXIFRAGACEÆ, and comprising about fifty-six species, natives of Temperate Europe, Asia, North America, and the Andes. It is a remarkable fact that while the Gooseberry and the Black and Red Currant occur wild on the North-West Himálaya, the European introduced fruits are but very rarely seen in cultivation, at the hill stations, nor have the indigenous forms been cultivated by the hill tribes (*Watt*).

R. 350

Margin notes

MEDICINE.
Root.
342

FOOD.
Fruit.
343

Leaves.
344

345

346

FODDER.
347

348

MEDICINE.
349

350

351

Ribes glaciale, *Wall.; Fl. Br. Ind., II., 410 ;* Saxifragaceæ.

Syn.—R. acuminatum, *Wall.*

Vern. —*Kukuliya, kala káliya, mángle* (=the red variety), *durbui, dongole* (=the black) (Atkinson), Him. Names ; *Robhay,* Bhutia.

References.—*Brandis, For. Fl.,* 214; *Gamble, Man. Timb.,* 173; *Wall., Cat.,* 6833; *Atkinson, N.-W. P., V.* (Foods), 44, 72, 73; *Watt, Cal. Exhb. Cat.:—VI.,* 156; *VII.,* 217; *Gazetteers:—Simla Dist.,* 11; *N.-W. P., X.,* 310, 714; *Agri.-Hort. Soc. Ind. Journ., XIII.,* 385.

Habitat.—A small shrub of the Temperate and Alpine Himálaya, from Bhotán to Kashmír, occurring at altitudes between 7,000 and 12,000 feet. Atkinson describes two varieties—the red, occurring rarely, and the black, which is common above 10,000 feet, the latter of which he says is the R. acuminatum of Wallich.

FOOD.
Fruit.

352

Flowers.

353

TIMBER.

354

355

Food.—The FRUIT of R. glaciale is a round, smooth, red (or black?) berry, as large as a common red currant but very sour and unpalatable. Strachey found it near Nabhi in Byáns, where it is very abundant and yields a fruit described by him as small and insipid. The FLOWERS appear in May and the fruit ripens in September-October (*Atkinson*).

Structure of the Wood.—White, compact, moderately hard. Weight about 63℔ per cubic foot.

R. Grossularia, *Linn.; Fl. Br. Ind., II., 410.*

THE ROUGH OR HAIRY GOOSEBERRY.

Syn.—R. himalensis, *Royle* ; R. alpestre, *Dcne.*

Vern.—*Galdam, lepcha, sirkuchi, baikunti* (Byans), Kumaon ; *Sirgochi,* Juhar ; *Amlánch, kánsi, pilsa, teila, súr-ka-chúp* (Upper Chenab & Lahoul), Pb.

References.—*Boiss., Fl. Orient., II.,* 815; *Brandis, For. Fl.,* 213; *Gamble, Man. Timb.,* 173; *Stewart, Pb. Pl.,* 102; *Kuram Valley Rept.,* 13, 17, 24, 57; *DC., Orig. Cult. Pl,* 276; *Atkinson, Him. Dist. (Vol. X., N.-W. P. Gas.),* 310, 714; *also Econ. Prod., N.-W. P., V.,* 72; *Royle, Ill. Him. Bot.,* 225; *Watt, Cal. Exhb. Cat., VI.,* 157; *Gazetteer, Simla Dist.,* 11; *Ind. Forester, XI.,* 3; *Agri.-Hort. Soc. Ind., Trans., VI.,* 235; *Journ., VI.* (Pro.), 58; *VII.* (Pro.), 34; *XIII.,* 386.

Habitat.—In India the Gooseberry is found wild in the Western Alpine Himálaya, from Kumáon to Kashmír, at altitudes between 9,000 and 12,000 feet. In its wild state it is distributed to North and Alpine Europe, the Atlas, and the mountain ranges of Greece and the Caucasus.

CULTIVA-
TION.

356

ATTEMPTS AT CULTIVATION.—Many attempts have been made by Europeans in India in the plains and on the hills both to cultivate introduced European varieties and to improve the indigenous stock, but hitherto all have proved unsuccessful. Thus **Firminger,** speaking of the cultivation of introduced varieties in the plains, says : "The climate is so utterly unsuited to it that it cannot even exist here." On the hills of India, where one would think it might have a better chance of growing, no introduced variety has been recorded to have thriven or borne fruit fully. Royle suggests that the reason for this may be the shortness of time that elapses between the season of their flowering and the outbreak of the rains. [It seems probable greater success would be attained than hitherto by grafting European stock on to indigenous roots. There are, however, large tracts of the Himálaya, such as in Kulu, where the Gooseberry should thrive and fruit well. In Pangi the wild bush is very plentiful, and, though the fruits are not good, the plant bears profusely.—*Ed., Dict. Econ. Prod.*]

FOOD.
Fruit.

357

DOMESTIC.

358

Food..—The wild gooseberry of the Himálaya produces a small, hairy, very sour FRUIT, which is most unpalatable, and is hardly ever eaten even by natives.

Domestic.—"In one or two villages of the Hariab District it is employed as a hedge plant" (*Aitchison*).

R. 358

The Black and Red Currant. (*W. R. Clark.*)	**RIBES rubrum.**

Ribes nigrum, *Linn. ; Fl. Br. Ind., II., 411.* $\quad$ 359

THE BLACK CURRANT.

> **Vern.**—*Pápar,* KUMAON; *Murádh, nábar, kadash niangna mandrí, hádar, belí, sháktekas,* PB.; *Askúta,* LAHOUL.
>
> **References.**—*Boiss., Fl. Orient., II., 815 ; Brandis, For. Fl., 215 ; Gamble, Man. Timb., 173 ; Atkinson, Econ. Prod., N.-W. P., V., 72 ; Firminger, Man. Gard. Ind., 267 ; Stewart, Pb. Pl., 102 ; DC., Orig. of Cult. Pl., 278 ; Watt, Cal. Exhb. Cat., VI., 157 ; Smith, Econ. Dict., 145 ; Falconer, Com. Forms, 27 ; Gazetteer, N.-W. P., X., 310, 714.*
>
> **Habitat.**—An erect unarmed shrub, which occurs on the Temperate Western Himálaya, from Kanawar to Kashmír, at altitudes between 7,000 and 12,000 feet. In distribution the black currant extends to North Asia and North Europe.

ATTEMPTS AT CULTIVATION.—With regard to the cultivation of the currant on the plains of India, **Firminger** writes : "Plants have often been raised from seed, as well as on one or two occasions been imported in ships that have brought ice from America. There does not seem the least probability of the plant ever being brought to succeed in any part of India. In Lower Bengal more particularly, it has been found quite impossible to keep it alive during the hot and rainy seasons. On the Nilgherries, currant trees manage to live, but do not thrive even there, and their cultivation is attended with very unsuccessful results. At Ferozepore, I raised a plant of the black currant from seed in the cold season and managed to preserve it during the heat and rains until the following cold season. But it perished then, as often happens with delicate plants upon the effort to start into growth at the approach of a season more congenial to them."

> **CULTIVATION. 360**

Food.—The FRUIT of the wild variety is said by **Dr. Stewart** and Major Garstin to be very like that of the cultivated black currant, and quite as large and very palatable. The flowers appear in July and the fruit ripens in August-September.

> **FOOD. Fruit. 361**

R. orientale, *Poir. ; Fl. Br. Ind., II., 410.* $\quad$ 362

> **Syn.**—R. LEPTOSTACHYUM, *Dcne.;* R. VILLOSUM, *Wall.;* R. GLUTINOSUM, *Jacq.*
>
> **Vern.**— *Gwáldakh, kagják* (KAGHAN), N.-W. P.; *Nangke, nyái phulánch* (the fruit=*nyangha*) (CHENAB), PB.; *Askúta, askútar,* LADAK ; *Yange,* SPITI ; (The fruit=) *Aksíswérai,* AFGH.
>
> **References.**—*Boiss., Fl. Orient., II., 817 ; Brandis, For. Fl., 214 ; Gamble, Man. Timb., 173 ; Stewart, Pb. Pl., 102 ; Aitchison, Fl. Kuram Valley, 57 ; Journ. Agri.-Hort. Soc. Ind., XIV., 52.*
>
> **Habitat.**—A shrub which grows to a height of 6 feet and is found in Kashmír and Baltistan, at altitudes between 8,000 and 12,000 feet. It is distributed to Cabul, Persia, Armenia, Asia Minor, and Greece.

Medicine.—The FRUIT ripens in October, and has a mawkish sweet taste. Aitchison states that the shrub is a very common one in the Hariab District, and that the BERRIES, taken one or two at a time, are considered by the natives an excellent purgative.

> **MEDICINE. Fruit. 363 Berries. 364**

R.-rubrum, *Linn.; Fl. Br. Ind., II., 411.* $\quad$ 365

THE RED CURRANT.

> **Syn.**—R. HIMALAYENSE, *Dcne.* (*not of Royle*).
>
> **Vern.**—*Dák, kagh dák, ráde, áns, phulánch, nangke hádar, khádrí, murádh, nábar, nábre,* PB.; *Gwaldakh,* KHAGAN ; *Wara wane,* TRANS-INDUS.
>
> **References.**—*DC., Prodr., III., 481 ; Orig. Cult. Pl., 277 ; Boiss., Fl. Orient., II., 816 ; Brandis, For. Fl., 215 ; Gamble, Man. Timb., 173 ; Stewart, Pb. Pl., 102 ; Aitchison, Fl., Kuram Valley, 13 ; Atkinson, N.-W. P. Gas., X., 310, 715 ; Econ. Prod., N.-W. P., Part V., 73 ; Gazetteers :—Dera Ismail Khan, 19 ; Bannu, 23 ; Ind. Gardener, 303*

R. 365

| RICINUS communis. | The Castor Oil Plant. |

Habitat.—An erect unarmed shrub of the Western Himálaya, from Kumáon to Kashmír, occurring at altitudes between 8,000 and 12,000 feet. It is distributed to Alpine Europe, the Caucasus, and Altai.

FOOD.
Berries.
366

Food.—Brandis notices that in Lahoul the BERRIES are yellow when unripe, but become black, although with the taste of red currants when ripe. Atkinson says the FRUIT is small and more acid than is agreeable.

Fruit.
367

Cultivated red currants have been introduced from England, and by Sir E. O. Buck have been made to fruit freely near Simla.

Rice, see **Oryza sativa,** *Linn.;* Vol. V., 498-654.

(*G. Watt.*)

368

RICINUS, *Linn.; Gen. Pl., III., 321.*

A genus of EUPHORBIACEÆ which possesses but one species, though under that there are numerous forms, which have been designated varieties and even species by some writers, although they would seem to possess no higher claim than that of cultivated races, many of which may, perhaps, be only sports from garden cultivation, perpetuated because of their beauty of foliage. In India there are two primary races—a perennial bushy state with large seeds and an annual condition with small seeds. The former yields in considerable quantity an inferior oil, employed for illumination and lubrication, and the latter a superior oil, the qualities of which constitute the medicinal oil of commerce.

[*2209;* EUPHORBIACEÆ.

369

Ricinus communis, *Linn.; Fl. Br. Ind., V., 457; Bot. Mag. No.*

CASTOR OIL PLANT, PALMA CHRISTI; HUILE DE CASTOR, RICIN DE PALMA-CHRISTI, *Fr.;* RICINUS SAMENÖL, *Germ.* Often called in Indian books of travel, &c., LAMP OIL.

Syn.—RICINUS INERMIS, *Jacq.;* R. LIVIDUS, *Jacq.;* R. SPECIOSUS, *Burm.;* R. SPECTABILIS, *Blume;* R. VIRIDIS, *Willd.;* CROTON SPINOSUS, *Linn.*

Vern.- *Arand, arend, erand* or *erend, rand, ind* (these and other names in Hindustan are generally made to terminate in "*i*" when the seeds are denoted), HIND.; *Bherendá,* BENG.; *Eradom,* SANTAL; *Eri,* ASSAM; *Areta, alha, orer,* NEPAL; *Rak-lop,* LEPCHA; *Renr, lenr, anrar, anda* (a plantation=*renrwári*), BEHAR; *Gab,* URIYA; *Grundi,* C. P.; *Nerinda,* GOND; *Arand, rendi, reri, bhatreri,* N.-W.-P.; *Ind-rendi,* KUMAON; *Anerú, harnauli, arand* or *arind, bedanjir,* PB.; *Arhand,* PUSHTU; *Bas-anjir, bus-anjir,* AFG.; *Edia, arend,* RAJ.; *Ayrun-kukri, hérán,* SIND; *Erund, ind, rund, yarand,* DECCAN; *Erendi,* BOMB.; *Erandi, yarandícha,* MAR.; *Diveligo, diveli,* GUZ.; *Amanakkam, sittamunuk,* TAM.; *Kottei, muttu,* COIMBATORE; *Amadam, amdi, sittamindi, ayendal* (*ámudapu, chittámudapu* (small-seeded medicinal form), *eramudapu,* TEL.; *Haralu, harlu,* KAN.; *Tonda,* MALASAR; *Avanakku,* MALAY.; *Kyeksu, kesú,* BURM.; *Endaru,* SING.; *Eranda, ruvuka, vátári* (=anti-rheumatic), *rakta éránda,* SANS.; *Khirvá* (DeCandolle mentions, see historic chapter *kerrua, kerroa,* & *charua* as Arabic names, but they do not appear to be so), ARAB.; *Bedánjir* or *bédanjír,* PERS.; *P'i-ma,* CHINESE; *Jarak,* SUMATRA; *Camiri,* AMBOYANA; *Kiki,* EGYPTIAN; Κίκι and κρότων, ANCIENT GREEK.

References.—DC., *Prodr., XV., ii., 1017;* Roxb., *Fl. Ind., Ed. C.B.C., 690;* Brandis, *For. Fl., 445;* Kurz, *For. Fl. Burm., II., 400;* Gamble, *Man. Timb., 363;* Dalz. & Gibs., *Bomb. Fl. Suppl., 78;* Stewart, *Pb. Pl. 197;* Rept. *Pl. Coll. Afgh. Del. Com., 108;* DC., *Orig. Cult. Pl., 422-25;* Rev. A. Campbell, *Rept. Econ. Pl., Chutia Nagpur, No. 3221;* Graham, *Cat. Bomb. Pl., 183;* Mason, *Burma and Its People, 492, 762;* Sir W. Elliot, *Fl. Andhr., 14, 44, 51, 151;* Sir W. Jones, *Treat. Pl. Ind., V., 57;* Rheede, *Hort. Mal.* (*Avanacu and Pandi-avanacu*), *II., t. 32, 57 & 60;* Rumphius, *Amb., IV., 92-97, t. 41;* Burmann, *Fl. Ind., 307, t. 63, f. 2;* Thesaurus, *Zey., 205-206;* Loureiro, *Fl. Cochin-China, II., 584;* Stocks, *Sind;* Forskahl, *Egypt, 75;* Pharm. *Ind., 201, 462;* British Pharm., *291;* Flück. & Hanb., *Pharmacog., 567;* U. S. Dispens., *15th Ed., 1031-36;* Fleming, *Med. Pl. & Drugs* (*Asiatic Reser., XI.*), *176;*

R. 369

Products of India. 507

The Castor Oil Plant.   (*G. Watt.*)   **RICINUS communis.**

Ainslie, Mat. Ind., I., 254; II., 472; O'Shaughnessy, Beng. Dispens., 556; Irvine, Mat. Med. Patna, 34, 90; Pereira, Mat. Med., II., 416; Ed. B. & R., 531; Moodeen Sheriff, Supp. Pharm. Ind., 214-16; U. C. Dutt, Mat. Med. Hind., 231, 297; S. Arjun, Cat. Bomb. Drugs, 126, 212; K. L. De, Indig. Drugs, Ind., 101; Murray, Pl. & Drugs, Sind, 33; Waring, Basar Med., 37; Bent. & Trim., Med. Pl., 237; Dymock, Mat. Med. W. Ind., 2nd Ed., 704-708; Smith, Contri. Mat. Med. China, 54-55; Pharm. Jour. & Trans., 2 ser. II., 419; VII., 229; VIII. (Groves), 250; also IV., 282; VII., 534; I., 350; Amer. Jour. Pharm., XXVI., 207; XXVII., 99; Chemical News, XXII., 229; Watts, Dict. Chemistry, I., 815; VII., 270; VIII., 414; Gmelin's Chemistry, XVII., 131; Birdwood, Bomb. Prod., 78, 288; Baden Powell, Pb. Pr., 375, 421; Drury, U. Pl. Ind., 365-367; Atkinson, Him. Dist. (X., N.-W. P. Gas.), 749, 772; Duthie & Fuller, Field & Garden Crops, II., 38-39; Useful Pl. Bomb. (XXV., Bomb. Gas.), 125, 220-221, 225; Forbes Watson, Indian Prod., 99, 152; Royle, Prod. Res., 28-34, 213; Liotard, Dyes, 68, 72, 75; Hawkes, Report on Oils of South India, published 1855, 6, 8, 16, 24-27; Cooke, Oils and Oilseeds, 68-70; McCann, Dyes and Tans, Beng., 30-35; Simmonds, Science and Commerce, their Infl. on Manuf., 542; Tropical Agriculture, 405-8; Shortt, Man. Ind. Agri., 241-245; Hoey, Trade and Manuf., N. Ind., 191 & 192; Milburn, Oriental Commerce (1825), 287; Wilson, External Trade of Bengal, from 1813 to 1826, 14-15; Buchanan-Hamilton, Jour. through Mysore, Canara, etc., I., 109, 229, 286, 380, 410; II., 225, 323, 384; III., 240, 326; Cavane. Dissert. Oleum Palmæ Christi, published 1764; Marsden, History of Sumatra, 92; Long, History of Jamaica, 712; Man. Madras Adm., I., 288; Nicholson, Man. Coimbatore, 225-26; Gribble, Man., Cuddapah; Moore, Man., Trichinopoly, 72; Behar Peasant Life by Grierson, 246; Bombay Admin. Rep., 1872-73, 357; Bombay, Man. Rev. Accts., 103; Settlement Report:—Panjáb (Peshawar), 13; Oudh (Hardoi), 15; Central Provinces (Upper Godavery), 36; Nursingpore, 52; Chindwara, 23; Nagpur, 271; Chanda, 81; Wardha, 63, 68; Madras (South Arcot), 109; Gazetteers:—Bombay, II., 63-64; V., 106; VIII., 189; XII., 153; XV., ii., 19; XVIII., 270; XVIII., ii., 44; XIX., 165; XXIV., 170-71; N.-W. P., I., 84, 153; IV., lxxvii; Orissa, II., 15; 134; 159, 181, App. IV., VI.; Mysore and Coorg, I., 65, II., 11; Ind. Forester, I., 37, 93, 97; V., 212, 214; X., 260; XII., App., 28; XIV., 370; Smith, Dict. Econ. Pl., 307; Encycl. B., V., 200; Balfour, Cyclop. Ind., I., 603.

Habitat.—Modern botanical writers seem agreed that **Ricinus** is not a native of India, and that the cultivation of the plant very probably spread from Africa, the country which can alone be said to possess it in a truly wild condition. **Sir J. D. Hooker** (*Flora British India*) says on this subject that **R. communis** is cultivated throughout India and naturalised near habitations; distributed throughout the Tropics generally, but probably indigenous to Africa. **M. DeCandolle** also holds that existing evidence is in favour of an African origin. **Flückiger & Hanbury** (*Pharmacography,* 567) affirm, on the other hand, that it is a native of India, and they apparently hold that opinion chiefly on account of the undoubted antiquity of the knowledge of the drug, as denoted by Sanskrit literature. **Bentley & Trimen** (*Medicinal Plants*) say that it is believed to be a native of India.

The writer has seen the plant growing in the scrubby jungles of the outer Himálaya at considerable distance from human influence, and in association with **Citrus medica** *var.* **acida**—the sour lime—and **Moringa pterygosperma**—the horse-radish tree   The two last mentioned plants are accepted as truly wild in these localities, although neither have very probably ever been seen to possess a stronger claim to that position botanically than their frequent associate—**Ricinus communis.** In the Sutlej valley north of Simla, there are many such groves, and in the valleys south-west of Simla, towards Kasauli, extensive jungles, approaching almost to forests, exist, in which the prevailing trees are the plants named,

Conf. with p. 54.

with, in addition, Ægle Marmelos—the bael tree—and Feronia elephantum—the elephant apple. Intermixed with these occur many undoubtedly
indigenous trees and shrubs such as Mallotus philippinensis, Euphorbia Roy-
leana, and Cassia Fistula, &c., but others of no less certain exotic nature,
such as Excæcaria sebifera, Opuntia Dillenii, Nicandra, &c., are equally
abundant. The extent to which such situations are, in fact, being permeated
with American weeds is rapidly becoming one of the striking features of
the vegetation of the outer Himálaya. It is therefore extremely difficult
to form any opinion as to whether Citrus medica, Moringa pterygosperma,
Ægle Marmelos, Feronia elephantum, and Ricinus communis are wild in such
localities, or are only escapes from cultivation. None of these plants have,
as far as the writer is aware, been recorded in any part of India as inhabiting primeval forests (or such as might be called so) in association
with indigenous plants only. In the plains of India the castor oil plant is
frequent on rubbish heaps and waste lands near human dwellings, but in
the vast expanse of India it nowhere manifests a stronger claim to being
accepted as indigenous than has been denoted in the above brief description of its occurrence in the outer Himálaya.

RESIN.
Plant.
370

Resin —Kurz says that this PLANT yields a white resin. No other writer
alludes to this subject, and the substance may, therefore, be said to be
quite unknown.

DYE & TAN.
371

Dye and Tan.—Castor oil is frequently employed by the Indian dyers
as an auxiliary in certain dye preparations as, for example, in dyeing with
Morinda —which see; also consult Liotard's *Memorandum on Dyes,
68, 72 75,* &c., &c. Its uses are well known to the Calico-printers of India
and Europe. The oil has also the reputation of being one of the best for
dressing tanned hides and skins. In Spons' *Encyclopædia (p. 1382)* it is
stated that "The common kinds are largely used by leather-dressers, principally, perhaps, for Morocco leather, but with equal success on all descriptions ; moreover, it repels rats and other vermine, and does not interfere with
subsequent polishing." In India cod oil is largely used for this purpose.
Its value as a preservative is well known. On this account it is employed
by the Natives of India in preserving their water buckets and agricultural
appliances, harness, &c., made of leather. Hence, in all probability, the
high caste Hindu prejudice, against it, the *chumar* of oils. (*Conf.* with
the articles on Leather, also Tans and Tanning.)

FIBRE.
Stems.
372

Bark.
373

Fibre. —While Ricinus communis does not itself yield fibre, it is largely
cultivated in Assam to feed the *eri* silk-worm. For an account of this subject the reader is referred to the article Silk. An excellent paper-pulp is,
however, said to be made from the STEMS with their BARK, the latter containing a fibre though not of sufficient value to justify its separation. As
some 500 maunds of stems are obtained from an acre of land, it seems probable that where grown in the vicinity of paper mills it would be more
profitable to dispose of the stems to the paper-maker than to use them
as fuel or thatching as is the present custom.

OIL.
Seeds.
374

Oil.—It is scarcely necessary to say that the SEEDS of this plant yield
the Castor-Oil of Commerce. The chapter below on that subject deals with
the chief features of interest both agriculturally and commercially, and it
need only be added, therefore, that there are two primary forms of the plant,
namely, the large and the small-seeded. The former yields the "lamp oil"
of Madras and other parts of India, and the latter affords the more valuable
oil, specially pure samples of which constitute the medicinal castor-oil.
The following account of the uses of castor-oil is from the pen of Mr.
O. N. Bryce. As it gives some facts of a practical nature in addition to
those alluded to above under Dye and Tan, it may be here given as an
introduction to the more detailed account which will be found below :—

R. 374

Products of India. 509

and Castor-Oil Cake. (*G. Watt.*) RICINUS communis

OIL.

" The uses of the castor plant are many; the oil is the only eligible one for lubricating all sorts of machinery, clocks, watches, &c.; it is the best lamp oil we have in India, and gives an excellent white light, far superior to that of mineral oils, petroleum, rapeseed, mustard, linseed, and all other oils, whether vegetable, animal, or mineral. This I state after comparison. The slowness with which the oil burns effects a saving of consumption ranging from ¼ to ½. Its freedom from danger as lamp oil is another recommendation. All Railways in India burn castor-oil. It gives out very little soot, almost imperceptible, which quality no other lamp oils possess. This is a great recommendation; most important of all, it is the cheapest lamp oil in existence, while it is the cheapest and best for the manufacture of all kinds of soaps, candles, pomatums, and perfumed oils. All the great perfumers of London and Páris use castor-oil for the manufacture of Golden oil, and its beneficial effects are due to the demulcent quality of the castor-oil, which keeps the head cool and the pores of the skin and roots of the hair soft and open. It is a medicinal oil, and the most largely used as a *mild demulcent laxative*. The use of cold-drawn oil gives a splendid light: no other oil can vie with this light, it being almost electrical in its brilliancy. When boiled it does not give such a splendid light: the reason for this is, in the boiling or sweating, or both processes, the electric light-giving portion of the oil gets dissipated to a certain extent. The shelling of seed and filtering processes are comparatively very costly. The oil drawn from unheated shelled seeds without boiling (*cold filtration* being substituted for boiling) is of the best quality. If cold-drawn oil can be manufactured sufficiently cheap to be within reach of all as *lamp oil*, it would be used by many. At present, however, it is selling at prices ranging from R40 to 50 per maund in India. The reason for this is that castor seed has first to be exported to Europe, then it is bought by the manufacturers from the traders in castor seed, and after manufacture of the oil, it is bottled, labelled, corked, capsuled, packed, and shipped to India. If manufactured here it could be sold with profit at R10 per maund." "The oil," continues **Mr. Bryce** in another part of his paper, " dissolves completely in alcohol, and this, incorporated with a solution of copal, the resinous exudation of the well known tree **Vateria indica** in alcohol, can be formed into a varnish which will be found on trial very useful in polishing all kinds of high class furniture, carriages, saloon cabins of ships, picture frames, oil-paintings, parchment, vellum, maps, and drawings, cloths and canvasses, and leathers of all kinds, &c., &c., &c." According to **Hawkes** the Railways of India use for lubrication a preparation of castor-oil agitated with nitric acid until it forms what he calls " a kind of palmine. "

CAKE.
375

CASTOR-OIL CAKE.—Though of very considerable value as a manure, the cake, obtained after expression of the oil, is regarded by the European cattle-breeders as highly injurious to cattle. One of the chief subjects for which the agricultural chemist's opinion is desired is, therefore, the freedom or otherwise of a sample of oil-cake from adulteration with castor. It is somewhat singular, however, that in many parts of India this opinion does not prevail. Thus **Dr. Buchanan-Hamilton** says of Mysore (a statement repeated in the *Gazetteer* of that Province) that a decoction prepared from the cake is given to buffaloes from an opinion that, like the leaves of the plant, it has the power of increasing the flow of milk. Since, however, a strong prejudice exists in Europe against the cake, it would seem desirable that the mixture of this cake with others should be discouraged, where such admixture intentionally or otherwise takes place. But while this is so, the cake has a distinct claim to consideration. It is highly valued as a manure, and, according to **Mr. O. N. Bryce**, a new use for it has

RICINUS communis.	Medicinal uses of

OIL.

been discovered of much value. **Mr. Bryce** says:—"The oil-cake has also found an excellent use, in addition to its being one of the most valuable manures we have, as a material for making gas which gives a superior light. At the Allahabad East Indian Railway station the gas lamps are lighted with the gas obtained from the castor-oil cake, which has hitherto been regarded either as manure or as a material of no use. The East Indian Railway Company obtain their supply of castor-oil cake from their castor-oil mills." For some years past the public have been familiar with the fact that the palace, public offices, and streets of Jeypore are lighted with gas prepared chiefly from a crude and cheap form of castor-oil grown for the purpose. But that the Railway Companies have it within their power to prepare from the by-product (castor-oil-cake), obtained from their oil presses, a gas by which the platforms of their larger stations could be lighted, is a fact of considerable interest. To non-residents in India this may not seem so interesting until they are reminded that in only a few of the larger cities of India is coal gas available by which railway stations could be lighted. In an article which appeared in the *Chemist and Druggist*, the following particulars are given regarding the yield of gas from this oil. " When castor-oil is not available other seed oils are used for making the gas. The price of the former is about 22s. per maund of 82℔, and of the other seed oils 20s. From 1 maund of castor-oil the gas manager obtains 1,250 cubic feet of candle gas. The cost of manufacture (exclusive of the oil) is about 8s. 6d. per 1,000 cubic feet." In connection with the account below of Panjáb castor-oil cultivation, it will be found that it is stated the cake is sold as fuel for stationary engines.

In the Annual Reports of the Agricultural Department of Bengal will be found frequent mention of the experiments conducted with castor-oil cake as a Manure. In the report for 1888-89, p. 19, it is stated that " castor-cake and bone-meal mixed together form a better fertilizer for sugarcane than either of these manures alone." In the same report it is said that for the soil of the Seebpore farm castor-cake proved a better manure than salt-petre for wheat. Its merits as a manure for potatoes have been shown by frequent tabular statements. For example, an acre of land manured with 1,920℔ of the cake gave 9,966℔ of potatoes, whereas an acre manured with 960℔ cake and 480℔ bone-meal gave 7,752℔ of potatoes; cowdung gave a return of 5,220℔, and an acre not manured at all only 5,100℔ of potatoes. On the other hand, while many writers say that castor-oil cake is used as a manure for betel—*pán*—in the Director's report for 1886, App. lvii., it is remarked that the cake used for that purpose is mustard-cake " castor-cake destroys the plants. " Many writers affirm that the value of castor-cake, as a manure, is in consequence of the high amount of phosphates which it contains. This would appear to be an incorrect explanation. All oil-cakes are richer in phosphates than most other vegetable manures, but castor-oil cake contains considerably less phosphates (according to **Professor Anderson,** *Elements of Agricultural Chemistry, p. 196*) than mustard and rape-cake, and infinitely less than poppy-cake. In **Morton's** *Cyclopædia of Agriculture* rape-seed cake is shown to contain 4·34 per cent. of phosphates; linseed-cake 3·83 per cent.; and hemp-cake 1·08 per cent. According to **Anderson** rape-seed cake possesses 3·87 per cent. of phosphates; poppy-cake 6·93 per cent.; cotton-seed cake 2·19 per cent.; and castor-seed cake 2·81 per cent. (*Conf.* with OIL-CAKE under **Oils** in this work, Vol. V., 476.)

MEDICINE.
376

Medicine.—One of the earliest Anglo-Indian Medical authorities to deal with the subject of castor oil was **Dr. John Fleming,** who, in the *Asiatick Researches,* Vol. XI. (1810), gave the following account:—

the Castor-Oil. (*G. Watt.*) **RICINUS communis.**

MEDICINE:
Seeds.
377

"This plant is cultivated, for both economical and medicinal purposes, over all *Hindostan.* The expressed oil of the SEEDS, so well known in *Europe,* under the name of *castor oil,* is more generally used, as a purgative, than any other medicine; and perhaps there is no other on which we may, with so much confidence, rely as a safe, and, at the same time, an active cathartic. It may be given, with propriety, in every case in which that class of remedies is required (unless when the most drastic are necessary) and to patients of every age and constitution; for though it seldom fails to produce the effect intended, it operates without heat or irritation.

"The oil should be expressed, in the manner directed by the *London* College, from the decorticated seeds, and without the assistance of heat. That which is obtained by boiling the seeds in water is injured both in smell and taste, and becomes sooner rancid than the oil procured by expression." **Dr. Fleming's** contemporary, and in some respects his predecessor, **Dr. Ainslie,** wrote of it : "The oil is highly prized as a purgative medicine by the Native practitioners of India, who conceive it to be particularly indicated in cases of costiveness, from it operating freely and without irritating; it is usually given daily, in small quantities, to new born children, for three weeks together; and is also considered as an invaluable medicine as an external application in various cutaneous affections." "The BARK of the root of the trees," says Ainslie, "is a powerful purgative, and when made into a ball about the size of a lime, in conjunction with chillies and tobacco leaves, is an excellent remedy for gripes in horses. In the Mysore country, where the castor-oil plant is much cultivated, two varieties are distinguished: our present article, which is the RICINUS COMMUNIS, FRUCTUS MINOR, and which is in Canarese *chicca hárálú:* and the larger sort, which is the RICINUS COMMUNIS, FRUCTUS MAJOR, and which in Canarese is *dódú hárálú.*" The distinction into these two forms will be found below in the chapter on CULTIVATION of Castor-oil, and especially in the passages from **Dr. Buchanan-Hamilton's** account of his *Journey through Mysore, &c.* But long before any of these early explorers in the field of Indian Economic Products, the distinction had been made into the large and the small-fruited forms of castor-oil. **Rheede,** the Dutch Governor of Malabar in 1636-1691, distinguished these two forms, and he was followed in this opinion by **Rumphius,** the Dutch Governor of Amboina (1627-1702), who gave a detailed account of the action of the OIL, the LEAVES, and the ROOT-BARK. It is thus somewhat remarkable that the early trade in castor-oil to Europe should have taken place from the West instead of the East Indies. This subject will be returned to again in the chapter below on the HISTORY of CASTOR-OIL; suffice it to add here that **Rumphius** deals fully with the medicinal properties of the plant which the subsequent writers—**Fleming, Ainslie, Buchanan-Hamilton,** and **Roxburgh**—only elaborated. Speaking of the uses of the leaves, for example, he says that by the Indians they are employed to arrest the secretion of milk in puerperal women. For this purpose, he tells us, they are reduced to a pulp and applied to the breasts, when in three days the flow of milk is interrupted (*conf.* with the remarks below on this subject). Rumphius also refers to the use of the oil, in external application and as a purgative internally. **Dr. U. O. Dutt** (*Hindu Mat. Med. of Sanskrit Writers*) tells us that "the root of **Ricinus communis** and the oil obtained from the seeds have been used in medicine by the Hindus from a very remote period. They are mentioned by **Susruta.**" It thus follows that the use of castor-oil was known and the plant very probably cultivated in India many centuries before the Christian era. The two varieties of that plant are, at all events, referred to in the *Susruta Ayurveda,* under their common names as

Bark.
378

RICINUS communis.	Medicinal Uses of

MEDICINE.

the "red" and the "white." "Their properties are said to be identical. Castor-oil is regarded as purgative and useful in costiveness, tympanitis, fever, inflammation, &c. It is much praised for its efficacy in chronic rheumatic affections, in which it is used in various combinations. One of its synonyms is *vátári* or anti-rheumatic. The root of the plant is also said to be particularly useful in the local varieties of rheumatism, such as lumbago, pleurodynia, and sciatica. As a purgative, castor-oil is recommended to be taken with cow's urine, or an infusion of ginger or a decoction of the combination called *Dasamula.*" Thus, according to the above abstract of Sanskrit medical opinions, all that we now know of the drug was known to the Hindus from a period so ancient that it is difficult to believe the plant was not then extensively cultivated, if indeed it was not a native of the country.

The following passage from **Dr. Dymock's** *Materia Medica of Western India* deals mainly with the views held regarding this drug by the Muhammadan doctors :—" Mohametan writers describe two kinds, red and white; the red is said to be the most active. They consider the oil a powerful resolvent and purgative of cold humors and prescribe it in palsy, asthma, colds, colic, flatulence, rheumatism, dropsy, and amenorrhœa; of the SEEDS, ten kernels rubbed down with honey are sufficient as a purge. A poultice of the crushed seeds is used to reduce gouty and rheumatic swellings and inflammation of the breasts of women during lactation. The leaves have similar properties but in a less degree. The fresh juice is used as an emetic in poisoning by opium and other narcotics; made into a poultice with barley meal it is applied to inflammatory affections of the eye. The root-bark is used as a purgative and alterative in chronic enlargements and skin diseases; it is also applied externally. In the Bombay Lying-in Hospital the leaves are applied to the breasts of women to stop the secretion of milk. In the Konkan the oil is applied to eruptions of the skin supposed to be due to heat of blood " (*Dymock, Mat. Med. W. Ind., p. 704*).

Many writers draw attention to the action of the seeds when eaten. Thus **Guy & Ferrier** (*Forensic Medicine*) say that the seeds "act on the stomach and intestines with a violence quite disproportionate to the action of the oil which they would yield on compression. Two or three seeds act as a drastic purgative; three seeds have destroyed the life of an adult male in 46 hours and twenty seeds that of a young lady in five days, with symptoms of violent irritation of the stomach and bowels, and an appearance as of one affected with malignant cholera." (*Conf.* with paragraph on Chemistry, 516-517.)

Juice.
379
Leaves,
380

Dr. Shortt says that "the JUICE of the LEAVES is given internally to increase the flow of milk. Cattle are fed with the leaves with the same object; most cows eat the leaves freely and readily. Externally the leaves are applied to wounds and bruises, when it acts as a detergent." It will be found that in the paragraph below, under the heading of SPECIAL OPINIONS, a similar statement is made. In fact, while most writers hold that a poultice of the leaves will arrest the flow of milk, others, like **Shortt**, make the opposite statement. For example, **Drury** says : " The leaves heated and applied to the breasts, and kept on for twelve or more hours, will not fail to bring milk after child-birth. The same applied to the abdomen will promote the menstrual discharge." Most writers affirm, however, that,

Flowers.
381

applied externally to the breasts, a poultice of the leaves arrests the flow of milk. **Bellew** says that the FLOWERS are often used as a laxative medicine. In several of the Bombay Gazetteers it is stated that the fresh leaves are applied in guinea-worm in the form of a poultice. The dry root is also mentioned as being regarded as a febrifuge. **Dr. Buchanan-Hamilton** says that the oil, in addition to being taken internally as a purgative by the

people of Mysore, is used by the Sudras and lower castes to anoint the head | MEDICINE.
when they labour under any complaint which they attribute to heat in the
system.

SPECIAL OPINIONS.— § The following opinions regarding local uses
of castor-oil in the provinces of India have been obligingly furnished for
this work by medical officers in correspondence with the Editor :—" The leaf
applied as a poultice to the breast will bring on copious discharge of milk"
(*Surgeon-Major Lionel Beech, Coconada.*) " Useful where it is desirable to
give purgatives in connection with diseases of the pelvic organs " (*Brigade-
Surgeon G. A. Watson, Allahabad*). " The heated leaves applied to the
breast acts as a lactagogue" (*Honorary Surgeon E. A. Morris*). " The leaves
are used in painful joints" (*Assistant Surgeon T. N. Ghose, Meerut*).
" In cases of abrasion of the conjunctiva, common castor-oil dropped into
the eye is an efficient means of removing irritation. Its efficiency in the
diarrhœa of childhood is too well known to call for any special remark "
(*Surgeon S. H. Browne, M.D., Hoshangabad, Central Provinces*).

THERAPEUTICS.—It is unnecessary in a work like the present to deal
with the medicinal properties and methods of administration recognised and
followed in Europe with a drug so well known as castor-oil. The objects
of the present article are more to exhibit the historic facts, the properties
that have a bearing on manufacture, the local and special uses of the drug
known to the people of India, and the agricultural and commercial facts
that may be of value to those interested in the trade in the seed and oil.
The reader is, therefore, referred to the *Pharmacopœia of India and
Great Britain*, to the *Pharmacographia* of **Fluckiger & Hanbury**, and
other such works for the purely medical aspects of the subject.

The following account by an American writer of the preparation and
general properties of the medicinal oil may, however, be given, as it reviews
in an able manner all the literature that exists on the subject:—

EXTRACTION OF THE MEDICINAL OIL.—" This may be extracted from | Extraction
the seeds in three ways : 1 by decoction ; 2, by expression ; and 3, by | of Oil.
the agency of alcohol or other solvent. | 382

" The process by decoction, which has been practised in the East and
West Indies, consists in bruising the seeds, previously deprived of their
husk, and then boiling them in water. The oil, rising to the surface, is
skimmed or strained off, and afterwards again boiled with a small quantity of water to dissipate the acrid principle. To increase the product it
is said that the seeds are sometimes roasted. The oil is thus rendered
brownish and acrid ; and the same result takes place in the second
boiling, if care is not taken to suspend the process soon after the water
has been evaporated. Hence it happens that the West India oil has
generally a brownish colour, an acrid taste, and irritating properties.

" The oil is obtained in this country by expression. The following, as
we have been informed, are the outlines of the process usually employed
by those who prepare it on a large scale : The seeds, having been
thoroughly cleansed from the dust and fragments of the capsules with
which they are mixed, are conveyed into a shallow iron reservoir, where
they are submitted to a gentle heat, insufficient to scorch or decompose
them, and not greater than can be readily borne by the hand. The object
of this step is to render the oil sufficiently liquid for easy expression. The
seeds are then introduced into a powerful hydraulic press. A whitish oily
liquid is thus obtained, which is transferred to clean iron boilers, supplied
with a considerable quantity of water. The mixture is boiled for some
time, and, the impurities being skimmed off as they rise to the surface, a
clear oil is at length left at the top of the water, the mucilage and starch

RICINUS communis.	Chemical Composition of

MEDICINE.

having been dissolved by this liquid, and the albumen coagulated by the heat. The latter ingredient forms a whitish layer between the oil and water. The clear oil is now carefully removed; and the process is completed by boiling with a minute proportion of water, and continuing the application of heat till aqueous vapour ceases to rise, and till a small portion of the liquid, taken out in a vial, continues perfectly transparent when it cools. The effect of this last operation is to clarify the oil, and to render it less irritating by driving off the acrid volatile matter. But much care is requisite not to push the heat too far; as the oil then acquires a brown hue, and an acrid peppery taste. After the completion of the process, the oil is put into barrels, and sent into the market. There is reason, however, to believe that much of the American oil is prepared by merely allowing it to stand some time after expression, and then drawing off the supernatant liquid. One bushel of good seeds yields five or six quarts, or about 25 per cent., of the best oil. If not carefully prepared, it is apt to deposit a sediment upon standing; and the apothecary may find it necessary to filter it through coarse paper before dispensing it. Perhaps this may be owing to the plan just alluded to of purifying the oil by rest and decantation. We have been told that the oil in barrels usually deposits in cold weather a copious whitish sediment, which is redissolved when the temperature rises. A large proportion of the drug consumed in the eastern section of the Union has been derived, by way of New Orleans, from Illinois and the neighbouring States, where it has been at times so abundant that it has been used for burning in lamps, and for lubricating machinery. According to the census reports there were manufactured in the United States in the year 1880, 893,802 gallons of castor-oil, valued at $790,741. The process for obtaining castor-oil by means of alcohol has been practised in France; but the product is said to become rancid more speedily than that procured in the ordinary mode. Such a preparation has been employed in Italy, and is asserted to be less disagreeable to the taste, and more effective, than the common oil obtained by expression. According to M. Parola, an ethero-alcoholic extract and an ethereal or alcoholic tincture of the seeds operate in much smaller doses than the oil, and with less disposition to irritate the bowels or to cause vomiting." (*See Am. Journ. of Med. Sci., N. S., XIII., 143.*)

Chemical Properties.
383

Chemical and other Properties of the Oil. — " Pure castor-oil is a thick, viscid, colourless liquid, with little or no odour, and a mild though somewhat nauseous taste, followed by a slight sense of acrimony. As found in the shops it is often tinged with yellow, and has an unpleasant smell, and parcels are sometimes, though rarely, met with of a brownish colour and hot acrid taste. It does not readily congeal by cold. When exposed to the air it slowly thickens, without becoming opaque." "When cooled it becomes thicker, generally depositing white granules, and at about—18° C. (0·4°F.) it congeals to a yellowish mass." It is heavier than most of the other fixed oils; its sp. gr. having been stated to be 0·969 at 55°F. or 0·950 to 0·970. It differs also from other fixed oils in being soluble in all proportions in cold absolute alcohol. Weaker alcohol, of the sp. gr. 0·8425, takes up about three-fifths of its weight. It has been supposed that adulterations with other fixed oils might thus be detected, as the latter are much less soluble in that fluid; but Pereira has shown that castor-oil has the property of rendering a portion of other fixed oils soluble in alcohol; so that the test cannot be relied on. (*P. J. Tr., IX, 498.*) Such adulterations, however, are seldom practised in this country. Castor-oil is soluble also in ether, and in its weight of glacial acetic acid. Its proximate composition has been repeatedly investigated. By saponification it yields several fatty acids, one of which appears to be palmitic

R. 383

Products of India. 515

the Castor-Oil Plant.   (*G. Watt.*)   **RICINUS communis.**

MEDICINE.

acid $C_{16} H_{32} O_2$, but the prevailing acid is *ricinoleic acid*, $C_{18} H_{34} O_2$, which is solid below $0°C.$, does not solidify in contact with the air by absorption of oxygen, and is not homologous with oleic or linoleic acid, neither of which is found in castor-oil. Castor-oil is thickened if warmed with one part of starch and five parts of nitric acid (sp. gr. 1·25), *ricine-laidin* being formed. From this *ricinelaidic acid* may be obtained in brilliant crystals. Ricinoleic acid is converted by caustic potassa into caprylic alcohol and sebacic acid, with disengagement of hydrogen; and the same products are obtained by the reaction of potassa with the oil itself. (*See Journ. de Pharm.*, *Août*, *1855, p. 113.*) Its purgative property is supposed by M.M. Bussy and Lecanu to belong essentially to the oil, and not to any distinct principle which it may hold in solution. According to M. O. Popp, castor-oil differs from all other fixed oils in turning the plane of polarized light to the right (*A. J. P., Aug. 1871, p. 354; from Arch. de Pharm., 1871*). Various substances containing colouring principles which they yield to castor-oil produce beautiful fluorescence in that fluid when heated with it; such as logwood, turmeric, camwood, &c. (*Charles Horner, P. J. Tr., Oct. 1874, p. 282*). "Castor-oil, which is acrid to the taste, may sometimes be rendered mild by boiling it with a small proportion of water. If turbid, it should be clarified by filtration through coarse paper. On exposure to the air, it is apt to become rancid, and is then unfit for use" (*U. S. Dispens., pp. 1033—1035*).

CHEMISTRY.
384

CHEMICAL COMPOSITION.—It is of importance that a fairly complete statement of the chemistry of the subject be placed before the reader, and for this purpose the following passages from the learned authors of the *Pharmacographia* may be given. It will be seen that in some respects the information here supplied repeats what is contained in the above quotation from the *United States Dispensatory*. The account from the *Pharmacographia* is in some respects, however, a more recent statement, and it, moreover, is of a more directly chemical nature. The former passages exhibit the bearing of the chemical properties of the drug on the process of extraction of the oil. "The most important constituent of the seed is the fixed oil, called *castor oil*, of which the peeled kernels afford at most half of their weight.

"The oil, if most carefully prepared from peeled and winnowed seeds by pressure without heat, has but a slightly acrid taste, and contains only a very small proportion of the still unknown drastic constituent of the seeds. Hence the seeds themselves, or an emulsion prepared with them, act much more strongly than a corresponding quantity of oil. Castor-oil extracted by absolute alcohol or by bisulphide of carbon, likewise purges much more vehemently than the pressed oil. The castor-oil of commerce has a sp. gr. of 0·96, usually a pale yellow tint, a viscid consistence, and very slight yet rather mawkish odour and taste. Exposed to cold it does not in general entirely solidify until the temperature reaches $18°C.$ In thin layers it dries up to a varnish-like film. Castor-oil is distinguished by its power of mixing in all proportions with glacial acetic acid or absolute alcohol. It is even soluble in four parts of spirit of wine (·838) at $15°$ C., and mixes without turbidity with an equal weight of the same solvent at $25°$ C. The commercial varieties of the oil, however, differ considerably in these as well as in some other respects."

"The optical properties of the oil demand further investigation, as we have found that some samples deviate the ray of polarized light to the right and others to the left. By saponification castor-oil yields several fatty acids, one of which appears to be *palmitic acid*. The prevailing acid (peculiar to the oil) is *ricinoleic acid*, $C_{18} H_{34} O_3$; it is solid below $0°C.$,

RICINUS communis.	Chemical Composition of the Castor Oil Plant.

CHEMISTRY.

does not solidify in contact with the air by absorption of oxygen, and is not homologous with oleic or linoleic acid, neither of which is found in castor-oil. Castor-oil is, nevertheless, thickened if six parts of it are warmed with one part of starch and five of nitric acid (sp. gr. 1·25), *ricinelaidin* being thus formed. From this *ricinelaidic* acid may be easily obtained in brilliant crystals."

"As to the albuminoid matter of the seed, **Feleury** (1865) obtained 3·23 per cent. of nitrogen, which would answer to about 20 per cent. of such substances. The same chemist further extracted 46·6 per cent. of fixed oil, 2·2 of sugar and mucilage, besides 18 per cent. of cellulose. **Tuson**, in 1864, by exhausting castor-oil seeds with boiling water, obtained from them an alkaloid which he named *ricinine*. He states that it crystallizes in rectangular prisms and tables, which, when heated, fuse, and, upon cooling, solidify as a crystalline mass; the crystals may even be sublimed. Ricinine dissolves readily in water or alcohol, less freely in ether or benzol. With mercuric chloride, it combines to form tufts of silky crystals. soluble in water or alcohol. **Werner** (1869) on repeating **Tuson's** process with 30℔ of Italian castor-oil seed, also obtained a crop of crystals, which in appearance and solubility had many of the characteristics ascribed to *ricinine*, but differed in the essential point that when incinerated they left a residuum of magnesia. **Werner** regarded them as a magnesium salt of a new acid. **Tuson** repudiates the suspicion that *ricinine* may be identical with **Werner's** magnesium compound. **E. S. Wayne** of Cincinnati (1874) found in the leaves of **Ricinus** a substance apparently identical with **Tuson's** *ricinine*; but he considers that it has no claim to be called an alkaloid. The testa of castor-oil seeds afforded us 10·7 per cent. of ash, one-tenth of which we found to consist of silica. The ash of the kernel, previously dried at 100° C., amounts to only 3·5 per cent." (*Flückiger & Hanbury, Pharmacog., pp. 569-570*).

FOOD.
Seeds
385

Food.—Repeated mention occurs in works on this subject of a "sweet" form of R. communis, the SEEDS of which are edible. This statement is, however, as often contradicted as made. Thus **Rumphius**, when commenting on the account of this plant as given by **Dapperus** (in his work on Persia), says that that author gives it the following vernacular names:—Arabic, *kikajon, santjone, alkaroa* and *kerva*: Persian, *kuntsut*. His brief description of the plant, continues **Rumphius**, while it agrees on the whole with the facts known regarding the Indian and Amboyan plant, differs in an important respect (in which **Dapperus** was probably deceived), in that he says that a special form of the plant yields a sweet and pleasant oil which is employed for culinary purposes. **Rumphius** doubts the existence of any form of **Ricinus** yielding edible seeds. At the same time this very opinion which **Rumphius** discredits is frequently upheld both by Indian, Chinese, and African writers. Thus, for example, **Smith** (*Contr. Mat. Med. China*) says "a species or variety of **Ricinus** is said to have a smooth fruit, and to be innocuous. It must have been from some such Euphorbiaceous plant that a castor-oil is said to be obtained by the Chinese and used in cooking food." It is commonly reported that a kind of castor-oil is largely used by the Africans for culinary purposes, and by the Negroes of the West Indies. **Bellew** states that a form of castor-oil is used for that purpose also at Ghuzin. **Aitchison** remarks that the seeds are, at Jhelum, put into curries. The writer has no personal acquaintance with a non-purgative form of **Ricinus**, but he has been assured on the very best authority that, as the oil escapes from the cold presses used at the East Indian Railway works, it is always for a short time quite inert, and that it acquires its characteristic odour and purgative property only after being kept. There would, however, be little matter for surprise in the discovery of a form of **Ricinus**

FOOD.

which did not possess the purgative property, or which possessed it in so slight an extent that habit and use rendered it possible to eat the seeds or the oil without any after-effects, just as many cultivated states of Cucurbitaceous fruits which, in their wild form, are highly purgative, have been made inert by cultivation. Aitchison says that in the Hari-rud valley and Khorasan the use of the oil as a purgative is quite unknown, the oil being used purely for illumination purposes.

FODDER.
Leaves.
386
Cake.
387

Fodder.—According to most writers cows are fond of the LEAVES, and the natives of Madras regard them as tending to increase the flow of milk (*Conf. with Vol. III., 418*). The CAKE, after the expression of the oil, it has already been remarked, is in Europe regarded as highly injurious to cattle. Stocks and other Indian writers say, however, that it is regularly given to Indian cattle, especially buffaloes, and no mention is made of its being injurious. (*Conf.* with concluding paragraph regarding Orissa 518.)

DOMESTIC.
Stems.
388
Husks.
389
Timber.
390

Domestic Uses.—The dried STEMS and HUSKS after the extraction of the seeds constitute a highly combustible fuel which is largely used in boiling sugar-cane juice. The TIMBER is one of the poorest in India, an ordinary storm destroying the branches freely; but on being cut these dry and harden, and thus attain a considerable degree of strength. In that state they are largely used in the construction of the rafters for thatching, and as wattle in the walls of mud huts. The chief recommendation, it is said, for this purpose is the reputed immunity of these stems and twigs from the attacks of white ants and other insects. Dr. Shortt, however, remarks that the living plant is more subject to the attacks of white ants than almost any other agricultural crop. The interior of the stem, he says, is often entirely destroyed by these insects, and filled up with mud in the construction of their covered passages. This apparent contradiction is somewhat like the statement that, taken internally, the leaves increase the flow of milk, while applied externally they arrest it. These two statements would seem to call for further and more careful investigation. Bees are said to be fond of the castor-oil plant, and in this fact Mr. Bryce sees another recommendation in favour of castor-oil cultivation, since apiculture might be combined with castor-oil plantations, thus yielding an abundant supply of wax and honey. Except in certain very restricted regions, however, the natives of India have shown no desire to cultivate bees, and it is probable it will take a good many centuries before they can be made to advance so far as to call in the aid of the bee in supplementing their daily wants. Indeed, in the plains of India generally, bees could scarcely thrive, since honey-yielding-plants do not exist at the very season when climatically these insects should be most active.

Oil-cake.
391

The OIL-CAKE is a highly valued manure and a useful fuel. These facts have been repeatedly alluded to: but it may be added that in Mysore and other parts of India a special fuel is prepared which consists of a certain proportion of the oil-cake or boiled seed mixed with cow dung.

CASTOR SEED AND OIL.

CASTOR SEED & OIL.
392

The information on the subject of Indian CASTOR-OIL and -SEED, hitherto placed at the disposal of the public, having been of the most imperfect nature, a special effort has been made to bring together in this article the chief facts and figures contained in the official correspondence of the Revenue and Agricultural Department of the Government of India. The writer ventures to think that, while it has been found the better course to reproduce the material placed at his disposal, to a large extent in its original form, instead of the issue of a compilation from all sources in the shape of a separate account for each province, that, never-the-

Conf. with
pp. 550—552.

SEED
and
OIL.

less much new and interesting information will be found in the following brief notices of the castor- oil and seed industry of the chief Provinces of India. The editor would here beg to apologise to the authors
of the several reports and correspondence consulted by him for the liberty
he has, in some instances, taken in arranging their paragraphs in accordance with the system of dealing with agricultural subjects pursued in this
work. He trusts that this action will be found to have in no case distorted or injured the value of the original papers. Indeed, in some
cases, so valuable are these papers that it is much to be regretted that
information (such as has been discovered in the Proceedings of the Government of India) under Madras, Bengal, and the Panjáb was not available
for the other Provinces. Fortunately for the North-West Provinces and
Oudh, an excellent account exists in **Messrs. Duthie & Fuller's** *Field and
Garden Crops*, which, for the convenience of persons specially interested
in this crop, has been reproduced almost in its entirety. In the case of
Bombay a brief note by the Director of Agriculture (found in an official
letter of instructions regarding the cultivation recommended in connection
with seeds supplied to Europe) has been given as an introduction to a series
of quotations from the Gazetteers. These, it is believed, convey the chief features of the castor-oil cultivation and manufacture of Western India. In
concluding these introductory remarks it may be added that, while a separate section of the present article has been set apart for an account of the
EXPRESSION OF THE OIL, under that heading will be given only such papers
as more directly deal with that subject. It was, however, found undesirable
to break up the provincial information to such an extent as would allow of
the complete relegation of the facts given to separate paragraph headings.
The reader who may desire information on the various methods of preparation and purification of the oil will, therefore, have to glance over the facts
given below provincially before he reads the separate articles on the subject
in which he is more particularly interested. And this may also be said
of the paragraphs headed **Habitat, History,** and **Races** or **Varieties.** Certain information has been unavoidably repeated, and the facts in the
various paragraphs overlap each other, but the authenticity of the sources
of information have thereby been preserved, a feature of this work on which
the editor places considerable stress.

HISTORY.
393

HISTORY OF CASTOR-OIL.

The brief remarks made above, under the paragraphs **Habitat,** as
also **Medicine,** will have conveyed the impression that the writer is by
no means satisfied that it has been proved that the castor-oil plant may
not have been a native of India as well as of Africa. It has been repeatedly
pointed out (in this work) that M. DeCandolle has adopted a triple system
of determining whether or not a plant can be regarded as a native of a
certain tract of country, *viz.*, the historic facts that bear on the question, the
concurrence of the names given to it being ancient in the language or languages of the region in question, and the still further confirmation obtained
from botanical science, in the plant belonging to a family or genus that possesses uncultivated wild representatives. Everything points to the fact that,
in most tropical countries, castor-oil has been cultivated from a remote antiquity, and if M. DeCandolle's mode of determining nativity be applied to
India as well as to Africa, it seems probable the claim will, in the aggregate
of evidence, be made out in as strong terms for the one country as for the
other. It is true that no botanist has distinctly said that he has seen **Ricinus**
in India under such circumstances as to justify the affirmation that it is a
native. It is the sole representative of the genus to which it has been
assigned, so that the evidence that might be adduced from the existence

| History of Castor-Oil. | (*G. Watt*) | **RICINUS communis.** |

HISTORY.

Conf. with pp. 508, 544.

of other wild members is not available. The Natural Order to which Ricinus belongs (EUPHORBIACEÆ) is, however, by no means imperfectly repre-sented in this country. Out of a total of 197 Genera, India possesses mem-bers of 77, and the Indian species number 623, out of a grand total for the whole world of 3,000. As already shown, the castor-oil plant has been observed in the valleys of the outer Himálaya under circumstances that, in the case of **Citrus medica, Ægle Marmelos,** etc., have been accepted as sufficient to establish these trees as indigenous to India, though the castor-oil has been refused that position. While, therefore, botanical evidence is of a more or less negative character, the facts afforded by philology and his-tory are considerably stronger than is generally supposed. One of the earliest Sanskrit medical works (the *Susruta A'yurveda*) possesses unmis-takable reference to castor-oil, and speaks of the uses of the fresh parts of the plant in such terms as to preclude the possibility of an imported drug being alluded to. The names given to it by most of the subsequent Sanskrit writers denote its properties, and its oldest name *eranda* has passed into the most diversified languages and dialects of India with a persistency seen with almost no other economic plant. If, therefore, a wide-spread diffusion of a name indicates dispersion of the plant from the region where the original form of that name took its birth, the castor-oil plant must have either been made known, throughout the length and breadth of India, by the Aryan invaders, or those enlightened people found the plant in India on their arrival, and Sanskritized its aboriginal name. Whichever view be taken historic and philological evidence prove an ancient existence or cultivation, and afford no proof of a derivation from Africa, since the names given to the plant in this country are quite unconnected with those of Africa, Arabia, Egypt, or of ancient Greece and Rome. Indeed, even to the pre-sent day, the medicinal virtue of the oil is not known or at least not appre-ciated in the tract of country through which the plant would have had to be conveyed if it reached the frontier or shores of India through African or Arabian influence. Aitchison (*Notes on the Products of Western Afghánis-tán and North-Eastern Persia*) says imported " cold drawn" castor oil is well known as a medicine and so are the seeds of a CROTON, but neither the local oil nor the seeds of this plant are employed as medicine. Indeed, the people would never admit that the oil could be of use as a drug except at Hosenabad. Household tapers, *maluk*, are made by crushing the seeds along with raw cotton wool until the oil is expressed, and then rolling the cotton laden with oil into the form of " tapers." It may also be added, as having a certain significance, that the Muhammadans of India do not now (or at all events only rarely) give the plant or its medicinal products names de-rived from the Persian or Arabic. That the plant is not and never could have been a native of the supposed home of the Aryan race goes without saying. It is essentially a tropic-loving species, and on the Himá-laya is rarely seen in cultivation (except as an annual in the gardens of European settlers on the temperate tracts), above 6,000 feet in altitude. It is more than likely, therefore, that **Ricinus communis** was a native of South-ern Asia as well as of Africa, though it may have disappeared from its Asiatic wild habitat before the arrival, on the shores of India, of the earliest European writers. Whether native to India or not it must have been cul-tivated in this country before the date of the earliest Sanskrit medical works, since the Aryans were not likely to have learned so accurate a know-ledge of its properties prior to their invasion of India ; but it is very gene-rally accepted that the Sanskrit Medical works of the Hindus were penned in India. It is of less importance, therefore, to the present issue whether or not the early Aryans knew of it (before their arrival in this country), than that it was known to, and probably cultivated by, the people of India

R. 393

RICINUS
communis.   History of Castor-Oil.

HISTORY.   at the time Susruta wrote his great medical work. That fact fixes the Indian
acquaintance with castor-oil many centuries before the Christian era. But
there is nothing to disprove the suggestion that their knowledge of the plant
and its medicinal virtues may not have a still greater antiquity, prior even
to the Aryan invasion, or, say, 4,000 years ago. If this view be not accepted
the difficulty exists in accounting for Sanskrit and other Indian names
which are in no way traceable to external languages. M. DeCandolle
(in his *Origin of Cultivated Plants*) gives much interesting information
as to the historic and botanic facts which throw light on the origin of castor-
oil cultivation in other parts of the world. As the information he has collected in certain respects has a direct bearing on the question of Indian
cultivation, it may not be out of place to give here his views : "In no
country," he says, "has the species been found wild with such certainty as
in Abyssinia, Sennaar, and the Kordofan. The expressions of authors and
collectors are distinct on this head. The castor-oil plant is common in
rocky places in the valley of Chire, near Goumalo, says Quartin Dillon ; it
is wild in those parts of Upper Sennaar which are flooded during the rains,
says Hartmann. I have a specimen from Kotschy, No. 243, gathered on
the northern slope of Mount Kohn, in the Kordofan. The indications of
travellers in Mozambique and on the coast of Guinea are not so clear, but
it is possible that the natural area of the species covers a great part of tropical Africa. As it is a useful species, and one very conspicuous and
easily propagated, the Negroes must have early diffused it. However, as
we draw near the Mediterranean, it is no longer said to be indigenous. In
Egypt, Schweinfurth and Ascherson say the species is only cultivated
and naturalized. Probably in Algeria, Sardinia, and Morocco, and even in
the Canaries, where it is principally found in the sand on the sea-shore, it
has been naturalized for centuries. I believe this to be the case with specimens brought from Djedda, in Arabia, by Schimper, which were gathered
near a cistern. Yet Forskal gathered the castor-oil plant in the mountains of Arabia Felix, which may signify a wild station. Boissier indicates it in Beluchistán and the south of Persia, but as ' sub-spontaneous,'
as in Syria, Anatolia, and Greece."

"Rheede speaks of the plant as cultivated in Malabar and growing in
the sand, but modern Anglo-Indian authors do not allow that it is wild.
Some make no mention of the species. A few speak of the facility with
which the species becomes naturalized from cultivation. Loureiro had
seen it in Cochin-China, and in China 'cultivated and uncultivated,'
which perhaps means escaped from cultivation. Lastly, for the Sunda
Islands, Rumphius is as usual one of the most interesting authorities. The
castor-oil plant, he says, grows especially in Java where it forms immense
fields and produces a great quantity of oil. At Amboyna, it is planted
here and there, near dwellings and in fields, rather for medicinal purposes.
The wild species grows in deserted gardens *(in desertis hortis)*; it is
doubtless sprung from the cultivated plant (*sine dubio degeneratio domestica*). In Japan the castor-oil plant grows among shrubs and on the
slopes of Mount Wuntzen, but Franchet & Savatier add, probably introduced. Lastly, Dr. Bretschneider mentions the species in his work of 1870,
p. 20; but what he says here, and in a letter of 1881, does not argue an
ancient cultivation in China."

"The species is cultivated in tropical America. It becomes easily
naturalized in clearings, on rubbish-heaps, etc., but no botanist has found
it in the conditions of a really indigenous plant. Its introduction must
have taken place soon after the discovery of America, for a common
name, *lamourou*, exists in the West India Islands; and Piso gives another
in Brazil, *nhambuguacu, figuero inferno* in Portuguese. I have received

R. 393

the largest number of specimens from Bahia ; none are accompanied by
the assertion that it is really indigenous."

" In Egypt and Western Asia the culture of the species dates from so remote an epoch that it has given rise to mistakes as to its origin. The ancient Egyptians practised it extensively, according to **Herodotus, Pliny, Diodorus,** &c. There can be no mistake as to the species, as its seeds have been found in the tombs. The Egyptian name was *kiki.* **Theophrastus,** and **Dioscorides** mention it, and it is retained in modern Greek, while the Arabs have a totally different name, *kerrua, kerroa, charua.* **Roxburgh** and **Piddington** quote a Sanskrit name, *eranda, erunda,* which has left descendants in the modern languages of India. Botanists do not say from what epoch of Sanskrit this name dates ; as the species belongs to hot climates, the Aryans cannot have known it before their arrival in India, that is, at a less ancient epoch than the Egyptian monuments. The extreme rapidity of the growth of the castor-oil plant has suggested different names in Asiatic languages, and that of *Wunderbaum* in German. The same circumstances and the analogy with the Egyptian name *kiki* have caused it to be supposed that the *kikajon* of the Old Testament, the growth, it is said, of a single night, was this plant.

" I pass a number of common names more or less absurd as *palma Christi, girasole,* in some parts of Italy, etc., but it is worth while to note the origin of the name *castor-oil,* as a proof of the English habit of accepting names without examination, and sometimes of distorting them It appears that in the last century this plant was largely cultivated in Jamaica, where it was once called *agno casto* by the Portuguese and the Spaniards, being confounded with **Vitex agnus castus** a totally different plant. From *casto* the English planters and London traders made *castor.*" (*DeCandolle, Origin of Cultivated Plants, p. 422.*)

The writer has in more than one place in this work urged that the mere existence of a Sanskrit name for a plant is not by itself proof of vast antiquity nor of the knowledge having been possessed by the Aryan people prior to their invasion of India. In the case of plants alluded to in the Sanskrit medical works it should not be forgotten that the authors wrote in and of India. Where the names given by these writers do not occur in earlier Sanskrit works or are not derived from roots in that language of an earlier date, they may be Sanskritized forms of the aboriginal names for the plants. This idea is confirmed when, as in the case of **Ricinus,** these names cannot be traced to other countries, and when also the plant is itself a tropical species which was not likely to have existed in the supposed home of the Aryan race. While holding this view the writer is disposed to think that **DeCandolle,** in the above passage, has placed an unduly small reliance on the evidence obtained from the Sanskrit records of **Ricinus,** more especially in the light of the remarkable prevalence in India of forms of the Sanskrit name in almost every language and dialect of this vast empire.

The history of the modern introduction of castor-oil into European medical practice is instructive and may be here given in the words of the *Pharmacographia :*—" The castor-oil plant was cultivated by **Albertus Magnus,** Bishop of Ratisbon, in the middle of the thirteenth century. It was well known as a garden plant in the time of **Turner** (1568), who mentions the oil as *Oleum cicinum vel ricininum.* **Gerarde,** at the end of the same century, was familiar with it under the name of *Ricinus* or *Kik.* The oil, he says, is called *Oleum cicinum* or *Oleum de Cherua,* and used externally in skin diseases. After this period the oil seems to have fallen into complete neglect and is not even noticed in the comprehensive and accurate *Pharmacologia* of **Dale** (1693). In the time of **Hill** (1751)

RICINUS communis.	Cultivation of the

and **Lewis** (1761) *Palma Christi* seeds were rarely found in the shops and the oil from them was scarcely known. In 1764 **Peter Canvane**, a physician who had practised many years in the West Indies, published a *Dissertation on the Oleum Palmæ Christi, sive Oleum Ricini ; or (as it is commonly called) Castor Oil,* in which he strongly recommended its use as a gentle purgative. This essay, which passed through two editions, and was translated into French, was followed by several others, thus thoroughly drawing attention to the value of the oil. Accordingly we find that the seeds of **Ricinus** were admitted to the London Pharmacopœia of 1788, and directions were then given for preparing oil from them. Woodville, in his *Medical Botany* (1790), speaks of the oil as having '*lately come* into frequent use.' At this period, and for several years subsequently, the small supplies of the seeds and oil required for European medicine were obtained from Jamaica. This oil was gradually displaced in the market by that produced in the East Indies : the rapidity with which the consumption increased may be inferred from the following figures, representing the value of the castor-oil shipped to Great Britain from Bengal in three several years, namely, 1813-14, £610; 1815-16, £1,269; 1819-20, £7,102."

The facts connected with the growth and present position of the Indian castor-oil trade will be found dealt with in considerable detail in the section of this article (below) devoted to **Trade** (*p. 554*), as also in the concluding remarks in the provincial section under **MADRAS** (*p. 528*).

CULTIVATION.

Some short time ago the editor received from a commercial firm the following series of questions regarding castor-oil :—

1st— What is the cost of picking or collecting the seeds per bushel ?

2nd— What percentage of oil will one bushel of seed yield ? (*conf.* with various passages below, especially that of N.-W. P., *p. 541*).

3rd— What is the average yield of seed per acre ? (4 to 6 cwt. average good crop – *conf.* with para. on the subject below, *pp.* 531, 534, 540, 542, 544).

4th – What is the average number of plants per acre ?

5th— What is your present market value of seed per bushel ?

6th— What is the name of the best castor-oil producing plant, and where can the seed be obtained ?

7th— How often will the plant bear seed each year in your climate?

8th— What is the average weight of one bushel of castor-oil seed ?

9th— Is there any use for the oil-cake obtained from the seed after pressing ? (*conf.* with remarks under Oil in paragraph above at *p. 510*).

If to these questions he added one or two more, such as on the area in each province under the crop, the total production, amount used locally, and surplus available for export, &c., &c., the category of unanswerable questions would be complete. At the present moment the literature of the subject is so imperfect that all that can be done is to bring together a few of the more important passages from the publications and correspondence which the writer has been enabled to consult. It may be hoped that this admission, together with the list of questions quoted above, may, however, call forth more precise particulars, so that a revision of the present compilation may, at some early date, be possible. The answers to many of the questions may, however, be learned by a perusal of the present compilation.

CULTIVATED RACES AND VARIETIES.

Before proceeding to give the series of quotations that exhibit the main facts known regarding the cultivation of the plant and the production in

Castor-Oil Plant in Madras.	(*G. Watt.*)	RICINUS communis.

India of seed and oil, it may, in this place, serve a useful purpose to review the botanical facts that exist regarding the cultivated forms of the species. Fluckiger, DeCandolle, and most other writers on this subject allude to the existence of a remarkable multiplicity of forms as a proof of ancient cultivation. Some of these forms (as, for example, the Indian agricultural conditions) are purely *races*, while others are not even entitled botanically to so important a position, but are at most only *sports* produced by the gardener's art, which have greatly added of late years to the existing stock of ornamental foliage plants. A third class of forms, however, are of a considerably higher value, some of which, by various botanists, have been assigned positions as *species*, but by others have been reduced to *varieties* of one species— **Ricinus communis**. This latter view is that taken by **Dr. J. Muller**, the botanist, who has hitherto perhaps devoted the greatest attention to the subject. In **DeCandolle's** *Prodromus* (*XV., Pt. 2, pp. 1016-21*) he has assigned all the forms met with in the world to sixteen varieties which he has demonstrated can have no higher claim, since their characteristics are scarcely hereditary, while the forms pass one into the other by numerous transitions, and thus prove that they constitute a group which are hardly more than cultivated states of one species.

From the earliest European, and even still more so from the ancient Sanskrit works down to the present day, all Indian writers are, however, agreed that there are two chief forms met with, but that under each of these there are numerous modifications in colour and shape of leaves and presence or absence of spinous appendages on the fruits. These two great types have received various names which denote, on the one hand, the size of the seed; on the other, the colour of the stems, petioles, and leaves, or perchance the stature and persistence of the plants. The one form is a tall bush or almost tree, a perennial grown as a hedge plant or to afford shade around fields in which more delicate crops are being cultivated. This yields a larger seed and abundance of an inferior oil which is generally expressed by a different process from that adopted with the other form. The second race is an annual plant, sometimes grown as a pure crop, though more frequently in rows or lines through a mixed field containing other crops. It yields a small seed, the better qualities of which, by an expensive and careful process, yields the superior qualities of the oil of commerce and of pharmacy. The oil obtained from the first race is largely used in India for illumination purposes and is hence often called "Lamp oil," but it also finds a place as a lubricating oil, and for many of the industries of Europe where the more expensive article would prove a superfluous extravagance. It is commonly stated that the bushy form of the species not only affords an oil which, through the method of preparation, possesses far less of the purgative property than that of the annual plant; but that apart from changes effected by the system of expression, the oil is naturally less purgative. Whether or not this be actually the case the writer cannot definitely say, but if it be so there would be little difficulty in believing that a development of that property might have resulted in the production of a form of **Ricinus** that afforded edible seeds and a sweet oil such as is frequently mentioned by writers on the subject.

I.—MADRAS.

Mr. **J. F. Price**, Acting Director of Revenue Settlements and Agriculture, Madras, furnished the following valuable statement of the castorseed and oil industry of that Presidency :—

VARIETIES.—There are two chief and distinct varieties of the castor-oil plant. These are known as the "*Large*" and "*Small*" kinds, and, though

Side notes:
CULTIVATION in Madras.

Races & varieties.

MADRAS.

Varieties. 396

RICINUS
communis.   Cultivation of the

CULTIVATION
in
Madras.

Varieties.

called by different names in various parts of the country, are everywhere distinctly the same. Besides these, however, there are the following : The "*Pyra*" peculiar to the Kistna District, and remarkable from having no lateral branches. This is always grown as a mixed crop. The "*Mulli-kottai,*" which is a variety found in the Coimbatore District. It is of small size and the capsules are smooth instead of being prickly. The Godavari and Kistna seeds are very much alike, being small and grey in colour. They find much favour in the Calcutta market owing to the quantity of oil which they yield. Salem is the largest in point of size and is red in colour. It is much esteemed in this Presidency. The Bellary seed is of medium size and is also red. The large-seeded form is distinguished both by the size of the plant and the seeds. It is always grown as a separate crop, and it is from this that lamp oil is principally obtained. The other variety, which is small, both as regards size and seeds, is almost always cultivated as a mixed crop, with millet and Bengal-gram, and the oil procured from it is used, chiefly, for medicinal purposes.

LARGE FORM.
Method of
Cultivation.
397

I.—Large form of Castor-oil plant.

Method of Cultivation.—The method of cultivation appears to be pretty much the same everywhere, though it varies in slight particulars in different districts. The annexed account is one taken from a paper drawn up by the Deputy Director of Revenue Settlement, South Arcot District, giving an account of the mode of cultivation of castor-oil in NORTH ARCOT, and this may be taken as typical for the whole Presidency.

Season.
398

SOIL : SEASON OF SOWING, &c.—The best ground for the cultivation of the castor-oil plant is red loam, that at the foot of hills being in most request. It also grows well on alluvial earths. The land is ploughed, in May or June, thrice with the ordinary native plough. A large flock of sheep is, if procurable, picketted on the field, for a few nights : if these cannot be obtained, 24 cart-loads of manure, equivalent to about 7 or 8 tons, are applied to each acre. After the manuring has taken place, the ground is ploughed again twice. In July or August and after the first showers have fallen, the sowing takes place. One plough forms a furrow. The man sowing follows and drops a seed into this at intervals of about a foot. A second plough follows the sower, covering the seed in. From 16 to 24℔ of seed are said to be required for one acre of land. This, however, seems too much ; 12 to 14℔ would probably be nearer the mark.

In the GODAVARI District, there is some difference in the mode of cultivation. There is no special manuring, the reason of this, apparently, being that much of the land upon which the castor-oil plant, which is said to be so exhausting that it is never put down for two successive years on the same ground, is grown, is rich deltaic soil.

Method of
Sowing.
399

METHODS OF SOWING.—The modes of sowing are threefold, *viz.,*—

1st—Dibbling holes, with a heavy pointed stick, some 1 or 1½ yards apart, and putting into each two seeds, on which a little water is poured, the hole being then filled up. This is considered the best method.

2nd—A plough is driven along the field which has previously been well turned up. Behind this follows another plough to the pole of which a hollow bamboo tube, with a perforated cocoanut shell fixed to the top of it, is attached. A man walking alongside this drops, at every yard, two seeds into the shell, which fall through the tube to the bottom of the furrow cut by the first plough and are then covered by the sowing plough. This is the course usually followed with the large variety.

3rd—Is a method identical with that adopted in North Arcot and already described. The quantity of seed employed when sowing according

R. 399

to the first mode is about 5℔, with the other two 7 or 8℔. This is much less wasteful than the system holding in North Arcot.

IRRIGATION : WEEDING.—The castor-oil crop is never irrigated, and it depends entirely upon what rain may fall. The seeds germinate in about a week, and a month after this, the land is ploughed twice or thrice to clear it of weeds and thin the crop. At this stage, the plant is not infrequently subject to the attacks of caterpillars. The remedies employed are sprinkling with ashes and removing the insects by hand.

Nothing more in the way of cultivation is subsequently done to the crop.

AFTER-TREATMENT.—In the fourth or fifth month after sowing, the flowering begins, and in the sixth the capsules are formed. The picking is taken in hand in the seventh month and terminates with the ninth, when the cattle eat off what leaves remain and the stems are cut as fuel.

MODE OF HARVESTING.—The seed-pods are gathered by hand and are stacked in the corner of a house and covered with straw. A weight is then placed on the heap in view to excluding the air. After a period of six days, the capsules become soft and the envelopes rotten. They are then exposed to the sun for two days and, when thus dried, are beaten with a heavy wooden mallet about 2 feet long and 1½ broad. This removes about half the seed.

The remaining capsules are again dried and the beating process is repeated, which results in the husking of the balance of the seeds. The whole is then cleaned and is ready for the next process—that of extracting the oil. The refuse obtained in shelling the beans is either put into the manure heap, or else is mixed with cowdung, beaten into cakes, dried and used as fuel.

II.—Small form of Castor-oil Plant.

METHOD OF CULTIVATION.—The foregoing account relates to the cultivation of the large variety. The small species is usually grown with dry crops, being sown by driving a plough at pretty wide intervals over the field directly after the other crops have been put down. The seeds are sown about a yard apart. The plant continues bearing as long as five years, but the outturn is less than that of the large description, though its value is said to be 50 per cent. higher. The oil of the small kind only is, as a rule, used for medicinal purposes.

YIELD AND COST OF CULTIVATION.—It has been very difficult to obtain anything approaching accurate information as regards cost of cultivation, outturn, and value of the produce per acre. This has partly arisen from carelessness and want of interest on the part of many of those from whom enquiries had to be made. I have, however, done the best that I could. It is highly probable that the outturn is understated and that the estimate of cost of cultivation is over the mark. It must be borne in mind, as regards the latter, that a large portion of the charges shown are the estimated value of work performed by the ryot himself and his family. The calculations are all based upon the assumption that the cultivator does nothing and that everything is carried out by means of hired labour. This of course is not the case In the cost of cultivation is included the assessment paid to Government. No allowance has, in any instance, been made by the Revenue officers, for the value of the leaves used as fodder or or of the stalks cut for fuel. These items cannot be very large and still they would amount to something.

YIELD.—The average estimated outturn per acre is 473℔ of cleaned seed, the highest 884℔ which is the outturn, when grown as a single crop, and the lowest 232℔ which is for castor grown with other crops.

CULTIVATION in Madras.

Large Form. Irrigation. 400

After-Treatment. 401

Harvesting. 402

SMALL FORM. Method of Cultivation. 403

Cost. 404

Yield. 405

R. 405

CULTIVATION
in
Madras.
Small Form.
Value, Cost
& Profit.
406

VALUE.—The average estimated value of the outturn per acre is R15-8-5, the highest is R28 and the lowest R5-13-4.

COST.—The average estimate of cost of cultivation is R10 per acre, the highest is R15-8-0 and the lowest R4-15-6.

PROFIT.—The average estimated profit per acre is R5-8-5, the highest is R17-10-6 and the lowest R0-13-10.

I am inclined to think that growing the castor-oil plant is a more profitable pursuit than would appear, from the figures now given, to be the case.

Extraction of
Oil.
407

EXTRACTION OF OIL.--There are three processes by which the oil is extracted :—

I.—The seed is roasted in a pot, pounded in a mortar and placed in four times its volume of water, which is kept boiling. The mixture is then frequently stirred with a wooden spoon. After a time the pot is removed from the fire and the oil skimmed off. The residue is then allowed to cool and, next day, is again boiled and skimmed. The oil thus procured is superior to that first obtained and is kept separate.

II.—The seed is first boiled and then dried in the sun for two or three days. It is then pounded and the further process is as in No. I. The oil thus manufactured is of superior quality and is used only for lamps and greasing cart wheels.

III.—The seed is soaked for a night in water and, next morning, ground up in the ordinary native oil-mill. The oil is removed by putting a piece of cloth into the pulp and then squeezing it into a pot. This oil is used for lamps and dyeing purposes.

The cake appears to be used only for fuel.

Test.
408

EXPERIMENTS TO TEST YIELD OF OIL.—Experiments were made at the Government Farm at Saidapet in view to ascertaining the relative merits of the various methods of expressing castor-oil. The quantity of beans operated upon, in each case, was 100℔. In experiment No. I, the oil was cold drawn, the seed being crushed in a screwpress with horizontal rollers and the resulting pulp put into gunnies and pressed. In experiment No. II, which was divided into A and B, extraction was effected by the ordinary native process of roasting and pounding the seed and boiling the result in water, the oil being, as it rose, skimmed off. The difference in results between A and B was chiefly the result of the over-roasting of the beans in the former case. In experiment No. III, the ordinary native mill was employed and the oil, after the pulp had been well pressed, was boiled. The oil obtained in No. I experiment is stated to have been very pure : in the others it was dirty.

The annexed statement exhibits the outturn in each experiment.

	No. I.	No. II.	No. III.
		A B	
Oil per cent. ℔	36·5	27·0 32·	30·43
Cake	36·8	} Not deter-	{ 43·48
Husk and Wastage	27·7	} mined.	{ 29·09
	100·		100·

It will be observed that the yield of oil obtained by cold drawing was considerably larger than by the other processes. The general native estimate of the yield of castor-oil is 25 per cent. on the quantity of beans used. This is probably pretty nearly the real figure. The experiments

R. 408

at the Government Farm were conducted under European supervision and were carefully carried out : this would account for the increase in yield.

Trade in Madras Castor-seed and Oil.

OIL.—Nearly all the oil manufactured in this Presidency (I believe myself correct in saying) is consumed locally, the main reason being that it is very difficult to get ships to carry oils and that none will do this with other goods. The trade of extracting the oil is a profitable one. It is to be regretted that it cannot be carried on in this country to a greater extent, as the cake which is a very excellent manure is lost to us when the seed is exported. Through the courtesy of a mercantile friend, upon whose statement I can place reliance, I am in a position to afford information which, though your department may be in possession of the major portion of it, may be of interest to mercantile men not connected with Madras.

THE TRADE IN CASTOR-SEED is one which has increased of late years in a very marked degree. In 1875-76, 22,737 cwts. went to foreign countries ; in 1883-84 the figures were 446,029. The coast exports which are chiefly to Bengal and which have not increased in the same degree, were, in 1875-76, 468,791 cwts. ; and in 1883-84, 586,481 cwts. The bulk of the export to foreign countries goes to Marseilles and Venice. There are also shipments for Sebastopol, Odessa, and Lisbon. During the last two years from 70,000 to 120,000 cwts. were shipped to London, partly for consumption in England and partly for re-shipment to the Continent and America. Coconada is the port from which there are the largest exports. The month in which the seed is available is July. Masulipatam and Kotapatam are also ports whence heavy shipments are made. Salem and Bellary furnish the major portion of what goes from Madras.

The remarkably rapid manner in which the use of kerosine oil is superseding that of the vegetable oils for lighting purposes is, so my friend is of opinion and, I think, with some reason, one of the causes of the great increase in the export of castor-seed. He writes : " The seed is exported only when there is a large excess over the internal requirements of the country, but in consequence of the great fall in the value of kerosine oil that excess is likely for the future to be large and more regular."

The above account of the Castor-oil of Southern India embraces the chief facts brought out in the official records, consulted by the Editor, but many of the district notices are instructive. For example, Mr. H. R. Farmer, Sub-Collector, Cuddapah, says that " In this part of the country there are three kinds of Castor-seed :—

(i) Chittamudalu : (ii) Chennamidalu : and (iii) Peddaamidalu.

" The first is a small seed rarely cultivated. The second is medium sized and chiefly cultivated in dry lands, and the third is the large seed usually grown in betel and sugarcane gardens. It is generally cultivated with other crops but sometimes alone. The plant flourishes in soft loamy soil and requires ordinary ploughing and manuring. After a fall of rain in July or August, the land is first ploughed and the seeds handsown in parallel rows. The rows and the seeds are at a distance of one foot from each other. The shoots spring up in 10 days, and the land is turned up with the plough after 20 days. This is believed to secure a better harvest. The plant begins to bear berries in about 4 months. These berries are collected with the hand as often as they are ripe at about the end of January. They are heaped up for a couple of days and then dried in the sun and the seeds are separated from the outer shell. An acre of land takes from 4 to 6 Cuddapah measures of seed.

**TRADE.
Oil.
409**

**Seed.
410**

**Cuddapah.
Varieties.
411**

**Method of
Cultivation.
412**

R. 412

RICINUS communis.	Cultivation of the

TRADE.

Yield & Percentage of Oil.

413

Uses of the Oil & Oilcake.

414

"The average outturn is stated to be about ¾ *putti* or nearly 4¾* cwt per acre.† Twelve measures of seeds (27℔) produce by the process of home manufacture 3 measures (7℔) of oil. A measure of oil generally weighs 144 *tolahs* or 3¾ ℔. Thus 27℔ of seeds produced (3 × 3¾) 11¼℔ of oil. The percentage can, therefore, be stated to be nearly 41. The oil is used for lamps and as an unguent. It is, of course, a well-known purgative. After extracting the oil what remains of the berries are used as manure for betel and sugarcane gardens."

The information published by **Dr. Shortt** (*Man. of Indian Agriculture*) is to a large extent a reprint of **Lieutenant Hawkes'** report in castor-oil, a publication which appeared in 1853 (see the reprint in full below, *p. 547*). In some passages **Dr. Shortt** would appear to be even incorrect as, for example, when he says (p. 243) that "the small-seeded variety grows into a large umbrageous tree 30 to 40 feet in height." He adds that he had on his estate trees that measured four feet in girth, one foot above the soil. "It is a handsome tree, and seeds freely, yeilding 15℔ of seed per tree per annum." Of the large-seeded form **Dr. Shortt** simply remarks that it yields the lamp oil of the bazárs "so expressively termed *veluku yennie*," but he does not inform us to what height it grows. His remarks regarding percentage, yield of oil, and production per acre being apparently based on personal experiments may be here quoted :—

Yield of Oil and of Seed.

415

"From my own personal experiments, one pound of oil is obtained from three pounds of oil seed. The process consists in first scorching the seeds in an earthern pan over a fire, and pounding them in a mortar to reduce them to flour. Two bottles of water are boiled, and into the boiling water, the pounded castor seeds are thrown and the mixture is well and constantly stirred with a wooden ladle for a time till the oil gradually rises to the surface, and is skimmed off. A fresh supply of boiling water is added to the mass in the vessel, and it is boiled for the second time to remove any remnants of oil it may contain. The whole of the oil obtained is now boiled to evaporate any water it may contain and the oil is then ready to be bottled for use. Care must be taken to see that the oil does not get burned in boiling.

The average outturn per acre of seed is about 5 cwt., and 12 measures of seeds equal to 27℔ produce 3 measures or 7℔ of oil. A measure of oil weighs 144 *tolahs* or 3¾℔, thus 27℔ of seeds produce 11¼℔ of oil and the percentage may be reckoned at 40 to 45 per cent."

MYSORE.

416

MYSORE & COORG.

Dr. Buchanan-Hamilton in his most instructive report of his journey through these provinces (published 1807) gives the main facts which have since been simply republished by most subsequent writers. The following two passages out of some seven or eight gives the chief ideas regarding the forms grown, methods of cultivation and yield. The process of preparing the oil narrated by **Dr. Buchanan-Hamilton** will be found in the paragraph below on that subject :—

"*Haralu* is the **Ricinus palma-christi** of **Linnæus**. In the *Ashta gram* two varieties of it are common; the *chica*, or little *harulu*, cultivated in gardens; and the *doda*, or great *harulu*, that is cultivated in the fields and the plant of which I am now to give an account. In the spring plough five times before the 8th of May. With the first good rain that happens afterwards, draw furrows all over the field at a cubit's distance,

---

* The *Madanapalle putti* contains 320 Cuddapah measures.
† One measure (holding 132 tolahs of second sort rice) of castor-seeds weighs 88 tolahs and 38⅔ tolahs are taken as equal to 1℔.

| Castor Seed and Oil. | (*G. Watt.*) | RICINUS communis. |

and, having put the seeds into these at a similar distance, cover them by drawing furrows close to the former. When the plants are eight inches high, hoe the intervals by drawing the *cuntay* first longitudinally, and then transversely. When the plants are a cubit and a half high, give the intervals a double ploughing. The plant requires no manure, and in eight months begins to produce ripe fruit. A bunch is known to be ripe by one or two of the capsules bursting; and then all those which are ripe are collected by breaking them off with the hand. They are afterwards put into a heap or large basket; and the bunches, as they ripen, are collected once a week, till the commencement of the next rainy season, when the plant dies. Once in three weeks or a month, when the heap collected is sufficiently large, the capsules are for three or four days spread out to the sun, and then beaten with a stick to make them burst. The seed is then picked out from the husks, and either made by the family into oil for domestic use, or sold to the oil makers." " It is cultivated on the two best qualities of land, and on the better kinds of *Marulu.* When the same piece of ground is reserved always for the cultivation of this plant, the succeeding crops are better than the first; when cultivated alternately with *ragy*, it seems neither to improve nor injure the soil for that grain."

Then again, he adds, further particulars of interest :—"*Harulu* is cultivated on a particular soil, which is reserved for the purpose, and consists of ash-coloured clay mixed with sand. There are here in common use three kinds of *harulu :* the *phola* or field; and the *doda* and *chittu*, which are cultivated in gardens. A red kind is also to be seen in gardens, where it is raised as an ornament. The *chitt'-harulu* produces the best oil. Next to it is the *phola* that is cultivated in the fields. In the course of a few days, any time in the three months following the vernal equinox, plough three times. With the next rain that happens, plough again, and at the same time drop the seeds in one furrow at the distance of one cubit and a half, and then cover them with the next furrow. A month afterwards hoe with the *cuntay*, so as to kill the weeds, and to throw the earth in ridges towards the roots of the plant. It ripens without further trouble. At the time the *harulu* is planted, seeds of the pulses called *avaray* and *tovaray* are commonly scattered throughout the field. In four months after this, the *harulu* begins to produce fruit, and for three months continues in full crop. For two months more it produces small quantities."

II.—BENGAL.

BENGAL. 417

Conf. with pp. 549-551.

Rajshahye. 418

In answer to a circular letter, issued by the Government of India in the Department of Revenue and Agriculture, calling for information regarding the cultivation of the Castor-oil Plant in Bengal, the following particulars were elicited from Divisional and District officers :—

1. Rajshahye and Cooch Behar Division.

CULTIVATION.—Lord H. Ulick Brown reported as follows :—" The plant is not much cultivated in this division. In RUNGPORE no castor seed is cultivated ; but the plants are sometimes found growing wild in the fields and jungles, no notice being taken of them by the cultivators. It is cultivated in the northern part of the Hurrypore estate in Dinagepore for the purpose of manufacturing *endí* cloth from silk obtained from cocoons fed on its leaves, and the oil is used for dyeing a cloth called *fotas* worn by the women of the lower classes in that district. Two sorts of castor plant are grown. The leaves of one are of a purple tinge, and of the other green. The soil upon which the plant is cultivated is of sand and clay mixed. The land requires constant ploughing and the seed is dropped into the furrows by a man following the plough.

| RICINUS communis. | Cultivation of the |

EXPRESSION OF OIL. "The method of extracting the oil is as follows : — The seeds are first roasted in a pan, then husked and crushed, next placed in an earthen vessel and boiled in water mixed with alkali; the oil then rises on the surface and is taken out and put into a separate vessel. The oil thus obtained is again boiled and purified."

Chota Nagpore.
419

2. Chota Nagpore Division.

The castor-oil plant is not regularly cultivated in any part of this division. It is grown to a small extent by agriculturists and others, chiefly on the borders of their *bari* or homestead lands. Oil is extracted from the seeds according to the usual native crude process. It is used for burning in lamps and for medicinal purposes.

Chittagong.
420

3. Chittagong Division.

The cultivation of castor-seed is very rare in this division.

Patna.
421

4. Patna Division.

Mr. F. M. Halliday, Commissioner of Behar, furnished the following information regarding the cultivation of Castor-oil in that Division :—

CULTIVATION.—"Three sorts of plants are grown in the Patna District— the *bhadoi*, also called *bugrarah ;* the *basanti* or popularly called *saluk reori* or *gohumi ;* and the *chanaki*.

"The *bhadoi* or *bugrarah*, so called from its being grown along with other *bhadoi* crops, is sown in May and begins to ripen in January and February. The seeds of this sort are larger than those of the *basanti* and *chanaki*, but contain a larger quantity of shell. The *basanti* or *gohumi*, or *saluk reori*, is sown in the beginning of September and in the latter end of August, and begins to ripen in March and April. This sort is largely sown in *dofasli* (double cropped) lands, and the seeds fetch a good price in the market on account of their containing a larger quantity of oil than the *bhadoi*. The third or *chanaki* is cultivated to a comparatively small extent, and is so called from the seed being thrown to a distance from the plants when the fruits ripen. Its quantity is good, but the bursting of the pods causes loss.

"In the District of Patna, the plants grow to a height of from 2 to 16 feet. Castor seed, more or less, is sown throughout the district, but as a rule only in the *dearahs*. On these river banks the yield is larger, and very little trouble is necessary for the cultivation. It is sown in *panka* (moist land of *dearahs*) and sandy loam fields left uncultivated, and ploughed throughout the year, yield a much larger quantity of castor seed than those in which it is grown with other crops. The seeds are never soaked in water, but are sown on dry land at a distance of 2 to 2½ feet apart in furrows made by the plough or irregularly. When the seeds spring up and grow to a height of about 1½ feet, the fields are ploughed again, and the process is repeated at an interval of 15 to 20 days. No irrigation is necessary where the soil is very dry. From 3 to 4 seers of seeds are required for the sowing of one *bigha* of land=27,225 square feet. The seeds are sown in May, August, and September, and harvested in January, February, March, and April.

Yield.
422

YIELD.—"The yield varies according to the nature and quality of the soil from 4 to 12 maunds per *bigha*. A portion of the produce is retained for local consumption, and the rest purchased by merchants and exported to Calcutta.

Expression of Oil.
423

EXPRESSION OF OIL.—"All the home processes described in the 'Note' issued by the Government of India (*see pp 549-551*) are resorted to in the Patna District. The oil thus obtained is very impure and is used in burning in native lamps.' It is very coarse and thick, and of a deep brownish colour. The general rate at which it is locally sold is 4 annas per seer. The refuse is used as manure for fields; 2½ seers of *basanti* and 3 seers of *bhadoi* seeds give a seer of oil by this process. The kernel is not separated from the outer shell in the extraction of oil. Hydraulic presses are not used at all, but screws manufactured by **Jessop & Co.**, and by **Madhub Bysack** are used in the Dinapore Sub-division, and the ordinary oil mill is also employed in the Behar Sub-division."

"In the SHAHABAD District the seed is not widely cultivated. It is chiefly sown to fill up spaces in sugarcane or *rahar* fields, and. rarely, to form a protective hedge to other crops. In the southern part of this district the plant growing on sandy soil yields seeds of smaller size, and to obtain large seeds irrigation is necessary once or twice. The outturn per acre, when grown with other crops, is estimated at about 3½ to 4½ cwt. The seed is not soaked in water before it is sown Bursting of the seed is the sign of its maturity, and as soon as one or two of the pods have burst, the cluster is plucked off by the hand,' placed under the shade for a few days, and then exposed to the sun for the separation of the outer husk."

" There are two different modes of extraction of oil observed in this district :—

"(1) The seed is roasted in an earthen pot and then put into a mortar and pounded. The shell is then separated and the kernel boiled in water, and the oil on coming to the surface is skimmed off.

"(2) The seed is neither broken nor husked,' but crushed entire and pounded by an apparatus called *dhenki ;* water is added whenever required.

" The yield of the oil varies from 25 to 30 per cent. of the seed.

" Of the sub-divisions of the MOZUFFERPORE District, that of Hajeepore grows the largest area of castor-oil. Here the plant is of two kinds—one grown in high lands and cultivated with other crops, and the other on the sandy soils of the *dearah* lands, and cultivated alone. The following is a description of the castor oil plant cultivated on high lands with other crops :—

" Height 8 to 12 feet. In well-moistened fields the plants throw out the stems from their very roots and yield fruit in each of these stems ; each bunch of fruit contains 50 to 100 shells. The colour of the seeds is shining black, dotted with red. The seeds are sown twice a year. In the month of *Asarh* (June and July), they are sown in spare lands along with *makai, rahar,* and cotton, and in the month of *Kartick* (October) with potatoes. No special preparation of the land is made, nor is any process of manuring the soil used. The seeds are sown dry, at a distance of a yard apart from one another. Ordinarily five seers of seeds are sown in a *bigha* of land. Those sown in the month of *Asarh* require no irrigation, that being the rainy season ; but those sown in *Kartick* require watering, which is done for the benefit of potatoes, and consequently serves the double purpose of irrigating both the crops. The fruit of both the sowings begins to ripen at one time in the month of *Chait* (March), and extends to the middle of *Jeth* (May). The time of plucking the bunches is uncertain. The cultivator's children keep a sharp eye on the bunches, and when they find a shell out of a bunch burst, and the seeds likely to fall out, they pluck off the cluster and keep it carefully in the shade inside the house. They do the same with other bunches till the plant becomes fruitless. The fruits are then placed inside a ditch prepared for the purpose. A little water, mixed with cowdung, is sprinkled over them, and they are then covered over with a mat or piece of gunny. They are kept so for three days, and then taken out and put in the sun, when the shells leave the

CULTIVATION
in
Bengal.

Shahabad.

seeds. They are then crushed, and the oil extracted. Seeds which are
preserved for sowing are cleaned in a different way. The bunches are
kept exposed to the sun for two or three days, and when thoroughly dry
they are well thrashed on a plank which causes the shells to separate
from the seeds. The cost of cultivation is not much, as only secondary
attention is paid by the cultivators to it. Five seers of seeds at the present
time cost about eight annas. The sowing of the seeds and the plucking
of the fruits are generally done by the wives and the children of the
cultivators. In large cultivations labourers are employed and are
generally paid in kind with a handful of grain. In some parts of the sub-
division they are paid in cash at the rate of one anna each for the labour
of the whole day. No attention is paid to the plants after the fruits have
been plucked, and they are generally broken down for fuel.

" The large-seeded form only is cultivated in this sub-division. The
one sown in the month of *Asarh* is called the '*banghi*,' and the other
sown in *Kartick*, the *pashtahi* or *jethna*. The outturn is from five to six
maunds per *bigha*. Sixteen seers of oil extracted from a maund of
seeds. The seeds sell at two to four rupees per maund, and are generally
purchased by merchants and traders from the cultivators direct. The sale
of oil depends on the outturn of the produce; when it is above the average,
it sells at five or six seers for the rupee, otherwise four to five seers. Petty
cultivators extract oil for their own use for burning in lamps.

" The plants growing on *dearah* lands are called '*dearah rendis*,'
height 5 to 6 feet, colour white, kernel small, and contains less oil than
that of the highland plants. These are cultivated alone in fields like
other crops and the fields are prepared similarly as those of *rabi* fields.
A hole is made in the ground, and two seeds are sown together with heads
down. Each hole is made 4 feet apart from the next hole. The sowings
begin in the month of *Assin*. A *bigha* of land takes five or six seers of
seeds, and in a field well-looked after and occasionally dug up gives an
outturn of four to five maunds. The weeding and clearing of the fields
begin in *Aghran*, or as soon as the seeds germinate and continue so till
the plants grow to a height of three to four feet. When the plants are
growing the intermediate earth is well dug up. Labour is paid at the rate
of one anna each with food, the cultivation lasting about four days. At
the beginning of digging, or *tumnying* in native phraseology, the plants
are made to head down by the operation of cutting off the small scattered
roots, leaving the principal or central root to nourish the plants. The heading down of the plants produces several branches, each one coming out
from the different junctions of the stems, and each branch brings forth three
to five clusters containing 25 to 30 fruits each. Should the plants be
allowed to grow straight they will have one bunch only. The other particulars are almost the same as reported above.

" In the Sitamarhi Sub-division the plant can be grown almost on any
kind of soil. The best land suited for it is sandy clay—sandy about 6
to 8 inches, and then good clay below for 4 to 5 feet deep. It grows
luxuriantly and gives an abundant outturn in new lands either just cleared
from forests, or which had the advantage of being overflooded during the
rains. The *dearah* lands are the best suited for it, where the yield is at
least 16 maunds an acre. Here it is grown in almost all sorts of soil, and
is sown twice during the year—once in *Magh* (January) and then in
Asarh (July). The field is first well prepared, much like a field for
wheat sowings. It requires ploughing up six times at a cost of R1-8.
In this part the seeds are not soaked in water, but sown in lines with hand
one yard apart; 12 to 14lb of seeds are required if it is sown alone, and its
cost is 8 annas. It requires no watering, but the fields are manured before

R. 424

Castor Oil Plant in Bengal.   (*G. Watt.*)   **RICINUS communis.**

they are ploughed up. When the plants grow about one hand high, the field is weeded at a cost of R1-4, and they grow up almost about waist high ; the top of the plant is torn off to make the plant give a number of branches and off-shoots. The more branches there are the more is the yield. The field at this stage is again ploughed to serve double purposes : firstly, to weed out any grass or other undergrowth ; and, secondly, to give bushes on the tree to bring out more off-shoots. The ploughs generally knock the plants down a little, and it is usually the case that the plants bring out off-shoots at every such bend or inclination. This costs here about 8 annas per acre. After this it requires no further care. It is to be noted that the seeds are sown with their mouth portion below and the other sides upwards. The total cost for cultivation is R3-12 per acre.

" The outturn in this part is about 10 maunds per acre, if there be no accidents to the plants. Frosts, lightning, and hailstones are very injurious to the crop. These make the ear of the corn dry, and the produce in consequence thereof is very meagre. The outturn ranges from 6 to 12 maunds per acre, and this is in accordance with the quantity and nature of the soil."

5. Dacca Division.

From Mr. **N. S. Alexander**, the Officiating Commissioner of the Division, the following facts were obtained :—In the DACCA and BACKER-GUNGE districts there is no regular cultivation of castor seed. In the Kishoregunge Sub-division of the MYMENSINGH District, besides the use of the seeds for oil, the leaves to some extent are used to feed silk-worms. The plant, it is stated, is grown here chiefly to form a fence round other crops. The sowing season extends in Kishoregunge and Jamalpore from April to July, and in Netrokonah from September to October ; the seeds are pounded and boiled and the oil skimmed off and used mostly for lighting, the whole process being most rude and unsophisticated.

The District Superintendent of Police, BACKERGUNGE District, reported to **Mr. Alexander** as follows :—

METHODS AND SEASONS OF CULTIVATION.—" It does not appear that castor seed is, in any part of the district, grown to any great extent, or for sale, and the oil is manufactured for lighting purposes only, except occasionally for rubbing on the body for rheumatism. There are two kinds cultivated, the '*punchi*' or small and '*boro*' or big seeds. There are two seasons for sowing, *viz.*, from August to November, according to the locality during the rains, and in March and April. The seed is not soaked before sowing, and it is sown at intervals of about 3 feet. If sown alone, about 2 seers are required for one *bigha* of land. When sown with other crops, such as sugarcane, cotton, and pulse, less seed is of course used. The seed sown in March and April begins to ripen in January, and that sown in the rains to be ready for plucking from March to the end of May. The plants of the larger variety are allowed to remain up to the third year ; but the quality and quantity of the crops deteriorate as the plants get older. The plants of the smaller kind are cut down early. The fruit of both kinds bursts, and the seeds are thrown to a distance ; so it is the custom to pluck them before they burst. They are then covered over for six or seven days, after which they are exposed to the sun for three or four days, and then separated from the shell.

YIELD.—" The yield, when sown alone, is said to amount to about 10 maunds per *bigha*, and when sown with other crops to about one and a half to two maunds. The former estimate appears to be high. It is generally cultivated on new lands thrown up round tanks and ditches when sown alone.

RICINUS
communis.   Cultivation of the

CULTIVATION
in
Bengal.

Dacca.

Extraction
of Oil.
428

EXTRACTION OF OIL.—" The process for extracting the oil is as follows :—The seeds are first partially roasted and then crushed in a
'*dhenki*,' or sometimes pounded by hand. *2nd*—They are then boiled,
and the oil, as it rises to the surface, is skimmed off.   Another process
is to boil the seeds first, then to expose to the sun till dry, and crush and
boil again, and skim as above. After the first boiling the oil is again
boiled to cleanse it of impurities, and is then stored for use. The proportion of oil to seed is said to be about 25 per cent."

Of the non-official gentlemen whom **Mr. Alexander** consulted, Nawab
Ahsanoollah, of Dacca, was the only one who afforded any information.
He reported as follows :—" In portions of Dacca and Mymensingh, specially on higher grounds, castor seed is generally seen grown with turmeric.
It is sown between June and July. The land requires deep ploughing and
thorough weeding. The seed is sown in rows a little elevated from the
general level of the field and a yard apart from each other. The elevated
rows are made with earth dug from the sides to serve as ditches for rain
water to flow through. This system of elevated rows brings to the surface
earth from below which aids the plants to grow luxuriantly. The seed wants
soaking for about 12 hours. If the land is very low and exposed to inundation, even after the seeds have germinated, the plants generally die away.
The soil preferable for it is sandy loam. The above method brings the best
outturn, yielding from 10 to 12 seers per each plant. The cost of cultivating
and preparing a field to receive castor seed is about R10 a *bígha*.
This amount generally covers the cost of growing and tending the plants
till the fruits are ripe. The fruits are generally plucked when the plants
are standing, which gives time for younger fruits to grow and ripen.
The fruits are thus gathered for three or four months together from November. When gathered they are dried in the sun when they burst of
themselves and give the seeds. The labour of collecting the seeds is
about R2 per each *bígha*. The plants, when dry, serve for fuel.

" In portions of Tipperah, where castor seed cultivation is very rare,
the foregoing method is not in vogue. . There they sow the seeds broadcast, though considerably apart from each other, and the plants therefore
are not at all luxuriant. They generally select the riversides, which are
occasionally inundated during the rains. The rivers overflowing leave
alluvial deposits, which make excellent soil when dry. The cost of cultivation, &c., here is about two-thirds that in Dacca, and the outturn is proportionately low. The seed also is of a smaller size."

Presidency.
429

6. Presidency Division.

From the reports sent in to **Mr. A. Smith**, Officiating Commissioner
of the Division, it appears that anything like a regular cultivation of castor
seeds on an extensive scale is unknown in any of the districts under his
charge. He was therefore unable to furnish a detailed report on the subject; but communicated the following general facts :—" From the reports
it appears that there are two varieties of the plant, white and red. The
seeds of the former are sown with other *rabi* crops, while the latter kind
grows spontaneously in the wet season. The Collector of JESSOR states
that the varieties which grow in his district are two, *viz.*, the *khudi* and
bagha, *i.e.*, small and large. The former grows spontaneously in the
jungle, and no care is taken of its growth. The latter is grown by cultivators around their fields so as to form a hedge to protect the crops from
cattle. The mode of cultivation, &c., is almost the same in the several
districts of this division. I therefore give below, the Commissioner states,
an extract from the Nuddea report which is somewhat detailed. The
seeds are generally sown along with chilli and turmeric in July; sandy

Castor Oil Plant in Bengal.	(*G. Watt.*)	**RICINUS communis.**

loam is said to be the best suited for the cultivation, of these plants' which do not require any special care beyond what is needed for chilli and turmeric with which it is cultivated. When the fruits are ripe, they burst of themselves or with little pressure, and give out the seeds of which there are three in each. The process of extracting oil from the seeds is thus described: the seeds are first boiled in water and dried up in the sun. When dry they are crushed under a *dhenki*. The powdered seeds are placed in an earthen *handi* containing water and boiled again; the oil floats on the surface, when it is taken out by a spoon and placed in a separate vessel and again boiled for the purpose of purification. A *bigha* of land, it is estimated, can yield 6 maunds of seed, from which about 3 maunds of oil can be extracted. The net profit per *bigha* is estimated at R20."

7. Burdwan Division.

In the districts of BURDWAN, BANKOORA, and BEERBHOOM the castor plant is not cultivated. Of the other districts, Mr. J. Beames, the Commissioner of the Burdwan Division, wrote as follows:—

Burdwan.
430
Bankoora.
431
Beerbhoom.
432
Midnapore.
433

"In MIDNAPORE this crop is extensively grown in pergunnahs Nayabasan, Cheara, and Khelar, which are situated on the rivers Subunrakha and Dolang. Sandy loam soil is most congenial to its growth, so the *chur* lands on the sides of these two rivers are specially suited to the cultivation of this crop. It is sown in September and beginning of October, and reaped in February and March. Before sowing the soil is ploughed thrice in the course of a month, and when the plants put out two or three sets of leaves, the soil is slightly stirred up with *kodalis*, after which nothing further is required, except gathering it. No irrigation whatever is needed. The average cost of cultivation per *bigha* is shown on the margin.

	R	a.	p.
Ploughing for three times before sowing .	1	2	0
Stirring for one time only . . .	0	6	0
Seeds, about 12 seers, for sowing .	0	12	0
Rent for land on which the crop is grown .	1	4	0
TOTAL .	3	8	0

"The outturn of the crop on a *bigha* of land is on an average 3 maunds, worth about R7-8. The *ryats* therefore get a profit of about R4 per *bigha*. The *ryats* always keep as much of the seed as is required for their home use, and the surplus is sold to the *mahajans*, who come for this purpose. They seldom prepare oil for sale. The process they adopt in making oil for their home use is as follows:—The seeds are slightly fried, and then crushed in *dhenkis*, and afterwards put in a *handi* with a sufficient quantity of water. It is then boiled, and when the oil floats on the surface of the water, they take it out and put it in an earthen pot.

"In HOOGHLY the plant is not cultivated as a regular crop. It grows wild on waste lands in almost every village in the district, and is planted only to form hedges around fields. It grows best on sandy soil. The seeds are sown along the line of the hedge; they germinate during the latter part of the rains, and grow well. In some villages the seed is collected and sold; in others it is collected when required for medicinal purposes. Each plant yields eight ounces of seed on an average.

"In HOWRAH the plant is grown to a very limited extent within the jurisdiction of thana Doomjoor for demarcating the boundaries of fields. Its produce is pounded in a mortar, and a small quantity of crude oil thus obtained is used for lighting purposes, but it is never sold or exported."

8. Bhagulpore Division and Santal Pergunnahs.

With the exception of the report from Mr. W. B. Oldham, Deputy Commissioner of the Sonthal Pergunnahs, Mr. G. N. Barlow, C.S.I., the Commissioner of the Division, was unable to furnish any useful information from the district reports; but added the following facts, apparently

CULTIVATION
in
Bengal.

Bhagulpore.

from his own experience and observation :—"The mode of cultivation followed in the several districts of this division is almost the same. In Monghyr the field is repeatedly ploughed some eight or nine times, and the clods broken up. The cultivators, following the plough, drop the seeds, one by one, into the furrows. As soon as this is completed, a flat piece of wood is dragged over the ground till the field is levelled. When the plants are big enough not to be injured by weeding, they are weeded two to four times. Alluvial *dearah* soil, intermixed with a small quantity of sand, is well adapted for its growth, but stiff clay is unsuitable.

"In Bhagulpore, too, the seed is best sown by dropping them in furrows about a cubit and half apart, and the seed to be about a cubit apart ; this crop is weeded twice generally :—1st, when the plants are one foot high : and, when they are two feet high.

"The suitable soil selected for cultivation is the high land called '*chúsra*' or ashy coloured soil on the *dearah*, and the '*ghuksia*' or light soil having sand on the surface and strong soil underneath.

Purneah.
437

"In Purneah the seeds are sown by hand, and they germinate in about nine days. On low lands the seeds are sown after once ploughing without manuring the land, but on high lands they are sown after four or five ploughings and the lands are also manured with cowdung. When the plants are about three feet high they are weeded once for all. High sandy land is selected for the cultivation.

Maldah.
438

"In Maldah the seeds are put into the ground generally on the top of mounds raised in a row round mulberry fields or other small crops at a distance of about a foot. High lands are generally selected for the cultivation.

"The annexed statement will show the cost of cultivation, average outturn per acre, and net profit :—

Name of District.	Cost of Cultivation and Collection of Crops.	Average out-turn.	Net profit.	REMARKS.
	R a. p.	Mds.	*R a. p.*	
Monghyr .	18 0 0	30	40 0 0	It will be seen that the cost of cultivation at Monghyr and Maldah is much greater than in Bhagulpore and Purneah. The outturn per acre in these two districts is also proportionately large. This is owing to the greater care taken for the cultivation, such as ploughing several times, manuring, watering, &c., &c.
Bhagulpore	3 14 0	8½	25 0 0	
Purneah .	5 10 0	12	16 0 0	
Maldah .	20 0 0	30	27 0 0	

"The varieties of castor seed that grow in the districts of Monghyr, Purneah, and Maldah are two in number, *viz.*, one big and the other small ; but in Bhagulpore five varieties are cultivated, *viz.*, (1) *ghawria*, or *moothia*, (2) *jhokia*, (3) *chanaki*, (4) *gohumah*, (5) *bhadaiah*. The first four species are sown in October and November and the crops gathered from March to May. The *bhadaiah* is sown in May and June with other *bhadai* crops, and the seed is gathered from November to January. The times for sowing and gathering the different species of crops are almost the same in every district.

Extraction
of Oil.
439

EXTRACTION OF OIL.—"As regards the extraction of oil there is nothing exceptional in the methods practised anywhere in this division."

Castor Oil Plant in Bengal.	(*G. Watt.*)	RICINUS communis

Mr. Oldham's report is as follows :—

FORMS GROWN.—"There are three varieties of castor seed in the subdivision. Their names are as follows :—

1st, *Chunaki* (small in size): 2nd, *Gohuma* (middle-sized): and 3rd, *Jagia* or *Ruksa* (big in size).

"There is also a sub-variety of *jagia* called *khaguria*, but it is not cultivated in this part of the district. *Chunaki* and *jagia* are sown in June or July, and *gohuma* in November or December.

METHODS OF CULTIVATION.—"The method of sowing *chunaki* castor seed is very simple, and no preparation of the lands for this seed is necessary. Holes are made in the ground about two inches deep and three inches in diameter, one seed being put into each hole, the holes being three to four feet apart. The seeds germinate and the young plants appear after a week. The crop does not require any further care. The cost of cultivation of *chunaki* is not more than an anna a *cottah*, including the value of seed, which is sown at the rate of 2 seers a standard *bigha*. It is sown on small plots of land attached to the houses of the cultivators. Its stems serve as living supports for creeping vegetable plants. The poor classes of people use the stems, which range from 10 to 12 feet in height, for thatching, and the leaves and oil are medicinally employed. The oil is used as a remedy for rheumatic pains, and the rubbing of oil on the affected part of a patient, and the application of the leaves, give immediate relief. The plants of *chunaki* are allowed to stand for two or three years, and the old plants produce fruits earlier, but their seeds are inferior to those of the new plants. The fruits of the *chunaki* begin to ripen in November or December, and continue till March. When ripe they burst, throwing the seeds to some distance. The bunches of fruits are plucked and placed, mixed with water or liquid dung, in small trenches or tubes. They are allowed to rot for two or three days, after which period they are taken out and put in the sun, and when dry the fruits are crushed by hand or flat wooden mallets, and the seeds cleaned by winnowing. The outturn of *chunaki* when the crop is good, is 10 seers a *cottah* or 5 maunds a *bigha*. The seeds are bright black, dotted with grey, and have a hard shell.

"*Jagia* is sown in *bari* lands, generally along with Indian-corn or *rahar*, in June or July, before the rains set in. The lands are ploughed two or three times, and the seeds scattered by a man or woman, who drops them one by one, 3 feet apart, in the furrows made by a plough. When the sowing is finished, the fields are made even by the application of a wooden flat timber, called *chowki*. By this process all seeds are covered with earth, and the soil of the field retains moisture. The plants of *jagia* are similar to those of *chunaki*, but they are not so strong as the latter. The crop does not require any care until the fruits begin to ripen in February or March, when they are collected and seeds prepared by the same process as followed in case of *chunaki* seeds. The cost of cultivation of this variety is R3 per *bigha*, and the yield 4 maunds. Six to seven seers are required for a standard *bigha* of land, but as it is generally sown along with other crops, the quantity does not exceed 3 seers per *bigha*. Seeds of *jagia* are of red colour and a little flat, while those of *khajuria* are a little round, resembling dates, from which it takes its name.

"The third or the best variety is *gohuma*. It is extensively cultivated in this sub-division, and resembles somewhat wheat in colour, and hence it is called *gohuma*. It is sown in *bari* lands, but thrives best in loamy soil. The fields are well prepared and seeds sown at the distance of one or two spans in furrows. *Gohuma* plants are watered once or twice when the soil becomes hard and when there is a means of irrigation, but the outturn

CULTIVATION in Bengal.	of the fields which have natural moisture is greater than that obtained from irrigated land. The banks of streams and rivers yield a good *gohuma* crop, superior to the produce of other lands both in quality and quantity.
Bhagulpore. Cost. 441	Cost.—"The cost of cultivation of the *gohuma* is R3-4 to R3-8 per *bigha*, including the labour of plucking the fruits and husking the seeds; and the outturn when the crop is good is 6 maunds to 6½ maunds per *bigha*. Five seers of seed is required for a standard *bigha* of land, if sown alone. The stem of this variety does not exceed 6 feet at the outside and is about an inch in diameter. *Gohuma* is sown in November and its fruits begin to ripen in March and continue till April. It is harvested in the same manner as *chunaki* and *jagia*.
Expression of Oil. 442	Expression of Oil.—"Oil of all castor seeds is extracted in this part of the country by a very simple process: seeds are partially roasted in a pan and pounded in a mortar without being husked. The stuff so prepared is mixed with water and placed in an earthen pot or jar over a fire, the quantity of water used being two or three inches above the level of the pounded seed. As the water evaporates the oil rises to the surface, when it is poured in a vessel. The jar is then taken down and the mixture allowed to cool, after which some cold water is added to the mixture, and the jar is then placed in the sun. The quantity of oil absorbed in the mixture then appears at the top and is removed by the hand. The oil thus obtained is boiled again in a separate pan, by which process any moisture and all other extraneous matter are got rid of. The oil becomes purified and fit for consumption in burning lamps. There is another process for the extraction of oil from castor seeds followed by the people of this subdivision. It differs from that already explained very slightly. In this process the seeds are boiled with water before being pounded. When they become soft they are placed in the sun, and pounded afterwards when they are dry.
Santal Pergunahs. 443	"*Chunaki* seed gives oil 35 per cent.; *gohuma* 37·5 per cent.; and *jagia* 31 per cent." In addition to the above, Mr. Oldham furnished the following brief report prepared by Mr. E. B. Harris, Sub-divisional Officer, Rajmahal, on the cultivation and manufacture of castor oil in his sub-division:— Cultivation.—"Very little castor is cultivated in this sub-division. The total quantity of land sown will be about 75 *bighas*. The cost of cultivating each *bigha* is from R6 to R9, and the production of each *bigha* is from 4 to 5 maunds, the price of which is from R16 to R20. "The land is ploughed two or three times, and the seed generally sown in September. About 10 seers of seed is necessary to sow one *bigha* of land.
Variety. 444 Extraction of Oil. 445	"Varieties.—Big fruits are sown and produced in this sub-division. "Extraction of Oil.—Cultivators here generally extract oil at home for burning in lamps in the following manner:—1st, Seeds roasted on heated sand put in a pan over a fire, which coagulates the albumen and liquifies the oil: 2nd, Slightly crushed and husked: 3rd, Pounded in mortar: 4th, Boiled in water: 5th, As the water evaporates the oil rises to the surface when it is taken out by a cup. The oil is again put on fire in a separate pot to soak the watery portion of the oil. The yield of the oil is about one-third of the fruit. The castor oil is sold by the poorer class of people for burning purposes only."
Orissa. 446	**9. Orissa Division.** The cultivation of castor seed is not very extensively carried on in any of the districts of this division, except in Cuttack, where it is grown to

| Castor Oil Plant in Bengal. | *(G. Watt.)* | **RICINUS communis.** |

some extent. The following report furnished by **Mr. W. R. Larminie**, Commissioner, will show the method of cultivation pursued, the cost of produce, and the general outturn of the crops in this division :—

VARIETY.—"There are two principal varieties of castor seed *jara*— one large and the other small. The former is called *burra* (big) *gaba* or *boda jura* and the latter *chuni* (small) or *kuji gaba. Burra gaba* has two sub-varieties, called *pata jara* and *kala jara*, and the smaller kind has also two, called *chuni* proper and *jahuri*. The large sort grows to the height of about 8 to 14 feet, and has purple leaves, with red veins and stems; while the smaller attains a height of from 3 to 6 feet, having green leaves and stems. *Bara jara* is generally grown along with other crops, such as *arhar*, turmeric, cotton, &c., on homestead and highlands, which produce two crops and have a light and sandy soil. The time of sowing extends from May to July, and that of reaping from January to March. The plants last for two or three years, if properly looked after. The *sana* or *kuji gaba* thrives best in sandy alluvial formations along the banks of rivers. The cultivation is also carried on on light soils, which yield two crops, on which early rice (*baili*) or other miscellaneous crops are raised; also on land newly cleared. The seed is sown from November to December, and the gathering of the crops extends from March to April. Pumpkins and melons are often grown with this variety.

MODE OF CULTIVATION.—"The land is ploughed, harrowed, and manured with cowdung several times before it is ready to receive the seed, which is sown either broadcast or in drills, the latter process being considered the better of the two. After the sowing is over, the land is once more harrowed. Prior to the sowing, the seeds are in some places soaked in water for six or seven days, the water being changed daily to facilitate germination; and where this is not done, and the land is dry, it has to be irrigated before the sowing commences. It is said that the crop is never a full one if the watering before sowing is omitted. When the plants grow up, the soil is frequently loosened with a spade, and also irrigated when necessary. Beyond this nothing more is required in the way of cultivation. Plants grown with other crops are treated in the same way as those with which they are mixed."

HARVESTING.—"When the capsules are ripe, they are plucked by the hand, and left in a heap for three to five days, a mixture of cowdung and water being daily sprinkled over them. They are then exposed to dry in the sun. Most of the capsules burst themselves, and those that do not are opened by means of a mallet or stick.

COST OF CULTIVATION, OUTTURN, &c.—"The cost of cultivation and the yield per acre, when the crop is grown separately, have been estimated to vary from R6 to R9, and 3 to 6¼ maunds, according to the nature of land on which the crop is grown. The average value of produce varies from R7·8 to R31, and the profit accruing to the cultivator is put at R2 to R16. As the cultivators themselves do the ploughing, sowing, &c., and find their own ploughs and bullocks, the cost is nominal, and the crop, where it is grown, may be said to be permanent.

EXTRACTION OF OIL.—"There are two processes for extracting oil from castor seeds—one is by boiling and the other by pressing. The first process is as follows :—The seed is partially roasted or boiled and dried in the sun, and then crushed into a pulpy consistent mass. This pounded stuff is then mixed with water and boiled immediately, or, after being stowed away for two or three days, until the oil rises to the surface, when it is scooped out or poured into another vessel. The oil thus obtained is again heated to drive off the watery particles by evaporation. The boiling of the pounded stuff is repeated for five or six times until the whole of

RICINUS communis.	Cultivation of the

CULTIVATION in Bengal. Orissa.

the oil is extracted, and the refuse emits a stench which is perceived from a distance. No oil-cake is obtained by this process. The ordinary process, however, is to press the seeds in an oil mill or *ghana*. By this process, although less oil is obtained, the refuse is saved for oil-cakes, and it is also less laborious and better suited for manufacture. The other process is adopted for extracting small quantities of oil only for home consumption, and the oil thus obtained is considered to be of superior quality. The average cost of fuel, labour, &c., in the first process is about annas 9·4, and that of extracting oil by the second process about annas 7·6 per maund of seed. The average yield of oil per maund by the two processes is about 12¾ seers (varying from 17 seers to 8⅓ seers) and 11⅜ seers (varying from 12¼ to 10 seers) respectively.

Conf. with p. 517.

"The poorer classes, such as peasants, *bauris*, &c., use the oil for cooking purposes and for anointing their bodies. It is also employed as a purgative and for various other medicinal purposes; also for lighting. The oil-cake is a good manure for different kinds of vegetables, and also serves for fodder."

III.—NORTH-WEST PROVINCES AND OUDH.

N.-W. PROVINCES & OUDH. Varieties. 452

VARIETIES.—"Two varieties are reported to be grown in the Azamgarh District, known respectively as *reri* and *bhatreri*. The former is the taller of the two, and is said to be invariably cut down after the first year, whilst *bhatreri* trees are allowed to remain for two or three years. The seeds of the *bhatreri* variety are reported to be richer in oil than those of the other variety.

Distribution. 453

DISTRIBUTION.—"Castor is grown to a greater or less extent in every district of the Provinces, but usually as a field border, and very rarely as a sole crop. The only division, indeed, in which the area it covers is large enough to deserve mention is Allahabad, where it is reported to be grown alone on between 1,200 and 1,300 acres, situated principally along the margin of the river Jumna. It is, on the other hand, a very common bordering to cotton and sugar-cane fields, and is not uncommonly grown on isolated patches of a few square yards in the neighbourhood of dwelling-houses, and used as a support for the creeping bean known as *sém* (**Dolichos Lablab**). It thrives on a rich soil, but curiously enough succeeds exceedingly well when sown along the top of the high mud-banks which commonly surround orchards and vegetable gardens. In this situation the young plants are protected from flooding, and their roots rapidly strike deep enough to acquire sufficient moisture.

Season. 454

SEASON.—"It is sown at the commencement of the monsoon or in the hot weather just before the rains break. The seeds are either sown behind the plough, being dropped at intervals of about 18 inches in every alternate furrow, or they are planted by hand. In the latter case, a little manure is commonly buried with them. The young plants are occasionally earthed up to prevent the accumulation of water round the bottom of the stem. The seeds ripen in March and April.

Harvesting. 455

HARVESTING.—"When ripe the seed pods are picked, and are either dried in the sun and broken by rolling (Azamgarh), or are buried in the ground and allowed to rot. The latter is the common practice in Doáb districts.

Method. 456

METHOD OF EXTRACTING THE OIL.—"The oil is extracted by boiling, and the operation is not performed by the professional oil-pressers (or *telis*) but by the gram-parchers (*bhurjis*). The seeds are first slightly roasted, then crushed in a mortar, and then boiled in water over a quick fire, when the oil rises to the surface and is skimmed off. As a rule, the seeds yield

Castor Oil Plant in Bombay.   (*G. Watt.*)   **RICINUS communis.**

a quarter of their weight of oil, but seeds of the *bhatreri* variety are said to yield as much as one-third of their weight.

OTHER USES.—" Young castor leaves are relished by cattle, and the dried stalks are utilised for thatching. Castor trees are commonly cut down after their first year, but it has been already mentioned that the *bhatreri* variety of Azamgarh is commonly allowed to stand for three or four years, when it yields a crop of seed each spring, and is finally cut down, not because its bearing powers are exhausted, but because it is a breeding ground for a hairy brown caterpillar which is supposed to bring ill-luck.

YIELD OF SEED.—" A well-grown castor plant will yield as much as 10 seers (=20℔) of seed in a season, but the plants which are grown round fields rarely give more than from ¾ seer to 1½ seer apiece. The yield of individual plants grown together as a single crop in a field is much less than this, since flowering is hindered by a loss of light and air, when the plants are not separated from each other by a considerable space.

" The castor is popularly ranked as the *chamár* amongst plants, and men of this caste are particularly afraid of a blow from the stalk of a castor plant" (*Fuller & Duthie's Field and Garden Crops, Part II., pp. 38-39*).

IV.—BOMBAY.

The Director of Land Records and Agriculture, in a correspondence on the subject of castor-oil, furnished the following brief note :—

CULTIVATION.—" There are two varieties—the large and the small seeded. The former is arboreous and perennial. It is grown either with irrigation or without, but, in the absence of irrigation, requires a deep loamy soil. This variety is very generally grown as a protective border around plots of sugar-cane, betel-vine, and other garden crops. It is sown at various times of the year. The oil extracted from the seed of this variety is darker and thicker than that obtained from the small-seeded kind. It is used for burning and dressing leather and similar purposes, but not as a medicine.

" The latter is an annual, grown without irrigation or manure as a rain crop sown at the commencement of the monsoon. It is sown at the rate of about 10℔ per acre. It is selected in parts of Gujarát as the first crop taken after a long fallow on light uplands. It is always grown in a distinct area. If grown in alluvial or garden land, it develops into a tree, like the large-seeded variety and shows a tendency to become perennial. Such land is, however, avoided, as in a deep soil it exhibits a strong tendency to fruit prematurely. A light shallow soil is therefore preferred. The medicinal oil is obtained from this variety. The percentage of oil is slightly larger, though the market price of both kinds is practically the same.

" The residue of the seed after extraction of the oil—castor-oil cake—is a valuable manure" (*Records of the Government of India, Revenue and Agricultural Department, B, July 1886, Nos. 36-37*).

In amplification of the above the following passages may be here given from the Gazetteers of Bombay :—

SURAT.—" The castor-oil plant, *divelo*, is very extensively cultivated in the southern parts of the district. In 1874-75, 39,200 acres were under castor-oil, or 9˙96 per cent of the total area under cultivation. The oil extracted from this plant is of the greatest purity, and is used for burning. It is also, even without refinement, adapted for medicinal purposes. As a rule, the castor-oil plant is not sown by itself, but with pulse" (*Vol. II., 63-64*).

CUTCH AND MAHI KANTHA.—" Castor-oil seed, *erandia*, an early crop, is generally sown in the same field as cotton. In parts of Abdasa and

CULTIVATION in N.-W. Provinces & Oudh.

Yield.
457

BOMBAY.
458

Surat.
459

Cutch.

RICINUS communis.	Cultivation of the

CULTIVATION in Bombay.

Vágad it is in a very few cases sown by itself. When sown together the proportion of castor seed to cotton seed varies from one-tenth to one-fifteenth " (*Vol. V.*, *106*).

Kathiawar.
460

KÁTHIÁWÁR.—" Castor seed, *erandi*, is a crop of small importance found in all parts of Káthiáwár. It grows in black soil and is both a hot weather and a rainy season crop. The hot weather castor plant is sown in March or April and reaped in November and December; the rainy weather castor plant is sown in June and reaped in October. The oil is locally used for lamps " (*Vol. VIII.*, *183*, *425*).

Khandesh.
461

KHÁNDESH. - " Castor-plant, *erandi*, an early crop sown in June and reaped between the middle of September and October, has in most parts of Khándesh two varieties, one annual and small seeded, the other perennial and tree-like with large seeds. Of the castor tree there are many sorts, which, wanting much water, are commonly planted on the boundaries and along the leading water channels of sugar-cane plantations. The castor plant is grown as an ordinary cold weather field crop. To extract the oil, the seeds are roasted, ground in a handmill, and boiled over a slow fire, the oil being carefully skimmed as it rises to the surface. The refuse forms an excellent manure for plantain trees, and the stems are useful in thatching roofs " (*Vol. XII.*, *152-53*).

Kanara.
462

KÁNARA.—" Of the castor-plant, *vudla* or *harlu*, two varieties, *chiti* or spotted and *dodda harlu* or large, are grown to a very small extent. From the large or *dodda* species medicinal castor oil is made; the spotted seed yields a greater quantity of oil which is commonly used as lamp-oil. The oil is extracted either by boiling or in a mill " (*Vol. XV.*, *pt. II.*, *19*).

Ahmadnagar.
463

AHMADNAGAR.—" Castor seed, *erandi*, is sown either in June or November, in black soil, sometimes round other crops and oftener in patches by itself. It grows without water or manure, and is harvested in November or February. The stem and flowers are red. It is not much grown and the oil is more used as a lamp-oil than as a medicine. The oil is extracted by husbandmen for home use by boiling the bruised bean and skimming the oil that rises to the surface. By this process four *shers* of seed yield one *sher* of oil. The leaf is used as an application for guinea-worm and the dried root as a fever-scarer. A larger variety with green stem and flowers, but otherwise the same as the smaller variety, is grown in gardens round other crops. Both varieties are perennial and grow to a considerable size. They are rarely allowed to remain on the ground for a second year " (*Vol. XVII.*, *270*).

Satara.
464

SATÁRA.—" Castor seed, *erandi*, is grown in black soil without water or manure. It is sown either in June or November, and is harvested in November or February. It is sometimes grown round other crops, and more often in patches by itself. It is not much grown, and is more used as a lamp-oil than as a medicine The people extract the oil for home use by boiling the bruised bean and skimming the oil as it rises to the surface. By this process 4℔ of the seed yield 1℔ of oil. The leaf is used as an application for guinea-worm, and the dried root as a febrifuge. A large variety of the castor plant, probably **Ricinus viridis**, is grown in gardens round other crops. Except that the stem and flowers of the large variety are green and those of the small variety are red, the two plants do not differ from each other. Both varieties are perennial and would grow to a considerable size if they were allowed to remain on the ground for a second year " (*Vol. XIX.*, *165*).

Kolhapur.
465

KOLHÁPUR.—" *Erandi*, with an area of 1,986 acres, is either grown by itself or with late *juari* and maize in red, black, and alluvial soils, and harvested in January. There are three varieties, *chitkya* or small-seeded, *dholkya* or large-seeded, and *ghaderandi*. The first two varieties are

CULTIVATION
in
Bombay.

Kolhapur.

grown as ordinary field crops, and the third, as it requires much water, is planted either near the leading channel of a sugar-cane field or around the field. The oil, which is more used for burning than as a medicine, is drawn by boiling the bruised beans and skimming from the surface. To raise the oil to the surface, cold water is poured on. The refuse is an excellent manure for plantain trees and the stems are used as fuel or thatch. The average acre outturn is 530℔ " (*Vol. XXIV., 170-71*).

V.—PANJÁB.

PANJAB.
466
Conf. with pp. 549-551.

In reply to the enquiry addressed by the Government of India to the Panjáb Government on the subject of Castor-seed and oil production, the late Colonel E. G. Wace, Commissioner of Settlements and Agriculture, replied that he had instituted an investigation in Karnal, Hoshiarpur, Multan, Amritsar, and Delhi districts on the subject, and had also invited the co-operation of the Sind, Panjáb and Delhi Railway Company (who manufacture for their own use a large quantity of the oil), but the replies from the district officers were mostly to the effect that the plant was not grown or if grown not for the purpose of the manufacture of the oil. The reply obtained from the Agent of the Railway Company forwarded, however, a report drawn up by Mr. Scrofton, the Company's Oil-mill Superintendent, which Colonel Wace submitted for the information of the Government of India. The report being thus the only statement available that furnishes facts with a more or less bearing on the Panjáb, is published in this place, though its great merit lies in the practical experience (thus made public) in the extraction of the oil which should cause it to be rather assigned a place in the section of this article which deals with that subject. Castor-oil is doubtless grown in the Panjáb to but a limited extent and nowhere as a field crop. The plant is, however, a nearly always constant associate of village sites, specially on the lower hills and outer ranges of the Himálaya. It seems highly likely, however, that in these situations it is rarely if ever grown with any care, being more spontaneous than cultivated, and, as suggested by Colonel Wace, it is not viewed as an oil-yielding plant. In the villages around Simla its leaves are viewed as valuable food for cattle and useful as a domestic medicine for external application. These facts would thus seem to lend some degree of support to the contention that in the scrubby jungles that occur in the unculturable glades on these hills the plant may possibly be indigenous (*Conf.* with the remarks under Habitat, p. 507).

The following is Mr. Scrofton's report :—

CULTIVATION.—"In the Panjáb where irrigation is available, and it is intended to raise crops by this method, the seed should be sown in March and April, otherwise after the first good fall of rain about the end of June, after the seed germinates, the plant grows very rapidly, and requires no special care except watering till it is about two feet high. The yield is best where the plant is grown in a sandy soil.

Cultivation.
467

DAMAGE FROM FROST.—"The castor plant is liable to serious damage if the frost is at all severe, and young plants in particular are totally destroyed.

Frost.
468

QUALITY OF SEED.—"The seed grown in the North-West Provinces and Panjáb are about the same size, but the seed grown in the Panjáb is much heavier and yields a better percentage of oil; the plants, after they are cut down, do not yield so good a crop.

Seed.
469

OUTTURN.—"The yield of a good crop is from 30 to 35 maunds of seed per acre. [This is very probably a high estimate.—*Ed., Dict. Econ. Prod.*]

Outturn.
470

RICINUS communis.	Pests which attack

CULTIVATION in the Panjab.

				Seers.
From one maund of Hathras seed the outturn of oil is	.	.	13
„ „ Shahjahanpur „ „	.	.	15½
„ „ Panjáb „ „	.	.	17

OIL.
471

OIL, MANUFACTURE OF, FOR LUBRICATING AND LIGHTING.—"The seed taken over is ready cleaned from stones and dust; it is first placed between two rollers in the upper storey and slightly crushed, and in falling down the husks separate from the kernel and are blown by a fan down a shoot into a bin, and the kernels fall straight down into another bin. The seed is then placed in a kettle and heated by steam; during the time it is heating, the seed is mixed up by a revolving stirrer fixed at bottom of the kettle and driven from the shafting: the seed is then taken out and wrapped in pieces of canvas or goats' hair blanketing, and then placed in either hair or woollen envelopes, which fit in to the trays in the hydraulic press; it is then pressed up to 1½ tons, the cake that comes out is mixed with a small quantity of seed, and ground up under two heavy stone edge runners. it is then heated in another ·kettle, and placed in stronger presses and pressed up to 2½ tons.

Oil-cake.
472

OIL-CAKE AS FUEL.—"The cake that comes out after the second pressing has the oil edges pared off and stacked; a portion of cake is mixed with the husks, and burnt as fuel, and the surplus is sold to private firms as fuel for stationary boilers.

Clarifying of.
473

OIL, CLARIFYING OF.—"The oil, after being pressed out, is conveyed by iron pipes from the hydraulic presses into iron tanks sunk level with the floor, from which it is pumped into kettles made of copper or cast iron (these hold about 5 maunds of oil each), after adding 3 seers of water to each maund of oil. It is then boiled by steam; the scum rising to the surface is skimmed off with a ladle, and the skimmings placed in a bucket; after boiling about 2½ hours the water evaporates, the steam-cock from the boiler is then shut off, and the blow-off cock under the kettle opened, and the steam let out; the oil is allowed to cool down; it is then ladled into zinc buckets, and carried to the filtering tank, and filtered through canvas into the tank; the sediment remaining at the bottom of the kettle, is placed with the skimming, and the oil drained out of it.

Storing of.
474

OIL, STORING OF.—"The filtering and storage tanks are connected by pipes, through which the oil is pumped from filtering tanks to the storage tanks. The oil for lighting purposes is stored at least twelve months before using; it is then equal to any class of vegetable oil for burning and giving a brilliant white light, and no offensive smell comes from it. Castor oil is the only vegetable oil used by this Company all the year round for train lighting, signals, &c." (*Records of the Government of India, Revenue and Agricultural Department, B Proceedings, March 1884, Nos. 16-28*).

PESTS.
475

PESTS WHICH ATTACK THE CASTOR OIL PLANT.

An extensive correspondence has taken place on this subject, chiefly in connection with Assam and Madras, where the most serious insect pest seems to have attained its most alarming proportions, *viz.*, a noctine moth, **Achœa melicerte,** *Drury.* For a brief notice of this moth see the article **Pests,** Vol. VI., p. 144, also *Indian Museum Notes on Economic Entomology,* 1889, p. 104, and in Balfour's *Agricultural Pests of India,* p. 100.
The following is a translation of a report by a native gentleman of Madras on this subject, which was communicated to the Government of India by the Director of Revenue Settlements and Agriculture:—
"The insects (called in Canarese *Thindala Húla*) damaging castor-oil plants near Nundibahoor are like caterpillars, but there are no hairs on the body. These insects when grown fat are 3 inches in length and 1

the **Castor Oil Plant**.	(*G. Watt.*)	**RICINUS communis.**

inch in diameter. It is not known by what name they are called in Telugu and English. These insects lay at first small eggs of the size of poppy seeds behind the leaves of castor-oil plants in *Magha Karthy*. The eggs crack in a few days and form insects of a hair thickness, which gradually grow large in four or five days, and eat away all the leaves, leaving only the stems. These insects, after they have become large, eat away the leaves and flowers of castor-oil plants on 2 or 3 acres of land in one night and leave bare branches. The plants very rarely revive afterwards and yield any produce; if they revive at all, they will not yield one-fourth of the produce. The *rayats* smoke the plants to rid them of the insects as soon as they appear, but they will not leave the plants. As it is impossible to smoke all the plants when seeds are sown on an extensive area, the *rayats* give up all hopes of this crop" (*Records of the Government of India, Revenue and Agricultural Department, B, December 1886, No. 1*).

In the above papers mention is also made of an opinion held by the *rayats* that a few good showers of rain drive away the pest.

Mr. J. Wood-Mason, Superintendent of the Indian Museum, communicated the following description of the caterpillar which may prove of service in the identification of the insect:— "The caterpillar is greatly elongated, especially in the two legless regions of the body, is slender, and tapers slightly to either extremity. Its first pair of abdominal legs, though greatly reduced in size, yet still retains much of its locomotor function, as is evidenced by the indifferent copies of the excellent native (Ceylon) drawings quoted above, and by the persistence, of the minute terminal retentive hooks with which it, like its successors, is provided: the caterpillar is, in fact, only a commencing semi-looper. Its colouration is striking: the head is black, symmetrically marked with sharply-defined yellow spots and stripes, as follows: A large oval spot on each side between eyes and occiput; another only about one-third the size of the first on each side of the vertex; a third pair right in front and less than half the size of these; and in front of, and internal to, these, again a fourth pair, mere dots, separated from one another and from the middle line only by a very narrow linear space of the black ground colour; a club-shaped stripe parallel to each of the arms of the epicranial suture; and between this and the anterior end of the lateral spot on each side a second stripe of similar shape, but only half the length; and a third pair of stripes converging posteriorly on the space included between the arms of the epicranial suture: the clypeus is also yellow. The dorsal region of the body is marked with three dark-brown longitudinal bands of equal width, one dorsal and two lateral. The dorsal band is marked along the middle with a dull red chain-like pattern coinciding with the outlines of the chambers of the subjacent heart, and it is separated from the dark lateral band of each side by a more or less distinct ochreous subdorsal band of somewhat less width. The two ochreous subdorsal bands are always paler and often even bright yellow at their two edges where they come into relation with the dorsal and lateral bands (their upper yellow line being in the latter case expanded or intensified at the anterior end of the fifth ring of the body into a yellow spot forming with a median dorsal spot, at the same level a transverse series of three yellow spots [in light-coloured specimens], and they bear on each body-ring a round black spot. The two dark lateral bands extend on to, and consequently have an irregular outline at, the leg bases, on each of which they bear a yellow spot, and they are ornamented nearer to their upper than to their lower margin by a continuous or discontinuous dull red line, which passes just below the spiracles. The two short conical dorsal horns of the eleventh ring are red like the dorsal chain-like marking, the thoracic legs,

RICINUS communis.	**Expression of Oil**

and the infra-spiracular line. Below, the animal is ochreous-brown with a median dark line expanded, between the legs into long oval spots and with a yellow line on each side concolorous with the outsides of the abdominal legs, the retentive hooks of which are dark brown. All the markings of the body are made up of a variously aggregated and a variously coloured speckling; and, it is probable, were suffused with a bluish tinge in the living insect.

"This caterpillar varies greatly in the depth of colour and in the distinctness and the continuity of the body-markings; light-coloured specimens with the bands and other markings all very distinct and completely continuous and dark-coloured specimens with the bands and markings all (with the exception of those of the head, which never vary) very indistinct and very discontinuous, occurring with almost all conceivable means between these two extremes of colouring, and indicating, it is probable, a high degree of protective adaptation to surroundings.

"The caterpillar pupates within the living leaf of its food-plant (the castor-oil plant, *Ricinus communis*), to which it has lately proved a destructive pest in the Bellary District, Madras" (*J. Wood-Mason, Esq., Curator, Indian Museum, Calcutta, in the Records of the Government of India*).

EXPRESSION OF OIL.

EXPRESSION
of
Oil.
476

In many of the passages quoted above reference has been made to this subject, notably the report furnished by **Mr. J. F. Price**, regarding the castor-seed and oil industry of Madras, and **Mr. Scrofton's** account of the preparation of the oil used by the North-Western State Railways of the Panjáb and Sind. In the Madras report will be found the results of special experiments conducted with the object of ascertaining the yield of oil. The earliest, and in some respects the best, statement hitherto published on this subject was that given by **Lieutenant (now Colonel) Hawkes** as the Madras Exhibition Jury Report of 1853. Three years later **Hawkes** brought out, in pamphlet form, his *Report upon the Oils of South India*—an improved issue of the Jury Reports. It may safely be said that the majority of subsequent writers have satisfied themselves by repeating **Hawkes'** account of Castor-oil (and in some cases without any acknowledgment), but have not attempted to give more recent facts or figures. Indeed, the extent to which this has been carried has retarded rather than advanced the Castor-oil trade of India. The statements thus originally made 35 years ago have been republished as the results of recent experiments. It has therefore been thought desirable to give here in its entirety (since not readily accessible) **Hawkes'** report on castor-oil, so that comparisons may be possible between his statements and those which it is believed are for the first time made public in this work. It would seem highly desirable that this republication of **Hawkes'** report may be accepted as a record, from which progression may be made, by future writers giving only fresh results and additional facts. The editor has before him perhaps a dozen books and reports in which **Hawkes'** facts and figures, and **Hawkes'** facts and figures only, are given, as an account of the castor-oil industry of India in various periods from 1853 to 1890. (*Conf. with Cooke's Oils and Oil-seeds; Drury's Useful Plants of India; Spons' Encycl., p. 1380; Balfour's Cyclopædia, Ricinus, 421; Shortt's Man. Indian Agri., etc., etc.*)

EXTRACTION
of
Oil.
Madras.
477

EXTRACTION OF OIL IN MADRAS.

In discussing the subject of Indian castor-oil Hawkes divides his account into 1st, the large-seeded, and 2nd, the small-seeded forms :—

Large Form.
478

1st—*The Large-seeded Form.*

"The oil, obtained from the large-seeded variety of the **Ricinus com-**

munis has obtained the above name from the fact of its being used almost solely for burning in the commonest lamps and for feeding torches. For this purpose the seeds are sometimes partially roasted, to coagulate the albumen and liquify the oil, and then pressed in the ordinary mill, or boiled with water, or the roasting process is omitted; in either case the colouring matter of the husks of the seeds and other impurities give the oil a dark colour, and if the roasting process is carried too far, a slightly empyreumatic odour is communicated. By carefully shelling the seed and rejecting all impurities, the Natives prepare a clear oil for medicinal purposes (by boiling) nearly equal to that extracted from the small-seeded variety. The price of this oil varies in different parts of the country from R1-10 to R3-13-6 per maund of 25℔. The average of nineteen large stations in all parts of the Presidency for the quarter ending 31st October 1854, was R2-8-6 per maund. It is chiefly used for burning in lamps, and from its viscidity and drying qualities only in those of the simplest description. The average export of this oil for the last six years has been 97,561 gallons per annum. Lamp-oil made into a kind of *palmine* by agitation with nitric acid, is largely used as a lubricating agent for Railway locomotives in India."

2nd—The Small-seeded Form.

"The small-seeded variety of the Ricinus communis is supposed to yield the best product, and is therefore universally employed in preparing the oil exported to Europe for medicinal purposes.

"The fresh seeds, after having been sifted and cleaned from dust, stones, and all extraneous matters are slightly crushed between two rollers, freed by hand from husks and coloured grains and enclosed in cleaned gunny. The packets of seed then receive a slight pressure in an oblong mould which gives a uniform shape and density to them. The 'Bricks', as they are technically called, are then placed alternately with plates of sheet iron in the ordinary screw or hydraulic press. On the application of a gradually increasing pressure, the oil exudes through the pores of the gunny and is received in clean tin pans. Water in the proportion of a pint to a gallon of oil being added the whole is boiled until the water has evaporated; the mucilage will be found to have subsided and encrusted the bottom of the pan, whilst the albumen solidified by heat forms a thin layer between the oil and the water. Great care must be taken to remove the pan from the fire the instant the whole of the water has evaporated, which may be known by the bubbles having ceased, for, if allowed to remain longer, the oil, which has hitherto been of the temperature of boiling water or 212°F., suddenly rises to that of oil or nearly 600° F., thereby heightening the colour and communicating an empyreumatic taste and odour. The oil is then filtered through blanket, flannel, or American drill and put into cans for exportation. It is usually of a light straw colour, sometimes approaching to a greenish tinge. The cleaned seeds yield from 47 to 50 per cent. of oil. The following is the result of experiments made at Madras and Calcutta to ascertain the percentage of oil in castor-oil seed (January 27th, 1853) :—

"Calcutta 1,400℔ of seed yielded kernels and raw oil as follows :—

	Kernels. ℔	Oil. ℔
1st sort	632	324
2nd „	184	87½
3rd „	164	76½

making a total of 980℔ of kernels and 488℔ of raw oil from 1,400℔ of seed.

RICINUS communis.	**Extraction of Oil**

EXTRACTION of Oil.

Madras.

Small Form.

" Madras 1,400℔ of seed yielded raw oil as follows :—

	℔
1st sort	318
2nd ,, ...	88
3rd ,, ...	74

making a total of 480℔ of oil from 1,400℔ of seed.

" The cost of the Madras oil is as follows :—

	R	a.	p.
1,400℔ of seed at R3-3 per bag of 164℔ . . .	27	3	4
Husking and selecting kernels and coolie hire . .	3	11	9
Crushing, moulding, pressing, boiling . .	2	7	1
Filtering and sundries	2	8	0
Overseer's pay, godown rent	1	6	2
300 empty quart bottles, corks, etc.	34	4	8
Cleaning and packing charges	4	8	0
TOTAL .	**76**	**1**	**0**

Deducting the price of the bottles this gives an average of 1 anna 4$\frac{11}{24}$ pies per quart first, second, and third sort oil. This oil is chiefly used as a mild purgative, and, by Natives, for anointing the head. Soap of good quality may be made from it, but the cost, and disagreeable smell which it communicates, preclude its general use. The average export from the year 1849-50 to 1852-53 was 11,325 gallons per annum.

" The method of extracting this oil by the boiling process is thus given by Ainslie :— The seeds are boiled for two hours in water, dried for three days in the sun, husked and pounded. They are then boiled in fresh water until the whole of the oil has risen to the surface. Castor oil, being entirely soluble in alcohol of sp. gr. ·825, any adulteration of it with other fixed oils may be ascertained by dissolving a sample in eight times its weight of spirit ; the fixed oil is not dissolved but floats on the surface. This, however, is not an infallible test."

MANUFAC-TURE.

Mysore.

480

MANUFACTURE OF OIL IN MYSORE AND COORG.

" The *haralu*, or castor oil, is made indifferently from either the large or small varieties of the **Ricinus**. It is the common lamp oil of the country, and is also used in medicine. What is made by boiling, as described below, is for family use; all that is made for sale is expressed in the mill. To form the cake seventy seers of the seed require only five seers, *kachcha* measure (1·39 ale quarts) of water, and give 60 seers (4·17 ale gallons) of oil; which after being taken out of the mill, must be boiled for half an hour, and then strained through a cloth. The cake is used as fuel.

" The following is the process for making castor oil for domestic use. The seed is parched in pots containing about a seer, which is somewhat more than a quart. It is then beaten in a mortar, by which process balls of it are formed. Of these, from four to sixteen seers are put into an earthen pot, with an equal quantity of boiling water, and boiled for five hours; during which care must be taken, by frequent stirring, to prevent the decoction from burning. The oil now floats on the surface, and is decanted off in another pot, in which it is boiled by itself for a quarter of an hour. It is then fit for use, and by the last boiling it is prevented from becoming rancid." The above passage has been taken from the *Mysore and Coorg Gazetteer* since its existence there conveys the idea that it is applicable to the present day. It is, however, almost word for word a reprint of Dr. Buchanan-Hamilton's account published in 1807.

EXTRACTION OF OIL IN SHAHABAD, BENGAL.

"The castor-oil plant has two or three marked varieties and is largely cultivated principally for home use; but a considerable quantity for the castor-oil factories of Dinapore. The native process only succeeds in making a very impure oil, which is so offensive for its smoky qualities in burning that it is not sought after by them for that purpose; but only for anointing leathern well ropes, shoes, etc., and being a cheap oil is largely used for the latter purpose. It is thick and viscid, and, extracted under the native process, soon turns rancid :—while by the European process it is next to cocoanut-oil one of the purest and best burning oils known. The plant requires scarcely any cultivation :—and in south Shahabad is oftener sown in the borders of a valuable field as an hedge than for any other purposes. It loves, however, a sandy loam, and will not grow in the clays. Its yield under the native process is about 33 per cent. of the impure oil above described :—and I believe a larger quantity, and I know a purer oil is extracted by the European process. Newly cleared jungle lands grow the castor plant abundantly, and its extended cultivation is only bounded by the demands in the market, as long as the rates are remunerative :—for although the sowing and tendence of the plant costs little trouble; yet the picking of the seed is a troublesome process, and it requires a much larger amount of room to come to perfection. The natives sow and uproot the plant yearly. I do not know why this should be, as it grows and yields abundantly the second and third years in hedges or other open places. When cultivated by itself the natives always sow the seed too close, and consequently the plant is comparatively small :—for attaining its full perfection, no place is better for it than a hedge or a bank" (*R. W. Bingham, Esq., Agri.-Horti. Soc. Ind. Journ. XII.*, *341*).

Lieutenant Hawkes' account and that given above from the Agri.-Horticultural Society's Journal are the earliest articles on this subject and those from which most writers have compiled. Though in some respects disconnected, the following passages may be here given as exhibiting the more recent facts that have been brought to light. The editor is fully conscious that he has failed to produce an original essay on the subject of castor-oil and has furnished instead mainly a series of quotations. He has been forced to adopt this course from the incomplete nature of available information. Had he attempted anything else he could have given the cost of production in one province not in another; the yield in some not in others; the value of the oil produced in certain districts but not in the majority; and the method of expression of oil in many but not in all districts. It therefore seemed the preferable course to allow the original writers to speak for themselves, and while the reader may have to work through many pages, instead of being furnished with definite paragraphs treating of distinct sections of the subject, if he takes the trouble to do so he will, it is hoped, discover information on the particular features of the castor-oil industry in which he is interested. Of no part of this article is an admission of imperfect treatment more necessary than in the present. To obtain a knowledge of the methods of extraction and yield of oil, the reader must consult all the foregoing pages as well as those under this chapter.

MANUFACTURE OF OIL IN CALCUTTA.

In order to obtain this much-to-be-desired information, the Government of India issued, to Local Governments and Agricultural Societies in July 1883, *A Note on Castor-seed*—which had been drawn up by Mr. T. N. Mukharji. In their forwarding letter (accompanying the *Note*) the admission was made by the Government of India that the "Note" was obviously

RICINUS communis.	Manufacture of Oil

incomplete as regards the methods and cost of cultivation, outturn and value of the crop, and the invitation was accordingly made for supplementary information. A large portion of the new material here published was, as the result of that enquiry, furnished by the local authorities, but it need not be repeated that even now the available particulars are in many respects still incomplete. The " Note " in its opening paragraphs gives some of the main facts of castor cultivation compiled chiefly from **Hawkes** and **Drury**, and it concludes by giving Hawkes' experiments in the yield of oil, etc., from 1,400℔ of Calcutta and a like quantity of Madras seed. **Mr. Mukharji**, however, contributed in addition the results of a personal enquiry. He tells us that " at the Calcutta market the big variety comes from Upper Bengal and the North-West Provinces, while the smaller seeds come chiefly from the Madras Presidency." This same fact has been commented on by other writers. " Most of the Indian oil," says **Spons'** *Encyclopædia,* " is extracted in and exported from Calcutta, the crushed seed being sent up from Madras for the purpose—an inexplicable proceeding." The commercial writer quoted by **Mr. Price** (see paragraph above under the heading Madras) alludes to this subject also. But **Mr. Mukharji** in dealing with the Calcutta trade gives perhaps the best account which has as yet appeared of the methods of expressing the oil in Bengal. He writes :—

" *Extraction of oil, home process.*—Cultivators generally extract oil at home for burning in lamps. The process is as follows :—

(1) Seeds partially roasted on heated sand put in a pan over a fire which coagulates the albumen and liquifies the oil.

(2) Slightly crushed by hand-mill and husked.

(3) Pounded in a mortar.

(4) Boiled in water, 2 quarts of water being added to 5℔ of pounded seed.

(5) As the water evaporates, the oil rises to the surface, when it is taken out with a spoon and put in a separate vessel.

(6) The oil thus obtained is boiled again, by which process the sediment, the remaining moisture, and all other extraneous matter is burnt up or precipitated, and the oil becomes more purified.

" The oil thus obtained is, however, still very impure, thick and viscid, and offensively smokes when burnt in lamps. It is also used to anoint shoes, water-bags used for raising water from wells, and other agricultural appliances made of leather. The yield of oil is 33 per cent. of seed.

" *Scientific process.*—The oil extracted with the aid of scientific appliances, hand-mill or a hydraulic press is of a superior quality.

" **Messrs. Khettra Mohan Bysack,** one of the leading castor-oil manufacturers in Calcutta, whose oils, marked KB1, KB2, and KB3 are largely exported to Europe and Australia, and who have won a first class order of merit at the Melbourne Exhibition, have furnished the following information on the subject :—They generally use the big Bengal and the small Madras seed. The price at Calcutta for both is about 6s. per cwt. The proportion of oil yielded is about 40 per cent. by the former and 37 per cent. by the latter, but the latter is said to yield the best kind suitable for medicinal purposes, while the oil from the former is chiefly used for burning in lamps and for other purposes. They manufacture four kinds of oil distinguished according to the degree of refinement :—

I.—Cold-drawn, the best kind, only used in medicine.

II.—Cold-drawn, No. 1, used in arts and manufacture.

III.—Cold-drawn, No. 2, less refined, used as above and also for burning in lamps.

IV.—Cold-drawn, No. 3, unrefined, used for burning and in machines.

"Hydraulic press is not used, as its unsuitability has been found by experience to lie in the difficulty to adjust the amount of power to be bestowed in the different stages of pressing. A hand-mill is used, which is said to have been invented by Messrs. Jessop & Co. at the early stage of Indian castor-oil trade with foreign countries. This machine has not been patented. The process followed by this firm is described below :—

"*I.*—The seeds are first cleaned with the hand ; women are employed in this work. They place a quantity of seed on a smooth board, to which they give one or two strokes with a flat wooden mallet which break the seeds to two or three pieces, thus rendering the separation of the husk easy. The broken seeds are then winnowed with a common basket winnower, which removes the husk from the kernel. The kernels are then dried in the sun and afterwards broken by a crushing machine. They are then put within small canvas or gunny bags and pressed in the hand machine, the oil falling into a pan placed underneath. The oil is collected in large galvanized iron vats and bleached by exposure to the sun, which also causes the sediment to precipitate. It is next boiled in order to evaporate any remaining moisture. Vegetable charcoal is added to it, and the oil is then thrice filtered through flannel or blotting paper. The oil thus obtained is of the purest quality, only used in medicine, and manufactured to order. No fire is applied during the pressing, and hence no irritating part of the seed finds its way to the oil. The yield is, however, 10 per cent. less than No. II.

"*II.*—The seeds are husked, crushed and pressed as before. At the time of pressing fire is put underneath the machine, the heat from which liquifies the oil and increases the yield, with which, however, a certain portion of the irritating or injurious part of the seeds is mixed, which is avoided by process No. I. It is then bleached and boiled as before, and filtered with the addition of animal and vegetable charcoal. This kind of oil is also not ordinarily manufactured.

"*III.*—Process as above but not filtered. Largely manufactured and exported, price 30s. per cwt.

"*IV.*—The seed is not husked by the hand, but by machine, and is therefore not quite free of husk, and the oil is accordingly not so clean. Other processes are as in No. III. The oil is not filtered. Price 25s. per cwt."

MANUFACTURE OF OIL IN BENGAL JAILS.

The following memorandum on the manufacture of Castor-oil, as carried on in the Rajshahye Central Jail (Bengal) by E. O. Bensley, Superintendent, may be given as an example of the process of oil manufacture followed at the Jails of India :—

"I can give no information regarding the cultivation of the castor plant, as the plant is not grown in this district for marketable purposes. A few plants are grown by villagers in the immediate vicinity of their homesteads, and the oil is extracted from the seed in the rudest fashion for domestic purposes. The Central Jail in this district has, however, for its special industry, the manufacture of castor-oil. Between 700 and 800 maunds are turned out monthly. Castor-seed for the purpose is boated down here from Bhagulpore, Patna, and Revelgunge, chiefly from the two latter markets.

"As the process of manufacture in this jail differs somewhat from that employed by the seed-crushers in Calcutta, I propose giving a detailed description of it under the following heads :—

(1) Cleaning and grading of the seed :
(2) Splitting of the seed :

MANUFAC-
TURE
in
Bengal.
Calcutta.

Jails.
483

RICINUS communis.

MANUFAC-
TURE
in
Bengal.

Jails.

   (3) Sunning the seed and winnowing it, so as to separate the shell from the kernel :
   (4) Crushing the kernel :
   (5) Putting into canvas bags and pressing in the screw-presses :
   (6) Boiling : and
   (7) Straining.

   "*First.*—The cleaning and grading of the seed are done by the female prisoners, who first remove from the seed all extraneous matter, such as dust, pebbles, other seeds, etc., and who then by means of sieves with different size meshes grade the seed into four sizes. The need for grading will-be apparent when the next stage is described.

   "*Second.*—The splitting of the shell is done with a machine which consists of two smooth iron rollers placed parallel to one another and working towards one another. It is worked by hand by a simple arrangement of cogged-wheels. One of the cylinders or rollers is fixed, the other is movable by a screw adjustment. By means of the latter contrivance the space between the cylinders can be regulated to the required distance. The space is increased or diminished according to the size of the seed about to be split. The great point is to give the seed a sufficient squeeze so as to split the shell without crushing any of the oily matter out of the kernel. It will now be understood why it is necessary to grade the seed, and experience has taught us that castor-seed can be placed in four grades according to size. A wooden box is placed above the cylinders to hold about four or five seers of seed at a time. This keeps the cylinders constantly supplied. The cylinders are each about two feet long, so the process of splitting goes on very rapidly. Seed-crushers in Calcutta use a mallet and board for the purpose of splitting the seed; but apart from this being a slow process, entailing a large employment of hand-labour, it has the disadvantage of frequently bruising the kernel and extracting the oil which is consequently lost.

   "*Third.*—The seed is passed on to the winnowers, who separate the husk from the kernel on large masonry platforms. Sunning is a very necessary step in castor-oil manufacture, not only to dry any moisture that there may be in the kernel, but to liquify and facilitate the exit of the oily matter.

   "*Fourth.*—The kernel is now taken to the crushing machine, which is similar to the splitting machine, excepting that the two cylinders work close together and are both immovable. In this the kernel is crushed.

   "*Fifth.*—The crushed kernel is now put into pieces of canvas about 15 inches by 12 inches, the sides of which are folded over so as to prevent escape of the kernel. These canvas bags are placed alternately with iron plates into the screw-press, the pressure from which is applied horizontally by means of two powerful screws. Each screw has a wheel fixed to it, in which are sockets for the insertion of lever-bars. Four men are generally employed in working the screws by means of lever-bars. As pressure is applied to the canvas bags, the oil oozes out of them into a trough placed below. At the back of the press a fire is kept up to facilitate the exudation of oil. Each canvas bag holds about half a seer of crushed kernel, and each *feed* of the press requires from 130 to 150 such canvas bags full.

   "*Sixth.*—The thick slimy oil thus obtained is passed into the hands of the boilers, who, mixing it with water in the proportion of 40 parts of oil to 5 or to 8 parts of water, boil it in large copper pans. The boiling of the oil is perhaps the most delicate stage of the whole process of the manufacture. To know exactly when to stop the boiling is a point of knowledge acquired only by great experience. Thermometers were used at one time, but the results were not satisfactory. Now the experienced boiler is guided by his eye, and by his sense of touch; directly he sees that the bubbling of the oil is about ceasing, and directly he finds that the sediment that lies in grains at the bottom of the pan has acquired a certain colour and certain crisp consistence, he knows the oil is sufficiently boiled, and he quickly extinguishes the fire.

"*Seventh.*—The oil is removed from the pans and passed through a bed of charcoal, and through six or eight folds of calico. It is now ready for use.

"The quality of oil made in this jail holds a position intermediate between the qualities known as No. 2 and No. 3 of the oil market. It is used exclusively for lubricating and burning purposes, and supplied to all Government Departments in Calcutta, including State and Guaranteed Railways. Its price varies from R10 to R10-4-0 per maund of 82℔ avoirdupois. Some cold-drawn medicinal oil is also made here and supplied to the Calcutta Medical Depôt. The process of manufacture is the same as described in the note on Castor-seed. The yield of oil from Bhagulpore seed varies from 36 to 37½ per cent., while that from Patna and Revelgunge hardly ever yields more than 34 to 34½ per cent. This percentage is taken not on crushed and clean seed, but on seed that comes straight from storing godowns with all extraneous matter still in it. The husk of the seed is used in the jail for boiling the oil and for keeping up the fires at the back of the oil-presses, and is also issued to the cooks as fuel for cooking the prisoners' food " (*Extract from the Records of the Government of India Revenue and Agricultural Department, B, September 1884, Nos. 27-8*).

TRADE IN CASTOR SEED AND OIL.

I. Foreign.—In the chapter above which deals with History of castor-oil reference has been made to the fact that England and most continental countries drew their supplies at first from the West Indies. The trade consisted almost entirely in the supply of the expressed oil for medicinal purposes. About the beginning of the present century attention was directed to India as a producing country, and the discovery that it possessed two kinds of the oil, one an inferior and much cheaper article than the other, rapidly led to the development of the industrial purposes for which the cheaper article was suitable. A new interest was thereby given to the subject which soon manifested itself in the demand for the seed to be expressed at oil mills in Europe. This was doubtless a natural result of increased demands which could not be satisfied through the heavy freight charges on the transport of oil. It is significant that neither the first (1813) nor the second (1825) edition of Milburn's *Oriental Commerce* makes any allusion to the export of castor-seed for India though the oil is dealt with. The passages and table of trade returns quoted below, however, occur in that work :—

"The following are the quantities of castor-oil imported, and sold at the East India sales, in the years, 1804 to 1808 inclusive, with sale amount, and the average price per ℔ :—

Year.	March Sale.		September Sale.		Total.		Average per ℔.		
	℔	£	℔	£	℔	£	£	s.	d.
1804 . . .	20,207	2,309	...	...	20,207	2,309	0	2	3
1805 . . .	4,603	258	15,627	1,944	20,230	2,202	0	2	5
1806 . . .	...	...	1,352	27	1,352	27	0	0	5
1807 . . .	4,727	774	8,200	1,302	12,927	2,076	0	3	3
1808 . . .	3,503	49	659	7	4,162	56	0	0	3

RICINUS communis.	Trade in Castor

| TRADE. Foreign. | "16 cwt. of castor-oil are allowed to a ton. The permanent duty is 9*d.* per ℔ and the temporary or war duty 3*d.*, making the whole 1*s.* per ℔." The returns published by H. H. Wilson (*Review of the External Commerce of Bengal for 1813 to 1828*) have already been quoted (*see Chapter on History*); but it may be said that they manifest a less capricious state of the trade than the above. The imports into Great Britain had attained in 1819-20 the value of £7,102. In the account of castor-oil given by Hawkes (re-printed above), the figures of the trade during 1850-55 are dealt with. The average exports of the oil for the past six years, says Hawkes, have been 97,561 gallons. He makes no mention of the exports of castor-seed, they being then too unimportant. The exports of the oil had thus expanded, it may be said, in 50 years from 2,000 to 100,000 gallons. Passing over a gap of some 30 years more the earliest detailed information available regarding castor-oil occurs in the trade returns for 1876-77. Previous to that year the figures of the trade were included under the heading "all other kinds" since they were too unimportant to necessitate separate treatment. Indeed the trade in castor-seed was not separately returned until the year 1877-78. In that year the exports in castor-oil were 1,411,216 gallons valued at R19,26,427 and of castor-seed 4,521 cwt. valued at R27,412. The largest quantity of the oil went to the United Kingdom from Bengal, next in importance being Australia with 429,268 gallons. Of the seed, Italy took the largest quantity, being followed by the United States; these two countries in fact took between them 95 per cent. of the total exports. The chief exporting province in 1877-78 was Bengal with nearly ¾℔ of the entire quantity. Thus it may not incorrectly be said that from 1850 to 1880 the exports in castor-oil expanded from 100,000 to nearly 3 million gallons, and that a new trade had at the same time started in the export of the seed to be pressed in Europe for the oil. These facts give some conception of the magnitude of the increased uses discovered for the oil. From the year above detailed (1877-78) it may be desirable to review the Foreign trade in the two chief sections, *viz.*, EXPORTS of Indian produce, and Imports of Foreign produce. *Exports of Indian Produce.*—The following table exhibits the exports of Indian Castor-oil and Castor-seed from 1878-79 to 1889-90 :— |

Years.	Castor-oil.		Castor-seed.	
	Gallons.	R	Cwt.	R
1878-79	2,119,757	31,53,969	74,214	5,00,056
1879-80	2,651,889	32,10,703	237,601	11,80,768
1880-81	2,890,803	31,04,701	76,461	4,33,858
1881-82	3,009,288	29,77,122	250,696	11,77,090
1882-83	2,571,588	25,42,347	222,156	10,02,517
1883-84	3,102,063	31,29,704	512,444	24,00,488
1884-85	3,207,045	30,84,602	476,396	21,03,379
1885-86	2,190,888	21,86,228	670,537	29,89,514
1886-87	2,676,012	27,18,370	610,893	28,86,182
1887-88	2,677,005	26,42,816	720,951	31,94,388
1888-89	2,092,913	20,31,467	585,769	31,28,741
1889-90	2,664,990	32,83,163	894,631	60,74,290

It will thus be seen that while the exports in castor-oil from India have been stationary during the past ten years, if indeed they could not rather be said to have declined, the exports in the seed have expanded within that period from an average of the three years previous to 1881-82 of 130,000 cwt. to 894,631 cwt., and in value from an average for the three previous years of

R. 484

Seed and Oil.	*(G. Watt.)*	**RICINUS communis.**

some 7 lakhs of rupees to 60 lakhs. While, therefore, it is customary to find writers pointing to the stationary nature of the Indian export trade in castor-oil, as a proof of the effect produced by the modern use of mineral oils as lubricants, it would seem that in any such contention the trade in castor-seed should be taken into account as well as of castor-oil. Assuming that a gallon of castor-oil weighs 9·9℔, the exports of castor-seed from India in 1889-90 would have afforded over 3,500,000 gallons of oil. It would thus appear that in that year, India actually furnished the world with a little over six million gallons. The more natural explanation of the stationary nature of the Indian export trade in the oil would thus seem to be the advantages of the oil expressed in Europe by more profitable methods, having its usual effect on a manufacture in which the gains of a small and impecunious dealer are largely, though temporarily supplemented by a short-sighted policy of adulteration. It may be said that the history of Indian jute manufacture establishes the value of this contention. So long as Native manufacturers were the only competitors against the new industry started in Scotland, in the production of jute bags, etc., India went to the wall. But no sooner had jute mills been opened out in India itself under European management and governed by European skill and capital, than Dundee felt that it had no longer a monopoly. There seems no doubt that the facilities for adulteration have more to say to India's backwardness in oil production than all other considerations. Most oils find a market on account of some definite property. If adulterated they are rendered useless, and since they cannot be again purified by any cheap process, the insecurity of possible adulteration must operate against the growth of a large trade in Native-made Indian oils. That castor-oil has felt the effect of the modern use of mineral oils as lubricants, there seems no doubt, but that the demand for the oil in the aggregate has not been seriously diminished, is equally true. The ultimate utilization of the large Indian supply is a matter of considerable speculation and doubt, but that new uses are being yearly found for it seems undeniable.

Area and Outturn. —It has been admitted, in the foregoing chapters, that statistics are not available to show the area in India annually under this crop. Indeed from its being largely grown as a hedge plant or in strips through fields of other crops the difficulty is very considerable in arriving at anything like a safe estimate. Some idea may, however, be formed from a calculation of the area necessary to produce the quantities shown in trade statistics. For example, if we assume that an acre of land devoted to castor-oil yields 5 cwt. of seed, and that the production of oil is 30 per cent. on the weight of seed (and these are low estimates since by European methods of expression the yield is often over 40 per cent.) the amounts of oil and seed exported from India in 1889-90 would have alone required roughly an acreage of 330,000 for their production. But the Indian consumption of the oil is very great so that as a haphazard guess it may be assumed that the total area in India devoted to the crop in 1889-90 was not far short of half a million acres. If this suggestion be for the present accepted as approximately correct, the total production of seed would have been 2,500,000 cwt. (at five cwt. to the acre,) or of 8,400,000 gallons of oil (at 30 per cent.). The Indian consumption of castor-oil by this calculation would in the year in question have been a little under two million gallons. The writer does not advance this estimate as anything more than an example of one way by which an idea of the acreage and production of oil may be arrived at. The data employed in the calculation may be corrected as found necessary, and the results stated anew. The most serious errors involved, are in the acceptance of any averages as trustworthy, and in the assumption of the area required to produce the amount

| RICINUS communis. | Trade in Castor Seed and Oil. |

consumed in India. That figure has been arrived at after careful consideration of numerous scattered commercial and agricultural returns, but it is offered only as a suggestion that may approximate to the truth.

Having now attempted to give some idea of the possible production, attention may be turned to a feature of the castor-oil trade regarding which, we fortunately possess more precise information, namely, the destination of the Indian supply in the foreign markets. Taking every third year since 1877-78 it may be said of the OIL that the supply to the United Kingdom has steadily declined, while that to Australia has correspondingly increased. After these two countries the balance of the Indian production goes to the Straits Settlements, Hong-kong, Mauritius, and Ceylon, in quantities ranging from 80,000 to 250,000 gallons, the order of importance being that given. The trade with the United Kingdom is striking. In 1880-81 it received out of the total exports of oil 1,775,074 gallons ; in 1883-84, 1,578,670 gallons ; in 1887-88, 1,066,102 gallons ; and in 1889-90, 782,550 gallons out of the totals shown in the table at page 554. Australia, on the other hand, took in 1880-81, 679,391 gallons, in 1883-84, 823,851 gallons in 1887-88, 892,940 gallons ; and in 1889-90, 1,005,780 gallons. The supplying provinces during these periods have also exhibited certain important changes ; the trade from Calcutta has fallen off and that from Madras greatly improved. Thus Bengal exported in 1880-81, 2,721,233 gallons ; in 1883-84, 2,681,847 ; in 1887-88, 2,079,856 gallons ; and in 1889-90, 2,148,171. Madras, on the other hand, exported in 1880-81, 107,520 gallons ; in 1883-84, 331,529 gallons ; in 1887-88, 569,293 gallons, and in 1889-90, 510,302 gallons. The trade from Bombay is now almost insignificant, though in 1880-81 it exported considerably more than Madras.

The trade in the seed manifests an almost opposite condition. France is by far the most important receiving country followed by the United Kingdom, Italy, Russia, and Belgium. The amounts sent to Egypt appear to go there in order to await their final destination to Mediterranean ports. The following table exhibits the exports in castor-seed for the periods dealt with above :—

INDIAN EXPORT TRADE IN CASTOR-SEED IN CWTS.

Receiving Countries.	1880-81.	1883-84.	1887-88.	1889-90.
Italy	47,760	106,747	91,448	81,867
United Kingdom . .	22,023	78,597	104,724	237,388
France	5,763	323,406	433,626	450,301
Egypt	...	...	61,995	...
Russia	...	...	...	62,958
Belgium . . .	...	...	...	43,785

INDIAN EXPORT TRADE IN CASTOR-SEED IN CWTS.

Exporting Provinces.	1880-81.	1883-84.	1887-88.	1889-90.
Bombay	59,691	63,341	247,974	565,764
Madras	16,257	446,029	472,806	325,741
Bengal . . .	513	3,074	171	3,124

Imports, Foreign Produce.—The imports are unimportant. During the year 1889-90, India received 625 gallons of oil chiefly from Ceylon, the

R. 484

	RIVEA
The Clove-scented Creeper. (*W. R. Clark*)	**hypocrateriformis.**

entire supply going to Madras. In seed the traffic is even less important in 1886-87 : India received 1,317 cwt., in 1887-88, 443 cwt., in 1888-89, 1 cwt., and in 1889-90, 4 cwt.

TRADE.

II. COASTWISE TRADE.—Madras is the chief exporting province having in 1889-90 furnished to other provinces and Native States 313,063 gallons of oil and 98,860 cwt. of seed. The bulk of the oil went to Burma (164,840 gallons), and to Bombay (141,122 gallons) and of the seed to Bengal (97,696 cwt.). The next most important province in meeting the coastwise traffic is Bombay. In 1889-90, it exported 16,183 gallons of oil mainly to Kattywar, and of seed 3,099 cwt. chiefly to Baroda. Bengal naturally follows as the next important province. It exported 7,319 gallons to Burma and Bombay, and of seed none to other provinces, though a small exchange took place between ports within the province.

Coastwise.
485

III. INTERNAL TRADE BY ROAD, RIVER AND RAIL.—Calcutta generally heads the list of exporting centres in oil. It furnished 24,094 maunds in 1888-89 to Bengal Province (17,676 maunds) and to Assam (6,198 maunds). Bombay Province stands as next important, and gave for the year in question 15,446 maunds, chiefly to the port town (10,720 maunds), and to Rájputana and Central India (4,072 maunds). The exports from the North-West Provinces in oil are unimportant and go chiefly to the Central Provinces (1,628 maunds), Rájputana and Central India (1,446 maunds), the Panjáb (1,174 maunds), and to Bengal (1,133 maunds). Bombay port town furnished to Rájputana and Central India (1,828 maunds), to Berar (1,341 maunds), and to the Central Provinces (1,078 maunds). The transactions in castor-seed may now be dealt with. Bengal is the chief exporting province, having furnished in 1888-89, 2,23,740 maunds chiefly to Calcutta. The Nizam's Territory is next important, having given 2,22,739 maunds mainly to Bombay town. The North-West Provinces and Oudh stand next in the list, having exported, 1,82,720 maunds to Calcutta. Bombay province furnished 1,42,823 maunds chiefly to Bombay port.

Internal.
486

The above review exhibits the chief provincial exchanges. The port towns drain from the provinces to meet the foreign trade, but in the case of Calcutta the demand for seed is largely to supply the necessities of local mills, since Calcutta is the chief exporting town for oil. The drain into the Central Provinces, Rájputana, and the Panjab manifests the fact already alluded to that these tracts cultivate the plant to a small extent only and insufficient for local demand.

(*W. R. Clark*)

Riedleia corchorifolia, *DC.;* see **Melochia corchorifolia,** *Linn.;* Vol. V., 225; STERCULIACEÆ.

Rinderpest, see **Oxen, Buffaloes, and allied species of Bovinæ,** Vol. V., 673.

RIVEA, *Chois.; Gen. Pl., II., 868.*

Rivea hypocrateriformis, *Chois.; Fl. Br. Ind., IV., 184;* CONVOL-
THE MIDNAPORE CLOVE-SCENTED CREEPER. [VULACEÆ.

487

Syn.—RIVEA BONA-NOX, *Roxb.;* R. FRAGRANS, *Nimmo;* R. ORNATA, *Aitch.;* CONVOLVULUS HYPOCRATERIFORMIS, *Lamk.;* C. CANDICANS, *Wall.;* LETTSOMIA UNIFLORA, *Roxb.;* L. BONA-NOX, *Roxb.;* ARGYREIA UNIFLORA and BONA-NOX, *Sweet.*

Vern.—*Kulmi luta,* BENG.; *Phanja,* BOMB.; *Búdthi kiray,* TAM.; *Bod dikúra, niru boddi, boddi,* TEL.

References.—*DC., Prod., IX., 326; Roxb., Fl. Ind., Ed. C.B.C., 166; Voigt, Hort. Sub. Cal., 351; Brandis, For. Fl., 342; Dalz. & Gibs.,*

R. 487

Bomb. Fl., 168 ; _Aitchison, Cat. Pb. & Sind Pl._, 100 ; _Sir W. Elliot, Fl. Andhr._, 29, 135 ; _Useful Pl. Bomb._ (_Vol. XXV., Bomb. Gas._), 202 ; _Gazetteers_:—_Bombay, V._, 27 ; _N.-W. P._ (_Bundelkhand_), _I._, 82 ; _IV._, lxxiv ; _Agri.-Horti. Soc. Ind. Journ._ (_Old Series_), _X._, 19 ; _Ind. Forester, III._, 237 ; _XII., App._, 17.

Habitat.—A twining shrub common in the dry forests of Western India, and from Lahore to Mysore; and found also by Griffith in Assam. The flowers open at night so that by most persons this would be regarded as a form of the Moon-flower. (See **Ipomœa bonanox**, Vol. IV., 483.)

Food.—The LEAVES and YOUNG SHOOTS are boiled by the Natives of India, with salt and chillies, and eaten as a vegetable.

Rivea ornata, _Chois.; Fl. Br. Ind., IV., 183._

Habitat.—From an economic point of view it is scarcely necessary to specially notice this species. It occurs in the Deccan Peninsula, and a variety, known as **Griffithi** is said to be found in Bengal. The latter very possibly is the plant alluded to by Gamble (_List of Climbers, etc., in Darjeeling, p._ 57) the SEEDS of which are eaten.

ROBINIA, _Linn. ; Gen. Pl., I._, 499.

Robinia pseudacacia, _Linn. ; Baron von Mueller, Select Extra-Tro-_ [_pical Plants, 8th Ed., p._ 422 ; LEGUMINOSÆ.

THE NORTH AMERICAN LOCUST ACACIA.

Habitat.—A large tree, which attains a height of 90 feet. It is indigenous to North America from Alleghany to Arkansas. It is very hardy and has been introduced into several European countries, and of late cultivated on some of the treeless southern slopes of the Himálaya.

Domestic Uses.—Baron F.lvon Mueller describes this tree as a most useful one. He says : —" The hard and durable wood is in use for a variety of purposes, and is particularly eligible for tree-nails, axle-trees, and turnery, strength greater than that of the British oak, weight lighter. The Natives used the wood for their bows. It is one of the best trees for renovating exhausted soil and for improving poor land. Recommended as one of the easiest grown of all trees on bare sand, though standing in need of twice as much mineral aliment as **Pinus sylvestris**, and nearly as much as poplars. It pushes through shifting sand its spreading roots, which may attain a length of 70 feet. The roots are poisonous."

With reference to the cultivation of the tree in India, the following remarks taken from the _Indian Agriculturist_ review the recent action taken by the Government of India in the Forest Department :—" The variety (of **Robinia**) introduced three or four years ago by Mr. Parsons of the Annandale Gardens at Simla resembles the **Acacia** so closely that its botanical name is the ' pseudo' or false Acacia. The Inspector General of Forests has recently written a note on the value of this and other varieties of the **Robinia** to the unprotected slopes of the Lower Himálayas, in which he has recommended the purchase of large quantities of seed from Italy where it has been largely utilised, more especially in the southern Alps for fixing embankments, landslips, and torrents. Its roots form a perfect carpet in the surface soil, from which innumerable suckers spring up forming, if treated as a coppice, a very dense thicket. Its propagation is easy, as it can be grown from seed which, in Europe, is usually sown in autumn, from root suckers, and from cuttings. The facility with which the **Robinia pseudacacia** has taken possession of one or two barren slopes in Simla, although possessing a hot southern aspect, seems to point to the fact that the cultivation of the tree on a larger scale is likely to be beneficial. We understand that the Government of India has ordered from the Italian Government seeds of the **Robinia** and other trees recommended

FOOD.
Leaves.
488
Young shoots.
489
490
FOOD.
Seeds.
491
492
DOMESTIC.
493

The Dog-rose. *(W. R. Clark.)*	**ROSA alba.**

by the Inspector General, which will be offered to the local Government for distribution through the Panjáb." The utilization of the Simla gardens as a nursery from which a large stock of seedlings could be distributed seems likely to prove highly beneficial to the Simla district.

[LILIACEÆ.

Rocambole, see Allium scoradosprasmum, *Linn. ;* Vol. I., 174;

Roccella, see Lichens, Vol. IV., 636.

Rock Crystals, see Quartz, Vol. VI., 378; also Carnelian, Vol. II., 167.

Rodents, see Furs, Vol. III., 458; Hares, Vol. IV., 202; Rats, Mice, Marmots, etc., Vol. VI.; also Skins, Vol. VI., Pt. II.

RODETIA, *Moq. ; Gen. Pl., III., 25.*

A genus of AMARANTACEÆ, described in the *Genera Plantarum* as very nearly allied to **Bosia,** to which latter Sir J. D. Hooker has, subsequently, in the *Flora of British India,* reduced it. As the only Indian species has, however, been left undescribed under **Bosia,** it is necessary to deal with it under its former name—Rodetia Amherstiana.

494

Rodetia Amherstiana, *Moq. ; Fl. Br. Ind., IV., 716 ;* AMARANTACEÆ.

Syn.—DEERINGIA AMHERSTIANA, *Wall. ; see* BOSIA AMHERSTIANA, *Hook. f.*
Vern.—*Bilga,* KOTI, PB.
References.—*DC., Prod., XIII., 2, 323 ; Gamble, Man. Timb., 302 ; Watt, Cal. Exhib. Cat., VI., 158 ; VII., 217 ; Gazetteer, N.-W. P., X., 316.*
Habitat.—A large straggling shrub of the Western Temperate Himálaya from Kashmír to Kumáon at altitudes between 4,000 and 7,000 feet.
Dye.—A black dye is obtained from the LEAVES (*Gamble*).
Food.—The YOUNG SHOOTS are fried in *ghí* and eaten. The Natives also eat the bright crimson BERRIES.
Structure of the Wood.—Grey and soft. Weight 41℔ per cubic foot

495

DYE.
Leaves.
496
FOOD.
Young Shoots
497
Berries.
498
TIMBER.
499

Rohan, see Soymida febrifuga, *Juss. ;* Vol. VI.; MELIACEÆ.

Rohu, a common fresh-water fish ; see Vol. III., 384.

Rondeletia tinctoria, *Roxb. ;* see Wendlandia tinctoria, *DC. ;*
[RUBIACEÆ.

Ropes, see Cordage, Vol. II., 566.

ROSA, *Linn. ; Gen. Pl., I., 625.*

A genus of erect or climbing shrubs belonging to the Natural Order ROSACEÆ, and comprising about thirty distinct species with very numerous cultivated sub-species and varieties. Nine species are described in the *Flora of British India* as occurring wild in India, but a very much larger number of cultivated forms also occur, the economic uses of which are much more important than those of the indigenous species. It is not within the scope of the present work to give a complete list of all the roses cultivated in India for ornamental purposes, but a description of the chief groups may be found in Brandis' *Forest Flora,* 199, and in the *Flora of British India,* II., 363. The writer of this article has, therefore, thought it best to give merely, in alphabetical order, as detailed an account as possible of the wild and cultivated roses of India, which are employed for economic purposes.

500

Rosa alba, *Linn. ; Fl. Br. Ind., II., 364 ;* ROSACEÆ.

Syn.—R. GLANDULIFERA, *Roxb. ;* nearly allied to R. CANINA, *Linn. ;* the common English Dog-rose (*Brandis*).
Vern.—*Guláb,* HIND.; *Swet, sheuti gulab,* BENG.; *Gulseoti,* PB.; *Gul,* BOMB.
References.—*Roxb., Fl. Ind., Ed. C.B.C., 407 ; U. C. Dutt, Mat. Med. Hind., 320 ; Baden Powell, Pb. Pr., 347 ; Gazetteer, Mysore & Coorg, I., 60, etc., etc.*

501

R. 501

ROSA damascena.	**The Cabbage and Damask Roses.**

OIL.
502

MEDICINE.
Flowers.
503
504

OIL.
505

MEDICINE.
Petals.
506

Syrup.
507
508

Habitat.—" A tall spreading shrub, indigenous probably in the Caucasian region, and possibly wild in Afghánistán and North-Western India" (*Brandis*). Roxburgh suggests a Chinese origin for this species, since he knew it had been brought thence to the Botanic Gardens at Calcutta; but all later writers have agreed in describing it as indigenous to the Caucasus. It is found in gardens throughout India, and cultivated partly for ornament, but also as an article of economic value.

Oil.—The greatest possible confusion exists as to the species of roses which are used in the preparation of *attar*. Probably **R. alba** is employed, but as it is not so to any extent, compared with several other species, the description of this manufacture will be reserved till the species are described which are known to be most used for that purpose.

Medicine.—The FLOWERS are employed as a cooling medicine in fevers and also in palpitation of the heart. Dose 6 *máshas* (*Baden Powell*).

Rosa centifolia, *Linn.; Fl. Br. Ind., II., 364.*

THE HUNDRED-LEAVED or CABBAGE ROSE.

Vern.—*Guláb, gulab-surdi*, HIND.; *Goláp*, BENG.; *Guláb, gul-i-surkh*, (stem=) *kubjak*, (conserve=) *gul-khand*, PB.; *Troja*, TAM.; *Roja*, TEL.; *Gulábi*, KAN.; *Paninir, mawar*, MALAY.; *Nesi poen hnin-si*, BURM.; *Vard*, ARAB.; *Guli surkh, gul*, PERS.

These vernacular names seem, in most cases, to be applied indiscriminately to other species and varieties of roses; indeed, no definite names for the various species exist, but all are known by the name *guláb*, with various prefixes which indicate the colour of the flower.

References.—*Roxb., Fl. Ind., Ed. C.B.C., 407; Brandis, For. Fl., 200; Stewart, Pb. Pl., 85; Aitchison, Cat. Pb. & Sind Pl., 57; Pharm. Ind., 82; Moodeen Sheriff, Supp. Pharm. Ind., 216; Ainslie, Mat. Ind., I., 345; O'Shaughnessy, Beng. Dispens., 326; S. Arjun, Bomb. Drugs, 52; Murray, Pl. & Drugs, Sind, 142; Mys. Cat. (Cal. Exhib.), 21; Butler, Top. & Stats., Oudh and Sultanpore, 36; Watt, Cal. Exhib. Cat., V., 242; Baden Powell, Pb. Pr., 346; Lisboa, U. Pl. Bomb., 390; Birdwood, Bomb. Pr., 33, 192, 193; Piesse, Perfumery, 190, 198; Kew Off. Guide to the Mus. of Ec. Bot., 61; Kew Off. Guide to Bot. Gardens and Arboretum, 141; Gazetteers:—Mysore & Coorg, I., 60; W. W. Hunter, Orissa, II., 178, App. VI.; Bombay, VII., 40.*

Habitat.—A native of the Caucasus and Assyria; cultivated in India. This, with **R. damascena** and **R. gallica**, constitutes the group of cultivated roses known as the **Galliceæ.** That group is most likely to be confused with the **Caninæ** the species of which are **R. indica, R. alba,** and **R. microphylla.**

Oil.—According to Ainslie this is the rose chiefly employed in Persia for the manufacture of *attar* and rose-water, but later writers are agreed in saying that, in India at any rate, the damask rose (**R. damascena**) is the one most used. In Bulgaria **R. centifolia** is, according to Flückiger & Hanbury, the species chiefly employed for that purpose.

Medicine.—The PETALS of this species are official in the Pharmacopœia of India. They possess a sweetish bitter, faintly astringent taste and a roseate odour. They are mildly laxative and are used in combination with senna, cassia pulp, and chebulic myrobalan. The confection is much employed as a vehicle for potent metallic preparations (*S. Arjun*). They are given in the form of a SYRUP as a laxative to infants.

R. damascena, *Mill.; Fl. Br. Ind., II., 364.*

THE DAMASK, BUSSORA, or PERSIAN ROSE.

Vern.—*Guláb, sudburg*, HIND. & BOMB.; *Guláppá, irojáppá*, TAM.; (the flower buds=) *Gulál-akali*, GUZ. & MAR.; *Gulál, gul, guláb*, AFGH.

For other vernacular names which are applied to this and the preceding species, see under **R. centifolia** and **R. gallica**—the other members of this group.

R. 508

The Damask Rose.   (*W. R. Clark.*)   **ROSA damascena.**

References.—*Voigt, Hort. Sub. Cal., 194; Brandis, For. Fl., 200; Kur⁸, For. Fl. Burm., I., 440; Dals. & Gibs., Bomb. Fl., 31 (Suppl.); Stewart, Pb. Pl., 85; Aitchison, Cat. Pb. and Sind Pl., 57;—Afgh. Del. Com., 62, 63;—Kuram Valley Flora, 9, 54;—Prod., Western Afghan. and N.-E. Persia, 176; Firminger, Man. of Gardening for India, 469; Wall, Cat., 684; Pharm. Ind., 82; O'Shaughnessy, Beng. Dispens., 327; Dymock, Mat. Med. W. Ind., 298—301; Dymock, Warden, & Hooper, Pharmacog., I., 574-578; Murray, Pl. & Drugs, Sind, 142; Butler, Top. and Stats., Oudh and Sultanpore, 36; Dr. Jackson in Journ. Asiat. Soc., Beng. (1839); Watt, Cal. Exhib. Cat., V., 242; Drury, U. Pl., 367; Birdwood, Bomb. Pr., 192; Royle, Ill. Him. Bot., 203; Piesse, Perfumery, 198; Balfour, Cyclop., III., 439, 440; Spons' Encycl., Vol. II., 1427; Kew Off. Guide to the Mus. of Ec. Bot., 61; Kew Off. Guide to Bot. Gardens and Arboretum, 141; Gazetteers:—Mysore & Coorg, I., 60; N.-W. P. (Bundelkhand), I., 81; (Agra Dist.), IV., lxxi.; Ind. Forester, XIV., 368.*

Habitat.—Although this is the commonest Indian garden rose, its native country is absolutely unknown. At Patna, Ghazipore, Amritsar, Lahore, and several other places in India, large areas of ground have been converted into rose gardens, in which this species is the one chiefly cultivated for the manufacture of *attar* and rose-water from its flowers.

History.—The preparation of *attar* of roses and rose-water seems to have been entirely unknown to the ancients; indeed, it is not till the close of the thirteenth century that we find any mention of the latter, while the first authentic description of an essential oil only occurs in a work by **Geronimo Rossi** of Ravenna, who wrote about the end of the sixteenth century. In India, *attar* of roses is said to have been first discovered by Nur-i-Jehan Begum, A.D. 1612, on the occasion of her marriage with the Emperor Jehanghir. A canal in the palace garden was filled with rose-water in honour of the event, and the princess, observing a scum on the surface caused it to be collected, and found it to be of admirable fragrance, on which account it received the name of Atar-Jehanghiri, *i.e.*, the perfume of Jehanghir. In English commerce *attar* of roses only began to be recognised in the beginning of this century. **Aitchison** says that "in Persia the rose is the flower of all flowers in beauty and scent, it, however, lasts too short a time, owing to the hotwinds, as they at once put an end to all its beauty. Wherever it is grown, or in however small quantity, the flowers are daily collected by the owners of each garden, and handed over to the distiller, who manufactures from them rose-water, *guláb.* Rose-water is a luxury which the very poorest of the Persian ladies and dandies cannot do without; in almost the smallest village it is to be procured." **HISTORY. 509**

Oil.—The essential OIL or OTTO (*attar*) OF ROSES, obtained by distillation of the FLOWERS, in many parts of India, and also imported from Persia and Turkey, is much valued by the natives both medicinally and as a perfume. **OIL. Otto. 510**

MANUFACTURE.—The *attar* of roses imported to Great Britain is chiefly produced in a small tract of country on the southern side of the Balkan Mountains, but smaller quantities, which are almost entirely consumed locally, are manufactured about Cannes and Nice in the south of France, at Medinet Fayum, south-west of Cairo, and in Tunis. The whole of the *attar* and rose-water manufactured in India is consumed in the country. **Manufacture. 511**

In Bulgaria **R. centifolia** is the species used for making the otto, or *attar*, but the damask rose is also employed to a certain extent. These shrubs are cultivated by peasants in gardens and open fields, in which they are planted in rows as hedges 3 or 4 feet high. The best localities are those occupying southern or south-eastern slopes. Plantations in high mountainous regions generally yield less, and the oil is of a quality that easily congeals. The flowers attain perfection in April and May, and are gathered before sun-

ROSA damascena.	The Damask Rose.

OIL.

Manufacture.

rise; those not wanted for immediate use are spread out in cellars, but are always used for distilling the same day. The apparatus is a copper still of the simplest description, connected with a straight tin tube, cooled by being passed through a tub fed by a stream of water. The charge for a still is 25 to 50℔ of roses, from which the calices are not removed. The first runnings are returned to the still; the second portion, which is received in glass flasks, is kept at a temperature not lower than 15°C. for a day or two, by which time most of the oil, bright and fluid, will have risen to the surface. From this, it is skimmed off by means of a small tin funnel having a fine orifice, and provided with a long handle. There are usually several stills together. The produce is extremely variable. According to Baur, it may be said to average 0·04 per cent. Another authority estimates the average yield as 0·037 per cent. (*Pharmacographia, 264*).

In India the rose principally cultivated for the manufacture of *attar* and rose-water, is the **R. damascena.** An interesting account of the methods by which rose-water and the otto are manufactured in Ghazipore, the principal seat of the industry in India, is given by **Dr. Jackson** in the *Journal of the Asiatic Society of Bengal*, and is reproduced in O'Shaughnessy's *Bengal Dispensatory.* The following abstract has been taken from the latter work :—About 300 *bighas* or 150 acres of land around the station of Ghazipore are laid out in small detached fields as rose gardens. These gardens are annually let out by the zamindars who own them at about R5 per *bigha* for the land and R25 per *bigha* for the rose bushes of which there are 1,000 in each *bigha*. The expense of cultivation amounts to about R8-8. If the season is good 1,000 rose bushes should yield one lakh of roses, and these are sold to distillers at a price varying from R40 to R70 per lakh. The cultivators themselves very rarely manufacture. The rose bushes come into flower at the beginning of March and continue throughout April. In the early morning the flowers are plucked and conveyed in large bags to the distillers.

The apparatus for distilling is of the simplest description : it consists of a large copper or iron boiler well tinned, capable of holding from 8 to 12 gallons, with a large body, a rather narrow neck, and a mouth about 8 inches in diameter, on the top of which is fixed the head of the still. This is merely an old *dekchi* with a hole in the bottom to receive the tube or water which is well luted in with flour and water. This tube is composed of two pieces of bamboo fastened together at an acute angle and covered in their whole length with a coating of string, over which mud is luted to prevent the vapour from escaping. The lower end of the tube is carried down into a long-necked vessel or receiver, called a *bhubka*. This is kept in a *handi* of water which, as it gets hot, is changed. The end of the tube in the *bhubka* is padded with cloth to keep in the vapour. The boiler is let into an earthen furnace, and after being charged with the roses and a sufficient quantity of water distillation is slowly proceeded with. A boiler of the size described will hold from eight to twelve thousand roses, on these from ten to eleven seers of water will be poured, and eight seers of rose-water is distilled. This after distillation is placed in a glass *carboy* and exposed to the sun for several days to become ripe, after which the mouth is stopped with cotton over which a covering of moist clay is put to prevent the scent from escaping.

To procure the *attar*, after the rose-water has been distilled, it is placed in a large metal basin, which is covered with wetted muslin to prevent dust and insects from getting in; this vessel is then let into the ground which has been previously wetted with water and allowed to remain thus during the whole night. During the cool a little film of *attar* forms on the

R. 511

The Damask Rose. (*W. R. Clark.*) ROSA damascena.

OIL.
Manufacture.

surface of the rose-water which is removed in the morning by means of a feather and placed in a small phial, and day after day, as the collection is made, it is placed for a short time in the sun, and after a sufficient quantity has been procured it is poured off clear into small phials. The first few days' distillation does not produce such fine *attar* as is obtained afterwards, since it is mixed with dust and particles of dirt from the still. From one lakh of roses, it is calculated that 180 grains or one tolah of *attar* is produced; more may be obtained if the roses are full sized and the nights cool to allow of the congelation. The natives never remove the calices of the roses, but place the whole in the still, as it comes from the gardens.

The rose-water should always be twice distilled, the water procured from the first distillation being used to pour over the roses for the second.

Description & Chemical Properties.—Otto of roses, when recently extracted, has a pale greenish hue, afterwards it becomes light yellow in colour. The colour of the *attar* obtained varies much in different years, even in roses grown on the same ground. Emerald-green, bright yellow, and even reddish *attars* are often seen. Its specific gravity is 0·87 to 0·89. It solidifies at 11 to 18°C. (varying in proportion to the amount of stearoptene it contains), and its boiling point is 228·8°. Rose oil is made up of a liquid constituent containing oxygen, to which it owes its perfume, and the solid hydrocarbon, stearoptene, above mentioned, which is entirely odourless. This latter is very variable in amount, being present in largest quantity in the oil produced in a cold northern climate like that of Great Britain. The liquid portion of the rose oil has not yet been obtained entirely free from stearoptene, but when most of the hydrocarbon is removed the oil is perfectly liquid at 0°C., although the presence of stearoptene in small amount is shown by its solidifying into a gelatinous mass when placed in a cooling mixture. The oil thus purified is said to have a very fine and powerful odour, and when dissolved in spirit does not give rise to any crystalline separation (*Flück. & Hanb., Pharmacographia; Pharmacographia Indica*).

CHEMISTRY.
512

Uses.—Otto of roses is of no medicinal value, but is sometimes used as a SCENT for ointments. Rose-water is sometimes made with it, but is not so good as that obtained by distillation. Its principal utilisation is in perfumery and the manufacture of snuff (*Flück. & Hanb., Pharm.*). In India rose-water and the *attar* are much used by the natives at their festivals and weddings. They are distributed to the guests as they arrive and sprinkled with profusion in the apartments. The WATER is also used as a medicine or as a vehicle for other mixtures.

USES.
Scent.
513

Water.
514

Adulteration.—Much of the rose-water in India is adulterated with water before being sold. Pure otto is hardly to be obtained; it is adulterated before being shipped for India, and on arrival is mixed with sandal-wood oil. In India an adulterated otto of roses is largely prepared by the addition of sandal-wood chips to the flowers before they are distilled. The writers of the *Pharmacographia Indica* say they have been unable to ascertain that rose oil is in India ever adulterated with Rusa grass oil (the Geranium oil of English commerce), and that the dealers in that article do not appear to know anything of its use for this purpose in Turkey.

ADULTERA-
TION.
515

Medicine.—The FLOWERS of R. damascena are also used medicinally. For this purpose the BUDS are preferred, as they are " more astringent than the expanded flowers; they are considered to be cold and dry, cephalic, cardiac, tonic, and aperient, removing bile and cold humours; externally applied the PETALS are used as an astringent. The STAMENS are thought to be hot, dry, and astringent " (*Dymock*).

MEDICINE.
Flowers.
516
Buds.
517
Petals.
518
Stamens.
519

36A

R. 519

ROSA damascena.	Trade in Damask Rose.

MEDICINE.

Scented Water.
520

SPECIAL OPINIONS. — § " Mild, laxative" (*Assistant Surgeon Nehal Sing, Saharunpore*). " The *Tukm-i-gul* or ' Rose seeds ' of the shops are really the stamens" (*W. Dymock, Bombay*). " The SCENTED WATER yielded on distillation is much used as a vehicle for other medicines. It seems to enhance the action of dilute sulphuric acid in summer diarrhœa. The following lotion is a useful cooling application to the head in fever :— Vinegar 1 part, rose-water 1 part, water 10 parts. A mixture containing 1 part of rose-water to 2 parts of water drunk *ad libitum* causes free running from the nose, it is therefore said to have a cooling effect, to allay thirst, and to relieve head symptoms in fevers. Much used in astringent gargles and injections, and collyria " (*Assistant Surgeon Jaswant Rai, Mooltan*).

TRADE.
521

Trade.—Flückiger & Handury (*Pharmacog., Ed. 1874*) state that "the commerce to India (in rose-water and *attar* of roses), though much declining, still exists; and in the year 1872-73, 20,100 gallons of rose-water, valued at R35,178 (£3,517), were imported into Bombay from the Persian Gulf." This information had been, it is stated, derived from *The Statement of the Trade and Navigation of the Presidency of Bombay for 1872-73*. The authors of the *Pharmacographia Indica* say : " The Indian market is supplied with dry roses from all parts of the table-land ; both buds and expanded flowers arrive together, and are valued at R4½ per Surat maund of 37½℔. The buds are separated and sold for R7 per maund. The expanded flowers are worth only R3 per maund, and are purchased for the preparation of *gulkand*. Rose-water, to the extent of 20,000 to 30,000 gallons annually, is imported into Bombay from the Persian Gulf; two qualities are met with—*yuk-atishi* (once distilled) and *du-atishi* (twice distilled), value R4 to R4½ per *carboy* of 20℔."

"Otto of Roses is imported from Persia and Turkey, and a small quantity is made in India."

In the *Trade and Navigation* returns for British India there are two, places under which, one or both of the above substances, might appear, *viz.*, " Essential Oils " and " Perfumery." Of the former the IMPORTS have (during the past five years) been 2,954 gallons for 1885-86, 2,622 gallons for 1886-87, 2,683 gallons for 1887-88, 2,826 gallons for 1888-89, and 2,756 gallons for 1889-90. The declared values of these imports were, for the years named, R44,803, R32,881, R44,847, R46,994, and R52,886. Of the latter (Perfumery) the quantities are not shown, but the value of the imports is given as having been in 1885-86 R5,57,111, 1886-87 R5,58,741, 1887-88 R3,07,407, 1888-89 R2,60,184, and 1889-90 R3,03,675. The EXPORTS from these foreign imports to foreign countries are not very important. During the past five years the average value of the essential oils sent away from this country came to only R1,700 and of the perfumery R64,500. Of the essential oile about two-thirds came from the Straits Settlements and China, and a little less than a third from the United Kingdom. It is probable, therefore, that these imports, of essential oils, do not include the *attar* of roses, and that that substance is entirely classed as a perfume. An examination of the sources from which the perfumes are derived reveals the fact that of the total imports in 1889-90 (R3,03,675 worth), China furnished R1,84,133, Persia R69,557, the United Kingdom R16,548, Turkey in Asia R13,071, and the balance from Aden, Arabia, Ceylon, etc. Of these imports Bombay took R2,24,665 worth, Bengal R39,817, Burma R23,598, Madras R8,271, and Sind R7,324. While it would be difficult to obtain the details of the total imports of perfumery into India, it is possible to give a statement of that section of the trade that took place with Bombay. But since it may almost be said that the imports into Bombay of rose-water constitute so large a proportion of the total for

Trade in Damask Roses. (*W. R. Clark.*) ROSA damascena.

TRADE.
Imports.
522

all India that they may be regarded as the total imports the loss of particulars regarding the other provinces need not be considered as of great moment.

The following analysis may be furnished of the IMPORTS into Bombay of preparations from roses, for the past three years, and to allow of comparison for the year 1872-73 as well; but the grand totals are the values of these and all other perfumes.

Articles and whence imported.		1872-73 Quantity.	1872-73 Value.	1888-89 Quantity.	1888-89 Value.	1889-90 Quantity.	1889-90 Value.	1890-91 Quantity.	1890-91 Value.
Attary, Persian—			R		R		R		R
From Turkey	Cwt.	11	159	...	...	...	...	47	700
" Red Sea	"	15	220	...	...	...	...	...	...
" Persian Gulf	"	625	9,367	1,290	20,264	1,252	17,603	1,347	21,708
" Arabia	"	...	...	...	...	...	...	12	220
TOTAL		651	9,746	1,290	20,264	1,252	17,603	1,406	22,628
Rose Flowers, Dried?—									
From Arabia	Cwt.	...	...	16	224	8	120	1,814	25,227
" Persia or Persian Gulf	"	729	7,289	1,615	23,458	1,562	21,383	4	54
" Turkey in Asia	"	...	...	...	...	...	...	...	...
TOTAL		729	7,289	1,631	23,682	1,570	21,503	1,818	25,281
Rose-water—									
From Red Sea	Gals.	29	51	...	...	...	...	...	...
" Turkey in Asia	"	...	...	22	47	62	86	60	125
" Persia or Persian Gulf	"	20,097	35,170	49,079	90,527	72,560	1,34,200	55,836	1,05,542
TOTAL		20,126	35,221	49,101	90,574	72,622	1,34,286	55,896	1,05,667
GRAND TOTAL OF ALL FORMS OF PERFUMERY.		...	1,43,794	...	1,83,205	...	2,24,555	...	1,95,105

ROSA gallica.	The Persian & French Roses.

TRADE.
Imports.

The reader will be able to judge of the importance of this trade from the table above, which exhibits the Bombay section. It will, for example, be seen that the trade has by no means manifested any tendency to decline, and that, instead of 20,000 to 30,000 gallons (stated by the authors of the *Pharmacographia Indica*) being the annual imports of rose-water into Bombay, the average of the past three years comes to 59,155 gallons. But there is, besides, a very considerable trade in importing dry roses which must be used up by the Indian drug-sellers and chiefly in the manufacture of rose-water. India has, in fact, a very considerable traffic in rose-water and *attar* of roses, and a certain amount of the produce of this local industry is known to be exported. The exact share has not, however, been recorded, so that the only particulars that can be learned of the trade are obtained from the study of the exports of Indian-made essential oils and perfumes (excluding Musk, for which see Vols. III., 58, and V., 307). The following table exhibits these EXPORTS during the past five years :—

Exports.
523

Exports of Indian Produce and Manufacture.

YEARS.	ESSENTIAL OILS.		PERFUMES.	
	Gallons.	R	Gallons.	R
1885-86	8,833	1,80,157		31,942
1886-87	12,523	1,82,129	Not	42,539
1887-88	15,622	2,23,660	recorded.	60,699
1888-89	15,270	2,67,797		48,442
1889-90	12,128	2,02,773		42,410

About R30,000 worth of the exports of essential oils go annually to the United Kingdom, the same amount to Egypt, and also to Arabia; and about R20,000 worth to France, to Turkey in Europe, and to China. It may also be said that $\frac{4}{5}$ths of these exports go from Bombay. Of the perfumes France, Zanzibar, the Straits Settlements, and Aden take the major portion, but the exports are chiefly made from Bengal not from Bombay. It has already been said that it is difficult to assign the share of these exports that should be regarded as *attar* of roses and rose-water. The other substance of importance included in these returns is Rusa oil ; see Vol. I., 250-252.—[*Ed., Dic. Econ. Prod.*]

524

Rosa Eglanteria, *Linn. ; Fl. Br. Ind., II., 366.*
THE YELLOW PERSIAN ROSE.

Syn.—R. LUTEA, *Mill.*

References.—*DC., Prodr., II., 607 ; Boiss., Fl. Orient., II., 671 ; Brandis, For. Fl., 201 ; Gamble, Man. Timb., 166 ; Aitchison, Kuram Valley Flora, 9, 15, 54 ; Stewart, Pb. Pl., 86 ; Baden Powell, Pb. Pr., 595.*

Habitat.—A shrub occurring in the drier parts of the inner Himálaya from Kishtwar westward, and in Western Tibet, at altitudes between 8,000 and 11,000 feet. It is distributed to Afghánistán, thence westward to Asia Minor, and northward to Siberia. Never apparently cultivated in India, except perhaps in Kashmír and Chumba near villages.

DOMESTIC
Hedge Plant.
525
526

Domestic.—Aitchison mentions that it is used as a HEDGE PLANT in the Hariab district.

R. gallica, *Linn. ; Fl. Br. Ind., II., 364.*
THE FRENCH or RED ROSE.

This may be taken as the characteristic species of the group to which the Cabbage rose, Moss rose, Province rose, and Damask rose, etc., belong.

Products of India. 567

The Manipur Wild Tea Rose. (*W. R. Clark.*) ROSA involucrata.

References.—*Brandis, For. Fl.,* 200 ; *Pharm. Ind.,* 81 ; *O'Shaughnessy, Beng. Dispens.,* 326 ; *Watt, Cal. Exhib. Cat., V.,* 242 ; *Kew Off. Guide to Bot. Gardens and Arboretum,* 141 ; *Gazetteer, Mysore & Coorg, I.,* 60.

Habitat.—A native of South and Central Europe and Asia Minor. Cultivated in India.

Medicine.—The dried PETALS are officinal in the Pharmacopœias both of Europe and of India. They are slightly tonic and astringent, but are chiefly employed, in the form of an acid infusion, a confection, or syrup, as vehicles for other medicines.

MEDICINE.
Petals.
527

Rosa gigantea, *Collett ; Linn., Soc. Jour., xxviii. (1890),* 55.
THE MANIPUR WILD TEA ROSE.

528

Syn.—R. XANTHOCARPA, *Watt, MSS.*

Habitat.—A profuse climber, found in Manipur and the Shan States, at altitudes of 4,000 to 5,000 feet. In Dr. Watt's herbarium there are many specimens of this elegant plant, regarding which various notes are given as to its properties and uses. It would appear to have very large yellow flowers (often 4 to 5 inches in diameter) which have the rich perfume of the Tea-scented Roses. Dr. Watt remarks, " This seems nearly allied to, but quite distinct from, R. indica, *Linn.*: and as it is unquestionably wild, may this not be the ancestral form from which the Tea Roses have been derived ? It was nowhere seen near villages, but was found to frequent the forests ; it perfumed the air with its rich fragrance during the month of April."

Food.—The large yellow FRUITS, which in some respects resemble small apples, both in external appearance and flavour, are regularly collected and eaten, and are even offered for sale on the fruit-sellers' baskets in the capital of Manipur, to which town they are conveyed from the forests on the Burma-Manipur frontier. It was this peculiarity of the fruit that suggested the name given to the plant by Dr. Watt, and under which his specimens No. 6320 were issued (in 1882) to the Herbaria of India and Europe.

FOOD.
Fruits.
529

Domestic.—The stout stems are largely employed by the Nagas for walking sticks and spear-shafts.

DOMESTIC.
530
531

R. indica, *Linn. ; Fl. Br. Ind., II.,* 364.
The Indian Roses of this group are strongly scented. The chief forms are the Tea Roses and the Hybrid Perpetuals.

Syn.—R. CHINENSIS, *Jacq.*; R. SINICA, *Linn.*; R. FRAGRANS, *Redouté ;* R. semperflorens, *Bot. Mag.,* t. 284.

Vern.—*Kanta, kat-guláb,* BENG. ; *Sádá guláb,* PB.

References.—*Roxb., Fl. Ind., Ed. C.B.C.,* 407 ; *Brandis, For. Fl.,* 200 ; *Firminger, Man. Gard. Ind.,* 477 ; *Baden Powell, Pb. Pr.,* 347 ; *Agri.- Horti. Soc. Ind. Journ. (New Series), IV., Sec. I.,* 169 ; *Gazetteers :— Mysore & Coorg, I.,* 60 ; *N.-W. P. (Agra Div.), IV.,* lxxi.

Habitat.—A native of China, but early introduced into India, where it has found a congenial home, and is much cultivated for ornamental pur- poses. The roses of this series are not, however, supposed to be so deli- cately perfumed as are those of the Gallicæ. They have, however, the advantage of flowering throughout the year.

R. involucrata, *Roxb. ; Fl. Br. Ind., II.,* 365.

532

Syn.—R. LYELLII, *Lindl.* ; R. PALUSTRIS, *Buch.*

References.—*Roxb., Fl. Ind., Ed. C.B.C.,* 407 ; *Don. Prodr. Nepal,* 235 ; *Wight, Ic.,* t. 234 ; *Kurz, For. Fl. Burm., I.,* 440 ; *Lindl., Mono. Roses,* 12, t. I.

Habitat.—By the sides of streams in the Gangetic plain ; along the lower Himálaya from Kumáon eastwards ; and from Rájputana to My- sore and Burma. The wild rose of tropical tracts of India.

R. 532

ROSA sericea.	The Musk Rose.

533

Rosa macrophylla, *Lindl. ; Fl. Br. Ind., II., 366.*

Syn.—R. HOFFMEISTERI & GULELLMI WALDEMARII, *Klotssch ;* R. HOO-KERIANA, *Bertol.*

Vern.—*Guláb, ban-guláb, HIND. ; Jikjik, akhiári, breri, shingári, ban-guláb, bankújrú, yal, trind, túmbi,* PB.

References.—*Brandis, For. Fl., 203 ; Gamble, Man. Timb., 167 ; Stewart, Pb. Pl., 86 ; Don, Prodr., 235 ; Baden Powell, Pb. Pr., 595 ; Gazetteers :—Mysore & Coorg, I., 60 ; N.-W. P. (Agra Div.), IV., lxxi. ; Him. Dist., X., 309 ; Ind. Forester :—IV., 91 ; IX., 197 ; XI., 3.*

Habitat.—A thorny, pink-flowered shrub, common on the Temperate Himálaya, from the Indus to Sikkim, at altitudes of between 8,000 and 10,000 feet. The more alpine forms are more spinose, but have larger and more sweetly-scented flowers than those of lower altitudes. They are also attacked by different species of fungi.

FOOD.
Fruit.
534

Food.—"Its FRUIT is said to be eaten, and is stated by Madden to become very sweet when black and rotten " (*Stewart*).

TIMBER.
535

Structure of the Wood.—Hard, compact, of slow growth. Weight 57℔ per cubic foot.

DOMESTIC.
Hedge Plant.
536

Domestic.—This rose is grown as a HEDGE PLANT in some parts of the Himálaya. In Kanawar, a PERFUME is said to be extracted from the flowers for export to the plains (*Stewart*).

Perfume.
537
538

R. moschata, *Mill. ; Fl. Br. Ind., II., 367.*

THE MUSK ROSE ; The NOISETTE ROSE is a hybrid between this and R. indica.

Syn.—R. RECURVA, *Roxb. ;* R. BRUNONII, *Lindl. ;* R. LINDLEYI, *Herb. Wall. ;* R. PUBESCENS, *Roxb. ;* R. GLANDULIFERA, *Herb. Roxb.*

Vern.—*Kúji, kujai, karer, kwia, kwiala,* HIND. ; *Phulwari, chal,* KASHMÍR ; *Karir, kajei, kiyo, gúngári, kiú, kahi guláb ghure,* PUSHTU ; *Gul-nasta-ran, gul-nastran,* AFGH.

References.—*Gamble, Man. Timb., 166 ; Stewart, Pb. Pl., 85 ; Aitchison, Fl. Kuram Valley, 10, 54 ; Botany, Afgh. Del. Com., 63 ; Econ. Prod., W. Afgh. & N.-E. Persia, 177 ; U. C. Dutt, Mat. Med. Hind., 306, 317 ; Gazetteers :—Mysore & Coorg, I., 60 ; N.-W. P. (Him. Dist.), X., 309 ; Kew Off. Guide to Bot. Gardens and Arboretum, 141 ; Journ. Agri.-Horti. Soc., 1871-74 ; IV., I, 171 ; Ind. Forester :—IV., 91 ; IX., 197 ; XIII., 52, 57 ; XIV., 368.*

Habitat.—A large, thorny, climbing shrub, found in the North-West Himálaya, from Afghánistán to Nepál, ascending to 11,000 feet. The most elegant as it is the most abundant of North-West Himálayan Roses. The neighbourhood of Simla is, in spring, literally perfumed with the odour of this elegant and copiously flowering rose.

TIMBER.
539

Structure of the Wood.—Moderately hard, porous, of slow growth.

DOMESTIC.
Wood.
540

Domestic.—The WOOD is used on the Himálaya for making walking sticks.   [*208, t. 42, f. 1.*

541

R. sericea, *Lindl. ; Fl. Br. Ind., II., 367 ; Royle, Ill. Him. Bot.,*

Syn—R. TETRAPETALA, *Royle ;* R. WALLICHII, *Trattin.*

Vern.—*Chapala,* KUMÁON.

References.—*Brandis, For. Fl., 202 ; Gamble, Man. Timb., 167 ; Don, Prodr., 236 ; Ind. Forester, January 1885, XI., 3 ; Gazetteer, N.-W. P. (Him. Dist.), X., 309.*

Habitat.—A thorny shrub found on the Temperate Himálaya, from the Sutlej to Bhotán and Manipur, at altitudes between 9,000 and 14,000 feet. Distributed to China. The Himálayan white rose with four petals, the lowers being borne on short lateral twigs all along the branches.

TIMBER.
542

Structure of the Wood.—Very hard, darkening on exposure.

R. 542

The Rosemary.	(*W. R. Clark.*)	ROTHIA trifoliata.

[*Bot., 208, t. 42, f. 2.*

Rosa Webbiana, *Wall.; Fl. Br. Ind., II., 366; Royle, Ill. Him.*
Nearly allied is the SCOTCH or BURNET ROSE, **R. spinosissima.**
Syn.—R. UNGUICULARIS, *Bertol.;* R. PIMPINELLIFOLIA, *Hook. f. & T.*
Vern.—*Khántián, shingári,* HAZARA; *Chúa,* LAHOUL; *Sia, sea,* LADAK & SPITI; *Sikanda, manyar, sháwali ringyál, kugína,* PB.
References—.*Brandis, For. Fl., 202; Gamble, Man. Timb., 166; Stewart, Pb. Pl., 86; Watt, Cal. Exhib. Cat., VI., 158; Gazetteer, N.-W. P. (Him. Dist.), X., 309.*
Habitat.—An erect pink-flowered shrub of the dry inner Himálaya from Kashmír to Kumáon and Western Tibet, occurring at altitudes between 3,000 and 13,500 feet.
Food.—The FRUIT is eaten by Natives.
Domestic.—In Spiti the STEMS and BRANCHES are piled up on the roofs of the houses and used as fuel.

543

FOOD.
Fruit.
544
DOMESTIC.
Stems.
545
Branches.
546

Rosaries, see Beads, Vol. I., 426.
Rose Bay, see Nerium odorum, *Soland.;* Vol. V., 348; APOCYNACEÆ.
Rose, Christmas, see Helleborus niger, *Linn.;* Vol. IV., 216;
[RANUNCULACEÆ.

Rose of Sharon, see Narcissus Tazetta, *Linn.;* Vol. V., 317;
[AMARYLLIDACEÆ.

ROSMARINUS, *Linn.; Gen. Pl., II., 1197.*
[LABIATÆ.

Rosmarinus officinalis, *Linn.; Boiss., Fl. Orient., IV., 636;*
ROSEMARY.
Vern.—*Rusmari,* HIND.; *Ukleel-ul-jilbul, hasalban-achsir,* ARAB.
References.—*Pharm. Ind., 167; O'Shaughnessy, Beng. Dispens., 488; Year-Book Pharm. (1875), 259; (1879), 467; Watt, Cal. Exhib. Cat., V., 242; VI., 158; Atkinson, Econ. Prod., N.-W. P., V., 41; Birdwood, Bomb. Prod., 66, 226; Smith, Dict. Econ. Pl., 356; Ind. Forester, XII., 59; Agri.-Horti. Soc. Ind. Trans., III., 198.*
Habitat.—A densely leafy shrub, 3 to 4 feet high, indigenous to the South of Europe, Asia Minor, and Northern Africa, and cultivated in India chiefly in the gardens of Europeans.
Perfumery.—Rosemary is employed as an adjunct in many perfumes, *e.g.,* Hungary water and Eau-de-Cologne. A CONSERVE and a LIQUEUR are also made from it.
Medicine.—The OIL and a SPIRIT prepared from it are officinal in the Indian and British Pharmacopœias. They have carminative properties, but are rarely exhibited internally. The preparations of Rosemary are chiefly employed externally as ingredients of stimulating liniments for baldness. Rosemary was held in high esteem by the ancients as a powerful stimulant to the nervous system.
Food.—It is enumerated by **Birdwood** in his List of Condiments and Spices.

547

PERFUMERY.
Conserve.
548
Liqueur.
549
MEDICINE.
Oil.
550
Spirit.
551
FOOD.
552

Rostellularia procumbens, *Nees;* see Justicia procumbens, *Linn.;*
[Vol. IV., 557; ACANTHACEÆ.

ROTHIA, *Pers.; Gen. Pl., I., 477.* [LEGUMINOSÆ.

Rothia trifoliata, *Pers.; Fl. Br. Ind., II., 63; Wight, Ic., t. 199;*
Syn.—LOTUS INDICUS, *Desr.;* TRIGONELLA INDICA, *Linn.;* DILLWYNIA TRIFOLIATA, *Roth.;* GLYCINE LEPTOCARPA, and HOSACKIA INDICA, *Grah.*
Vern.—*Nurreypitten keeray,* TAM.; *Nucka kura,* TEL.
References.—*DC., Prodr., II., 382; Roxb., Fl. Ind., Ed. C.B.C., 588; Us. Pl. Bomb., 197; Gazetteers:—N.-W. P., IV., lxx.; Bundelkhand, I., 80; Mysore & Coorg, I., 59; Ind. Forester, III., 237.*

553

RUBIA cordifolia.	The Indian Madder.

FOOD.
Leaves.
554
Pods.
555

Habitat.—A copiously branched diffuse annual found on the tropical plains of India, from Bundelkhand to Ceylon ; distributed to Australia.

Food.—The LEAVES and PODS are boiled and eaten as a vegetable by the Natives, especially in times of famine.

Rouge, see Carthamus tinctorius, *Linn. ;* Vol. II., 193 ; COMPOSITÆ.

ROUREA, *Aubl. ; Gen. Pl., I., 432.*

556

Rourea commutata, *Planch. ; Fl. Br. Ind., II., 47 ;* CONNARACEÆ.

Syn.—CNESTIS MONADELPHIA, *Roxb., Fl. Ind., Ed. C.B.C., 388.*
With reference to this species Sir J. D. Hooker (*Fl. Br. Ind.*) remarks that it is so similar in most respects to R. santaloides, that he scarcely doubts its being an Eastern form of that plant, distinguished by the nervation of the leaflets which are more usually acute at the base. He further says he has seen some Tavoy specimens which are almost intermediate.

Vern.—*Súkurtothi,* SYLHET ; *Kowatothi,* CHITTAGONG HILL TRACTS.
Reference.—*Kurz, For. Fl. Burm., I., 325.*

FOOD.
Aril.
557

Habitat.—A large scandent evergreen shrub, found in Assam, Sylhet, Chittagong, and Burma, and distributed to China.

Food.—The Natives of Assam and Chittagong eat the fresh ARIL of the ripe seeds (*Roxb.*).

558

R. santaloides, *W. & A. ; Fl. Br. Ind., II., 47.*

Syn.—CONNARUS SANTALOIDES, *Vahl. ;* C. MONOCARPUS, *W. & A.* (not of *Linn.*).
Vern.—*Vardárá,* BOMB., MAHR. ; *Kerindi-wel,* CINGH.
References.—*DC., Prodr., II., 85 ; Beddome, Fl. Sylv. Anal. Gen., lxxxi ; Gamble, Man. Timb., 114 ; Dals. & Gibs., Bomb. Fl., 53 ; Burman, Thes. Zeylan., 199, t. 89 ; Dymock, Mat. Med. W. Ind., 208 ; S. Arjun, Bomb. Drugs, 36 ; Watt, Cal. Exhib. Cat., V., 243 ; Gazetteer, Bomb., XV., 431.*

MEDICINE.
Root.
559
Stem.
560

Habitat.—A climbing shrub, found in the Western Peninsula from the Koncan to Travancore and in Ceylon.

Medicine.—The ROOT and STEM are employed as a bitter tonic and prescribed in rheumatism, scurvy, diabetes, and pulmonary complaints. "Many fanciful virtues are attributed to it by the Natives. It is believed to promote the growth of a *fœtus in utero,* the development of which has been arrested " (*S. Arjun*).

ROYLEA, *Wall. ; Gen. Pl., II., 1212.*

561

Roylea elegans, *Wall. ; Fl. Br. Ind., IV., 679 ;* LABIATÆ.

Syn.—PHLOMIS CALYCINA, *Roxb., Fl. Ind., Ed. C.B.C., 462 ;* BALLOTA CINEREA, *Don.*
Vern.—*Patkarru,* HIND. ; *Tit-patti, kauri,* KUMÁON ; *Kaur, kauri,* PB.
References.—*DC., Prodr., XII., 516 ; Gamble, Man. Timb., 301 ; Stewart, Pb. Pl., 172 ; Watt, Cal. Exhib. Cat., V., 243 ; VII., 218 ; Baden Powell, Pb. Pr., 595 ; Atkinson, Him. Dist., 315, 749.*
Habitat.—A tall hoary undershrub, found in the sub-tropical Western Himálaya from Kashmír to Kumáon, at altitudes between 2,000 and 5,000 feet.

MEDICINE.
Leaves.
562

Medicine.—An infusion of the LEAVES is drunk for contusions produced by blows, and about Kumáon the same preparation is used as a bitter tonic and febrifuge (*Stewart*).

TIMBER.
563

Structure of the Wood.—White and hard. Weight 52℔ per cubic foot.

RUBIA, *Linn. ; Gen. Pl., II., 149.*

[*128, bis. f. I.;* also *Ic., t. 187 ;* RUBIACEÆ.

564

Rubia cordifolia, *Linn. ; Fl. Br. Ind., III., 202 ; Wight, Ill., t.*
THE INDIAN MADDER.

R. 564

| The Indian Madder. | (*W. R. Clark.*) | RUBIA cordifolia. |

Syn.—R. Munjista, *Roxb.*; R. Mungisth, *Desv.*; R. javana, *DC.*; R. alata, *Wall.*; R. purpurea, *Dcne.*; R. scandens, *Zoll. & Morr.*; R. chinensis, *Regl. & Maack.*; R. mitis, *Miq.*

Vern.—*Manjit, manjith, majith,* Hind.; *Manjistha, majith, manjit,* Beng.; *Majathi, majetti,* Assam; *Enhu, chenhu,* Naga; *Ryhoi,* Khasia Hills; *Moyúm,* Manipur; *Sóth,* Bhutia; *Vhyem,* Lepcha; *Btsod,* Thibetan; *Manjistá, manjisthá,* Uriya; *Majethi, munjit,* Kumáon; *Dandú, faharghás,* Kashmir; *Kúkarphali, tiúrú, manjit, khúri, sheni, rúna, mitú, majit, múnsat, rúnang,* Pb; *Manjit,* Deccan; *Manjit, madar,* Bomb.; *Manjéshta,* Mar.; *Manjitti, shevelli,* Tam.; *Támravalli, manjishtige, manjishta, tige, chiranji,* Tel.; *Manjushta, manjustha,* Kan.; *Man-chetti,* Malay.; *Manjista, velmadata,* Sing.; *Manjistha, manjishthá, kála-méshiká,* Sans.; *Fóvvah,* Arab.; *Rúnás,* Pers.

References.—*DC., Prodr., IV.,* 588; *Roxb., Fl. Ind., Ed. C.B.C.,* 125, 416; *Voigt, Hort. Sub. Cal.,* 399; *Kurz, For. Fl. Burm., II.,* 5; *Gamble, Man. Timb.,* 219; *Cat., Trees, Shrubs, etc., Darjiling,* 49; *Dals. & Gibs., Bomb. Fl.,* 121; *Stewart, Pb. Pl.,* 115; *Kuram Valley Rept., Pt. I.,* 65; *Graham, Cat. Bomb. Pl.,* 93; *Mason, Burma and Its People,* 512, 787; *Sir W. Elliot, Fl. Andh.,* 41, 112, 173; *Pharm. Ind.,* 118; *Fleming, Med. Pl. & Drugs (Asiatic Resear., XI.),* 177; *Ainslie, Mat. Ind., I.,* 202; *II.,* 182; *O'Shaughnessy, Beng. Dispens.,* 378; *Irvine, Mat. Med. Patna,* 67; *Medical Topog., Ajmir,* 146; *Moodeen Sheriff, Supp. Pharm. Ind.,* 218; *U. C. Dutt, Mat. Med. Hindus,* 178, 309; *Sakharam Arjun, Cat. Bomb. Drugs,* 71; *Murray, Pl. & Drugs, Sind,* 195; *Dymock, Mat. Med. W. Ind., 2nd Ed.,* 415; *Dymock, Warden & Hooper, Pharmacog. Ind., Pt. III.,* 231; *Year-Book Pharm.,* 1874, 193; 1879, 214; *Watts' Dict. Chemistry, III.,* 1061; *Birdwood, Bomb. Prod.,* 299; *Baden Powell, Pb. Pr.,* 354, 442; *Drury, U. Pl. Ind.,* 369; *Atkinson, Him. Dist. (X., N.-W. P. Gaz.),* 749, 773; *Useful Pl. Bomb. (XXV., Bomb. Gaz.),* 246; *Forbes Watson, Indian Prod.,* 124; *Econ. Prod. N.-W. Prov., Pt. III. (Dyes and Tans),* 19; *Liotard, Dyes,* 54, 55, 89, 95; *Wardle, Dye Report,* 36, 55; *McCann, Dyes and Tans, Beng.,* 45-50; *Mr. H. Z. Darrah, Note on Cotton in Assam,* 26; *Kew Off. Guide to the Mus. of Ec. Bot.,* 82; *Kew Off. Guide to Bot. Gardens and Arboretum,* 72; *Milburn, Oriental Commerce* (1813), *II.,* 218; (1825), 293; *Man. Madras Adm., I.,* 360; *The Report of the Land Revenue Settlement Report of Mandlah Dist.,* 88; *Gazetteers\:—Bombay, VIII.,* 183; *XV.,* 436; *N.-W. P., X.,* 311; *Orissa, II.,* 160, 179; *Mysore & Coorg, I.,* 61, 436; *Agri. Horti. Soc. :— Ind. Trans. VI.:(Pro.),* 60; *Journals, II., Sel.,* 295; *III., Pro.,* 50, 65, 66, 71, 72; *IV., Sel.,* 207-209; *V.,* 27-36; *VIII.,* 183-185, *Pro.,* 196; *IX.,* 296; *X.,* 151-157, *Sel.* 22, 23; *XI.,* 316-324; *Pro.,* 71 (*Pro. for.* 1860) 20, 21; *XII.,* 397; *XIII.,* 390; *New Series, I.,* 77; *VI.,* 193-197; *VII.,* 361; *Ind. Forester, I.,* 97; *XIV.,* 394; *Spons'. Encycl.,* 863; *Encyclop. Brit., XV.,* 176; *Balfour, Cyclop. Ind., III.,* 447; *Ure, Dict. Inds. Arts & Man., III.,* 3.

Habitat.—A herbaceous creeper with perennial roots, which is met with throughout the hilly districts of India from the North-West Himálaya eastwards, and southwards to Ceylon. On the Himálaya, it occurs up to an altitude of 8,000 feet. It is distributed throughout North-Eastern Asia from Dahuria to Java and Japan, and is found also in Tropical Africa. It is a very variable plant, but **Dr. Watt** (*Calc. Exhib. Cat., Pt. II.,* 54) remarks :—"There are two easily recognised primary forms, met with in India." These he distinguishes as *var. 1,* **cordifolia** *proper,* the diagnostic characters of which are that the leaves are five costate, rarely three, the veins are impressed and the surface of the leaf rough or hispid. "This," **Dr. Watt** continues, "is the form chiefly met with on the Himálaya, appearing near the Chenab and extending eastward to Sikkim and Bhotán, to the Khasia and Naga Hills, Burma, South India, and Ceylon. It seems nowhere to be cultivated, but is largely collected as a wild dye-stuff and carried to the plains to be sold. The root and lower or ground twigs are the dye-yielding portions. This form I regard as inferior in

R. 564

RUBIA
cordifolia.

dye-property to the next, although it is the one generally used in India and sold as Madder." The other form which Dr. Watt distinguishes he calls *var. 2*, khasiana, and says its diagnostic marks are, leaves three, rarely five costate, often almost with a solitary midrib; the surface of these leaves smooth, not hispid, and the veins not impressed. "This form," Dr. Watt continues, "is the richest in Madder dye-principle. It is occasionally met with in Sikkim, but attains its greatest development eastward in the Khásia and Naga Hills. It seems nowhere to be met with to the west of Sikkim. I repeatedly collected this form and compared it with the true R. cordifolia, thinking that it would probably be found to possess characters sufficient to justify its entire separation from that species if not its identification with R. Manjista, *Roxb*. But while arriving at the conclusion that it was probably only a variety of R. cordifolia, I satisfied myself as to its superior dye-yielding property. I had been struck with the perfection of the red dye with which the Nagas colour the hair decorations of their spears, etc., and I at first concluded that this was the plant from which they obtained it. I was soon after convinced, however, that neither of these supplied the favourite red, but a third plant which I was shewn, namely, R. sikkimensis, *Kurz*. Before proceeding to discuss this interesting discovery, I venture to repeat my conviction that *var.* khasiana is a far richer dye-yielding plant than the ordinary R. cordifolia. I am inclined to suspect that the experiments, which were once made with the view to discover whether R. cordifolia in a cultivated form could compete with the European Madder, may have failed because the inferior variety was experimentally cultivated. If it happened that a consignment of *var.* khasiana reached Europe, it is likely that its richness in dye-property suggested the idea that the cultivation of R. cordifolia would be as profitable as that of R. tinctorum, and that disappointment followed from experimenting with the ordinary North-West Himálayan form."

It may be here added in conclusion that Sir J. D. Hooker, in the *Flora of British India*, describes three forms of this species, one, apparently the khasiana dealt with above, the second from the Western Peninsula (cordifolia *proper* of Dr. Watts' classification), and the third collected by Griffiths on the Western Himálaya; but Sir Joseph considers none of them sufficiently constant or marked to be worthy of the name of varieties. This observation is doubtless correct from the botanical standpoint, but the superiority in dye property is an economic feature of great importance. In the Herbarium khasiana may at once be recognised by the fact that it stains the paper of several sheets above and below it, while cordifolia rarely does so.

DYE.
Root.
565

Dye.— Manjít root or East Indian Madder, obtained for the most part from this species, is much employed by the Natives of India in dyeing coarse cotton fabrics or the thread which is woven into such fabrics, various shades of scarlet, coffee brown or mauve. The colour obtained from *manjít* is brighter though more fleeting than that of European Madder. Few attempts appear to have been made to introduce the cultivation of the latter article into India. The East Indian Madder of commerce "consists of a short stalk from which numerous cylindrical roots about the size of a quill diverge; these are covered by a thin brownish suber, which peels off in flakes, disclosing a red brown bark, marked by longitudinal furrows. The taste is sweetish at first, afterwards acrid and bitter" (*Mat. Med. W. Ind.*, 416).

CULTIVATION
566

CULTIVATION.

Rubia cordifolia does not now at any rate appear to be cultivated in India. Stewart, indeed, mentions that he was told distinctly that it was

Madder or Munjit.	(*W. R. Clark.*)	**RUBIA cordifolia.**

sometimes cultivated in Kashmír, but says he thinks that reference was made to the cultivation of R. tinctorum. In a letter, however, from Dr. Campbell to the Agri.-Horticultural Society of India, published in Volume V. (1844) of their Journal, he says that the long tendrils of R. cordifolia are near Darjíling cut into pieces of about a foot long, laid in the ground, which is partially delved, while the tree stumps are allowed to remain. The plant grows over the stumps, sending out shoots from every joint. In a further communication, he informs the Society that it is grown from seed also which is sown broadcast, and when the land is not thickly studded with tree stumps, that poles are put in, for the plant to twine around. The only other reference to the cultivation of this species occurs in McOann's *Dyes and Tans of Bengal*, where that writer says he was informed by the Deputy Commissioner of Darjíling that R. cordifolia used to be extensively cultivated in that district. It appears from the above that, although the *Manjít* was cultivated in Dr. Campbell's time, the wild plant only is now used for dyeing purposes.

Chemical Composition.—The Chemistry of *Manjít* was first thoroughly investigated by Dr. Stenhouse who published a paper on the subject in the Proceedings of the Royal Society of London. He found that the colouring matters in *Manjít* were not identical with those of the common madder since it contained no *alizarin* (see under R. tinctorum), and its colouring matter was a mixture of *purpurin* and an orange dye which he called *munjistin*. This has a formula ($C_8H_6O_3$) and is thus nearly related in composition to *purpurin* ($C_9H_6O_3$) and to *alizarin* ($C_{10}H_6O_3$).

The tinctorial power of *Manjít* was first examined in 1835 by Professor Runge, who stated that it contained twice as much colouring matter as ordinary madder. Subsequent experiments, however, have proved that this result was incorrect, and that actually the colouring power of *manjít* is considerably less than that of the madder obtained from R. tinctorum, and the experiments of Mr. John Thom, a Lancashire Cotton Printer, conclusively proved that the *garancine* (a commercial preparation of madder) from *manjít* has only about half the tinctorial power of that made from the best madder, *viz.*, Naples roots. "These, however," he says, "yield only 30 to 33 per cent. of *garancín*, whereas *manjít* yields 52 to 55 per cent." "The inferiority of *manjít* as a dye stuff results from its containing the comparatively feeble colouring matters *purpurin* and *munjistin*, of the latter of which only a small part is available, while its presence in large quantity appears to be positively injurious; so much so indeed that *manjít*, garancin, freed by boiling water from the greater part of the *munjistin* which it contains, yields much richer shades with alumina mordants than before."

METHODS OF DYEING.—Dyeing with *Manjít* seems to be practised in much the same manner all over India, the colour being produced by steeping the cloth in an infusion of the wood or roots of the plant and afterwards being mordanted with a solution of alum. Various slight modifications of this method are, however, adopted in the different parts of the country.

Bengal.—The stem of the plant is preferred to the root, although both are used. The wood is first dried, then crushed and pounded, and thereafter generally boiled with water, but sometimes merely left to steep for two or three hours in cold water. Alum seems to be usually employed as a mordant. The following district reports, taken from McOann's *Dyes and Tans of Bengal*, will serve to show the various modifications of the process in the different parts of the province. In Cuttack the method employed is "The *Manjistha* is bruised and boiled with water; on cooling it is strained and the tincture mixed with a little alum. The cloth is steeped in this, squeezed well, and then taken out and dried. This is repeated three or four times, until the required colour is attained. To dye 60 yards of cloth a yard wide in this

RUBIA **cordifolia.**	Cultivation of the Indian

.CULTIVATION
in
Bengal.

Method of
dyeing.

way requires 1½ seers of *Manjistha* and 1 chittack of alum." In Midnapur "Madder is cut into very small chips, which are carefully washed and boiled in water for six hours : the water has then become very red. The silk or cloth to be dyed red, is boiled for 10 minutes in water made alkaline by the addition of some ash, generally that of burnt plantain leaves, and is then steeped in a solution of alum in water, and then drenched several times in the red dye, above obtained." In the Maldah district the method is as follows :—" Six seers of iron dross (*jhama*) are placed in a vessel into which is poured the liquor resulting from boiling four seers of rice water (*kanji*) with ½ seer of parched Indian corn (*makai*), water is added and the vessel is left out in the sun daily until its contents acquire a sweetish taste which takes place in about fifteen days. In another vessel one seer of powdered *Manjistha* wood is boiled in 9 seers of water, and the dye is ready when the water has been boiled down to one-third of its original quantity. In this one chittack of alum is boiled before its use. In a third vessel *tori* water is prepared by boiling the fruit of the *tori* plant " (**Cæsalpinia Sappan**) " in water. The material to be dyed is first wetted in the *tori* water, which acts as a mordant, and then dried in the sun. It is then dipped in the *Manjistha* dye and afterwards in the decoction first mentioned, upon which it becomes purple." " In the Darjeeling District, no mordant seems to be used. **Dr. Schlich** remarks :—' To prepare the dye, the plant is well dried, then pounded and boiled; as a rule, especially among the Lepchas, it is mixed with lac. The cloth to be dyed is then steeped in it. Elsewhere, however, it is stated that *hana* and *bahera* (myrabolans) are used as mordants in dyeing with *manjit*, the articles to be dyed being steeped in a solut on obtained by pounding and boiling these before using the *Manjit*' "(*McCann, Dyes and Tans of Bengal*).

ASSAM.
570

Assam.—The process usually followed is to dye the thread which is afterwards to be woven into cloth. It is described as follows :—" The dye itself is made by cutting a few dried sticks of *Majetti* into pieces, and steeping them in cold water for some hours, a red infusion results. The thread or cloth to be dyed is first boiled in water containing the pounded bark of the *Jam* tree (**Eugenia Jambolana**), and afterwards dried in the sun. It is then steeped in the *Majetti* dye, to which is added a mixture of goat's milk, lime, and *kolkhar* (a potash extracted from plantain ashes) in equal quantity; afterwards the whole is boiled gently for about 15 minutes. The cloth or thread is then removed and dried in the shade."

N.-W.
PROVINCES.
571

North-West Provinces.—Sir E. O. Buck, in his *Dyes and Tans of the North-West Provinces*, describes the method pursued in dyeing with *Manjit* as follows :—" *Manjit* is too expensive a dye to be much used : dying with it is almost confined to the towns of Faruckhabad and Bareilly. The roots are mixed with a little sweet oil and powdered; afterwards ground in a hand-mill and strained with water. The cloth to be dyed is boiled with the powder until of the shade required, the colour being fixed with alum. Occasionally cloth seems to be dyed by being steeped in a cold infusion of the powdered roots in water." " *Manjit* is also used in calico-printing." Elsewhere the same writer says that the mordants used are alum and the red earth called *géru*, which is chemically an impure oxide of iron, that the alkalies are employed to brighten the colour, while a preliminary washing in tannic astringents is sometimes practised as well.

PANJAB.
572

Panjáb.—Mr. Baden Powell, in his *Panjáb Products*, says that the greater part of the madder used in the Panjáb is that imported by the Lohani Afgháns from the hills of North Baluchistán, Kabul, and Khorassan, and is brought in large quantities to Multan by the Shikarpore and South Afghán Povindahs, and to Peshawar through the Kaibar and northern passes by Kabul merchants, so that in all probability much of the madder used in

Madder or Manjit.	(*W. R. Clark.*)	**RUBIA cordifolia.**

the Panjáb is madder proper (the product of **R. tinctorum** which is said to abound in those countries) and not *Manjít*. The process of dyeing, as practised by native dyers in the Panjáb, is :—" The fabrics to be dyed are first steeped in a decoction of *máin*, the galls of the tamarisk and then submitted to the madder solution hot. It is fixed by alum as a mordant, the galls seem to impart to the cloth a facility for taking the colour." This is a deep, full red, which is quite permanent, but is not brilliant, and cannot be called a scarlet. **Mr. Baden Powell** continues :—" It is most remarkable, however, that the beautiful and permanent dye known as ' Turkey red ' and which does deserve the name of scarlet, is a dye of Indian origin, but notwithstanding the fact that this process originated in India, it does not now appear to be either remembered or practised, at all events in Hindustan and Upper India."

Rájputana—In Ajmere a similar process is practised to that followed in the Panjáb. No details from other parts of Rájputana are available.

Bombay.—*Manjít* dyeing is practised more or less all over the Bombay Presidency, but one of the principal seats of the industry is at Diu in Gujerat. The process there pursued is fully described in Narayan Daji's *Art of Dyeing in Western India,* from which the following abstract has been made. The fabric to be dyed undergoes a preliminary soaking for about twelve hours in a solution of carbonate of soda, after which it is taken out and dried. It is then steeped and worked up in a mixture containing goat's dung, sweet oil (from **Sesamum indicum**), and carbonate of soda, made into an emulsion with water. This process is repeated four times a day for three days, after which, for about six days more, the fabric is alternately steeped in pure water and dried. The cloth by this process acquires a peculiar whiteness and silk-like softness, and much of the future success depends on its proper treatment at this stage. The mature and immature fruits of **Terminalia Chebula,** the leaves, flowers, and bark of Cassia Tamala, the galls of **Quercus infectoria,** and some coarse sugar, are next finely powdered and stirred up with cold water. In this the prepared fabric is steeped, and then taken out and dried. The cloth is next alumed. An infusion of the bark of **Symplocos racemosa** is made, alum is dissolved in it, and the fabric steeped in the mixture, taken out, and again dried. It is next steeped in cold water, squeezed and beaten about in all directions. The washing may be done in sea water at first ; but must be finished in fresh water. The colour is next imparted. The madder dye-beck consists of a large copper vessel which is fitted over a fire-place. Water is put into the vessel and made to boil, upon which a powder, made of *manjít* root, the flowers of **Woodfordia floribunda,** and tamarisk galls, is added, and the whole is boiled up together. The fabric is next added to and boiled in the mixture for three hours. During the process, the decoction becomes almost colourless in appearance, while the red colour is imparted to the cloth. That is afterwards well worked up, squeezed, then washed in clean water, and dried. The dyed fabric is finished and made fit for packing by first steeping it in a watery solution of goat's dung and leaving it to soak for one night. On the following morning it is spread out on the ground, its upper surface is sprinkled over with water, and it is left to dry in the sun. This process is repeated the next day on the other side of the fabric. Lastly, it is again washed in clean water, dried in the sun, and packed.

The compound permanent tints produced from madder are usually four, and are known at Diu as (1) " *Gulábi* " or Pink, (2) " *Jambals* " or purple, (3) " *Kalo* " or Rust black, (4) " *Achhokalo* " or " *Roz* " (the Portuguese name) light black. To produce the first of these the fabric is treated as in ordinary madder-dyeing, but to impart the colour one seer of alum and *lodhra* (**Symplocos racemosa**) solution is taken, and to it one seer of water

RUBIA cordifolia.

CULTIVATION
in
Madras.
575

MEDICINE.
576

TRADE.
577

Trade in Madder or Manjit.

and 1¼ seers of sugar are added, and the cloth is steeped in this and then dried before being dipped in the madder. The darker shades are imparted by treatment with a black dye, made of impure oxide of iron, before the material is boiled in the *manjít*.

Madras.—In this province also *manjit* is used for dyeing purposes, but only locally, and that not to any great extent, as dyeing with the *chay-root* (**Oldenlandia umbellata**) almost entirely supersedes the use of madder in the production of various shades of red. No separate accounts of the methods of dyeing with *manjít*, as practised by the Natives of Madras, seem to be available.

Medicine.—In Muhammadan works on medicine, *Manjít* is described as having deobstruent properties, and is prescribed for use in paralysis, jaundice, obstructions in the urinary passages, and in amenorrhœa. The fruit is said to be useful in hepatic obstructions, and a paste, made by rubbing up the roots with honey, is mentioned as a valuable application for the removal of freckles and other discolorations of the skin. The whole plant is said to be alexipharmic, and it is frequently hung up in houses to avert the evil eye or tied to the necks of animals with the same object (*Dymock, Mat. Med. W. Ind.*). In Hindu medicine it is used chiefly as a colouring agent. and all medicated oils are first prepared for use by being boiled with the root. It is also regarded as astringent and a valuable external application in external inflammations, ulcers, and skin diseases. A preparation of *manjít* and liquorice root rubbed into a paste, with the sour liquid produced by the acetous fermentation of paddy, is applied over fractures to reduce swelling and inflammation (*U. C. Dutt, Mat. Med. Hind.*). Ainslie observes that "the *hakims* are in the habit of prescribing an infusion of it as a grateful and deobstruent drink in cases of scanty lochial discharge after lying-in" (*Mat. Ind., II., 182*). The *Pharmacopœia of India* remarks that Dr. George Playfair, in a note appended to his *Telíf Sheríf*, states that if taken in doses of about three drachms several times daily, it powerfully affects the nervous system, inducing temporary delirium with a powerful determination to the uterine system. In reality, *manjít* has probably little, if any, medicinal virtues, but as no thorough investigation has been made of the action of the small amount of acrid and resinous material it contains, it ie impossible definitely to declare it valueless as a therapeutic agent.

TRADE.

Till within the last few years a considerable import trade in *manjít* and madder existed, but the substitution of artificially-prepared alizarine and other aniline dyes has caused it to decline greatly. In 1885-86, R2,02,038 worth of madder and *manjít* was imported into India, while in the year 1889-90 the value of the total import of these substances was only R29,488. The value of the annual imports of alizarine and aniline dy has, on the other hand, almost quadrupled itself during that period. The principal exporting country is Persia, and Bombay is the only importing town. The exports of *manjit* and madder are comparatively small. In 1889-90 they amounted in value to R6,987 in all. Of this Bombay exported 321 cwt., valued at R3,407, Sind 310 cwt., value R3,198, and Madras 22 cwt., valued at R382. The last mentioned was probably not madder at all but *chay-root*, which is sometimes called Indian Madder. The countries to which the *manjít* and madder were sent were the East Coast of Africa, Persia, and Turkey in Asia. Madder from Sind, which is in all probability the true madder (**Rubia tinctorum**) and is imported into Sind from Baluchistán and Afghánistán, fetches R17 a cwt., about double the price of Indian and Persian Madder.

R. 577

Rubia sikkimensis, *Kurz; Fl. Br. Ind., III., 203.*

Vern.—*Moyúm,* MANIPUR.

References.—*Kurz, in Journ. As. Soc., 1874, II., 188; Watt, Cal. Exhib. Cat., II., 55; Major Trotter, Report on Manipur Dyes (1883); Ind. Forester, XIV., 394.*

Habitat.—A stout, handsome creeper, found on the Eastern Himálaya, in Sikkim and Bhotán, at an altitude of from 2,000 to 5,000 feet, in the Mishmi Mountains of Upper Assam, and in Manipur and the Naga Hills

Dye.—According to Dr. Watt this and not the preceding species is the principal source of the brilliant red dye used by the hill tribes of Naga Hills and Manipur. In the Calcutta Exhibition Catalogue, he gives a long account of the method of dyeing practised by these people, which, since it is the account of an eye-witness of the process and throws an entirely new light on the question of the origin of part at least of the Indian *Manjit,* seems to the writer worthy of being republished here. Dr. Watt writes:—

"Apparently, the Lepchas of Sikkim do not know that this plant yields Madder dye; but I suspect that the thick heavy roots, (many times thicker than the roots and twigs of **R. cordifolia**) which are sold in the bazárs, belong largely to this species, though probably used as an adulterant. This opinion seems to be strengthened by the fact that until 1874 the plant here dealt with was not named or even known to exist. Specimens had of course been collected, but they escaped attention, having remained for many years in the larger Herbaria unpublished. In the Naga Hills and in Manipur this species alone supplies the brilliant red dye used by the hill tribes to colour their cloths, hair decorations for spears, shields and earrings, rings, etc., as well as to colour their cane and bamboo plaited work.

"The process of extracting the dye is curious. It was shown to me after considerable trouble. A woman came one morning to the Residency, Manipur, bringing with her—*1st,* Two or three bundles of the root and stem of **R. sikkimensis,** *Kurz; 2nd,* a slab of the bark of **Quercus fenestrata,** *Roxb.; 3rd,* a bundle of twigs and leaves of **Symplocos racemosa,** *Roxb.; 4th,* a packet of seed and a specimen of the plant yielding these seeds, which I identified as **Leucas cephalotes,** *Spreng.,*—a Labiate plant common in fields throughout India, more especially Bengal. I have been told that the seed yields an oil which is used for illuminating purposes, but I can find no mention of this fact in works on Indian Economic Botany—; *5th,* two skeins of cotton thread, one of which was of a yellow colour and had been prepared beforehand by a process which I was to see applied to the second one,—it had been steeped in some mordant or metallic salt; *6th,* two earthen vessels; and *7th,* a small basket.

"I was told that it was necessary first to prepare the second skein of cotton, so as to give it time to dry in order that it also might, if possible, be dyed. The woman sat down and set fire to the bundle of twigs and leaves of **Symplocos racemosa.** When completely burned to ashes these were carefully collected and placed in the corner of the basket and a little water sprinkled over and allowed to soak for a few minutes, then more water was sprinkled, until ultimately a yellowish liquid began to strain through and trickle into one of the earthen vessels. This liquid tasted bitter and no doubt contained some alkali salt which I have not as yet had time to identify chemically. When enough liquid had thus been obtained the second or unprepared skein of cotton was placed in the vessel and boiled for some time; after which it was removed, wrung out, and hung up to dry.

"The second process was then proceeded with. The woman and her assistants commenced to pound the chips of **Rubia,** using about equal

578

DYE.
579

The European Madder.

DYE.

proportions of root and stem. When this had been done the powder was mixed with a handful of the seeds of **Leucas** and intimately combined and rubbed together by the hand on a stone. This mixture was then placed in the other earthen vessel and boiled with about three proportions of water to one of the mixed powder. When boiling, the prepared skein of cotton was plunged into the solution, which was now of a deep red colour. It was turned round and round in the boiling liquid upon the extremity of a small twig held in the hand, and when dyed to the required depth it was removed and allowed to strain off the surplus liquid. Thereafter it was washed several times and hung out to dry.

"I asked what was the use of the Oak-bark, and was told that it was for deepening the colour from red to brown of the darkest possible shade. A few pieces were thrown in, and the skein of cotton prepared in my presence was treated as before, when a beautiful red-brown colour was the result.

"I have gone into detail on the process of dyeing from **R. sikkimensis,** because I am assured by many distinguished authorities that it has been reported as not yielding Madder dye, and because the process described seems to be known to the hill tribes of Assam and the Naga Hills only. I trust that this preliminary account may suggest the lines upon which a more thorough investigation should be instituted by the authorities in Assam.

"I suspect that the bulk of the Madder plant of Assam will be found to be derived from **R. sikkimensis,** instead of from **R. cordifolia,** and that a considerable proportion of the Madder exported from Sikkim may be derived from this plant also.

"Since writing the above I have had the pleasure to receive from my friend **Major Trotter,** Political Agent, Manipur, a most interesting account of the dyes and process of dyeing in practice in that State. I wrote specially asking that he should investigate the subject of the beautiful madder red in order to confirm my own observations. Greatly to my delight I received a most interesting series of specimens, amongst which were some 30 good specimens of **Rubia sikkimensis,** putting an end to any doubt as to this plant being the source of the Naga red instead of the equally abundant **R. cordifolia.** Instead of **Leucas cephalotes,** however, **Major Trotter** sends me the seeds of **Perilla ocimoides,** *Linn.,* another of the LABIATÆ as the dye-auxiliary. Perhaps both plants are used, the action being similar to the use of oils in the extraction of other dyes."

580

Rubia tinctorum, *Linn.; Fl. Br. Ind., II., 203.*

THE EUROPEAN MADDER ; ALIZARI, GARANCE, *Fr.;* FARBEROTHE, *Germ.* ; MEE, *Dutch* ; MARIONA, *Rus.* ; ROBBIA, *It.*

Vern.—*Bacho,* PB.; *Manyunth,* SIND.; *Rodang, rodan,* AFGH. & PERS.

References.—*DC. Prodr., IV., 589; Boiss., Fl. Orient., III., 17; Stewart, Pb. Pl., 117; DC., Origin Cult. Pl., 41; Aitchison, Cat. Pb. & Sind Pl., 71; Botany Afg. Del. Com., 73; Notes on the Prod. of West Afghan and N.-E. Persia, 177; O'Shaughnessy, Beng. Dispens., 378; Dymock, Mat. Med. W. Ind., 415; Dymock, Warden, & Hooper Pharmacog., 231; Year-Book Pharm., 1874, 193; Watts, Dict. of Chem., III., 741; Davies, Trade and Resources N.-W. Frontier, ccclxxvi., 40; Watt, Cal. Exhib. Cat., II., 57; VI., 158; Crookes, Dyeing, 228; Hummel, Dyeing, Text fab., 344; Smith, Dic., 254; Encycl. Brit., XV., 176; Ure, Dic. Indus., Arts and Manu., III., 1; Rep. Inland Trade, Br. Burma, 1884-85, Ap. II.; Agri.-Hort. Soc. Ind.:—Transactions, II., 269-270; III., 13; Pro., 242; IV., 170; Jour., II., Sel., 295; IV., 239, 240; Sel. 45, 57-64, 154; Pro., xxvii., xxxii., xc.; V., 27; Sel., 27, 33; X., 151-157; Pro. 171; XI., Pro., 77; (Pro. for 1860), 20, 21, 70, 110, 111, 114, 115; XII., Pro. (for 1862), 27; XIII., 158, 159; (Pro. for 1863), 51, 52*

Habitat.—A climbing herb with perennial root, cultivated in Kashmír, Sind, and throughout Afghanistán; distributed westward from Persia to Spain, both cultivated and semi-cultivated, or wild. With reference to the indigenous habitat of the true madder of commerce, DeCandolle writes :— "The madder is certainly wild in Italy, Greece, the Crimea, Asia Minor, Syria, Persia, Armenia, and near Lenkoran. As we advance westward in the south of Europe, the wild indigenous nature of the plant becomes more and more doubtful. There is uncertainty even in France. In the north and east, the plant appears to be 'naturalized in hedges and on walls' or 'sub-spontaneous,' escaped from former cultivation. In Provence and Languedoc, it is more spontaneous or wild, but here also it may have spread from a somewhat extensive cultivation. In the Iberian peninsula it is mentioned as 'sub-spontaneous.' It is the same in the north of Africa. Evidently the natural, ancient, and undoubted habitation is western temperate Asia, and the south-east of Europe. It does not appear that the plant has been found beyond the Caspian Sea in the land formerly occupied by the Indo-Europeans, but this region is still little known. The species only exists in India as a cultivated plant and has no Sanskrit names" (*Origin of Cult. Pl.*).

Dye.—The dried and ground ROOTS of Rubia tinctorum was formerly one of the most valued and most largely employed dye-stuffs both in Europe and Asia. It has, however, now been almost entirely replaced in European countries by the coal-tar derivatives—Alizarin and allied colouring matters.

DYE.
Roots.
581

History.—Madder cultivation has been carried on from a very remote era, and the plant was known to the Greeks and Romans under the names of *Erythrodanon* and *Rubia*. Its cultivation in India does not, however, appear to date from a very early period, as no Sanskrit names for the true madder seem to exist, nor are there any records in the works of Sanskrit authors of its cultivation or use, although several references occur to the employment of R. cordifolia both in medicine and in the arts.

HISTORY.
582

Cultivation.—In Europe madder is chiefly cultivated in marshy land, and given very careful tillage. The land is deeply ploughed and well manured with rich well-rotted horse-dung or farm-yard manure. In Avignon the seed is sown in March and the young plants are carefully tended and kept free from weeds. Experience having shown that the older the root the richer it is in colouring matters, the plant is left in the soil for at least eighteen months, but oftener for a longer time, care being taken to promote the development of the root by covering the portions above ground with soil. The plant blooms in July and the seed is collected in August. The plant is cut down with sickles, the stem and leaves being used as fodder for cattle. In November, after having been in the ground eighteen or thirty months, the root is dug out by means of a specially constructed plough. In Holland madder is not grown from seed, but shoots are planted in the month of May, and put in well-tilled ground in rows about two feet or more apart. The roots are dug up with long, narrow-bladed spades. In India the cultivation of the true madder plant is carried on chiefly in Kashmír and in some parts of Sind. Few references exist as to the method in which the cultivation is carried on, but in Kanawar it is said to be propagated by buds or sets from the rhizome, and that the roots are occasionally dug up after two years, but are usually allowed to remain in the ground for five or six seasons. Aitchison, in his latest work on the Products of Western Afghánistán and North-Eastern Persia, states that it is "cultivated throughout that country in orchards under the shade of trees, and where irrigation is plentiful; the plant takes three years before it can yield the proper size of roots to be considered a good marketable commodity. The cultivation of the plant in

CULTIVATION
583

CULTIVATION

orchards is said rather to improve the bearing of the trees than cause them injury; the fact is, to get a good crop of madder, the soil requires to be well manured and most liberally irrigated, thus the fruit trees benefit as well as the madder. The finest is said to be cultivated at Anar-dara, Koin, and Yezd, from whence the roots are imported in immense quantities to Herat. At Herat a good deal is also produced, but not of such a fine quality. From Herat it is re-exported in all directions, a great deal to Afghán proper and India, besides in some bulk to Turkistán." In the earlier volumes of the Proceedings of the Agri.-Horticultural Society of India considerable interest seems to have been evinced in the cultivation of Madder in India; French Madder seed was obtained from Avignon and was distributed, together with full directions as to its cultivation, to various members of the Society in Upper India. The seeds were said to have germinated, but no further report as to their ultimate successful cultivation or the reverse seems to have been made, and the artificial production of alizarine from coal-tar seems to have effectually blighted the hopes, at first entertained, that the cultivation of the true madder in India might ultimately prove a successful industry and oust the supplies of French and Dutch Madder from the English market.

CHEMISTRY.
584

Chemical Composition.—Of the substances that exist in ordinary madder, the most important are those called alizarin and purpurin, which are in fact the essential red-colouring matter of the dye stuff. Alizarin may be extracted from ordinary madder by the action of solvents, and obtained in crystals without sublimation, hence its existence preformed may be inferred. Its exact chemical composition is not definitely established, but **Wolf & Strecker's** formula ($C_{10} H_6 O_3$) is most in accordance with its relations to chloronaphthalic acid ($C_{10} H_5 C_{10} H_5 Cl O_3$), both these compounds being converted by the action of nitric acid into phthalic acid. Alizarin is very little soluble in water, more soluble in alcohol, but nearly insoluble in aluminium salts. Its alkaline solutions have a beautiful violet or purple colour, and it forms lakes of various colours with the earthy and heavy metallic oxides.

Purpurin is extracted from madder by the same processes as alizarin, and separated from it by its superior solubility in alum liquor. It has been regarded by some chemists as not distinct from the former, but may be distinguished by its easy solubility in alum liquor, with which it forms a solution of a beautiful pink colour with yellow fluorescence; secondly, by the colour of its alkaline solutions, which are cherry red or bright red, whereas alizarin forms violet solutions with alkalis. Its optical properties are also distinct. From Schunck's experiments it appears that the final result of dyeing with madder and its preparations is simply the combination of alizarin with the various mordants employed. Purpurin almost entirely disappears from all madder colours which have been subjected to a long course of after-treatment. Madder root also contains certain yellow colouring materials, but these are useless, if not positively injurious. in the process of dyeing. Resinous matters, extractives, and sugar, probably glucose, are the other organic constituents of madder root (*Watts Dict. of Chemistry*).

Method
of Dyeing.
585

Methods of Dyeing.—The methods of Madder dyeing pursued in Europe do not come within the scope of the present work, while the Indian methods employed with the roots of R. **tinctorum** do not differ in any particular from those used with the indigenous plant R. **cordifolia.**

FOOD.
Plant.
586

Food.—In India as in Europe the PLANT is used as a fodder for cattle, and in Sind camels especially are said to be fond of it. The bones of animals fed on it become red, as also do the claws and beaks of birds, probably on account of the affinity of the colouring matter for phosphate of lime.

RUBUS, *Linn.; Gen. Pl., I., 616.*

A genus of creeping herbs or erect or sarmentose shrubs belonging to the Natural Order ROSACEÆ. Some of the species are so difficult of discrimination that no fewer th n 500 have been described, of which only 100 are allowed in the *Genera Plantarum* and *Flora of British India.* Of these 41 species, with numerous sub-species and varieties, are described in the latter work, as indigenous to India. Many of these yield a fruit, which is eaten locally by the Natives, and is in some cases collected and brought to the bazaars of hill stations for sale to the European residents. Except as an article of food, the different species of **Rubus** are of little economic value.

587

Rubus biflorus, *Ham.; Fl. Br. Ind., II., 338;* ROSACEÆ.

588

Vern.—*Chánck, kantanch, khariára,* KASHMIR; *Akhreri, karer, ankren, bumbal, insra, bátang, kalkalín (dher,* near Simla), Pᴮ.

References.—DC., *Prodr., II., 558; Brandis, For. Fl., 198; Gamble, Man. Timb., 165; Stewart, Pb. Pl., 86; Atkinson, N.-W. P. Foods, 68; Gasetteer, N.-W. P. (Him. Dist.), X., 309; Ind. Forester, IX., 197; Balfour, Cyclop., III., 448; Kew Off. Guide to Bot. Gardens and Arboretum, 53.*

Habitat.—A strong, rambling shrub, with its stem covered by a white pulverulent epidermal layer; found on the Temperate Himálaya from Sirmore to Sikkim and Bhután, at altitudes between 7,000 and 9,000 feet.

Food.—Its FRUIT is drooping, sub-globose, in colour red or deep orange, sweet to taste. It is eaten by Natives.

FOOD.
Fruit.
589
590

R. ellipticus, *Smith; Fl. Br. Ind, II, 336; Wight, Ic., t. 230.*

Syn—R. ROTUNDIFOLIUS, *Wall.;* R. GOWRY PHUL (and GOWREEPHUL), *Roxb.;* R. FLAVUS, *Ham.;* R. SESSIFOLIUS, *Miq. Var. 1st,* HIRTA= R. HIRTUS, *Roxb.;* R. WALLICHIANUS, *W. & Arn;* R. AFFINIS, *Madden; Var. 2nd,* DENUDATA=R. ROTUNDIFOLIUS, *Wall.*

Vern.—*Tolu aselu, escalu, cesi,* NEPAL; *Kashyem,* LEPCHA; *Esar, hish alu, hisalu, jogiya-hisálu,* KUMÁON & GARHWAL; *Gauri-phul, hisára* KASHMIR; *Akhi, ankri, kunáchi, gurácha, pukána,* Pᴮ.

References.—DC., *Prodr., II., 563; Roxb., Fl. Ind., Ed. C.B.C., 408; Brandis, For. Fl., 197; Kurz, For. Fl. Burm. I., 438; Gamble, Man. Timb., 165, 166; Darjiling List, 35; Stewart, Pb. Pl., 86, 87; Baden Powell, Pb. Pr., 596; Atkinson, Econ. Prod., N.-W. P., Foods, 44, 68; Ind. Forester:—II., 25; IV., 198; VIII., 26, 404; IX., 197; XIII., 57; Madras Man. Adminis., I., 314; Report of the Shan States, by Mr. Aplin, 1887-88; Agri-Horti. Soc. Ind.:—Trans., IV., 105, 141; VI., 247; Jour., XIII., 386; Balfour, Cyclop., III., 448; Gasetteers:—Simla District, 12; N.-W. P., X., 309.*

Habitat.—A tall, sub-erect bush, met with in the temperate and subtropical Himálaya, from the Indus to Sikkim and Bhotán, at altitudes between 4,000 and 7,000 feet. It is found also on the Khasia Hills, Manipur, and Burma, also in the Western Peninsula of India, from Kanara southwards, and in the central parts of Ceylon.

Food.—The FRUIT is yellow and has the flavour of a raspberry; in the Himálaya it is commonly eaten either raw or made into a preserve, and is certainly one of the best wild fruits of India (*Gamble*). It is offered for sale at most hill stations, such as at Simla, and is regularly shown at the Horticultural Society's shows of that station.

FOOD.
Fruit.
591

Structure of the Wood.—Moderately hard, light brown, with very broad medullary rays.

TIMBER.
592

R. fruticosus, *Linn.; var.* DISCOLOR; *Fl. Br. Ind., II, 337.*

THE BLACKBERRY or BRAMBLE.

593

Syn.—R. DISCOLOR, *Weihe & Nees.*

Vern.—*Alish, shálidag gánch, akhi, kanachi, chench, pakhána,* Pᴮ. *Karwárei,* TRANS-INDUS.

References.—DC., *Prodr., II., 560; Brandis, For. Fl., 197; Gamble, Man. Timb., 165; Stewart, Pb. Pl., 87; Boiss., Fl. Orient., II., 695; Baden*

R. 593

| RUBUS lasiocarpus. | The Raspberry. |

Powell, Pb. Pr., 596 ; Gazetteer, N.-W. P. Him. Dist., X., 309 ; Agri.-Horti. Soc. Ind., XIV., 511 ; Ind. Forester, XIII., 68 ; Smith, Dic., 60.

Habitat.—A large shrub, which occurs on the western temperate Himá-laya, from Marri to Jamu, at altitudes between 3,000 and 7,000 feet. It is distributed to Afghánistán and westward to the Atlantic.

FOOD.
Fruit.
594
595

Food.—Produces pink flowers, and many small, black, more or less hemispherical fleshy FRUITS, which are edible.

Rubus idœus, *Linn.; DC., Prodr., II., 558.*

THE RASPBERRY.

References.—*Firminger, Man. Gard. Ind., 249 ; O'Shaughnessy, Beng. Dispens., 325 ; Year-Book Pharm. (1873), 442 ; Balfour, Cyclop. Ind., II., 449 ; Smith, Dict., 346 ; Agri.-Horti. Soc. Ind. Trans., VII. (Pro.), 91.*

Habitat.—A deciduous shrub, with perennial creeping roots and bien-nial stems, technically termed *canes;* it is found wild in Europe as far north as 70° Lat., and southward ; appears to have been abundant on Mount Ida in Asia Minor. The true raspberry of Europe has never been introduced to any great extent into India, an omission one can scarcely understand, since there seems little doubt but that it would succeed well in the Himálaya, and might even pass into a semi-wild state. It would certainly prove an invaluable addition to the very few fruits of an indigenous or even introduced nature met with at hill stations.

596

R. lanatus, *Wall; Fl. Br. Ind., II., 331.*

Syn.—R. TILIACEUS, *Herb., Str., & Winterb.*

Vern.—*Hisálu,* KUMÁON.

Reference.—*Atkinson, N.-W. P. Foods, 44, 66.*

FOOD.
Fruit.
597
598

Habitat.—A rambling creeper, which occurs wild in the Central and Western Himálaya, from Kumáon to Nepál, at altitudes between 5,000 and 8,000 feet.

Food.—It yields in the spring an insipid edible FRUIT.

R. lasiocarpus, *Smith; Fl. Br. Ind., II., 339; Wight, Ic., t. 232.*

Syn.—R. DISTANS, *Don.*; R. HORSFIELDII, *Miq.*; R. ALBESCENS and R. ROSÆFLORUS, *Roxb.*; R. ROTUNDIFOLIUS, *Royle*; R. MYSORENSIS and INDICUS, *Heyne*; R. PARVIFOLIUS, *Moon*; R. FURFURACEUS, *Wall*; R. BIJUGUS, *Focke.*

A large number of varieties of this species are described in the *Flora of British India,* all of which are apparently very constant in their characters, but difficult of definition. The fruit of **Rubus lasiocarpus** constitutes almost invariably the so-called blackberry, which is sold in the bazaars of most of the hill stations, and not that of **R. fruticosus.**

Vern.—*Kala aselu,* NEPAL; *Kajutalam,* LEPCHA; *Kaleri-hisálu,* N.-W. P.; *Kálawar, kála-hisalu,* KUMÁON; *Kandiári, kharmuch, surganch, túlanch, oche,* KASHMIR; *Gunácha, pagúnai, pakána, pakáni, gurácha, kandiára, kharmach,* PB.; *Surganch, niú, kalliachi, klenchu, galka, kalga,* PUSHTU; *Mansakhta,* TRANS-INDUS; *Gunacha, pukana,* HAZARA.

References.—*DC., Prodr., II., 558 ; Roxb., Fl. Ind., Ed. C.B.C., 409 ; Brandis, For. Fl., 198 ; Kurz, For. Fl. Burm., I., 439 ; Gamble, Man. Timb., 166 ; Cat., Trees, Shrubs, &c., Darjíling, 36 ; Dals. & Gibs., Bomb. Fl., 89 ; Stewart, Pb. Pl., 87 ; Graham, Cat. Bomb. Pl., 64 ; Rum-phius, Amb., V., 88, f. 47, I ; Firminger, Man. Gard. Ind., 249 ; Watt, Cal. Exhib. Cat., VI., 159 ; VII., 218 ; Birdwood, Bomb. Prod., 151 ; Baden-Powell, Pb. Pr., 596 ; Atkinson, N.-W. P. Foods, 44, 67, 68 ; Useful Pl. Bomb. (XXV., Bomb. Gas.), 155 ; Report on the Shar States by Mr. Aplin, 1887-88 ; Gazetteers:—N.-W. P. (Him. Dist.), X., 309 ; Mysore & Coorg, I., 56 ; III., 17 ; Agri.-Horti. Soc., Ind. Jour. XIV., 45, 61 ; Ind. Forester, II., 25 ; VIII., 26 ; Balfour, Cyclop. Ind, III., 449.*

Habitat.—A large, rambling, very variable plant, met with in the Tem-perate Himálaya, from Kashmir to Sikkim, at altitudes between 5,000 and

R. 598

Products of India. 583

Indian Wild Raspberries. (*W. R. Clark.*) | RUBUS niveus.

10,000 feet, also on the Khásia Hills, in the Western Peninsula, on the higher Gháts from Kanara southwards, in Burma, and in Ceylon. It is distributed to Java.

Food.—The FRUIT, which is red, orange, or of a glaucous blue black colour, is somewhat dry, but very palatable; large quantities of it are imported into the bazárs of hill stations for sale to Europeans. Firminger describes it as similar in flavour to the common English Blackberry, but vastly superior, and says that by judicious cultivation it might be rendered very productive (*Man. Gard. Ind.*).

Domestic.—Balfour recommends the use of this species as a HEDGE PLANT.

Rubus lineatus, *Reinw. ; Fl. Br. Ind., II., 333.*
> Syn.—R. PULCHERRIMUS, *Hook.*
> Vern.—*Gempé aselu,* NEPAL.
> References.—*Gamble, Man. Timb., 166; Darjiling List, 36; Watt, Cal. Exhib. Cat., VI., 159; VII., 218.*

Habitat.—A strong, sub-erect herb, found in the Sikkim Himálaya, at altitudes between 6,000 and 9,000 feet, and distributed to Java.

Food.—FRUIT red, edible.

Structure of the Wood.—Yellowish brown, resembling that of R. ellipticus.

Domestic.—STEMS used to make fences (*Gamble*).

R. moluccanus, *Linn. ; Fl. Br. Ind., II., 330 ; Wight, Ic., t..225.*
> Syn.—R. RUGOSUS, *Smith* ; R. ALCEÆFOLIUS, *Poir.* ; R. MICROPETALUS, MACROCARPUS, and FAIRHOLMIANUS, *Gard.* ; R. CORDIFOLIUS, *Don.* ; R. REFLEXUS, *Ker* ; R. HAMILTONIANUS, *Seringe.*
> Vern.—*Bipem-kanta,* NEPAL ; *Sufokji,* LEPCHA ; *Katsol,* KUMÁON.
> References.—*Roxb., Fl. Ind., Ed. C.B.C., 408; DC. Prodr., II., 566; Brandis, For. Fl., 197 ; Kurz, For. Fl. Burm., I., 437 ; Gamble, Man. Timb., 165 ; Darjiling List, 35 ; Dals. & Gibs., Bomb. Fl., 89 ; Graham, Bomb. Fl., 64 ; Atkinson, Econ. Prod., N.-W. P., Part V., Foods, 44, 66, 68 ; Indian Forester :—II., 25 ; III., 178 ; IV., 241 ; VI., 239 ; 309 ; Gasetteer, Mysore and Coorg, III., 17.*

Habitat.—A large, scrambling shrub, common in many parts of the Central and Eastern Tropical and Temperate Himálaya, from Kumáon to Sikkim, at altitudes between 3,000 and 7,000 feet. It occurs also in Assam and on the Khásia Hills at altitudes between 3,000 and 5,000 feet. It is found also in the Eastern Peninsula, in the Western Peninsula, on the Gháts from Bombay southward, and in Çeylon and Burma. It is distributed to the islands of the Malayan Archipelago.

Medicine.—Although no mention is made in the works of modern Indian writers, of any medicinal virtues supposed to be possessed by this plant, Rumphius states that, in his time, it was largely used as a medicine by the Malayans, who considered the FRUIT a valuable remedy for the nocturnal micturition of children, and the LEAVES a powerful emenagogue and abortefacient.

Food.—The FRUIT is red and edible. It makes very fair jam (*Atkinson*).

R. niveus, *Wall ; Fl. Br. Ind., II., 335.*
> Syn.—R. GRACILIS, *Roxb.*
> With reference to the synonymy of this species Sir J. D. Hooker says, he finds it "impossible to arrange satisfactorily the forms of this most puzzling plant," and gives a numerous list of varieties to which the reader is referred (*V. S., l.c.*).
> Vern.—*Pila hisálu,* N.-W. P. ; *Kalga,* PB.

FOOD.
Fruit.
599

DOMESTIC.
Hedge Plant.
600

601

FOOD.
Fruit.
602
TIMBER.
603
DOMESTIC.
Stems.
604

605

MEDICINE.
Fruit.
606

Leaves.
607
FOOD.
Fruit.
6c8

609

R. 609

RUBY.	The Rubies of India.

References.—*Roxb., Fl. Ind., Ed. C.B.C., 409 ; Gamble Man. Timb., 165 ; Watt, Cal. Exhib. Cat., VI., 160 ; Atkinson, N.-W. P. Foods, 44, 67, 68 ; Gazetteer, N.-W. P. (Him. Dist.), X., 309.*

Habitat.—A large, rambling shrub, met with on the Temperate Himálaya, from Kashmír to Bhután, at altitudes between 6,000 and 10,000 feet on the west and 5,000 to 11,500 feet on the east.

FOOD.
Fruit.
610

Food.—Atkinson states that this species yields a very succulent, pleasantly-tasted reddish or yellowish brown FRUIT.

611

Rubus nutans, *Wall. ; Fl. Br Ind., II., 334.*

Vern.—*Sinjang,* (the berries=) *sinjang-lho,* BHUTIA.

References.—*Trans. Linn. Soc., XX., 45 ; Watt, Cal. Exhib. Cat., VI., 160 ; Atkinson, Econ. Prod., N.-W. P., Part V., 44, 68 ; Ind. For., XI., 3.*

Habitat.—An unarmed shrub, met with in the Temperate Himálaya, from Garhwal to Kumáon, at altitudes between 8,000 and 10,000 feet.

FOOD.
Fruit.
612

Food.—Its FRUIT consists of scarlet drupes, which are edible, and have a pleasant sub-acid flavour.

613

R. paniculatus, *Smith ; Fl. Br. Ind., II., 329.*

Syn.—R. TILIACEUS, *Smith.*

Vern.—*Kála hisálu,* KUMÁON ; *Anchu, patharola,* GARHWAL ; *Kála-akhi, púlla,* PB. ; *Numing rik,* LEPCHA.

References.—*Brandis, For. Fl., 196 ; Gamble, Man. Timb., 165, xix ; Darjiling List, 36 ; Stewart, Pb. Pl., 87 ; Atkinson, Econ. Prod., N.-W. P., Part V., 44, 66 ; Gazetteer, N.-W. P., X., 309 ; Stewart in Journ. Agri.-Horti. Soc. Ind., XIV., 19.*

Habitat.—A very rambling climber, which has all the parts, except the upper surface of the leaves, covered with a dense tomentum ; found on the Temperate Himálaya from Hazara to Sikkim, at altitudes between 3,000 and 8,000 feet, and in the Khásia mountains between 4,000 and 5,000 feet.

FOOD.
Fruit.
614
TIMBER.
615

Food.—The FRUIT consists of numerous large round black drupes. It is edible but insipid in flavour.

Structure of the Wood.—Soft and porous with very large medullary rays.

616

R. rosæfolius, *Smith ; Fl. Br. Ind., II., 341 ; Hooker, Ic. Pl., t. 349*

Syn.—R. PINNATUS, *Willd. ;* R. ASPER, *Don. ;* R. SIKKIMENSIS, *O. Kunze ;* R. PANICULATUS, *Clarke.*

References.—*Roxb., Fl. Ind., Ed. C.B.C., 409 ; Brandis, For. Fl., 198 ; Kurz, For. Fl. Burm., I., 439 ; Gamble, Man. Timb., 165 ; Darjiling List, 36 ; Firminger, Man. Gard. Ind., 249, 485 ; Atkinson, Econ. Prod., N.-W. P., Pt. V, 68 ; Gazetteer, N.-W. P., X., 309 ; Ind. For., IX., 197 ; Kew Bulletin (1889), 24.*

Habitat.—A small shrub, found on the Temperate Himálaya, from Kumáon to Sikkim, at altitudes between 3,000 and 7,000 feet. It occurs also in the Khásia Hills on the hills of Ava and Martaban, and is distributed to Java. It is naturalised and cultivated in the tropics and warm temperate regions, and in cultivation has often double flowers.

FOOD.
Fruit.
617
618

Food.—The FRUIT is large, red, and edible, and is frequently collected and sold in the Darjíling bazár.

RUBY.

This name is applied by lapidaries and jewellers to two distinct minerals— the true or oriental Ruby and the Spinel-ruby. The former may be called a red variety of Corundum and distinguished from the different forms of the latter, by its composition, hardness, and crystalline form. In composition, the true ruby is merely the clear, red-coloured, crystalline form of alumina or aluminic oxide (Al_2O_3), in which the colour is due to the presence of metallic oxides, chiefly those of chromium or of iron or of both. The spinel-ruby, on the other hand, is an aluminate of magnesium, and may be expressed in chemical formula as

R. 618

Countries in which Rubies are found. (*W. R. Clark.*)	RUBY.

Mg O, Al$_2$ O$_3$. In hardness, the ruby is inferior only to the diamond, a fact which affords the simplest test, to determine whether a stone is a true ruby or a spinel; the sharp edge of a corundum crystal will scratch either a spinel or garnet, but has no effect on a ruby. The oriental ruby crystallises in the rhombohedral system, whereas the crystalline form of the spinel and garnet belong to the cubical series. This fact can, however, only rarely be traced, as the oriental ruby usually occurs in small pebbles or rounded fragments in which the crystalline form may be altogether lost. When a true ruby is, however, viewed through the dichroiscope, the colour is seen to be resolved into a carmine and an aurora-red and red slightly inclining to orange, whereas cubical crystals such as those of the spinel are not dichroidal, and do not, therefore, exhibit a similar breaking up of their colours. The Ruby receives the name "Oriental" from the fact of the finest red and violet varieties being obtained from Ceylon, Ava and other parts of the East. (*Encycl. Brit.*; *Watts, Chemistry*).

Ruby, *Ball, in Man. Geol. Ind., III., 427, 622.*   619

    Vern.—*Lal, yáqút, surkhi,* HIND.; *Manak* (the true ruby), *lábri* (the spinel), *tambra* (the garnet), PB.; *Kembu, kallu, mánikkam,* TAM.; *Kempu rai,* TEL.; *Yákút rumáni* (a first class oriental ruby), *labri* (the garnet), *kyouk-nees ballamya* (applied both to the true and the spinel-ruby), BURM.; *Manikya,* SANS.; *Balaksh, yaqut,* ARAB.

    References.—*Mason, Burma and Its People, 573, 577-579, 731, 732; Irvine, Mat. Med. Patna, 61; Watts, Dict. Chemistry, II., 86; V., 132, 400; Manual, Geology of India, Pt. II., 708; IV. (Mallet Mineralogy, 42, 51; Baden Powell, Pb. Pr., 48; Pb. Manuf., 202; Forbes Watson, Indust. Survey Ind., 413; Ain-i-Akbari (Blochmann's Trans.), I., 414; Linschoten, Voyage to East Indies (Ed. Burnell, Tiele & Yule), I., 80, 97; II., 140, 156; Milburn, Oriental Commerce (1825), 207, 317; Davies, Trade and Resources, N.-W. Boundary, India, clxxxi.; Marco Polo, Travels (Yule's Transl.), I., 149, 152; Tavernier, Travels in India (Ball's Transl.), I., 383, 399; II., 100, 102, 103, 123, 127, 128, 129, 449, 465, 472; Yule, Mission to Ava, 347; Man. Madras Adm., II., 82; Settlement Report, Central Provinces, Chanda Dist., 4; Admin. Rept. Burma (1874-75), 92; (1882-83), 56; Tropical Agriculturist (May 1st, 1888), 737; (June 1889), 845; (July 1889), 2, 49; Ind. Agriculturist, July 10th, 1886; Oct. 5th, 1889; Journ Asiat. Soc. Bengal, I., 353; II., 75; VIII., 372; Indian Economist, V., 14; Madras Journ. of Lit. and Science, IX., 121; Encyclop. Brit. XXI., 47; Balfour, Cyclop. Ind., III., 449.*

SOURCES.

    (*r*) *Southern India.*—According to Balfour, fine rubies have, from time to time, been discovered in many of the corundum localities in Southern India, particularly in the gneiss at Viralimodos and Sholasigamany. They occur also, though rarely, in the Trichingode *taluk,* and at Mattapollaye.

    (*2*) *Ceylon.*—Rubies are occasionally found in Ceylon, chiefly at Badulla and Saffragani, and also, it is said, at Matura. They usually occur with sapphires, but are rarer than these gems, and are not usually of good colour.

    (*3*) *Afghánistán.*—Rubies have been brought from Jagdalak or Gandamak in Afghánistán; indeed, Mr. Streeter mentions having possessed a ruby of 10½ carats from the latter locality, but most of the stones reputed to be Afghan rubies are merely spinels.

    (*4*) *Badakhshan.*—The delicate rose pink variety of spinel known as *balas* ruby was worked for centuries in Badakhshan. The mines were situated on the river Shighnan, a tributary of the Oxus. In the time of Marco Polo these mines were the chief source of the spinel-ruby, and were wholly in the hands of the King of Balkh, who kept up the value of the stones by permitting only a limited number to be exported. When Murad Beg of Kunduz conquered Badakhshan, he found the out-

SOURCES.
Southern India. 620
Ceylon. 621
Afghanistan. 622
Badakhshan. 623

RUBY.	History of the

SOURCES.

turn of the mines so poor that he abandoned working them. In 1866, the reigning Mir had one of them re-opened, but without much result (*Ball*). A short account of the workings was given in 1837 by a traveller in those regions who attempted to visit the mines, but was hindered by the authorities from doing so. The matrix, he states, is a red sandstone, or a limestone, largely impregnated with magnesia; most. probably it is the latter, since that appears to be the usual matrix. The galleries are numerous and easily cut, but the frequent influx of water caused much trouble when the mines were worked.

Upper Burma.
624

(5) *Upper Burma.*—The chief sources, however, both of the oriental and of the spinel-ruby are the mines of Upper Burma. Mr. Ball in an appendix to his edition of *Tavernier's Travels*, gives a long and interesting account of these mines, which, as it contains practically all that is known on the subject, may be reproduced here. " The principal ruby mines of Burma are situated in three valleys which are known by the names of their chief villages respectively, namely, Mogok (or Mogout), Kathé and Kyat-pyen. The elevated tract including these valleys is situated at a distance of about 90 miles north-north-west from Mandalay, and is at elevations of from 4,000 to 5,500 feet above the sea. The ruby tract, as now defined by the most recent scientific examination, occupies an area of 66 square miles, but mining is at present limited to an area of about 45 square miles. A totally distinct ruby tract is situated in the marble hills at Sagyin, which is only 16 miles from Mandalay. So far as is known, it is of comparatively little importance, the rubies and other gems which are found there being of inferior quality. Other localities, about 15 miles to the north and north-east of Sagyin, are reported to produce rubies, but nothing certain is known about them.

HISTORY.
625

HISTORY.

" The ruby mines of Burma were first made known by European travellers towards the end of the fifteenth century. In the sixteenth century there are more definite references by Portuguese travellers, but they are not of much practical importance. Tavernier gives an account of the mines and their produce from hearsay, from which it would appear that the reputation they then bore was not very high, or he would probably have made an effort to visit them. The yield, he says, did not exceed 100,000 *écus* (say £22,500 per annum), and he found it profitable to carry rubies from Europe to Asia for sale. The principal authorities of the present century previous to the conquest of Upper Burma are Mr. Crawford, the Pere Guisseppe d' Amato, who visited the mines about 1833, and Dr. Oldham, who visited Ava and collected information about the year 1855, when with Sir Arthur Phayre's Mission, Mr. Bredmeyer, who was in the service of the King and visited the mines in the year 1868, and Mr. Spears and Captain Strover of the British Burma Commission, both of whom have placed on record their observations. From these authorities we learn that the rubies which were found were generally small, not averaging more than a quarter of a *rati*, and that the large stones were generally smuggled away, but few of them reaching the King. It was supposed that the Chinese and Tartar merchants who visited Mogok and Kyat-pyen conveyed most of them out of Burma. The large rubies were generally flawed, and Mr. Spears states that he never saw one exceeding half a rupee in weight, *i.e.*, about 22 carats. The King's revenue derivable from the monopoly was variously stated by these authorities at from £12,500 to £15,000. The more recent information now available confirms these estimates. The figures stated on official authority are R90,000 to R1,00,000, the highest sum being R1,50,000, paid in one year.

R. 625

| Ruby Mines of Burma. | (*W. R. Clark.*) | RUBY. |

Besides which, however, was the reservation of stones above a certain size, but it seems to be generally admitted that few large stones were found, and of these a proportion, in spite of severe punishments for concealment, never reached the King; there is no basis then for an estimate of the total revenue which he received from the mines.

"If one may judge from the appearance of the rubies forming part of the treasure taken at Mandalay, and which are now exhibited at the South Kensington Museum, valuable stones were rare, as, except a few of the smaller ones, none seems to be perfect. As is well known, recent accounts by experts have represented the prospects of the mines in a much more favourable light, and the true value will probably be ere long ascertained by the energetic operations of a Company conducted on scientific principles."

Besides rubies, various other varieties of corundum have been found, such as sapphires, oriental emeralds, oriental amethyst, oriental topaz and white sapphires. Spinels of various colours are also abundantly met with.

Occurrence.
626

MODE OF OCCURRENCE AND SOURCE OF THE GEMS.—Mr. Ball continues : " Although it has, for some time, been known that the rubies of Sagyin were derived from crystalline limestones or marble, the source of the gems in the principal region at Mogok, Kyatpyen, and Kathé was not actually ascertained till lately when these localities were visited by Mr. Barrington Brown. It was known that they were for the most part actually obtained in derivative gravels, and it had been inferred that the so-called clefts and lodes of a report which appeared before this examination were really fissures in limestones, where the stones had accumulated as the result of the solution of the limestone and by gravitation into these recesses. Mr. Brown has shown that the geological formation consists of recent deposits of hill wash and alluvium, and old crystalline limestones, schists, pigmatite, and other metamorphic rocks. In order to explain the relationship which exists between these formations and the rubies, it will be convenient to describe the various systems of mining by which the mode of occurrence will be made apparent. The mines as worked by the natives may be divided into four classes as follows :—

    I.—*Twinlones,* or pits sunk in the alluvium of the valleys.
    II.—*Mewdwins,* or open cuttings in the hill wash over which water is led.
    III.—*Loodwins,* or workings in caves and fissures.
    IV.—*Quarries* in a bed of coarse calcspar in the limestone, which appears to be the true original matrix of the gems.

The *twinlones* are square pits, which are sunk in the alluvium of the valleys down to the gem-bearing gravels, which occur at varying depths. These pits have to be timbered to support the sides, and, as far as possible, exclude water, which, however, finds access, and the first operation every day is to bale out the water which has accumulated during the night. The gravel is hoisted out in baskets by means of bamboo poles similar to those which are used in India for raising water from wells. The gravel is then washed in shallow baskets made of closely-woven bamboo, and the rubies as they are picked out are placed in a bamboo tube full of water and are sorted at the close of the day's work. The larger pits are generally cleared out in about ten days, and the smaller in half the time; when working in one is finished the timber is removed and another pit is started.

" *Mewdwins.*—These are open cuttings on the slopes of the hills to which water is conducted, often from a considerable distance, and discharged with as great a head as possible on the ruby clay and sand, which is shovelled under it by the miners. The lighter portions are carried down by the stream, the boulders removed by hand, and the residue placed in the sluices and washed, where it is caught by *riffles*, from whence it is

HISTORY.

removed and washed in baskets as in the preceding process. The circumstances appear to be such as would suit a more scientific application of
hydraulic methods than are known to the Natives.

Loodwins.—These are natural caves and fissures in the limestone rock,
in the floors and crevices of which the rubies have accumulated in consequence of the solution by water of the limestone matrix. In the ordinary
sense of the term these are not mines, *i.e.*, the miners do not excavate the
rock, but merely scramble through the natural passages and tunnels
to the spots where the loam containing the rubies is found—this they
either carry to the surface in baskets, or it is hoisted up by means of balance
poles—and it is then washed at the surface at the nearest watercourse. From such caves the finest rubies ever found have been obtained,
and from one in the Pingu Hill, near Kyatpyen, Mr. Brown states that, after
the detritus had been passed, of every basketful of the ruby clay which
was raised half consisted of rubies.

A certain Royal mine of this character is said to have produced a ruby
as large as a walnut, and in another the rubies were found in association with the bones of some extinct animal of very large size. This
description opens up a somewhat wide vista of speculation, and one
can hardly resist the temptation of prophesying as to the wonderful
discoveries which may be made when adits and shafts are driven to afford
access to these natural caves and fissures in the mass of the marble hills.
In such safe receptacles it is not unreasonable to suppose that stones which
have suffered but little from attrition and fracture may be found, and that
there the greatest prizes will be obtained.

Quarries.
627

Quarries.—To the north of Mogok village, at a distance of about three
quarters of a mile, a bed of calcspar in the limestone, which is 20 feet wide,
produces rubies, but in order to obtain them the use of powder has to be
employed as well as a hammer, and when chipped out the gems are more
or less fractured; but good stones have been obtained. Whether any
method can be devised of avoiding the injury resulting from the use of
explosives is at present doubtful. It is not easy to suggest how a firm
rock, such as this calcspar, could be mined without recourse being had to
violent methods of some kind.

The rose pink rubellite (a variety of tourmaline) is obtained on the
margin of the Mobychoung river, 15 miles south of Mogok and 3 miles
from Mamlong. The mines in the alluvium are worked by a rude hydraulic
system, and the produce is sent to China, large pieces obtaining a good
price.

Under the arrangements which have been made with the New Burma
Ruby Mine Company, the rights and interests of the miners have apparently been very fully safeguarded, but whether the miners on their part
will refrain from smuggling and comply with the regulations, and disclose
their more valuable finds and submit them to taxation, remains to be seen.
The total production of rubies in 1887, when the country was disturbed,
amounted to only R42,486 worth, but in the first two months of 1888
R21,883 worth had been obtained. Stones of from 5 to 20 carats' weight
were sold during this period, and the highest price obtained for one was
R500.

The mode of occurrence of the rubies in calcspar is, I believe, somewhat
unusual, though spinel is known to be found in calcareous rocks. It is
generally the case that the corundum minerals are found in mica schists ;
such is stated to be the case in Zanskar in the Himálayas, and also in
Ceylon."

Domestic.
628

Domestic.—According to the ideas of native jewellers, rubbies come
next in value to diamonds. They must be hard and transparent (*shafóf*).

R. 628

DOMESTIC.

The most esteemed kind is the *yákút rumáni* "whose colour is like the seed of a pomegranate." Like most other jewels, rubies have, in the East, some fancied talismanic or medicinal virtues attached to them. Thus, a ruby worn on the finger is supposed to protect the wearer from the night-mare in his sleep and from evil dreams, and in many parts of India a bracelet formed of nine gems, of which the ruby is one, is supposed to protect the wearer from the evil eye (*Baden Powell*).

The largest true ruby known in Europe is said to be one of the size of a small hen's egg, which was presented by Gustavus III. of Sweden to the Empress of Russia when he visited St. Petersburg. Rubies, however, of larger size have been described by **Tavernier** and other oriental travel-lers, but it is probable that in many instances spinels have been mistaken for real rubies. There seems little doubt that the great historic ruby set in the Maltese cross in front of the State crown of England is a spinel (*Encycl Brit.*). In Europe, the oriental ruby has long been regarded as of higher value than any other precious stone. **Mr. Streeter**, in his *Precious Stones and Gems*, states that a ruby of perfect colour weighing 5 carats, is worth, at the present day, ten times as much as a diamond of equal weight, and, as the weight of the stone increases, its value rises rapidly, so that rubies of exceptional size command enormous prices. There is consequently much temptation to replace the true stone by spinel, garnet, or even paste. An excellent imitation of the colouring of the true ruby is obtained by oxide of chromium, and the paste made up of silicate of alumina is almost as hard as rock crystal, so that it is often very difficult to distinguish the spurious from the genuine article.

TRADE.

TRADE.
629

For an account of the production of, and trade in, rubies in British Burma previous to the year 1889, the reader is referred to **Mr. V. Ball's** statements briefly reviewed above. According to a statement recently published by the Revenue and Agricultural Department, which shows the quantities and values of gems produced in each British Province and Native State of India during the calendar year 1889, amounted to 65,628·5 carats of rubies valued at R33,848 and 4,496 carats of spinel valued at R259 produced in the Ruby Mines District of British Burma. There is no record of their production in any other Province or State. It is under-stood, however, that the Company that took up the concession to work these mines is by no means satisfied with the results they have attained.

Rue, see **Ruta graveolens,** *Linn.;* p. 594; RUTACEÆ.

RUELLIA, *Linn.; Gen. Pl., II., 1077.*

630

A genus of Acanthaceous plants, comprising 150 species, found in the tropics and warm temperate regions of both hemispheres. Nine species are described in the *Flora of British India* as belonging to this genus, but many others, formerly regarded as **Ruellias,** and some of them of economic value, have been reduced to other genera. Few of the species now left in this genus are of much use.

Ruellia indigotica, *Fortune,* see Strobilanthes flaccidifolius, *Nees.;* Vol. [VI., Pt. II; also Indigo, Vol. IV, 451; ACANTHACEÆ.

R. longifolia, *T. Anders.; Fl. Br. Ind., IV., 412.*

631

Syn.—DIPTERACANTHUS LONGIFOLIUS, *Stocks.*
Vern.—*Surata,* C. P.
References—*Boiss., Fl. Orient., IV., 519; Settlement Rep., Chanda Dist., C. P., App. VI.*
Habitat.—Found on the plains in Sind, the Central Provinces, and dis-tributed to Balúchistán.

RUMEX dentatus.	·Ferment used by Santals. (*W. R. Clark.*)

FOOD.
Leaves.
632

Food.—The LEAVES are said to be eaten as a vegetable in the Central Provinces.

[*Ind. Or., t. 282.*

633

Ruellia prostrata, *Lamk.; Fl. Br. Ind., IV., 411 ; Beddome, Ic. Pl.*

Syn.—R. WIGHTIANA, *Wall.* ; R. REPENS, *Heyne;* DIPTERACANTHUS PROS-TRATUS, *Nees.*

Var. dejecta=RUELLIA RINGENS, *Roxb.;* R. REPENS, *Wall.* chiefly ; R. DECCANENSIS, *Grah.;* DIPTERACANTHUS DEJECTUS, *Nees;* D. PROSTRA-TUS, *Griff.*

Vern. —*Upu-dali,* MALYAL.; *Nilpuruk,* CING.

References.—*DC., Prodr., XI., 124; Roxb., Fl. Ind., Ed. C.B.C., 473; Grah., Cat. Bomb. Pl., 162, 185 ; Ainslie, Mat. Ind., II., 482.*

Habitat.—A perennial herb of the Deccan Peninsula, which is distributed to north Behar. The variety **dejecta** is, however, much more prevalent, being found throughout India from the Panjáb to Assam and Ceylon.

MEDICINE.
Juice.
634

Medicine.—The JUICE of the leaves, boiled with a little salt, is supposed on the Malabar coast to correct a depraved state of the humours (*Rheede*). "They are sometimes given with *pundum* or liquid copal as a remedy for gonorrhœa " (*Ainslie*).

635

R. suffruticosa, *Roxb.; Fl. Br. Ind., IV., 413.*

Syn.—R. REPENS, *Wall.;* DIPTERACANTHUS SIBUA, *Nees.*

Vern.— *Chaulia,* SANTAL.

References.—*Roxb., Fl. Ind., Ed. C.B.C., 476; DC., Prodr., XI., 121 ; Revd. A. Campbell, Econ. Pl., Chutia Nagpur, No. 8422; Gasetteer, N.-W. P. (Agra Dist.), IV., lxxvi.*

Habitat.—A small, erect undershrub, found in Lower Bengal (throughout Chutia Nagpur), and in the upper Gangetic plain.

MEDICINE.
Root.
636
DOMESTIC.
Root.
637

Medicine.—Revd. A. Campbell states that the ROOT is used medicinally by the Santals in gonorrhœa, syphilis, and renal affections generally.

Domestic.—The ROOT is employed to cause fermentation in the grain from which the Santals make their *haudi* or beer.

Rum, see Spirits, under the article Narcotics, Vol. V., 332 ; also the article Sugar.

638

RUMEX, *Linn. ; Gen. Pl., III., 100.*

A genus of perennial herbs or annuals, rarely shrubby, belonging to the Natural Order POLYGONACEÆ. About 100 species are described, but Sir J. D. Hooker, in the *Flora of British India*, states that the genus wants revision and a great reduction of species, and that most of the Indian ones may be referred to European. Besides being used for the purposes hereafter described, several species of Rumex are eaten by the Natives of India as pot-herbs, and Aitchison, in his work on the economic products of Afghánistán, says that the stem of a species of Rumex is used by the natives of that country as a tinder.

639

Rumex dentatus, *Linn.; Fl. Br. Ind., V., 59 ;* POLYGONACEÆ.

Syn.—R. ROXBURGHIANUS, *Wall. partim* ; R. KLOTZSCHIANUS, *Meissn.;* R. OBTUSIFOLIUS, *Herb. Ham.* Sir J. D. Hooker remarks that except for its shorter pedicels it would pass for R. OBTUSIFOLIUS, *Linn.*

References.—*Boiss., Fl. Orient., IV., 1013 ; Aitchison, Bot. Afgh. Del. Com , 106; Murray, Pl. & Drugs of Sind, 98.*

Habitat.—An annual herb, found on the plains of India from Assam and Sylhet to the Indus; it ascends the Himálaya to an altitude of 1,000 feet. It occurs also in Sind and the Koncan.

DYE.
Root.
640

Dye.—Murray, speaking of R. obtusifolius, *Linn.*, but probably referring to this plant, says that in Sind the ROOTS are used as a dye by the Natives.

R. 640

| A Substitute for Rhubarb. | (*W. R. Clark.*) | RUMEX nepalensis. |

Medicine.—The ROOT is said to be used as an astringent application in cutaneous disorders.

Rumex hastatus, *Don; Fl. Br. Ind., V., 60.*

Vern.—*Amlora,* KUMÁON; *Khattimal, katambal, ámi, malori-ghá, amla, amlora,* PB.

References.—*DC., Prodr., XIV., 1, 72; Gamble, Man. Timb., 303; Stewart, Pb. Pl., 187; Gazetteers:—Panjáb, Simla Dist., 12; N.-W. P., X., 316; Stewart, in Journ Agri.-Horti. Soc. Ind., XIV., 9.*

Habitat.—An undershrub of the North-West Himálaya, which occurs chiefly on dry hill-sides from Kumáon to Kashmír, at altitudes between 1,000 and 8,000 feet, and distributed to Afghánistán. Through its exceptional abundance it has originated the name Almora for a North-West Himálayan hill station. When abundant and in fruit, many parts of the outer Himálaya are often made to take the tints of the moors of Scotland.

Food.—The LEAVES have a pleasant acid taste, and are eaten raw by the Natives (*Stewart*).

Structure of the Wood.—Light red, moderately hard, with broad medullary rays.

642

FOOD.
Leaves.
643
TIMBER.
644
645

R. maritimus, *Linn.; Fl. Br. Ind., V., 59.*

Syn.—R. PALUSTRIS, *Sm.;* R. WALLICHII, *Meissn.;* R. WALLICHIANUS *Meissn.;* R. ACUTUS, *Roxb.;* R. ROXBURGHIANUS, *Wall.;* R. COMOSUS & SETACEUS, *Ham.*

Vern.—*Júl-palum,* HIND.; *Bun-palung,* BENG.; *Jungli-palak, húlá obúl, sagúkei, khattikan,* (Bazar seed=) *bij band,* PB.

References,—*DC., Prodr., XIV., 1, 59; Boiss., Fl. Orient., IV., 1014; Roxb., Fl. Ind., Ed. C.B.C., 309; Stewart, Pb. Pl., 187; Dymock, Mat. Med. W. Ind., 661; Gazetteers:—N.-W. P., I., 84; IV., lxxvii; X., 316, 749.*

Habitat.—An annual plant, common in marshes in Assam, Silhet, Cachar, and Bengal, and in the plains of Northern India. In the Panjáb Himálaya it is found in similar localities up to 12,000 feet in altitude. It is distributed to Europe, Asia, North Africa, and North and South America.

Medicine.—The PLANT has cooling properties and is often eaten by Natives as a pot-herb especially in the warm weather. The LEAVES are applied to burns, and the SEEDS are, according to Atkinson, the *bij band* of the bazárs, or are, in Stewart's opinion, often sold as such, although not the genuine article (see **Sida cordifolia,** *Linn.*). They are used in native medicine as an aphrodisiac.

MEDICINE.
Plant.
646
Leaves &
Seeds.
647

R. nepalensis, *Spreng.; Fl. Br. Ind., V., 60; Wight, Ic., t. 1810.*

Syn.—R. ROXBURGHIANUS, *Schultes f. (non Wall.);* R. RAMULOSUS, *Meissn.;* R. HAMATUS, *Trevir.;* R. TUBEROSUS, *Roxb.*

References.—*DC., Prodr., XIV., 1, 55; Boiss. Fl. Orient., IV., 1011; Aitchison, Fl. Kuram Valley, 91; Irvine, Mat. Med. Patna, 91; Gazetteer, N.-W. P., X., 316; Stewart, in Journ. Agri.-Horti. Soc. Ind., XIV., 9.*

Habitat.—A tall, branched perennial found on the Temperate Himálaya from Bhután to Kashmír, at altitudes between 4,000 and 9,000 feet. It occurs also on the Gháts of the Western Peninsula from the Koncan to the Nilghiris, and is distributed westward to Asia Minor and South Africa and eastward to Java.

Medicine.—The tuberous ROOTS are said to be sold in the bazárs of Bengal under the name of *Rewund chini* as a substitute for rhubarb. They are given in constipation in doses of 10 gr. to 120 gr. (*Irvine*).

648

MEDICINE.
Roots.
649

R. 649

RUNGIA parviflora.	The Sorrel.

650

Rumex vesicarius, *Linn.; Fl. Br. Ind., V., 61.*

SORREL, BLADDER DOCK.

Vern.—*Chúka, ambari, chukeká sák,* HIND.; *Chúka, chúka-pálang, chak, chúk,* BENG.; *Súkha sag,* ASSAM; *Ambut chúka,* C.P.; *Chúka, chúka-pálak, chúka-pálang,* N.-W. P.; *Triwakka, khatbíri, khattítan, katta mítha, salúni,* PB.; *Chók, chóka, talúni,* PUSHTU; *Chúka,* SIND; *Ambari, chúk-ka,* DECCAN; *Chúka,* BOMB.; *Shakkán-kirai,* TAM.; *Chukka kúra,* TEL.; *Kala-khen-boun,* BURM.; *Súri,* SING.; *Chukra, amlavetasa, shutavedhí,* SANS.; *Humbíjít, hamáz, humarbostaní,* ARAB.; *Turshah, tursak, túrshumuk,* PERS.

References.—*DC., Prodr., XIV., 1, 70; Roxb., Fl. Ind., Ed. C.B.C., 309; Voigt, Hort. Sub. Cal., 326; Stewart, Pb. Pl., 187; Aitchison, Cat. Pb. and Sind Pl., 125; Graham, Cat. Bomb. Pl., 172; Mason, Burma and Its People, 780; Sir W. Elliot, Fl. Andhr., 45; Ainslie, Mat. Ind., I., 398; Irvine, Mat. Med. Patna, 21, 24; Medical Topog. Ajmir, 130, 132; Moodeen Sheriff, Supp. Pharm. Ind., 218; U. C. Dutt, Mat. Med. Hindus, 209, 295; Sakharam Arjun, Cat. Bomb. Drugs, 114; Murray, Pl. & Drugs, Sind, 99; Dymock, Mat. Med. W. Ind., 2nd Ed., 658, 659; Birdwood, Bomb. Prod., 174; Baden Powell, Pb. Pr., 372; Atkinson, Him. Dist. (X., N.-W. P. Gas.), 316, 708; Econ. Prod. N.-W. P., Part V., 37, 42; Useful Pl. Bomb. (XXV., Bomb. Gas.), 170; Settlement Report, Central Provinces, Chanda Dist., 82; Gazetteers:—N.-W. P., IV., lxxvii; Mysore and Coorg, I., 65; Agri.-Hortí. Soc. Ind., XIV., 6; Ind. Gard., 199; Balfour, Cyclop. Ind., III., 454.*

Habitat.—An annual monœcious species, 6 to 12 inches high; indigenous to the Western Panjáb, the Salt Range and Trans-Indus Hills. Found in most other parts of India either in a state of cultivation or as a garden-escape.

MEDICINE.
Juice.
651
Whole Herb.
652
Bruised Leaves.
653
Seeds.
654
FOOD.
Plant.
655

Medicine.—The JUICE of the plant is considered by natives to be cooling, aperient, and, to a certain extent, diuretic (*Ainslie*). It is used to allay the pain of toothache, and from its astringent properties it is supposed to check nausea. The WHOLE HERB is given internally to allay burning at the pit of the stomach and to improve the appetite. Externally a pulp composed of the BRUISED LEAVES is applied to the skin to allay the pain of the bites of reptiles and the stings of scorpions. The SEEDS are said to have similar properties, and are, besides, prescribed roasted in dysentery. The ROOT also is medicinal (*Dymock*).

Food.—The PLANT is cultivated as a vegetable almost throughout India, and is used by the natives both in the raw and cooked state. It is usually grown in patches near wells, and may be procured almost all the year round.

RUNGIA, *Nees; Gen. Pl., II., 1120.*

656

Rungia parviflora, *Nees, var.* pectinata; *Fl. Br. Ind., IV., 550;*
[*Wight, Ic., t. 1547;* ACANTHACEÆ.

Syn.—R. PARVIFLORA, *Nees;* R. POLYGONOIDES, *Nees;* JUSTICIA PECTINATA, *Linn.;* J. PARVIFLORA, *Retz.;* J. INFRACTA, *Vahl.*

Vern.—*Tavashú múrunghie, púnakapúndú,* TAM.; *Pindi kúnda,* TEL.; *Bir lopong arak',* SANTAL; *Pindi,* SANS. The names *Pitpáprá, pittapá-padá,* although more properly applied to Fumitory (**Fumaria officinalis**), are also used in the bazárs of Bombay as the names for this and the following species, both of which are used as substitutes for that drug.

References.—*Roxb., Fl. Ind., Ed. C.B.C., 45; Voigt, Hort. Sub. Cal., 491; Graham, Cat. Bomb. Pl., 165; Dals. & Gibs., Bomb. Fl., 195; Aitchison, Cat. Panjáb & Sind Pl., 113; Ainslie, Mat. Ind., II., 412; Dymock, Mat. Med. W. Ind., 594; S. Arjun, Bomb. Drugs, 108; Gazetteers, N.-W. P., IV., lxxvi., X., 315.*

Habitat.—The variety **pectinata**, occurs universally throughout India, from the Himálaya to Ceylon and Pegu. The other forms occur in the Deccan and Burma.

| Garden Rue. | (*W. R. Clark.*) | RUTA graveolens. |

Medicine.—Besides being employed as an adulterant of Fumitory, this species has itself distinct medicinal uses among the Natives of India. The JUICE of the small and somewhat fleshy LEAVES is considered cooling and aperient and is prescribed for children suffering from small-pox in doses of a tablespoonful or two twice daily. The bruised LEAVES are applied to contusions to relieve pain and diminish swelling (*Ainslie*). Among the Santals the ROOT is given as a medicine in fevers (*Rev. A. Campbell*).

MEDICINE.
Juice.
657
Leaves.
658
Root.
659
660

Rungia repens, *Nees; Fl. Br. Ind., IV., 549; Wight, Ic., t. 465.*

Syn.—JUSTICIA REPENS, *Linn.;* DICLIPTERA REPENS, *Rœm.*

Vern.—*Kodaga saleh,* TAM.; *Salundyi,* CING.

References.—*Roxb., Fl. Ind., Ed. C.B.C., 44; Graham, Cat. Bomb. Pl., 165; Dals. & Gibs., Bomb. Fl., 196; Ainslie, Mat. Ind., II., 158; Drury, U. Pl. Ind., 369; S. Arjun, Bomb. Drugs, 109; Burm., Thes. Zeylan., 7, t. 3, f. 2.*

Habitat.—A procumbent, rooting, ramous weed, common throughout India, from the Panjáb and Bengal to Ceylon.

Medicine.—The LEAVES resemble both in smell and taste those of thyme; while fresh they are bruised, mixed with castor-oil, and applied to the scalp in cases of tinea capitis (*Ainslie*). The whole PLANT, dried and pulverised, is given in doses of from 4 to 12 drachms in fevers and coughs, and is also considered a vermifuge (*Drury*).

MEDICINE.
Leaves.
661
Plant.
662

Rusa oil, see Andropogon Schœnanthus, *Linn.;* Vol. I., *249;*
[GRAMINEÆ.

RUTA, *Linn.; Gen. Pl., I., 286.*
[485; RUTACEÆ.

Ruta graveolens, *Linn., var.* **angustifolia;** *Fl. Br. Ind., I.,*
GARDEN RUE.

663

Syn.—R. ANGUSTIFOLIA, *Pers.;* R. CHALEPENSIS, *Wall.*

Vern.—*Sadáb, pismarum, satari,* HIND.; *Ispund, ermul,* BENG.; *Marúya,* URIYA; *Sudáb, katmal,* PB.; *Sadáf, pismarum, satari,* DECCAN; *Satáp,* BOMB.; *Satápa,* MAR. & GUZ.; *Arvada,* TAM.; *Sadápa, arudu,* TEL.; *Nágadali-sappu,* KAN.; *Aruda,* SING.; *Sadápaka, somalata,* SANS.; *Arúda, féjan,* ARAB.; *Sudáb,* PERS.

References.—*Roxb., Fl. Ind., Ed. C.B.C., 362; Voigt, Hort. Sub. Cal., 182; Dals. & Gibs., Bomb. Fl., 17; Stewart, Pb. Pl., 38; Graham, Cat. Bomb. Pl., 36; Sir W. Elliot, Fl. Andhr., 16, 165; W. & A., Prodr., 146; Folkard, Plant-Lore and Legends, 104, 531; Pharm. Ind., 39, 40; British Pharm., 292; U. S. Dispens., 15th Ed., 1038; Ainslie, Mat. Ind., I., 351; O'Shaughnessy, Beng. Dispens., 260; Irvine, Mat. Med. Patna, 40; Moodeen Sheriff, Supp. Pharm. Ind., 218; Mat. Med. S. Ind. (in MSS.), 74; Sakharam Arjun, Cat. Bomb. Drugs, 29; Murray, Pl. & Drugs, Sind, 89; Bent. & Trim., Med. Pl., I., Pl. 44; Dymock, Mat. Med. W. Ind., 2nd Ed., 122-24; Dymock, Warden & Hooper, Pharmacog. Ind., I., 249; Year-Book Pharm., 1874, 263, 623; 1875, 260; 1879, 467; Birdwood, Bomb. Prod., 17; Baden Powell, Pb. Pr., 335; Man. Cuddapah Dist., Madras, 199; Gazetteer, Mysore and Coorg, I., 58; Hunter, Orissa, II., App. VI., 181; Balfour, Cyclop. Ind., III., 495; Smith, Dict., 357.*

Habitat.—A small, branching under-shrub, 2 or 3 feet high, cultivated in Indian gardens for the medicinal properties of its leaves and seeds; distributed westward to the Canaries.

Oil.—By distillation with water the fresh HERB yields a small quantity of volatile OIL. This has a pale yellow colour when fresh, but becomes brown by keeping. Its odour is strong and disagreeable, and it has an acrid and nauseous taste. It has a specific gravity of 0·837 at 18°C., boils at 228°—230°C., and solidifies between +1° and 2° into shining crystalline laminæ. Oil of Rue is chiefly a mixture of a hydrocarbon with an aldehyde

OIL.
Herb.
664

R. 664

or ketone belonging to the series $C_n H_{2n} O$ (*Bentley & Trimen; U. S. Dispensatory*).

MEDICINE.
665

Medicine.—Rue was held in high estimation by the ancient Greeks and Romans, who considered it a valuable resolvent, diuretic, and emmenagogue, and attributed to it also many fanciful virtues. Thus **Aristotle**, in his *History of Animals*, states that weasels, before fighting with serpents, rub themselves against this plant as a protection against venom, and **Mithridates** is said to have used rue as one of the ingredients of his famous antidote against poisons. During the middle ages the PLANT was hung

Plant.
666

round the neck as a charm against vertigo and epilepsy; it was considered emblematic of good luck, and a protection against sorcery. Muhammadan medical writers, both Arabian and Indian, class the drug among "attenuants, vesicants, and stimulants." Its properties are said to be hot and dry in the third degree; it strengthens the mental powers, acts as a tonic and digestive, and increases the urinary and menstrual excretions. It is described as a powerful antaphrodisiac and an abortefacient in pregnant women. The Hindus received the drug from the West, and with its cultivation

Leaves.
667

brought all the superstitions regarding it. They use the dried LEAVES as a fumigatory for children suffering from catarrh; powdered and in combination with aromatics, they give them as a remedy for dyspepsia, and with the fresh ones they make a tincture which they use as an external remedy in the first stages of paralysis. They consider rue in all its forms injurious to pregnant women. In the Panjáb the leaves are taken by the Natives as a remedy for rheumatic pains. The HERB and the OIL act as stimulants

Herb.
668
Oil.
669

chiefly of the uterine and nervous systems. Rue has also been regarded as an anthelmintic. In large doses it is an acro-narcotic poison. When fresh its topical action is acrid, and if much handled it produces redness, swelling, and even vesication.

It may be given internally in hysteria, amenorrhæa, epilepsy, flatulent colic, etc., and externally may be used as a rubefacient. The oil is the

Rue Tea.
670

best form for administration, but RUE TEA is a popular remedy. The dose of the powdered leaves is 10 grs to ʒi, of the oil one to four minims (*Dymock; Murray; Bentley & Trimen*).

DOMESTIC.
671

Domestic.—It was formerly employed in Europe as a condiment, and in the East is sometimes placed in beds to keep off insects.

672

Ruta tuberculata, *Forsk.; Fl. Br. Ind., I., 485.*

References.—*Boiss, Fl. Orient., I., 939; Murray, Pl. & Drugs of Sind, 89.*
Habitat.—Common on the hills of Sind, and extending thence westward to Egypt and Algeria. It is used for the same purposes as **R. graveolens.**

Rye, see **Secale cereale,** *Linn.;* Vol. VI.; GRAMINEÆ.

R. 672

Printed in the United States
By Bookmasters